AAPG Treatise of Petroleum Geology

Reprint Series

The American Association of Petroleum Geologists
gratefully acknowledges and appreciates the leadership and support
of the AAPG Foundation in the development of the
Treatise of Petroleum Geology.

Structural Concepts and Techniques III

Detached Deformation

Compiled by
Norman H. Foster
and
Edward A. Beaumont

Treatise of Petroleum Geology
Reprint Series, No. 11

Published by
The American Association of Petroleum Geologists
Tulsa, Oklahoma 74101, U.S.A.

Library of Congress Cataloging-in-Publication Data

Structural concepts and techniques / compiled by Norman H. Foster and
Edward A. Beaumont.

3v cm. -- (Treatise of petroleum geology reprint series ; no.
9-11)
Bibliography: p.
Contents: 1. Basic concepts, folding, and structural techniques --
2. Basement-involved deformation. -- 3. Detached deformation.
ISBN 0-89181-408-6 (v. 1) : $34.00. -- $18.00 (v. 1, v. 2, v. 3: pbk.)
ISBN 0-89181-409-4 (v. 2)
ISBN 0-89181-410-8 (v. 3)
1. Geology, Structural. 2. Folds (Geology) 3. Petroleum --
Geology. I. Foster, Norman H. II. Beaumont, E. A. (Edward A.)
III. American Association of Petroleum Geologists. IV. Series:
Treatise of petroleum geology reprint series ; no. 9, 10, 11, etc.
QE601.S8637 1988 39231
CIP

551.8 -- dc19

American Association of Petroleum Geologists Foundation
Treatise of Petroleum Geology Fund*

Major Corporate Contributors
($25,000 or more)

Chevron Corporation
Mobil Oil Corporation
Pennzoil Exploration and Production Company
Sun Exploration and Production Company

Other Corporate Contributors
($5,000 to $25,000)

The McGee Foundation, Inc.
Texaco Philanthropic Foundation
Union Pacific Resources Company

Major Individual Contributors
($1,000 or more)

C. Hayden Atchison
Richard R. Bloomer
A. S. Bonner, Jr.
David G. Campbell
Herbert G. Davis
Lewis G. Fearing
James A. Gibbs
George R. Gibson
William E. Gipson
Robert D. Gunn
Cecil V. Hagen
William A. Heck
Harrison C. Jamison
Thomas N. Jordan, Jr.
Hugh M. Looney
John W. Mason
George B. McBride
Dean A. McGee
John R. McMillan
Rudolf B. Siegert
Jack C. Threet
Charles Weiner
Harry Westmoreland
James E. Wilson, Jr.

The Foundation also gratefully acknowledges the many who have supported this endeavor with additional contributions, which now total $12,240.50.

*Contributions received as of January 23, 1989.

Treatise of Petroleum Geology
Advisory Board

INTRODUCTION

This reprint volume belongs to a series that is part of the *Treatise of Petroleum Geology*. The *Treatise of Petroleum Geology* was conceived during a discussion we had at the 1984 AAPG Annual Meeting in San Antonio, Texas. When our discussion ended, we had decided to write a state-of-the-art textbook in petroleum geology, directed not at the student, but at the practicing petroleum geologist. The project to put together one textbook gradually evolved into a series of three different publications: the Reprint Series, the Atlas of Oil and Gas Fields, and the Handbook of Petroleum Geology; collectively these publications are known as the *Treatise of Petroleum Geology*. With the help of the Treatise of Petroleum Geology Advisory Board, we designed this set of publications to represent the cutting edge in petroleum exploration knowledge and application. The Reprint Series provides previously published landmark literature; the Atlas collects detailed field studies to illustrate the various ways oil and gas are trapped; and the Handbook is a professional explorationist's guide to the latest knowledge in the various areas of petroleum geology and related fields.

The papers in the various volumes of the Reprint Series complement the different chapters of the Handbook. Papers were selected on the basis of their usefulness today in petroleum exploration and development. Many "classic papers" that led to our present state of knowledge have not been included because of space limitations. In some cases, it was difficult to decide in which Reprint volume a particular paper should be published because that paper covers several topics. We suggest, therefore, that interested readers become familiar with all the Reprint volumes if they are looking for a particular paper.

We have divided the topic of structural concepts and techniques into three volumes. The first volume contains papers that discuss Basic Concepts, Folding, and Structural Techniques. The first paper in Basic Concepts is the classic 1979 paper by Harding and Lowell, "Structural styles, their plate-tectonic habitats, and hydrocarbon traps in petroleum provinces." We have used Harding and Lowell's classification of structural styles to group papers in volumes II and III. Basic Concepts also includes papers on stress analysis and pore-pressure effects. Folding includes papers describing folding processes and geometries. Structural Techniques is a collection of practical papers included to help the petroleum geologists solve structural problems encountered in exploration and development. These papers relate ways to visualize and predict the three-dimensional arrangement and location of strata in the subsurface.

Volume II contains papers related to Basement-Involved Deformation. These papers are subdivided into Extensional, Compressional, and Strike-Slip Deformation. Extensional Deformation includes papers discussing crustal rifting and normal faulting. Compressional Deformation includes papers discussing foreland deformation. In most cases, these papers use the Rocky Mountains as an example. Strike-Slip Deformation includes papers discussing strike-slip or wrench fault deformational processes and the consequent effects these processes have on folding, faulting, basin formation, and sedimentation.

Volume III, Detached Deformation, is subdivided into three sections: Extensional Deformation, Compressional Deformation, and Salt Tectonics. Extensional Deformation contains papers that discuss listric normal faulting and growth faulting (a species of listric normal faulting). Compressional Deformation contains papers on the processes and mechanics of low-angle thrust faulting. Salt Tectonics is a collection of papers on salt movement and salt dissolution, and their subsequent effect on the structural geometry of associated strata.

Edward A. Beaumont
Tulsa, Oklahoma

Norman H. Foster
Denver, Colorado

Table of Contents

Structural Concepts and Techniques III

Extensional Deformation

Compressional Deformation

Salt Tectonics

Table of Contents

STRUCTURAL CONCEPTS AND TECHNIQUES I

Basic Concepts

Folding

Structural Techniques

TABLE OF CONTENTS

STRUCTURAL CONCEPTS AND TECHNIQUES II

BASEMENT-INVOLVED DEFORMATION

EXTENSIONAL DEFORMATION

COMPRESSIONAL DEFORMATION

STRIKE-SLIP DEFORMATION

Extensional Deformation

OCEANOLOGICA ACTA, 1981, N° SP

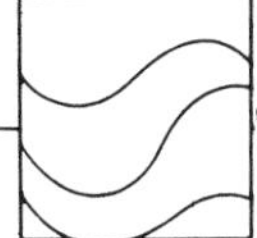

Listric normal faults
Rifting
Subsidence
Stretching

Failles normales listriques
Distension
Subsidence
Étirement

Listric normal faults

A. W. Bally [a], D. Bernoulli [b], G. A. Davis [c], L. Montadert [d]
[a] Shell Oil Company, Houston, Texas, USA.
[b] Universität, Basel, Switzerland.
[c] University of Southern California, Los Angeles, California, USA.
[d] Institut Français du Pétrole, 1 et 4, avenue de Bois-Préau, BP n° 311, 92506 Rueil-Malmaison, France.

ABSTRACT

The importance of listric normal faults in the formation of sedimentary basins is becoming increasingly more obvious. Based on reflection seismic sections and surface observations, the following genetic types may be differentiated :
— listric normal faults involving the basement that are associated with some crustal attenuation. Such faults occur during the formation of rifts that often precede the formation of passive continental margins (e.g., Gulf of Biscay, Galicia Bank) ;
— superficial soft-sediment listric normal faulting related to deltaic systems and/or to drifting sequences associated with the subsidence of passive continental margins (e.g., Gulf of Mexico) ;
— listric normal faulting associated with the genesis of accretionary wedges of active continental margins (e.g., Colombia) ;
— syn- and post-orogenic faulting associated with the stretching and shearing of orogenic systems and parts of their foreland (e.g., Great Basin).
The role of normal faulting in the evolution of "geosynclines" and folded belts (e.g., the Alps) is better understood in the light of observations on continental margins and late orogenic basins.

Oceanol. Acta, 1981. Proceedings 26th International Geological Congress, Geology of continental margins symposium, Paris, July 7-17, 1980, 87-101.

RÉSUMÉ

Les failles normales listriques

L'importance des failles normales listriques dans la formation des bassins sédimentaires est de plus en plus reconnue. A partir des coupes de réflection sismique et des observations de terrain, il est possible de distinguer les types génétiques suivants ;

— les failles normales listriques affectant le substratum associées avec un certain amincissement de la croûte. De telles failles apparaissent lors de la formation des rifts, qui précède souvent celle des marges continentales passives (ex. : Golfe de Gascogne, Bancs de Galice) ;

— les failles normales listriques dans les sédiments mous des systèmes deltaïques et/ou des séquences mobiles associées à la subsidence des marges continentales passives (ex. : Golfe de Mexico) ;

— les failles normales listriques associées à la genèse des prismes d'accrétion des marges continentales actives (ex. : Colombie) ;

— les failles syn- et post-orogéniques associées à l'étirement et au cisaillement des systèmes orogéniques et d'une partie de leur avant-pays (ex. : Great Basin).

Grâce aux observations sur les marges continentales et les bassins post-orogéniques, il est possible de mieux comprendre le rôle des failles normales listriques dans l'évolution des « géosynclinaux » et des chaînes plissées (ex. : les Alpes).

Oceanol. Acta, 1981. Actes 26e Congrès International de Géologie, colloque géologie des marges continentales, Paris, 7-17 juil. 1980, 87-101.

INTRODUCTION

Eduard Suess, in his "Face of the Earth" (1904-1924, Vol. IV, p. 536, 542 and 582), introduced the concept of listric faults. He described upward concave curved fault surfaces from the coal districts of Northern France, Belgium and Germany, and noted that in a cross-section such faults appear shovel-like (Greek listron, shovel). For many years the term was adopted by German geologists working in coal mining districts, but otherwise gained little acceptance outside Central Europe. Note that Suess was specifically dealing with listric thrust faults, some of which were in fact refolded by later deformation. However, listric normal faults have also been illustrated — although not so named — from the coal districts of Central France (Pruvost, 1960-1963), as well as from Northern France (Bouroz, Stievenard, 1958). Cloos (1936) also preferred to limit his illustrations of the term to thrust faults, although in his clay experiments he showed listric normal fault systems with considerable detail. Kirchmayer and Mohr (1963) gave a detailed review of the terminology of curvilinear and curviplanar structural elements and listed a number of examples of listric faults from coal-bearing sequences, from non-coal-bearing rocks and rock deformation from experiments. These authors also suggested that surface geologic studies do not easily reveal the listric nature of faults, because outcrop conditions often prevent adequate geometric control of fault planes. A number of recent publications, however, illustrate listric normal growth faults from outcrops [e.g., from Spitzbergen, Edwards (1976) ; from Ireland, Rider (1978) ; Crans *et al.* (1980) ; and from Wales, Woodland and Evans (1964)]. See, also, Figure 1 from the Pennine Alps and Figure 2 from Haiti.

Quite independently from all the previously mentioned work, petroleum geologists of the Gulf Coast of the USA during the late 1930's recognized growth faults, that is, faults that display stratigraphic thickening on their downthrown side and have increased throw with depth. Typically, such faults flatten with depth and, therefore, are listric normal growth faults (Ocamb, 1961 ; Bruce, 1973 ; Busch, 1975). Experimental work by H. Cloos (1936) did illustrate listric normal faults in connection with the simulation of graben-like structures, and E. Cloos (1968) made similar experiments to illustrate listric growth faults of the Gulf Coast type.

A number of potential fault surfaces suggested by the theoretical studies of Hubbert (1951) and Hafner (1951) are in fact curved or listric surfaces. A theory of growth faulting in a deltaic environment is offered by Crans *et al.* (1980). Thus, it is surprising to find that most textbooks in structural geology pay little or no attention to the curved nature and the downward flattening tendency of fault surfaces. In recent years, the more widespread use and publication of reflection seismic lines suggests to us that a dominant percentage of all faults are listric or that they were at least initiated as listric faults.

Figure 1
Triassic marbles from Kaltwasser Pass, east of Simplon Pass, Italian-Swiss Alps. Note that dolomitic blocks have been rotated in domino-like fashion, with essentially planar faults separating them. The ductility contrast is such that listric faults are not formed ; instead, the more ductile calcareous material serves as cushion and fills the space between the rotated dolomitic blocks by ductile flow. Extension may be expressed as a function of the dip of the rotated beds.

The seismic documentation of listric thrust faults is rather well known from décollement-controlled fold belts (e.g., Bally *et al.*, 1966 ; Beck, Lehner, 1974 ; 1975 ; Gordy *et al.*, 1975 ; Royse *et al.*, 1975 ; Hamilton, 1977 ; 1979) and is, therefore, not the subject of this paper.

Instead, in this essay we will try to highlight the nature of the evidence favoring listric normal faulting in widely differing settings. This is useful in the context of a discussion of passive continental margins, because the documentation of the listric geometry of normal faults is critical to an understanding of the formation of still intact passive margins, as well as of older passive margins that can now be observed as deformed remnants in ancient folded belts. As McKenzie (1978), Royden *et al.* (1980), Christie and Sclater (1980), and Le Pichon and Sibuet (1980 manuscript) have all pointed out, lithospheric stretching may well be responsible for the initiation of passive margins, and at least the stretching in the brittle crust may be accomplished by listric normal faulting.

Much like listric thrust faults, the shape and complexity of listric normal faults is controlled by ductility contrasts within the rock sequence that is intersected by them. In this context, it is important to differentiate synsedimentary soft

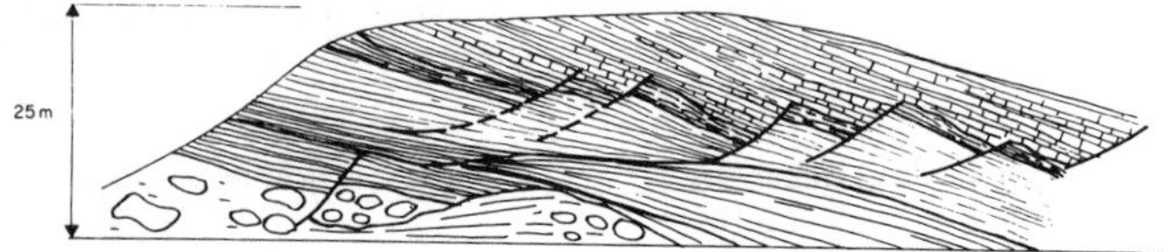

Figure 2
Listric normal faults in calcareous soft sediments from Haiti (late Miocene, NN II, Jaemel road).

sediment listric normal faulting and other listric faults in consolidated rocks that rise to the surface from listric normal fault structures that involve the underlying basement. Only in the latter case may attenuation of the crust be suggested.

The term listric fault is purely descriptive. We like to differentiate listric thrust faults from listric normal faults, and among them we separate listric growth faults from those that do not show any measurable growth through time. Finally, we note that many faults have been refolded, and thus it may be useful to determine whether the listric nature of a fault is primary, or due to secondary refolding.

LISTRIC NORMAL FAULTS INVOLVING THE BASEMENT OF PASSIVE MARGINS

Passive margins typically display a lower section of rifted structures that often — but by no means always — is separated by a "breakup" unconformity (Falvey, 1974) from overlying less disturbed sequences that subsided during spreading and cooling of the adjacent ocean. These latter, "drifting" sequences are often characterized by soft-sediment gravity type tectonics that do not involve the underlying basement. Let us first discuss basement faulting associated with the rifting phase.

Lowell and Genik (1972) and Lowell *et al.* (1975) postulated crustal attenuation for the Southern Red Sea by listric normal faulting affecting the whole thickness of the crust. These authors correctly point out that in matching conjugate passive margins across oceans, it is important to allow for the amount of extension in the underlying faulted basement. Unfortunately, however, the seismic sections provided by Lowell *et al.* (1975) were not adequate to document the listric nature of the Red Sea faults.

In the nearby Afar triangle, Morton and Black (1974) set up a model that recognizes the tilted fault blocks in the area, but avoids curved faults. Instead, rigid blocks are slipping along inclined straight faults into a ductile substratum, and progressive thinning takes place as new sets of inclined normal faults offset the earlier fault blocks. The authors relate the tilting of the beds to the amount of stretching of the crust. As the paper presents only a model, no documentation of the shape of the fault plane is provided.

The recent studies by de Charpal *et al.* (1978) and montadert *et al.* (1979 *a* and *b*) provide the first direct evidence for listric normal faulting that involves the basement of a passive margin. The basement of the Northern Bay of Biscay is of Variscan age and in part overlain by early Mesozoic basinal sequences that were initiated during an earlier tensional event (Triassic). Major rifting and extension in a marine regime occurred during early Cretaceous times and ceased by Aptian time. Blocks were rotated during the rifting phase by about 20 to 30° along listric normal faults. These faults can be mapped seismically in the Northern Bay of Biscay (Fig. 3 and 4), and Figure 5 shows a subhorizontal reflection (after correction for velocity pull-ups) which appears to be the sole of the listric normal fault system. To show that this situation is not limited to the Gulf of Biscay, similar features are shown on a line west of Galicia Bank (Fig. 6).

Refraction velocities (Avedik, Howard, 1979) indicated on Figure 4 suggest that the sole fault separates an upper faulted brittle crust from a lower crust with velocities exceeding 6.3 km/sec. In other words, as emphasized by Montadert *et al.* (1979 *a*), the base of the sole fault is a mechanical discontinuity within the crust, which, during the rifting period, was about 6-8 km below sea level. The same authors cautioned in an alternate interpretation that the basal flat reflector shown on Figures 5 and 6 may also represent the Moho itself, in which case the Moho would be the main decoupling level for the tilted fault blocks.

Montadert *et al.* used a reconstruction based on migrated seismic profiles. They determined for their preferred interpretation that the upper continental crust was reduced by about 30% from an unusually thin crust of about 6-8 km to a new thickness of about 4-5 km ; thus, the upper brittle crust was, according to the authors, attenuated by listric normal faulting, while it is presumed that the lower crust of about 3 km thickness was attenuated by creep. Le Pichon and Sibuet (1980) tested the stretching model of McKenzie (1978) on the data of the Gulf of Biscay and conclude that "the amount of brittle stretching in the upper 8 km of continental crust reaches a maximum value of about 3 and is equal to the stretching required to thin the continental crust and presumably the lithosphere".

We conclude that listric normal faults involving the basement of passive margins can be mapped by reflection seismic surveys. So far the correlation and identification of

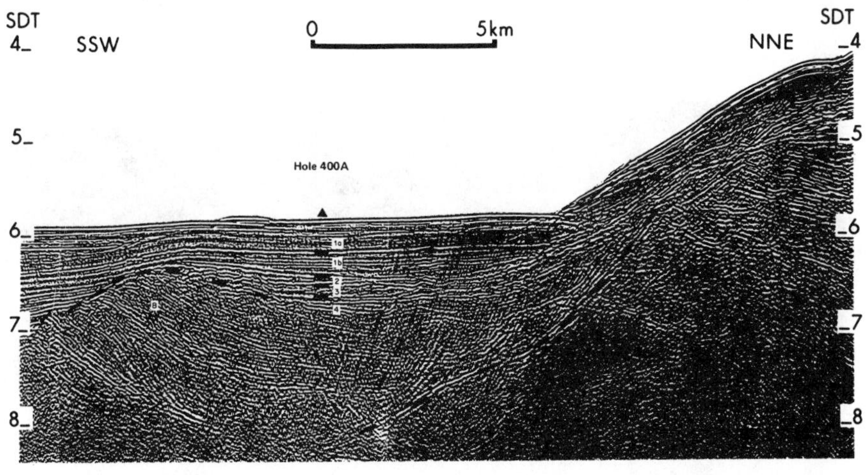

Figure 3
Northern Gulf of Biscay, section across IPOD Hole 400. The acoustic basement (B) consists of layered sedimentary rocks of probable Jurassic and early Mesozoic age. These formerly continuous layers were faulted and tilted during the early Cretaceous rifting phase. Vertical scale in seconds, two-way travel time. Unit. 1-A : Quaternary to late Pliocene ; 1-B : Pliocene to early Miocene, 2 : Miocene to late Paleocene ; 3 : Maastrichtian-Campanian ; 4 : late Albian, late Aptian. After Montadert et al., *1979 a.*

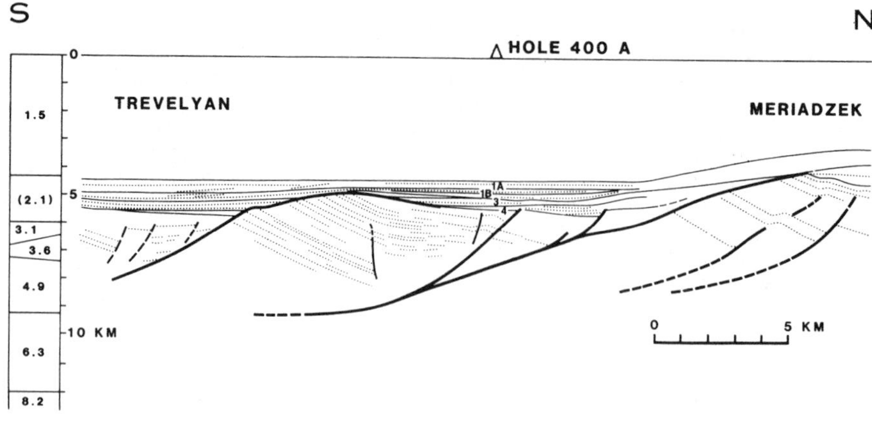

Figure 4

Northern Gulf of Biscay, line drawing of the section of Figure 3 converted to depth and migrated. Horizontal and vertical scale, the same. note listric normal fault bounding the tilted blocks. near the base of the listric faults is a horizontal reflector corresponding to the interface between 4.9 km/sec. and 6.3 km/sec. layers, as defined by seismic refraction (Avedik, Howard, 1979). Moho discontinuity is at 12 km. After Montadert et al., *1979 a.*

reflections between the tilted blocks has been elusive and, consequently, estimates of the amounts of stretching remain crude. It is hoped that in coming years more attempts will be made to regionally map faults within the basement of passive margins and to tie such work with refraction and wide angle reflection studies and calibration wells. Listric normal faults provide an elegant way for extension of the upper brittle crust, but the process which leads to the attenuation of the more ductile underlying lower crust and upper mantle will be more difficult to characterize and to document.

LISTRIC NORMAL FAULTS LIMITED TO DEFORMATION OF SEDIMENTS OF PASSIVE MARGINS

Thick sedimentary sequences overlie the rifted portions of many mature passive margins. These sequences subsided while the passive margin "drifted" away from the mid-ocean ridge. Structural deformation in these sedimentary sequences is often dominated by deformation along listric normal faults and by associated diapiric phenomena. Such deformation is spectacularly dramatized in areas of rapid deltaic deposition such as the Niger Delta (Delteil *et al.*, 1976 ; Weber, Daukoru, 1976 ; Lehner, de Ruiter, 1977) or the Gulf of Mexico, where mobile salt plays an additional role in the genesis of growth faults. In this paper we will limit ourselves to examples from the Gulf of Mexico.

The Gulf Coast Tertiary may be viewed as a thick deltaic clastic wedge prograding on a substratum of high pore pressure shales that in turn is underlain by a Mesozoic carbonate sequence deposited on an unstable basal evaporitic sequence. The crust of the Gulf of Mexico thins rapidly under the Gulf Coast ; where drilled on land — in areas of normal continental crustal thickness — a Paleozoic basement is observed, with evidence of superposed Pennsylvanian and Triassic rifting events. Overlying this basement, a lower set of listric normal faults and "roller" structures is associated with the basal evaporitic sequence, within which the listric faults appear to sole out. Such faults and structures are illustrated on Figures 7 and 8.

Another set of listric faults is associated with the Tertiary wedge. These are the well known Tertiary growth faults which have been illustrated by a number of authors from the Gulf Coast (e.g., Bruce, 1973 ; Busch, 1975 ; Roux, 1977 ; Bally, 1980). A net of more or less slope-parallel anastomosing growth fault systems affects all Cenozoic deposits of the Gulf. The shape of most master faults is listric, and they and their associated fault systems "sole out" or flatten within the overpressured shale section of the Gulf Coast. Hardin and Hardin (1961), Busch (1975), Curtis (1970 ; 1980), Curtis and Picou (1978) among others have described the interaction of sedimentation and growth faulting. The listric normal growth faults of the Gulf of Mexico often interact with diapiric structures that in some cases involve salt and in others, overpressured shales. Clearly, gravity tectonics dominate the scene.

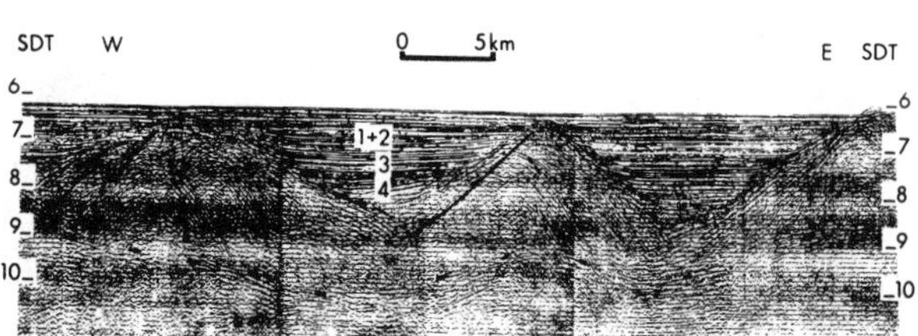

Figure 5

Northern Bay of Biscay, seismic profile south of Goban Spur showing tilted blocks with listric faults. Note the horizontal reflector below the tilted blocks. It is observed on the deepest part of the margin. On Trevelyan, it corresponds to the boundary between a 4.9 km/sec. layer and a 6.3 km/sec. layer only 3 km thick. The Moho is about 12.5 km below sea level. Profile CM 16 processed. After Montadert et al., *1979 a.*

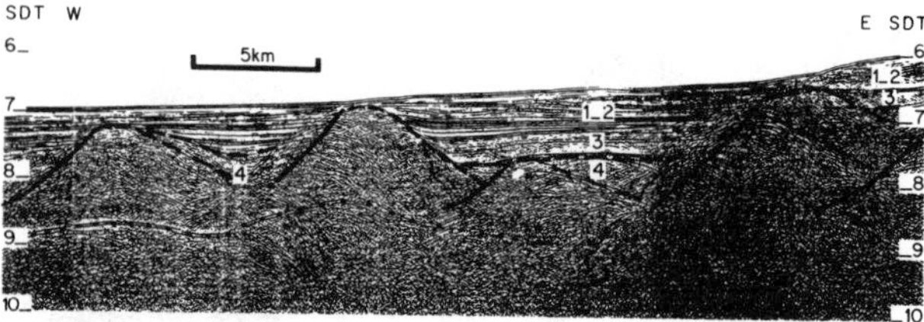

Figure 6

Seismic profile immediately west of Galicia Bank, showing tilted blocks and listric faults and a horizontal reflector below, as on Figure 4 Seismic formations 1-4 are the same as defined in northern Biscay. Profile IFP-CNEXO-CEPM, processed. After Montadert et al., *1979 a.*

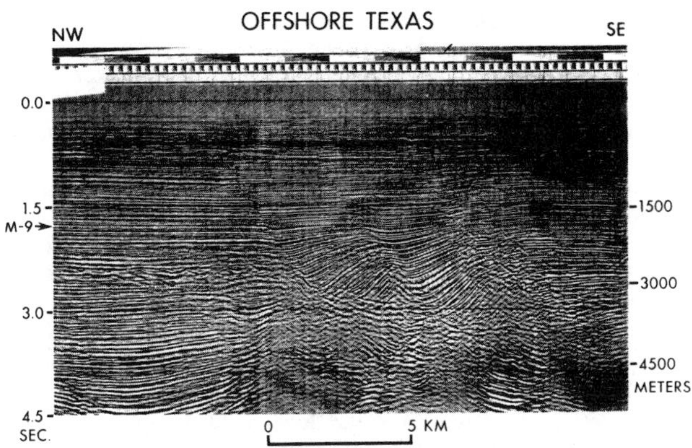

Figure 7
Offshore Texas, reflection seismic section. Note extensive listric normal growth faults in Miocene section. M indicates Miocene marker beds. After Bally, 1980.

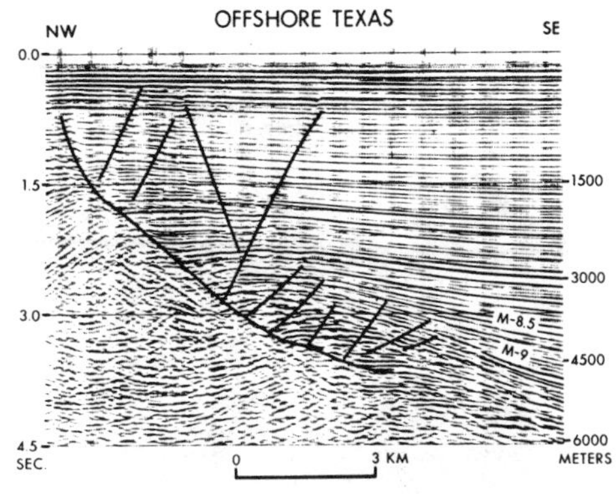

Figure 8
Offshore Texas, reflection seismic profile, showing the interaction of growth faults and diapiric structures. M-9 is a Miocene marker bed. After Bally, 1980.

Soft-sediment growth faults are widespread on many passive margins. Unfortunately, published documentation is less commonly available. While the growth fault systems associated with deltaic sequences always are spectacular, lesser but nevertheless obvious listric growth faults also occur in areas with smaller rates of sedimentation and areas that are not underlain by salt or thick shale sequences. Our undocumented suspicion is that in almost all cases one is looking at gravity tectonics in a shelf-slope setting with overpressured shales. In summary, it may be stated that soft-sediment listric normal growth faults are due to gravity tectonics on passive margins. They appear to be limited to the sedimentary section and do not involve the underlying basement. Such listric normal faults are often also associated with salt or clay diapirs.

Poor quality seismic data commonly do not permit tracing fault systems at depth, and thus the question whether a listric growth fault intercepts the basement or whether the fault is restricted to sediments cannot be answered until high quality reflection data are obtained. If the critical evidence is lacking, it may be tempting to infer basement controlled faulting (often with limited or no mappable growth) for the early rifting phase of a margin. On the other hand, soft-sediment listric growth faulting appears to be preferably associated with the subsidence and slope progradation of the subsiding "drifting" sequence.

The reader is reminded that listric normal growth faulting that also does not involve the basement occurs within the accretionary wedges associated with subduction zones (see example from Colombia by Beck and Lehner, 1975).

LATE SYNOROGENIC AND/OR POSTOROGENIC LISTRIC NORMAL FAULTS ASSOCIATED WITH FOLDED BELTS

Extensional faulting linked by strike-slip fault systems often fragment folded belts late, during, or soon after their tectogenesis. Complex intramontane sedimentary basins form in this manner (i.e., the Vienna basin, Grill *et al.*, 1968 ; Kroell, Wiesender, 1972 ; Kroell, Wessely, 1973 ; and Mahel, 1974 ; the Pannonian basin, Horvath, Stegena, 1977 ; and Sclater *et al.*, 1980 *a* and *b* ; and the wider Cenozoic Basin and Range province of the Western Cordillera of North America).

Listric normal faulting associated with the stretching and collapse of folded belts may well represent the rifting event that initiates continental separation and the formation of two opposing conjugate passive margins. Thus, some of the late Paleozoic basins and the widespread Triassic graben systems on both sides of the Central Atlantic appear related to post-collisional extension of the backbone of the late Paleozoic folded belts. Contrast them to the more isolated graben systems of the Red Sea or the African-South American conjugate passive margins, where a propagating crack causes the formation of rift systems that are superposed on a much older and stabilized Precambrian craton.

Here we will only discuss selected aspects of listric normal faulting in the Western Cordillera of North America, because there, although far from being satisfactory, the complex phenomenology of listric normal faulting appears to be somewhat better documented than in other folded belts.

Mid- and young Cenozoic extensional fault and strike-slip fault systems extend from Central British Columbia into the Basin and Range province of the US and Northern Mexico. Hamilton and Myers (1966) have provided an early and unusually farsighted synthesis of the Cenozoic tectonics of the US portion of the Cordillera. Davis and Burchfiel (1973) and Liggett and Ehrenspeck (1974), among others, have shown that normal fault-controlled horst and graben systems may be linked by strike-slip or transform fault zones (Fig. 9).

Extensive summaries of the geology and geophysics of the Basin and Range are given by Thompson and Burke (1974), Newman and Goode (1979), Smith and Eaton (1978), and Armentrout, Cole and Terbest (1979) ; more concise overviews are provided by Eaton (1979 ; 1980), Stewart (1978), Stewart and Carlson (1978), and Davis (1980).

In the western United States normal faults were first described by G. K. Gilbert (1875) as dominating the Cenozoic structure of the Great Basin area. It was not until 70 years later, however, that systematic changes in the dips

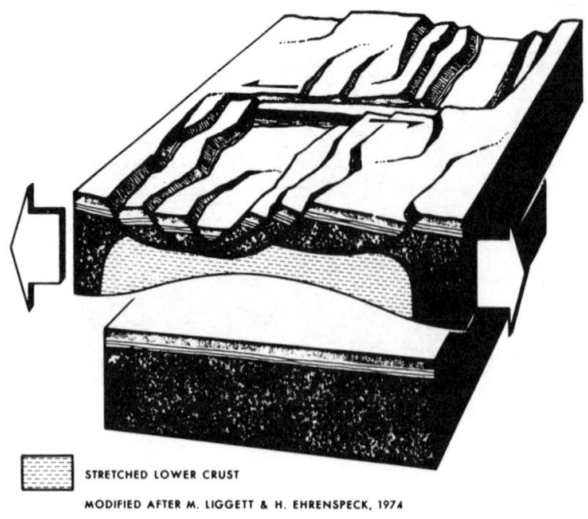

Figure 9
Sketch illustrating strike-slip faulting, listric normal faulting and crustal attenuation, modified after Liggett and Ehrenspeck, 1974.

of some basin and range normal faults with depth were recognized. Chester Longwell (1945) reported that normal faults in the Desert Range and Grand Wash Cliffs areas of southern Nevada had much lower dips (10 to 30°) than might be expected from their occurrence in the Basin and Range province. Furthermore, Longwell stated (1945, p. 111) that he could perceive an "upward-concavity of all fault surfaces that are exposed extensively enough for accurate appraisal of their form". Cautiously extrapolating from his observations in Southern Nevada, Longwell predicted (p. 117) that it would not be surprising to him if it was later found that many of the basin and range faults typically "represented in published sections as plane surfaces, are actually curved in cross-section". Little immediate attention was paid to Longwell's paper, but Moore (1960), Hamilton and Myers (1966), and Armstrong (1972) were among those who later stressed the probable importance to Great Basin geology of both low-angle and listric normal faults.

Confirmation of Longwell's prediction has been slow to come from conventional field studies in the Great Basin, primarily because of the geologist's inability to "see" to depth along the major range-front faults of the region, covered or masked as they are by basinal sedimentary fills. However, recent improvements in seismic reflection technology and increased seismic exploration by the petroleum industry of the Great Basin area have helped to document Longwell's insight into the geometry of basin and range faulting.

Surface geologic evidence for the existence of listric normal faults in the western United States has come from several kinds of observations, as amplified below — among them : 1) direct observations in non-basin areas, e.g., Longwell (1933 ; 1945) ; 2) the geometric interrelations between tilted or rotated hanging-wall strata and normal faults ; 3) subsurface geologic data from wells, boreholes, etc, e.g., Proffett (1977) ; and 4) interpretation of a curviplanar fault geometry at depth based on the arcuate geometry of many range-front faults as seen in plan view (Moore, 1960).

Hamblin (1965) measured a downward decrease in dip along normal faults at the western edge of the Colorado Plateau, where topographic relief is great. He noted that faults such as the Hurricane fault of Utah and Arizona were essentially vertical or dipped very steeply at the top of the plateau, but had average westward dips of 60° along the Colorado River in the Grand Canyon. Proffett (1977) cites rates of flattening at depth for the Hurricane fault of Hamblin (1965), the faults described by Longwell (1945), and faults studied by him in the Yerington area, Nevada, of 1.3-2°/100 m, 0.7-1.3°/100 m, and 0.3-0.7°/100 m, respectively. He concludes from the diverse geologic settings of these three examples that the rate decreases with increasing depth in the crust.

The reverse drag of hanging-wall strata into the Colorado Plateau border faults was explained by Hamblin (1965) as a geometric consequence of the downward decrease in dip of these faults. The tilting of hanging-wall strata into dip orientations opposite to that of bounding normal faults has sometimes been considered as evidence for a listric geometry of faulting, but Thompson (1971) has correctly noted that such rotation can also occur during simultaneous block faulting and tilting on *multiple* and sub-parallel planar faults. However, his mechanism obviously does not apply if footwall strata below the faults in question do not exhibit rotation. Perhaps the most convincing evidence for listric normal faulting, using the geometry of hanging-wall strata, comes from those instances of growth faulting where such strata display a progressive decrease in dip upwards through the stratigraphic section. Relations of this type have been described by Wallace (1979) in the West Humboldt Range and by Proffett (1977) in the Yerington area, both in Nevada, and by Frost (1979) in the Whipple Mountains, southeastern California, and the adjacent Aubrey Hills, Arizona. These authors all interpret dip relationships as indicating continued or intermittent sedimentation during rotational displacement of downthrown blocks along listric normal faults.

Proffett (1977) describes in detail and interprets a complex system of listric growth faults in which earlier, relatively steep listric normal fault planes are intersected by subsequent generations of listric normal faults that rotate the earlier faults into flat positions of much shallower dip. This process leads to an east-west extension of more than 100 percent, with a crustal thinning of the same order supported by seismic refraction data. Dips of tertiary hanging-wall fanglomerate in the West Humboldt Range are as high as 30 to 35° to the east. Overlying basalt caps eroded fanglomerate and dips 8-10° eastward ; still higher basaltic terrace gravels dip only 1 or 2° (Wallace, 1979). Southwestward-dipping strata in the Aubrey Hills, on the east bank of Lake Havasu between California and Arizona, are truncated at depth by the northeast-dipping Havasu Springs fault. E. Frost (pers. comm., 1979) reports that Oligo-Miocene hanging-wall strata dip as steeply as 85°, whereas Mio-Pliocene beds at the top of the progressively shallowing section dip only 15°.

Anderson (1971), Wright and Troxel (1973), and Davis *et al.*, (1979) have mapped the downward flattening of closely spaced normal faults in separate areas of the southern Cordillera (Great Basin, eastern Mohave Desert, Fig. 9), but their faults flatten into structurally shallow and subhorizontal detachment surfaces (or have been inferred to do so) and should not be confused with the major crustal breaks that outline the mountain ranges of the Great Basin. In fact, the relation of these shallow detachment surfaces to major range-front faults is one of the remaining enigmas of basin and range tectonics.

In recent years growing attention has been focused on such low-angle detachment faulting, particularly of the type associated with the metamorphic core complexes of the western United States (see G. H. Davis, Coney, 1979 ; and G. A. Davis, 1980). Metamorphic core complexes are described as domal uplifts of metamorphic rocks that are separated from overlying rocks by *décollement* zones of Tertiary age "marking strikingly sharp thermal and strain gradients" (G. H. Davis, Coney, 1979). The detachment zones appear to be formed in an extensional regime, and they are part of the general phenomenon of "younger over older thrust faults", or "denudational faults", that have been described for many years from Nevada (e.g., Misch, 1960 ; 1971). Armstrong (1972) has determined for the Snake Range of Nevada that these Tertiary extensional faults result from thinning of supracrustal rocks by listric normal faulting (denudational tectonics). An example of this type of faulting from the Whipple Mountains of southeastern California is diagrammatically shown in Figure 10.

The interpretation of the detachment fault complexes of the Cordillera is highly debated at the present time. G. H. Davis and Coney (1979) view them in the context of the ductile stretching of the basement (megaboudinage), with the layered cover flowing passively during metamorphism and, at higher structural levels, exhibiting brittle detachments and associated normal faulting. G. A. Davis *et al.* (1979) from studies in the lower Colorado River area, rule out the possibility there that detachment faulting occurred synchronously with metamorphic and mylonitic flow in underlying basement rocks. This conclusion is based on geochronologic and geologic evidence that shows that regional mylonization of lower-plate rocks significantly predated the Miocene detachment faults, and that the major detachment surface discordantly overlies underlying tilted, folded and brittley deformed mylonitic gneisses. In fact, these gneisses were locally eroded prior to the Oligocene (?) and Miocene deposition of Tertiary sediments later affected by normal and low-angle faulting. Clasts of what are now lower-plate mylonitic rocks occur throughout the allochthonous Oligocene (?)-Miocene fanglomerates of the area. Profound North-East directed extension within upper-plate rocks was unaccompanied in the Whipple Mountains region by coeval extension in lower-plate rocks. This relation raises the possibility that the Whipple allochthon is an upper crustal landslide sheet moved northeastward under the influence of gravity, or, alternatively, that lower-plate extension commensurate with that observed in the allochthon occurred by unknown mechanisms in areas northeast of those studied to date.

It follows from the previous descriptions that surface geologic observations at best provide only indirect information on the geometry of the normal faults of the Basin and Range province and their possible relation to outcropping "younger over older" *décollement* zones. Proffett (1977) projected surface information into the subsurface using shallow boreholes. Ultimately, however, only reflection seismic studies may help to document the fault configurations controlling the Tertiary basins of the Western Cordillera. The first reflection seismic evidence for postorogenic listric normal faulting in the Western Cordillera was given by Bally *et al.* (1966). The Flathead normal fault of southeastern British Columbia appears to offset the Lewis overthrust by more than 6 km (down to the West). Reflection data do not permit mapping the fault itself ; however, a continuous reflection from the Mesozoic at shallow depths precludes a steep over-all attitude of the fault and demands

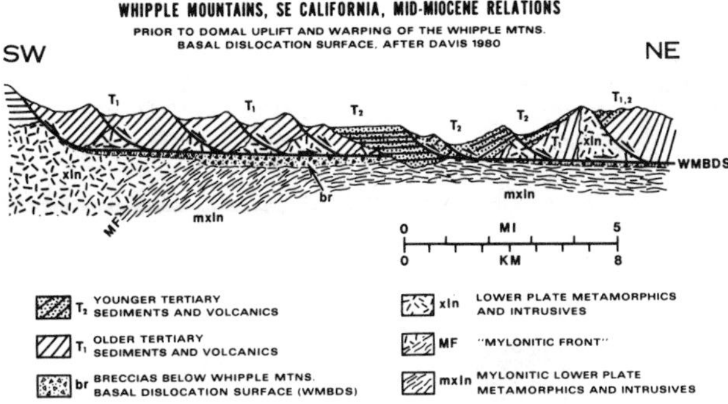

Figure 10
Diagrammatic cross-section across the Whipple Mountains, southeastern California, illustrating Middle Miocene geological relations prior to domal uplift and warping of the Whipple Mountains basal detachment surface (WMBDS). The cross-section illustrates two phases of rotational normal fault displacement along the basal detachment surface. T_1 : older Tertiary sedimentary and volcanic rocks ; T_2 : younger Tertiary sedimentary and volcanic rocks deposited across the detachment surface prior to their involvement in renewed fault displacement ; MF is a "mylonitic front", the abrupt non-fault contact between undifferentiated lower plate metamorphic and intrusive rocks (xln) and their largely mylonitic equivalents (mxln) ; br : breccias developed below the basal detachment surface. From Davis et al., *1979.*

an interpretation as a listric normal fault that probably merges in — and causes "backslippage" of — the older Lewis overthrust. Based on much poorer reflection seismic data, Bally *et al.* (1966) also postulated that the southern Rocky Mountain trench was part of a system of listric normal faults that separated eastward tilted blocks in Precambrian-Beltian sediments. Similar listric normal fault systems may also be responsible for the Tertiary basins of northern Montana. Note that in all examples from southeastern British Columbia, the listric normal faults are inferred to merge with pre-existing thrust faults and into a major solefault system that presumably overlies the crystalline basement (see also Myers, Hamilton, 1964).

More examples of reflection lines suggesting listric normal faults merging into and controlled by pre-existing thrust faults are offered by Royse *et al.*. (1975) from the Wyoming fold belt. An unusual set of lines (Fig. 11) from the Sevier Desert of Central Utah was published by MacDonald (1976). With the exception of one well, these sections are not calibrated, and the author offers only a limited interpretation of the data. Nevertheless, the sections convincingly show a reflection from a detachment zone, into which the overlying normal faults flatten. Whether the flat *décollement* zone represents an earlier overthrust — as the author suggests — that was used by the late listric fault system, or whether the whole system is only postorogenic in origin cannot be determined from the data.

Effimoff and Pinezich (1980) published reflection seismic sections from northeastern Nevada (e.g., Fig. 12 *a, b* and 13 *a, b*). Their sections display half-grabens tilted toward the east and bounded on the east by a major listric normal fault system. The flat sole of the listric fault is only rarely displayed by a reflection, but the listric nature of the fault is supported by the rotation of hanging-wall Tertiary beds into the fault plane.

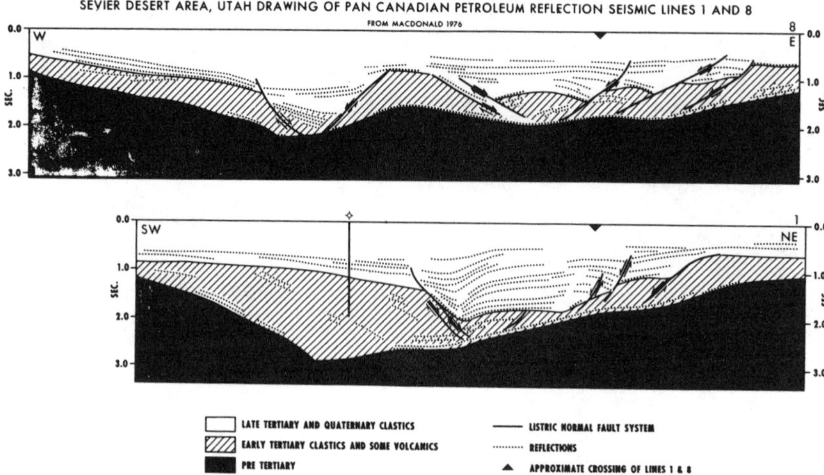

Figure 11
Drawing of seismic line by Pan Canadian Petroleum across the Sevier Desert in Utah, modified after an interpretation published by MacDonald, 1976.

Typically, the lower portions of the sections appear to be more intensely faulted, suggesting an early phase of intensive faulting followed by later tilting into the main fault system. In the mountains adjacent to the seismic lines shown by Effimoff and Pinezich, major low angle *décollement* zones (denudational faults) occur that separate chaotically disturbed Paleozoic strata from underlying less disturbed Paleozoic or Precambrian beds. The seismic lines of Effimoff and Pinezich suggest that the outcropping shallow *décollement* systems of the mountains may merge into the listric normal fault systems that underlie the adjacent basins. The relation may be far from simple, because the shallow surface *décollement* often appears to be offset by the steeper listric boundary fault of the valley. Thus, the surface *décollement* zones may well be the outcropping soles of earlier, originally deeper, listric normal fault systems that were uplifted and later modified by landslide-like gravity tectonics.

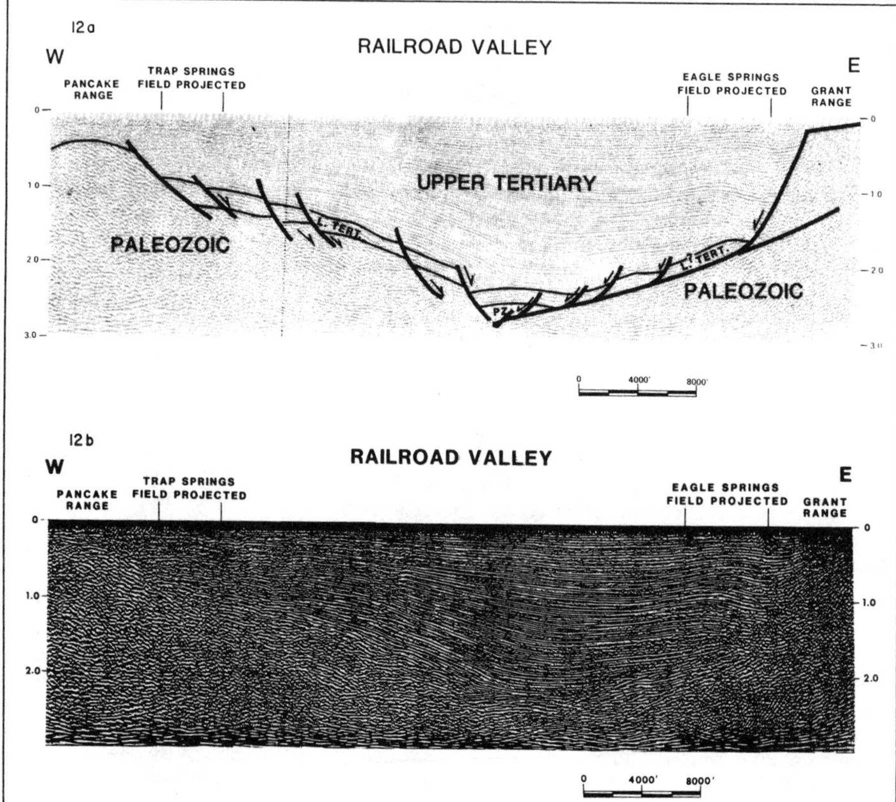

Figure 12
Seismic (migrated) profile across Railroad Valley, Nevada. a : interpreted line ; b : raw data of the same line. Effimoff and Pinezich, 1980.

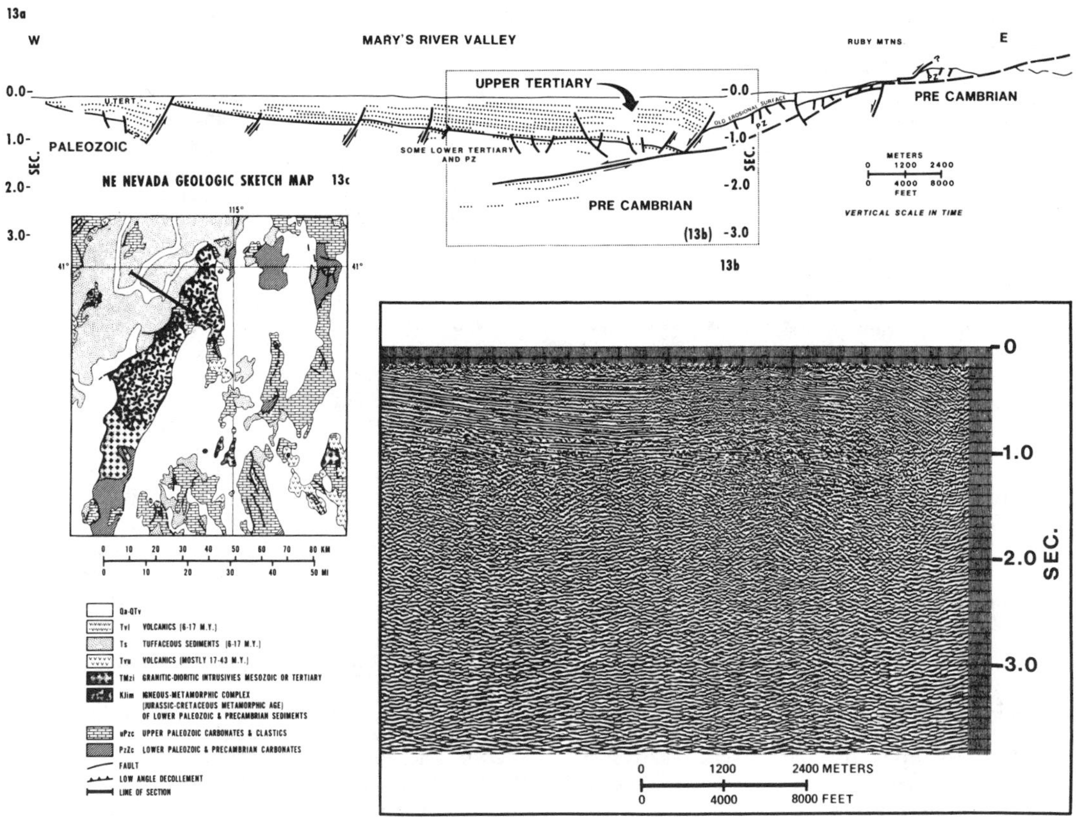

Figure 13
a : drawing of a seismic profile migrated across Mary's River Valley. Published by Effimoff and Pinezich, 1980. b : closeup of eastern segment of seismic line showing presumed continuation at depth of surface décollement *or Ruby Range. After Effimoff and Pinezich, 1980.*

The estimates for extension in the Basin and Range made by different authors range from 10 to 100% (for a review, see Davis, 1980) and are based mainly on judgments derived from surface geology. Although the estimates are probably in the right order, better estimates could be obtained from regional reflection seismic surveys. According to different authors, the depth of normal faulting or their "soling out" occurs in the range of 5-17 km. A preferred interpretation is that extension in the brittle upper crust in taken up by ductile stretching in the lower crust (Thompson, 1959 ; 1966 ; Hamilton, Myers, 1966 ; Stewart, 1971 ; 1978 ; Proffett, 1977). The transition between brittle and ductile crust may well be represented by a crustal low velocity zone which in the eastern Great Basin occurs between 5 and 15 km depth (Smith, 1978 ; and Eaton, 1980).

We conclude that large areas within the Cordillera of North America, from British Columbia to Mexico, have been extended by evolving systems of listric normal faults, linked to each other by strike-slip-transform faults or diffuse zones of transcurrent strain. These fault systems are responsible for the formation of basins initiated in late Paleogene time which, however, ceased to be active at different times in different segments of the Cordillera. The amount of stretching suggested by the normal faults is probably accompanied by corresponding ductile extension in the lower crust.

Most authors have emphasized the stretching aspects of the Basin and Range province. Geomorphological evidence, as well as the uplift of deep-seated metamorphic rocks during the Tertiary, suggest that the crust of the western Cordillera may also have been arched between the Interior Plains and the Sierra Nevada. In other words, it has become important to differentiate the effects of super-regional — possibly thermal — arching from the structural and thermal consequences of simple lithospheric stretching.

LISTRIC NORMAL FAULTS IN REMNANTS OF ANCIENT PASSIVE MARGINS OCCURRING IN DEFORMED FOLD BELTS

Miogeosynclinal sequences in folded belts are today generally recognized as remnants of former passive margins caught in the collisional drama of mountain building. Within this context, it is particularly important to develop criteria that allow differentiation of the following :

— listric normal faulting related to the rifting phase preceding the opening of an ocean. Such normal faults may be postorogenic faults of an earlier cycle, or else directly related to thermal events that affected much older crust ;

— soft-sediment listric normal faulting often related to the drifting phase preceding orogenic deformation : this is suggested by evidence from Atlantic-type passive margins, but has not yet been described from ancient folded belts ;

— late orogenic to postorogenic listric normal faulting imposed on the deformed passive margin sequences after they were deformed and incorporated into the folded belt (i.e., the basin and range faulting previously described).

To illustrate some of the problems, we review examples from the Alps. That the preorogenic Mesozoic evolution of the Alpine "geosyncline" was dominated by tensional block-faulting rather than by "embryonic" (Argand, 1916) compressional movements became clear in the late fifties and early sixties (e.g., Trümpy, 1960 ; 1975 ; Schindler, 1959). Particularly, the importance of normal faulting in the formation of early Jurassic basins was established by a wealth of sedimentological and paleotectonic observations (Wiedenmayer, 1963 ; Bernoulli, 1964) and has since been demonstrated in various regions of the Alpine-Mediterranean belt. The interpretation of ophiolites as remnants of Mesozoic oceanic lithosphere and an understanding of their significance for palinspastic restorations led to the recognition of former continental margins (Laubscher, 1969), and the early Jurassic period of normal faulting was identified with a phase of rifting preceding spreading in the Tethyan ocean (Bernoulli, Jenkyns, 1974).

The occurrence of early faulting in the Alpine belt is often inferred from circumstantial evidence. Where the existence of pre-Alpine faults is firmly established, their original geometry is usually more or less disturbed by later orogenic movements. In most cases the faults were reactivated during Alpine orogeny and *décollement* planes of sedimentary and shallow basement nappes may also have followed pre-existing listric surfaces. Particularly, the most distal areas of continental margins and their passage zone to oceanic crust have been the site of decoupling and most intense deformation, and documentation of the early history is most fragmentary in these areas.

Listric faulting as a mechanism for crustal thinning is inferred, usually in an intuitive manner, in many palinspastic reconstructions of Tethyan continental margins (e.g., Bosellini, 1973 ; Sturani, 1973). Helwig (1976), using Lowell's and Genik's (1972) evolutionary model of the Red Sea as a geotectonic model for the Triassic-Jurassic Alpine Tethys, has also postulated the occurrence of listric faulting to account for preorogenic crustal thinning required to maintain material balance during crustal shortening without subduction of continental crust Whether or not we admit lithospheric subduction of sometimes already attenuated continental crust (Bally, Snelson, 1980), a mechanism is required to account for the isostatic subsidence of continental margins that formed before their tectogenic involvement in mountain building.

The paleotectonic evolution of many Tethyan margins closely parallels that of the undeformed rifted margins of the Bay of Biscay (de Charpal *et al.*, 1978 ; Montadert *et al.*, 1979 *a* and *b*) and of Iberia (Groupe Galice, 1979 ; Graciansky *et al.*, 1979). In contrast to many rift systems that developed on an ancient subaerial surface (e.g., Basin and Range) rifting occurred in both areas in a pre-existing marine Mesozoic basin — in the case of the central and eastern Tethys discordantly across thick carbonate sequences that were marginal to the earlier, Triassic, Paleotethys ocean (Laubscher, Bernoulli, 1977). As a consequence, there are hardly any siliciclastic sediments associated with the early Jurassic phase of rifting, and evaporite deposits of Jurassic age are conspicuously lacking along the rift zone. Traces of volcanic activity are also extremely scarce along the Jurassic passive margins of the Tethys.

Inside the Alpine belt of the Mediterranean area, the southern Alps of northern Italy probably preserve the most complete and undisturbed record of a Jurassic to early Cretaceous passive continental margin of the Tethyan ocean. The Mesozoic sequences of the area are deformed by folding and some minor thrusts ; however, there are no large-scale nappe structures hampering paleotectonic restorations (Fig. 14 *a*). Particularly, the signature of north-south-trending Jurassic fault scarps is clearly recognizable across the east-west-striking Alpine structural grain.

Rifting in the South Alpine margin began, after precursory movements in the Triassic, during the early Jurassic. The synsedimentary nature of normal faults is clearly established by rapid changes of facies and formational thickness of the syn-rift sediments across the fault zones and by the existence of pronounced fault scarps that were the source areas for gravity flow deposits and carbonate turbidites in

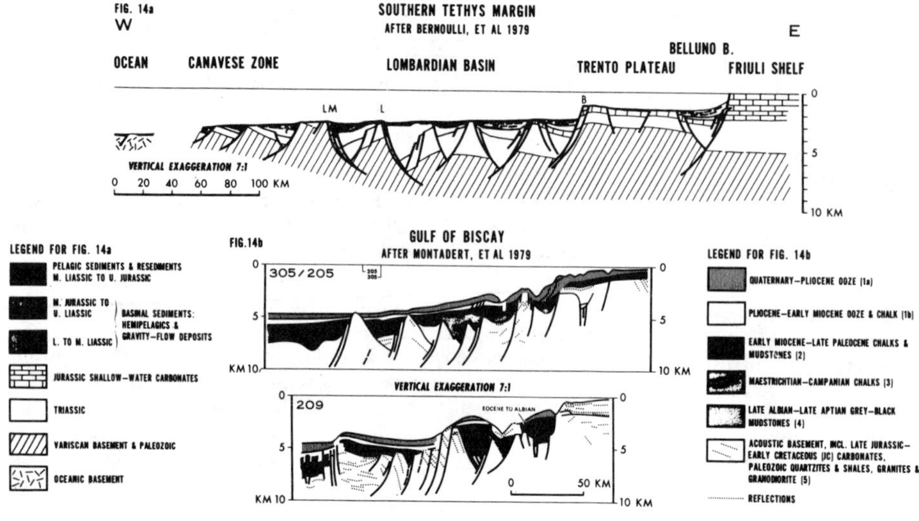

Figure 14

a : Palinspastic cross-section through the southern continental margin of the Tethys in the late Jurassic. After Bernoulli et al., *1979 b, modified. LM : Lago Maggiore fault ; MN : Monte Nudo trough ; L : Lugano fault ; G : Generoso trough ; B : Ballino-Garda escarpment. Vertical exaggeration, approximately 7 times. b : Depth sections across the north Biscay margin, based on seismic reflection sections. For explanation of numbers, see Figure 3. Vertical exaggeration, approximately 7 times.*

the adjacent basins (Bernoulli, 1964 ; Castellarin, 1972). Subsidence rates were highest during this early phase of disintegration of the margin and varied widely between the different fault blocks. Some of the blocks that were submerged only in the course of the late Liassic to early Middle Jurassic became submarine highs and plateaus (e.g., Trento plateau), where only limited amounts of pelagic sediments accumulated. With the onset of spreading and the formation of oceanic crust in the Ligurian-Piemontese ocean in the late early to middle Jurassic, subsidence rates decreased and were more evenly distributed over the margin.

Thus, through time the submerged distal continental margin (Lombardian zone) became increasingly starved, and only pelagic sediments were slowly deposited, their facies being determined mainly by prolonged subsidence, increasing water depth and basin-wide paleoenvironmental changes (Bernoulli *et al.*, 1979 *a* ; Bossellini, 1980).

A fruitful analogy is offered by the examples of the Gulf of Biscay (Fig. 14 *b*).

Evidence for *listric* faulting during the early breakdown of the margin can only be inferred and comes mainly from the sedimentary evolution of the margin. In the Lombardian basin, the syn-rift sediments are essentially composed of well-bedded, current-deposited spongolithic cherty limestones and associated gravity flow deposits (Lombardian siliceous limestones, Medolo group) deposited in fault-bounded troughs. Tilting of fault blocks as a consequence of listric faulting is suggested by the asymmetry of certain troughs reflected by the pattern of formational thicknesses. Deposition rates match approximately the rates of differential subsidence and results in approximately horizontal layering at the end of the rifting phase (compare Fig. 14 *a* from Bernoulli *et al.*, 1979 *b*, with Fig. 14 *b* from Montadert *et al.*, 1979 *a*). There is no unconformity at the base of the basinal syn-rift sequences, but lensing out of packages of strata and local unconformities are ubiquitous within the sequence. Locally, in outcrops, stacks of strata are observed that have been rotated along synsedimentary listric growth faults (Fig. 15). The formation of tilted fault blocks during sedimentation is also suggested by troughs and half-grabens bounded along one side by steep fault scarps, documented by coarse proximal resediments, and by a much smoother topography along the other side (e.g., the middle to late Liassic Monte Nudo basin, Fig. 14 *a*, and Kälin, Trümpy, 1977).

On top of the tilted blocks, particular facies of shallower water are observed in the lower Jurassic (e.g., Broccatello formation of the Lugano swell). These shallow-water areas were restricted in size and characterized by early subaerial and later submarine erosion, and by local angular unconformities (Wiedenmayer, 1963 ; Casati, Gaetani, 1968). Particularly spectacular outcrops of unconformities that may illustrate tilting associated with the rifting phase of the southern Tethys have been exquisitely illustrated by H. Eugster (Fig. 16, and Cadisch *et al.*, 1968) from the Engadine dolomites of Switzerland.

The depositional geometry of the syn-rift sediments of the southern Alps compares well with that of the corresponding formations of the Iberian (Formation 4, Groupe Galice, 1979) and of the Armorican margin (Montadert *et al.*, 1979 *a* and *b*). In the southern Alps, the early Jurassic basins measured some 25 to 40 km across ; this is in accordance with the observations along the Iberian and Armorican margins, where fault blocks from a few to 30 km across are observed. Likewise, the throw of some individual fault zones is on the same order with a maximum of 3 to 4 km ; this corresponds to the throw reconstructed for the early Jurassic Lugano fault (Bernoulli, 1964, Fig. 18, our 14 *a*). In the Iberian and Armorican margins, a polarity of the listric faults towards the axis of the rift zone is observed. In the southern Alps, it appears that rifting started in the central zone of the Lombardian basin with the step-wise foundering of new fault-blocks to the east and west during the Middle to Late Liassic (Fig. 14 *a*). In the Late Liassic to Middle Jurassic, the axis of rifting was finally shifted somewhat, and break-up and spreading occurred a hundred kilometers to the north and west.

In contrast to undeformed passive margins, the geometry of the presumably listric faults cannot be deciphered in the southern Alps. However, the size and aerial extension of the larger fault blocks suggests that the major fault zones

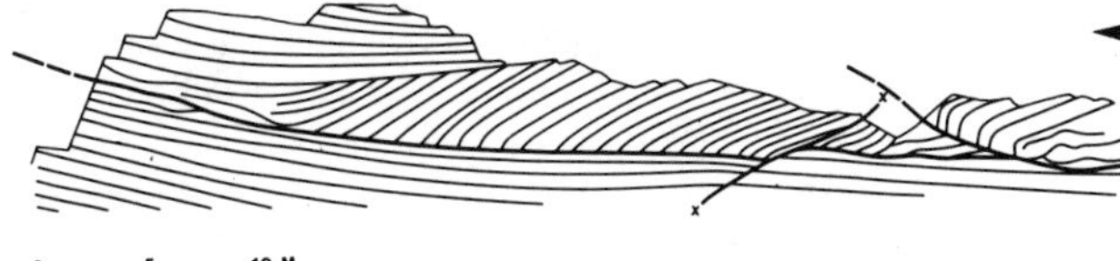

Figure 15
Synsedimentary growth fault in syn-rift sediments, Generoso trough. Lombardian basin. Lombardian siliceous limestones, lower Liassic, road from Caneggio to Bruzella, southern Switzerland. From Bernoulli, 1964.

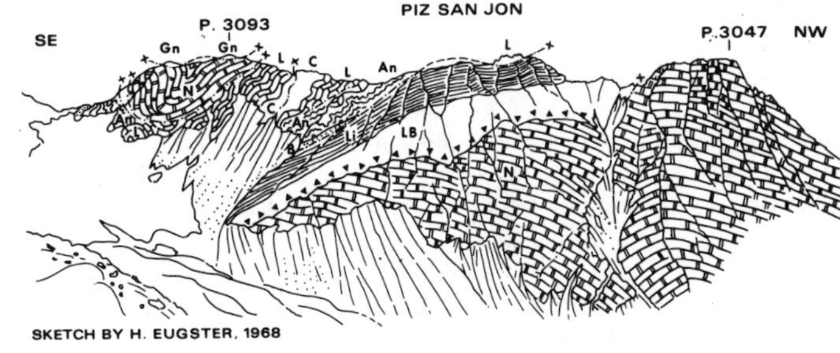

Figure 16
Piz San Jon (Lower Engadin, Switzerland) seen from the east, as sketched by H. Eugster (see Cadisch et al., *1968). Note the pronounced angular unconformity, overlain by Liassic breccias. The dip of the Norian dolomites may be due to rotation of beds along listric faults associated with the rifting phase of the southern margin of the Tethys. Gn : Gneiss of the Oetztal nappe ; Li : Liassic shales ; LB : Liassic breccias overlying angular unconformity ; N : Norian Hauptdolomite ; C : Carnian Raibler schichten ; L : Ladinian dolomites ; An : Anisian Muschelkalk ; B : Permowerfenian Buntsandstein ; x : thrust faults.*

sole within the pre-Triassic basement. In the case of the Lugano fault (Fig. 14 *a*), its present day geometry could indicate a flattening of the fault zone with depth. This north-south-trending fault zone cannot be followed directly into the underlying crystalline basement, but merges with an east-west-trending fault zone that cuts obliquely through Triassic and shallow crystalline basement rocks (see Spicher, 1980).

From middle Liassic to middle Jurassic times, the topography created by rifting in the Lombardian basin was nearly levelled by basinal, hemipelagic and gravity-flow deposits. Only along the platform margin of the Friuli shelf and along the escarpment bounding the Trento platform to the west did important submarine scarps persist throughout the Jurassic and the Cretaceous. In the Lombardian basin, however, local stratigraphic gaps, condensed facies and restricted areas of shallower pelagic facies suggest the persistence of a subdued morphology during post-rift regional subsidence.

The post-rift history of the Tethyan continental margins is characterized by prolonged subsidence probably consistent with exponential thermal decay (Bernoulli *et al.*, 1979 *a* ; Winterer, Bosellini, 1980) as observed in undeformed continental margins (Montadert *et al.*, 1979 *a* and *b*). Water depth at the end of rifting in the Lombardian zone was in the order of 1.000 m or more and about 2.500 m at the end of the Jurassic (Winterer, Bosellini, 1980). During the early Cretaceous the distal continental margin sank to a few kilometers water depth, as shown by the encroachment of deep oceanic facies onto the most oceanward parts of the continental margin. The occurrence of the early Cretaceous black shale formations, believed to be deposited below calcite compensation depth, in tectonic units derived from the most distal parts of the continental margin (Scisti di Levone, Canavese Zone ; Palombini in lowermost Austroalpine nappes ; Trümpy, 1975) suggest water depth in the order of 4 km. Isostatic sinking of the starved distal continental margin to this depth strongly suggests crustal thinning and cooling ; there are, however, so far only limited petrologic data to support this. Perhaps the occurrence of potassium-argon ages of biotite of 170 to 180 Ma in the lower crust of the southern Alps (McDowell, Schmid, 1968) could suggest a phase of crustal thinning and cooling during an early Jurassic phase of rifting. Also, the postulated pre-Alpine high position of the ultramafic rocks of the Ivrea zone and the extreme thinness of some Alpine basement nappes (e.g., Carungas nappe, lower Austroalpine) could possibly be explained in terms of Mesozoic crustal thinning, controlled by listric normal faults.

The paleotectonic evolution outlined for the southern Alps is in no way unique for continental margins of the Tethyan ocean. Although the geological documentation is in most cases very fragmentary, the sedimentary evolution suggests a history of rifting and subsidence (e.g., Bernoulli *et al.*, 1979 *a* ; Graciansky *et al.* ; 1979).

CONCLUSIONS

We have reviewed selected examples of listric normal faults to emphasize the importance of this fault type in the formation of passive margins and orogenic systems.

The documentation of listric normal faults is often elusive. Only reflection seismic data permit either mapping directly or else inferring more or less convincingly the faults themselves. On the other hand, surface data — and in some cases subsurface data based on wells or else from mines — provide a much more detailed perspective of the interaction of sedimentation with the formation of listric normal fault systems. Much work needs to be done to reconcile surface data with reflection seismic data and to come up with satisfactory documentation. It is particularly important to obtain better data on the flat soles of listric normal faults. Only rarely (as in the Bay be Biscay and the Galicia banks) are these displayed on seismic lines. The inference that the *décollement* zones of the metamorphic core complexes of the North American Cordillera may in some way represent the soles of listric normal fault systems that were uplifted to surface is still debated, but is clearly valid in some ranges within the orogen. The problem in the Cordillera is further complicated by the possible reactivation of older thrust fault sole systems by later listric normal faults ("backslippage").

Listric normal fault systems are probably caused by stretching of the upper brittle crust, and they appear to be linked by strike-slip or transform-type faults. The suggestion is that listric normal fault systems tend to bottom out in the middle of an attenuated continental crust (say, 10 km), in a position that may well coincide with some of the low velocity layers that have been reported by seismologists (Mueller, 1977).

The amount of stretching of the upper brittle crust associated with listric normal faulting probably cannot be deduced with any accuracy from extension observed at or near the surface unless regional seismic reflection data and/or drill holes permit linking the surface observation with the deeper subsurface. Rough estimates made in areas such as the Bay of Biscay or the Basin and Range suggest that the stretching of the upper brittle crust had to be accompanied by ductile stretching of the underlying basement to explain the crustal attenuation actually observed on limited refraction data. While it is theoretically appealing to postulate such stretching (or else possible creep) in the lower crust, it is much more difficult to document its occurrence. Such documentation would have to be in the form of geophysical measurements that unambiguously describe the rock properties of the lower crust (and maybe the upper mantle) in an environment that is today under extension. An entirely different type of documentation could come from outcrop studies of the now allochthonous crust originally underlying passive margins that today are involved in folded belts. Such studies would need to date basement fabrics that were generated during the rifting-stretching phase of the former passive margins. Prime candidates for such studies are the Eastern Alpine thrust sheets and the large crystalline thrust sheets of the Himalayas.

In all cases future studies focusing on crustal attenuation and listric normal faulting will have to be based on the integration of geophysical observations with surface and subsurface geological data.

Acknowledgements

Field studies by G. Davis were supported by NSF Earth Science Grants EAR 77-09695 and GA-43309. O. Kälin cooperated in the restoration of the Alpine section, and the Swiss National Science Foundation supported the work of D. Bernoulli. The authors thank the French Petroleum Institute and Shell Oil for support in the preparation of the manuscript. We thank particularly H. Scott for preparing a number of the illustrations and J. Cartwright for typing the manuscript.

REFERENCES

Anderson R. E., 1971. Thin-skinned distension in Tertiary rocks of southeastern Nevada, *Geol. Soc. Am. Bull.*, **82,** 43-58.

Argand E., 1916. Sur l'arc des Alpes occidentales, *Eclogae Geol. Helv.*, **14,** 145-191.

Armentrout J. M., Cole M. R., Terbest H., 1979. *Cenozoic paleogeography of the Western United States, Pacific Coast Symposium* 3, Pac. Sec. of Soc. Econ. Paleo. Min., Los Angeles.

Armstrong R. L., 1972. Low-angle (denudation) faults, hinterland of the Sevier orogenic belt, eastern and western Utah, *Geol. Soc. Am. Bull.*, **83,** 1729-1754.

Avedik F., Howard D., 1979. Preliminary results of a seismic refraction study in the Meriadzek-Trevelyan area, Bay of Biscay, in : *Initial reports of the Deep Sea Drilling Project*, edited by L. Montadert and D. G. Roberts, **48,** 1015-1024, US Government Printing Office, Washington.

Bally A. W., 1980. Thoughts on the tectonics of folded belts, *Geol. Soc. London Spec. Pub.*, **9,** 20 p.

Bally A. W., Snelson S., 1980. Realms of subsidence, *Can Soc. Pet. Geol. Mem.*, **6,** 9-94.

Bally A. W., Gordy P. L., Stewart G. A., 1966. Structure, seismic data, and orogenic evolution of Southern Canadian Rocky Mountains, *Can. Pet. Geol. Bull.*, **14,** 337-381.

Beck R. H., Lehner P., 1974. Oceans, new frontier in exploration, *Am. Assoc. Pet. Geol. Bull.*, **58,** 376-395.

Beck R. H., Lehner P., with collab. of **Diebold P., Bakker G., Doust H.,** 1975. New geophysical data on key problems on global tectonics, *Proc. Ninth World Petrol. Congress, Tokyo, Applied Science, Essex*, **2,** 3-17.

Bernoulli D., 1964. Zur Geologie des Monte Generoso (Lombardische Alpen), *Beitr. Geol. Karte Schweiz.*, N.S., **118,** 1-134.

Bernoulli D., Jenkyns H. C., 1974. Alpine, Mediterranean and central Atlantic Mesozoic facies in relation to the early evolution of the Tethys, in : *Modern and Ancient geosynclinal sedimentation*, edited by R. H. Dott and R. H. Shaver, *Soc. Econ. Paleo. Min., Spec. Pub.*, **19,** 129-160.

Bernoulli D., Kälin O., Patacca E., 1979 *a*. A sunken continental margin of the Mesozoic Tethys : the northern and central Apennines, *Assoc. Sedimentol. Fr., Pub. Spec.*, **1,** 197-210.

Bernoulli D., Caron C., Homewood P., Kälin O., Van Stuijvenberg J., 1979 *b*. Evolution of continental margins in the Alps, *Schweiz Mineral. Petrogr. Mitt.*, **59,** 165-170.

Bosellini A., 1973. Modello geodinamico e paleotettonico delle Alpi meridionali durante il Giurassico-Cretacico, *Accad. Naz. Lincei, anno 370, Quaderno N.*, **183,** 163-205.

Bouroz A., Stievenard M., 1958 (fide **Crans** *et al.*, 1980). La structure de gisement des charbons gras du Pas-de-Calais et la notion de faille Renmaux, *Soc. Géol. du Nord, Annales* **LXXVII,** 146 *et seq.*

Bruce C. H., 1973. Pressured shale and related sediment deformation-mechanisms for development of regional contemporaneous faults, *Am. Assoc. Pet. Geol. Bull.*, **57,** 878-886.

Busch D. A., 1975. Influence of growth faulting on sedimentation and prospect evaluation, *Am. Assoc. Pet. Geol. Bull.*, **59,** 217-230.

Cadisch J., Eugster H., Wenk E., 1968. *Erläuterungen Blatt Scuol-Schuls-Tarasp Schw. Geol. Kommission*, Kümmerly and Frey AG, Bern.

Casati P., Gaetani M., 1968. Lacune nel Triassico Superiore e nel Giurassico del Canto Alto-Monte di nese (Prealpi Bergamasche), *Boll. Soc. Geol. Ital.*, **87,** 719-731.

Castellarin A., 1972. Evoluzione paleotettonica sinsedimentaria del limite tra piattaforma veneta e bacino lombardo al nord di Riva del Garda, *G. Geol.*, **38,** 11-212.

de Charpal O., Guennoc P., Montadert L., Roberts D. G., 1978. Rifting, crustal attenuation and subsidence in the Bay of Biscay, *Nature*, **275,** 5682, 706-711.

Christie P. A. F., Sclater J. G., 1980. An extensional origin for the Buchan and Witchground Graben in the North Sea, *Nature*, **283,** 729-732.

Cloos E., 1968. Experimental analysis of Gulf Coast fracture patterns, *Am. Assoc. Pet. Geol. Bull.*, **52,** 420-444.

Cloos H., 1936. *Einführung in die Geologie*, Bornträger, Berlin, 503 p.

Crans W., Mandl G., Haremboure J., 1980. On the theory of growth faulting : a geomechanical delta model based on gravity sliding, *J. Pet. Geol.*, **2,** 3, 265-307.

Curtis D. M., 1970. Miocene deltaic sedimentation, Louisiana Gulf Coast, in : *Deltaic sedimentation Modern and Ancient*, edited by James P. Morgan, *Soc. Econ. Paleontol. Min., Spec. Pub.*, **15,** 293-308.

Curtis D. M., in press. Sources of oils in Gulf Coast Tertiary, *J. Sediment. Pet.*

Curtis D. M., Picou E. B. Jr., 1978. Gulf Coast Cenozoic : a model for the application of stratigraphic concepts to exploration, on passive margins, *Trans. Gulf Coast Assoc. Geol. Soc.*, **XXVIII,** 103-120.

Davis G. A., 1980. Problems of intraplate extensional tectonics, Western United States, in : *Continental tectonics, studies in geophysics*, Natl. Acad. Sci., 84-93 ; and 1979, in Newman G. W. and Goode H. D.

Davis G. A., Burchfiel B. C., 1973. Garlock Fault : an intracontinental transform structure, southern California, *Geol. Soc. Am. Bull.*, **84,** 1407-1422.

Davis G. A., Anderson J. L., Frost E. G., 1979. Regional Miocene detachment faulting and Early Tertiary (?) mylonitic terranes in the Colorado River trough, southeastern California and western Arizona, in : *Geological excursions in the Southern California area*, edited by P. L. Abbott, San Diego State University, Dept. of Geological Sciences, 73-108.

Davis G. H., Coney P. J., 1979. Geological development of the Cordilleran metamorphic core complexes, *Geology*, **7,** 120-124.

Delteil J. R., Rivier F., Montadert L., Apostolescu V., Didier J., Goslin M., Patriat P. H., 1976. Structure and sedimentation of the continental margin of the Gulf of Benin, in : *Continental margins of Atlantic type*, Ann. Acad. Brasil, **48,** 51-66.

Eaton G. P., 1979. Règional geophysics, Cenozoic tectonics, and geological resources of the Basin and Range province and adjoining regions, in : *Basin and Range Symposium and Great Basin Field Conference*, edited by G. W. Newman and H. D. Goode, Rocky Mtn. Assoc. of Geol. and Utah Geol. Assoc., 11-40.

Eaton G. P., 1980. Geophysical and geological characteristics of the crust of the Basin and Range province, in : *Continental tectonics, studies in geophysics*, Natl Acad. Sci., 96-113.

Edwards M. B., 1976. Growth faults in Upper Triassic deltaic sediments, Svalbard, *Am. Assoc. Pet. Geol. Bull.*, **60,** 314-355.

Effimoff I., Pinezich A. A., 1980, in prep. Tertiary structural development of selected valleys based on seismic data-Basin and Range province, northeastern Nevada, *Pap. R. Soc. Meet., London.*

Falvey D. A., 1974. The development of continental margins in plate tectonic theory, *J. Aust. Petrol. Explor. Assoc.*, 59-106.

Frost E. G., 1979. Growth fault character of Tertiary detachment faulting, Whipple Mountains, San Bernardino County, California, and Buckskin Mountains, Yuma County, Arizona, *Geol. Soc. Am. Abstr. Progr.*, **11,** 7, 429.

Gilbert G. K., 1875. Report on the geology of portions of Nevada, Utah, California, and Arizona, examined in the years 1871 and 1872, *US Geogr. Geol. Surv. W. 100th Mer. Rep.*, **3.**

Gordy P. F., Frey F. R., Ollerenshaw N. C., 1975. Structural geology of the foothills between Savanna Creek and Panther River, SW Alberta, Canada, Guidebook Can. Soc. Petrol. Geol., *Can. Soc. Explor. Geophys., Explor. Update* 1975, Calgary.

de Graciansky P. Ch., Bourbon M., Chenet P. Y., de Charpal O., Lemoine M., 1979. Genèse et évolution comparées de deux marges continentales passives : marge ibérique de l'Océan Atlantique et marge européenne de la Tethys dans les Alpes occidentales, *Bull. Soc. Geol. Fr.*, **7,** 21, 665-674.

Grill R., Kapounek J., Kupper H., Pape A., Plöchinger B., Prey S., Tollmann A., 1968. Guide to excursion 33c Austria Neogene basins and sedimentary units of the Eastern Alps near Vienna, Int. Geol. Congress, XXIII Session, *Geol. Surv. Austria Vienna*, **III,** 1-75.

Groupe Galice, 1979. The continental margin off Galicia and Portugal : acoustical stratigraphy, dredge stratigraphy and structural evolution, in : *Initial reports of the Deep Sea Drilling Project,* edited by J. C. Sibuet, W. B. F. Ryan *et al.,* US Government Printing Office, Washington, **47,** 2, 633-662.

Hafner W., 1951. Stress distributions and faulting, *Geol. Soc. Am. Bull.,* **62,** 373-398.

Hamblin W. K., 1965. Origin of "reverse drag" on the downthrown side of normal faults, *Geol. Soc. Am. Bull.,* **76,** 1145-1164.

Hamilton W., 1977. Subduction in the Indonesian region, in *Island arcs, Deep Sea trenches and back-arc basins,* edited by W. Talwani and W. C. Pittman, Maurice Ewing Series, Am. Geol. Un., **1,** 15-31.

Hamilton W., 1979. Tectonics of the Indonesian region, *US Geol. Surv. Prof. Pap.,* **1078,** 345 p.

Hamilton W., Myers W. B., 1966. Cenozoic tectonics of the western United States, *Rev. Geophys.,* **4,** 509-549.

Hardin F. R., Hardin G. C. Jr., 1961. Contemporaneous normal faults of Gulf Coast and their relation to flexures, *Am. Assoc. Pet. Geol. Bull.,* **45,** 238-248.

Helwig J., 1976. Shortening of continental crust in orogenic belts and plate tectonics, *Nature,* **260,** 5554, 768-770.

Horvath F., Stegena L., 1977. The Pannonian Basin : a Mediterranean interarc basin, in : *Structural history of the Mediterranean basins,* edited by Biju-Duval and L. Montadert, Editions Technip, 133-142.

Hubbert M. K., 1951. Mechanical basis for certain familiar geologic structures, *Geol. Soc. Am. Bull.,* **62,** 355-372.

Kälin O., Trümpy D., 1977. Sedimentation and Palaotektonik in den westlichen Südalpen : Zur triasisch-jurassischen Geschichte des Monte Nudo-Beckens, *Eclogae Geol. Helv.,* **70,** 2, 295-350.

Kirchmayer M., Mohr K., 1963. Zur Terminologie krummflächiger and krummliniger Gefügeelemente, *Bergbauwissenschaften 10,* **H. 16,** 378-385 ; or 1964, in *Felsemechanik und Ingenieurgeologie,* Springer Verlag Wien, 106-114.

Kroell A., Wessely G., 1973. Neue Ergebnisse beim Tiefenaufschluss im Wiener Becken, *Erdoel-Erdgas Z.,* **89,** 400-413.

Kroell A., Wiesender H., 1972. The origin of oil and gas deposits in the Vienna Basin (Austria), *24th Int. Geol. Congr. Section,* **5,** 153-160.

Laubscher H. P., 1969. Mountain building, *Tectonophysics,* **7,** 551-563.

Laubscher H., Bernoulli D., 1977. Mediterranean and Tethys, in : *The Ocean Basins and Margins, 4A : The Eastern Mediterranean,* edited by A. E. M. Nairn, W. H. Kanes and F. G. Stehli, Plenum Publ. Corp., New York, 1-28.

Lehner P., de Ruiter P. A. C., 1977. Structural history of the Atlantic margin of Africa, *Am. Assoc. Pet. Geol. Bull.,* **61,** 961-981.

Le Pichon X., Sibuet J.-C., 1980. Passive margins : a model of formation, submitted to *J. Geophys. Res.*

Liggett M. A., Ehrenspeck H. E., 1974. Pahranagat shear system, Lincoln County, Nevada, Argus Exploration Co., *Rep. Invest. NASA,* **5,** 21809, 6 p.

Longwell C. R., 1933. Rotated faults in the Desert Range, southern Nevada (Abstr.), *Geol. Soc. Am. Bull.,* **44,** 93.

Longwell D. R., 1945. Low-angle normal faults in the Basin and Range province, *Am. Geophys. Union Trans.,* **26,** 107-118.

Lowell J. D., Genik G. J., 1972. Sea-floor spreading and structural evolution of Southern Red Sea, *Am. Assoc. Pet. Geol. Bull.,* **56,** 247-259.

Lowell J. D., Genik G. J., Nelson T. H., Tucker P. M., 1975. Petroleum and plate tectonics of the southern Red Sea, in : *Petroleum and global tectonics,* edited by A. G. Fischer and S. Judson, Princeton Univ. Press, Princeton, N.J., 129-153.

MacDonald R. E., 1976. Tertiary tectonics and sedimentary rocks along the transition : Basin and Range province to Plateau and Thrust Belt province, Utah, in : *Geology of the Cordilleran Hingeline,* edited by G. J. Hill, Rocky Mtn Assoc. Geol., Denver, 281-317.

Mahel, 1974. *Tectonics of the Carpathian-Balkan regions. Explanation to the tectonic map of the Carpathian-Balkan regions and their foreland,* Geol. Inst. of Dionyz Stur Bratislava, 454 p.

McDowell F. W., Schmid R., 1968. Potassium-argon ages from the Valle d'Ossola section of the Ivrea-Verbano zone (northern Italy), *Schweiz. Mineral. Petrogr. Mitt.,* **48,** 205-210.

McKenzie D., 1978. Some remarks on the development of sedimentary basins, *Earth Planet. Sci. Lett.,* **40,** 25-32.

Misch P., 1960. Regional structural reconnaissance in central-northeast Nevada and some adjacent areas : observations and interpretations, in : *Guidebook to the geology of East-Central Nevada, Eleventh Annual Field Conference,* edited by J. W. Boettcher and W. W. Sloan Jr., Intermt Assoc. Pet. Geol., Salt Lake City, 17-42.

Misch P., 1971. Geotectonic implications of Mesozoic *décollement* thrusting in parts of eastern Great Basin, *Geol. Soc. Am. Abstr. Progr.,* **3,** 164-166.

Montadert L., Roberts D. G., de Charpal O., Guennoc P., 1979 *a.* Rifting and subsidence of the northern continental margin of the Bay of Biscay, in : *Initial reports of the Deep Sea Drilling Project,* US Government Printing Office, Washington, **48,** 1025-1060.

Montadert L., de Charpal O., Roberts D. G., Guennoc P., Sibuet J.-C., 1979 *b.* Northeast Atlantic passive margins : rifting and subsidence process, in : *Deep drilling results in the Atlantic Ocean : continental margins and paleo-environment,* edited by M. Talwani, W. W. Hay and W. B. F. Ryan, Maurice Ewing Series 3, Am. Geophys. Union.

Moore J. G., 1960. Curvature of normal faults in the Basin and Range province of the western United States, *US Geol. Surv. Prof. Pap.,* **400,** B, 409-411.

Morton W. H., Black R., 1974. Crustal attenuation in Afar, in : *Afar depression in Ethiopia,* edited by A. Pilger and A. Roesler, E. Schweizerbart, Stuttgart, 55-65.

Mueller S., 1977. A new model of the continental crust, *Am. Geophys. Un. Monogr.,* **20,** 289-317.

Myers W. B., Hamilton W., 1964. Deformation accompanying the Hegben Lake earthquake of August 17, 1955, *US Geol. Surv. Prof. Pap.,* **435,** I, 55-68.

Newman G. W., Goode H. D., 1979. *Basin and Range Symposium and Great Basin Field Conference,* Rockey Mtn Assoc. of Geol. and Utah Geol. Assoc., Denver.

Ocamb R. D., 1961. Growth faults of South Louisiana, *Gulf Coast Assoc. Geol. Soc. Trans.,* **11,** 139-173.

Proffett J. M. Jr., 1977. Cenozoic geology of the Yerington district, Nevada, and implications for the nature and origin of Basin and Range faulting, *Geol. Soc. Am. Bull.,* **88,** 247-266.

Pruvost P., 1960-1963. *Les jeux propres du socle révélés par l'histoire de certains bassins houillers à la périphérie du domaine alpin,* Livre à la Mémoire du Pr. P. Fallot, Tome II, Soc. Geol. Fr., 1-17.

Rider M. H., 1978. Growth faults in western Ireland, *Am. Assoc. Pet. Geol. Bull.,* **62,** 2191-2213.

Roux W. F. Jr., 1977. *The development of growth fault structures,* Am. Assoc. Pet. Geol. Struct. Geol. Sch. Course Notes, 33 p.

Royder L., Sclater J. G., Von Herzen R. P., 1980. Continental margin subsidence and heat flow : important parameters in formation of petroleum hydrocarbons, *Am. Assoc. Pet. Geol. Bull.,* **64,** 173-187.

Royse F. Jr., Warner M. A., Reese D. C., 1975. Thrust belt structural geometry and related stratigraphic problems, Wyoming-Idaho-northern Utah, in : *Symposium on deep drilling frontiers in the central Rocky Mountains,* Rocky Mtn Assoc. Geol., Denver, 41-54.

Schindler C. M., 1959. *Zur Geologie des Glärnisch,* Beitr. Geol. Karte der Schweiz, N.F. 107.

Sclater J. G., Royden L., Horvath F., Burchfiel B. C., Semken S., Stegena L., 1980 *a* (in prep.). *The formation of the intra-Carpathian Basins as determined from subsidence data.*

Sclater J. G., Royden L., Semken S., Burchfiel B. C., Stegena L., Horvath F., 1980 *b* (in prep.). *Continental stretching : an explanation of the late Tertiary subsidence of the Pannonian Basin.*

Smith R. B., 1978. Intra-plate tectonics of interior of Western Cordillera, *Geol. Soc. Am. Mem.,* **152,** 111-141.

Smith R. B., Eaton G. P., 1978. *Cenozoic tectonics and regional geophysics of the western cordillera,* Mem. 152, Geol. Soc. Am., Denver.

Spicher A., 1980. *Carte géologique de la Suisse*, 1 : 500 000, 2e éd., Comm. geol. Suisse.

Stewart J. H., 1971. Basin and Range structure : a system of horsts and grabens produced by deep-seated extension, *Geol. Soc. Am. Bull.*, **82**, 1019-1044.

Stewart J. H., 1978. Basin-range structure in western North America : a review, *Geol. Soc. Am. Mem.*, **152**, 1-31.

Stewart J. H., Carlson J. E. (compil.), 1978. Geologic map of Nevada, 1 : 500 000, US Geol. Surv., Reston, VA.

Sturani C., 1973. Considerazioni sui rapporti tra Appennino settentrionale ed Alpi occidentali, *Accad. Naz. Lincei, anno 370, Quaderno N.*, **183**, 119-142.

Suess E., 1904-1924, *The face of the earth*, I-V, Oxford.

Thompson G. A., 1959. Gravity measurements between Hazen and Austin, Nevada : a study of basin-range structures, *J. Geophys. Res.*, **64**, 217-229.

Thompson G. A., 1966. The rift system of the western United States, in : *The world rift system*, edited by T. N. Irvine, *Geol. Surv. Can. Pap.*, **66**, 14, 280-290.

Thompson G. A., 1971. Thin skin distension in Tertiary rocks of southeastern Nevada : discussion, *Geol. Soc. Am. Bull.*, **82**, 3529-3532.

Thompson G. A., Burke D. B., 1974. Regional geophysics of the Basin and Range province, *Ann. Rev. Earth Planet. Sci.*, **2**, 213-238.

Trümpy R., 1960. Paleotectonic evolution of the Central and Western Alps, *Geol. Soc. Am. Bull.*, **71**, 843-908.

Trümpy R., 1975. Penninic-Austroalpine boundary in the Swiss Alps : a presumed former continental margin and its problems, *Am. J. Sci.*, **275-A**, 209-238.

Wallace R. E., 1979 (preprint). Earthquakes and the pre-fractured state of the western part of the North American continent, *Conference on Intra-continental Earthquakes*, Ohrid, Yugoslavia, 11 p.

Weber K. J., Daukoru E., 1976. Petroleum geology of the Niger Delta, *Ninth World Petrol. Congress Tokyo 2*, 209-221, Applied Science Publishers, Essex.

Wiedenmayer F., 1963. Obere Trias bis mittlerer Lias zwischen Saltrio und Tremona (Lombardische Lapen), *Eclogae Geol. Helv.*, **56**, 529-640.

Winterer E. L., Bosellini A., 1980. Subsidence and sedimentation on a Jurassic passive continental margin (Southern Alps, Italy), manuscript submitted to *Am. Assoc. Pet. Geol. Bull.*

Woodland A. W., Evans W. G., 1964. The country around Pontypridd and Maesteg, in : *Mem. Geol. Survey Great Britain : the geology of the South Wales coalfield, Pt. IV*, 238-273, H.M. Stationery Office, London.

Wright L. A., Troxel B. W., 1973. Shallow-fault interpretation of Basin and Range structure, southwestern Great Basin, in : *Gravity and tectonics*, edited by K. A. de Jong and R. Scholten, John Wiley, New York, 397-407, 502 p.

The American Association of Petroleum Geologists Bulletin
V. 68, No. 7 (July 1984), P. 801-815, 32 Figs.

Listric Normal Faults: An Illustrated Summary[1]

JOHN W. SHELTON[2]

ABSTRACT

Normal faults are commonly listric, that is, the dip flattens with depth. Movement along this type of fault is instrumental in formation of several types of structural traps (e.g., rollover anticlines and upthrown-fault-block closures). Some listric faults are restricted to sedimentary rocks, whereas others offset basement rocks. Theoretical data, rock-mechanical and simulated model experiments, and foundation-engineering tests and failures suggest that this type of fault may occur where brittle rocks overlie ductile rocks in an extensional regime. In some places the ductile section may be thin and bounded sharply at its top. Also, the extensional regime may be locally derived within a broader stress regime of another type, as evidenced by transtension associated with strike-slip movement and arched strata in a compressive setting.

The flattening of the fault reflects an increase in ductility of the rocks with depth and, in some cases, deformation of the fault due to compaction or tilting of the upthrown block. The dip angle may vary along the strike of the fault in response to changes in throw. In cross section, a listric fault may consist of relatively short, en echelon fault segments. This geometry may be particularly characteristic of growth faults. Sedimentary faults may sole in ductile strata, or they may represent the brittle part of a fault-flow system. Fault patterns commonly are characterized by bifurcation, some of which may occur near the ends of individual faults comprising a zone.

Although unequivocal recognition of listric normal faults requires unusually extensive outcrop data, close subsurface control, or high-quality seismic data, their presence is suggested indirectly by such features as increasing dip with depth toward the controlling fault ("reverse drag"), thick progradational sandstone overlying ductile strata, and in some cases arcuate fault patterns, basins, or uplifts.

Listric normal faults form during rifting, drifting, and evolution of passive continental margins with concomitant basinal development. Listric faults confined to the sedimentary prism are common features on passive margins, especially in progradational, post-evaporite sequences. The basement is offset by listric faults as a fundamental element in the development of other types of basins, including those which formed during postorogenic extension. They also occur as secondary extensional features in an overall compressive stress regime due to plate convergence and during transform or strike-slip faulting.

INTRODUCTION

A listric fault is characterized by a decreasing angle of dip with depth. It, therefore, is a curved surface which is concave upward. Apparently the concept was introduced by Edward Suess in the early part of this century (Bally et al, 1981) as part of his description of faults in coal mines in northern France.

Listric thrust faults have been recognized for a long time as a basic feature of thin-skinned tectonics, with decollements. Now, as deep faults soling in the ductile crust, they are also considered an integral part of suturing during plate convergence (e.g., Thompson, 1976). Although listric normal faults have been recognized as updip (or upslope) segments of gravitational slides (e.g., Reeves, 1925, 1946; Hubbert and Rubey, 1959; Wise, 1963), most commonly they have been regarded as a special feature of syndepositional faults in strongly subsident basins containing thick shale (with or without salt) below progradational sandstone sections. This general opinion probably derives from the abundance of sedimentary faults in the northern Gulf Coast basin (Texas and Louisiana) and the common knowledge of "rotational slips" and associated failures in foundation engineering (Figure 1). Apparently little significance was given to the early work of Davis (1925) and Longwell (1933, 1945), who described listric normal faults offsetting crystalline and/or basement rocks in the western United States; to the theoretical treatment of Hafner (1951), who showed curved stress trajectories including conditions for listric normal faults; or to the work of Carey (1958), who described listric normal faults as a major feature in development of rift valleys. It seems reasonable, therefore, to regard listric geometry as a common feature of both thrust and normal faults displacing sedimentary and/or basement rocks.

Wernicke and Burchfiel (1982) have grouped normal faults into two categories: rotational and nonrotational. The rotational category is divided into (a) those with rotation of beds along listric faults, and (b) those with rotation of beds and faults along planar or listric faults. Nonrotational faults have no rotation of structures along planar faults.

[1]Manuscript received, May 18, 1983; accepted, December 7, 1983.

[2] ERICO, Inc., Tulsa, Oklahoma 74172.

For documentary data which have not been published, the writer is indebted to Shell Oil Co. for materials from the Gulf Coast and to G. W. Hart for data and interpretations in the Arkoma basin. Laura F. Serpa and R. E. Denison provided information on basement-involved faults and detached sediments, respectively.

Appreciation is gratefully expressed to many colleagues and acquaintances who for more than 2 decades have stimulated thought on this subject of listric (rotational) normal faults. Kaspar Arbenz kindly reviewed the original manuscript and made numerous helpful suggestions. Appreciation is also expressed to AAPG Editors M. K. Horn and Richard Steinmetz, Science Director Edward A. Beaumont, and reviewers for their valuable comments. Yet the author must assume responsibility for errors or any aberration in accepted thought.

David E. Brooker drafted the illustrations, and Sherry Hempel, Mildred P. Lee, and Dianne O'Malley prepared the typescript. O'Malley also assisted in compilation of the references. S. W. Carey kindly provided the reference noted herein to his work.

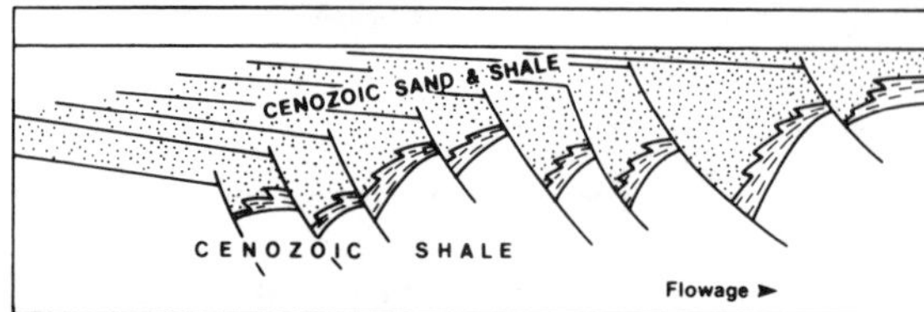

Figure 6—Schematic cross section across Texas part of northern Gulf of Mexico basin, with normal faulting due to flowage of ductile shale. After Bruce (1973).

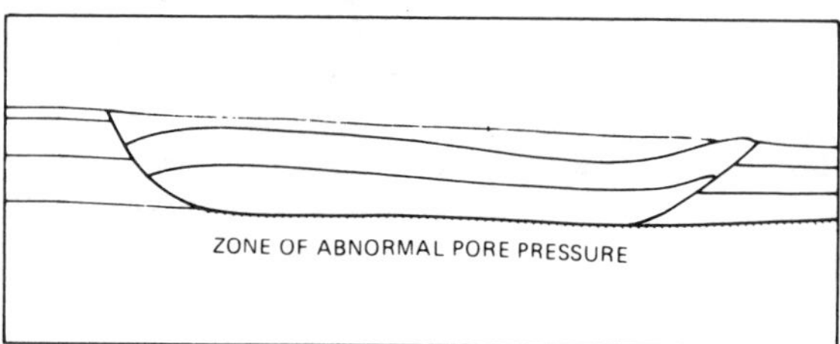

Figure 7—Normal faulting due to gravitational sliding in response to dip of strata with abnormal pore pressure. After Hubbert and Rubey (1959).

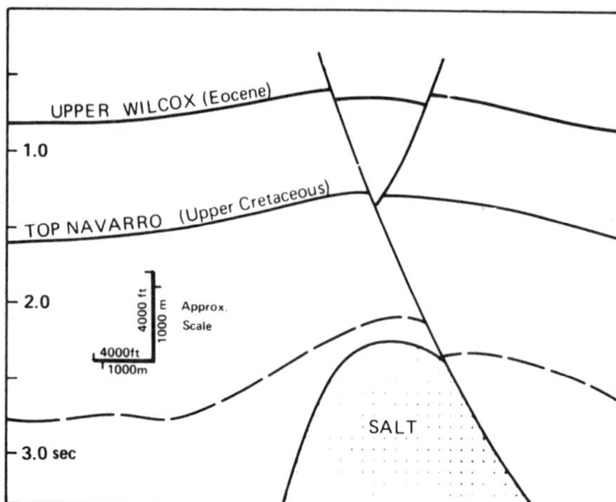

Figure 8—Seismic cross section of Pescadito dome, Webb County, Texas, showing normal faults due to extension in strata overlying salt. After Halbouty (1979).

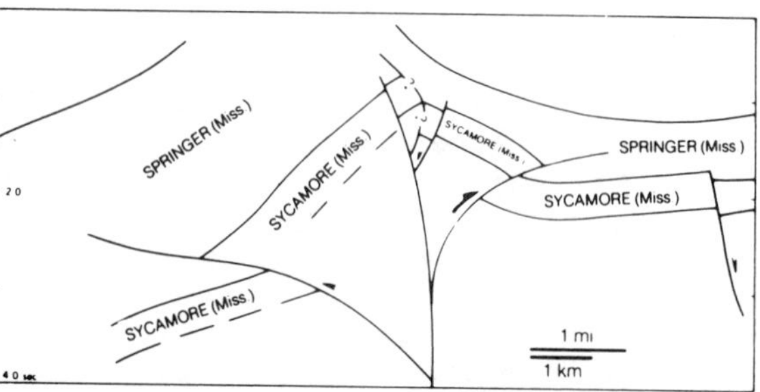

Figure 9—Seismic cross section across wrench fault zone in Ardmore basin, Oklahoma, with flower structure which contains minor listric normal faulting due to extension of Mississippian and older strata. After Harding and Lowell (1979).

wrench faulting (Harding and Lowell, 1979) (Figure 9); (b) axial collapse associated with subduction (Beck et al, 1975) (Figure 10); (c) uplift and/or arching of earlier formed foldbelt.

LISTRIC NORMAL FAULTS

The flattening of the dip of a normal fault with depth may reflect one or more environmental conditions or processes at depth. The first group is conditions that are inherent, that is, conditions that contributed to formation of listric faults.

1. Increase in ductility in sedimentary prism, generally involving thick overpressured shale and/or salt, with extension of "overburden" due to flowage or decollement of "substrate."
2. Increase in ductility in crust (with extension of ductile "substrate").

The second group includes processes that operated after formation of the fault. This includes deformation of fault by:

1. Compaction of shale in footwall (Figure 11).
2. Arching during uplift initiated in rocks below the fault (e.g., due to salt or igneous intrusion; Figure 8).
3. Increased tilting (with rotation about an axis parallel with strike of fault) of entire upthrown fault block reflecting movement along subjacent "underlying" fault (Roux, 1977) (Figure 12).

Theoretical and experimental data together with case histories from foundation engineering suggest that the listric feature may be a basic element of some normal faults. Included in the theoretical and experimental data demonstrating, or allowing inference of, listricity in extensional conditions are results of the following works:

1. One set of stress trajectories derived by Hafner (1951).

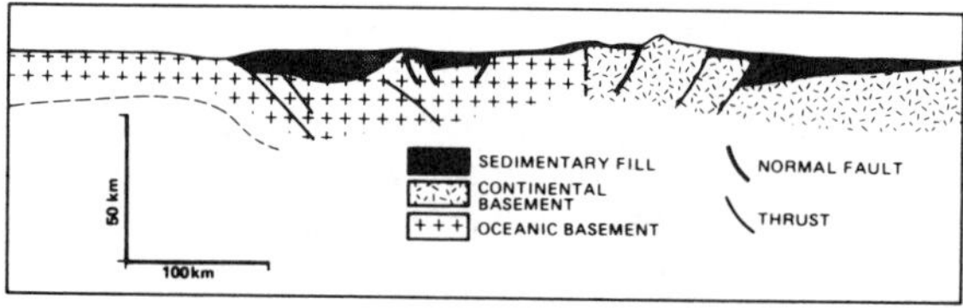

Figure 10—Cross section of Andean orogene, showing measure of bilateral symmetry, with two outer zones of compression and axial zone(s) of block faulting, interpreted to include listric normal faults. After Beck et al (1975).

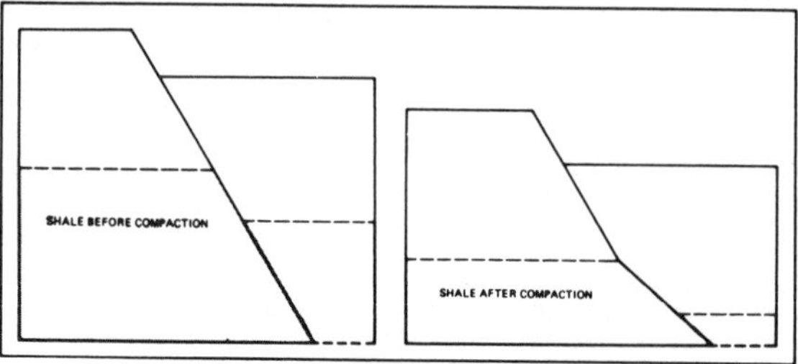

Figure 11—Flattening of fault due to compaction of shale.

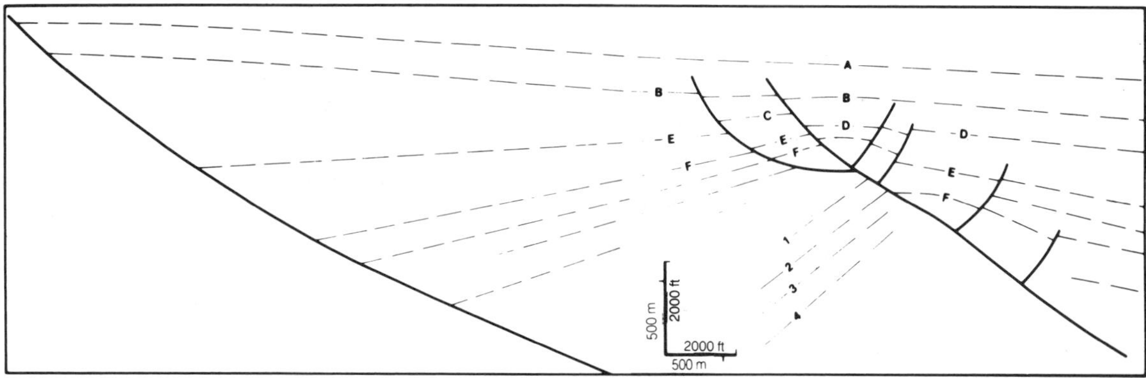

Figure 12—Seismic cross section of local structure in offshore Texas part of northern Gulf Coast basin, showing deformed fault due to rotation of upthrown block reflected by attitude of strata 1-4. After Roux (1977).

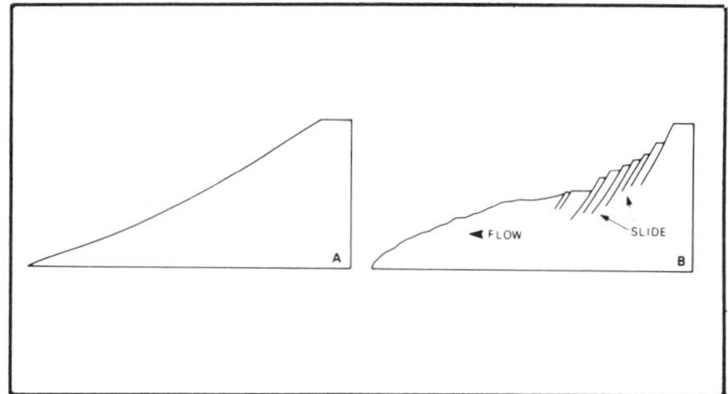

Figure 13—A. Profile of slope before failure. B. Cross section of clayey deposits after slide flow. After Longwell and Flint (1962, p. 142-143). This type of failure, occurring where clays are very sensitive, is thought to be analogous to fracture (fault) and flow (incipient diapir) relationship of numerous sedimentary listric faults.

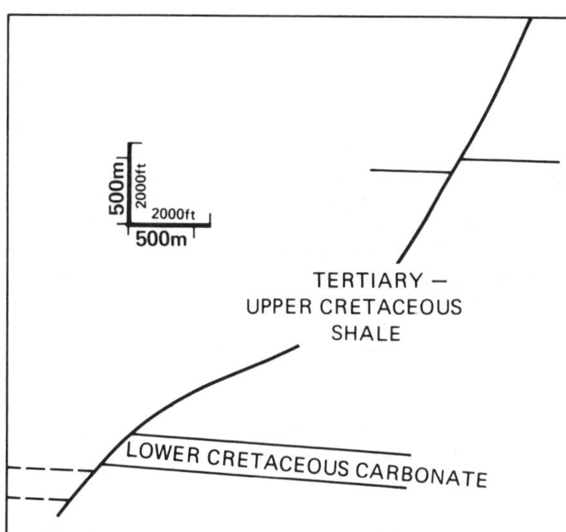

Figure 14—Cross section through Fashing field, south-central Texas, showing listric normal fault in Tertiary and Upper Cretaceous shale. Dip of fault steepens with further depth in Lower Cretaceous carbonates. After Murray (1961).

2. Theoretical model for shale tectonics by Odé (Crans et al, 1980).

3. Geomechanical model by Crans et al (1980).

4. Theoretical considerations by Muehlberger (1961) predicting decreasing dip angle of fracture surface with increase in confining pressure.

5. Rock-mechanical experiments by Handin and Hager (1957) showing decreasing angle of internal friction with increase in confining pressure, and by von Kármán (Handin and Hager, 1957) showing decrease in angle between fracture surface and least principal stress axis with increase in pressure.

6. Rock-mechanical experiment by Heard (1966) showing listric normal fault with distributed flow under high confining pressure and high temperature.

7. Model experiment by H. Cloos (1930) simulating grabens and rifts.

8. Model experiment by Rettger (1935) simulating faulting due to gravitational gliding.

9. Model experiment by P. Diebold (Crans et al, 1980) simulating faulting due to loading.

10. Model experiments by E. Cloos (1968) simulating growth faults.

GEOMETRY

Dip

The dip of a listric fault flattens with depth, but it either "dies out" in ductile rocks that deform by flowage or it becomes a decollement zone. There is a strong tendency to consider the latter as the dominant disposition of a listric fault. Where gentle regional dip exists, creep may contribute to development of decollement zones or sole ductile faults. Yet, the relationship of listric faults to shale and salt diapirs suggests that a fault-flow system (Figure 13), which is analogous to the slide flow in soil mechanics, may be very common. In terms used in foundation engineering, base failures, where the ductile substrate flows in response to asymmetric loading, are probably more analogous to listric sedimentary faults than slope failures, which are not necessarily bounded by listric surfaces. Flowage associated with base failure may be regarded as a form of

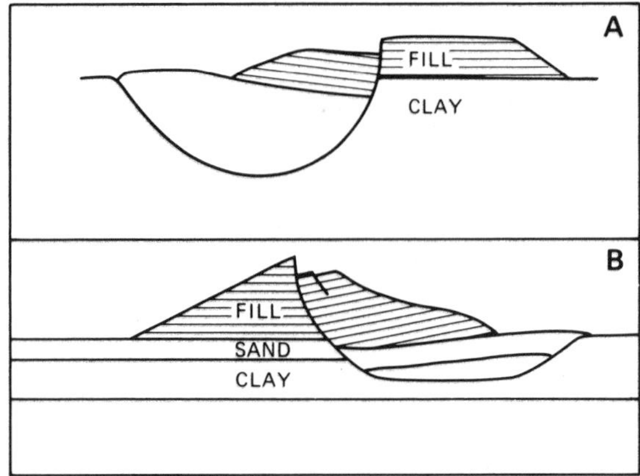

Figure 1—Foundation failures resembling configurations of faults in sedimentary rocks. A. Rotational slip in foundation due to localized loading of uniform clay. B. Base failure due to loading of foundation with thin clay. After Terzaghi and Peck (1948).

Wernicke and Burchfiel indicate that large-scale displacement on low-angle listric normal faults results in a series of tilted planar-fault blocks, forming "extensional allochthons."

Both normal and thrust listric faults, along with planar faults, are of major significance to the explorationist because they are an important element in the formation of traps in faulted strata. Presently a commonly held opinion is that listric normal and thrust faults may be sequentially related (or even coincident) in some areas that undergo changes in tectonic regime. For example, listric thrust faults may be reactivated as normal faults when an earlier formed orogenic belt is subjected to extension (Bally et al, 1966), and, conversely, normal faults may be reactivated as thrusts during the evolution of a continental margin from a passive to active phase (Cohen, 1982). Further, the location of thrusts with displacement during the active phase (after basinal subsidence) may be predetermined by buried normal faults that formed during the earlier passive phase (during basinal subsidence). Listric normal faults are probably important elements in the development of many basins. Downward dip-slip movement of faulted strata in the hanging wall of a listric normal fault may result in "reverse drag" in half grabens or "rollover" (dip-direction reversal), with formation of an anticlinal feature (Figure 2). Absolute movement, with rotation of an upthrown block, may result in a tilted fault block with reverse drag. Significant variations in displacement along the strike of a fault present conditions for closure against it (Figure 3). The closure may also result from differential rotation (along the strike of a fault) of an entire block which itself is downthrown with respect to a subjacent "underlying" fault (Figure 3C), or by changes in stratigraphic thicknesses along the strike of the fault. The detailed geometry of the faults provides subtle trapping potential. For example, lateral branching or overlapping ends of faults are possible elements of subsidiary traps. Also, movement along individual faults of a fault zone may result in several traps rather than one larger trap.

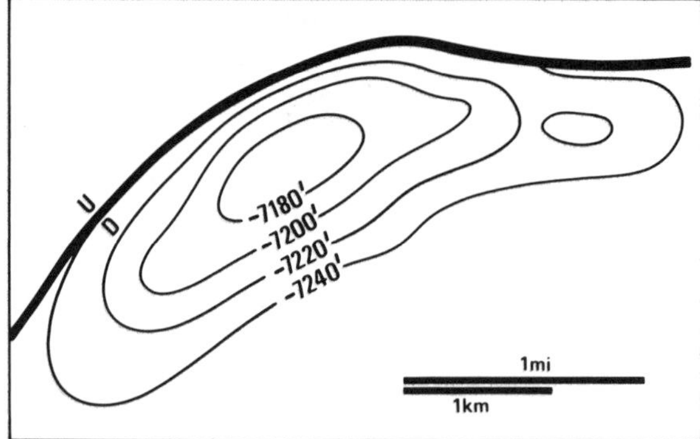

Figure 2—Structural map of top of Wilcox Group (Eocene) in South Bancroft field, Beauregard Parish, Louisiana, showing rollover anticline. After Murray (1961).

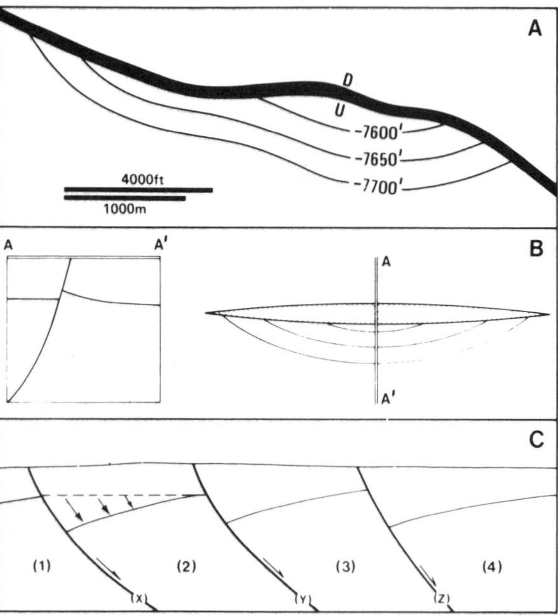

Figure 3—A. Structural map of Lower Cretaceous marker in Pleasanton field, Atascosa County, Texas, depicting tilted-fault-block trap. After Murray (1961). B. Hypothetical fault closure due to absolute movement of upthrown block. C. Fault blocks with potential for trap in each upthrown block due to rotation of that block (e.g., dip in block 2 is due to rotational movement along fault X).

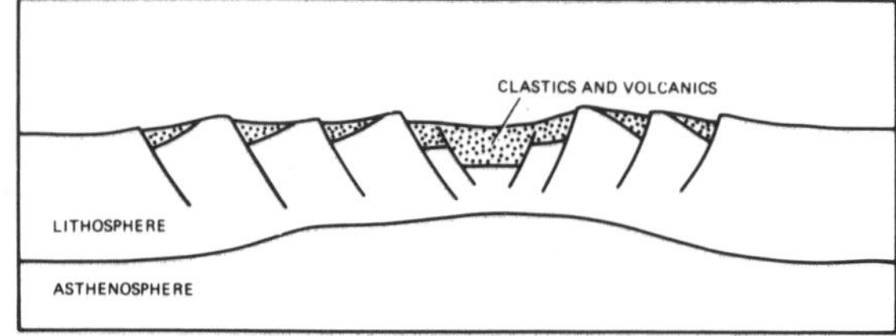

Figure 4—Schematic cross section of major rift zones rupturing a continent—before possible drifting and formation of passive continental margins. After Dewey and Bird (1970).

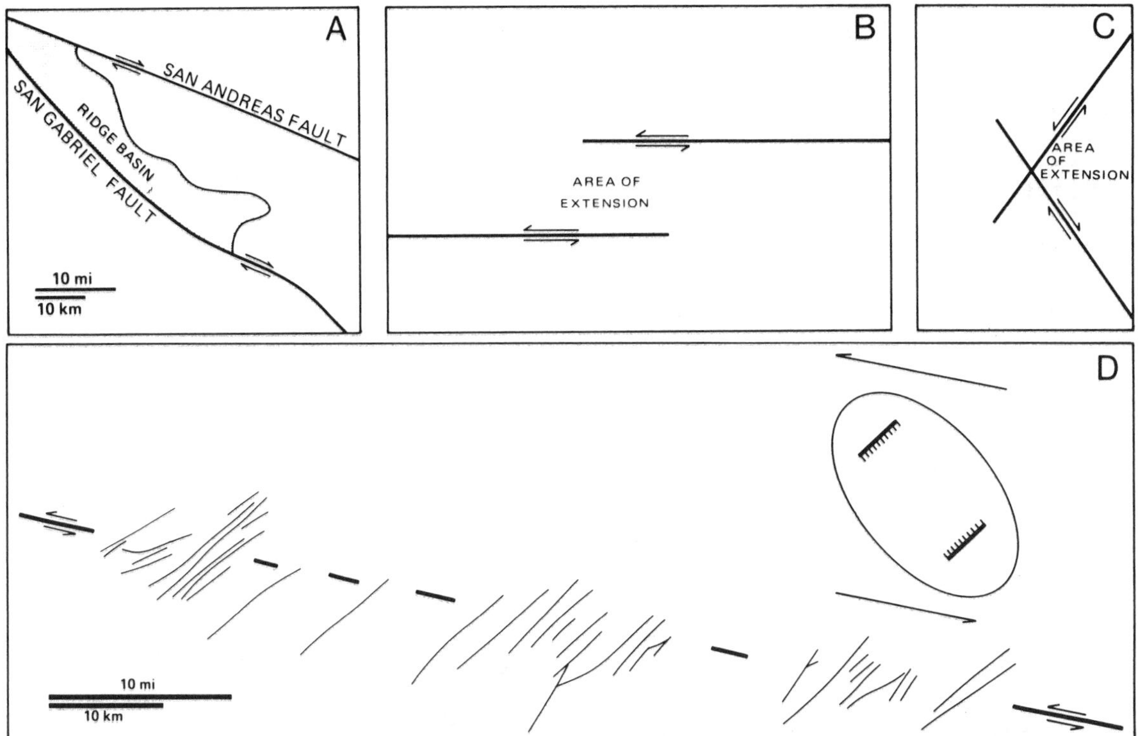

Figure 5—Areas of extension with normal faults resulting from wrenching. A. Extension with formation of Ridge basin due to divergent wrenching between San Andreas and San Gabriel faults (after Wilcox et al, 1973). B. Extension due to movement along parallel en echelon wrench or transform faults. C. Extension due to conjugate wrench faults (based on discussion in Wilcox et al, 1973). D. En echelon normal faults of Lake Basin fault zone in south-central Montana due to left-lateral faulting (after Fanshawe and Alpha, 1954; Harding, 1974). Faults of this type may be riedel shears along which there is some dip-slip component.

This review is restricted to normal faults, with description of (1) faults along which the apparent relative displacement of the hanging wall was down with respect to the footwall and (2) faults which formed in a local or regional stress regime wherein the maximum principal stress, σ_1, is interpreted to have been vertical or near vertical. In many places movement of strata along listric faults is dip-slip and rotational, with the axis of rotation being parallel with the strike of the fault. Under conditions where the primary feature is a strike-slip fault, the dip-slip component of the total displacement across the fault may also be comparatively "small," and movement along the listric fault may vary significantly from dip slip. The scale of the "small" displacement, of course, may be more than 1,000 m (3,300 ft).

In this paper, concepts are presented before examples; the topics, in order, are: causes of normal faulting and of listric normal faults, geometry, propagation, growth faults, evidence for listric faults, and occurrences. The primary references are Bally et al (1981) and Bally (1983). The former is a resumé of listric normal faults in various geologic settings, in particular, passive continental margins and orogenic systems. The latter, which is a pictorial atlas of seismic sections illustrating various structural styles, contains outstanding examples of listric normal faults from several extensional provinces.

CAUSES OF NORMAL FAULTING

Normal faults occur in response to extension, which may be crustal extension, sedimentary-section extension, or basement and/or sedimentary-section extension.

1. Crustal extension results from (a) divergent plate movement, expressed by rifts (Figure 4); (b) arching by thermal expansion (e.g., development of a plume); and (c) transtension accompanying divergent wrenching (wrench or transform faulting) and movement along parallel to subparallel en echelon faults or "plates" or along conjugate wrench faults (Wilcox et al, 1973; Harding and Lowell, 1979; Burchfiel and Royden, 1982) (Figure 5).

2. Sedimentary-section extension results from (a) flowage of ductile substrate (shale and/or salt) (e.g., Bruce, 1973; Woodbury et al, 1973; Humphris, 1978) (Figure 6); (b) increase in stratal dip and resultant gravitational sliding (e.g., Hubbert and Rubey, 1959) (Figure 7); (c) bending, or arching, during uplift (e.g., associated with salt or igneous intrusion; see Figure 8), and flexural or concentric folding associated with compressional folding; (d) strike-slip faulting (possible normal separation along at least part of the length of a fault which may be a riedel shear; Figure 5D).

3. Basement and/or sedimentary-section extension results from (a) uplift during transpression accompanying

lateral "extrusion" that results in extension and subsidence in the area of loading. Ductility of the substrate generally reflects overpressure in shale and/or plasticity of salt.

Some faults flatten at depth through a shale and steepen below it (e.g., Murray, 1961) (Figure 14). That relationship is generally attributable to shale compaction, but in places it may reflect a lower original angle of dip through the more ductile shale. Roux (1977) presents a very persuasive case that shale compaction does significantly reduce the dip and throw. However, the listric nature of faults in relatively brittle rocks and the listric nature of rotational slips in foundation failures indicate that flattening of dip is an inherent feature of many normal faults.

Dip of the steeper part of a listric normal fault commonly has been recorded as approximately 60°, following the theoretical/experimental work of Anderson (1942) and Hubbert (1951) and from subsurface data. However, some faults are near vertical at the surface or in the near surface.

According to de Sitter (1964), the dip of a fault is generally steeper near its ends (along strike) and flatter along the middle section, where throw is commonly greater.

Small-scale faults in Svalbard are zones which show en echelon patterns in cross sectional view (Edwards, 1976) (Figure 15A). This configuration is shown also in some of the experiments by H. Cloos (e.g., 1930) and E. Cloos (1968) (Figure 15B). It seems reasonable that some listric faults may in fact be zones composed of shorter faults, some of which are listric (Figure 15C).

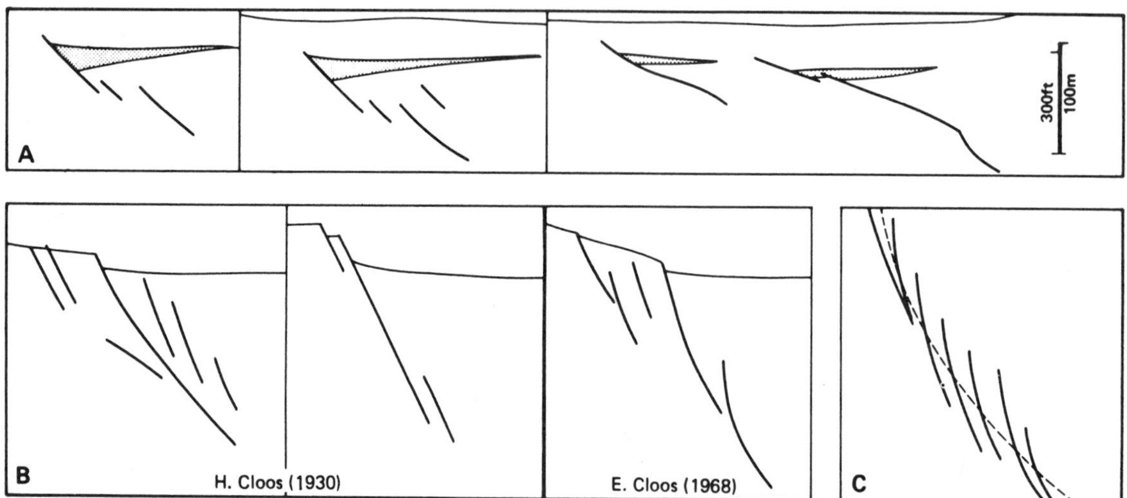

Figure 15—Configuration of listric normal faults in cross section. A. Triassic growth faults which are discontinuous (en echelon in part). After Edwards (1976). B. Faults produced experimentally in small-scale clay models. After H. Cloos (1930) and E. Cloos (1968). C. Proposed pattern of discontinuous en echelon faults comprising listric normal fault zone.

Plan View

Local and regional normal fault patterns are well known from outcrops in the Middle East, east Africa, and the western United States (e.g., de Sitter, 1964; Hamblin, 1965; Holmes, 1965; Anderson, 1971; Robson, 1971; Stewart, 1971), and from subsurface data in basins such as the Gulf of Mexico, North Sea, and coastal Nigeria (e.g., Weber and Daukoru, 1976; Evamy et al, 1978; Galloway et al, 1982) (Figure 16). Several miscellaneous features are noted below.

Arcuate sedimentary faults are probably common in deltaic strata, whereas essentially straight fault traces may be common in nondeltaic deposits. This difference may reflect "point" loading in the former and "line" loading in the latter. In map view, tilted upthrown fault blocks are generally convex in the direction of tilt (Moore, 1960). It is common for faults to branch or splay toward their ends or to show subsidiary faults there (de Sitter, 1964) (Figure

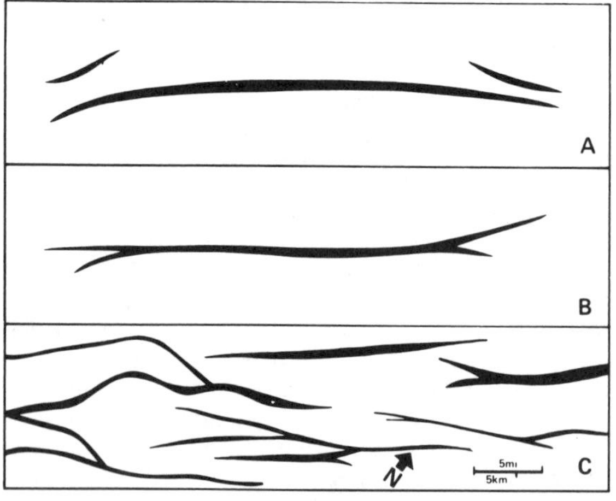

Figure 16—Map-view patterns of normal faults. A, Hypothetical subsidiary faults (after de Sitter, 1964); B, branches (or splays) developed near ends of major faults. C. Oligocene fault pattern in part of Texas Gulf Coast.

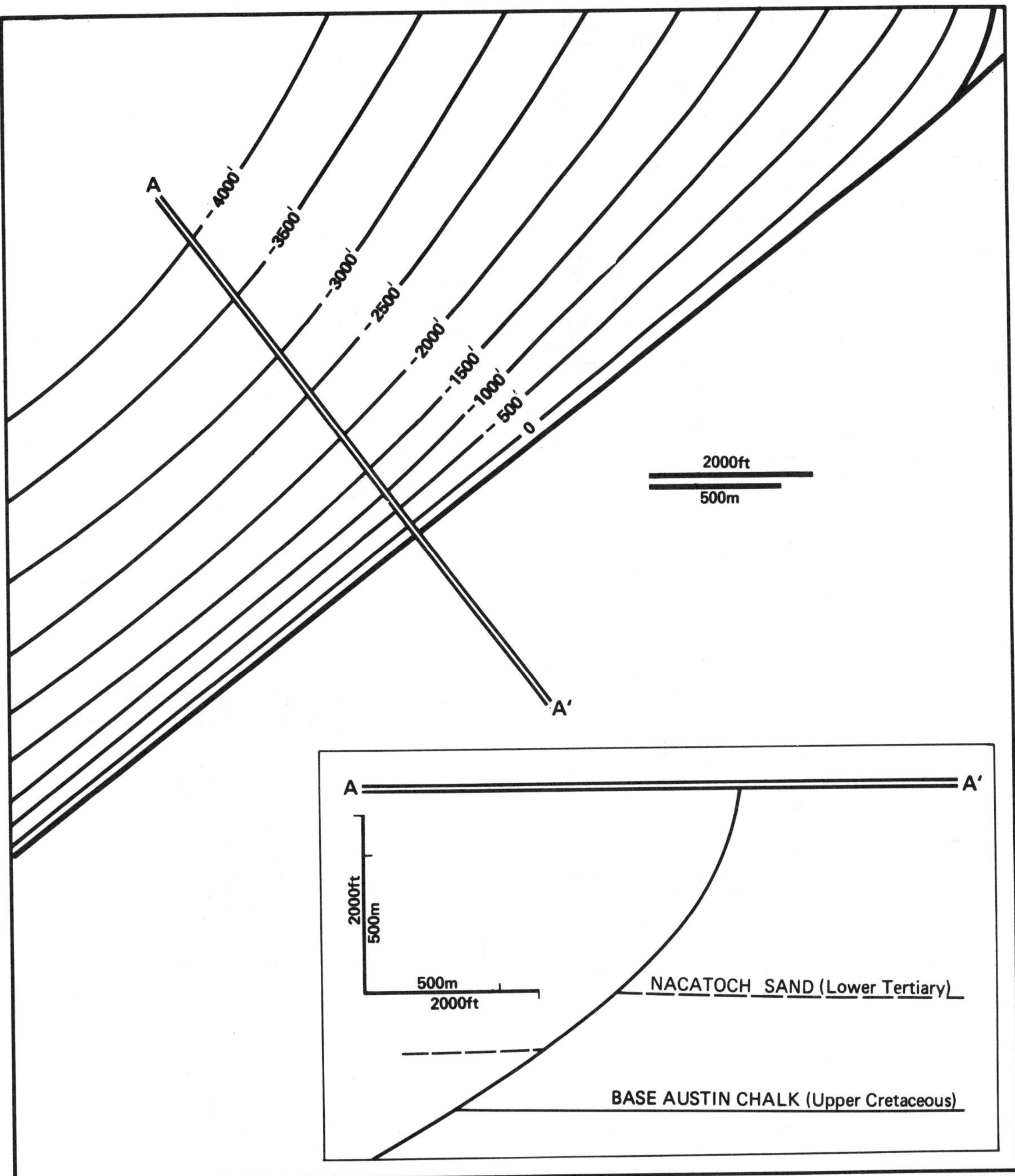

Figure 17—Contour map and cross section of listric normal fault at Lone Star field, northeast Texas. Fault is near vertical at surface, which is strongly erosional. After Bunn (1951). Nacatoch Sand dips away from fault in both blocks, suggesting that it was part of anticlinal feature at time faulting was initiated.

16). Although a fault zone may be very extensive, individual faults within it may be very limited in length, and contiguous (or successive) faults along strike may show some overlap of their lengths. Also, in the Gulf of Mexico basin where the age of major fault zones decreases gulfward (generally basinward), a particular zone may contain older faults immediately gulfward of younger faults.

Propagation

Many faults are initiated in a local area and extend, propagate, or grow laterally (de Sitter, 1964). With propagation of a fault, one or both blocks may move in such a way that the greatest displacement (shift) of the block(s) is at or near the fault—a circumstance that results in reverse drag in the active block(s), including possible rollover in the downthrown block where the direction of dip of the fault is the same as regional dip. In an active upthrown block, the most favorable area for trap development would correspond to the section of fault with most displacement (Figure 3B).

Roux (1977) suggests that the throw of a fault not only decreases laterally toward the ends but also both upward and downward. This type of fault configuration is an indication that the horizontal (plan view) fault pattern in some cases may assist in estimating the general vertical geometry. This relation is inferred also from the work of Moore (1960), who correlated the direction of convexity of uplifted (upthrown) fault blocks with the direction of fault-block tilt, although Stewart (1978) has noted that other situations are common.

Propagation of a fault upward and/or downward is generally inferred to be along a continuous surface. However, Roux (1977) and Crans et al (1980) suggest that fault propagation is not that simple and that many individual faults may compose a major fault as generally mapped. The en echelon pattern has appeal because the propagation of an existing listric fault upward or downward may require an unrealistic geometry if that fault is one continuous surface. For example, the Lone Star fault (Bunn, 1951) (Figure 17) and the Mount Enterprise fault zone (Jackson, 1982) in east Texas show very steep dips at the present surface, and their upward extensions to the original surface before erosion would require an unusually large component with near vertical dip. The same problem exists where faults are characterized by many episodes of movement during deposition. Therefore, it is suggested that en echelon faults in cross section may comprise a zone which is generally shown as one continuous listric fault.

Roux (1977) has documented relatively steep branch (horsetail) faults which formed after compaction and flattening of master faults (Figure 18A). This type of branch may reflect reactivation of a fault which was initiated at or near the surface. Where faulting is initiated at depth, a branch fault may form in the downthrown block (Figure 18B).

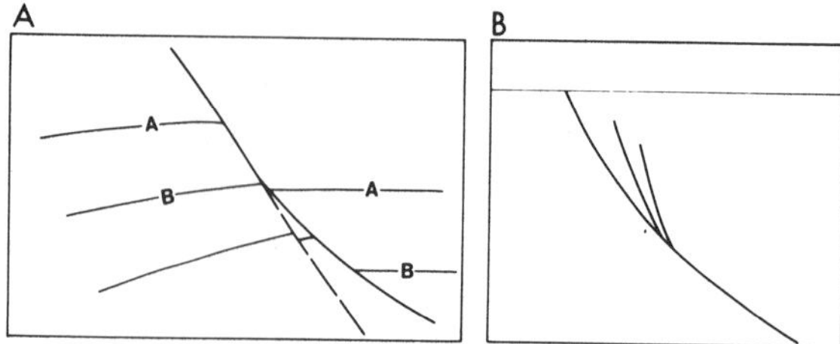

Figure 18—Branching (horsetailing) of normal faults in cross section. A. Late, downward horsetail fault in upthrown block of listric growth fault in Upper Tertiary of offshore Texas. After Roux (1977). B. Hypothetical upward horsetail in downthrown block of major fault.

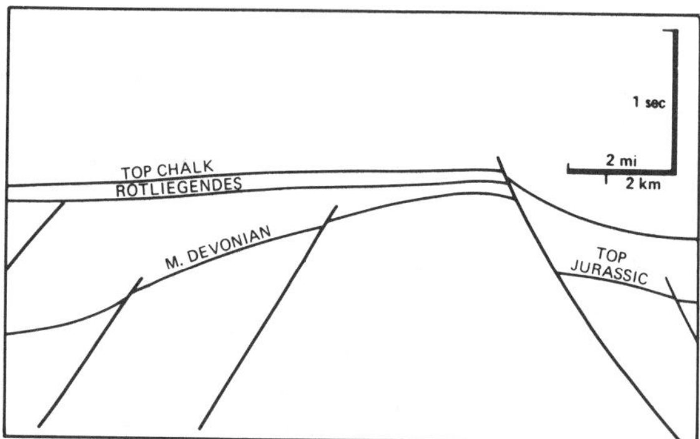

Figure 19—Seismic cross section of Argyll field, North Sea, with listric fault bounding riftlike Central graben on west. After Pennington (1975).

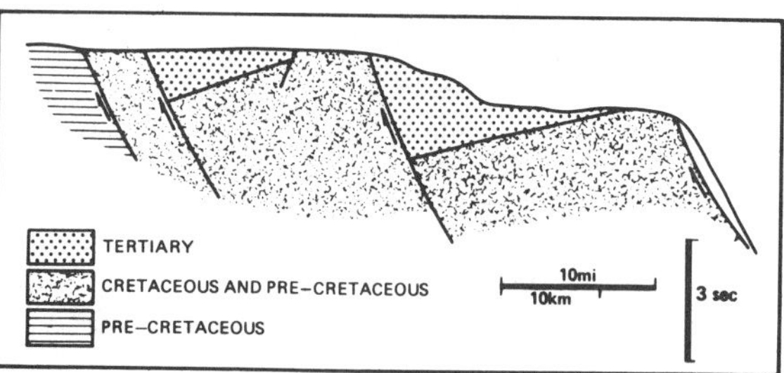

Figure 20—Schematic cross section of southern part of Bay of Biscay showing listric faults which formed during rifting. After Boillot et al (1971).

Growth Faults

If growth faults are defined as those which were active during deposition, almost all normal faults are growth faults because the downthrown block is a likely depositional site. Listric growth faults seemingly are regarded by most workers as a basic feature of regions where the faults are considered to be sedimentary, but it is now reasonable to conclude that listric growth faults are common even where the faults offset the basement (Figures 19-23). It seems possible that a listric growth fault may represent a zone of smaller listric faults or a zone of en echelon faults (Figure 24). A reason for that suggestion is that an original listric fault surface, if extended during significant growth (deposition) as a single continuous surface, would probably not retain a realistic shape for a normal fault.

Evidence for Listric Faults

The best types of evidence include abundant data on the position of fault surfaces in the subsurface, generally in an oil or gas field, unusual outcrop data where local relief allows delineation of the fault surface with depth, and seismic definition of the fault (Bally, 1983). Indirect, but not conclusive, evidence suggesting listric faults includes:

1. Sharply arcuate fault patterns.
2. Sharply arcuate uplifts or basins.
3. Increase in stratal dip in hanging walls with depth together with increase in dip toward controlling growth

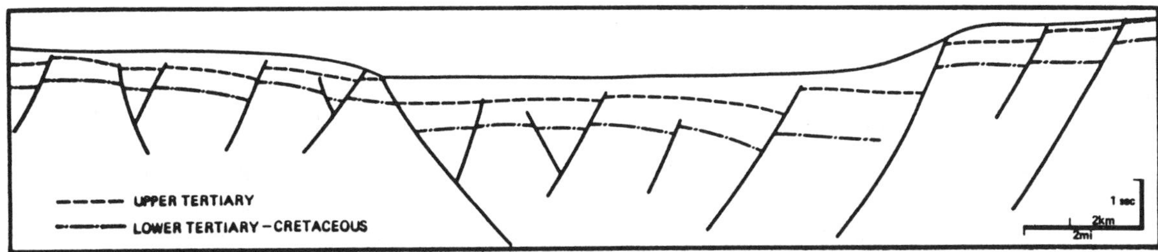

Figure 21—Seismic cross section in central Mediterranean region illustrating listric normal faults which formed during rifting.

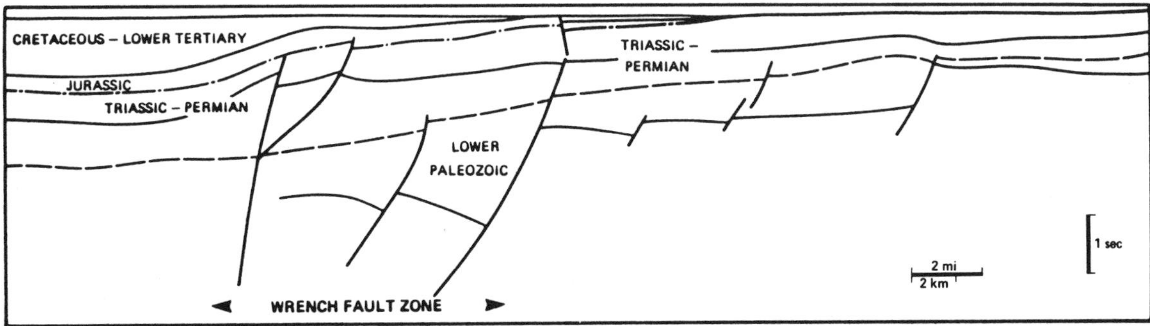

Figure 22—Seismic cross section across Tornquist-Teisseyre wrench fault zone separating Danish-Polish basin (left) from Fennoscandian shield (right). Fault zone in cross section contains lower Paleozoic listric normal growth faults.

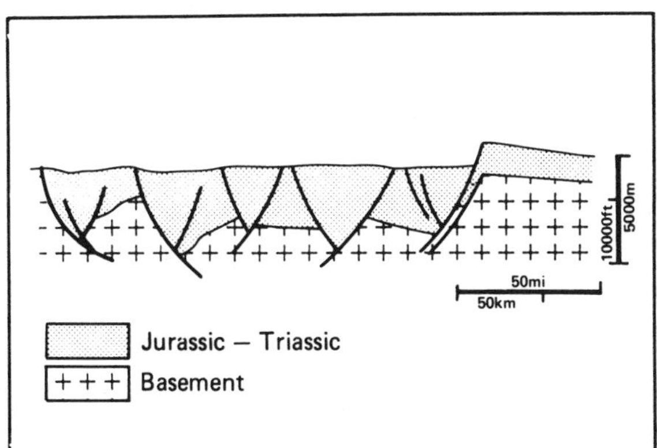

Figure 23—Palinspastic paleostructural cross section of eastern Italian Alps showing development of basin due to movement along listric normal faults. After Bernoulli et al (1979).

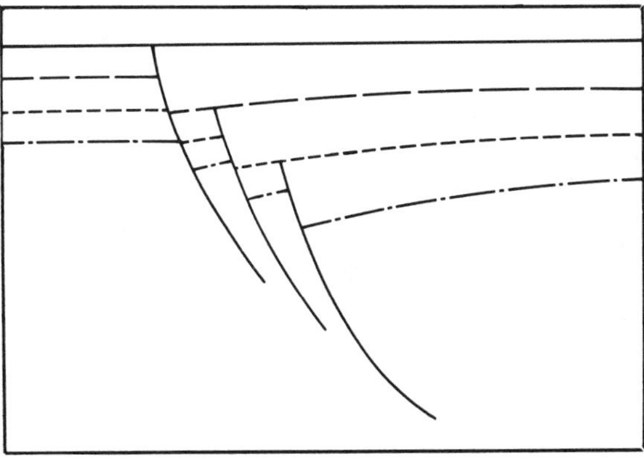

Figure 24—Hypothetical configuration of growth fault in cross section showing propagation of individual listric faults within zone.

fault (reverse drag).

4. Reverse drag in hanging wall where footwall strata show no evidence of rotation about an axis parallel to the strike of the fault. If strata in the subsurface were inclined before faulting, the attitude of these beds with respect to the fault may not be a criterion for a listric fault. For example, the Nacatoch in Figure 17 dips away from the listric fault. Also, it should be noted that absolute movement of the downthrown block from geometric considerations could result in reverse drag along a planar fault with significant lateral changes in throw (Figure 25).

5. Differential tilt between imbricate fault blocks (successively steeper dips in the dip direction of the faults) (Wernicke and Burchfiel, 1982).

6. Progradational stratigraphic succession, with thick ductile shale below brittle sandstone.

Planar faults rather than listric faults may form where the affected strata are entirely brittle (i.e., fault dies out above any ductile rocks) or, in some cases, where the fault has not been deformed.

OCCURRENCES

In terms of global tectonics, listric normal faults occur:

1. In rifts within various geologic settings. Some may

precede formation of passive continental margins (e.g., Bally et al, 1981; Harding, 1983).

2. On passive continental margins during drifting as they form and subside.

3. As sedimentary faults related to subsidence of passive continental margins (e.g., Bruce, 1973; Bally et al, 1981), with "base failure" (involving overpressured shale and a salt) or gravity sliding.

4. In deformed basins, including those which formed along passive margins (miogeosynclines) (Bally et al, 1981)

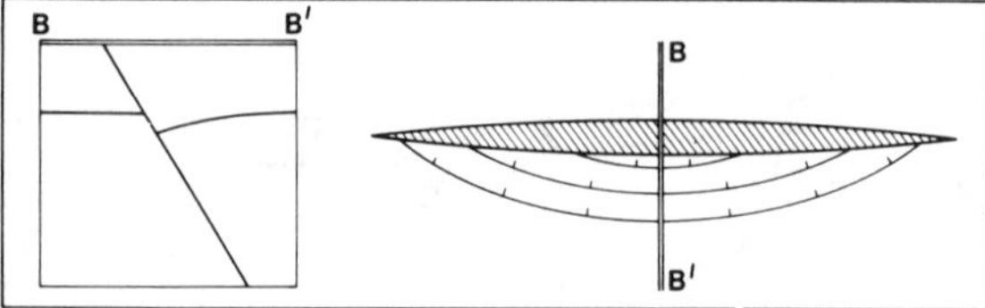

Figure 25—Hypothetical structure map and cross section showing reverse drag along a planar normal fault due to movement of downthrown block of areally restricted fault.

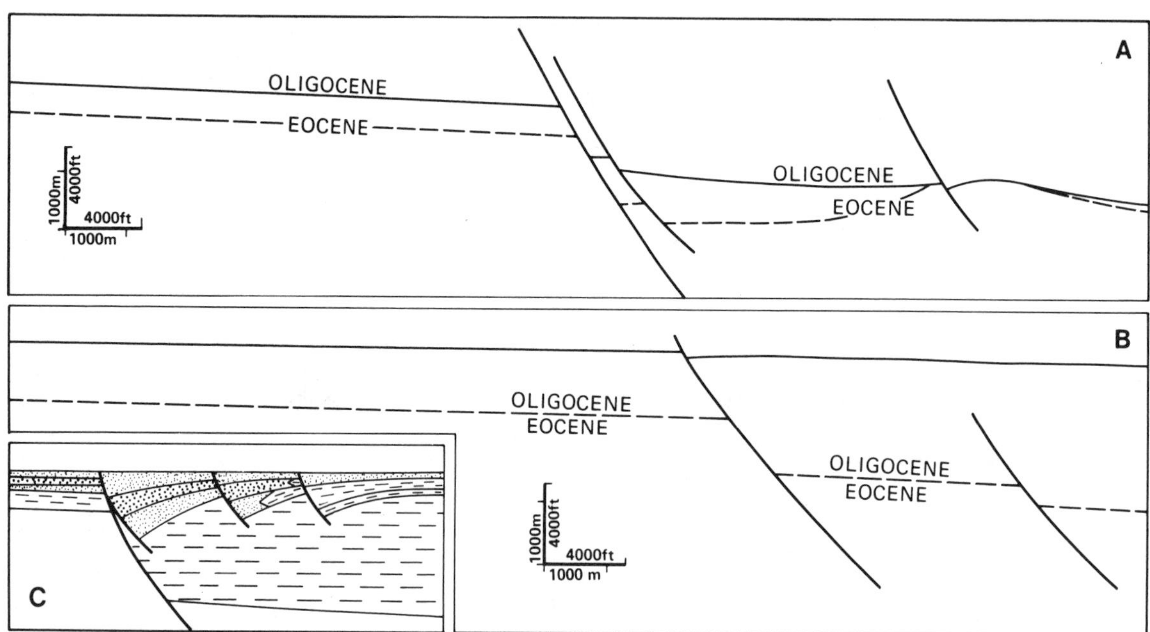

Figure 26—Cross sections A and B across Vicksburg flexure. Significant subregional displacement across this fault zone, with interpretive listric faults, apparently reflects basinal development. C. Schematic cross section showing possible relationship between major fault in basinal development (like that reflected by Vicksburg flexure) and sedimentary faults. After unpublished Shell Oil Co. report.

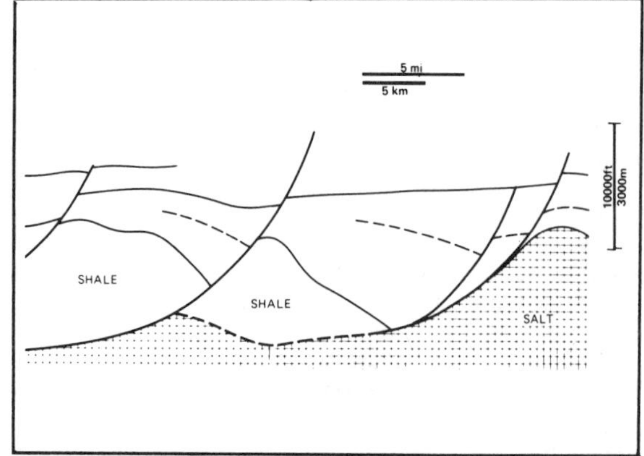

Figure 27—Seismic cross section of listric sedimentary growth faults due to salt and shale flowage. Area is outer continental shelf in offshore southeast Texas and Louisiana where diapiric uplifts are semicontinuous. After Woodbury et al (1973).

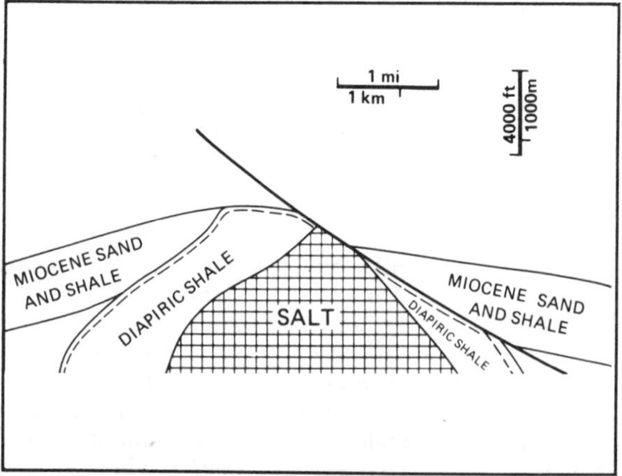

Figure 28—Cross section of Valentine salt dome, La Fourche Parish, Louisiana, with listric normal fault associated with diapiric shale and salt. After Halbouty (1979).

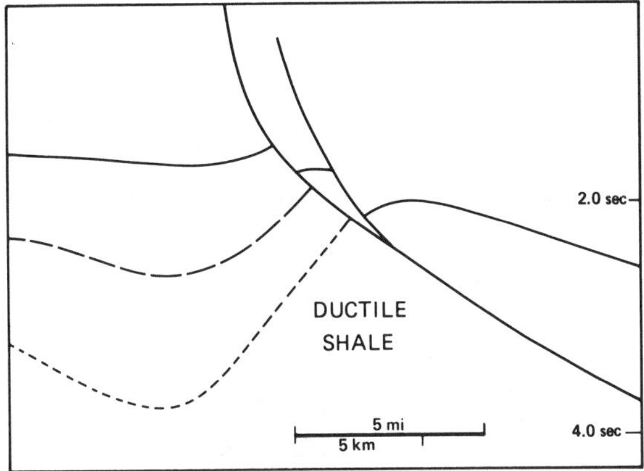

Figure 29—Seismic cross section of offshore Texas showing listric normal faults above ductile shale, which probably is incipient diapir. After Bruce (1973).

tral Mediterranean (Figure 21); Gulf of Suez (Lowell and Genik, 1972; Lowell et al, 1975; interpretation of Robson, 1971, by Harding and Lowell, 1979; Harding, 1983); Lake Superior Precambrian rift (Weiblen and Morey, 1980); and Rio Grande rift (Brown et al, 1983).

Passive Continental Margins

Falvey (1974) suggested that rifted stratigraphic sections and basement on passive margins commonly underlie less deformed sequences which formed during "drifting." Although the subsidence of passive margins undoubtedly reflects to some extent isostatic adjustment to the load of the sedimentary prism (Dietz, 1963; Hsü, 1965; Bott, 1978), subsidence to significant depths is thought to be by movement along basement faults (e.g., Shelton, 1968). It is suggested that these faults may be listric (Figure 26), that they are similar to and possible outgrowths of faults bounding rifts at earlier stages, and that they may be ulti-

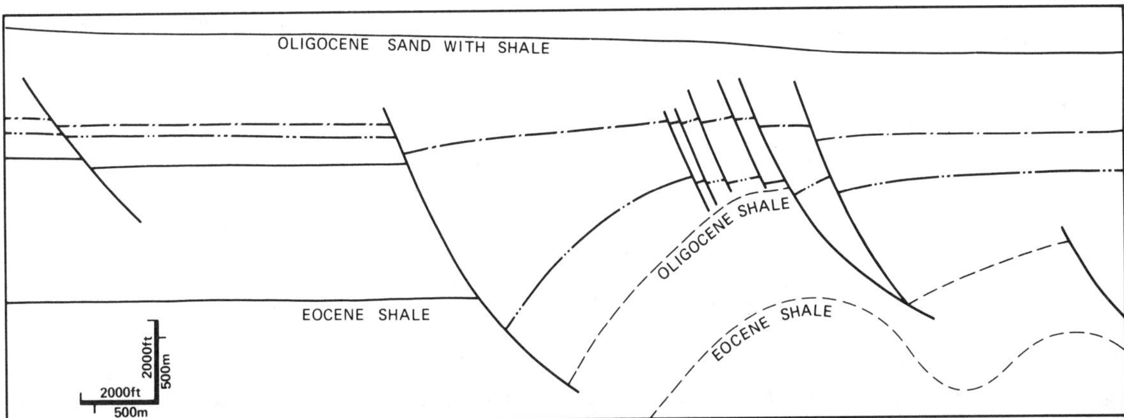

Figure 30—Cross section of North Maude Traylor field, Jackson and Calhoun Counties, Texas. Listric normal fault is related to flowage of ductile Eocene shale. After unpublished Shell Oil Co. report.

or as foredeeps.

5. As late-orogenic and postorogenic faults after earlier formation of foldbelts—very similar to rifts (Bally et al, 1981).
6. In axial zones of orogenes on active continental margins (Beck et al, 1975).
7. Along transform fault boundaries as a result of transtension or in extended upper part of transpressional (flower) structures.

Rifts

Examples of rifts, excluding postorogenic faults, where listric normal faults are fairly well documented or where interpretation of them from available data is reasonable, include North Sea (e.g., Bowen, 1975; Pennington, 1975; Evans and Parkinson, 1983; Harding, 1983) (Figure 19); Bay of Biscay (Boillot et al, 1971; de Charpal et al, 1978; Montadert et al, 1979) (Figure 20); offshore eastern United States (Sheridan, 1974; 1977; Crutcher, 1983); cen-

mately responsible for sedimentary faults where ductile strata are thick owing to movement along this type of fault during deposition (Figure 26C). Basement-involved faults in this type of setting are illustrated by Morgan and Dowdall (1983) from Baltimore Canyon Trough, and by Petrobras (1983) from Potiguar basin, offshore Brazil.

Sedimentary Faults on Passive Margins

Deformation of progradational sedimentary sequences on passive margins may be dominated by half grabens, reverse drag, and rollover related to listric normal faults, which commonly are associated with overpressured or diapiric shale or salt diapirs, and which were active during deposition. The best known areas for this type of deformation with listric growth faults are the northern Gulf of Mexico (Figures 6, 27-30) and the Niger Delta (e.g., Hardin and Hardin, 1961; Ocamb, 1961; Bruce, 1973; Busch, 1975; Lehner and de Ruiter, 1977; Curtis and Picou, 1978; Evamy et al, 1978; Roux, 1979). Other areas are Sarawak

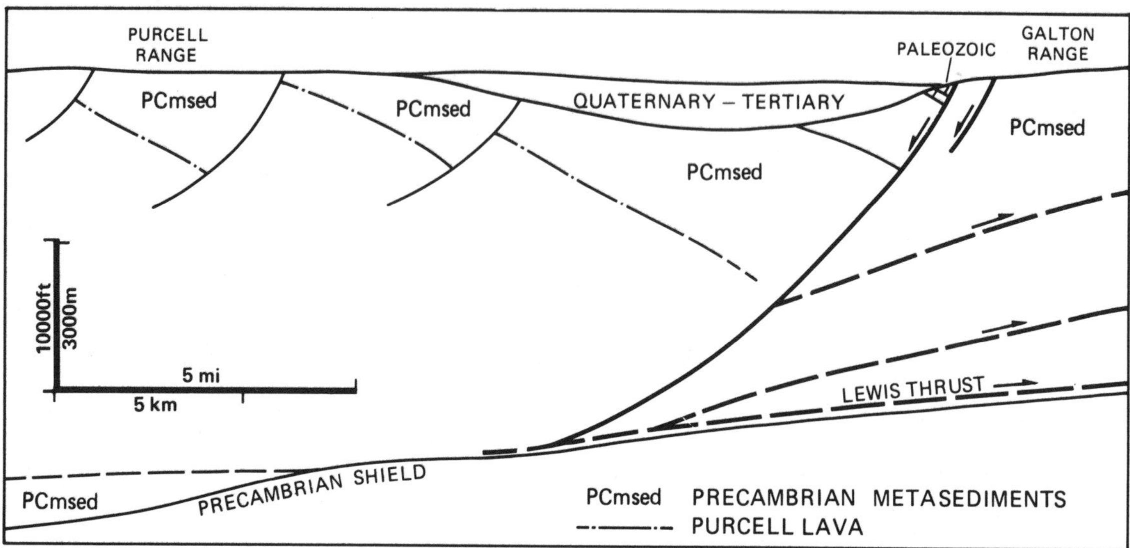

Figure 31—Seismic cross section across Rocky Mountain trench in southwestern Canada showing listric normal fault which developed by opposite movement (backslippage) along earlier listric thrust fault. After Bally et al (1966).

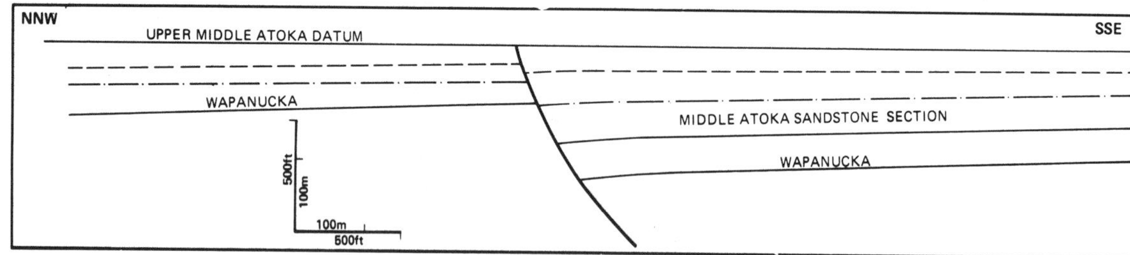

Figure 32—Paleostructural cross section of part of Arkoma basin, Oklahoma, with Atokan (Pennsylvanian) listric normal growth fault. After G.W. Hart (1978; personal communication, 1983).

and Sabah, east Malaysia (Schuab and Jackson, 1958; Scherer, 1980), offshore Brazil (Brown and Fisher, 1977), offshore eastern North America (Jansa and Wade, 1975; Sheridan, 1977), and North Sea (Gibbs, 1983).

Late-Orogenic to Postorogenic Faults

Balley et al (1981) described examples of listric normal faults from western North America which may be interrelated. These include (1) faults that may represent opposite movement ("backslippage") along preexisting listric thrust faults (Bally et al, 1966; Royse et al, 1975; Allmendinger et al, 1983) (Figure 31); (2) rifts, horsts and grabens, and half grabens; and (3) two types of faults in a mountain and valley system, with (a) an older decollement zone exposed in the mountain and (b) younger listric faults of the valley which may offset the decollement and contribute further to basinal development in the valley.

This region is where some of the early studies demonstrated tilted fault blocks and low-angle listric normal faults (Davis, 1925; Longwell, 1933, 1945). Additional studies documenting listric normal faults in this region include those of Mackin (1960), Moore (1960), Osmond (1960), Hamblin (1965), Hamilton and Myers (1966), Anderson (1971), Armstrong (1972), MacDonald (1976), Proffett (1977), Bally et al (1981), and Robison (1983). It is suggested that listric normal faults in this tectonic setting may be common in many orogenic belts. Listric faults may bound some grabens and half grabens in which the Triassic of the Appalachian system is present (Barrell, 1915; King, 1959, p. 50). As noted previously, faults in this type of setting may be parts of rift systems; correspondingly, the Triassic would be related to rifting which preceded opening of the Atlantic.

Axial Zones of Orogenes

In regard to subduction zones of island arcs, Beck et al (1975) have proposed that Pacific island arcs show a central collapse zone. The Andean orogene also contains an axial zone of block faulting and collapse (Figure 10). This zone apparently has a causal relationship to a young volcanic belt.

Transform Boundaries and Strike-Slip Faults

Normal faults may form as a result of transtension associated with lateral movement—transform and/or strike-slip fault zones (Wilcox et al, 1973) (Figure 5). These normal faults may be listric (e.g., Southwest Lone Grove field, southern Oklahoma; Westheimer and Schweers, 1956). Those superimposed on a more fundamental crustal wrench or transform fault zone may be large-scale features (Figure 22).

In areas of transpression, normal faults in the extended part of the uplifted flower structure may possibly be listric (Figure 9), similar to those associated with compressional folding or extension over salt or igneous intrusives.

Deformed Basins

Faults in these settings are essentially of the same types as those which formed during rifting and drifting. Bernoulli et al (1979) have mapped Mesozoic listric normal growth faults in the eastern Italian Alps (Figure 23). These faults affect a dominantly carbonate section. Woodward (1961) and Wagner (1976) have described growth faults affecting the lower Paleozoic section in the Appalachian basin; they may well be listric.

The Arkoma basin of Oklahoma and Arkansas was a foredeep during the Atokan (Pennsylvanian), when both basement faults and listric sedimentary faults were active (G. W. Hart, personal communication, 1983) (Figure 32). The latter type is very similar to those of the Gulf of Mexico basin.

CONCLUSIONS

Listric normal faults may be an integral part of basinal development, and formation of several types of structures, with potential for entrapment of hydrocarbons, results from movement along them. Listric normal faults occur in the various geologic settings reflecting extensional stress regimes that are crustal and/or relatively superficial (restricted to the sedimentary prism). Their formation is enhanced by, or perhaps requires, a ductile "substrate."

The detailed three-dimensional geometry of listric faults may be expressive of subsidiary structures with some exploration or development potential. For example, short en echelon faults in cross section may constitute a zone, and define tilted fault blocks which are areally and stratigraphically restricted. Further, splays or subsidiary faults, commonly developed near the ends of major faults, present conditions for potential traps.

REFERENCES CITED

Allmendinger, R. W., J. W. Sharp, D. Von Tish, L. Serpa, L. Brown, S. Kaufman, J. Oliver, and R. B. Smith, 1983, Cenozoic and Mesozoic structure of the eastern Basin and Range province, Utah, from COCORP seismic-reflection data: Geology, v. 11, p. 532-536.

Anderson, E. M., 1942, The dynamics of faulting and dyke formation with applications to Great Britain: Edinburgh, Oliver and Boyd, 191 p.

Anderson, R. E., 1971, Thin-skinned distension in Tertiary rocks of southeastern Nevada: GSA Bulletin, v. 82, p. 43-58.

Armstrong, R. L., 1972, Low-angle (denudation) faults, hinterland of the Sevier orogenic belt, eastern Nevada and western Utah: GSA Bulletin, v. 83, p. 1729-1754.

Bally, A. W., ed., 1983, Seismic expression of structural styles: AAPG Studies in Geology 15, vs. 1-2.

——— D. Bernoulli, G. A. Davis, and L. Montadert, 1981, Listric normal faults: Oceanologica Acta, 26th International Geological Congress, Paris, 1980, p. 87-101.

——— P. L. Gordy, and G. A. Stewart, 1966, Structure, seismic data and orogenic evolution of southern Canadian Rocky Mountains: Bulletin of Canadian Petroleum Geology, v. 14, p. 337-381.

Beck, R. H., P. Lehner, P. Diebold, G. Bakker, and H. Doust, 1975, New geophysical data on key problems on global tectonics: 9th World Petroleum Congress Proceedings, Tokyo, v. 2, p. 3-17. London, Applied Science Publishers Ltd.

Bernoulli, D., C. Caron, P. Homewood, O. Käglin, and J. Van Stuijvenberg, 1979, Evolution of continental margins in the Alps: Schweizerische Mineralogische und Petrographische Mitteilungen, v. 59, p. 165-170.

Boillot, G., P. A. Dupeuble, M. Lamboy, et al, 1971, Structure et histoire geologique de la marge continentale au nord d l'Espagne (entre 4° et 9°W), *in* Histoire structurale du Golfe de Gascogne: Institut François du Petrole, Collected Colloquial Seminars, no. 22, v. 22, p. V6-1–V6-52.

Bott, M. H. P., 1978, Subsidence mechanisms at passive continental margins, *in* Geological and geophysical investigations of continental margins: AAPG Memoir 29, p. 3-9.

Bowen, J. M., 1975, The Brent field, *in* A.W. Woodland, ed., Petroleum and the continental shelf of north-west Europe: New York, John Wiley and Sons, p. 353-362.

Brown, L. D., S. Kaufman, and J. E. Oliver, 1983, COCORP seismic traverse across the Rio Grande rift, *in* Seismic expression of structural styles: AAPG Studies in Geology 15, v. 2, p. 2.2.1-1–2.2.1-6.

Brown, L. F., Jr., and W. L. Fisher, 1977, Seismic-stratigraphic interpretation of depositional systems—examples from Brazilian rift and pull-apart basins, *in* Seismic stratigraphy—applications to hydrocarbon exploration: AAPG Memoir 26, p. 213-248.

Bruce, C. H., 1973, Pressured shale and related sediment deformation—mechanisms for development of regional contemporaneous faults: AAPG Bulletin, v. 57, p. 878-886.

Bunn, J. R., 1951, Lone Star field, *in* F. A. Herold, ed., Occurrence of oil and gas in northeast Texas: University of Texas Publication 5116, p. 195-200.

Burchfiel, B. C., and L. Royden, 1982, Carpathian foreland fold and thrust belt and its relation to Pannonian and other basins: AAPG Bulletin, v. 66, p. 1179-1195.

Busch, D. A., 1975, Influence of growth faulting on sedimentation and prospect evaluation: AAPG Bulletin, v. 59, p. 217-230.

Carey, S. W., 1958, The tectonic approach to continental drift, *in* S. W. Carey, ed., Continental drift—a symposium: Hobart, Australia, University of Tasmania, p. 173-363.

Cloos, E., 1968, Experimental analysis of Gulf Coast fracture patterns: AAPG Bulletin, v. 52, p. 420-444.

Cloos, H., 1930, Zur experimentellen Tektonik: Naturwissenschaften, v. 18, p. 741-747.

Cohen, C. R., 1982, Model for a passive to active continental margin transition: implications for hydrocarbons exploration: AAPG Bulletin, v. 66, p. 708-718.

Crans, W., G. Mandl, and J. Haremboure, 1980, On the theory of growth faulting—a geomechanical delta model based on gravity sliding: Journal of Petroleum Geology, v. 2, p. 265-307.

Crutcher, T. D., 1983, Southeast Georgia embayment, *in* Seismic expression of structural styles: AAPG Studies in Geology 15, v. 2, p. 2.2.3-27–2.2.3-29.

Curtis, D. M., and E. B. Picou, Jr., 1978, Gulf Coast Cenozoic—a model for the application of stratigraphic concepts to exploration on passive margins: Gulf Coast Association of Geological Societies Transactions, v. 28, p. 103-120.

Davis, G. A., J. L. Anderson, and E. G. Frost, 1979, Regional Miocene detachment faulting and early Tertiary(?) mylonitic terranes in the Colorado River trough, southeastern California and western Arizona, *in* P. L. Abbott, ed., Geological excursions in the southern California area: San Diego State University, Department of Geological Sciences, p. 73-108.

Davis, W. M., 1925, The Basin Range problem: Proceedings of the National Academy of Sciences, v. 11, p. 387-392.

de Charpal, O., P. Guennoc, L. Montadert, and D. G. Roberts, 1978, Rifting, crustal attenuation and subsidence in the Bay of Biscay: Nature, v. 275, p. 706-711.

de Sitter, L. U., 1964, Structural geology, second edition: New York, McGraw-Hill Book Co., p. 119-137.

Dewey, J. F., and J. M. Bird, 1970, Mountain belts and the new global tectonics: Journal of Geophysical Research, v. 75, p. 2625-2647.

Dietz, R. S., 1963, Collapsing continental rises—an actualistic concept of geosynclines and mountain building: Journal of Geology, v. 71, p. 314-333.

Edwards, M. B., 1976, Growth faults in Upper Triassic deltaic sediments, Svalbard: AAPG Bulletin, v. 60, p. 341-355.

Evamy, B. D., J. Haremboure, P. Kammerling, W. A. Knaap, F. A. Molloy, and P. H. Rowlands, 1978, Hydrocarbon habitat of Tertiary Niger Delta: AAPG Bulletin, v. 62, p. 1-39.

Evans, A. C., and D. N. Parkinson, 1983, A half-graben and tilted fault block structure in the northern North Sea, *in* Seismic expression of structural styles: AAPG Studies in Geology 15, v. 2, p.2.2.2-7-2.2.2-11.

Falvey, D. A., 1974, The development of continental margins in plate tectonic theory: APEA Journal, v. 14, p. 59-106.

Fanshawe, J. R., and A. G. Alpha, 1954, Tectonic map of portions of southern Montana and northern Wyoming: Billings Geological Society 5th Annual Field Conference Guidebook.

Galloway, W. E., D. K. Hobday, and K. Magara, 1982, Frio Formation of Texas Gulf coastal plain—depositional systems, structural framework, and hydrocarbon distribution: AAPG Bulletin, v. 66, p. 649-688.

Gibbs, A. D., 1983, Secondary detachment above basement faults in North Sea—Clyde field growth fault (abs.): AAPG Bulletin, v. 67, p. 469.

Hafner, W., 1951, Stress distributions and faulting: GSA Bulletin, v. 62, p. 373-398.

Halbouty, M. T., 1979, Salt domes, Gulf region, United States and Mexico, second edition: Houston, Gulf Publishing Co., p. 47-87.

Hamblin, W. K., 1965, Origin of "reverse drag" on the downthrown side of normal faults: GSA Bulletin, v. 76, p. 1145-1164.

Hamilton, W., and W. B. Myers, 1966, Cenozoic tectonics of the western United States: Review of Geophysics, v. 4, p. 509-549.

Handin, J., and R. V. Hager, 1957, Experimental deformation of sedimentary rocks under confining pressure: tests at room temperature on dry samples: AAPG Bulletin, v. 41, p. 1-50.

Hardin, F. R., and G. C. Hardin, 1961, Contemporaneous normal faults of Gulf Coast and their relation to flexures: AAPG Bulletin, v. 45, p. 238-248.

Harding, T. P., 1974, Petroleum traps associated with wrench faults: AAPG Bulletin, v. 58, p. 1290-1304.

——— 1983, Graben hydrocarbon plays and structural styles: Geologie en Mijnbouw, v. 62, p. 3-23.

——— and J. D. Lowell, 1979, Structural styles, their plate-tectonic habitats, and hydrocarbon traps in petroleum provinces: AAPG Bulletin, v. 63, p. 1016-1058.

Hart, G. W., 1978, Structural influence of growth faults on middle Atokan sandstone distribution in the Oklahoma portion of the Arkoma basin (abs.): GSA Abstracts with Programs, v. 10, p. 7.

Heard, H. C., 1966, Illustration on cover: Journal of Geological Education, v. 14, no. 5.

Holmes, A., 1965, Principles of physical geology, second edition: New York, Nelson and Sons/Ronald Press Co., p. 1044-1108 p.

Hsü, K. J., 1965, Isostosy, crustal thinning, mantle changes, and the disappearance of ancient land masses: American Journal of Science, v. 263, p. 97-109.

Hubbert, M. K., 1951, Mechanical basis for certain familiar geologic structures: GSA Bulletin, v. 62, p. 355-372.

——— and W. W. Rubey, 1959, Role of fluid pressure in mechanics of overthrust faulting: I. mechanics of fluid-filled porous solids and its application to overthrust faulting: GSA Bulletin, v. 70, p. 115-166.

Humphris, C. C., 1978, Salt movement on continental slope, northern Gulf of Mexico: AAPG Studies in Geology 7, p. 69-86.

Jackson, M. P. A., 1982, Fault tectonics of the East Texas basin: University of Texas Bureau of Economic Geology Geological Circular 82.4, 31 p.

Jansa, L. F., and J. A. Wade, 1975, Geology of the continental margin off Nova Scotia and Newfoundland, *in* W. J. M. van der Linden and J. A. Wade, eds., Offshore geology of eastern Canada, v. 2—regional geology: Geological Survey of Canada Paper 74-30, p. 51-106.

King, P. B., 1959, The evolution of North America: Princeton, New Jersey, Princeton University Press, 190 p.

Lehner, P., and P. A. C. de Ruiter, 1977, Structural history of the Atlantic margin of Africa: AAPG Bulletin, v. 61, p. 961-981.

Longwell, C. R., 1933, Rotated faults in the Desert Range, southern Nevada (abs.): GSA Bulletin, v. 44, p. 93.

——— 1945, Low-angle normal faults in the Basin and Range province: Transactions of the American Geophysical Union, v. 26, p. 107-118.

——— and R. E. Flint, 1962, Introduction to physical geology, second edition: New York, John Wiley and Sons, 504 p.

Lowell, J. D., and G. J. Genik, 1972, Sea-floor spreading and structural evolution of southern Red Sea: AAPG Bulletin, v. 56, p. 247-259.

——— ——— T. H. Nelson, and P. M. Tucker, 1975, Petroleum and plate tectonics of the southern Red Sea, *in* A. G. Fischer and S. Judson, eds., Petroleum and global tectonics: Princeton, New Jersey, Princeton University Press, p. 129-153.

MacDonald, R. E., 1976, Tertiary tectonics and sedimentary rocks along the transition: Basin and Range province to Plateau and Thrust Belt province, Utah, *in* G. J. Hill, ed., Symposium on the geology of the Cordilleran hingeline: Rocky Mountains Association of Geologists, p. 281-317.

Mackin, J. H., 1960, Structural significance of Tertiary volcanic rocks in southwestern Utah: American Journal of Science, v. 258, p. 81-131.

Montadert, L., D. G. Roberts, O. de Charpal, and P. Guennoc, 1979, Rifting and subsidence of the northern continental margin of the Bay of Biscay: Initial Reports of the Deep Sea Drilling Project, v. 48, p. 1025-1060.

Moore, J. G., 1960, Curvature of normal faults in the Basin and Range province of the western United State*r*: U. S. Geological Survey Professional Paper 400-B, p. 409-411.

Morgan, L., and W. Dowdall, 1983, The Atlantic continental margin, *in* Seismic expression of structural styles: AAPG Studies in Geology 15, v. 2, p. 2.2.3-30-2.2.3-35.

Muehlberger, W. R., 1961, Conjugate joint sets of small dihedral angle: Journal of Geology, v. 69, p. 211-219.

Murray, G. E., 1961, Geology of the Atlantic and Gulf Coastal province of North America: New York, Harper and Brothers, p. 167-201.

Ocamb, R. D., 1961, Growth faults of south Louisiana: Gulf Coast Association of Geological Societies Transactions, v. 11, p. 139-173.

Osmond, J. C., 1960, Tectonic history of the Basin and Range province in Utah and Nevada: Mining Engineering, v. 12, p. 251-265.

Pennington, J. J., 1975, Geology of Argyll field, *in* A. W. Woodland, ed., Petroleum and the continental shelf of north-west Europe: New York, John Wiley and Sons, p. 285-297.

Petrobras, 1983, Potiguar basin, Brazil, *in* Seismic expression of structural styles: AAPG Studies in Geology 15, v. 2, p. 2.2.3-59-2.2.3-65.

Proffett, J. M., Jr., 1977, Cenozoic geology of the Yerington district, Nevada, and implications for the nature and origin of Basin and Range faulting: GSA Bulletin, v. 88, p. 247-266.

Reeves, F., 1925, Shallow folding and faulting around the Bearpaw Mountains: American Journal of Science, 5th series, v. 10, p. 187-200.

——— 1946, Origin and mechanics of thrust faults adjacent to the Bearpaw Mountains, Montana: GSA Bulletin, v. 57, p. 1033-1047.

Rettger, R. E., 1935, Experiments on soft rock deformation: AAPG Bulletin, v. 19, p. 271-292.

Robison, B. A., 1983, Low-angle normal faulting, Marys River valley, Nevada, *in* Seismic expression of structural styles: AAPG Studies in Geology 15, v. 2, p. 2.2.2-12-2.2.2-16.

Robson, D. A., 1971, The structure of the Gulf of Suez (Clysmic) rift with special reference to the eastern side: Quarterly Journal of the Geological Society of London, v. 127, p. 247-276.

Roux, W. F., Jr., 1979, The development of growth fault structures: AAPG Structural Geology School Course Notes, 33 p.

Royse, F., M. A. Warner, and D. C. Reese, 1975, Thrust belt structural geometry and related stratigraphic problems, Wyoming-Idaho-northern Utah, *in* Symposium on deep drilling frontiers of the central Rocky Mountains: Rocky Mountains Association of Geologists, p. 41-54.

Schaub, H. P., and A. Jackson, 1958, The northwestern oil basin of Borneo, *in* L. G. Weeks, ed., Habitat of oil: AAPG, p. 1330-1336.

Scherer, F.C., 1980, Exploration in east Malaysia over past decade, *in* Giant oil and gas fields of the decade 1968-1978: AAPG Memoir 30, p. 423-440.

Shelton, J. W., 1968, Role of contemporaneous faults during basinal subsidence: AAPG Bulletin, v. 52, p. 399-413.
Sheridan, R. E., 1974, Atlantic continental margin of North America, *in* C. A. Burk and C. L. Drake, eds., The geology of continental margins: New York, Springer-Verlag, p. 391-408.
——— 1977, Passive margin structural styles: AAPG Structural Geology School Course Notes, 11 p.
Stewart, J. H., 1971, Basin and Range structure—a system of horsts and grabens produced by deep-seated extension: GSA Bulletin, v. 82, p. 1019-1044.
——— 1978, Basin-range structure in western North America, *in* R. B. Smith et al, eds., Cenozoic tectonics and regional geophysics of the western Cordillera: GSA Memoir 152, p. 1-31.
Terzaghi, K., and R. B. Peck, 1948, Soil mechanics, in engineering practice: New York, John Wiley and Sons, p. 394-406.
Thompson, T. L., 1976, Plate tectonics in oil and gas exploration of continental margins: AAPG Bulletin, v. 60, p. 1463-1501.
Wagner, W. R., 1976, Growth faults in Cambrian and Lower Ordovician rocks of western Pennsylvania: AAPG Bulletin, v. 60, p. 414-427.
Weber, K. J., and E. Daukoru, 1975, Petroleum geology of the Niger Delta: 9th World Petroleum Congress Proceedings, Tokyo, v. 2, p. 209-221. London, Applied Science Publishers Ltd.
Weiblen, P. W., and G. B. Morey, 1980, A summary of the stratigraphy, petrology and structure of the Duluth complex: American Journal of Science, v. 280-A, p. 88-133.
Wernicke, B., and B. C. Burchfiel, 1982, Modes of extensional tectonics: Journal of Structural Geology, v. 4, p. 105-115.
Westheimer, J. M., and F. P. Schweers, 1956, Southwest Lone Grove field, Carter County, Oklahoma, *in* Petroleum geology of southern Oklahoma, vol. I: AAPG, p. 144-153.
Wilcox, R. E., T. P. Harding, and D. R. Seely, 1973, Basic wrench tectonics: AAPG Bulletin, v. 57, p. 74-96.
Wise, D. U., 1963, Keystone faulting and gravity sliding driven by basement uplift of Owl Creek Mountains, Wyoming: AAPG Bulletin, v. 47, p. 586-598.
Woodbury, H. O., I. B. Murray, P. J. Pickford, and W. H. Akers, 1973, Pliocene and Pleistocene depocenters, outer continental shelf, Louisiana and Texas: AAPG Bulletin, v. 57, p. 2428-2439.
Woodward, H. P., 1961, Reappraisal of Appalachian geology: AAPG Bulletin, v. 45, p. 1625-1633.
Ziegler, P. A., 1982, Geological atlas of western and central Europe: Amsterdam, Elsevier, 130 p.

THE AMERICAN ASSOCIATION OF PETROLEUM GEOLOGISTS BULLETIN
VOL. 52, NO. 3 (MARCH, 1968), P. 420-444, 38 FIGS.

EXPERIMENTAL ANALYSIS OF GULF COAST FRACTURE PATTERNS[1]

ERNST CLOOS[2]
Baltimore, Maryland 21218

ABSTRACT

Gulf Coast faults are normal faults with the exception of those around salt domes. Normal faults must accommodate both a horizontal and vertical component of displacement. The horizontal component increases the distance between two points on opposite sides of the fault surface: The vertical component is larger and must be accommodated simultaneously by the same mechanism. Experiments show that horizontal and vertical components can be derived from one motion in practically horizontal surfaces.

The accumulation of all horizontal components in the Gulf Coast embayment makes considerable horizontal displacement of the sedimentary blanket necessary.

The regional Gulf Coast fault pattern and its many local variations are therefore thought to be caused by regional gravity creep of the sedimentary blanket into the basin. As creep takes place the sliding sediments break away from the stationary ones forming a marginal graben, and, nearer the coast, asymmetrical down-to-basin faults with reverse drag, and antithetic faults.

Creep has taken place for a long time as is proved by the fact that there was sedimentation after and during faulting in association with many growth faults. The first faults may well have been the peripheral ones as they occurred first in the experiments. Faults within the basin are later, as shown by the facts that they were buried under sediments, grew upward into them, and transect younger formations.

Local domes show local fault patterns, but at many places the regional fault pattern suppressed local ones. Experimentation suggests that a general mechanism can explain both regional and local phenomena rather simply by one general cause which is modified by specific local conditions within the large and heterogeneous Gulf Coast embayment.

Introduction

Faults and fault patterns in the coastal plain bordering the Gulf of Mexico have been described comprehensively by Murray (1961, chaps. 4 and 5). With very few exceptions "normal (gravity-tensional) faults predominate" (Murray, p. 167; see Fig. 1, this paper).

Some of these normal faults are caused by regional stresses which result in regional fault patterns, and others are due to local stresses such as those generated by the rise of salt domes. Interference between these two patterns is common.

The faults can be classified also according to special characteristics:

1. Normal faults which limit a graben such as are prominent in the inner zone of the Gulf Coast embayment: Balcones-Luling, Mexia-Talco, South Arkansas, and Pickens-Gilbertown fault systems;
2. Down-to-basin faults with reverse drag (sags, rollovers) are most common near the coast;
3. Antithetic faults;
4. "Growth faults" (Ocamb, 1961) or "contemporaneous faults" (Hardin and Hardin, 1961) with increased sedimentary thickness on the downthrown side; and
5. Faults related to salt domes.

There are gradations and overlapping characteristics among the different types of faults as well as combinations, which make it difficult and artificial to separate them. Therefore I have tried to design models for typical situations such as a graben, a down-to-basin fault with reverse drag and antithetic faults, and a salt dome with radial faults. After typical situations were reproduced I tried to imitate the regional pattern of Gulf Coast faulting in one model, with moderate success.

Experimental Method, Purpose, and Procedure

The experimental method used here has been described by H. Cloos (1928, 1930, 1931) and E. Cloos (1955) and is based on the fact that soft clay fractures when subjected to stress.

[1] Manuscript received, April 10, 1967; accepted, July 31, 1967.

[2] Professor of Geology, The Johns Hopkins University, and consultant, Esso Production Research Company.

The late H. N. Fisk, former Professor of Geology at Louisiana State University and later Chief of the Geological Research Section, Humble Oil and Refining Company, now Esso Production Research Company, Houston, suggested an experimental study of Gulf Coast faulting and especially down-to-basin faults. The work was done at the Geological Laboratory, The Johns Hopkins University. I am grateful to the Esso Production Research Company for permission to publish this paper and for much stimulating discussion at the Esso laboratory.

I am grateful also for much needed editorial help by readers and critics. Illustrations were prepared by Josephine Spemann, Ulf Wiedemann, and Charles Weber.

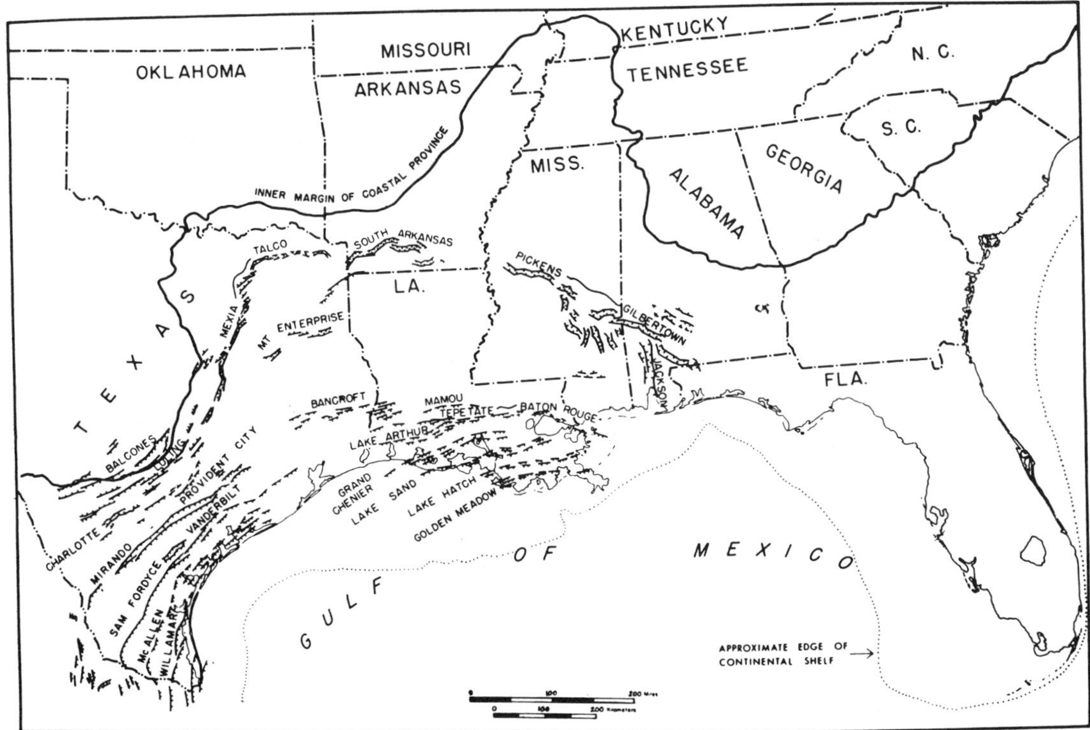

FIG. 1.—Diagrammatic representation of principal strike-fault systems in northern Gulf Coast (from Murray, 1961, Fig. 4.1; published with permission of Harper & Row, New York).

It is easy to imitate fracture patterns in clay models. The value of such imitation lies not in quantitative determinations of parameters but in the imitation of a movement pattern or stress pattern which is thought to be the cause of fractures or fracture patterns, or in the correction of erroneous notions which do not explain an observed pattern. In the following pages experiments are described for each geological situation which was studied. Each experiment was repeated many times and improved as errors were eliminated. For details on the experimental method, see E. Cloos (1955).

General Considerations

Because Gulf Coast faults are normal faults their displacements have a vertical and a horizontal component. Figure 2A shows this on a cross section and Figure 2B shows a map view of the horizontal component and the fault trace. The horizontal component means an extension between two points on opposite sides of the fault plane. On curved faults the horizontal component is less at upper levels and larger at lower and gently dipping levels. This fact is important in the interpretation of curved normal faults and reverse drag as shown subsequently.

In most cross sections of Gulf Coast faults the horizontal component is underemphasized in comparison with the vertical component by reason of the gross vertical exaggerations which distort the geological profile (see Murray, 1961, Figs. 4.22b, 4.25, 4.27, 4.31, and others). Exaggerations are

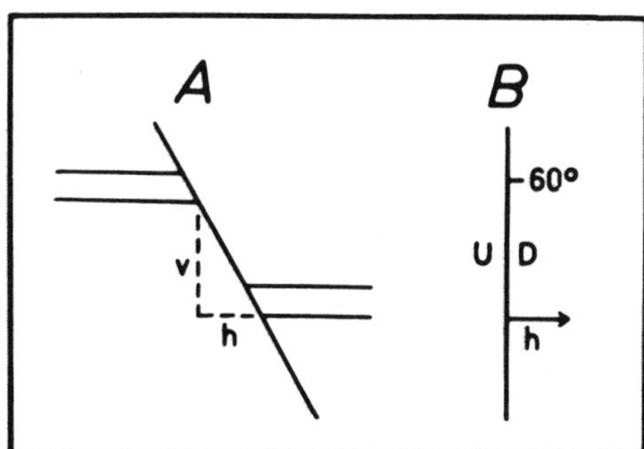

FIG. 2.—A, Profile of normal or gravity fault with horizontal components *h* and vertical component *v*. B, Map view of 2A.

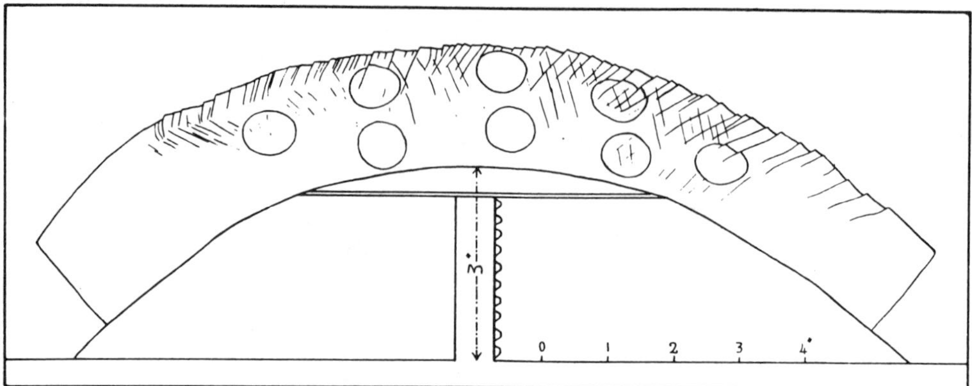

FIG. 3.—In bent beam, extension is distributed through length of beam with faults spaced evenly. Vertical uplift of clay model is carried by stiff but flexible metal sheet.

commonly used because sections with a scale of 1:1 do not permit showing stratigraphic details, although such sections do convey a far more accurate picture.

GRABEN STRUCTURES

As shown in Figure 1 (Murray, 1961, Fig. 4.1) the area which contains the "principal strike fault systems in northern Gulf coastal province" is bounded on the west, north, and east by the Balcones-Luling, Mexia-Talco, South Arkansas, Pickens, and Gilbertown graben systems. The Mexia-Talco graben system consists of smaller *en échelon* grabens but the remaining systems are almost uninterrupted. The Gulf Coast graben system was discussed recently by Walthall and Walper (1967), who compared it with African and European systems (p. 102) and concluded that these grabens are the result of post-orogenic movements of the Ouachita belt.

A graben bounded by two or more normal faults is an area of extension, as is generally recognized. The term "extension" does not imply that the faults are due either to tension or to compression. Hubbert (1951, p. 367) has shown that at depths of more than 30 ft faults in Gulf Coast sediments are the result of compressive stresses. However, there is no question but that an increase of distance normal to the strike of faults accompanies normal faulting. Consequently models were built in which the clay was extended.

If the extension is distributed about equally across the entire length of the model, normal faults will form that strike normal to the direction of extension and dip from 40° to 70° (*see also* Hubbert, 1951). In profile the dip of the faults will be either to the left or right.

Such a setup was described by H. Cloos (1930, p. 741) and E. Cloos (1955, p. 246) and can be produced readily by modeling the clay on a one-way stretch belt, or on a rubber bladder, or by bending a beam by doming as shown in Figure 3.

A graben is a special case of such extension and can be produced experimentally by placing a clay block over two metal sheets or pieces of plywood which then are pulled apart. The clay block is not extended across its entire length but only above the joint between the base plates; then shear fractures appear at the separation point and grow upward. A graben forms between the fractures as the center block sags downward.

This experiment has been repeated many times and most recently it was pictured by Badgley (1965, p. 168, Fig. 5-17). H. Cloos (1953, Pl. XLIV) pictures the same experiment in a drawing. The extension here was produced by doming (see also H. Cloos, 1949, Fig. 56, facing p. 225). A similar graben caused by lateral pull of two separating base plates is shown by H. Cloos (1936, p. 265). Finally the same experiments are illustrated by Hills (1963, p. 187, Figs. VII-29, A and B).

REVERSE DRAG

To imitate reverse drag on down-to-basin faults, I began experimentation with a setup which provided maximum symmetry: a clay model, pulled equally in opposite directions, with a gap opening between the base plates of brick or wood blocks as shown in Figure 4. This experi-

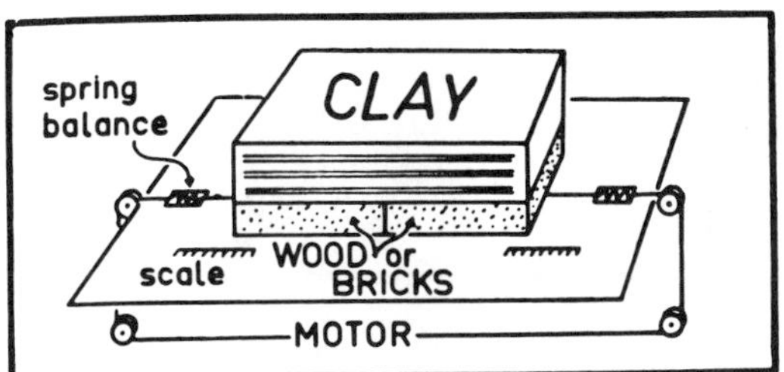

FIG. 4.—Diagrammatic setup of experiment for symmetrical pull and movement.

ment also has been described by H. Cloos (1930, p. 744) but not for the same reason. In that experiment he tried only to produce a graben; my experiment was aimed at maximum symmetry in a graben.

Symmetrical pull was maintained by spring balances on which the pull could be measured within 0.5 lb or less. The rate of movement was observed on a scale which showed the displacements of the moving base plates. Even with equalized pull, movements could become unequal due to differences in friction between the base and the table top, if, for example, one side "stuck" due to spilled clay.

Figure 5 shows the front of a model on blocks before movement began. "Bedding" is scored on the front surface of the model; the clay is homogeneous.

Figure 6 shows the formation of a symmetrical graben. Bedding remains horizontal in the center and is dragged slightly along the faults. Displacement is equal at the two faults. On the left side one fault is prominent; on the right side antithetic faults appear in a zone.

The horizontal displacements of the blocks are equal within 3 mm of the 20-cm mark in the center. The pull on two scales was 16 lb on each side when fracturing began. Fractures first formed at the bottom where the blocks separated. Before they reached the surface plastic deformation occurred in the upper half of the model, and fracturing in the lower half. The curvature of the faults (in cross section) was caused by plastic sag in the upper part of the model before fracturing, thereby steepening the position of the shear planes.

Clay-particle movements were traced from successive photographs, as shown in Figure 7. Particles were identified and traced on acetate sheets and their paths are shown as dashed lines. In the footwalls particles moved horizontally away from the center at equal rates. Near the top of the model, particles first sagged due to plastic deformation and then moved horizontally away from the center. Within the graben the particles moved downward; the subsiding block spread and the horizontal distance between reference points increased slightly. The faults also moved laterally and thus opened the gap into which the block dropped. The movements were symmetrical within the limits of the model: away from the center in the footwalls and downward with a slight tendency toward the left.

FIG. 5.—Clay model on wooden blocks before movement. Scale in cm.

FIG. 6.—Symmetrical graben after symmetrical pull. Clay graben has slid down between blocks. Displacement is 20 mm on right side and 22 mm on left side. Scale in cm.

For a symmetrical graben caused by one-sided pull, a second model was designed identical with the first except that only the right block was pulled while the left block remained stationary. This experiment showed identical fracture patterns except for small details, regardless of whether the pull was toward left or right.

Figure 8 appears symmetrical and seems identical with the first experiment (Fig. 6). A difference is detected only when particle movements are traced (Fig. 9).

The left boundary is an inclined plane. The subsiding graben block slides along that plane, and the entire block can move only parallel with that boundary surface. It can do this only if the right block makes room for it by moving horizontally toward the right. There are no suggestions of downbending. Thus, it is evident that the movement alone is not the cause of asymmetries.

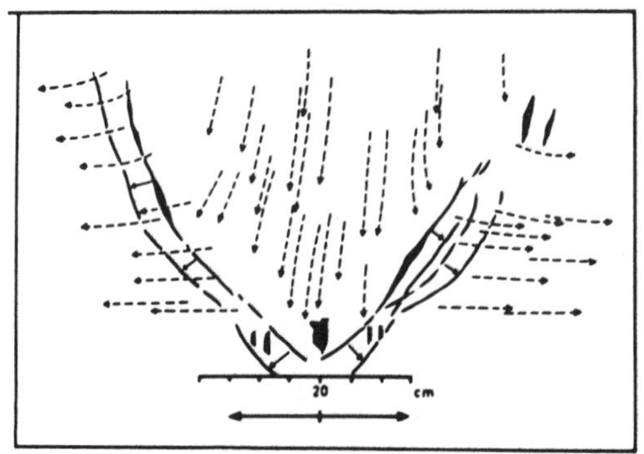

FIG. 7.—Migration of reference points during deformation. Traced from successive photographs.

A symmetrical graben can be modeled also without the gap into which the block drops. The experimental setup is the same except that the base of the graben block impinges on the table top and this results in a zone of complex deformation. In such a model the graben block commonly breaks up into a group of subsidiary grabens and horsts, and the structure does not resemble the down-to-basin faults of the Gulf Coast (see H. Cloos, 1930; 1936, p. 265; 1949, Fig. 56; 1953, Pl. XLIV; Badgley, 1965, p. 168; Hills, 1963, p. 187).

DOWN-TO-BASIN FAULTS

All down-to-basin faults in the Gulf Coast are essentially asymmetrical graben structures in which the gulfward side has become a downbend (reverse drag, rollover) with synthetic and antithetic faults, and the updip side is bounded by a master fault that dips toward the basin. At many places stratigraphic thicknesses on the downthrown side are greater than on the upthrown side. Such faults have been called "growth faults" (Ocamb, 1961) or "contemporaneous faults" (Hardin and Hardin, 1961).

The dip of the master fault seems to lessen with depth (Murray, 1961, p. 182). This has been observed directly in the field by Hamblin (1965) who discussed most ably the phenomena of reverse drag and asymmetric faults.

A good example of a down-to-basin fault with reverse drag from the Isthmus of Tehuantepec is

FIG. 8.—Symmetrical graben formed by asymmetrical pull toward right. Final stage: bedding remains horizontal, drag on faults is normal; dip slip on faults is equal. Scale in cm.

shown in Figure 10 (Castellot and Caletti, 1958). Here the Gulf of Mexico is on the north and the structure is a mirror image of Texas and Louisiana down-to-basin faults. In Figure 10 there is no well control (vertical lines) for the lower two horizons, but the upper four controlled layers show the reverse drag very well.

It has been shown already that the direction of tectonic transport is not the cause of the asymmetry of down-to-basin faults. To find the reason for the asymmetry a model was built on two overlapping metal sheets which move across the table top (Fig. 11).

Five stages of the experiment are shown in Figures 12–16. The right sheet overlaps the left one, as shown by arrows; the shaft of the upper arrow ends at the edge of the upper sheet. The left arrow indicates only the direction of motion of the lower sheet. The rate of motion is equal for the two sheets within the limits of the experiment.

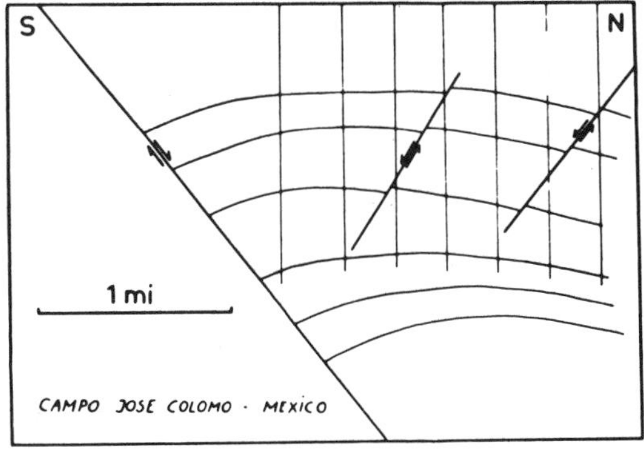

FIG. 10.—Reverse drag in generalized cross section of José Colomo field, Isthmus of Tehuantepec, State of Tabasco, Mexico (after Castellot and Caletti, 1958, Fig. 2). Vertical lines are wells. Lower two layers are uncontrolled, but controlled layers show structure.

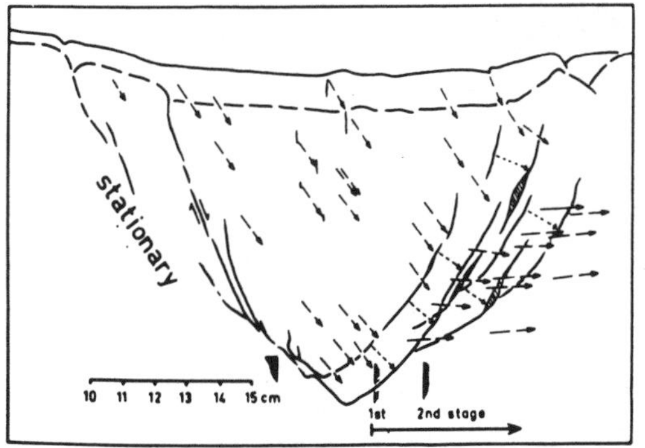

FIG. 9.—Migration of reference points during unequally pulled experiment. If pull is toward left, picture would become mirror image.

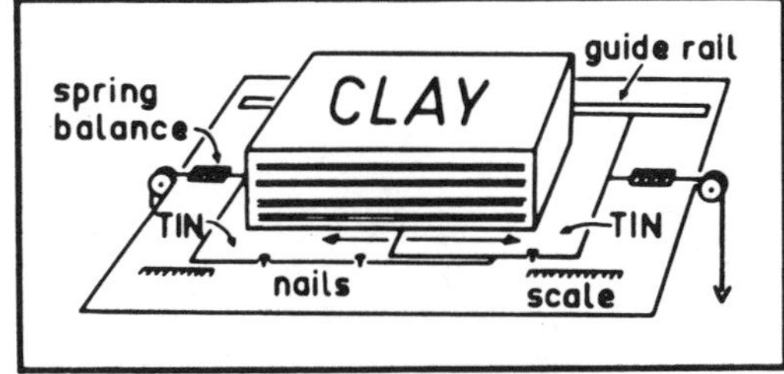

FIG. 11.—Experimental setup for asymmetrical fault pattern with symmetrical pull. Metal sheets overlap. Nails and metal rail prevent rotation of metal sheets. Spring balances permit equalizing pull. Movements are measured as displacements of metal sheets on table top.

FIG. 12.—Clay model on overlapping metal sheets before deformation. Left end of upper arrow is at edge of right, upper sheet.

After only a few millimeters of lateral displacement, fractures appear. At stage 2 (Fig. 13) a sharp normal fault begins to form at the edge of the upper sheet and reaches about a third of the way to the top of the model. The four lowest layers are faulted, the displacement decreasing upward. The fault also steepens upward. A second, smaller, left-dipping, synthetic fault appears on the right. On the left side, a small synthetic fault reaches upward through the lowest layer and several antithetic faults accompany the slight reverse drag. Above the faulted area in the upper two-thirds of the model the bedding has sagged symmetrically above the graben.

In Figure 14 faults reach the surface, but the upper two layers are not displaced. Fractures have just begun to form near the top. Near the base displacements on the right have grown larger, and the small fault on the right extends higher although not to the surface. Antithetic faults at the left have become more extensive; the fault zone has become wider. Downbending is asymmetrical toward the right in the lower three layers.

At this stage a master fault at the right, a fault zone at the left, and reverse drag in the lower half of the model are well established. The master fault begins at the edge of the upper sheet and steepens toward the top of the model.

In Figure 15 the two sides of the graben are distinctly different. Downbending, especially in the lower half of the model, is very pronounced and is accomplished in part by plastic deformation and in part by rotation of blocks on antithetic (as related to the downbend) left-dipping faults in a zone that grows wider as deformation continues. A few synthetic right-dipping faults also appear near the top in the left fault zone; horst and graben patterns occur in the upper half of the model.

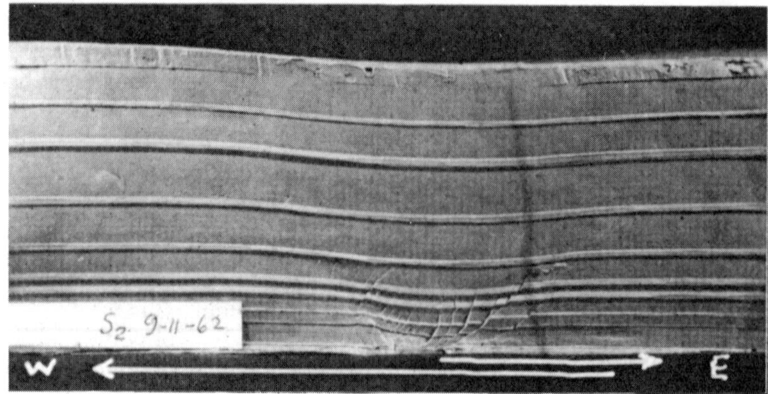

FIG. 13.—Asymmetrical graben after deformation begins. Faults appear first at base and downbending develops in lowest layers.

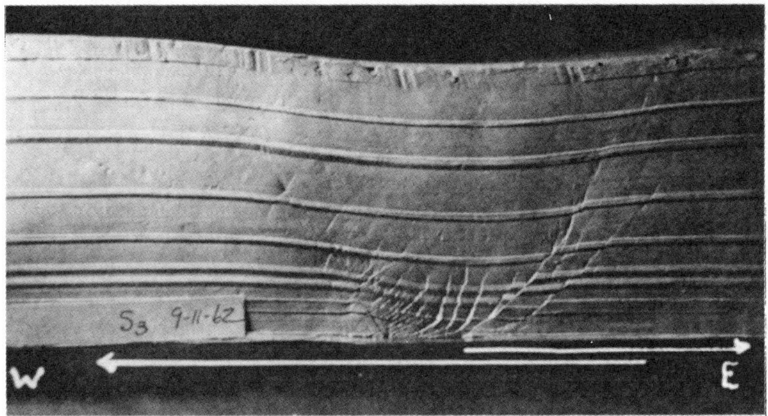

FIG. 14.—Faults at right and fault zone at left have reached surface. Asymmetry grows more pronounced. Downbending grows upward and includes upper layers.

Figure 16 is the final stage of the experiment. The block on the right has moved as a unit. The fault at the left has widened to include the entire area of downbending. Antithetic faults prevail; blocks have rotated clockwise, intensifying the downbending. In the upper part of the model faults dip left and right accommodating the flexure. Next to the master fault, bedding in the graben is still almost horizontal.

The movements of particles (Fig. 17) indicate the horizontal outward movement of the two blocks toward the left and right away from the central area. Within the graben the particle paths are asymmetrical. They move downward only nearest to the master fault. In the left half of the graben particles first move down and then toward the left as, for example, points A and F.

Line A-B-C dipped 4° at the beginning and line A′-B′-C′ dipped 24° at the end of the experiment. During the same time interval the line extended from 26 mm to 38 mm. This illustrates the development of a downbend toward the master fault and of an extension which can be accommodated only by plastic flow and additional faulting. As the downbend develops the antithetic faults flatten. Line F-G shows the same relationship; an almost horizontal line becomes tilted and longer as the points migrate from F and G to F′ and G′. Figure

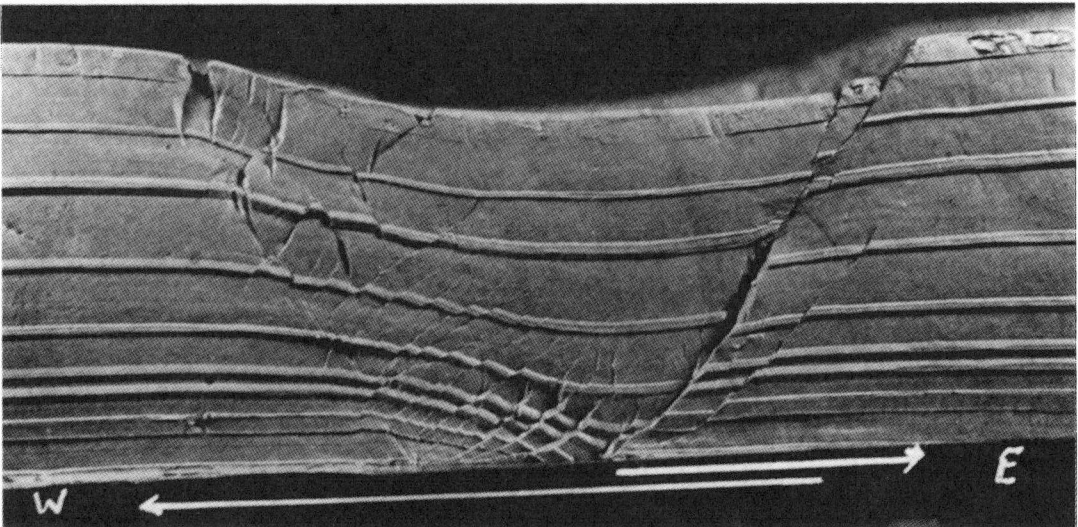

FIG. 15.—Master fault at right separates right block from asymmetrical graben. Left fault zone is wide and accommodates downbending in lower layers, where blocks between antithetic faults rotate clockwise.

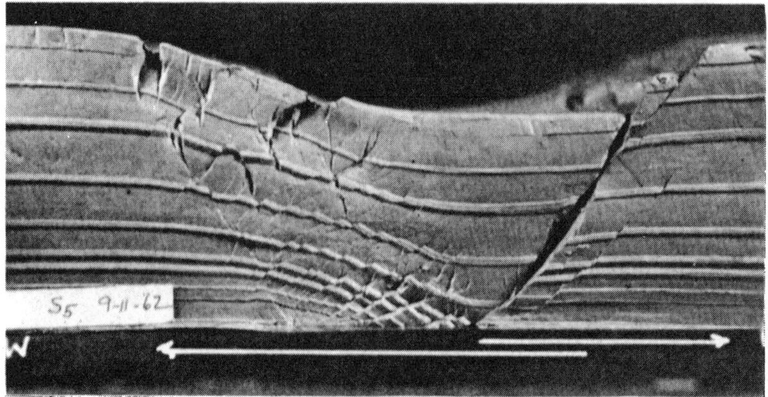

Fig. 16.—Antithetic faults dip left and blocks rotate in downbend. Master fault is sharp and prominent, flattening downward.

18 pictures an essentially similar experiment in three dimensions. The right block was moved toward the right. A master fault is covered with striae; the downbend shows mostly antithetic faults.

Asymmetry is not a function of the motion of the two blocks, but the underlapping sheet removes the lower layers of the subsiding graben as they slide down the master fault plane onto the sheet. An identical asymmetry, with similar faulting and downbending, is obtained if pull is in one direction only. In such cases, particles move only away from the master fault which remains stationary.

Discussion of Down-to-Basin Fault Experiment

The experiment essentially imitates what is seen in field observations of down-to-basin faulting: a master fault, reverse drag, and antithetic faults. In addition, there is some thickening on the downthrown side, and the dip of the beds above the glide plane steepens. The master fault dips more gently at depth and, due to rotation, there seems to be less displacement above than below.

If one visualizes the right side (master fault) of Figure 16 as stationary and the left side as moving away from it, the structural pattern is produced by gliding on a surface alone without need for salt ridges (Quarles, 1953), faulted anticlines, basement participation, differential compaction, or any other mechanism. Furthermore, the master fault is curved as has been observed by Hamblin (1965). The experiment strongly supports the slipping-plane hypothesis as outlined by

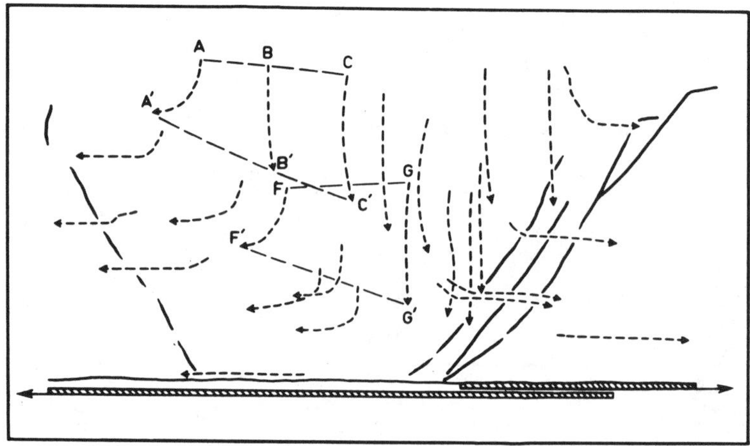

Fig. 17.—Migration of reference points in asymmetrical graben experiment (Figs. 11–16). Line A-C rotates into position A'-C', line F-G becomes F'-G'. Traced from successive photographs.

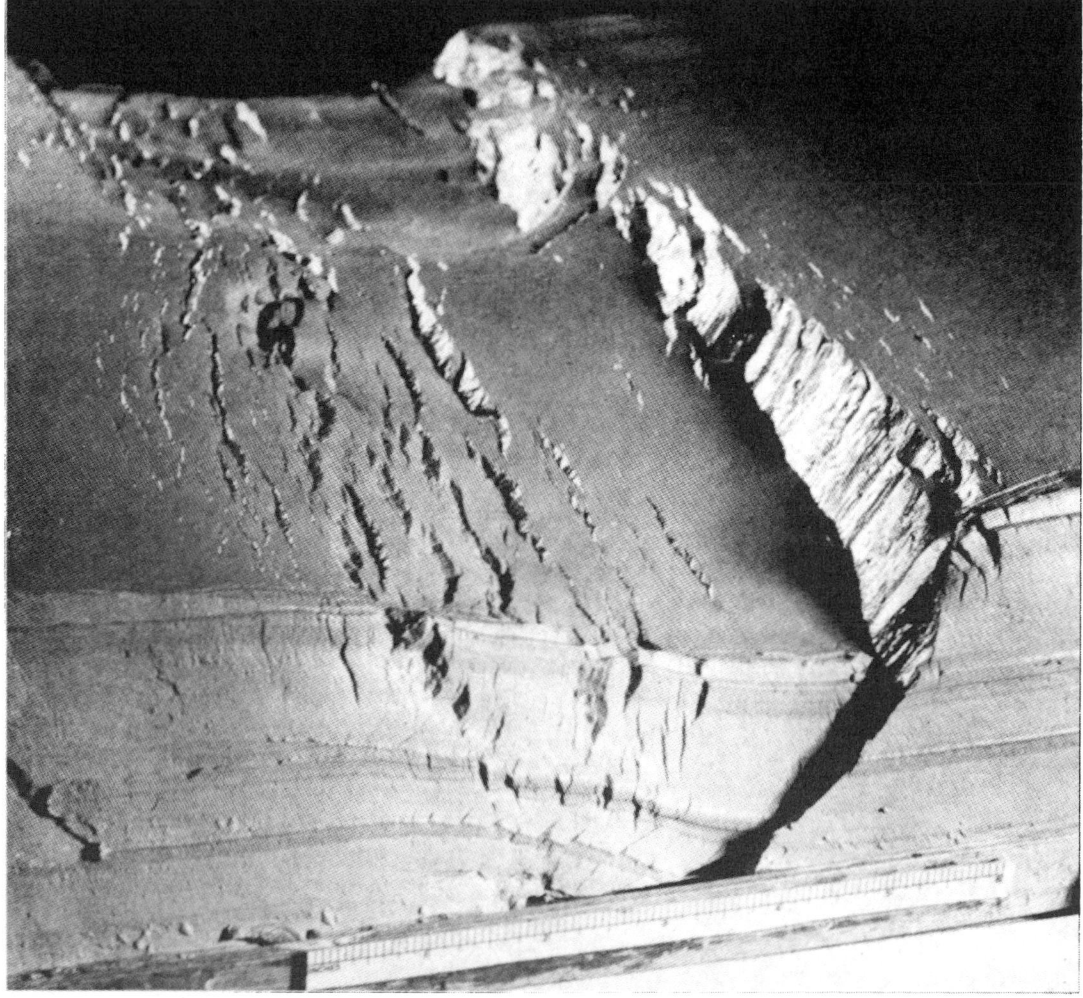

FIG. 18.—Reverse drag, master fault with striae, antithetic faults in oblique photograph. Right block moved right; left block remained stationary.

Quarles (1953, p. 498–500).

A creep mechanism can produce all the elements which have been observed on down-to-basin faults with reverse drag but only if the material at the base of the graben creeps first down the master fault and then along the glide plane. When faults first form in the symmetrical graben, both the updip and the down-to-basin shear planes exist, but the latter will develop into a master fault and the updip plane degenerates into a downbend with tension and a series of either synthetic or antithetic faults, or both. Where the material does not creep downdip and no rotation occurs, the updip fault, as well as the downdip fault, can become the master fault. This happens in the peripheral graben and is discussed in a subsequent section. It also happens in a few places near the coast, and occurred in the experiments described below.

The experiments suggest that reverse drag should be more common and more intense on down-to-basin faults and rarer or much less intense on updip faults.

DOWN-TO-BASIN FAULTS AND SEDIMENTATION (GROWTH FAULTS)

In experimentation with clay, fractures are observed on free surfaces of the model. These surfaces are most sensitive and easily destroyed. The experimental material therefore was changed from clay to sand. The addition of sand to a sand model does not disturb existing structures. Also, glass walls provide a clean and even cross section. Sand moves along glass with little friction, and

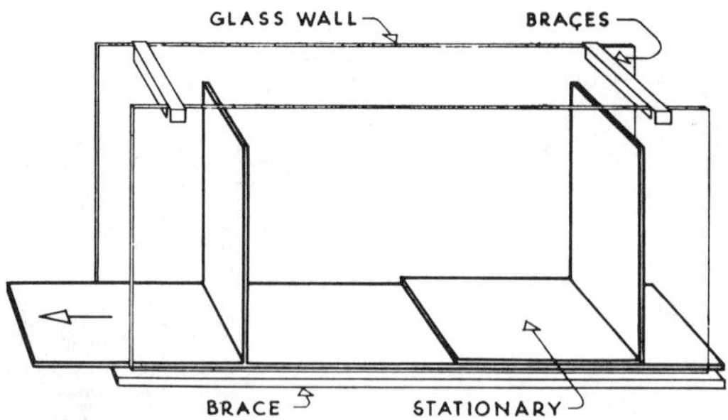

FIG. 19.—Setup for growth fault experiment. Partitions and floors are ¼-in. plywood. Glass plates are ¼-in. plate glass. Length of box 52 in., width 7 in. (schematic, not to scale).

the glass influences the fault pattern very little.

A long box was constructed of two plate glass walls, with a stationary end, and a movable end (Fig. 19). The movable end (left) was rigidly connected to a movable floor which reached almost the length of the box. The stationary end (right) was connected to a stationary floor above the movable one and extended to the center of the box. Sand used ranged from very fine to 2 mm in diameter.

Sedimentation was simulated by filling sand into the graben as it formed so that the slope of the sand surface did not intersect the fault which continued to grow upward as deformation continued. Sand was replenished as soon as the fault became visible as a feeble line at the surface.

The sand experiment served two purposes: (1) to permit sedimentation during faulting, and (2) to observe asymmetrical down-to-basin faulting. However, for the latter purpose, clay is very much better.

In the first stage of the experiment (Fig. 20) faulting has affected all but the topmost layer. The second layer from the top, below the dark line, has been faulted, and more sand was added to it during faulting; the layer is thicker in the graben than on the two sides. The top layer also is faulted and was filled in after considerable faulting. The right (left-dipping) fault is steeper than the left one. Most of the layers were lowered vertically and remained horizontal between the two major faults. The lowest white layer and the darker one below it show pronounced asymmetry and downbending.

In the second stage (Fig. 21) all layers are faulted and the right fault is sharp and does not change dip. The left side is more asymmetrical and several faults appear in the left half of the graben. The dip of the left fault is gentler than that of the right fault.

The positions of both major faults (as they were in stage 2; Fig. 21) were marked by black lines on the glass (Fig. 22) to show changes of dip or position which resulted from further movement. The right fault remained stationary. The left fault moved toward the left with the entire moving block and several smaller faults appeared within the graben. As movement continued, there was downbending toward the right on several normal faults, which form complicated horst and graben patterns. Sedimentary thicknesses are very uneven, but layers in the graben generally are thicker (Fig. 23) due to the addition of sand during faulting.

At the end of the experiment the graben was not refilled, in order to demonstrate the intersection of the angle of repose of the sand with the fault plane and also with the uppermost key bed. The right side of Figure 23 shows the intersection: the fault dips 72° to the left, and receded

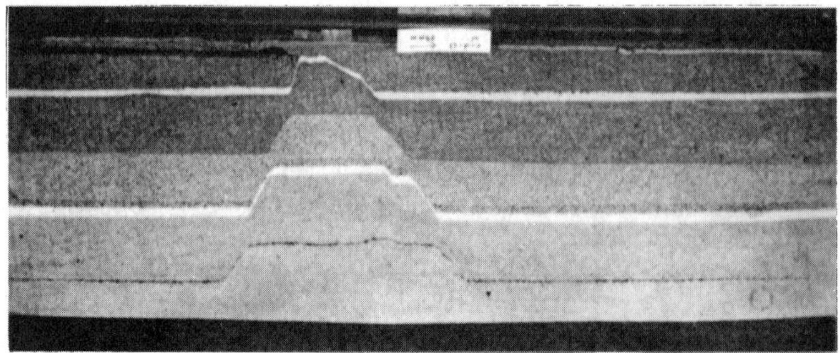

FIG. 20.—Asymmetrical graben in sand. Lower layers were faulted; top layers were filled after faulting. Right fault is steeper and sharp without change in dip; sedimentation in graben is thicker than on "highlands."

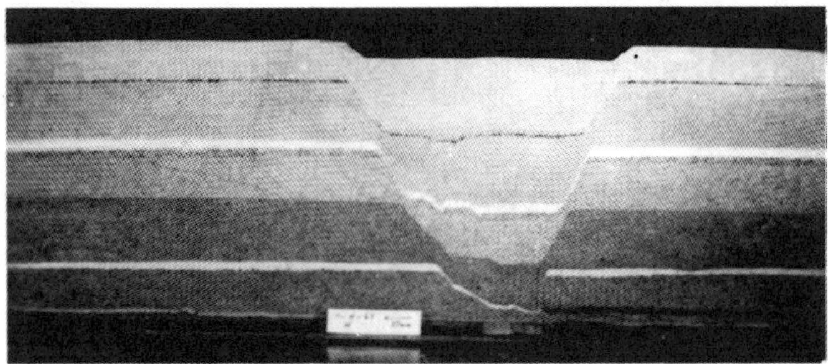

Fig. 21.—Faulting continued. Top layer has been faulted. Fault grows upward without change in dip.

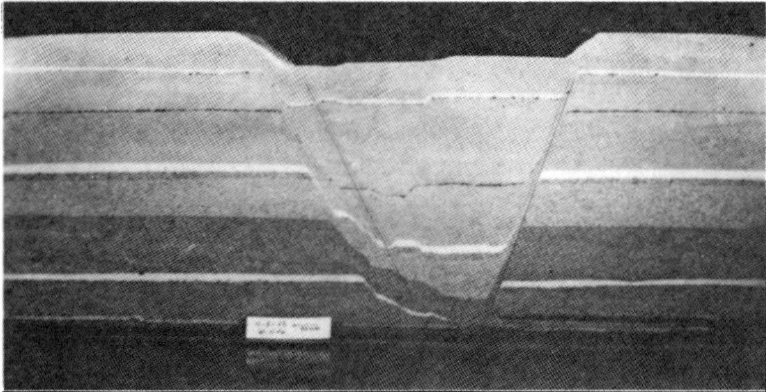

Fig. 22.—Two layers have been added and are faulted. Fault positions at stage of Figure 21 are black lines on glass. Right fault is stationary; left fault migrates toward left with entire block. As graben widens, new faults appear.

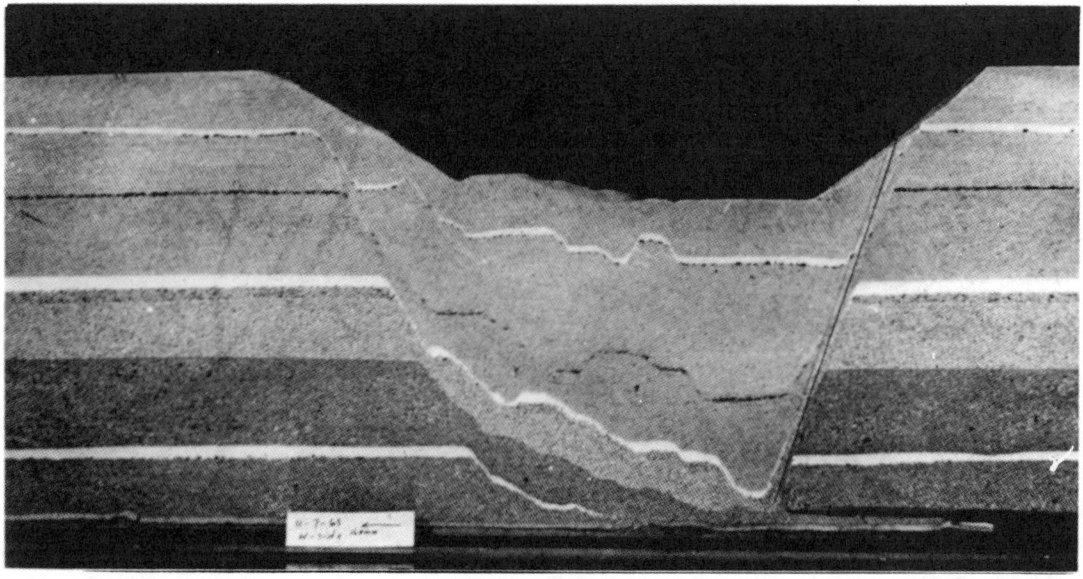

Fig. 23.—Final stage of faulted sand layers. Total displacement is 160 mm. Angle of repose of dry sand is very much less than that of fault dip.

somewhat toward the right as deformation progressed. The layers are horizontal up to the fault and break off abruptly at the fault. The sand surface dips about 36° toward the left. The dip depends on packing of the sand, its grain size, and moisture content. Limestone dust makes vertical bluffs, as does the upper white layer of white ground plastic ("microthene") shown in Figure 23.

The intersection of the fault plane with the angle of repose of the sand is, of course, quite artificial and scarcely is representative of sedimentation in basins receiving sediment. In a sedimentary basin the profile of the sedimentation surface has an angle measured in minutes. Sedimentation continues at a slow rate covering faults as soon as they reach the surface. Thus, even a fault with major displacements at depth is not apparent at the surface. The experiment suggests what the faulting process might be. It also suggests that the fault continues upward without deflection or change in dip into the younger sediments as they become buried.

If the entire graben area subsides, thicknesses should increase abruptly across the master fault and then may decrease again on the other side. The largest displacement should be along the down-to-basin fault. At the same time, the displacement should increase with increasing depth for those layers which are affected for the longest time.

In the sand experiment, particle paths (Fig. 24) are very similar to those of the clay experiment. On the right along the master fault, movement parallels the fault and is guided by it. Particles in the stationary right block do not move; in the left block the movement is horizontal toward the left. Between these two blocks are areas where particles move down parallel with the master fault and then turn into a horizontal path when downward motion is arrested and new faults open within the graben. The curved paths are more angular than in clay because sand is not plastic and cannot bend. The effect is the same however in all experiments. Points A to D, which are in one bed, first move along parallel paths. Then point A turns toward the left but the others continue downward. This extends distance A-D, and line A-D begins to dip toward the right. As this happens more faults appear in order to accommodate the extension. This phenomenon is also a downbend, even if it is broken somewhat due to the physical properties of the material used.

DISCUSSION OF GROWTH-FAULT EXPERIMENT

The experiment shows that the faults continue upward into new and previously unfaulted layers which are added above the fault. The dip-slip therefore grows downward but it also did this in the clay model where no sediments were added.

The increase in thickness consists of two components: (1) real increase from deposition, and, (2) steepened dip which, when drilled through, will show apparent increases in thickness of beds. It is impossible to appraise the relative impor-

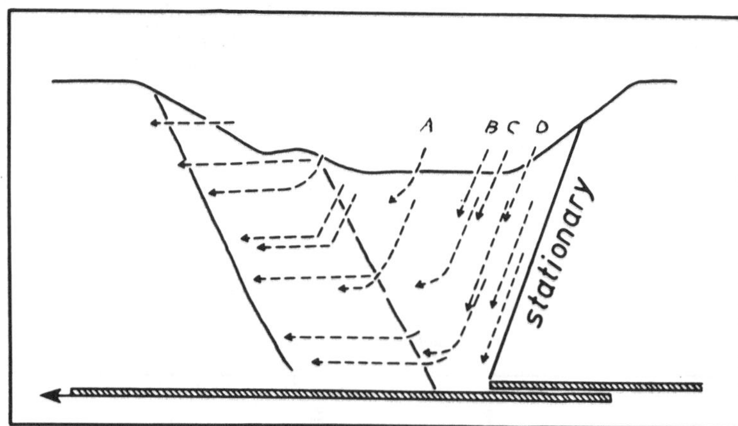

FIG. 24.—Sand-particle paths during deformation (Figs. 21–23). Stationary block is on right with sharp fault dipping left. Left block moves on lower floor.

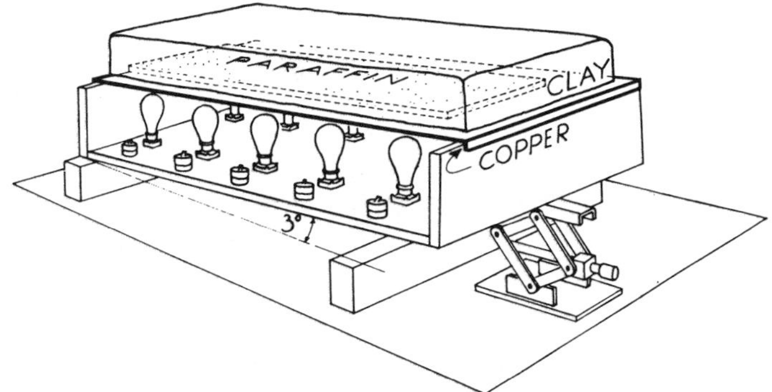

Fig. 25.—Setup for gravitational creep experiment with wet clay (diagrammatic).

tance of these two factors without more information. However, company files must contain abundant information.

In addition to growing faults, the sand-box experiment shows a crude downbend with subsidiary faults. Furthermore, it shows again that the direction of pull has nothing to do with the asymmetry of the resulting structures.

Regional Fault Pattern

The preceding discussion shows that down-to-basin faults with reverse drag can be explained rather simply and can be imitated experimentally. It was tempting to try an imitation of the entire semicircular Gulf Coast pattern and graben zones. Many experiments were made until a fair imitation was achieved.

First, a tank was constructed 12 × 36 in. and 1 in. thick with small holes in the top on 1-in. centers. A clay model was built on that surface and the tank inclined 1°. When water pressure was turned on at about 10 lb the entire clay model slid into the sink within moments and essentially in one slab.

The greatest disadvantage of this arrangement is the fact that, when surface tension in clay experiments is relieved by a film of water, shear fractures do not appear but tension fractures do and the model becomes worthless when it breaks up into slabs and irregular blocks.

After much experimentation, the setup of Figure 25 was constructed. It consists of a wooden box above which a copper plate can be heated by rows of light bulbs inside the box. A layer of paraffin on the copper sheet can be modeled to any shape and thickness and the clay placed above it.

The model is tilted a few degrees and the melting paraffin becomes the lubricant which permits the clay to creep downslope. Fracture patterns form on the surface and the sides of the model. After the desired stage of faulting is reached the lights are turned off and the paraffin hardens. The fractures which cut through to the base of the model are preserved in the paraffin and can be correlated with the surface fractures after the clay is removed.

Figure 26 shows two down-to-basin faults which formed in clay during creep. The angle of inclination of the copper sheet is 3.5°. Reverse dip is toward the right and toward the major faults. Antithetic faults facilitate the downbending as creep proceeds toward the left.

Gulf Coast Fault Pattern

An attempt was made to duplicate the entire Gulf Coast pattern in one experiment in which creep of a large area was directed away from a stationary land mass of approximately crescent shape. As a pilot experiment, a clay model was built on a metal plate which was pulled. Later the model was built on paraffin and allowed to creep (Figs. 26, 27). Figures 28–32 show the initial experiment, but the pattern is the same in the gravity-induced creep experiment (Fig. 27). Figures 28 and 29 are photographs of the model from above. Figure 28 is illuminated at a low angle from the north (top), and Figure 29, from the south.

For comparison, faults from the Gulf Coast structure map by Murray (1961, Fig. 4.1) are superimposed on the model photograph (Fig. 30). I think that the coincidence is striking. Many well-known structures are represented in the model and some additional ones appear. Some faults dip

FIG. 26.—Downdip faults in wet clay model. Downbend is clockwise (toward left) with some updip faults visible on surface. Scale in cm.

the wrong way—*i.e.,* a down-to-basin fault in nature is represented by an updip fault in the model. It is nevertheless a fault which indicates a horizontal component of extension. An equivalent of the South Arkansas graben is not present in the model, but if motion had continued it might have appeared. (It would appear late in this model because the clay did not adhere to the metal sheet at the base of the model and therefore lagged behind.) The strike of the faults is the same on the map and in the model, and the east Texas to south Mississippi embayment is represented as well as the down-to-basin fault zones which strike parallel with the coast line.

If the directions of the horizontal components are plotted for both the map and the model the

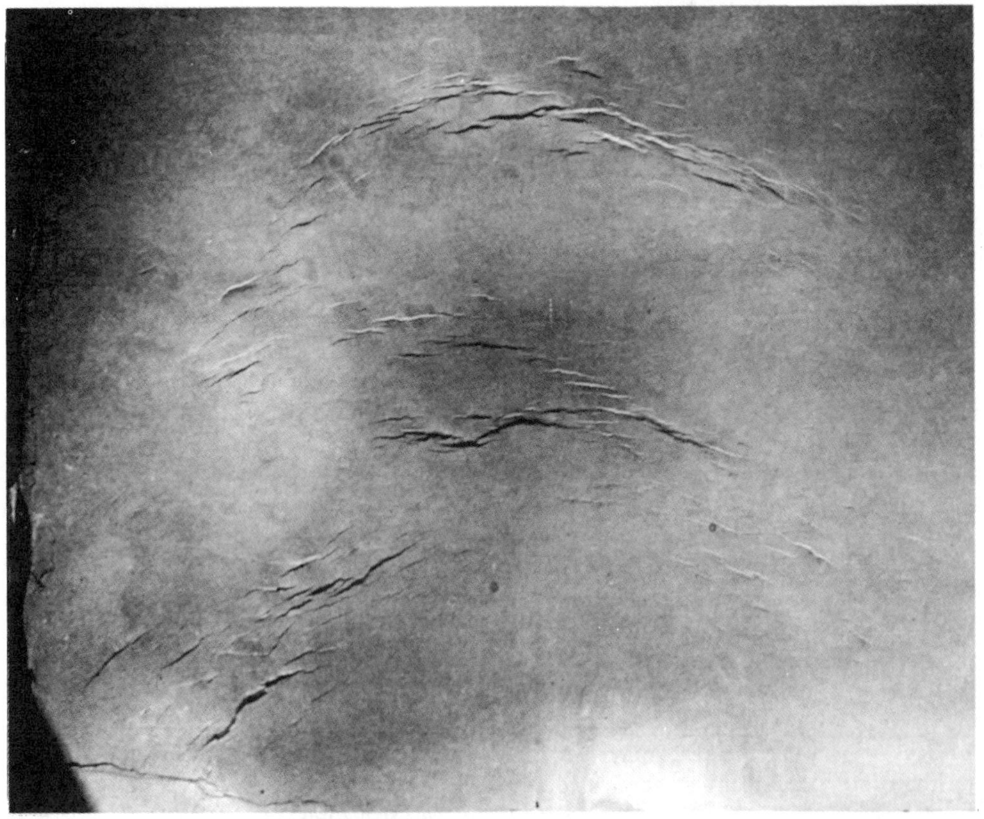

FIG. 27.—Fault pattern on top surface of clay model caused by gravitational creep.

resemblance becomes still more striking. Figure 31 shows the horizontal components added to Murray's map. No attempt was made to indicate the component quantitatively by the length of the arrows. Figure 32 is a similar plot for the model. The difference between Figures 31 and 32 is only in the details. The Mississippi delta area is not shown in the model (Fig. 32), and faults in the model are shown in the Gulf far from shore where they are known to exist.

Discussion of Experiments

The experiments were performed in the following sequence: symmetrical grabens were done many years ago by both H. and E. Cloos. Asymmetry was studied systematically on the suggestion of H. N. Fisk, especially with reference to reverse drag because of the obvious interest of the petroleum geologist in possible closures near down-to-basin faults. The presence of down-to-basin faults clearly proved the existence of a horizontal motion which probably caused both the horizontal and vertical components. Next, sand was added to simulate sedimentation and the asymmetrical faults grew upward into new sediments. Then, an attempt was made to imitate the entire Gulf Coast pattern in one experiment. The results showed similarity with the map pattern in spite of (1) the rather primitive nature of the model, and (2) certain unlikely comparisons: (a) there is no metal sheet being pulled beneath the Gulf Coast and (b) the Gulf Coast fault pattern comprises a huge heterogeneous area with many irregularities in which faults appeared in sequence during a long period of time. However, the experiment shows a horizontal component that results in a pattern similar to nature with rather similar faults. The final effort was aimed at gravity gliding on paraffin which is far less rigid than the metal sheet experiment and therefore may be somewhat more similar to nature.

Even if the method is crude, the resulting patterns are remarkably similar to those which actually exist and suggest that a horizontal component may well be the explanation of the map pattern.

The imitation of the entire Gulf Coast fault pattern in one experiment does not imply that there was only one act of motion during a limited time. The peripheral graben faults appeared early, the faults nearer the coast appeared later, and some of these are still active. Even in the very short time of that experiment (2–3 hr) not all faults occurred simultaneously but in succession. The first ones appeared after 1 hr, the last ones after 3 hrs.

The sequence in the experiment is similar to that in nature: the marginal faults appear first and the interior ones later, because the clay sheet first tears loose at its margins and then collapses when the fault arcs reach the interior.

The experiments suggest that the regional fracture pattern of the Gulf Coast area was caused by creep of the sedimentary blanket toward the lowest point of the basin. It seems justifiable to conclude that the nearly identical patterns of the experiments and the Gulf Coast are due to similar movements and stress patterns.

The creep mechanism does not require movements of separate basement blocks, or salt intrusions, or flexures and folds for the localization of faults; on the contrary, the mechanism functions better without them. Of course, it cannot be assumed that the basement surface is as smooth as a table top, nor is the Gulf Coast basin fill as homogeneous as the clay in the model. Major obstructions probably do not exist, however, because they would prevent creep. Well-known obstructions, such as the Llano-Burnet uplift, determine the shape of the area which creeps and in turn the pattern of Gulf Coast faults.

Creep begins at a very delicately balanced stage; *i.e.,* when the gravity component barely exceeds the cohesive strength of the sedimentary column, the friction on the glide plane, and the friction on the sides of the basin. Creep is continuous, as shown by contemporaneous sedimentation and faulting, the increase of downward displacements, and the appearance of structures at several stages during the geologic history.

Continuous creep does not necessitate simultaneous faulting of equal order of magnitude at all points. In the model a fault appears here, another one there. Then the displacement dies out and a new fault opens between the first two. Later one of the two faults shows more displacement and the other two seem to become temporarily dormant, *et cetera.*

In the Gulf Coast area, facies changes, thickness changes, transgressions, and innumerable other inhomogeneities will influence the formation of faults and fault zones but the regional fracture

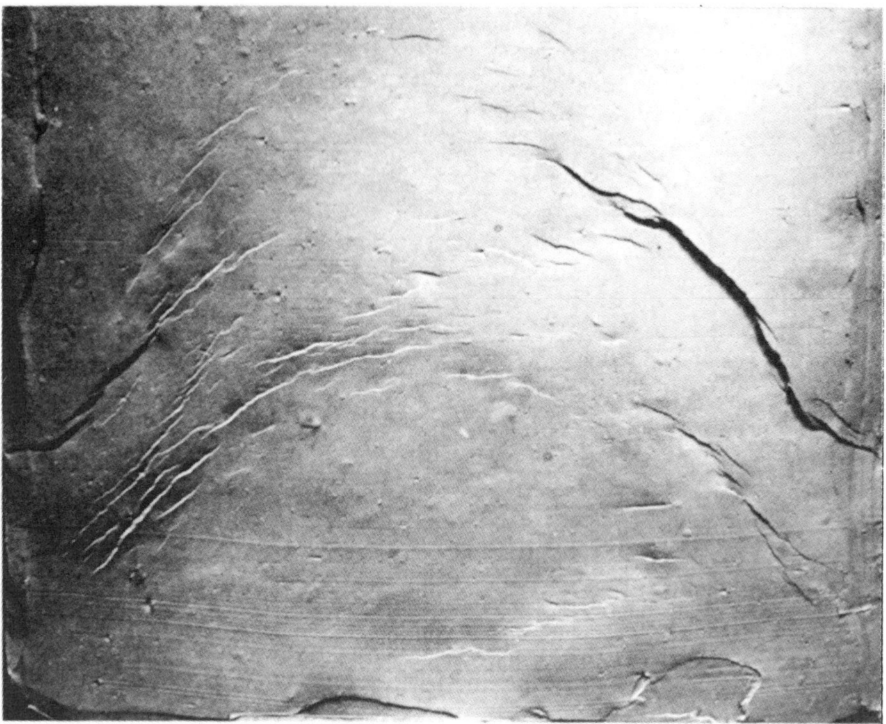

Fig. 28.—Fault pattern on model surface illuminated at low angle from north. Deep shadows are on south-dipping faults; light fault scarps dip north.

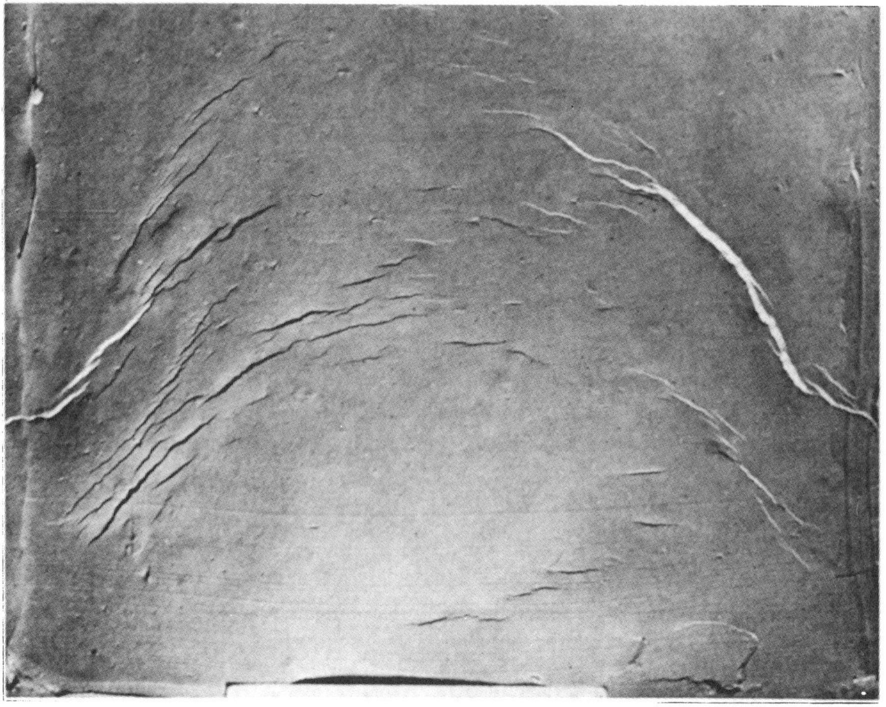

Fig. 29.—Same surface as Figure 28, illuminated from south.

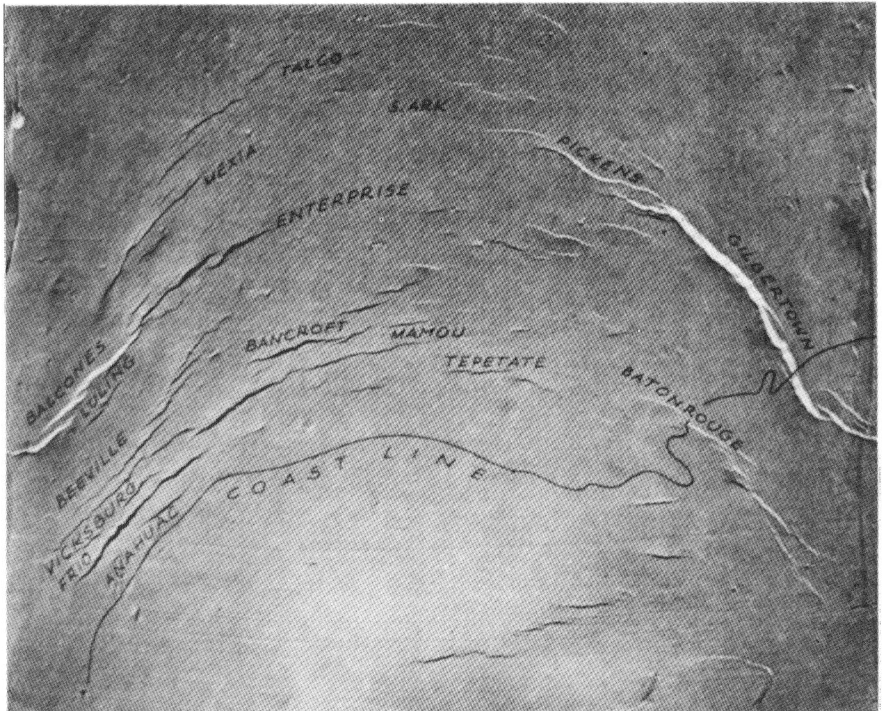

FIG. 30.—Map of Gulf Coast fault pattern superimposed on model (Fig. 29).

pattern transgresses all these and the direction of the horizontal component remains constant.

The peripheral graben zone (Balcones-Mexia-Talco-South Arkansas-Pickens-Gilbertown) follows the "contact" between stationary margins of the Gulf Coast and the central area which creeps away from it. This resembles the familiar "Bergschrund" crevasse where a glacier pulls away from a mountain ice field. It is explained easily as that zone along which the creeping sediments are torn loose from the stationary area. This zone coincides in the Gulf Coast with the approximate subsurface updip edge of the salt, thus suggesting that the salt may well facilitate the creep.

This tear-off zone also is typically not asymmetrical but a more or less symmetrical graben formed between a downdip and an updip fault. Asymmetry is not developed because there is not much creep and the material at the base of the graben is not removed basinward.

The Mexia-Talco *en échelon* grabens are in the position of extension fractures in a zone between a stationary and a moving block as observed at many places on a large and small scale. The existence of such *en échelon* grabens indicates the position of the boundary between the creeping basin fill and the stationary Llano uplift. On the east side of the basin the boundary of the moving sheet may well flare southeastward and the Pickens-Gilbertown system is therefore not an *en échelon* structure.

INTERFERENCE OF REGIONAL AND LOCAL COMPONENTS

Radial fault patterns are well known in more or less circular domes (Murray, 1961, p. 217, E). The faults are normal faults and prove extension either above an up-arched dome or in its vicinity. In elongate domes or anticlinal updomings, fractures normally are parallel with the long axis of the dome because extension across the dome is greater than extension parallel with the long axes. There are, however, exceptions in true folds in which the limbs have been compressed. This rarely applies in Gulf Coast domes and anticlines. Fractures above circular and elongate domes have been demonstrated experimentally (E. Cloos, 1955, p. 247–253, Pls. 5, 6). In the following section, these fracture patterns are called "local" because they are related to a specific local structure

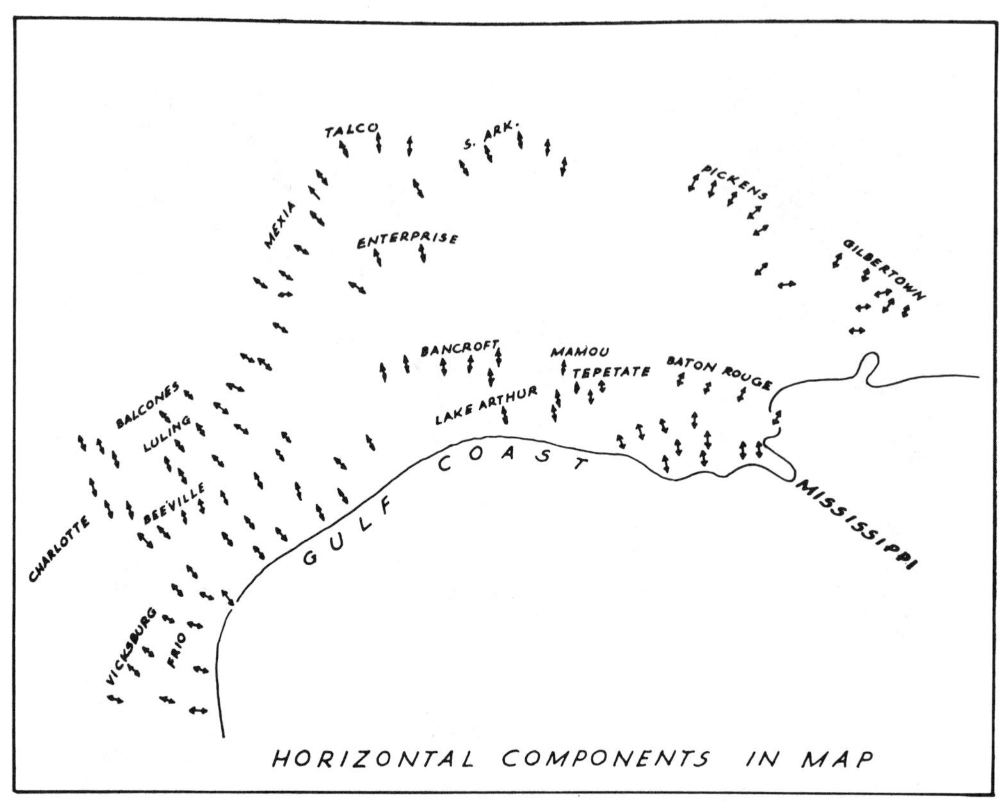

FIG. 31.—Horizontal extension component of Gulf Coast faults, based on Murray (1961, Fig. 21). Arrows show direction; length of arrow meaningless.

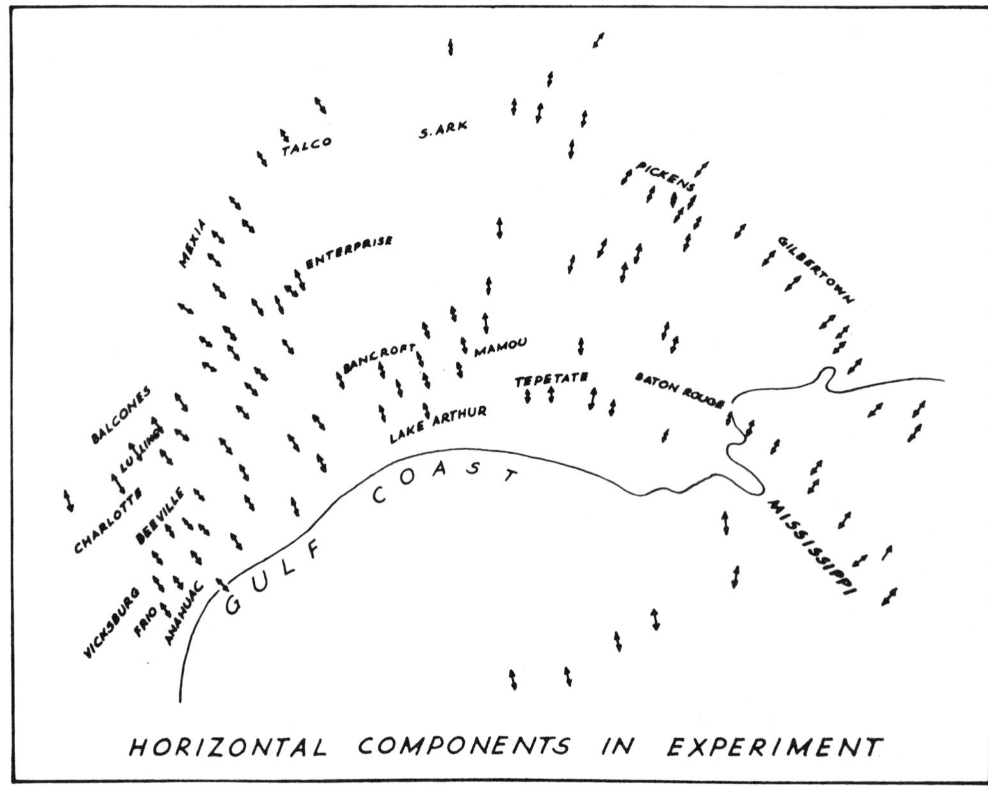

FIG. 32.—Extension components of faults in model, Figures 28–30. Arrows indicate directions, not quantities.

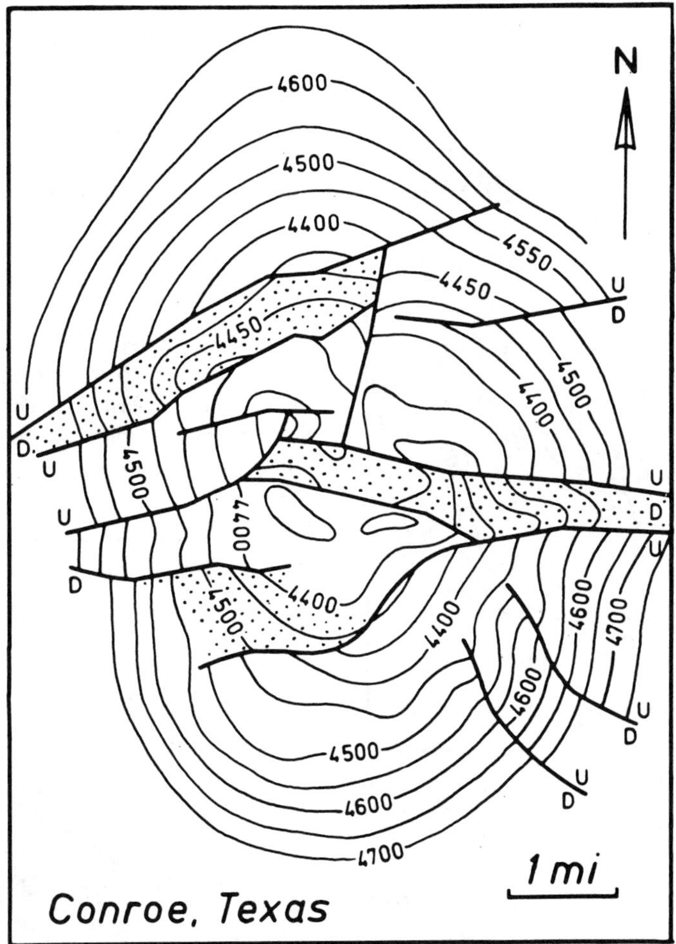

FIG. 33.—Conroe field, Montgomery County, Texas (after Carlos, 1953, p. 104). Local components should result in radial faults. Superimposed regional trend is east-west. Compare with Figure 34.

(dome, anticline). The Gulf Coast pattern is called "regional" because it is uniform through large areas. Accordingly, it seems proper to speak of a "local component of horizontal extension," and a "regional component of horizontal extension." This regional component is well shown in Figures 31 and 32.

In elongate anticlinal structures such as salt ridges the two components may coincide and regional faults then parallel the long axis and the local faults. In that case the two components are not distinguished easily.

Where the local and regional components do not coincide, the regional one is triggered by the local one or the local component is subdued by the regional one.

There are many examples of a regional component superimposed on the local component; some of these are shown in Figures 33–37. In each of these examples, the patterns should not be only regional. Obviously the local patterns were suppressed.

Another outstanding example of a dominant regional component is the salt dome at Reitbrook near Hamburg, Germany (Behrmann, 1949, p. 200). This dome is cited frequently as an example of normal faulting above a salt dome. The dome is almost circular or somewhat elongate from northwest (Fig. 38) to southeast. Many normal faults show extension in the direction of the long axis and trend northwest and east-west. Their strikes intersect at angles of 20°–30°. If the

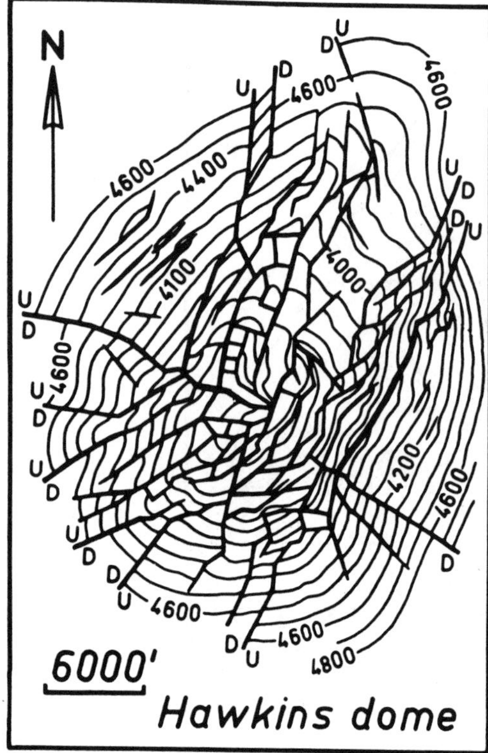

Fig. 34.—Hawkins field, Wood County, Texas (after Wendlandt, 1951). Radial faults are response to local updoming; regional component barely noticeable.

doming alone had caused the faulting, the faults should be radial as at Hawkins (Fig. 34) or other domes.

The lack of evidence at Reitbrook for a local component of extension and the presence of the northeast-trending faults strongly suggest the existence of a regional component oriented northwest-southeast. Behrmann (1949) suggested that the dome had risen above a much broader domed area which provided this regional component. It may not be accidental that long salt ridges northwest of the area also trend northeast and roughly across the dome. The salt ridges and faults above the dome seem to be contradictory unless salt withdrawal toward the northwest provided the regional component. To assume northwest creep analogous to the Gulf Coast creep may seem speculative in the absence of other evidence.

Discussion of Previous Views

Gravity sliding, gravity flow, or the "regional component" has been mentioned repeatedly as a possible cause for the pattern of Gulf Coast faults and reverse drag on down-to-basin faults.

Wallace (1952, p. 63) described a "strong regional grain" or a "regional tension developed as a result of general flexing of the north flank of the geosyncline which has continued over a long period of depositional history." He also recognized a "dominant factor in controlling both regional faulting and those faults which cross deep-seated domes" and the fact that "down-to-the-south major faults on structures may be indistinguishable from regional faults." Local faults on domes or other structures have been recognized and called "abnormal." The map by Wallace shows many examples of regional and local faults and their interference. The only real difference

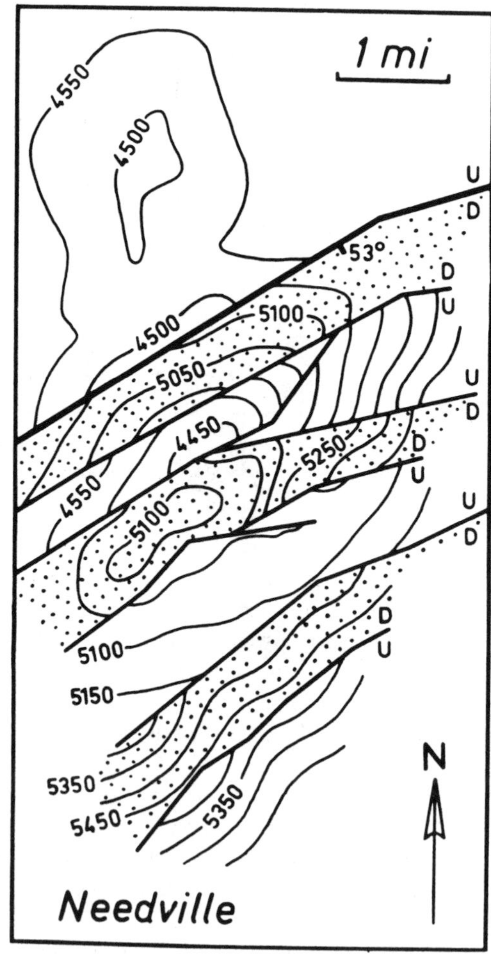

Fig. 35.—Needville field, Fort Bend County, Texas (after Greenman and Gustafson, 1953, p. 140). Regional trend is superimposed on local dome.

between the conclusions of Wallace and this paper is in suggesting that the cause of "regional" faulting is either downwarp (Wallace) or gravity creep (E. Cloos).

Quarles (1953, p. 490) discussed one of the reasons for faulting thus: ". . . the complete column of sediments slips down toward the Gulf on bedding planes and produces a fault where the sliding beds essentially break off the stationary ones." In the subsequent discussion he favored participation of salt ridges in faulting but discussed the "slipping plane hypothesis" (p. 498).

Bornhauser (1958, p. 352) discussed the gravity-flow theory: " . . . folding of sedimentary beds is the result of flowage of incompetent beds

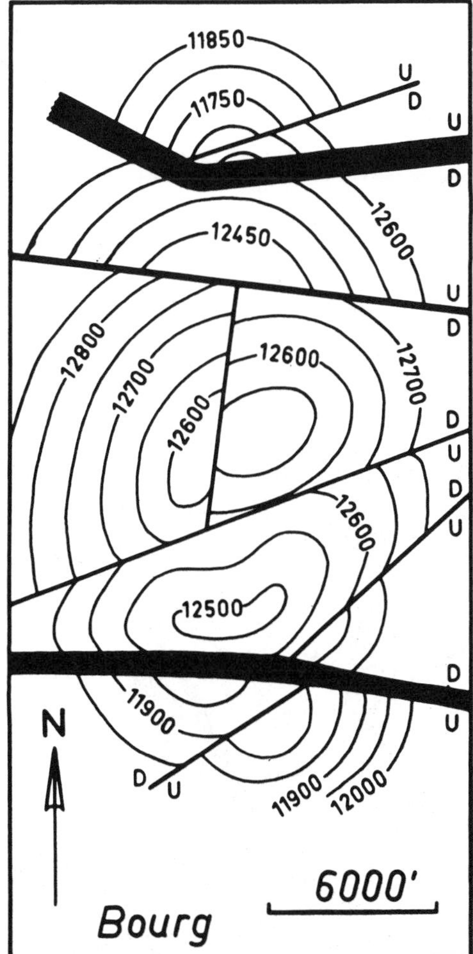

Fig. 37.—Bourg field, Lafourche and Terrebonne Parishes, Louisiana (after DeHart, 1955, Fig. 5). Local north-south anticlinal structure with superimposed regional faults.

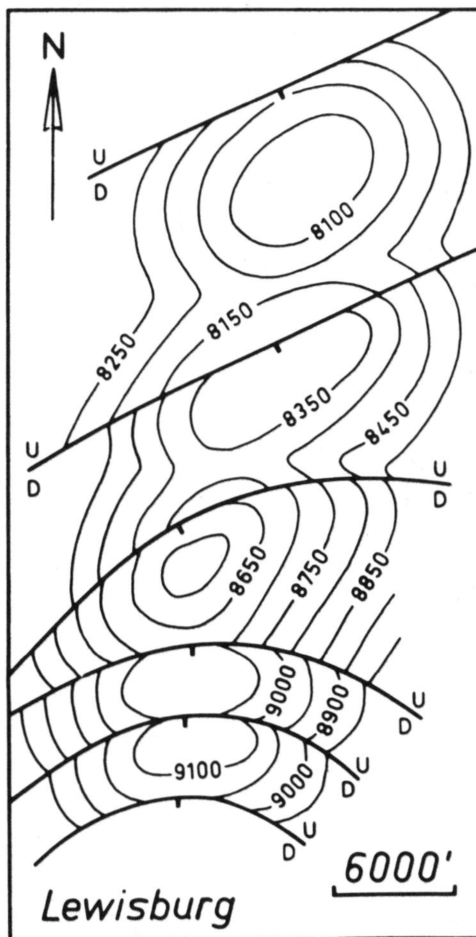

Fig. 36.—Lewisburg field, Acadia and St. Landry Parishes, Louisiana (after Ocamb and Grigg, 1954, Fig. 4). Local north-south-trending anticline with regional faults superimposed.

under load down a sloping surface." He referred to M. King Hubbert (1937, 1945) and suggested evaporites as the locus of possible glide horizons. The flow takes place in waves and, when it takes place, tensional stresses may cause shearing. Bornhauser thought that faulting usually is associated with gravity-flow folding and the process of downfolding provided an explanation for the mechanics involved in this faulting; his reverse drag is linked to folding and folds. My experiments show that folds are not necessary in reverse drag and that gravity flow or creep may be the cause of the fracture pattern. Bornhauser (1958, p. 357) also used the term "master fault"

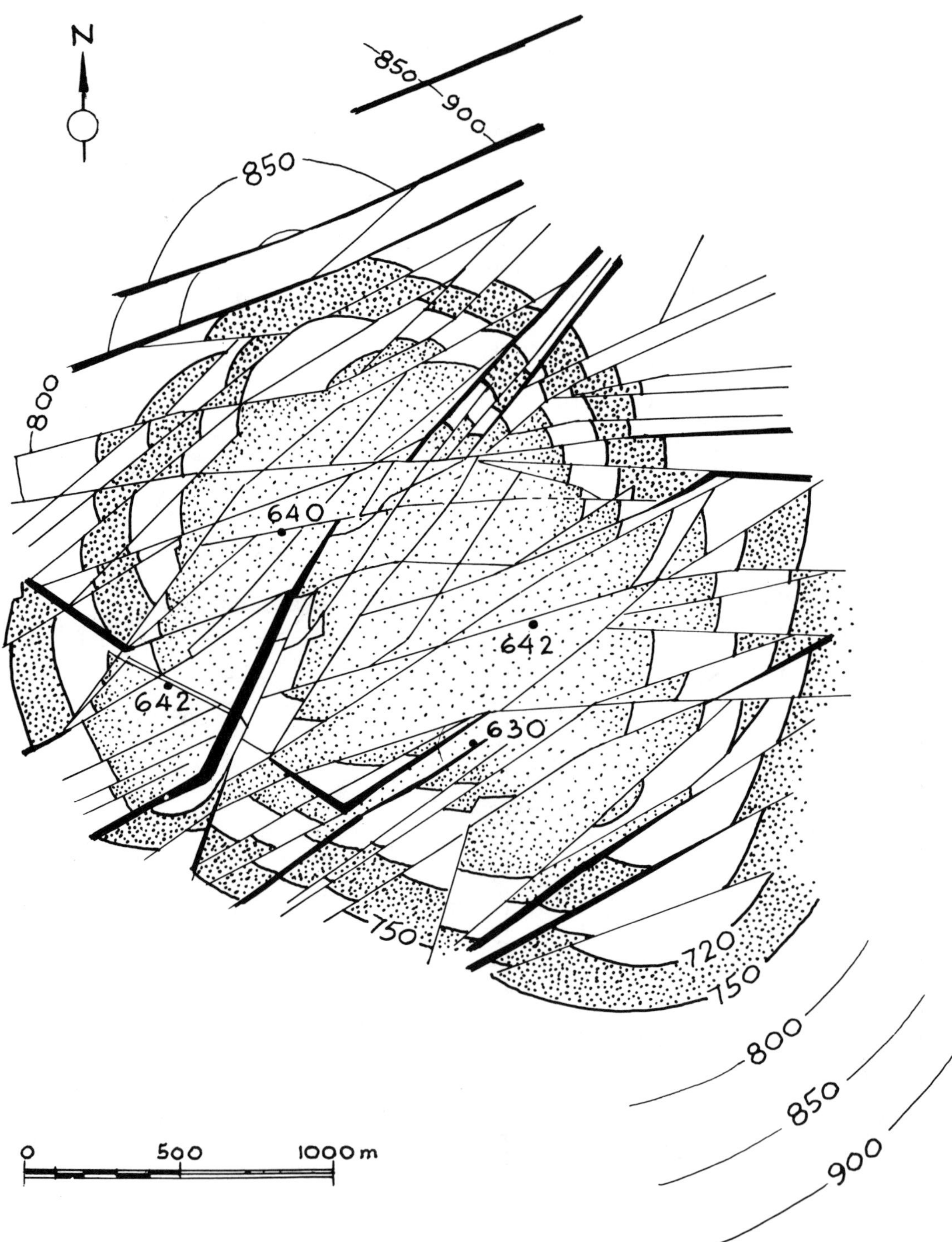

FIG. 38.—Reitbrook dome, near Hamburg, Germany. Nearly circular dome should show radial fracture pattern. "Regional" trend seems to prevail suggesting strong NW-SE component of extension (after Behrmann, 1949, p. 200, Fig. 7).

which seems to be a good term for the fault along which reverse drag is observed.

Hamblin (1965) discussed most recently "reverse drag" on normal faults and showed that the phenomenon is not restricted to Gulf Coast faults. He has mapped reverse drag in the field and gave several examples. Reverse drag is, according to Hamblin, the result of displacement on curved faults.

My experiments confirm the curved-fault principle but show that the curved fault may not be the only explanation. The fault may have to lead into a slip plane which is capable of accommodating the horizontal component.

Hubbert and Rubey (1959) discussed the possibility of lateral movements of rock masses during overthrust faulting and they showed that the presence of high fluid pressures reduce friction at the base of a moving sheet to such an extent that overthrusting becomes likely. They showed also that abnormally high fluid pressures exist in the Gulf Coast area.

Furthermore Rubey and Hubbert (1959, Fig. 9, p. 194) showed a diagrammatic cross section of a zone of abnormal fluid pressure and the beginning of a bedding-plane glide surface on the flank of a geosyncline. The authors stated (p. 194, 195):

> Under these conditions, at some depth where the critical relationship between the fluid pressure–overburden ratio and the lateral component of stresses acting on the area is exceeded, a thick section of sedimentary rock would part from its foundation and start to move. The critical relationship between pressure ratio and lateral stresses might be exceeded as the result of unusually rapid deposition, of an increased horizontal compression, or of a slight steepening of the flank of the geosyncline. . . .
>
> In this diagram (Fig. 9) the thrust plane is shown as breaking loose along a normal fault or faults at its rear or upslope margin and sliding downslope into the geosynclinal trough. . . .
>
> The writers do not intend to emphasize unduly the gravitational sliding hypothesis and Figure 9 might just as well have shown no downslope sliding, but a regional compression or horizontal push from the rear. . . . The general principles illustrated in Figure 9 apply without regard to the nature of the forces that cause movement of the thrust plate. The effect of high interstitial fluid pressures is virtually the same—that is, they would greatly facilitate gliding on bedding–plane surfaces—whether the lateral stresses that cause the movement owe their origin to gravitational forces, regional compression, or some other source.

It is quite evident that slip planes, creep, curved faults, and a horizontal component have been thought of by several authors as possible causes of the Gulf Coast fault pattern. The last hypothesis has been worked out most elaborately and seems to be the most convincing.

Conclusions

The experiments presented here demonstrate that the regional Gulf Coast fault pattern may well be the result of gravity creep of the basin fill. Such creep also would provide the mechanism for asymmetries in down-to-basin faulting: curved master faults, reverse drag, antithetic faults, steepening of dips with increasing depth, and even some thickening in the downthrown block.

Creep would provide also the necessary extension for the marginal fault system or tear-off rift at the edge of the basin. If creep began in that zone it would explain its symmetry, *en échelon* graben system, and absence of reverse drag. The assumption of creep is supported strongly by the prevalence of reverse drag on down-to-basin faults and its rarity on updip faults.

Finally, a regional component must cause the suppression of local faults and the prevalence of regional ones in structures that would be expected to show a local fault pattern.

Local patterns are abundant, however, and are without "overprint." Whether regional or local patterns dominate may well be determined by the rate of deformation.

If creep is assumed, no other explanatory mechanisms are needed and the entire Gulf Coast pattern can be explained in rather simple terms by a universal cause.

References Cited

Badgley, Peter C., 1965, Structural and tectonic principles: New York, Harper & Row, 521 p.

Behrmann, R. B., 1949, Geologie und Lagerstätte des Ölfeldes Reitbrook bei Hamburg, *in* Erdöl und Tektonik in Nordwestdeutschland: Hannover-Celle, Amt für Bodenforschung, p. 190–221.

Bornhauser, Max, 1958, Gulf Coast tectonics: Am. Assoc. Petroleum Geologists Bull., v. 42, p. 339–370.

Carlos, P. F., 1953, Conroe field, Montgomery County, Texas, *in* McNaughton, D. A., ed., Guidebook, field trip routes, oil fields, geology: Houston, AAPG-SEPM-SEG, Joint Ann. Mtg., p. 104–109.

Castellot, A. E., and R. P. Caletti, 1958, Exploración de los horizontes profundos del campo José Colomo: Asoc. Mexicana Geólogos Petroleros Bol., v. 10, p. 409–420.

Cloos, Ernst, 1955, Experimental analysis of fracture patterns: Geol. Soc. America Bull., v. 66, p. 241–256.

Cloos, Hans, 1928, Über antithetische Bewegungen: Geol. Rundschau, Bd. 19, p. 246–251.
——— 1930, Zur experimentellen Tektonik: Naturwissenschaften, Jahrg. 18, Hft. 34, p. 741–747.
——— 1931, Zur experimentallen Tektonik, Brüche und Falten: Naturwissenschaften, Jahrg. 19, Hft. 11, p. 242–247.
——— 1936, Einführung in die Geologie: Berlin, Gebrüder Bornträger, 503 p.
——— 1949, Gespräch mit der Erde: 2d ed., Munich (München), R. Piper & Co. Verlag, 389 p.
——— 1953, Conversation with the earth: New York, Alfred A. Knopf, 413 p.
DeHart, B. H., Jr., 1955, The Bourg field area, Lafourche and Terrebonne Parishes, Louisiana: Gulf Coast Assoc. Geol. Socs. Trans., v. 5, p. 113–123.
Greenman, W. E., and E. E. Gustafson, 1953, Needville field, Fort Bend County, Texas, *in* McNaughton, D. A., ed., Guidebook, field trip routes, oil fields, geology: Houston, AAPG-SEPM-SEG, Joint Ann. Mtg., p. 138–140.
Hamblin, W. K., 1965, Origin of "reverse drag" on the downthrown side of normal faults: Geol. Soc. America Bull., v. 76, p. 1145–1164.
Hardin, Frank R., and George C. Hardin, Jr., 1961, Contemporaneous normal faults of Gulf Coast and their relation to flexures: Am. Assoc. Petroleum Geologists Bull., v. 45, p. 238–248.
Hills, E. Sherbon, 1963, Elements of structural geology: New York, John Wiley & Sons, 483 p.
Hubbert, M. King, 1937, Theory of scale models as applied to the study of geologic structures: Geol. Soc. America Bull., v. 48, p. 1459–1520.
——— 1945, Strength of the earth: Am. Assoc. Petroleum Geologists Bull., v. 29, p. 1630–1653.
——— 1951, Mechanical basis for certain familiar geologic structures: Geol. Soc. America Bull., v. 62, p. 355–372.
——— and William W. Rubey, 1959, Role of fluid pressure in mechanics of overthrust faulting; I. Mechanics of fluid-filled porous solids and its application to overthrust faulting: Geol. Soc. America Bull., v. 70, p. 115–166.
Murray, Grover E., 1961, Geology of the Atlantic and Gulf coastal province of North America: New York, Harper and Bros., 692 p.
Ocamb, R. D., 1961, Growth faults of south Louisiana: Gulf Coast Assoc. Geol. Socs. Trans., v. 11, p. 139–175.
——— and R. P. Grigg, Jr., 1954, The Lewisburg field area, Acadia and St. Landry Parishes, Louisiana: Gulf Coast Assoc. Geol. Socs. Trans., v. 4, p. 183–200.
Quarles, Miller, Jr., 1953, Salt ridge hypothesis on origin of Texas Gulf Coast type of faulting: Am. Assoc. Petroleum Geologists Bull., v. 37, p. 489–508.
Rubey, William W., and M. King Hubbert, 1959, Role of fluid pressure in mechanics of overthrust faulting; II. Overthrust belt in geosynclinal area of western Wyoming in light of fluid-pressure hypothesis: Geol. Soc. America Bull., v. 70, p. 167–206.
Wallace, W. E., 1952, South Louisiana fault trends: Gulf Coast Assoc. Geol. Socs. Trans., v. 2, p. 63–67.
Walthall, Bennie H., and Jack L. Walper, 1967, Peripheral Gulf rifting in northeast Texas: Am. Assoc. Petroleum Geologists Bull., v. 51, p. 102–110.
Wendlandt, E. A., 1951, Hawkins field, Wood County, Texas, *in* Herald, F. A., ed., Occurrence of oil and gas in northeast Texas: Texas Univ. Bur. Econ. Geology Pub. 5116, p. 153–158.

Reprinted by permission of Scientific Press Ltd. from *Journal of Petroleum Geology*, v. 2, no. 3 (1980), p. 265-307.

ON THE THEORY OF GROWTH FAULTING*: A GEOMECHANICAL DELTA MODEL BASED ON GRAVITY SLIDING

W. Crans, G. Mandl and J. Haremboure

A geomechanical delta model is presented that explains and permits quantitative reproduction of the main features associated with growth faulting. The model is based on a soil-plasticity analysis of gravity sliding of overpressured clays and silts on very gentle delta slopes. In this analysis, the packets involved in gravity sliding—called 'units'—are well quantified. The delta body may behave as a stack of such units ('multi-unit' delta model), which can behave differently depending on such parameters as sedimentation rates, changes in lithology and compactional behaviour. Two prominent and essentially different structural expressions of the model are discussed: (a) the regularly spaced growth-fault pattern, without recognisable toe regions; (b) the complete slide structure with a well-developed toe region. The shapes of the roll-over structures (particularly the positions of the crests), the thickening of the layers near the growth faults, antithetic faulting and horsetailing of growth faults can be derived from the model without the need to invoke any deeper seated mechanism in the substratum. The basic assumptions of this geomechanical model are supported by observations in Ireland and examples from literature. It may contribute to the reconstruction of the hydrocarbon migration history in a delta by accounting for the synsedimentary development of stresses and fault structures that may control hydrocarbon migration. Recently, geomechanical aspects came in focus again for the assessment of seismic amplitude anomalies in the search for hydrocarbons (Crans and Berkhout, 1979).

INTRODUCTION

Purpose and scope

Growth faults and the structures associated with them provide the characteristic hydrocarbon traps in areas of continuous deltaic deposition (Gulf Coast Province, Niger Delta). They result from contemporaneous failure of the prograding delta slope. Although it is generally agreed that these slope failures are caused by the weight of the sediment on the slope, opinions differ with respect to the constraining conditions that control the development of synsedimentary faults.

Various controlling conditions have been postulated, such as steepening of the sea-bottom slope due to spontaneous upheaval of deep-seated clay masses, local overloading of the slope by excessive sedimentation, or the slip of slope sediment along a prescribed slip path. Studies based on these hypothetical

* *This research work was carried out in the Koninklijke/Shell Exploratie en Produktie Laboratorium, Rijswijk, The Netherlands, during 1968-1972 by: Dr. W. Crans, Exploration Consultant, GeoQuest International, Inc., 4605 Post Oak Place, Suite 130, Houston, Texas 77027; Dr. G. Mandl, Koninklijke/Shell, Exploratie en Produktie Laboratorium, Rijswijk, The Netherlands; J. Haremboure, Shell U.K. Exploration & Production Ltd., UEE/9 Shell Centre, London SW1 7NA, U.K. Some basic features have been published previously in short notes by Crans, Mandl and Shippam (1973) and Mandl and Crans (1979).*

constraints have been mainly concerned with the geometric and kinematic implications for the slide structures; little attention has been paid to the mechanical properties of the fluid-filled sandy and clayey materials involved.

A satisfactory understanding and a quantitative description of the processes by which gravity generates growth faults and associated structures cannot be obtained without considering mechanical parameters and laws. In particular, knowledge of the stress field induced in the sediment body by its own weight is essential, as neither the role of abnormal pore pressure in gravitational sliding nor the deformations and various slide structures produced inside the delta slope can be adequately dealt with without a detailed analysis of the stress field.

The present investigation is therefore aimed at achieving an understanding of the mechanics of synsedimentary faulting and gravitational sliding. The main objective is to develop a theoretical mechanical model that explains and connects relevant features of the sliding process by means of quantitative relationships that can be used to interpret and integrate geological data into a picture of the subsurface geometry. If it is to serve as an interpretative tool, the model should be flexible enough to allow the parameters of geological significance to be varied.

Basic to any mechanical model is a theoretical specification of the material behaviour, i.e. a relation between the stresses and the deformations they produce inside the particular type of material considered. The loose or slightly consolidated sediment found in overpressured delta slopes may be characterised as a frictional granular material that opposes deformation by frictional and cohesive forces between the individual particles. Its behaviour is thought to be accounted for best by the theory of 'soil plasticity'—a variant of the theory of 'ideally plastic' materials—which is therefore used consistently in developing our mechanical model.

To facilitate the mathematical analysis, simplifications have to be introduced regarding the outer configurations of the sediment body and its environment. In approximating the reality of deltaic provinces, we assumed that the fluid-filled sediment is deposited on a very gentle slope in layers that are thin compared to their length and lateral extent. The base of the stack of sediment layers is assumed to remain planar and inclined at a small constant angle. Likewise, the submarine surface of the sloping sediment is supposed to remain planar; possible irregularities being levelled out by sedimentation and currents. It seems further justified, in view of the high length/thickness ratios observed in reality, to treat the sediment layer as infinitely long in the mechanical analysis. The most important simplification, however, is obtained by treating the problem in vertical planes parallel to the dip of the slope. This simplification, which is motivated by the large lateral extent of the sediment body, implies that the stresses do not noticeably change in strike direction and that the plastic deformation of the sliding sediment mass is the result of material displacement in dip-parallel sections only ('plane strain').

Within the framework of these simplifications, the mechanical delta model describes how the stress field develops in the growing sediment body, how pore-fluid overpressures initiate slope failure, and how a first slide path develops. After the first sliding 'unit' has formed on the slope, continuing accretion of sediment and changes in abnormal fluid pressure alter the stress field. Consequently, new structures form and old ones change. The model, in accounting quantitatively for these processes, describes in particular the formation of new sliding units ('Multi-Unit Delta Model') and the growth, deformation and 'splintering' of synsedimentary faults.

In addition, the model allows computer simulation of the development of anticlinal ('roll-over') structures adjacent to growth faults and toe-thrusts. Characteristic types of these (often hydrocarbon-bearing) structures near growth faults are computed and related to the controlling mechanical parameters. Antithetic faulting in these roll-over regions is also accounted for in a realistic manner.

The present paper reviews the model and its main features without going into theoretical and mathematical detail. Essential features of the model are compared with outcrops of deltaic sediments observed in the Upper Carboniferous of Ireland.

Main conclusions

The theory developed leads to an idealised model—the Multi-Unit Delta Model—which may be useful for the interpretation of subsurface data, e.g. in the Niger Delta. It permits the following conclusions:

A. Growth faulting in deltaic regions can be explained solely as the result of gravity sliding of overpressured clay and silts. Clay uplifts, for instance, may affect the structures, but are in no way necessary for the initiation of growth faulting in such regions.

1. Overpressured clay and silt layers can slide under their own weight (gravity sliding) on a very gentle slope ($\angle 3^\circ$) along a surface-parallel plane, here called the 'basal slip-plane'. Shearing along the basal slip-plane causes a reduction in basal friction, offsetting the balance of gravitational and frictional forces in the sliding mass. Static equilibrium is restored by a surface-parallel 'buttressing' force activated inside the sediment itself. This force causes the creeping sediment to compact in dip-parallel direction, thus providing the space necessary for the slide to proceed.
2. The thickness (Z_{pe}) of a sliding sediment layer, the so-called 'unit', is well defined; it is proportional to the initial depth of paleo-overpressure (Z_{ig}) and approximately inversely proportional to the delta slope δ.
3. The delta body is composed of a stack of such units [Multi-Unit Delta Model (Fig. 8)].
4. The units are bounded at their upper end by rotational normal faults (growth-faults) with dips of 45°-65° relative to the (momentary) paleo-surface.
5. From its formation, a growth fault terminates in the surface-parallel basal slip-plane, with which the maximum paleo principal stress makes an angle of about 30° (i.e., it is not vertical).
6. The curvature of the growth fault and the genesis of the basal slip-plane in a homogenous clay layer are due mainly to overpressure (*)*.

B. The structural expression of the Multi-Unit Delta Model lies between two extremes:

I. *The open-ended (or incomplete) slide structure*, without recognisable toe region, appearing when the overpressure relaxation rate is relatively large compared with the sedimentation rate (Fig. 8). On account of its appearance, this manifestation is here called the *'regularly-spaced growth-fault pattern'*.

1. The spacing of growth-fault families (i.e. growth faults terminating in the same basal slip-plane) is proportional to the thickness of a unit (Z_{pe}) and approximately inversely proportional to the square of the delta slope (δ).
2. Therefore, a reduction in the spacing between growth-fault families is likely to be associated with an increase in the depositional slope and/or a higher paleo-overpressure level.
3. The slide-velocity profile (*) for this manifestation of the multi-unit delta is determined mainly by the lateral compaction of the units (see A.1. above).

II. The complete slide structure**, consisting of a roll-over region, a seemingly undeformed region and a toe region (adjacent to a low-angle gravity thrust), appears when the sedimentation rate during build-up of the formation is large compared with the overpressure relaxation rate (Fig. 17).

1. For the growth faults bounding the upper ends of the units, conclusion A.4 still holds. At their lower ends, the units are bounded by thrust (toe) faults with dips that must lie between 25° and 45° with respect to the (momentary) paleo-surface.
2. In this manifestation of the Multi-Unit Delta Model, the distribution of inflowing sediment along the delta slope can considerably affect the slide-velocity profile. This was not the case for the regularly spaced growth-fault pattern (see conclusion BI.3).

C. Two main causes of antithetic faulting in the roll-over region can be distinguished:

1. A discontinuity in the slide-velocity profile at the level of a plane of weakness, produced by an inhomogeneity in the sediment packet (Fig. 21). Antithetic faults caused in this way can be synsedi-

---

**For terms followed by a bracketed asterisk, please refer to 'Glossary of concepts used'.*

***The complete slide structure can be considered as a degeneration of the regularly spaced growth-fault pattern, since the roll-over regions of successive units coalesce. Likewise, it is assumed that the toe regions of successive units coalesce. Field evidence for this theoretical construction has still to be found.*

mentary and concave downwards when the formation is still overpressured.

2. A kink-like transition of a growth fault into the associated basal slip-plane (Fig. 20b).

   As the unit is deposited, the growth fault curves smoothly into the basal slip-plane (Fig. 20a). During subsequent overpressure relaxation the transition of the growth fault into the basal slip deforms into a 'kink'. The antithetic faults associated with this transition are post-sedimentary and become less curved as the degree of overpressuring diminishes. Moreover, they are related to horsetailing of the growth faults (see conclusions E.1).

D. The shape of the roll-over structures is determined merely by the slide-velocity profile, so that many different roll-over structures can correspond to the same stress state (the same holds for a possible toe region).

   1. The position and shape of the crestal surface in the roll-over structure depend on the slide-velocity profile (Fig. 15) and the ratio of the (average) slide velocity and deposition rate.
   2. As a result of the plastic deformation in a roll-over structure, the oldest sediment beds deposited during fault movement need not show noticeable thickening. Beds deposited before the onset of growth faulting may even show thinning near the growth fault as a result of plastic deformation (Fig. 12d).

E. During the relaxation of overpressure, the potential normal faults become 'straighter', their shape being determined by the momentary overpressure profile (*), which in turn is determined by the fluid drainage in the formation. Therefore, if the growth fault is still active, straight faults branch off from the initially curved growth fault during the relaxation process (Fig. 25).

   1. The straightening of a growth fault during overpressure relaxation manifests itself as 'horsetailing' (Figs. 25, 26).
   2. The 'horsetail' pattern is a record of the history of overpressure relaxation (Figs. 25, 26).
   3. Where a growth fault shows no curvature after correction for compaction, it has been formed under hydropressured conditions. Faults of similar age in the same region should also be plane.

Glossary of concepts used

The plasticity theory deals with permanent —*plastic*—deformation of loaded material. By definition, the (yield) *strength* of a material is the stress at which plastic deformation occurs. For *ideally plastic* material, this strength is independent of the plastic deformation; in practice, hardening or softening of the material often occurs after some plastic deformation. Generally, the strength of a material is characterised by a yield criterion. Regions in the material where the stress state satisfies the yield criterion are in the *plastic state.*

The stresses in the plastic state have to satisfy the equilibrium equations, the yield criterion, and the stress boundary conditions.

In *soil mechanics, Coulomb's yield criterion* is commonly used to characterise soil strength. According to this criterion, the material fails by slipping along planar elements—*slip planes* —along which the shear stress has reached a critical value that depends on the normal stress on the plane involved (*frictional* material) and the cohesion of the material. One can prove that the slip elements form *potential faults.*

The yield criterion for the stresses in an ideally plastic material is quadratic in form, so that there are generally two different solutions for the stress state— the *active*—and *passive-plastic states*—that satisfy the set of stress equations. A *roll-over region* (adjacent to a growth fault) is a field example of an active plastic state, and a *toe region* (adjacent to low-angle gravity thrust) of a passive plastic state.

In the entire (passive-plastic) toe region of our model, the surface-parallel normal stress (surface-parallel pressure) is larger than the subvertical normal stress. The subvertical normal stress, on the other hand, is the largest compressive stress in the major part of the (active-plastic) roll-over region. If both types of plastic region are present, the growth-fault family upslope and the toe-fault family downslope are connected by a surface-parallel *basal slip-plane,* so that the *sliding path is closed* (Fig. 7a).

It follows from the theory that a roll-over region must accompany a toe region, but it is quite possible for a roll-over region to exist alone if the surface-parallel pressure does not reach the limiting value for the passive-plastic state and, consequently, the surface-parallel basal slip plane dies out downslope in the sediment. Such a structure, which consists only of an (active-plastic) roll-over and a nonplastic (*stiffened*) region, has been called an *open-ended slide structure*.

Gravity sliding of a long, gently sloping fluid-filled sediment layer under its own weight is made possible by overpressure alone. The shapes of the potential faults are determined by the *overpressure profile*, i.e. the pressure in excess of hydrostatic fluid pressure as a function of depth.

Growth-fault structures have been calculated for *'leaky' overpressured* formations, i.e. formations in which the downward fluid-pressure gradient below the initial depth of overpressure Z_{ig} is equal to the lithostatic gradient (Fig. 5a). An opposite of leaky overpressure is *sealed overpressure* for which the fluid-pressure gradient is hydrostatic in the overpressured region but there is a discontinuity in fluid pressure across the sealing layer in the formation (Fig. 5b).

The plastic strains in the loaded region are obtained from the *flow rules* or *velocity equations* and from the boundary conditions for the velocities. In the model given, both the roll-over and the toe regions are *statically determined*, i.e. the stress state in these regions is fully determined by the boundary conditions for the stresses. It follows from plasticity theory that the boundary conditions for the plastic strains in such a case can be chosen independently of those for the stresses; (in the theory of elasticity, for instance, this is impossible, since the stresses and strains are related linearly by Hooke's law).

As boundary conditions for the strains in both the roll-over and toe regions one can therefore introduce, in the first instance, arbitrary *slide-velocity profiles* that describe variations in the surface-parallel slide velocity as a function of depth. However, the plastic deformation fields in the plastic regions have to satisfy a restrictive requirement (compatability condition), in that the shearing rates along the slip planes must be in the direction of the prevailing shear stresses.

In some parts of the geochemical analysis of the sliding sediment layer, the rheological properties of the soil have been extended to include Bingham behaviour. In Bingham materials the stresses can exceed the yield stresses, the excess stresses causing viscous fluid flow.

REVIEW OF GROWTH-FAULTING THEORIES

In view of the significance of growth faults and structural deformation related to growth faulting for oil exploration, it is not surprising that there is extensive literature on the subject. At present one can distinguish three main schools of thought on the cause of growth faulting in deltaic regions, which attribute it to: (i) the development of shale waves (ridges) in the shaly part of the formation, (ii) local excess loading of the delta slope, or (iii) free gravitational sliding.

Shale waves and growth faulting

The idea that growth faulting is caused by the development of shale waves was strongly supported by Odé (IR, 1962)*, who assumed that overpressured clay or shale layers behave, in the first instance, as salt layers. It is a well-known phenomenon that a density inversion occurs if such a salt layer is buried beneath sediment layers, leading to gravitational instability.

According to Odé, the most likely requirements for gravitational instability in clay or shale are (following this train of thought) the following:

(i) the existence of thick and uniform clay or shale layers of large horizontal extent;

(ii) comparatively rapid burial of these shale layers under denser sediment (Fig. 1).

Under these conditions gravitational instability may occur sooner or later. That is to say, clay will tend to rise in crests separated by deepening troughs, and sediment will start sliding downwards along the steepening flanks of the clay 'waves'. Figure 1 shows that the spacing of the growth fault produced in this way is directly related to the wavelength of the shale 'waves'.

Merki (IR, 1971) described the structural development of the Cenozoic Niger Delta employing a similar concept of a causal relationship between the growth of shale ridges and the formation of growth faults. One can, of course, raise objections against this explanation of growth-fault genesis, firstly because gravitational instabilities in salt and shale layers behave differently. This objection has already been mentioned by Odé himself (IR, 1962) and

**IR indicates an internal SHELL report.*

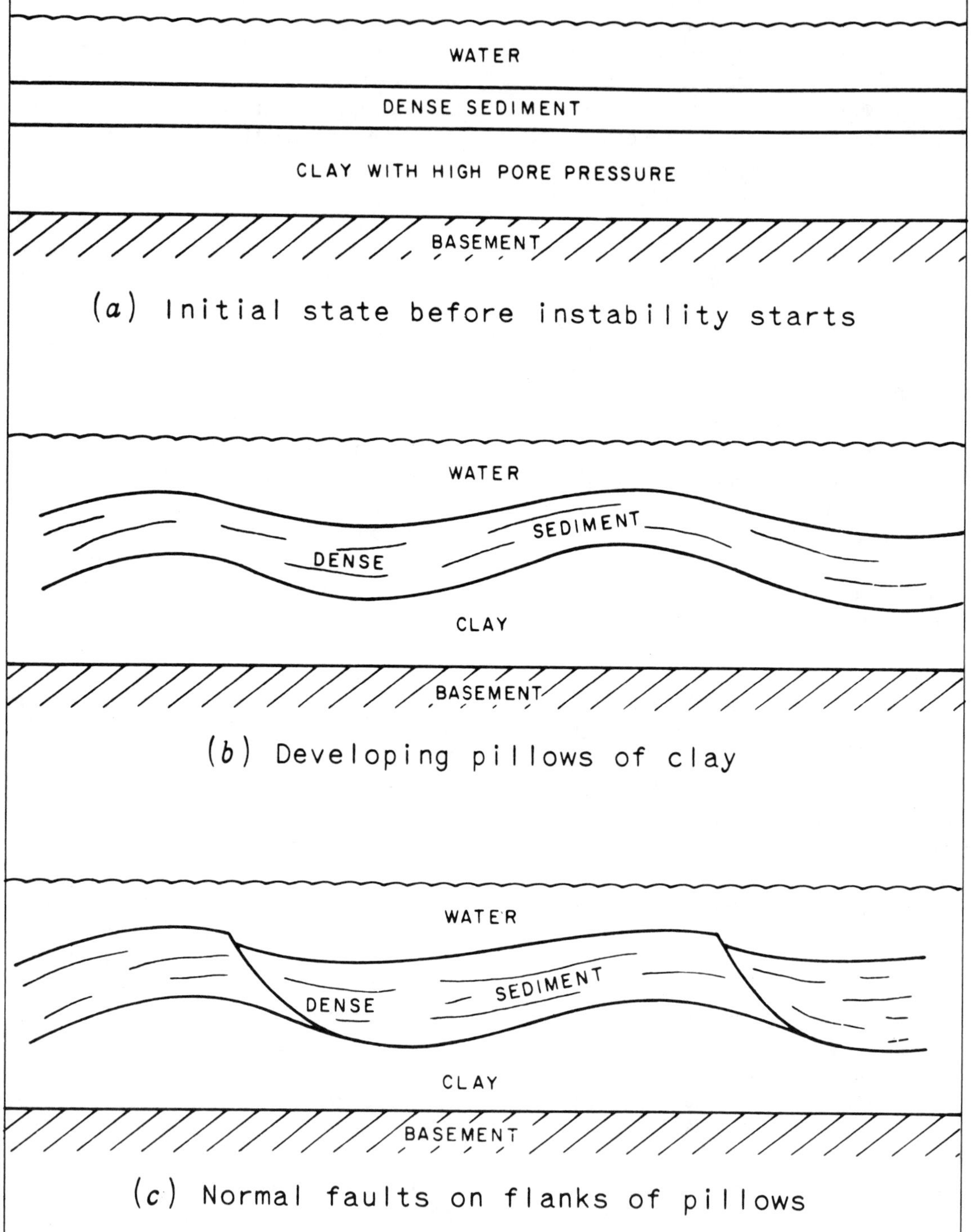

(a) Initial state before instability starts

(b) Developing pillows of clay

(c) Normal faults on flanks of pillows

DEVELOPMENT OF NORMAL FAULTS ON FLANKS OF CLAY PILLOWS
(after Odé, 1962)

Fig. 1

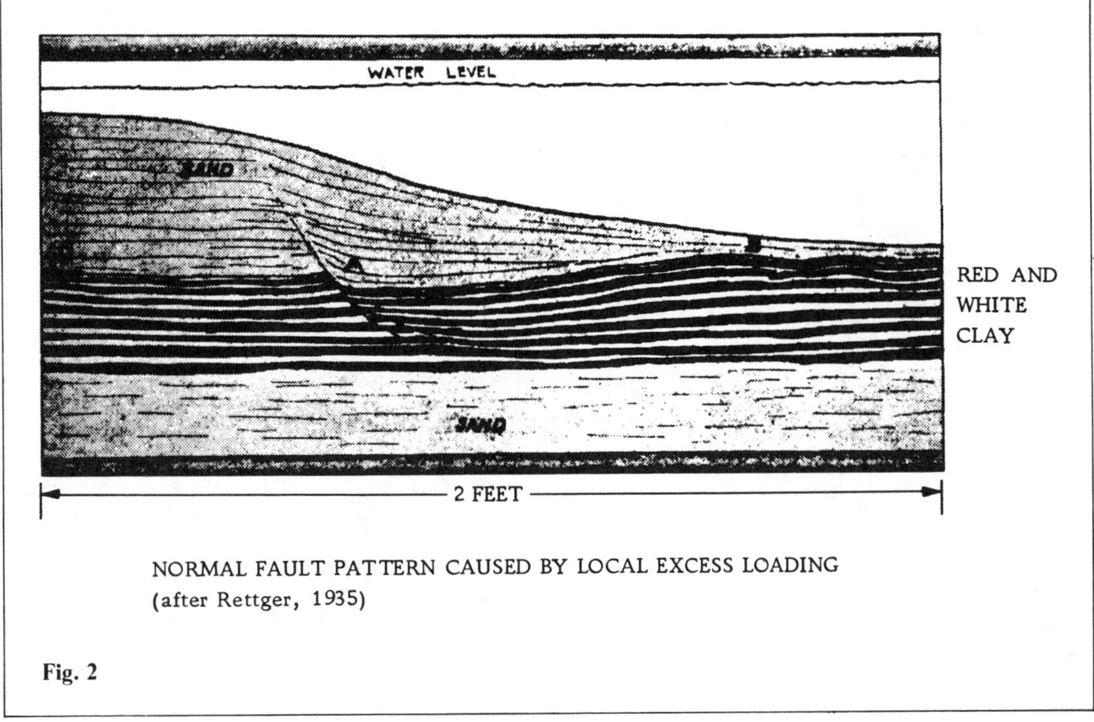

NORMAL FAULT PATTERN CAUSED BY LOCAL EXCESS LOADING (after Rettger, 1935)

Fig. 2

by Crans & Spijer (IR, 1971). Secondly, as Lohse (IR, 1958) suggests, in slightly dipping, overpressured clays and shale layers, gravity sliding can occur without being preceded by doming. The mechanism of low-angle gravity sliding has been further investigated by Hubbert & Rubey (1959) using Terzaghi's concept of land-sliding [Terzaghi (1950, 1956)]. Odé (IR, 1962) in trying to relate growth faulting to gravity sliding along the steepening flanks of the shale waves, expanded upon these concepts.

It will be shown in this report series that gravity sliding can occur in rather thin sediment packets on very gentle slopes ($\angle 3°$) long before the shale or clay layer has been buried by a heavier sediment layer. Naturally, at a later stage, clay waves may affect the shape, but hardly the spacing, of growth faults. Hence, within the framework of the present model, the bulges and clay diapirs observed in the field are interpreted as having been created later than the faults (see below, under "open ended slide structure").

Local excess loading of the delta slope

The idea that growth faulting is caused by local excess loading of the delta slope (or in other words, 'differential loading') is supported by model experiments of Rettger (1935) and Diebold (IR, 1963) (Figs. 2 and 3, respectively). Figure 2 shows, apart from normal faults at point A, a toe region at point B, which may account for the upheavals.

If one attributes growth faulting to differential loading, it is difficult to find a theoretical explanation for the spacing of the faults, which in the case of shale-waves concept was conveniently associated with the dominant wavelength of the instabilities (see above, under "Shale waves and growth faulting"). Compared with the Multi-Unit Delta Model, excessive local loading would also initiate growth faulting at a much later stage of sediment deposition.

A variation on the differential-loading theory is that developed by van den Abeele (IR, 1968), in which a strength- and a depositional-profile for the delta are distinguished. The former is a mechanically stable profile, while the latter corresponds to the prevailing sedimentation distribution. According to van den Abeele, the depositional profile in most deltas is steeper than the strength profile, producing an unstable situation that leads to gravity sliding.

Free gravitational sliding of the sediment packet.

The third possibility is to explain growth faulting as the result of free gravity sliding towards the basin. The concept of a free gravity slide is older than the theory of shale waves, and is best illustrated by the model experiments of Cloos (1968). By melting a

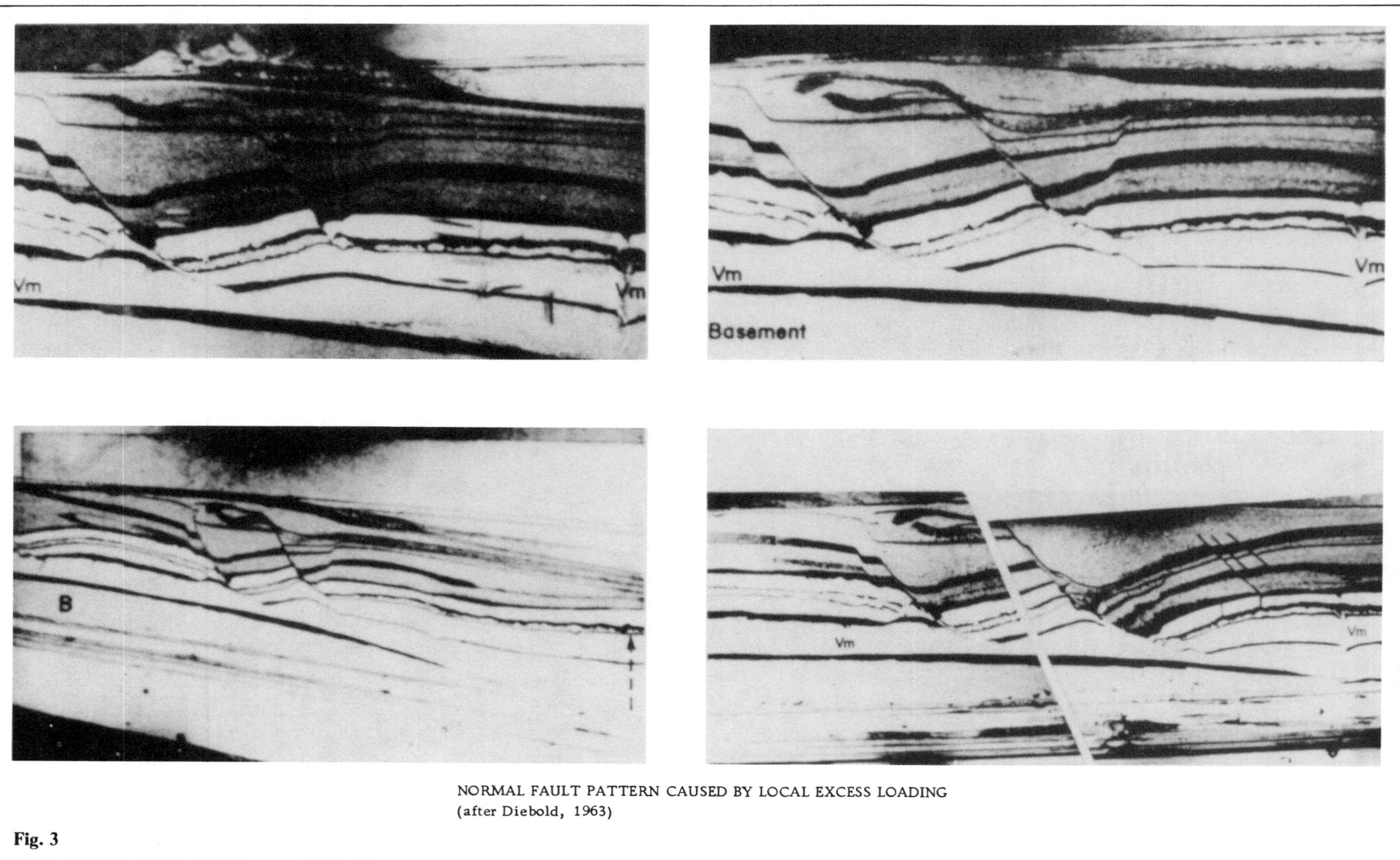

NORMAL FAULT PATTERN CAUSED BY LOCAL EXCESS LOADING
(after Diebold, 1963)

Fig. 3

paraffin layer under a sloping clay packet, he introduced an underlying plane of weakness along which gravity sliding could be initiated (Fig. 4a). The deformation pattern adjacent to the fault could also be followed (Fig. 4b). Cloos tries to explain the growth-fault structures in the Gulf Coast by this mechanism [Fig. 4c; see also Dickey *et al.*, (1968)].

The growth-faulting theories and model experiments mentioned above deal largely with the kinematic aspects of the problem and do not go beyond a qualitative discussion as far as the deformational behaviour of the material itself is concerned. So far, little has been done on the quantitative formulation of the growth-fault process, which has led some geologists to draw erroneous conclusions. For instance, knowing nothing else of the process of growth faulting, it is reasonable to assume that the maximum principal stress is due to gravity, and therefore practically vertical in orientation. Consequently, growth faults should have a dip of approximately 60°, as follows from Coulomb's slip concept. In reality much smaller dip angles are observed, and this has been attributed to vertical compaction in a report by Roux (IR, 1968), which presents a well-documented qualitative survey of growth faulting.

Livingstone (IR, 1968), on the other hand, has proved that in some cases the flattening cannot have been caused by compaction, so that other factors must have been responsible. Livingstone also questioned the assumption that the principal maximum stress should be vertical at depth. Some investigators have suggested that growth faulting could be produced by gravitational sliding, and that the dip of faults of this type could be a function of overpressure.

The Delta Model described in the present report shows that surface-parallel slip planes exist from the beginning of growth faulting and that at the depth of such a slip plane, the maximum principal stress deviates by approximately 60° from the vertical.

THE MULTI-UNIT DELTA MODEL

Overpressure and strength parameters

When loose sediment is deposited on a gently inclined, plane delta slope ($\angle 3°$), its weight has a surface-parallel component that tends to pull it down the slope. As long as this force is balanced by the reactive shear stress along the slope, the sediment will stay on the slope in stable equilibrium. As sedimentation continues and the height of the sediment layer increases, both the 'pulling' weight component and the reactive shear stress increase proportionally. The equilibrium will remain stable as long as the slope-parallel shear stress remains below the limit determined by the shear strength of the sediment. Since loose or slightly consolidated sediment owes its shearing strength to friction at the contacts between the individual particles, the sediment will 'strengthen' as the interparticle contact pressures rise. If the pore pressure remains hydrostatic, i.e. balancing the full weight of the water column only, the interparticle contacts have to carry the full increase in the submerged sediment weight. The shearing strength along the slope will increase accordingly in proportion to the sediment height, and stability will be maintained for any thickness of the hydropressured layer under the conditions mentioned. In general, however, the pore pressure increases beyond the hydrostatic level as a result of incomplete drainage during compaction of the fine-grained sediment. Under deltaic conditions, this overpressuring can start at very shallow depth (of the order of 10m).

Overpressure can vary with depth in different ways, but here 'leaky' overpressure is mainly considered (Fig. 5a). For this somewhat idealised type of overpressure, the subvertical effective stress $\bar{\sigma}_{zz}$ remains constant below the initial depth of overpressure $z = Z_{ig}$, while the fluid-pressure gradient $(\frac{\delta p}{\delta z})$ is equal to the lithostatic one. Such overpressure is only possible when fluid flow 'leaks' through the sediment.

An essentially different type of overpressure is 'sealed' overpressure (Fig. 5b), (occasionally considered in this report series), where both the fluid-pressure gradient and the effective-stress gradient in the overpressured region are equal to those in the hydropressured region, but there is a jump Δp in the fluid pressure across the sealing layer.

In the case of 'leaky' overpressure, plastic slip becomes possible along a surface-parallel plane at a depth Z_{pe}, the basal slip-plane*. The critical depth Z_{pe} is determined by the

**In the case of 'sealed' overpressure, gravity sliding becomes possible along the sealing layer if the pressure jump Δp is large enough.*

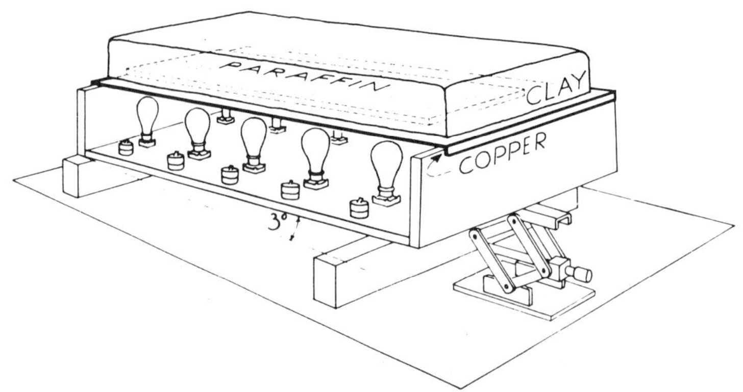

Fig. 4a. Gravitational creep experiment with wet clay (diagrammatic)

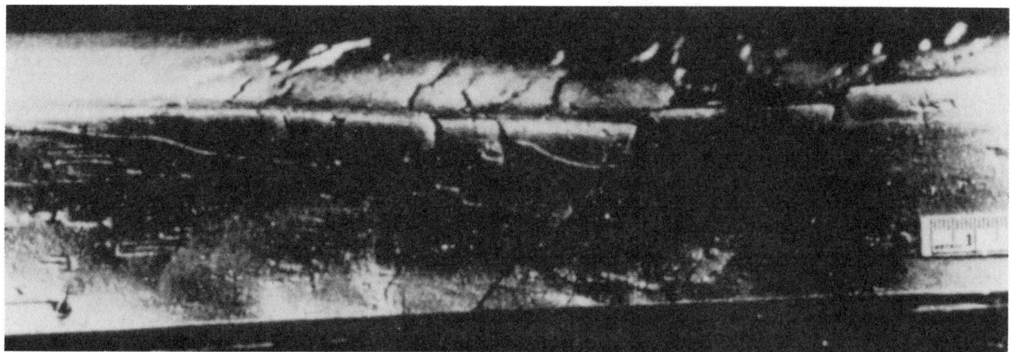

Fig. 4b. Downdip faults in wet clay model. Downbend is clockwise (towards left) with some updip faults visible on surface

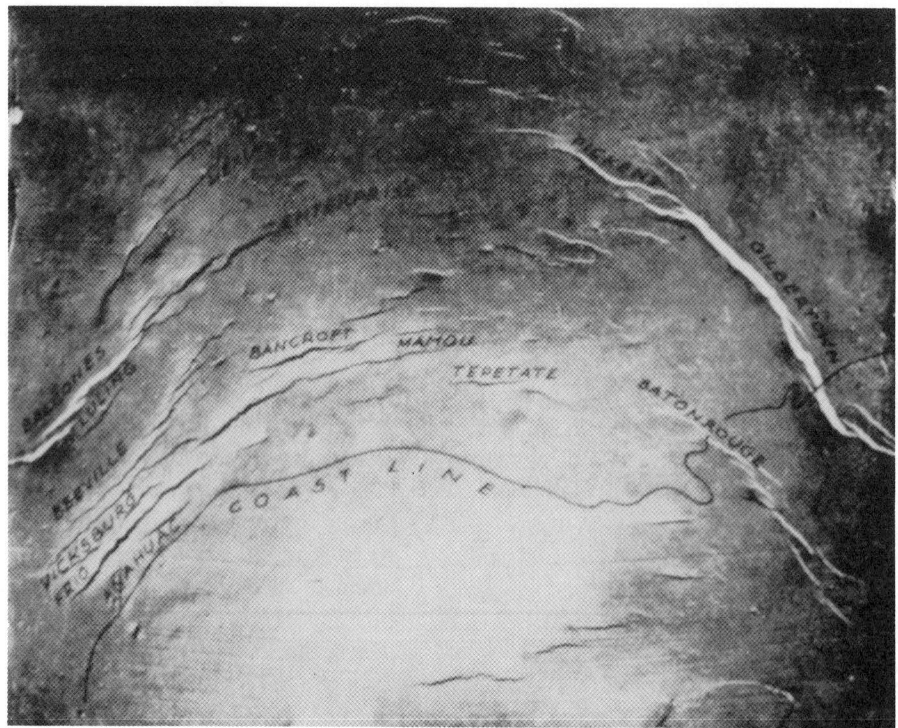

Fig. 4c. Map of Gulf Coast fault pattern superimposed on model

Fig. 4 FAULT PATTERN CAUSED BY FREE GRAVITATIONAL SLIDE (after E. Cloos, 1968)

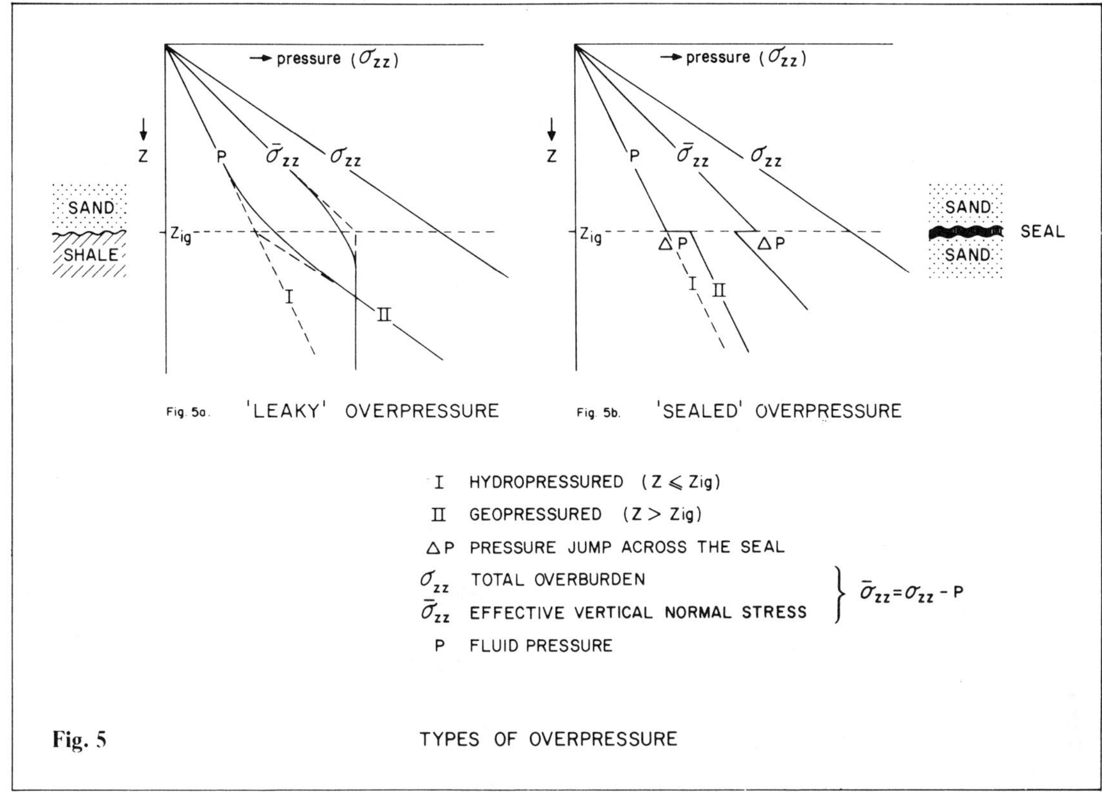

Fig. 5 TYPES OF OVERPRESSURE

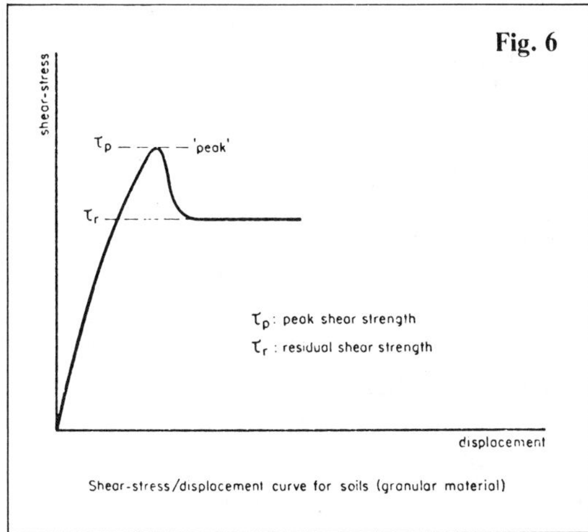

Fig. 6

Shear-stress/displacement curve for soils (granular material)

expression:

$$Z_{pe} = \frac{c}{\gamma \sin\delta} + Z_{ig}\frac{\tan\rho}{\tan\delta} \quad (Z_{ig} \angle Z_{pe}) \quad (1)$$

where the cohesion c and the angle of internal friction ρ are the parameters determining the strength of the sediment, and γ is the submerged specific weight.

The 'unit'

Reduction of basal friction

At the moment when the sediment layer with the critical thickness Z_{pe} (the 'unit') is about to start sliding down the slope, the driving weight component is still completely balanced by the friction along the base of the layer; after onset of sliding, a small amount of slip reduces the shearing resistance at the basal slip-plane (a phenomenon well-known from many laboratory soil tests and field observations). The reduction in basal friction is expressed by $r_o = \tau_r / \tau_p$, where τ_p is the peak and τ_r the residual value of the shearing resistance (see the shear-stress/displacement curve of Fig. 6).

As a result of this reduction in basal friction, the subhorizontal effective normal stress ($\bar{\sigma}_{xx}$ in Figs. 7b & c) has to increase down-slope to support the unbalanced weight component of the slowly gliding layer, which will now be subjected to an extra lateral compaction. The up-slope (active plastic) part of the sliding sheet will slide along a concave (upwards) slip surface S_aB_a (Fig. 7c).

It should be understood, that, in general, sliding proceeds slowly and only gradually affects larger parts of the slope. After sliding has commenced, somewhere at the base of the unit, the local reduction in basal friction will gradually spread down-slope in pace with the slope-parallel compaction caused by the unbalanced weight component. Naturally, such slow controlled slope failure is only possible when a 'buttressing' stress builds up inside the

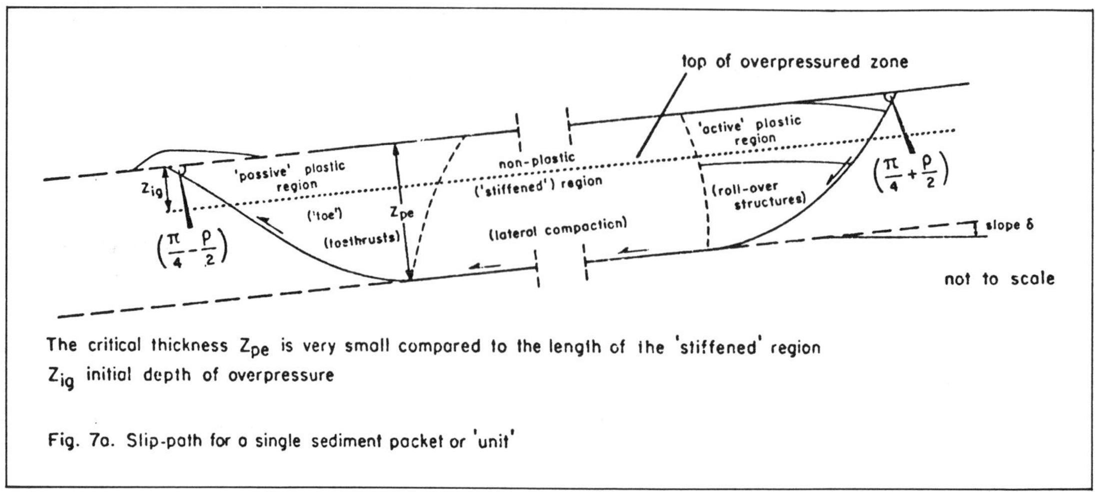

The critical thickness Z_{pe} is very small compared to the length of the 'stiffened' region

Z_{ig} initial depth of overpressure

Fig. 7a. Slip-path for a single sediment pocket or 'unit'

layer to support the 'loosened' part and prevent an 'avalanche-type' slide.

Plastic material has the remarkable property of being capable, within certain limits, of building up such a reactive stress. Soil-plasticity theory accounts for this fact by distinguishing two limiting states: the active and passive plastic states. An infinite sloping layer may therefore be in a 'hanging' (active) plastic state, or in a 'self-buttressing' (passive) plastic state; the only difference being in the magnitude of the surface-parallel effective normal stress ($\bar{\sigma}_{xx}$) in the two cases (Fig. 7a).

Up-slope from where sliding starts, the unit is in the 'hanging' state, and the surface-parallel normal stress is well-defined. Because of the reduction in basal friction, the surface-parallel normal stress increases down-slope. When the conditions necessary for sliding have been maintained for long enough, it will attain—at a certain distance down-slope—the limiting value associated with the buttressing (passive plastic) state (Fig. 7a). There, the slide path will terminate in a low-angle gravity thrust or 'toe' fault bounding the toe region of the slide.

Between the active plastic region up slope and the passive plastic 'toe' down-slope, the sediment cannot be in the plastic state: it has 'stiffened'.

Slip-line fields

Associated with the stress state in a plastic region is a slip-line field, i.e. the traces of surfaces along which the yield condition is satisfied. In plastic flow of soils, actual material slip may occur along these surfaces, this being one of the outstanding features of plasticity theory which accounts for both continuous and slip-like deformation.

Since, generally, in every point of the plastic region, two planar elements exist, the slip-line field is built up of two families of slip lines* (Figs. 7b & c).

Figure 7c shows the computed slip-line fields for both the active and passive plastic regions under overpressured conditions. The slip lines that curve smoothly into the base of the unit (basal slip-plane) are the traces of potential (growth or toe) faults, since they obviously can provide a sliding path. In the active plastic region, a growth fault and its associated roll-over region will develop, while a toe region will form in the passive plastic region. The basal slip-plane $z = Z_{pe}$ is an envelope for a set of slip lines both in the active and in the passive plastic regions. For comparison, the slip-line fields corresponding to the active and passive plastic states of a hydropressured layer have also been given in Figure 7b. The slip lines asymptotically approach straight lines for greater depth and therefore do not initially constitute potential slide paths.

Length of the stiffened region

In the case of the existence of a passive plastic toe region, active and passive plastic regions cannot coexist in a unit along the slope unless separated by a stiffened buffer region. The three regions together form a complete sliding unit bounded by a closed slide path (*) (see Fig. 7a).

The maximal length L of the stiffened region

*Following common soil-plasticity usage, the direction along the slip lines including the smaller positive angle with the maximum principal stress $\bar{\sigma}_1$ is denoted as the α-direction, and the other slip-line element as the β-direction (Figs. 7b & 7c).

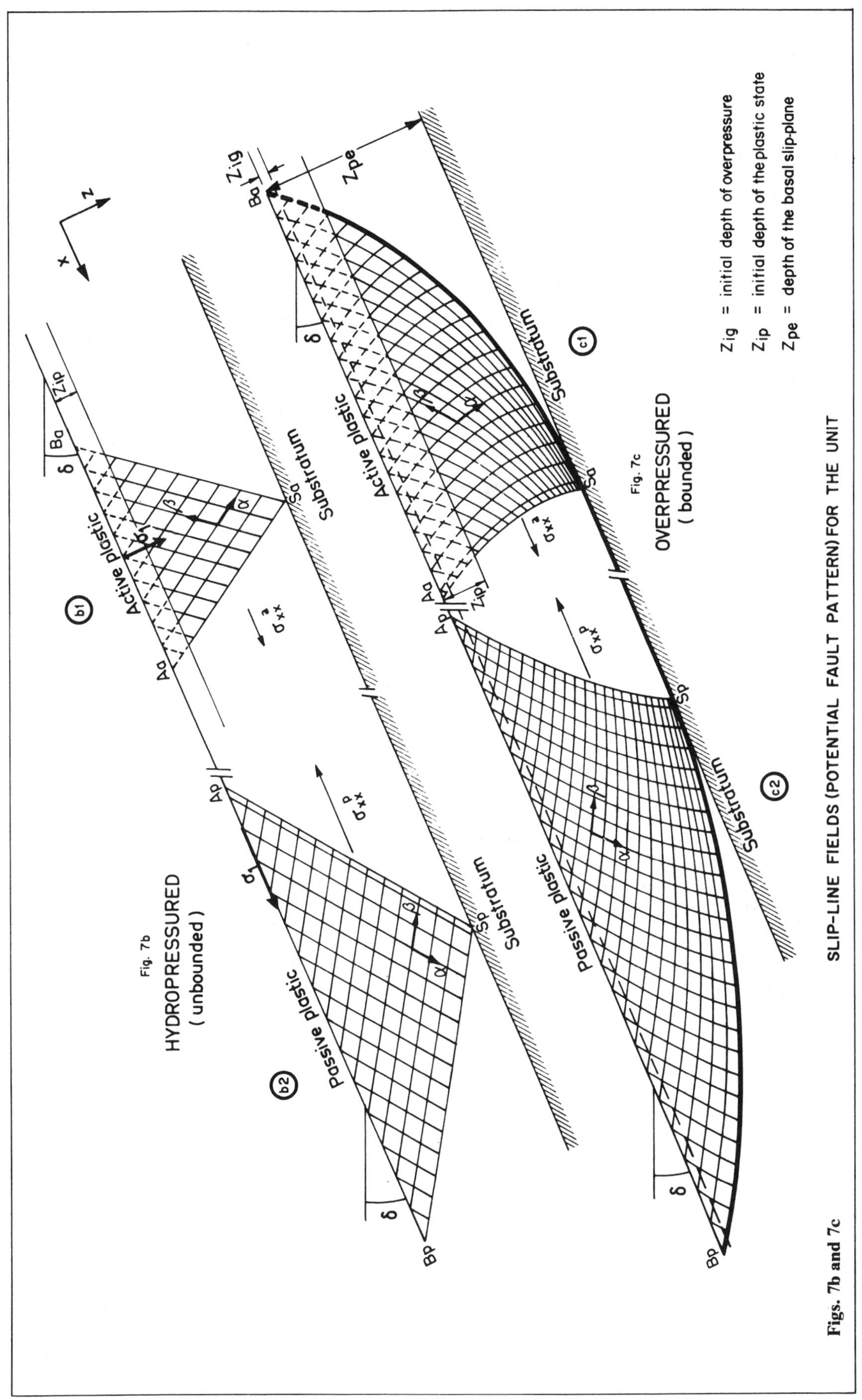

Figs. 7b and 7c SLIP-LINE FIELDS (POTENTIAL FAULT PATTERN) FOR THE UNIT is determined by the 'buttressing' stress that the layer itself can provide, i.e. the difference in the limiting values of the surface-parallel effective stresses ($\bar{\sigma}^{p}_{xx} - \bar{\sigma}^{a}_{xx}$ in Fig. 7b) corresponding to the active and passive plastic states, and is a function of the following variables;

a. Reduction in basal friction (cf. p. 273): $r_o = \tau_r / \tau_p$.

b. Relaxation of overpressure:
As sedimentation continues, the overpressure along the basal slip-plane may relax, and this will tend to restore the original basal friction by increasing the effective subvertical stress $\bar{\sigma}_{zz}(Z_{pe})$. This strengthening of the basal slip-plane is accounted for by the parameter

$r_\sigma = \bar{\sigma}^{m}_{zz}(t, Z_{pe}) / \bar{\sigma}^{i}_{zz}(Z_{pe})$, where $\bar{\sigma}^{m}_{zz}(t, Z_{pe})$ is the instantaneous value of $\bar{\sigma}_{zz}(Z_{pe})$ and $\bar{\sigma}^{i}_{zz}$ is its initial value at the moment of first slip along the basal slip-plane. The influence that reduction and re-strengthening of basal friction have upon the length L of the stiffened region in the unit is accounted for in the theory by the combined parameter $R(t) = r_o r_\sigma$.

c. Sedimentation rate:
Due to sediment accretion, the top of the unit becomes loaded by a shear stress that acts down-slope. This load too must be balanced by the surface-parallel buttressing stress. Therefore, sediment accretion—described by the relative thickening $\xi(t)$ of the unit—will tend to reduce the maximal length L of the stiffened part of the unit.

Taking into account these parameters, the maximal length of the stiffened part of the unit can be approximated by:

$$L = \frac{\pi Z_{pe} |1 + \xi(t)|^2}{|1 - R(t) + \xi(t)| \cos\rho} |1 - \Phi| \quad (2)$$

where

$$\Phi = \frac{2}{\pi}\left\{\xi(1+\xi)^{-2}\sqrt{1+2\xi} + \arcsin \xi(1+\xi)^{-1}\xi\right\}.$$

The quantity Φ may be neglected when the sediment load is small ($\xi \ll 1$).

In the event that the denominator tends to zero, the length L of the stiffened region approaches infinity, which means that no toe region will be formed. Equation (2) shows that such a situation will occur when the relaxation of overpressure—represented by the expression $|1 - R(t)| < 0$—is balanced by the sedimentation represented by the parameter $\xi(t)$*. On the other hand, if the sedimentation rate is relatively large (so that $\xi(t)$ becomes large) the length L becomes finite and a toe region is likely to be formed in the unit.

A negative denominator indicates that the movement along the basal slip-plane has ceased. In this event, the effective stress $\bar{\sigma}_{zz}$ normal to the basal slip-plane becomes so large that the condition for plastic slip is no longer fulfilled.

It may be recalled that the presence of a toe region is not necessary for growth faulting. If there is no toe region ('open-ended slide structure'), the space necessary for a slip movement along the basal slip plane is created by lateral compaction of the unit. This lateral compaction is caused by the buttressing, lateral effective normal stress ($\bar{\sigma}_{xx} - \sigma^{a}_{xx}$ in Figs. 7b & c), which increases in the x-direction and is caused by the reduction in basal friction. Any surface depressions caused by this lateral compaction process are filled by extra sediment from, for instance, mud currents.

Genesis of the multi-unit delta

In a deltaic environment, the first unit formed will be buried under sediment. Assuming that the delta progrades without a change in sea level, the roll-over structure adjacent to the growth fault will be buried beneath sediment that becomes continuously coarser, causing the overpressure profile to change. The change in overpressure profile during the sedimentation process will depend on, among other things, the rate of sedimentation.

The slip movement along the growth fault S_aB_a (Fig. 7c) will continue as long as the conditions are such that plastic slip along the basal slip-plane remains possible. When that is the case, the roll-over region remains active plastic, allowing plastic slip and deformations to proceed.

However, the change in overpressure profile will considerably affect the shape of the growth fault S_aB_a, which can lead to bifurcation of the growth fault at the basal slip-plane. In Figure 23a, computer-calculated growth faults are

**Note that equation (2) was derived under the assumption of a uniform overpressure profile. In reality, however, the overpressure top in the lower part of the delta slope will be located at greater depth. Consequently a toe region cannot be formed in all those cases where the value L (predicted by eq. (2)) is large enough for the associated basal slip-plane to meet hydropressure conditions somewhere down-slope.*

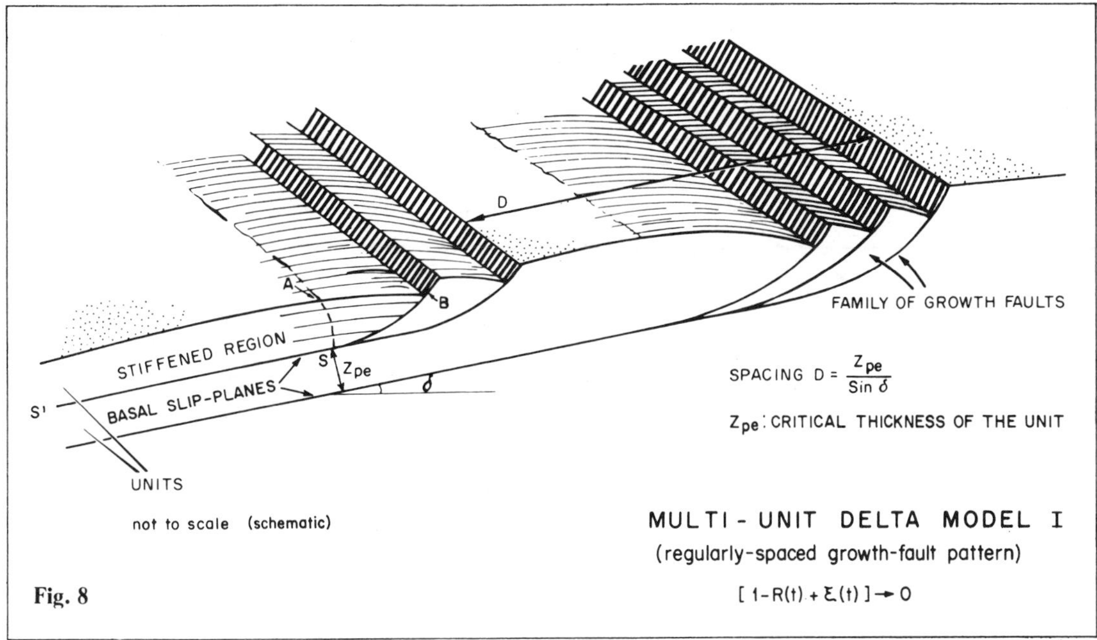

Fig. 8

presented corresponding to different levels of 'leaky' overpressure (cf. synsedimentary development, under "Development of growth faults", below). This figure shows that the growth fault will change its position and shape continuously during the sedimentation process, while the basal slip plane will remain unchanged. The reduction in friction after some slip along the growth faults is therefore not taken into account when dealing with the stress field in the roll-over region.

The stiffened part of the first unit serves in the initial state as a 'rigid' substratum for the freshly deposited material. Clearly, this new sediment loads the underlying unit and, depending on its overpressure state, the first unit may become plastic again. This development of the stress state in the unit during continuous sedimentation and overpressure relaxation has been extensively considered in unpublished parts of these studies, where it is proved that, under certain conditions, the unit can indeed become plastic again, but that a new surface-parallel slip-plane cannot form again in the unit. Consequently, the unit maintains its integrity and can serve as substratum for further sediment accretion. When the new sediment layer has reached a critical thickness, a second unit caves in and starts to slide without disturbing the underlying unit. This process may be repeated, building up the delta body as a stack of units.

Manifestations of the multi-unit delta

Two main structural expressions of the multi-unit delta model can be distinguished, depending on whether or not there is a toe region present:

a. The open-ended (or incomplete) slide structure without a recognisable toe region (Fig. 8). According to equation (2) for the length of the stiffened region, such a structure occurs when re-strengthening of the basal slip plane by overpressure relaxation compensates for the down-slope drag of the accreting sediment load.
b. The complete slide structure showing a well-defined toe region (Fig. 17). This structure occurs when the rate of sedimentation is large compared with the overpressure relaxation rate.

Open-ended slide structure (Fig. 8)

In this case the sediment units slide individually on top of each other; each parallel slip plane terminating upslope in a single growth fault or in a 'family' of growth faults. In this way, a regularly spaced growth-fault pattern appears in the delta body.

The subvertical spacing of successive surface-parallel basal slip planes Z_{pe} is well defined [cf. eqn. (1)]. It is further assumed that the branching-off from the surface-parallel slip plane is initiated by an inhomogeneity in the overpressured material, e.g. the change in lithology (Fig. 9a). Consequently, the horizontal spacing D between successive families of growth faults is essentially controlled by the unit thickness Z_{pe}, according to the equation

$$D = Z_{pe}/\sin\delta \qquad (3)$$

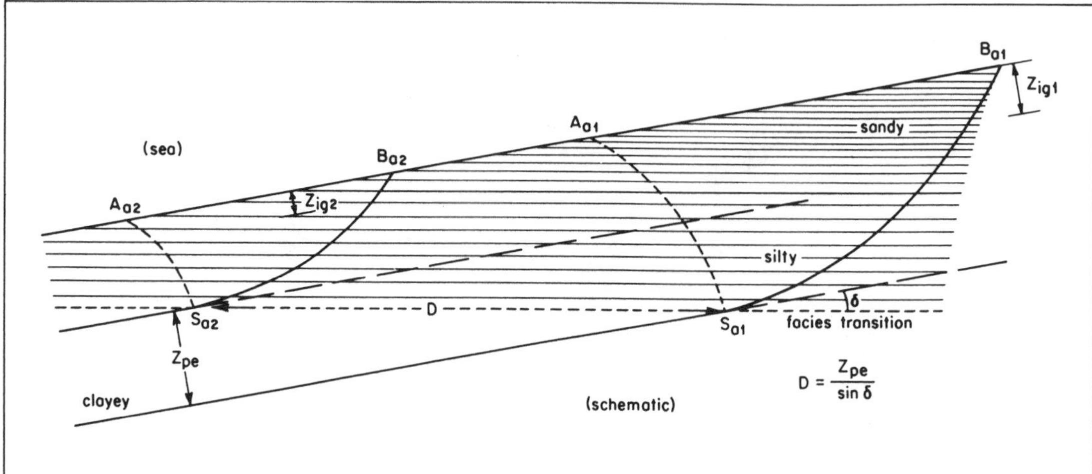

Fig. 9a. Branching-off of growth faults ($S_{a1}B_{a1}$ & $S_{a2}B_{a2}$) caused by a facies transition. In this regularly spaced growth-fault pattern, each unit terminates in a seperate roll-over region ($S_{a1}A_{a1}B_{a1}$ & $S_{a2}A_{a2}B_{a2}$)

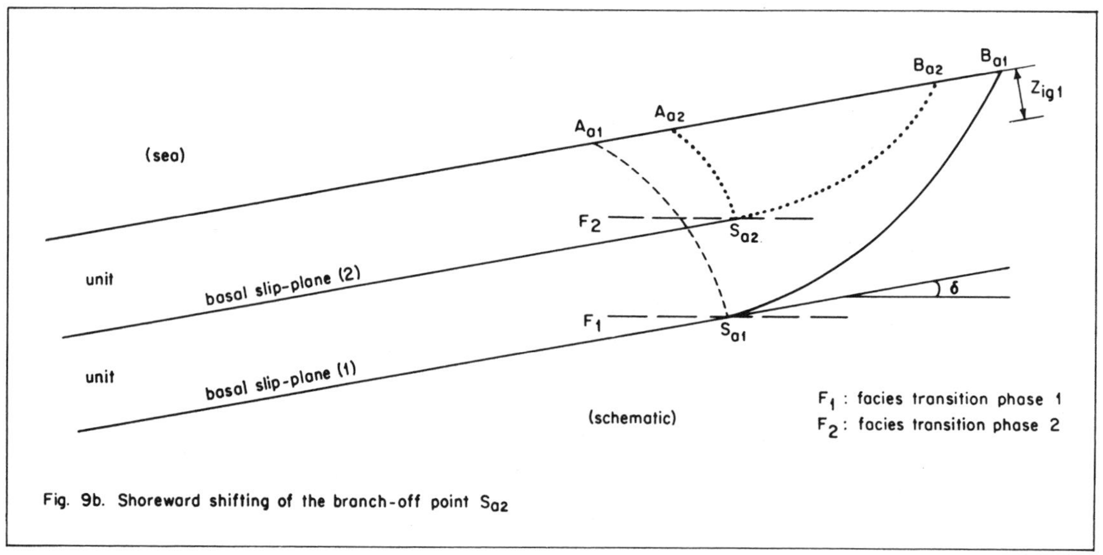

Fig. 9b. Shoreward shifting of the branch-off point S_{a2}

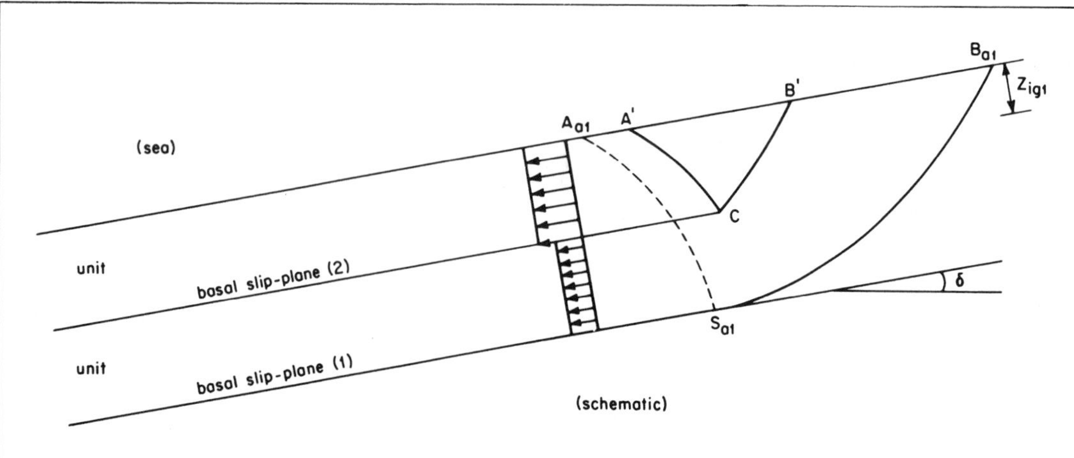

Fig. 9c. Roll-over structure with new fault system (antithetic fault CA' and growth fault CB') due to a velocity discontinuity along the second basal slip-plane

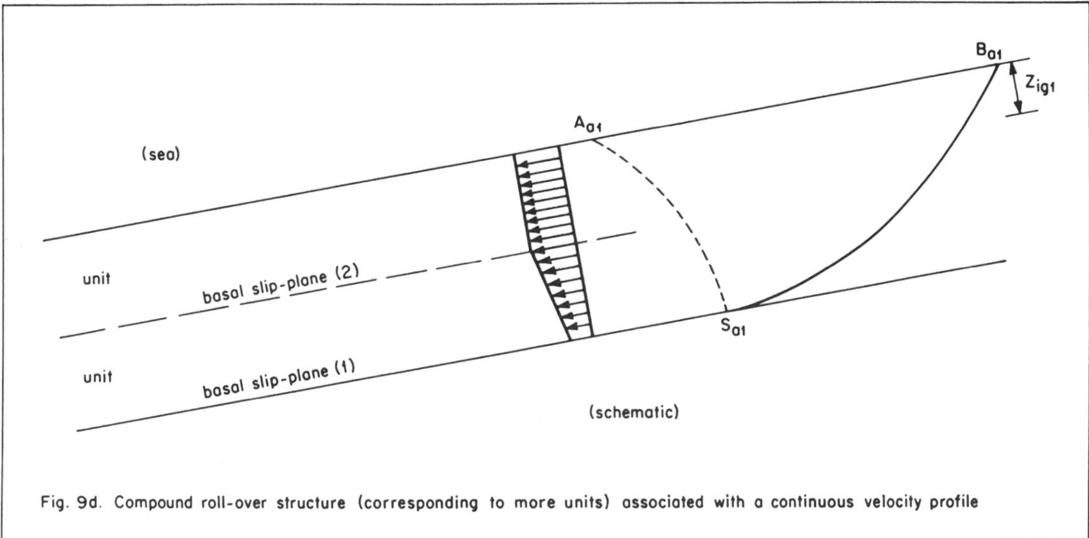

Fig. 9d. Compound roll-over structure (corresponding to more units) associated with a continuous velocity profile

Substituting equation (1) into (3), the spacing of the families of growth faults becomes proportional to δ^{-2} for small slope angles. A steepening of the paleo-slope, caused for example by basement subsidence or deformation of the whole sediment packet (e.g. by density inversion), must therefore be reflected in a closer spacing of the growth faults.

It should be stressed that, in this structural expression of the Multi-Unit Delta Model, the space necessary for the movement along the growth faults is assumed to be created solely by expulsion of fluid from the lower, shaly part of the formation, i.e. by lateral compaction of the units.

Deformation pattern of the roll-over region

The deformation pattern in the active plastic region can be found by applying the proper flow rules of soil mechanics (*) to the slip-line field and introducing certain boundary conditions. The flow rules provide a velocity field with which the plastic deformation of the sediments—in particular the formation of roll-over structures—can be calculated and plotted by the computer. The profile of the slide velocity down-slope enters the calculation as the most essential boundary condition that determines the shape of the roll-over structures.

An example of a velocity field calculated for the active plastic region of the single unit is given in Fig. 10a*. Since the unit starts to slide only when the critical thickness Z_{pe} has been reached, the slide movement is post-sedimentary for the markers in the unit (Fig. 10b). Consequently, there is no expansion of the layers; in fact, on the contrary, the plastic deformation can cause some thinning of the layer. In calculating the deformation of the active plastic region of the unit, it was assumed that the boundary slide velocity was uniform for the entire sediment packet, since the stiffened part of the unit moves as rigid body.

In Fig. 10c, a more advanced stage of syn-sedimentary deformation is presented, showing the well-known roll-over anticline and the expansion of the layers in the downthrown block. This calculation was based on a velocity field geometrically similar to that in the active plastic region of the single unit. This approximation seems justified because of the close similarity between the active plastic roll-over regions in Figures 10c and 15d**.

By varying one of the boundary conditions while keeping the others constant, the effect of such variations can be studied. Some results showing the effect of variations in sliding rate are given in Figures 11, 12 and 13*. They illustrate the development of three growth-fault structures that shared the same sedimentation rate v_d, but were formed at different uniform sliding rates v_s. For the uniform slide-velocity profile, the ratio $\Lambda = v_s/v_d$ appears to control the shape of the roll-over structure and its crestal surface c, as well as the relative thickening of the beds in the downthrown block.

The relative thickening ($\Delta T = (T_d - T_u)/T_u$) measured along the fault surface in cross-

**Figures 10, 11, 12 and 13 are calculated with the ordinary or Zagaynov velocity equation.*
***See also next section for further comments on the procedure followed.*

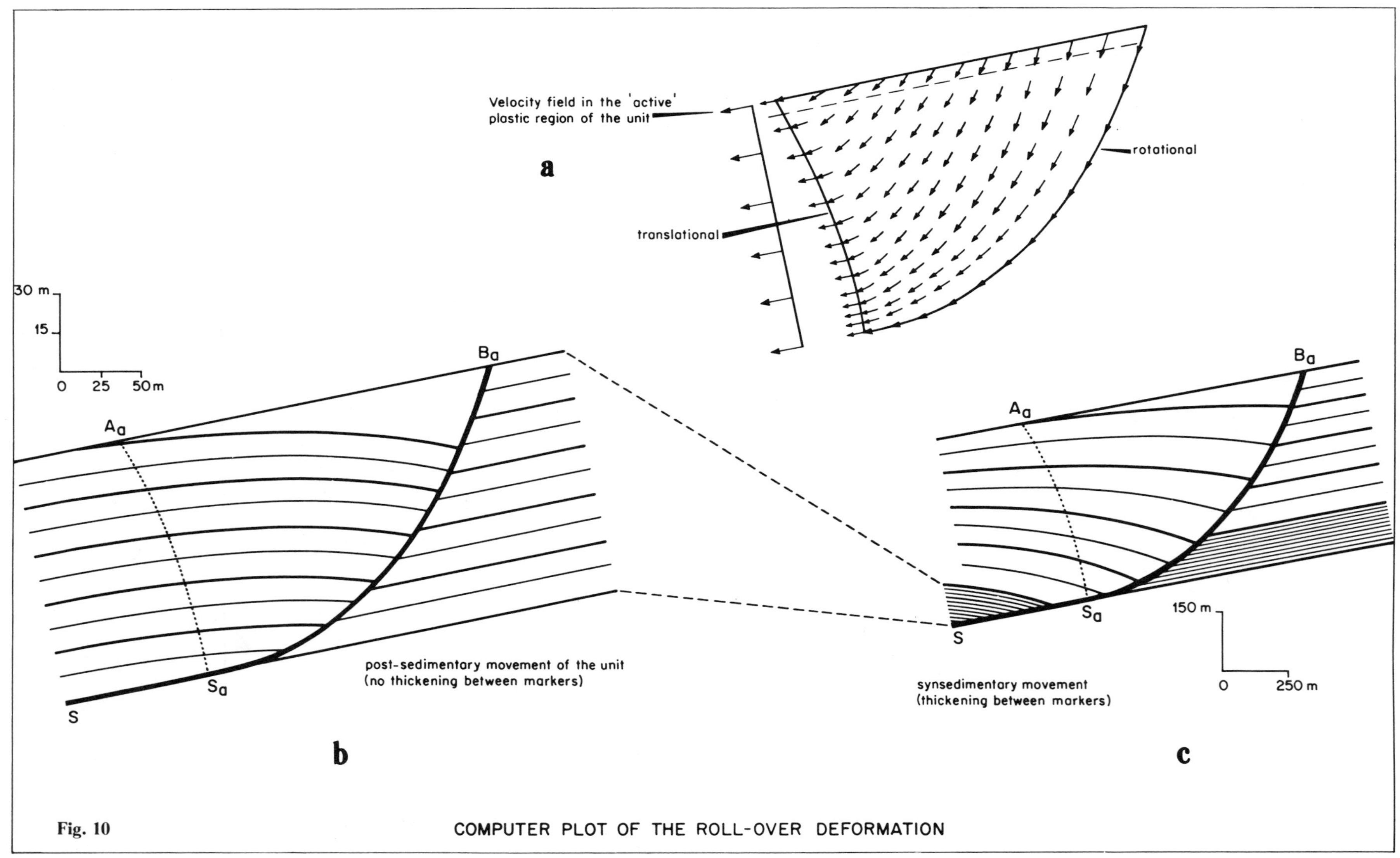

Fig. 10 COMPUTER PLOT OF THE ROLL-OVER DEFORMATION

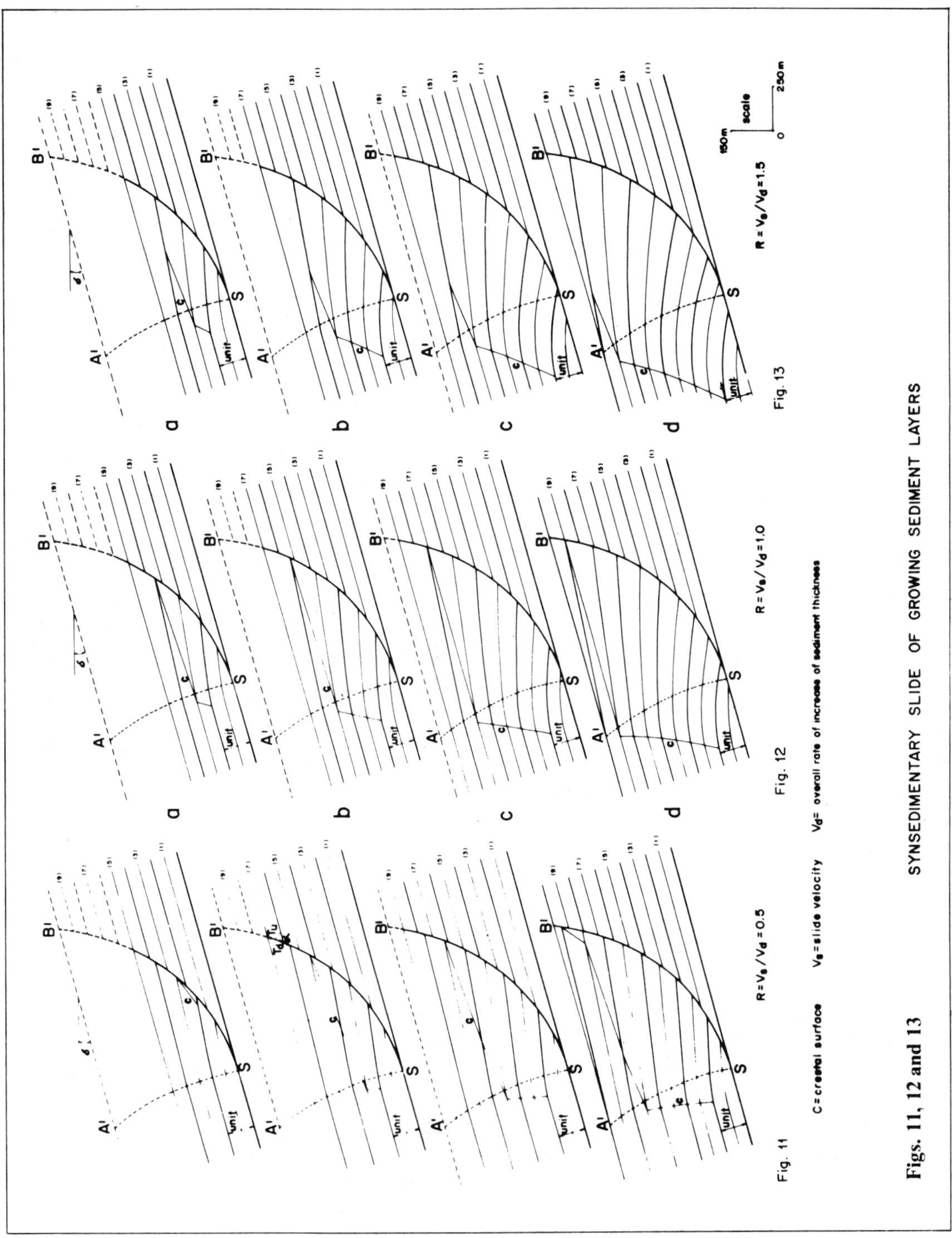

Figs. 11, 12 and 13 SYNSEDIMENTARY SLIDE OF GROWING SEDIMENT LAYERS

section (see Fig. 11b) is related to Λ by

$$\Delta T = \frac{f_d \cdot v_s \cdot \Delta t}{f_u \cdot v_d \cdot \Delta t} = f \cdot \Lambda \qquad (4)$$

where the function $f = f_d/f_u$ follows from the computed velocity field and depends on the position of the bed on the fault surface. The much-used growth index GI is related to ΔT by GI = $\Delta T + 1$. Some authors, e.g. Pfau (IR, 1965), have correlated large growth indices with the occurrence of hydrocarbons.

Variations in the slide-velocity profile

As has already been stated, the slide-velocity profile is mainly controlled by the lateral compaction behaviour in the case of the open-ended slide structure (regularly spaced growth-fault pattern). Clearly, when the total sediment packet consists of material with different (lateral) compaction behaviours, the slide-velocity profile will deviate from the uniform* one.

To obtain an idea of the consequences of different slide-velocity profiles, a pattern card has been compiled (Fig. 15) in which the profiles are roughly subdivided into three classes:

a. Overflow: the maximum slide velocity lies in the upper part of the sediment packet.
b. Underflow: the maximum slide velocity lies in the lower part of the packet.
c. Intermediate flow: slide-velocity profiles not falling within the above classes.

It is clear from the pattern card (Fig. 15) that the roll-over structure is certainly not determined by the stress state alone, but largely by the slide-velocity profile**. The overflow-type profiles (Figs. 15a, b, c) generate the familiar roll-over anticlines with right limb curved concave downward and crestal surface (c) concave upward. The underflow-type profiles (Figs. 15g, h, i), on the other hand, produce roll-over structures that show a drag along the growth fault (i.e. upward concave right limbs).

The synsedimentary roll-over structures shown here have been calculated using the velocity fields associated with the final state of the delta-body build-up. In other words, the appropriate portions of the final velocity field have been substituted for the real synsedimentary velocity fields. This procedure is not rigorously correct and serves merely to illustrate the various types of roll-over structure that can be associated with the same stress field. The correct procedure***, which accounts for the synsedimentary changes in the effective stress field step-by-step, produces a family of growth faults rather than a single fault, as shown in Figure 23a. This partly obscures the correct roll-over structure.

Apparent shale wave due to growth faulting

When a growth fault remains active during prograding of the delta, i.e. during deposition of increasingly sandy deposits, the continuous slip will bring a sandy section in the downthrown block into lateral contact with (i) a clayey section further down dip or (ii) a clayey section in the upthrown block (Fig. 14). Since the clay is under-compacted, the above juxtaposition will result in differential loading, the sandy section being denser than the shale on either side. Consequently, the shale will tend to form uplifts, which are thus a result of faulting rather than its cause (see also "Shale deformation and growth faulting", under "Field Observations", below).

Disturbances in the regular spacing of growth faults

The mechanism suggested above for the branching-off of the growth fault from the basal slip-pane implies an upward coarsening of the sediment. When the facies boundaries are more or less horizontal, as can be expected in a regularly prograding delta, a regular growth-fault pattern will develop (Fig. 9a).

Irregularities in the facies distribution will of course affect the regularity of the pattern. Slight irregularities will result from, for instance, the variations in the lithologic sequence related to minor shore-line fluctuations. A more marked transgressive phase (F_2 in Fig. 9b), causing a notable shoreward shift of the facies boundaries involved, will bring about a comparable shift in the branching-off points of the growth faults in the respective sediment intervals. Consequently, the branching-off points may come to be situated in the (active plastic) roll-over region adjacent to the preceding growth fault (Fig. 9b). Such superposition of roll-over regions will cause mutual interference which can give rise to different structural patterns.

When, for instance, the basal slip-plane of a younger unit extends into the active plastic region of a preceding one, it will act as a plane of weakness and can create a discontinuity in the slide-velocity profile. This will lead to the generation of new growth faults and antithetic faults (Fig. 9c), as discussed below.

**For the complete slide structure, factors other than differential lateral compaction lead to a nonuniform slide-velocity profile.*

***Although the shape of slide-velocity profiles may, in the first instance, be chosen quite arbitrarily, the resulting velocity field in the roll-over region must comply with an energy condition (see Glossary, above). After calculation of the velocity field, inspection revealed that some slide-velocity profiles did not generate a compatible velocity field. The slide-velocity profile can, of course, be corrected to give a compatible velocity field. The dotted lines in some slide-velocity profiles in Figure 15 are the contours of the original slide-velocity profiles, while the unbroken lines are the corrected ones.*

****It is possible to execute both procedures with the computer program for the calculation of roll-over structures given in another internal report.*

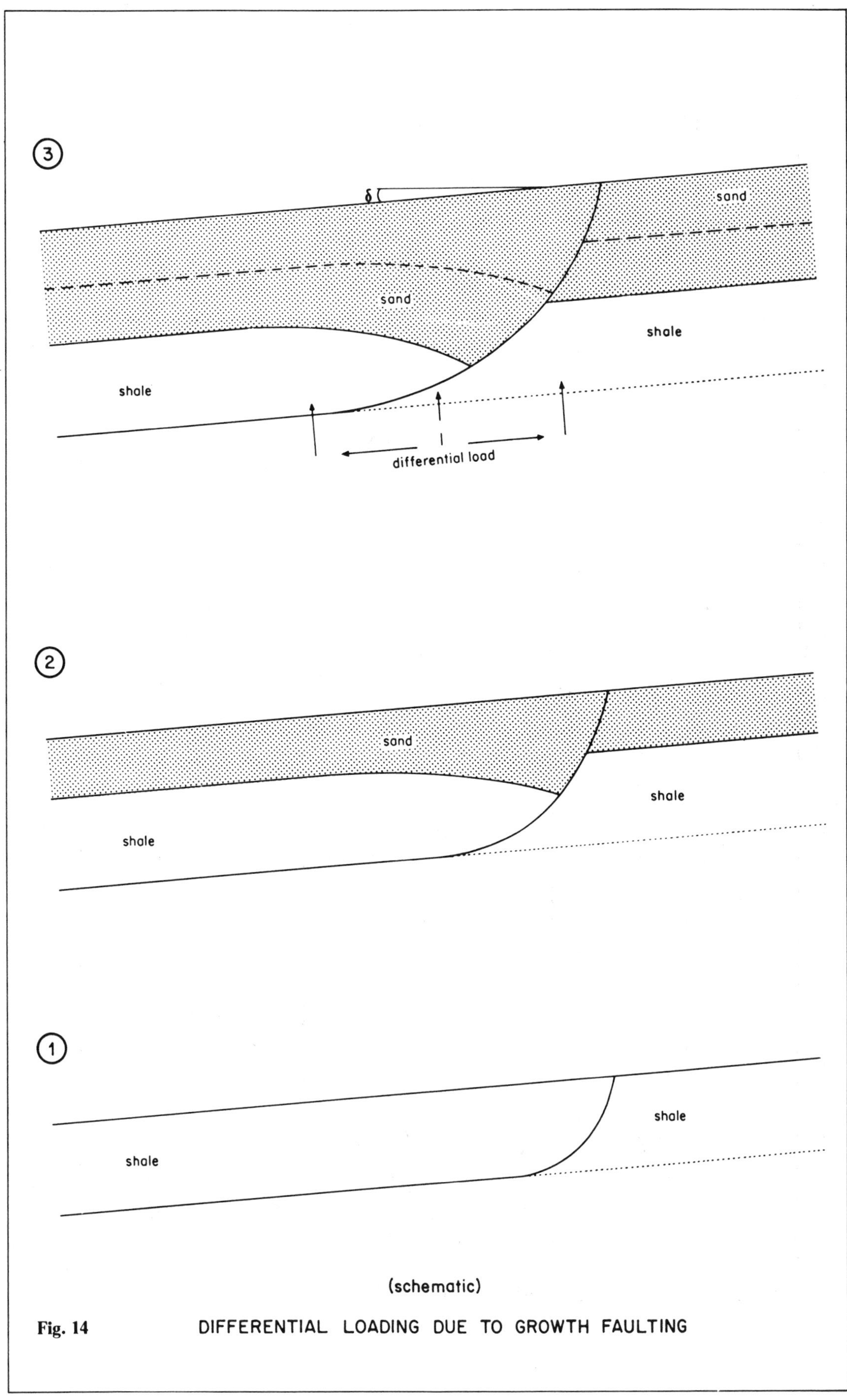

Fig. 14 DIFFERENTIAL LOADING DUE TO GROWTH FAULTING

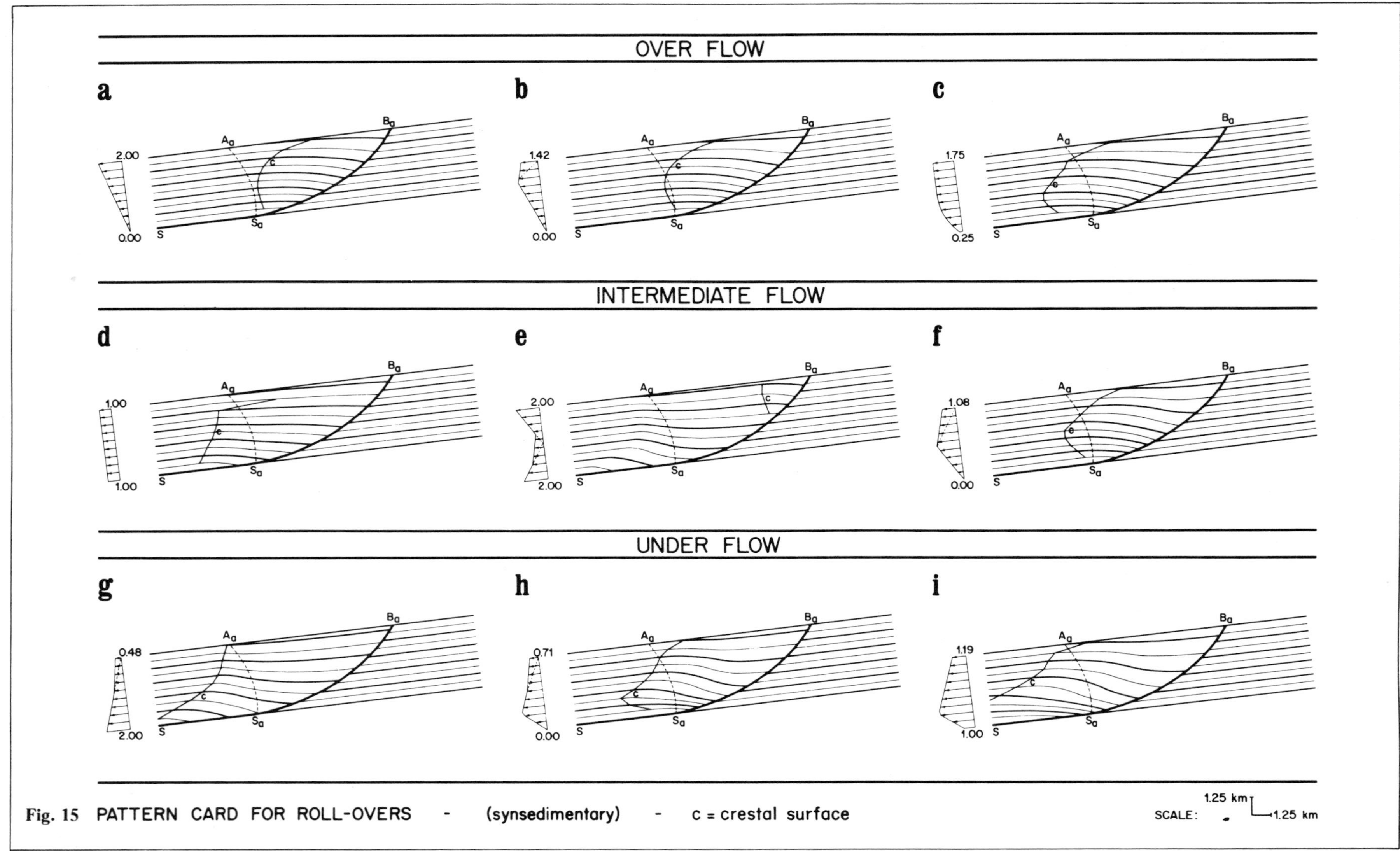

Fig. 15 PATTERN CARD FOR ROLL-OVERS - (synsedimentary) - c = crestal surface

Another possibility is that the slide-velocity profile does not suffer a jump across the base of the younger unit (Fig. 9d), in which case the profile does not induce a new fault (the geometry of the bounding growth fault $S_{a1}A_{a1}B_{a1}$ could eventually be affected by a change in the overpressure profile). The active plastic region $S_{a1}A_{a1}B_{a1}$ corresponding to more than one unit is called a 'compound' roll-over structure. The pattern card (Fig. 15) can also be applied to compound roll-over structures such as these.

Complete slide structure

It was stated above that a toe region is likely to be generated in a unit when the sedimentation rate is relatively high compared with the overpressure relaxation rate. Clearly, because of the upward movement in a toe region, underlying toes will interfere with overlying units. Although a whole range of interference modes is feasible, we only consider the case in which the toe regions amalgamate to form one large, compound plastic region (Fig. 17). Moreover, the existence of one compound active plastic region (see last subsection) is assumed. Although the situation shown in Figure 17 occurs under special conditions, it is presented here as an illustration of a full degeneration of the regularly spaced growth fault pattern.

At first sight, the multi-unit character has completely disappeared in this structural expression. However, the region between the compound plastic regions can be considered as consisting of a stack of units that have been stiffened initially.

One can prove that, for a relatively high sedimentation rate, liquefaction can occur in the lower part of a unit near the basal slip-plane, since the water-saturated clays can behave as a Bingham material. Clearly, since the thickness of the units is small compared with the thickness of the total sediment packet, various slide-velocity profiles can be built-up by the interaction of solid viscous interlayers. Since the stress fields in both the roll-over and toe regions are statically determined, i.e. independent of boundary displacement, theoretically any slide-velocity profile can be introduced.

In practice, the slide velocity will be determined by several factors. In the complete slide structure, for instance, the distribution of sedimentation in the roll-over region upslope can considerably affect the slide-velocity profile. In the open-ended slide structure (regularly spaced growth-fault pattern), mentioned above, it is determined by the lateral compaction of the units.

Before discussing the complete slide structure, the influence of different slide velocity profiles on the passive plastic region is first examined by computation. To illustrate the consequences of the different slide-velocity profiles on gravity sliding, a pattern card for post-sedimentary toe structures is included (Fig. 16). The slide-velocity profiles used originally for the roll-over structures (Fig. 15) were used again here, but it was necessary to correct some of them to avoid energy-incompatible velocity fields*.

Figure 16 again shows that not only the deformation pattern, but also the fault pattern, is determined by the slide-velocity profile rather than by the stress state (the fault zones near S_pA_p are fitted by hand on the basis of computer plots).

In addition to the pattern card for the post-sedimentary gravity slides, the synsedimentary formation of a complete slide structure with compound active and passive plastic regions was calculated (Figs. 18a-c). The same slide-velocity profile (parabolic; see Fig. 18a) was used for the calculation of both plastic regions, neglecting lateral compaction.

Considering the final state of the complete, synsedimentary slide structure (Fig. 18c), one sees that the compound active plastic region

---

**Most toe structures in Figure 16 have been marked with an α, and two (Figs. 16e & g) with a β. α indicates that the velocity field (on which the deformation pattern in a toe region has been based) has been calculated according to the α-formulation of the boundary-value problem for the velocity field. In this formulation, the slide-velocity profile prescribes a velocity component along the α-boundary slip line S_pA_p. One can also formulate the velocity boundary problem by letting the slide-velocity profile dictate a velocity component along the β-boundary slip line S_pB_p; this is called the β-formulation.*

This duality in formulating the velocity boundary problem requires consideration beyond the bounds of this paper. For the toe region, the α-formulation of the velocity boundary value problem is most appropriate if the calculated velocity field—associated with a certain type of slide-velocity profile—turns out to be compatible with the stress field. This is true for the toe structures in Figure 16, except Figures 16e & g. For the latter, only the β-formulation generates a compatible velocity field.

The pattern card for roll-over structures (Fig. 15) was based exclusively on the β-formulation of the velocity boundary-value problem, since one can show that the β-formulation is adequate for the roll-over region.

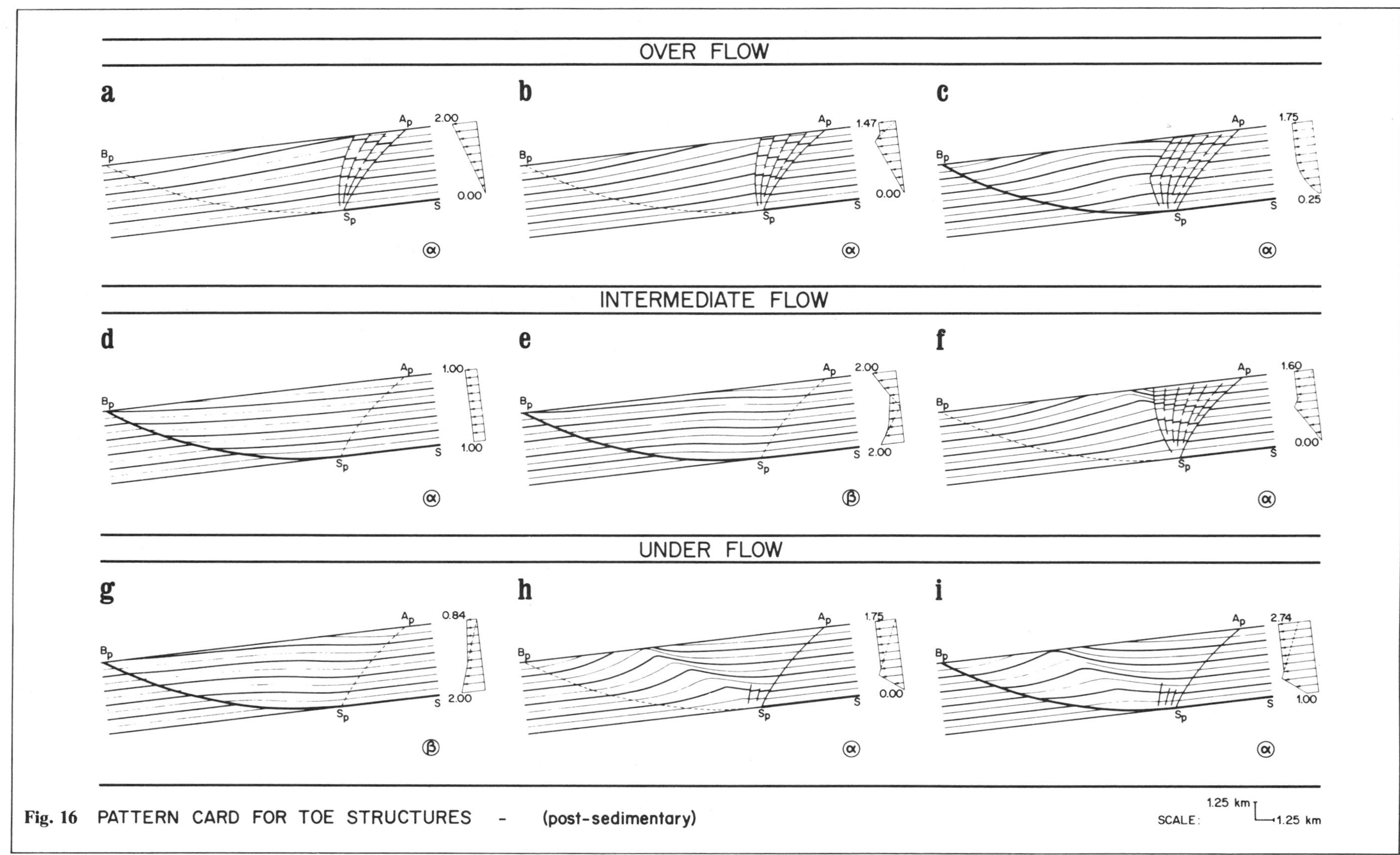

Fig. 16 PATTERN CARD FOR TOE STRUCTURES - (post-sedimentary)

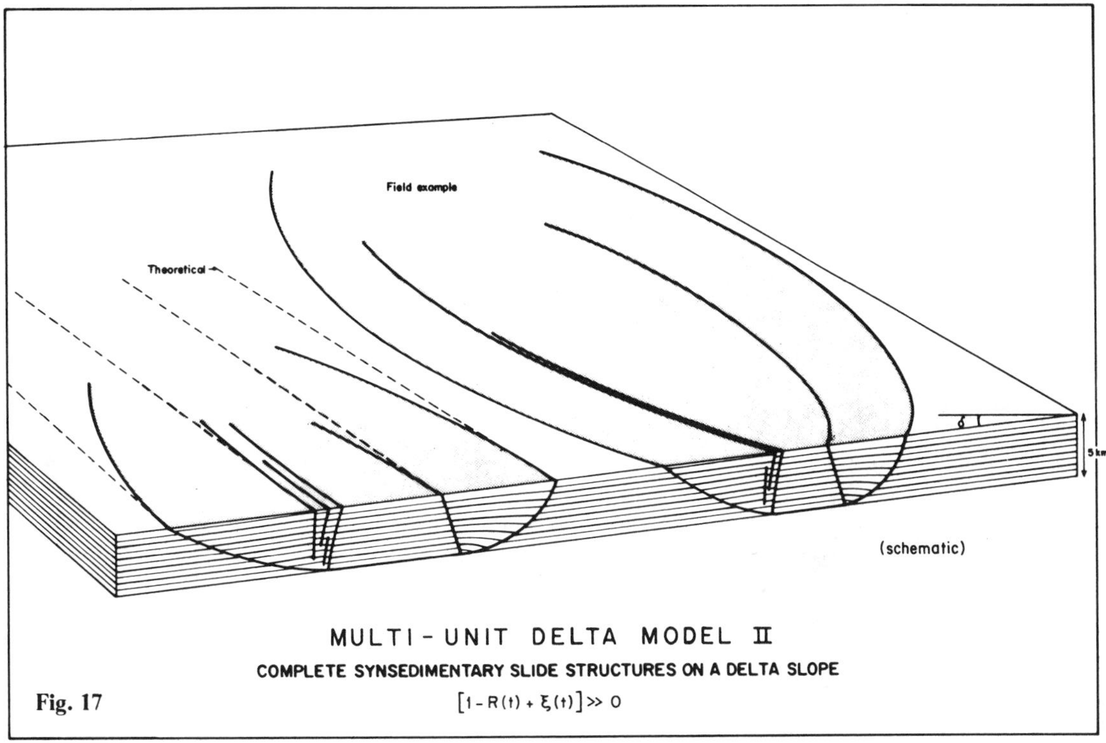

MULTI - UNIT DELTA MODEL II
COMPLETE SYNSEDIMENTARY SLIDE STRUCTURES ON A DELTA SLOPE
$[1 - R(t) + \xi(t)] \gg 0$

Fig. 17

has the well-known roll-over structure. The toe region, however, shows an unexpected deformation pattern for thrust-like regions. The thrust movement causes material to move upwards along the fault S_pA_p, resulting in a thinning near the fault S_pA_p (Fig. 18c). This gives the surprising impression that the layers thicken towards the thrust faults S_pB_p, so that this fault also looks, at first glance, like a growth fault.

In a repetition of such complete synsedimentary slide structures on a slope, as shown in Figure 17, the combination of the region of an upslope structure and the growth-fault region of a neighbouring one gives the impression of a back-to-back structure of two growth faults. In complete slide structures, such structures can be recognised by the fact that the fault S_pB_p is a low-angle thrust appearing at the surface with an angle $\angle 45^\circ$, while the growth fault S_aB_a is a normal fault appearing at the surface with an angle $\urcorner 45^\circ$.

Imbricated toe structures

In a toe region associated with one or more sliding units, successions of toe thrusts may develop. An example of such an 'imbricated' structure is shown by the seismic profile of Figure 19, taken from literature [Lewis (1971)]. A mechanism that might account for these structures within the framework of our model is the one that determines the distance between the (active plastic) roll-over region and the (passive plastic) toe region. According to equation (2), overpressure relaxation R(t) tends to increase the distance L, while sediment accretion $\xi(t)$ has the opposite effect. Hence, when overpressure relaxation increases faster than sediment accretion, old toe thrusts may be immobilised and new ones activated further down slope. When the rate of accretion exceeds the rate at which overpressure is reduced, the toe region moves closer to the active plastic region, and successive toe thrusts may develop up slope. All of these toe thrusts would branch off from the same basal slip-plane.

Antithetic faulting

A smooth (unfaulted) roll-over structure is formed when the growth fault S_aB_a curves smoothly into the surface-parallel basal slip-plane S_aS (for instance, Fig. 10). When, however, the transition of the fault S_aB_a into the basal slip-plane degenerates into a kink, other potential fault planes will be activated. The two cases are compared in Figure 20. Figure 20a shows the velocity field for the case in which the transition of the growth fault S_aB_a into the basal slip-plane is smooth. The velocity on S_aA_a is then surface-parallel. Figure 20b shows the velocity field for a kink-like transition of the growth fault into the slip plane. Clearly, the velocity on S_aA_a can no longer be surface-parallel everywhere. Since everywhere to the

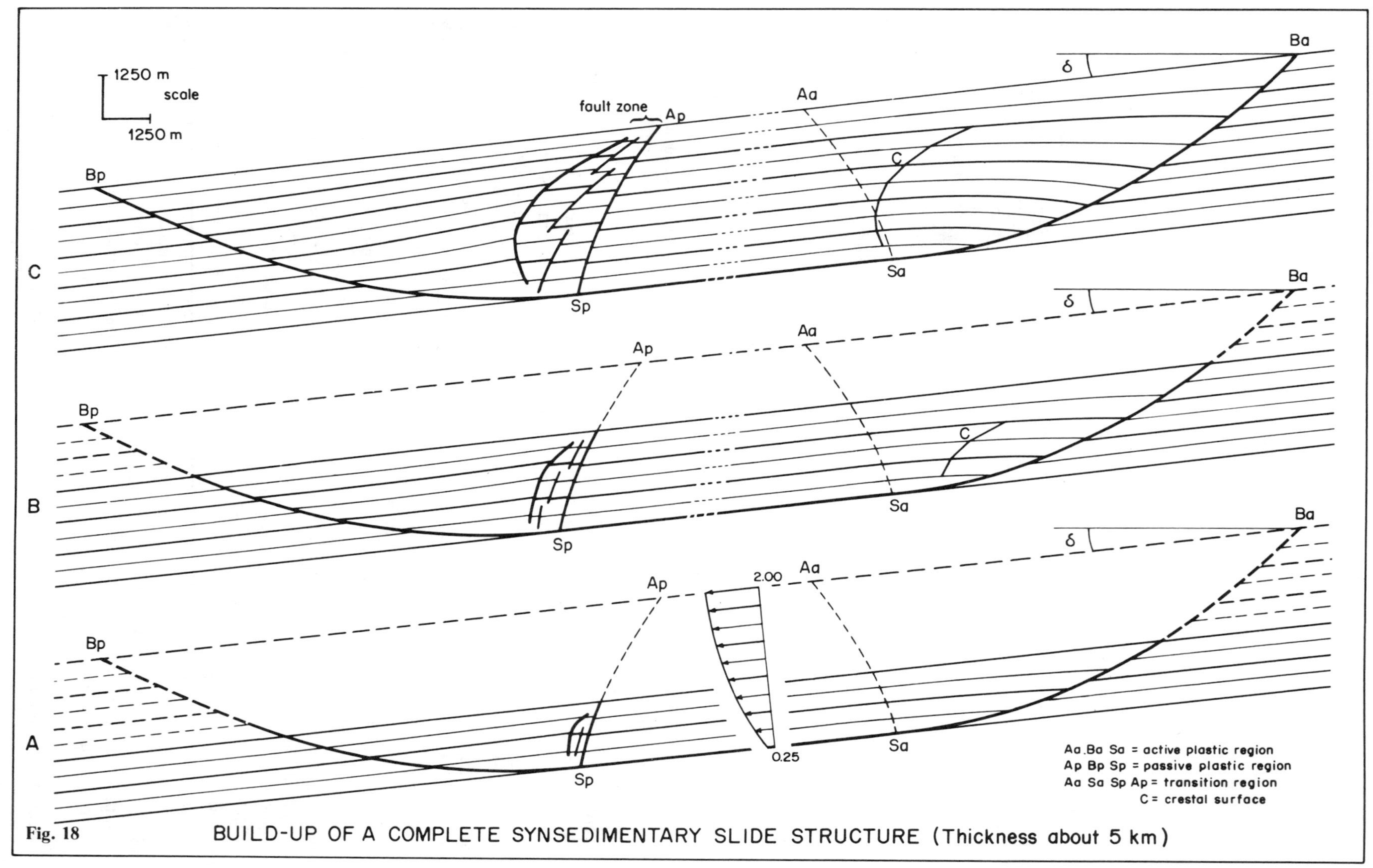

Fig. 18 BUILD-UP OF A COMPLETE SYNSEDIMENTARY SLIDE STRUCTURE (Thickness about 5 km)

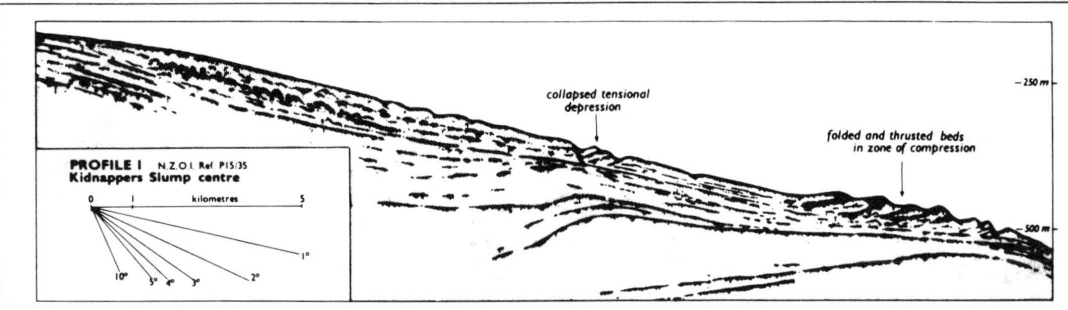

SEISMIC SECTION OF A COMPLETE SLIDE STRUCTURE ON A CONTINENTAL SLOPE, SHOWING AN IMBRICATED TOE REGION (after Lewis, 1971)

Fig. 19

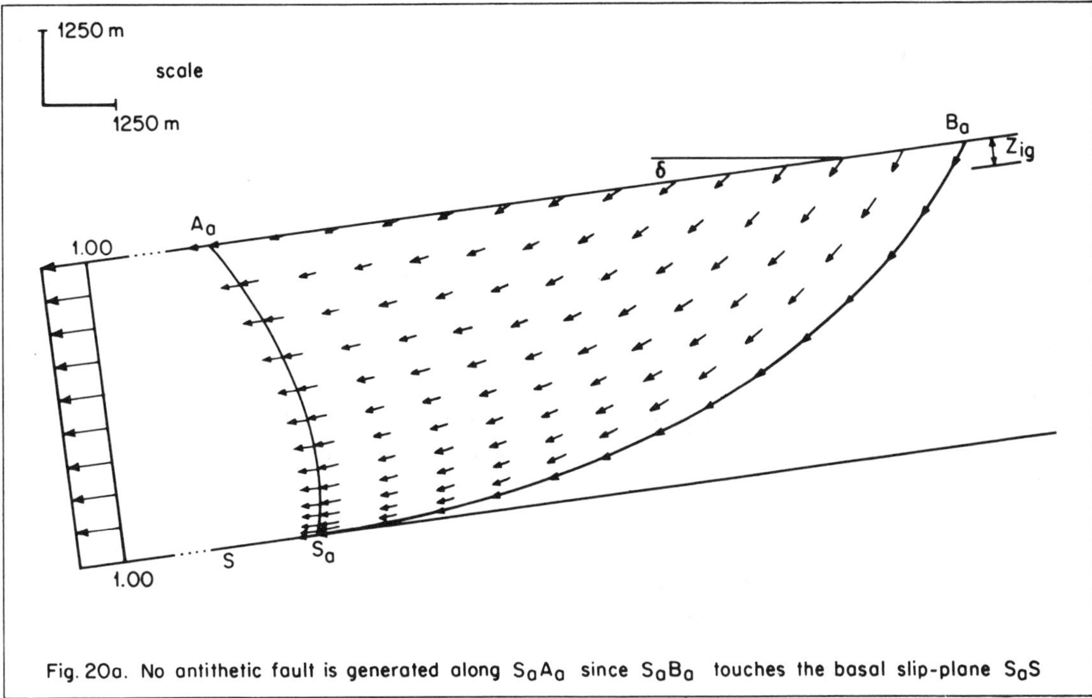

Fig. 20a. No antithetic fault is generated along S_aA_a since S_aB_a touches the basal slip-plane S_aS

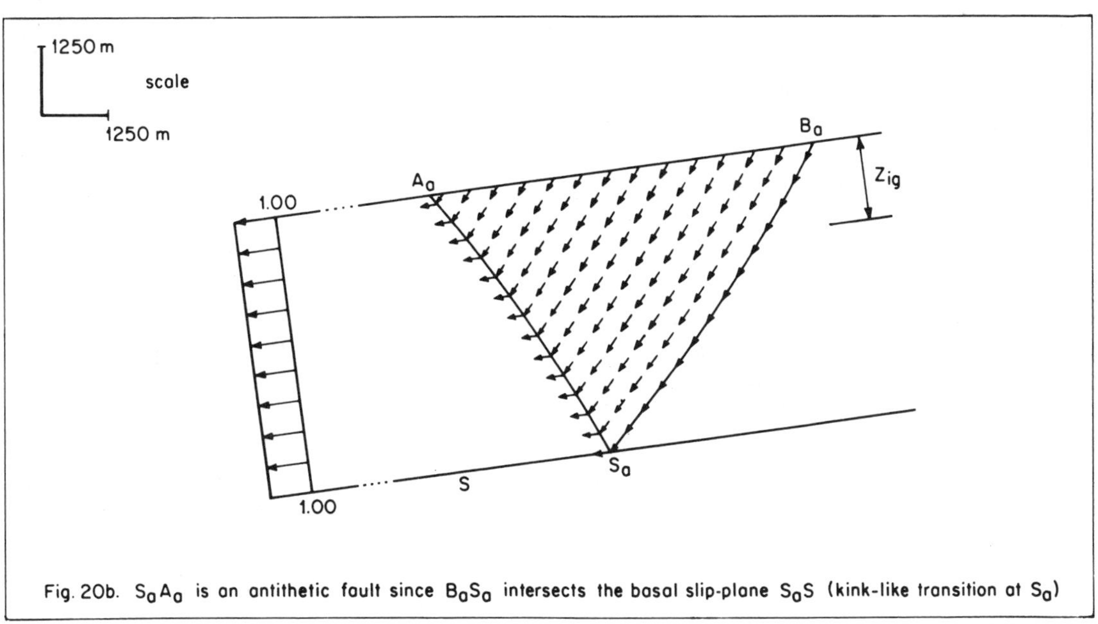

Fig. 20b. S_aA_a is an antithetic fault since B_aS_a intersects the basal slip-plane S_aS (kink-like transition at S_a)

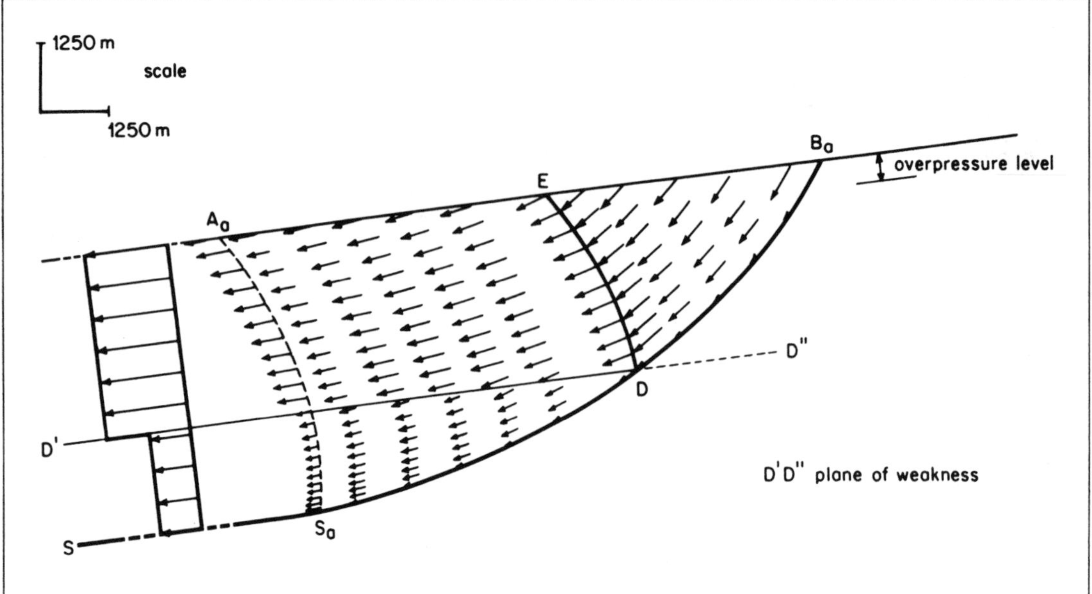

Fig. 21. Antithetic faulting (DE) associated with a discontinuity in the slide-velocity profile (plane of weakness terminates on the growth fault S_aB_a)

left of S_aA_a the sediment packet slides down in a surface-parallel direction, there must be a velocity discontinuity along S_aA_a; this means that this slip line is activated as an antithetic fault.

It is clear from the above reasoning that the throw along the antithetic fault depends on the curvature of the growth fault at S_a, which is directly related to the fluid pressure at that level. In contrast to the growth fault S_aB_a, which can be considered as being attached to the fixed 'upthrown' block, the antithetic fault moves with the sliding material. Consequently, a whole set of antithetic faults may be generated.

A special case of 'kink transition' arises when a surface-parallel plane of weakness intersects the sediment packet. Such a plane of weakness (D″D′ in Fig. 21) may be associated with a marked change in lithology, e.g. relatively thick shales on top of a sandy sequence. Its introduction does not disturb the prevailing stress state, and the shape of the growth fault S_aB_a is therefore unchanged. However, the material lying above the plane of weakness D″D may slide down faster than the underlying material, in which case the slide-velocity profile will show a discontinuity at D″D′, as sketched in Figure 21.

Naturally, the point of generation of an antithetic fault depends on the extent and shape of the 'plane' of weakness inside the active plastic region. In Figure 21 it is assumed that the growth fault S_aB_a cuts the plane of weakness D″D′ at point D, so that extra slip along D′D requires mobilisation of an antithetic fault ED, as in the case of Figure 19b.

As sliding continues, DD′ also becomes warped (Fig. 22) and surface-parallel sliding is no longer possible. Any slipping of the overlying sediment will therefore probably be restricted to the part D′C down slope of the crestal point C of the plane of weakness. Accordingly, active antithetic faulting will now take place along $A_a'C$ in Figure 22*.

The configuration shown in Figure 22 might give the erroneous impression that the fault in C has resulted from uplifting. As shown in our example, however, such crestal faulting can be generated by partial down-warping of an active

*From the reasoning above, the configuration of Figure 21 holds approximately while the throw along the growth fault is still small; that in Figure 22 applies for well-developed roll-over structures.

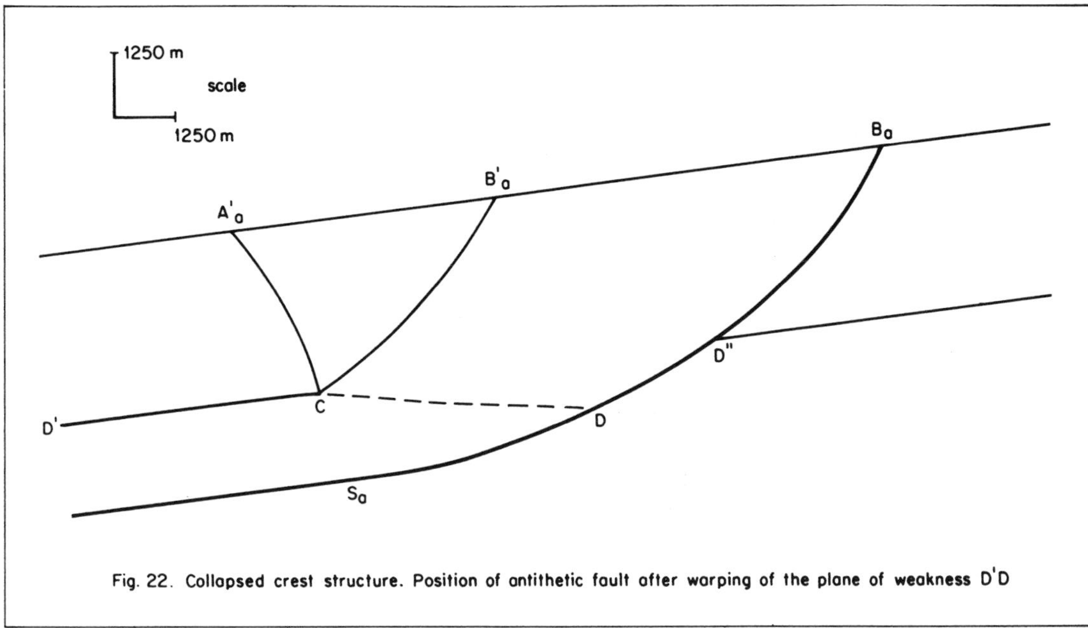

Fig. 22. Collapsed crest structure. Position of antithetic fault after warping of the plane of weakness D'D

plane of weakness, along which purely lateral sliding occurs.

It follows from our model that the antithetic faults are concave downwards if they are generated under overpressured conditions. If generated under hydropressured conditions, they are almost straight, but compaction may render them concave upwards at a later stage.

Development of growth-fault zones ('splintering', 'horsetailing')

During build-up of the sediment packet from a unit thickness to its final thickness, the fluid-pressure profile will change continuously (see under "Genesis of the multi-unit delta"). The same is true after deposition because of the compaction process. Since the fluid-pressure profile directly affects the attitude of the growth fault during both the deposition of the delta body and the post-sedimentary compaction period, growth faults will change in shape. New branches will split off and sliding motion will proceed along them. One has to distinguish between syn- and post-sedimentary 'splintering' of growth faults.

Synsedimentary development

With the computer program for the roll-over structure, one can simulate the formation of families or bundles of growth faults. One can also introduce different instantaneous overpressures and eventually, different slide-velocity profiles for every step during the building-up of the total sediment packet.

An example of a computer-simulated synsedimentary fault zone is presented in Figure 23. A synsedimentary rise in the top of the 'leaky' overpressure (Z_{ig}) (see Fig. 23b) was assumed, guaranteeing that the growth faults $S_aB_a^1$, $S_aB_a^2$, ... , $S_aB_a^n$ curve smoothly at the fixed point S_a into the same basal slip-plane S_aS'. In simulating the roll-over structure, successive parts of the slide-velocity profile shown in Figure 23a were employed.

For the sake of comparison, the final slide-velocity profile was used to compute Figure 23c by employing increasing portions of the final velocity field to approximate the syn-sedimentary formation of the roll-over structures.

Post-sedimentary development ('horsetailing')

As already mentioned, the curvature of the growth faults, and in particular the genesis of the surface-parallel basal slip-plane, are a direct consequence of overpressure. It is easily seen that relaxation of overpressure tends to straighten the lower part of a growth fault. Therefore, when fault movement continues during the relaxation process, typical fault configurations can appear depending on the mode of fluid-pressure relaxation.

To demonstrate how the changes in growth geometry depend on the fluid-pressure history, we consider a state of 'leaky' overpressures (curves OD_1C in Figs. 24a & b), which gradually approaches the hydrostatic state (I) as compaction proceeds. As a mathematical analysis

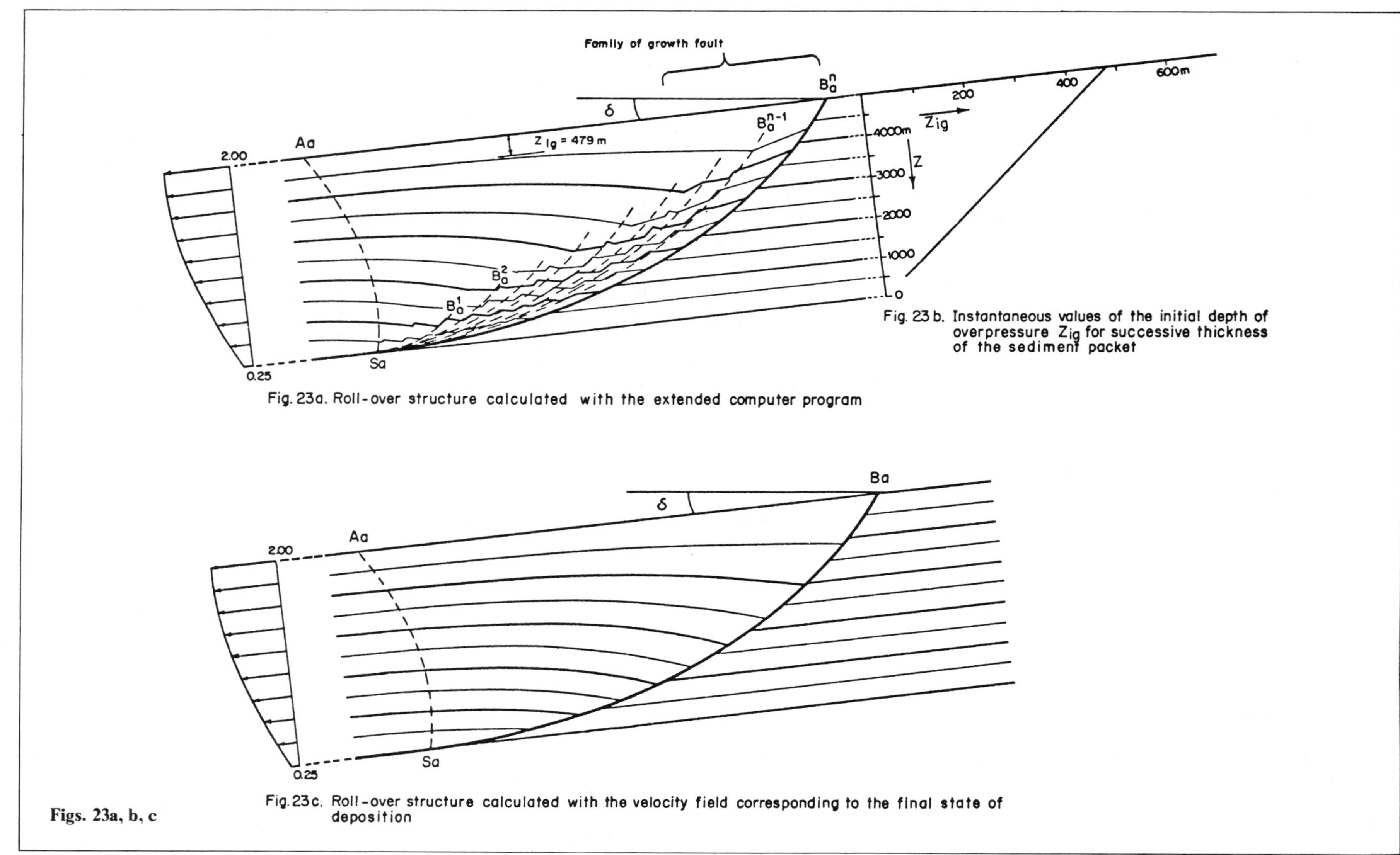

Fig. 23a. Roll-over structure calculated with the extended computer program

Fig. 23 b. Instantaneous values of the initial depth of overpressure Z_{ig} for successive thickness of the sediment packet

Fig. 23c. Roll-over structure calculated with the velocity field corresponding to the final state of deposition

Figs. 23a, b, c

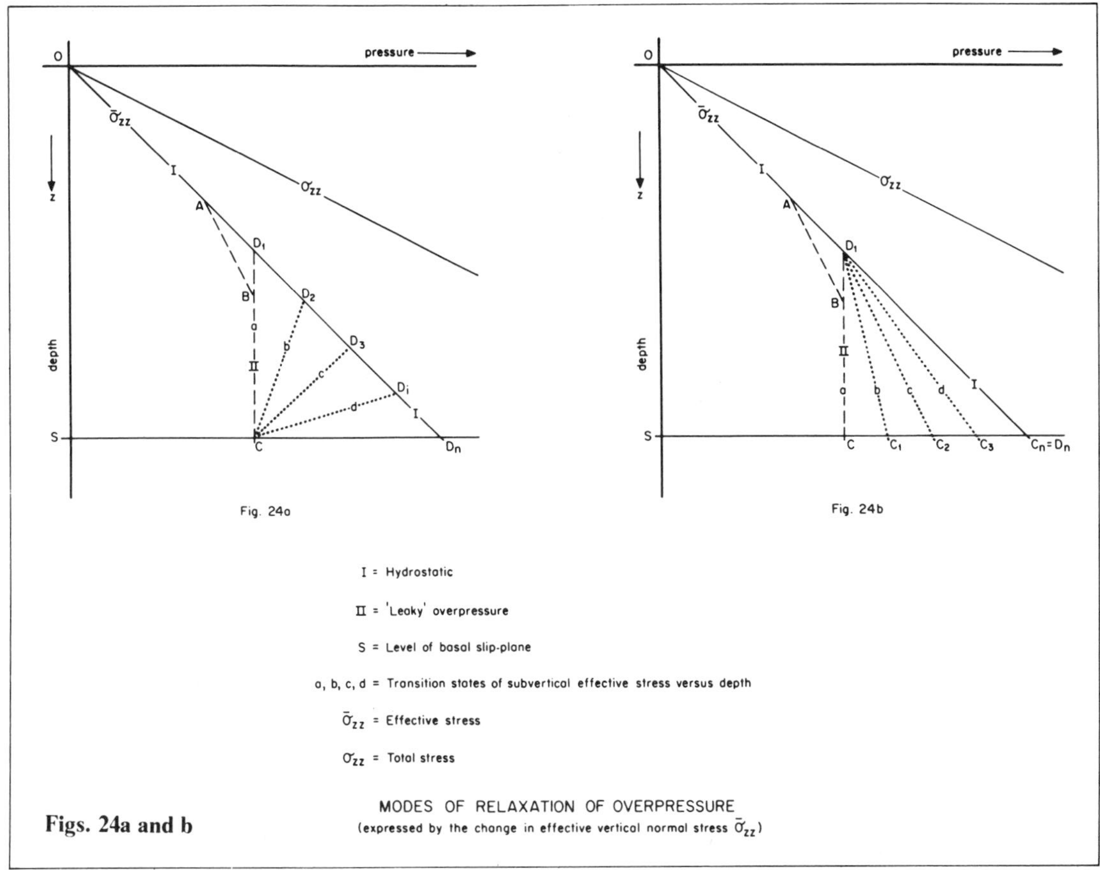

Figs. 24a and b

of this transition process is outside the scope of the present model, we compare only the consequences of two contrasting (hypothetical) pressure histories.

The first mode of overpressure relaxation is associated with drainage from the top of the overpressured sediment packet. During the time interval considered, the overpressure, and hence the subvertical effective stress $\bar{\sigma}_{zz}$, is supposed to remain constant near the basal slip-plane (cf. point C in Fig. 24a), while the top of the overpressured zone gradually falls (cf. D_1, D_2, ..., D_n in Fig. 24a). Interpolating linearly between the overpressure top and the constant pressure at the basal slip-plane, the distribution of the subvertical effective stress $\bar{\sigma}_{zz}$ passes through the hypothetical transition states OD_2C, OD_3C, ..., OD_nC indicated in Figure 24a.

Associated with these transition states are potential faults of well-defined shape. Their exact location, however, is the subject of some speculation. If we assume that all subsequently activated growth faults branch off from the basal slip-plane at the same point, the resulting fault pattern (shown in Fig. 25a) will resemble an 'upward horsetail'. This convergence of subsequent faults at the end of the basal slip-plane may be associated with the facies change that has already determined the location of the first fault S_0B_0, which corresponds to curve OABC in Figures 24a & b.

Another possibility, the 'downward horse-tailing' shown in Figure 25b, seems less likely since it requires extension of the originally activated part SS_0 of the basal slip-plane in an upslope direction.

The second mode of overpressure relaxation considered is illustrated in Figure 24b. It corresponds to drainage from below, which might be a result from fluid flow along the basal slip-plane. The corresponding up- and downward-converging fault bundles are shown in Figures 26a & b.

It is interesting to note that the two upward directed 'horsetail' fault patterns (Figs. 25a and 26a) are very similar, although associated with different drainage processes. The two down-ward-diverging fault patterns on the other hand, show a remarkable qualitative difference that should allow a decision as to the type of overpressure relaxation that has been active. Drainage from the top leads to faults that osculate about the youngest (B_0S_6 in Fig. 25b),

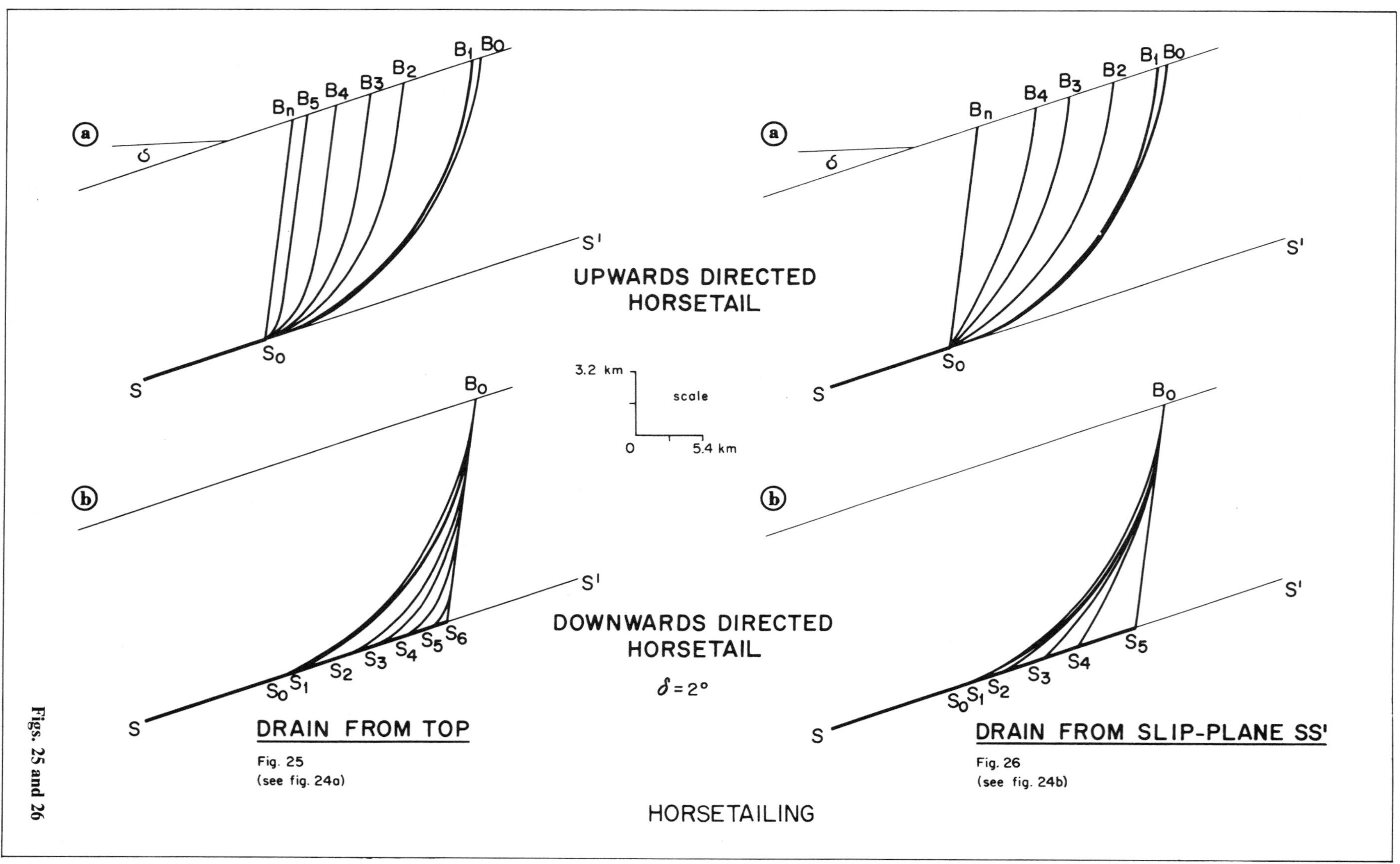

Figs. 25 and 26

while drainage from below produces faults that envelope the oldest growth fault (B_0S_0 in Fig. 26b).

Interpretations of field cases frequently show the 'horsetail' type faults of Figure 26b, which would suggest a history of overpressure relaxation of the type shown in Figure 24b.

It was mentioned in the preceding section that kink-like transition of a growth fault into a mobile basal slip-plane gives rise to antithetic faults. Since this is indeed the case in Figure 26, antithetic faults will occur at point S_0 in Figure 26a, or at points $S_2, S_3, \ldots, S_5$ in Figure 26b (these antithetic faults have not been drawn in Fig. 26 for the sake of clarity).

As pressure relaxation proceeds, both the subsequent growth faults and the associated antithetic faults become straighter. Moreover, as the splintering of growth faults is, in general, a post-sedimentary process, little or no bed growth will occur along lower parts of the 'splinter' faults.

It should be possible to distinguish antithetic faults of the post-sedimentary type from those generated by an extra plane of weakness (cf. Fig. 21) during build-up of the sediment packet. The latter type of antithetic fault, which is synsedimentary and usually embedded in strongly overpressured material, is therefore concave downwards.

Needless to say, the age of genesis of antithetic faults has an important bearing on the migration and accumulation of hydrocarbons in the roll-over structures.

Structural variations along the strike of a major growth fault

In the field, the more important growth faults usually line up and form sub-parallel regional trends. In many cases, however, the structural picture shows strong variation along the strike.

Such lateral variations are not described directly by the present model, which provides only a two-dimensional picture. It is, however, reasonable to assume that they should be related to lateral variations in the significant parameters, e.g. the depth and slope angle of the basal slip-plane and the rate of sliding along it. These parameters depend, in turn, on variations in sedimentation rate, facies and overpressure profile, if we exclude possible movements of the substratum. The present model can, however, show the qualitative effect of these parameters [see also Crans & Spijer (IR, 1971), p.4].

The Multi-Unit Delta Model and hydrocarbon migration

Two aspects in particular of the Multi-Unit Delta Model are important for the migration of hydrocarbons: (i) the stress state in the formation according to the model, and (ii) the fault pattern [see also Crans & Spijer (IR, 1971)].

In general, it may be assumed that an important part of the hydrocarbons generated in deltaic complexes has to migrate through overpressured shales to reach the sandy potential reservoirs. A mechanism that has been suggested for migration through such shales assumes that hydrocarbons may be injected into the shales when their pressure is sufficiently higher than the pore-fluid pressure. They would then be able to form micro-cracks, which on average are perpendicular to the smallest principal-stress axis (parallel to the maximum principal stress), and hence in a stable shale mass hydrocarbons would move vertically upwards.

However, from the present model of growth faulting it follows that, at least during faulting, the maximum principal stress in a large part of the sliding stack of units tends to be parallel to the sediment slope. The above migration mechanism suggests that hydrocarbons that have migrated upwards into such a sliding sediment pile may continue to move upwards in a direction more or less parallel to the sediment surface.

Growth faults and basal slip-planes are expected to control hydrocarbon migration, firstly because they form capillary barriers against transverse hydrocarbon flow, and secondly because they may facilitate flow along the fault.

It has been established by means of numerous experiments in a ring-shear aparatus [Mandl, De Jong and Maltha (1977)] that, even without the juxtaposition of different sediments (common to growth faults), a fault will gain a certain capillary sealing capacity from the textural changes that the sediment undergoes inside the shear zone. Therefore, not only growth faults, but also basal slip-planes and antithetic faults of very moderate throw, will oppose transverse hydrocarbon flow. At the same time dilatancy of densely packed sediment in the shear zones may provide conduits of higher permeability.

In addition, it should be recalled that the stresses inside the growing sediment body gradually change during sedimentation before the state is reached where surface-parallel slip

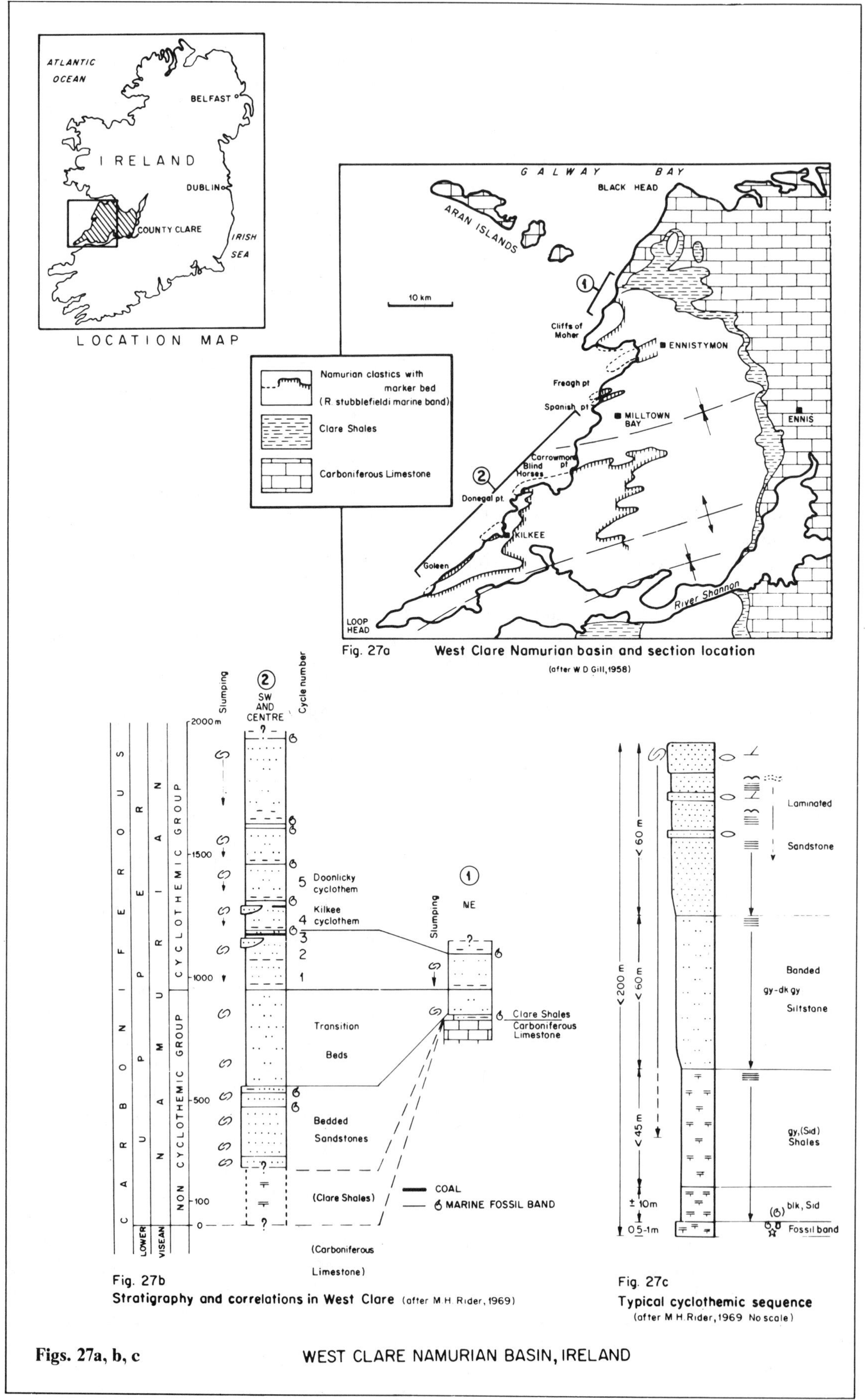

Fig. 27a West Clare Namurian basin and section location (after W D Gill, 1958)

Fig. 27b Stratigraphy and correlations in West Clare (after M H Rider, 1969)

Fig. 27c Typical cyclothemic sequence (after M H Rider, 1969 No scale)

Figs. 27a, b, c WEST CLARE NAMURIAN BASIN, IRELAND

can commence. Therefore, sets of parallel shear planes develop with successively decreasing dip angles in association with these stress changes and precede the formation of a basal slip-plane. In general, slip along these precursory slip elements will be negligible. Nevertheless, one may speculate that these features concentrated near the basal slip-plane may produce a permeability anisotropy that favours flow along its upper side.

It may be anticipated that the effect of the fault system on hydrocarbon migration will be different for open-ended and complete-slide structures. The open-ended slide structure is characterised by regularly spaced growth faults (Fig. 8), which together with the basal slip-planes divide the delta slope into compartments, each having its own compaction history*. Hydrocarbon migration from the deeper shaly beds of the delta deposits to the sandy upper part of the roll-over regions will be confined to individual compartments.

Antithetic faults may form important obstacles to upward and lateral hydrocarbon migration and it is therefore important to recognise the time of genesis of the antithetic faults in relation to the period of migration. The distinction between the early, downward concave, synsedimentary antithetic faults related to a discontinuity in the slide-velocity profile and the post-depositional, almost straight, antithetic faults that accompany overpressure relaxation can be helpful in this respect.

When the slide structure is complete, both the (active plastic) roll-over region and the (passive plastic) toe region (Fig. 17) are potential hydrocarbon traps. In this expression of the Multi-Unit Delta Model, the sediment packet involved in the slide movement is completely enveloped by the main fault system, which assumedly is transversely sealing. Consequently, it would seem unlikely that hydrocarbons can enter the complete slide structure after a toe has formed.

FIELD OBSERVATIONS IN THE CARBONIFEROUS OF IRELAND

In the development of a mechanical model of the kind described above, certain simplifying assumptions have to be made. It is essential to verify the validity of such assumptions by comparison of the model with field and subsurface examples.

Synsedimentary faulting in outcrops of deltaic sediment in the Upper Carboniferous of the West Clare Namurian Basin (Rider, 1969 and 1978) provided such a comparison.

1. Geological setting

The Upper Carboniferous deposits outcrop in a narrow strip along the Atlantic coast of County Clare, from south of Galway Bay to the estuary of the River Shannon (Fig. 27a).

The Visean Carboniferous Limestone is overlain by a 400-1,900m thick Namurian sequence. These sediments consist of shales and siltstones, with subordinate sandstones and several thin coal seams. They display an outstanding variety of slump features [see also: Brindley & Gill (1959), and Gill & Kuenen (1957)].

The Namurian is subdivided (Fig. 27) into a lower, Noncyclothemic Group, interpreted as clastic shelf and slope deposit, and an upper Cyclothemic Group of which almost 1,000m of strata are exposed.

This upper Group consists of a number of cyclothems of varying thickness, which are typically shale (with marine fossils at the base)—siltstone-sandstone sequences (Fig. 27c). They are interpreted as having been deposited in a deltaic environment (Rider, 1969 and 1978). In this Cyclothemic Group, the synsedimentary deformations are mainly manifested as faulting.

The Carboniferous sediments were folded later in a vast synclinorium (Fig. 27a) during the Variscan orogeny (Gill, 1962).

2. Growth faulting

The range of one growth fault (Fig. 28a) is confined to the sediments of one cyclothem. In other words, the deformation of the sediments of a cyclothem took place independently of that of the underlying cyclothems (Fig. 29b). This observation is in agreement with the model.

The fault is usually a discrete slip plane, concave towards the downthrown compartment (Figs. 28a, 28e, 30a), then becoming more or less parallel to the bedding planes of the overlying sediments (Figs. 28a, 28i, 30b and 30c). This geometry is comparable to that of the delta model: a curved slip plane in the

* *This is supported by the discontinuity in overpressure that is frequently observed when drilling through a growth fault [see also Dickey et al. (1968)].*

'active' plastic region with a rotational velocity field, becoming a surface-parallel slip plane in the nonplastic region, where the velocity field is purely translational (Fig. 10a).

The sedimentary 'packet' found above the fault plane consists mainly of sandstones with subordinate siltstones (Fig. 28a) and is usually 30-80m thick. Towards the distal part of the fault, one finds fewer and fewer sandstones underlain by more and more siltstones.

The sandstone beds are described as displaying a roll-over geometry near the fault [Rider (1969 and 1978) & Fig. 30a]. However, this structure is often not clear because:

(a) an individual bed may show frequent variations in thickness close to the steep part of the fault (Fig. 28f), or
(b) within the same zone, in addition to the roll-over, there is a comparatively large normal drag along the fault. Although this drag is not often represented on sections of the large growth faults, it is known to occur in Nigeria, for instance, but is often masked on seismic sections by diffractions from the fault.

Sand volcanoes (Gill & Kuenen, 1957, and our Fig. 28d) can be encountered near the fault, at the top of the sedimentary 'packet'. Their formation may have been favoured by local liquefaction due to shearing along the slip planes.

Other deformations, which are believed to have occurred later than the bulk of the movements along the fault, are also found; particularly normal faulting, which creates a small horst-and-graben structure parallel to the main fault trend (Figs. 28c, 28g). Similar features are known in the Gulf Coast large-scale growth faulting.

Further away from the steep part of the fault, the sediments above the surface-parallel portion of the slip plane appear completely undisturbed. This might be interpreted as an indication of the existence of a 'stiffened' region. Within this region, occasional horizons appear to be deformed over large distances as a result of sliding phenomena of secondary importance which do not affect the adjacent strata (Fig. 28b). Apparently these beds have a lower shearing strength and were in a contained plastic state.

The sediments below the fault plane usually consist of siltstones underlain by the basal shales (Fig. 28a). These siltstones sometimes show strong deformation in the immediate vicinity of the fault, while the sediments above the fault are undisturbed. For instance, in the region where the fault plane is still curved, numerous shear drag folds associated with many shear planes may be encountered just below the main slip plane (Fig. 28h). This deformation by shearing is typical of plasticity. In the distal area, where the slip plane is subhorizontal, chaotically deformed siltstones are reported in several cases (Rider, 1969).

No clear-cut evidence of a 'toe' region was found. One can only observe that, towards the distal part of the fault, the siltstones overlying it become more disturbed and even chaotic (Figs. 28a , 28j).

The sediments on both sides of the fault are overlain by beds that are flat and undisturbed (Fig. 28). Before their deposition, topographical relief created by the fault movement had been smoothed out.

It is worth noting that the measured palaeocurrent directions (Figs. 28c, 29a) are roughly perpendicular to the fault trend and oriented towards the downthrown side, as one might expect for a synsedimentary movement along the fault.

An idea of the magnitude and spacing of the faults can be gained from some outcrops, as illustrated in Figures 29a and 29b.

3. Shale deformation and growth faulting

Some small (less than 20m high) mud lumps are known in County Clare. Like the faults, they are each confined to a single cyclothem. The shale is often highly sheared. A number of small-scale synsedimentary deformations, due to the growth of the diapiric shales, are visible in the overburden. These mud lumps are found in a limited area only, and are sometimes underlain by a slip plane (Fig. 31a).

The evidence indicates that, in the majority of cases, such mud lumps are not responsible for the formation of the growth faults. Only in one instance (Fig. 31b) does a synsedimentary fault that seems to bound such a lump appear to have been active during the growth of the mud lump.

4. Comparison of the synsedimentary faulting in Ireland with the Multi-Unit Delta Model

The above-described features of the synsedimentary faults in the West Clare Namurian indicate a plastic (*) behaviour of the sediments, which supports one of the basic assumptions made in the mechanical model.

The existence of overpressures, which make low-angle sliding possible in the model, has not

THIS PAGE INTENTIONALLY BLANK

b. View of an extensive layer of balled sandstones, within the 'undisturbed' packet of slumped sediments.
Note: - the sandstone slump balls
- the flow structure in the finer material serving as matrix
- the undisturbed beds above and below the disturbed layer
(Blind Horse Cove)

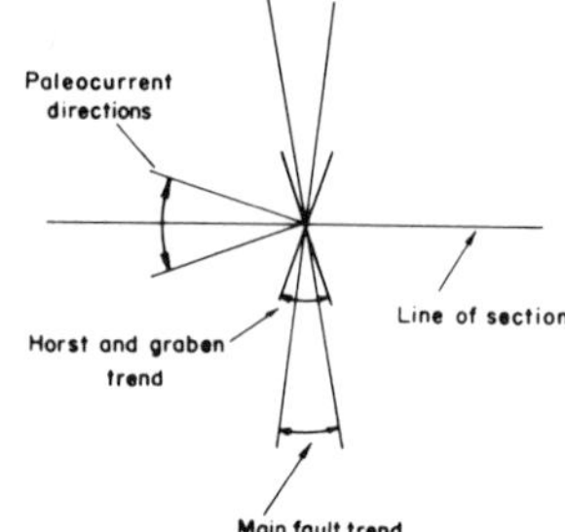

c. Directional elements of a growth fault

a. Synthetic cross section showing most of the features that can be seen associated with syn-sedimentary faulting in West Clare (adapted from Gill and Rider according to field observations)

Not to scale

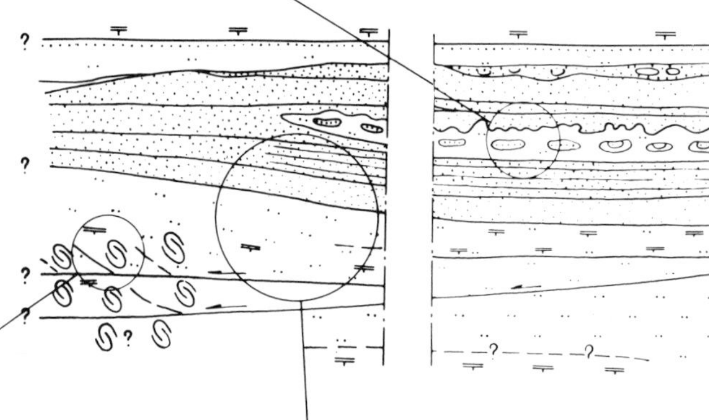

j. View showing the intense disturbances that can occur in the lower part of the slumped 'packet,' away from the steep part of the fault
(Donegal Point)

i. Subhorizontal slip plane (x), several hundreds of metres away from the steep part of the fault
(North of Spanish Point)

Fig. 28 MAIN FEATURES OF THE GROWTH

d. Sand volcano. Note the successive layers of ejecta and the vent.
(Goleen)

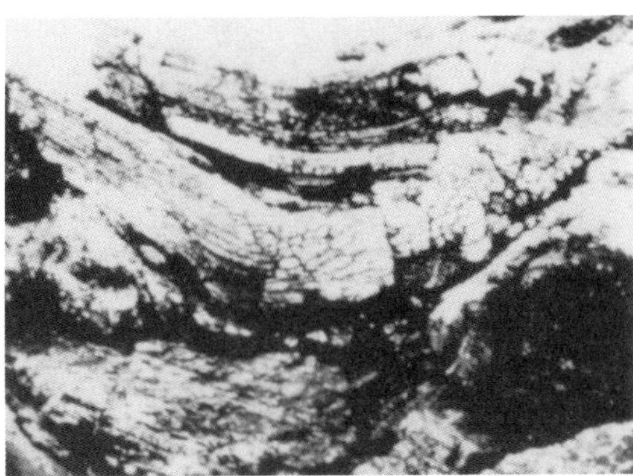

e. The upper steep part of the fault (x)
Note: - the steepness of the fault plane (upper right)
- the difference in lithology: sandstones above the slip plane, siltstones below
- the incipient brecciation in some sandstone beds
- the peculiar deviation of the sandstone beds from the simple roll-over geometry
(Blind Horse Cove)

f. The upper, steep part of the fault (x)
Note: - the difference in lithology across the fault: sandstones on the left, siltstones on the right
- the thickness variations in the sandstone beds
- the flattening out of the slip plane (lower left)
(Donegal Point)

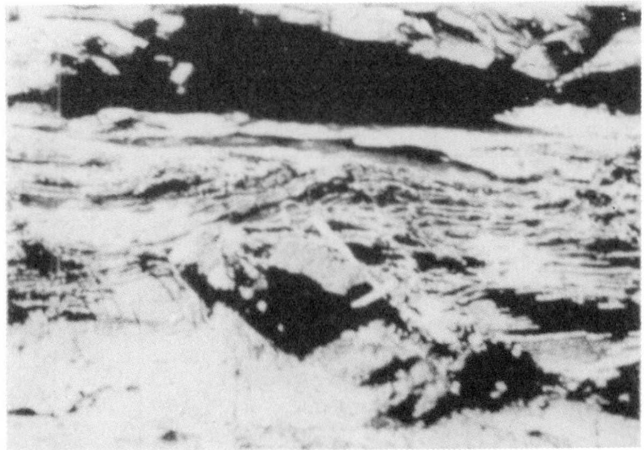

h. Deformation of the siltstone beds below the main slip plane (x). Note the numerous shear folds and associated shear planes (s)
(Blind Horse Cove)

g. View showing normal faulting determining locally a horst and graben structure in the slumped 'packet' above the slip plane (x)
(Carrowmore Point)

FAULTS IN THE WEST CLARE NAMURIAN, IRELAND

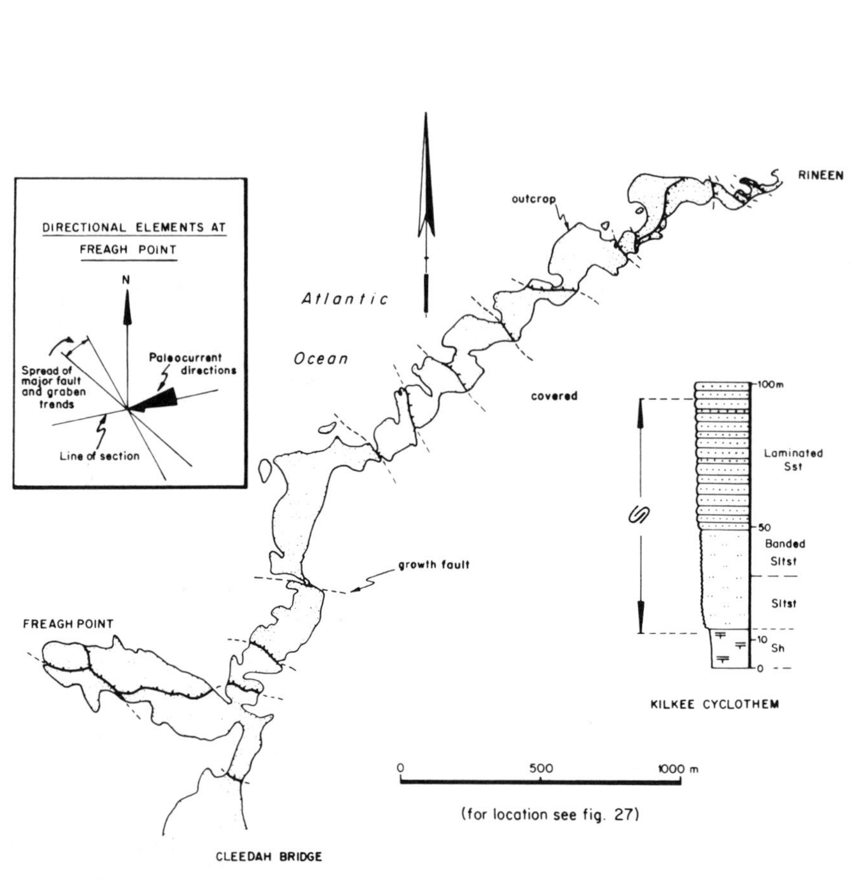

Fig. 29a. Arcuate shape and approximate parallelism of the faults in plan view
(map and other data after M.H. Rider, 1969)

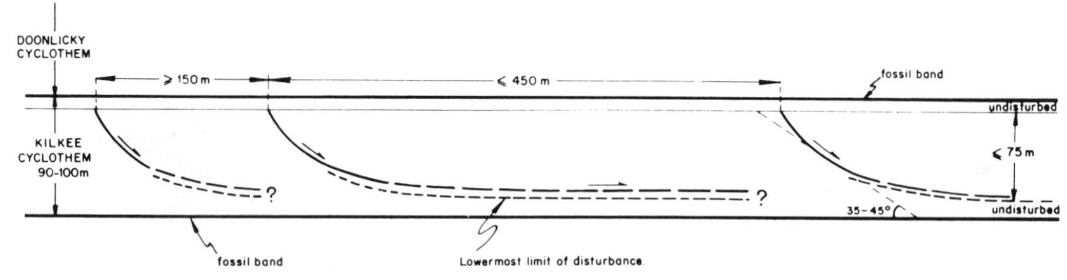

Fig. 29b. Diagrammatic section showing the spread in spacing of the growth faults
(for location see fig. 27)

SPATIAL RELATIONSHIPS OF GROWTH FAULTS IN THE WEST CLARE NAMURIAN, IRELAND

Fig. 29a and b

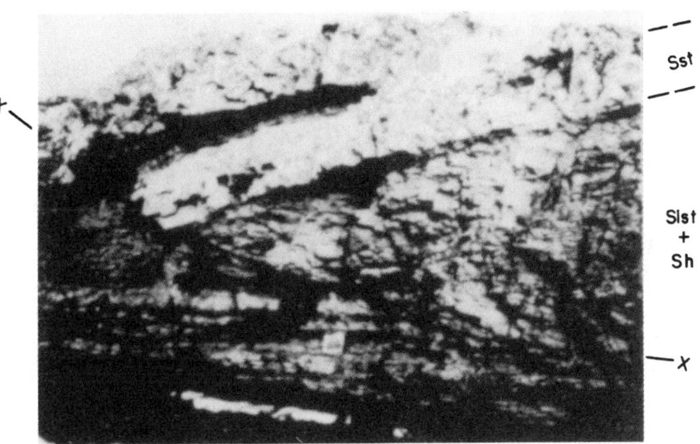

Fig. 30a. General view. Note the roll-over structure of the sandstone beds and the slip plane (x).

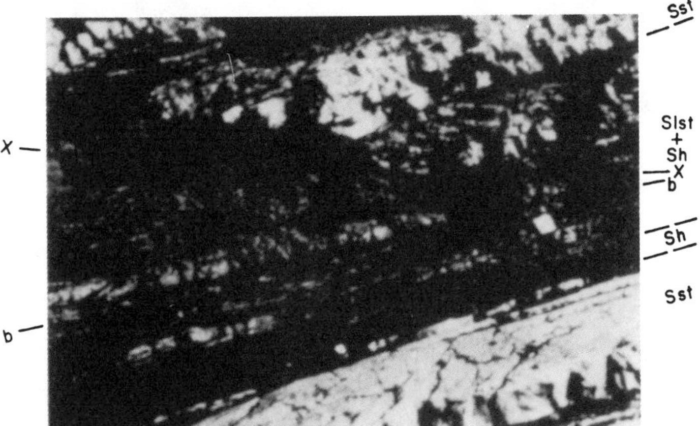

Fig. 30b and c. Close-ups showing the flattening of the slip plane (x) into a bedding plane (b) within a sequence of alternating siltstones and shales.

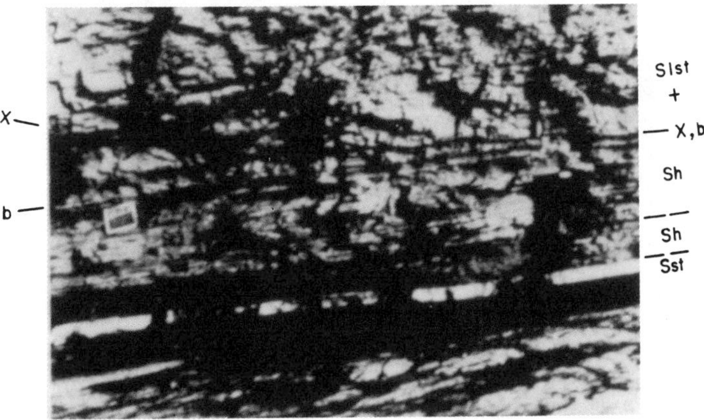

SMALL PENECONTEMPORANEOUS FAULT AT GOLEEN, WEST CLARE NAMURIAN BASIN, IRELAND (for location see fig. 27).

Fig. 30

Fig. 31a. Shale diapir near Kilkee (for location see fig. 27). Note its asymmetry, the deformation in the overburden and the underlying slip plane (x).

Fig. 31b. Synsedimentary faulting (x) at Goleen (for location see fig. 27). Note the synclinal deformation and thickening of the beds on the downthrown side (left), the anticlinal deformation and thinning of the beds on the upthrown side (right), below which the presence of a shale diapir might be inferred.

Figs. 31a and b

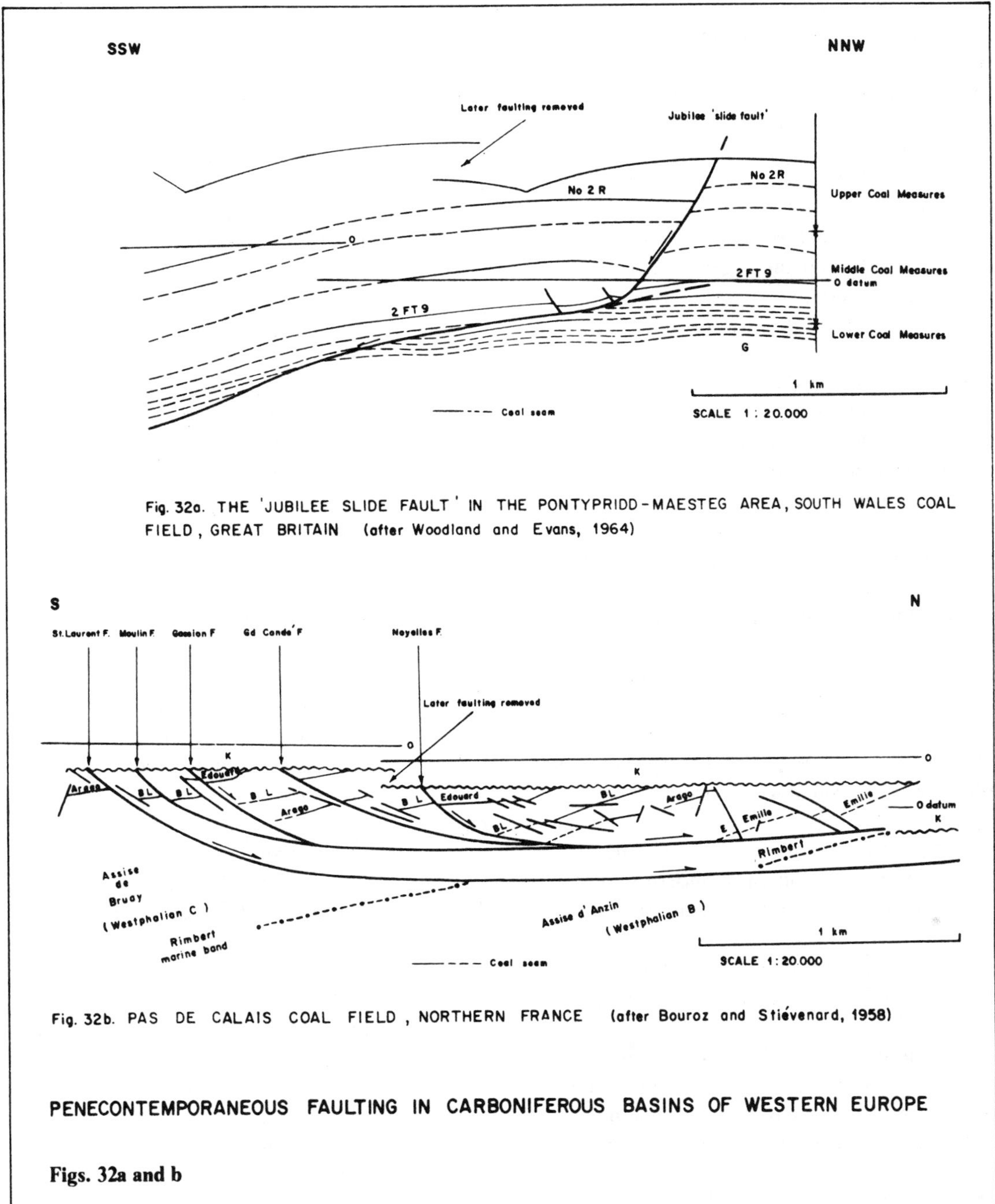

Fig. 32a. THE 'JUBILEE SLIDE FAULT' IN THE PONTYPRIDD-MAESTEG AREA, SOUTH WALES COAL FIELD, GREAT BRITAIN (after Woodland and Evans, 1964)

Fig. 32b. PAS DE CALAIS COAL FIELD, NORTHERN FRANCE (after Bouroz and Stiévenard, 1958)

PENECONTEMPORANEOUS FAULTING IN CARBONIFEROUS BASINS OF WESTERN EUROPE

Figs. 32a and b

been proved directly in County Clare. However, it should be remembered that overpressures can occur at very shallow depths and that gravity sliding requires less overpressures if it takes place at somewhat greater slope angles, as might have been the case in Western Ireland. Moreover, the existence of sand volcanoes and mud lumps may be taken as an indication that overpressure existed at least at certain stages of development.

Field evidence also supports the main theory of the model; namely, that slip and faulting can occur on a fairly gentle slope as soon as a critical sediment thickness is reached, and do not require movements in the substratum for their initiation. Another aspect of the model, namely that of a stack of sedimentary sheets gliding independently on top of each other, is verified by the existence of independent faulting in the successive cyclothems.

There are, however, a number of differences between this field example and the large deltaic complexes in line with which the model was established. In the well-known deltaic oil provinces the growth faults are thought to have been initiated in marine shales. They were active for a considerable length of time, during which they affected a large part of the sedimentary sequence. Moreover, successive growth faults may have been active simultaneously.

In Ireland, where synsedimentary faults are also found in deltaic sequences, they are of a much smaller size and affect a much smaller part of the sedimentary sequence (30-80m). Here, the basal slip plane is generally situated in siltstones overlying marine shales. The faults are each restricted to one sedimentary cycle.

The circumstance that sedimentation took place in a number of distinct cycles and probably at a lower rate than in Nigeria, for example, may be responsible for the fact that each fault was active for a much shorter period.

Nevertheless, it is permitted to use the features observed in County Clare as an example of this type of synsedimentary faulting, since the mechanical principles involved are the same in both cases, regardless of the size of the features studied.

5. Other examples of penecontemporaneous faulting in the Carboniferous basins of Western Europe

Conditions in coal mines are generally well suited to the detection of subhorizontal slip planes, which are difficult to recognise in wells or on most seismic sections. Two examples from the literature [Bouroz & Stievenard, (1958), Woodland & Evans (1964)] are shown in Figure 32.

As in the case of the Namurian of Ireland, the growth episode seems rather restricted, and the basal slip plane is not in a thick marine shale section, but in a coal-bearing, paralic sequence. However, this interval is predominantly clayey in South Wales (interval 2ft 9in Fig. 32a: 36% clay and subordinate siltstones, 5% sand in lenses, 9% coal). In the Pas de Calais basin (Fig. 32b), there are some indications that, as in County Clare, the paleoslope was steeper than in the large recent deltas. In both cases, the dimensions of the faulting are large by comparison with the Irish example and approach those encountered in Nigeria: one fault affects several hundred metres of sediments.

Acknowledgment

The authors are indebted to Shell Research B.V. for permission to publish this work and to GeoQuest International, Inc., Houston, Texas, U.S.A., for final editing.

LIST OF SYMBOLS

c	cohesion
D	spacing of the growth-fault families
L	maximum length of the stiffened region in a unit
p	fluid pressure (pore pressure)
Δp	fluid-pressure drop across a sealing layer
r_o	$=\tau_p/\tau_r$, reduction factor of basal friction
r_σ	$=\bar{\sigma}^m_{zz}(Z_{pe})/\bar{\sigma}^i_{zz}(Z_{pe})$ relative change in the (sub) vertical effective normal stress $\bar{\sigma}_{zz}$ before and after sliding
R	$= r_o \cdot r_\sigma$
T_d, T_u	distance between two subsequent marker lines—measured along the growth fault—in the downthrown and upthrown blocks
v_d	deposition rate
$v_s(z)$	slide velocity
x, y, z	co-ordinates in the sloping layer x: down-slope direction z: perpendicular to the sloping surface
Z_{ig}	initial depth of overpressure (geopressure)
Z_{ip}	initial depth of the plastic state
Z_{pe}	depth of the basal slip-plane; critical thickness of the unit
α-, β-axes	positive direction along the α- and β-slip lines including angles of $(\theta+\mu)$ and $(\theta-\mu+\pi)$ with x-direction
γ	submerged specific weight
δ	slope angle of the sediment layer
θ	angle between the first effective principal stress $\bar{\sigma}_1$ and the x-direction
Λ	$= v_s/v_d$
μ	$\pi/4-\rho/2$
$\xi(t)$	relative thickening of the unit (to a first approximation proportional to the sedimentation rate)
ρ	angle of internal friction
$\bar{\underline{\sigma}}$	effective stress tensor
σ_{ii}	i^{th} normal component of the total stress tensor
σ_{ij}	shear component of the total stress tensor ($i \neq j$)
σ_i	i^{th} principal stress
τ_p, τ_r	the maximum and residual values of the shear strength

REFERENCES

BOUROZ, A., and STIEVENARD, M., 1958. La structure de gisement des charbons gras du Pas de Calais et la notion de faille Renmaux. *Soc. géol du Nord. Ahnales* LXXVII, pp. 146 *et seq.*

BRINDLEY, J. C., and GILL, W. D., 1958. Summer field meeting in southern Ireland, 1957. *Proc. Geol. Assoc.*, **69**, pp. 241-261.

CRANS, W., MANDL, G., and SHIPPAM, G., 1973. Geomechanische Modelle Tektonische Struktüren. Sonderforshungsbereich **77**, Felsmechanik, Universitât Karlsruhe. Jahrsb. 1973.

CRANS, W., and BERKHOUT, A. J., 1979. On the probabilistic assessment of seismic amplitude anomalies. Paper presented at S.E.G., Fall Convention, New Orleans, La., U.S.A. (preprint).

CLOOS, E., 1968. Experimental analysis of Gulf Coast fracture patterns. *Amer. Assoc. Petrol. Geol. Bull.*, **52**, pp. 420-444.

DICKEY, P. A., SHRIRAM, C. R., and PAINE, W. R., 1968. Abnormal pressures in deep wells of southwestern Louisiana. *Science*, **160**, no. 3838, pp. 609 *et seq.*

GILL, W. D., 1962. The Variscan fold belt in Ireland. *In:* "Some Aspects of the Variscan Fold Belt", (Ed. K. Coe), Manchester University Press.

GILL, W. D., and KUENEN, P. H., 1957. Sand volcanoes on slumps in the Carboniferous of County Clare, Ireland. *Quart. Journ. Geol. Soc., London*, **113**, pp. 441-460.

LEWIS, K. B., 1971. Slumping on a continental slope inclined at 1°-4°. *Sedimentology*, **16**, pp. 97-110.

MANDL, G., and CRANS, W., 1979. Gravitational glidings in deltas. Proc. Symposium on "Thrust and Nappe Tectonics". Imperial College, London. (Publication pending).

MANDL, G., DE JONG, L. N. J., and MALTHA, A., 1977. Shear zones in granular material. *Rock Mechanics*, **9**, pp. 95-144.

RETTGER, R. E., 1935. Experiments on soft rock deformation. *Bull. Amer. Assoc. Petrol. Geol.*, **19**, pp. 271-292.

RIDER, M. H., 1969. Sedimentological studies in the West Clare Namurian, Ireland, and in the Mississippi River delta. Ph.D. Thesis, Imperial College, London.

——, 1978. Growth faults in Carboniferous of Western Ireland. *Amer. Assoc. Petrol. Geol. Bull.*, **62**, 11, pp. 2191-2213.

TERZAGHI, K., 1950. Mechanism of landslides. *Geol. Soc. America*. (Berkey Volume). Application of Geology to Engineering Practice, pp. 83-123.

——, 1956. Varieties of submarine slope failures. Eighth Texas Conference on Soil Mechanics and Foundation Engineering. Sept. 1956, Paper 3.

WOODLAND, A. W., and EVANS, W. G., 1964. The country around Pontypridd and Maesteg. Pp. 238-273. *In: Mem. Geol. Survey Great Britain: the Geology of the South Wales Coalfield*. Pt. IV, H.M. Stationery Office, London.

Reprinted by permission of the Geological Society of America from W. K. Hamblin, *Geological Society of America Bulletin*, v. 76 (1965), p. 1145-1163.

W. K. HAMBLIN *Dept. Geology, Brigham Young University, Provo, Utah*

Origin of "Reverse Drag" on the Downthrown Side of Normal Faults

Abstract: "Reverse drag," also called "downbending" or "turnover," was first recognized by Powell in the Colorado Plateau and subsequently was found to be an important structure associated with "down-to-basin" faulting in the Gulf Coast region. Many conflicting hypotheses have been proposed to explain this structure, most of which are based upon subsurface studies or observations in local areas.

Detailed field studies of reverse drag made in the western Colorado Plateau reveal that the flexure extends practically the entire length of most normal faults in the region. It is characterized by a broad, asymmetrical arc on the downthrown block, approximately 1 mile wide, with maximum dips of more than 30 degrees near the fault plane. Normal drag is common adjacent to the fault on both the upthrown and downthrown blocks. Reverse drag has been formed repeatedly during recurrent movement along the Hurricane and Grand Wash faults, clearly indicating that it is genetically related to faulting. The magnitude of the flexure is roughly proportional to displacement, and the trend of the fold closely parallels the trend of the fault. In many places reverse drag passes both vertically and laterally into antithetic faults.

Observations in the Grand Canyon reveal that the dip of the faults with which reverse drag is associated decreases with depth.

It is concluded that reverse drag results from an alternate response to the same forces that produce antithetic faults and develops because of curvature of the fault plane at depth. Normal movement along a curved fault plane, in effect, tends to pull the blocks apart as well as to displace them vertically. Adjustments to fill the incipient gap by rupture produces antithetic faults, whereas failure by flexing develops reverse drag.

CONTENTS

Geological Society of America Bulletin, v. 76, p. 1145–1164, 17 figs., 4 pls., October 1965

INTRODUCTION

One of the seemingly minor but interesting observations made by Powell in his exploratory trip down the Colorado River was that beds on the downthrown block of some normal faults in the Grand Canyon region dip into the fault plane, in a manner exactly opposite to the flexure produced by drag (Powell, 1875, p. 184). This structure was subsequently found to be common throughout much of the Gulf Coast area as well as the Colorado Plateau, and has been referred to as "reverse-drag," "downbedding," and "turn-over." Its origin and tectonic significance, however, are not well understood. Previous studies concerned with it have largely been restricted to short descriptions made in conjunction with investigations of a particular oil field, or to general geologic studies of local areas. Many geologists, therefore, have considered this structure to be an anomalous feature resulting from some abnormal combination of local tectonic events. Even in the Gulf Coast, where the greatest interest in this structure has been shown, there is little agreement concerning the possible mechanism by which reverse-drag flexures develop. This is due, at least in part, to the fact that these flexures are not exposed in that area and can be studied only by subsurface methods.

The present paper represents a detailed field study of reverse drag in the area of the Colorado Plateau, where it was first recognized and is best exposed. Its purpose is to present additional field data concerning the characteristics of reverse-drag flexures in an effort to determine the origin and significance of this type of structural deformation.

ACKNOWLEDGMENTS

The writer wishes to express his appreciation to R. C. Moore, W. Fisher, and J. Sorauf for many critical discussions concerning this problem, and to Charles Bondurant and Lorin Dutson for assistance in the field. M. K. Hubbert and J. H. Mackin read the manuscript and offered many helpful suggestions. The National Science Foundation through Grant G-13357 generously supported all phases of the investigation.

PREVIOUS HYPOTHESES

Faulted Monocline

Dutton (1882, p. 115) was apparently the first to describe reverse-drag flexures carefully and to formulate a theory explaining their origin. He observed a persistent downward flexing of the beds on the downthrown side of the Hurricane fault in the western part of the Grand Canyon, and a corresponding, although less well-defined, upward flexure on the upthrown block. Similar features along other faults in the region were also mentioned by Dutton but were not described in detail.

Dutton concluded that reverse-drag flexures resulted from faulting of earlier monoclines in which displacement had been opposite to that of faulting. The flexures were thus thought to antedate the faults and represent a distinct period of deformation which had lowered the strata to the east. This theory was later accepted by Davis (1901) and Johnson (1909, p. 158), who believed that the monoclines in the Grand Canyon district were relatively old structures, along whose lines later faulting had occurred with opposite throw.

Nobel (1914, p. 76) in his study of the Muav Canyon fault in the central part of the Grand Canyon recognized recurrent movement along Precambrian faults and considered reverse drag to be the result of a faulted monocline, although his structure sections show that the strata on the upthrown block are horizontal.

Koons (1945, p. 165–166) and C. E. Bondurant (1963, M.A. thesis, Univ. of Kansas) later reaffirmed the work of Dutton, and to many workers in the Colorado Plateau the theory of a faulted monocline is considered to be well established.

Compression After Faulting

In the eastern United States, Foye (1924, p. 240) briefly described abnormal dips near the eastern boundary fault of the Connecticut Triassic basin. These flexures are apparently similar to those found in the Colorado Plateau, although they are not nearly so well expressed. Foye believed that a period of compression occurred after faulting and that a monoclinal flexure developed in the sedimentary strata adjacent to the crystalline upland. The reverse drag would thus postdate the faulting.

More recently B. Mears (1950, Ph.D. dissert., Columbia Univ.), after an extensive review of the literatuıe concerning reverse drag, concluded that the flexures associated with faulting in Oak Creek Canyon, Arizona, also resulted from local compression during later phases of deformation.

Double Movement

Sears (1925, p. 23) believed that reverse-drag flexures are truly drag phenomena and result from double movement along the fault in opposite directions. Concerning the flexures in the Gallup-Zuni basin of New Mexico he states that "... at first the upthrown side rose higher than it is at present and afterward it settled slightly to its present level; the second movement, though of smaller throw than the first, was accompanied by greater pressure, which caused the drag on the beds of the downthrown side."

Spooner (1929, p. 217–222) described reverse drag in the Homer dome in Louisiana and also considered the flexure to be the result of displacement in opposite sense on an earlier fault. He attributed the downbending to upward movement on the downthrown block after the upthrown side had become stabilized.

Elastic Rebound

Brucks (1929, p. 265–268) believed that reverse-drag flexures associated with the Luling fault in Texas resulted from elastic rebound. This hypothesis states that prior to actual slippage along the fracture, the rocks would behave elastically under long-continued shearing stresses. After rupture, the warped strata would rebound to a position of no strain when faulting occurred. This explanation has also been invoked by a number of German geologists in an attempt to explain reverse-drag flexures in Europe and Africa (George W. R. Knetsch, personal communication). A similar phenomenon is discussed by Reid (1911, p. 413–444) and Benioff (1938, p. 77–84) in relation to strike-slip faults in California.

Sag

Various workers have concluded that complex conditions involved in the hypotheses described previously are unnecessary and that reverse-drag flexures result from sagging of the downthrown block near the fault plane. Most proponents of the sag hypothesis arrived at their conclusions independently from studies of widely separated regions. There is, therefore, little agreement as to the structural conditions that cause the downthrown block to sag near the fault plane.

Gardner (1941, p. 258–260) studied the Hurricane fault near the town of Hurricane, Utah, where late Cenozoic volcanoes are abundant. He suggested that the sag along the Hurricane fault may be related to the eruption of large masses of lava in the area. Storm (1945, p. 1338) believed that reverse drag in the Gulf Coast results from slumping but does not explain the factors which would cause the beds to sag near the fault plane. Perkins (1961, p. 181) suggests that slump along the Gulf Coast faults may be due in part to gravity flowage.

Many geologists in the Gulf Coast area believe that sagging along the downthrown side of the faults resulted from a combination of bedding-plane faults and medium- to low-angle normal faults. A thick sedimentary section could thus slip down the regional dip on a lubricated medium such as a salt layer or an incompetent shale bed (Russell, 1957, p. 69; Quarles, 1953, p. 498). Beds near the fault would thus sag down to fill the hypothetical gap left after the section slid down the bedding plane. This has become known as the "slipping-plane hypothesis" (Quarles, 1953, p. 498) and is considered by many to adequately explain reverse drag in the Gulf Coast. The basic elements of the hypothesis were first illustrated by Lotze (1930, p. 222), who observed antithetic faults and reverse-drag flexures in the Moringer district of Germany. He concluded that the fault blocks separated under tension and developed a fissure-like cavity. The strata on the hanging wall would not be able to support their own weight and would collapse to fill the void. Later, Kamen-Kaye (1953, p. 2178–2179) explained how curvature of low-angle faults at Las Mercedes, Venezuela, would produce a greater than regional dip near the fault plane and a progressive increase in the vertical displacement of the beds from the bottom to the top.

Salt-Ridge Hypothesis

After carefully considering the prevailing explanations of Gulf Coast faulting, Quarles (1953) rejected the sag hypothesis and suggested that reverse drag is formed from a faulted arch over a salt ridge. He (1953, p. 490)

"... assumes that salt from a sedimentary layer has intruded the overlying beds in the form of a long ridge or dike and forms a deep-seated anticline by arching the strata in a manner similar to the formation of a salt dome.... Arching upward without inward movement at the sides would tend to stretch and break the strata and the weight of the overlying section would cause the beds to sink into the adjacent depression along a medium- to low-angle fault."

Faulted Anticlines

Strahler (1948) studied the geomorphology and structure of the West Kaibab fault zone, Arizona, and concluded that the downbending of strata adjacent to faults in the Kaibab region represents faulted limbs of low anticlines or domes. He believed that, inasmuch as the anticlines are limited by faults, faulting and bending must have been contemporaneous, and he concluded that there is no evidence in the Kaibab region for two periods of deformation as proposed by Dutton, Noble, and Koons. A somewhat similar explanation is presented by Bornhauser (1958, p. 355) for reverse-drag flexures in the Gulf Coast.

Differential Compaction

Honea (1956, p. 53) concluded that reverse-drag flexures in the Gulf Coast could result from processes described in any of the theories previously proposed. He also concluded that one theory may stand out as a major factor in one area, but be ineffective in another. In addition, he proposes that "... differential compaction contemporaneous with deposition figured strongly in the formation of this structure in the Gulf Coast region." Shirley (1960, p. 80), studying the structure of the Rayne Field, also concluded that compaction could cause reverse-drag flexures. He stated that

"... faulting was occurring at the same time that major sedimentary deposition was taking place. The presence of this active fault caused an increase in sediment on the downthrown side of it by changing the sedimentary factors of the area. The subsequent increase in sediment on the downthrown side of the fault produced a sedimentary overburden which caused compaction of the sediment adjacent to the downthrown trace of the fault. This had the effect of creating a dip into the fault."

OBSERVATIONS IN THE WESTERN PART OF THE COLORADO PLATEAU

General Statement

Excellent exposures in the western part of the Colorado Plateau present a unique opportunity to obtain an accurate and detailed concept of reverse-drag flexures. On the Esplanade of the Grand Canyon, for example, the structure is not only completely exposed on the plateau but excellent cross sections are found in the deep canyon walls below (Pl. 1). Details of the three-dimensional aspect of the flexures and associated faults are thus unusually clear. Northward, on the Shivwits, Uinkaret, and Kanab plateaus, the conditions are only slightly less favorable. Moreover, numerous late Cenozoic basalts have crossed many faults at various times and have subsequently been displaced by recurrent movement. Reverse-drag flexures developed in the basalts are too young to be obscured by erosional debris and are generally completely exposed.

The basic approach used in this study was detailed mapping because many structural relationships, which are generally obscure in other areas, are clearly expressed in the Grand Canyon region and provide important documentation concerning the origin and significance of this seemingly anomalous feature. Most of the Hurricane fault zone, where reverse drag is exceptionally well expressed, was mapped on a scale of 3 inches to 1 mile. Elsewhere mapping was generally limited to 1 inch to 1 mile and to detailed observations of local areas. Special attention was focused on the following: (1) areal extent and spatial relationships between flexures and faults, (2) geometric details of the flexures, (3) age of flexing *vs.* age of faulting, (4) characteristics of the faults, and (5) structures associated with reverse drag.

Areal Extent

Detailed mapping in the western part of the Colorado Plateau indicates that reverse drag is practically coextensive with normal faulting in the area (Fig. 1). For example, the flexure can be traced as a continous structural feature for more than 160 miles along the Hurricane fault, and extends throughout the entire length of many associated minor faults. In addition, reverse drag is expressed in the late Cenozoic basalts throughout the central half of the Grand

Wash fault and is exposed at various intervals along the southern part of the Sevier and West Kaibab faults. Erosional debris covers much of the downthrown block along the northern Sevier and West Kaibab faults, so reverse-drag exposures are available and reverse drag was not found is at the southern end of the Hurricane fault near Peach Springs, Arizona, and along associated minor faults to the west where the displacement is less than 100 feet.

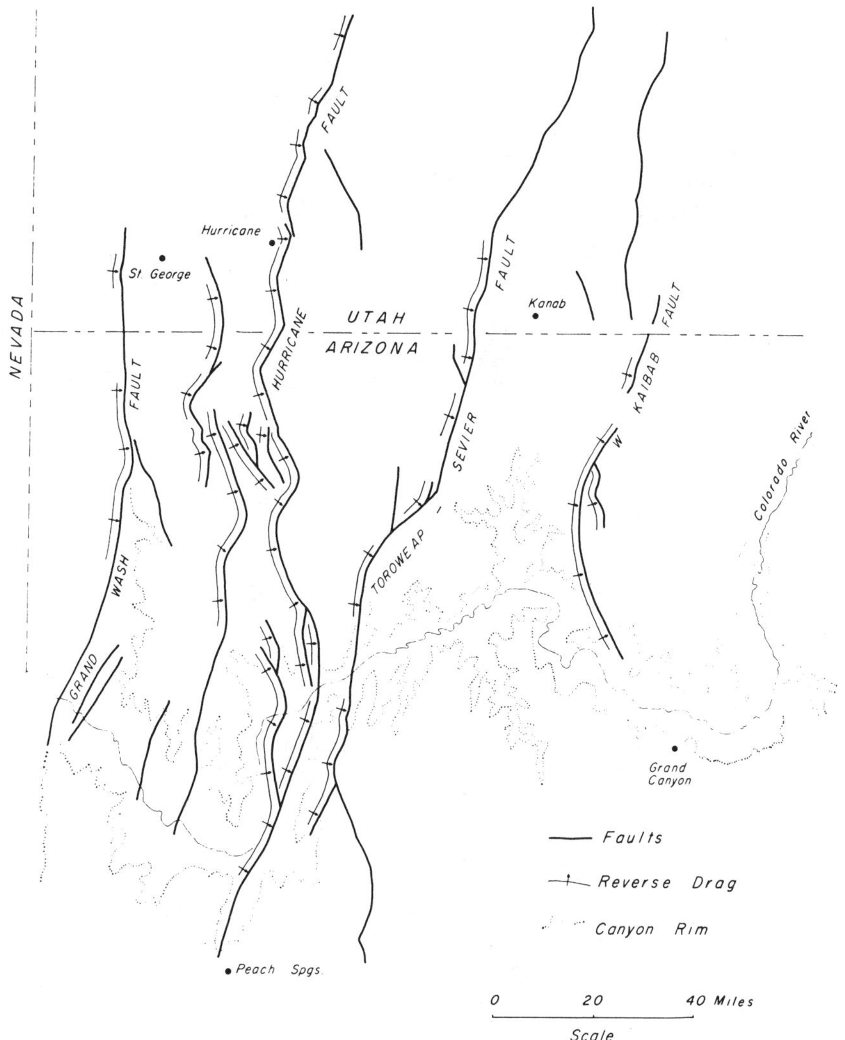

Figure 1. Map showing major faults and reverse-drag flexures in the western Grand Canyon region

flexures in those areas could not be demonstrated. Available evidence, however, strongly suggests that flexures continue under the alluvial cover. It is apparent from Figure 1 that reverse drag is much more common than has heretofore been realized and that it is not an anomalous feature found only in a few exceptional areas. The only area where adequate

These data, together with scattered reports of reverse drag elsewhere in the Colorado Plateau (Sears, 1925, p. 23; B. Mears, 1950, Ph.D. dissert., Columbia Univ.); the Northern Rockies (Rogers and Lee, 1923, p. 52); the Triassic basins of the eastern United States (Foye, 1924, p. 240; G. deV. Kline, 1960, Ph.D. dissert., Yale Univ.); Ontario (Kay, 1942, p. 610); Ger-

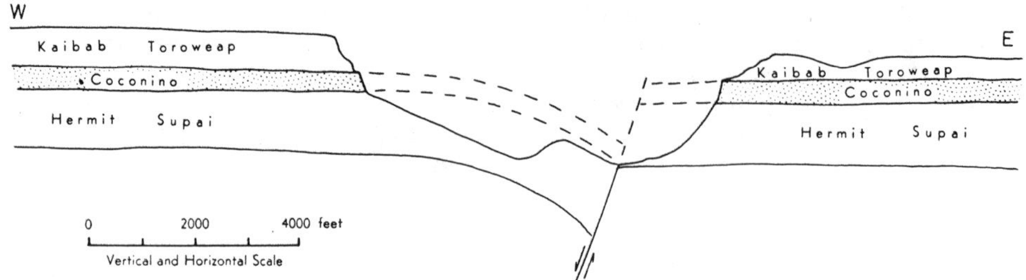

Figure 2. Section showing reverse-drag flexure along the Muav Canyon fault, Arizona. (*Modified after* Noble, 1914, Pl. 1)

many (G. W. R. Knetsch, personal communications, 1963; Lotze, 1930); Venezuela (Kamen-Kaye, 1953, p. 2178); and the Gulf Coast (Quarles, 1953) indicate that reverse drag is not an abnormal structure, but is a common feature associated with normal faulting in many different areas. Forces responsible for this structure are, therefore, not exceptional nor freakishly influenced by local conditions, but often come into action during formation of normal faults. This structure exists in many other regions in which normal faulting is common, but is covered by erosional debris from the upthrown block.

Geometry of Reverse-Drag Flexures

Characteristics of the geometry of reverse-drag flexures in the Grand Canyon region are illustrated in Figures 2–7 and Plate 2. The flexure is a simple structure in which strata on the downthrown block are bent down toward the fault plane in a large, graceful, asymmetrical arc. Along most major faults, where displacements range from 1000 to 2000 feet, the fold is approximately 1 mile wide, with maximum dips of more than 30 degrees near the fault plane. Displacement caused by flexing alone may thus be as much as 500 feet. Flexures along minor faults are correspondingly smaller, and little or no evidence of reverse drag can be detected where the throw is less than 100 feet. Throughout the extent of a given fault the radius of curvature of the flexure near the fault plane is relatively constant except where there is significant variation in the amount of displacement.

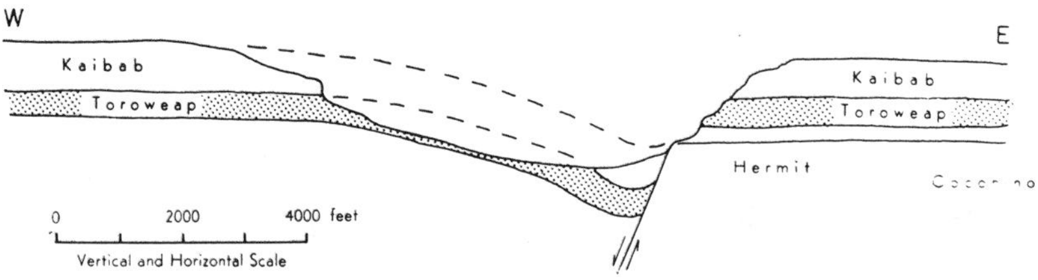

Figure 3. Section showing reverse drag along the Big Springs fault near Big Springs, Arizona. (*Modified after* Strahler, 1948, p. 520)

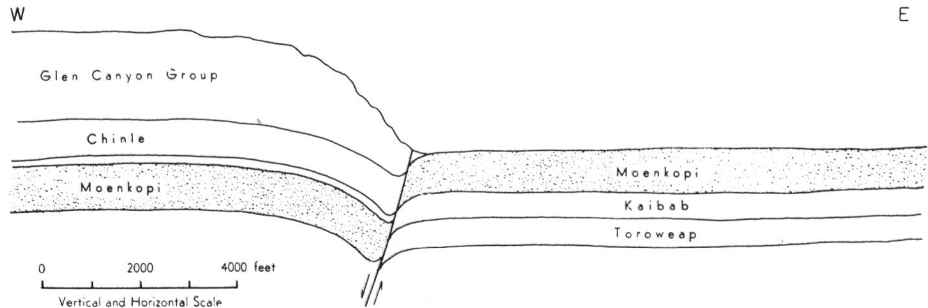

Figure 4. Section showing reverse drag along the Sevier fault at Pipe Springs National Monument, Arizona. The scarp in this area is an obsequent fault-line scarp carved on the resistant Navajo Sandstone. (*Modified after* Gregory, 1950, p. 147)

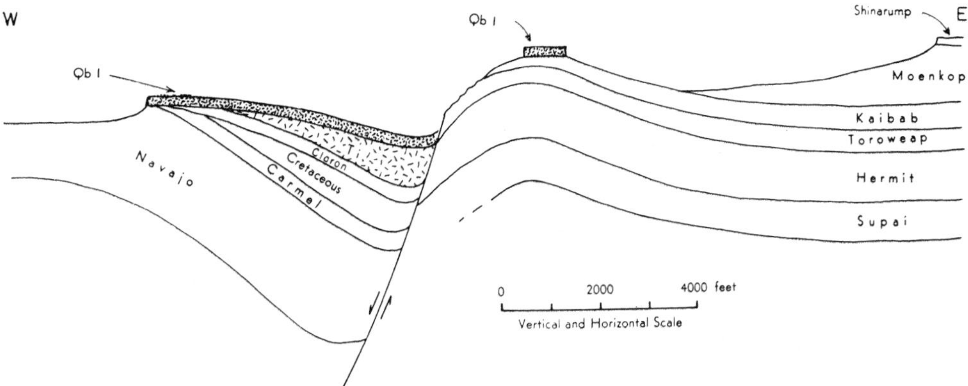

Figure 5. Section across the Hurricane fault approximately 10 miles north of Hurricane, Utah, showing reverse-drag flexures in Mesozoic and Cenozoic sediments and older Quaternary basalts (Qbl)

Normal drag is commonly found within a few hundred feet of the fault plane on both the upthrown and downthrown blocks, with dips ranging up to 75 degrees (Figs. 4–7, and 9). Best exposures are preserved on upthrown blocks along the Hurricane and Grand Wash faults, where normal drag can be traced from the top to the base of the cliffs, a height commonly greater than 1000 feet. Elsewhere, exposures are sufficient to indicate that some degree of normal drag is typical along the faults where reverse drag is present.

The axis of reverse-drag flexures closely parallels the fault trace and may be straight, curved, or zig-zag, according to the pattern of the fault system (Fig. 1). Moreover, the flexure dies out as the fault dies out and never continues beyond the extent of the fault trace.

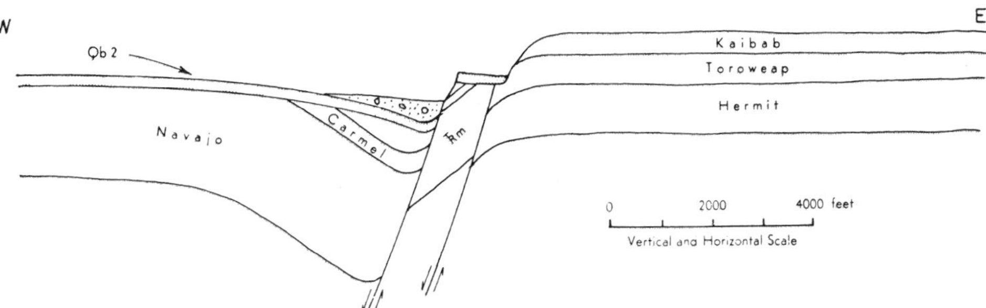

Figure 6. Section across the Hurricane fault along the Virgin River just north of Hurricane, Utah, showing a large reverse-drag flexure in Mesozoic sediments and a smaller flexure in the younger Quaternary basalts (Qb2)

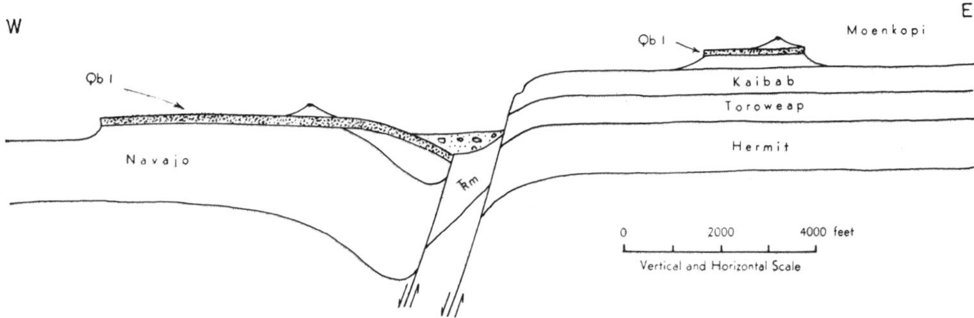

Figure 7. Section across the Hurricane fault approximately 8 miles south of Hurricane, Utah, showing variations in reverse-drag flexures in the Navajo Sandstone and older Quaternary basalts (Qbl)

Characteristics of Flexures on the Upthrown Block

In most areas beds on the upthrown block, beyond the influence of normal drag, are essentially horizontal or conform to the regional dip in the area. Slight upward flexing of the strata on the upthrown block, however, has been reported in some areas (Dutton, 1882, p. 114; Koons, 1945, p. 165–166; B. Mears, 1950, Ph.D. dissert., Columbia Univ.), and is well expressed along the Hurricane fault north of the town of Hurricane and near the Utah-Arizona border, as well as locally on the south side of the Grand Canyon (Figs. 5, 7, and 9; Pl. 2). Elsewhere in the western Colorado Plateau strata on the upthrown block show no indication of flexing considered to be reverse drag.

The extent of the area affected by upward flexing on the upthrown block is highly erratic. In some areas abnormal dips extend only a few hundred yards beyond the fault; elsewhere the flexure continues more than 3 miles. Dips as high as 40 degrees have been measured north of the town of Hurricane, Utah, but in most areas maximum dip on the upthrown block is less than 7 degrees. The magnitude of dip is variable, not only from one region to another but also within a local area. Along the Hurricane fault, near the Utah-Arizona border, dips on the upthrown block, adjacent to the fault, range from 20 degrees to the east to horizontal within a distance of 6 miles, although displacement of the fault remains essentially constant.

The highly erratic distribution of flexures on the upthrown block stand out in marked contrast to the continous and much more extensive flexures of reverse drag on the downthrown block. The two flexures appear to be independent and not genetically related.

The most striking evidence of the complete independence of flexing on the upthrown and downthrown blocks is found in the attitudes of late Cenozoic basalts which have crossed the Hurricane or Grand Wash faults at various times and have been subsequently displaced by recurrent movement along the faults (Figs. 5–7; Pl. 3). In both areas reverse drag is well developed in basalts on the downthrown block whereas these same flows on the upthrown side have remained horizontal.

Detailed mapping on both the upthrown and downthrown blocks of the Hurricane fault indicates that flexures on the upthrown block result from prefaulting warps and have no genetic relationship to displacements on the faults. Locally, where Laramide and Tertiary structures are colinear, some flexing on the downthrown block is obviously the result of prefaulting folds (Fig. 8). This is especially evident north of the town of Hurricane, where steep dips on the upthrown block extend northward to Cedar City, Utah. Available evidence indicates that the Kanarra fold described by Gregory and Williams (1947) and illustrated by Mackin (1960, p. 125) is the northern extension of the east limb of the Virgin anticline on the upthrown block of the Hurricane fault. Near the Utah-Arizona border a similar relationship exists where "anomalous" dips on the upthrown block represent prefaulting monoclines and other flexures. Mapping by Gregory (1950), Cook (1960), and the writer clearly indicates a series of broad, shallow, northeast-plunging folds cut by the Hurricane fault.

It is thus clear that although some flexing considered to be reverse drag is the result of faulted Laramide folds, reverse drag develops in many areas only on the downthrown block without a corresponding flexure forming on the upthrown block. The presence of reverse drag on both the upthrown and downthrown blocks in some areas appears to be a coincidence of later faulting on older folds.

Age of Flexing

A basic element in the reverse-drag problem is the age of flexing with respect to faulting. Some hypotheses make the flexures older than the faulting (faulted monocline, faulted folds, and salt ridge) whereas others state that faulting and flexing were contemporaneous (double movement, elastic rebound, and some sag hypotheses). Still others propose that flexures are definitely postfaulting (compression after faulting, some sag hypotheses, and differential compaction).

Studies along the Hurricane and Grand Wash faults shed much light upon this problem, for in a number of places late Cenozoic basalts have crossed the faults at various times and have subsequently been displaced by recurrent movement.

The area in the vicinity of Hurricane, Utah, was studied in detail by Gardner (1941), who recognized four major periods of deformation. Figure 5 is a modification of Gardner's cross section showing the structural relationships near Black Ridge, approximately 10 miles north of the town of Hurricane, Utah. The complex flexing in this area resulted from at least three sep-

REVERSE-DRAG FLEXURE EXPOSED ON THE ESPLANADE IN THE WESTERN PART OF THE GRAND CANYON, ARIZONA

View looking south along a branch of the Hurricane fault

HAMBLIN, PLATE 1
Geological Society of America Bulletin, volume 76

REVERSE-DRAG FLEXURE ALONG THE HURRICANE FAULT SOUTH OF THE COLORADO RIVER, ARIZONA

View looking south near Granite Canyon. Note that the strata on the upthrown block dip to the east in the foreground but are horizontal in the background, whereas the reverse drag on the downthrown block extends along the entire length of the fault.

HAMBLIN, PLATE 2

HURRICANE FAULT, SHOWING REVERSE-DRAG FLEXURES IN THE QUATERNARY BASALT AND UNDERLYING NAVAJO SANDSTONE, UTAH

View is to the north. Note that the reverse drag in the Navajo Sandstone is greater than that in the younger basalt, and the strata on the upthrown block are essentially horizontal. The settlement in the background is Hurricane, Utah.

HAMBLIN, PLATE 3
Geological Society of America Bulletin, volume 76

Figure 1. Hurricane fault surface near the Utah–Arizona border, showing vertical dip

Figure 2. The surface of a branch of the Hurricane fault near the bottom of the Grand Canyon, showing the relatively low dip of the fault plane at depth

TWO EXPOSURES OF THE HURRICANE FAULT SURFACE, ARIZONA AND UTAH

HAMBLIN, PLATE 4
Geological Society of America Bulletin, volume 76

arate disturbances. The prominent eastward dip of Paleozoic and Mesozoic strata on both the upthrown and downthrown blocks represent the northward extension of the east limb of the Virgin anticline (Fig. 8). Maximum dips range up to 40 degrees. This flexure antedates the fault ward dip of the Paleozoic strata near the fault is the result of normal drag along the fault plane.

Southward, in the vicinity of the town of Hurricane, Utah, a similar relationship is of particular interest, for it clearly shows that the

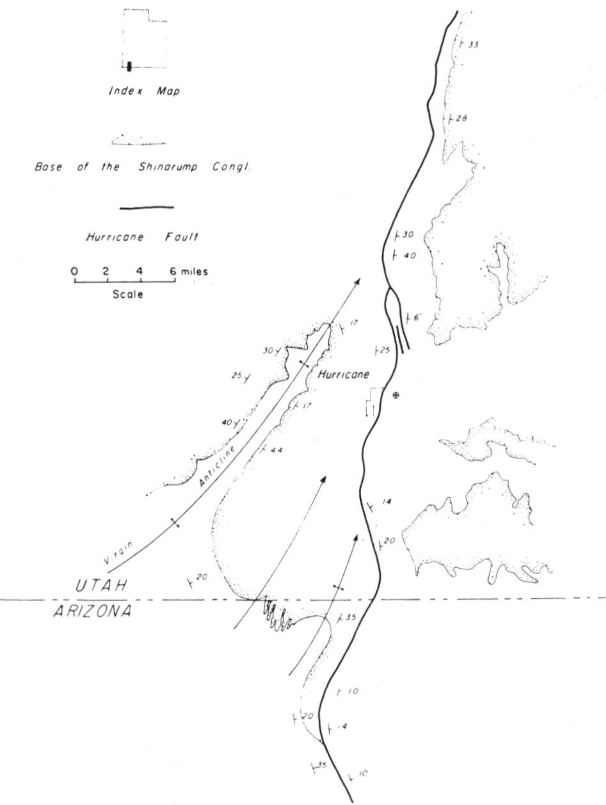

Figure 8. Map showing the relationship between the Virgin anticline and related flexures on the downthrown block and "anomalous" eastward dips on the upthrown block. Note that the beds on the upthrown block dip from 10° to 40° E. adjacent to the anticline, whereas the beds above the town of Hurricane adjacent to the shallow syncline are horizontal.

and is completely independent of it although locally the two structures are parallel. The dip of the Claron Formation is less than that in the underlying strata and represents a second period of deformation. This flexure is interpreted to be reverse drag developed by the first movement along the Hurricane fault. The third period of deformation is recorded in the Quaternary basalt (Qbl) which has been displaced 2000 feet. Reverse drag is well expressed by the dip slope on the downthrown basalt, but remnants on the upthrown block are horizontal. Strong westmagnitude of reverse drag becomes progressively greater in the older beds and that flexing developed during each period of significant movement. A section across the Hurricane fault along the Virgin River illustrating this relationship is shown in Figure 6. Here, displacements along the Hurricane fault zone developed along two separate fault planes less than 1000 feet apart. Flexures in the Mesozoic strata on the downthrown block are part of the east limb of the Virgin anticline modified near the fault plane by reverse drag. Recent movement along the western fault plane is well documented by younger basalt (Qb2) which crossed the fault line and lapped up against the eastward-retreating fault scarp. Recurrent movement along the western fault plane subsequently sheared off the lava flow and dropped the western block 200 feet. A slight reverse-drag flexure can be seen in the flows on the downthrown block along the gorge cut by the Virgin River. Alluvium several tens of feet thick has accumulated in the resulting depression adjacent to the fault plane. The remnant of the flow clinging to the scarp on the upthrown block is essentially horizontal and can be traced more than 2 miles along the Hurricane Cliffs above the town of Hurricane. Paleozoic strata on the upthrown block are also horizontal except for large drag flexures found locally adjacent to the fault plane. The tilted slice of Moenkopi shale between the fault planes probably represents a remnant of a normal drag flexure developed adjacent to the earlier eastern fault plane.

Additional evidence indicating that reverse drag develops contemporaneously with faulting is found approximately 8 miles south of Hurricane. The structural relationships are shown in a structural cross section (Fig. 7) and an oblique areal photograph (Pl. 3).

The regional dip of the Navajo Sandstone, cropping out below the lava, is to the northeast into a broad, shallow syncline adjacent to the Virgin anticline (Fig. 8). Prominent reverse-drag flexures are clearly exposed in the Navajo Sandstone with maximum dips up to 36 degrees. The early Quaternary basalt (Qbl) is much older than the lavas shown in Figures 5 and 6 for it flowed more than 1 mile across the fault line upon an erosional surface developed on Moenkopi Formation and Navajo Sandstone. Recurrent movement has displaced this flow at least 1400 feet. A prominent reverse-drag flexure developed in the basalt on the downthrown block, with maximum dips less than in the underlying Navajo Formation. Numerous remnants of the basalt on the upthrown block are all horizontal. Normal drag is found locally on the upthrown block in the Kaibab and Toroweap limestones, but is not conspicuous. Isolated exposures of westward-dipping Moenkopi shale adjacent to the fault scarp indicate movement along at least two separate fault planes.

The structural relationships shown in Figures 5–7 reveal the following sequence of tectonic events (Fig. 9). (1) Compressional forces developed the Virgin anticline and adjacent flexures during the Laramide orogeny but did not greatly deform the strata in the area shown in the cross sections. (2) Tertiary faulting along two separate fault planes subsequently displaced the sedimentary sequence approximately 5000 feet and preserved a slice of Moenkopi shale between the two fault blocks (Fig. 9, A and B). Reverse-drag flexures formed on the downthrown block at this time and the area was subjected to a long period of erosion. (3) The earliest lavas flowed across the fault trace on the smooth erosional surface and were subsequently displaced 1200 feet (Fig. 9, C and D). This period of deformation developed reverse drag in the basalts and accentuated the drag previously formed in the sedimentary sequence. (4) The scarp formed by the second period of deformation was eroded back a few hundred feet, followed by extrusion of lavas which crossed the fault line and lapped up against the receding cliffs (Fig. 9, E). (5) A subsequent movement then sheared off the edge of the younger transgressive flow (Qb2) and dropped the western block 200–300 feet (Fig. 9, F). This developed a small reverse-drag flexure in the youngest flow and further accentuated the flexures developed during the previous periods of deformation.

A similar but less complicated relationship can be seen in the central part of the Grand Wash fault where only one flow crossed the fault line and has been displaced approximately 500 feet. Like exposures found in the Hurricane fault zone the numerous remnants of basalt on the upthrown block are essentially horizontal, but reverse drag has produced dips on the downthrown block ranging up to 16 degrees.

Exposures in the Grand Wash area are especially significant in that they clearly illustrate that reverse drag in the basalts is not the result of primary dip. Flow structures, volcanic necks, dikes, and the geometry of the flow units together with paleocurrent structures in the interflow sediments indicate that the principal source of lava was at the north end of Grand Wash and that the lava flowed southward toward the Colorado River. Near the fault plane, however, the lava dips eastward toward the fault as much as 16 degrees whereas the remnants on the upthrown dip slightly westward or conform to the regional slope to the south. The reverse drag in the basalt, therefore, must represent a postdepositional flexure and not primary dip to the east.

It thus seems clear that reverse drag can develop on the downthrown block exclusively, and forms during each significant movement along the fault plane. All exposures studied indicate that the magnitude of the flexure is roughly proportional to the displacement, except in areas disturbed by prefaulting warps. Moreover, exposures of Cenozoic basalts near the town of Hurricane, Utah, definitely prove that flexing may occur in increments.

Characteristics of the Faults

Faults in the western part of the Colorado Plateau with which reverse drag is associated are characterized by a sinuous pattern with large segments trending alternately N. 10° W. and N. 20° E. (Fig. 1). Displacement ranges from less than 100 feet to more than 10,000 feet. Extensive exposures of the actual fault plane show well-preserved striations directly down dip. In addition, the lavas which have crossed the fault plane and have subsequently been displaced show no indications of strike-slip displacement. Zones of breccia ranging up to 3 feet thick and normal drag adjacent to the fault plane are common.

Most scarps are composite fault-line scarps resulting from differential erosion of the nonresistant Moenkopi shale, together with segments of a true fault scarp produced by recurrent movement along the fault planes. In major faults, such as the Hurricane, displacement has occurred along several fault planes in a zone approximately ¼ mile wide. There is excellent evidence that repeated movement has occurred along the fault zone throughout much of late Cenozoic time, as recognized by Huntington and Goldthwait (1904) and Gardner (1941).

High on the Shivwits Plateau, literally thousands of square yards of the actual plane of the Hurricane fault may be found at numerous localities, extending from the Grand Canyon rim northward past the town of Hurricane. Similar but less extensive outcrops are also found along the Grand Wash fault and on many of the minor faults in the area. All these exposures show that the fault planes at an average elevation of 4000 feet are essentially vertical or dip at a very high angle to the west (Pl. 4, fig. 1)

Equally good exposures of the actual fault surface of the Hurricane and associated minor faults were found deep within the Grand Canyon, several thousand feet below the plateau surface. These were studied from Peach Springs Wash northward approximately 20 miles, and on the north side of the canyon below the Esplanade. In contrast to the nearly vertical attitude high on the plateau, the average dip of the fault planes deep within the canyon is 60° W., and in some areas dips as low as 55 degrees were measured (Pl. 4, fig. 2). Unfortunately, the best exposures of the Hurricane fault on the north side of the Colorado River at Whetmore Wash show a zone of complex fractures in which the dip of the main plane of displacement could not be determined. All measurements made elsewhere in the canyon on the Hurricane and associated faults indicate dips less than 70 degrees, suggesting a gradual decrease in dip of the fault plane at depth.

Associated Structures

Small antithetic faults[1] are commonly associated with reverse-drag flexures in the western Colorado Plateau and are best developed where displacements on the master fault exceed 1000 feet. Antithetic faults are generally less than 2 miles long and strike northeast or northwest following the trend of the major conjugate fracture system in the area. The average dip is 60° E. and displacements are generally less than 200 feet. In many places minor normal faults, dipping in the same direction as the master fault, parallel the antithetic faults so that small grabens are developed on the crest of the reverse-drag flexure.

Antithetic faults are best expressed in the late Cenozoic basalts north of the town of Hurricane, where well-defined east-facing scarps rise 30–50 feet above the surface of the surrounding flows. Southward in Arizona, where the Hurricane fault zone is more complex, antithetic faults are more abundant and have greater displacements. They are, however, difficult to distinguish from branches of the master fault system because many fault scarps have been obliterated by erosion.

On the Esplanade of the Grand Canyon, exposures of branch faults of the Hurricane fault system show that reverse-drag flexures may pass both vertically and laterally into antithetic faults. This relationship is especially clear in the fault system between Parashant and Andrus canyons where reverse drag is well expressed in the Supai Formation on the Esplanade, but high on the rim of the canyon the fold gives way to antithetic faults in the Kaibab and Toroweap formations. This seems to be a significant relationship, for if antithetic faults grade laterally and vertically into reverse-drag flexures, the

---

[1] The term "antithetic fault" is used in the sense proposed by Cloos (1928), and refers to faults of minor displacement which dip toward the major fault plane.

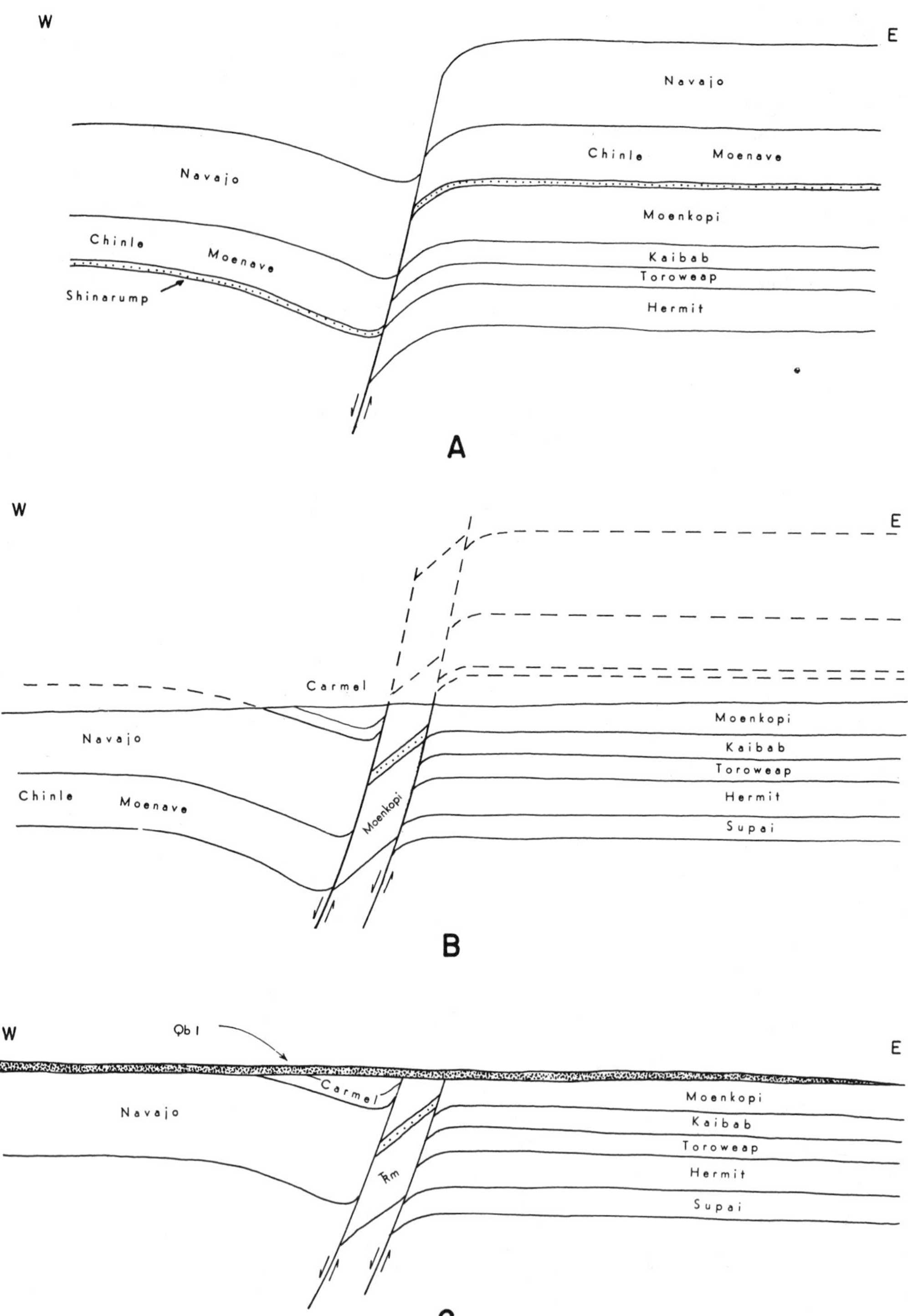

Figure 9. Sections showing the sequence of events in the development of the structural relationships near the town of Hurricane, Utah. A, early faulting along the Hurricane fault. Development of initial reverse drag flexure. B, recurrent movement along a second fault plane. Reverse drag accentuated and a slice of Moenkopi preserved between fault blocks. Total displacement approximately 5000 feet. C,

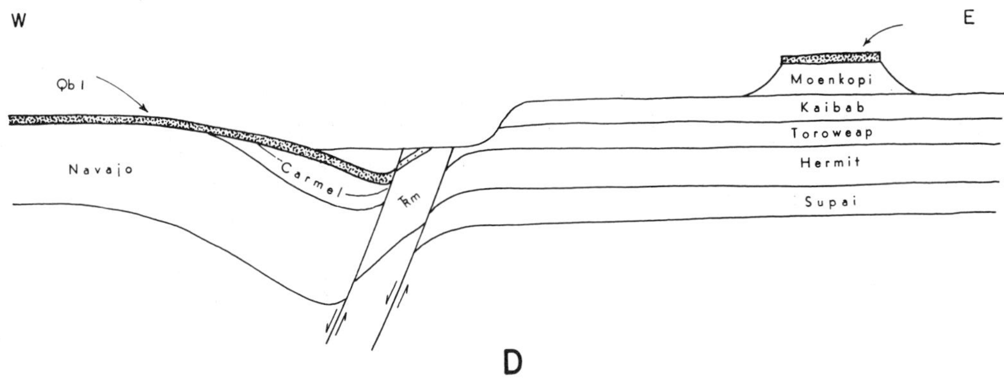

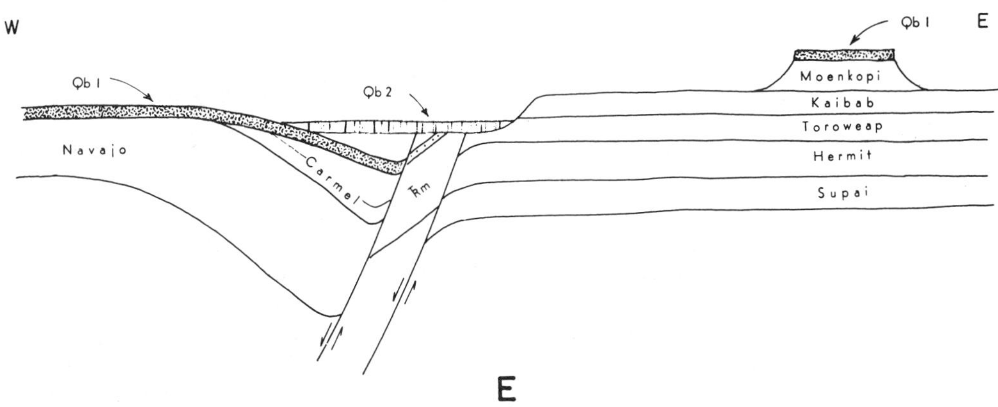

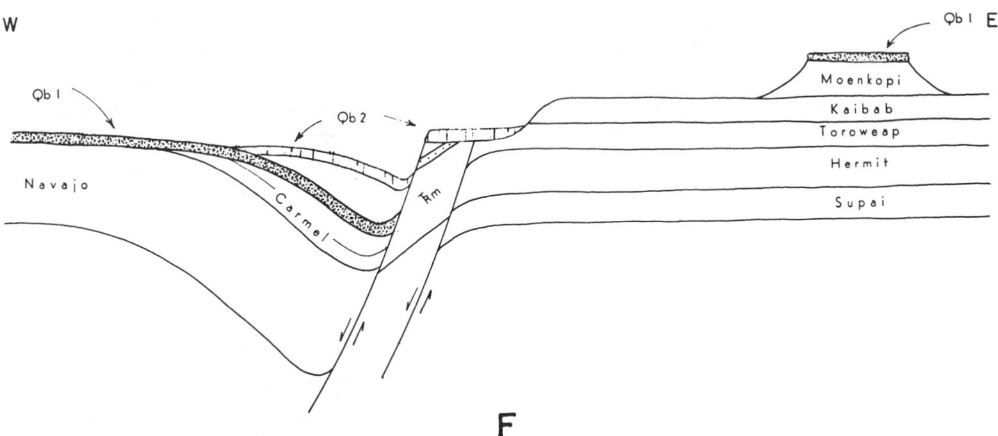

extrusion of earliest basalt across fault line. D, recurrent movement. Displacement of basalt 1200 feet. Reverse drag developed in basalt and accentuated in older sediments. E, extrusion of younger lavas across fault line. F, recurrent movement. Reverse drag developed in younger basalts and accentuated in older rocks.

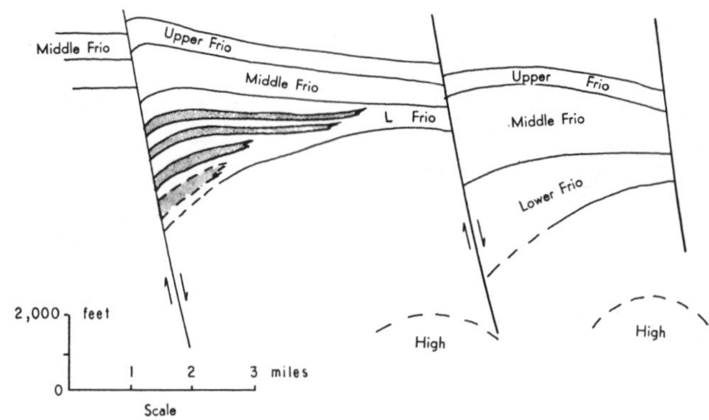

Figure 10. Section across the Francitas area, Jackson County, Texas, showing the characteristics of reverse drag in the Frio sediments of the Gulf Coast. (*After* Walters, 1959, p. 57)

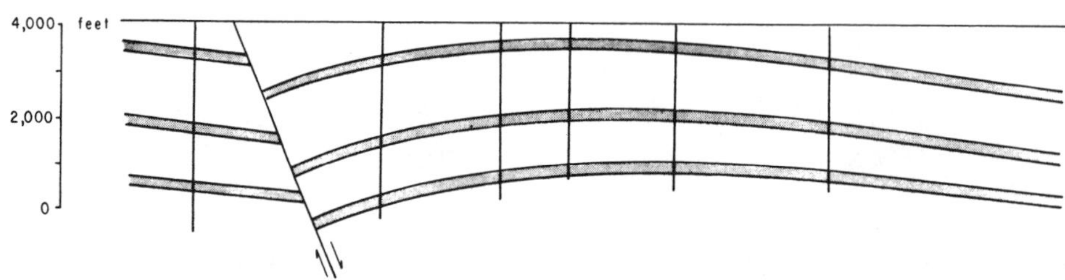

Figure 11. Section across the Aqua Dulce–Stratton Field, Texas, showing typical reverse drag in the Gulf Coast area. (*Modified after* Honea, 1956, p. 52)

two structural features probably resulted in response to the same tectonic force.

OTHER EXAMPLES OF REVERSE DRAG

Cross sections showing typical examples of reverse drag reported from other parts of the world are shown in Figures 10–16. Most of the detailed information comes from the Gulf Coast, where extensive drilling has demonstrated reverse-drag flexures adjacent to most major faults in the region. A comparison of Figures 10–13 with Figures 2–7 and Plates 2 and 3 indicates that the geometry of the flexures in the Gulf Coast is similar to that in the Colorado Plateau. The fold consists of a broad asymmetrical arc approximately 1 mile wide with maximum dips adjacent to the fault. It is not known if normal drag is common adjacent to the fault plane but antithetic faults are associated with the flexure in many areas (Quarles, 1953, p. 501). Reverse drag in the Gulf Coast closely follows the faults and in some areas the flexure has been traced laterally up to 240 miles (Honea,

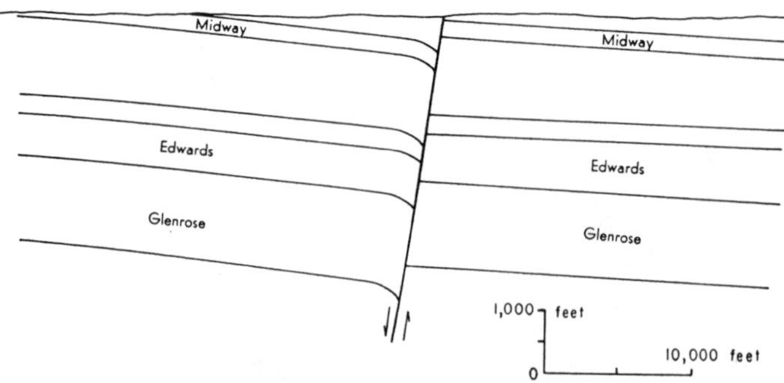

Figure 12. Section across the Luling field, Texas, showing reverse-drag flexures along the Luling fault. (*After* Brucks, 1929, p. 263)

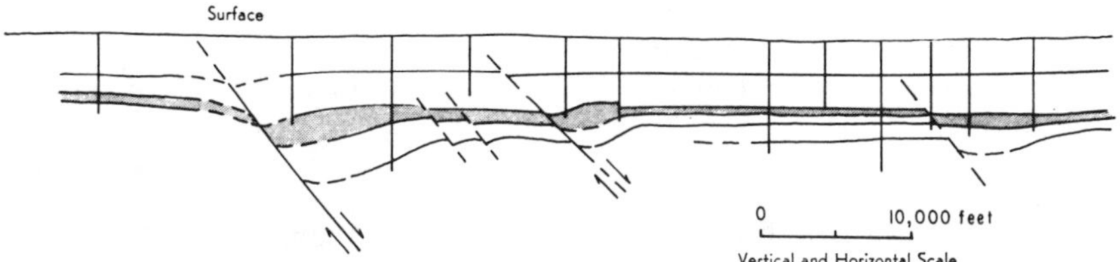

Figure 13. Section across the Rincon y Turria fields, Starr County, Texas, showing the general characteristics of reverse drag on the down-thrown block and "normal" drag adjacent to the fault plane. (*After* Bornhauser, 1958, p. 156)

1956, p. 51). Moreover, the faults with which reverse drag is associated in the Gulf Coast have the same general characteristics as those in the Colorado Plateau, except that the faults in the Gulf Coast have a more gentle dip. The majority of the faults appear to be steepest near the surface and decrease somewhat or flatten at depth (Murray, 1961, p. 163). Displacement varies up to several thousand feet. Many writers have emphasized the fact that sediments are noticeably thicker on the downthrown side of the fault and have interpreted this to be indicative of recurrent movement along the fault during sedimentation (Hardin and Hardin, 1961, p. 242).

Reverse drag along the border faults of the Triassic basins in eastern North America has been reported by Foye(1924), Longwell (1922), and Kline (1960, Ph.D. dissert., Yale Univ.), but details of the flexures in this area are generally obscured. Along the eastern boundary fault in Connecticut, Foye (1922) reports that the regional dip of 10–15 degrees increases rapidly over a distance of ½ mile to 40–50 degrees near the fault line. Cross sections by

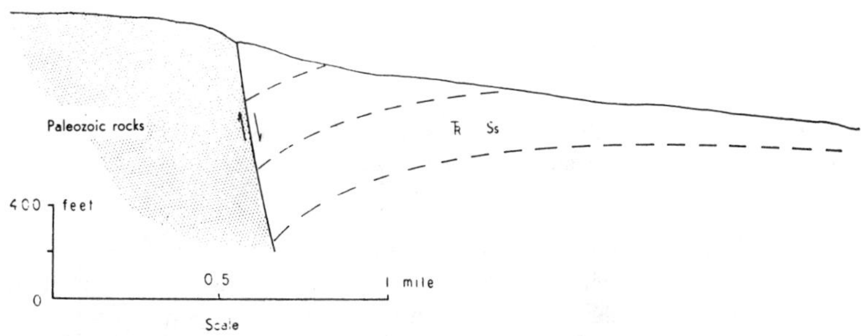

Figure 14. Section across the north border fault of the Triassic basin in Cobequid Bay, Nova Scotia, showing reverse drag in the Triassic sediments. (*After* Klein, 1960 dissert.)

Kline (1960 dissert.) show approximately the same relationship in Nova Scotia (Fig. 14).

Similar flexing has been reported from the Ottawa-Bonnechere graben by Kay (1942), where Lower Paleozoic rocks dip into the fault plane at angles ranging up to 50 degrees (Fig. 15). Paucity of exposures in this area prevents complete delineation of the structure but from the descriptions presented by Kay the flexures

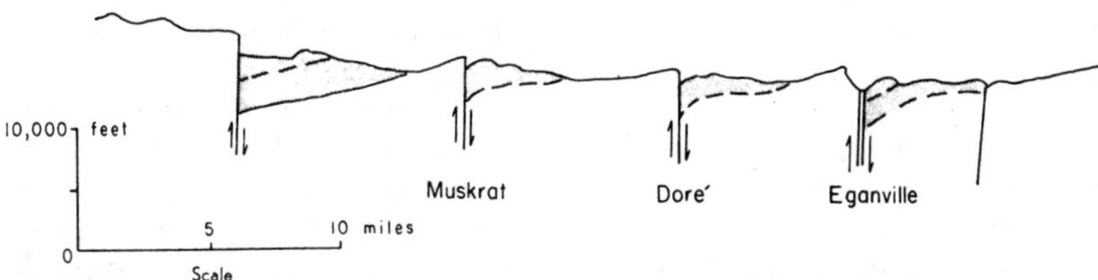

Figure 15. Section across the Ottawa–Bonnechere graben, Canada, showing reverse drag in the Paleozoic sediments adjacent to the major faults. (*After* Kay, 1942, Pl. 2)

adjacent to the Muskrat, Dori, and Eganville faults are similar in most respects to reverse drag in the Gulf Coast and the Colorado Plateau.

Edwards and others (1944, p. 23) describe reverse-drag flexures associated with faults in the Wonthaggi Coalfield, Victoria, Australia. Flexures on both the upthrown and downthrown blocks were observed and referred to as "rise to the downthrow" and "fall to the upthrow." A cross section showing these structures in the Dudley Basin (Fig. 16) indicates that the flexures on the downthrown blocks are remarkably similar to the reverse drag on the Colorado Plateau. Work with the coal seams of the region indicates that this change in dip is a reliable indication of an approaching fault and the size of the flexure is roughly proportional to the fault displacement. Antithetic faults are commonly associated with reverse drag and are developed near the crest of the flexures. In some areas the flexure is brought about by minor movements ranging from a few inches to several feet along a series of parallel antithetic faults. Reference to the cross section (Fig. 16) indicates that the flexures on the upthrown block are limited in extent and size and are apparently not coextensive with the more pronounced flexing on the downthrown block.

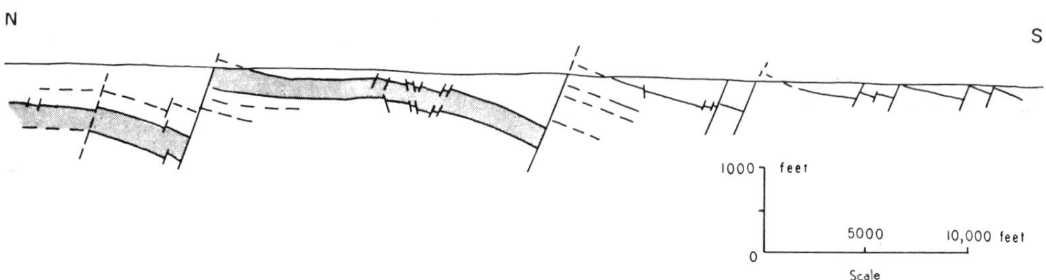

Figure 16. Section across several faults in the Wonthaggi coal basin, Victoria, Australia, showing reverse drag and associated antithetic faults. (*After* Edwards and others, 1944)

It is apparent from Figures 10–16 that the reverse-drag flexures in other parts of the world are remarkably similar to those in the western Colorado Plateau.

SUMMARY

A summary of the significant observations made during this study together with their implications is as follows:

(1) Reverse drag is a common structure associated with most normal faults in the western Colorado Plateau. It is also found in many other regions and is therefore not the result of some abnormal combination of local tectonic events.

(2) Reverse drag is a large asymmetrical fold ranging up to more than 1 mile in width with maximum dips of more than 30 degrees near the fault plane. In the western Colorado Plateau normal drag is common adjacent to the fault plane on both the upthrown and downthrown blocks. The flexure, therefore, does not represent a drag phenomenon.

(3) Reverse drag is essentially coextensive with associated normal faults. In the Colorado Plateau the flexure dies out where the displacement of the fault decreases below 100 feet.

(4) The magnitude of reverse drag is roughly proportional to the amount of displacement on the fault. This relationship is especially clear in the late Cenozoic basalts that have crossed the Hurricane and the Grand Wash faults at various times and have been subsequently displaced by recurrent movement.

(5) Reverse drag has developed with each significant movement along the fault plane.

The relationships mentioned in (3), (4), and (5) strongly suggest that the fault and flexures are genetically related and were formed at the same time by a single process. Each significant movement along the fault plane develops a corresponding reverse-drag flexure.

(6) Reverse-drag flexures that developed on the downthrown block in the Cenozoic basalts without a corresponding flexure forming on the upthrown block indicate that the two flexures are independent.

(7) Flexures referred to as reverse drag on the upthrown block are notably erratic in their distribution, magnitude, and areal extent. Dips are invariably less than those in the reverse-drag flexures on the downthrown block. There is thus no apparent relationship between the "anomalous" flexures on the upthrown block and the fault. In many places the flexures on

the upthrown block can be demonstrated to be the result of Laramide folds cut by Tertiary faults.

(8) Reverse drag may pass both vertically and laterally into antithetic faults. This strongly suggests that the two structural features result from the same tectonic forces.

(9) Extensive outcrops of the actual fault surfaces indicate a decrease in the dip of the fault plane at depth, suggesting that the fault plane is a curved surface.

CONCLUSIONS

The results of this study indicate that reverse-drag flexures are not abnormal structures, but are common features associated with many normal faults in widely separated tectonic provinces. Hypotheses attempting to explain the origin of these structures by some unique characteristic of local areas (salt-ridge hypothesis, sagging due to the withdrawal of magma, compaction, and slipping-plane hypothesis) or by an abnormal combination of local tectonic events (faulted monocline, compression after faulting, and double movement) may apply locally but cannot represent the general case. All available evidence points to the conclusion that reverse drag develops from some fundamental mechanism associated with normal faulting.

Such a mechanism is suggested by the observations made in the Grand Canyon which indicate that the dip of the faults decreases with depth and the fault plane is actually a curved surface. Large exposures of low-angle normal faults in the Basin and Range Province have also been described by Longwell (1945). Sound theoretical support for these observations has been established by McGee (1883), Prandtl (1923), Evison (1959, p. 172–177), and Kanizay (1962, p. 12–16), who have attempted to demonstrate mathematically that the field of slip lines (fault planes) curve. According to their calculations normal faults should be essentially horizontal at great depths and should curve to a vertical attitude at the surface. Normal movement along a curved fault plane would tend to pull the blocks apart as well as displace them vertically so that an incipient gap would form between the upthrown and downthrown blocks. Subsidence or collapse to fill the potential void would generally develop antithetic faults and associated minor grabens (Fig. 17A). If, however, the physical conditions were such that failure occurred by flexing instead of rupture, the strata adjacent to the fault plane would bend down to fill the potential void and form a reverse-drag flexure (Fig. 17B). It is concluded, therefore, that reverse drag represents an alternate response to the same tectonic forces that produce antithetic faults and develops because of curvature of the fault plane at depth.

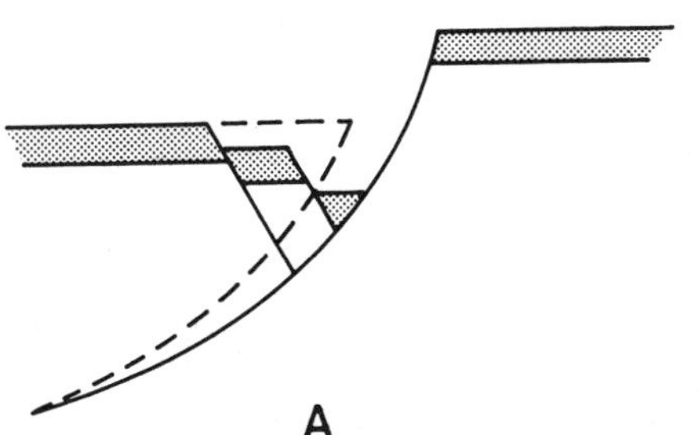

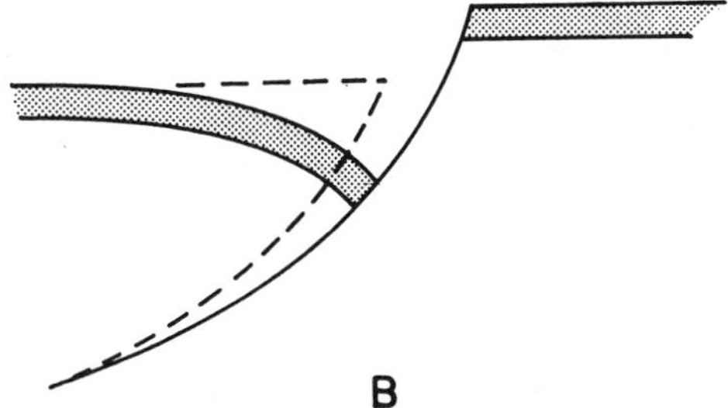

Figure 17. Diagrams illustrating the mechanism by which reverse drag may be produced. Normal movement along a curved fault plane would tend to pull the blocks apart as well as displace them vertically. Subsidence to fill the incipient gap may develop antithetic faults (A) or reverse drag (B).

In many respects this theory is similar to the slipping-plane hypothesis reported from the Gulf Coast (Quarles, 1953), but it does not require special conditions such as a bedding-plane fault. This hypothesis has been proposed by several workers who apparently arrived at their conclusions independently. To the writer's knowledge Lotz (1930) was the first to describe the mechanism, and later Kamen-Kaye (1953) carefully illustrated how this process would explain the greater than regional dips near faults at Las Mercedes, Venezuela.

This hypothesis appears to be in harmony with all the facts observed in the western Colorado Plateau as well as other regions where reverse drag has been reported. It fully explains how reverse drag may develop in many different areas and why the structure has been formed repeatedly by each significant movement along the fault plane. It comports with the geometry of the flexures and the relationship between the magnitude of the fold and the displacement of the faults. Moreover it explains the close association between reverse drag and antithetic faults.

If it is correct, reverse drag is probably much more widespread than has heretofore been realized and would be expected to form wherever normal faulting is produced by tension.

REFERENCES CITED

Benioff, H., 1938, The determination of the extent of faulting with application to the Long Beach earthquake: Seismol. Soc. America Bull., v. 28, p. 77–84

Bornhauser, Max, 1958, Gulf Coast tectonics: Am. Assoc. Petroleum Geologists Bull., v. 42, p. 339–370

Brucks, E. W., 1929, Structure of typical oil fields of the United States: Am. Assoc. Petroleum Geologists, v. 1, p. 261–267

Cloos, H., 1928, Über antithetisch Bewegungen: Geol. Rundschau, v. 19, p. 246–251

Cooke, E. F., 1960, Geologic atlas of Washington County, Utah: Utah Geol. and Mineralog. Survey, Bull. 70, 119 p.

Davis, W. M., 1901, An excursion to the Plateau Province of Utah and Arizona: Harvard Univ. Museum Comp. Zoo. Bull., v. 42, p. 1–50

Dutton, Clarence E., 1882, Tertiary history of the Grand Canyon district: U. S. Geol. Survey, Mon. 2, 264 p.

Edwards, A. B., and others, 1944, The geology of the Wouthaggi Coalfield, Victoria: Proc. Aust. Inst. Min. Met., N. S., no. 134, p. 1–54

Evison, F. E., 1959, On the growth of continents by plastic flow under gravity: Geophys. Jour. Royal Astronomical Soc., v. 3, no. 2, p. 155–189

Foye, Wilbur G., 1924, Abnormal dips near the eastern boundary fault of the Connecticut Triassic: Science, new series, v. 59, p. 240

Gardner, Louis S., 1941, The Hurricane fault in southwestern Utah and northwestern Arizona: Am. Jour. Sci., v. 239, no. 4, p. 258

Gregory, H. E., 1950, Geology and geography of the Zion Park region, Utah and Arizona: U. S. Geol. Survey Prof. Paper 220, 200 p.

Hardin, F. R., and Hardin, G. C., 1961, Contemporaneous normal faults of Gulf Coast and their relation to flexures: Am. Assoc. Petroleum Geologists Bull., v. 45, p. 238–248

Honea, John W., 1956, Sam Fordyce-Vanderbilt fault system of southwest Texas: Gulf Coast Assoc. Geol. Soc. Trans., v. VI., p. 51

Huntington, E., and Goldthwait, J. W., 1904, The Hurricane fault in the Toqueville district, Utah: Harvard Univ. Museum Comp. Zoo. Bull., v. 42, p. 200–259

Johnson, D. W., 1909, A geological excursion in the Grand Canyon district: Boston Soc. Nat. History Proc., v. 34, p. 135–162

Kamen-Kaye, Maurice, 1953, Curvature of low-angle faults at Las Mercedes, Venezuela: Am. Assoc. Petroleum Geologists Bull., v. 37, p. 2178–2195

Kanizay, S. P., 1962, Mohr's theory of strength and Prandtl's compression cell in relation to vertical tectonics: U. S. Geol. Survey Prof. Paper 414-B, p. 1–16

Kay, G. Marshall, 1942, Ottawa-Bonnechere graben and Lake Ontario homocline: Geol. Soc. America Bull., v. 53, p. 585–646

Koons, E. D., 1945, Geology of the Uinkaret Plateau, northern Arizona: Geol. Soc. America Bull., v. 56, p. 151–180

Longwell, C. A., 1922, Notes on the structure of the Triassic rocks in southern Connecticut: Am. Jour. Sci., v. 204, p. 223–236

—— 1945, Low-angle normal faults in the Basin and Range Province: Am. Geophys. Union Trans., v. 24, p. 107–118

Lotze, F., 1930, Der Westrand des Leinetalgrabens Zwischen Hardegsen und Moringen: Abh. d. Preuss. Geol. L. A., N. F., no. 116, p. 195–237

McGee, W. J., 1883, On the origin of hade of normal faults: Am. Jour. Sci., 3rd ser., v. 26, p. 294–298

Murray, G. E., 1961, Geology of the Atlantic and Gulf Coast Province of North America: New York, Harper & Bros., 692 p.

Noble, L. F., 1914, The Shinumo quadrangle, Grand Canyon district, Arizona: U. S. Geol. Survey Bull. 549, no. 548–553, p. 76

Perkins, Hunt, 1961, Fault closure-type fields, southeastern Louisiana: Gulf. Coast Assoc. Geol. Soc. Trans., v. II, p. 177–196

Powell, J. W., 1875, Exploration of the Colorado River of the West: Washington, D. C., Govt. Printing Office, 291 p.

Prandtl, L., 1923, Spannungsverteilung in plastischen Körpen: Delft, 1st Internat. Cong. on Applied Mechanics Proc., p. 41–54

Quarles, Miller, Jr., 1953, Salt-ridge hypothesis on origin of Texas Gulf Coast type of faulting: Am. Assoc. Petroleum Geologists Bull., v. 37, p. 489–508

Reid, H. F., 1911, The elastic rebound theory of earthquakes: Univ. Calif. Dept. Geol., Bull. 6, p. 413–444

Rogers, G. S., and Lee, Wallace, 1923, Geology of the Tullock Creek Coal Field: U. S. Geol. Survey Bull. 749, 181 p.

Russell, Wm. L., 1957, Faulting and superficial structures in east-central Texas: Gulf Coast Assoc. Geol. Soc. Trans., v. VII, p. 65–71

Sears, Julian D., 1925, Geology and coal resources of the Gallup-Zuni Basin, New Mexico: U. S. Geol. Survey Bull. 767, 52 p.

Shirley, J. W., 1960, Structure and stratigraphy of Rayne Field: Gulf Coast Assoc. Geol. Soc. Trans., v. X, p. 78–85

Spooner, W. C., 1929, Homer Oil Field, Claiborne Parish, Louisiana, structure of typical American oil fields: Am. Assoc. Petroleum Geologists Bull., v. 1, p. 196–228

Storm, L. W., 1945, Résumé of facts and opinions on sedimentation in Gulf Coast region of Texas and Louisiana: Am. Assoc. Petroleum Geologists Bull., v. 29, p. 1304–1335

Strahler, Arthur N., 1948, Geomorphology and structure of the West Kaibab fault zone and Kaibab Plateau, Arizona: Geol. Soc. America Bull., v. 59, p. 513–640

Walters, J. E., 1959, Effect of structural movement on sedimentation in the Pleasant-Francitas area, Matagorda and Jackson counties, Texas: Gulf Coast Assoc. Geol. Soc. Trans., v. IX, p. 51–58

Manuscript Received by the Society July 13, 1964

The American Association of Petroleum Geologists Bulletin
V. 57, No. 5 (May 1973), P. 878-886, 10 Figs.

Pressured Shale and Related Sediment Deformation: Mechanism for Development of Regional Contemporaneous Faults[1]

CLEMONT H. BRUCE[2]
Houston, Texas 77046

Abstract Regional contemporaneous faults of the Texas coastal area are formed on the seaward flanks of deeply buried linear shale masses characterized by low bulk density and high fluid pressure. From seismic data, these masses, commonly tens of miles in length, have been observed to range in size up to 25 mi in width and 10,000 ft vertically. These features, aligned subparallel with the coast, represent residual masses of undercompacted sediment between sandstone-shale depoaxes in which greater compaction has occurred. Most regional contemporaneous fault systems in the Texas coastal area consist of comparatively simple down-to-basin faults that formed during times of shoreline regression, when periods of fault development were relatively short. In cross-sectional view, faults in these systems flatten and converge at depth to planes related to fluid pressure and form the seaward flanks of underlying shale masses. Data indicate that faults formed during regressive phases of deposition were developed primarily as the result of differential compaction of adjacent sedimentary masses. These faults die out at depth near the depoaxes of the sandstone-shale sections.

Where subsidence exceeded the rate of deposition, gravitational faults developed where basinward sea-floor inclination was established in the area of deposition. Some of these faults became bedding-plane type when the inclination of basinward-dipping beds equaled the critical slope angle for gravitational slide. Fault patterns developed in this manner are comparatively complex and consist of one or more gravitational faults with numerous antithetic faults and related rotational blocks.

Postdepositional faults are common on the landward flanks of deeply buried linear shale masses. Many of these faults dip seaward and intersect the underlying low-density shale at relatively steep angles.

Conclusions derived from these observations support the concept of regional contemporaneous fault development through sedimentary processes where thick masses of shale are present and where deep-seated tectonic effects are minimal.

Introduction

Regional contemporaneous faults are structural features that develop on the shoreward sides of major depocenters during deposition. They are characterized by thickening of sediment on the basinward, downthrown side, where rollover (reverse drag) is common. Carver (1968) enumerated possible causes of regional contemporaneous faults, the most significant of which are basement tectonics, deep salt or shale movement, slump across flexures, slump at the shelf edge, differential compaction, response to crustal loading, or combinations of these factors.

In southern Texas, where thick masses of Tertiary shale are present and where salt diapirism is not dominant, extensive contemporaneous fault systems are developed that extend for several miles subparalleled with the coast. Data from deep wells that have encountered low-density (high-pressure) shale, together with information gained through improved seismic methods, now provide information which indicates shale to be a dominant factor in the formation of these fault systems. A diagrammatic cross section (Fig. 1) illustrates how Tertiary sedimentary rocks are related to the underlying pre-Tertiary section in southern Texas. In this area, the Tertiary section is composed primarily of sandstone in the west, which grades into shale in a seaward direction. Contemporaneous faults, indicated by arrows, are shown to be developed where relatively abrupt changes from sandstone to shale occur. The heavy black line extended across this section to separate Tertiary sandstone above from the underlying shale illustrates a "ridge and valley" effect that becomes more pronounced in a seaward direction. These "ridges" represent residual masses of undercompacted shale between sandstone-shale depoaxes along which greater compaction has occurred. These linear shale bodies, commonly referred to as "shale masses," are the subject of this discussion.

Shale Masses

The term "shale mass" as used in this paper is the same as that defined by Musgrave and Hicks (1968), in that it is a body of shale that is not less than 500 ft in the smallest dimension. The shale

[1] Manuscript received, June 9, 1972; accepted, October 21, 1972.

[2] Mobil Oil Corporation.

The writer thanks many co-workers within Mobil Oil Corporation for helpful advice and constructive criticism. Special thanks are extended to D. R. Palmore, J. H. Parsley, and R. W. Aldrich for seismic interpretation. I thank the Gulf Coast Association of Geological Societies and its Past-President, Donald R. Boyd, for permission to reprint this revised paper from GCAGS Transactions, Volume 22.

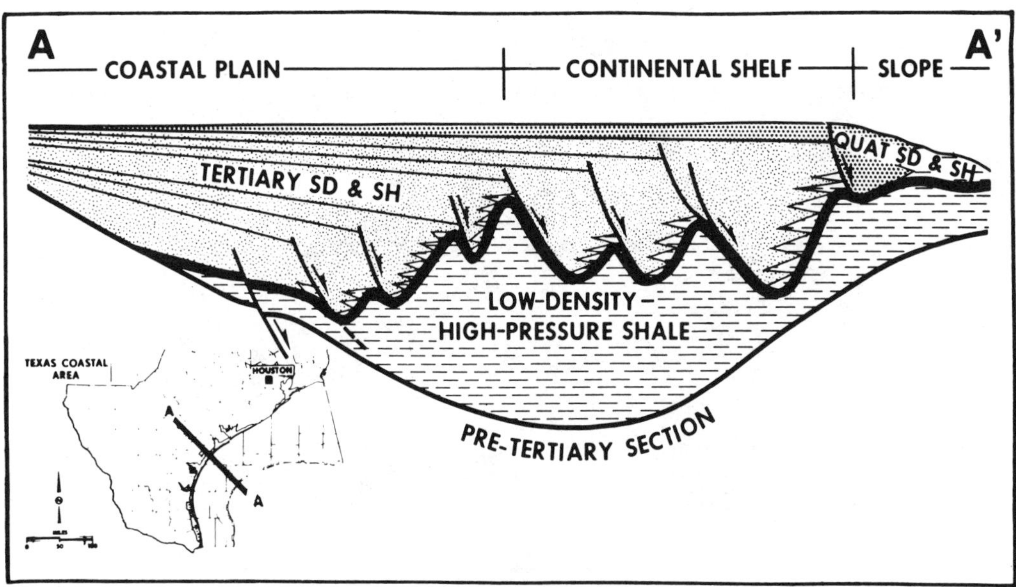

FIG. 1—Diagrammatic cross section across Texas part of northern Gulf of Mexico basin.

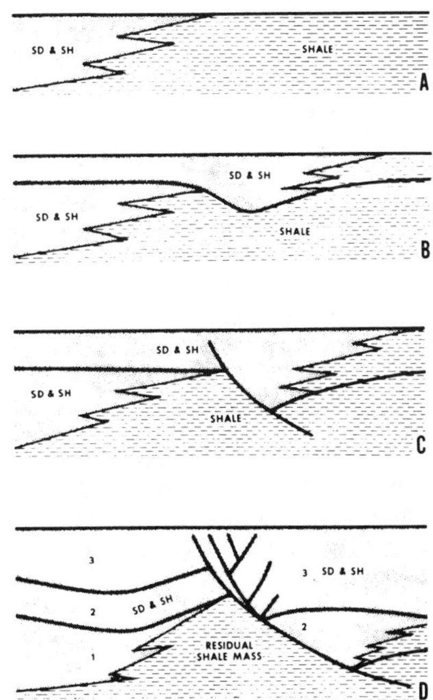

FIG. 2—Diagrammatic illustration showing four stages in development of residual shale mass.

can be in the original state of deposition or may be in any stage of deformation. The simplest forms of shale masses are flat lying and are developed by deposition (Fig. 2A). However, because of the instability of newly deposited clay materials, shale masses of this simple form are uncommon. Initial development of residual shale masses of regional extent occurs where regressive deposition of sandy sediment on a plastic clay surface results in compaction and downwarping of the unconsolidated clay under the heavier sandy section (Fig. 2B). Through continued deposition (Fig. 2C), growth faults develop and remain active as long as the depositional axis is maintained along the same line. Figure 2D shows mature development of a residual shale mass and illustrates the relation of the mass to facies on the landward side and to structure on the seaward side. These relations also are illustrated by a line of seismic data (Fig. 3) that crosses a major contemporaneous fault system where the residual shale mass is more than 10,000 ft vertically and 10 mi across.

The principal process involved in the development of residual shale masses similar to those found in southern Texas is considered to be differential compaction resulting from differential loading and shale diagenesis, both of which con-

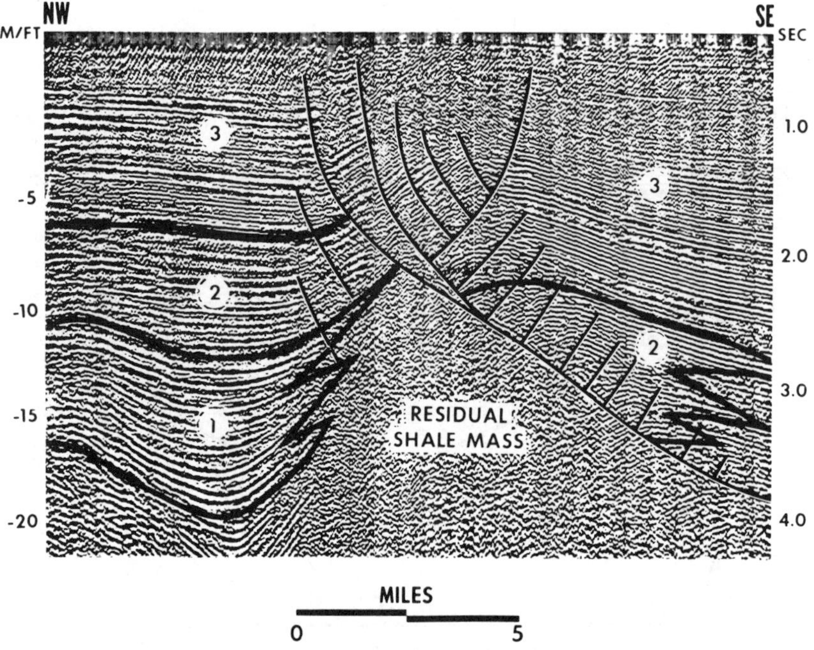

Fig. 3—Seismic illustration of residual shale mass. Numbers on either side of fault system indicate depositional sequence of sand and shale units.

trol subsurface fluid movement. In the initial stages of residual shale-mass development (Fig. 2A, B), water is free to escape from both the shale and the adjacent sandstone-shale sections. At first the water loss is greater from the shales, because of the relatively higher water content. However, as subsidence continues, water loss from the shale decreases progressively, as a result of decreasing permeability, until a critical depth is reached where orderly expulsion of water from the shale section is restricted and abnormally high pore pressure is developed within the shale mass. If subsidence continues to depths where fluid pressure within the shale approaches total overburden pressure, compaction ceases.

Some fault patterns observed on seismic records indicate that masses of this type tend to expand. It is probable, although the subject is not considered in this study, that under these conditions the shale may become diapiric locally along the linear shale mass. During the process of abnormal pressure development within the shale mass, the flanking sections of interbedded sandstones and shales continue to compact normally with water loss through the permeable sandstone layers. Water expulsion through the sandstone layers continues as long as the interlayered permeable strata are in communication with the surface.

Additional compaction occurs within the flanking sediments when the critical temperature-pressure level is reached at which montmorillonite is altered to illite. During this process, large volumes of interlayer water are transferred from the montmorillonite into the pore-water system of the host sedimentary section, as suggested by Powers (1967) and Burst (1969). This released interlayer water also can escape through permeable sandstone layers present within the sedimentary section. The amount of water in motion during this stage of clay-mineral diagenesis is considered by Burst (1969) to be 10-15 percent of the compacted bulk volume of the shale involved. This second-stage water loss does not affect the pressured shale section because permeable strata are not present within the mass to permit the escape of fluids. The extended period of water loss from the interbedded sandstone-shale section reduces the thickness of the flanking sediments, leaving them draped on the landward side of the pressured shale mass, and further accentuates the faulting that has developed on the seaward side (Fig. 2D, 3).

Fault Mechanisms

The processes responsible for development of regional contemporaneous faults similar to those

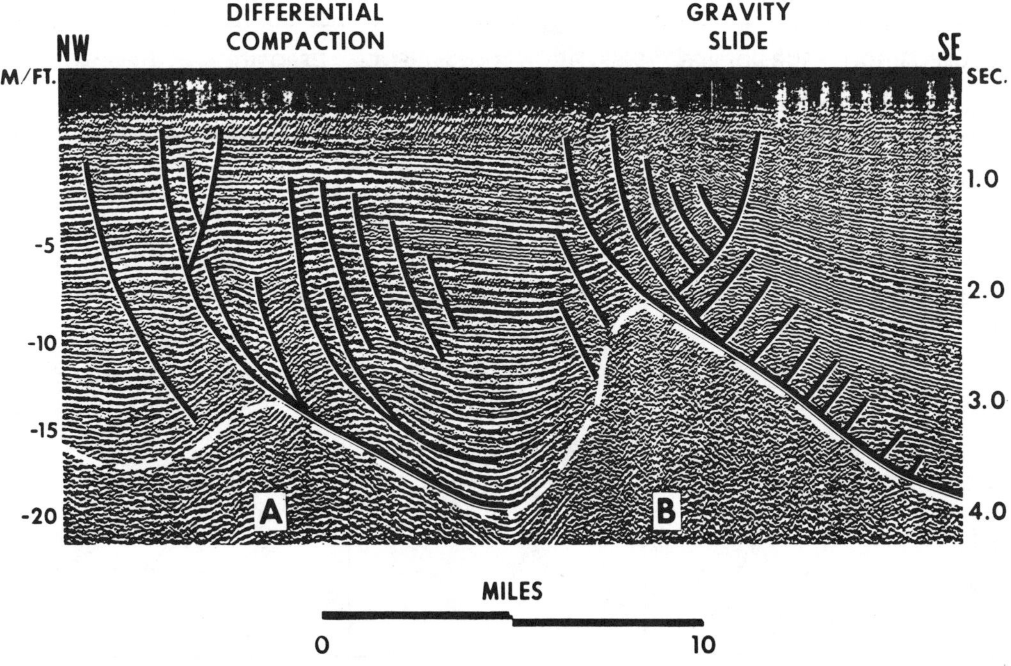

FIG. 4—Seismic illustration showing differences between fault systems formed by differential compaction and gravity slide. Dashed white line shows configuration of shale masses.

present in southern Texas are considered to be differential compaction and gravity slide. Examples are shown on a line of seismic data (Fig. 4) which includes the fault system shown in Figure 3 and a second fault system located approximately 10 mi northwest. The heavy dashed white line extending across this seismic section illustrates how these two fault systems are interpreted to be related to underlying shale masses, labeled "A" and "B." Although fault patterns within these systems are different, faults in both systems flatten and converge at depth to form the southeast (seaward) flanks of the underlying masses. Progressive flattening at depth of major basinward-dipping faults is common in Tertiary sedimentary rocks of the Texas coastal area, regardless of mechanism, and is considered to be related directly to progressive increases in subsurface fluid pressure with depth.

Where gravitational slide faults are developed, the mechanism is considered to follow in general the fluid-pressure hypothesis presented by Hubbert and Rubey (1959). In this hypothesis they propose that development of high fluid pressure greatly reduces internal friction, thus facilitating the formation of low-angle gravitational faults. Faulting associated with shale mass "B" is a typical example of faults formed through gravitational sliding. Fault systems formed in this manner are characterized by one or more major faults that are downthrown basinward and many antithetic (adjustment) faults. Together, these faults produce extremely complex systems.

Faults formed by differential compaction, in contrast, consist predominantly of normal faults that are downthrown basinward. Carver (1968) considered differential compaction faults to represent simple shear failure in sediments that became more compacted on the downthrown side. Faults formed in this manner flatten at depth and die out at near the depositional axis of the adjacent syncline. The fault system associated with shale mass "A" illustrates, in cross-sectional view, faults which die out at depth near the synclinal axis between shale masses "A" and "B." The relatively short distance between these masses confined the sandstone-shale section between them in such a manner that major faulting could occur only by dehydration and compaction. Across several shallow faults, present between the depths of 5,000 and 10,000 ft near the center of the syncline, no displacement is observed below the depth of 10,000 ft. Faults of this type are best explained by differential water loss from adjacent blocks during the process of compaction.

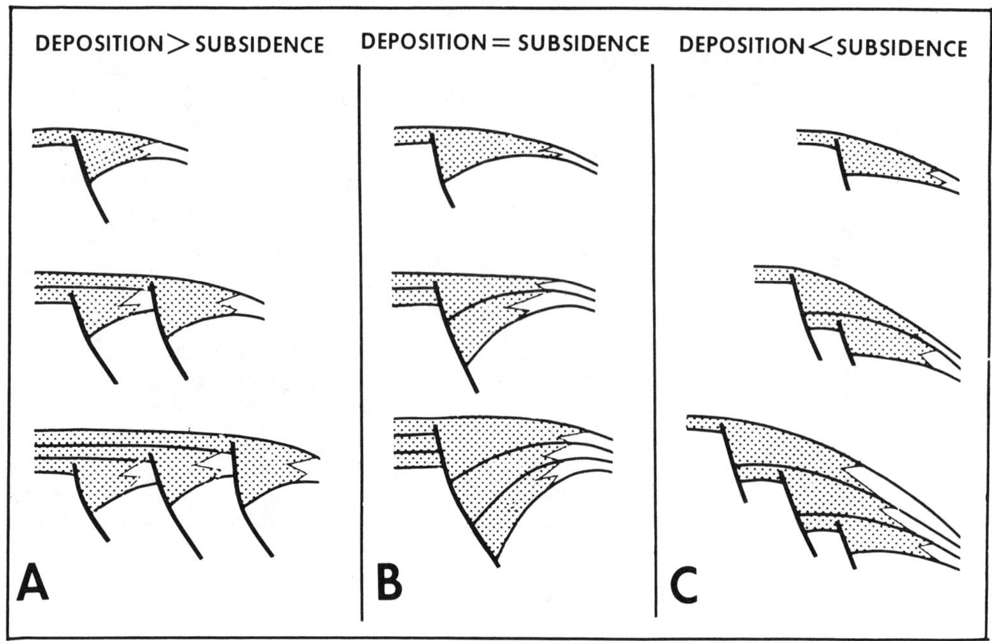

Fig. 5—Diagrammatic illustration showing development of three types of contemporaneous fault systems.

Fault Types

Tertiary deformation in southern Texas was controlled primarily by sediment overburden distribution, rather than deep-seated tectonic activity. Three basic types of regional contemporaneous faults are recognized (Fig. 5), with differentiation based on rates of deposition of sandy sediment upon unconsolidated clay surfaces. Two of these types are considered to be associated with sea floors which were relatively flat at time of deposition, and the third appears to be formed in areas of slope environments where sea floor subsidence exceeded the rate of deposition.

The first example (Fig. 5A) represents faults formed during a regressive sequence of deposition (progradation locally), when the amount of sediment available for deposition was greater than the space available for accumulation. Under these conditions each successive depoaxis was formed seaward from that of the adjacent underlying unit. Antiregional dip, developed adjacent to the downthrown sides of these faults, varies in relation to the amount of sediment deposited. In areas where still-stand depositional conditions prevailed, the rate of faulting was sufficient to accommodate all incoming sediment (Fig. 5B). In these areas, strong antiregional dip developed that increased with depth and time. Contemporaneous faults, formed during still-stand and regressive phases of deposition, are common in southern Texas and are considered to have developed primarily through differential compaction associated with relatively flat sea floors.

Faults formed during transgressive phases of deposition are present in southern Texas; however, they are less common than the other two forms. Where subsidence exceeded the rate of deposition (Fig. 5C), the sea floor is considered to have been inclined basinward at an angle related to the rate of subsidence. The primary cause of sea-floor subsidence and tilting was not dependent on differential compaction and differential loading, as described for faults formed during regressive and still-stand phases of deposition, but was controlled by forces below or outside of the area of deposition. These forces may have been related to either salt movement or basement tectonics. Other manifestations of contemporaneous faulting can be explained when sea-floor inclination and basinward formational dips are considered with rates of deposition. The most significant of these are gravity-slide faults, many of which become bedding-plane types at depth.

Faults of bedding-plane type begin with normal displacement (Fig. 6A) and become bedding-plane faults where, at depth, the dip of the fault plane becomes the same as the dip of the basinward tilted beds (Fig. 6C). After bedding-plane characteristics have been developed, all major subsequent displacement is along one or more planes parallel with the bedding. Quarles (1953) has shown that faulting of this type produces a hypothetical gap in the normal part of the fault zone above the point where the fault becomes bedding-plane type. This hypothetical gap, which is relatively wide in the upper section, does not develop in nature but is filled by collapsed material from both sides of the fault zone. An interpreted seismic section (Fig. 7) illustrates in cross-sectional view a complex fault system that became bedding-plane type. At this location the collapse zone is approximately 7 mi wide near the depth of 5,000 ft; however, at greater depths, the width of the fault zone decreases progressively, reaching the point of nonrecognition near the 15,000 ft level. Reconstructed sections (Fig. 8) prepared from these data indicate the presence of three types of faulting which account for the complicated nature of the system. A phase of faulting associated with still-stand deposition (Fig. 8A) was followed by faulting that formed during a transgressive sequence of deposition (Fig. 8B). During later deposition (Fig. 8C), sea-floor inclination and abnormal subsurface fluid pressure were sufficient to become factors in fault development. All subsequent major down-to-basin faults are shown to flatten and to converge at depth into faults of gravity-slide type. Figure 8D duplicates present structural conditions shown in Figure 7. These illustrations demonstrate, in part, why faults formed during transgressive phases of deposition are relatively difficult to recognize when compared with contemporaneous faults of the other types. Seismic and well data indicate that faults formed during periods of transgressive deposition generally flatten and converge at relatively shallow depths, and that later displacement results in development of other faults which complicate preexisting structure and make recognition of mechanisms difficult.

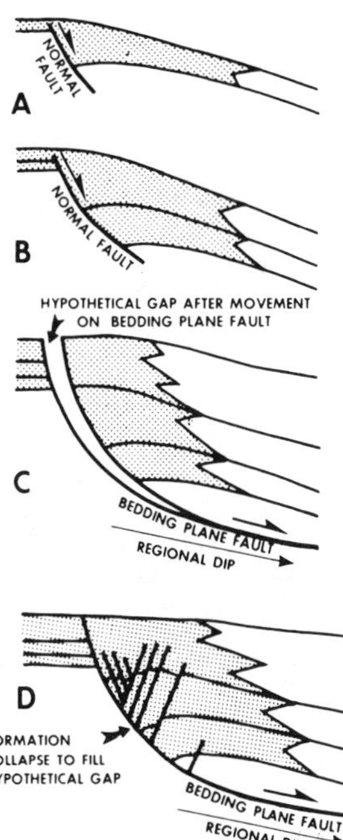

Fig. 6—Diagrammatic illustration showing development of bedding-plane fault.

Faults formed on the landward sides of linear shale masses are usually postdepositional and have relatively small amounts of displacement. Generally, these faults dip basinward and show little indication of flattening at depth. They appear to enter the low-density shale at relatively steep angles. Differences in fault types, relative to landward or seaward locations on linear shale masses, are illustrated on a line of seismic data (Fig. 9) that covers a distance of approximately 30 mi and crosses a broad shale mass with a well-developed contemporaneous fault system. The configuration of the low-density shale mass, as interpreted from seismic, gravimetric, and well data, is represented on the section by a heavy dashed white line. These data indicate a mass with dimensions greater than 10,000 ft vertically and 25 mi across. Contemporaneous faulting on the southeast (seaward) side of the mass is considered to have formed during a phase of still-stand deposition with considerable thickening of sediments on the downthrown side. The angle of faulting is relatively steep above the crest of the underlying shale mass, being 50-60°, but at depth the dip of the fault plane flattens to approximately 10° where it coincides with the seaward slope of the low-density (high-pressure) shale surface. This faulting cannot be traced beyond the adjacent synclinal axis, which is shown to be at a depth of approximately 20,000 ft. On the northwest (landward) part of the shale mass the faults are mostly postdepositional and have relatively small amounts of displacement. They show little indication of flattening at depth, but appear to penetrate the low-density shale at nearly right angles. Fault patterns observed here are common to other fault systems in southern Texas where faulting occurred in regressive and still-stand phases of deposition.

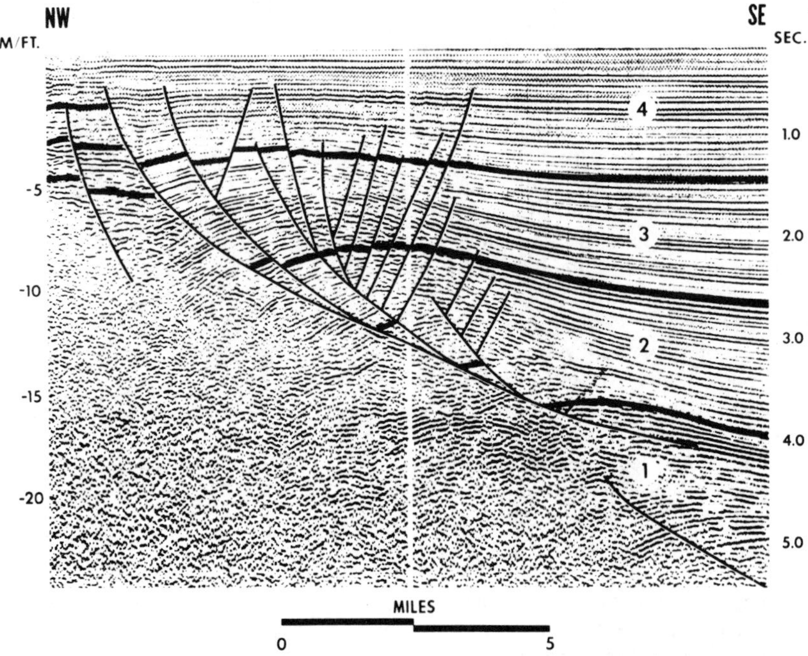

FIG. 7—Seismic illustration of combination differential compaction and bedding-plane fault system. (AQUAPULSE-Courtesy Western Geophysical Company.)

Fluid Pressure-Fault Angle Relation

Flattening at depth is normal for contemporaneous faults in the Texas coastal area. Subsurface data indicate fault flattening to be coincident with progressive increases in fluid pressure with depth. An example of the fluid pressure-fault angle relation is illustrated by a line of seismic data (Fig. 10) that crosses a major contemporaneous fault system where well information is sufficient to establish the relation down to 20,000 ft. The normal vertical pressure gradient for the Gulf Coast Tertiary sedimentary section is considered to be 1.0 psi/ft of depth. The two pressure components are 0.465 psi/ft for fluid and 0.535 psi/ft for sediment. Well data indicate fluid pressure along the line of this section to be normal down to approximately 5,000 ft. The bottomhole fluid-pressure gradients, as determined from drilling mud weights at total depth, are recorded for each of the four tests shown on the section. These data indicate that, near the depth of 15,000 ft, the fluid pressure/overburden ratio increases to more than 0.90 psi/ft. It can also be observed that the angle of faulting reflects this pressure increase, in that a progressive change in dip occurs from 60° in the upper normally pressured section to 15° where fluid pressure reaches the higher value. Similar fluid pressure-fault angle

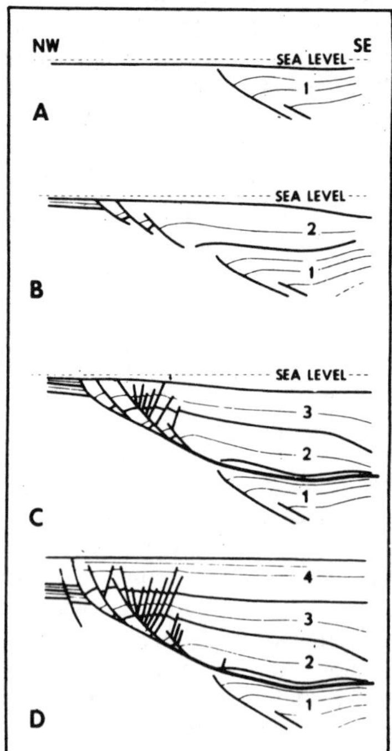

FIG. 8—Diagrammatic illustration showing development of combination differential compaction and bedding-plane fault system.

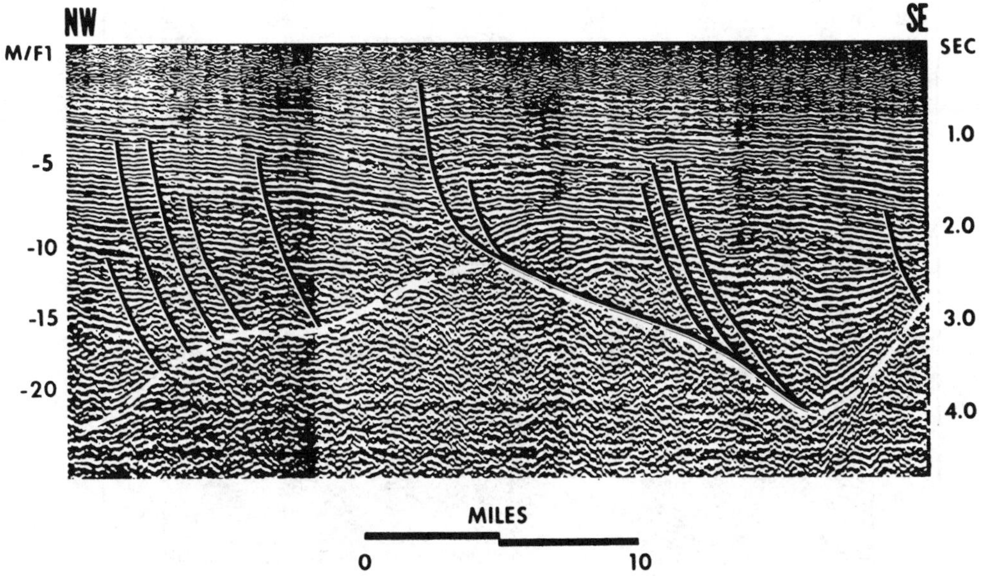

FIG. 9—Seismic illustration showing differences in faults on landward and seaward sides of shale mass. Dashed white line shows configuration of shale mass.

relations have been observed along other contemporaneous fault systems in Texas. The depth at which flattening occurs, and the degree of flattening attained, vary between areas depending on the physical and chemical properties of the shale involved. The principal factors controlling fluid pressure are the number and thickness of sandstone layers in communication with the surface and the clay-mineral composition of the shales.

SUMMARY AND CONCLUSIONS

In southern Texas, where salt diapirism is minimal, seismic data and well information indicate low-density (high-pressure) shale to be a dominant factor in the formation of regional contemporaneous fault systems. Observations concerning fault development in this area, where thick shale is dominant, are as follows:

1. Regional contemporaneous fault systems are formed on the seaward flanks of underlying shale masses, where all down-to-basin faults flatten and converge at depth. These fault systems may be formed either by differential compaction or gravitational sliding.

2. Faults formed in sediments deposited on the landward flanks of underlying shale masses are primarily postdepositional. These faults, which involve little or no sedimentary thickening, do not converge at depth but intersect the abnormally pressured shale at relatively steep angles.

3. Regional contemporaneous faults associated with regressive and still-stand phases of deposition were the dominant types formed during Tertiary time. Subsurface data indicate that fault systems formed under these conditions developed during relatively short periods of time. These fault systems, with comparatively simple patterns, appear to have formed primarily by differential compaction with some associated gravity adjustments.

4. Complex fault systems are found where gravitational sliding occurred. Faults formed in this manner are in areas of rapid subsidence, where sea-floor inclination was relatively steep (slope environments). Differential compaction also occurred within these systems.

Conclusions derived from these observations support the concept of the establishment of regional contemporaneous faulting through sedimentary processes. The physical and chemical properties of Gulf Coast Tertiary shale were such that abnormal fluid-pressure development occurred within the shale at relatively shallow depths. During periods of regressive deposition, linear masses of abnormally pressured shale developed between depoaxes of sandy sediments in which greater compaction occurred. The role of shale in

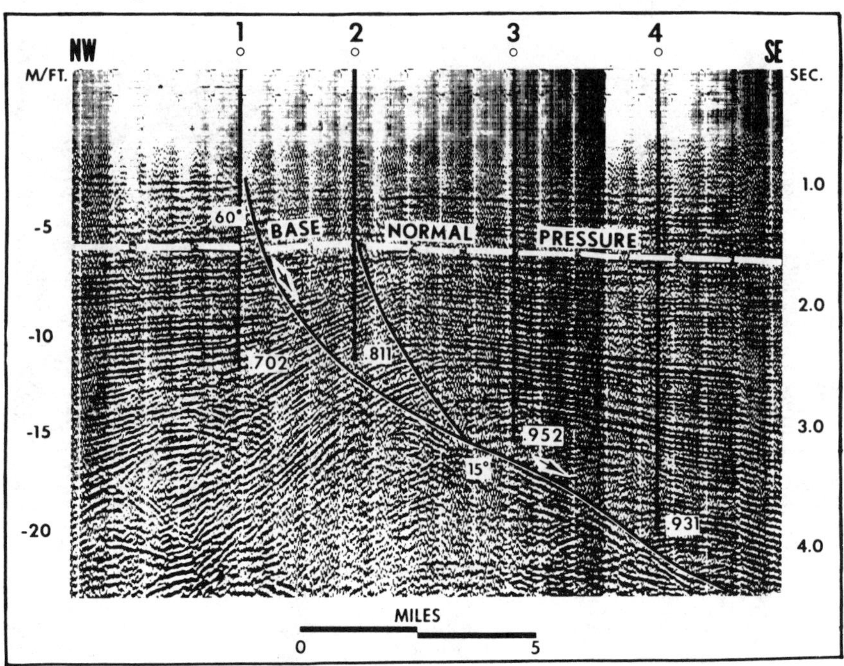

Fig. 10—Seismic illustration with well locations showing fluid pressure/overburden ratio at total depth adjacent to fault plane.

the development of regional contemporaneous faults formed in this manner is generally passive, in that structural uplift is minimal. These mechanisms for fault origin and growth satisfy the observed characteristics of regional contemporaneous fault systems found in the Texas coastal area, where thick masses of shale overlie the pre-Tertiary section and where salt diapirism is not dominant.

References Cited

Bruce, C. H., 1972, Pressured shale and related sediment deformation: a mechanism for development of regional contemporaneous faults: Gulf Coast Assoc. Geol. Socs. Trans., v. 22, p. 23-31.

Burst, J. F., 1969, Diagenesis of Gulf Coast clayey sediments and its possible relation to petroleum migration: Am. Assoc. Petroleum Geologists Bull., v. 53, no. 1, p. 73-93.

Carver, R. E., 1968, Differential compaction as a cause of regional contemporaneous faults: Am. Assoc. Petroleum Geologists Bull., v. 52, no. 3, p. 414-419.

Hubbert, M. K., and W. W. Rubey, 1959, Role of fluid pressure in mechanics of overthrust faulting. 1. Mechanics of fluid-filled porous solids and its application to overthrust faulting: Geol. Soc. America Bull., v. 70, no. 2, p. 115-166.

Musgrave, A. W., and W. G. Hicks, 1968, Outlining shale masses by geophysical methods, *in* Diapirism and diapirs: Am. Assoc. Petroleum Geologists Mem. 8, p. 122-136.

Powers, M. C., 1967, Fluid-release mechanisms in compacting marine mudrocks and their importance in oil exploration: Am. Assoc. Petroleum Geologists Bull., v. 51, no. 7, p. 1240-1254.

Quarles, M. W., Jr., 1953, Salt-ridge hypothesis on origin of Texas Gulf Coast type of faulting: Am. Assoc. Petroleum Geologists Bull., v. 37, no. 3, p. 489-508.

Compressional Deformation

BULLETIN OF THE AMERICAN ASSOCIATION OF PETROLEUM GEOLOGISTS
VOL. 18, NO. 12 (DECEMBER, 1934), PP. 1584-1596, 9 FIGS.

MECHANICS OF LOW-ANGLE OVERTHRUST FAULTING AS ILLUSTRATED BY CUMBERLAND THRUST BLOCK, VIRGINIA, KENTUCKY, AND TENNESSEE[1]

JOHN L. RICH[2]
Cincinnati, Ohio

ABSTRACT

The structural relations of the Cumberland overthrust block are such as would occur if gliding on the thrust plane took place parallel with the bedding along certain shale beds in such a way that the thrust plane followed a lower shale bed for some distance, then sheared diagonally up across the intervening beds to a higher shale, followed that for several miles, and again sheared across the bedding to the surface.

Reasons are given for the belief that subsidiary faults and folds within the block are superficial and do not extend below the thrust plane. This possibility should be borne in mind when exploration of such structures for oil or gas is contemplated.

Study of the Cumberland block throws new light on the broader problems of the nature of folding and faulting in the sedimentary rocks bordering great mountain ranges and on the function of friction in setting limits to the distance through which overthrust blocks can be moved.

INTRODUCTION

It is now generally recognized that low-angle overthrust faults are common and that the movement on some of them has amounted to several miles, but in most cases the manner in which such thrusts are produced, the limits of movement set by friction, and the function of the nature and attitude of the rocks in guiding and controlling the faulting have not been clearly understood.

A clue to the solution of one such problem has been revealed by a study of the Cumberland fault block, a rectangular overthrust mass about 125 miles long and 25–30 miles wide, located at the common corners of the states of Virginia, Kentucky, and Tennessee. This area, which marks the outer limit of the Appalachian overthrusting in that district, shows the essential mechanics of the thrust faulting because the process has not been carried far enough to obscure the record.

CUMBERLAND FAULT BLOCK

The Cumberland overthrust block has been described in consider-

[1] Presented before the Association at the Houston meeting, March 24, 1933. Manuscript received, May 4, 1934.

[2] Department of geology, University of Cincinnati.

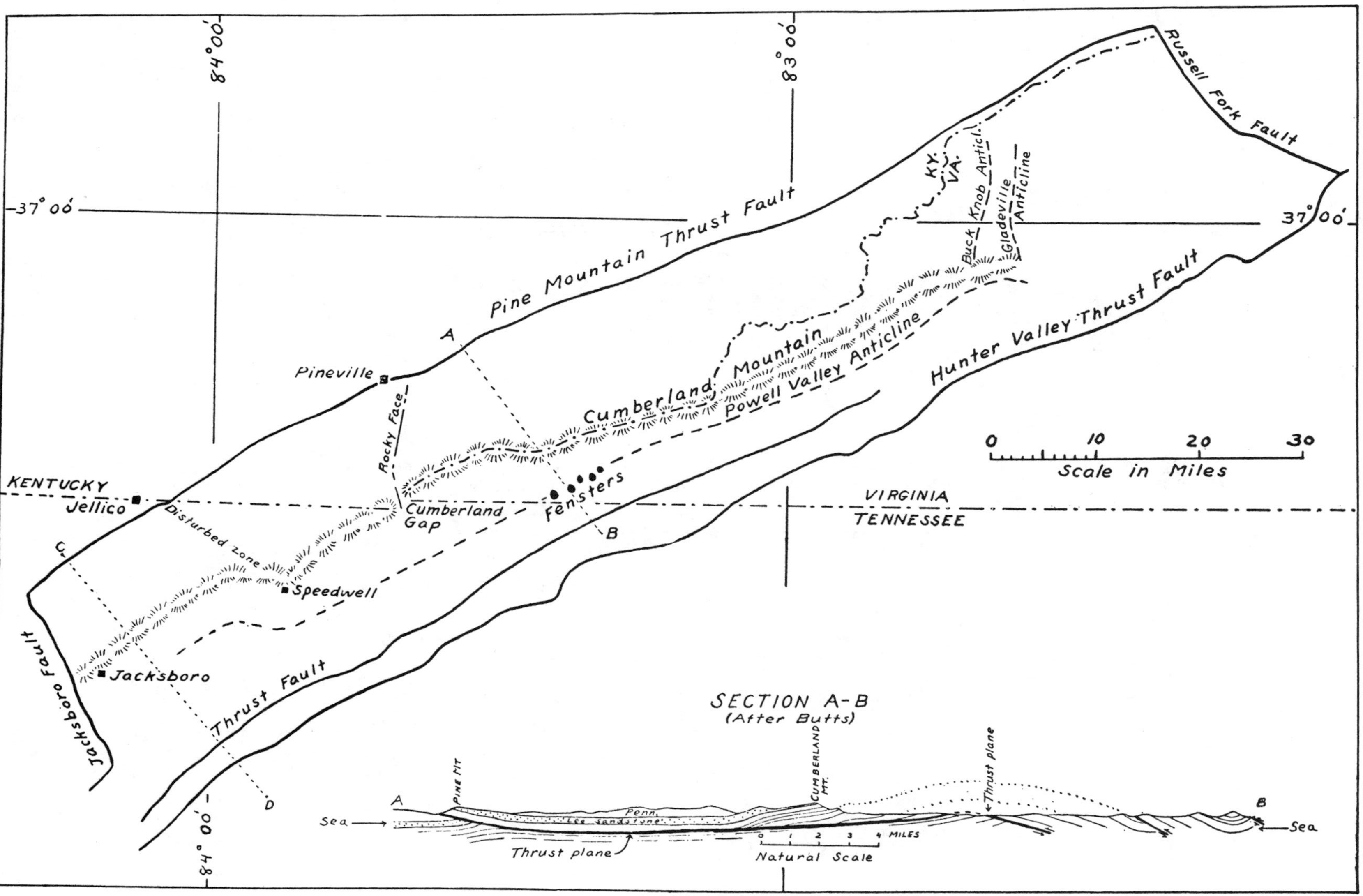

FIG. 1.—Cumberland overthrust block.

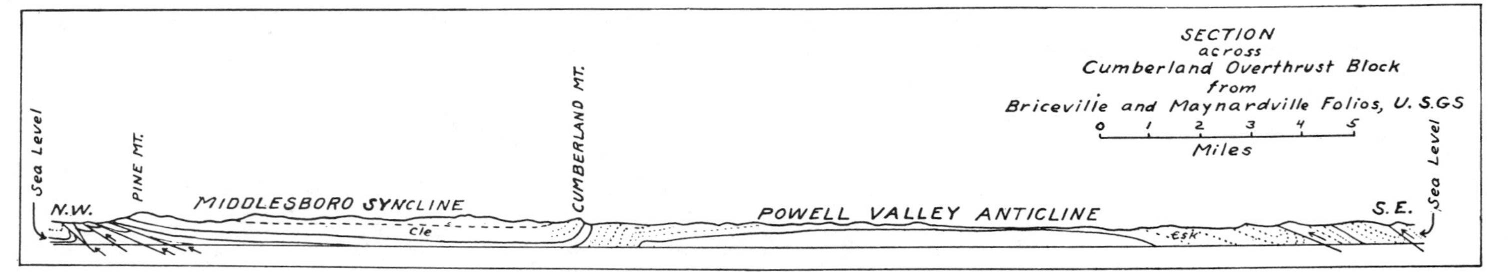

FIG. 2.—Topographic and geologic cross section of western part of Cumberland overthrust block along line *CD* (Fig. 1), showing flat-topped Powell Valley anticline and flat-bottomed Middlesboro syncline; also showing earlier, now disproved, conception of relations of Pine Mountain thrust fault. From Briceville and Maynardville folios, United States Geological Survey.

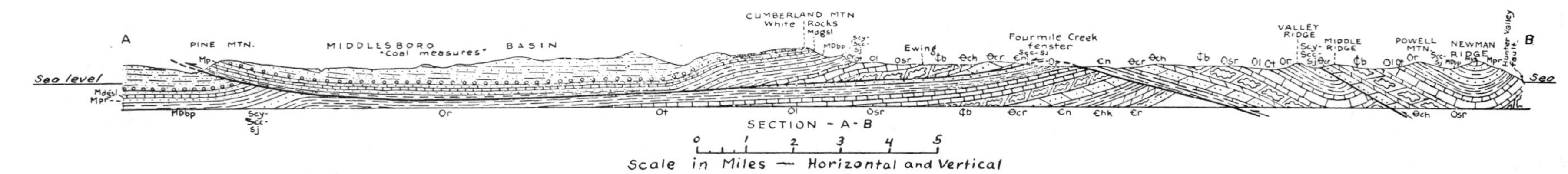

FIG. 3.—Cross section of Cumberland overthrust block showing fensters and newer conception of relations of Pine Mountain fault. Courtesy of Charles Butts and Virginia Geological Survey.

able detail by Wentworth,[3] and certain of its features in still more detail by Butts.[4]

The block (Fig. 1) is bounded on the northwest by the Pine Mountain thrust fault; on the southwest by the Jacksboro tear fault, upthrown toward the southwest; on the northeast by the Russell Fork tear fault, along which there has been some overthrusting, but not so much as on the Jacksboro fault; and on the southeast by the Hunter Valley thrust fault. The latter fault plays no part in the mechanics of the block—it merely marks it off from other fault blocks farther southeast.

STRUCTURE WITHIN BLOCK

The structure within the Cumberland block is illustrated by the two accompanying cross sections (Fig. 2 and Fig. 3). Figure 2 is taken from the Briceville folio (No. 33) of the United States Geological Survey, and Figure 3, from Butts' paper referred to in the foregoing paragraphs.

From these two sections it is apparent that the Cumberland block is divided lengthwise into two major units—a broad, flat-bottomed syncline called the Middlesboro syncline or Middlesboro basin, and a broad, flat-topped anticline, called the Powell Valley anticline—separated by the sharp Cumberland Mountain monocline.

The topographic features of the block are expressions of these structures and of the varying resistances of the rocks brought to the surface by them. Pine Mountain and Cumberland Mountain are monoclinal ridges formed by the outcrop of the resistant basal sandstone (part of the Lee formation) of the Pennsylvanian system. The Middlesboro synclinal basin between them is topographically expressed as a mountainous region of maturely dissected Pennsylvanian rocks and the Powell Valley anticline is expressed as an anticlinal lowland—part of the Great Valley of Tennessee—developed on the relatively non-resistant Ordovician limestones.

The Cumberland block is broken crosswise by three lines of disturbance (Fig. 1), trending roughly parallel with one or another of the tear faults at its ends and therefore bearing essentially shearing-angle relations to the outlines of the block as a whole. The significance of these cross-disturbances is discussed on a following page. Other structural disturbances within the block include at least two strike

[3] Chester K. Wentworth, "Russell Fork Fault of Southern Virginia," *Jour. Geol.*, Vol. 29 (1921), pp. 351–69.

[4] Charles Butts, "Fensters in the Cumberland Overthrust Block in Southwestern Virginia," *Virginia Geol. Survey Bull. 28* (1927), pp. 1–12.

faults near the top of the Cumberland Mountain monocline. One is on Brush Mountain about 7 miles northeast of Cumberland Gap,[5] and the other detaches Powell Mountain from Cumberland Mountain about 3 miles southwest of Cumberland Gap. No reference to the last-mentioned fault has been found, and it was not examined in detail in the field, so it is not known whether it is a normal or a thrust fault. These strike faults may have significance as adjustment phenomena—tensional or otherwise—to the movements which accompanied the overthrusting of the Cumberland block.

INTERPRETATION OF STRUCTURE

The older interpretation of the structure of the Cumberland block is indicated by the cross section (Fig. 2) taken from the Briceville and Maynardville folios. That interpretation shows a thrust fault extending indefinitely downward from the base of Pine Mountain. It presents a difficult problem as to the mechanics of the structure for it offers no explanation of the sharp Cumberland monocline or of the flatness of the Powell Valley anticline.

The interpretation of the structure of the block was fundamentally changed by the discovery of several fensters, or "windows," on the Powell Valley anticline[6] which revealed the fact that beneath the relatively flat-lying Ordovician rocks of the Powell Valley anticline is a thrust plane, below which Silurian rocks are exposed. Later drilling in these fensters proved the presence of the normal Ordovician section below the Silurian.

Butts interpreted the presence of the Silurian rocks beneath the Ordovician at the fensters as indicating that the thrust plane revealed in the fensters must be the same as that which crops out at the base of Pine Mountain and that the thrust plane must, therefore, underlie the whole of the Cumberland block at shallow depth (Fig. 3). Butts suggested that the thrust may have followed the Chattanooga shale, on or within which gliding would have been relatively easy. He estimates that the forward movement on the thrust at this point has been about 7 miles.

Butts' interpretation is believed to be correct and is accepted as a basis for the discussion which follows.

If, in accordance with conventional ideas, the thrust plane is thought of as originally a plane surface, it would have required considerable warping since the thrusting occurred to have produced its

[5] G. H. Ashley and L. C. Glenn, "Geology and Mineral Resources of the Cumberland Gap Coal Field, Kentucky," *U. S. Geol. Survey Prof. Paper 49* (1906).

[6] Charles Butts, *op. cit.*

present attitude, with the northerly dip into the Middlesboro syncline. In addition to the pronounced apparent warping of the thrust plane shown on Butts' cross section (Fig. 3), an interesting and significant feature is the fact that in the area between the fensters and Cumberland Mountain, the strata dip northward into the thrust plane. This feature finds a ready explanation in, and is a necessary consequence of, the mechanics of thrusting outlined in the following section. Besides, the explanation to be proposed accounts for the present attitude and northward dip of the thrust plane, as shown in Figure 3, without the necessity of any warping of the thrust plane since it was formed.

EXPLANATION OF CUMBERLAND THRUST

The peculiar structural features of the Cumberland block, namely, the flat-topped Powell Valley anticline, the sharp monoclinal structure at Cumberland Mountain, the flat-bottomed Middlesboro syncline between Cumberland Mountain and Pine Mountain, and the

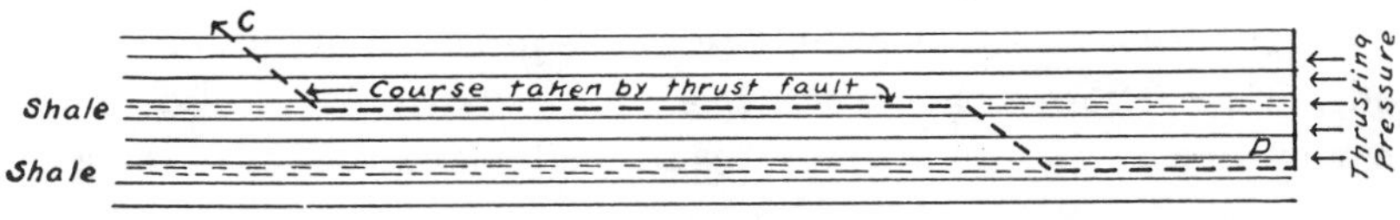

FIG. 4.—Diagram representing course of incipient thrust plane in series of sedimentary rocks. Plane follows beds of easy gliding, such as shales, and breaks diagonally across more brittle beds from one shale to another, and, finally, up to surface.

apparent warping of the thrust plane, all fit into a consistent picture when the thrusting is considered as having taken place along and across the bedding of the sedimentary rocks along some such path

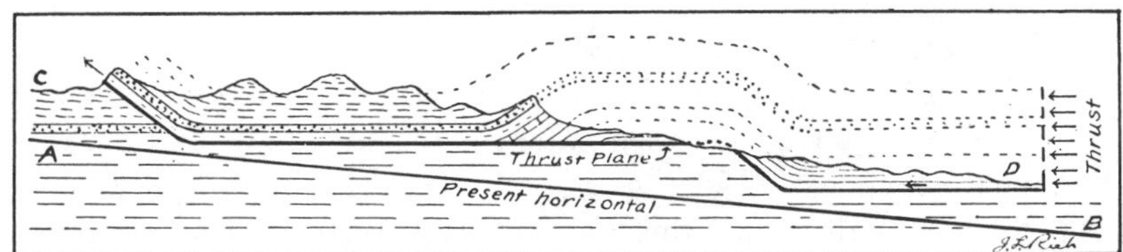

FIG. 5.—Diagram showing result of movement on thrust plane such as that shown in Figure 4. Tilting so that horizontal is represented by line *AB* and erosion down to line *CD* gives structure and topography to be compared with present cross section of Cumberland block (Fig. 3).

as that represented by the dotted line, *CD*, of Figure 4, and the whole region as having been subsequently tilted slightly toward the north.

The thrust plane may be pictured as following some zone of easy gliding such as the lower shale of Figure 4 until frictional resistance became too great; then shearing diagonally up across the bedding to another shale; following that for several miles, and finally shearing across the bedding to the surface.

Movement along such a thrust plane would inevitably produce almost an exact counterpart of the present structure of the Cumberland block, as may be seen by comparing the actual cross section (Fig. 3) with the diagram (Fig. 5)[7] and the model (Fig. 6). The model was made by piling sheets of thin paper into a form like that of Figure 4, with a sheet of paper inserted along the course of the potential thrust plane, and then pushing the paper from one side while the other was held rigid. To reproduce, almost exactly, from the diagram or model, the present structure of the Cumberland block, it is only necessary to tilt the diagram or model so that a horizontal line corresponds with the line *AB* of Figure 5 and then to erode the surface down to a line such as *CD*. The fensters then appear along the line where the thrust plane passes into the upper gliding bed after having broken across the bedding from the one below.

Examination of the diagram (Fig. 5), and the photograph of the model (Fig. 6) shows that whether a broad, flat-topped anticline or a narrow one with a rounded crest is formed where the thrust breaks

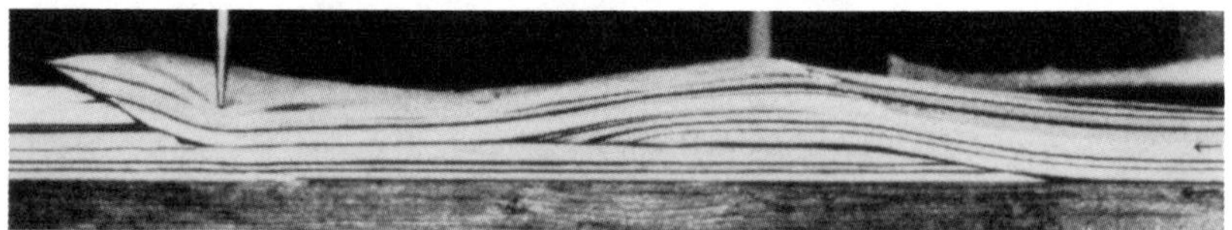

FIG. 6.—Model made by piling paper sheets as in Figure 4 and thrusting toward left. Note similarity to structure of Cumberland block.

upward from the lower gliding plane depends on the amount of thrusting after the break occurs. If the forward movement is slight, a narrow anticline is formed, whereas greater movement produces a broad, flat-topped anticline.[8]

The Powell Valley anticline seems to illustrate both conditions in different parts of its course. At the west end it is broad and flat-

[7] The diagrams, Figures 4 and 5, indicate the thrust pressure as applied entirely at one end. Without going further into the matter, attention is called to the fact that this representation is only diagrammatic, and that it seems possible, and certain features make it seem even probable, that the actual force which caused the slipping on the thrust plane was applied from above as well as from the side through the medium of an overthrust sheet now removed by erosion.

[8] This relation is easily seen by constructing a model like that shown in Figure 6 and varying the amount of thrust.

topped; toward the east it gradually narrows and assumes a rounded crest; and ends abruptly before the end of the block is reached. In accordance with the proposed interpretation, this condition indicates that at the west end, where the thrust broke across from the lower to the upper gliding bed, the strata were pushed forward for several miles; at the east end of the present Powell Valley anticline the forward movement on the lower gliding plane was sufficient to form a narrow anticline; and still farther east the thrusting occurred only along the upper of the two gliding zones, so that no anticline was formed.

Confidence in the foregoing interpretation of the nature of the thrusting is heightened by the discovery, in the southeast corner of the map in the Maynardville folio, of a structure section, reproduced as Figure 7, showing a beveled remnant of a structure such as that shown on Butts' cross section (Fig. 3), at and immediately north of

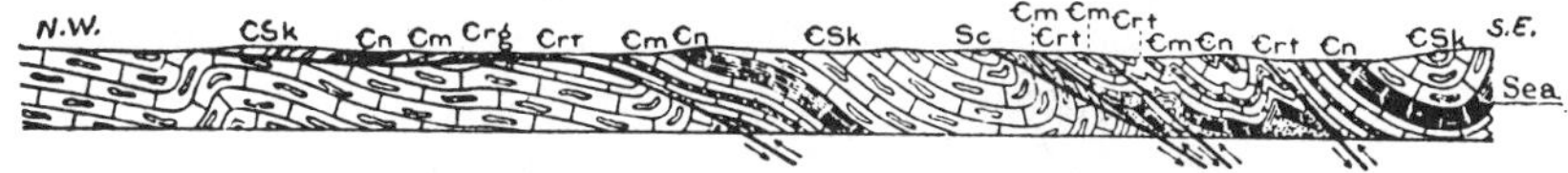

FIG. 7.—Reproduction of structure section from southeast corner of Maynardville folio. Note shape of thrust plane at left and manner in which strata dip into thrust plane along flat part of its course. Compare with cross section of Cumberland block (Fig. 3) in area northwest of fensters.

the fensters, and also on the model (Fig. 6), where the thrust breaks from the lower gliding plane up to the higher. This structure section proves that the type of thrust plane and thrust movement indicated for the Cumberland block is not unique, but occurs in connection with other faults of the Appalachian valley.

CROSS FRACTURES IN CUMBERLAND BLOCK

As has already been mentioned (Fig. 1), the Cumberland block is broken across by at least three lines of disturbance—one between Speedwell and Jellico, another between Cumberland Gap and Pineville, and a third extending from the eastern end of the Powell Valley anticline near Little Stone Gap northward toward Pine Mountain at Pound Gap.[9]

These lines of disturbance cross the block in directions which make the same angles with the Pine Mountain fault and with the block as a whole as the tear faults at its ends—essentially shearing

[9] See map accompanying the report of J. Brian Eby and others, "The Geology and Mineral Resources of Wise County and the Coal-Bearing Portion of Scott County, Virginia," *Virginia Geol. Survey Bull. 24* (1923).

angles. The most southwesterly of these lines of disturbance extends diagonally northwestward across the block from the vicinity of Speedwell toward Jellico. No adequate description of the details of its structure has been found; but, as inferred from a meager description[10] and from the geological map of Tennessee, it seems to be a line of rather sharp monoclinal dip northeast, so that the portion of the block southwest of this line stands somewhat higher structurally than that on the northeast. The trend of the disturbed zone appears to be about northwest. The zone has the appearance of an incipient tear fault separating an area of somewhat greater thrust movement in the Cumberland block west of it from one of less movement on the east.

The middle one of the three cross zones of disturbance is the Rocky Face fault zone extending from Cumberland Gap northward toward Pineville, and trending approximately parallel with the Jacksboro tear fault at the end of the block. It has been described by Ashley and Glenn[11] and certain of its implications discussed by Rich.[12] At Rocky Face it is essentially a buckled half dome on the east side of a fault upthrust toward the northwest.

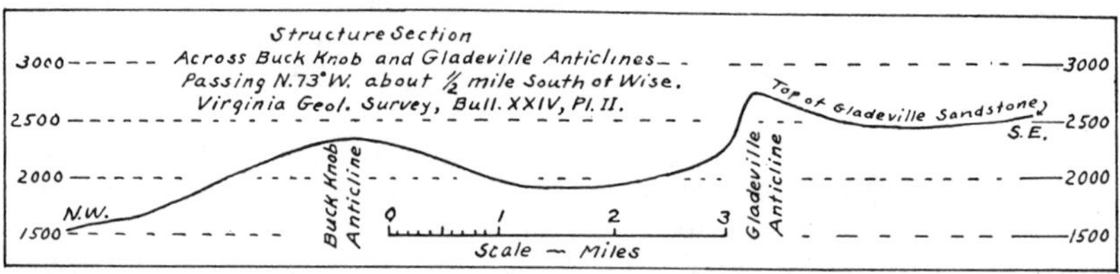

FIG. 8.—Profile across Buck Knob and Gladeville anticlines showing asymmetry of latter.

The third disturbed zone crosses the block northward from the point where the Powell Valley anticline dies out. It has been described and mapped by structure contours.[13] It consists of two parallel anticlines, the Buck Knob anticline and the Gladeville anticline, whose forms and relations to each other are shown in cross section in Figure 8. The eastern of these, the Gladeville anticline, is very unsymmetrical, being steep on the west side. It is marked by two sharp subsidiary domes along its crest.

[10] L. C. Glenn, "Northern Tennessee Coal Field," *Tennessee Geol. Survey Bull. 33-B* (1925).

[11] G. H. Ashley and L. C. Glenn, *op. cit.*

[12] John L. Rich, "Physiography and Structure at Cumberland Gap," *Bull. Geol. Soc. America*, Vol. 44 (1933), pp. 1219–36.

[13] J. Brian Eby and others, *op. cit.*

All these cross fractures are believed to be superficial features of the Cumberland thrust block, representing a tendency toward breaking into segments as the block was pushed northwestward, and it is believed that all these cross fractures and folds will be found to end downward at the thrust plane underlying the block.

It seems that the formation of the Buck Knob and Gladeville anticlines can be accounted for by the excess northwestward shoving of that portion of the block east of the end of the Powell Valley anti-

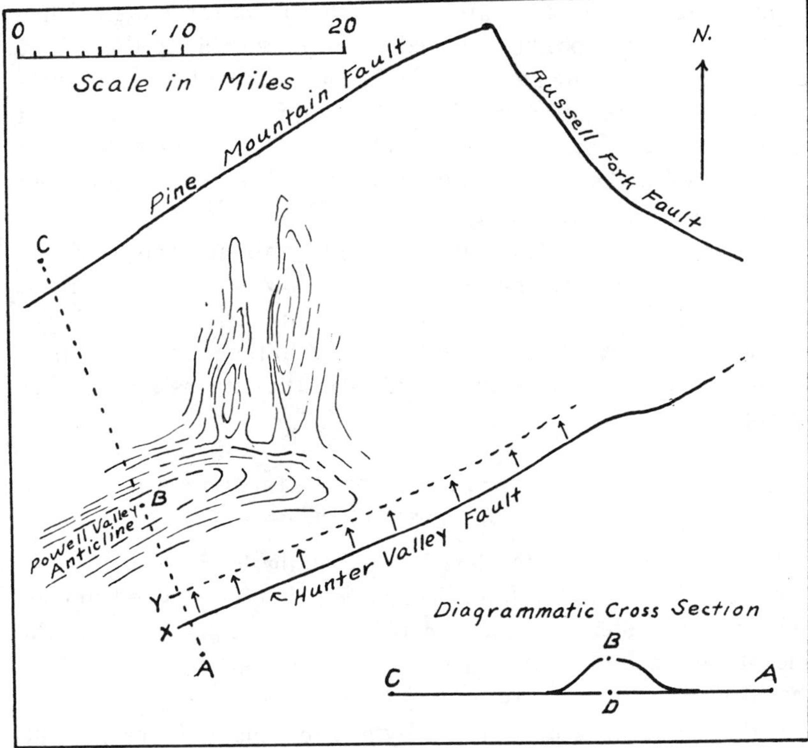

FIG. 9.—Sketch map of eastern end of Cumberland block to illustrate discussion of origin of Buck Knob and Gladeville anticlines. General form of anticlines indicated by sketch structure contours.

cline. Assuming that the Powell Valley anticline was formed during the thrusting, and as a result of the shearing up of the thrust plane from a lower bed to the horizon of the Chattanooga shale, and further assuming a uniform push from the southeast and a movement which may be diagrammatically represented by the distance XY (Fig. 9); then the northwestward movement of a point north of the anticline must have been less than XY by an amount equal to the difference between the length of a line such as *ABC* over the top of the anticline, and a straight line *ADC* (see cross section, Fig. 9), whereas east of the end of the Powell Valley anticline the northwestward movement would have been the full amount, *XY*.

The beds east of the end of the anticline, being crowded northwestward in this way farther than those on the west, must have tended to tear loose along a shearing plane or incipient tear fault, forming a block of the same general shape as the Cumberland block as a whole. In view of the trend of the incipient tear fault, the excess northward movement of the area east of the end of the Powell Valley anticline must have produced a wedging effect similar to that which caused the upthrusting along the Jacksboro tear fault at the southwestern end of the Cumberland block. In this case, the movement seems not to have gone far enough to form an actual break, but only enough to form the highly unsymmetrical Gladeville anticline and the Buck Knob anticline in front of it.

The extreme sharpness of the Gladeville anticline and its asymmetry suggest strongly that it may pass downward into a thrust fault. The anticline bears many points of similarity to the Rocky Face fault zone near Cumberland Gap, even to the presence of sharp local domes along its crest, but it differs in that it has not been so far developed.

PROBABLE SHALLOWNESS OF LOCAL STRUCTURES IN CUMBERLAND BLOCK

The exposure of the fensters in the Cumberland block (Fig. 1) seems to make inevitable Butts' conclusion that the entire Cumberland block is underlain by a low-angle thrust plane and that the Pine Mountain fault marks the line along which the thrust plane sheared upward to the surface.

If the whole block is underlain by a thrust plane, it seems probable that any local anticlines or synclines appearing within the block are only superficial and do not extend down below the thrust plane which, as nearly as can be judged at present, follows the approximate horizon of the Chattanooga shale. There seems to be no reason to expect that such features as the Rocky Face half-dome or the Buck Knob or Gladeville anticlines extend below the thrust, especially since, as in the latter case, the presence of such features in the rocks above the thrust seems to be required by the dynamics of the thrusting.

BROADER ASPECTS—POSSIBLE CORRELATION OF PINE MOUNTAIN FAULT WITH SEQUATCHIE VALLEY ANTICLINE

Analogy with the features displayed by the Cumberland overthrust block strongly suggests that the Sequatchie Valley anticline[14] is the expression of similar low-angle thrust faulting carried not so far as in the Cumberland block. The Sequatchie anticline, where eroded most deeply, reveals thrust faulting in its lower part; its distance from the western margin of the Appalachian Valley belt of strong shingle-type thrust faulting is comparable with that of the Pine Mountain fault; and the anticline west of Lookout Mountain in the southern part of the area occupies a position analogous to that of the Powell Valley anticline.

It is not difficult to picture a somewhat greater original movement along the thrust plane bringing the thrusts now exposed in the deeper parts of the Sequatchie anticline up to the surface (before erosion), shoving Walden Plateau farther westward, breaking it loose along a diagonal tear fault somewhere north of Pikeville, Tennessee, and producing an overthrust block almost identical with the Cumberland block.

BROADER BEARING ON GENERAL PROBLEM OF OVERTHRUST FAULTING

Revelation of the way in which thrust faults may follow for long distances along the bedding of strata, such as shales, on which movement is easy, completely changes the basis of calculations of the possible distance the rocks may move on overthrust faults. In the past such calculations have been based on the coefficient of friction of dry granite. The friction involved in movement along the bedding of clay shales, probably wet with water at the time of movement, is something of an entirely different order than that for granite.

Wentworth, for example, made a calculation of the force required to move the Cumberland block,[15] but he used the coefficient of friction of dry granite. His results would have been radically different had they been based on movement along the bedding of shales.

Recognition of the importance of thrusting along bedding planes throws a new light on many problems in thrust faulting. For example, the striking fact that in crossing the Appalachian valley southeast-

[14] The Sequatchie Valley anticline, as conspicuously shown on the geologic maps of Tennessee and Alabama, forms a remarkable anticlinal valley essentially in line and trend with the Pine Mountain fault, though separated from it by a gap of nearly 50 miles. The portion of the Allegheny Plateau between Sequatchie Valley and the Appalachian Valley province is known as Walden Plateau.

[15] Chester K. Wentworth, *op. cit.*

ward from the Cumberland block toward the great up- and overthrust mass of the crystalline Appalachians one passes half a dozen or more large thrust faults which bring up repeatedly, shingle-fashion, almost the whole of the unmetamorphosed Paleozoic sedimentary series without once bringing up the crystalline basement, makes it appear likely that the thrusting in that part of the Appalachian valley is entirely confined to the sediments, which have been sheared off from the underlying crystalline basement, pushed forward, and piled up in shingle-fashion by a great plunger moving from the southeast—presumably the pre-Cambrian mass of the crystalline Appalachians—which came up diagonally along a shear plane to the base of the sediments, then rode forward horizontally, pushing and piling up the sediments before it. It is not unlikely that the pre-Cambrian basement beneath the Appalachian valley would be found to be undisturbed.

THIS PAGE INTENTIONALLY BLANK

Erratum to the following paper:

The last full sentence on the bottom paragraph of page 367 (page 183 of this volume) should read: "Figure 12 shows a plot of age determinations from southeast British Columbia by Gabrielse and Reesor (1964) which is compared with linear plot of the first 129 random numbers, having values 350 or less as suggested by Ross (1966)."

This plot was inserted on Figure 12 to call attention to the criticism by Ross (1966) regarding histogram analysis of isotopic age data.

Reprinted by permission of the Canadian Society of Petroleum Geologists from *Bulletin of Canadian Petroleum Geology*, v. 14, no. 3 (1966), p. 337-381.

BULLETIN OF CANADIAN PETROLEUM GEOLOGY
VOL. 14, NO. 3, (SEPT., 1966), P. 337-381

STRUCTURE, SEISMIC DATA, AND OROGENIC EVOLUTION OF SOUTHERN CANADIAN ROCKY MOUNTAINS[1]

A. W. BALLY,[2] P. L. GORDY,[2] and G. A. STEWART[2]
Calgary and Edmonton, Alberta

ABSTRACT

In the Rocky Mountain Foothills, major oil and gas accumulations occur in the folded and faulted leading edges of thrust sheets involving Paleozoic carbonates. These structures underlie a complex of imbrications involving Mesozoic clastic rocks. In this area the integration of seismic and geologic data leads to the definition of prospects and also illustrates concepts fundamental to an understanding of mountain building.

Reflection data show that for its entire width of about 80 miles, the Rocky Mountain fold belt is underlain by the gently westward dipping extension of the crystalline Precambrian Shield. Shortening, exceeding 100 miles in Paleozoic beds, takes place along décollement zones and curved thrust faults which flatten at depth (listric thrust faults). Late Mesozoic and early Tertiary thrusting was followed by late Tertiary normal faulting. Reflection data suggest that these normal faults, which are steep at the surface, also flatten at depth (listric normal faults) and may merge with older thrust faults.

Reflection sections show that at depth the structural style on both sides of the Rocky Mountain Trench is similar and they suggest a continuation of the westward dipping basement beneath, and well to the west of the Trench. Therefore the Trench and the associated post-orogenic Tertiary basins are probably related to a system of shallow, listric, normal faults that are responsible for the location and direction of this morphologic feature.

A palinspastic reconstruction based on seismic and subsurface data is essential background for discussions concerning the relations between the Rocky Mountains and the igneous and metamorphic western half of the Cordillera. More generally, relations between continental drift and the formation of the Western Cordillera are placed in perspective using such reconstructions.

The seismic reflection data shown provide insight into the structure of the crust down to depths of ten kilometers, and effectively bridge the gap between surface geology and deep crustal refraction data.

CONTENTS

---

[1]Manuscript received August 2, 1966; essence of paper presented to meeting of Royal Society of Canada, Vancouver, B.C., June 8, 1965.

[2]Shell Canada Ltd.

Plates (in pocket)

Introduction

Some one hundred years ago the British Government issued a "Blue Book" reporting the results of Captain Palliser's expedition to the Canadian west in which James Hector (1863) published the first geologic observations on the Canadian Rocky Mountains. Today the Southern Canadian Rockies rank among the best known mountain ranges of the world, both geologically and geophysically.

Pioneer work by Dawson, McConnell, Willis and Daly, to name a few, established basic stratigraphy and led to the recognition of widespread overthrust and folding phenomena. This period was followed by mapping, done mainly by officers of the Geological Survey of Canada, a project that is still in progress. The discovery of gas, and later of oil, at Turner Valley stimulated the search for additional hydrocarbon accumulations. Petroleum geologists have published many valuable contributions to the geology of the area. Excellent summaries of the state of knowledge of the time are given in the "Western Canada Sedimentary Basin Symposium" (Clark, 1954), North and Henderson (1954a), Hume (1957), Fox (1959), and Shaw (1963). Recently, the stratigraphy of western Canada has been compiled in the "Geological History of Western Canada" (A.S.P.G., 1964b).

During the early Forties the first geophysical surveys (gravity and seismic) were undertaken in selected areas. They led to discovery of the Jumping Pound, Sarcee and Pincher Creek gas fields. A stimulating synthesis of geologic and geophysical data was published by Link (1949). During the Fifties extensive regional seismic surveys were undertaken, culminating in discovery of the Waterton, Wildcat Hills, West Jumping Pound and other, not yet fully evaluated gas fields. During this period only one field, Savanna Creek, was discovered using surface geologic methods.

It soon became apparent that seismic data were not only valuable from an economic point of view, but that they contributed greatly to a better understanding of regional structure and problems related to mountain building. The importance of widespread décollement phenomena was clearly demonstrated. Fox (1959), Shaw (1963) and others published regional sections which evidently were based on seismic information, but only recently has Keating (1966) published some of the supporting data.

Because of the lack of geophysical documentation relating to the geology of the Rockies and Foothills of Alberta and southeastern British Columbia, the authors consider it desirable to publish basic seismic reflection data at this time. The seismic sections presented provide illustrations of the internal structure of a typical thrusted and folded mountain belt and permit dealing with the "shortening" aspects of mountain building in a reasonably quantitative manner.

This paper is based on the fundamental contributions of many predecessors. It is not possible to credit all who have contributed to a better understanding of the geology of the Rockies but we must single out the work of the Geological Survey of Canada, particularly its fine surface maps and Memoirs. Publications of the Alberta Society of Petroleum Geologists have also added much to a geological understanding of the Rockies.

We thank Shell Canada Limited for releasing the seismic information and allowing publication of this paper.

A judicious amalgamation of geophysical and geological talents and know-how is the key to an adequate understanding of most geologic problems. We cannot name all the numerous past and present colleagues at Shell who helped us to arrive at a better understanding of the geology and geophysics of the Rockies and Foothills, but we wish to thank our closest associates: among the geophysicists, A. Junger, G. Robertson, D. W. Smith and F. Van Goor, and among the geologists, J. M. Alston, J. E. Davidson, G. I. Lewis and O. L. Slind.

We also extend our thanks to G. E. Merritt who critically reviewed the manuscript and to C. G. Devenyi who is responsible for the majority of the drawings in this report.

Regional Framework

Three major fold belts rim the North American craton. They are the Appalachian-Ouachita system on the southeast and east, the Innuitian system on the north and the Western Cordillera on the west. The first two, folded during Late Paleozoic time, were later eroded deeply and partly buried, whereas the Western Cordillera, folded during Late Mesozoic and Tertiary time, is better exposed.

Each system can be divided into an inner igneous belt and an outer non-igneous fold belt. The former was characterized by eugeosynclinal sedimentation followed by a breakup into epi-eugeosynclines accompanied by the emplacement of igneous intrusions and metamorphism. The outer fold belt evolved through a miogeosynclinal and exogeosynclinal

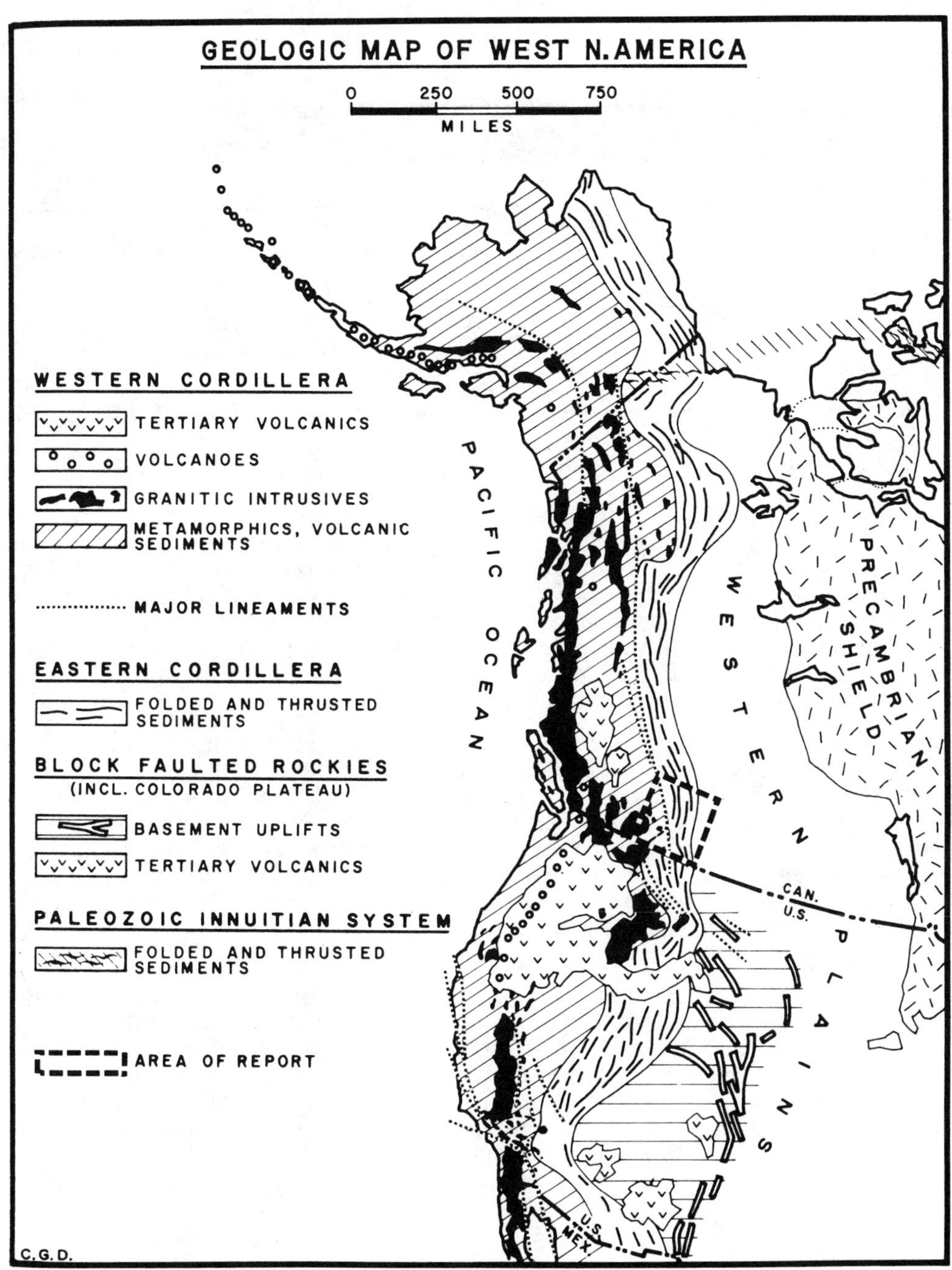

FIG. 1.—Geologic map of western North America, showing location of report-area.

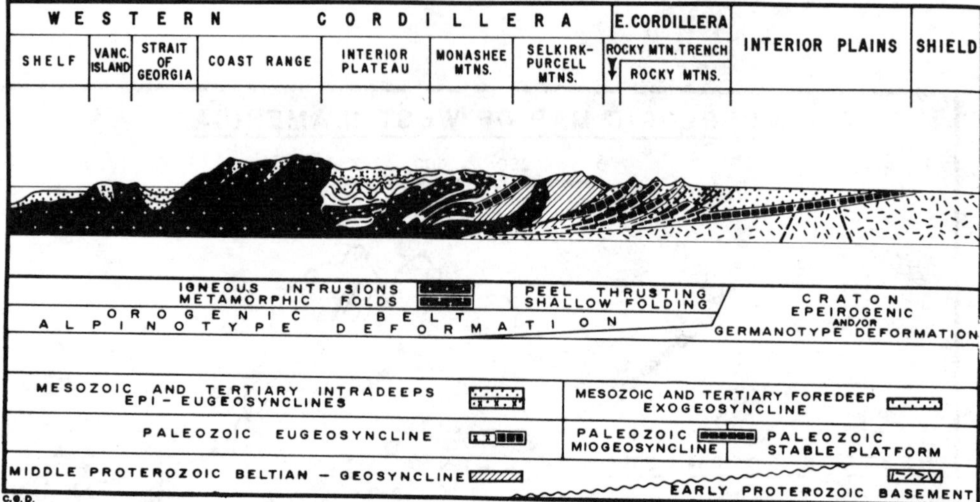

FIG. 2.—Sketch showing relationship of geosynclinal, structural and physiographic units.

phase into a fold belt characterized by peel-thrusting, that is, décollement folding and thrust faulting. Large block-faulted (germanotype) uplifts involving crystalline basement and the overlying cratonic cover form the transition zone between the outer miogeosynclinal fold belt and the adjacent shield. The application of these subdivisions to the Canadian Cordillera is shown on a map (Fig. 1) and sketch section (Fig. 2). The sketch shows major physiographic subdivisions at the top, and structural styles and geosynclinal phases below.

Any attempt to understand the phenomenon of mountain building has to relate the geology of these belts to a unified and evolutionary concept. This paper is an attempt to improve the foundations for such a concept. We plan to document the internal structure of a typical outer, non-igneous fold belt. Most of this paper is devoted to the evidence for widespread décollement over a gently westward dipping shield-type basement. Some conclusions are particularly relevant to an understanding of the evolution of the adjacent igneous belt.

It must be emphasized that the Canadian Rocky Mountains (or Canadian Rockies) strike into the folded belts of Montana, Wyoming, Idaho, Utah and Nevada. The Rocky Mountains and foothills of central Montana, Wyoming, Colorado and New Mexico form a different structural province with predominant germanotype block-faulting involving the Precambrian basement. The Mackenzie and Franklin Mountains of the Northwest Territories are a branch of the Canadian Rockies also characterized by variations on the theme of décollement folding and thrust faulting.

SUBDIVISIONS OF EASTERN CORDILLERA

The major geologic and physiographic subdivisions shown on the geologic map (Fig. 3) and the provinces map (Fig. 4) are based largely on proposals by Bostock (1948) and North and Henderson (1954a). From east to west, the main subdivisions are:

1. *The Interior Plains,* underlain by a relatively undisturbed sequence of Paleozoic, Mesozoic and Cenozoic sediments directly overlying a gently westward dipping Canadian Shield.

2. *The Foothills,* whose structural skeleton is formed by relatively large and flat thrust sheets involving Paleozoic carbonates, with some frontal imbrications. Intensive imbrications of Mesozoic clastic rocks drape this Paleozoic "skeleton." The whole is underlain by the relatively undisturbed westward continuation of the Canadian Shield. A large portion of this paper is devoted to the evidence for this.

3. *The Front Ranges,* formed by thrust sheets, stacked in imbricate fashion, mainly involving Paleozoic carbonates and Precambrian (Beltian) carbonate and clastic rocks. This province is also underlain by a basically undisturbed westward extension of the Canadian Shield.

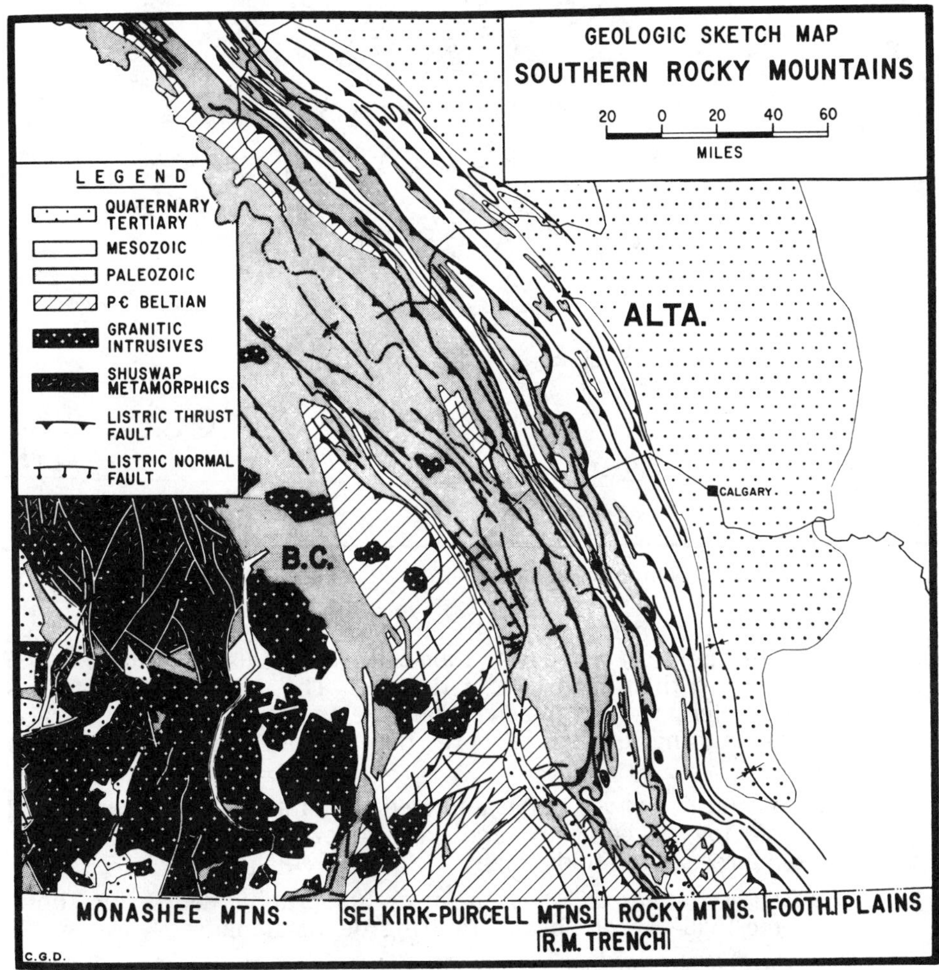

FIG. 3.—Geologic sketch map, Southern Rocky Mountains.

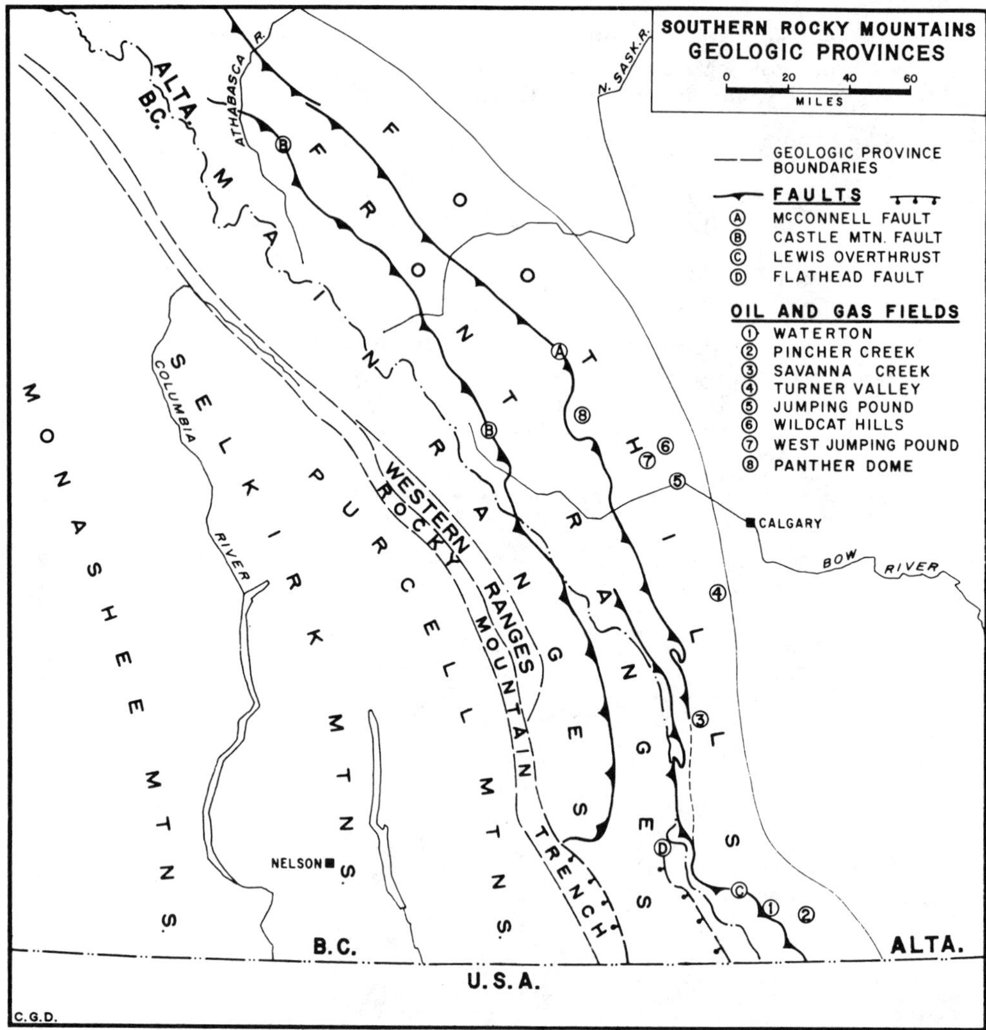

Fig. 4.—Southern Rocky Mountains, geologic provinces.

4. *The eastern Main Ranges,* with elements composed largely of Precambrian (Beltian) and lower Paleozoic sediments, seemingly much less deformed than in the Front Ranges. The *western Main Ranges* are characterized by intensive cleavage in lower Paleozoic strata. Normal faulting occurs in both the eastern and western Main Ranges.
5. *The Western Ranges,* characterized by intensive cleavage and thrusting toward both east and west. Normal faulting is common and some recumbent folds occur.
6. *The Rocky Mountain Trench,* a linear topographic depression showing some evidence of normal faulting on its flanks.
7. *The Purcell Mountains,* essentially a large anticlinorium complicated by thrust faulting, involving thick sequences of Proterozoic

and lower Paleozoic sediments, intruded by large, more or less discordant granitic batholiths.

8. *The Selkirk-Monashee Mountains,* characterized by extensive gneiss complexes (Shuswap and Valhalla), by more or less discordant granitic intrusives (Nelson batholith) and by eu- and epieugeosynclinal Paleozoic and Mesozoic clastic rocks and volcanics.

Simplified Stratigraphy

The amount of stratigraphic information on the area is overwhelming; fortunately an outstanding synthesis is available in the "Geological History of Western Canada" (A.S.P.G., 1964b). Table 1 is a table of formations expanded to indicate lithologies, thicknesses, velocity distributions, and the stratigraphic position of two major reflection events. The area described on Table 1 is shown on Figure 5.

For purposes of regional structural analysis the stratigraphy can be simplified as follows.

Two types of Precambrian rock occur: (1) Metamorphic and igneous rocks of the western extension of the Canadian Shield formed during the Hudsonian orogeny 1,640-1,820 m.y. ago (Burwash *et al.,* 1964), and (2) a thick series in excess of 45,000 feet of Beltian (Proterozoic) quartzites, argillites, and carbonates, with basic intrusives and extrusives. The Beltian disconformably overlies the Canadian Shield. An unconformity separates the Beltian Purcell Series from the younger Proterozoic Windermere Series.

The Paleozoic succession consists primarily of carbonates and some shales with thicknesses generally increasing towards the west. Three groups of unconformities occur, one below the Cambrian, another between the Middle Ordovician and the base of the upper Middle Devonian, and the third between the Pennsylvanian and the base of the Lower Cretaceous. They indicate gentle tilting movements but no effects of true alpinotype orogenic movements can be observed.

The lower Mesozoic succession (Triassic and Jurassic) is incomplete due to the above mentioned group of unconformities, caused by tilting movements during late Paleozoic and early Mesozoic time. Where present, shales, some coarse clastic rocks, and rare carbonates characterize the succession.

Uppermost Jurassic, Cretaceous, and Tertiary sediments consist of conglomerates, sandstones, and shales of marine and continental character, showing some affinities to Alpine molasse sediments. With the possible exception of some beds in the Kootenay Formation, no equivalent to "flysch-like" deposits occurs in the Canadian Rocky Mountains.

Summary of Seismic Procedures

The structural style of the Foothills and Rocky Mountains can be understood by the integration of geological and seismic data. Both reflection and refraction seismic techniques are used in the Foothills.

TABLE 1.—Table of formations.

		FORMATION	LITHOLOGY	THICKNESS	AV. VELOCITY (MAX. DEVIATION) IN FEET/ SEC.	RANGE OF VELOCITIES	REMARKS
TERTIARY CLASTICS	PLEISTOCENE-RECENT		GRAVEL, SAND, SILT, GLACIAL DRIFT	0-5000'	11,000 (1000)	10,000-12,000	
		UNCONFORMITY					
	MIOCENE	ST. EUGENE (CONFINED TO CRANBROOK AREA)	GRAVEL, SAND, SILT				
		UNCONFORMITY					
	EOCENE AND OLIGOCENE	KISHENENA (FLATHEAD VALLEY)	NON-MARINE MUDSTONE, MARL, SILTSTONE, SANDSTONE, CONGLOMERATE	6000'+			
		UNCONFORMITY					
	PALEOCENE	PASKAPOO - PORCUPINE HILLS	NON-MARINE SANDSTONE, SHALE, COAL, BASAL CONGLOMERATE	4000'+			
		UNCONFORMITY					
MESOZOIC CLASTICS EXOGEOSYNCLINAL (FOREDEEP)	PALEOCENE AND/OR UPPER CRETACEOUS	WILLOW CREEK	NON-MARINE SANDSTONE, SHALE, MUDSTONE	350-2700'		12,000-14,000	
	UPPER CRETACEOUS	EDMONTON (BLOOD RESERVE, ST. MARY RIVER)	NON-MARINE SANDSTONE, SHALE, COAL, BASAL CONGLOMERATE	1000-3100'	11,400 (400)		
		BEARPAW	MARINE BLACK SHALE	0-60'			
		BELLY RIVER	NON-MARINE SANDSTONE, MUDSTONE, SHALE	1200-4000'	12,600 (1700)		
		WAPIABI	MARINE SHALE, SILTSTONE	1100-1800'	12,800 (1000)		
		CARDIUM	MARINE SANDSTONE, SILTSTONE, SHALE	30-450'	13,400) (2400)		
		BLACKSTONE	MARINE SHALE, SILTSTONE, BASAL GRIT	400-1000'	13,100 (1900)		
	LOWER CRETACEOUS	CROWSNEST VOLCANICS (CONFINED TO CROWSNEST AREA)	VOLCANIC AGGLOMERATE, TUFF	0-1800'		13,000-15,000	
		INTRUSIVES (CONFINED TO FLATHEAD AREA)	TRACHYTE, SYENITE 95-112 M.Y. OLD				
		BLAIRMORE	NON-MARINE SANDSTONE, SILTSTONE, SHALE, BASAL CONGLOMERATE	1000-6500'	14,700 (1900)		BASAL MESOZOIC REFLECTION EVENT "NEAR MISS."
		UNCONFORMITY					
	LOWER CRETACEOUS & UPPER JURASSIC	KOOTENAY	NON-MARINE SANDSTONE, SILTSTONE, SHALE, COAL	0-4000'	14,800 (2600)		
	JURASSIC	FERNIE	MARINE SHALE, SILTSTONE, LIMESTONE, SANDSTONE	100-1000'+	13,700 (3000)		
		UNCONFORMITY					
	TRIASSIC	WHITEHORSE	DOLOMITE, SANDSTONE	0-800'			
		SULPHUR MOUNTAIN	LAMINATED ARGILLACEOUS, SILTSTONE, SANDSTONE	0-1000'			
		UNCONFORMITY					
PALEOZOIC CARBONATES MIOGEOSYNCLINAL	PERMIAN	ISHBEL	QUARTZITIC SANDSTONE, SILTSTONE, CHERT	0-2000'		20,000-21,000	
		UNCONFORMITY					
	PENNSYLVANIAN	SPRAY LAKES GROUP: KANANASKIS	MARINE CHERTY DOLOMITE	0-170'			
		SPRAY LAKES GROUP: TUNNEL MTN. (CONTAINS MINOR DISCONFORMITIES	MARINE DOLOMITIC SILTSTONE, SANDSTONE	0-1800'			
	MISSISSIPPIAN	RUNDLE GROUP: ETHERINGTON	MARINE LIMESTONE, SILTY DOLOMITE, ANHYDRITE	0-850'	20,400 (2500)		
		RUNDLE GROUP: MOUNT HEAD	MARINE THIN-BEDDED LIMESTONE, SILTY DOLOMITE	0-1000'			
		RUNDLE GROUP: LIVINGSTONE (TURNER VALLEY-SHUNDA-PEKISKO)	MARINE CRINOIDAL, CHERTY LIMESTONE	800-1400'			
		BANFF	MARINE, DARK ARGILLACEOUS, CHERTY LIMESTONE	500-1050'			
		EXSHAW	MARINE, BLACK SHALE	10-40'			
		UNCONFORMITY					
	UPPER DEVONIAN	PALLISER	MARINE, MASSIVE LIMESTONE, DOLOMITE	900-1200'	20,000-21,000		
		ALEXO	MARINE, SILTY LIMESTONE, DOLOMITE, SILTSTONE	20-600'			
		FAIRHOLME	MARINE LIMESTONE, SHALE, DOLOMITE, DOLOMITIZED REEFS	950-1500'			
		UNCONFORMITY					
	LOWER AND/OR MIDDLE DEVONIAN	BASAL DEVONIAN CLASTICS	SILTY & SANDY DOLOMITE, RED BEDS	0-120'			
		UNCONFORMITY					
	ORDOVICIAN	MONS & SARBACH (FRONT RANGES OF BOW VALLEY & NORTH)	LIMESTONE, DOLOMITE, PUTTY-COLOURED SHALE	0-1500'			
	UPPER CAMBRIAN	LYNX (ELKO) (FRONT RANGES)	MARINE DOLOMITE, SILTY DOLOMITE, SHALE, LIMESTONE	0-1800'+			
	MIDDLE CAMBRIAN	ARCTOMYS (FRONT RANGES)	SILTSTONE. SILTY DOLOMITE, SHALE (SHALLOW WATER)	0-200'+			
		PIKA (FRONT RANGES)	MARINE LIMESTONE, DOLOMITE	0-320'			
		ELDON	MARINE LIMESTONE, DOLOMITE	700-1000'			CAMBRIAN REFLECTION EVENT "NEAR BASEMENT"
		STEPHEN	MARINE LIMESTONE, SHALE	100-400'			
		CATHEDRAL (BURTON)	MARINE DOLOMITE, LIMESTONE	200-1000'			
		MOUNT WHITE (EAGER)	MARINE SHALE, SANDSTONE, QUARTZITE, LIMESTONE	500-1500'			
	LOWER CAMBRIAN	ST. PIRAN - GOG (CRANBROOK)	QUARTZITE, SHALE	20-8000'		17,000-18,000	
		UNCONFORMITY					
BELTIAN	MIDDLE-LOWER LATE PROTEROZOIC	UPPER PURCELL (CRANBROOK WATERTON AREA) NOT PRESENT IN FOOTHILLS	ARGILLITE, DOLOMITE, QUARTZITE, PURCELL LAVA	8000-12,000'	17,000 - 18,000		
		LOWER PURCELL (CRANBROOK WATERTON AREA) NOT PRESENT IN FOOTHILLS BASE OF PURCELL NOT OBSERVED	QUARTZITE, ARGILLITE, LIMESTONE, DOLOMITE	8000-20,000'			
		UNCONFORMITY					
BASEMENT	EARLY PROTEROZOIC	BASEMENT	IGN. & METAMORPHIC "SHIELD" TYPE ROCKS, CONSOLIDATED DURING HUDSONIAN (1600-1900 M.Y.)				

C.G.D.

Generally, fair quality reflection data are obtained by conventional methods, that is, continuous profiling, split shots, single holes, 20 to 80-pound charges with geophone spreads of about 1,800-2,400 feet. For reasons of accessibility and costs the lines generally follow the stream valleys (the majority of which are in the dip direction). Operational procedures have been summarized recently by Keating (1966).

The reflection-record sections used in the Plates were prepared from field records corrected for elevation and weathering effects. They are variable-area recorded (VAR) time sections, and are compressed somewhat vertically. As a purpose of this paper is to show good representative data, some sections have been pieced together from projected line segments (Fig. 5).

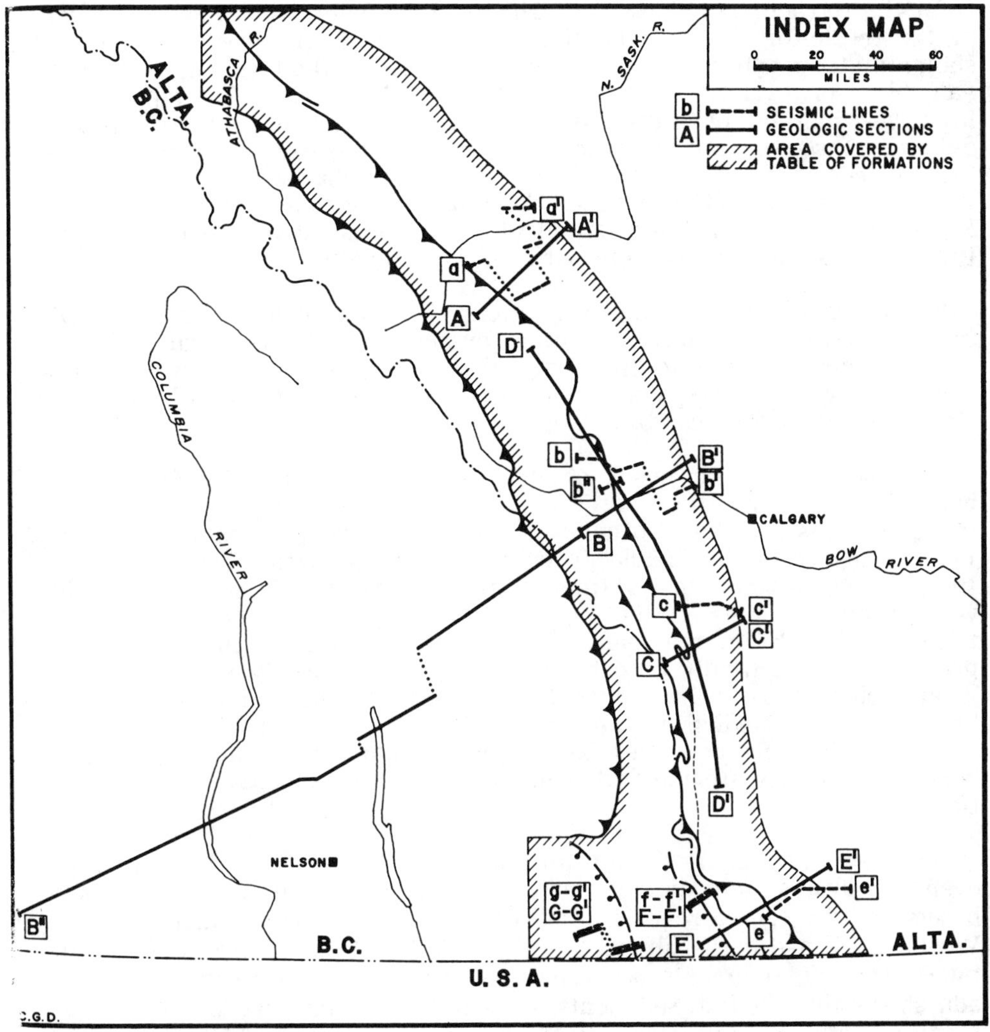

Fig. 5.—Index map, locations of geologic and seismic sections.

Although a large velocity contrast occurs at the Mississippian-Mesozoic interface, a corresponding reflection is rarely observed. Instead, an excellent reflection comes from the Lower Blairmore-Jurassic interval (see also Keating, 1966). Where Mesozoic and Paleozoic strata are structurally conformable, the correct Mississippian pick is a late phase of this Lower Blairmore-Jurassic band. However, where Lower Mesozoic rocks are thrusted over younger Mesozoics, the question of conformity is in many cases uncertain from reflection data alone and is usually established from refraction data.

A strong, persistent arrival from the Mississippian is usually obtained from late-arrival refraction shooting (second event refraction technique of Keating, 1966). The shooting distance chosen is greater than the critical distance (the minimum distance a refraction may be recorded from the refractor) but less than the distance where the time of arrival of the refraction would equal the arrival time of the direct wave from shotpoint to geophone. The in-line refraction method (along strike) is valuable for velocity identification of the Paleozoic top. The broadside method of second arrival refraction profiling is useful for defining Mississippian structure and establishing the location of the leading edges of Paleozoic thrust sheets (Richards, 1959; Blundun, 1956). While refraction outlines the Paleozoic skeleton, the mapping of Mesozoic structure and estimation of the internal structure of multiple thrust sheets can be done only with reflection data.

Numerous steep dips, and even intersecting dips evident on the VAR sections, obviously require migration for accuracy of structural interpretation. However, due to the complex geometry of seismic trajectories and velocity variations, only an approximation of the true position for steeply dipping events can be achieved.

The most consistent event obtained in Foothills shooting is a wide-band reflection from the Cambrian. This event is calibrated by wells in the Plains east of the Foothills. It has two important characteristics: (1) Usually it is identifiable (even in bad-record areas), and (2) it is the only event that shows time variations due to all the changes in velocity and thickness of the Mesozoic and Paleozoic strata from surface to basement. From regional studies the underlying basement is interpreted to be essentially autochthonous, dipping smoothly to the west at about 200 feet per mile. With this assumption one can use the Cambrian event to predict the amount of Paleozoic section above regional, once the Mesozoic velocities have been estimated. Examples of Cambrian time uplifts are illustrated in the description of characteristic sections.

We generally assume that in an autochthonous position the Cambrian event is always immediately underlain by the crystalline Precambrian basement and therefore often call the event a "near-basement event." We have discarded the alternative that the Cambrian event is underlain by Proterozoic sediments because experience elsewhere in western Canada shows that Beltian sediments are seismically characterized by prominent reflections. An exception to this will be described in the section discussing the west side of the Rocky Mountain Trench.

Description of Characteristic Sections

The location of the sections is indicated on Figure 5. Note again that the seismic sections are spliced together to provide continuity. The geologic sections are located subparallel to the seismic lines and incorporate pertinent surface, subsurface and seismic data. Although the sections are largely self-explanatory, the following comments provide additional background and emphasize pertinent points. Relevant reference material to each of the sections is listed chronologically in Part C of the References.

A. *Geologic Section A-A′ and Seismic Section a-a′, Ram River-Stolberg Area* (*Plates* 1, 2)

The Cambrian event is outstanding on seismic section a-a′. The event has not been identified nearby, but numerous penetrations into the carbonate sequence by wells (e.g. Phillips Ancona) and the high quality of the seismic data leave little doubt as to its near-basement Cambrian origin.

Proceeding from east to west, a velocity uplift of the Cambrian event due to the stacked carbonate imbrications of the Stolberg structure can be seen. A second velocity uplift farther west is due to the carbonates of the outcropping Brazeau thrust sheet. Note there is no sizable velocity uplift corresponding to the Ram River feature, suggesting that the total amount of carbonates overlying the basement event does not change because of "redistribution" of the carbonates from the Brazeau sheet to the Ram sheet.

The geologic section displays many characteristic features: (1) The basement dips gently to the west, after correction of velocity effects on the Cambian event. (2) Faults tend to be steep at surface but they flatten out at depth, where they ultimately merge into bedding planes (listric thrust faults). (3) In a westward direction faults always cut deeper into the section. (4) The Cambrian is involved in the thrusting and forms the base of the Brazeau thrust sheet. The westward root of this sheet is clearly indicated on the seismic section. The Brazeau sheet has been penetrated by a number of wells which show that Cambrian overlies the main fault (see also Imperial Cal. Stan. Nordegg 6-17, on Shaw's (1963) cross-section). (5) On surface, a stratigraphic displacement in the order of 10,000 feet along the front of the Brazeau sheet largely conceals the established thrusting distance of at least 20 miles. This phenomenon occurs elsewhere in the Foothills and can be explained by splitting of the main thrust fault into a number of smaller thrusts that distribute the total displacement.

B. *Geologic Section B-B′ and Seismic Section b-b′, Bow Valley Area* (*Plates* 1, 3, 4)

If a prototype structure section for the Canadian Rockies and Foothills had to be selected, the authors would probably choose this section. Again, three characteristic seismic reflection sections have been spliced together and the geologic section attempts to synthesize all available information. Because the seismic and geologic sections are not located on the same line, different wells are shown as calibration. Many other wells have been drilled in the area, all of which support the interpretation shown on section B-B′.

The nearest basement well is California Standard Parkland 4-12, about 60 miles southeast; regional seismic data leave little doubt that the Cambrian event represents the section immediately overlying the crystalline basement. On seismic section b-b′ the Cambrian event can be traced well into the Front Ranges of the Rockies. Small velocity uplifts occur below the Jumping Pound and West Jumping Pound structures, but no sizable velocity uplift is recognized beneath the leading edges of the two western thrust sheets. We assume this is due to a thin wedge-like leading edge which causes the velocity uplift to be very gradual. After correction only an increased tilt of the Cambrian event can be seen.

In the area of sections B-B′ and b-b′ the carbonate skeleton consists of a lower sheet, an intermediate sheet which outcrops about 10 miles south at Moose Mountain, and an uppermost sheet, the McConnell thrust.

The lower sheet can be split into three parts: (1) The gas-bearing structures of Jumping Pound and West Jumping Pound-Morley (frontal offshoots underlain by a common bedding fault and showing cumulative thrust slip of three to four miles). (2) An intermediate flat part of normal stratigraphic thickness that evidently conceals the displacement of the frontal offshoots in the form of a basal Mississippian bedding-plane fault. (3) A rearward lens-like feature at the place where the bedding-plane fault cuts into deeper Paleozoic strata, thus forming a broad anticline.

Douglas (1950) described similar features observed at the surface and interpreted them following a mechanism suggested by Rich (1934) and others for the Appalachians. The mechanism is shown on the two lower sections of Figure 13. Because the transition between the rearward and the middle sector is so critical we include a detailed closeup of another seismic section, shot in the valley of the south branch of Ghost River (Pl. 4).

Overlying the Paleozoic carbonate skeleton is an envelope of complex Mesozoic imbrications. Most should be considered as being sheared off the top of the lower sheet by the higher Moose Mountain sheet. To construct a consistent cross-section we must then assume that the cumulative length of any Mesozoic stratum (e.g. the Cardium) should correspond to at least the cumulative length of the underlying Paleozoic top. One important restriction must be placed on this statement. Some of the Mesozoic imbrications may be sheared off the top of a Paleozoic sheet located farther west than the western margin of the section. That this is in some instances the case is suggested on our section B-B′ east of the Morley well where we show thrust displacements of the Cardium Formation that greatly exceed the slip shown for the underlying frontal offshoots. This implies that the Mooose Mountain sheet sheared Mesozoic beds off their carbonate substratum before the more easterly lower thrust sheet was formed. Formation of the lower element warped or folded the overlying Moose Mountain and McConnell sheets and simultaneously the faults underlying the frontal offshoots folded and faulted the bedding-plane fault connecting the Mesozoic imbrications. This is illustrated on Figure 6. If there are differences in strike between the upper and lower fault systems one will observe a map picture whereby a younger

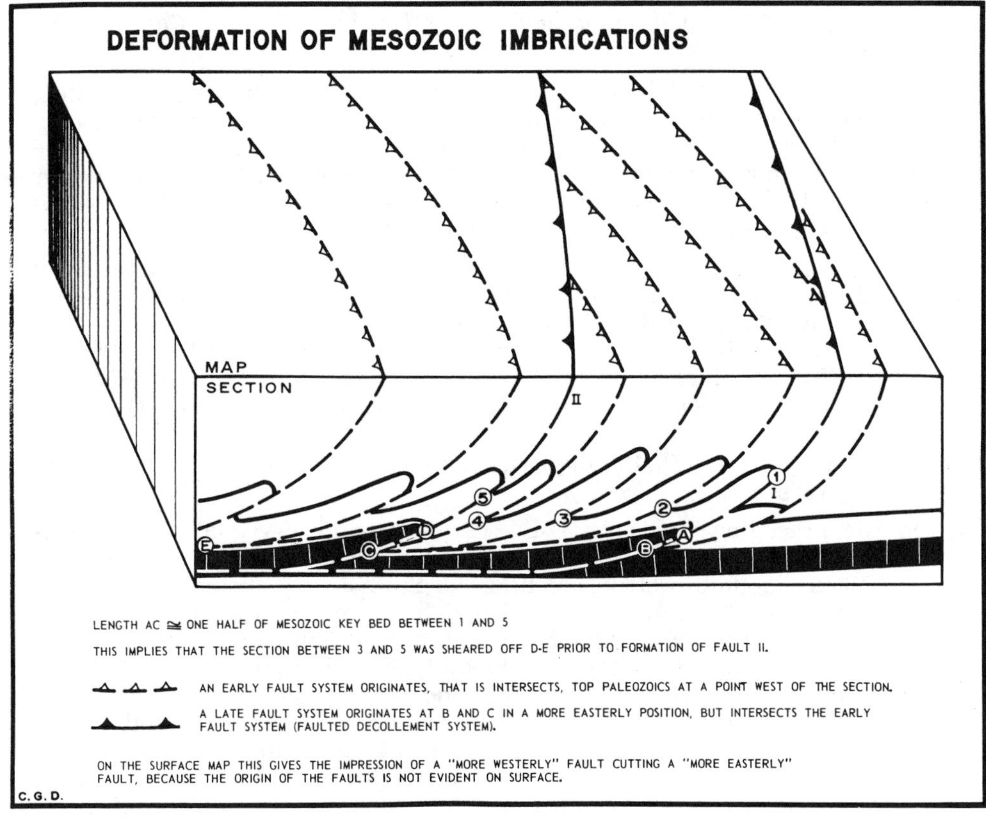

FIG. 6.--Deformation of Mesozoic imbrications.

fault cuts older faults, suggesting that a more westerly element was formed later and intersected more easterly elements. In reality we are dealing with a deeper and more easterly fault system that broke through an earlier formed system.

When section B-B′ is related to the longitudinal section (Pl. 8), the lens-like western portion of the lower sheet can be seen to split into several imbrications below the Moose Mountain culmination to the south, but to remain intact in the more northerly Panther culmination. In both localities the existence of the lower feature was predicted and subsequently confirmed by wells.

A comparison between section A-A′ and section B-B′ shows differences in structural style. In the Ram-Stolberg section fewer imbrications involving Mesozoic clastic rocks occur because the Paleozoic sheets do not lie "flush" on top of each other and therefore much of the Mesozoic sequence is undisturbed and preserved below the Brazeau thrust sheet. This enables us to better understand surface maps. It appars that areas marked by a multitude of Mesozoic imbrications indicate the presence of carbonate sheets stacked on top of each other without intervening Mesozoic clastic strata.

C. *Geologic Section C-C′ and Seismic Section c-c′, Turner Valley-Highwood Area* (*Plates* 5, 6)

This section, located about 50 miles southeast of the one preceding, illustrates the structure of the Turner Valley oil field and the neighbouring Highwood and Sullivan Creek structures, which are located to the west and are non-productive. The geologic section C-C′ extends farther west than the seismic section in order to tie to the Front Ranges and provide some continuity with longitudinal section D-D′.

The sections illustrate a classic oil exploration area with particularly good seismic data. Compared with the two preceding sections, we can observe elements of both structural styles, that is, thrust sheets that override the relatively intact Mesozoic sequence, and thrust sheets that sheared off a considerable amount of Mesozoics from their substratum (Sullivan Creek structure and Mesozoic imbrications to the east).

D. *Geologic Section D-D′ Longitudinal Section Scalp Creek to Oldman River* (*Plate* 8)

On the preceding C-C′ and B-B′ sections we observed a gently west-dipping basement and in the west two structural highs (the projection of the Moose Mountain structure and the Sullivan Creek feature). Both structures are on an anticlinal trend which can be traced from the Scalp Creek area via Panther River-Moose Mountain-Sullivan Creek into the area of the Savanna Creek gas field and beyond. The anticlinal trend is characterized by axial culminations and depressions designated on the longitudinal section. Extensive drilling has been done on top and on the flanks of these culminations and some of the more important wells are shown on the section. The section is based on numerous seismic lines, many of which show the presence of the undisturbed Cambrian event.

It is important to recognize regional axial culminations and depressions in order to utilize the information for axial projections into cross-sections. In the Alps and in many other areas of alpinotype deformation, axial projections from culminations into the neighbouring depressions have frequently been the key to an understanding of complex structural problems. It is, however, often difficult to explain the nature of the axial culminations themselves. Frequently based on analogy with the Alps, it has been concluded that major culminations represent autochthonous basement highs.

Section D-D′ gives insight into the nature of axial culminations and depressions in the Rockies. Clearly, there are no basement highs underlying the culminations, and basement depressions are also absent. Proper correction for velocity effects has removed anomalies evident on the raw data, leaving only a possibility for minor features smaller than the resolution power of reflection seismic, i.e. about 1,000 feet, by present seismic methods. We therefore conclude that the occurrence of axial culminations is due either to deeper imbrications and thrust sheets or to local variations in thickness of the stratigraphic sequence involved in the thrust sheet or else to a combination of both. Although seismic and drilling have convincingly eliminated the possibility of major basement highs, it is often difficult to conclude which of the above alternatives is likely because the internal structure of stacks of carbonate imbrications is rather difficult to unravel on reflection lines. Erdman (1950) pointed out that axial culminations and depressions tend to line up. The authors also have qualitatively plotted culminations and depressions on a regional map and found they tend to line up in predominantly northeasterly directions. Because this was not done in a strictly quantitative manner it was of course difficult to decide which culmination is minor and which is major. Also, to be meaningful, plotting should be done on a palinspastic base.

E. *Geologic Section E-E′ and Seismic Section e-e′, Lewis Overthrust Area* (*Plates* 5, 7)

Figure 7 illustrates the surface geology in a simplified manner. Two major gas fields, Waterton and Pincher Creek, are shown on the sections. On seismic section e-e′ the Cambrian event can be followed with reasonable continuity under the Waterton structure and the Lewis overthrust. Adjacent to the east end of this section, another seismic line published by Robertson (1963) revealed the presence of an intra-basement reflection dipping from 7.5 km depth in the east to 14 km depth in the west.

The Pincher Creek structure is rather obscure on this section, due to the small throw on the fault (between 500 and 1,000 feet). The structure is better recognized on refraction data. The Waterton structure is readily recognizable by its large velocity uplift and the anticlinal form of the composite Mesozoic envelope. The internal structure of this stack of Paleozoic imbrications is difficult to unravel seismically but is evident in numerous wells that have penetrated this feature.

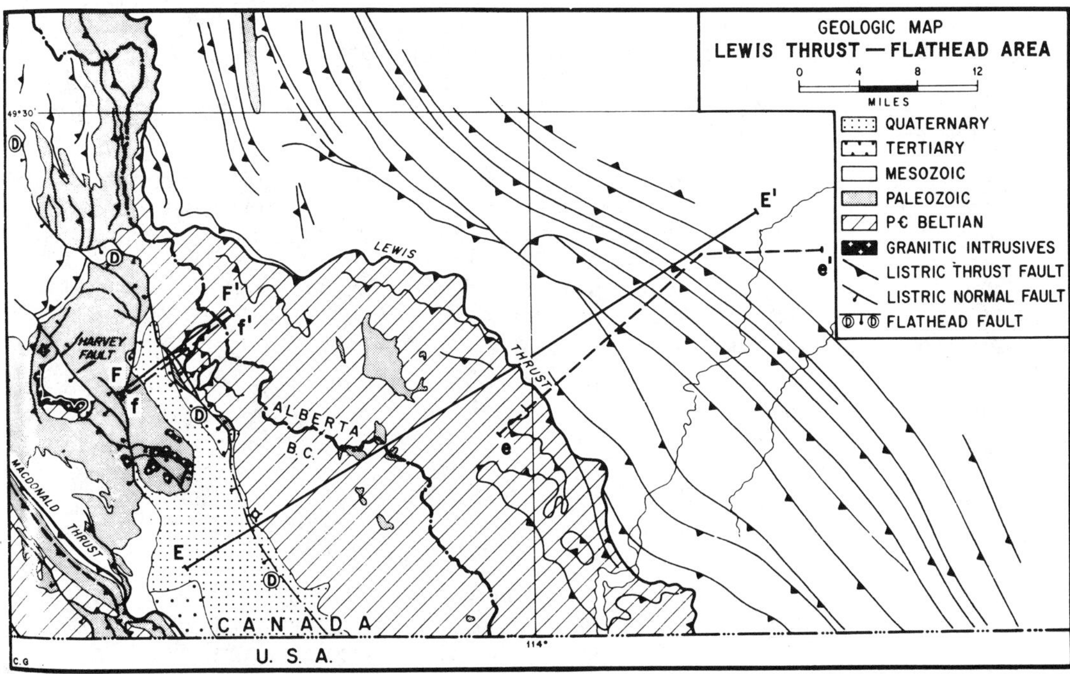

FIG. 7.—Geologic map of Lewis thrust-Flathead area.

Discordance of 15° in the strike between the Mesozoic imbrications and the underlying Pincher Creek structure has been discussed by Erdman *et al.* (1953) and Fox (1959). Erdman *et al.* suggest that the Paleozoic rocks of the Pincher Creek structure were deformed in early Laramide time and later overridden by Mesozoic imbrications (see also Gallup, 1955). Fox prefers the reverse sequence of events, that is, the Lewis overthrust mass formed first, moving and rotating Cretaceous imbrications in front of it, with the Pincher Creek structure formed after this early phase.

Our cross-section shows that the imbrications overlying the Pincher Creek structures are sheared off the top of the complex Waterton structure and strike parallel to that feature. Whether the Pincher Creek structure was formed prior to, or after, the deformation of the overlying Mesozoic imbrications, cannot be decided with the data available, but we feel that evidence elsewhere in the Foothills and Rockies strongly favours a deformation progression from west to east as suggested by Fox, and illustrated in a generalized manner on Figures 6 and 13.

F. *Geologic Section F-F′ and Seismic Section f-f′, Flathead Area* (*Plate* 9)

To the west, the Precambrian masses of the Lewis overthrust are limited by a normal fault dipping about 40° to the west, which at places on surface shows stratigraphic throws exceeding 10,000 feet and dies out to the north (see Flathead fault on Fig. 7). Associated with the fault is a narrow fault basin filled with continental clastic rocks of early Oligocene to late Eocene age (Russell, 1964). These beds contain pebbles derived from the neighbouring mountains. The beds are deformed in proximity to the fault and generally dip into the fault. Hence the clastics were deposited after emplacement of the Lewis overthrust and the normal fault formed during, but mostly after, deposition of the Tertiary beds (Pardee, 1950).

On surface (Fig. 7), the Lewis overthrust is offset by the Flathead fault and regional mapping shows clearly that the Lewis overthrust lies at depth under the mountains to the west. Many interpretations have been suggested for the Flathead fault. They cover the spectrum from normal faults to east-dipping reverse faults and overthrusts; recently, Oswald (1964) and Dahlstrom, Daniel, and Henderson (1962) published a sequence of interpretations illustrating low-angle normal faults.

With many other geologists we think that the Flathead fault and the adjacent Tertiary basin are part of a large province characterized by the faulted Tertiary basins in Montana. The Rocky Mountain Trench may be an extension of this province, which to the south probably extends into the Basin and Range province of Nevada. The seismic documentation of the Flathead fault shown will aid our understanding of the Tertiary structural evolution of this whole province.

In places the Flathead fault is a complex zone containing elements of Beltian and Paleozoic rocks (Price, 1965). Seismic section f-f′ and geologic section F-F′ illustrate a strong "basal Mesozoic" event which can be followed with continuity and which dips to the west. Refraction data calibrated by drilling show that this event is underlain by Paleozoic carbonates, and that it has not been displaced by the Flathead fault as far west as data are available. It must be concluded that the Flathead normal fault flattens at shallow depth and, as shown, we believe it merges with the Lewis overthrust. This suggests the Lewis overthrust has been "stretched" or "bottlenecked" after its emplacement, an event also related to the formation of the Tertiary basins.

The conclusion is that the Flathead fault is a listric normal fault formed after emplacement of the Lewis overthrust by "back-slippage" along a pre-existing thrust during a phase of post-orogenic uplifting. The predominance of east dips in the Tertiary beds west of the fault is perhaps due to rotation along a listric normal fault. A similar situation, largely supported by surface geology and drilling, was reported by Dahlstrom *et al.* (1962).

G. *Geologic Section G-G′ and Seismic Section g-g′, Rocky Mountain Trench* (*Plates* 10, 11)

The Rocky Mountain Trench is a linear topographic depression extending well over 1,000 miles, generally regarded as separating the eastern from the western Cordillera. Structures on both sides of the Trench are commonly truncated and in many places faulting is associated with its margins. Within it, Tertiary and younger sediments predominate.

Some mapping and a fair amount of speculative material have been published on the Trench. Its origin is still debated and opinions range from formation due to thrust faulting, normal faulting, and transcurrent faulting, to glaciation and stream erosion.

Our seismic data are located in the southern portion of the Rocky Mountain Trench, in an area that has recently been remapped. Figure 8 is based essentially on this work. Leech (1959) points out that this area is not necessarily characteristic of other portions of the Trench, and this implies that conclusions based on information here should not be generalized too eagerly. At the same time this portion of the Trench is the only one mapped and studied in reasonable detail, whereas the remainder of the Trench is less well known.

In our area both sides of the Trench are flanked by mountains underlain by generally east-dipping Beltian strata. The Trench is characterized by block-faulting of a type also observed in neighbouring Montana; the faults appear to be antithetic normal faults separating blocks tilted towards the east. The valley of the Trench contains some outcrops of Upper Paleozoic carbonates, and Miocene sands, silts and gravels occur in one area. Fluvial and glacial deposits fill several bedrock depressions in the Trench with thicknesses in the order of 1,500 to 5,000 feet (Lamb and Smith, 1962; Thompson, 1962).

Seismic data (section g-g′) were obtained on both sides of the Trench; however, reflection data across its floor are too poor to be of use. By Foothills standards the quality of the reflections is poor, but the few data obtained are of considerable interest.

On the east side of the Trench we see a continuation of the Cambrian event which can be followed with continuity from the Plains across the Foothills and the Fernie Basin. At about 2.0 seconds we observe some flat bands of energy which are characteristic for the "seismic-structural

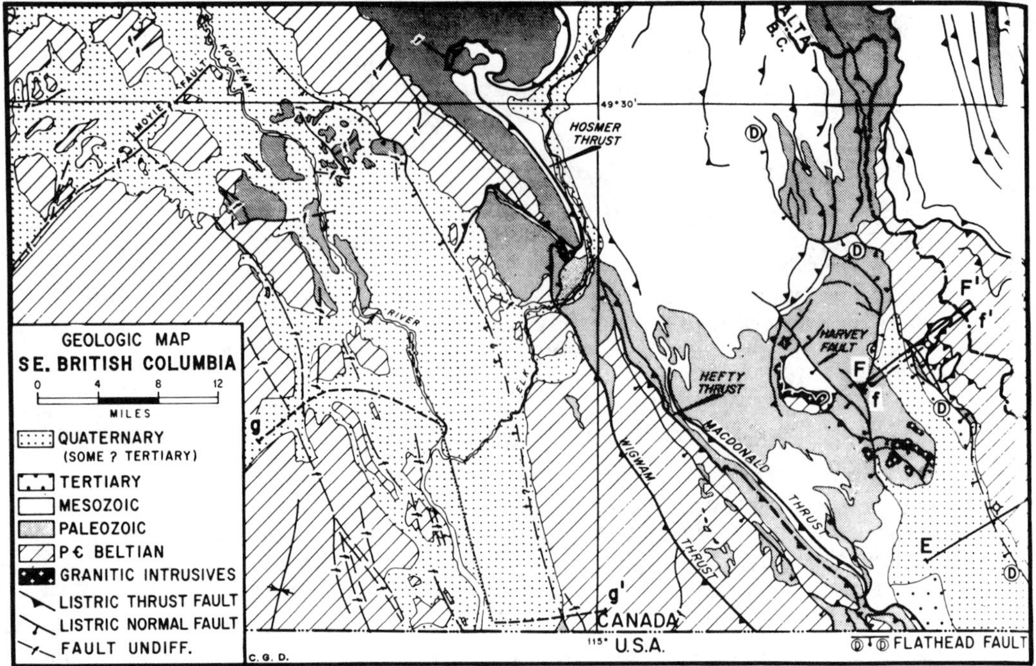

FIG. 8.—Geologic map of southern British Columbia.

style" of the Rocky Mountains. Regional data indicate that the extension of the major thrust faults, located east of the Trench and shown on the east half of section G-G′, project into this zone of flat reflections.

On the west side of the Trench we observe an event that falls in the projected extension of the "Cambrian" event; but we also observe an indication of another reflection that diverges from it. We like to speculate that the wedge contained between both events may be the autochthonous eastern wedge-edge of the Belt Series. The lower event therefore cannot be here approximated with the Cambrian as previously assumed for all Foothills sections, but it would still represent the top of the "shield-type basement." If we assume average overlying velocities of 17,000 feet per second, the top of the Precambrian Shield would lie at a depth of about 36,000 feet. An alternative interpretation would suggest that the divergent dip is the root zone of a major thrust, e.g. the Lewis thrust.

On the west side we again observe some flat and some converging dips, suggesting a layered structure. Converging dips are in many instances symptomatic of leading edges of thrust sheets (i.e. the intersection of beds with thrust faults).

Because our conclusions will be subject to controversy we list below what we consider to be reasonably well established geophysical facts concerning this part of the Trench:

1. At depth, the structure on both sides of the Trench appears to be essentially similar and is layered.

2. The Cambrian event of the Foothills can be traced to the east side of the Trench.

3. On the west side of the Trench we recognize a reflection that is on the regional projection of the Cambrian event of the Foothills. This reflection appears to branch out to the west into two events.

We now combine these points with the following geological surface observations:

1. Major normal faults occur on the east side of the Trench; some also are observed on the west side.

2. Farther north in the Cranbrook-Fort Steele area, a complex system of transverse faults (Moyie-Dibble Creek faults) crosses the Trench and can be seen on both sides. Although the exact relations between these faults are obscure, the suggestion is that major strike-slip movements have not occurred along the Trench.

3. The observed normal faults form part of a regional system that includes the faults on flanks of the Tertiary basins of Montana and the Flathead fault. The latter is therefore an adequate analog and could serve as a model to explain normal faulting in the Rocky Mountain Trench.

In addition, based on reconstructions discussed in the following paragraph (see also Plate 12), it must be pointed out that until late Late Cretaceous time the area of the Trench was occupied on both sides by sedimentary sequences that are presently "stacked up" in the Front Ranges of Alberta.

Consideration of these geophysical and geological points suggests the propositions (1) that the Trench was formed in Tertiary time and after the main thrusting phase, (2) that the Trench is underlain by an undisturbed gently westward dipping basement, and (3) that location and strike of the Trench is dictated by a complex system of curved low-angle normal faults or "listric normal faults." The eastward tilting of the associated fault blocks is explained in terms of reverse drag (Hamblin, 1965). We agree with Leech (1965) that at least the southern portion of the Trench is structurally controlled and that stream erosion and depositional filling "were governed directly by topography consequent upon block faulting."

Regional Considerations and Palinspastic Restoration

The basement contours shown on Figure 9 are based on seismic control combined with subsurface information from the Plains, the Foothills and the eastern Front Ranges. For the westerly zones of the Rocky Mountains control is limited to the area south of Latitude 50°N. and based only on a few regional reflection lines. Accessibility considerations and the location of National Parks preclude obtaining more data to the north. There is, however, little doubt in our minds that a gently westward dipping basement is characteristic for the Rocky Mountains of Alberta and British Columbia.

The most intriguing aspect of the data presented in this paper is the seismic definition of a Cambrian event and the underlying basement top below the Rocky Mountains, which together with the surface geology set clearly defined boundaries for the construction of structural cross-sections. This in turn drastically limits speculations concerning the deep structure of the Rocky Mountains. It also enables us to construct reasonably accurate palinspastic restorations.

On Plate 12 each imbrication and thrust sheet is shown in its present and restored position. Offsets in the line of section are disregarded, and for ease of construction the restoration is tied to the present basement gradient, whereas to the west no specific datum was assumed. Note that nowhere in the structure section or in its restoration is the Paleozoic section completely missing. We assume that no major structural units

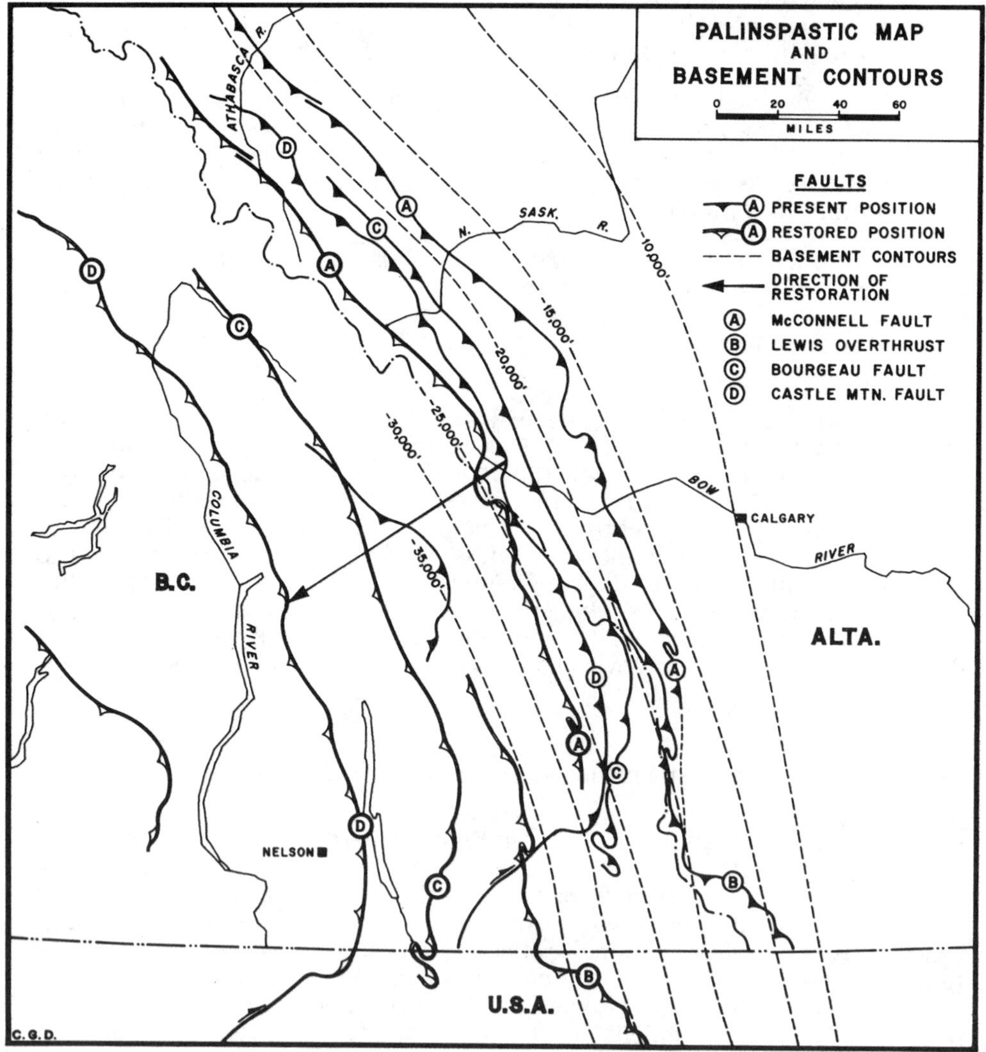

FIG. 9.—Palinspastic map and basement contours, Southern Rocky Mountains.

have been removed by erosion; therefore, our construction is considered conservative.

It appears that shortening in the Rocky Mountains is in the order of 50 per cent, or about 100 miles for the Main Ranges, and about 120 miles for the eastern Purcell Mountains. Comparable amounts of shortening have been postulated by others (North and Henderson, 1954a; Shaw, 1963).

It is instructive to show the present geology of the areas that apparently once were occupied by the Paleozoic and Mesozoic sediments now "piled up" in the Rocky Mountains (Fig. 8). The western half of structural cross-section B′-B″ across the Selkirk and Purcell Mountains is based on maps and sections published by the Geological Survey of Canada. We are particularly grateful to J. E. Reesor who provided manuscript maps and cross-sections across the Valhalla complex and who reviewed the structure section. His comments induced us to modify our previous interpretations. We also wish to thank G. B. Leech, J. D. Aitken, G. G. L. Henderson and C. D. A. Dahlstrom for their comments. Responsibility for the interpretation is of course ours, but their critical review helped us to avoid extravagant interpretations in areas unfamiliar to us.

The restoration shown on Plate 12 assumes that the Alberta Plains at the east end of the section have remained geographically fixed, and that the Rocky Mountains, as well as the granite and gneiss masses of interior British Columbia, all were folded and thrust from the west, onto and over the neighbouring craton. However, geologists familiar with the Purcell and Selkirk Mountains are reluctant to assume extensive allochthoneity of gneiss domes and discordant batholiths (see also Reesor, 1965). Our own interpretation can be supported only with reference to analogs, such as the Penninic zone of the Alps.

The restoration implies that the space presently occupied by igneous rocks, yielding Jurassic, Cretaceous and Tertiary radiometric ages (Fig. 12), was occupied by Paleozoic and Mesozoic geosynclinal sequences until the end of the Cretaceous. We therefore postulate that the igneous rocks, which probably contain metamorphic equivalents of Precambrian to Jurassic sediments, were moved into their present position during Late Cretaceous and Tertiary time. In this context it is interesting to note that K-Ar radiometric ages from interior British Columbia appear to correlate with the ages of deposition of sandstone sequences in the Foothills and Plains of southern Alberta, indicating that metamorphic events coincide roughly with the deposition of coarse clastics to the east. In a much broader sense, our restoration agrees with Shaw's (1963) suggestion that the Pacific shelf edge may have been extended westward by an amount exceeding 100 miles.

There is, however, an alternative to the proposed restoration. Rather than assuming a fixed position for the Plains, it could be postulated that the inner igneous portion of the Cordillera was geographically fixed during Mesozoic and Tertiary time. According to this mechanism, the Precambrian Shield-type basement was dragged under the batholiths and gneiss domes of British Columbia, which in part would represent the remobilized part of the downfolded craton. Charlesworth (1959) expressed a similar view in proposing shield underthrusting.

Because the craton represents the main portion of the North American continent, this alternative in effect assumes continental drift during Mesozoic and Tertiary times. Such a proposition is strengthened by the observation that the formation and subsidence of the Atlantic shelf occurred during Jurassic to Tertiary time, that is, simultaneous with deformation indicated for the Western Cordillera.

Mechanically, the two alternatives appear to be similar and describe relative movements (underthrust versus overthrust). A decision is difficult and not within the scope of this paper. It must be considered however, that any synthesis of Cordilleran mountain building must also reconcile observations made on the Pacific side of this mountain belt.[1] Recent studies show that the western Cordillera is characterized by extensive thrusting towards the west involving igneous and metamorphic rocks, as well as sediments of Mesozoic and younger age. The considerable shortening observed at the surface on both sides of the Cordillera is likely to be coupled to shortening within deeper portions of the continental crust.

The palinspastic position of major Paleozoic structural elements of the Foothills and Rockies is shown on Figure 9. Because of differential décollement movement, this reconstruction is not necessarily valid for the overlying Mesozoic sequences.

Inspection of a structural map (e.g. Fig. 3) of the Foothills indicates that it is not possible to restore units along directions perpendicular to their present strike, without crowding the restored units in the west. The best palinspastic restoration results from selection of a particular arbitrary direction (roughly N.E.-S.W.) perpendicular to the overall strike of the Rocky Mountains. However, this procedure is probably basically incorrect because its application to similar mountain ranges with greater plan curvature, such as the Alps, the Carpathians and the Himalayas, would lead to sizable difficulties. Therefore, our reconstruction should be viewed as an approximation that is adequate because it is based on seismic data, but, less adequate because it uses a constructional procedure that is not fully acceptable.

Summary of Structural Evolution of Southern Canadian Rocky Mountains

A. INTRODUCTION

In discussing various aspects of the structural evolution of the Rocky Mountains several authors have suggested a succession of structural events, mainly inspired by the interrelations of folds and faults. Stratigraphic evidence has been used only to indicate that formations involved in the structural deformation give a lower time limit to the process of deformation, and most discussions revolve around these main questions:

1. Does folding precede thrust faulting?
2. Are "folded" faults formed during or after the faulting process?
3. Does deformation proceed from west to east or from east to west?
4. Does normal faulting precede or follow thrust faulting?

[1]Readers are referred to "Tectonic History and Mineral Deposits of the Western Cordillera," (Canadian Inst. Mining and Metallurgy, Spec. Vol. 8, 353 p.), published after this paper was in press. Ed.

Another group of authors has attempted to reconstruct the history of the Rocky Mountains by using mainly stratigraphic and paleogeographic deductions. The bulk of the stratigraphic evidence indicates that structural deformation has proceeded from west to east, and that therefore the easternmost elements of the Cordillera were the last formed. Thus, many authors assume earlier deformation in the western eugeosynclinal belt. Documentation for such movements during Paleozoic time is extremely meagre, but there is no doubt that the western geosyncline was deformed during early Mesozoic time. Often overlooked is the fact that the type of deformation in the western eugeosynclinal belt is characterized by intrusive and metamorphic events, in contrast to the peel-thrusting style exhibited in the Rocky Mountains proper. The igneous and metamorphic events are frequently dated by radiometric methods and labeled as "orogenic" events. Problems relating to this practice have been reviewed recently by Gilluly (1966).

A few authors (e.g. Alden, 1932; Pardee, 1950) have used a geomorphic approach to unravel the late structural evolution of the Rocky Mountains. This is rather astonishing, because only the latest phase of deformation triggered the erosional processes that led to formation of a spectacular mountain range.

Any discussion of structural events faces semantic problems concerned with the questions: What is a geosyncline? What is an orogeny? What is a foredeep? Reference to our preferred usage of some terms on Figure 2, Figures 10 to 14, and Plate 13 may avoid terminological problems with the following brief summary of the sequence of structural events related to the evolution of the Southern Canadian Rocky Mountains.

B. GEOSYNCLINAL PHASE (PLATE 13)

Plates 12 and 13 illustrate the restored Rocky Mountain geosyncline. The boundary between stable platform sediments and geosynclinal sediments is arbitrarily drawn at the Foothills boundary where the carbonate skeleton is involved in thrust faulting. The boundary between eu- and miogeosyncline straddles the Purcell Mountains, which contain the first major granitic intrusives (Fig. 2).

Relations with the western extension of the eugeosyncline are obscure because the record has been destroyed by numerous granitic intrusions. What little evidence there is suggests some continuity with the eugeosynclinal sequences of Prince of Wales Islands in Alaska and the Orcas Island Group in the Strait of Georgia.

It appears that rocks of the miogeosyncline were deposited on the peneplained extension of the Canadian Shield, which had been consolidated during the Hudsonian orogeny (about 1,700 m.y. ago). We interpret this basement also to underlie the eugeosynclinal portion of the Cordilleran geosyncline. This is suggested by Precambrian age determinations reported from California (Wasserburg *et al.,* 1959) and more generally by the fact that some better known eugeosynclinal systems (e.g. Alps and Appalachians) are clearly underlain by a sialic, shield-type, pre-geosynclinal basement.

Miogeosynclinal subsidence began in Middle Proterozoic time (Helikian of Stockwell, 1965) with Purcell strata and lasted some 1,100 m. years

The most intriguing aspect of the data presented in this paper is the seismic definition of a Cambrian event and the underlying basement top below the Rocky Mountains, which together with the surface geology set clearly defined boundaries for the construction of structural cross-sections. This in turn drastically limits speculations concerning the deep structure of the Rocky Mountains. It also enables us to construct reasonably accurate palinspastic restorations.

On Plate 12 each imbrication and thrust sheet is shown in its present and restored position. Offsets in the line of section are disregarded, and for ease of construction the restoration is tied to the present basement gradient, whereas to the west no specific datum was assumed. Note that nowhere in the structure section or in its restoration is the Paleozoic section completely missing. We assume that no major structural units

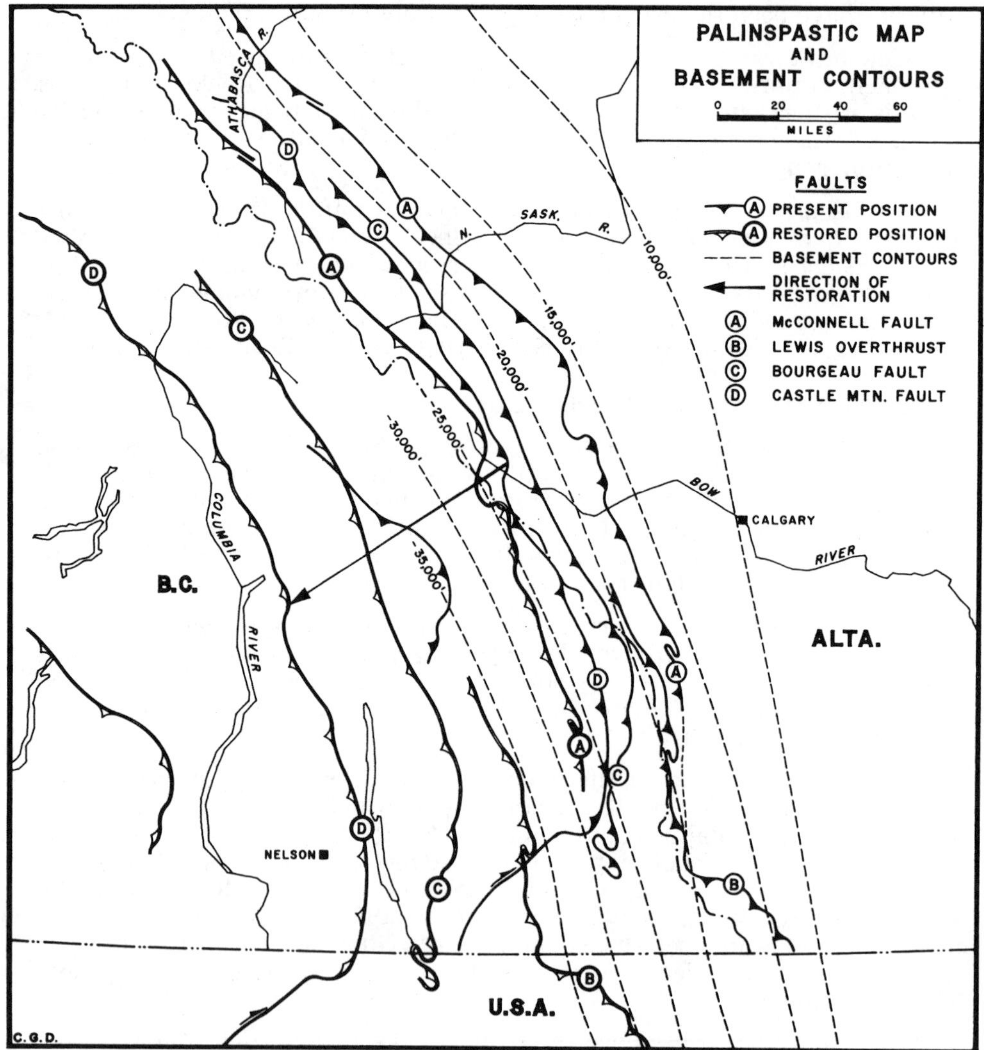

FIG. 9.—Palinspastic map and basement contours, Southern Rocky Mountains.

have been removed by erosion; therefore, our construction is considered conservative.

It appears that shortening in the Rocky Mountains is in the order of 50 per cent, or about 100 miles for the Main Ranges, and about 120 miles for the eastern Purcell Mountains. Comparable amounts of shortening have been postulated by others (North and Henderson, 1954a; Shaw, 1963).

It is instructive to show the present geology of the areas that apparently once were occupied by the Paleozoic and Mesozoic sediments now "piled up" in the Rocky Mountains (Fig. 8). The western half of structural cross-section B′-B″ across the Selkirk and Purcell Mountains is based on maps and sections published by the Geological Survey of Canada. We are particularly grateful to J. E. Reesor who provided manuscript maps and cross-sections across the Valhalla complex and who reviewed the structure section. His comments induced us to modify our previous interpretations. We also wish to thank G. B. Leech, J. D. Aitken, G. G. L. Henderson and C. D. A. Dahlstrom for their comments. Responsibility for the interpretation is of course ours, but their critical review helped us to avoid extravagant interpretations in areas unfamiliar to us.

The restoration shown on Plate 12 assumes that the Alberta Plains at the east end of the section have remained geographically fixed, and that the Rocky Mountains, as well as the granite and gneiss masses of interior British Columbia, all were folded and thrust from the west, onto and over the neighbouring craton. However, geologists familiar with the Purcell and Selkirk Mountains are reluctant to assume extensive allochthoneity of gneiss domes and discordant batholiths (see also Reesor, 1965). Our own interpretation can be supported only with reference to analogs, such as the Penninic zone of the Alps.

The restoration implies that the space presently occupied by igneous rocks, yielding Jurassic, Cretaceous and Tertiary radiometric ages (Fig. 12), was occupied by Paleozoic and Mesozoic geosynclinal sequences until the end of the Cretaceous. We therefore postulate that the igneous rocks, which probably contain metamorphic equivalents of Precambrian to Jurassic sediments, were moved into their present position during Late Cretaceous and Tertiary time. In this context it is interesting to note that K-Ar radiometric ages from interior British Columbia appear to correlate with the ages of deposition of sandstone sequences in the Foothills and Plains of southern Alberta, indicating that metamorphic events coincide roughly with the deposition of coarse clastics to the east. In a much broader sense, our restoration agrees with Shaw's (1963) suggestion that the Pacific shelf edge may have been extended westward by an amount exceeding 100 miles.

There is, however, an alternative to the proposed restoration. Rather than assuming a fixed position for the Plains, it could be postulated that the inner igneous portion of the Cordillera was geographically fixed during Mesozoic and Tertiary time. According to this mechanism, the Precambrian Shield-type basement was dragged under the batholiths and gneiss domes of British Columbia, which in part would represent the remobilized part of the downfolded craton. Charlesworth (1959) expressed a similar view in proposing shield underthrusting.

to the end of the Paleozoic. This enormous span of overall subsidence was interrupted by epirogenic events identified by interregional unconformities that subdivide the geosyncline into sequences (Sloss, 1963; Webb, 1964).

Westward from the stable cratonic platform of the western Plains, the stratigraphic sequences become more complete and unconformities less pronounced. The origin of these unconformities is obscure, but we believe they are due to warping of the entire North American continent related to early deformation of the Appalachian and Innuitian fold belts. There is, however, no reason to link unconformities with major mountain building events in the western Cordillera. A brief description of the stratigraphic sequences and their deformation follows.

1. *Purcell Series*

   This series appears to be an unconformity-bounded sequence, and following Sloss' practice it could perhaps be designated by a name of Indian origin (e.g. Skookumchuk sequence). It represents a time span of at least 600 m.y., as indicated by "common lead" ages of about 1,300-1,400 m.y. (Lang *et al.,* 1960; Wanless *et al.,* 1958) and 700-800 m.y. indicated by K-Ar ages of granitic intrusions of probable pre-Windermere age (Leech, 1962b). During this time 30,000 to 40,000 feet of clastic and carbonate rocks were deposited in a sequence that is frequently compared with the Tertiary of the Gulf Coast. However, the apparent rate of deposition of the Purcell is much lower.

2. *Windermere Unconformity*

   This unconformity has been described by Walker (1926) and Reesor (1957a), who point out that deformation was limited to the formation of gentle open folds which, however, in places led to erosion of up to 20,000 feet of Purcell rocks. A major geanticline was formed in the Purcell Mountains. Leech (1962b) suggests that the above mentioned granitic intrusion and deformation preceding deposition of the Windermere system were related. This in effect means that the Grenville orogeny affected the Rocky Mountains (East Kootenay orogeny of White, 1959). Although the evidence indicates the occurrence of a major structural event, we question the advisability of calling it an orogeny, because subsidence resumed subsequently. What little can be seen does not suggest the presence of a major folded belt with its intensive folding and widespread granitic intrusion.

3. *Sauk Sequence (Windermere-Cambrian-Lower Ordovician)*

   This sequence includes more than 4,000 feet of Windermere clastic rocks, Lower Cambrian quartzites and about 16,000 feet of Middle and Upper Cambrian and Lower Ordovician carbonates. The inclusion of the Windermere system in this sequence may be debatable because it is generally assumed that the Lower Cambrian is underlain by an unconformity. However, Windermere rocks are commonly conformably overlain by the Lower Cambrian, and it appears that the Lower Cambrian onlaps a sub-Windermere erosional surface.

An anomaly occurs in the Mount Forster section (Pl. 12, sec. B′-B″) where Upper Cambrian directly overlies Windermere beds (Walker, 1926; Reesor, 1957b) and only rudiments of the Sauk sequence (about 450 feet of Upper Cambrian Jubilee and McKay) are preserved, whereas in the surrounding area the sequence is represented by thicknesses exceeding 15,000 feet (see also North, 1964). Whether this "Mount Forster high" is of local or regional significance remains a question.

4. *Middle Ordovician Unconformity*

Probably only a minor unconformity underlies the Middle Ordovician Mount Wilson Quartzite in the Southern Canadian Rocky Mountains (Norford, 1964) and there is little evidence for any significant structural deformation preceding deposition of the overlying sequence.

5. *Tippecanoe Sequence (Ordovician-Silurian-Lower Devonian)*

Due to extensive pre-Middle Devonian erosion, the record of this sequence is poorly preserved. About 1,500 feet of Upper Ordovician-Silurian Beaverfoot-Brisco carbonates overlie the Mount Wilson Quartzite.

6. *Sub-Middle Devonian Unconformity*

Next to the sub-Windermere unconformity, this is by far the most important break in deposition in the Rocky Mountain geosyncline. The significance of this unconformity can best be grasped by a study of published subcrop maps (A.S.P.G., 1964b). Prominent subsurface features in the Western Plains such as the Peace River Arch, the Western Alberta Arch, and the land-mass of Montania had been already formed by Middle Devonian time. Uplift causing these features possibly started towards the end of the Silurian Period and lasted through Early Devonian and early Middle Devonian time, although evidence for this in the Southern Rocky Mountains is meagre. As previously mentioned, we visualize tilting and warping of large areas of the continent being related to major events in the Appalachian and Innuitian fold belts.

It is conceivable that the intrusion of the Ice River syenites (K-Ar dates between 330 and 392 m.y., Baadsgaard *et al.*, 1961; Rapson, 1963) is related to deformation preceding deposition of the Middle Devonian-Mississippian sequence. Here again however, we are reluctant to equate such an intrusion with the occurrence of a major orogenic event. Clearly, in the Southern Canadian Rocky Mountains there is no evidence for the existence of such an event, with its associated intensive folding, faulting and extensive granitic intrusion.

7. *Kaskaskia Sequence (Middle Devonian, Upper Devonian, Mississippian and Pennsylvanian)*

This sequence contains all the major Paleozoic hydrocarbon accumulations of western Canada. A gradual transgression began with red beds, clastic rocks and evaporites of the Middle Devonian Elk Point Group, followed by the overall transgressive reefal sequences of the Middle and Upper Devonian. The latter contains a minor erosional unconformity at the base of the Fammenian (McLaren and Mountjoy, 1962). During Mississippian time the sequence changed to a major regressive cycle, culminating with siltstones, sandstones and carbonates during Chester and Early and Middle Pennsylvanian time, which concluded the geosynclinal phase.

In British Columbia, Pennsylvanian and Permian beds unconformably overlie intensely folded and metamorphosed rocks of the Shuswap complex and related formations, a relationship which led White (1959) to postulate the Cariboo orogeny. This deformation could be an extension of the Antler orogenic zone of Nevada. It suggests the end of the geosynclinal regime and the beginning of orogenic movements heralding formation of the Rocky Mountains. However, evidence in southern British Columbia for such an orogeny is limited and, until more information becomes available, we prefer not to emphasize its importance.

To sum up, it can be stated that geosynclinal sedimentation persisted over a period exceeding 1,100 m. years. We recognize two major breaks, the first preceding deposition of the Windermere system, and the second preceding the Middle Devonian. In our opinion neither break can be viewed as manifesting true orogenic disturbances because in the Rocky Mountains they are not related to the full development of a folded belt. More likely, the unconformities are related to warping of the North American continent in response to orogenic phases in the Grenville, Appalachian and/or Innuitian systems. Only towards the very end of the geosynclinal phase (i.e. Pennsylvanian) can we observe meagre indications of orogenic activity in interior British Columbia.

C. PERMIAN TO MIDDLE JURASSIC WEDGE OF MULTIPLE UNCONFORMITIES (FIGURE 10)

In discussing stratigraphic data from a structural point of view, it is often convenient to differentiate periods and areas of increased tectonic activity from periods and areas of tectonic quiescence. Such separation leads to a simple subdivision of a stratigraphic time interval as illustrated on Figure 10. The time interval A-A′ is represented by the extremes of a complete hiatus and the complete sequence. Between the two is a transition zone of partial sequences and hiati, which we designate as a "wedge of multiple unconformities." The term refers to the phenomenon of one major unconformity branching into wedges of lesser or short stratigraphic sequences and minor unconformities. On a map, the zero isopachs of the short sequences, which in most cases represent the intersection of two unconformities, will tend to be regionally subparallel if the warping or tilting movements systematically affected the same area and are of related origin. Note also that stratigraphic units predating the wedge of multiple unconformities subcrop in the realm of the complete hiatus. Depending on the nature of local uplifts these subcrops are not necessarily parallel to the strike of the adjacent fold belt.

This wedge of multiple unconformities may be viewed as the transition of an intracratonic hiatus into a geosynclinal sequence which lasted for about 100 m. years.

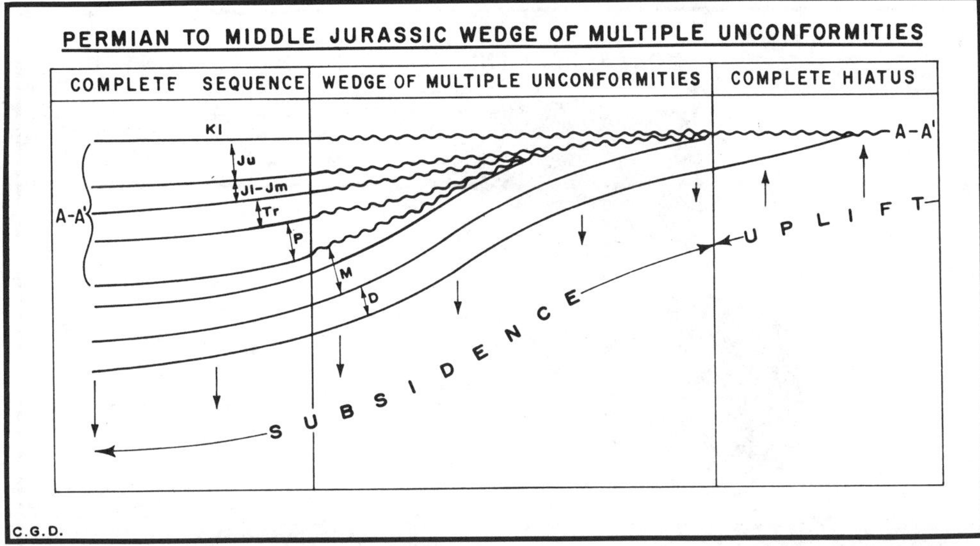

FIG. 10.—Permian to Middle Jurassic wedge of multiple unconformities.

The interval from Permian to Jurassic is represented by a wedge of multiple unconformities. In the Alberta Plains, the Lower Cretaceous laps on to a sub-Cretaceous subcrop. In a westerly direction this unconformity branches out and separates, respectively, relatively thin Permian, Triassic and Lower and Middle Jurassic beds. This suggests a tilting movement, with uplift in the east and subsidence in the west, a concept corroborated by evidence for an eastern source of the sandstones found in the Triassic and Lower-Middle Jurassic sequences.

In the interior of British Columbia, about 20,000 feet of Permian Cache Creek beds occur, consisting of basal greenstones, overlying cherts, argillites and tuffs, and an upper section of carbonates. In places this succession, which may be of Pennsylvanian age, is reported (White, 1959) to unconformably overlie older folded and metamorphosed beds. The intrusion of ultramafics characterizes the Triassic, and during Triassic and Jurassic time thick marine and continental volcanic sequences, in places exceeding 20,000 feet, were deposited in basins separated by major structural highs, which in turn were the source for clastic rocks deposited in the adjacent basins (see also Wheeler, 1965).

In summary, the Permian to Middle Jurassic interval marks the beginning of genuine orogenic deformation in the eugeosyncline, with the intrusion of ultramafics and granites, metamorphism, and the formation of land masses that rimmed epi-eugeosynclinal basins. However, the breakup of the geosyncline had not yet directly affected sedimentation in the Rocky Mountains, although it caused westward tilting.

D. MIGRATING FOREDEEP (FIGURE 11)

During this phase and for the first time, an enormous influx of clastic material derived from the west was deposited as a clastic wedge over the miogeosynclinal sequence of the Rocky Mountains and the stable platform of the Plains (exogeosyncline of Kay, 1951; clastic wedge of King, 1959).

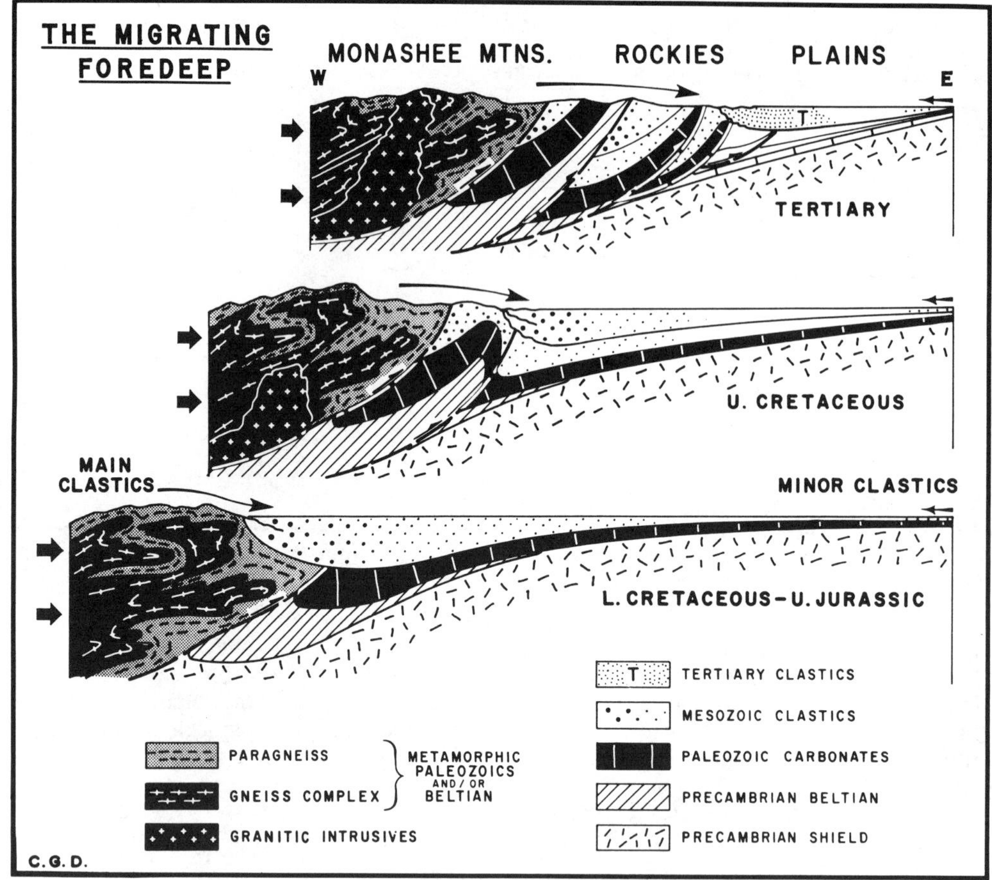

FIG. 11—The migrating foredeep.

1. *Upper Jurassic Kootenay-Nikanassin transition sequence*

   This sequence can be viewed as part of the preceding wedge of multiple unconformities because it wedges out in the Plains and appears to be bounded by two unconformities in the Foothills and Plains. In the Front Ranges however, contact between Kootenay and Lower Cretaceous Blairmore appears to be transitional. More important than this transitional contact, is the appearance of thick sequences of coarse clastic rocks (up to 4,000 feet in the Fernie area) clearly derived from a western source area (Newmarch, 1953; Springer *et al.*, 1964; Rapson, 1964).

2. *Lower Cretaceous*

   This is an essentially transgressive sequence composed mainly of alluvial and deltaic coarse clastic sediments derived from the west. Published maps suggest depocentres in the form of coalescing fans which shifted along the mountain front. Interpretations of multiple minor transgressions and regressions are probably due to failure to recognize lateral shift of these depocentres. In the southern Foothills and Front Ranges, the top of the Lower Cretaceous is marked by the Crowsnest volcanics, a sequence of volcanic breccias, flows, and tuff beds yielding a K-Ar age determination of 96 m.y. (Folinsbee *et al.,* 1957). Nearby alkaline intrusives of the Flathead area yield ages of between 61 and 126 m.y. (Gordy and Edwards, 1962; Leech *et al.,* 1963). The igneous bodies in this area are considered to be allochthonous and the intrusions are not associated with major contemporaneous dislocations in the Flathead area.

3. *Upper Cretaceous*

The base of the Upper Cretaceous marks the climax of the transgression that started in late Jurassic time and roughly coincides with the worldwide Cenomanian transgression. Only during the Late Cretaceous (Campanian-Maestrichtian) do we recognize a widespread regression. Here again we visualize a major source area in the west, as shown by the occurrence of thick sequences (exceeding 5,000 feet) of clastic rocks in the Foothills and Front Ranges. During early Late Cretaceous time minor coarse clastic rocks were deposited in the area, although an important depocentre (Dunvegan) occurred in northeastern British Columbia. Later, the deposition of sandstones resumed in the southern Foothills, as shown by thick alluvial sequences of Belly River and Edmonton strata.

4. *Paleocene*

Although Paleocene clastic rocks could be grouped with the Upper Cretaceous because they mark the climax of the regressive phase, they deserve special attention in a discussion dating structural events. Paleocene beds are involved in the deformation of the Fooothills, e.g. the Willow Creek Formation in the southern Foothills (Douglas, 1950), Paleocene beds in the Nordegg area (A.S.P.G., 1958a), and the Foothills east of Jasper National Park (Lang, 1947; Irish, 1965). In the southern Foothills, Douglas (1950) described an erosional unconformity separating Willow Creek (uppermost Cretaceous-Paleocene) from the Porcupine Hills Formation (Paleocene). The latter does not appear to be involved in the Foothills deformation. The Porcupine Hills Formation is presently correlated with the Paleocene Paskapoo Formation, which farther north is involved in deformation (Tozer, 1956; Taylor *et al.,* 1964). Should the Porcupine Hills Formation be younger than the Foothills Paskapoo, one could fix the formation of the easternmost Foothills as an intra-Paleocene event, as suggested by Bossort (1957). Such speculations, however, can be substantiated only by detailed paleontologic and palynologic correlations. In any event, the Paskapoo marks the end of the foredeep phase.

Deposition of the Mesozoic clastic wedge of the Rocky Mountains reflects orogenic, metamorphic and igneous events in interior British Columbia. Figure 12 shows a plot of age determinations from southeast British Columbia by Gabrielse and Reesor (1964) and another plot of the same data by Ross (1966). Schematically sketched are the major sandstone-

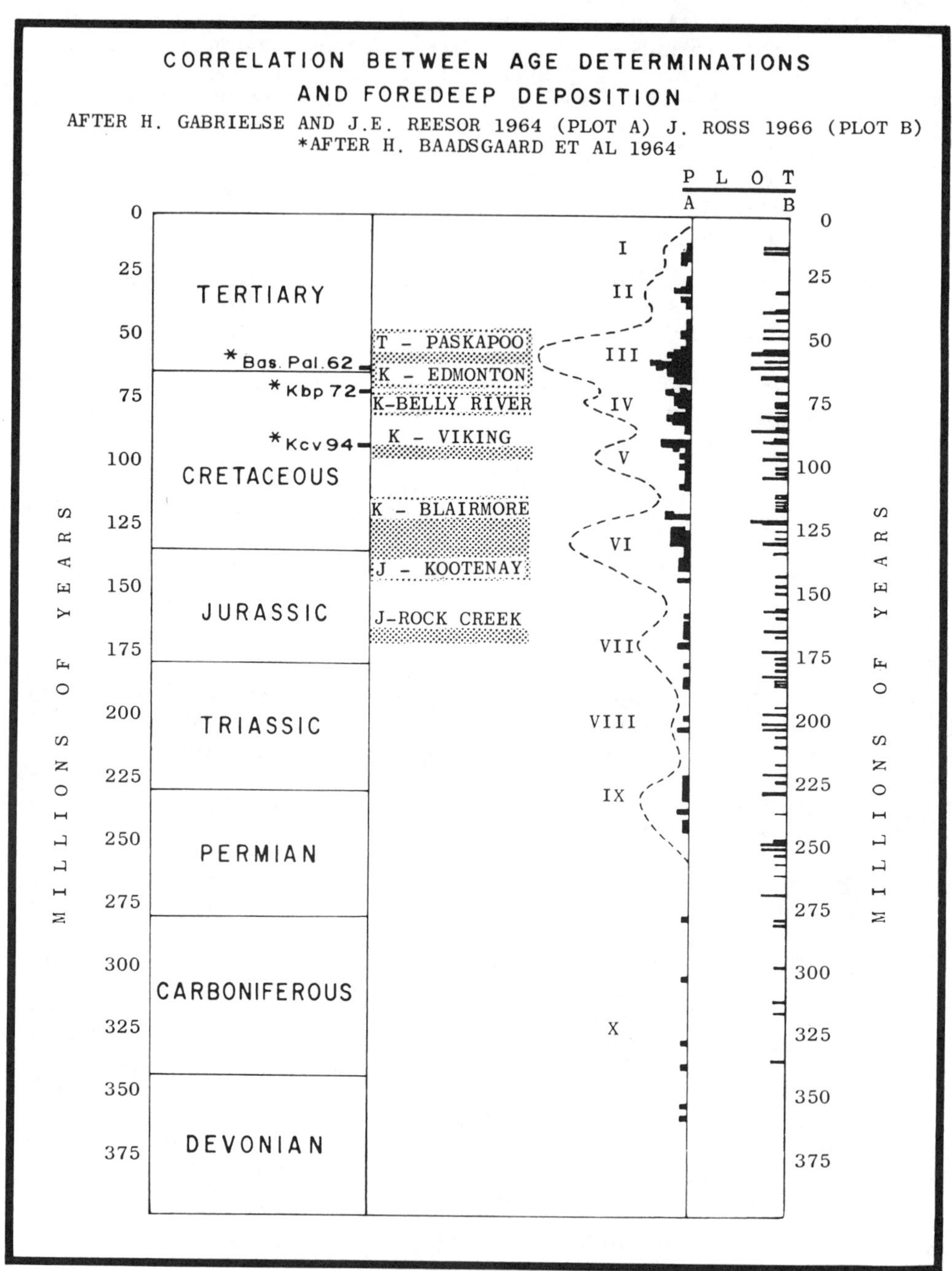

FIG. 12.—Correlation between age determinations and foredeep deposition.

conglomerate accumulations of the Foothills and Plains and their calibration by the age determinations of Baadsgaard *et al.* (1964). We are aware of the pitfalls in the use of radiometric dates, but are nevertheless intrigued by the apparent correlation of metamorphic-igneous events in British Columbia with the occurrence of coarse clastic rocks in the adjacent foredeep.

The specific nature of our correlation may be obscure and debatable. We feel, however, that clastics of the foredeep were derived largely from highlands in the west formed as a consequence of granitic intrusion and metamorphism (Nelson uplift of Rudkin, 1964; see also Wheeler, 1965). Whether this "high" remained autochthonous or moved eastward with the deformation of the Rockies has been discussed previously. The occurrence of igneous pebbles yielding K-Ar whole rock ages from 113 to 174 m.y. from the Lower Cretaceous (Albian) McDougall-Segur conglomerate further supports this concept (Norris *et al.,* 1965).

On a palinspastic base, the main depocentres of the coarse clastic rocks appear to have migrated eastward, i.e. the main axis of the foredeep migrated eastward. Thick, coarse clastic material of the Kootenay Formation occurs in the Front Ranges, whereas in the more easterly Foothills, coarse clastic rocks of the Lower Cretaceous overlie the thin distal end of the underlying Kootenay Formation. Similarly thick sandstone sequences of the Upper Cretaceous Belly River and Edmonton Formations overlie the thin distal end of the Lower Cretaceous foredeep.

On Figure 11, the westward side of the younger foredeep is shown unconformably onlapping its deformed predecessor. There is no positive support for this but we prefer this view because it avoids the extreme postulate, that the Rocky Mountains all were formed during one short phase bracketed between the end of the Paleocene and early Oligocene. We believe the Foothills were deformed during Eocene time and, because they are linked with a common sole fault to the Front Ranges and Main Ranges, that deformation also affected these units. It appears more reasonable to assume that deformation of the Front Ranges followed deposition of the Upper Cretaceous, and that the Main Ranges were conceivably deformed much earlier, perhaps after deposition of the Lower Cretaceous.

In summation, a foredeep sequence developed in the Rocky Mountains during most of Mesozoic time. The clastic rocks were derived from Paleozoic and metamorphic rocks outcropping in highlands in central British Columbia. As deformation progressed, some of these clastic rocks were reworked into new and younger foredeeps. The axis of the assymetrical basins migrated eastward. The underlying basement of the Rocky Mountains kept subsiding during this period.

Our interpretation of the sequence of structural events is similar to that proposed by Armstrong and Oriel (1965) for the Idaho-Wyoming thrust belt, a province quite analogous to, and on regional strike with, the Alberta fold belt. These authors interpret onlap relations of specific Mesozoic foredeep sequences on faults, and a progression of deformation from west to east.

E. MAIN OROGENIC PHASE (FIGURE 13)

We suggested that during Mesozoic time deformation proceeded from west to east, concomitantly with the formation and migration of a foredeep. It may then seem to be contradictory to speak of any "main" phase of deformation. Inspection of section B′-B″ (Pl. 12) shows that about 25 miles of shortening occurred in the Foothills after Paleocene time. To the west, and at depth, this amount is carried by the same group of sole faults that carried the earlier displacements of the Front and Main Ranges. These 25 miles represent approximately 25 per cent of the total amount of shortening of the Rocky Mountains that occurred between Paleocene and Late Eocene-Oligocene time. The remaining 75 per cent all could have occurred during the same time, but we prefer to assume that it occurred during Mesozoic time. The main orogenic phase is thus taken to mean the formation of the Foothills.

Within this phase, the sequence of deformation can be unraveled only by structural deductions. We prefer the following sequence (Fig. 13):

Preliminary Phase: Gentle folding. This has been demonstrated in a few cases (Henderson and Dahlstrom, 1959; Mountjoy, 1960), but we believe that folding does not necessarily precede the formation of thrust faults and therefore suggest that this phenomenon is of minor significance.

Phase 1: Step-like faulting as illustrated by Douglas (1950) and shown on Figure 13, or straight low-angle faulting. All transitions

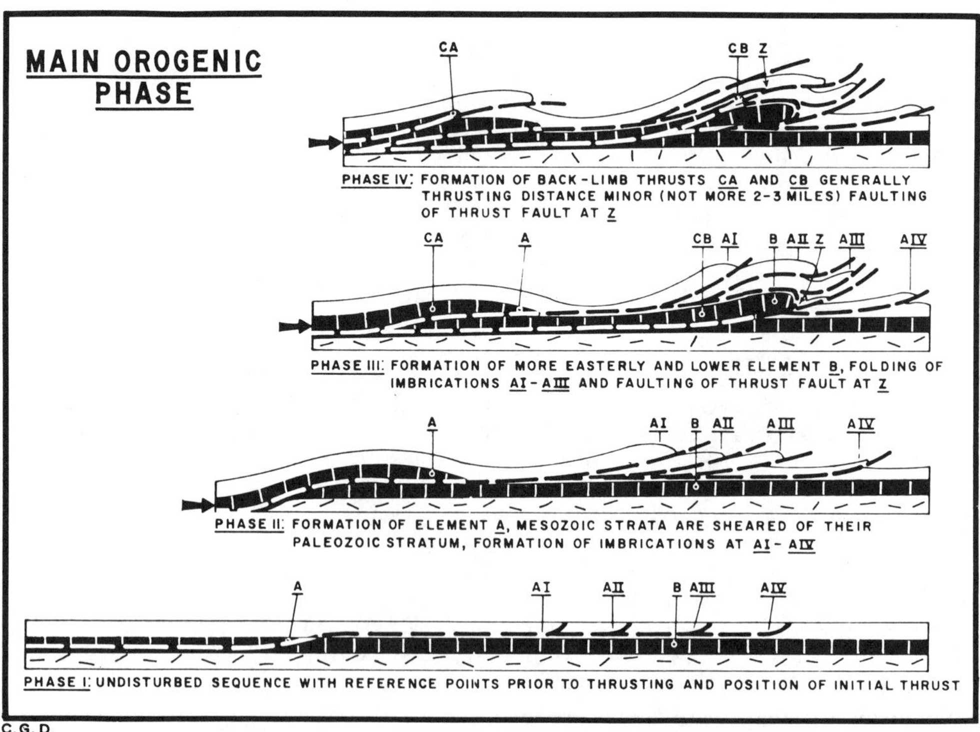

FIG. 13.—Main orogenic phase.

between the two types are conceivable. The regional distribution of the original fault traces can be seen on the restoration (Pl. 12).

Phase 2: Overthrusting and, in the case of step-like faulting, simultaneous warping and some drag folding at the leading edge of the thrust sheet. In the Foothills this process frequently caused the Mesozoic formation underlying the thrust sheet to be sheared off its Paleozoic base.

Phase 3: Folding of thrust sheets and imbrications formed during Phase 2, by an underlying element formed later, and in the same manner as the previously formed overlying sheet. The displacement of the more easterly lower sheet is generally less than that of the upper sheet. This phase leads to the formation of folded faults (e.g., Scott, 1951; Douglas, 1950, and others). The same phase also leads to the formation of faulted listric thrusts as illustrated on Figure 6.

Phase 4: Back-limb thrusting as suggested by Douglas (1950). These thrusts, which are associated with underlying highs, are in our opinion of a minor magnitude, and have minor displacements.

In conclusion, we interpret that deformation of the Eastern Cordillera proceeded from west to east, that is, that the higher and more westerly thrust sheets were formed prior to the lower and more easterly elements. As in the case of back-limb thrusting, we can conceive of minor exceptions. In many cases, where a reversal of this overall sequence of events is indicated, we believe that we are dealing with listric thrusts that have been faulted by deeper, more easterly and younger faults (Fig. 6).

F. MORPHOROGENIC UPLIFT (FIGURE 14)

This last phase of deformation lead to formation of the present-day mountains (Fig. 14). It involved a large regional uplift of the whole of western Canada, with intensive uplifting in the Rockies and simultaneous formation of intramontane basins by downfaulting along listric normal faults. In the Plains, uplift decreased in an easterly direction as indicated by the position of young terrace systems. During all preceding phases of deformation the basement underlying the Rockies was subsiding and tilted more and more to the west. The morphorogenic uplift thus was a readjustment movement which, however, was not nearly sufficient to flatten the westerly dip of the basement. Alden (1932), Russell (1951, 1954), Pardee (1950) and Cook (1960) have described phenomena related to this phase. Most supporting observations have been made in Montana and farther south.

Cook (1960) rightly points out that mountains formed during and before the main orogenic phase (Laramide Revolution) usually had their bases at elevations little higher than sea level. It appears that both mountains and plains were nearly base-levelled after deposition of the Upper Eocene and Oligocene gravels (Swift Current, Cypress Hills of Saskatchewan). These gravels contain components of Beltian sediments which probably originated from northwestern Montana (Vonhof, 1965).

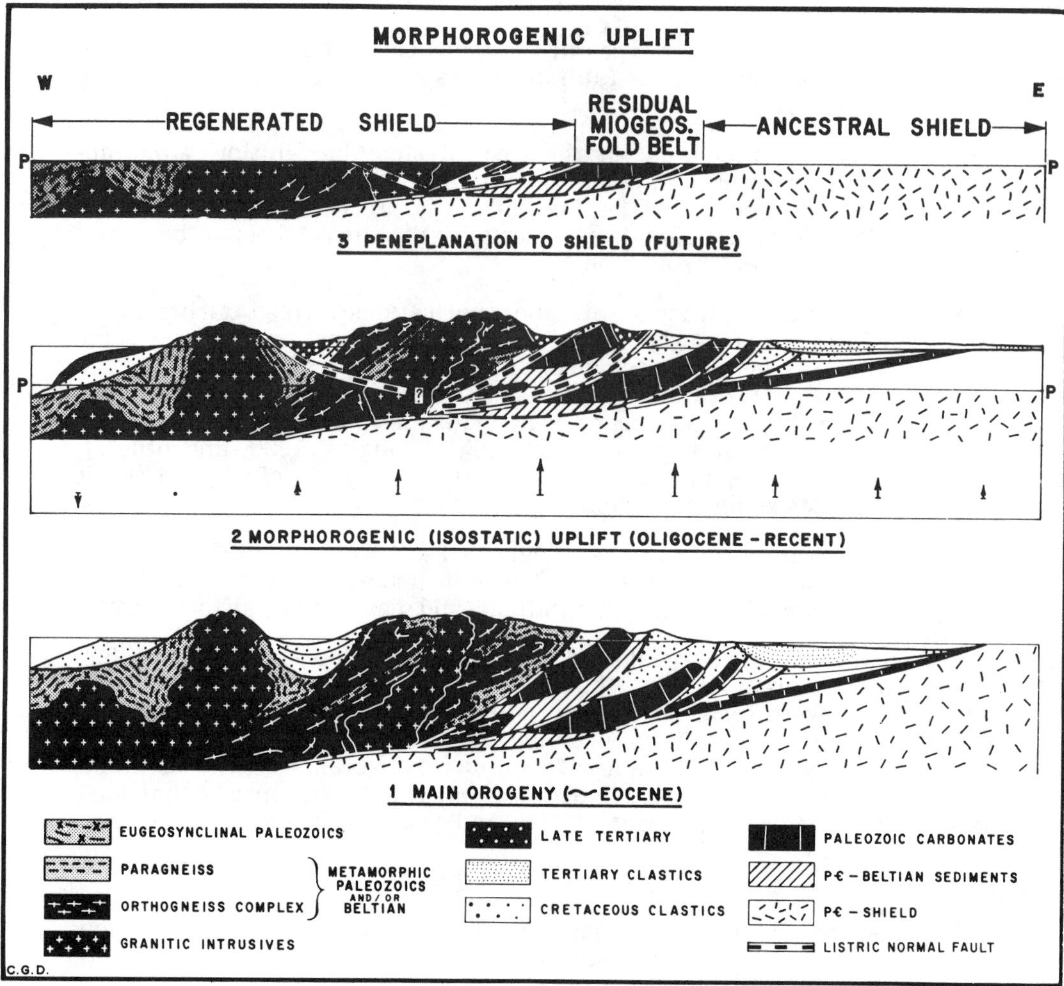

FIG. 14.—Morphorogenic uplift.

Alden (1932, p. 6) projects the Cypress Plain into the mountains of Glacier National Park, assuming a gradient similar to the one exhibited by underlying, better-preserved terraces. "Such a gradient rising in a westerly direction from the head of the Cypress Hills at about 4,850 feet above sea level would reach the mountains at about 8,300 feet—that is, about 750 feet below the crest of the Chief Mountain—and, if continued westward at 100 feet to the mile, would overtop all but the highest peaks east of and along the Continental Divide in Glacier National Park." With reference to the later physiographic history, the same author states: "The development of the Cypress Plain was followed by differential uplift and dissection to depths ranging from 700 to 1,500 feet on the plains and possibly 2,000 to 3,000 feet in the mountains."; and with respect to recent and late Pleistocene uplift, "Certain phenomena [described in detail in Alden's publication] suggest that the regional uplifts continued through Pleistocene into Recent time. The depths of Recent stream erosion range from 50 to 200 feet, and in general the streams appear to be still actively cutting downward."

All this means that since Oligocene time the mountains have been uplifted more than the Plains and that the total amount of differential uplift is considerable. Seismic data show that during this uplift the Rocky Mountains and Plains were linked by a common westward dipping basement that remained intact. The formation of the Rocky Mountain Trench and possibly other valleys and intramontane basins (e.g. Flathead basin) are related to the same deformation, being related to rotational movement along essentially post-Oligocene, listric normal faults.

The timing and morphologic expression of the process is best summarized by quoting from Pardee's (1950) abstract which reports on the formation of basins in northwestern Montana:

"By Oligocene time the region had been generally reduced to a surface of moderate to slight relief. During the Oligocene and Miocene the drainage became sluggish or ponded . . . and in these [areas] accumulated the Tertiary 'lake beds' . . . In the late Miocene or early Pliocene the surface comprised areas of older rocks that . . . had been eroded to slight or moderate relief; and areas of the 'lake beds' that formed gently sloping or level plains . . . this surface is called . . . the Late Tertiary peneplain."

"Further leveling . . . was interrupted by a general re-elevation of the region accompanied by greatly accelerated local crustal movements that relatively elevated the present mountains. These movements continued intermittently and with decreasing intensity through the Pliocene and, except for small displacements on some of the faults as late as the Recent epoch, ceased in early or middle Pleistocene. They are thought to constitute a distinct late stage of the Cenozoic mountain building."

"During the halt in the uplift of the mountains, wide stream valleys as much as 1500 feet deep were eroded in the elevated and deformed peneplain. In the basins during this pause, called the Old Valley cycle, the 'lake beds' were reduced to gently sloping plains collectively referred to as No. 1 Bench. With renewed uplift the more vigorous streams deepened their channels across the mountain blocks as fast as the surface rose and thus excavated narrow inner valleys or gorges. In this, the Present cycle of erosion, No. 1 Bench of the 'lake bed' areas was, in most of the basins, dissected to a series of terraces."

We believe that this description applies also to the Southern Canadian Rockies, where the morphology and Tertiary-Pleistocene stratigraphy have not yet been studied in detail. The effect of this last phase on the morphology and drainage history of the Cordillera must have been profound. Our restoration shows that during Mesozoic and early Tertiary time, the drainage pattern of the Central Cordillera of interior British Columbia was oriented eastward toward the foredeep, a hypothesis that might be supported by a regional morphologic study finding remnants of late Cretaceous and Tertiary stream systems with drainage direction towards the Plains. The present drainage to the Pacific Ocean and the extensive longitudinally oriented valleys probably originated during the late morphorogenic uplift with its associated systems of listric normal faults.

Figure 14 also shows a speculative interpretation that assumes the present Rockies will eventually be peneplained to a much deeper level (marked P-P). We then would see an ancestral shield, a regenerated portion of the shield, and a remnant fold belt in juxtaposition. This concept may aid in understanding tectonic maps of the Canadian Shield (e.g. Stockwell, 1965), where we observe the juxtaposition of metamorphic provinces of different age, which in places are separated by remnants of fold belts of relatively more miogeosynclinal aspect (e.g. Schefferville area or Belcher Islands).

REFERENCES

A. *Publications of Alberta Society of Petroleum Geologists*

A.S.P.G., 1953, Third Annual Field Conference and Symposium: *Editor,* J. C. Sproule, 230 p.

——, 1954, Fourth Annual Field Conference, Banff-Golden-Radium; *Editor,* J. C. Scott, 182 p.

——, 1955, Fifth Annual Field Conference, Jasper National Park; 189 p.

——, 1956, Sixth Annual Field Conference, Bow Valley; *Editor,* F. G. Fox, 153 p.

——, 1957, Seventh Annual Field Conference, Waterton; *Editors,* E. W. Jennings and C. R. Hemphill, 180 p.

——, 1958a, Eighth Annual Field Conference, Nordegg; *Editor,* C. R. Hemphill, 203 p.

——, 1958b, Annotated Bibliography of the Sedimentary Basin of Alberta and of adjacent parts of British Columbia and Northwest Territories (1845-1955); 499 p.

——, 1959, Ninth Annual Field Conference, Moose Mountain-Drumheller; *Editor,* G. H. Austin, 196 p.

——, 1960, Photogeological Interpretation and Compilation, Exshaw-Golden (2 sheets); *Editor,* G. G. L. Henderson.

——, 1961, Turner Valley-Savanna Creek-Kananaskis Road Log; *by* J. C. Scott and W. J. Hennessey, 19 p.

——, 1962a, Twelfth Annual Field Conference, Coleman-Cranbrook-Radium: *Editors,* R. L. Manz and H. Mogensen, v. 10, no. 7, Alberta Soc. Petroleum Geologists Jour., p. 333-453.

——, 1962b, Crowsnest-Cranbrook,-Windermere Road Log; *compiled by* P. L. Gordy and C. J. Bruce, 23 p.

——, 1963, Ghost River Area Route Log; *compiled by* C. B. Geisler, I. R. Halladay and G. A. Wilson, 36 p.

——, 1964a, Fourteenth Annual Field Conference, Flathead Valley; *Editor,* A. J. Goodman, v. 12, Spec. Issue, Bull. Canadian Petroleum Geology, 276 p.

——, 1964b, Geological History of Western Canada; *Editors,* R. G. McCrossan and R. P. Glaister, 232 p.

——, 1965a, Fifteenth Annual Field Conference, Cypress Hills; *Editor,* R. L. Zell, Pt. 1 (Technical papers), 288 p.

——, 1965b, Fifteenth Annual Field Conference, Cypress Hills; *Editor,* I. Weihmann, Pt. 2 (Road log and maps), 22 p.

B. *Selected References*

Aitken, J. D., 1963, Ghost River type section: Bull. Canadian Petroleum Geology, v. 11, no. 3, p. 267-287.

——, 1966, Sub-Fairholme Devonian rocks of the eastern Front Ranges, southern Rocky Mountains, Alberta: Geol. Surv. Canada Paper 64-33.

Alden, W. C., 1932, Physiography and glacial geology of eastern Montana and adjacent areas: U.S. Geol. Survey Prof. Paper 174.

Armstrong, F. C., and Oriel, S. S., 1965, Tectonic development of Idaho-Wyoming thrust belt: Am. Assoc. Petroleum Geologists Bull., v. 49, no. 11, p. 1847-1866.

Armstrong, J. E., 1959, Physiography of the Rocky Mountain Trench: Canadian Inst. Mining Metallurgy Bull., v. 52, no. 565, p. 318-321.

Baadsgaard, H., Folinsbee, R. E., and Lipson, J., 1961, Potassium-Argon dates of biotites from Cordilleran granites: Geol. Soc. America Bull., v. 72, no. 5, p. 689-702.

Baadsgaard, H., Cumming, G. L., Folinsbee, R. E., and Godfrey, J. D., 1964, Limitations of radiometric dating: *in* Osborne, F. F., *Editor*, Geochronology in Canada, Royal Soc. Canada Special Pub. 8, p. 20-38.

Beach, H. H., 1943, Moose Mountain and Morley map-areas, Alberta: Geol. Surv. Canada Mem. 236, 74 p.

Beveridge, A. J., and Folinsbee, R. E., 1956, Dating Cordilleran orogenies: Royal Soc. Canada Trans., ser. 3, v. 2, sec. 4, p. 19-43.

Billings, M. P., 1938, Physiographic relation of Lewis overthrust in northern Montana: Am. Jour. Sci., v. 35, p. 260-272.

Blundun, G. J., 1956, The refraction seismograph in the Alberta Foothills: Geophysics, v. 11, no. 3, p. 828-838.

Bossort, D. O., 1957, Relationship of the Porcupine Hills to early Laramide movements: *in* Alberta Soc. Petroleum Geologists 7th Ann. Field Conf., p. 46-51.

Bostock, H. S., 1948, Physiography of the Canadian Cordillera, with special reference to the area north of the 55th parallel: Geol. Surv. Canada Mem. 247, 106 p.

Burwash, R. A., Baadsgaard, H., Peterman, Z. E., and Hunt, G. H., 1964, Precambrian, (Chap. 2): *in* McCrossan, R. G., and Glaister, R. P., *Editors*, Geological history of western Canada, Calgary, Alberta Soc. Petroleum Geologists, 232 p.

Cairnes, C. E., 1934, Slocan mining camp, British Columbia: Geol. Surv. Canada Mem. 173, 137 p.

Charlesworth, H. A. K., 1959, Some suggestions on the structural development of the Rocky Mountains in Canada: Alberta Soc. Petroleum Geologists Jour., v. 7, no. 11, p. 249-256.

Clapp, C. H., 1932, Geology of a portion of the Rocky Mountains of northern Montana: Montana Bureau Mines Geol. Mem. 4.

Clark, L. M., 1948, Geology of Rocky Mountain Front Ranges near Bow River, Alberta: *in* Clark, L. M., Chm., Alberta Symposium: Am. Assoc. Petroleum Geologists Bull., v. 33, no. 4, p. 614-633.

——, 1954, Cross-section through the Clarke Range of the Rocky Mountains of southern Alberta and southern British Columbia: *in* Alberta Soc. Petroleum Geologists, 4th Ann. Field Conf., p. 105-109.

——, *Editor*, 1954, Western Canada sedimentary basin; a symposium: Tulsa, Am. Assoc. Petroleum Geologists, 521 p.

——, 1964, Cross-section of Flathead Valley in vicinity of Sage Creek, British Columbia: Bull. Canadian Petroleum Geology, v. 12, Spec. Issue, p. 345-349.

Cook, H. J., 1960, New concepts of Late Tertiary crustal deformations in the Rocky Mountain region of North America: Int. Geol. Congress 21st Session, Norden, Pt. XII, p. 198-212.

Crabb, J. J., 1962, Coal mining in southeast British Columbia and historical outline: Alberta Soc. Petroleum Geologists Jour., v. 10, no. 7, p. 335-340.

Crickmay, C. H., 1964, The Rocky Mountain Trench: A problem: Canadian Jour. Earth Sciences, v. 1, p. 184-205.

Dahlstrom, C. D. A., 1960, Concentric folding: *in* Edmonton Geol. Soc. 2nd Ann. Field Trip Guide Book, p. 82-84.

——, Daniel, R. E., and Henderson, G. G. L., 1962, The Lewis thrust at Fording Mountain: Alberta Soc. Petroleum Geologists Jour., v. 10, no. 7, p. 373-395.

Daly, R. A., 1912, Geology of the North American Cordillera at the Forty-ninth Parallel: Geol. Surv. Canada Mem. 38, pts. 1, 2, 3, 857 p.

——, 1915, A geological reconnaissance between Golden and Kamloops, B.C., along the Canadian Pacific Railway: Geol. Surv. Canada Mem. 68, 260 p.

Dawson, G. M., 1875, Report on the geology and resources of the region in the vicinity of the Forty-ninth Parallel from the Lake of the Woods to the Rocky Mountains: Montreal, British North America Boundary Commission, p. 379.

——, 1886, Preliminary report on the physical and geological features of the portion of the Rocky Mountains between Latitudes 49 d. and 51 d. 30 ft.: Geol. Surv. Canada Ann. Rept. 1, p. 169.

——, and McConnell, R. G., 1885, Report of the region in the vicinity of the Bow and Belly Rivers, Northwest Territory: Geol. Surv. Canada Rept. Prog. 1882-1884, p. 168.

Douglas, R. J. W., 1950, Callum Creek, Langford Creek, and Gap map areas, Alberta: Geol. Surv. Canada Mem. 255, 124 p.

——, 1951, Preliminary map, Pincher Creek, Alberta: Geol. Surv. Canada Paper 51-22.

——, 1952, Preliminary map Waterton, Alberta: Geol. Surv. Canada Paper 52-10.

——, 1956, Nordegg, Alberta: Geol. Surv. Canada Paper 55-34.

——, 1958, Mount Head map-area, Alberta: Geol. Surv. Canada Mem. 291, 241 p.

Edmonton Geological Society, 1959, Field Trip Guide Book, Cadomin area.

——, 1960, Second Annual Field Trip Guide Book, Rock Lake.

——, 1962, Fourth Annual Field Trip Guide Book, Peace River: *Editor,* E. E. Pelzer, 121 p.

——, 1963, Fifth Annual Field Trip Guide Book, Sunwapta Pass Area; *Editor,* D. E. Jackson.

——, 1964, Sixth Annual Field Trip Guide Book, Medicine and Maligne Lakes: *Editor,* R. Green, 68 p.

——, 1965, Seventh Annual Field Trip, David Thompson Highway.

Erdman, O. A., 1946, Cripple Creek, Alberta: Geol. Surv. Canada Paper 46-22.

——, 1950, Alexo and Saunders map-areas, Alberta: Geol. Surv. Canada Mem. 254, 100 p.

——, Belot, R. E., and Slemko, W., 1953, Pincher Creek area, Alberta: *in* Alberta Soc. Petroleum Geologists 3rd Ann. Field Conf. and Symposium, p. 139-157.

Evans, C. S., 1930, Some stratigraphic sections in the Foothills region, between Bow and North Saskatchewan Rivers, Alberta: Geol. Surv. Canada Summ. Rept. 1929, pt. B, p. 25-35.

——, 1933, Brisco-Dogtooth map-area, British Columbia: Geol. Surv. Canada Summ. Rept. 1932, pt. A II, p. 106-187.

Faust, L. Y., 1951, Seismic velocity as a function of depth and geologic time: Soc. Explor. Geophysicists, v. 18, no. 2, p. 271-288.

Fitzgerald, E. L., 1962, Structure of the McConnell thrust sheet in the Ghost River area, Alberta: Alberta Soc. Petroleum Geologists Jour., v. 10, no. 10, p. 553-574.

Folinsbee, R. E., Ritchie, W. D., and Stansberry, G. F., 1957, The Crowsnest volcanics and Cretaceous geochronology: *in* Alberta Soc. Petroleum Geologists 7th Ann. Field Conf., p. 20-26.

Fox, F. G., 1959, Structures and accumulation of hydrocarbons in southern Alberta Foothills, Alberta, Canada: Am. Assoc. Petroleum Geologists Bull, v. 43, no. 5, p. 992-1025.

Gabrielse, H., and Reesor, J. E., 1964, Geochronology of plutonic rocks in two areas of the Canadian Cordillera: *in* Osborne, F. F., *Editor,* Geochronology in Canada, Roy. Soc. Canada, Special Pub. 8.

Gallup, W. B., 1954, Geology of Turner Valley oil and gas field, Alberta, Canada: *in* Clark, L. M., *Editor,* Western Canada sedimentary basin; symposium: Tulsa, Am. Assoc. Petroleum Geologists, p. 397-414; originally published 1951.

——, 1955, Geology of the Pincher Creek gas and naptha field and its regional implications: Am. Assoc. Petroleum Geologists Rocky Mountain Section, Geol. Record, Petroleum Information Pub., p. 153-164, Denver; Billings Geol. Soc. 6th Ann. Field Conf., p. 150-159.

——,, 1957, Relation of Laramide movements to the Cretaceous and Tertiary sediments of western Canada: Alberta Soc. Petroleum Geologists Jour., v. 5, no. 6, p. 125-126.

Gilluly, J., 1963, The tectonic evolution of the western United States: Quart. Jour. Geol. Soc., v. 119, pt. 2, p. 133-174.

——, 1966, Orogeny and geochronology: Am. Jour. Sci., v. 264, p. 97-111.

Gordy, P L., and Edwards, G., 1962, Age of the Howell Creek intrusives: Alberta Soc. Petroleum Geologists Jour., v. 10, no. 7, p. 369-372.

Gussow, W. C., 1960, The Pre-Devonian unconformity in North America: Int. Geol. Congress, 21st Session, Norden, Part XIX, p. 158-163.

——, and Hunt, C. W., 1959, Age of Ice River complex, Yoho National Park, British Columbia: Alberta Soc. Petroleum Geologists Jour., v. 7, no. 3, p. 62.

Hage, C. O., 1942, Folded thrust faults in Alberta Foothills, west Turner Valley: Royal Soc. Canada Trans., ser. 3, v. 36, sec. 4, p. 67-68.

——, 1943a, Beaver mines: Geol. Surv. Canada Map 739A.

——, 1943b, Dyson Creek map-area, Alberta: Geol. Surv. Canada Paper 43-5.

——, 1946, Dyson Creek: Geol. Surv. Canada Map 827A.

Hake, B. F., Addison, C. C. and Willis, R., 1942, Folded thrust faults in the Foothills of Alberta: Geol. Soc. America Bull., v. 53, p.

Hamblin, W. K., 1965, Origin of "reverse drag" on the downthrown side of normal faults: Geol. Soc. America Bull., v. 76, no. 10, p. 1145-1164.

Hector, J., 1863, Journals of the exploration of British North America: London.

Henderson, G. G. L., 1954a, Geology of the Stanford Range: British Columbia Dept. of Mines Bull. 35.

——, 1954b, Southern Rocky Mountains, tectonic compilation map: Alberta Soc. Petroleum Geologists, Map no. 1.

——, 1959, A summary of the regional structure and stratigraphy of the Rocky Mountain Trench: Canadian Inst. Mining Metallurgy Bull., v. 52, no. 565, p. 322-327.

——, and Dahlstrom, C. D. A., 1959, First-order nappe in Canadian Rockies: Am. Assoc. Petroleum Geologists Bull., v. 43, no. 3, p. 641-653.

Holland, S. S., 1959, Symposium on the Rocky Mountain Trench: Introduction: Canadian Inst. Mining Metallurgy Bull., v. 52, no. 565, p. 318.

Hume, G. S., 1931, Geology, Turner Valley sheet, southwest quarter, Alberta: Geol. Surv. Canada Map 261A.

——, 1931, Geology Turner Valley sheet, northwest quarter, Alberta: Geol. Surv. Canada Map 262A.

——, 1932, Geology, Jumping Pound sheet, Alberta: Geol. Surv. Canada Map 277A

——, 1933, Waterton Lakes-Flathead Valley area, Alberta and British Columbia: Geol. Surv. Canada Summary Rept. 1932, pt. B, p. 1-20, 1933; and British Columbia Bur. Mines Ann. Rept. 1932, p. 164-167.

——, 1936, The west half of Wildcat Hills map-area, Alberta: Geol. Surv. Canada Mem. 188, 15 p.

——, 1938, The stratigraphy and structure of southern Turner Valley, Alberta: Geol. Surv. Canada Paper 38-22.

——, 1941a, Jumping Pound, Alberta: Geol. Surv. Canada Map 653A.

——, 1941b, A folded fault in the Pekisko area, Foothills of Alberta: Royal Soc. Canada Trans., ser. 3, v. 35, sec. 4, p. 87-92.

——, 1943, Preliminary maps, Stinson Creek, Alberta: Geol. Surv. Canada Paper 43-8.

——, 1957, Fault structures in the Foothills and eastern Rocky Mountains of southern Alberta: Geol. Soc. America Bull., v. 68, no. 4, p. 395-412.

Irish, E. J. W., 1965, Geology of the Rocky Mountain Foothills, Alberta (between latitudes 53° 15′ and 54° 15′): Geol. Surv. Canada Mem. 334, 241 p.

Jones, A. G., 1959, Vernon map-area, British Columbia: Geol. Surv. Canada Mem. 296, 186 p.

Jones, P. B., 1964, Structures of the Howell Creek area: Bull. Canadian Petroleum Geology, v. 12, Spec. Issue, p. 350-362.

Kay, M., 1951, North American geosynclines: Geol. Soc. America Mem. 48, 143 p.

Keating, L. F., 1966, Exploration in the Canadian Rockies and Foothills: Canadian Jour. Earth Sciences, v. 3, no. 5.

King, P. B., 1959, The evolution of North America: Princeton, N.J., Princeton University Press, 190 p.

Kirkham, V. R. D., 1939, The Moyie-Lenia overthrust fault: Jour. Geology, v. 38, no. 4, p. 364-374.

Lamb, A. T., and Smith, D. W., 1962, Refraction profiles over the southern Rocky Mountain Trench area of British Columbia: Alberta Soc. Petroleum Geologists Jour., v. 10, no. 7, p. 428-437.

Lang, A. H., 1947, Brûlé and Entrance map-areas, Alberta: Geol. Surv. Canada Mem. 244, 65 p.

Leech, G. B., 1959, The southern part of the Rocky Mountain Trench: Canadian Inst. Mining Metallurgy Bull., v. 52, no. 565, p. 327-333.

——, 1960, Fernie, west half, Kootenay District, British Columbia: Geol. Surv. Canada Map 11-1960.

——, 1962, Structure of the Bull River Valley near Latitude 49° 35′: Alberta Soc. Petroleum Geologists Jour., v. 10, no. 7, p. 396-407.

——, 1962b, Metamorphism and granitic intrusions of Precambrian age in southeastern British Columbia: Geol. Surv. Canada Paper 62-13.

——, 1965, Discussion of "The Rocky Mountain Trench: A Problem:" Canadian Jour. Earth Sciences, v. 2, p. 405-410.

——, Lowdon, J. A., Stockwell, C. H., and Wanless, R K., 1963, Age determinations and geological studies: Geol. Surv. Canada Paper 63-17.

Link, T. A., 1935, Types of Foothills structure of Alberta, Canada: Am. Assoc. Petroleum Geologists Bull., v. 19, no. 10, p. 1427-1471.

——, 1949, Interpretations of Foothills structures, Alberta, Canada: Am. Assoc. Petroleum Geologists Bull., v. 33, no. 9, p. 1475-1501.

——, 1953, History of geological interpretation of the Turner Valley structure and Alberta Foothills, Canada: *in* Alberta Soc. Petroleum Geologists 3rd Ann. Field Conf. and Symposium, p. 117-133.

Little, H. W., 1960, Nelson map-area, west half, British Columbia: Geol. Surv. Canada Mem. 308, 205 p.

Long, A., Silverman, A. J., and Kulp, J. L., 1960, Isotopic composition of lead and Precambrian mineralization of the Coeur D'Alene District, Idaho: Econ. Geology, v. 55, p. 645-658.

MacKay, B. R., 1943, Preliminary map, Foothills belt of central Alberta: Geol. Surv. Canada Paper 43-3.

Martin, R., 1956, Jumping Pound gas field: *in* Alberta Soc. Petroleum Geologists 6th Ann. Field Conf., p. 125-140.

McConnell, R. G., 1887, Report on the geological structure of a portion of the Rocky Mountains: Geol. Surv. Canada Summ. Rept., 1886, v. 2, p. 41.

McLaren, D. J., and Mountjoy, E. W., 1962, Alexo equivalents in the Jasper region, Alberta: Geol. Surv. Canada Paper 62-23.

Mountjoy, E. W., 1960, Structure and stratigraphy of the Miette and adjacent areas, eastern Jasper National Park, Alberta: unpub. Ph.D. thesis, Univ. Toronto.

Nelson, S. J., Glaister, R. P., and McCrossan, R. G., 1964, Introduction (Chap. 1): *in* McCrossan, R. G., and Glaister, R. P., *Editors*, Geological history of western Canada, Calgary, Alberta Soc. Petroleum Geologists, 232 p.

Newmarch, C. B., 1953, Geology of the Crowsnest coal basin with special reference to the Fernie Area: British Columbia Dept. Mines Bull. 33.

Norford, B. S., 1964, Ordovician-Silurian (Chap. 4, Pt. 2): *in* McCrossan, R. G., and Glaister, R. P., *Editors*, Geological history of western Canada, Calgary, Alberta Soc. Petroleum Geologists, 232 p.

Norris, D. K., 1961, An interstratal peel on Maverick Hill, Alberta: Alberta Soc. Petroleum Geologists Jour., v. 9, p. 188-191.

——, 1964, Microtectonics of the Kootenay Formation near Fernie, British Columbia: Bull. Canadian Petroleum Geology, v. 12, Spec. Issue, p. 383-398.

——, 1964, The Lower Cretaceous of the southeastern Canadian Cordillera: Bull. Canadian Petroleum Geology, v. 12, Spec. Issue, p. 512-535.

——, Stevens, R. D., and Wanless, R. K., 1965, K-Ar age of igneous pebbles in the McDougall-Segur Conglomerate, southeastern Canadian Cordillera: Geol. Surv. Canada Paper 65-26.

North, F. K., 1964, Cambrian (Chap. 3, Pt. 2): *in* McCrossan, R.G., and Glaister, R. P., *Editors*, Geological history of western Canada, Calgary, Alberta Soc. Petroleum Geologists, 232 p.

——, and Henderson, G. G. L., 1954a, Summary of the geology of the southern Rocky Mountains of Canada: *in* Alberta Soc. Petroleum Geologists 4th Ann. Field Conf., p. 15-81.

——, ——, 1954b, The Rocky Mountain Trench: *in* Alberta Soc. Petroleum Geolists 4th Ann. Field Conf., p. 82-100.

Oswald, D. H., 1964, The Howell Creek structure: Bull. Canadian Petroleum Geology, v. 12, Spec. Issue, p. 363-377.

Pardee, J. T., 1950, Late Cenozoic block faulting in western Montana: Geol. Soc. America Bull., v. 61, no. 4, p. 359-406.

Park, C. G., and Cannon, R. S., 1943, Geology and ore deposits of the Metaline Quadrangle, Washington: U.S. Geol. Surv. Professional Paper 202.

Patterson, J. R., and Storey, T. R., 1960, Caledonian earth movements in western Canada: Int. Geol. Congr., 21st Session, Norden, Pt. XIX, p. 150-157.

Penner, D. G., 1957, Turner Valley oil and gas field: Alberta Soc. Petroleum Geologists 7th Ann. Field Conf., p. 131-137.

Price, R. A., 1958, Structure and stratigraphy of the Flathead North map area (east half), British Columbia and Alberta: unpub. Ph.D. thesis, Princeton Univ., 363 p.

——, 1961, Fernie map-area, east half, Alberta and British Columbia: Geol. Surv. Canada Paper 61-24.

——, 1962, Geologic structure of the central part of the Rocky Mountains in the vicinity of Crowsnest Pass: Alberta Soc. Petroleum Geologists Jour., v. 10, no. 7, p. 341-351.

——, 1964, The Precambrian Purcell System in the Rocky Mountains of southern Alberta and British Columbia: Bull. Canadian Petroleum Geology, v. 12, Spec. Issue, p. 399-426.

——, 1965, Flathead map-area, British Columbia and Alberta: Geol. Surv. Canada Mem. 336, 221 p.

Rapson, J. E., 1963, Age and aspects of metamorphism associated with the Ice River complex, British Columbia: Bull. Canadian Petroleum Geology, v. 11, no. 2, p. 116-124.

——, 1964, Lithology and petrography of transitional Jurassic-Cretaceous clastic rocks, southern Rocky Mountains: Bull. Canadian Petroleum Geology, v. 12, Spec. Issue, p. 556-586.

Reesor, J. E., 1957a, The Proterozoic of the Cordillera in southeastern Alberta: *in* Gill, J. E., *Editor*, Proterozoic in Canada, Royal Soc. Canada, Special Pub. 2.

——, 1957b, Lardeau (east half), Kootenay District, British Columbia: Geol. Surv. Canada Map 12-1957.

——, 1958, Dewar Creek map-area with special emphasis on the White Creek batholith, British Columbia: Geol. Surv. Canada Mem. 292, 78 p.

——, 1965, Structural evolution and plutonism in Valhalla gneiss complex, British Columbia: Geol. Surv. Canada Bull. 129, 128 p.

Rice, H. M. A., 1937, Cranbrook map-area, British Columbia: Geol. Surv. Canada Mem. 207, 67 p.

——, 1941, Nelson map-area, east half, British Columbia: Geol. Surv. Canada Mem. 228, 86 p.

Rich, J. L., 1934, Mechanics of low-angle overthrust faulting illustrated by Cumberland thrust block, Virginia, Kentucky and Tennessee: Am. Assoc. Petroleum Geologists Bull., v. 18, no. 12, p. 1584-1596.

Richards, T. C., 1959, Broadside refraction shooting: Geophysics, v. 24, no. 4, p. 725-748.

——, and Walker, D. J., 1959, Measurement of the thickness of the earth's crust in the Alberta plains of western Canada: Geophysics, v. 24, p. 262-284.

Robertson, G., 1963, Intrabasement reflections in southwestern Alberta: Geophysics, v. 28, no. 5, pt. 2, p. 910-915.

Ross, C. P., 1959, Geology of Glacier National Park and the Flathead Region: U.S. Geol. Surv. Prof. Paper 296.

Ross, J. V., 1966, A note on histogram analysis of isotopic age data: Canadian Jour. Earth Sciences, v. 3, no. 2, p. 259-262.

Rudkin, R. A., 1964, Lower Cretaceous (Chap. 11): *in* McCrossan, R. G., and Glaister, R. P., *Editors*, Geological history of western Canada, Calgary, Alberta Soc. Petroleum Geologists, 232 p.

Russell, L. S., 1951, Age of the front-range deformation in the North American Cordillera: Royal Soc. Canada Trans., ser. 3, v. 45, sec. 4, p. 47-69.

——, 1954, The Eocene-Oligocene transition as a time of major orogeny in western North America: Royal Soc. Canada Trans., sec. 4, ser. 3, v. 48, sec. 4, p. 65-69.

——, 1954, Mammalian fauna of the Kishenehn Formation, southeastern British Columbia: Nat. Museum Canada Bull. 132.

——, 1964, Kishenehn Formation: Bull. Canadian Petroleum Geology, v. 12, Spec. Issue, p. 536-543.

Scott, J. C., 1951, Folded faults in Rocky Mountain Foothills of Alberta: Am. Assoc. Petroleum Geologists Bull., v. 35, no. 11, p. 2310-2347.

——, 1953, Savanna Creek structure: *in* Alberta Soc. Petroleum Geologists 3rd Ann. Field Conf. and Symposium, p. 134-138.

——, Hennessey, W. H., and Lamon, R. S., 1957, Savanna Creek gas field, Alberta: *in* Alberta Soc. Petroleum Geologists 7th Ann. Field Conf., p. 113-130.

Shaw, E. W., 1963, Canadian Rockies — orientation in time and space: *in* Childs, O. E., *Editor,* Backbone of the Americas, Tulsa, Am. Assoc. Petroleum Geologists Mem. 2, p. 231-242.

Stockwell, C. H., 1965, Tectonic map of the Canadian Shield: Geol. Surv. Canada Map 4-1965.

Sloss, L. L., 1963, Sequences in the cratonic interior of North America: Geol. Soc. America Bull., v. 74, no. 2, p. 93-114.

Springer, G. D., MacDonald, W. D., and Crockford, M. B. B., 1964, Jurassic (Chap. 10): *in* McCrossan, R. G., and Glaister, R. P., *Editors,* Geological history of western Canada, Calgary, Alberta Soc. Petroleum Geologists, 232 p.

Taylor, R. S., Mathews, W. H., and Kupsch, W. O., 1964, Tertiary (Chap. 13): *in* McCrossan, R. G., and Glaister, R. P., *Editors,* Geological history of western Canada, Calgary, Alberta Soc. Petroleum Geologists, 232 p.

Thompson, T. L., 1962, Origin of the Rocky Mountain Trench in southeastern British Columbia by Cenozoic block faulting: Alberta Soc. Petroleum Geologists Jour., v. 10, no. 7, p. 408-427.

Tozer, E. T., 1953, The Cretaceous-Tertiary transition in southwestern Alberta: *in* Alberta Soc. Petroleum Geologists 3rd Ann. Field Conf. and Symposium, p. 23-31.

——, 1956, Uppermost Cretaceous and Paleocene non-marine Molluscan faunas of western Alberta: Geol. Surv. Canada Mem. 280, 125 p.

Vonhof, J. A., 1965, The Cypress Hills Formation and its reworked deposits in southwestern Saskatchewan: *in* Alberta Soc. Petroleum Geologists 15th Ann. Field Conf., pt. 1, p. 142-161.

Walker, J. F., 1926, Geology and mineral deposits of Windermere map-area, British Columbia: Geol. Surv. Canada Mem. 148, 69 p.

Wanless, R. V., and Leech, G. B., 1958, Lead isotope studies of the Sullivan and other deposits in the East Kootenay District, southeast British Columbia: Am. Geophysical Union Trans., v. 39, p. 535.

Warren, P. S., 1951, The Rocky Mountain geosyncline in Canada: Royal Soc. Canada Trans., v. 45, 3d ser., sec. 4, p. 1-10.

——, and Stelck, C. R., 1958, Continental margins, western Canada: Alberta Soc. Petroleum Geologists Jour., v. 6, no. 2, p. 29-42.

Wasserburg, G. J., Wetherill, G. W., and Wright, L. A., 1959, Ages in the Precambrian terrane of Death Valley, California: Jour. Geology, v. 67, no. 6, p. 702-708.

Webb, J. .B., 1951, Geological history of plains of western Canada: Am. Assoc. Petroleum Geologists Bull., v. 35, p. 2291-2315; reprinted with revisions, 1954, *in* Western Canada sedimentary basin; a Symposium, p. 3-28.

——, 1964, Historical summary (Chap. 16): *in* McCrossan, R. G., and Glaister, R. P., *Editors,* Geological history of western Canada, Calgary, Alberta Soc. Petroleum Geologists, 232 p.

Wheeler, J. O., 1961, Rogers Pass map-area, British Columbia: Geol. Surv. Canada Prel. Map 4-1961.

——, 1965, The Tectonic evolution of the southern Canadian Cordillera (Abstract): Royal Soc. Canada, June Meeting, Vancouver, p. 13-14.

White, W. H., 1959, Cordilleran tectonics in British Columbia: Am. Assoc. Petroleum Geologists Bull., v. 43, no. 1, p. 60-100.

Willis, B., 1902, Stratigraphy and structure, Lewis and Livingstone Ranges, Montana: Geol. Soc. America Bull., v. 13, p. 305-352.

Wright-Broughton, C., 1960, Photogeological interpretation and compilation Exshaw-Golden: (2 sheets), Alberta Soc. Petroleum Geologists.

C. *Subject Index of Selected References* (in approximate chronologic order)

REGIONAL SECTION A-A′

MacKay (1943); Erdman (1946, 1950); Link (1949); Douglas (1956); A.S.P.G. (1958a); Shaw (1963).

REGIONAL SECTION B-B′

McConnell (1887); Hume (1932, 1936, 1941a); Evans (1930); Beach (1943); Clark (1948); A.S.P.G. (1954, 1956, 1959, 1960); North and Henderson (1954a); Fox (1959); Wright-Broughton (1960); Fitzgerald (1962).

REGIONAL SECTION B′-B″ WESTERN ROCKY MOUNTAINS AND BRITISH COLUMBIA ONLY

Daly (1915); Walker (1926); Evans (1933); Cairnes (1934); Henderson (1954a); North and Henderson (1954b); A.S.P.G. (1954, 1956, 1960, 1962b); Jones (1959); Reesor (1957b, 1965); Little (1960); Wheeler (1961, 1965).

REGIONAL SECTION C-C′

Hume (1931, 1938, 1941b); Hage (1942, 1943b, 1946); Gallup (1954); A.S.P.G. (1953, 1959, 1961); Link (1953); Penner (1957); Douglas (1958); Keating (1966).

REGIONAL SECTIONS E-E′ AND F-F′

Dawson (1875, 1886); Willis (1902); Daly (1912); Clapp (1932); Hume (1933); Billings (1938); Hage (1943a); Pardee (1950); Douglas (1951, 1952); A.S.P.G. (1953, 1957, 1962a, 1964a); Erdman *et al.* (1953); Clark (1954, 1964); Gallup (1955); Ross (1959); Price (1958, 1962, 1965); Fox (1959); Robertson (1963); Jones (1964); Oswald (1964); Russell (1964); Keating (1966).

ROCKY MOUNTAIN TRENCH AND SECTION G-G′

Rice (1937); North and Henderson (1954b); Holland (1959); Leech (1959, 1960, 1962, 1965); Thompson (1962); A.S.P.G. (1962a); Lamb and Smith (1962); Crickmay (1964).

STRUCTURAL EVOLUTION MAINLY BASED ON STRUCTURAL CONSIDERATIONS

McConnell (1887); Willis (1902); Daly (1912); Hume (1933, 1941b, 1957); Hake *et al.* (1942); Hage (1942); Douglas (1950, 1958); Erdman (1950); Clark (1954); Gallup (1954, 1955); Link (1953); Scott (1953); Henderson and Dahlstrom (1959); Mountjoy (1960); Dahlstrom *et al.* (1962); Oswald (1964).

STRUCTURAL EVOLUTION BASED ON STRATIGRAPHIC AND MORPHOLOGIC CONSIDERATIONS

Daly (1912); Warren (1951); Pardee (1950); Russell (1951, 1954); Webb (1951, 1964); Alden (1932); Reesor (1957a); Warren and Stelck (1958); White (1959); Patterson and Storey (1960); Gussow (1960); A.S.P.G. (1964b); Wheeler (1965).

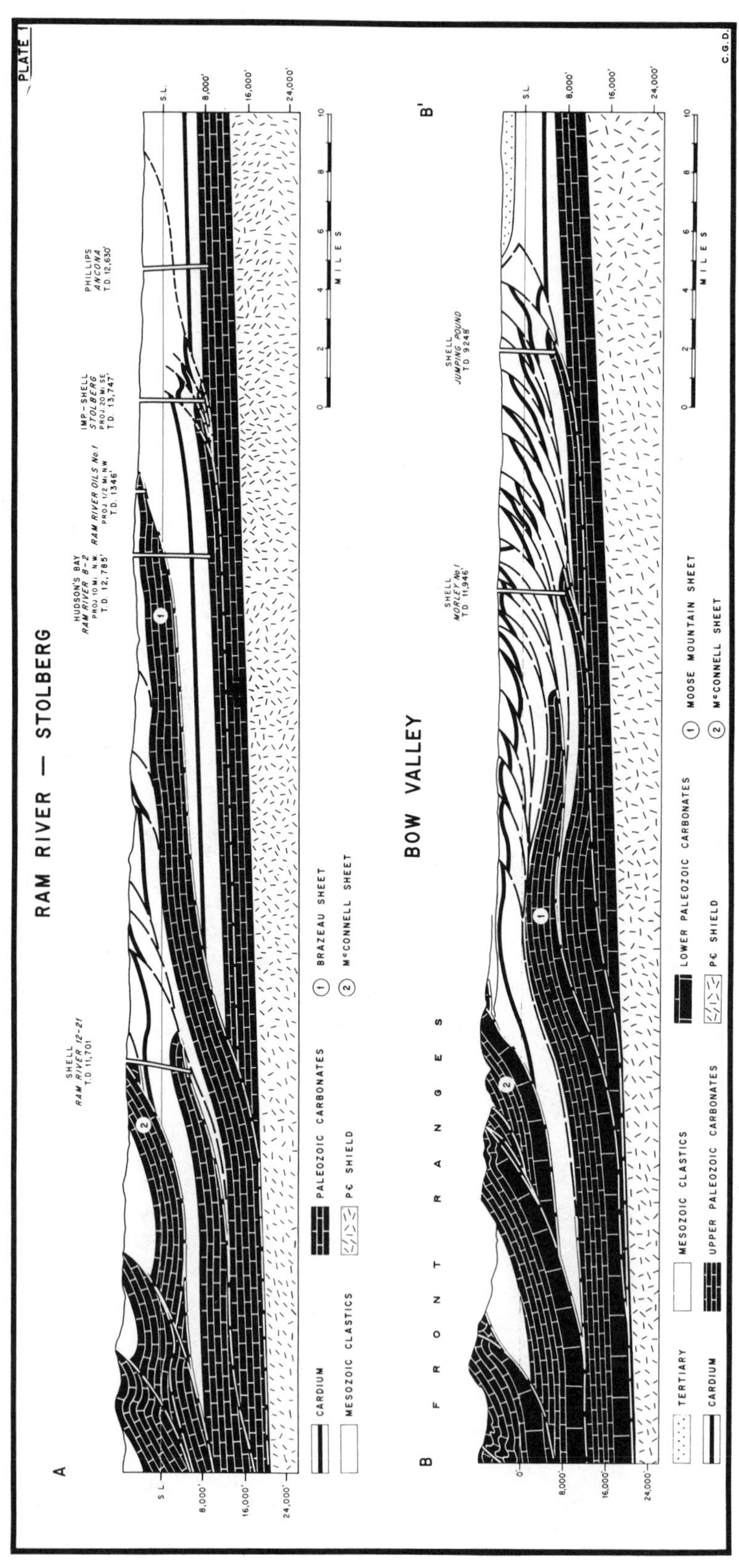

PLATE 1
RAM RIVER — STOLBERG
A
SHELL
RAM RIVER 12-21
T.D. 11,701'
HUDSON'S BAY
RAM RIVER 8-2
PROJ. 10 MI. N.W.
T.D. 12,785'
RAM RIVER OILS No. 1
PROJ. 1/2 MI. N.W.
T.D. 1346'
IMP-SHELL
STOLBERG
PROJ. 20 MI. S.E.
T.D. 13,747'
PHILLIPS
ANCONA
T.D. 12,630'
S.L.
8,000'
16,000'
24,000'
MILES
CARDIUM
MESOZOIC CLASTICS
PALEOZOIC CARBONATES
P€ SHIELD
1 BRAZEAU SHEET
2 McCONNELL SHEET
BOW VALLEY
B
B'
FRONT RANGES
SHELL
MORLEY No. 1
T.D. 11,946'
SHELL
JUMPING POUND
T.D. 9248'
TERTIARY
CARDIUM
MESOZOIC CLASTICS
UPPER PALEOZOIC CARBONATES
LOWER PALEOZOIC CARBONATES
P€ SHIELD
1 MOOSE MOUNTAIN SHEET
2 McCONNELL SHEET
C.G.D.

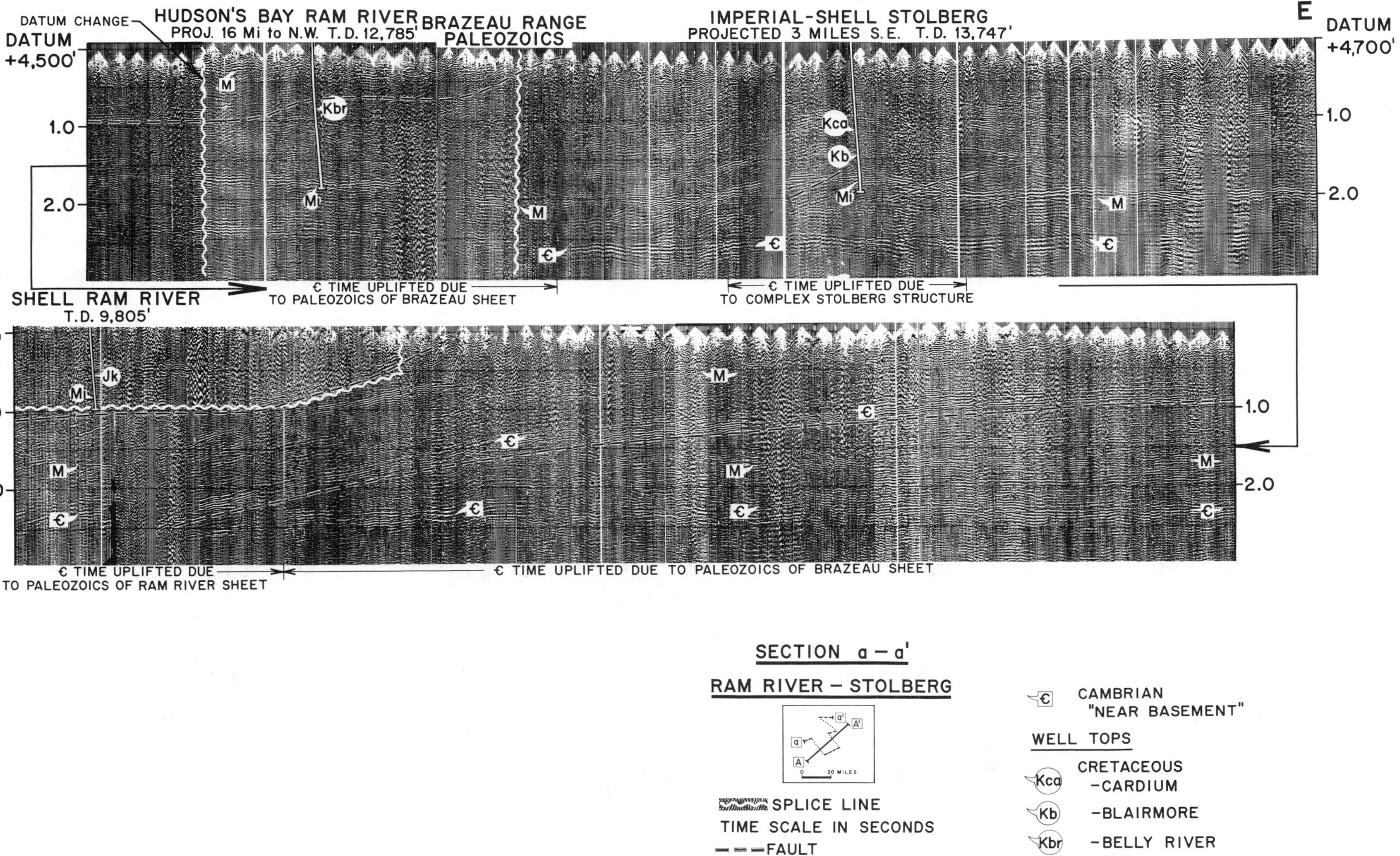
DATUM CHANGE
HUDSON'S BAY RAM RIVER
PROJ. 16 Mi to N.W. T.D. 12,785'
BRAZEAU RANGE
PALEOZOICS
IMPERIAL-SHELL STOLBERG
PROJECTED 3 MILES S.E. T.D. 13,747'
E
DATUM
+4,700'
DATUM
+4,500'
1.0
2.0
M
Kbr
Mi
Kca
Kb
€
€ TIME UPLIFTED DUE
TO PALEOZOICS OF BRAZEAU SHEET
€ TIME UPLIFTED DUE
TO COMPLEX STOLBERG STRUCTURE
W
SHELL RAM RIVER
T.D. 9,805'
Jk
€ TIME UPLIFTED DUE
TO PALEOZOICS OF RAM RIVER SHEET
€ TIME UPLIFTED DUE TO PALEOZOICS OF BRAZEAU SHEET
SECTION a – a'
RAM RIVER – STOLBERG
a
a'
A
A'
0
20 MILES
SPLICE LINE
TIME SCALE IN SECONDS
FAULT
REFLECTION EVENTS
BASAL MESOZOIC
"NEAR MISSISSIPPIAN"
CAMBRIAN
"NEAR BASEMENT"
WELL TOPS
CRETACEOUS
-CARDIUM
-BLAIRMORE
-BELLY RIVER
JURASSIC
-KOOTENAY
MISSISSIPPIAN
C.G.D.

PLATE 3

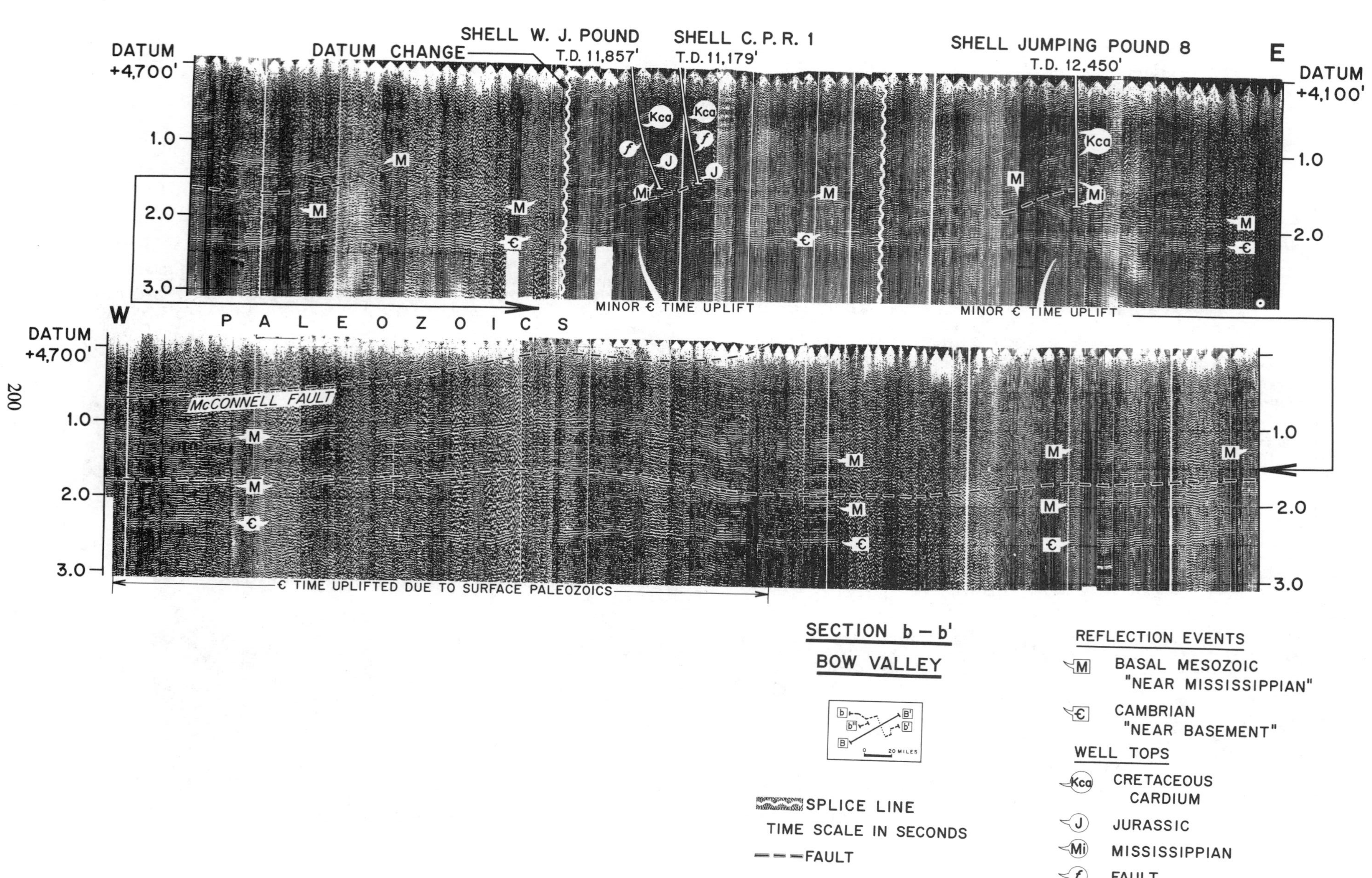

DATUM +4,700'
DATUM CHANGE
SHELL W. J. POUND
T.D. 11,857'
SHELL C. P. R. 1
T.D. 11,179'
SHELL JUMPING POUND 8
T.D. 12,450'
E
DATUM +4,100'
1.0
2.0
3.0
MINOR Є TIME UPLIFT
MINOR Є TIME UPLIFT
W
P A L E O Z O I C S
DATUM +4,700'
McCONNELL FAULT
Є TIME UPLIFTED DUE TO SURFACE PALEOZOICS
SECTION b – b'
BOW VALLEY
20 MILES
REFLECTION EVENTS
M BASAL MESOZOIC "NEAR MISSISSIPPIAN"
Є CAMBRIAN "NEAR BASEMENT"
WELL TOPS
Kca CRETACEOUS CARDIUM
J JURASSIC
Mi MISSISSIPPIAN
f FAULT
SPLICE LINE
TIME SCALE IN SECONDS
FAULT
C.G.D.

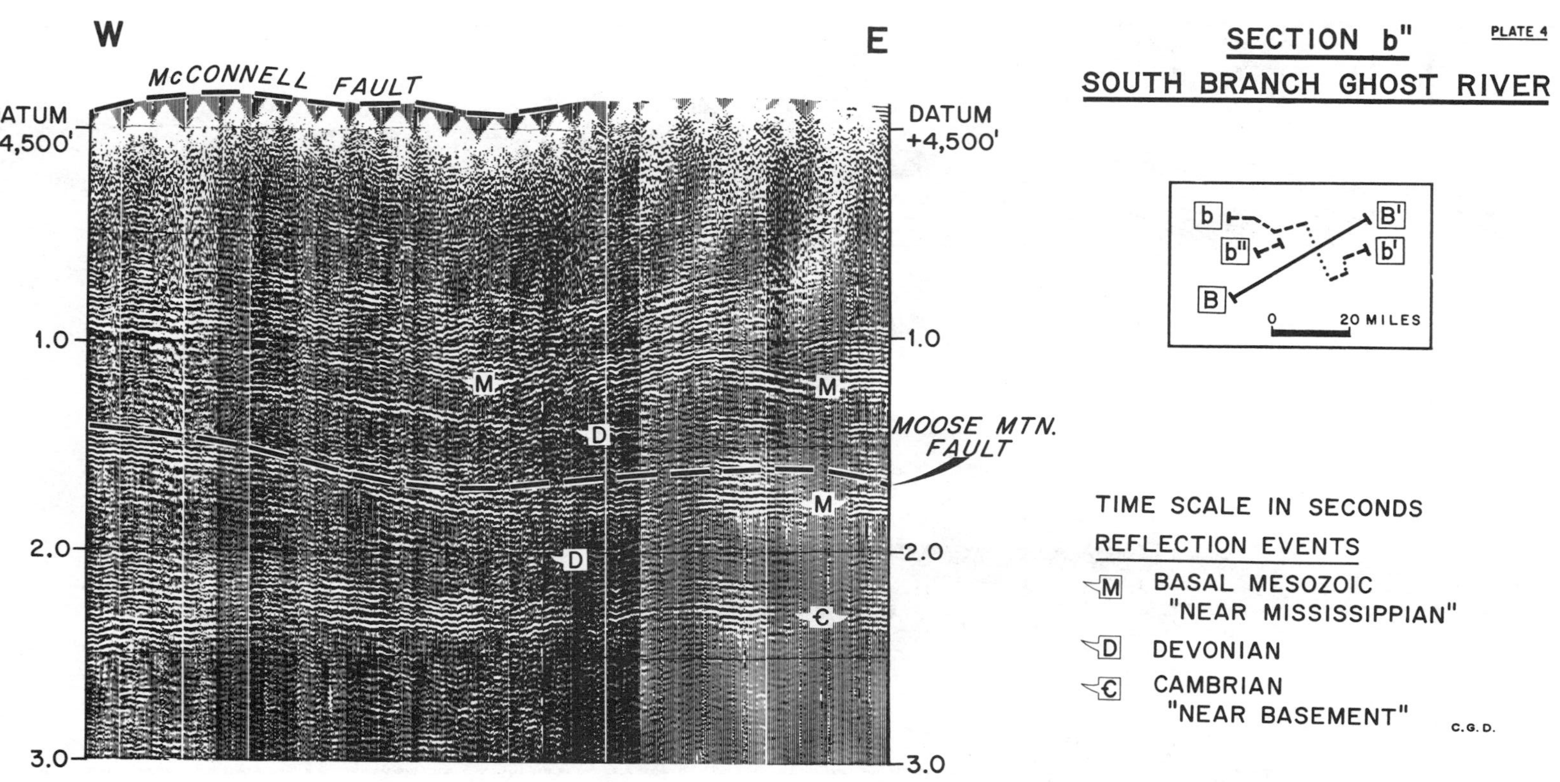
W
E
McCONNELL FAULT
DATUM +4,500'
1.0
2.0
3.0
M
D
MOOSE MTN. FAULT
C
SECTION b"
PLATE 4
SOUTH BRANCH GHOST RIVER
b
b"
B
B'
b'
0
20 MILES
TIME SCALE IN SECONDS
REFLECTION EVENTS
M BASAL MESOZOIC "NEAR MISSISSIPPIAN"
D DEVONIAN
C CAMBRIAN "NEAR BASEMENT"
C.G.D.

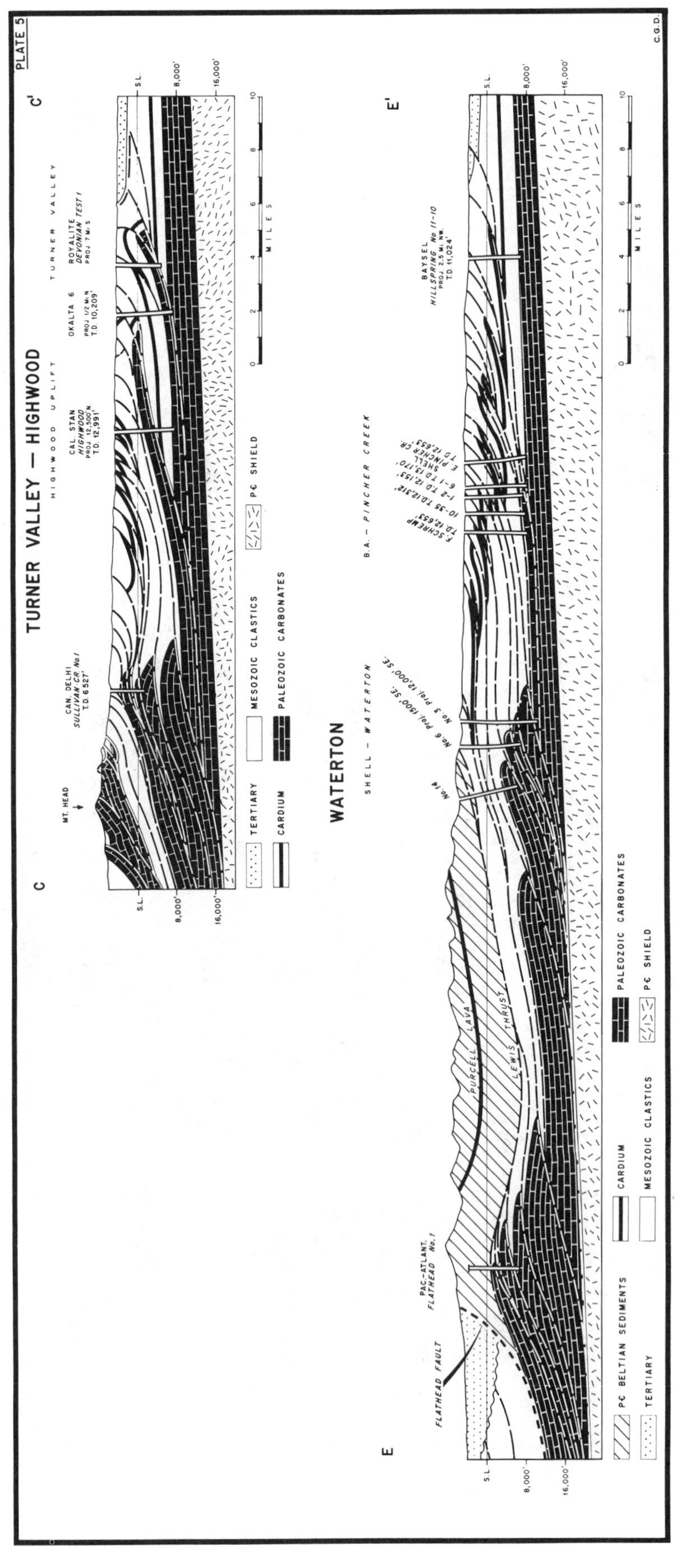

PLATE 5
TURNER VALLEY — HIGHWOOD
C
C'
HIGHWOOD UPLIFT
TURNER VALLEY
MT. HEAD
S.L.
8,000'
16,000'
TERTIARY
CARDIUM
MESOZOIC CLASTICS
PALEOZOIC CARBONATES
P€ SHIELD
MILES
WATERTON
E
E'
SHELL - WATERTON
B.A. - PINCHER CREEK
FLATHEAD FAULT
PURCELL LAVA
LEWIS THRUST
P€ BELTIAN SEDIMENTS
C.G.D.

PLATE 6

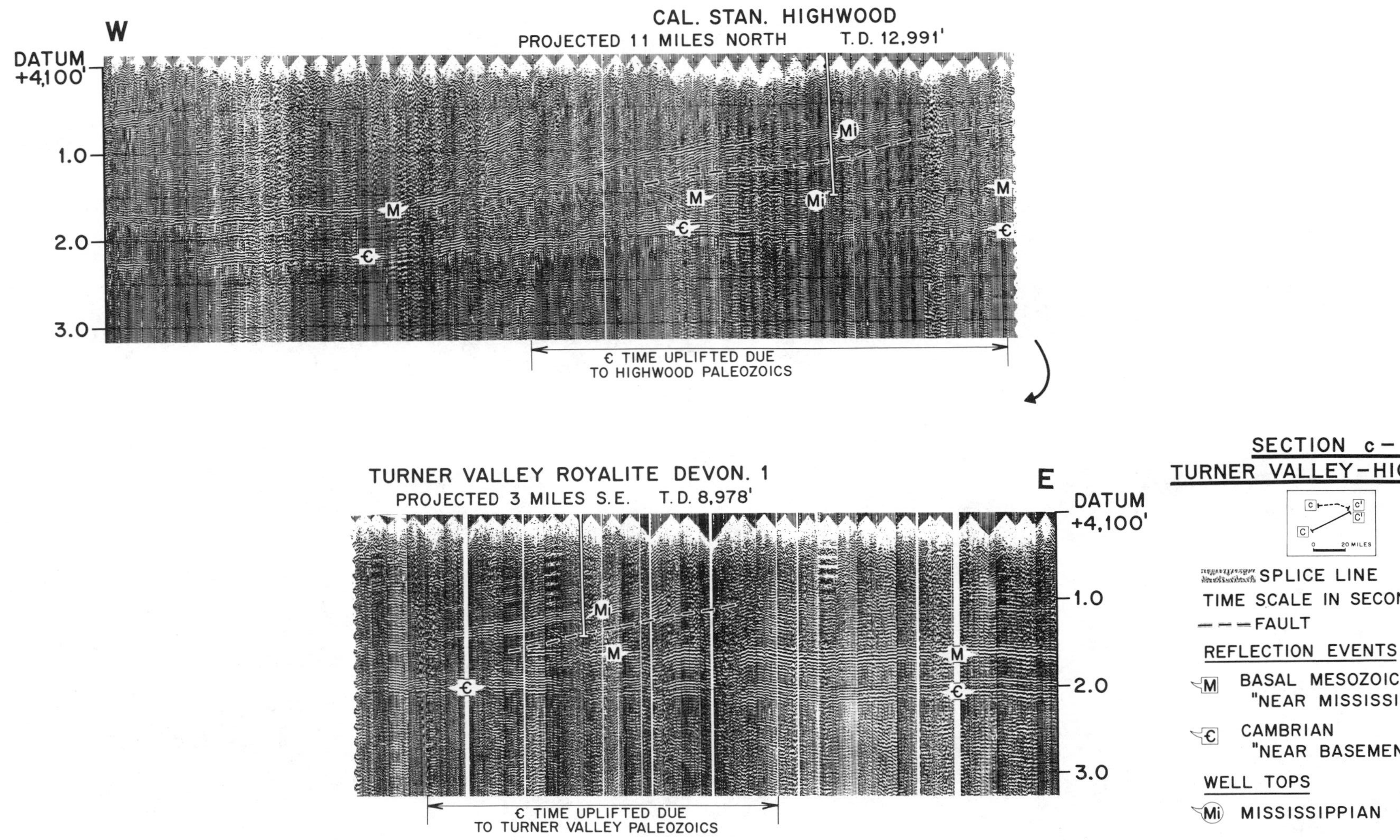

W
CAL. STAN. HIGHWOOD
PROJECTED 11 MILES NORTH T.D. 12,991'
DATUM +4,100'
1.0
2.0
3.0
€ TIME UPLIFTED DUE TO HIGHWOOD PALEOZOICS
TURNER VALLEY ROYALITE DEVON. 1
PROJECTED 3 MILES S.E. T.D. 8,978'
E
€ TIME UPLIFTED DUE TO TURNER VALLEY PALEOZOICS
SECTION c – c'
TURNER VALLEY–HIGHWOOD
0 20 MILES
SPLICE LINE
TIME SCALE IN SECONDS
FAULT
REFLECTION EVENTS
M BASAL MESOZOIC "NEAR MISSISSIPPIAN"
€ CAMBRIAN "NEAR BASEMENT"
WELL TOPS
Mi MISSISSIPPIAN
C.G.D.

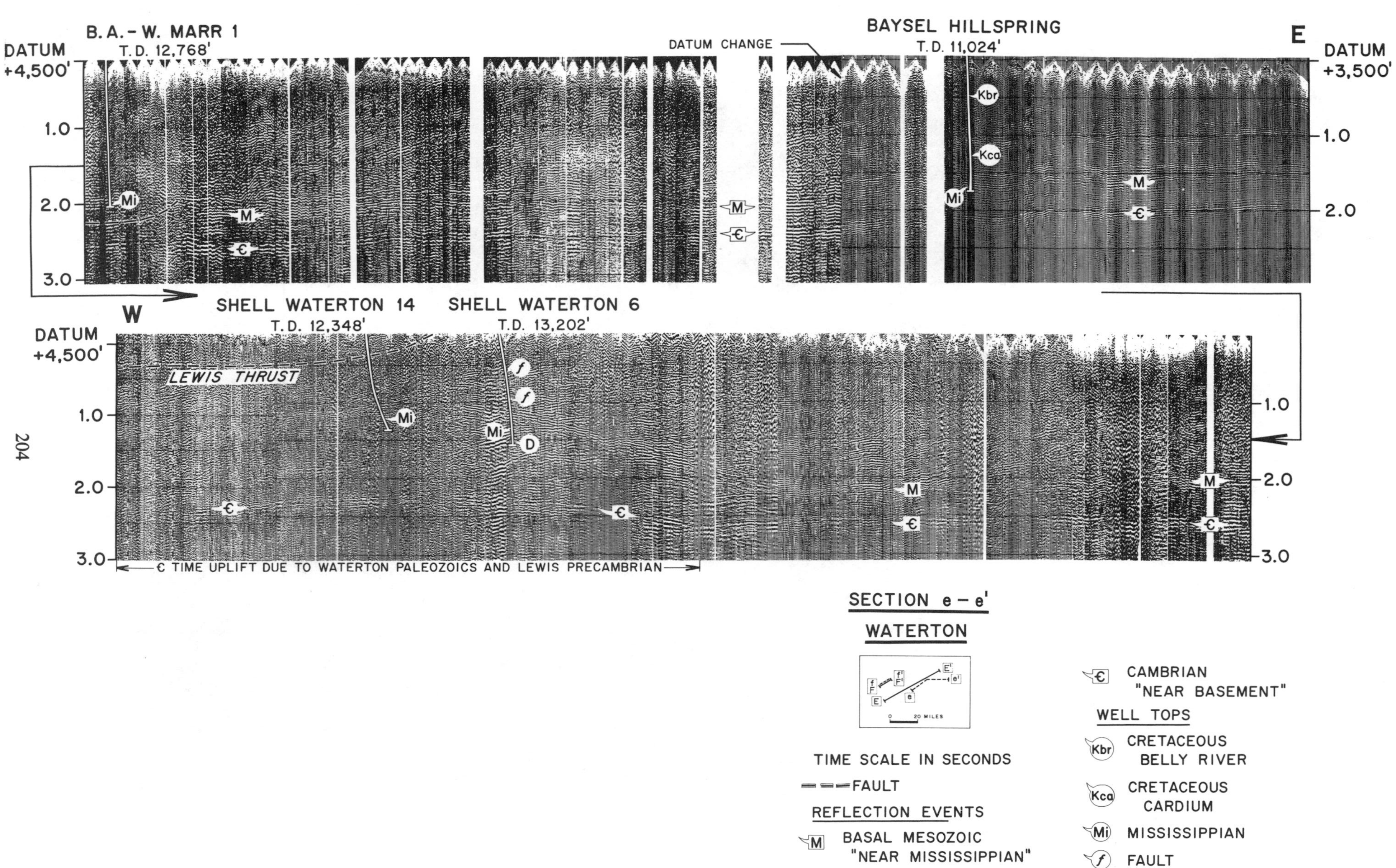
B.A.-W. MARR 1
T.D. 12,768'
DATUM +4,500'
1.0
2.0
3.0
DATUM CHANGE
BAYSEL HILLSPRING
T.D. 11,024'
E
DATUM +3,500'
Kbr
Kca
Mi
M
€
W
SHELL WATERTON 14
T.D. 12,348'
SHELL WATERTON 6
T.D. 13,202'
LEWIS THRUST
f
D
€ TIME UPLIFT DUE TO WATERTON PALEOZOICS AND LEWIS PRECAMBRIAN
SECTION e – e'
WATERTON
0
20 MILES
TIME SCALE IN SECONDS
FAULT
REFLECTION EVENTS
BASAL MESOZOIC "NEAR MISSISSIPPIAN"
CAMBRIAN "NEAR BASEMENT"
WELL TOPS
CRETACEOUS BELLY RIVER
CRETACEOUS CARDIUM
MISSISSIPPIAN
FAULT

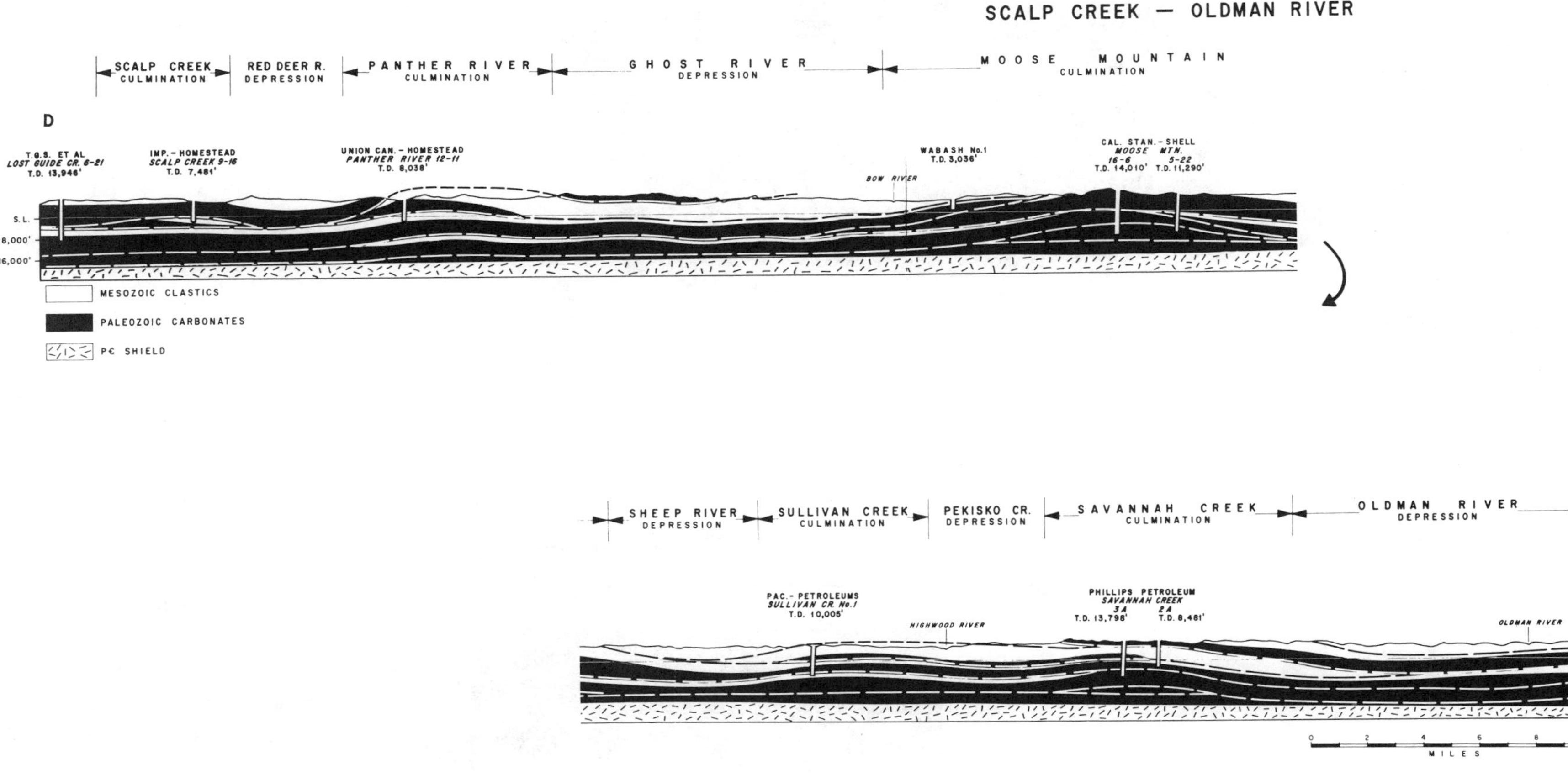
SCALP CREEK — OLDMAN RIVER
SCALP CREEK CULMINATION
RED DEER R. DEPRESSION
PANTHER RIVER CULMINATION
GHOST RIVER DEPRESSION
MOOSE MOUNTAIN CULMINATION
D
T.G.S. ET AL LOST GUIDE CR. 6-21 T.D. 13,946'
IMP.-HOMESTEAD SCALP CREEK 9-16 T.D. 7,481'
UNION CAN.-HOMESTEAD PANTHER RIVER 12-11 T.D. 8,038'
WABASH No.1 T.D. 3,036'
BOW RIVER
CAL. STAN.-SHELL MOOSE MTN. 16-6 5-22 T.D. 14,010' T.D. 11,290'
S.L.
8,000'
16,000'
MESOZOIC CLASTICS
PALEOZOIC CARBONATES
P€ SHIELD
SHEEP RIVER DEPRESSION
SULLIVAN CREEK CULMINATION
PEKISKO CR. DEPRESSION
SAVANNAH CREEK CULMINATION
OLDMAN RIVER DEPRESSION
D'
PAC.-PETROLEUMS SULLIVAN CR. No.1 T.D. 10,005'
HIGHWOOD RIVER
PHILLIPS PETROLEUM SAVANNAH CREEK 3A 2A T.D. 13,796' T.D. 8,481'
OLDMAN RIVER
S L
8,000'
16,000'
0 2 4 6 8 10
MILES
C.G.D.

PLATE 9

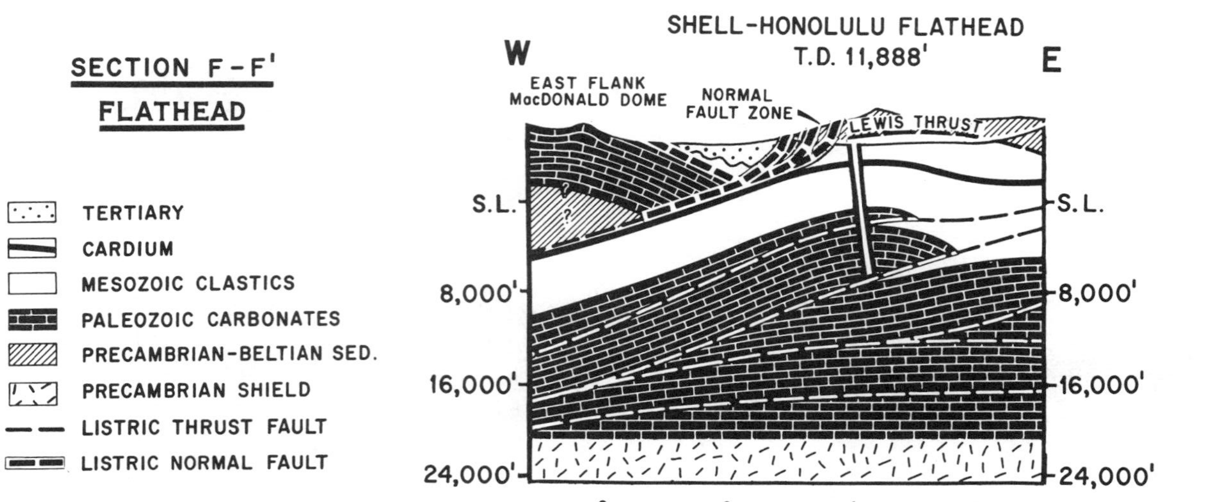

SECTION F - F'
FLATHEAD
TERTIARY
CARDIUM
MESOZOIC CLASTICS
PALEOZOIC CARBONATES
PRECAMBRIAN-BELTIAN SED.
PRECAMBRIAN SHIELD
LISTRIC THRUST FAULT
LISTRIC NORMAL FAULT
SHELL-HONOLULU FLATHEAD
T.D. 11,888'
W
E
EAST FLANK
MacDONALD DOME
NORMAL
FAULT ZONE
LEWIS THRUST
S.L.
8,000'
16,000'
24,000'
0
2
4
6
MILES

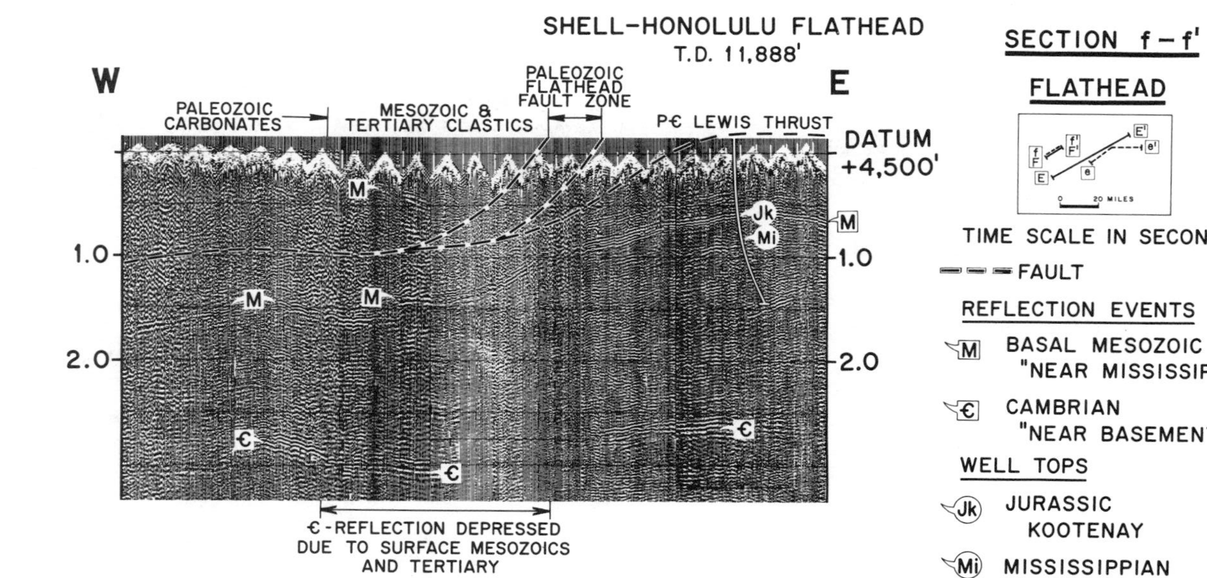

SHELL-HONOLULU FLATHEAD
T.D. 11,888'
W
E
PALEOZOIC
CARBONATES
MESOZOIC &
TERTIARY CLASTICS
PALEOZOIC
FLATHEAD
FAULT ZONE
P€ LEWIS THRUST
DATUM
+4,500'
1.0
2.0
M
Jk
Mi
€
€-REFLECTION DEPRESSED
DUE TO SURFACE MESOZOICS
AND TERTIARY
SECTION f - f'
FLATHEAD
20 MILES
TIME SCALE IN SECONDS
FAULT
REFLECTION EVENTS
BASAL MESOZOIC
"NEAR MISSISSIPPIAN"
CAMBRIAN
"NEAR BASEMENT"
WELL TOPS
JURASSIC
KOOTENAY
MISSISSIPPIAN

C.G.D.

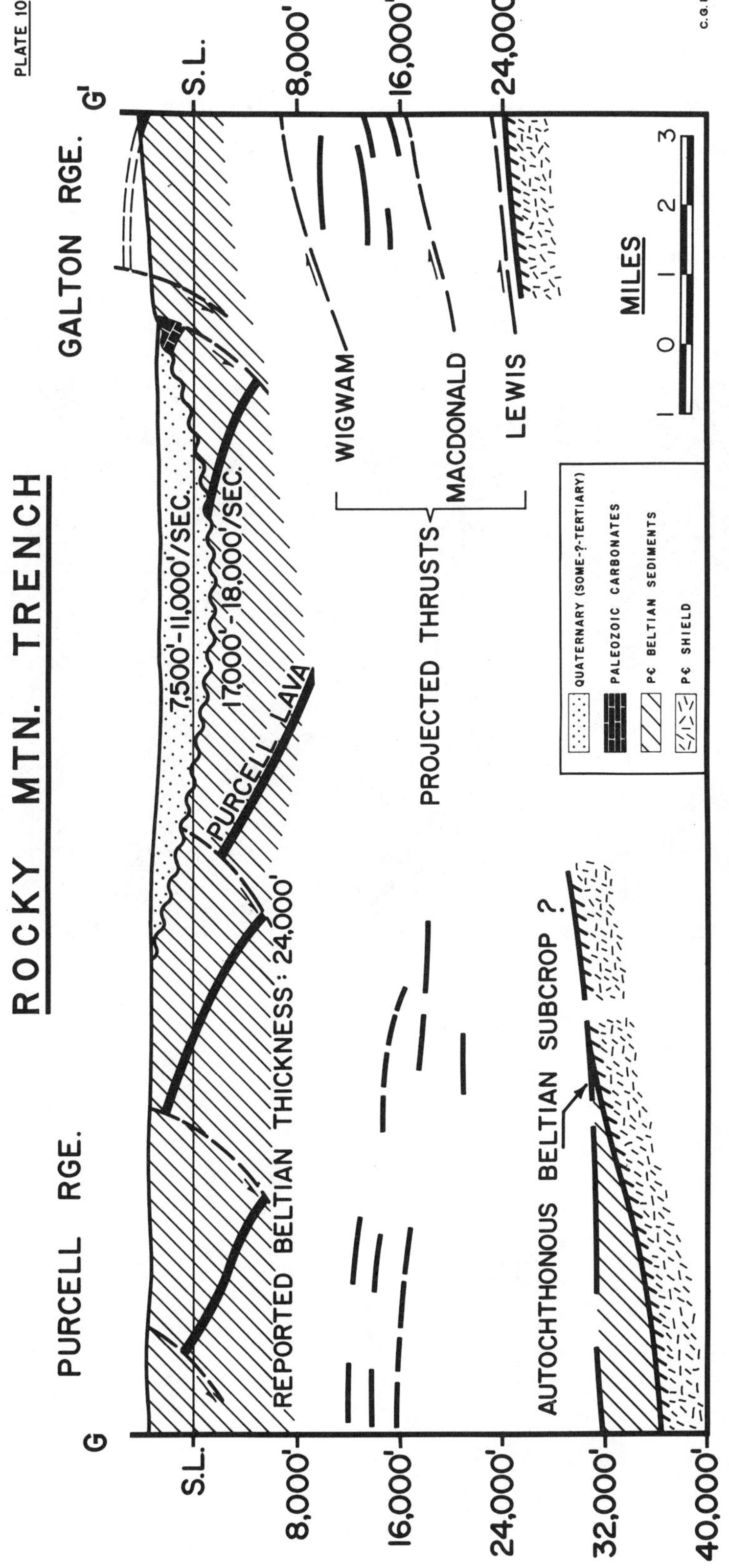
PLATE 10
ROCKY MTN. TRENCH
PURCELL RGE.
GALTON RGE.
G
G'
S.L.
8,000'
16,000'
24,000'
32,000'
40,000'
7,500'-11,000'/SEC.
17,000'-18,000'/SEC.
PURCELL LAVA
REPORTED BELTIAN THICKNESS: 24,000'
PROJECTED THRUSTS
WIGWAM
MACDONALD
LEWIS
AUTOCHTHONOUS BELTIAN SUBCROP ?
QUATERNARY (SOME-?-TERTIARY)
PALEOZOIC CARBONATES
P€ BELTIAN SEDIMENTS
P€ SHIELD
MILES
1 0 1 2 3
C.G.D.

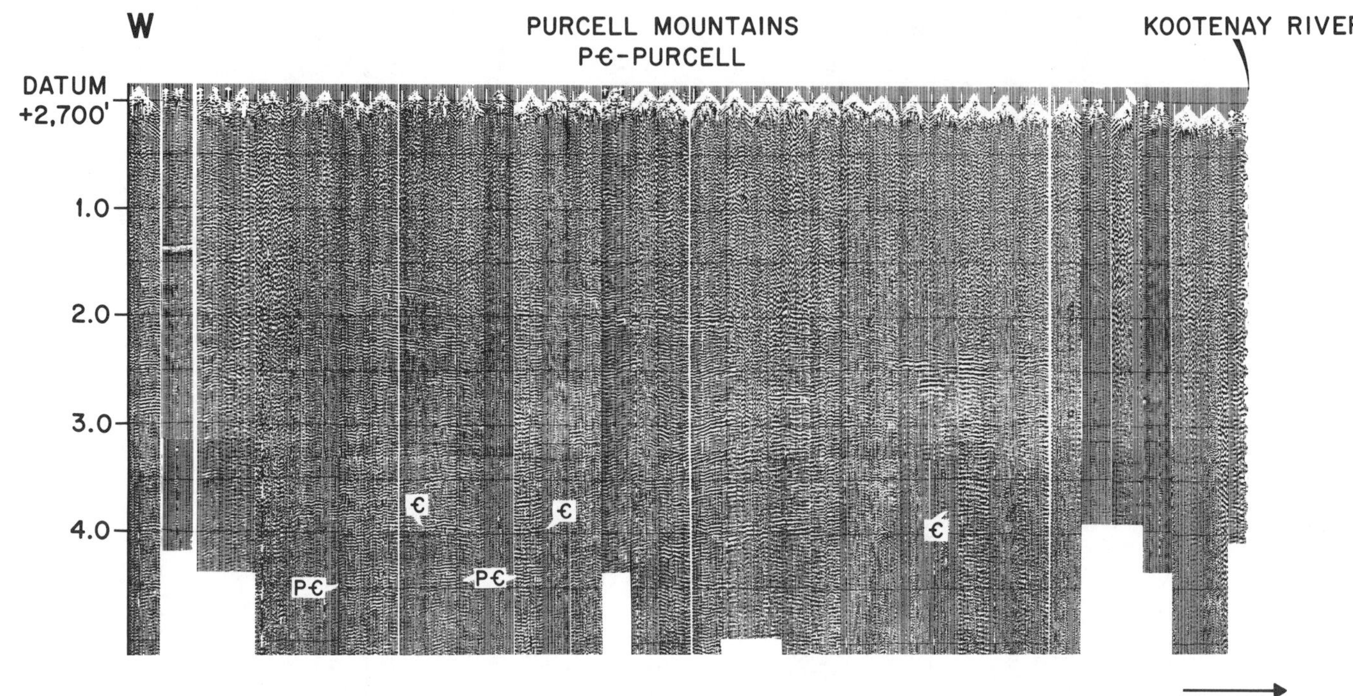
W
PURCELL MOUNTAINS
P€-PURCELL
KOOTENAY RIVER
DATUM
+2,700'
1.0
2.0
3.0
4.0
€
€
€
P€
P€

PLATE 11

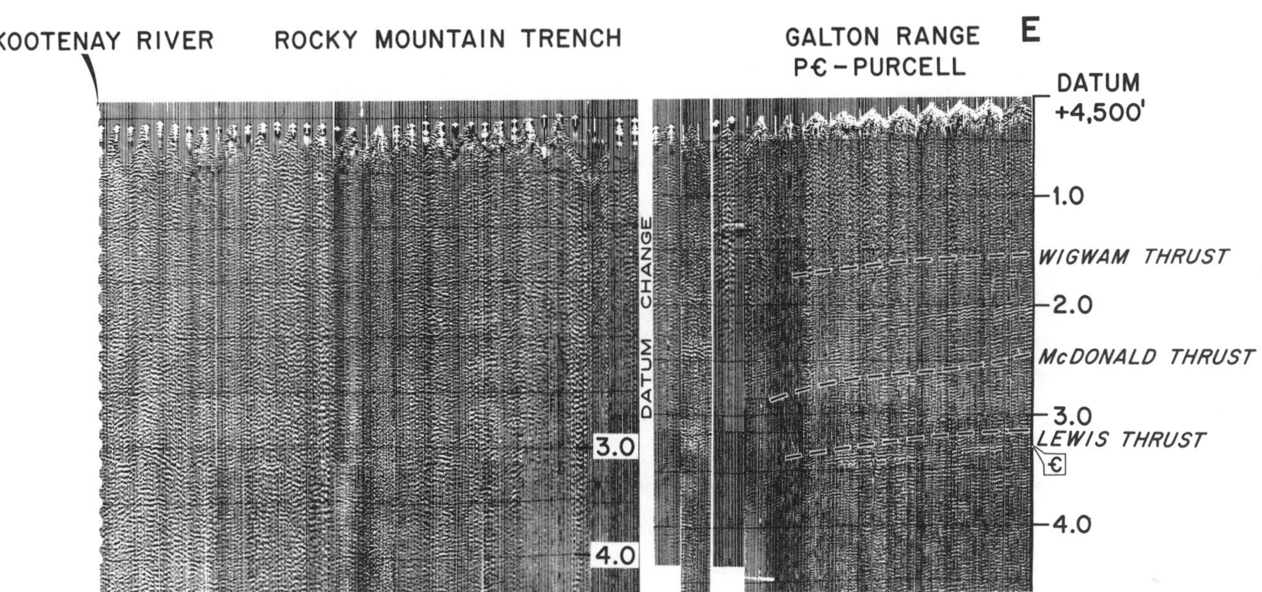

SECTION g – g'

ROCKY MOUNTAIN TRENCH

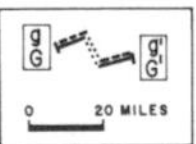

SPLICE LINE

TIME SCALE IN SECONDS

REFLECTION EVENTS

€ CAMBRIAN "NEAR BASEMENT"

P€ PRECAMBRIAN

C.G.D.

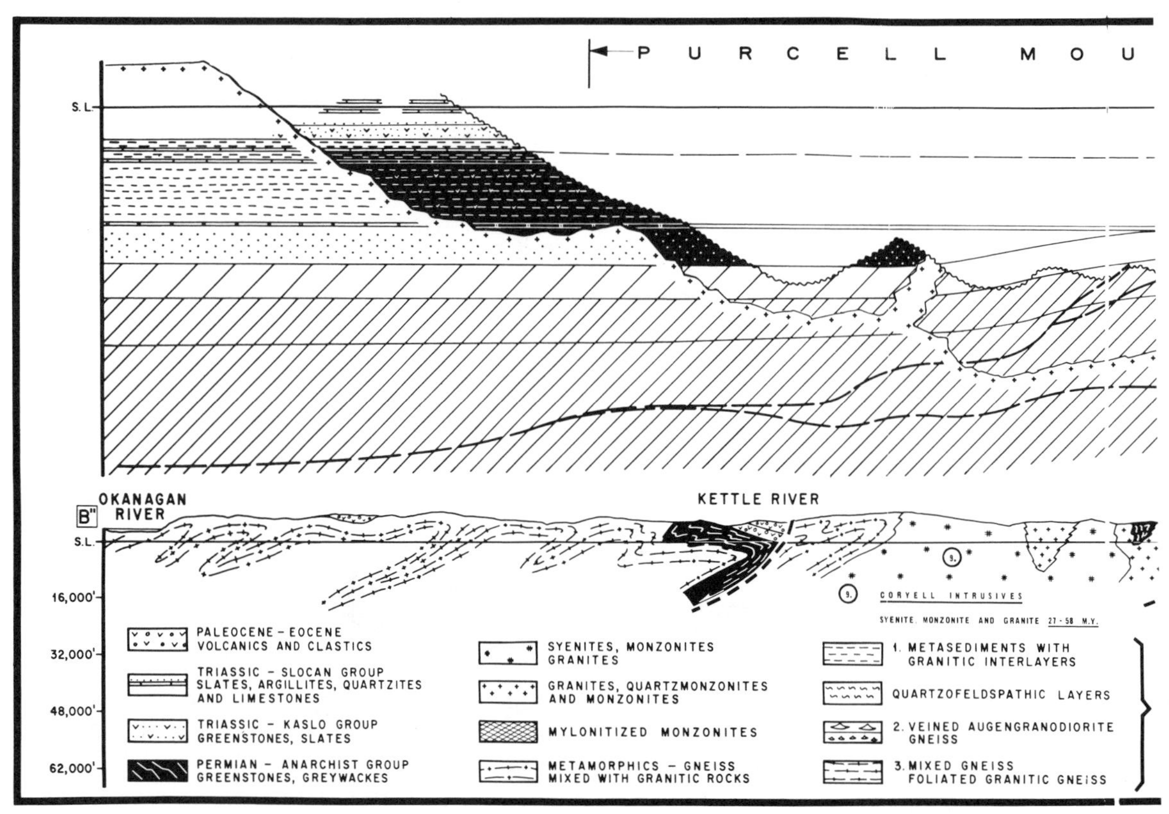
PURCELL MOU
S.L.
B" OKANAGAN RIVER
KETTLE RIVER
S.L.
16,000'
32,000'
48,000'
62,000'
CORYELL INTRUSIVES
SYENITE, MONZONITE AND GRANITE 27-58 M.Y.
PALEOCENE - EOCENE VOLCANICS AND CLASTICS
TRIASSIC - SLOCAN GROUP SLATES, ARGILLITES, QUARTZITES AND LIMESTONES
TRIASSIC - KASLO GROUP GREENSTONES, SLATES
PERMIAN - ANARCHIST GROUP GREENSTONES, GREYWACKES
SYENITES, MONZONITES GRANITES
GRANITES, QUARTZMONZONITES AND MONZONITES
MYLONITIZED MONZONITES
METAMORPHICS - GNEISS MIXED WITH GRANITIC ROCKS
1. METASEDIMENTS WITH GRANITIC INTERLAYERS
QUARTZOFELDSPATHIC LAYERS
2. VEINED AUGENGRANODIORITE GNEISS
3. MIXED GNEISS FOLIATED GRANITIC GNEISS

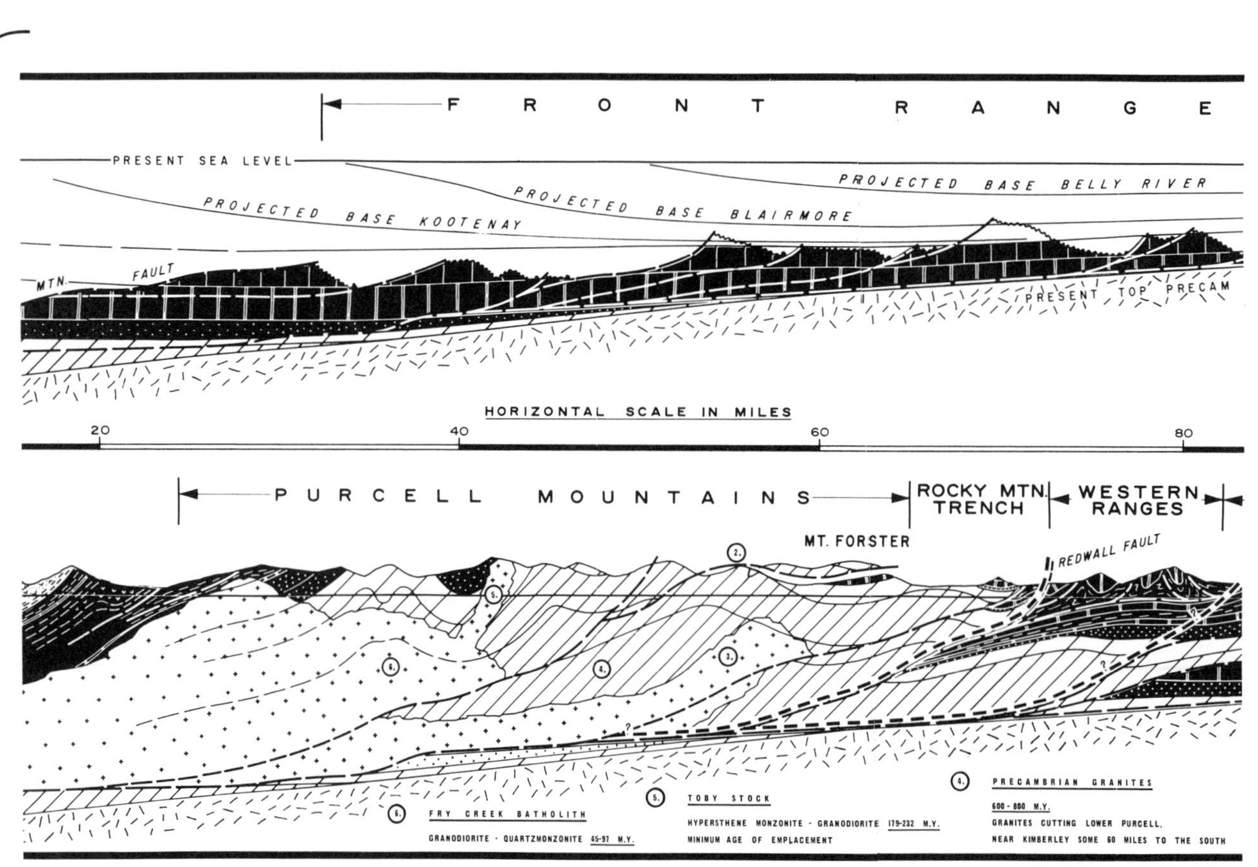
FRONT RANGE
PRESENT SEA LEVEL
PROJECTED BASE BELLY RIVER
PROJECTED BASE BLAIRMORE
PROJECTED BASE KOOTENAY
MTN. FAULT
PRESENT TOP PRECAM
HORIZONTAL SCALE IN MILES
20
40
60
80
PURCELL MOUNTAINS
ROCKY MTN. TRENCH
WESTERN RANGES
MT. FORSTER
REDWALL FAULT
FRY CREEK BATHOLITH
GRANODIORITE - QUARTZMONZONITE 45-97 M.Y.
TOBY STOCK
HYPERSTHENE MONZONITE - GRANODIORITE 179-232 M.Y.
MINIMUM AGE OF EMPLACEMENT
PRECAMBRIAN GRANITES
600-800 M.Y.
GRANITES CUTTING LOWER PURCELL,
NEAR KIMBERLEY SOME 60 MILES TO THE SOUTH

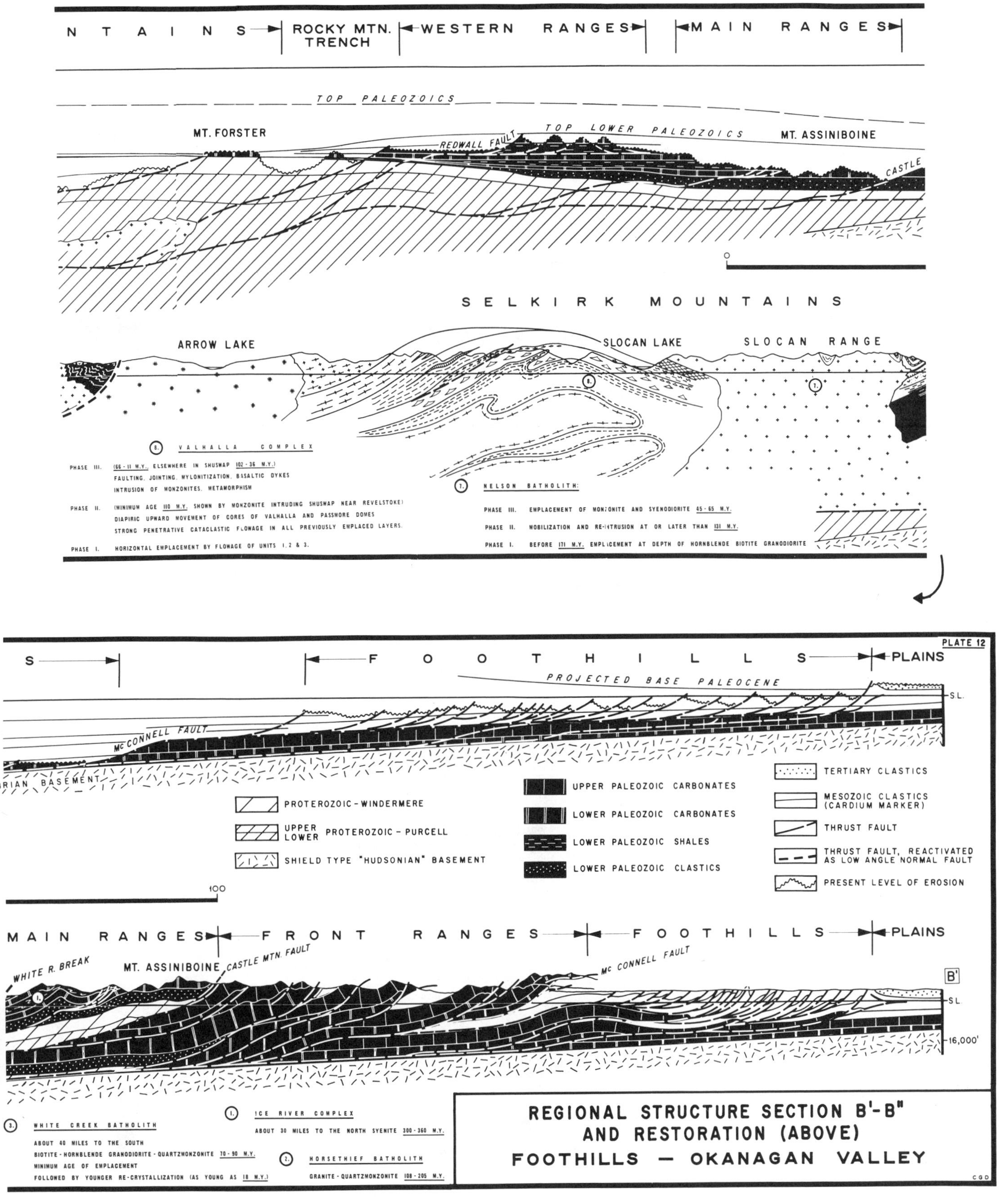

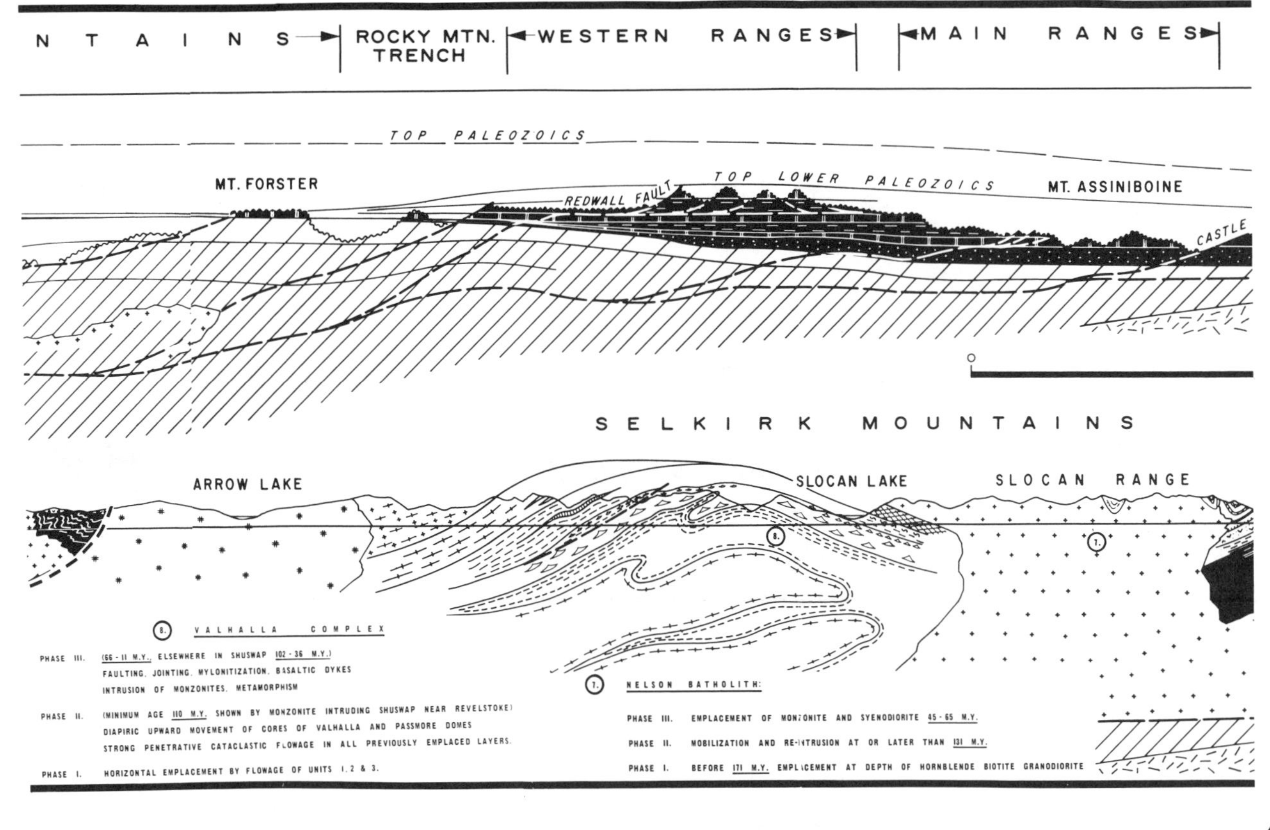

PLATE 12

S — F O O T H I L L S — PLAINS

PROJECTED BASE PALEOCENE

SL.

Mc CONNELL FAULT

BRIAN BASEMENT

PROTEROZOIC - WINDERMERE

UPPER / LOWER PROTEROZOIC - PURCELL

SHIELD TYPE "HUDSONIAN" BASEMENT

UPPER PALEOZOIC CARBONATES

LOWER PALEOZOIC CARBONATES

LOWER PALEOZOIC SHALES

LOWER PALEOZOIC CLASTICS

TERTIARY CLASTICS

MESOZOIC CLASTICS (CARDIUM MARKER)

THRUST FAULT

THRUST FAULT, REACTIVATED AS LOW ANGLE NORMAL FAULT

PRESENT LEVEL OF EROSION

100

MAIN RANGES — FRONT RANGES — FOOTHILLS — PLAINS

WHITE R. BREAK

MT. ASSINIBOINE

CASTLE MTN. FAULT

Mc CONNELL FAULT

B'

SL.

16,000'

3. WHITE CREEK BATHOLITH

ABOUT 40 MILES TO THE SOUTH
BIOTITE - HORNBLENDE GRANODIORITE - QUARTZMONZONITE 70 - 90 M.Y.
MINIMUM AGE OF EMPLACEMENT
FOLLOWED BY YOUNGER RE-CRYSTALLIZATION (AS YOUNG AS 18 M.Y.)

1. ICE RIVER COMPLEX

ABOUT 30 MILES TO THE NORTH SYENITE 300 - 360 M.Y.

2. HORSETHIEF BATHOLITH

GRANITE - QUARTZMONZONITE 108 - 205 M.Y.

REGIONAL STRUCTURE SECTION B'-B"
AND RESTORATION (ABOVE)
FOOTHILLS — OKANAGAN VALLEY

CGO

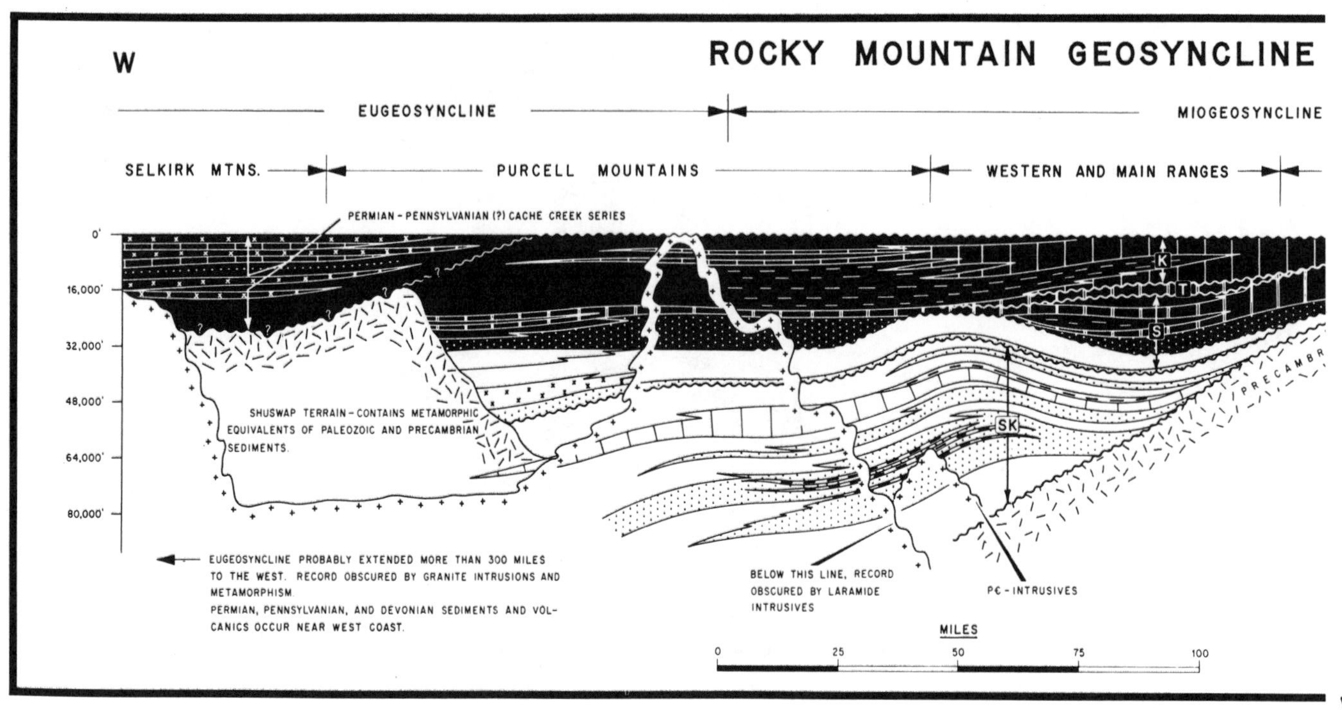
W
ROCKY MOUNTAIN GEOSYNCLINE
EUGEOSYNCLINE
MIOGEOSYNCLINE
SELKIRK MTNS.
PURCELL MOUNTAINS
WESTERN AND MAIN RANGES
PERMIAN - PENNSYLVANIAN (?) CACHE CREEK SERIES
0'
16,000'
32,000'
48,000'
64,000'
80,000'
K
T
S
SK
SHUSWAP TERRAIN - CONTAINS METAMORPHIC EQUIVALENTS OF PALEOZOIC AND PRECAMBRIAN SEDIMENTS.
PRECAMBR
EUGEOSYNCLINE PROBABLY EXTENDED MORE THAN 300 MILES TO THE WEST. RECORD OBSCURED BY GRANITE INTRUSIONS AND METAMORPHISM.
PERMIAN, PENNSYLVANIAN, AND DEVONIAN SEDIMENTS AND VOLCANICS OCCUR NEAR WEST COAST.
BELOW THIS LINE, RECORD OBSCURED BY LARAMIDE INTRUSIVES
P€ - INTRUSIVES
MILES
0
25
50
75
100

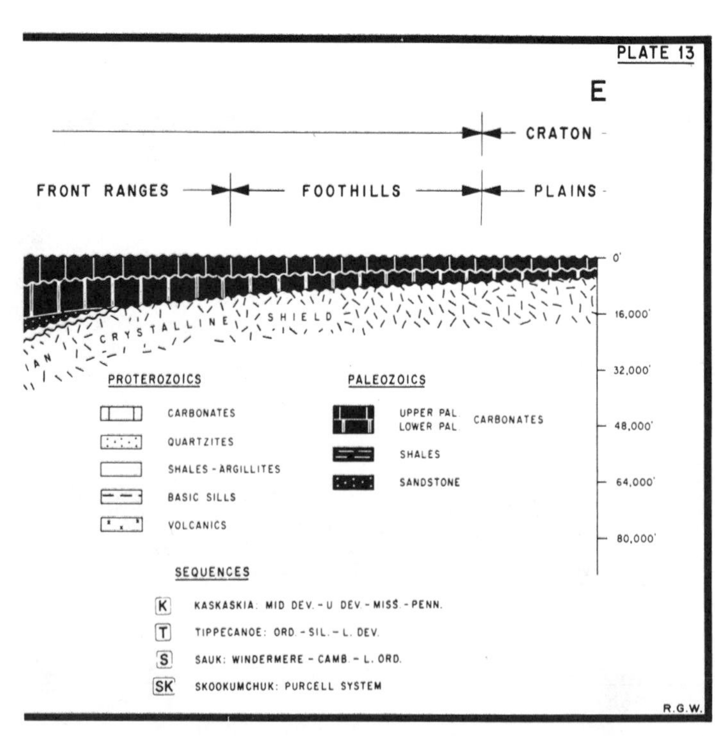
PLATE 13
E
CRATON
FRONT RANGES
FOOTHILLS
PLAINS
0'
16,000'
32,000'
48,000'
64,000'
80,000'
CRYSTALLINE SHIELD
PROTEROZOICS
CARBONATES
QUARTZITES
SHALES - ARGILLITES
BASIC SILLS
VOLCANICS
PALEOZOICS
UPPER PAL
LOWER PAL
CARBONATES
SHALES
SANDSTONE
SEQUENCES
K KASKASKIA: MID DEV. - U DEV. - MISS. - PENN.
T TIPPECANOE: ORD. - SIL. - L. DEV.
S SAUK: WINDERMERE - CAMB. - L. ORD.
SK SKOOKUMCHUK: PURCELL SYSTEM
R.G.W.

Reprinted by permission of the Rocky Mountain Association of Geologists from D. W. Bolyard, ed., *Deep Drilling Frontiers of the Central Rocky Mountains*, 1975, p. 41-54.

THRUST BELT STRUCTURAL GEOMETRY AND RELATED STRATIGRAPHIC PROBLEMS WYOMING–IDAHO–NORTHERN UTAH

by
F. Royse, Jr.,[1], M. A. Warner[1] and D. L. Reese[1]

INTRODUCTION

The thrust belt of western Wyoming, eastern Idaho and northeastern Utah (Fig. 1) has not been drilled for oil and gas to the same extent as have most other areas in the Rockies with preserved thick Phanerozoic sedimentary rock sequences. The thrust belt province is considered by many geologists as having too long and complicated a history of multiple deformation, uplift and erosion to harbor large oil and gas reserves even though known sedimentary rock thicknesses and facies seem favorable for past hydrocarbon generation and migration. Uncertainty in prediction of subsurface structural form, stratigraphy and timing are considered major exploration obstacles.

Much of this uncertainty can be alleviated by proper integration of structural and stratigraphic principles with available surface, seismic, aeromagnetic, paleontologic and well data. Recent studies of this sort have been published by Bally and others (1966), Dahlstrom (1970) and Price and Mountjoy (1970) for the south Canadian thrust province. In a comparable way, sufficient geologic observations have been made in the Idaho-Wyoming thrust belt by numerous geologists and geophysicists to establish with confidence certain models of basic structural types as typical of the province. These are: (1) concentric folds, (2) decollement, (3) reverse faults, (4) "tear" faults, (5) younger normal faults. Use of these as models along with proper geometric constraints permits an interpretation of thrust belt structure which is consistent with available geologic information.

In addition, awareness of the age and areal distribution of stratigraphic features such as synorogenic conglomerates and angular unconformities allows a stepwise palinspastic restoration of the structural and stratigraphic history.

A comprehensive report on thrust belt geology is not the intent, nor is it within the scope, of this paper. Rather, regional cross-sections and selected examples and discussions of typical features are presented to illustrate how the present structural form and stratigraphic history of parts of the thrust belt may be interpreted. The structural discussion is one of geometry rather than genesis, although one often implies certain aspects of the other.

GENERAL STRUCTURAL FRAMEWORK

Distribution of major thrust faults and younger extensional faults is illustrated on Figure 1. This, plus two regional structural cross-sections, X-X′ and Y-Y′ (Plates I and II, pocket) are an interpretation of the gross structural form of the thrust belt.[2]

A westward thickening wedge of Upper Precambrian, Paleozoic and Mesozoic sediments was compressed into a zone about 65 miles wide, roughly half its original width. East-west shortening was achieved by doubling the sedimentary section through motion on low angle reverse (thrust) faults and contemporaneous associated concentric folding. This period of compressional deformation began in latest Jurassic and continued through early Eocene, Later, normal (extensional) faulting occurred from Eocene to present (Armstrong and Oriel, 1965).

Seismic, aeromagnetic (Fig. 2) and surface data indicate the crystalline basement is not deformed over most of the region except for broad warping and is, therefore, structurally detached from the sedimentary cover by a regional *decollement.* The crystalline basement is, however, involved in thrusting along the western thrust belt margin in the central Wasatch Range north of Salt Lake City (Fig. 1) (Eardley, 1944). Bell (1952) described the outcrop of crystalline basement (Farmington Canyon complex) as a series of thrust slices or plates more than two miles thick. The central Wasatch Range is an uplift which is part of the Sevier orogenic belt described by R. L. Armstrong (1968). Uplift is pre-Evanston Formation (pre-Paleocene) in age according to mapping by Mullens and Laraway (1964) and is probably post Crawford thrusting since it apparently deforms this thrust system. Such an age indicates synchroniety between uplift and motion on the Absaroka thrust system (Plate II, in pocket) which is Santonian(?) through Maestrichtian.

STRUCTURAL GEOMETRY OF THRUST FAULTS

Empirical rules have been devised for interpreting thrust belt structural form below the level of direct observation. These have evolved over the years primarily from data gathered in the Canadian Rockies and Appalachians. These rules, modified from those set forth by Dahlstrom (1970) for the eastern Canadian thrust belt, are illustrated

[1]Chevron Oil Co., Denver, Colorado.

[2]Data for these illustrations were collected from numerous publications as well as proprietary files. Most valuable among published data sources are surface maps by the U.S. Geological Survey. The basic ideas pertaining to structural interpretation stem from articles by Rich (1934), Wilson and Stearns (1958), Bally and others (1966), Dahlstrom (1970) and conversation with many geologists, especially C. D. A. Dahlstrom, Peter Verrall and W. G. Brown.

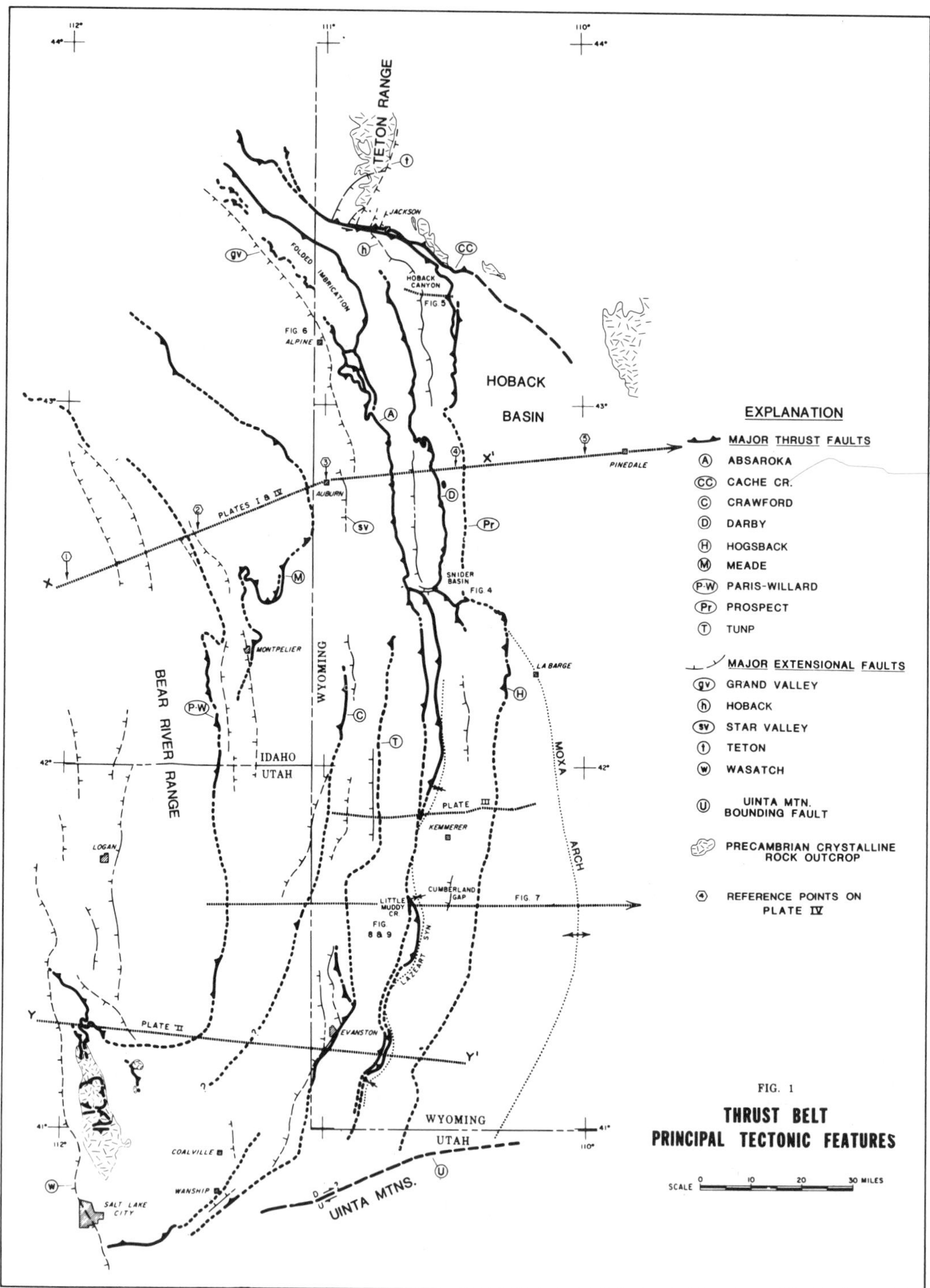

Fig. 1 — Index map showing principal tectonic features of thrust belt and location of figures (text) and plates (pocket).

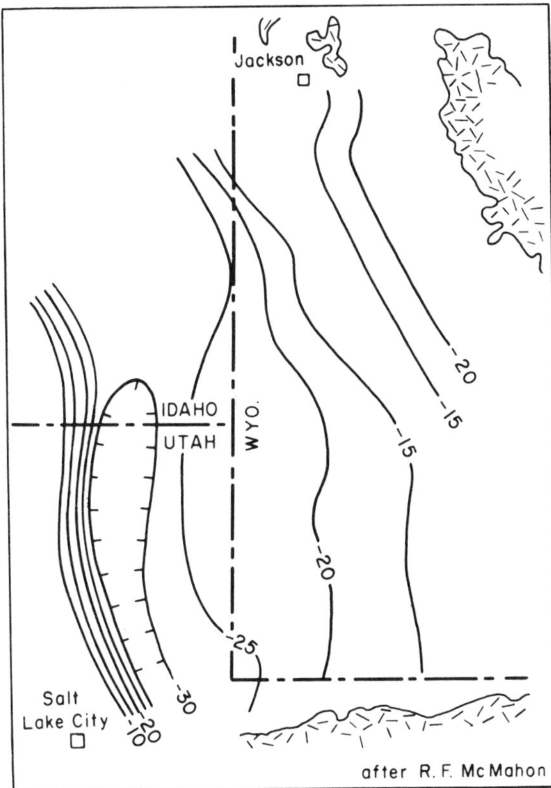

Fig. 2 — Generalized structure map, top of magnetic basement. Sea level datum. Contour int. = 5000 ft.

schematically in Figure 3 and can be summarized as follows: (1) folds have essentially concentric geometry, (2) thrust faults cut up section in the direction of tectonic transport, (3) thrust faults tend to parallel bedding in incompetent rocks and be oblique to bedding in relatively competent rocks, and (4) *major* thrust faults are younger in the direction of tectonic transport. Surface, seismic, and well data indicate that these concepts are applicable toward solving structural problems in the Idaho-Wyoming thrust belt. The fact that such rules can be devised at all indicates that a consistent, fundamental, long continuing deformational process has been at work.

Of great importance in predicting subsurface structure in this region is the fact that deformation is "brittle" in nature, and neither plastic flow or cleavage folding occurs to a significant degree. Concentric folding is the rule, and thickness and surface area of beds do not change. These features make it necessary and relatively easy to employ the concept of volumetric balance when constructing structural cross-sections (Dahlstrom, 1969). If correctly drawn, profiles of deformed sediments can be restored to an undeformed state without creating large unexplained voids, excess rocks, or improbable detachments. Sections should be drawn parallel to the direction of tectonic transport where plunge out of the plane of section is not very steep. The regional structural sections X-X′ and Y-Y′ (Plates I and II, in pocket) are examples of this interpretational technique and illustrate the principles outlined above.

These sections are, of course, interpretive; but they are not conjecture nor are they casually drawn. Seismic data, when combined with surface and sparse well control, enables us to structurally define the subsurface in the eastern thrust belt with some confidence. Two features most critical to subsurface interpretation which can be seen on seismic sections are: (1) the position and length of specific footwall stratigraphic cutoffs and (2) the general shape of the major thrust fault plane. From this information the amount of fault displacement and a correlation between surface hanging wall structure and subsurface fault plane shape can be made. Westward, where seismic data is not available, or is not interpretable because of poor record quality, erosion has exposed enough of the footwall structure of major thrusts to allow application of the principles established to the east.

The seismic time section (Plate III, in pocket) illustrates the tendency for major thrust faults to have a steplike profile. The Absaroka fault trace cuts steeply eastward through Paleozoic and Lower Mesozoic beds, flattens and nearly parallels bedding for several miles in the Cretaceous, then cuts abruptly up section to the surface. The Hogsback thrust shows similar form. Such steplike profiles are evident on many other seismic lines.

To the west, the Meade thrust (Plate I, in pocket) has been eroded deeply enough to reveal an outcropping steplike profile which has been subsequently deformed, probably by motion on the younger Absaroka thrust plane below. Movement on such a steplike surface has certain structural consequences for beds in the hanging wall which are illustrated on the idealized fault diagram (Fig. 3) (Gretener, 1972; Rich, 1934). The "typical" synclinal hanging wall feature shown on the diagram is a product of the transport and rotation of nearly flatlying beds over a steplike fault profile with long intervening flat traces. The gross synclinal character of many of the mountain ranges such as the Salt River Range, Wyoming Range, Hoback Range and Bear River Range probably derive such a steplike profile in the underlying bounding thrust.

It follows from such analysis that the map position of major steps (ramps) in the fault trace would be under and slightly west of the linear zone of tightly folded and imbricately faulted anticlinal structures of variable asymmetry which are observed to form the western borders of these ranges. This zone of locally intense deformation appears to result from rocks being ramped and rotated over a major step in the footwall. The long persistent west dip

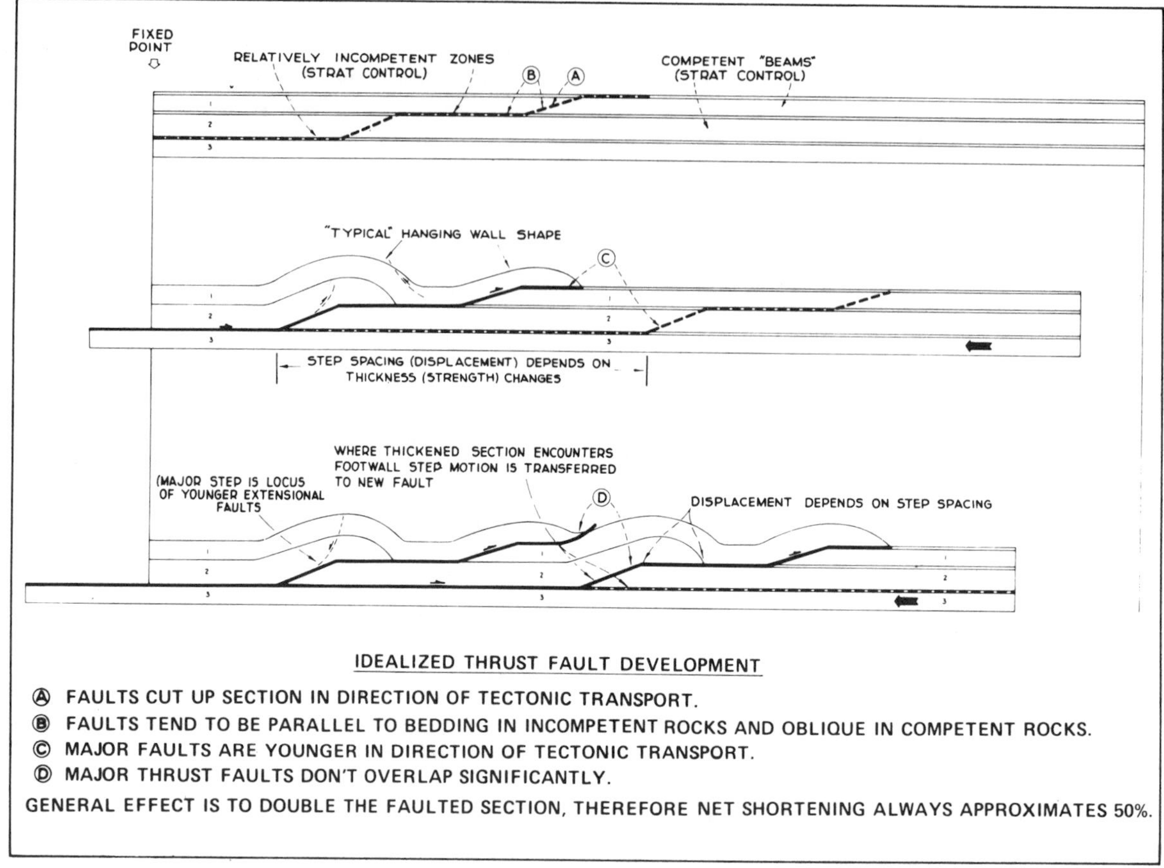

Fig. 3 — Schematic diagram illustrating empirical rules regarding thrust faults.

panels of Cretaceous and Jurassic rocks such as that in Greys River Valley west of the Wyoming Range (Plate I, in pocket), and that comprising the west flank of the Meridian Anticline (Plate III) are also clues to the subsurface position of major footwall steps.

An application of the principle of volumetric balance can be observed on section X-X′ (Plate I, in pocket). Note that the long hanging wall cutoff in lower Mississippian rocks shown for the Meade thrust is matched in the footwall farther west; the total shortening is known from outcrop preservation of both hanging wall and footwall cutoffs of Nugget sandstone (Cressman, 1964). The interpretation is admittedly speculative; however, one can be reasonably sure that a long footwall section of lower Mississippian rocks must exist in the subsurface to the west.

The main detachments occur at different stratigraphic positions. Apparently, even crystalline basement rocks (gneiss, schist, granite) contain detachment zones in the central Wasatch Range. Detachment positions must be zones of relative weakness (Gretener, 1972, p. 601). Although weak zones are generally equated with shales, other rock types may become weak under certain overburden pressure and temperature conditions. What may be a detachment zone in one area may not be in another. Well, seismic and surface control indicate the basal detachment lies within Cambrian shale and carbonate near the top of the crystalline Precambrian rocks over most of the subject area. Other preferred detachment positions are in the incompetent Jurassic Twin Creek Formation and basal Mississippian rocks in the west, and within the upper Cretaceous shale in the east. Still other detachment positions occur locally, consequently complicating the hanging wall structure; for example the Prospect thrust rides in the Triassic Ankareh Formation for about six miles.

Mackin's (1950) method of interpreting geologic maps by downdip (downplunge) viewing is extremely useful in the thrust belt. The wild looking subsurface interpretation shown on the northern regional cross-section X-X′ (Plate I, in pocket) in the vicinity of the Salt River Range is merely a southward (downplunge) projection of the surface form mapped by Rubey (1958). The extreme fold-fault shortening of Cambrian, Ordovician and Devonian

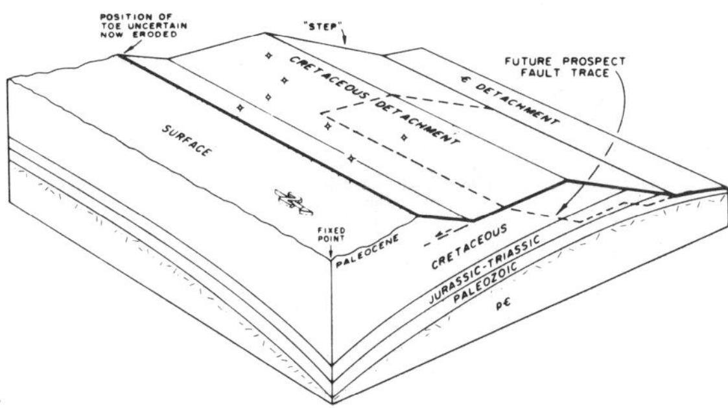

A. - EARLY DARBY FAULT PLANE

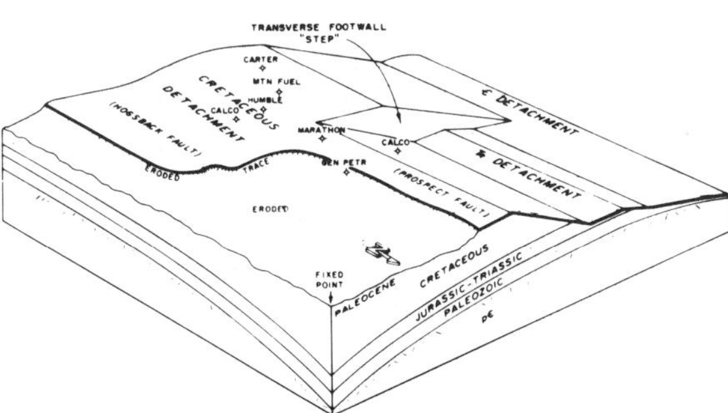

B. - COMPOUND PROSPECT - DARBY (HOGSBACK) FAULT PLANE

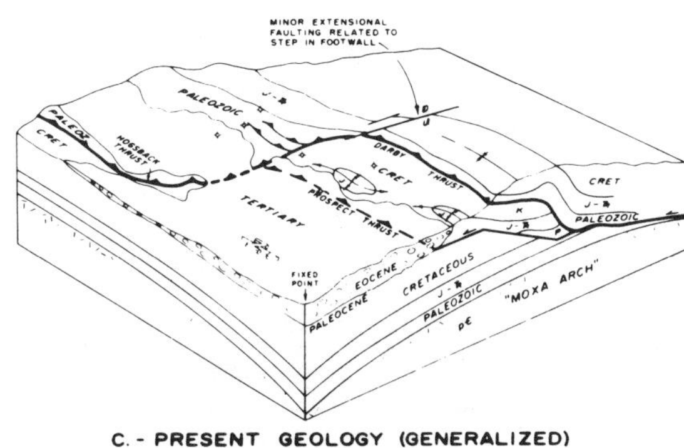

C. - PRESENT GEOLOGY (GENERALIZED)

Fig. 4 — Block diagrams showing evolution of linkage between the Darby, Prospect and Hogsback thrust faults; and the resulting hanging wall geology.

rocks, and the lack of such shortening in overlying Mississippian rocks as shown on the surface map require a detachment between them. This form, or something comparable, must project southward under the Permo-Pennsylvanian rocks to the line of section or some discontinuity like a transverse fault must intervene. There is no evidence for such a transverse separation.

Downdip viewing of geologic maps, when combined with well and seismic control, is also useful in interpreting the structure of the Snider Basin area northwest of LaBarge, Wyoming (Fig. 1) where there has been some controversy over the nature of the linkage between the Darby, Prospect and Hogsback thrusts.[3] Viewing a geologic map of this area from the east (down the dip of the thrusts) gives one the impression that the Darby thrust is uplifted and folded by the Prospect fault. Well and seismic control indicate that this is what occurred (Fig. 4). Motion on the Prospect fault has folded an older Darby fault plane and rocks on the Darby hanging wall. Actually, the effects of the Prospect fault motion on Darby hanging wall rocks can be seen for several miles west of Snider Basin to the vicinity of the older Absaroka thrust trace. Fault motion south of the intersection between Darby and Prospect thrusts was on a single fault plane, the Hogsback. The Hogsback thrust, then, becomes a composite plane of earlier Darby and later Prospect thrust motion. In longitudinal profile (north-south) the Prospect thrust steps stratigraphically up section to the south and joins the older Darby fault. The steep south plunge of the Prospect hanging wall anticline at Snider Basin results from eastward disappearance of this step where both Darby and Prospect thrust planes join and ride in the same upper Cretaceous detachment horizon. Descriptions of this type of thrust fault linkage in the Canadian Rockies are discussed by Dahlstrom (1970, p. 375). He refers to them as a type of "tear" fault which is an integral part of the boundary of the deformed hanging wall panel, and as being distinctive from those tears which are wholly within one plate. Hanging wall structural plunge like the steep south plunge on the Snider Basin anticline results from motion along this type of "tear." Surface maps indicate that similar thrust fault intersections, where motion on a younger thrust below deforms and links with an overlying thrust across a transverse step ("tear"), apparently occurs at several places along the trace of the Absaroka fault zone; specifically, five miles west of Snider Basin (Absaroka-Commissary thrusts), southeast of Alpine, Wyoming (Absaroka-Murphy-Firetrail thrusts), and just north of Little Muddy Creek south of Kemmerer, Wyoming.

The problem of naming faults which have composite motion through linkage of two or more faults might be handled by assigning each fault a name, and assigning a general name to the whole linking system. Thus we would have the Prospect, Darby and Hogsback thrusts, and may call the system the Darby thrust zone, since the name "Darby" has much historical usage. This is similar to the procedure followed by Armstrong and Cressman (1963) in naming the Bannock thrust zone. We interpret thrust faults of the Idaho-Wyoming thrust belt to group, from west to east, into a Bannock zone (Willard-Paris

[3]Armstrong and Oriel (1965) named the large fault which crops out 7.5 miles west of LaBarge, Wyoming, the Hogsback thrust. It had previously been called "Darby" by some workers.

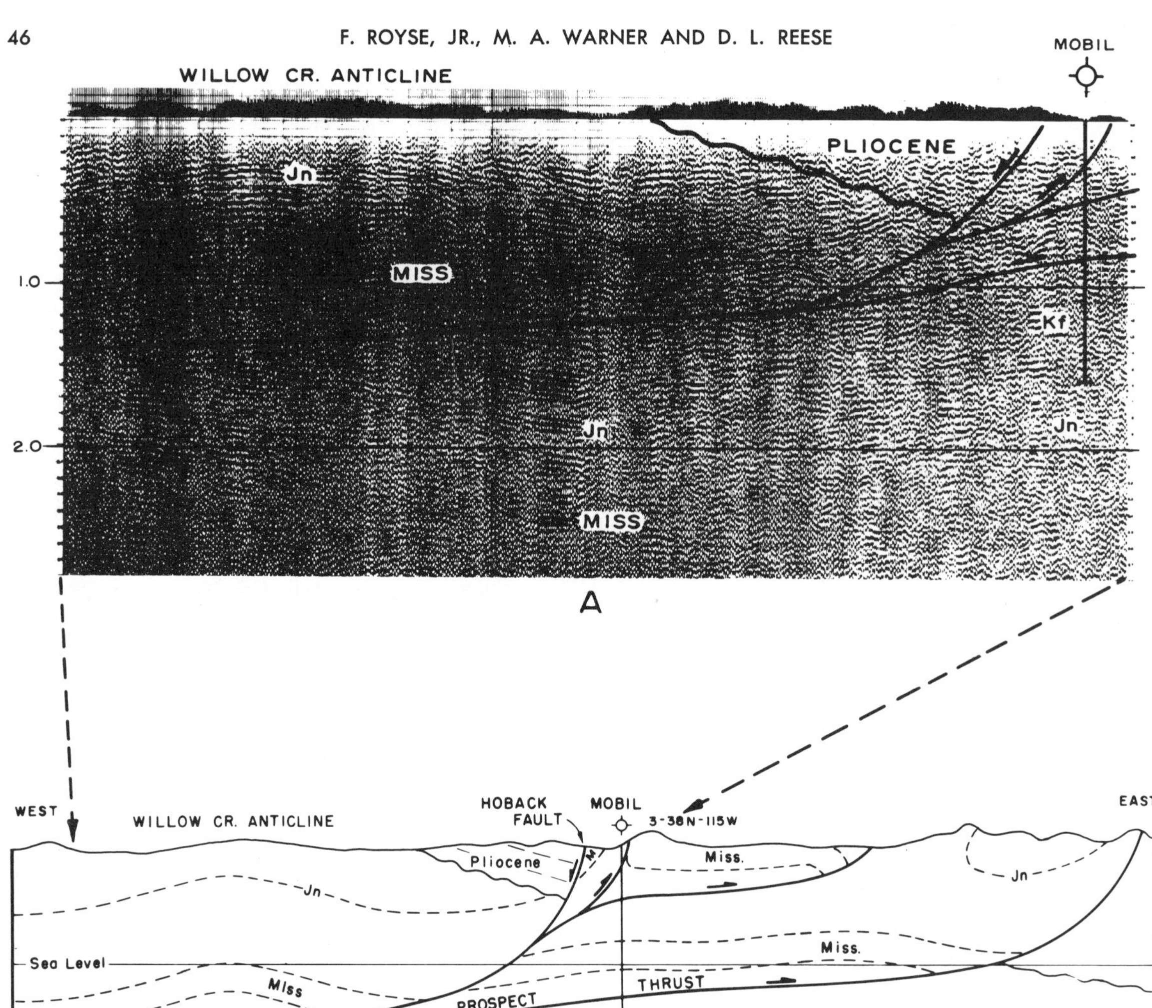

Fig. 5 — Hoback Canyon area, Wyoming. A. Seismic time section; B. Generalized structural section. Shows rotation of Pliocene into Hoback normal fault. Normal fault is restricted to hanging wall of Prospect thrust.

thrusts), a Meade-Crawford zone, an Absaroka zone, and a Darby zone. There are stratigraphic as well as structural reasons for such zonation which will be discussed later.

STRUCTURAL GEOMETRY OF EXTENSIONAL FAULTS

The thrust belt has been undergoing extension from Eocene to present along a series of major normal faults (Armstrong and Oriel, 1965) (Fig. 1). Many of the north-south valleys, such as Grand Valley, are half-grabens, bounded on the east by normal faults. Published interpretations generally depict the normal faults as being high angle and cutting through the entire sedimentary section and crystalline basement. Some evidence indicates that many of the normal faults flatten with depth and sole into older underlying thrust planes. According to this idea, portions of the major "thrusts" have composite movement;

early compressive motion (thrust faulting) and later extensional motion (normal faulting). The position of the major normal faults appears to be related to "step" geometry in the underlying thrust sheet. There is a coincidence between the position of the young normal faults and older thrust fault plane steps. An example of this type of fault is the Flathead fault in the thrust belt of southeastern British Columbia and northwestern Montana (Bally and others, 1966; Dahlstrom, 1970; Labrecque and Shaw, 1973). The Flathead is high angle where exposed at the surface but seismic and well data indicate the fault flattens with depth, and joins but does not cut the underlying Lewis thrust. Several pieces of evidence suggest the normal faults of the Wyoming-Idaho-Utah thrust belt have the same characteristics as the Flathead fault.

The Hoback fault south of Jackson, Wyoming (Fig. 1) is a major extensional fault that is probably restricted to the hanging wall of the Prospect thrust. The Hoback fault can be mapped for a distance of 35 miles. The trace has an arcuate shape which roughly parallels the trace of the Prospect thrust. Immediately southwest of Jackson, the surface traces of the Hoback and Prospect faults converge and probably merge under the Snake River Plain west of Jackson (Fig. 1). There is no indication that the younger Hoback fault cuts the Prospect thrust. In the vicinity of Hoback Canyon, two miles north of the line of section in Figure 5, the Hoback fault has a net slip of approximately 6,000′. Net slip can be determined using the thickness of the Pliocene Camp Davis Formation which was deposited on the downthrown side of the Hoback fault. The Camp Davis Formation has an outcrop width of 10,500′ and dips eastward into the fault at an angle of 30°-35°. Assuming the formation was deposited nearly horizontal, and has since been rotated into the fault, the amount of rotation requires 6,000′ of net slip. Because east dipping antithetic faults are absent, the consistent east dip of the Pliocene into the Hoback fault plane (reverse drag) is evidence the dip of the fault plane lessens at depth.

The Mobil Camp Davis Unit well was drilled through the hanging wall of the Prospect thrust and documents the stratigraphic position and sequence in the footwall. Seismic information ties the Mobil well and extends for several miles to the west-northwest (Fig. 5). The well establishes the position of the thrust at the east end of the seismic line, and the position of the thrust can be readily picked on the seismic data at the west end of the line. Additionally, the projected position of the Mississippian below the Mobil well is at 2.4± seconds while the strong Mississippian reflection at the west end of the line is at 2.0± seconds. It would be difficult to downdrop the Mississippian 6,000′ (750± milliseconds) on the Hoback fault and honor the seismic and well data.

The position of the Hoback fault appears to have been localized by a step in an imbricate thrust in the hanging wall of the Prospect thrust. The Mobil well drilled through two imbricate thrusts (Fig. 5); a minor one near the surface and a major imbricate with 3-4 miles of shortening at 2,700 feet. Both imbricate thrusts are exposed in the Hoback Canyon east of the well. The 2,700-foot imbricate branches from the main Prospect thrust west of the well, steps up-section through the competent Paleozoic section, then rides for a considerable distance in the Triassic. The footwall shoulder of the step coincides with the position of the Hoback fault.

The Grand Valley fault(s) near Alpine, Wyoming (Fig. 6) is another example of a major extensional fault which we feel is restricted to the hanging wall of the Absaroka thrust. The trace has an arcuate shape and is parallel to the trace of the Absaroka. A thick (10,000± foot) Pliocene sequence on the downthrown block has been rotated and dips eastward into the fault. The position of the fault coincides with the interpreted position of a major step in the underlying Absaroka thrust.

Additional examples are depicted on the structural cross-sections. North section X-X′ (Plate I, in pocket) shows extensional faulting west of the Wyoming Range to be restricted to the hanging wall of the Darby thrust and west of the Salt River Range to be restricted to the hanging

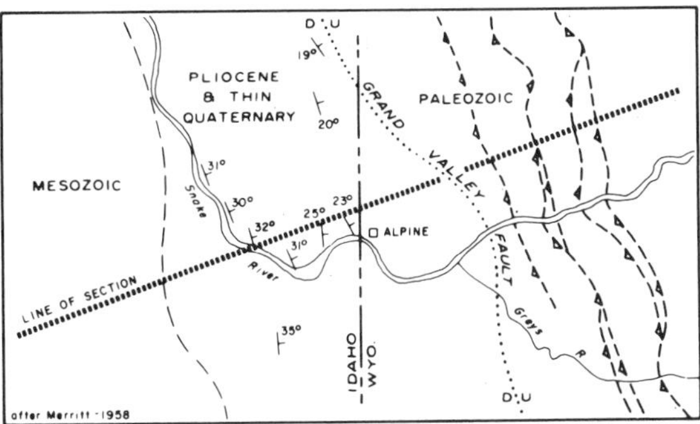

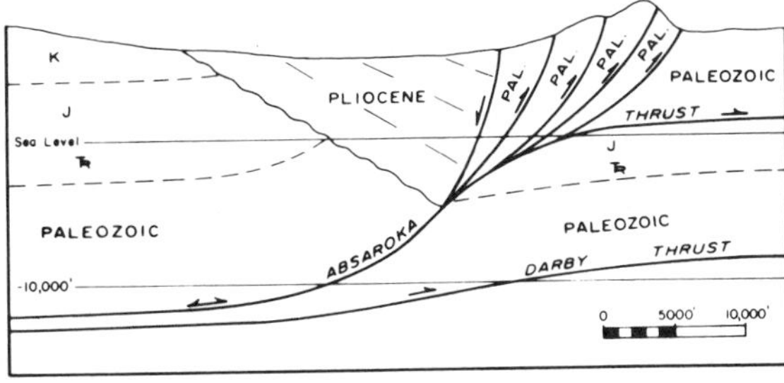

Fig. 6 — Alpine area, Idaho-Wyoming. Generalized geologic map and structural section. Shows rotation of thick Pliocene sequence into Grand Valley normal fault. Position of normal fault is controlled by a "step" in the underlying Absaroka thrust. Curvature of the normal fault plane is required when maintaining bed length in the Pliocene sequence.

wall of the Absaroka thrust. Section Y-Y′ (Plate II, in pocket) shows extensional faulting near the Utah-Wyoming state line which is restricted to the hanging wall of the Absaroka thrust. The characteristics in map view of such faults are: (1) they roughly parallel the major thrusts with which they are associated, (2) they may merge with but do not cut major thrust faults, and (3) the post-thrusting sediments on the downthrown (west) side are rotated so that they dip into the fault plane (reverse drag).

RESTORED CROSS-SECTION

A structural restoration (palinspastic) of the thrust belt along section X-X′ is shown on Plate IV (in pocket). Construction of such an illustration requires a great deal of faith in the basic thrust fault model presented herein, as well as a certain amount of recklessness. Illustrations of this sort are necessary, however, because proper stratigraphic mapping requires that control points be restored to relative pre-deformation positions.

An eastward progression in age of thrust deformation is generally accepted, and has been discussed in some detail by Oriel and Armstrong (1966). Additional stratigraphic data which bears on age of thrusting is discussed later. The connection between deformation (uplift) and sedimentation in space and time is summarized on Plate IV by restoring the section X-X′ stepwise to the end of each general period of major thrust motion. The sections are drawn as if the crystalline basement is moving west with the sedimentary section above being piled up on thrust faults and folds. The west end of the section is geographically fixed in the thrusted section. Reference points are indicated by numbers which finally become present day 30-minute longitude positions. Thicknesses of synorogenic deposits for each period are taken from regional isopach maps, restored to proper pre-thrust position, and flattened on a sea level datum.

The upper section shows thickness profiles of pre-thrusting (pre-Cretaceous) rocks completely restored. According to this interpretation, the crystalline basement surface was dipping approximately 6° westward. The wedge of sediments (including Precambrian) to be involved in thrusting was about 70,000 feet thick in the west and 11,000 feet thick on the east in this line of section. Thrusting was initiated on a surface which was buried far below sea level. Initial uplift of the site of the Bear River and Portneuf Ranges is believed to be a result of thrust faulting instead of uplift being the cause of thrust faulting as in conventional gravity slide models. Shortening in the brittle sedimentary section may be compensated westward by plastic flow in the basement when these rocks reach the depth (pressure) and temperature required for such behavior (Armstrong, 1974).

An eastward migrating foredeep related to Paris and Meade thrust faulting is indicated by the distribution of Cretaceous synorogenic deposits which predate the Ericson Formation. Post Ericson sediment distribution is more varied; local depositional basins began to form as the easterly migrating thrust belt impinged upon basement involved fold-fault uplifts on the Wyoming foreland province. The creation of the Moxa Arch, a broad basement involved fold, is synchronous with Absaroka thrusting. The Prospect-Darby thrust system overrides the arch north of LaBarge, Wyoming.

A total horizontal shortening of approximately 50% of the original width involved has been calculated in independent studies of different parts of the thrust belt province. Restoration of section X-X′ (Plate IV) shows 65 miles of horizontal shortening of the Mississippian from an original width of 130 miles or 50% (shortening indicated for the Wind River Mountains is not included). The southern regional section Y-Y′ (Plate II) shows about 52 miles of shortening of Mississippian from an original 111-mile width, or 47%. Monley (1971, p. 525) indicates 60 miles of shortening in a regional section intermediate between X-X′ and Y-Y′. Rubey and Hubbert (1959, p. 190) estimate about 85 miles of shortening of an original 175 miles, or 48%, for the Idaho-Wyoming thrust belt along latitude 43°. Bally and others (1966, p. 359) and Price and Mountjoy (1970, p. 16) estimate close to 50% shortening for the Canadian thrust belt. The tendency seems to be to double the thrusted section, thereby halving the width. Following the concept that thrust faulting will occur in a manner which requires the least work, it is apparently "easier" for a new fault to form in front of an older one than it is to ramp an already thrust-thickened section over a step and effectively triple the section (Fig. 3, Item D). The fact that major thrusts do not overlap in map view relates to this situation.

SYNTECTONIC DEPOSITS

The sedimentary record indicates that thrust faulting and concomitant uplift were episodic in the time period from Late Jurassic-Early Cretaceous to Early Eocene. Although the dominant movement during overthrusting was horizontal, a significant vertical component of motion was also involved. Syntectonic sedimentary deposits were formed concurrent with the creation and denudation of highlands on the rising thrust plates. Dating the preserved remnants of these syntectonic deposits not only gives evidence of the time of thrusting but also provides a key to facies mapping of Cretaceous sediments in the thrust belt.

Syntectonic conglomerates recognized in the southern part of the Idaho-Wyoming thrust belt include the Ephraim Conglomerate, the Echo Canyon Conglomerate, an unnamed unit at Little Muddy Creek (Fig. 1) and the Evanston Formation. The stratigraphic position of these conglomeratic units and their relationship to several major

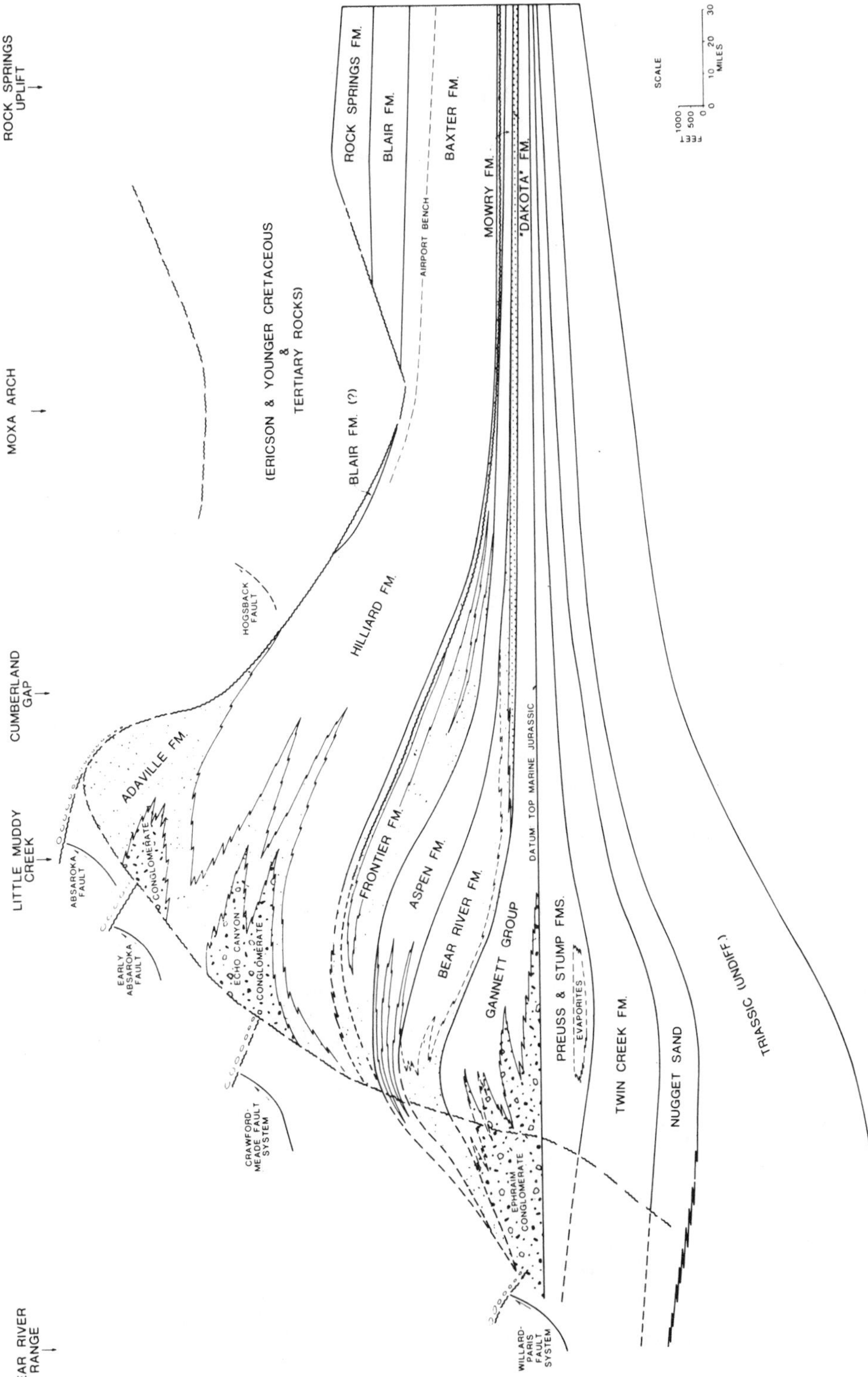

Fig. 7 — Stratigraphic diagram of restored pre-Evanston mesozoic rocks in western Wyoming and northern Utah showing relationship between thrust faulting and sedimentation. Restoration is based on palinspastic isopach and facies maps of individual formations. Line of section shown on Fig. 1.

thrust faults are shown in Figure 7. The temporal and spatial relations between deposition of the Ephraim Conglomerate and movement on the Paris thrust were summarized by Armstrong and Cressman (1963). Movement on the Paris thrust apparently started in latest Jurassic or earliest Cretaceous time. The nature of the clasts and the inverted stratigraphy shown in the Echo Canyon Conglomerate as described by Mullens (1971) indicate that it is a syntectonic unit that was probably deposited concurrent with movement on the southern extension of the Crawford fault. Data for dating the Echo Canyon are meager. It is unquestionably younger than the Ephraim Conglomerate and older than the Evanston Formation. Williams and Madsen (1959) considered fossils from the lower part of the Echo Canyon Conglomerate to be indicative of a late(?) Niobrara age. Preliminary palynology studies at Chevron Oil suggest (but by no means prove) that the Echo Canyon may be somewhat older than late Niobrara and equivalent to the middle part of the Hilliard Shale (Fig. 7). The tectonic implications of the Hams Fork Conglomerate Member of the Evanston Formation were pointed out by Oriel and Tracey (1970). The Hams Fork Conglomerate was deposited during and immediately after the latest major movement on the Absaroka thrust fault in Late Cretaceous Lance time (Rubey et al., 1961; Oriel and Tracey, 1970).

WANSHIP QUESTION

The Wanship Formation, recognized in a small area in northern Utah, is not included in the group of syntectonic deposits listed above because we believe that it is not related to a separate and distinct phase of thrusting. A structurally significant unconformity within the Cretaceous sequence in the area east of Wanship has been referred to by several authors (Peterson and others, 1953; Williams, 1955; Williams and Madsen, 1959; Crittenden, 1974). A unit consisting of about 2,100 feet of interbedded sandstone and shale with some lenticular coal beds and a prominent basal conglomerate overlies older Cretaceous and Jurassic formations with marked angular discordance that approaches 90° in places. Sharp folds and thrust faults in the older strata are truncated at the base of the conglomerate.

The unit above the unconformity was referred to as the Wanship Formation by Williams (1955). Williams and Madsen (1959) extended the usage of the term into the Coalville area, 12 miles north, by correlating the basal Wanship conglomerate with a conglomerate near the middle of a sequence previously included in the Frontier Formation at Coalville. By making this correlation they were able to fix the time of deposition of the basal conglomerate in the Wanship area as near the Carlile-Niobrara (Turonian-Coniacian) time boundary because the conglomerate in the Frontier Formation at Coalville was bracketed between marine deposits dated as early Carlile (mid-Turonian) below and early Niobrara (early Coniacian) above. Thus it could be shown on the basis of this correlation that a major tectonic event involving folding and thrust faulting took place in the Wanship area in middle to late Carlile time (late Turonian).

We believe that the correlation proposed by Williams and Madsen (1959) is in error and that the two conglomerate units are not synchronous. The age of the conglomerate in the Frontier Formation in the Coalville area is not in question. The paleontological data seem sufficient to demonstrate that this unit was deposited during the period between mid-Turonian and early Coniacian. However, the palynomorph suite of the Wanship Formation in the area east of Wanship contains the distinctive form *Aquilapollenites* which has not been found in rocks older than early Campanian (H. L. Ott, Chevron Oil Company report). The minimum age difference between the two conglomerates appears to span the better part of two full stages (Coniacian and Santonian) during which as much as 10,000 feet of sediments were deposited in other areas of the thrust belt. The Wanship Formation in the Wanship area is probably equivalent to the Evanston Formation and was deposited during or shortly after movement on the Absaroka thrust system as suggested by Crittenden (1974).

LITTLE MUDDY CREEK AREA

A thick synorogenic conglomerate that is significant to the interpretation of the structural and depositional history of the thrust belt is preserved in a highly deformed group of rocks along the frontal zone of the Absaroka fault system north of Little Muddy Creek in the SW¼ T. 19 N., R. 117 W. Vietti (1974) measured and described the upper 1,300 feet of the conglomeratic unit in an incomplete section where its total thickness appears to be about 2,000 feet. Rounded boulders with long dimensions of 7 feet or more are present in a massive conglomerate near the middle third of the conglomeratic unit. The coarser clastics grade into sandstones and mudstones both above and below.

Vertical changes in composition and abundance of clasts in the conglomeratic unit demonstrate an excellent example of inverted stratigraphy. Clasts derived from such distinctive formations as the Aspen, Bear River, Ephraim, Twin Creek and Thaynes can be identified. Many of the larger boulders are sandstone and appear to have been derived from the Nugget Formation. Clasts derived from the Aspen Formation are present near the base of the conglomeratic unit and the first appearance of clasts from formations older than the Aspen occurs at successively higher stratigraphic positions in an order opposite that in which the parent rocks were deposited. The sequence clearly indicates deposition concurrent with uplift of the source area on which progressively older formations were being bared to erosion.

Both Veatch (1907) and Walker (1950) mapped the conglomeratic unit as part of the Beckwith Formation, which, as originally defined, included the sequence of rocks now recognized as the Preuss and Stump Formations and

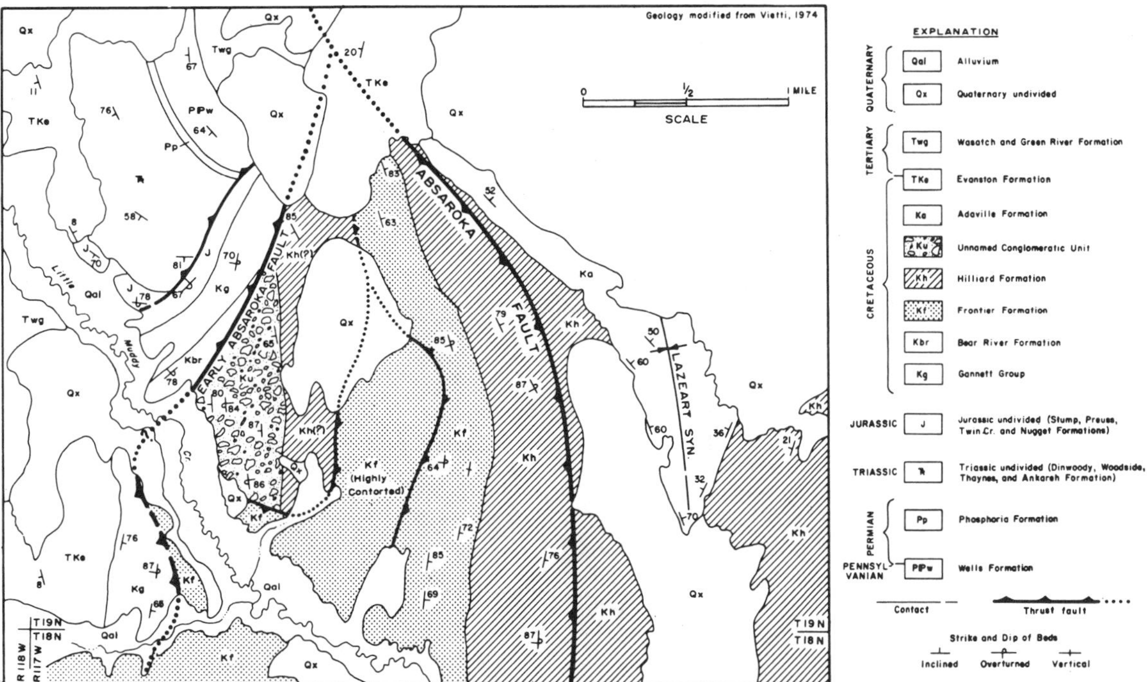

Fig. 8 — Generalized geologic map of Little Muddy Creek area, T19N, R117W, Lincoln County, Wyoming. Structural details in highly contorted areas have been omitted.

the Gannett Group. Kelly (1969) in discussing the complex structure of the area referred to "a steep unnamed fanglomerate sequence that appears to be post-Adaville and pre-Evanston in age." The conglomerate is obviously younger than the Beckwith (or Gannett) because it contains fragments derived from the Bear River and Aspen Formations which are younger than the Gannett Group. Palynomorphs recovered from samples collected below, within and just above the conglomeratic unit by Vietti (1974) and Chevron Oil Company geologists provide a basis for dating this unit and indicate that it is probably equivalent to part of the upper Hilliard or lower Adaville sequence exposed on the Hogsback thrust plate three miles to the east (Fig. 8).

The conglomeratic unit was water laid in an environment near sea level. Shales within the upper part of the unit and others immediately above its top contain marine microfossils. Oyster shells and sharks' teeth are found in sandstones in the same stratigraphic interval. The large size of the sandstone boulders and their lack of resistance to abrasion attest to a relative short distance of sedimentary transport from a nearby uplift.

Two faults with significant stratigraphic displacement of 5,000 feet or more have been mapped as parts of the Absaroka fault system in the Little Muddy Creek area. To avoid a lengthy discussion over which, if either, of these two faults should be called Absaroka, we shall use the terms Early Absaroka fault and Absaroka fault in an informal

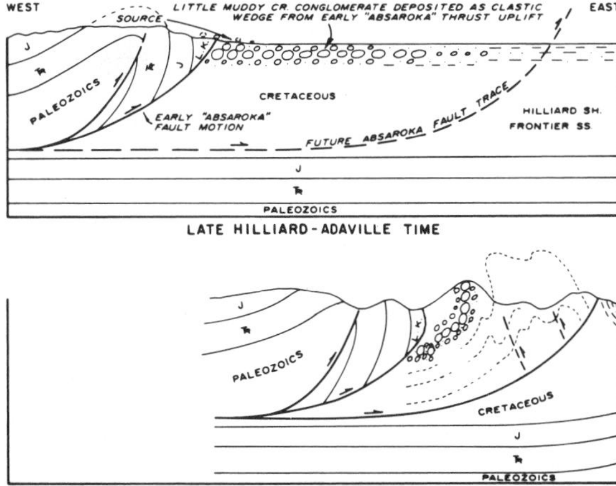

Fig. 9 — Diagrammatic cross sections in Little Muddy Creek area showing sequence of thrusting in Absaroka fault system. Upper diagram depicts a time shortly after movement on early Absaroka fault and deposition of the conglomeratic unit. Lower diagram illustrates the present structure after movement on Absaroka fault in latest Cretaceous.

designation on the geologic map (Fig. 8). The Early Absaroka fault places strata of the Lower Cretaceous Bear River Formation and Gannett Group in contact with the Upper Cretaceous conglomeratic unit. The stratigraphic displacement is on the order of 13,000 to 15,000 feet. The eastern fault, labeled Absaroka fault on Figure 8, has a stratigraphic displacement of 5,000 to 6,000 feet where it brings the Frontier Formation to within a few hundred feet of the base of the Adaville Formation. The rate of convergence of the two faults suggests that they probably join a short distance north of the place where they disappear beneath the cover of younger rocks. Subsurface data from the area north of Little Muddy Creek reveal only one significant fault in the Absaroka system. Both the Amerada No. 1 Fossil Unit and Hoxsey No. 1 Government wells in Section 23, T. 21 N., R. 117 W., penetrate the Absaroka plate but encounter only one major fault. Likewise, only one significant fault is interpreted on seismic lines located between these wells and the outcrops at Little Muddy Creek.

Neither of the two faults in the Little Muddy Creek area can be precisely dated but the field evidence shows that motion occurred in two separate stages. The diagrammatic cross-sections in Figure 9 show our interpretation of these events. Major thrusting on the Early Absaroka fault in late Hilliard (mid-Santonian) time created an uplift which supplied the coarse clastics in the conglomeratic unit. Horizontal movement at this time was probably on the order of 17 miles. Later movement on the Absaroka thrust below and in front of the Early Absaroka fault caused the intense folding and back thrusting now evident in the conglomeratic unit and older strata on the Absaroka hanging wall. Because the latter fault cuts the west flank of the Lazeart syncline which involves formations as young as the Adaville and is overlain by the Evanston Formation, motion here seems to have occurred during latest Cretaceous. From field evidence in the Hams Fork area, 25 miles north of Little Muddy Creek, Oriel and Armstrong (1966) concluded that major movement on the Absaroka fault took place during Late Cretaceous Lance (Maestrichtian) time.

From this, it appears likely that in the area north of Little Muddy Creek where only one fault is evident in the Absaroka system, motion may have occurred in at least two phases on the same fault plane. At Little Muddy Creek and to the south, motion was expressed on two separate faults in the frontal zone. Westward these two faults join in a single major detachment on which the total horizontal motion is in excess of 20 mi.

STRATIGRAPHIC RECONSTRUCTION

Meaningful stratigraphic analysis of a geologic province like the Idaho-Wyoming thrust belt cannot be accomplished without taking the effects of thrust faulting into consideration. These effects are twofold in nature. The original facies and thickness patterns of all of the stratigraphic units involved in thrust faulting have been compressed to some degree by shortening across the thrust belt. This shortening may amount to as much as 65 miles in Jurassic or older rocks. In addition, sediment source areas on uplifted fault plates had an important primary influence on the facies patterns of stratigraphic units deposited during the period of active thrust faulting from latest Jurassic to early Eocene time.

The map of the Bear River Formation and equivalents (Fig. 10) is a generalized reconstruction of the original isopach and facies patterns of one stratigraphic unit involved in the thrust faulting. Structural cross-sections of the type shown in Plates I and II were used to estimate the distance of transport on the various thrust faults. Stratigraphic control points have been restored to their original position relative to fixed points on the unfaulted foreland of the Green River Basin by moving them an amount and direction opposite the tectonic transport. In doing this the political boundaries were not moved from their present position relative to the unfaulted foreland. The map pattern is based on data derived from 110 control points of which 36 are surface sections and 74 are boreholes. The amount and direction that some representative points have been moved from their present ground position are shown by the small arrows in Figure 10.

When viewed in this manner, it is obvious that deposition of the Bear River Formation and equivalent strata centered in a structural downwarp in front of the Willard-Paris fault system. Much of the clastic material was derived from source areas on the hanging wall of this early fault plate. A large part of the deposition took place under marine conditions in an embayment that extended southwest across the Idaho-Wyoming boundary into northern Utah. The original pre-fault position of the type area of the Bear River Formation was near the southwest margin of the embayment. Numerous references to reports on the fossils found in the type area by Stanton (1892), White (1895), Veatch (1907) and Yen (1951) have fostered the idea that the Bear River Formation is nonmarine. However, the abundance of marine microfossils in shale samples collected at a number of localities north of the type area during this study demonstrate that much, if not all, of the shale is actually marine.

To the east and southeast of the area of thick shale deposition, the Bear River Formation passes into strata commonly included in the Thermopolis and Dakota Formations. Northwest of the shale area, equivalent strata are included in the Smiths, Thomas Fork and Cokeville Formations as recently defined by Rubey (1973). Interpretation of the original thickness and facies is more questionable in the western part of the map area. Not only is the sedimentary record fragmentary in this area but also a precise time correlation between the formations defined by Rubey and the Bear River Formation of the type area is yet to be established.

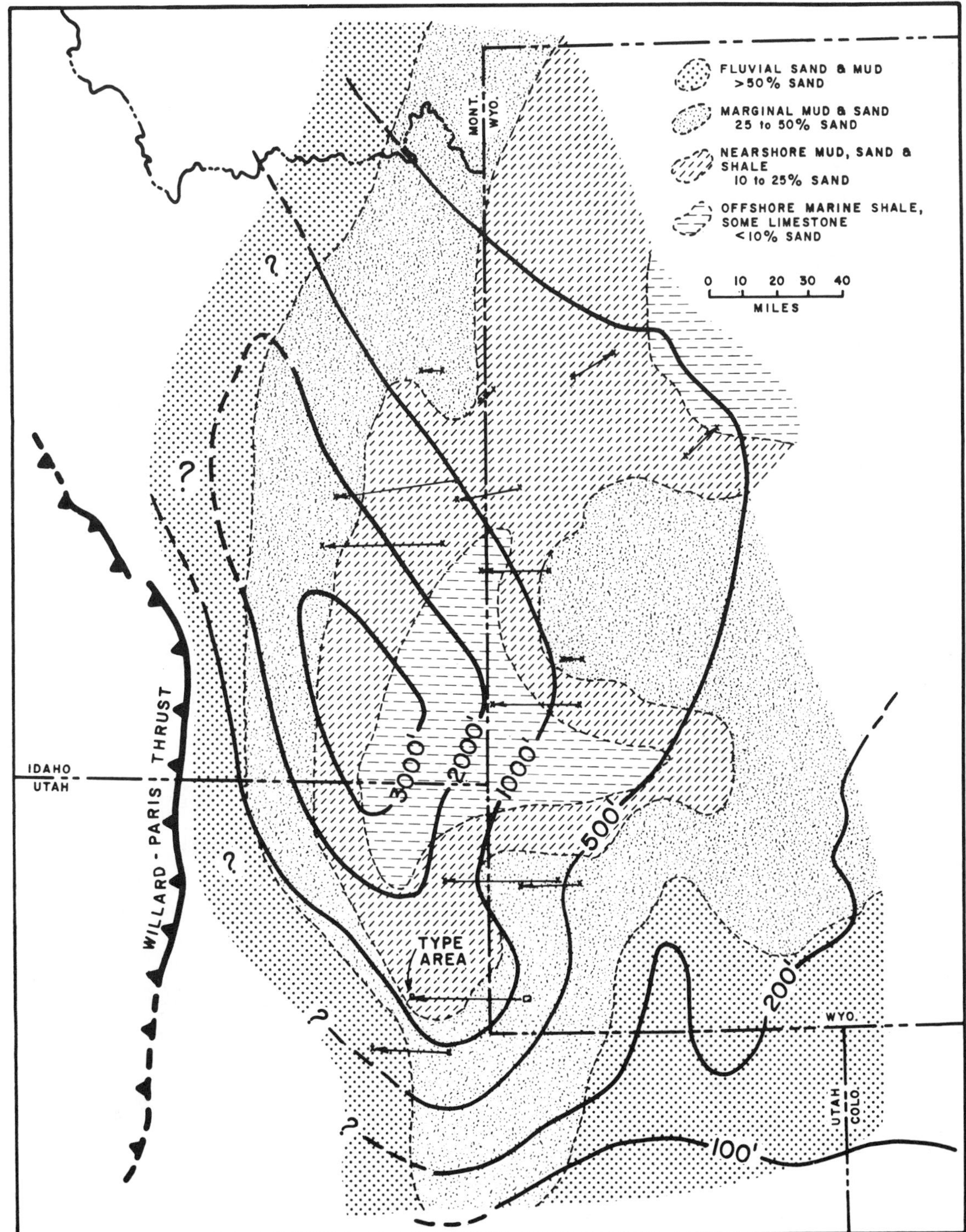

Fig. 10 — Palinspastic isopach and generalized facies map of the Bear River Formation and equivalent strata. Arrows show distance and direction some representative control paints have been moved from their present ground position to compensate for tectonic transport.

SUMMARY STATEMENT

Despite the structural complexity and scarcity of subsurface information in the thrust belt province of western Wyoming, southeastern Idaho and northern Utah, a geologic interpretation can be developed by invoking structural concepts and geometric constraints which have been found applicable in similar areas. When such concepts and constraints are applied to the wealth of information supplied by surface mapping and the less abundant information from seismic lines, other geophysical data and scattered bore holes, the resulting structural interpretation is reasonable and, therefore, can be accepted with some confidence. The structural model thus developed is vital to reliable stratigraphic mapping because of the stratigraphic disruptions brought about by thrust faulting. The original facies patterns of stratigraphic units involved in the thrusting can be mapped by using the structural model to unstack the thrust plates and restore the stratigraphic control points to their relative prefault positions. Conversely, stratigraphic information such as dating and correlating syntectonic sedimentary deposits must be considered in order to build an accurate structural model.

REFERENCES

Armstrong, F. C. and E. R. Cressman, 1963, The Bannock thrust zone, southeastern Idaho: U.S. Geol. Survey Prof. Paper 374-J, 22 p.

________________, and S. S. Oriel, 1965, Tectonic development of Idaho-Wyoming thrust belt; Am. Assoc. Petroleum Geologists Bull., v. 49, n. 11, p. 1847-1866.

Armstrong, R. L., 1968, Sevier orogenic belt in Nevada and Utah: Geol. Soc. America Bull., v. 79, n. 4, p. 429-458.

________________, and H. J. B. Dick, 1974, A model for the development of thin over-thrust sheets of crystalline rock: Geology, v. 2, n. 1, p. 35-40.

Bally, A. W., P. L. Gordy and G. A. Stewart, 1966, Structure, seismic data and orogenic evolution of southern Canadian Rocky Mountains: Bull. of Canadian Petroleum Geology, v. 14, n. 3, p. 337-381.

Bell, G. L., 1952, Geology of the northern Farmington Mountains, *in* Guidebook to the geology of the central Wasatch Mountains, Utah: Utah. Geol. Soc. Guidebook n. 8, p. 38.50.

Cressman, E. R., 1964, Geology of the Georgetown Canyon-Snowdrift Mountain area, southeastern Idaho: U.S. Geol. Survey Bull., 1153, 105. p.

Crittenden, M. D., Jr., 1974, Regional extent and age of thrusts near Rockport Reservoir and relation to possible exploration targets in northern Utah: Am. Assn. Petroleum Geologists Bull., v. 58, n. 12, p. 2428-2435.

Dahlstrom, C. D. A., 1969, Balanced cross-sections: Canadian Jour. of Earth Sciences, v. 6, n. 4, p. 743-757.

________________, 1970, Structural geology in the eastern margin of the Canadian Rocky Mountains: Bull. Canadian Petroleum Geology, v. 18, n. 3, p. 332-406.

Eardley, A. J., 1944, Geology of the north-central Wasatch Mountains, Utah: Geol. Soc. America Bull., v. 55, p. 819-894.

Gretener, P. E., 1972, Thoughts on overthrust faulting in a layered sequence: Bull. Canadian Petroleum Geology, v. 20, n. 3, p. 583-607.

Kelley, V. C., 1969, Westward overturning near the Absaroka thrust, Uinta County, Wyoming: Geol. Soc. Amer. Abstracts with programs, v. 5, p. 38-39.

Labrecque, J. E., and E. W. Shaw, 1973, Restoration of Basin and Range faulting across the Howell Creek Window and Flathead Valley of southeastern British Columbia: Bull: Canadian Petroleum Geology, v. 21, n. 1, p. 117-122.

Mackin, J. H., 1950, The down structure method of viewing geologic maps: Jour. Geology, v. 58, n. 1, p. 55-72.

Merritt, Z. S., 1958, Geologic map of Alpine area, Idaho-Wyoming, *from* Univ. of Wyo. Masters Thesis.

Monley, L. F., 1971, Petroleum potential of Idaho-Wyoming overthrust belt: *in* Future petroleum provinces of the United States — their geology and potential, Am Assoc. Petroleum Geologists Mem. 15, v. 1, p. 509-529.

Mullens, T. E. and W. H. Laraway, 1964, Geology of the Devil's Slide Quadrangle, Morgan and Summit Counties, Utah: U.S. Geol. Survey Mineral Investigations Map, MF-290.

________________, 1971, Reconnaissance study of the Wasatch, Evanston and Echo Canyon Formations in part of northern Utah: U.S. Geol. Survey Bull. 1311-D, 31 p.

Oriel, S. S. and F. C. Armstrong, 1966, Times of thrusting in Idaho-Wyoming thrust belt: Reply; Am. Assoc. Petroleum Geologists Bull., v. 50, n. 12, p. 2614-2621.

________________, and J. I. Tracey, Jr., 1970, Uppermost Cretaceous and Tertiary stratigraphy of Fossil Basin, southwestern Wyoming: U.S. Geol. Survey Prof. Paper 635, 53 p.

Peterson, R. H., D. J. Gauger and R. R. Lankford, 1953, Microfossils of the Upper Cretaceous of northeastern Utah and southwestern Wyoming: Utah Geol and Mineralog. Survey Bull. 47, 158 p.

Price, R. A. and E. W. Mountjoy, 1970, Geologic structure of the Canadian Rocky Mountains between Bow and Athabasca Rivers — a progress report: Geol. Assoc. of Canada Sp. Paper No. 6, p. 7-25.

Rich, J. L., 1934, Mechanics of low-angle overthrust faulting illustrated by Cumberland thrust block, Virginia, Kentucky and Tennessee: Am. Assoc. Petroleum Geologists Bull., v. 18, n. 12, p. 1584-1596.

Rubey, W. W., 1958, Geologic Map of the Bedford Quadrangle, Wyoming: U.S. Geol. Survey Geol. Quad. Map. GQ-109.

________________, 1973, New Cretaceous formations in the western Wyoming thrust belt: U.S. Geol. Survey Bull. 1372-I, 35 p.

________________, and M. K. Hubbert, 1959, Role of fluid pressure in mechanics of overthrust faulting, part II, Bull. Geol. Soc. Am., v. 70, n. 2, p. 167-206.

________________, J. I. Tracey and S. S. Oriel, 1961, Age of the Evanston Formation, western Wyoming, *in* Short papers on the geologic and hydrologic sciences: U.S. Geol. Survey Prof. Paper 424-B, p. B153-B154.

Stanton, T. W., 1892, The stratigraphic position of the Bear River Formation; Amer. Jour. Sci., v. 43, p. 98-115.

Veatch, A. C., 1907, Geography and geology of a portion of southwestern Wyoming: U.S. Geol. Survey Prof. Paper 56, 178 p.

Vietti, J. S., 1974, Structural geology of the Ryckman Creek Anticline area, Lincoln and Uinta Counties, Wyoming: Unpub. M. S. Thesis, Univ. of Wyoming.

Walker, C. L., 1950, Geologic map and structure sections Cumberland Reservoir — Little Muddy Creek area, Lincoln County, Wyoming, *in* Southwest Wyoming: Wyoming Geol. Assoc. 5th Ann. Field Conf., Guidebook, between P. 24 and 25.

White, C. A., 1895, The Bear River Formation and its characteristic fauna: U.S. Geol. Survey Bull., 128, 108 p.

Williams, N. C., 1955, Laramide history of the Wasatch-western Uinta Mountains area, Utah, *in* Green River Basin: Wyoming Geol. Assoc. 10th Ann. Field Conf. Guidebook, p. 127-219.

________________, and J. H. Madsen, Jr., 1959, Late Cretaceous stratigraphy of the Coalville area, Utah, *in* Guidebook to the geology of the Wasatch and Uinta Mountains transition area: Intermountain Assoc. Petroleum Geologists 10th Ann. Field Conf., p. 122-125.

Wilson, C. W., Jr. and R. G. Stearns, 1958, Structure of the Cumberland Plateau, Tennessee: Geol. Soc. America Bull., v. 69, n. 10, p. 1283-1296.

Yen, T. C., 1951, Fresh water mollusks of Late Cretaceous age from Montana and Wyoming: U.S. Geol. Survey Prof. Paper 233-A, p. 1-20.

Reprinted by permission of the Wyoming Geological Association and the Canadian Society of Petroleum Geologists from E. L. (Roy) Heisey et al., eds., *Rocky Mountain Thrust Belt Geology and Resources:* Wyoming Geological Association 29th Annual Field Conference, 1977, p. 407-439. Originally published in *Bulletin of Canadian Petroleum Geology*, v. 18, no. 3 (1970), p. 332-406.

STRUCTURAL GEOLOGY IN THE EASTERN MARGIN OF THE CANADIAN ROCKY MOUNTAINS [1]

CLINTON D. A. DAHLSTROM[2]

ABSTRACT

Hyrdocarbon accumulations in the "foothills" at the eastern margin of the Canadian Rocky Mountains are structurally trapped. Exploration for them entails predicting the deep geometric configuration of potential reservoir beds in imperfectly understood areas. This prediction is commonly derived from analogies with the most appropriate of the region's typical structures, a pragmatic approach which is effective because the "foothills" contain a limited suite of relatively simple structural types:

1. Concentric folds (with their attendant decollement)
2. Low-angle thrust faults (commonly folded)
3. Tear faults (usually transverse)
4. Late normal faults (commonly listric)

The assemblage in a particular area is also a function of the degree of deformation and of a lithology of the deformed rocks. Intensity of deformation increases from east to west. Regional stratigraphic changes alter the major lithologic units, while local isopach or facies changes alter the distribution of incompetent rocks within units.

The structural styles are all "thin-skinned," as the underlying Hudsonian basement is not involved.

This review article summarizes current knowledge of the geometry of "foothills" structures, their mode of occurrence, and some of the empirical rules for their interpretation.

INTRODUCTION

The object of most geological research is to find an explanation – to answer the question, "why is it?.. Obviously there must be an earlier stage, wherein the facts that need explanation are defined in aswer to the question, what is it?" The purpose of industrial exploration geology is to find a natural occurrence of economic value – that is, to answer the question, "where is it?" Prerequisite to both research and industrial work is an adequate answer to the "what" question. From a hundred years of academic studies and fifty years of industrial studies, a great deal has been learned about the geology of the Canadian Rocky Mountains, and enough of this knowledge has been published to provide a reasonably documented answer to the "what" question as it applies to the structure and stratigraphy of the southern and eastern parts of the Canadian Rockies. This paper is a review of some aspects of that structural data. It is supplementary to major contributions on tectonics by Armstrong and Oriel (1965), Bally *et al.* (1966), Price and Mountjoy (1970), Shaw (1963), Wheeler (1966 and 1970) and White (1959 and 1966); and on structural geology by Bally *et al.* (1966), Fox (1959 and 1969) and Keating (1966).

This review does not provide a comprehensive summary description of the areal structural geology of the Canadian Rocky Mountains. Rather it is a discussion of the limited association of structural types which comprise the structural pattern. These types are classified and an attempt is made to define the geometric rules that govern their behavior and to establish the relationships between them. Deliberately the discussion has been restricted to the geometric level of what can be observed to exist and what can reasonably be inferred to exist, which is the "what is it?" level of geologic work wherein one establishes the models that the academic must attempt to explain and the explorationist must attempt to exploit. It is hoped the paper will be useful to explorationists who, by the nature of their calling, are required to analyze fragmentary data and, at the earliest possible moment, to recognize the nature, shape and orientation of potentially productive structures. Necessarily they are obliged to use mental models to bridge between control points. This paper is intended to provide such models and would have been of considerable value to explorationists in the Canadian Rockies fifteen years ago. Now, of course, it is rather late, because there is a perverse but fundamental rule of exploration which decrees that enough data to provide a sound statistical basis for extrapolation by analogy is available only in areas which are so well known that there is no longer much need for that particular kind of data extension. However, as the marginal parts of orogenic belts are alike, perhaps these models can be exported to areas where exploration is less mature.

[1]Manuscript reviewed June 1970.

[2]Chevron Resources Co., San Francisco, Calif. The writer's acquaintance with the geology of the Canadian Rocky Mountains is a consequence of employment by Chevron Standard Limited in its mountain exploration program from 1955 to 1962, and its in-house educational program for geologists from 1965 to 1969. The writer is indebted to Chevron Standard for this opportunity to learn and for permission to publish this paper. Messrs. R.A. Stuart and P. Verrall read the manuscript and provided valuable criticisms and suggestions. The aerial photographs were taken by K.T. Hyde of Calgary for Chevron Standard. The illustrations were prepared in the drafting department of Chevron Standard under the supervision of Messrs. I.G. McLeod and A.C. Collin.

[3]This paper is reprinted from the Bulletin of Canadian Petroleum Geology (vol. 18, no. 3, Sept. 1970, pp. 332-406) with the permission of the Society. The author has edited, but not rewritten, this version to retain the discussion of structural principles and to eliminate the sections on speculative application of these principles to specific western Canadian examples.

As mentioned at the outset, this is a review article synthesizing data and ideas from a multitude of published sources. No systematic attempt is made tö trace concepts back to the originators: the references are to the most recent or complete author on a specific subject.

GENERAL GEOLOGY

The structural province characterized by the Canadian Rocky Mountains extends from the Osburn fault system and the Idaho batholith in Montana northward for a thousand miles to the Liard River, where it meets the Mackenzie Mountains province. To the east there is a platform of relatively undisturbed Mesozoic and Paleozoic rocks resting on a Hudsonian basement that is the downdip westward extension of the Canadian Shield. The Canadian Rockies province is both a structural and a stratigraphic entity.

Stratigraphy

The stratigraphic succession is broken into five units by four major uncomformities, one within the Proterozoic and the others respectively at the bases of the Cambrian, Devonian and Jurassic successions (Fig. 1). These units are distinctly wedge-shaped, thick in the west and thin in the east – on account of onlap at the base, depositional thinning within the unit, and erosional truncation at the top.

The lowest part of the exposed succession is the Proterozoic Purcell Series, which is thought to be more than 40,000 ft thick. It is primarily quartzite and argillite but there are a few carbonates in the upper part. The series thins rapidly to the east.

The Proterozoic Windermere clastic succession unconformably overlies the Purcell. It is truncated by an unconformity, which has produced a zero edge trending northerly across the southern mountain belt. Windermere rocks are exposed on the west side of the Purcell anticlinorium near the U.S. border, in the Main Ranges at Banff, and in the Front Ranges at Jasper.

Between the top of the Proterozoic and the base of the Upper Devonian are two distinctly different successions. In the southern part of the Canadian Rocky Mountains there are only a few hundred feet of Cambrian limestone and shale above a thin basal conglomerate, whereas to the north there are 15,000 to 20,000 ft of Silurian, Ordovician and Cambrian beds. The basal unit of this thick section is a through-going thick quartzite. The overlying units are primarily carbonate in the east with an abrupt facies change to shale in the west. The post-Laramide boundary between the thin and thick sections is a thrust fault, the Hosmer thrust. The Hosmer thrust sheet has a thick lower Paleozoic section, while the underlying thrust sheets have the thin Cambrian succession.

The Mississippian and Upper Devonian rocks are remarkably uniform shelf carbonates. There may have been a western shale facies (now largely eroded) with a northerly trend. Parallel to the facies trend in the southern Rockies the carbonate units are remarkably persistent to the extent that the cliff-forming Palliser, for instance, is a readily recognizable formation for 600 miles north from the U.S. border until it enters the western shaly facies.

The Jurassic to Paleocene rocks are a clastic succession of thick coarse sand wedges in the west giving way eastward to marine shales. Because the stratigraphic strike is roughly parallel to the structural strike, there is again remarkable north-south continuity of individual formations in the southern 500 miles of the Canadian Rockies. Jurassic rocks overlie Mississippian on the basal unconformity at the eastern edge of the Foothills, but westward this simple major unconformity splits into several lesser ones with intervening wedges of Triassic,

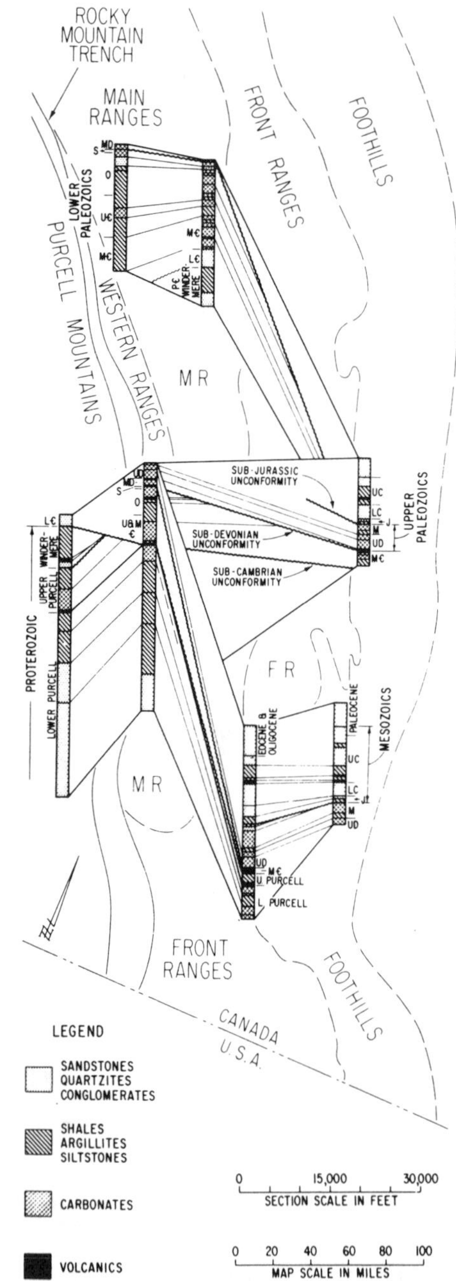

Fig. 1. Major units within the stratigraphy of the Canadian Rocky Mountains.

Permian and Pennsylvanian rocks. Some of the lesser unconformities in turn disappear westward into successions of continuous deposition. Conceivably, there may have been a complete Jurassic-to-Mississippian succession in the west, but the record has been stripped away by erosion.

The Geological History of Western Canada (McCrossan and Glaister, *eds.*, 1964) contains an extensive and authoritative account of the stratigraphy.

Physiographic Subdivisions

The Canadian Rockies have been split into a number of physiographic subdivisions which generally coincide with structural subprovinces (North and Henderson, 1954) (Fig. 2). Adjacent to the orogenically undisturbed rocks of the Great Plains on the east are the Foothills, which consist of deformed

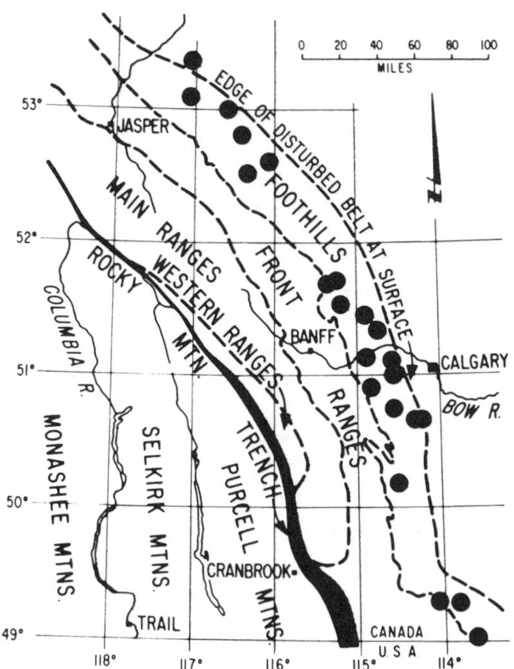

Fig. 2. Physiographic subdividions of the Canadian Rocky Mountains and location of discovered hydrocarbon accumulations.

Mesozoic clastic rocks. In the southern Foothills the deformation is primarily thrust faulting but in the northern Foothills only folding is visible at the surface. Within the Foothills a topographic distinction is sometimes made between the low relief produced by the shaly rocks of the upper Cretaceous (Outer Foothills) and the high relief produced by the resistant sandstones of the Lower Cretaceous (Inner Foothills).

Immediately to the west are the Front Ranges, which are formed of a relatively few major thrust sheets composed of Upper Paleozoic carbonates. The sheets are bounded by faults that are tens of miles long and have several miles of displacement. Individual sheets are folded and cut by subsidiary thrusts and tear faults.

The Main Ranges consist of lower Paleozoic and Proterozoic rocks thrust eastward over the adjoining Front Ranges. Within the eastern Main Ranges, where the Lower Paleozoics are in carbonate facies, the structures are broad, open folds with a relatively few thrusts and normal (oblique tear?) faults. In the western Main Ranges, however, where the Lower Paleozoics are in shale facies, the structures are tighter and the rocks are cleaved.

The Western Ranges are characterized by very tightly folded Lower Paleozoics of the shaly facies. Dips are steep, rocks are cleaved and the symmetry is commonly directed toward the west rather than the east as in the other structural divisions. Faults are steep and may appear to be normal, thrust or strike-slip. Much of the observed structure is surficial, being restricted to the rocks above the thick quartzite of the Lower Paleozoics.

The Rocky Mountain Trench is a broad linear valley marking the western boundary of the Canadian Rockies. Its floor is obscured by alluvium, so that its structure is in doubt.

The western side of the Trench is formed by the Purcell Mountains, a thrust-faulted, north-plunging anticlinorium of Windermere and Purcell rocks. The linear valley of the Purcell Trench separates the Purcell anticlinorium from the complex metamorphic and intrusive terrane in the Selkirk and Monashee Mountains to the west.

Hydrocarbon Accumulations

The typical Foothills hydrocarbon accumulation is a gas field in thrust-faulted Mississippian carbonate rocks. Field outlines are determined by structures that were formed during the latter part of the Laramide tectonic event. Presumably they are Oligocene in age, as there is no convincing evidence to indicate that individual structures grew over a protracted period. The principal reservoir rocks are porous Mississippian, in the upper part of the carbonate succession at or near the regional unconformity with the overlying Mesozoic rocks. The source rocks have not been defined, with the result that some regard the hydrocarbons as indigenous to the Mississippian while other contend that they moved downward from Mesozoic or upward from Devonian rocks. In any event there are pronounced disparities in age between source, reservoir and trap, making it likely that most Foothills fields are secondary accumulations produced by the redistribution of earlier ones. (The original accumulations might have been stratigraphic traps on the regional dip provided by basin subsidence and by epeirogenic uplift.) Turner Valley is the only oil field; the others are gas with varying amounts of hydrogen sulphide and condensate, which implies a complicated maturation and migration history. From this brief recitation of the characteristics of Foothills hydrocarbons, it is apparent that the explanation for their accumulation is a complex problem involving stratigraphy, tectonics, maturation, migration and structural geology.

Such a problem would be difficult enough if data were available, but much of the necessary information is lacking. The object of the initial cycle of Foothills exploration was to find structural closures in porous rocks and to drill them in the expectation that a rewarding number would contain hydrocarbons. Straightforward exploration of this type entailed mapping the stratigraphy and structure by surface, subsurface and geophysical methods. This work has been going on for forty years, and enough of the basic information is now in the public domain to permit generalizations about tectonics, sedimentation and structural geology. As a surprisingly large number of the locations drilled for purely structural reasons were found to have hydrocarbons, some of the essential data for maturation and migration studies presumably exists. It has not, however, been published, so it is currently impractical to go much beyond the concepts of migration and accumulation expounded by Gallup (1954), Gussow (1955) or Fox (1959).

All of the fields found to date have been in the Foothills belt. Their locations are shown in Figure 2, while the schematic cross-section of Figure 3 shows the typical structures. The fields, with the reserves ascribed to them by the Conservation Board, are listed below the appropriate structural type. Most of these fields have been described by Fox (1959).

Basic Concepts

Fundamental to and inherent in much of the discussion to follow are three basic geometric concepts – regional, no basement involvement, and structural families.

The regional idea developed in plains subsurface geology, where the gently west-dipping "layer cake" sequence permits the calculation of a "regional" rate of dip that can be used as the basis for extrapolation in map or cross-section. This extrapolation can be carried into the structurally deformed area, because much thrust-fault displacement occurs along planes parallel to the bedding and does not produce duplication. In such instances the stratigraphic succession and thicknesses are unaltered, so that regional gradients are still applicable. Consequently, in the marginal part of the deformed belt it is only at the forward edge of the thrust sheet that the beds are shoved "above regional," and westward they "return to regional" (see top of Paleozoics in Fig. 3). Eventually the amount of deformation beneath a specific horizon does increase to the point where the horizon must remain above regional (see base of Tertiary clastics in Fig. 3).

No Foothills well reached the Hudsonian crystalline basement, but a few nearby basement tests coupled with more easterly control, establish a regional configuration for basement itself. This can be extended by using "return to regional" points on other horizons, since it is evident that if the Mississippian, for instance, is at regional, all the horizons below it (including basement) are at regional as well. From such surface and subsurface data it is possible to establish that the basement continues at regional to the edge of the Front Ranges. Beyond that point seismic data obtained by Shell (Bally *et al.*, 1966) have shown an identifiable near-basement Cambrian event in the southern Rockes which maintains its regional dip as far west as the Rocky Mountain Trench. This is regarded as proof that the Hudsonian basement is not involved in the Laramide folding and thrusting that characterize the Canadian Rockies. The structures are "thin skinned."

For a particular structural environment there is a characteristic structural assemblage (or family) involving a restricted number of structural forms. (A "family" is comprised of a relative few basic forms and other types are lacking. For example, the foregoing discussion has excluded from the Canadian Rocky Mountain structural province, thrust faults or folds that displace or deform the Hudsonian basement – an exclusion that is supported by the fact that none exists on

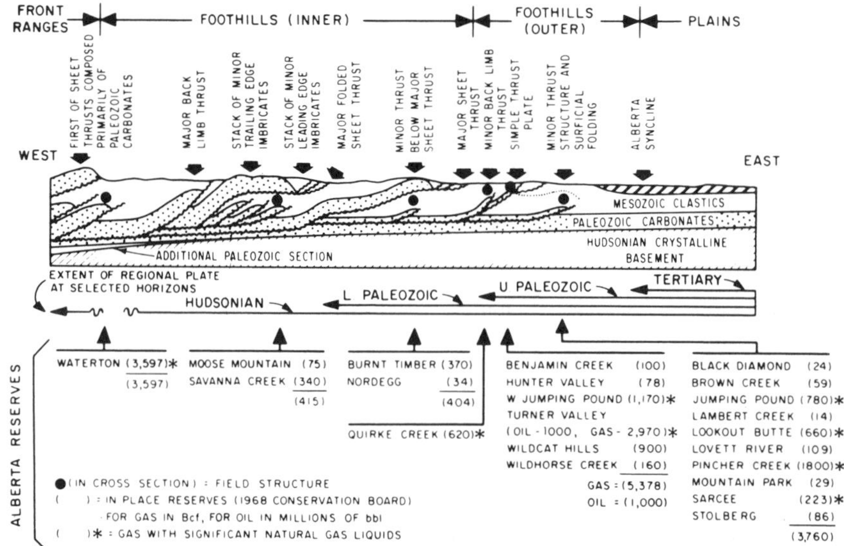

Fig. 3. Schematic cross-section of the Foothills to show common structures, field types and reserves (after R.E. Daniel, Chevron).

the outcrop.) The idea of structural familes is implicit in many of the structural definitions that we use, such as those for concentric and similar folding, which derive from a generally acknowledged observation that certain kinds of deformation are mutually exclusive. In a broader sense, the thought that a structural province can contain only a limited suite or family of structures is inherent in Bucher's (1933) concept of "natural laws" and DeSitter's (1956) treatment of "comparative structural geology." This paper will concern itself primarily with the group of structures that occurs in the Foothills and Front Ranges of the Canadian Rocky Mountains.

The structural assemblage, that is the "Foothills family" consists of: 1. low-angle thrust faults (commonly folded); 2. tear faults; 3. concentric folds, and their attendant detachment; and 4. late normal faults (commonly low-angle). These structures come in a wide variety of sizes and shapes, and can be associated with one another in different combinations to produce what sometimes appears to be a bewildering variety of structural styles. The number of fundamental forms, however, is limited and each form behaves according to rather simple geometric rules.

If the foregoing statement of the family concept is accepted, the interpretation of even a very complicated field problem becomes possible because the number of acceptable solutions is limited. If the concept is not accepted, the number of possible structural components that must be considered increases, and the number of apparently acceptable solutions may become very large indeed. The family concept is useful in ensuring that the models one uses in geological thinking are appropriate. As an extreme example, consider Wyoming, where the germanotype structures on the craton (e.g., Wind River Mountains) adjoin or even impinge upon the alpinotype structures of the Wyoming thrust belt. Here are two structural families in the same area. It could be quite misleading to use well-exposed structures in the Wind River Mountains as a model to help solve Wyoming thrust-belt problems, because these are two different structural environments and the structures within them are likewise dissimilar. Instead of going a few miles into the Wind River Mountains for a model, one would be far better advised to go hundreds of miles to the Canadian Rockies or the Appalachians, or thousands of miles to the Andes or the Zagros, to get models that belonged to the same basic structural family.

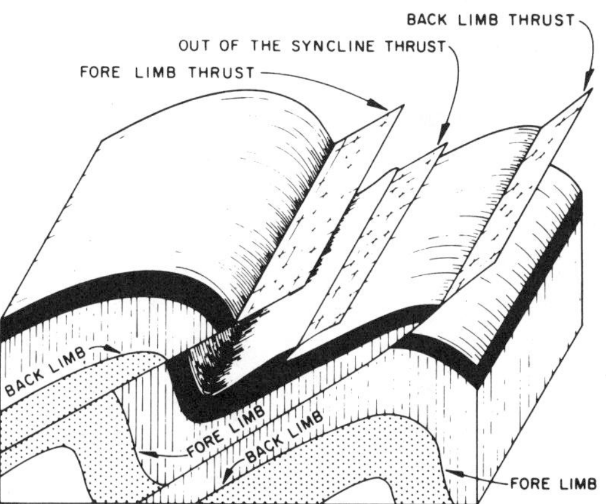

Fig. 4. Thrust-fault terminology describing the geometric fold-fault relationships.

THRUST FAULTS

Common usage among Canadian Rockies geologists equates "thrust fault" with "low-angle reverse fault." Regionally, the faults are low-angle but locally, on account of folding or imbrication, they may be very steeply dipping.

Classification

Thrust-fault terminology has developed to provide an informal classification of thrust faults on the basis of size, shape and relationship to other thrusts or to folds. A thrust terrane consists of a series of rock slabs stacked shingle-fashion one above the other. Each of these slabs is a "thrust sheet" or "thrust plate" which is bounded above and below by a "major thrust fault." Major thrusts are characterized by stratigraphic throws of several thousands of feet and lateral continuity of tens of miles. Normally the thrust plate is named from the lower boundary thrust. Within the plate itself there are substantial "subsidiary thrusts" and, near the updip edge, numerous smaller "imbricate thrusts." Characteristically, the subsidiary and imbricate thrusts dip rather steeply in their upper parts, but flatten in dip as they approach and join the underlying major or "sole fault." This concave upward or "listric" form is characteristic of thrust faults.

The listric form was originally suspected from a comparison of the steep fault dips in the Mesozoics of the Foothills with the gentler dips of the major faults in the Paleozoics of the Front Ranges. If one assumes that the basic structural pattern is the same in both areas, then the Foothills become the model for the upper half and the Front Ranges the model for the lower half of a general cross-section. If this basic assumption is valid, then the thrusts must be listric, an inference which seismic and well data now substantiate.

In a terrane of asymmetric folds it is common to designate thrusts cutting or paralleling the long, gently dipping limb as "back limb," and those cutting the short, steep limb as "fore limb" (Fig. 4). Since the thrust faults are gentler in dip than the axial planes of the folds, they must "migrate" in structural position from back limb to fore limb as they cut up-section through a fold or series of folds (Figs. 4 and 5). The terms "fore limb" and "back limb" are particularly useful when a stratigraphic horizon, such as the top of the Paleozoics, is specified.

In asymmetric folding, there is a substantial amount of interbed slippage between the upper and lower units on the long, gentle limbs. This relative movement (Hake, Addison and Willis, 1942) frequently gives rise to numerous small, low-angle "out of the syncline" thrusts or, under other stratigraphic conditions, to numerous drag folds. Imbrication at the rear or trailing edge of a thrust sheet can also produce similar features (Fig. 22).

The same terms are used in size domains ranging from the regional to the local, which introduces serious semantic dif-

Fig. 5. Updip migration of the Turtle Mountain thrust from back-limb to fore-limb to back-limb positions. The Frank Slide in the Crowsnest Pass, Alberta is in the foreground.

ficulties unless one is careful to consider the speaker's frame of reference. For example, it is clear from seismic data that the Hudsonian basement is not involved in the marginal deformation of the Canadian Rocky Mountains, which requires the whole plexus of faults that one sees at surface to join at depth in some fashion. In the regional context, then, even the largest "sheet thrusts" are "imbricates" to a basal "sole thrust" somewhere above the basement (Fig. 3).

Basic Rules of Thrust Geometry

Thrust faults alter the shape but not the amount of a rock volume by reducing area and increasing thickness according to the principle of least work. In a relatively undeformed sedimentary succession, the form of the thrust is profoundly affected by the distribution of strong and weak layers. These effects can be summarized in two simple geometric rules: 1. Thrusts cut up-section in the direction of tectonic transport (ordinarily up-dip). 2. Thrusts tend to be parallel to the bedding in incompetent rocks and oblique to the bedding in competent rocks.

In consequence of this geometric behavior thrusts need not alter the stratigraphic succession, but if they do, the overall thickness is increased as a result of bed-duplication (Fig. 6). Thrust faults do not cause bed omission, they do not thin the stratigraphic succession, and they do not thrust younger beds over older despite myriads of sections purporting to show such relationships. Some special circumstances where anomalies develop are shown in Figure. 7.

The applicability of Rich's (1934) "stair-step" hypothesis of thrust faulting to the Alberta Foothills was first fully appreciated and expounded by R.J.W. Douglas (1950). His cross-sections showed that the thrusts, in both the hanging wall and footwall, were usually within and parallel to the bedding of incompetent units and less often oblique to the competent units. Reconstruction of these cross-sections to the incipient fault stage indicated that incompetent shales and coaly units had acted as fault-gathering slippage zones (Fig 8). Statistical analysis of the stratigraphic position of thrust faults as determined from surface mapping (Fig. 9) supports these conclusions. It also shows that in thick incompetent units the thrusts tend to be near the contact with competent units. Individual thrusts do not necessarily utilize every fault-gathering zone in cutting through the section. Some faults, particularly the larger ones, transect lesser fault-gathering zones rather abruptly (Fig. 9).

To this point the discussion has pertained to transverse cross-sections parallel to the direction of tectonic transport, with a tacit assumption that in longitudinal cross-sections the fault remains at the same stratigraphic horizon. This is an

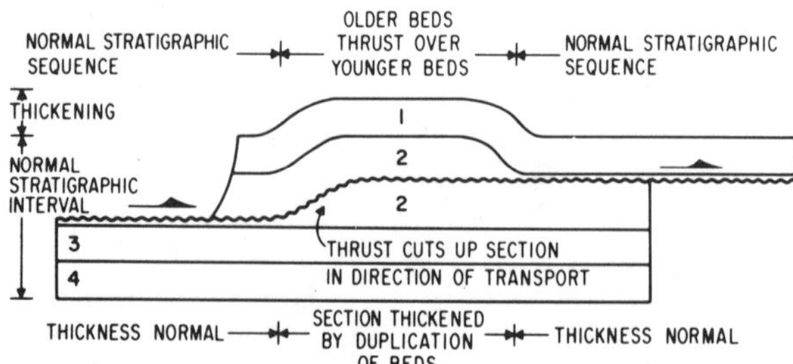

Fig. 6. The ideal thrust fault.

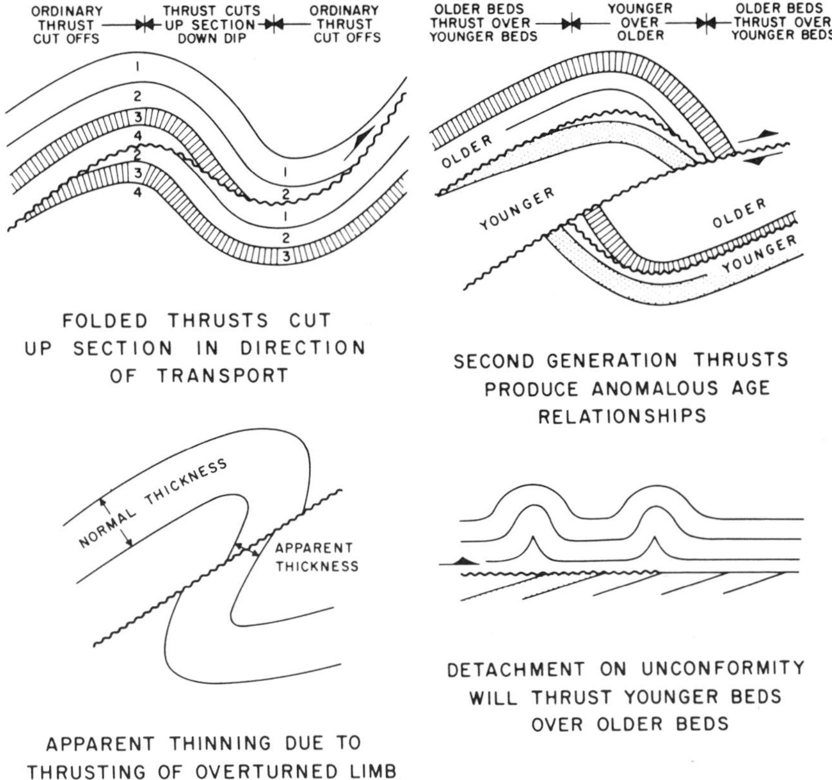

Fig. 7. Apparently anomalous age and thickness relationships produced by thrust faults.

acceptable assumption ordinarily, but important exceptions do occur. The Lewis thrust, for instance, to the north and to the south of the Waterton-Glacier salient at the U.S. border, cuts abruptly up-section along strike (Fig. 10). Though rare, this type of behaviour is important because it suggests that the basal sole fault above basement could behave in the same way.

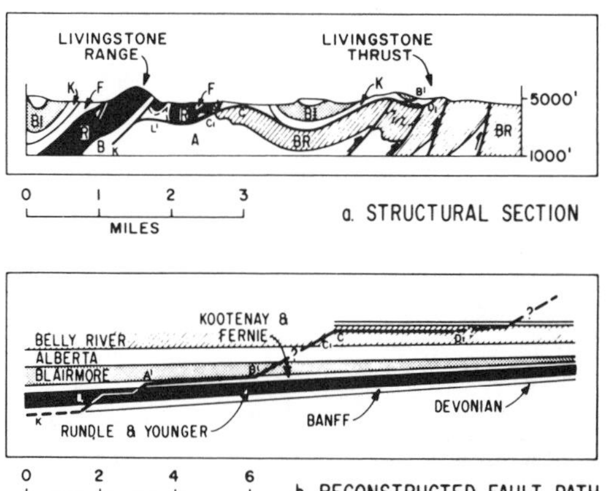

Fig. 8. Restoration of the cross-sectional trace of the Livingstone thrust to the incipient fault stage (from Douglas 1950).

Normally, one assumes that the sole fault will remain at approximately the same horizon along strike to both the north and south. The behaviour of the Lewis demonstrates that on occasion the sole fault (or basal detachment) can move rather abruptly up- or down-section along strike.

A useful way of describing such behaviour is by reference to the "cut-off," which is the intersection, in either the hanging wall or the footwall, of a specific bed with the thrust plane. Normally, the cut-offs on a thrust fault are roughly parallel to the strike of the faults, but the Lewis example demonstrates that cut-offs can be oblique or transverse. Bends of up to ninety degrees in the footwall cut-offs are possible, but cut-off lines do not close on themselves to become circular because thrust faults cut up-section in the direction of transport. Decapitated anticlines or synclines are not part of the Foothills structural association. This is an observational fact as well as a geometric requirement.

Folded Thrusts

All the basic types of thrust fault can be folded. Consequently, the Canadian Rocky Mountains contain folded examples of major sheet, fore-limb, back-limb, out-of-the-syncline and imbricate thrusts.

The McConnell fault is one of the largest of the Front Range sheet thrusts. At the Southesk River (Fig. 11) its hang-

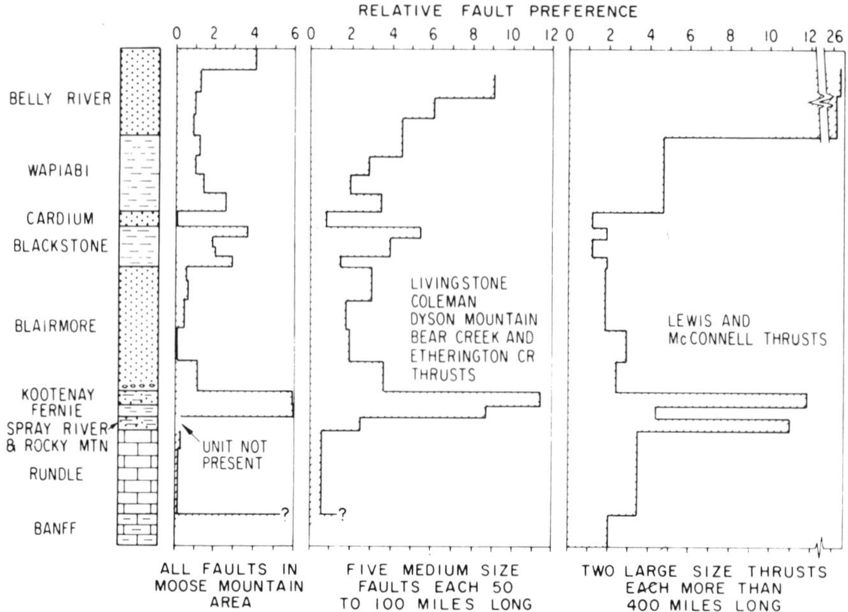

Fig. 9. Fault preference versus stratigraphy for Foothills and Front Range thrusts. Fault preference is the mapped length of thrust faulting within a specific stratigraphic unit divided by the thickness of that unit multiplied by an arbitarary constant. It is a relative measure of the tendency of faults to gather preferentially in specific units (after R.E. Daniel and J.O. Hayes, Chevron).

ing wall is parallel to the bedding in a lower unit of the Cambrian in the west and an upper unit in the east. The step from lower to upper horizon is accomplished by cutting abruptly across the middle unit, much as in the ideal instance portrayed in Figure 6. These changes in the thickness of Cambrian above the thrust produce an open anticlinal-synclinal pair at Fairholme level, although the fault plane directly below displays only a simple terrace. Futher east the fault plane is parallel to the bedding in the hanging wall, so that the fault plane and the hanging wall rocks show the same amount of folding. The McConnell example (Fig. 11) and the smaller-scale Appalachian example (Fig. 12) show the typical pattern, wherein thrusts are generally parallel to the bedding except for short segments in which the bedding is transected at a pronounced angle (Fig. 6). This "step fault" behaviour can change the thickness of the hanging wall and footwall sheets rather abruptly. When this happens, there are substantial differences in structural form between the fault plane itself and marker horizons in the overlying and underlying sheets.

Despite such discrepancies, there is commonly a relationship between a major folded sheet-thrust and a faulted anticlinal structure in the underlying plate. Coaxial relationships of the kind shown at Savanna Creek (Fig. 13) occur often enough to suggest strongly a cause-and-effect relationship that is only possible if the lower structure is the younger. This poses a problem, because the later thrusting is usually low-angle and shortens the lower plate more than the folding by uplift shortens the upper plate. Such an imbalance of shortening is geometrically unacceptable. Therefore, the upper plate must have moved in conjunction with the lower – either piggy-back, if the young fault is deep, or by reactivating part of the old fault, if the young fault is shallow (Fig. 14).

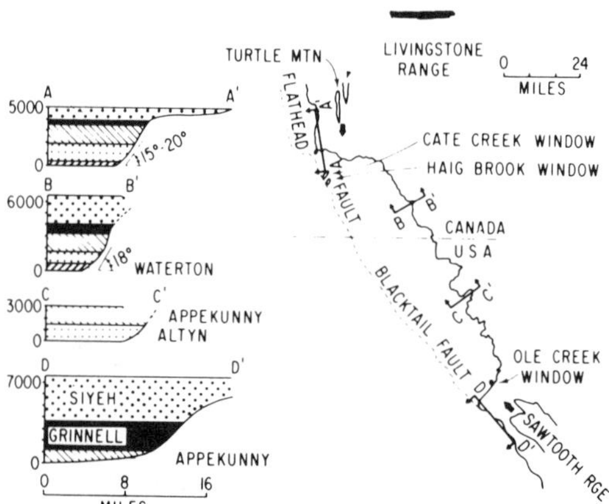

Fig. 10. Lewis thrust cutting up-section in both transverse and longitudinal sections (after Childers, 1964).

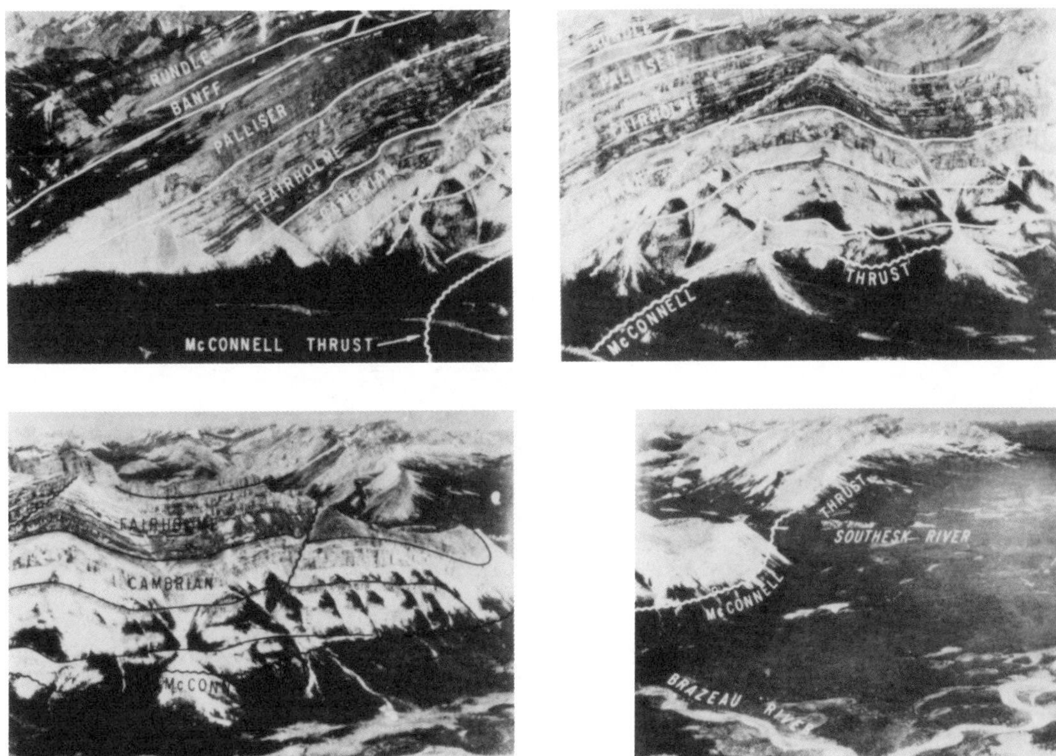

Fig. 11. Folding of the McConnell thrust at Brazeau River, Alberta. West is on the left as the photos look north. The four photos span a cross-sectional width of six miles. Vertical scale may be judged from the Fairholme which is approximately 1,400 ft thick.

The mechanism of folding a thrust by uplift from below is independent of the structure in the overlying plate. Consequently, the folded thrust could be a sheet thrust like the McConnell (Fig. 11) or a folded back-limb fault (Figs. 15 and 16) or, if it were sufficiently extensive, a folded out-of-the-syncline fault.

That uplift from below is not the only way of folding thrusts is shown by the folded back-limb fault at Dizzy Creek (Fig. 17). The Brazeau Range is a very large asymmetric anticline underlain by a major thrust-fault (Fig. 49). The Dizzy Creek fault arises on the gentle back-limb, is folded over the crest and dips very steeply down the fore-limb. If the fold in the fault plane is considered to be concentric (Fig. 31), it is apparent that it has to detach within the Cambrian at roughly the stratigraphic position of the major Brazeau thrust. Moreover, the dips are much too steep to have been produced by uplift from below. Both lines of evidence indicate that this back-limb fault was folded during the formation of the Brazeau anticline itself, and not by subsequent uplift. If it originated as a back-limb fault, there is a further implication that the crestal axis of the Brazeau anticline migrated relatively westward with time. A simple model can be made by pinning a tablecloth with one hand and pushing with the other to form an asymmetric fold of increasing amplitude. As the length of the fore-limb grows, the fold axis retains its geographic position but migrates relative to reference points on the folded surface. The Dizzy Creek folded fault suggests that this is what happened on the Brazeau Range anticline.

The examples cited (Figs. 16 and 17) show two different mechanisms whereby thrusts can be folded. Thrusts can be tightly folded if they were formed early enough and in the

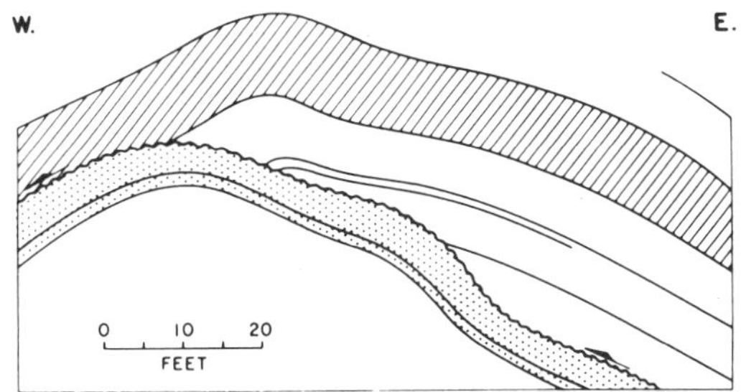

Fig. 12. Folded low angle thrust fault in the Appalachians of Maryland. Note the difference in shape between the anticline in the hanging wall and that in the footwall produced by "step-faulting." (After Cloos, 1964 with his permission.)

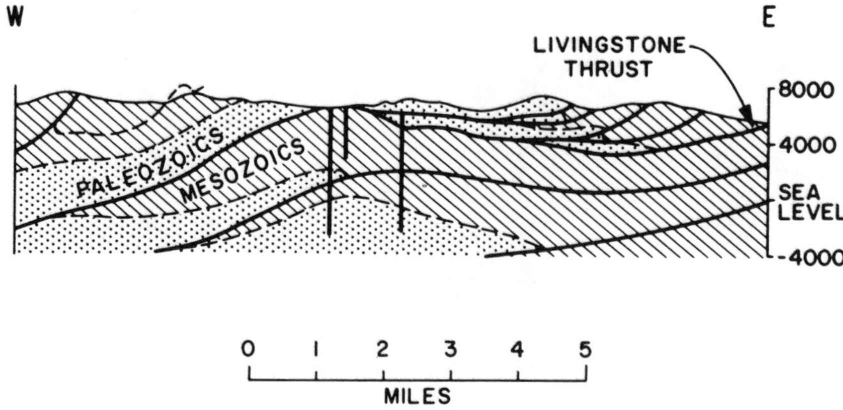

Fig. 13. Cross-section through the Savanna Creek gas field, Alberta, showing a folded sheet thrust and folded imbricates. Note the coaxial relationship between the fold in the Livingstone thrust plane and the underlying structure (after J.C. Scott, 1953 with permission of the American Association of Petroleum Geologists).

proper locations to be involved in subsequent concentric fold deformation, or they can be openly folded through uplift by younger structures below. Although the examples are back-limb faults the mechanisms are applicable to other types as well.

Folded out-of-the-syncline faults are smaller-scale structures to which the uplift-from-below mechanism is less applicable. In the Canyon Creek cross-section (Fig. 18), the thrust planes are concentrically folded, so it is clear that folding is subsequent to the faulting (Fig. 19a). It is equally clear from the principles of concentric folding (Fig. 31) that the lowest

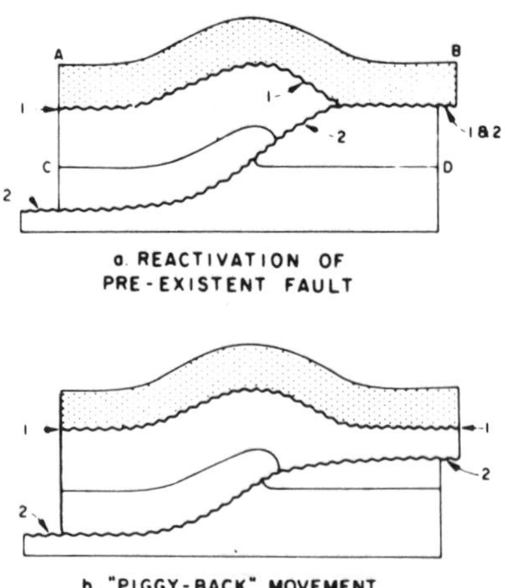

Fig. 14. Two ways of accommodating the differential in shortening between upper and lower plates when faults are folded by uplift from below. Fault 1 is first, fault 2 is second. Bedding lengths AB and CD are equal. The extra shortening between C and D requires that the upper plate move on a fault.

thrust plane in the section functioned as a detachment, and that the two anticlines shown in the cross-section do not exist below it. Out-of-the-syncline faulting is a method of moving the upper beds away from the syncline axis – a mechanism that also produces drag folds. In this instance it is possible that folding out of the out-of-the-syncline faults was a second stage of the same process that produces the faults initially. Under other conditions of stress and stratigraphy the sequence might be reversed, with drag-folding followed by thrusting.

By definition, a folded thrust demonstrates that some thrusting had taken place before the process that folded the fault became significant. The segment of the fault plane that is folded becomes inoperative, but it is possible for related thrusts to continue movement during and after the folding. This can be recognized in a simple folded pocket of imbricates where the lower ones in the stack are folded more than the upper (Fig. 19b). This is not conclusive evidence, however, because one can postulate an alternative explanation wherein the imbricates were all formed before folding. Because each imbricate is separated from its neighbor by a wedge-shaped panel of rock, subsequent folding would cause the inter-limb angle of the folds to become more open upwards (Fig. 20). This alternative explanation is tenable for situations where the angle between adjacent imbricates is small, as in the ideal example (Fig. 20) or at Savanna Creek (Fig. 13). However, if the angle between faults is large and the intervening wedge thick, as between the Dyson Creek fault and its imbricate at Moose Mountain (Fig. 16), then the imbricates become improbably steep in the restored section. Perhaps a more definitive piece of evidence is that the point of origin of the imbricates (point 0 in Fig. 20) is generally on the crestal part of the anticline (Figs. 13 and 16), which implies a casual relationship between the two phenomena. When this is the case, it implies that the anticline existed before imbrication.

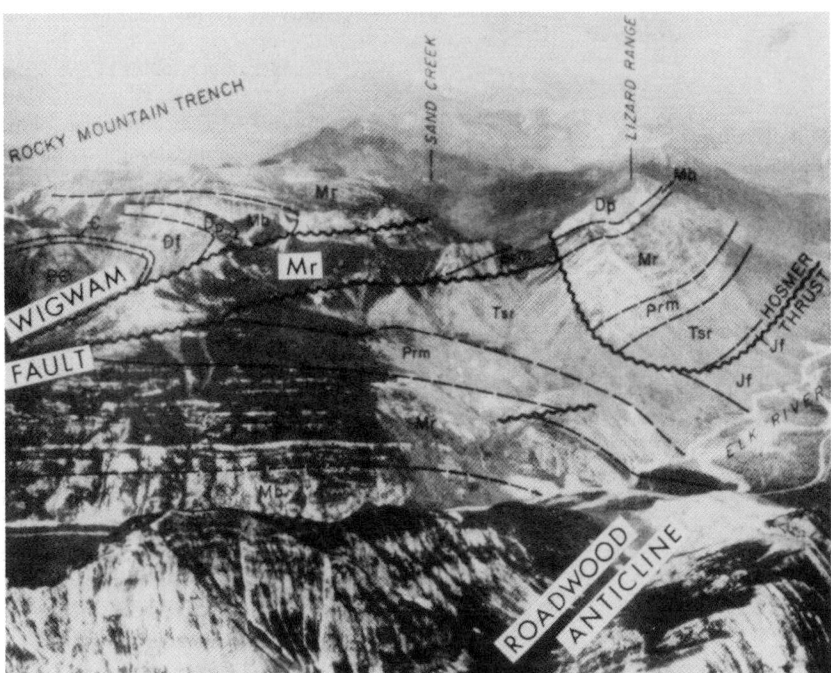

Fig. 15. The thrust underlying the Lizard Range of southeastern British Columbia is a folded fore-limb fault which becomes a back-limb fault to the west. Note that the Wigwam imbricates join the lizard thrust as in Figure 13 (a). Symbols refer to age and formation (e.g., Mr is the Mississippian Rundle Formation).

In summary then, folded thrusts can be a consequence of: 1. original shape (Fig. 6 and the basal sole thrust in Fig. 3); 2. folding during thrusting (the imbricates in Figs. 16 and 19a); 3. folding after thrusting, a. by normal folding process (Figs. 17 and 18), b. by uplift from below (the sole fault in

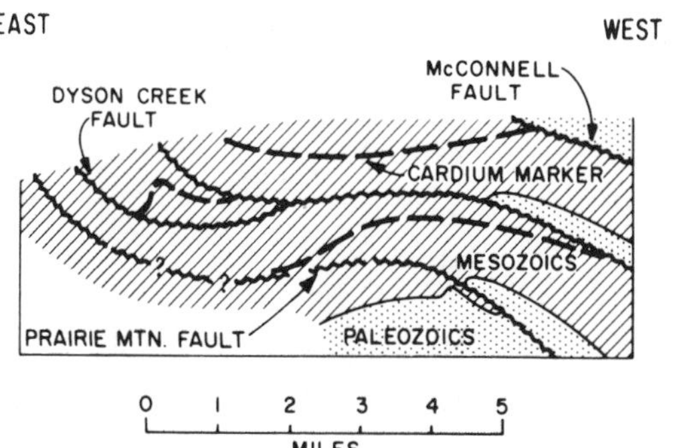

Fig. 16. A composite cross-section of the south end of the Moose Mountain culmination obtained by plunge projection. Note that east is on the left. The Prairie Mountain and Dyson Creek faults are back-limb thrusts on the Moose Mountain anticline (after Helen Chernoff, Chevron).

Figs. 13 and 14); 4. surficial effects.

The stair-step behaviour of thrust faults produces an original nonplanar fault surface which has the form of open asymmetric monoclinal folds. Although the fault plane does have a "folded" shape it is a primary shape, not one imposed by subsequent deformation. The degree of this original fault-plane warping depends on whether the thrusts develop before any folding (Fig. 6) or after substantial folding (Fig. 4). Either way, the amount of original fault-plane warp is small by comparison with the very steep dips that sometimes exist in faults that were subsequently folded. The existence of such steeply dipping folded-fault planes implies folding above a detachment, from which one can conclude that a) the folding was relatively early, b) the fold structures are restricted to a single plate and, c) the fold structures do not extend beyond the sole fault at the base of the fault plate. When the faults are folded into broad open structures, they represent folding by late uplift and the causative structure will be within the underlying sheet.

Near-surface deformation can also alter fault shapes by toreva faulting, by semi-plastic yielding of the footwall rocks, by overriding of the hanging wall on an erosion surface, etc. Since these are special local phenomena of restricted extent, they have not been discussed.

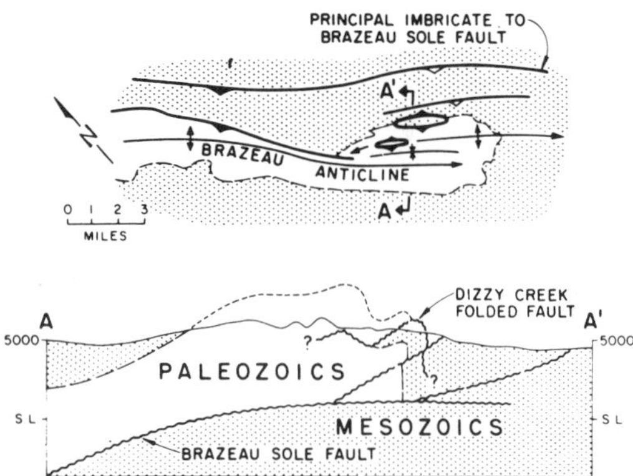

Fig. 17. Folded back-limb fault at Dizzy Creek on the Brazeau Range, Alberta (after O.A. Erdmann, 1950).

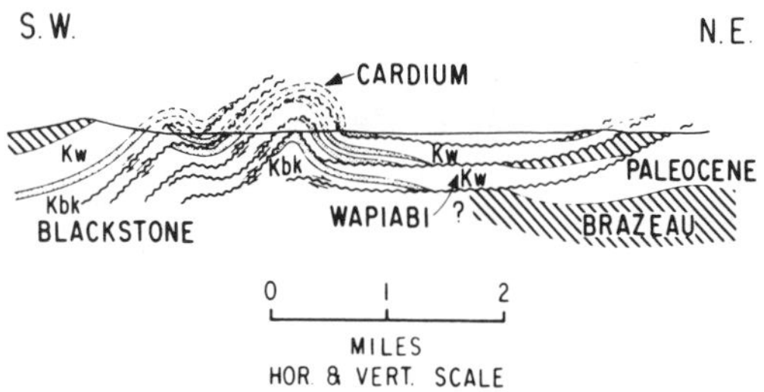

Fig. 18. Folded out of the syncline faults at Canyon Creek, Alberta (after J.C. Scott, 1951 with permission of the American Association of Petroleum Geologists).

Sequence of Fault Development

A question that is debated among Foothills geologists is whether there is a regular age sequence in fault development and, if there is, whether it begins on the foreland (east) or hinterland (west) side of the belt. This question should be examined from two standpoints — first, as it pertains to individual fault plates; second, in the regional sense.

By observationally derived definition, fault plates are bounded on their undersides by sole faults with substantial throw and lateral continuity. Instead of one fault there are usually several, of which it is customary to designate the largest and lowest as the sole fault and the minor upper ones as imbricates. Above and below this zone of imbrication the hanging-wall and footwall plates are usually simple. Under these circumstances it is possible to determine the age sequence in faulting by the distribution of stratigraphic throw. Ordinarily, most of the stratigraphic throw is on the lowest fault, indicating that it is the oldest and that faults within the imbricate zone have developed sequentially from lowest to highest (Fig. 21b). Since the reader will suspect some circularity and proof by definition in the foregoing argument (which can be explained but not briefly), it is fortunate that the same east-to-west conclusion has already been independently indicated by the folded imbricate stacks (Figs. 16 and 19).

The foregoing pertains to the distal or leading edge of a thrust sheet. There is also a proximal or trailing edge, which is seldom exposed but is known imperfectly from seismic and drill data. Deformation of the trailing edge of a sheet is presumed to result from the "bulldozer" action of the overlying thrust sheet having produced a stack of imbricates in the footwall, with attendant folding of the overlying thrust sheet and, occassionally, out-of-the-syncline updip faulting or drag folding (Fig. 22).

A trailing-edge imbricate stack is well exposed in the Waterton area near the U.S. border, where there is a complicated duplex fault zone (Fig. 23) at the base of the Lewis thrust sheet. The Lewis thrust is the lower boundary (the floor thrust) and the Mount Crandell thrust the upper boundary (the roof thrust) for a suite of more steeply dipping minor thrusts that thicken and shorten the intervening panel of rock. From the stratigraphy in the footwall and hanging wall of both the Lewis and Mount Crandell thrusts, it is apparent that both have cut gradually up-section through the Proterozoic sedimentary succession, which eliminates the possibility that the intervening minor faults antedate the Lewis and Mount Crandell thrusts and were truncated by them. Necessarily, the sequence is, first, a lower detachment which was, or subsequently became, the Lewis thrust; second, the Mount Crandell thrust; third, the minor faults. Some of the minor faults were folded by movement on other minor faults below them, which shows that the progression was from west to east (Fig. 22). Thirty-five miles northwest, the base of the Lewis thrust sheet is exposed in the Cate Creek and Haig Brook windows (Fig. 10). Although the duplex fault zone is thinner, the same basic situation exists (Fig. 24), so that the Waterton

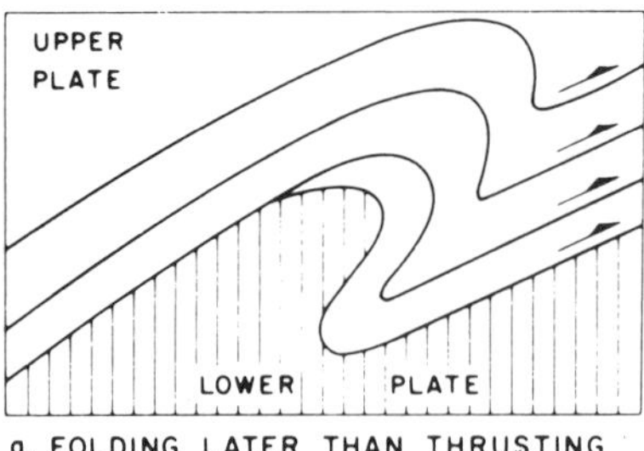

Fig. 19. Two basic patterns in folded stacks of imbricate thrusts (after J. Gilluly, 1960 with permission of the American Association of Petroleum Geologists).

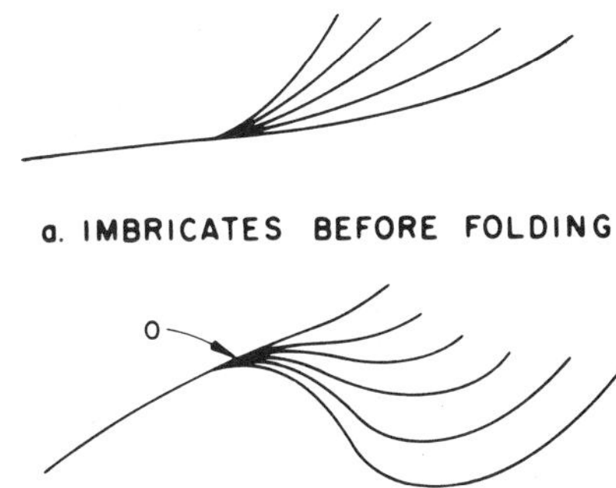

Fig. 20. Folding of an existent imbricate stack.

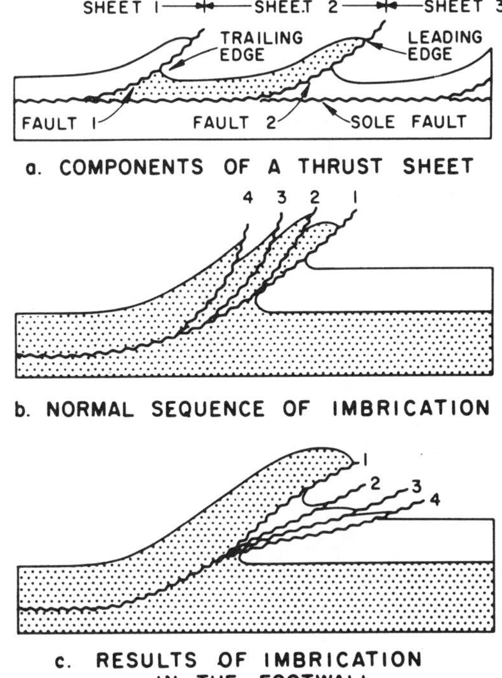

Fig. 21. Imbrication at the leading edge of a thrust sheet.

example does represent a class of structure and not an isolated instance. Structures at the trailing edge of a thrust sheet involve two major thrusts, the upper and lower boundaries – which naturally leads to the question of their relative ages and the sequence of faulting in the regional sense.

There are two lines of evidence which show that the regional sequence of faulting is from the hinterland (west) progressively toward the foreland (east). One, which has already been discussed, derives from the observation that anticlines in major folded sheet-thrusts are commonly underlain by faulted anticlinal structures in the underlying plates. As in the footwall imbricate stacks, it is obvious that the faulted lower structures developed after the upper sheet-thrust was emplaced. The same basic conclusion has been demonstrated by Bally *et al.* (1966). In the Foothills they show that the shortening, which is the ratio of original depositional length to deformed length, may be twice as much in the Mesozoics as in the underlying Paleozoics (Fig. 25). Therefore, there is a basal sole thrust or detachment in the fault-gathering zone at the base of the Mesozoic succession. Since this basal detachment is itself folded and offset by faults which cut the underlying Paleozoics, it is necessary to postulate two-stage deformation where-

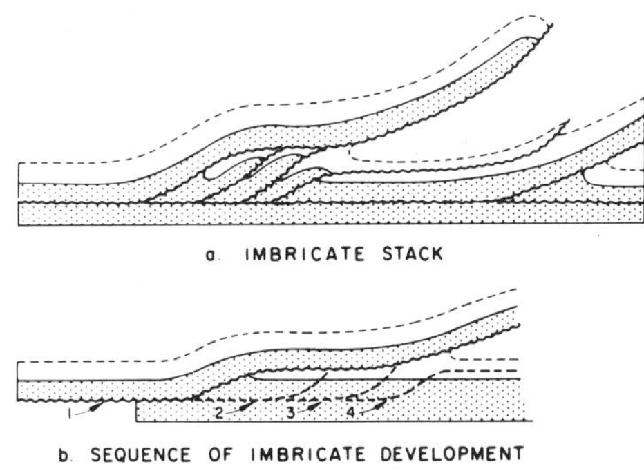

Fig. 22. Imbrication at the trailing edge of a thrust sheet.

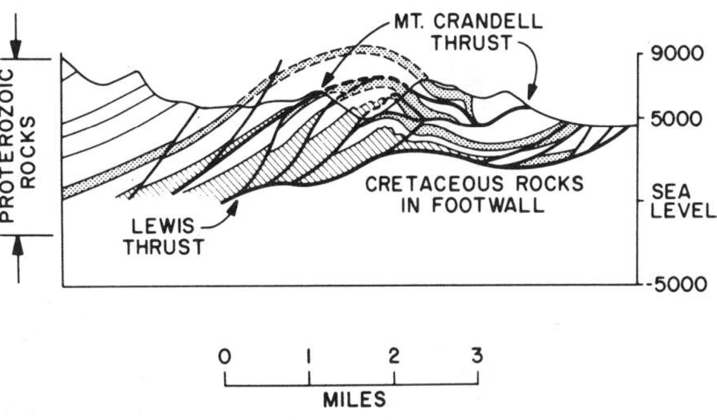

Fig. 23. Cross-section of the Lewis-Mount Crandell thrust zone at Waterton Lakes, Alberta (after Douglas, 1952).

in the Mesozoics were deformed first separately from and then in conjunction with the underlying Paleozoics. This is a west (upper) to east (lower) regional progression.

The local and regional sequences are not wholly compatible. At the local scale of a single thrust sheet, the evidence at the leading edge shows that the lowest fault in an imbricate stack is the oldest and the sequence is from east to west. At the trailing edge of an individual thrust sheet there are also imbricate stacks, and there the upper imbricate is probably first and the sequence is from west to east. The sequence in the trailing-edge imbricate stack is the same as that in the regional, but the leading-edge imbricates show the opposite sequence – an anomaly which can be either real or a misinterpretation. Generally the anomaly is believed to be a real one, and it is so considered in the next few paragraphs. (The possibility of another interpretation is discussed in the section on detachments.)

Much discussion about the sequence of fault development in the Foothills has hinged on an unspoken assumption that the sequence in the imbricates at the leading edge of a thrust sheet is a small-scale model of the sequence that must pertain in the regional relationships between thrust plates. This is not necessarily true, because it ignores the scale factor. If this assumption were applied literally to folding, one could use a hand-specimen-size drag fold as the model for structural extrapolations about geanticlines of regional dimensions. Obviously, this goes too far. Because some fundamental structural forms do occur in a spectrum of sizes spanning several orders of magnitude, it is very easy, very normal, and often very wrong, to assume that exactly the same interpretative rules will apply throughout the entire size spectrum. With structural geology being something less than a fully-developed

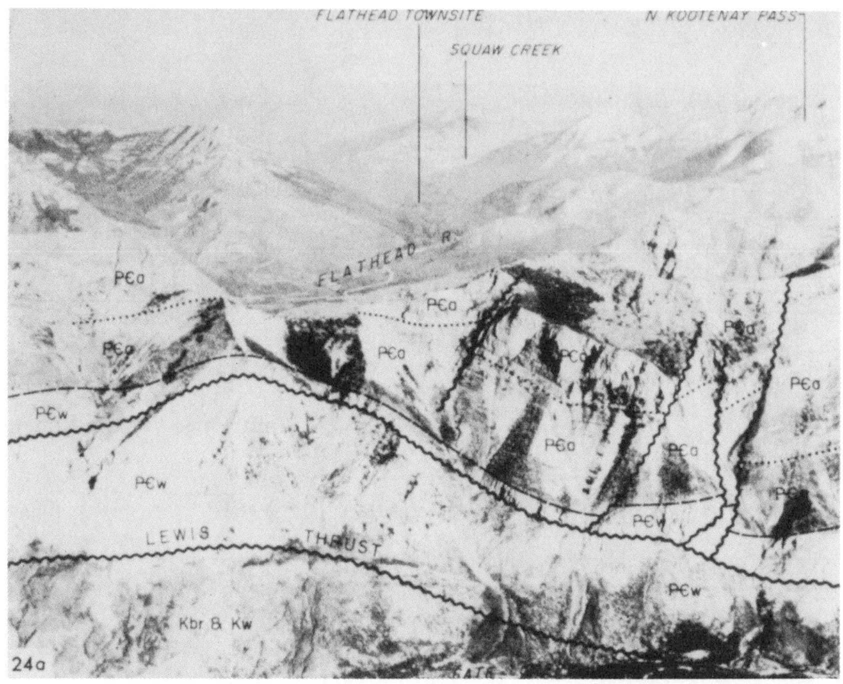

science, there is no way that one could have anticipated in advance that the local age sequence at the foreward edge of an individual fault plane would be the inverse of the regional sequence. However, if the field information indicates a difference, one is obliged to accept the possibility, because the difference may be an illustration of the fact that scale does influence geologic behaviour.

Another source of confusion has been a failure to discriminate between the fundamental and the incidental. The foregoing discussion of thrust sequence has been specifically directed toward the time at which the principal movement took place, and specifically unconcerned with the time of the very first or very last movement. This is not easy to manage, because individual thrust faults can move several times during their histories. This has been illustrated for the Lewis thrust zone (Fig. 23), for the roof fault in a folded imbricate zone (Figs. 22 and 23) and for the forward part of a folded sheet-thrust (Fig. 14); it will also be shown for tear displacements (Fig. 45). Some of these movements are clearly subsequent to the main thrust movement and should be considered as incidental. Moreover, with a west-to-east regional sequence, the whole deformed section must move relatively eastward on the underlying sole fault as each new structure forms in front of the migrating deformed zone. That the pre-existent movement planes in the deformed mass would remain totally inert during this transportation is inherently unlikely: some unsystematic incidental movement is to be expected.

Fig. 24. The duplex Lewis thrust zone in the Cate Creek window, southern British Columbia. An overall view of the whole zone and the structure adjacent to the floor thrust and the roof thrust are shown in separate photos to take advantage of the best exposures. The floor thrust is the Lewis but it is not known whether the roof thrust is the Mount Crandell fault as in Figure 23. Stratigraphic markers adjacent to the roof and floor thrusts show that the boundary faults were parallel to the bedding before the intervening imbricates developed.

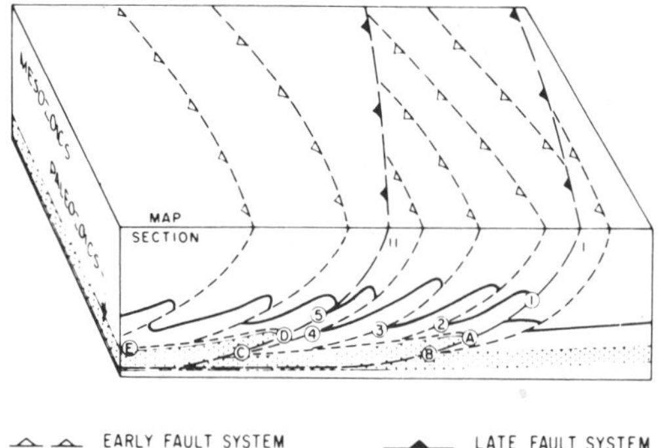

Fig. 25. Deformation of early Mesozoic imbrications originating in the underlying Paleozoics. See text for discussion (from Bally 1966).

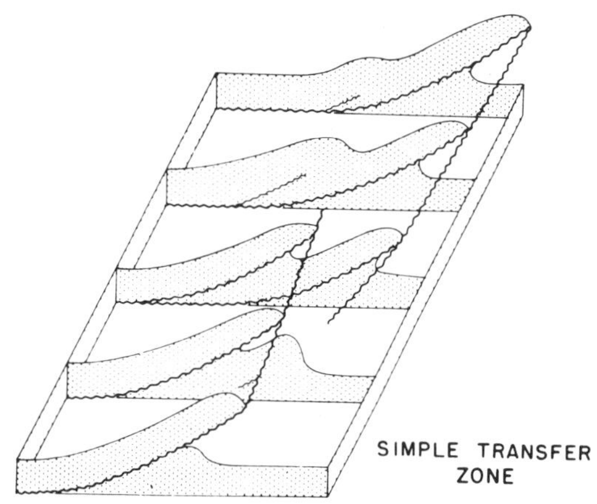

Fig. 26. Simple transfer zone between *en echelon* thrusts.

Distribution of Fault Displacement

Thrust-faulting is a mechanism equivalent in many respects to concentric folding, for shortening and thickening a packet of rock. Individual faults, like individual folds, do not continue forever. As they disappear, the amount of shortening of the packet must diminish or their function must be assumed by another structure, either fold or fault. As a general observation, the extent and amount of shortening within the Foothills as a whole is more nearly constant than that of individual structures. This implies the existence of a basic transfer mechanism that enables other structures to take over the shortening functions of those that die out. Since the Hudsonian basement is not involved in Foothills structures, there is a throughgoing sole fault that separates the basement from the overlying deformed rocks. The principal thrust-faults join the underlying sole thrust, and all the deformation above basement involves sliding upon it. This throughgoing sole fault is the mechanism for transferring displacement from one fault to its *en echelon* equivalent. The simplest form of transfer zone is a lap joint, wherein the upper thrust becomes first a back-limb fault and finally a drag fold on the underlying fault plate. The lower thrust disappears by losing displacement until it becomes a fore-limb thrust on a minor anticline which plunges underneath the upper thrust (Fig. 26). It is, of course, necessary that the principal structures within a transfer develop simultaneously. Therefore, the common assumption that fore-limb faults develop before back-limb faults is not true in transfer zones. Although transfer zones are generally complicated, it is usually possible to decide which faults are paired and thereby to correlate faults. Figure 27 shows the correlation in the Foothills of Alberta. Northward, correlation ceases to be effective as folding becomes the dominant method of shortening in the near-surface rocks.

Shortening necessarily implies thickening, but in a thrust terrane the thickening may be localized, not pervasive (Fig. 6). Since basement is throughgoing and gently dipping, differential structural thickening must produce structural culminations. The corresponding depressions represent areas of less thickening. If the stratigraphy and the shortening are consistent, this lack of thickening simply means that much of the thrust displacement is parallel to bedding and therefore does not cause bed duplication. Intuitively, one would expect that geometric or lithological conditions would introduce some pattern to the distribution of culminations and depressions, but none is evident on the regional scale (Fig. 28). Individual examples will be discussed later, when it will be seen that there are explanations for specific plunge phenomena. However, each one seems to be a special instance and generalities useful for prediction have not yet been recognized.

Relationship of Faulting and Folding

Most of the shortening in the southern part of the Foothills belt is accomplished by thrust-faulting, whereas in the north concentric folding predominates. Although one type of structure may be dominant, the other is never entirely lacking. The question of which is the initial phase of deformation has been debated at length by Foothills geologists. There are, of course, three possibilities: faulting first, folding first, or essential contemporaneity. The thrusting-first hypothesis postulates that the original stair-step fault shape is the prime cause of folding. The folding-first hypothesis derives from the fact that concentric folding requires a decollement parallel to the bedding, and that the stair-step pattern could result from the concentric folding of a rock packet with several detachment horizons. Regrettably, the end products of both processes have approximately the same geometry, so there is no way of demonstrating origin from a cross-section (Fig. 29). The third hypothesis, contemporaneity, says that the duration of the initial stage of pure faulting or pure folding is so fleeting as to be inconsequential, and that both thrusts and folds were functioning during deformation.

The data that might be used to discriminate between the two basic alternatives pertains to: 1. folded thrusts; 2. cut-off angles; and 3. displacement exchanges. By definition, a folded thrust demonstrates some folding subsequent to thrusting. As previously discussed, in some instances, this folding is normal concentric (flexural slip) folding, wherein the thrust planes

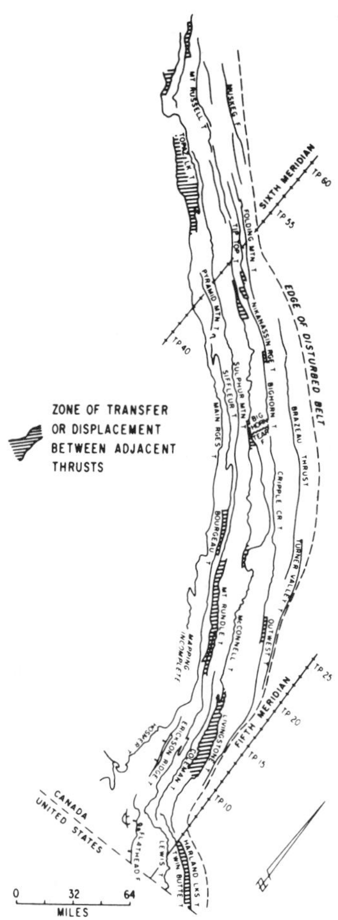

Fig. 27. Relationships between major thrusts in the eastern part of the Canadian Rocky Mountains.

are rather passive elements comparable to the adjacent bedding planes. In other instances the thrust is "folded" by uplift from below. In either event the evidence pertains to what happened after the fault was formed, since it can scarcely testify to any events before it was formed.

The original cut-off angles visualized in the stair-step hypothesis must be relatively shallow, to provide a low-friction sliding surface (Fig. 6), but the cut-off angles on fore-limb faults are often ninety degrees or even overturned (Fig. 15). Such evidence suggests (but does not prove) folding prior to thrusting, because interbed slippage caused by post-thrust folding can increase the cut-off angles drastically.

Displacement exchange is a more promising criterion. Field cross-sections at Turner Valley (Fig. 30) show that in this structure the shortening in the deep horizons is accomplished primarily by faulting, whereas in the upper horizons a comparable amount of shortening is provided by folding. The balance between diminishing fault throw and increasing vertical-fold limb shows that the fault was propagating itself upward in a growing fold. In such structures, and at a specific marker horizon, the sequence is clearly: fold first, fault second. Most areas lack the well control needed to establish the Turner Valley-type relationships in cross-section; however, these conditions can be identified from surface geology in two ways: The simple, direct and convincing way is to look at thrust terminations by down-plunge viewing (Mackin 1950), to see what kind of cross-section is represented. Commonly, they are like the Turner Valley example. The other way is to compare the axial planes of adjacent anticline-syncline pairs. If the fold-shortening is greater in the upper beds than in the lower, then the steep limb is longer and there must be a convergence of axial planes, with the axial plane of the anticline being much steeper than that of the adjacent syncline (Fig. 30). With due caution, this relationship can be used to infer the existence of a compensating thrust at depth. (The need for

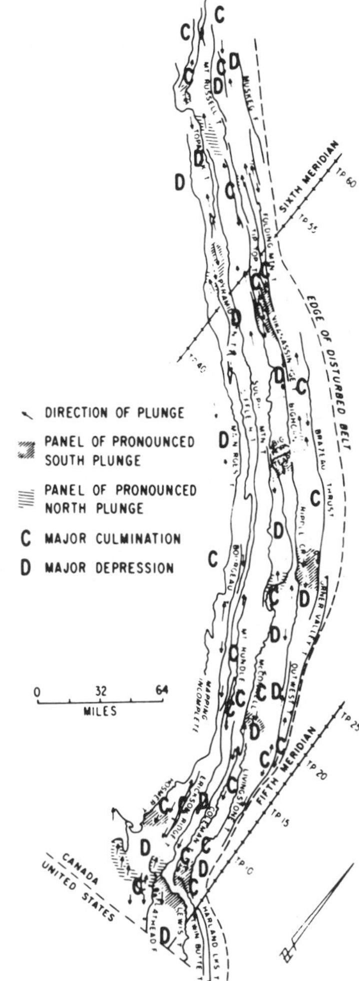

Fig. 28. Culminations and depressions in the eastern part of the Canadian Rocky Mountains.

caution is indicated by Fig. 40a). Applying these two criteria to disturbed-belt geological maps provides enough examples to suggest that it is common for thrusts to propagate upward at the expense of pre-existent folds.

The only definitive evidence available indicates that folding of a specific horizon at a particular place precedes thrusting of that same horizon at that same place; hence the writer prefers to use a "folding first" working hypothesis. As indicated in Figure 29, however, the geometric end product in both hanging wall and footwall is similar for both mechansims. In particular, the step-shape of the footwall is the same, and the hanging wall must necessarily accommodate itself to this shape as it moves forward. The significant difference, then, is whether the principal folding precedes this accommodation or is a consequence of it. In view of these similarities, the fault-first mechanism, like the wave theory of light, is often a convenient fiction for expository purposes, particularly in instances where the primary concern is the nature of the fault plane. In other instances, where the character of the folding or the amount of thrusting is critical, the fiction can be quite misleading.

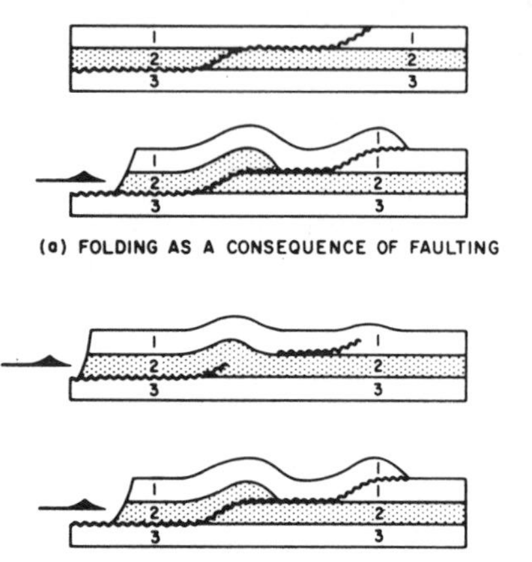

Fig. 29. Alternative sequences of folding and faulting (after P. Verrall, Chevron).

CONCENTRIC FOLDS

Principles

Concentric folding is characteristic of low-strain deformation at relatively low temperature and pressure. Under these conditions there is no flowage and the thickness of any individual bed is constant whether it is on the crest, flank or trough of a fold. This is a field-observed characteristic of concentric folds. Though not entirely necessary, it is usually assumed that there was no significant change in rock volume during deformation. If the constant-volume assumption and the constant-thickness observation are accepted, one must conclude that the surface area of an individual bed remains constant during deformation. For practical purposes, this means that the bed lengths of individual beds in transverse cross-section remain constant during deformation. In the simple linked folded succession in Figure 31 there is an inverse relationship between the circumferences of the synclines and anticlines, which permits bed-length balance to be maintained through a substantial thickness of section. However, this balance fails beyond the centres of curvature, where the uppermost beds are crowded into the synclinal trough and the lowermost into the anticlinal trough. The resultant crenulations and/or thrusting diminish the fold amplitude, thereby establishing upper and lower boundaries to the folded panel. These boundaries are, and must be, marked by detachments or decollement which segregate the folded panel from the overlying and underlying rocks (Fig. 31).

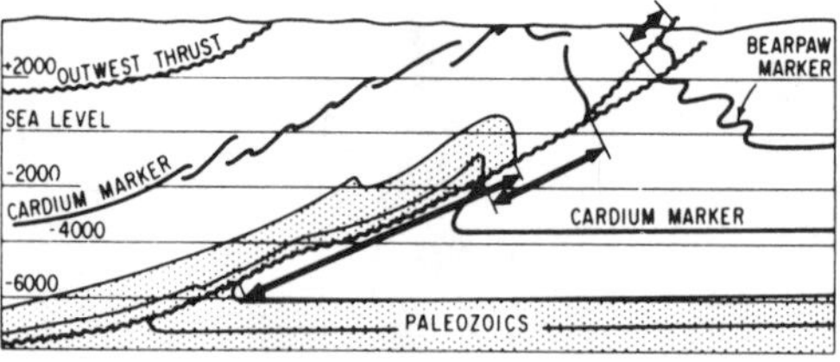

Fig. 30. Diminishing fault displacement compensated by increasing fold shortening in the Turner Valley structure, Alberta (based on a section by Gallup, 1951 with permission of the American Association of Petroleum Geologists).

Concentric folding can develop only in bedded rocks having internal slippage zones that permit interbed movement during folding. The pattern of folding depends on several variables. First is the character of the unit in the succession that is more competent than the adjacent rocks. This unit becomes the "dominant member" in folding and it has been demonstrated mathematically (Biot, 1961) experimentally (Donath and Parker, 1964) and observationally (Currie *et al.*, 1962) that the thickness of this dominant member determines the wave length of the folding. Amplitude is of course controlled by strain, not stratigraphy. Second, incompetent units are needed to provide detachment horizons. Third, at top and bottom, between the throughgoing detachment horizons and the discontinuous detachment that marks the outer limit of the simple folded section, there must be a relatively incompetent thin-bedded sequence that can be deformed to fill the tight fold cores (Fig. 32). These three are separate, stratigraphically dependent variables which can control the end product of concentric folding, but there are also several other factors such as temperature, pressure, and rate and amount of strain. This number of variables ordinarily precludes predictions of structural style from a knowledge of the stratigraphy alone. Nevertheless, the basic geometry is simple, and it is quite possible to understand and make predictions about concentric folds once the patterns have been mapped and identified.

Under the temperature, pressure and strain conditions that produced the Foothills and Front Range structures, four rather different fold types develop: 1. simple curved folds; 2. chevron folds; 3. pseudo-similar folds; and 4. box folds.

Simply linked fold pairs formed from smooth circular arcs or involute curvatures are more common than one might

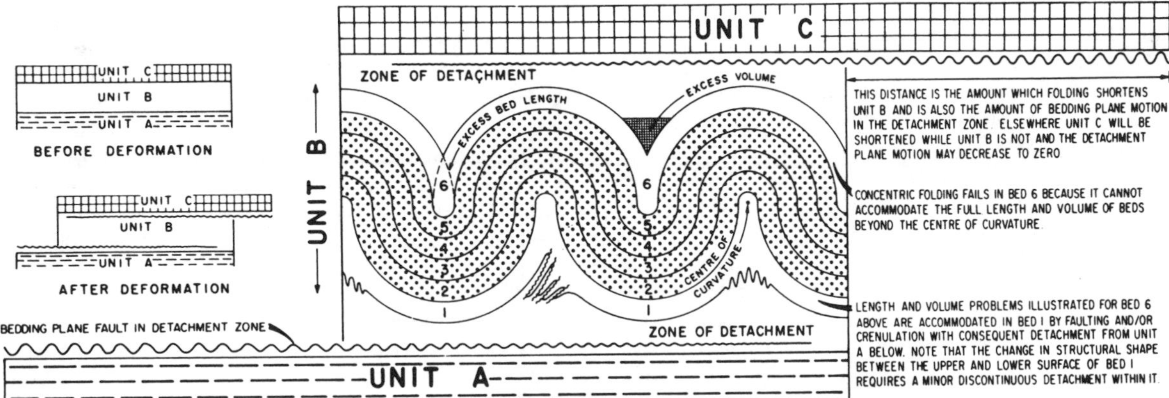

Fig. 31. The ideal concentric fold (from Dahlstrom, 1960).

expect. The prerequisite seems to be a distinctly defined dominant member that is substantially more competent than the overlying and underlying units. The appearance of a shaly facies near the top of the Paleozoic carbonates often provides such conditions in the Front Ranges (Fig. 33).

Chevron folds have planar fold limbs and angular crests or troughs. The axial planes of the folds are practically kink planes. In such folds the limbs are parallel and of constant thickness, hence they are concentric folds, although they have the similar fold ability to retain their shape unchanged from one horizon to another. Chevron folds can develop only in a thin-bedded succession of fairly competent beds, separated from one another by thin but effective slippage horizons (Fig. 34). Unless there is axial faulting along the "kink" plane, there must be flowage or crumpling at the apices of cheveron folds, even though in thin beds it is on such a small scale that it is not apparent.

A similar mechanism can function on a larger scale to provide "pseudo-similar" folding if the incompetent units are appropriately distributed for the development of partial detachments. In an anticline, the upper contact of the incompetent unit can have a very sharp radius of curvature and the lower contact a much more open curvature. This change in curvature permits the concentric pattern to resume (Fig. 35). Although thickening in the troughs and crests is by concentric processes, the end result is almost the same as though it were flowage. Grossly, these folds behave as similar folds because they retain their basic (but not their exact) form for vertical distances substantially beyond what one would expect from normal concentric folds. Such folding requires a specific combination of competent and incompetent units and therefore disappears at depth when the appropriate lithology ends.

Box folds are a basic concentric-fold form wherein the flanks and crests (or troughs) of folds are tabular rather than curved units (Figs. 36 and 37). In the simplest form of box fold the junctions between crest, flank, and trough are all "kink" planes. Predicting depth to detachment is relatively straightforward because the axial planes of the two anticlinal "shoulders" intersect at the point where the tabular crest disappears. In the synclinal areas the stratigraphic succession is normal down to the detachment – information that can be used to establish the fold amplitude and to predict detachment elevation. The resulting estimate can often be verified from the intersection of the axial planes of adjacent synclinal "trough corners," which is generally near but below the detachment horizon. Such predictions must be treated with some reservation because, as in psuedo-similar folding, it is possible for the radius of curvature to change along the axial planes if minor detachments develop within the section.

In concentric folds of either the concentric-arc or box type there are vertical changes in fold form which make the size and complexity of the exposed anticlines and synclines dependent on whether the erosional surface is near the upper or the lower detachment horizon. Figure 38 shows the change in

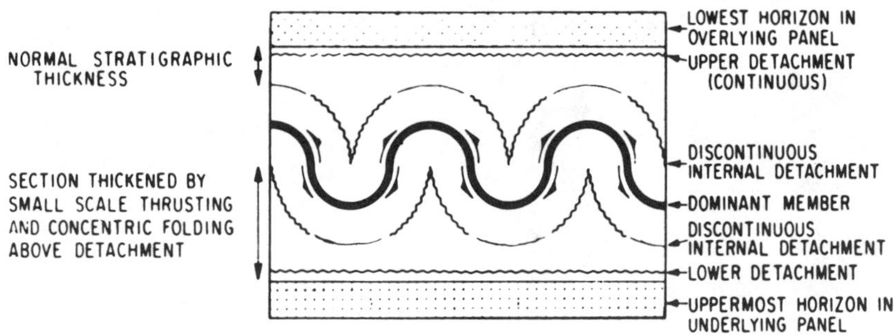

Fig. 32. Components of an ideal concentric-fold panel.

tectonic style with erosional depth in box-fold terrain. From Figure 31 it is evident that the same type of change also occur in folds with simple curved shapes. Several exploratory wells have inadvertently demonstrated the validity of these generalizations by drilling inordinate thicknesses of steeply dipping beds in the cores of "style ejectif" anticlines.

The foregoing discussion was restricted to the inherent features of concentric folding without considering the incidental features within a specific fold, although some of these can be of modest oil-field dimension. Though small, the crumpled cores of anticlines and troughs of synclines are not incidental features – they are inherent features which are a necessary consequence of concentric folding. Although the scale is small the deformation within these complex zones is by concentric folding, and by thrusting according to normal geometric rules. If inter-bed slippage within the dominant member and its associatied beds were perfect, no incidental minor structures would develop, but slippage is frequently inhibited, thereby producing subsidiary areas of shortening and extension (Fig. 39). The extensional features usually recognized are the normal faults (longitudinal, transverse or diagonal), which characterize the crestal part of an anticline above the dominant member. Presumably, comparable features should exist in the synclines, but these are seldom explored or mapped in detail. In synclinal areas above or in anticlines below the dominant member, it is common to find contractional features – either drag folds or small-scale thrust faults.

Within the southern Foothills, where thrusting is dominant, there is distinct asymmetry with the axial planes of folds almost always inclined toward the west. In the northern Foothills, where folding is the fundamental mode of shortening, there is no pronounced asymmetry and axial planes are apt to be inclined either way (Fitzgerald, 1968). Why this should be is not known. The writer would suggest that it depends on the glide horizon, and that lack of symmetry indicates a very effective glide horizon (e.g., evaporites). A parallel might be ocean-water waves, which, in deep water, are oscillatory with essentially vertical axial planes, and become translational with inclined axial planes only when the near shore bottom effect becomes important.

Detachments

In concentric folding, a detachment above and below the folded section is a geometric necessity. The upper detachment is not normally recognized because it is either hidden below a panel of simple dips or removed by erosion. The lower detachment does exist in the Canadian Rocky Mountains. The commonest exposed type (Fig. 40a) is one where the folds end concentrically against a detachment which is also a major thrust fault. This geometry could represent concentric folding above a detachment that was already a thrust or above a detachment that subsequently became a thrust. The writer uses the latter as a working hypothesis, postulating that the folds developed from west to east and that the thrust developed in response to the geometric necessity for a detachment. The change from a lower level of detachment to the next higher would have been accomplished subsequently by cutting upsection in the Rich "stair-step" fashion. As shown previously (Fig. 29), there are alternative explanations which others find equally satisfying.

The outer edge of the disturbed belt marks the boundary between the shortened and structurally thickened section of

Fig. 33. Concentric anticline in upper Paleozoic carbonate rocks at Cabin Creek, Alberta.

Fig. 34. Chevron folding in thin-bedded Ordovician limestone in the Western Ranges of eastern British Columbia (photo by G.G.L. Henderson).

Fig. 35. Pseudo-similar folding produced by concentric folding with multiple partial detachments.

the Foothills and the normal stratigraphic succession in the Plains. The regional dip in the Plains is westward, and the general dip in the Foothills is westward. These two areas of general west dip are connected by a short panel of east dip, producing a syncline-anticline pair commonly called the "front fold" and the Alberta syncline (Fig. 40b). The front fold was drilled in a number of places before it was discovered that the folding and faulting were restricted to the Mesozoics and that the upper Paleozoic objectives were at regional elevation. Seismic data show that the Tertiary units on the west flank of the Alberta syncline have an unbroken east dip, which implies a discontinuity with the underlying folded and faulted Mesozoics. This is the upper-detachment boundary of the deformed panel. Most of the east dip on this detachment is the result of uplift from below as the deformed Mesozoics were moved relatively eastward ("underthrust") beneath the detachment.

The other basic type of detachment in the Foothills is the sole fault to which the lesser thrusts join (Fig. 40c). The concentric-fold detachment and the sole-fault detachment are end members of a continuum. Where folding is dominant, as it is in the northern Foothills beyond the Peace River, the fold detachment is the basic model. Where thrusting is dominant, as it is in the southern Foothills near the U.S. border, the sole fault is the basic model. The distinction between them depends on whether the movement plane is parallel to bedding (a detachment) or trangresses bedding (a thrust). As suggested by Figures 40a and c, this distinction can be rather arbitrary and unrewarding. Moreover, it is probable that some detachments subsequently become major thrust faults. This is suggested by major thrust faults which are also the lower-detachment surfaces for concentrically folded panels (Dahlstrom, 1969). These examples are particularly convincing when the symmetry of the folds above the more steeply dipping part of the thrust is out of context with the regional style (Fig. 40a). Particularly good examples exist at Jasper, where some of the folds have eastward-dipping axial planes which appear to have been rotated from their original vertical or westward inclination by movement up a listric west-dipping thrust plane (Mountjoy, 1959).

If one is willing to accept the idea that folds can develop originally above a subhorizontal detachment and subsequently be moved up an inclined thrust fault with consequent rotation, then one should consider whether the same idea can be applied to the imbricate stack at the leading edge of a thrust sheet. The accepted idea, and the one used in the discussion to this point, is that the imbricates are formed *in situ* near the surface, and that the sequence of formation is east first and west last (Fig. 21b). This explanation is somewhat unlikely, because it requires the dip of the imbricates to become progressively

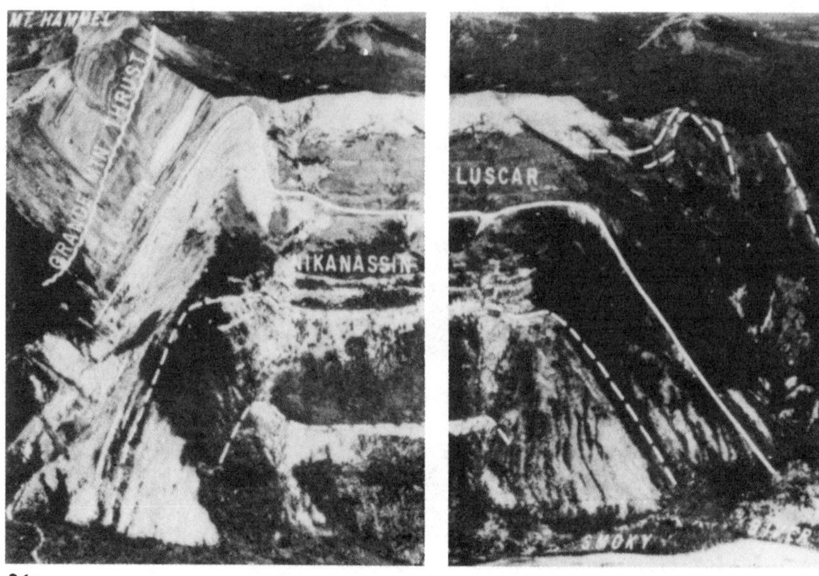

36a

36b

Fig. 36. Partial detachments on a box-fold anticline on the Smoky River, Alberta, and in a syncline at Kinuseo Creek, British Columbia. "Ears" on box-fold shoulders are a common phenomenon which are presumably tied to complementary features in the adjacent syncline of the type illustrated. In these partial detachments the core beds are isoclinally folded with a minimum of thrusting and crumpling.

steeper, rather than flatter as in pressure-box experiments. As pointed out earlier, the sequence of development of the leading-edge imbricates seems to be the inverse of that in the trailing-edge imbricate stacks and in the regional. While this is not impossible, it is a suspicious circumstance. Serious consideration should be given to the possibility that the imbricates form above a subhorizontal detachment in the normal west-to-east progression, and are then translated "en mass" up the lowest and last-formed imbricate. If this were true, then the sequence in the leading- and trailing-edge imbricate stacks and in the regional would all be consistently west (hinterland or upper) to east (foreland or lower).

A similar but not identical mechanism has been used to explain some normal faults as early formed thrusts that have rotated by continued folding above an underlying detachment (Fig. 53b). Balkwill (1969) has applied this mechanism regionally to explain the anomalous symmetry of the Western Ranges. When applied to back-limb faults, which ordinarily have steeper dips away from the anticlinal axis, it implies that the steeper faults were formed first. According to this idea the uppermost back-limb fault would be first, the lowest last, and the back-limb faulting would have gone on simultaneously with folding (Fig. 53b).

En Echelon Folds

EN ECHELON folds may be linked in a zig-zag or an elliptical pattern (Campbell 1958) (Fig 41). The elliptical linkage is restricted to special situations where the folds occur on a significantly tilted plane, such as drag folds on one flank of a major anticline. For *en echelon* folds which together comprise a gross anticlinal structure, the linkage is necessarily zig-zag. Each individual fold within such an anticlinal trend is a culmination, and the linkages usually occur in the intervening depression.

En echelon linkages in the Foothills are of four basic types (Fig. 42). The basic form is the simple zig-zag pattern, with two anticlines of opposed plunges separated by a doubly plunging syncline (Fig. 42b). Being relatively tight, these intervening synclines cannot persist very far upward in the section, and disappear abruptly upward on a local detachment horizon. In beds above the detachment horizon the *en echelon* separation is reflected by a small deflection of the anticlinal axis and a gentle saddle in the plunge (Fig. 42c). A simple variation of this pattern occurs when the synclinal linkage is replaced by a thrust which is a back-limb on one anticline and fore-limb on the other. These faults are usually west-dipping (Fig. 42a) in accord with normal foothills symmetry.

The three patterns which have been discussed are probably part of one basic model, the differences being related to depth of erosion. According to the dominant-member concept, the wave length of a fold is a function of the thickness of the controlling unit, but in simple zig-zag *en echelon* folding (Fig. 42b) there is a progressive decrease in wave length to zero. Since one fold is increasing in wave length as the other decreases, the changes cannot be related to thickness changes in the dominant member, which raises the geometric problems inherent in bending a two-foot bed into a one-foot fold. Another way of looking at the problem is to consider the depth of detachment, which is related to the size and curvature of the fold. Since more section is involved in a broad open fold than in a small tight one, it is apparent that the reduction of wave length in *en echelon* folding indicates that the depth to detachment decreases as the fold dies out. But lines of argument can be reconciled by the development of thrust faults to accommodate the reduced radius of curvature in the *en echelon* parts of the folds. Under these conditions the fault pattern is exposed only where erosion has cut fairly deeply into a fold; at a higher level a simple fold pair appears, and higher yet there is only a slight depression and a deflection of the fold axis (Figs. 42 and 43).

The fourth basic pattern occurs at the plunge end of some major culminations where the *en echelon* folds have the same, rather than opposed, directions of plunge (Figs. 42d). The two folds are separated by a folded west-dipping thrust, with a displacement that increases rapidly down-plunge. The underlying anticline is subsequent in age, and slightly oblique in direction, to the fold in the overlying sheet. The end result is a grossly anticlinal feature which consists of several anticlines stacked one above the other.

TEAR FAULTS

Kinds of Tear Faults

A tear fault is a species of strike-slip fault which terminates both upwards and downwards against movement planes that may be detachments or thrust faults or low-angle normal faults. In folded and/or thrust-filled terrain there are two basic types of tear fault: 1. tears wholly within one thrust sheet or packet of thrust sheets, which may be either (a) transverse (primary or secondary) or (b) oblique to the regional trend; 2. tears that are an integral part of the bounding surface of the deformed sheet.

Tear faults within a deformed sheet permit abrupt changes in the pattern of deformation through differential movement between component parts of the sheet. In primary tears, the amount of shortening on either side is consistent but the mechanisms are different, so that shortening may be provided by folding on one side of the tear and thrusting on the other (Fig. 44a). Such compensated differences in rock-shortening mechanisms demonstrate that the tear fault is an integral part of the structural fabric, which developed in the very early stages of deformation.

Secondary transverse tear faults provide a mechanism for transferring displacement between pairs of existent thrust faults. Such tears transect the intervening thrust sheet, thus permitting adjacent parts of the same thrust sheet to have markedly disparate displacement as shown in Figure 45. Formation of these tear faults is subsequent to that of the thrust planes and they are therefore secondary phenomena.

Two directions of oblique tear-faulting are possible – one with dextral, the other with sinistral offset. In either orientation the displacement nearly always lengthens the fold axes, making it possible to predict the direction of slip from the strike of the tear (Fig. 44b). Both fault directions can occur in the same area, but usually only one set develops. Oblique tear-faults are rather late in the deformational sequence, so that they offset fold axes.

Tear faults within a sheet or packet of sheets do exist in the Canadian Rocky Mountains but they are relatively rare. Most are primary transverse tears, and very few secondary oblique tears have been recognized. This scarcity of oblique tears could be because there is very little of the arcuation of the regional trend into the pronounced salients and re-entrants that, in other orogens, causes longitudinal extension of fold axes and thrust sheets (Fig. 46).

The second type, where the tear fault is an integral part of the boundary of the deformed panel, is not common in the Foothills either. However, those that do exist produce rather perplexing structures, particularly where the tear fault is in the unexposed part of the lower boundary of a thrust sheet. A simple example is shown in Figure 47, where the basal thrust is a bedding-plane fault, first at a lower horizon and then at a higher one. These extensive planar components are connected by three segments where the thrust cuts rapidly up-section, and by two vertical tear faults (Fig. 47a). The combination of these components gives the upper plate a tonguelike area protruding in front of the main body of the thrust sheet. When the upper plate moves forward as a unit, it produces three areas of bed-duplication, separated from one another by areas without structural thickening (Fig. 47b). As geologists mapping Foothills terrain are accustomed to looking for thrust faults, the basic criteria for recognizing faulting has become repetition of section. There is, of course, no repetition on the strike-slip segment of the tear faults, which makes them very hard to recognize in low-dip terrain. It is quite probable that on the usual geological map an area like that in Figure 47b would be represented by three unconnected thrusts. If the level of erosion is higher in the section, then the tear faults may not be exposed at all. The only clue to their presence would be anomalous panels of opposed-fold plunge along the trace of the buried tear fault (Fig. 47c). Such tear-related fold plunge may be recognized by its linearity and frequently by its steepness which can be in the 30 to 40 degree range, substantially steeper than normal in ordinary concentric folding. Zones of abnormal plunge in the Alberta Foothills which may represent such tear faultings are shown in Fig. 46.

The foregoing illustrations have shown tears as vertical faults, which is usual for oblique or transverse tear-faults within a thrust sheet. Movement on these tears is necessarily parallel to the intersection of the tear with the upper and lower boundary planes. The second kind of fault, where the tear is an integral part of the thrust-sheet boundary, is not so simple. Such tears are the paths whereby thrust faults move longitudinally from one zone of bedding-plane thrusting to another. These "steps" are more or less transverse to the strike of the fault. They could be steep or vertical, but there is no compelling mechanical reason why they must be steep. In an earlier discussion of the behaviour of thrusts with respect to bedding,

Fig. 37. Photo a – An anticlinal box fold with a simple isoclinal core at Cascade Mountain, Alberta. Note that part of the complementary syncline is preserved near the ridge crest. The exposed cross-section is somewhat oblique to the fold axis, so the width of the fold is exaggerated.

Photo b – An anticlinal box fold on the Murray River, British Columbia. At the upper contact of the sandstone unit in the southeastern part of the mosaic, the fold is a simple box anticline. In the basal part of the sandstone unit, the complex, crumpled core is exposed at river level. Compare with Figures 31 and 38.

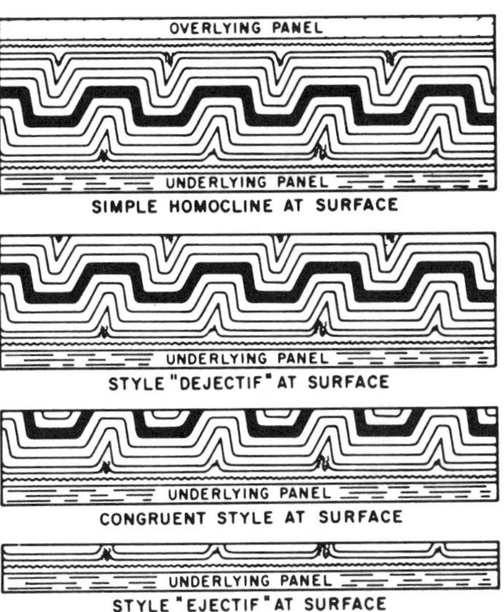

Fig. 38. Changes in the tectonic style as a simple panel of box folds is eroded to various depths.

it was shown that the Lewis thrust at one location cuts up-section along strike for a short distance at an angle of 18 degrees (Fig. 10). This is a kind of tear fault (Gwinn, 1964). When the dip of the tear fault is this low, there is no longer any necessity for the slip to be parallel to the intersection between thrust and tear. Displacement may be oblique to the tear fault. In Figure 48c and e, displacement is obliquely toward the tear fault, producing duplication of beds and an anticlinal structure parallel to the fault. Had the displacement been obliquely away from the tear fault, then there would have been omission of strata and the tear fault would have been reflected by a syncline (Fig. 48d). Tear-fault induced anticlines or synclines can be recognized by their trends, which are distinctly oblique or transverse to the normal structural strike.

NORMAL FAULTS

Kinds of Normal Faults

Normal faulting is an integral but minor part of the latter stages of the movement pattern in the marginal portion of an orogenic belt. Since the thrusting and folding which characterize this regime are contractional phenomena, geometrically devoted to reducing the areal extent of the sedimentary blanket, it is evident that normal faulting, which is an extensional phenomenon with the opposite effect of increasing the areal extent, must necessarily be of limited extent. The larger normal faults can be of two types (Fig. 52): 1. high-angle normal faults cutting both the deformed sedimentary section and the underlying basement; and 2. low-angle listric faults apparently restricted to the sedimentary section.

The high-angle normal faults are the kind usually inferred to bound the ranges in the Basin and Range province of California and Nevada. They also exist in the Wyoming equivalent of the Canadian Rockies, where the germanotype structures of

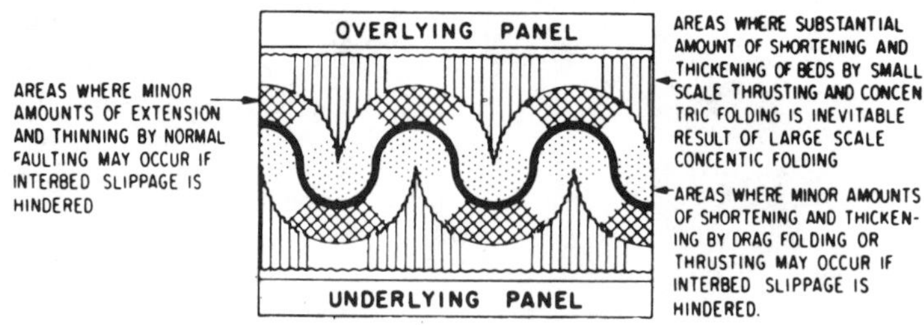

Fig. 39. Distribution of minor normal faults, thrust faults and folds in a concentrically folded panel.

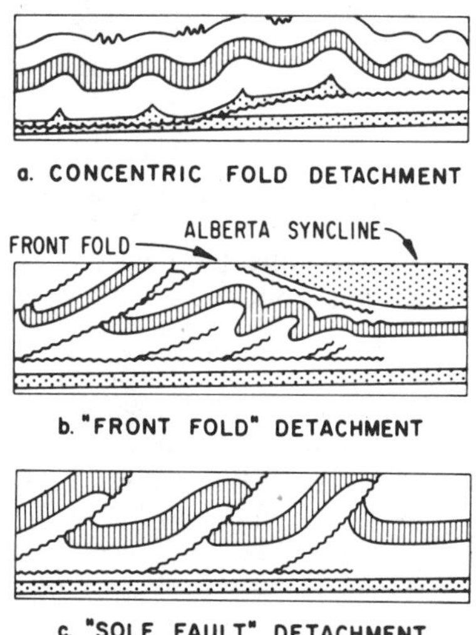

Fig. 40. Typical detachments in the Canadian Rocky Mountains.

the unstable platform impinge upon the alpinotype structures of the Wyoming thrust belt. There the existence, the form and the age of high-angle normal faults is not in contention, although it is not clear whether they are an integral but latter stage in the orogenic cycle that produced the thrust belt, or represent a subsequent episode of tectonic extension that is independent of the orogenic belt in both time and space. No faults of this type have been demonstrated in the Canadian Rocky Mountains.

Late extensional faulting, where the deformed sedimentary rocks are extended by low-angle normal faults, has only recently been suspected (Dahlstrom *et al.*, 1964) and demonstrated (Bally *et al.*, 1966) in the Foothills and Front Ranges. Although the area does not have many normal faults, it is possible that all those that do exist, including the ones in the southern part of the Rocky Mountain Trench, may be of this class. It this were true it would have a pronounced effect on interpretation, because these faults have some unique features.

There are also two kinds of normal faults that can exist within the orogenic belt that are not significant parts of the pattern: 1. pre-existent foreland faults; and 2. minor features related to folds and thrusts.

Deformation in an orogenic belt migrates from the hinterland toward the foreland, and in so doing may engulf structures of the pre-existent foreland structural family. One such characteristic structure is sporadic, basement-involved, steep normal faults related to epeirogenic warping. Less common are surficial "slump"-type glide structures within deltaic depocenters. Since these foreland features are extraneous to the orogen, they are apt to have an apparently random distribution within it. Consequently, such structures are apt to pass unrecognized unless they produce glaring facies changes or affect the nature and distribution of folding and thrust or tear faulting. Although it is apparent that such inherited structures should exist, and although many problems of Foothills geology have been temporarily shelved by using the "pre-existent fault" explanation, solidly documented examples are not available as yet.

Relatively minor normal faults which develop in consequence of the thrust and folding processes are fairly common. Faults with normal displacement can originate as thrusts in an inclined stress field but those in the Canadian Rocky Mountains have generally developed as ordinary thrusts and been subsequently rotated to their present orientation by continued folding and/or thrusting (Fig. 53). Minor adjustments to inadequate inter-bed slippage in concentric folding (Fig. 39) or thrust displacement oriented somewhat obliquely to tear-fault strike (Fig. 48) can also produce real, though usually insignificant, normal faulting.

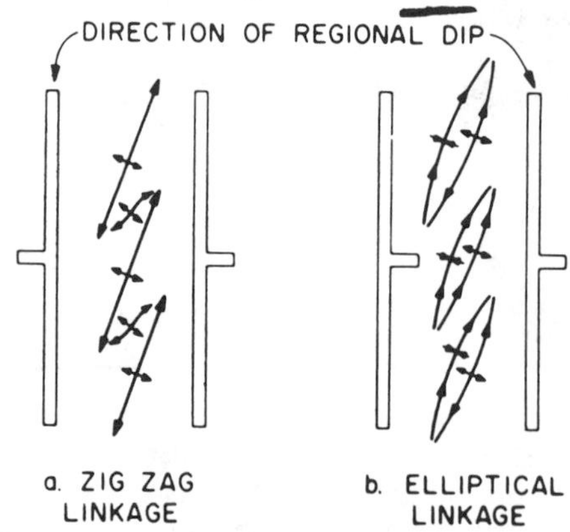

Fig. 41. Two types of *en echelon* fold linkage (after Campbell, 1958 with permission of Economic Geology).

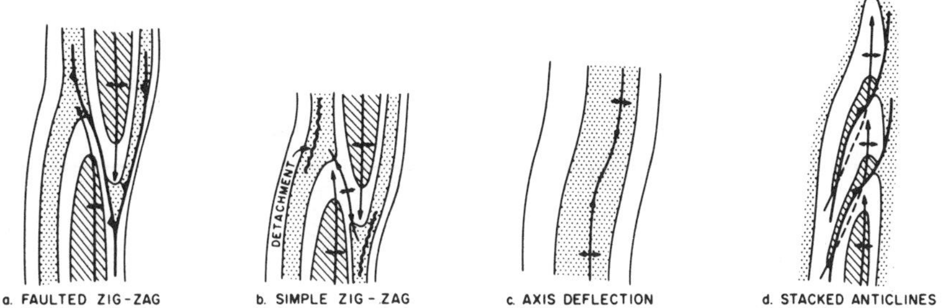

Fig. 42. Basic *en echelon* patterns in the Foothills.

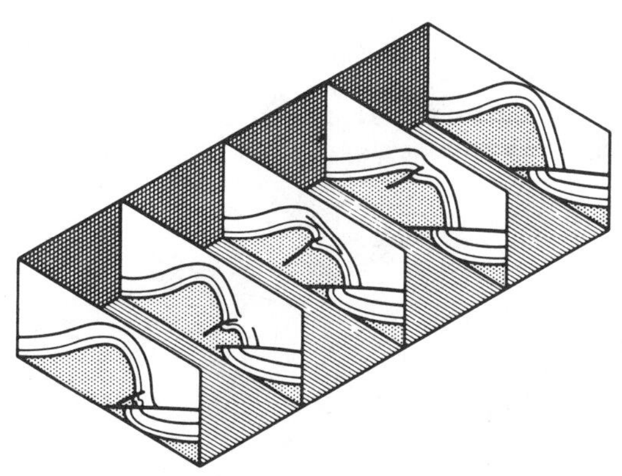

Fig. 43. Block diagram of an *en echelon* fold pair. Depending on the depth of erosion, this model would provide the map patterns in 42 a, b or c.

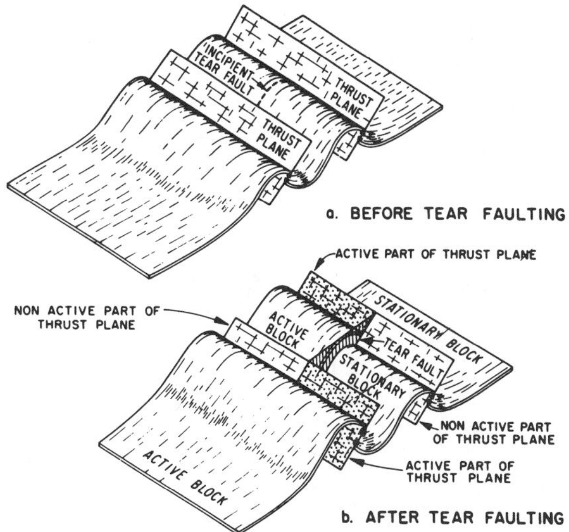

Fig. 45. Secondary transverse tear faults.

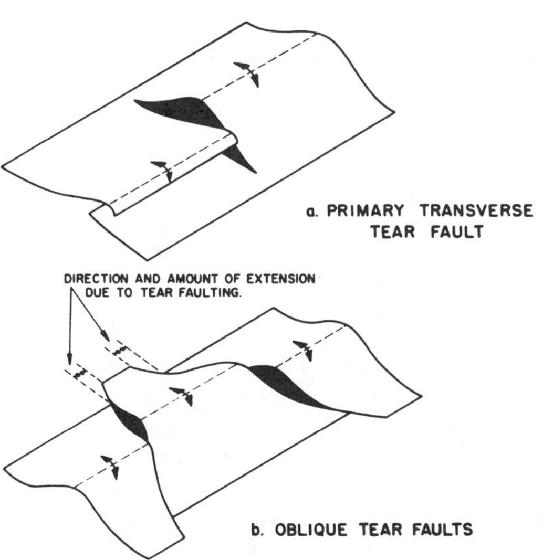

Fig. 44. Transverse and oblique tear faults within a single sheet.

Low-angle Normal Faults

A common practice is to describe any low-angle fault as a thrust, which makes it impossible to discriminate effectively between extensional features that make the sedimentary packet longer and thinner than it was originally, and the contractional features that make the sequence shorter and thicker. In this paper the distinction will be made by using two terms: low-angle normal fault, and thrust fault. Although these structures do have features in common, they are distinctly different phenomena, as shown in Figure 54. The most significant difference is in the possible stratigraphic succession: In thrust terrain the section is either normal or partially duplicated; in low-angle normal-fault terrain the section can be of normal thickness or it can be thinned, perhaps drastically, by omission of units. The second significant difference is in the structures that result from the accommodation of the upper sheet to the listric form of the fault: In thrusts the listric form induces a synclinal bend, with consequent interbed slippage that may cause secondary back-limb or out-of-the-syncline thrusting; the listric form of the low-angle normal fault, on the other hand, produces a "roll-over," which is a pronounced

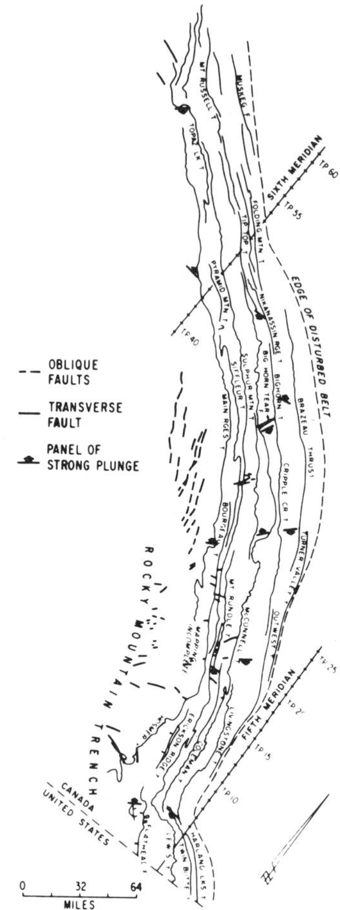

Fig. 46. Transverse and oblique structures in the eastern part of the Canadian Rocky Mountains.

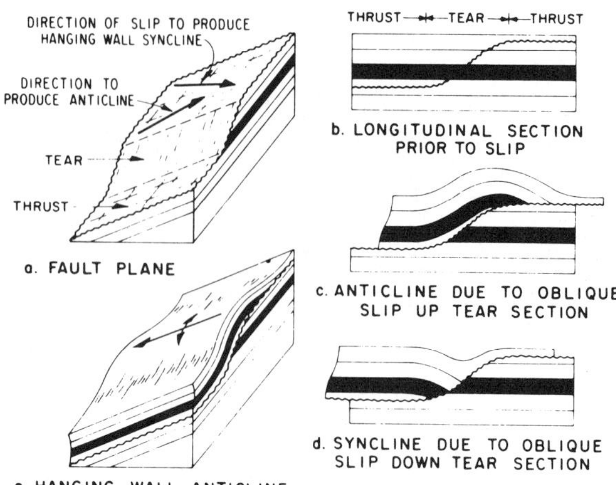

Fig. 48. Structures related to gently dipping tear faults with oblique slip.

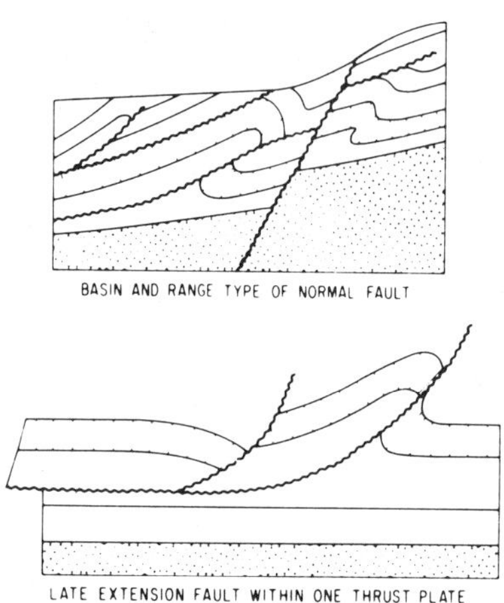

Fig. 49. Major types of late normal faults.

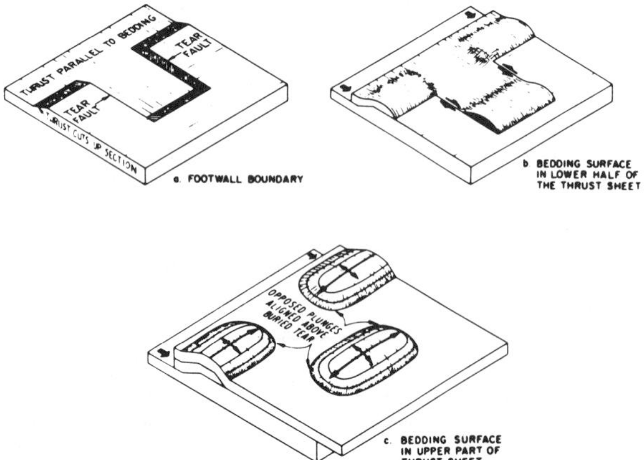

Fig. 47. Tear faults as an integral part of a thrust-sheet boundary.

panel of dip toward the fault plane (Hamblin 1965). A rollover anticline is a species of fold but is nevertheless an extensional and not a contractional feature. It is also a two-stage structure, in that the regional dip on the back limb may have been in existence for a long time before the rollover that produced the fore limb took palce.

Low-angle normal faults with large horizontal displacements and substantial omissions of section have been described by numerous authors from Nevada, western Utah and eastern California. Because of the mechanical difficulties inherent in major sheets like the ones diagrammed in Figure 51, it has been common practice to explain these low-angle normal faults as locally derived land-slide sheets, rather than to consider them as a fundamental part of the tectonic pattern. However, Hunt and Mabey (1966) (Fig. 52) have described a major folded imbricate low-angle normal fault in Death Valley which has many of the characteristics of Fig. 51c.

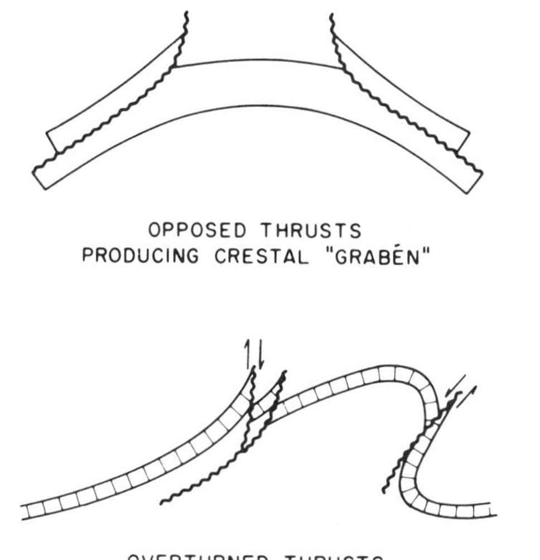

Fig. 50. Pseudo-normal faults in thrust-fault terrain.

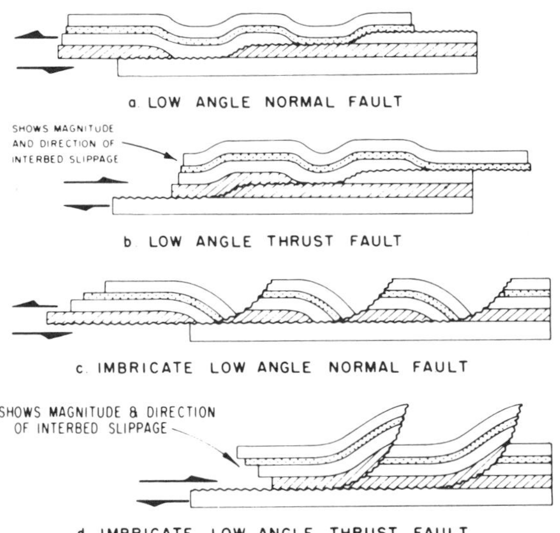

Fig. 51. Comparison of apparent stratigraphic sequences and structures produced by low-angle normal and thrust faults.

Low-angle normal faults in the Canadian Rocky Mountains developed in rocks already cut by thrust faults that provided ready made subhorizontal movement planes. If a low-angle normal fault follows one of these pre-existent thrusts, then the fault plane becomes the locus of two directions of movement – first, thrust movement, wherein the upper plate moves relatively eastward, and subsequently normal fault movement, wherein the upper plate moves relatively westward. If the faults were near the surface, the resulting rollover into the normal fault would produce a tectonic depression, which would be filled with fluvial debris from neighboring areas to provide a species of "growth fault" comparable in many respects to those of the Gulf Coast (Fig. 53).

The compound nature of the fault plane produces some unusual relationships. The footwall of the fault (A^1 B^1 D in Fig. 53c) is unaltered, but the upper plate is extended by the normal faulting to produce a fault gap (A^1 B^1 at base of unit 3). In this fault gap younger beds from the upper plate are brought into contact with the fault plane. If the compound nature of the fault plane is appreciated these age relationships are quite comprehensible but, if the fault is interpreted as an ordinary thrust fault, it appears to cut first up-section and then abruptly down-section in the direction of tectonic transport. Since this is contrary to the usual pattern of thrust behaviour, the interpretation would be judged invalid and an incorrect alternative would be evolved.

RELATIONSHIP BETWEEN STRUCTURE AND STRATIGRAPHY

The local effect of stratigraphy upon simple folding and faulting has been discussed above. The location of incompetent horizons determines the paths of thrust faults in cutting through a section. Where strain was by folding rather than thrusting, the location of the incompetent units determines where detachments occur, and the thickness of the competent units determines whether the folds between detachments will be large or small.

On a somewhat larger scale, abrupt changes in lithology can produce dramatic changes in structural style. In the Main Ranges of the Rockies there is an abrupt change in the Cambrian from a 10,000-ft succession of competent massive carbonates to an equivalent thickness of incompetent thin-bedded shaly carbonates (Fig. 1). The pronounced disparity in competency permitted a substantially different response to folding and faulting (Cook, 1967). In the shaly rocks to the west, folds are smaller and complex whereas, in the massive carbonates to the east, they are broad and open (Fig. 54). The facies boundary is the locus of faulting, both thrust and normal. There is also a change in fundamental style of folding. In the competent carbonates folding is concentric, but at the facies boundary cleavage appears and is thereafter ubiquitous to the west. It appears that units in the shaly facies are thicker than the stratigraphically equivalent units in the carbonate facies, which in view of normal compactibility ratios, is inherently improbable. The apparent thickening is probably a secondary feature related to the cleavage. The local distribution of cleavage is controlled by the facies boundary. To the east, in the Foothills, Front Ranges and Eastern Main Ranges, there is no cleavage; westward, however, there is an increase in the degree of strain, with cleavage and similar folding appearing first in the incompetent rocks and eventually in virtually all lithologies. At the facies boundary, the incompetent rocks are cleaved and the competent rocks are not, although pressure, temperature and stress conditions on the two facies must have been practically identical. Since other factors were comparable, the difference in response must have been due to lithology.

The broad regional effects of stratigraphy upon structure are necessarily a large-scale phenomenon, where the effects are not as clearly defined and the conclusions more subject to predilections of the observers. Local effects can often be demonstrated beyond reasonable doubt but regional effects seldom can. Nevertheless, most geologists would agree with the basic observation that the size of a structure depends on the thickness of section that is deformed. Folds in a thick section are necessarily larger than those involving a thin section (Goquel, 1962). From this it follows that deformation of a relatively thick section can produce a structural culumination. This conclusion can also be derived from the standard assumption that the amount of shortening along an orogenic belt does not change abruptly, so that the ratio of original deposited bed length to final horizontal deformed length is roughly the same for contiguous areas. If shortening is consistent, then any significant longitudinal stratigraphic thickening must be translated into a longitudinal structural thickening; that is, into a structural culmination. This is not the only, nor even the commonest way of producing a structural culmination. As previously discussed, tear faults and bedding-plane thrusts permit the transfer of thrust displacement without structural thickening, which makes it possible for culminations and depressions to develop in areas of constant shortening and contant stratigraphic thickness. There is also the possibility, as yet unrecognized in the Canadian Rocky Mountains, of differential post-orogenic warping. Nevertheless, a possible explanation for a regional culmination is a stratigraphically thick section.

In transverse sections across the whole of the deformed belt, it is necessary that the amount of shortening in the upper units be the same as that in the lower. The Laramide deformation must have shortened the lower Paleozoics as much as it shortened the Mesozoics, which, taken in conjunction with the fact that virually all the units thicken westward, makes it geometrically necessary that the exposed rocks become progressively older toward the west. When this trend reverses and

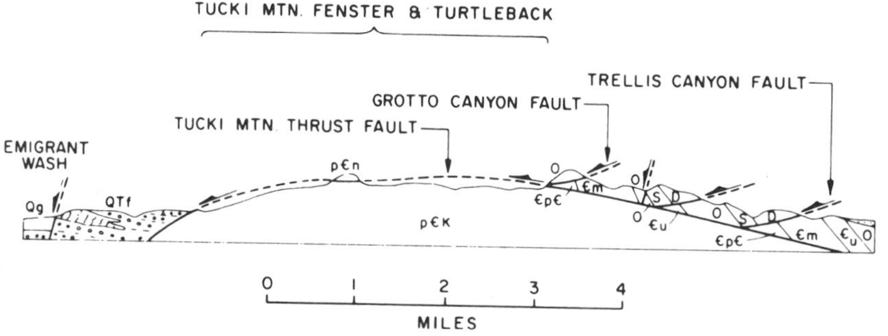

Fig. 52. The Tucki Mountain fault, a folded, imbricate, low-angle normal fault in Death Valley, California. The Symbols refer to age (e.g., S is Silurian). (From Hunt and Mabey, 1966.)

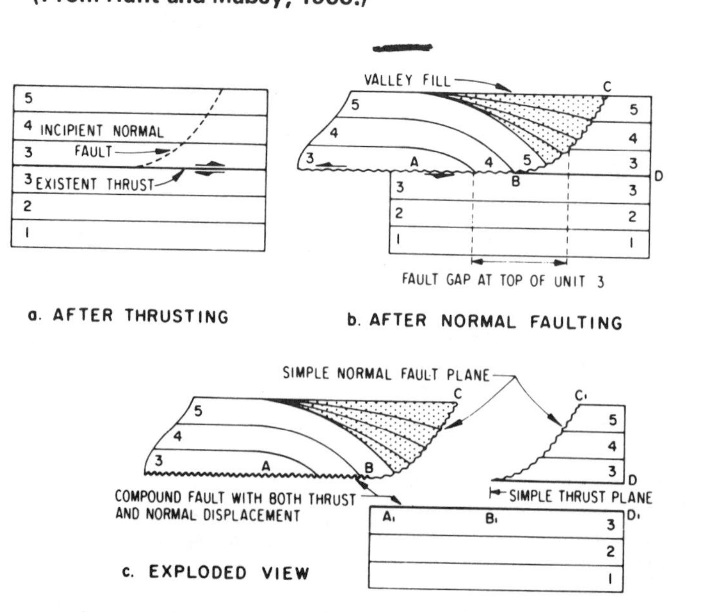

Fig. 53. Low-angle normal faults in thrust-faulted terrain.

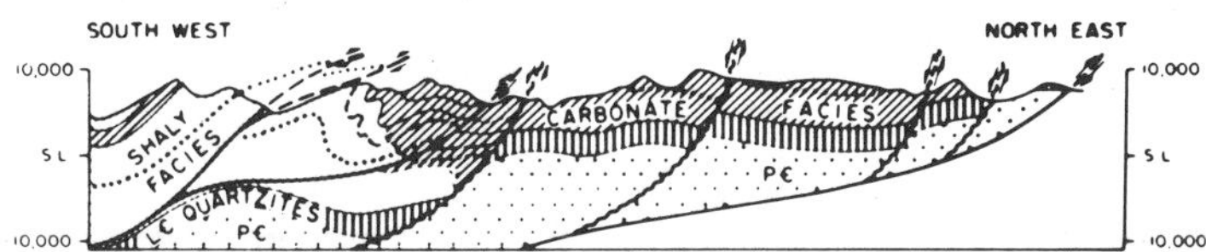

Fig. 54. Structural cross-section to show the change in structural style resulting from a major facies change from massive to shaly carbonates (after Cook, 1967 with his permission).

substantial areas of younger rocks are exposed to the west, then one should suspect that the section has thinned or that regional displacement has taken place parallel to bedding.

CONCLUSIONS

Specific structural environments contain a limited suite of structures. All the classes of structure need not be represented in a single area. In each structural class, as in other groupings of natural features, there are deviations from the norm in shape and particularly in size. The size range may span several orders of magnitude. The spectrum of tectonic styles which can develop from variations in size, form and structural combinations often obscures the underlying simplicities, particularly in the early stages of investigation.

As work proceeds and individual structures are understood it becomes much easier to separate what is fundamental from what is incidental. The first step is to recognize those components of the pattern that are recurrent, and to synthesize the basic characteristics of the individual components into normative models. In the marginal belt of the Canadian Rockies this means recognizing thrust faults, concentric folds, tear faults and late normal faults as the basic forms. Having identified the components, the second step is to devise geometric generalizations that describe the observed behaviour of each structural class. Thrust faults for instance, parallel the bedding in incompetent rocks and, in the direction of transport, cut up-section rather abruptly through the competent rocks. The third step is to determine the relationships between the components – for example, between concentric folds and detachments.

At this stage it becomes practical to use the established models as the basis for extrapolation into poorly known areas. One of the advantages of this method of analysis is that it depends primarily on observations and geometry and is relatively free of genetic implications. The structural classification, the structural models and the rules of geometric behavior discussed in the foregoing pages are not contingent on a causative force – they are the same whether they were produced by underthrust, gravity glide or gremlins. In areas where basic data are abundant, as they are in the Canadian Rocky Mountains, it is often possible to answer the "what is it?" question on purely local geometric grounds. In areas of lesser control, where models cannot be developed by local observation, it is necessary to go to foreign examples of the same structural province to obtain the needed models.

Answering the "what is it?" question is not an end in itself. It is only the first step toward solving the academic's "why is it?" problem, or the explorationist's "where is it?" problem. For the academic it is particularly important that the "what" problem be solved as far as possible without genetic assumptions. Premature introduction of genetic considerations ordinarily begs the question by assuming in advance what one is attempting to determine. The explorationist is commonly working beyond his data to the point where he is more concerned with reasonable risks (where is it, probably?) than with absolute truths. Consequently, he is obliged to extend his limited data with genetic assumptions. This is a right and proper thing to do, but it must be based on as solid a foundation as possible. The geometric analysis should be carried to its practical limit before a genetic extension is erected upon it. For both the academic and the explorationist, the geometric solution is only the first step in good geological work but, being the first, it is the foundation and therefore important.

REFERENCES

Armstrong, F.C. and Oriel, S.S., 1965, Tectonic development of Idaho-Wyoming thrust belt: Bull. Am. Assoc. Petroleum Geologists, v. 49, no. 11, p. 1847-1866.

Balkwill, H.R., 1969, Structural analysis of the Western Ranges, Rocky Mountains near Golden, British Columbia: Ph.D. thesis (unpubl.), Univ. of Texas at Austin.

Bally, A.W., Gordy, P.L. and Stewart, G.A., 1966, Structure, seismic data,and orogenic evolution of southern Canadian Rocky Mountains: Bull. Can. Petroleum Geology, v. 14, no. 3, p. 337-381.

Barnes, W.C., 1963, Geology of the northeast Whitefish Range, northwest Montana: Ph.D. thesis (unpubl.), Princeton Univ.

Beach, H.H., 1943, Moose Mountain and Morely map-areas, Alberta: Geol. Surv. Canada, Mem. 236, 74 p.

Biot, M.A., 1961, Theory of folding of stratified viscoelastic media and its implications in tectonics and orogenesis: Geol. Soc. America Bull. v. 72, no. 11, p. 1595-1620.

Bucher, W.H., 1933, The deformation of the earth's crust: Princeton, N.J., Princeton Univ. Press, 518 p.

Campbell, J.D., 1958, En echelon folding: Econ. Geology, v. 53, no. 6, p. 448-472.

Childers, M.O., 1964, Structure around Glacier National Park, Montana: Bull. Can. Petroleum Geology, v. 12, Field Conference Guide Book issue (August 1964), p. 378-382.

Cloos, E., 1964, Wedging, bedding plane slips and gravity tectonics in the Appalachians, in "Tectonics of the Southern Appalachians:" Virginia Polytechnic Inst., Dept. of Geol. Sci., Mem. 1, p. 63-70.

Cook, D.G., 1967, Structural style influenced by a Cambrian regional facies change in the Mount Stephen-Mount Dennis area, Alberta-British Columbia: Ph.D. thesis, (unpubl.), Queen's Univ.

Currie, J.B., Patnode, H.W. and Trump, R.P., 1962, Development of folds in sedimentary strata: Geol. Soc. America Bull., v. 73, no. 6, p. 655-674.

Dahlstrom, C.D.A., 1960, Concentric folding, in 2nd Ann. Field Trip Guide Book: Edmonton Geol. Soc. p. 82-84.

_____, Daniels, R.E. and Henderson, G.G.L., 1962, The Lewis thrust at Fording Mountain, British Columbia: Alberta Soc. Petroleum Geologists Jour., V. 10, no. 7, p. 373-395.

_____, 1969, The upper detachment in concentric folding: Bull. Can. Petroleum Geology, v. 17, no. 3, p. 326-346.

de Sitter, L.U., 1956, Structural geology: New York, McGraw-Hill Book Co., 552 p.

Donath, F.A. and Parker, R.B., 1964, Folds and folding: Geol. Soc. America Bull., v. 75, no. 1, p. 45-62.

Douglas, R.J.W., 1950, Callum Creek, Langford Creek, and Gap map areas, Alberta: Geol. Surv. Canada, Mem. 255, 124 p.

_____, 1952, Preliminary map, Waterton, Alberta: Geol. Surv. Canada, p. 52-10.

Erdman, O.A., 1950, Alexo and Saunders map-areas, Alberta: Geol. Surv. Canada, Mem. 254, 100 p.

Fitzgerald, E.L., 1968, Structure of British Columbia foothills, Canada: Bull. Am. Assoc. Petroleum Geologists, v. 52, no. 4, p. 641-664.

Fox, F.G., 1959, Structures and accumulation of hydrocarbons in southern Alberta foothills, Alberta, Canada: Bull. Am. Assoc. Petroleum Geologists, v. 43, no. 5, p. 992-1025.

_____, 1969, Some principles governing interpretation of structures in the Rocky Mountain orogenic belt: Geol. Soc. London, Spec. Vol, "Time and place in orogeny."

Gallup, W.B., 1951, Geology of Turner Valley oil and gas field, Alberta, Canada: Bull. Am. Assoc. Petroleum Geologists, v. 35, no. 4, p. 797-821.

Gilluly, J., 1960, A folded thrust in Nevada – inferences as to time relations between folding and faulting: Am. Jour. Sci. v. 258-A (Bradley Volume), p. 68-79.

Goguel, J., 1962, Tectonics (transl. 1952 edn.): San Francisco, Freeman and Company, 384 p.

Gussow, W.C., 1955, Oil and gas accumulations on the Sweetgrass arch, in Guide Book, 6th Ann. Conf. Billings Geol. Soc., p. 220-224.

Gwinn, V.E., 1964, Thin skinned tectonics in the Plateau and northwestern Valley and Ridge Provinces of the central Appalachians: Geol. Soc. America Bull., v. 75, no. 9, p. 863-900.

Hage, C.O., 1946, Dyson Creek: Geol. Surv. Canada, Map 827A.

Hake, B.F., Addison, C.C. and Willis, R., 1942, Folded thrust faults in foothills of Alberta: Geol. Soc. America Bull., v. 53, no. 2, p. 291-334.

Hamblin, W.K., 1965, Origin of "reverse drag" on the downthrown side of normal faults: Geol. Soc. America Bull., v. 76, no. 10, p. 1145-1164.

Henderson, G.G.L. and Dahlstrom, C.D.A., 1959, First-order nappe in Canadian Rockies: Bull. Am. Assoc. Petroleum Geologists, v. 43, no. 3, p. 641-653.

Hunt, C.B. and Mabey, D.R., 1966, Stratigraphy and structure, Death Valley, California: U.S. Geol. Surv. Prof. Paper 494-A.

Irish, E.J.W., 1965, Geology of the Rocky Mountain Foothills, Alberta (between latitudes 53° 15' and 54° 15'): Geol. Surv. Canada, Mem. 334, 241 p.

Jones, P.B., 1964, Structures of the Howell Creek area: Bull. Can. Petroleum Geology, v. 12, Field Conf. Guide Book Issue (August 1964), p. 350-362.

_____, 1969, The Tertiary Kishenehn Formation, British Columbia: Bull. Can. Petroleum Geology, v. 17, no. 2, p. 234-246.

_____, 1969, Tectonic windows in the Lewis thrust, southeastern British Columbia: Bull. Can. Petroleum Geology, v. 17, no. 2, p. 247-251.

_____, 1970, Folded faults and sequence of thrusting in Alberta foothills; Bull. Am. Assoc. Petroleum Geologists (in press).

Johns, W.M., 1959, Geologic investigation in the Kootenai-Flathead area, northwest Montana, 1. Western Lincoln county: Montana Bur. Mines Geol., Bull. 12.

_____, 1960, Geologic investigation in the Kootenai-Flathead area, northwest Montana, 2. Southeastern Lincoln County: Montana Bur. Mines Geol., Bull. 17.

_____, 1961, Geologic investigation in the Kootenai-Flathead area, northwest Montana, 3. Northern Lincoln County: Montana Bur. Mines Geol., Bull. 23.

_____, 1962, Progress Report 4, Geologic investigations in the Kootenai-Flathead area, northwest Montana, southwestern Flathead county: Montana Bur. Mines Geol., Bull. 29.

_____, Smith, A.G., Barnes, W.C., Gilmour, E.H. and Page, W.D., 1963, Progress Report 5, Geologic investigation in the Kootenai-Flathead area, northwest Montana, western Flathead county and part of Lincoln county: Montana Bur. Mines Geol., Bull. 36.

_____, 1964, Progress Report on geologic investigation in the Kootenai-Flathead area, northwest Montana, 6. Southeastern Flathead county and northern Lake county: Montana Bur. Mines Geol., Bull. 42.

Keating, L.F., 1966, Exploration in the Canadian Rockies and foothills: Can. Jour. Earth Sci., v. 3, no. 5, p. 713-723.

Leech, G.B., 1959, The southern part of the Rocky Mountain Trench: Can. Inst. Min. Metall. Bull., v. 52, no. 565, p. 327-333.

_____, 1960, Fernie, west half, Kootenay District, British Columbia: Geol. Surv. Canada, Map 11-1960.

Mackin, J.H., 1950, The down structure method of viewing geologic maps: Jour. Geology, v. 58, no. 1, p. 55-72.

McCrossan, R.G. and Glaister, R.P. (editors), 1964, Geological history of western Canada: Calgary, Alberta Soc. Petroleum Geologists, 232 p.

Mountjoy, E.W., 1959, Miette, Alberta: Geol. Surv. Canada, Map 40-1959.

_____, 1960, Structure and stratigraphy of the Miette and adjacent area, eastern Jasper National Park, Alberta: Ph.D. dissertation (unpub.), Univ. Toronto.

North, F.K. and Henderson, G.G.L., 1954, Summary of the geology of the southern Rocky Mountains of Canada, in 4th Ann. Field Conf. Guide Book: Alberta Soc. Petroleum Geologists, p. 15-81.

Oswald, D.H., 1964, The Howell Creek structure: Bull. Can. Petroleum Geology, v. 12, Field Conf. Guide Book Issue (August 1964), p. 363-377.

Price, R.A., 1962, Fernie map-area, east half, Alberta and British Columbia: Geol. Surv. Canada, Paper 61-24.

_____, 1964, Flexual-slip folds in the Rocky Mountains, southern Alberta and British Columbia: Geol. Surv. Canada, Reprint 78, 16 p.

_____, 1965, Flathead map-area, British Columbia and Alberta: Geol. Surv. Canada, Mem. 336, 221 p.

_____ and Mountjoy, E.W., 1970, Geologic structure of the Canadian Rocky Mountains between Bow and Athabaska Rivers, a progress report: Geo., Assoc. Canada, Spec. Paper no. 6, p. 7-26.

Ressor, J.E., 1957, The Proterozoic of the Cordillera in southeastern Alberta, in Gill, J.E., ed., Proterozoic in Canada: Royal Soc. Canada, Spec. Pub. 2.

_____, 1957, Lardeau (east half), Kootenay District, British Columbia: Geol. Surv. Canada, Map 12-1957.

_____, 1958, Dewar Creek map-area, with special emphasis on the White Creek batholith, British Columbia: Geol. Surv. Canada, Mem. 292, 78 p.

Rice, H.M.A., 1937, Cranbrook map-area, British Columbia: Geol. Survey Canada, Mem. 207, 67 p.

Rich, J.L., 1934, Mechanics of low-angle overthrust faulting illustrated by Cumberland thrust block, Virginia, Kentucky and Tennessee: Bull. Am. Assoc. Petroleum Geologists, v. 18, no. 12, p. 1584-1596.

Ross, C.P., 1959, Geology of Glacier National Park and the Flathead Region: U.S. Geol. Surv., Prof. Paper 296.

Scott, J.C., 1951, Folded Faults in Rocky Mountain foothills of Alberta, Canada: Bull. Am. Assoc. Petroleum Geologists, v. 35, no. 11, p. 2316-2347.

Shaw, E.W., 1963, Canadian Rockies – orientation in time and space in Childs, O.E., ed., Backbone of the Americas: Tulsa, Am. Assoc. Petroleum Geologists, Mem. 2, p. 231-242.

Stockwell, C.H., 1950, The use of plunge in the construction of cross sections of folds: Proc. Geol. Assoc. Canada, v. 3, no. 2, p. 97-121.

Thompson, T.L., 1962, Origin of the Rocky Mountain Trench in southeastern British Columbia by Cenozoic block faulting: Alberta Soc. Petroleum Geologists Jour., v. 10, no. 7, p. 408-427.

Verrall, P., 1968, Observations on geological structures between the Bow and North Saskatchewan Rivers, in 16th Ann. Field Conf. Guide Book: Alberta Soc. Petroleum Geologists, p. 107-118 and map.

Wanless, R.K., Loveridge, W.D. and Mursky, G., 1968, A geochronological study of the White Creek batholith, southeastern British Columbia: Can. Jour. Earth Sci., v. 15, no. 3, pt. 1, p. 375-386.

Wheeler, J.O., 1966, Eastern tectonic belt of western Cordillera in British Columbia, in A symposium on the tectonic history and mineral deposits of the western Cordillera in British Columbia and neighbouring parts of the United States: Can. Inst. Min. Metall., Spec. Vol. no. 8, p. 24-45.

_____, 1970, Summary and discussion in Structure of the southern Canadian Cordillera: Geol. Soc. Canada, Spec. Paper no. 6, p. 155-166.

White, W.H., 1959, Cordilleran tectonics in British Columbia: Bull. Am. Assoc. Petroleum Geologists, v. 43, no. 1, p. 60-100.

_____, 1966, Summary of tectonic history, in A symposium on the tectonic history and mineral deposits of the western Cordillera in British Columbia and neighbouring parts of the United States: Can. Inst. Min. Metall., Spec. Vol. no. 8, pp. 185-190.

Wilson, C.W. Jr. and Stearns, R.G., 1958, Structure of the Cumberland Plateau, Tennessee: Geol. Soc. America Bull., v. 69, no. 10, p. 1283-1296.

The American Association of Petroleum Geologists Bulletin
V. 66, No. 9 (SEPTEMBER 1982), P. 1196-1230, 34 Figs., 2 Tables

Thrust Systems[1]

STEVEN E. BOYER[2] and DAVID ELLIOTT[3]

ABSTRACT

A general geometric framework underlies the structure, evolution, and mechanical processes associated with thrust faulting. The main purpose of this paper is to review and extend this geometric framework. A certain family of lines must exist where thrust surfaces join along branch lines or end at tip lines. Starting from a description of these lines and individual thrust faults, we examine how they join into thrust systems, either as imbricate fans or duplexes. These thrust systems have distinctive map patterns commonly observed near culminations and windows. Many of the culminations have origins tied in with a particular thrust system. The order in which the fault slices form has a marked effect on the geometry of the thrust system. These systems must be identified to understand the provenance of the synorogenic sediments.

Part of a thrust belt may be dominated by one particularly large thrust sheet. In front and beneath these dominant sheets, there is a characteristic sequence of thrust systems with a regular pattern to the involvement of basement.

This overall geometric framework provides new insight into some classic areas, illustrated by a balanced cross section through the Mountain City and Grandfather Mountain windows, in the southern Appalachians and another from the Jura to the Pennines (in the western Alps).

INTRODUCTION

For almost any work in thrust belts, it is essential to establish the three-dimensional relations between faults. These geometric relations between thrusts arise, for example, in the following typical questions. Does my interpretation change if this thrust fault joins that one? Is it possible to establish the time sequence of faulting from the geometry? Is the thrust pattern associated in some way with the origin of my culmination? What does this map pattern of thrusts imply for the cross section?

In this paper we try to draw together part of the geometric principles that we find useful in deciphering three-dimensional thrust structures. This framework has reasonably wide application, and our examples come from the North American Cordillera, the Appalachians, the Caledonides, and the Alps. Although some of these geometric concepts are fairly old, they have seldom been reviewed; we have tried to establish their origin and say something about how the ideas evolved.

There are two general categories of tools useful in thrust belt analysis. The first is balanced cross sections, or cross sections that are both restorable to the undeformed state and conform to certain specified standards of admissibility (Dahlstrom, 1969; Elliott, in prep.). The same set of data may have several different solutions, all of which allow restoration and are otherwise admissible.

The second main category, and the subject of this paper, is concerned with the interconnections or relationships among faults. We shall see that the solutions of thrust belt problems fall into certain categories or types. Choice of the correct type of solution is commonly the most important single step in the interpretation. When we specify how the various parts connect to each other we may refer to the "logic" of the fault network, in the same way we speak of the logic of an electronic device. We will use a building-block approach, starting with the geometric elements and a very few thrusts, and then look at some characteristic patterns of larger numbers of thrusts.

GEOMETRIC ELEMENTS

A thrust sheet is a volume of rock bound below by a thrust fault. A useful convention is to name the thrust sheet after this underlying or leading thrust fault. This name usually continues to the trailing thrust, which joins the leading thrust along a branch line (Fig. 1). A sheet may have a distinctive stratigraphy, state of strain, or metamorphic grade, and on this basis individual sheets are often correlated long distances. There is a long tradition in the European Alps of focusing attention on the thrust sheet volumes (nappes), and we shall discuss this later in the paper.

The intersection of a thrust surface with a stratigraphic horizon is a cutoff line (Douglass, 1958, p. 132). A cutoff line could also be called an "edge" because it indicates the intersection of two surfaces. Dahlstrom's (1970, p. 352) terms "leading edge" and "trailing edge" thrust surfaces are in this sense self-contradictory, so we have shortened them to "leading and trailing thrust surfaces." J. K. Arbenz remarked (1981, personal commun.), "In

[1]Manuscript received, June 9, 1981; accepted, March 19, 1982.

[2]Sohio Petroleum Co., 633 17th St., Suite 2200, Denver, Colorado 80202.

[3]Johns Hopkins University, Department of Earth and Planetary Sciences, Baltimore, Maryland 21218. [*Editor's Note:* David Elliott passed away August 30, 1982, in Zurich, Switzerland.]

This paper is based partly on a Ph. D. thesis (Boyer 1976, 1978) and partly on a series of lectures presented at Johns Hopkins Univ., Calgary, Alberta; and Lausanne, Switzerland (Elliott 1977, 1978). This work was supported by the National Science Foundation, Earth Sciences Section, Grants 7723209 and 7926118. Field work by Boyer was also aided by The Geological Society of America, Sigma Xi, and The Johns Hopkins University Balk Fund. Elliott acknowledges the Donors of the Petroleum Research Fund, administered by the American Chemical Society, for partial support of this research. For their excellent assistance we are grateful to our reviewers, J. K. Arbenz, R. Groshong, and the students of structural geology at Johns Hopkins.

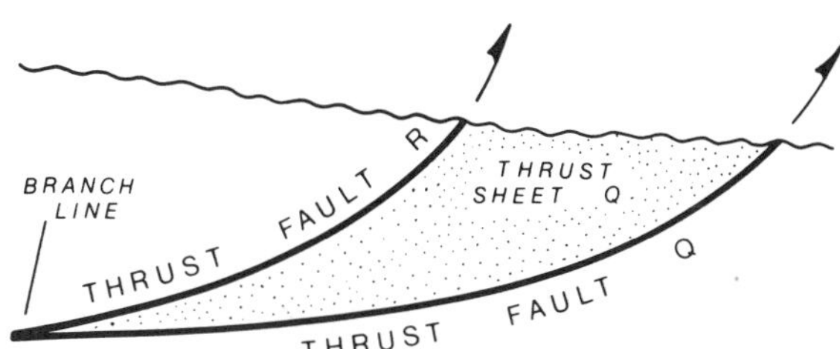

FIG. 1—Cross section through thrust sheet (Q), which is volume of rock above leading fault (Q) and below trailing fault (R).

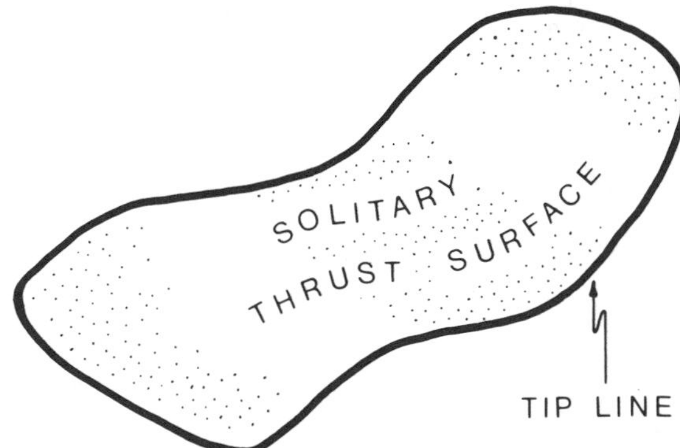

FIG. 2—Three-dimensional view of solitary thrust surface. This is buried, or blind, and its perimeter is everywhere a tip line.

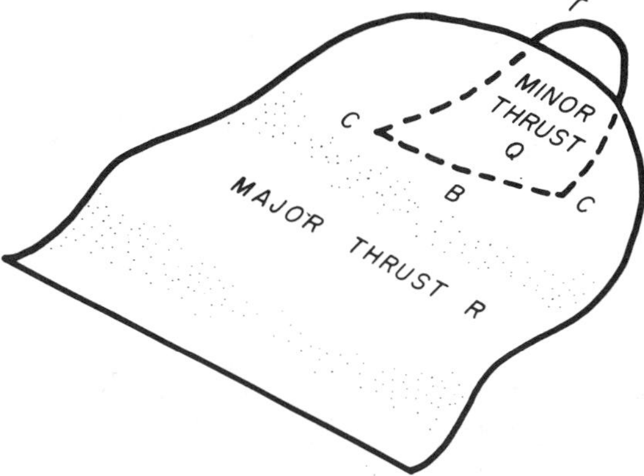

FIG. 3—Minor thrust surface (Q) branches off major thrust (R). Both thrusts meeting along branch line (B). Blind minor thrust (Q) has tip line (T) which meets branch line (B) at two corners (C).

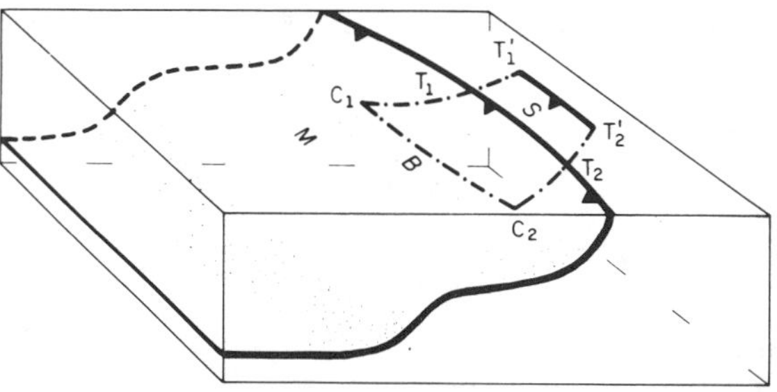

FIG. 4—Erosion cuts branching pair of thrust surfaces which might have once resembled Figure 3. Map pattern is main thrust (M) with isolated splay (S), which has one branch (B), two tip lines (T_1, T_2) that outcrop (T_1, T_2) at termination on map of fault trace.

industry usage, the term 'leading edge' refers to the line of intersection of a thrust surface with a specific stratigraphic horizon of that particular thrust sheet (e.g., the leading edge of the Madison). In fault closures the leading edge becomes the line along which the stratigraphic top seal passes to the fault seal."

So far we have concentrated on blind thrusts,[4] or those in which the tip line does not reach the ground surface (Fig. 2). These thrusts are particularly common near the frontal margin of the thrust belt. Many thrusts start blind and later turn upward to meet the synorogenic erosion surface.

Erosion through a tip or branch line produces the tip or branch points shown on maps. These are particularly important because near these special points we may observe many of the physical processes that produce thrusts (Elliott, 1976, p. 299). There are several examples illustrating the use of branch lines in the Moine thrust belt in Elliott and Johnson (1980, Fig. 17).

A splay rejoins the main fault once. Splays can crop out at the synorogenic erosion surface or arise by later erosion through an originally blind branching thrust (see Fig. 3). We can have an isolated splay (Fig. 4), where erosion cuts the tip line into two, and a diverging splay where erosion cuts the branch and tip lines once (Fig. 5). Maps show that on approaching their lateral termination, major thrusts often turn into a network of diverging splays. In a rejoining splay the tip line is fully removed and the map surface cuts the branch line twice (Fig. 6). Slightly more complex is a connecting splay, where the branch line is strung together along two different fault intersections (Fig. 7).

It is possible for an individual thrust surface to be completely surrounded by branch lines (Fig. 8), because a pair of fault surfaces may diverge after branching and then, farther up, converge to meet again at the branch line. Usually, but not necessarily, the thrust rejoins along strike as well as updip, so that the tip line is eliminated. A horse is a pod of rock completely bound by two or more such fault surfaces. This is an old and useful term (Dennis, 1967, p. 89) which we here extend beyond its original usage for pods bound by normal faults. Horses, unlike the other kinds of branching thrusts, are unlikely to meet a synorogenic erosion surface. Horses can be cut from either the hanging wall or the footwall of major thrusts and may consist entirely of inverted rocks (Fig. 9). They frequently decorate the edges of major thrusts and are particularly helpful in the field to identify thrusts that put shale upon shale. They may also provide stratigraphic information from beneath a major thrust, infor-

[4]Blind thrusts were used and described in Calgary lectures in 1977 by Elliott. We feel earlier usage exists; but were unable to locate the references.

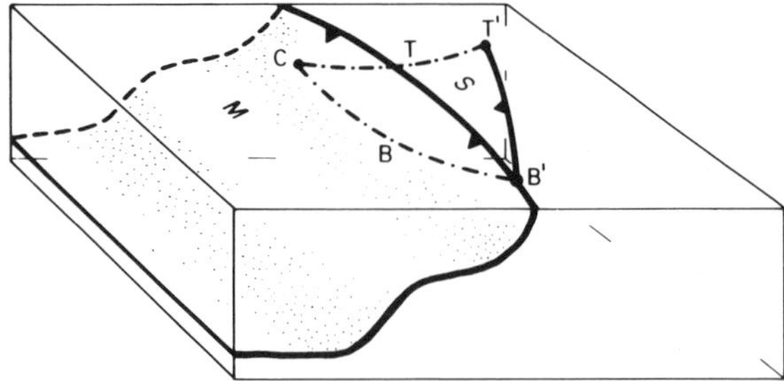

FIG. 5—Block diagram showing diverging splay (S), which has only one tip line (T) with a map termination (T′) and one branch line (B) that intersects erosion surface at B′. There is one corner C.

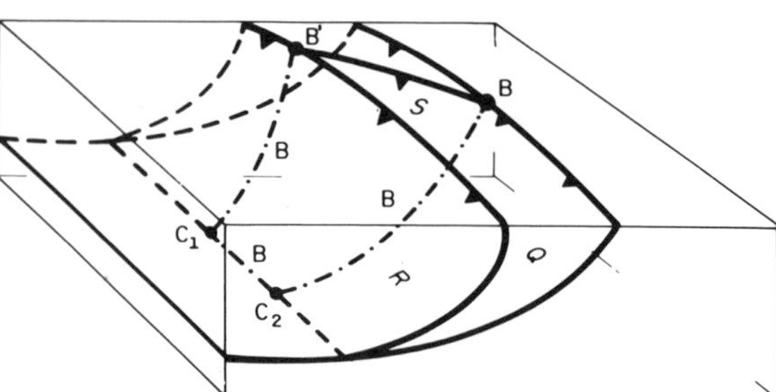

FIG. 7—Two major faults (Q, R) with connecting splay (S). Two branch lines (B) have surface terminations (B, B′) and one branch line at depth has two corners (C_1, C_2).

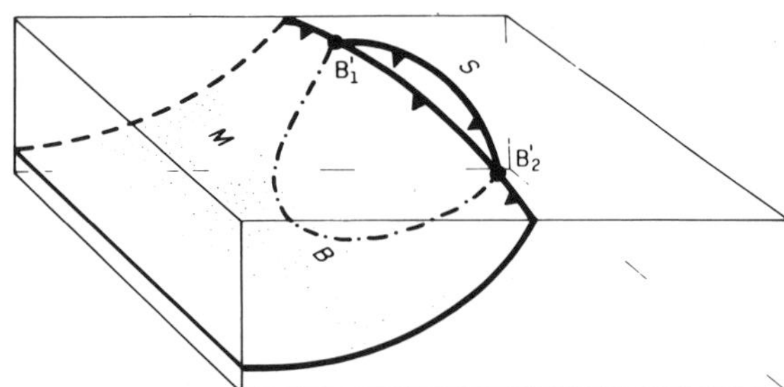

FIG. 6—Rejoining splay with one branch line (B) which intersects map at two branch points (B_1, B_2).

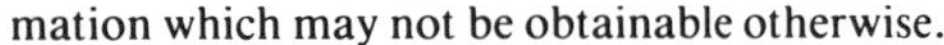

FIG. 8—Above: Horse in volume of rock surrounded by fault surfaces. Two fault surfaces meet at single closed branch line (B) with two cusps (U). Below: diagram illustrates half of horse, cut along line of section XX′.

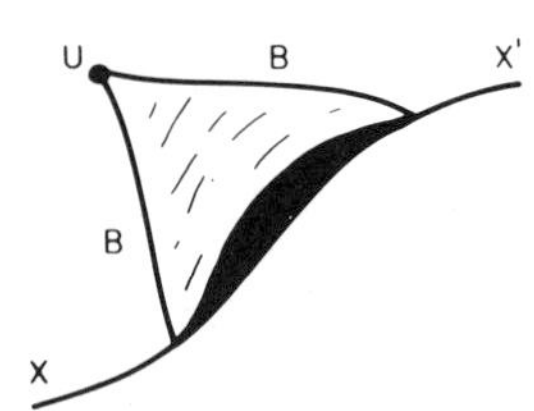

mation which may not be obtainable otherwise.

One must try to describe the three-dimensional geometry relative to the synorogenic erosion surface, not the current one and this can be difficult. For example, how do we distinguish a rejoining splay from an eroded horse on a map? If a rejoining splay has roughly parallel and gently plunging branch lines then cross sections can show two-dimensional pods that resemble horses (Fig. 10), how do we tell them apart?

Transfer Zones and Connectivity

A stratigraphic formation could be cut into interlocking pieces in such a way that if it were fully excavated and you started at one side of the map, you could walk around the ends of the faults on a tortuous path to the other side without having to jump across any faults (Douglas, 1958, p. 131). In this case each thrust sheet is related to its neighbor by an unfaulted envelope that acts as a transfer zone (Dahlstrom, 1970, p. 358). This description of interlocking thrust sheets is correct only for isolated and diverging splays, each of which has at least one tip line. It is not the case within rejoining and connecting splays or horses where each faulted part of the formation is surrounded by fault surfaces.

The extent to which thrust sheets are connected to each other depends on the relative length of tip and branch lines. Because all thrust surfaces are generated by tip lines (Elliott, 1976, p. 298) the degree of connectivity depends on the duration and intensity of activity in that part of the thrust belt. Other things being equal, horses are more common in the more internal, older, and deeper parts of thrust belts, whereas blind thrusts and isolated and diverging splays are more common in the younger, external, and shallower portions.

IMBRICATE THRUST SYSTEMS

So far we have concentrated on the geometric relations between two or three connected faults. Several nearby faults may join up in closely related branching array known as a thrust *system* or family (Rodgers, 1953, p. 130). Now we shall look at patterns that arise at thrust systems where a substantial number of faults are the same general shape and size.

If each thrust in a system repeats the size and shape of the neighboring thrust so that the thrust sheets overlap like roof tiles, all dipping in the same general direction, we have an imbricate system. This important special type of thrust system was recognized in 1883, when Suess defined schuppen structure as the repetition of strata by a

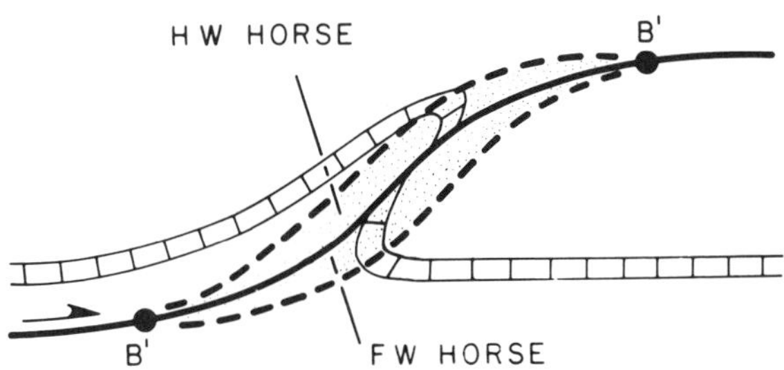

FIG. 9—Cross section through incipient horses. New fractures (dashed) may cut horses from either footwall (FW) or hanging wall (HW) of major thrust surface. Note that horses in this figure would consist entirely of inverted rocks.

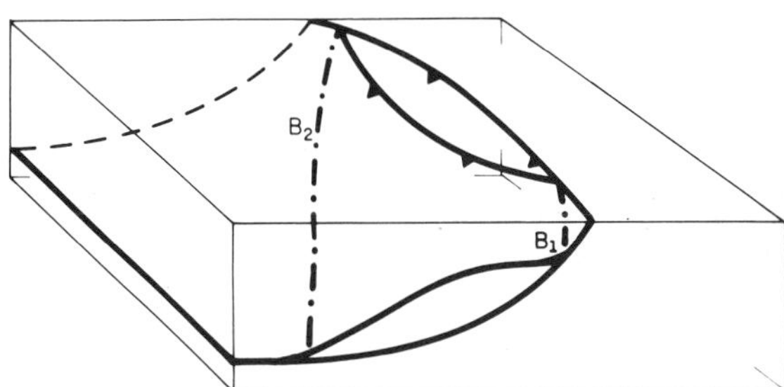

FIG. 10—Two fault surfaces meeting along two branch lines (B_1, B_2), whose map pattern resembles diverging splay and whose cross section resembles horse.

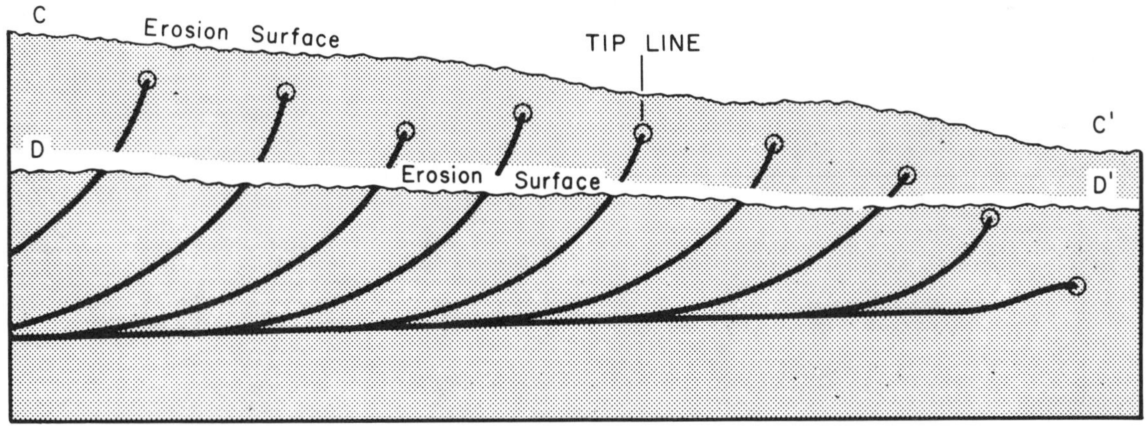

FIG. 11—Cross section of imbricate fan at two different levels of erosion. Each thrust sheet is an upward-opening crescentic slice, and all curve asymptotically downward to a common basal sole thrust. If most faults cut synorogenic erosion surface DD′ we have an emergent imbricate fan. Alternatively, it is possible that tip lines do not reach synorogenic erosion surface CC′, producing blind imbricate fan. Note that subsequent erosion (CC′ down to DD′) may obliterate any means of distinguishing two kinds of imbricate fans.

series of parallel and evenly spaced overlapping faults (see Suess, 1904, p. 112). As the concept evolved, several other terms came into use. The literal French equivalent is "structure ecailleuse" (Gosselet, 1885). Both the French and German expressions mean a scaly or flaky structure. However, de Margerie and Heim (1888) used "structure imbriquee" as the French equivalent of Suess's term. Hobbs in 1893 introduced "weatherboard structure," but later (1894) accepted a suggestion by Bernard Hobson to use imbricate structure as the English equivalent. In modern usage, schuppen zone is synonymous with imbricate zone but is less frequently employed.

Imbricate structures are an efficient means to shorten and thicken a sequence. Relative movement ". . . is trivial, as concerns adjacent members, but may in the aggregate lead to impressive telescoping of the affected zone" (Bailey, 1938, p. 607).

A sole thrust is the lower common thrust in an imbricate system (Dennis, 1967, p. 139). In an imbricate fan, a swarm of curved triangular thrust slices are asymptotically shaped downward to the sole thrust and spread out upward like an open fan (Fig. 11). Emergent imbricate fans where the faults reach the erosion surface are most common, but blind faults could produce blind imbricate fans. On approaching the synorogenic erosion surface the thrusts often increase their dip, and our impression is that they can meet the ground surface at about 60°. Imbricate fans dominate this near surface level.

An imbricate fan in which a thrust with maximum slip is at the front is a leading imbricate fan. If the thrust with maximum slip is at the back, it is a trailing imbricate fan (Fig. 12). These two terms were modified from Dahlstrom (1970, p. 352).

It is also possible to construct a thrust system with an imbricate family of subsidiary contraction faults asymptotically curving downward to a sole or floor thrust and upward to a roof thrust. This thrust system is known as a duplex (Fig. 12).

DUPLEXES

The term "duplex" first appeared in a paper by Dahlstrom (1970, p. 352), but he does not take credit for originating the duplex concept or coining the term, maintaining that these ideas were already present in the

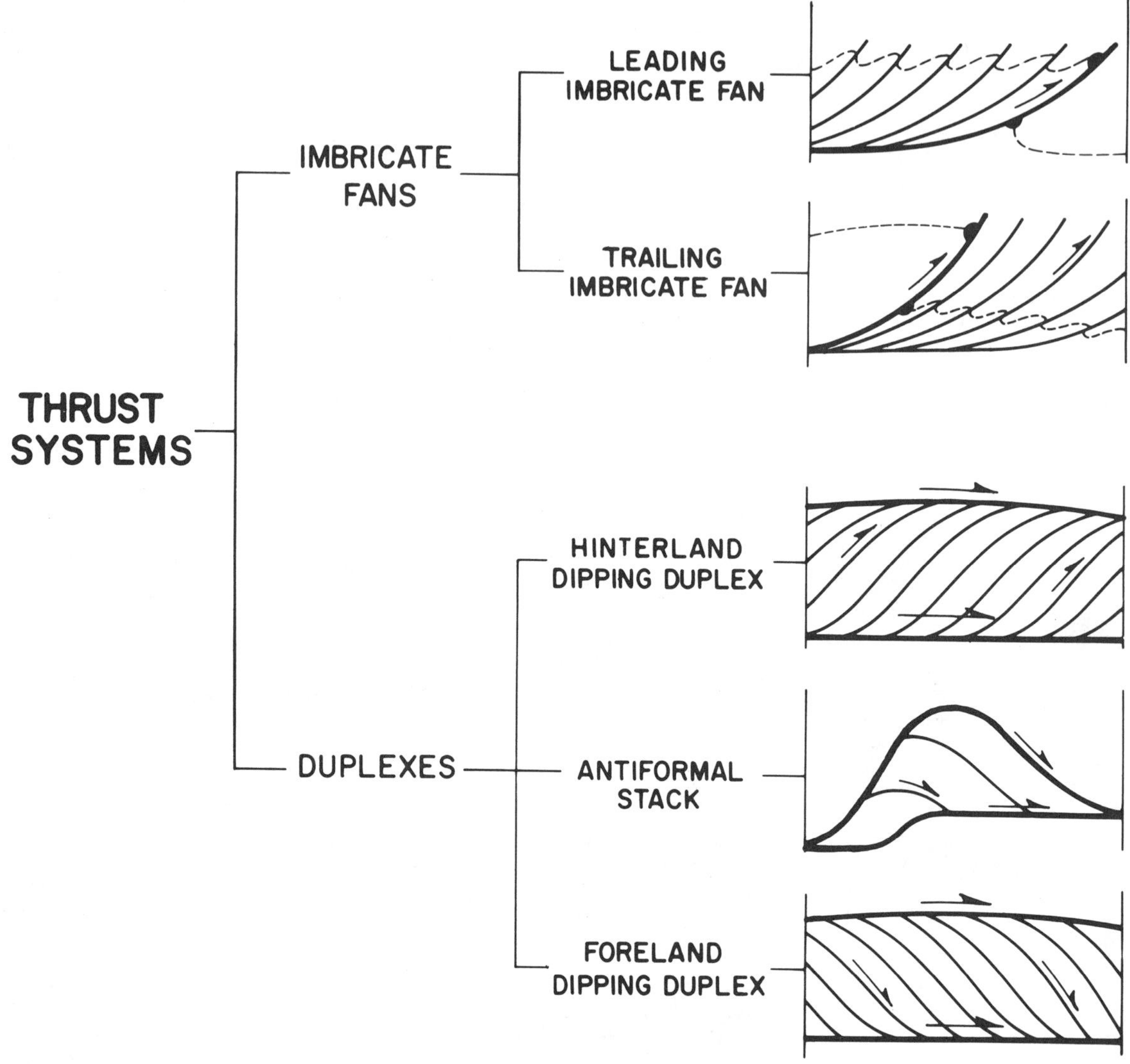

FIG. 12—Classification of different systems of thrusts; most are imbricate.

minds and writing of several people in Calgary (D. A. Dahlstrom, 1978, personal commun.). The evolution of the concept is best appreciated by studying some historically important examples.

Moine Thrust System

A 50-year controversy over the existence of the Moine thrust resulted in geologic mapping at about 1:10,000 scale, which was completed between 1883 and 1896 (Fig. 13). Early in this project, B. N. Peach interpreted the exposures on the east shore of Loch Eriboll as a duplex (McIntyre, 1954, p. 206). Shortly afterward, in an exceptionally exposed area with 0.6 mi (1 km) of local relief, a spectacular duplex zone (Foinaven duplex; Table 1) repeating a distinctive Lower Cambrian quartzite was discovered by Cadell (Peach et al, 1907, p. 491, Fig. 25) (Fig. 14). Throughout this early Scottish work the usage of "schuppen" and "imbricate zone" corresponded to what we would now call a duplex.

Lewis Thrust System

The Lewis thrust is one of the largest in the North American Cordillera (Table 2; Fig. 15). The Precambrian Belt sequence within the thrust sheet, spectacularly exposed in Glacier and Waterton Parks, shows remarkably little internal deformation, but going down through the sheet one may suddenly enter a duplex, a few hundred meters thick, whose floor is the Lewis thrust surface. This is particularly clear in the Willis (1902) illustration of the Chief Mountain klippe (Fig. 16). Douglas (1952) mapped a duplex near Mount Crandell (Fig. 17), which was cited by Dahlstrom (1970, p. 354) as his type exam-

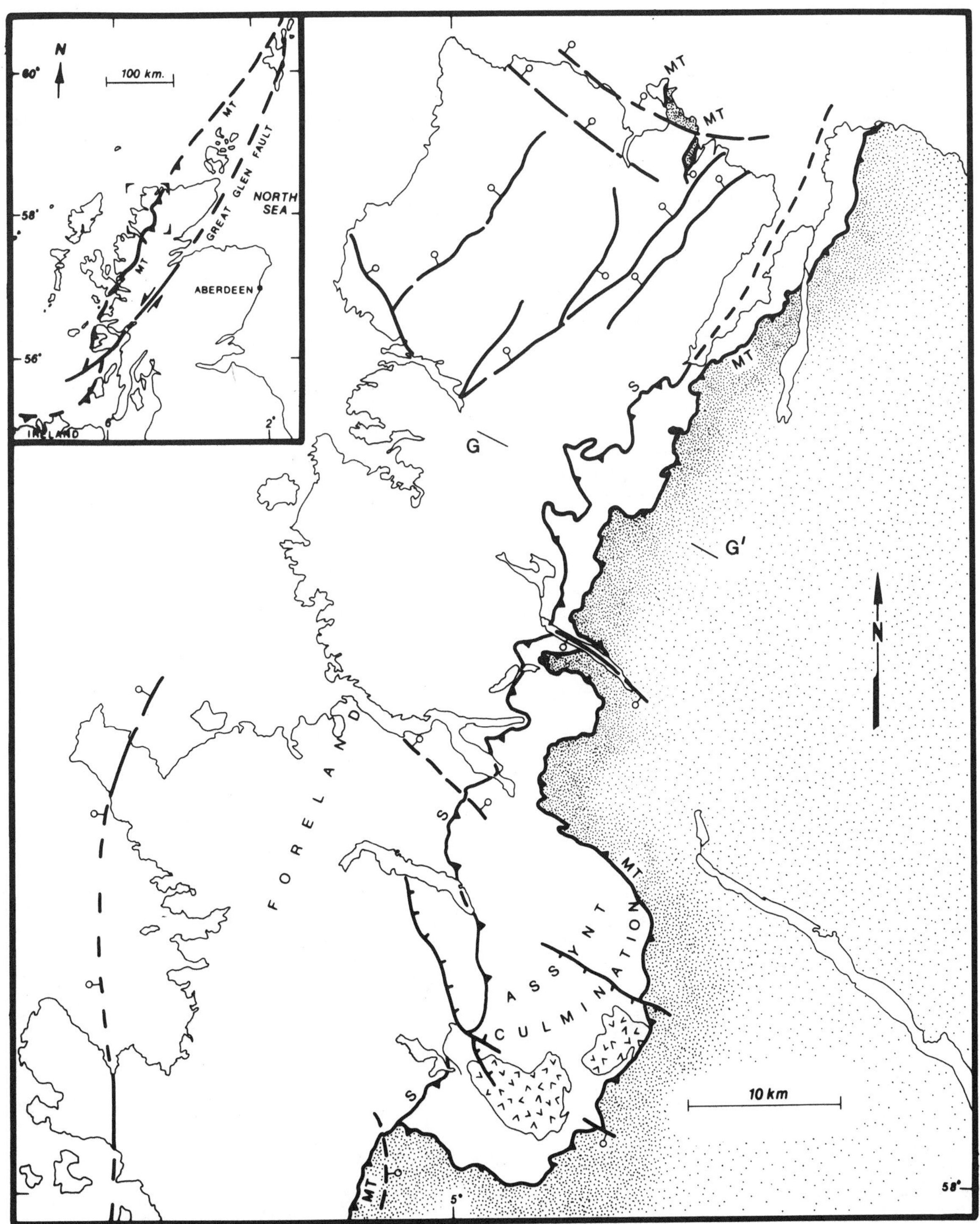

FIG. 13—Map of northern part of Moine thrust zone. MT is Moine thrust, S is Sole thrust, Moine schists stippled. Insert shows 500 mi (800 km) length of thrust zone.

ple. The Cate Creek and Haig Brook windows (Fig. 15) through the Lewis thrust sheet are particularly spectacular examples of duplexes (Dahlstrom, 1970, Fig. 24; Fermor and Price, 1976). It appears that substantial portions of the Lewis thrust are the floor of a duplex.

Klippen and reentrants demonstrate that the Lewis thrust once extended northeast of its present outcrop, possibly as much as 6 to 12 mi (10 to 20 km) (Fig. 15). Consequently, many thrusts that crop out in Cretaceous rocks, such as near the headwaters of the Carbondale River and Marias Pass, are part of a duplex whose roof was the Lewis thrust. Possibly, however, the Paleozoic rocks of the Livingstone and Sawtooth Ranges were brought up on emergent imbricate fans. Along strike, these Paleozoic rocks plunge beneath the Lewis thrust and might become part of the duplex containing the Waterton field (Gordy et al, 1977, Fig. 11b), but this is not clear, for other sections show the Waterton field as an imbricate fan (Gordy et al, 1977, Fig. 14).

A duplex is an imbricate family of horses—a "herd of horses"—but in the same way that a horse can change along strike into a splay (Fig. 10), it is possible for a duplex to change along strike into an imbricate fan.

The subject of along-strike variations in duplexes is discussed later, but it is important to emphasize here the lateral changes in the horses that make up a duplex. This change in shape of a horse, such as shown in Figure 8, is a result of oblique and lateral ramping giving the distinctive scoop shape of roof, subsidiary, or floor thrusts. The huge horses of Paleozoic carbonate rock underlying the Alberta foothills show just this sort of behavior, and are responsible for the doubly plunging culminations that affect the higher sheets. This is outstandingly illustrated by the longitudinal cross sections based on seismic and drill-hole data in Bally et al (1966, Plate 8). The Moine thrust belt also shows along-strike variation of a major duplex. In this case it is exposed at the surface for direct observation because of a regional dip of 10 to 15° which was imposed after thrusting (Elliott and Johnson, 1980, Fig. 6).

Dimensions and Internal Geometry

Duplexes have characteristic internal features (Fig. 18). Beds within a horse often trace out an elongate anticline-syncline fold pair, and bedding near the central inflection point roughly parallels the subsidiary faults. Above and below the duplex the bedding may be rela-

Table 1. Dimensions of Duplexes

Duplex	Contraction Ratio (L'/L_0)	Number of Horses (N)	Approx. Angle Subsidiary to Floor Thrust	Reference
Foinaven duplex in Moine thrust zone	0.29	34	40°	Fig. 14
Windows duplex, southern Appalachians	0.36	21	30-45°	Fig. 29
Lewis thrust (floor of duplex), North American Cordillera				
Chief Mtn.	—	2	23°	Fig. 16
Mt. Crandell	0.57	6	33°	Fig. 17
Cate Creek	0.58	2	31°	Dahlstrom (1970)
Haig Brook	0.6	12	27°	Fermor and Price (1976)
Central Appalachian Valley and Ridge	0.54	4	33°	Fig. 26
Idealized model				
Duplex constructed with kink folds	0.50	—	30°	Fig. 18

Table 2. Some Dominant Thrust Sheets

Dominant Thrust Sheet (Location)	Approximate Width Downdip (km)	Lithology
Moine (Northwest Scotland)	100	Precambrian and Cambrian siliciclastics refolded and metamorphosed to medium and high grades (Figs. 13, 14).
Semail Ophiolite (Oman)	100	Oceanic crust and upper mantle thrust over Arabian continental margin.
Austro-Alpine (European Alps)	150	Mesozoic continental margin unconformable on folded Paleozoic nonmarine sediments and quartzo-feldspathic crystalline complex of medium and high grade (Figs. 31, 34).
Lewis (North American Cordillera of U.S. and Canada)	100 to 200	Paleozoic shelf carbonates unconformable over late Precambrian shelf and slope sequence (Figs. 15, 17).
Jotun (Southwest Norway)	180	Crystalline Precambrian basement of two pyroxene granulites (Hossack et al, 1982).

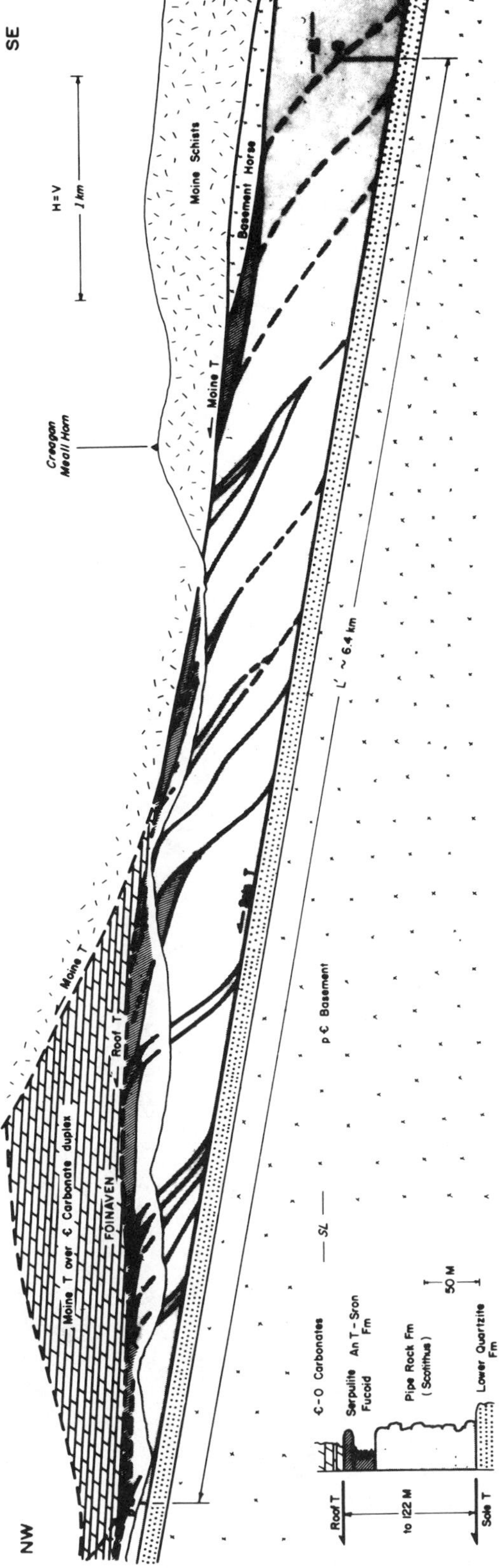

FIG. 14—Cross section of the Foinaven duplex from Moine thrust zone, line of section GG′ on Figure 13. Duplex has area PQ of 2.65 km² and is made up almost entirely of Cambrian Pipe Rock with a little An t-Sron. After Elliott and Johnson (1980, Fig. 4).

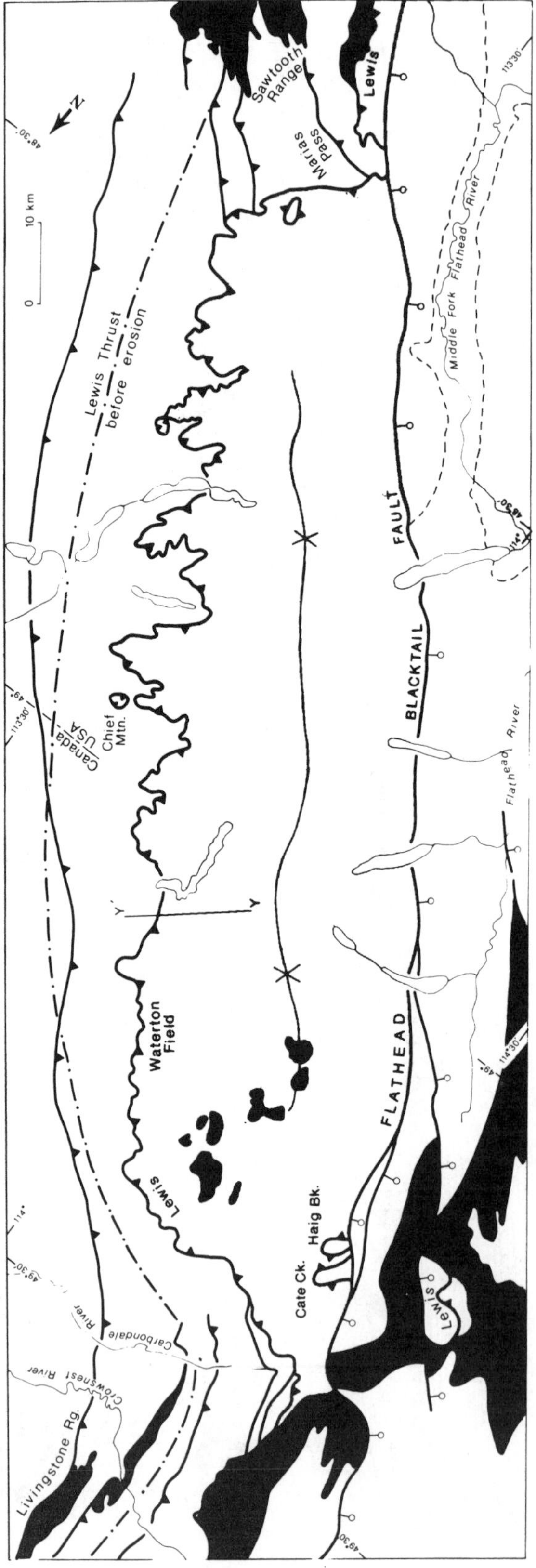

FIG. 15—Map of central portion of Lewis thrust sheet. Proterozoic is light pattern area, Paleozoic is dark area.

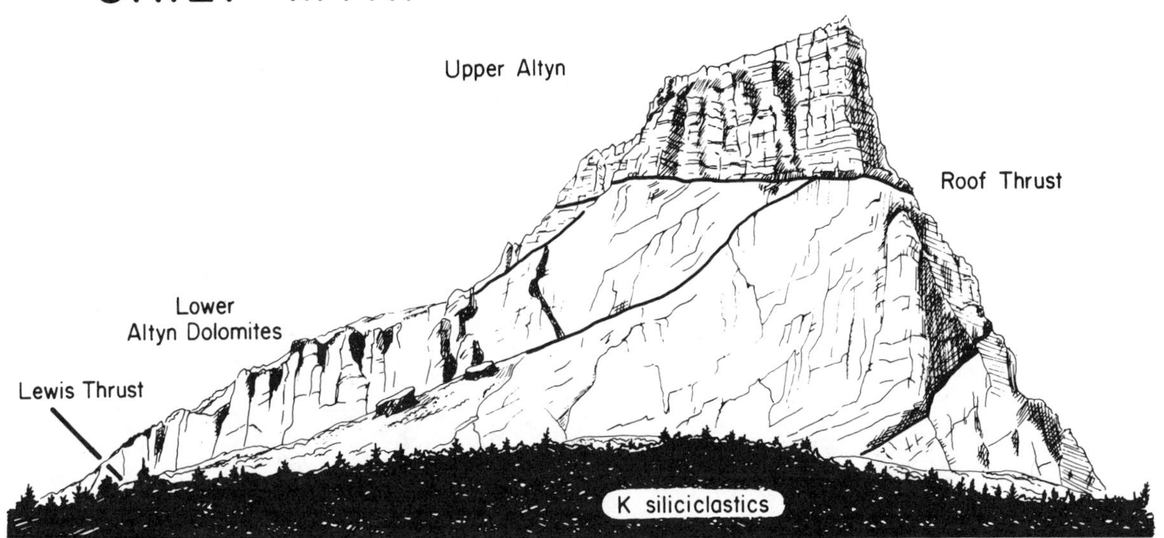

FIG. 16—Chief Mountain klippe, looking north, is eroded fragment of duplex in Precambrian Belt Supergroup whose floor is Lewis thrust. All lie on Upper Cretaceous siliciclastics. After Willis (1902, Fig. 5) and Dyson (*in* Nevin, 1949, Fig. 101).

tively undisturbed, and for long distances a particular stratigraphic unit may compose the hanging wall of the roof or the footwall of the floor.

The fold pairs within any one duplex have a somewhat similar shape and size, and it is often possible to draw enveloping surfaces through the whole sequence of horses. These enveloping surfaces are at very low angles to the floor thrust, seldom exceeding 15°.

While constructing current and restored cross sections through duplexes we repeatedly make a number of simple calculations (equations 1 to 7). Capitals indicate measurements over the complete thickness or length of a duplex, and small letters refer to one horse. The easiest measurements are the current duplex length (L′), structural thickness (H′), cross-section area (A), initial stratigraphic thickness of the formations making up the horses (t), and the current angle (β') between floor thrust and the central portion of the subsiding faults (Fig. 18).

We assume plane strain, with the cross-section area

$$A = H'L' = tL_o, \qquad (1)$$

where L_o is the initial length of the duplex. The overall shortening distance (S), accomplished by formation of the duplex, is

$$S = L_o - L', \qquad (2)$$

and from equation 1

$$S = A/t - L'. \qquad (3)$$

If the bed length (ℓ) within each horse is unaltered by deformation, then the total initial bed length within the duplex is

$$L_o = \Sigma\ell = N\ell_a, \qquad (4)$$

where N is the number of horses with an average bedding length of ℓ_a.

The perpendicular distance (h) between subsidiary faults bounding a horse may approximate the initial stratigraphic thickness (Fig. 18), h = t, so that the spacing between subsidiary faults, measured parallel with the floor thrust, is

$$p' = t/\sin\beta'. \qquad (5)$$

Now the total number of horses in the duplex is

$$N = L'/p', \qquad (6)$$

and the number of faults per unit length

$$N/L' = \ell/p'. \qquad (7)$$

A duplex beneath the Moine thrust provides an example of these calculations (Fig. 14). The duplex has a cross-section area (A) of 2.65 $10^6 m^2$ and a current length (L′) of 6.38 km. The average structural thickness (H′) is 415 m, although in the central and southeast part the duplex is thicker (H′ is 510 m). The duplex consists almost entirely of quartzite, although some shale is present toward the top, and the initial stratigraphic thickness (t) of the formations making up the duplex is 122 m. The initial length (L_o) is equivalent to 21.7 km from equation 1, and from equation 3 the shortening (S) is 15.3 km.

Only 20 faults, clearly observed on cliffs, were plotted

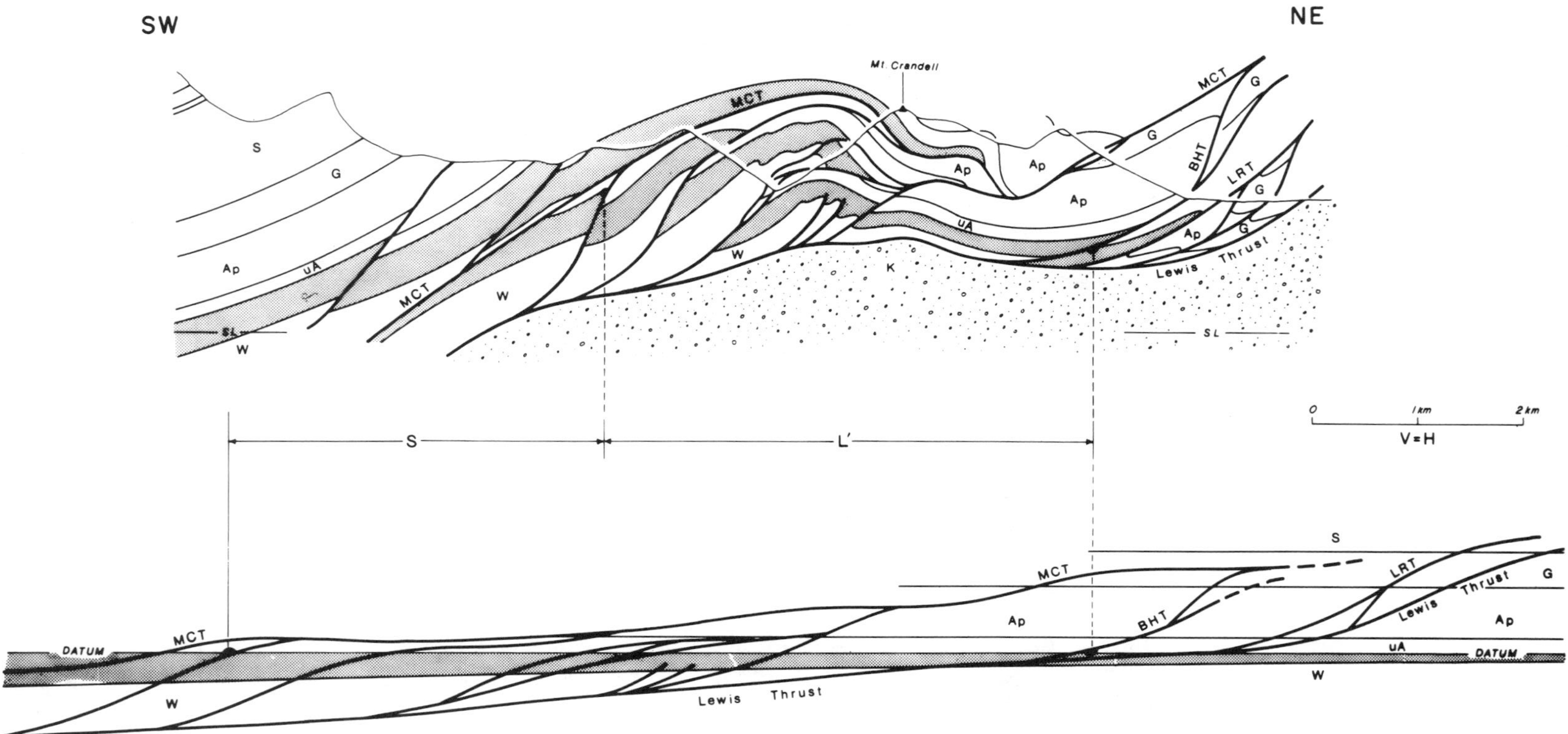

FIG. 17—Precambrian Belt Supergroup, comprising Waterton (W), lower Altyn (fine stipple), mid and upper Altyn (uA), Appekunny (Ap), Grinnell (G), and Siyeh (S), is thrust over Cretaceous siliciclastics (K, with pebble pattern) by Lewis thrust. Mount Crandell thrust (MCT) is roof and Lewis thrust is floor to duplex, and folded horse just northwest of Mount Crandell suggests that duplex developed toward foreland. Cross section is balanced (with current distance L′ between points recording a shortening of S), and is based on excellent control provided by over 2 km of local relief. Modified from Douglas (1952).

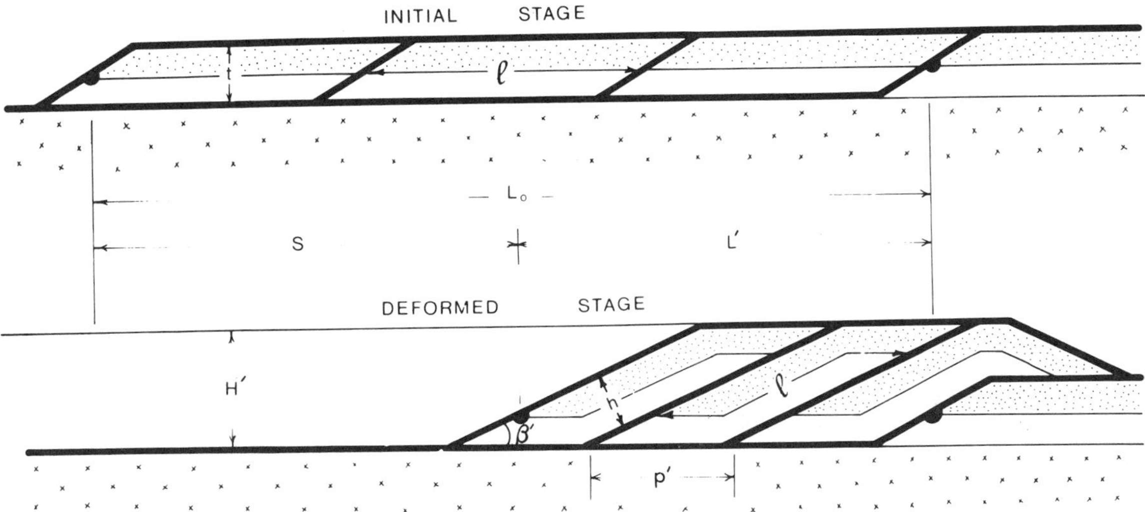

FIG. 18—Initial and deformed stages in formation of duplex, showing quantities L_o, L', S, t, H', ℓ, h, p', β' used in equations 1 through 8. Note elongate "S" folds in horses. Modified from Boyer (1978).

on the section. Many more subsidiary faults must be present, but how can this number be estimated? The angle β' between subsidiary faults and floor averages around 40°, but ranges between 25 and 55° and is smallest at the northwest end of the section.

The spacing between subsidiary faults must also vary (from equation 5) between 290 and 150 m, with a mean spacing (p') equal to 190 m. Consequently, the total number of faults (using equation 6) varies between 22 and 45, with the mean (N) being 34.

The shortening and structural thickening can vary from place to place. If the duplex involves a similar stratigraphic thickness and uniform mean bedding length (ℓ_n) in the horses, it follows from equations 4, 5, and 6 that

$$\left\{\frac{L'}{L_o}\right\} = \frac{p'}{\ell_n} = \left\{\frac{t}{\ell_n}\right\} \frac{1}{\sin \beta}. \qquad (8)$$

Therefore, as the shortening and thickening decrease, the ratio (L'/L_o) increases, and the angle β' must decrease.

As the Foinaven duplex gets thinner, the angle β' becomes smaller and the local shortening must diminish. By comparing different duplexes (as in Table 1), we see that β' varies inversely with the contraction ratio (L'/L_o).

Sequence of Development

There are two sides to any thrust belt. On one side is the internal zone or hinterland, and on the other, beyond the external zone and the margin of the thrust belt, is the foreland. When we require a sense of direction toward the hinterland or foreland, we use the terms "hindward" or "forward."

In this section, we review several different geometric arguments for the successive development toward the foreland, or a forward progression, of the subsidiary faults in duplexes. The oldest argument is best illustrated by an example, the Mount Crandell duplex (Fig. 17), where a higher horse is folded over a lower one proving the forward development (Dahlstrom, 1970, p. 352). Independently, Perry (1971, p. 195) deduced the same time sequence in a central Appalachian duplex.

The next arguments are based on a series of simple graphic experiments developed by Boyer (1978). First, we construct an idealized model based on typical dimensions and angles of observed duplexes (Table 1) assuming plane strain, constant bed lengths, and kink folds (Fig. 19). In the initial stage a major thrust with slip S_o has climbed upward in the section from a lower to an upper glide horizon, cutting rather steeply through a more competent sequence and making a footwall ramp. A new crack propagates from the base of the ramp, continues in the lower glide zone for some distance, and then cuts upsection to the higher glide horizon where it rejoins the preexisting major thrust. During the next time interval (Fig. 19, stage 1), this new fracture (S_2) slips, the overlying fault segment remains fixed, yet behind and in front of the new horse the major thrust slips by $S_o + S_1$. In other words, as slip is transferred to the new and lower fault a portion of the major thrust is deactivated and rides passively within the growing thrust sheet. The new horse, the inactive portion of the major thrust, and the rest of the overlying thrust sheet are folded, possibly in kinklike fashion, over the footwall ramp. Movement is again transferred to a new branch fault and the process is repeated (Fig. 19, stages 2, 3).

After the development of several subsidiary faults the duplex achieves its characteristic features: elongate folds within the individual imbricate horses with bedding locally parallel with the subsidiary faults, relatively undisturbed bedding above the roof thrust, and the same stratigraphic unit in the hanging wall along the roof thrust for a great distance. Compare the features in Figure 19 with those in the Mount Crandell duplex (Fig. 17).

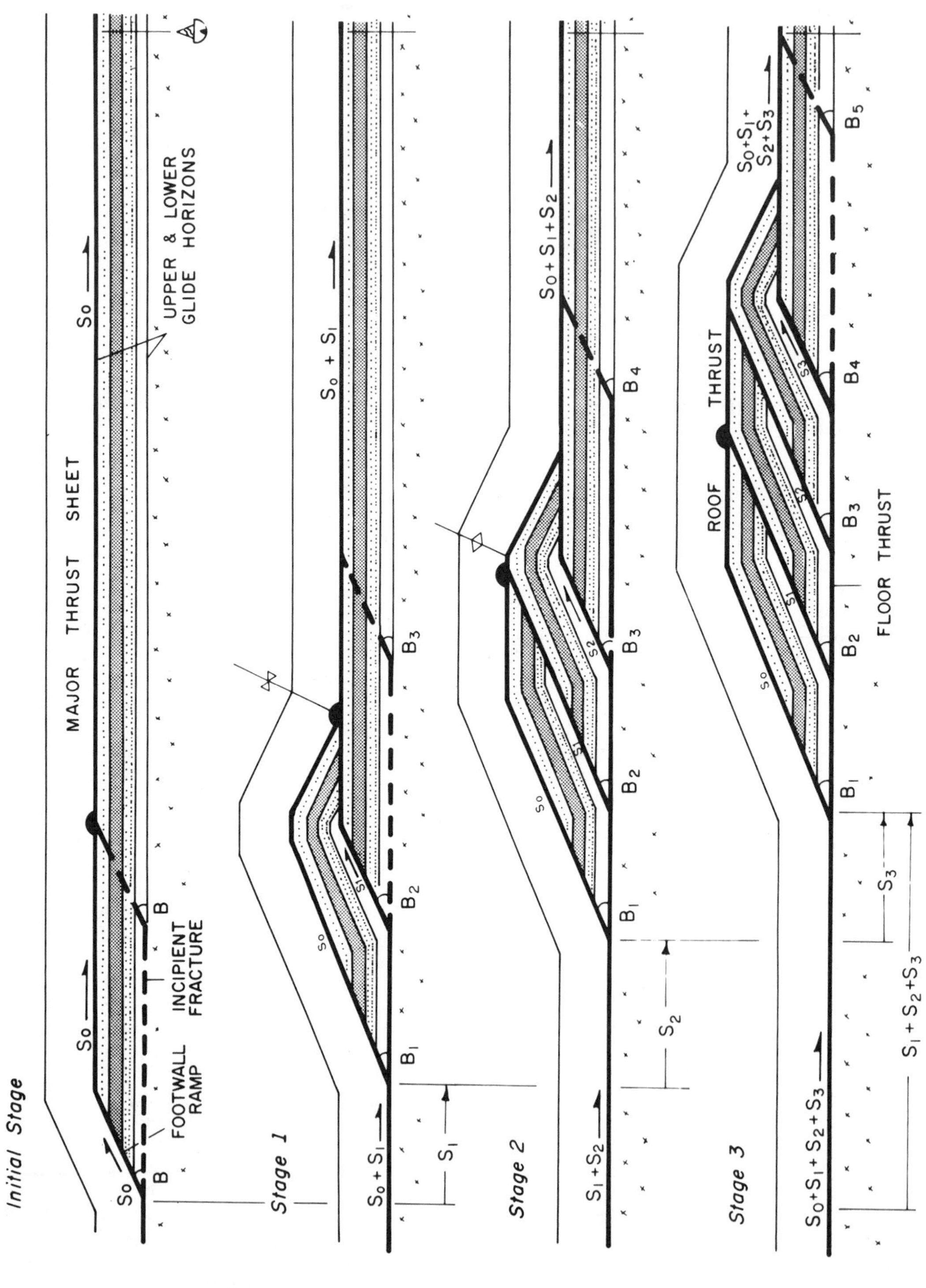

FIG. 19—Progressive collapse of footwall ramp builds up duplex. This is measured graphical experiment, assuming plane strain and kink folding, with angles and ratios of dimensions typical of natural examples (Table 1). Roof thrust sheet undergoes complex sequence of folding and unfolding, seen by following black half dot. Modified from Boyer (1978).

1

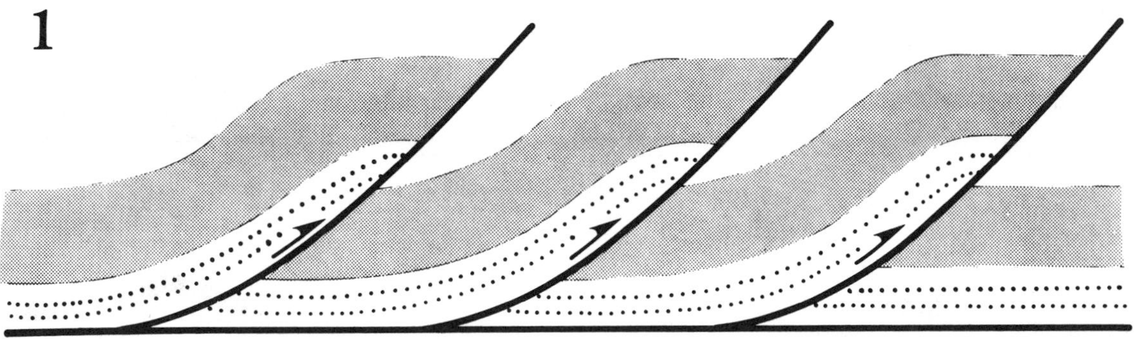

2

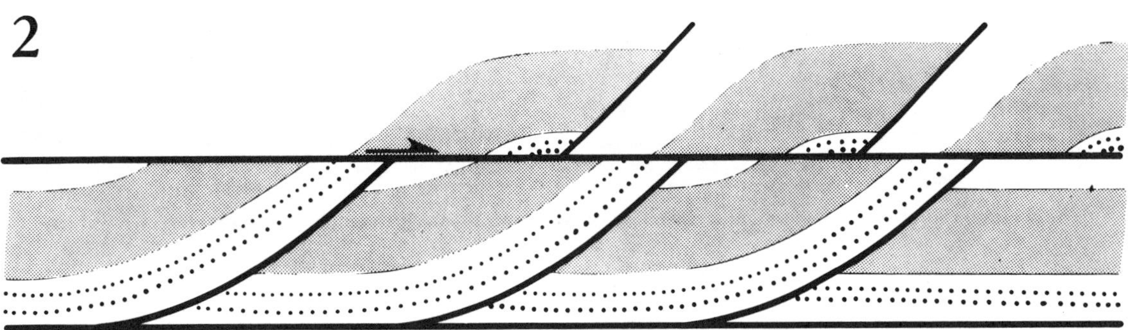

FIG. 20—Alternative method for developing duplexes. Earlier formed imbricate faults in stage 1 (above) are truncated by younger and higher thrust in stage 2 (below) (Boyer, 1978).

The faulted stratigraphic section is doubled in the area of each footwall ramp. As the thickened section enters the next ramp, the leading limb of the anticline is returned to the horizontal, and the fold is partially opened (solid half-dot in Fig. 19). At this point a new branch fault forms in this idealized model.

Duplexes are a mechanism for slip transfer from one glide horizon at depth to another at shallower levels. In the direction of movement, slip decreases along the floor and increases along the roof, and total slip at any point along the floor or roof thrust is dependent on the number of horses that lie between that point and the head of the duplex. Slip transfer and the creation of new horses causes structural thickening, duplex growth, and addition of mass to the moving thrust complex.

Let us now try some graphic experiments using alternative sequences of development. Because hanging-wall rocks of the roof thrust are often flat-lying or only gently folded, one might suggest that the high-angle subsidiary faults formed first by branching upward from the floor thrusts, then were truncated by the roof thrust. The resulting stratigraphic relations, however, are unlike those of any duplexes described to date (Fig. 20).

Could a duplex form by sequential branches that become younger toward the hinterland, a hindward progression? Here also the geometric features are not found in natural duplexes, although the imbricate fan is a possible structure (Fig. 21).

If the roof thrust were partially removed by erosion, one might be able to see the layout of the entire internal structure in the eroded duplex. For example, a number of southern Appalachian imbricate thrusts converge upward to the Rome thrust (Rodgers, 1970, p. 57), and the map pattern is probably an oblique section through a duplex. A map would show the roof thrust running across the subsidiary thrusts and appearing to "truncate" them (Fig. 22). This appearance of "truncation" or "overlap" in the Moine thrust belt led numerous authors to the erroneous belief that the higher Moine thrust was the last to form (Elliott and Johnson, 1980, p. 90). Repeated examples of this typical geometric pattern of apparent truncation in the southern Appalachians are interpreted, we think incorrectly, as a hindward sequence of imbrication (Rodgers, 1970, p. 178; Milici, 1975; Harris and Milici, 1977, p. 14). In Montana, precisely similar relations exist along the Lewis thrust and are misinterpreted as a hindward thrust sequence (e.g., Mudge and Earhart, 1980, p. 14).

But the problem is not trivial. We see that as each subsidiary thrust in turn joins the roof a steadily increasing portion of the roof is immobilized. Thus, there is an apparent paradox: a major roof thrust moved at different times along its map trace, yet everywhere it may bear the same name.

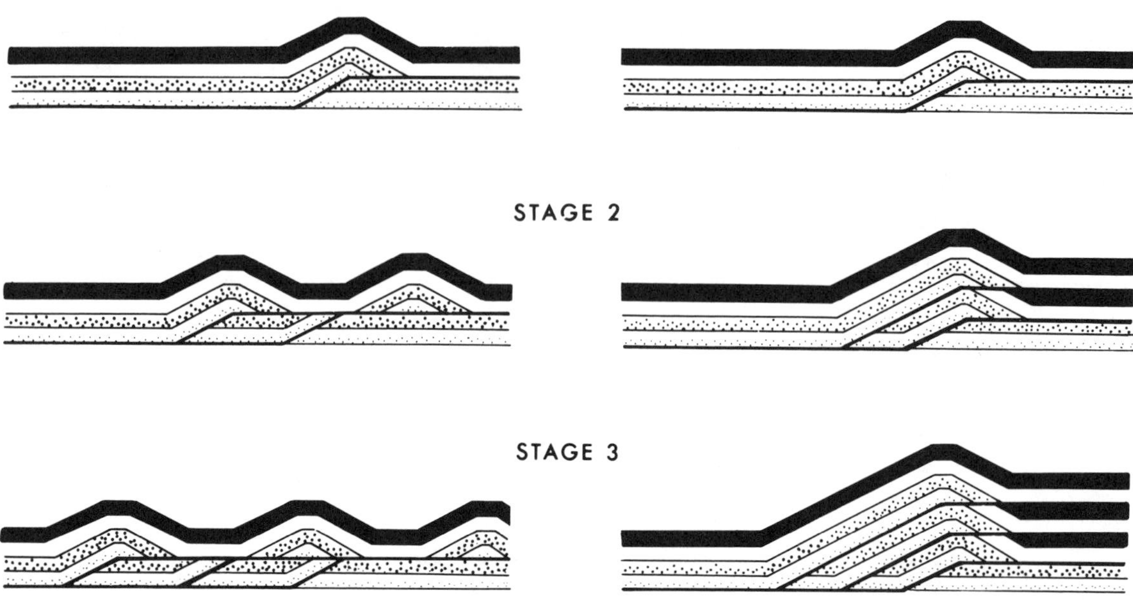

FIG. 21—Two alternative methods of hindward imbrication shown by measured graphical experiments (Boyer, 1978).

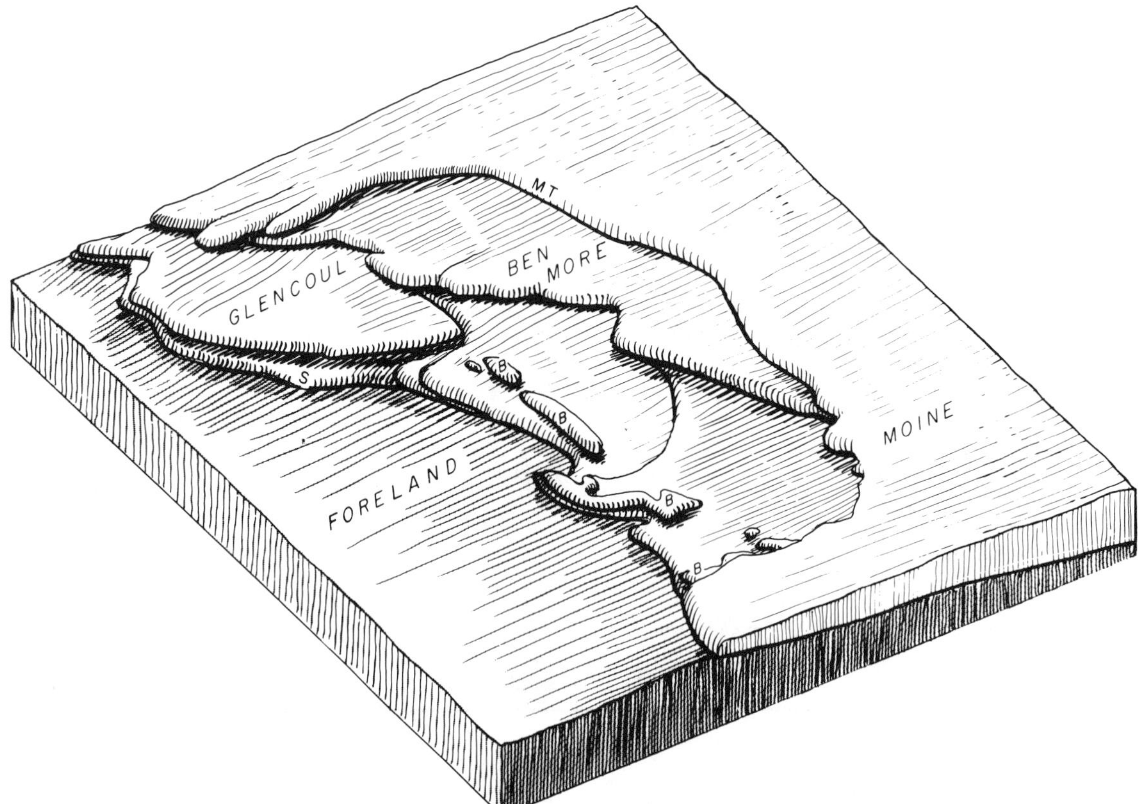

FIG. 22—Block diagram of Assynt culmination in Moine thrust zone (Fig. 13). This is oblique section through duplex. Diagram follows geologic interpretation of Elliott and Johnson (1980) and was inspired by plaster model on display at University of Edinburgh, and drawing by H. Cloos (1936, Fig. 290).

Antiformal Stacks and Forward-Dipping Duplexes

In a hindward-dipping duplex—the only kind discussed so far (Fig. 19)—the effect of shortening is to move the branch lines somewhat closer together. If the slip on the subsidiary faults is roughly equal to the length of the horse, then adjacent branch lines will bunch up and the horses will lie on top of one another. This occurs from time to time in an otherwise normal duplex, like the upper horses in the Mount Crandall duplex (Fig. 17). If the bunching of branch lines is widespread, the shingle-like imbricate pattern is destroyed. Instead, we have an antiformal stack of horses where each higher horse is folded about the lower ones. This folding dies out downward and provides unambiguous evidence of the sequence in which the horses were accumulated into the stack, as we see in the Dundonnell antiformal stack (Fig. 23). This small structure in the Moine thrust belt is historically important, for it was here that field geologists first worked out the piggyback or forward sequence of thrusting (Cadell and Horne, in Peach et al, 1907). However, fault "truncation" (discussed in the previous section) gave exactly the opposite time sequence, and this view prevailed until recently (Elliot and Johnson, 1980, p. 90-93).

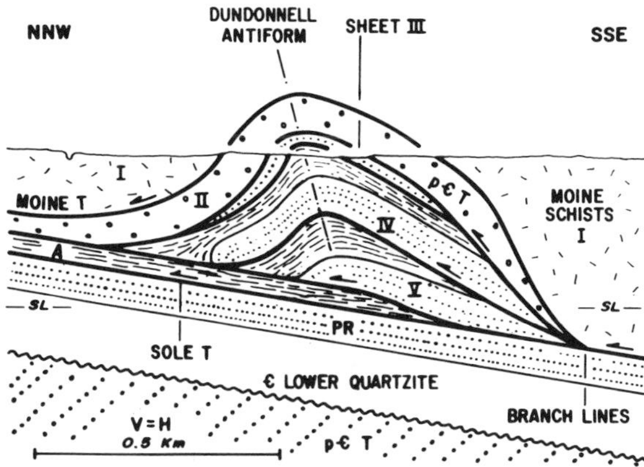

FIG. 23—Balanced cross section constructed normal to fold axis of Dundonnell antiformal stack, but at high angle to movement direction (see Elliott and Johnson, 1980, p. 90). Pebble pattern (pЄT) is Torridon group fluvial sandstones and conglomerates. Stippled pattern (PR) is Cambrian Pipe Rock formation, a quartzite. Dashed pattern A, is Cambrian An t-Sron formation, with an important shale. The sheets are numbered I to V in their order of formation.

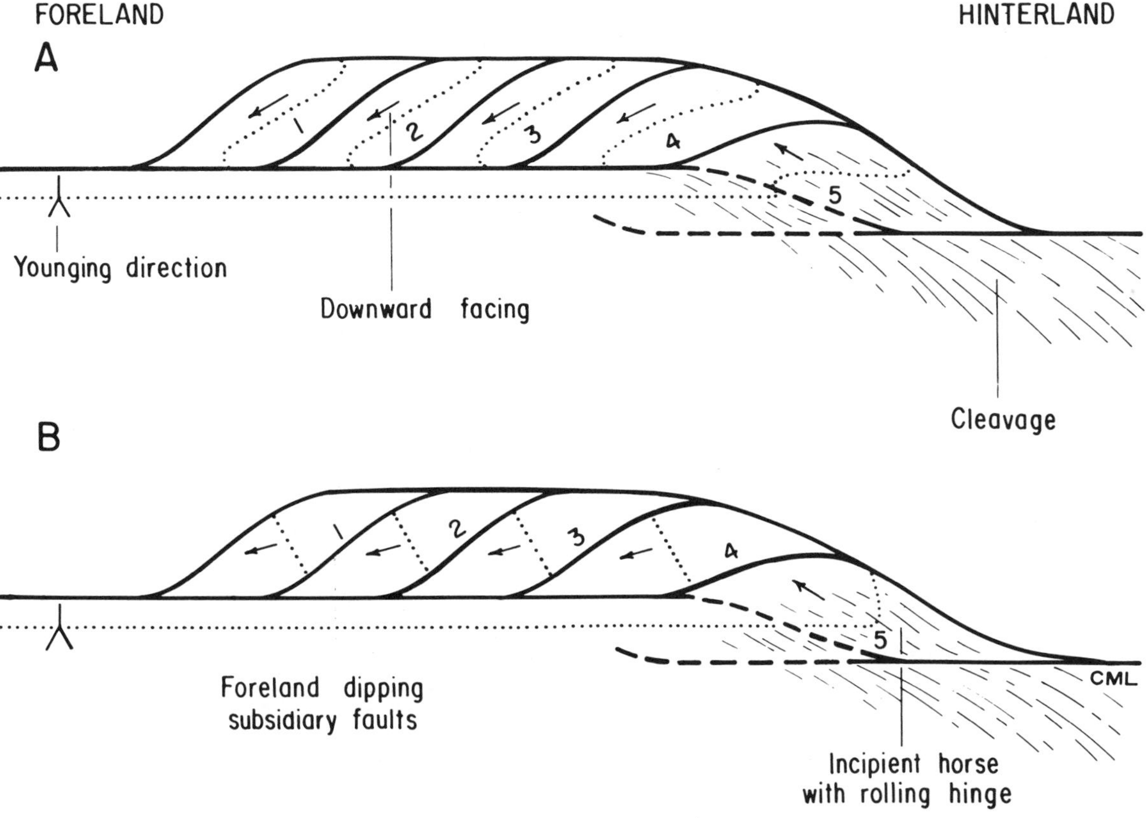

FIG. 24—Formation of forward-dipping duplexes. Horses numbered in order of formation; youngest horses (5) are not yet completely developed. Formations within horses may be predominantly right-way up (A) or upside-down (B), structures are often downward facing, but always forward facing.

In the Dundonnell structure, several of the higher and originally more internal branch lines are now over and forward of the underlying ones. This process can develop further when each branch line passes over and beyond the underlying one. We now recover once more the imbricate pattern (Fig. 24). The dip of the subsidiary faults suggests the term, forward-dipping duplexes, and the roof now contains the branch lines that started on the floor thrust. Forward-dipping duplexes were first proposed by N. Woodward (personal commun., 1981) for the southern part of the Mountain City Window.

Consider a horse just in the process of being plucked from a ramp and with cleavage developing in the surrounding rocks (Fig. 24). Stratigraphy, cross-bedding, or other primary features will usually provide the way-up or younging direction, and the cleavage/younging orientation within each horse fixes the structural facing of the horse from Shackleton's rule (Shackleton, 1958). Forward-dipping duplexes will usually consist of downward-facing horses (Fig. 24), even when the beds are upside down and originated at a footwall syncline with a rolling hinge (Fig. 24). In a similar fashion, hindward-dipping duplexes are usually composed of upward-facing horses, but both hindward and forward-dipping duplexes always have a forward structural facing. We have assumed here a particularly simple relation

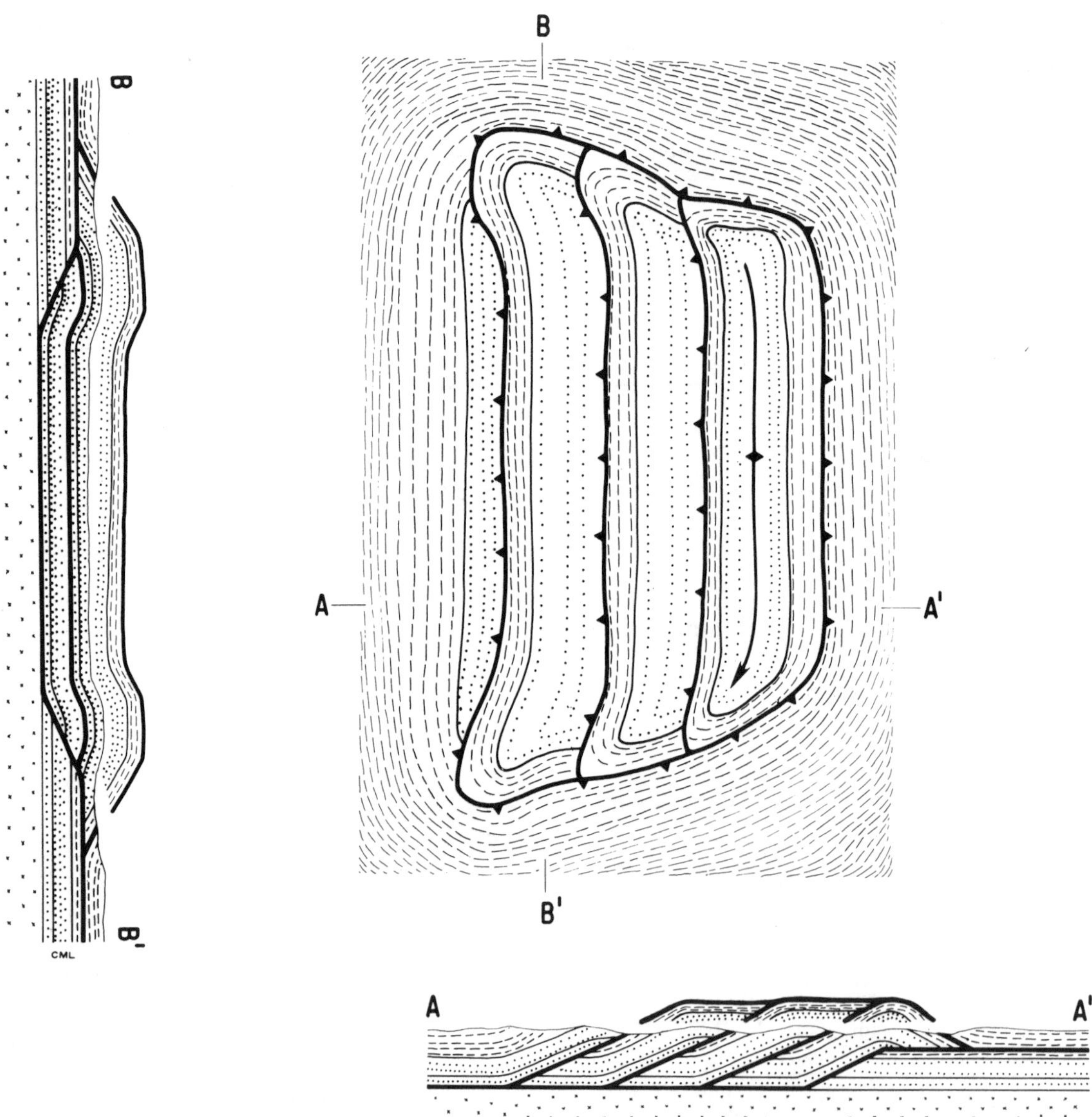

FIG. 25—One type of window into duplex. Anticline occurs at foreland side of window and overlies ramp in floor thrust. Other anticlines occur within each horse. Lateral termination of window may be lateral ramp in sole or floor thrust, with roof and floor thrusts rejoining along strike as well as across strike (Boyer, 1978).

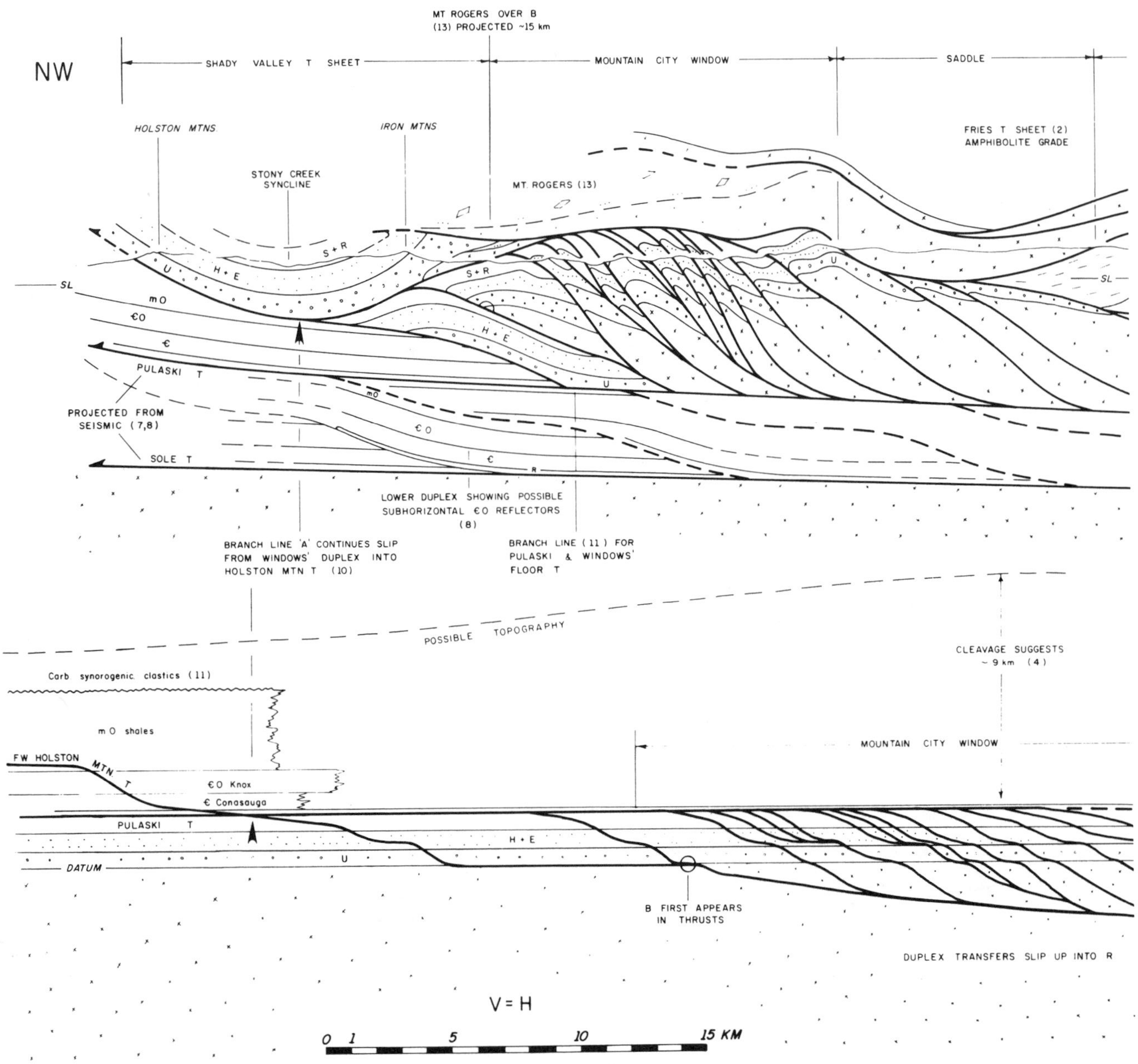

FIG. 29—Cross section through Blue Ridge in southern Appalachians along line AA′ of Fig. 28. The section is balanced below the Fries thrust.

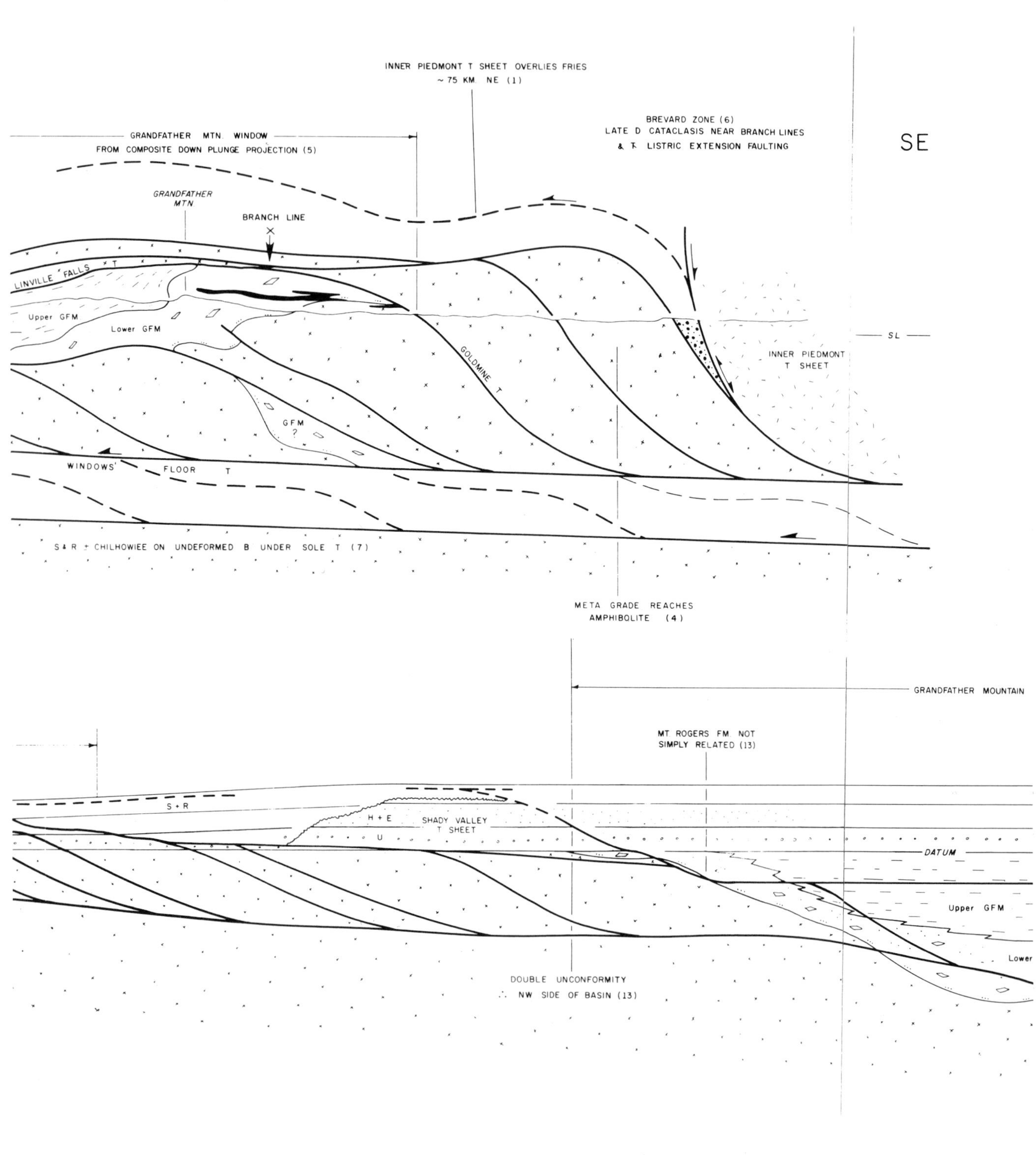
INNER PIEDMONT T SHEET OVERLIES FRIES
~ 75 KM. NE (1)
GRANDFATHER MTN. WINDOW
FROM COMPOSITE DOWN PLUNGE PROJECTION (5)
BREVARD ZONE (6)
LATE D CATACLASIS NEAR BRANCH LINES
& T LISTRIC EXTENSION FAULTING
SE
GRANDFATHER MTN
BRANCH LINE
LINVILLE FALLS T
Upper GFM
Lower GFM
GOLDMINE T
SL
INNER PIEDMONT
T SHEET
GFM
?
WINDOWS' FLOOR T
S & R ± CHILHOWIEE ON UNDEFORMED B UNDER SOLE T (7)
META GRADE REACHES
AMPHIBOLITE (4)
GRANDFATHER MOUNTAIN
MT ROGERS FM NOT
SIMPLY RELATED (13)
S + R
H + E
SHADY VALLEY
T SHEET
U
DATUM
Upper GFM
Lower
DOUBLE UNCONFORMITY
∴ NW SIDE OF BASIN (13)

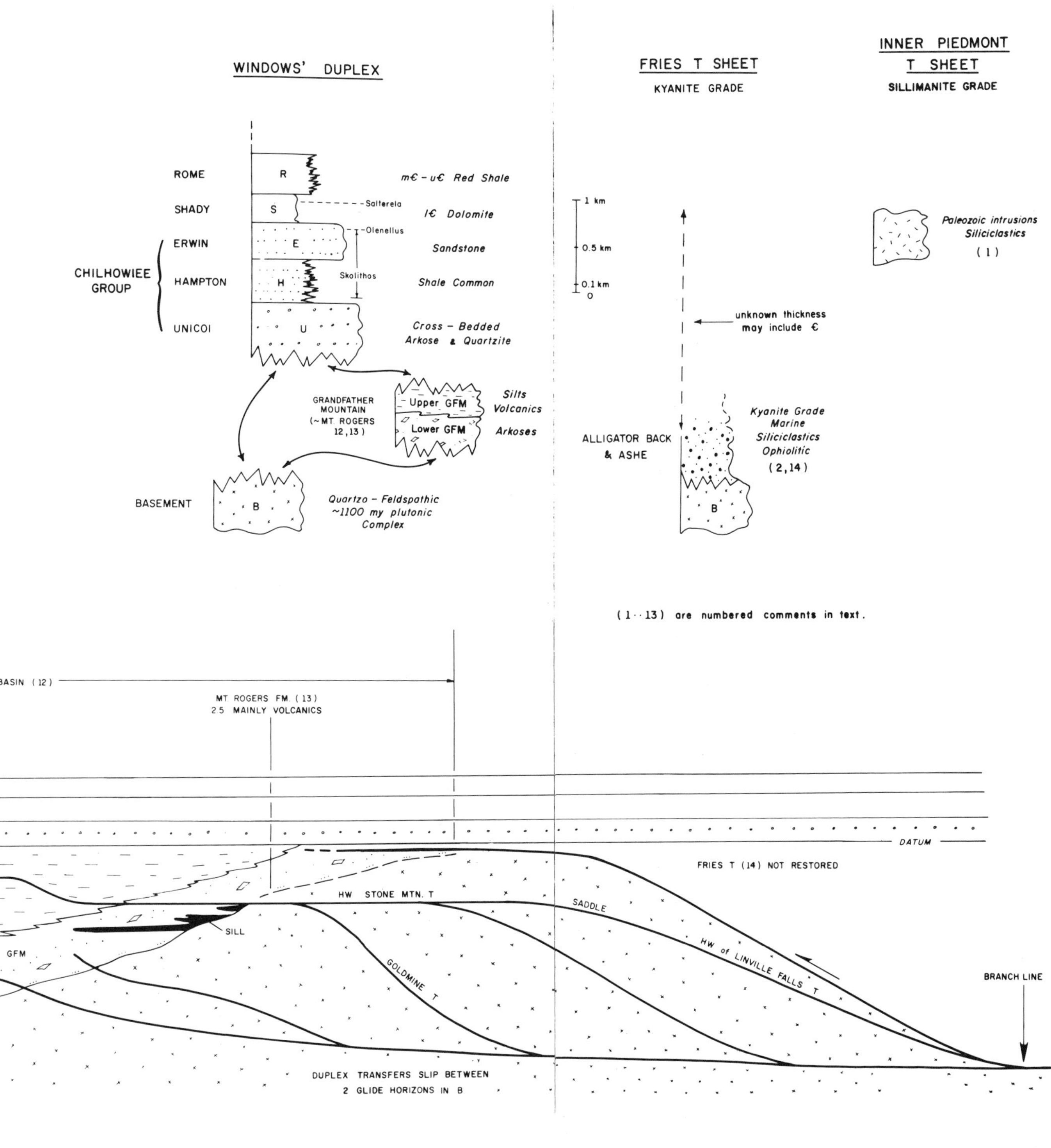
WINDOWS' DUPLEX
FRIES T SHEET
KYANITE GRADE
INNER PIEDMONT T SHEET
SILLIMANITE GRADE
ROME
SHADY
ERWIN
HAMPTON
UNICOI
CHILHOWIEE GROUP
R
S
E
H
U
Salterela
Olenellus
Skolithos
m€ – u€ Red Shale
l€ Dolomite
Sandstone
Shale Common
Cross – Bedded Arkose & Quartzite
1 km
0.5 km
0.1 km
0
unknown thickness may include €
Paleozoic intrusions Siliciclastics
(1)
GRANDFATHER MOUNTAIN (~MT. ROGERS 12,13)
Upper GFM
Lower GFM
Silts
Volcanics
Arkoses
ALLIGATOR BACK & ASHE
Kyanite Grade Marine Siliciclastics Ophiolitic
(2,14)
BASEMENT
B
Quartzo – Feldspathic ~1100 my plutonic Complex
(1··13) are numbered comments in text.
BASIN (12)
MT ROGERS FM (13)
2.5 MAINLY VOLCANICS
DATUM
FRIES T (14) NOT RESTORED
HW STONE MTN. T
SADDLE
SILL
GFM
GOLDMINE T
HW of LINVILLE FALLS T
BRANCH LINE
DUPLEX TRANSFERS SLIP BETWEEN
2 GLIDE HORIZONS IN B

THIS PAGE INTENTIONALLY BLANK

between formation of cleavage and the duplex, and of course much more complex situations can exist.

Culminations and Windows

Windows may form by simple differential erosions into undisturbed planar faults, but usually they form by erosion through culminations. Frequently, a culmination in a thrust surface is produced after the thrust was emplaced, and this folding could, in principle, run from top to bottom through the entire cross section including basement—an interpretation widely favored in older treatments. However, balanced cross sections and seismic exploration reveal that the folding itself is a consequence of thrust processes and is restricted to the rocks above a basal sole thrust. (This will be assumed throughout the rest of this paper.) Although the basal sole thrust is unfolded, it is not necessarily smooth; in fact it usually shows lateral and fronted ramps that create the overlying folds and culminations. The Moine thrust belt has several exposed examples showing this typical behavior, but in many other thrust belts the sole thrust is not available to direct observation (Elliott and Johnson 1980).

Culminations and Folds in Roofs of Duplexes

A culmination produced by a duplex with one frontal and two lateral ramps is idealized in Figure 25. In this model, the culmination has a flat top where the roof and floor thrusts are parallel and planar.

Erosion through this type of culmination produces a characteristic type of duplex window, which Tollman (1968, p. 46) called a "window of dislodged slices" (schurflingsfenster). The distinctive features on a map are: (1) a pattern of several subsidiary faults that join the roof thrust at the window margin, (2) doubly plunging anticlines within each horse, and (3) anticline crests and subsidiary faults are roughly parallel (Boyer, 1976, 1978).

There are several other ways in which a duplex can cause a culmination, such as dissimilar initial fault trajectories, changes in stratigraphic thickness, or variable contraction in the duplex. Excellent examples occur in places in the Moine thrust belt where contoured duplex thicknesses show a series of domes, basins, and saddles, rather like peas in a pod (Elliott and Johnson, 1980, Fig. 33).

If the slip on subsidiary faults is substantially less than the horizontal distance (p′) between horses, then not only is the overlying roof thrust immobilized, but it is also folded along with the new horse as it moves forward (Fig. 26). This produces a pattern "corrugated iron" and is one cause of the distinctive Appalachian Valley and Ridge folding and topography. The forward development of the duplex is accompanied by a sequential folding of the roof thrust, so that the family of congruent major folds are of progressively younger ages toward the foreland.

So far we have examined some of the geometric ways in which a duplex can form a culmination, but why should such a duplex and culmination form in one place and not in another? One reason would be a change in the ductility of the formations. Along-strike changes in carbonate facies toward bioherm buildups would make the stratig-

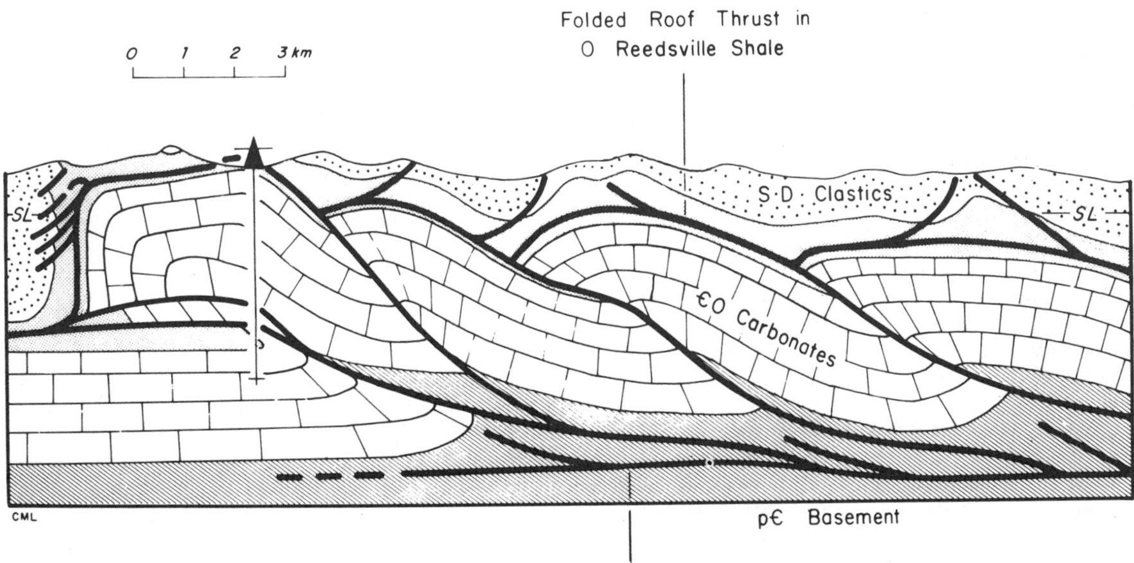

FIG. 26—Regular folds of central Appalachian Valley and Ridge are often underlain by duplex involving thick competent Cambrian-Ordovician carbonates. Small displacements on subsidiary faults which cut through large ramp produce periodically folded roof thrust. Sole is comparatively smooth in Lower Cambrian Waynesboro shales. Cross section in West Virginia after Perry (1978, Fig. 10).

raphy much less ductile and favor development of a duplex and culmination. In sandstones, small changes from carbonate to quartz cement could have the same effect. In the Moine thrust belt, local development of igneous intrusions slightly metamorphosed a shale and carbonate sequence, which during subsequent thrusting resulted in the Assynt culmination (Fig. 22).

Eyelid Windows

Now that we have looked at the possible causes of a culmination and the fault patterns within the window, let us examine the map patterns of the thrusts that frame the window.

An eyelid window is caused by the folding and erosion of an imbricate fan in such a manner that the window frame consists of several different thrust sheets (Fig. 27). The term "eyelid window" has a somewhat convoluted history. Sander (1921, p. 193, 212) originally suggested the name "scissors-window" (schernfenster), and later compared the pattern of framing thrust sheets to the overlapping blades in an iris diaphragm. Although Sander emphasized that scissors-window was a purely descriptive term, Tollman (1968, p. 47) pointed out that at one time these windows were thought to form by the opposed motion of the framing thrust sheets, like the blades in a pair of scissors. Oriel (1950, p. 46) suggested the more appropriate English name "eyelid window," and although the term was then in current use, he could find no published examples of its use.

In a discussion of the Goat Ridge eyelid window in Nevada, Gilluly (1960; Gilluly and Gates, 1965) argued that thrusts start folding as they form and are then abandoned for newer, higher, and as yet unfolded thrust surfaces (Fig. 21, right). He claimed that eyelid windows are diagnostic features of this hindward thrust progression.

Alternatively, we could have an imbricate fan that is then folded into a culmination—itself possibly owing to an underlying ramp or duplex. Erosion would produce an eyelid window, but one in which the thrusts all developed in forward progression. It seems that identical map patterns of eyelid windows could arise with either sequence of thrusting, and clearly each field example must be care fully analyzed.

The Lower Engadine eyelid window was the first one described (Sander, 1921). The Hot Springs eyelid window in the southern Appalachians is a more symmetrical example (Oriel, 1950).

Southern Appalachian Blue Ridge

Some 155 mi (250 km) southwest of our study area in the southern Appalachians, an important seismic line shows that the Blue Ridge and much of the Piedmont is a thin but southeast-thickening slab, 3 to 9 mi (5 to 15 km) thick, thrust over a Cambro-Ordovician sequence (Cook et al, 1979, 1980). We believe this basic geometry holds true over our region; and indeed some of this basic picture was anticipated from surface geologic evidence (e.g., Harris and Milici, 1977; Roeder et al, 1978). Regarding the Blue Ridge thrust complex itself, Cook et al (1979, p. 565) remarked that ". . . few seismic features are seen." Fortunately, the geologic map of the Blue Ridge provides exceptionally useful information for building a vertical cross section (Fig. 28). A key is two large windows, each on the order of 10^3 km^2—among the largest windows so far discovered in North America.

A balanced cross section (Fig. 29), in which each fault slice in the deformed state is restorable into the original stratigraphic framework, gives a picture of the fault trajectories just before movement occurred. Each formation within a fault slice has equal area in both the restored and deformed states, but in the lower grade portion of the cross section, to the northwest, where finite strains and fabric are less intense, formation thicknesses are conserved, and formation lengths remain unchanged.

What is the sequence in which the thrusting occurred? J. Rodgers feels that within most of the Valley and Ridge province the general progression is forward, but around the Grandfather Mountain and Mountain City windows he accepts the suggestion of Gilluly (1960) that the thrusts developed in hindward sequence (Rodgers, 1970, p. 170, 178). Hatcher (1978, p. 294-295) proposed a compromise in which thrusting progresses forward in two waves, so that thrusts in the first wave are cut off at the back by younger ones. We shall try and show that there are no difficulties in building up the Blue Ridge complex with a forward thrust progression.

To explain our cross section (Fig. 29), we will briefly list some conclusions in point form; these points have the same numbers on the cross section. Although we provide only partial references, it is important to note the superb work by the U. S. Geological Survey in this area (King and Ferguson, 1960; Bryant and Reed, 1970; Rankin 1970).

1. The Inner Piedmont thrust complex, composed of sillimanite-grade immature siliciclastics intruded by Paleozoic plutons, is the most internal tectonic unit and continues well to the southeast of this line of section. About 47 mi (75 km) northeast is the Smith River thrust sheet, which clearly overlies the Fries and is equivalent to part of the Inner Piedmont (Conley, 1978). This thrust complex may once have covered the Fries sheet on this line of section, but is now found only south of the Brevard zone because of deep erosion and Triassic faulting. It is the highest and largest single tectonic unit, so in a sense we can describe it as a dominant thrust sheet—a topic which we discuss later.

2. The Fries thrust sheet carries rocks that were folded several times, affected by a pervasive kyanite grade metamorphism of Ordovician age, and intruded by Silurian-Devonian granites. The Fries thrust surface juxtaposes this entire amphibolite-grade complex onto a substantially lower grade footwall (Rankin et al, 1973). We cannot yet restore the Fries or Inner Piedmont thrust complexes.

Our thrust sequence would start, then, with the Fries moving over our area of interest carrying the Inner Piedmont thrust complex on top of it.

3. The Linville Falls thrust moved next, accompanied by formation of a penetrative cleavage. In basement rocks this fabric is cataclastic and occasionally mylonite

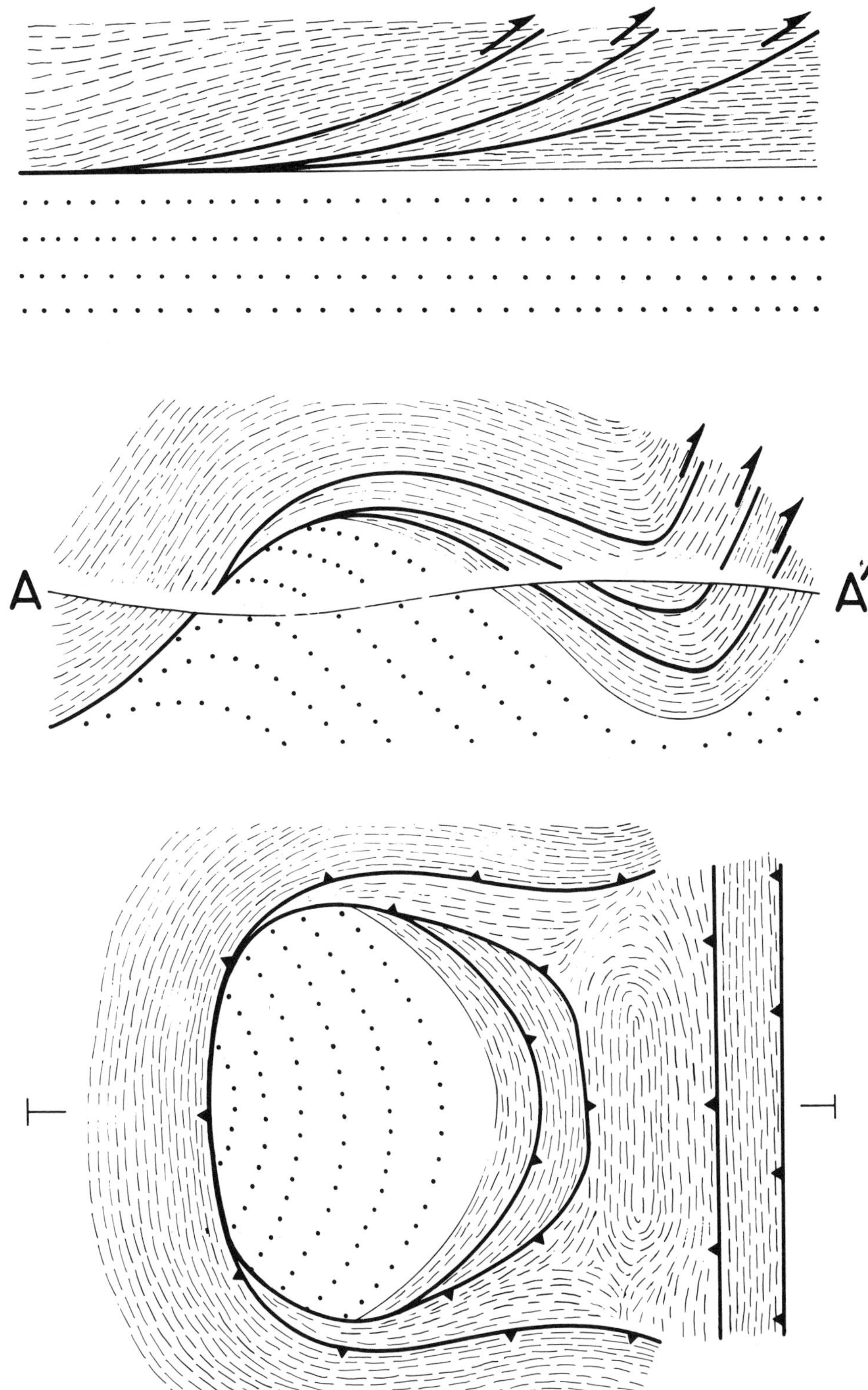

FIG. 27—Characteristic map pattern of an eyelid window (below and cross section AA′ (center). First we create an imbricate fan (center cross section), followed by sufficient but not excessive erosion (Boyer, 1978).

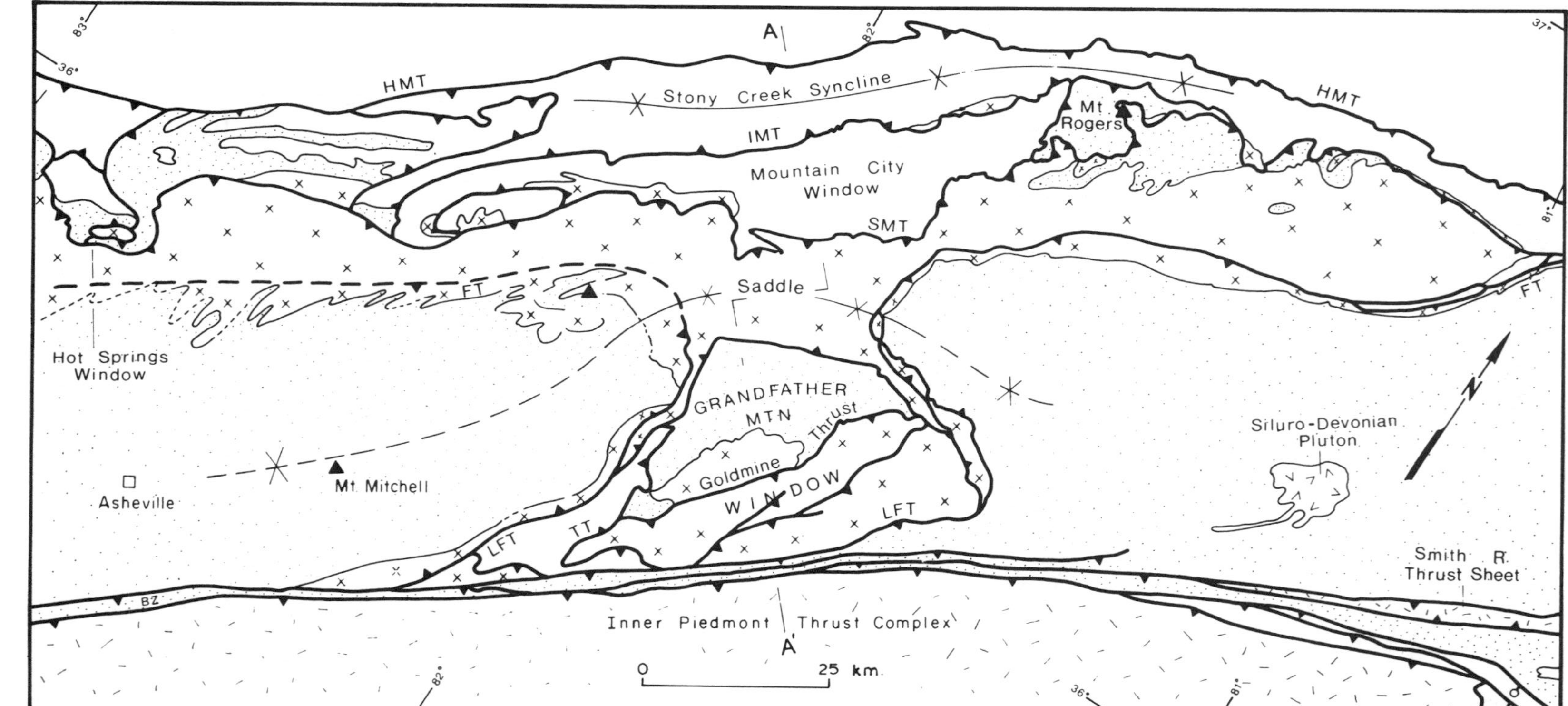

FIG. 28—Geologic map of Blue Ridge area in southern Appalachians of Tennessee, Virginia, and North Carolina. Holston Mountain thrust (HMT), Iron Mountain thrust (IMT), Stone Mountain thrust (SMT), and Linville Falls thrust (LFT) are all part of the imbricate system whose sole is intermittently exposed for over 47 mi (75 km) across strike. Precambrian basement is dashes or crosses, late Precambrian siliciclastics are stippled, Chilhowee group siliciclastics and Cambrian-Ordovician shales and carbonates are without pattern. Only major thrusts are shown, and no cross faults. Brevard fault zone is indicated by BZ. Based on Rankin et al (1973).

develops. The fabric curves asymptotically to the Linville Falls thrust and the strain and fabric are stronger near it.

Once the Linville Falls thrust was moving, a family of imbricate subsidiary faults cut upward to join the Linville Falls as a roof thrust and curved asymptotically downward to a floor thrust. This duplex continued to develop toward the northwest, with the roof's name changing to Stone Mountain thrust. We interpret the Mountain City and Grandfather Mountain windows as two culminations of the same duplex, the Windows duplex.

4. The main cleavage was accompanied by syntectonic metamorphism, which reaches into amphibolite-grade in the most southeastern horse within the Grandfather Mountain window, but drops off steadily toward the northeast and becomes greenschist. None of the horses brought up lower crustal or upper mantle rocks, suggesting that the initial trajectory of the floor thrust was subhorizontal (Bryant and Reed, 1970, p. 219). The northwestward decline in metamorphism is accompanied by a decrease of cleavage intensity, which becomes very weak and sporadic at the northwest side of the Mountain City window.

The Windows duplex had an initial length 78 mi (125 km), but its current length, between branch lines A and X, is 28 mi (45 km) which indicates a shortening of 50 mi (80 km).

5. The structure of the Grandfather Mountain window is based upon composite down-plunge projections (Boyer, 1976, 1978). Possibly the floor thrust climbs section along strike and updip to produce the Grandfather Mountain culmination. This interpretation is supported by the map pattern of large late-kink folds, which essentially define the culmination and whose minor structures are a crenulation cleavage, and by minor kink folds that deform the earlier main cleavage. Variable contraction is also a factor; for example, contraction may decrease beneath the saddle and then increase again within the Mountain City window. A third possible cause of the culmination is a still lower culmination in the floor thrust of the Mountain City window.

Erosion subsequently produced the duplex window (cf. Figs. 28 and 25). Also, the pattern of the Fries and Linville Falls thrusts, (the window frame), creates an eyelid window (cf. Figs. 28 and 27).

6. The Linville Falls, Fries, and several other thrusts beneath the Inner Piedmont sheet all merge toward branch lines in the Brevard zone. This merger is accompanied by an increase in the southeast dip of bedding and schistosity toward this narrow band of intense cataclastic deformation. Isotopic evidence summarized by Conley (1978, p. 1122) suggests that the principal time of activity of the Brevard zone, and therefore the branch lines for the major thrusts that lie within it, is Late Devonian.

The Brevard zone is not only a thrust structure, for Triassic faults can be traced into it (Fig. 28), and there is a scatter of Triassic mica ages from the mylonites (Butler and Dunn, 1968; Rankin, 1975, p. 329). The Brevard zone started as the site of major Late Devonian thrusting and was followed by Triassic listric extension faulting. There are good analogs to this pattern in the Cordillera (e.g., Bally et al, 1966, plate 5; Royse et al, 1975, p. 46).

7. Repeatedly the Valley and Ridge thrusts have hanging walls within the lower part of the Rome Formation, and the total overall shortening is roughly 75 mi (120 km) (Roeder et al, 1978, p. 23). This requires that a thrust, whose footwall is basal Rome, persist for at least 75 mi (120 km) beneath the Blue Ridge and Piedmont. The depth to this sole thrust agrees with that cited by Roeder et al (1978, p. 11) and with a seismic line shot 37 mi (60 km) southwest.

8. This seismic line, described by Harris and Milici (1977, Line TC-2), shows a major 2.5 mi (4 km) thick duplex involving Cambrian-Ordovician formations with the Pulaski thrust as roof and the sole thrust as floor. This lower duplex dips under the Windows duplex.

9. The floor thrust for the Windows Duplex we shall call the Windows floor thrust. Although we show the Windows thrust as a smooth and almost planar surface, it would be quite normal for this roof (of the lower duplex) to show low-amplitude folding.

Clark et al (1978) identified this thrust as overlying subhorizontal seismic reflections caused by to Cambrian-Ordovician sediments. These sediments were traced by Cook et al (1979, 1980) beneath the Blue Ridge and then for 93 mi (150 km) beyond the Brevard zone.

Cook et al (1979, p. 563; 1980, Fig. 3) called the Windows floor thrust the sole, and claimed that the underlying Cambrian-Ordovician sequence lies in place on its basement. We think that the geometric argument in point 7 proves this interpretation impossible.

10. At branch line A, the Windows floor and roof thrusts join the Holston Mountain thrust whose footwall cuts as high as mid-Ordovician shale. Some and possibly all of the 19 mi (30 km) of shortening in the Windows duplex may have transferred to slip on the Holston Mountain thrust, which therefore might have moved as the Windows duplex formed. If the Windows duplex started shortening in Late Devonian (see point 6) at 5 mm/year, then the total shortening of 50 mi (80 km) required 16 m.y. and was completed early in the Late Mississippian.

11. At a lower branch line, the Pulaski thrust merges with the Windows floor thrust, which then continues southeast across the section. The youngest formations currently overlain by the Pulaski are Early Mississippian. Because the Pulaski and Windows floor thrusts have a combined order of magnitude slip of 124 mi (200 km), any Carboniferous formation once deposited in those thrust sheets might have been in thick synorogenic facies and unlike any that crops out farther northwest.

12. The Grandfather Mountain formation consists of fluvio-deltaic sediments derived from both northwest and southeast sides of a steep and probably fault-bounded basin containing both felsic and basaltic flows and sills. The upper part of the basin is truncated by the Stone Mountain-Linville Falls thrust system, and in some way the stratigraphic sequence is continued into the hanging wall as the Mt. Rogers formation, one-half of which is volcanic.

13. One key to restoration of this complex thrust system is the presence 9 mi (15 km) northeast, in the Shady

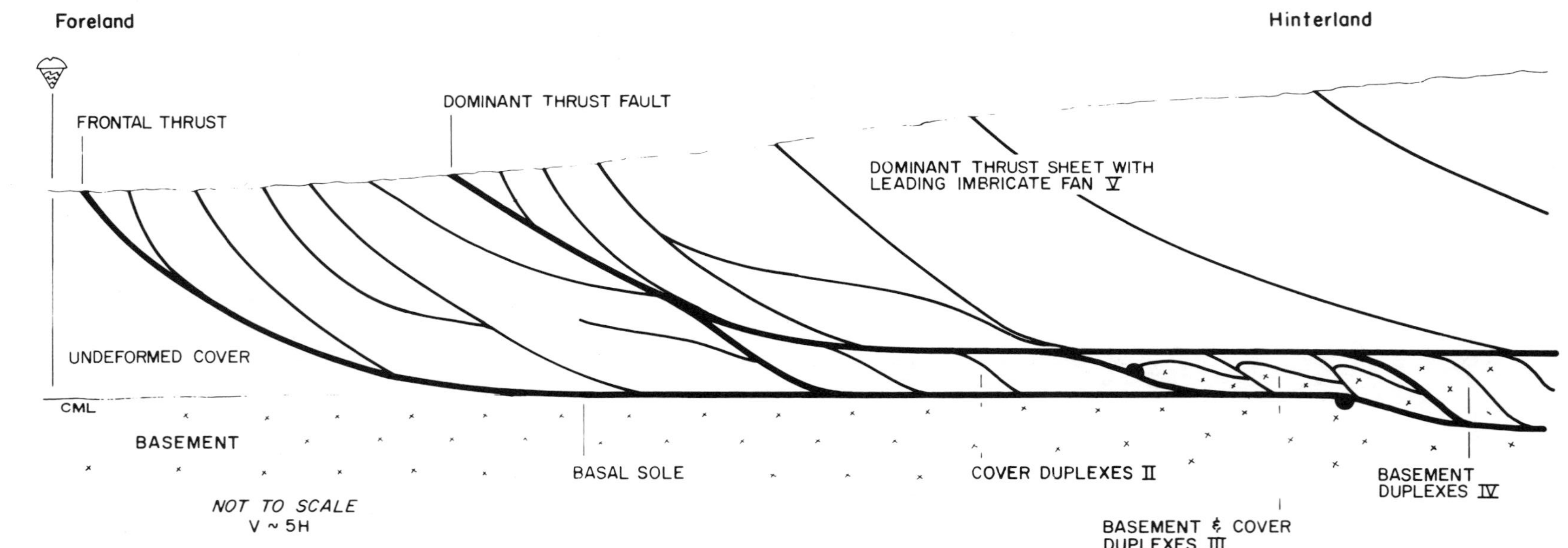

FIG. 30—Idealized sketch of thrust belt with dominant sheet shows five zones, each with characteristic thrust system and degree of basement involvement. Thrust faults at zone boundaries are bold. Vertically exaggerated and not to scale.

Valley thrust sheet, of a double unconformity—the intersection of the Unicoi/basement and Mt. Rogers/basement unconformities— suggesting a restoration to the northwest side of the Grandfather Mountain basin. A series of thrusts in the Mt. Rogers area, above and northeast of the Mountain City window, indicate a basin whose floor had considerable relief, an irregular shape, and sequences that could come from either side. This series of thrusts is too far (12 to 25 mi; 20 to 40 km) off this line of section for reliable projection.

14. Although it is not yet possible to restore the Fries thrust sheet, we do have constraints on its paleogeographic site. Formations recognized near the cross section (Ashe and Alligator Back) might have equivalents to the northeast (Lynchburg, Catoctin, and Candler or Chilhowie), indicating a range in age of deposition from late Precambrian to early Paleozoic (Rankin, et al, 1973).

Plate Tectonics and Blue Ridge Thrust System

The rocks of the Fries sheet suggest a marine siliciclastic sequence which is occasionally ophiolitic; one might infer a slope or possibly a rise sequence deposited on a newly rifted continental margin (Rankin, 1975, p. 315). In the northwest part of the Fries thrust sheet, the stratigraphic base is basement, so that at least the western sediments in the Fries sheet were deposited on continental crust.

Since we see the southeast side of its basin, the Grandfather Mountain formation could never have been laterally continuous into the Ash and Alligator Back formations, as thought by Hatcher (1978, p. 294, Fig. 4). It would appear that the Ash and Alligator Back sequences were deposited near the northwestern continental margin of a much larger basin, one whose adjacent ocean had an unknown southeast extent. Evidence of this ocean has now vanished, presumably into a thrust complex beneath the Inner Piedmont sheet, and several authors suggest that the Brevard zone started as a southeast-dipping subduction zone, which became a suture marking a continent/microcontinent collision (Odom and Fulligar, 1973; Rankin, 1975, p. 320; Hatcher, 1978, p. 295; Cook et al, 1979, p. 566, 1980, p. 213).

Is it true that there must be a subduction zone everywhere we see oceanic crust disappearing down a major thrust? The west side of Japan has active thrust faults that are on the opposite side of the islands from the subduction zone and that have an opposite sense of dip (Uyeda, 1977, p. 5). Mesozoic examples, such as the southern Andes (Dalziel, 1981), suggest that such a marginal basin can close right up, with the arc or microcontinent driven over the marginal basin and onto the continental foreland.

It would seem possible that the dominant Inner Piedmont sheet, either as an arc or a microcontinent, was thrust over the marginal basin and continental slope (Fries), and then everything thrust over the continental shelf (Linville Falls–Stone Mountain sheet and the Windows duplex), yet none of these thrusts would necessarily be traceable into the subduction zone.

THRUST BELTS WITH A DOMINANT SHEET

A portion of an orogenic belt may have an overlying dominant thrust sheet whose displacement is much larger than any of the others and whose motion dominates the evolution of this region.

We compare several dominant sheets in Table 2. Because of their great amount of slip and consequent uplift, many contain a substantial portion of coarse-grained crystalline rocks, such as an old quartzofeldspathic basement or an oceanic cumulate and mantle sequence. However, this is not a requirement, for example, the fronted part of the far-traveled Lewis thrust sheet consists of fine-grained clastics and carbonates of the Precambrian Belt group. At least part of the rocks now in front and beneath dominant thrusts are a stratified cover assemblage. Many of these assemblages unconformably overlie a basement of continental crust, often an old crystalline gneissic complex that we refer to as foreland basement. The stratified cover in some cases may include an Atlantic-type continental-margin assemblage, complete all the way to oceanic sequences. We have already discussed one dominant thrust sheet in the Blue Ridge thrust complex, and we shall examine one in the Alps. A generalized and vertically exaggerated cross section through a thrust belt with a dominant sheet shows five different zones (Fig. 30).

1. Closest to the foreland is a zone with cover telescoped into an emergent trailing imbricate fan above a basal sole thrust.

2. For a considerable distance beneath the dominant sheet, the cover is telescoped and thickened by duplexes over a deep-lying basal sole, below which foreland basement is still not involved. It is important to note that simply because crystalline rocks are found at the surface in a thrust sheet there is no reason to assume that foreland basement is necessarily involved in the thrust belt directly below.

3. Eventually a zone is reached (Fig. 30, III) well beneath the dominant sheet where the foreland basement does become involved with the cover in duplex thrust systems. Onset of this zone is controlled by metamorphic temperature in the basement and probably is a little less than 300°C (Voll, 1976).

If this temperature is reached well within the basement, the thrust may start as a glide horizon along this isotherm, and the fault must then cut upsection though lower temperature basement, probably as a ramp, before it reaches the cover.

4. At a deep level, a zone is reached in which the dominant thrust sheet directly overlies duplexes of foreland basement and the originally intervening cover rocks have been stripped off (Fig. 30, IV).

The thrust systems in zones II, III, and IV (Fig. 30) are duplexes, whose variation in thickness is an important cause of culminations and windows.

5. The dominant sheet is itself composed of numerous thrusts, all of which moved piggyback, and none of which had as great a slip as the dominant thrust fault. In other words, the dominant sheet is a leading imbricate fan whose sole is the dominant thrust fault.

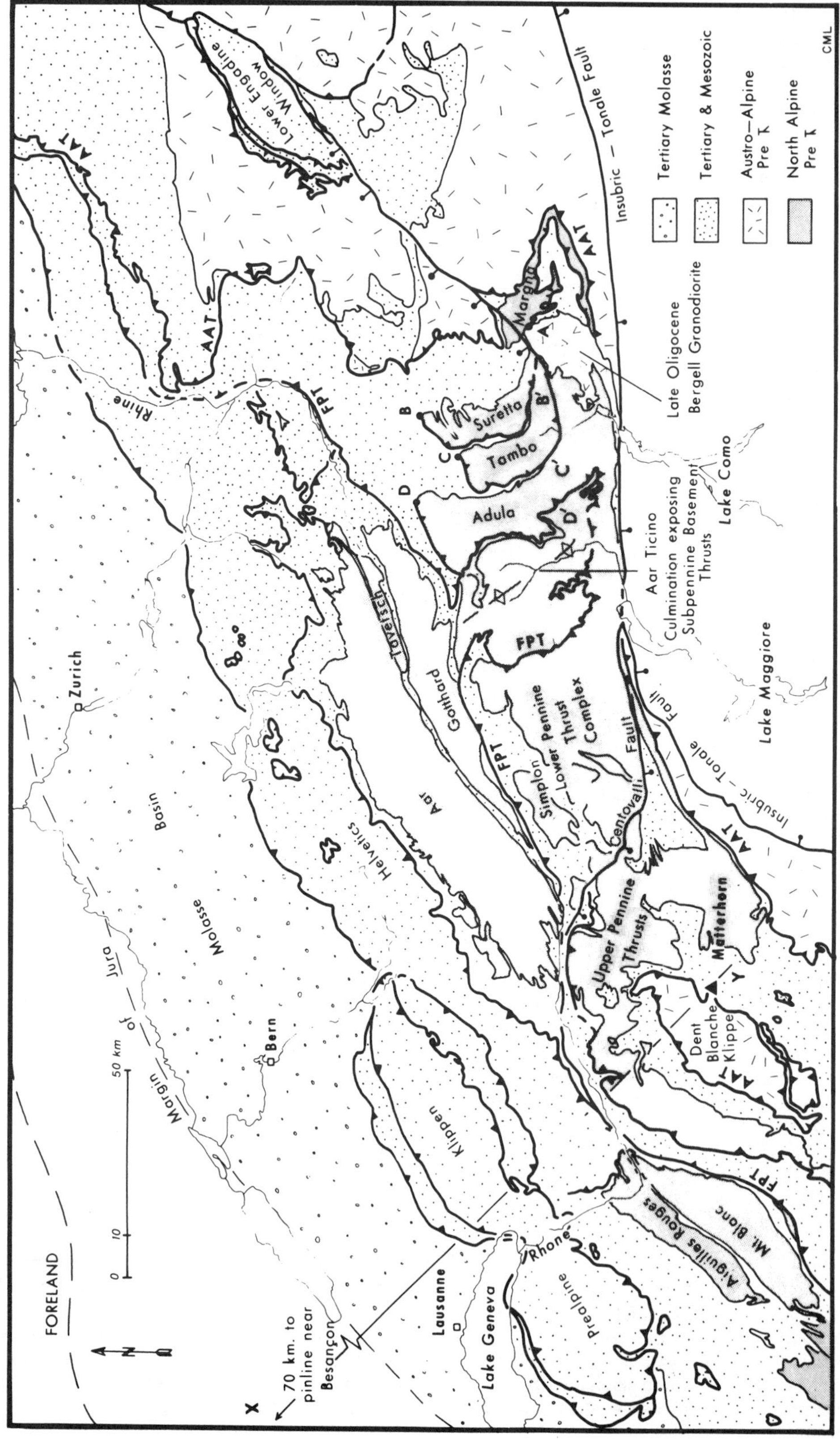

FIG. 31—Map of central Alps (modified from Spicher, 1972; Milnes, 1974).

ALPINE THRUST SYSTEMS

Helped by a concentration of work in a region of excellent exposure and high relief, thrust structures in the Alps were described at an early date and quickly became a widely accepted model. It is clear that a number of concepts developed in the Alps are precursors to the methods described here.

The emphasis and method of approach to Alpine nappes have traditionally been rather different from that outlined in this paper. To some extent, this Alpine approach is used in all the various chains about the Mediterranean, eastern Europe, the USSR, and the Himalayas.

Tollman (1968, p. 35) pointed out that a principal objective in the Alpine tradition is a description of the nappe body or thrust sheet volume, and that this is frequently accompanied by an incomplete characterization of the thrust surface. We will try to show that geometric reasoning based on thrust surfaces, tip and branch lines, and the various thrust systems, provides both an economical description of the structural geometry and a somewhat different perspective on the evolution of this classic belt.

Consider a cross section across the Jura, Prealps, and Helvetics to the Pennines, as shown on a map (Fig. 31). We rely heavily on outstanding reviews of Alpine geology by Trumpy (1973, 1980). The surface geology follows a recent cross section by Baud et al (1978), but our section is heavily modified and changed at depth. Below and northwest of the Frontal Pennine thrust these changes result in a cross section restorable by plane strain or balanced (Fig. 32). Letters identify corresponding points in the deformed and restored sections, and both letters and numbers on the cross section refer to the following points discussed in the text.

Pre-Triassic rocks (1) are principally a quartzo-feldspathic crystalline basement, but also include substantial Permo-Carboniferous sediments. Mesozoic rocks (2) start with a shallow-water carbonate-evaporate Triassic sequence which indicates that the underlying siliceous basement was continental crust at that time. The Jurassic and Cretaceous include a complete Atlantic-type rifted margin sequence, varying from a carbonate shelf, below and north-northwest of the Frontal Pennine thrust, to various deep-sea deposits with ophiolites.

The Tertiary-Cretaceous boundary is the horizontal datum in the restored section. The Tertiary rocks vary from flysch (3) to molasse (4) and their age and facies reflect a growing thrust belt, advancing toward the North Alpine cratonic foreland (Trumpy, 1980, p. 50).

We will describe the balanced part of the cross section in the rough chronological order of a forward-progressing thrust system. The initial stage starts with the Frontal Pennine thrust, already carrying, in piggyback fashion, a whole sequence of originally more internal sediments and thrust sheets. The Frontal Pennine thrust, including the Prealpine sheets, glided on a planar footwall of Eocene flysch, which overlies the undeformed Helvetic carbonates (5, from e to n). The most internal of this flysch was stripped off the underlying carbonate. Does this flysch now form a forward-dipping duplex of downward-facing structures, including the Ultrahelvetic thrusts (near 20)?

Southeast of (e) the underlying faults all join the Frontal Pennine thrust, which is thus the roof of enormous antiformal stacks and duplexes. Restoring the Mesozoic gives a length (f, h, k, l, m, n) of 68 mi (110 km) along the Tertiary-Cretaceous boundary (9), which was once overlain by Tertiary flysch. This flysch now beneath the Frontal Pennine thrust has a cross section area of 170 km^2. If most of this flysch came from northwest of (n) on the restored section, then its initial thickness (t) is 1.6 km (10).

The Helvetics then started to develop and fold the overlying Frontal Pennine thrust. Earlier and higher thrusts such as the Wildhorn (6) and the Diableret (7), were in turn folded by growth of the underlying and younger Morcles thrust sheet (8). We suggest that the Helvetic thrust system is an antiformal stack, 30 times larger but in a sense geometrically comparable with the Dundonnell structure (Fig. 23). Radiometric ages of the low-grade syntectonic metamorphism in the Helvetic stack are Oligocene (Trumpy, 1980, p. 60).

Each Helvetic subsidiary thrust in turn passed on its slip to part of the roof thrust in front and then became immobilized, moving the Prealpine thrust complex farther and farther northwest. In the late Oligocene (~25 m.y.), the Frontal Pennine thrust cut upward through the lower molasse to the erosion surface, and shed pebbles into the nearby Molasse basin (13). Eventually, Helvetic thrusts in western Switzerland overrode lower Miocene Subalpine molasse. After the Prealpine complex was established over its final foot wall rocks, in late early Miocene to early middle Miocene time (13-16 m.y.), the entire Frontal Pennine thrust system and its Helvetic underpinnings were abandoned.

In the next stage, slip started at a lower level, a decollement in Triassic evaporites beneath the Molasse basin. Imbrications from this basal sole thrust emerged in the Molasse basin northwest of the Prealps, where the youngest Molasse was deposited in late Miocene time (~6 m.y.). In the Subalpine region (14), the Molasse is underlain by a more intensely imbricated Mesozoic (area cdfg). The initial thickness of the Mesozoic thins down to the southeast.

An emergent imbricate fan gradually moved across the Molasse basin (15, between a and c), and the Molasse sediment was cannabalized and recycled. The current 28 mi (45 km) width records at least 6 mi (10 km) of shortening if we assume a roughly average thickness $t \sim 3.1$ km for the Mesozoic.

The pinline (16) is chosen in the foreland, northwest of Besancon. The current distance from the pinline to point (a) is 43 mi (70 km) (17), the entire width of the Jura. However, the Jura underwent a shortening of 15 to 19 mi (25-30 km) (Chauve, 1975, p. 23), giving an original length between pinline and point (a) of 53 mi (85 km). Orogenic deformation in the Jura ended before early late Pliocene (before ~2.5 m.y.).

Where does all the displacement go that accumulated on the basal sole beneath the Jura and Molasse basin?

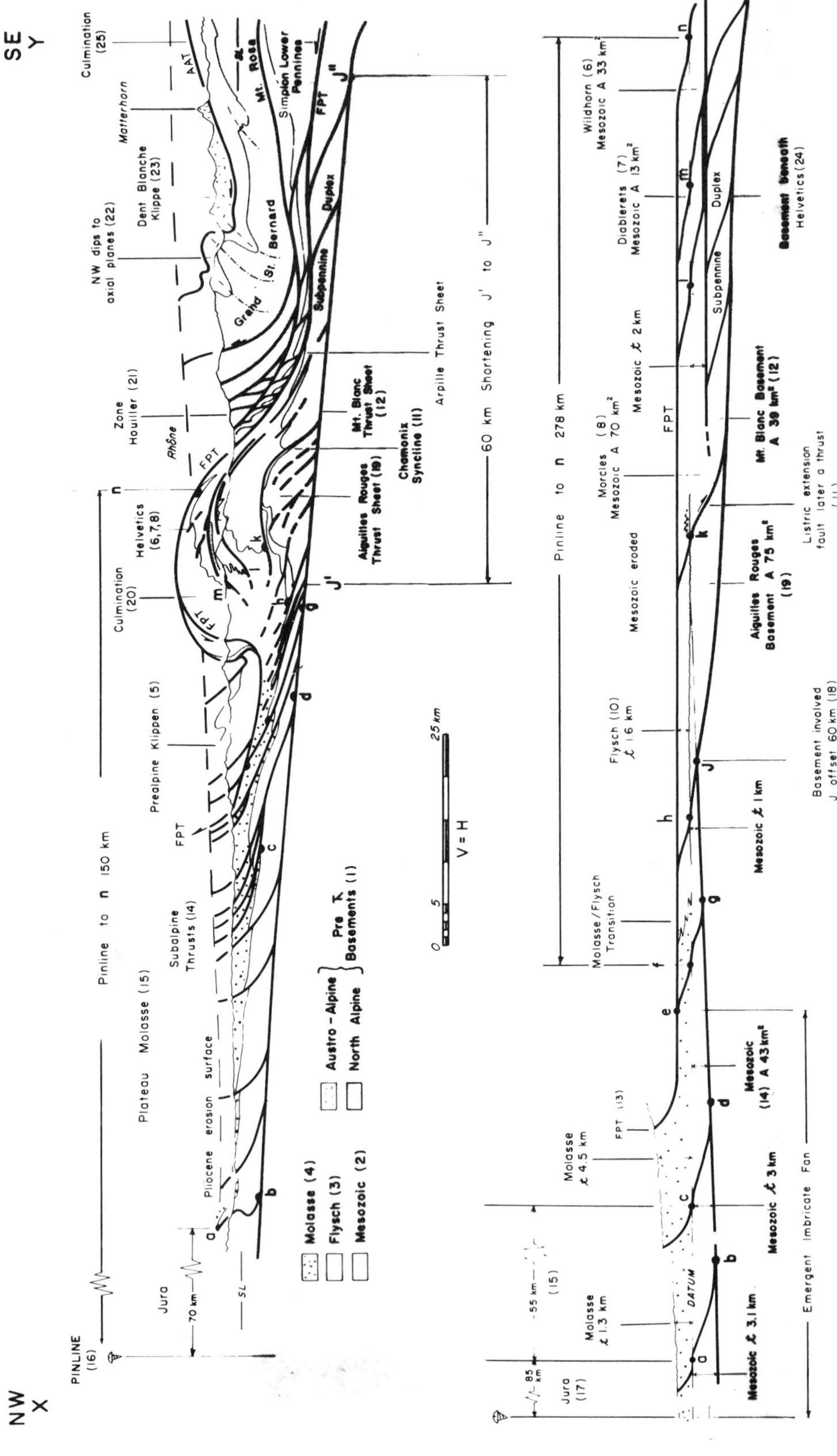

FIG. 32—Cross section through Alps along line XY′ (Fig. 31). Letters and numbers on section are location of points discussed in text.

One solution, implicit in most published cross sections, is to label the external basement massifs "parautochthonous" and to move most of the displacement over them. We favor an alternative solution where the displacement passes beneath the external basement massifs. In the undeformed state, the footwall and hanging wall cutoff must coincide (18). By keeping corresponding areas (Jheg) equal in restored and deformed state cross sections, the contact of basement and cover (J″) must be 37 mi (60 km) farther southeast than is J′ (cf. Hsu, 1979).

The Aiguilles Rouges massif (19) then appears as one subsidiary slice of a duplex involving basement and is most easily seen on the restored section. The tapering Mesozoic carbonate shelf sequence is completely removed from part of this massif (Badoux, 1972).

The Mesozoic in the Morcles nappe thickens abruptly (8), and the upper Eocene section consist of northwest-derived clasts, including large blocks of basement (Badoux, 1972). This suggests that a southeast side-down listric extension fault (11) was active until late Eocene time, a type of faulting widespread at that time (Trumpy, 1960, p. 882). We suggest that such a listric extension fault was subsequently used as a thrust and that the Chamonix syncline, separating the Aiguilles Rouges and Mount Blanc massifs, could mark its position. If so, then the thrust beneath the Mount Blanc massif (12) might have followed the subhorizontal basement decollement of the earlier listric extension fault.

The Frontal Pennine thrust goes over all the Helvetic sheets and then dips down to pass beneath the enormous klippe that makes up the Prealps. This major culmination (20) is due to a combination of four causes. First, horses of Ultrahelvetic flysch from farther southeast accumulated between the Prealps and Helvetics. Second, the Helvetic sheets have both their Mesozoic and Tertiary portions directly above one another, causing an antiformal stack. Third, as basement on the hanging wall moved onto the flat basal sole thrust, the external massifs arose as immense hanging wall anticlines ("Powell Valley-type," as in the Pine Mountain thrust sheet) folding the overlying thrusts. However, the external massifs themselves are a composite array of subsidiary faults and our fourth cause is a particularly intense local internal shortening and thickening. Some readers may recall the theory of "basement wedges" (Fig. 33), summarized by Umbgrove, 1950, p. 66, which had a following between the 1920s and 1950s, but is now seldom referred to. These basement wedges have the same effect—in our terms—as differential shortening in the upper part of a duplex.

It may be possible to date this event. Rare pebbles from the Helvetic nappes are first found in the upper Miocene molasse, but we attach a different significance to these pebbles than do previous writers. If the Helvetic thrust sheets are all horses, then they were not exposed to erosion during their development and emplacement. The pebbles do not record active Helvetic thrusting; instead, they mark when erosion through the stack of thrusts was sufficiently deep to reach the Helvetics, and must date growth of the external massifs.

The shortening recorded beneath the Frontal Pennine thrust is 75 mi (120 km), from the pinline to point (n). This value is probably much smaller than the shortening above and southeast of the Frontal Pennine thrust, and, although we cannot restore this part of the section, we wish to mention some interesting conjectures.

The imbricate slices making up the Prealps correlate with the region above the Frontal Pennine and below the Austro-Alpine thrust (21). First is a complex of three zones (Sion-Courmayeur, Houillere, and Great St. Bernard) each representing smaller scale duplexes. Possibly, then, there is a transition between emergent imbricate slices in the Prealps to buried multiple duplexes between the Frontal Pennine and Great St. Bernard thrusts.

Because of the structural relief of the Aar-Ticino culmination, the structure of the internal zones is clearer to the east (Fig. 31). The classic interpretation is to look down-plunge to the east-northeast at the map pattern. There were several strong phases of folding after the thrusts formed (Milnes, 1974; Milnes and Pfiffner, 1980), and our reconstruction is a very rough attempt to view the geometry, if these folds were removed (Fig. 34)

The region is dominated by about 10 major thrust sheets composed mainly of basement, but partly separated from each other by thin septa of Mesozoic cover. The axis of the Aar-Ticino culmination exposes the deepest level, a series of "Subpennine" thrust sheets (Milnes, 1974). We shall call this the Subpennine duplex and its roof is the Frontal Pennine thrust.

The overlying Pennine thrusts (Adula, Tambo, Suretta, Margna) are shown on published maps (e.g., Spicher, 1972) to end at points A, B, C, D (Fig. 31). We interpret these points as hanging wall cutoffs of the basement cover contacts, and suggest that the thrusts continue cutting upsection in the Mesozoic, eventually joining the Austro-Alpine thrust (Fig. 34). If so, then the Pennines are a duplex with the Frontal Pennine thrust as floor and the Austro-Alpine thrust as roof. These branch and cutoff lines are of considerable interest, for different arrangements are possible, and lead to quite different paleogeographic reconstructions.

Farther west on our cross section (Fig. 32), restoration of the Mesozoic (24) shows that the basement that once lay under the Helvetics can only partly be seen at map level. We suggest that this basement forms an extension of the Subpennine basement duplex. Above is the Simplon Lower Pennine basement duplex, which may be similar to those seen more clearly to the east (Fig. 34). Major culminations and depressions in the overlying Great St. Bernard and Austro-Alpine thrusts (23) could be due to variable thickness of the two basement duplexes (25).

The Austro-Alpine thrust sheet (AAT on Figs. 31, 32) has long been recognized as having a dominant role, and was vividly characterized up as a master overriding thrust sheet behaving like a giant earth-moving machine, a "Traineau ecraseur" (Termier, 1904). The five different zones beneath and in front of this dominant sheet are fairly clear (cf. Fig. 30 to Figs. 32 and 34), but we do not imply that the Austro-Alpine thrust sheet once covered, for example, the Prealps (Fig. 32). It is possible that some of the duplex systems beneath the Austro-Alpine thrust

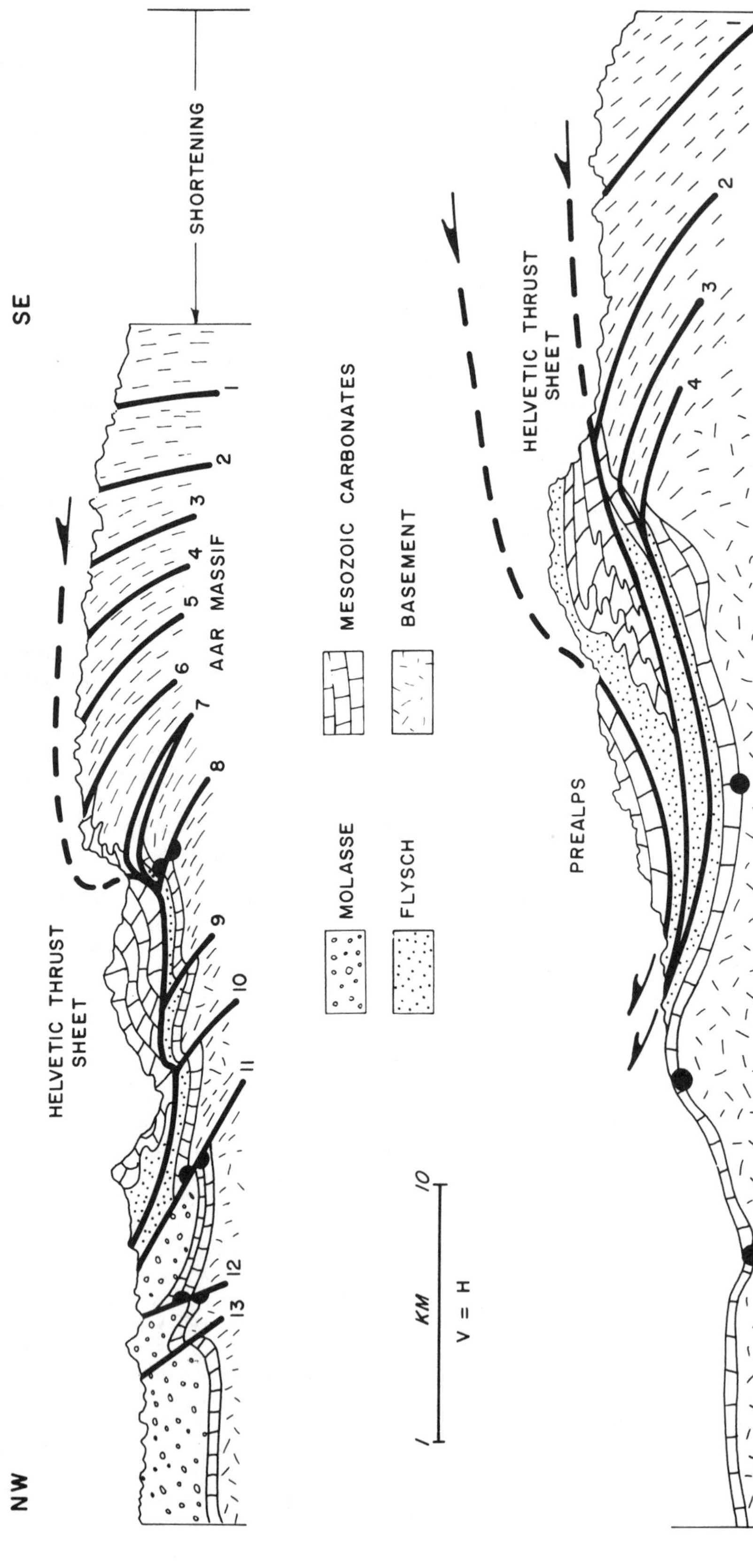

FIG. 33—Top: cross section from Molasse basin to Aar massif. Bottom: earlier stage in basin evolution (after Gunzler-Seiffert, *in* Umbgrove, 1950, Fig. 47; Holmes, 1978, Fig. 3.21). This figure illustrates that old idea of basement wedges may be equivalent to upper part of a duplex. Piggyback emplacement of Prealpine klippe and Helvetic thrust sheets by successive imbrication of underlying crystalline wedges are numbered in order of formation. Telescoping of slices steepens older wedges.

became emergent to the west, similar to the Lewis thrust and its underlying structures as discussed earlier.

ROOT ZONES

The root of a thrust surface on a map is where the thrust finally dives from view for the last time. When a number of thrusts do this in the same area, it is called a root zone. Part of the interest in root zones stems from plate tectonics, for if the root zone functioned as a subduction zone it is necessary, although not sufficient, that oceanic lithosphere once existed in the footwall. If, in addition, the two margins of the root zone were once continental lithosphere, then the root zone may also be a suture. Root zones have long been of interest, particularly in two regions already discussed: the internal Alpine root zone, just north of the Insubric-Tonale fault (Fig. 31); and the Appalachian Brevard zone (Fig. 28). Both these root zones are very narrow, often less than 6 mi (10 km) wide, and consist largely of cataclastic basement rock; the similarity of the two root zones is brought out by Burchfiel and Livingston (1967).

The classical Alpine picture is ". . . a belt of subvertical strata representing a former wide zone from which the Nappe has been squeezed out" (Rutten, 1969, p. 206). This theory of roots has three different aspects. First, the theory implies that development if the root zone caused the subvertical dips. But the steep dip could be imposed later than the motion of the thrusts whose roots make up the zone; Alpine field work that demonstrates this is reviewed by Milnes (1974). Further, a group of thrusts is more likely to intersect the ground surface if it is affected by a steeper than normal dip, so the correlation of steep dip with root zones may not be so much causal as statistical.

The second implication of the classical theory is that a root is the "inner margin of the former home of a nappe" or, in other words, a root coincides with a footwall cutoff. The slip distance of a thrust, measured in cross section, is the arc length between the hanging wall and footwall cutoffs. Often the footwall cutoff cannot be found on the map, and then the minimum possible slip is the distance from the hanging wall cutoff back to an assumed footwall cutoff just below the ground surface in the root. Classical theory assumes that these minimum slip estimates are actual values and that all the thrusts that root in the zone have footwall cutoffs at the same place—clearly an artifact of oversimplified assumptions.

The third part of the classical picture explains the narrow width of root zones as the effect of very large flattening strains imposed on originally much thicker thrust sheets. However, a thrust sheet is bound above and below by fault surfaces, and at the approach to the trailing branch line the thrust sheet must thin (Fig. 1). Consequently, root zones which represent these originally thin portions of thrust sheets should show frequent outcrops of branch lines, and this seems to be the case in both the southern Appalachians (Fig. 28) and the Alps (Fig. 31).

Why are root zones composed largely of cataclastic basement rocks, with the foliation parallel with the zone boundaries? Recall that the lowest stratigraphy in a thrust sheet is at a trailing branch line (Fig. 1). Consequently this portion of the sheet is more likely to consist of coarsely crystalline basement rocks. The region close to a branch line will be subject to the deformation induced by both the overlying and underlying thrust surfaces. Rocks near branch lines must be more highly deformed than elsewhere in a thrust sheet, having cleavage or schistosity congruent to both thrust surfaces. If the temperature is sufficiently high ($\sim 300°C$), cataclastic fabrics will form and will be most spectacular in the originally coarse basement.

Horses and duplexes should be particularly common in these deep internal zones, and the density of branch lines is high. Note that a "braided" map pattern is diagnostic of a steeply dipping duplex.

We mentioned earlier that several of the key physical processes acting during thrusting occur along lines. It is probable that the narrow dimensions, abundance of basement, and highly deformed mylonites so typical of root zones are directly related to the operation of branch lines at moderate metamorphic temperatures.

CONCLUSION

A thread which runs through this paper is the distinction between thrust sheet volumes, thrust fault surfaces, tip and branch lines, and corners (points). Interestingly, this is analogous to the sort of reasoning one uses to investigate crystals, with their grain volumes, grain boundaries, dislocation lines, and point defects.

We have looked at only a small part of the geometric framework that provides the basic underpinnings of fault systems. Left out, for example, are the relations between listric extension faults and thrusts—relations that must be extremely important throughout the Basin and Range province of the western United States.

Finally, some remarks on the plate-tectonic setting, the interpretation of which may depend on a correct reading of the thrust systems. It is, for example, popular to tie in the creation and emplacement of dominant thrust sheets directly with an arc, microcontinent, or continent in collision with another continent. Frequently the dominant thrust fault is equated with a plate boundary and subduction zone, but, as in the southern Appalachian Blue Ridge, it often is possible to place the subduction zone at a different position and even to give it an opposite dip. Clearly these deliberations are delicate and complex, and we feel that by untangling the large scale plate tectonics from descriptions of the smaller scale thrust systems both points of view will gain clarity and precision.

From a plate tectonic perspective, there is one major effect of thrust systems, which is well illustrated by our examples in the southern Appalachian Blue Ridge and the western Alps. We started with a tapered Atlantic-type continental margin of normal (19 to 22 mi; 30 to 35 km) thickness on the foreland craton. This continental crust thinned and eventually ended against oceanic crust. Later, the region became part of a convergent margin, and the old tapering wedge of continental crust was greatly shortened and thickened, up to about twice normal thickness. A principal cause of this crustal thicken-

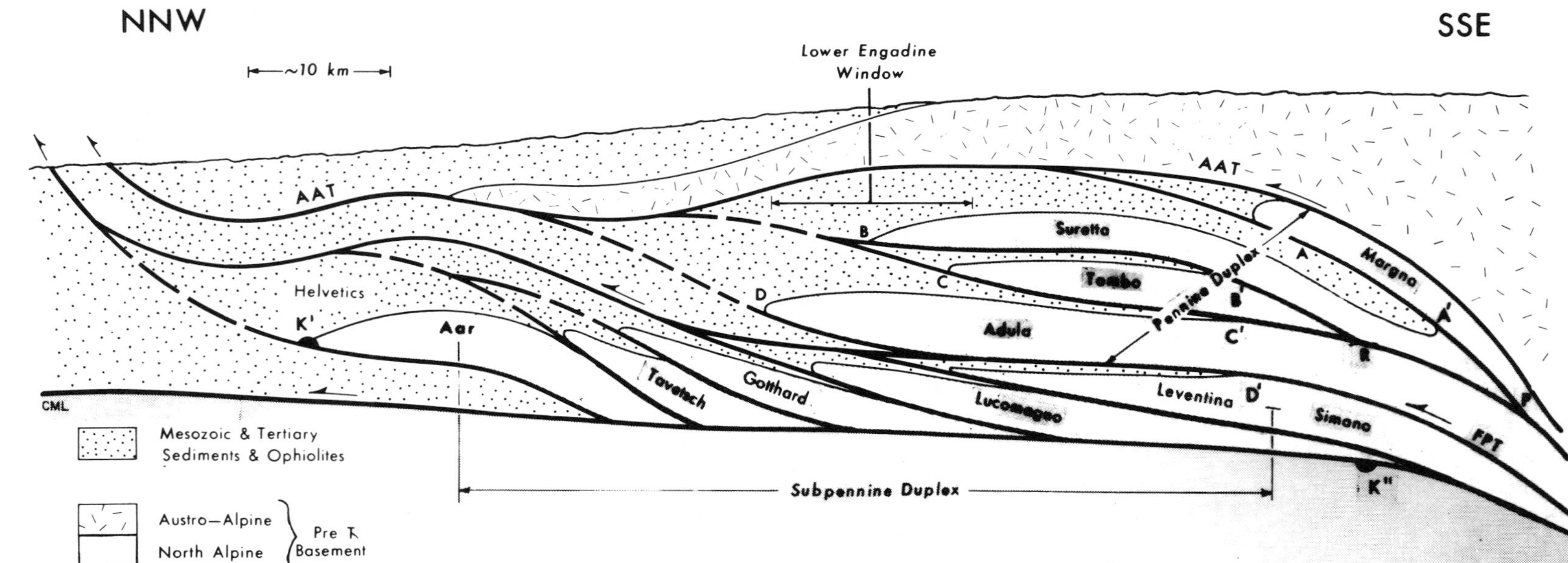

FIG. 34—Approximate down-plunge view in central Alps, west of Aar-Ticino culmination, suggesting duplex thrust systems. Only major faults which cut basement shown, but cover also shows intense but smaller scale families of duplexes. Effects of late folds removed, and scale is approximate. Letters identify points on map (Fig. 31).

ing is activity of the various kinds of thrust systems, and particularly by means of duplexes. Nor is this process restricted to shallow depths, for there are indications in the Alps and the Himalayas that thrust thickening persists down to the lower levels of the continental crust.

REFERENCES CITED

Armstrong, F. C., and S. S. Oriel, 1965, Tectonic development of Idaho-Wyoming thrust belt: AAPG Bull., v. 49, p. 1847-1866.

Badoux, H., 1972, Tectonique de la Nappe de Morcles entre Rhone et Lizerne: Materiaux Carte Geol. Suisse, new series, v. 143.

Bailey, E. B., 1938, Eddies in mountain structure: Geol. Soc. London Quart. Jour., v. 94, p. 607-625.

Bally, A. W., P. L. Gordy, and G. A. Stewart, 1966, Structure, seismic data, and orogenic evolution of southern Canadian Rocky Mountains: Bull. Canadian Petroleum Geology, v. 14, p. 337-381.

Baud, A., et al, 1978, Cross section Besancon-Ivrea zone, scale 1:200,000, Geologie Structurale, Programme 3ᵉ cycle: Pub. Lecture Notes, Inst. Geology, Univ. Lausanne.

Boyer, S. E., 1976, Formation of the Grandfather Mountain window, North Carolina, by duplex thrusting: Geol. Soc. America Abs. with Programs, v. 8, p. 788-789.

——— 1978, Structure and origin of Grandfather Mountain window, North Carolina: PhD thesis, Johns Hopkins Univ.

Bryant, B., and J. C. Reed, 1970, Structural and metamorphic history of the southern Blue Ridge, *in* G. W. Fisher et al, eds., Studies of Appalachian geology; central and southern: New York, Interscience, 460 p.

Burchfiel, B. C., and J. L. Livingston, 1967, Brevard zone compared to Alpine root zones: Am. Jour. Sci., v. 265, p. 241-256.

Butler, J. R., and D. E. Dunn, 1968, Geology of the Sauratown Mountains anticlinorium and vicinity: *in* Guidebook, Geol. Soc. America SE Section, Durham, N.C., Southeastern Geology Spec. Pub. 1, p. 19-47.

Chauve, P., 1975, Jura, *in* "Guides Geologiques Regionaux" Series: Paris, Masson and Cie, 216 p.

Clark, B. H., J. K. Costain, and L. Glover, 1978, Structural and seismic reflection studies of the Brevard ductile deformation zone near Rosman, North Carolina: Am. Jour. Sci., v. 278, p. 419-441.

Cloos, H., 1936, Einfuhrung in die Geologie: Berlin, Gebruder Borntraeger, 503 p.

Coleman, R. G., 1971, Plate tectonic emplacement of upper mantle peridotites along continental edges: Jour. Geophys. Research, v. 76, p. 1212-1222.

Conley, J. F., 1978, Geology of the Piedmont of Virginia—interpretations and problems, *in* Contributions to Virginia Geology—III: Virginia Div. Min. Resources, 154 p.

Cook, F. A., et al, 1979, Thin-skinned tectonics in the crystalline southern Appalachians; COCORP seismic-reflection profiling of the Blue Ridge and Piedmont: Geology, v. 7, p. 563-567.

——— 1980, The Brevard fault: a subsidiary thrust fault to the southern Appalachian sole thrust: Caledonides in the USA, Proc. I.G.C.P. 1979 Mtg.: Virginia Polytech. Inst. & State Univ., Dept. Geology, Mem. 2, 329 p.

Dahlstrom, C. D. A., 1969, Balanced cross-sections: Canadian Jour. Earth Sci., v. 6, p. 743-757.

——— 1970, Structural geology in the eastern margin of the Canadian Rocky Mountains: Bull. Canadian Petroleum Geology, v. 18, p. 332-406.

Dalziel, I. W. D., 1981, Back-arc extension in the Southern Andes—a review and critical reappraisal: Royal Soc. London Philos. Trans., ser. A, v. 300, p. 319-335.

de Margerie, E., and A. Heim, 1888, Les dislocations de l'ecorce terrestre: Zurich, J. Wurster, 154 p.

Dennis, J. G., 1967, International tectonic dictionary: AAPG Mem. 7, 196 p.

Douglas, R. J. W., 1952, Preliminary map, Waterton, Alberta: Canada Geol. Survey Paper 51-22.

——— 1958, Mount Head map area, Alberta: Canada Geol. Survey, Mem. 291, 241 p.

Elliott, D., 1976, The energy balance and deformation mechanisms of thrust sheets: Royal Soc. London Philos. Trans., ser. A, v. 283, p. 289-312.

——— 1977, Some aspects of the geometry and mechanics of thrust belts: Canadian Soc. Petroleum Geology, 8th Ann. Seminar Pub. Notes, Continuing Education Dept., Univ. Calgary, v. 1, 2.

——— 1978, Thrust sheets, *in* Geologie Structurale, Programme 3ᵉ cycle, 2ᵉᵐᵉ semaine: Published Lecture Notes, Inst. Geology, Univ. Lausanne.

———, in prep., Restorable cross sections.

——— and M. R. W. Johnson, 1980, The structural evolution of the northern part of the Moine thrust zone: Royal Soc. Edinburgh Trans. Earth Sci., v. 71, p. 69-96.

Fermor, P. R., and R. A. Price, 1976, Imbricate structures in the Lewis thrust sheet around Cate Creek and Haig Brook windows, southeast British Columbia: Canada Geol. Survey Paper 76-1B, p. 7-10.

Gilluly, J., 1960, A folded thrust in Nevada—inferences as to time relations between folding and faulting: Am. Jour. Sci., v. 258-A, p. 68-79.

——— and O. Gates, 1965, Tectonic and igneous geology of the northern Shoshone Range, Nevada: U.S. Geol. Survey Prof. Paper 465.

Gordy, P. L., F. R. Frey, and D. K. Norris, 1977, Geological guide for the CSPG 1977 Waterton-Glacier Park field conference: Canadian Soc. Petroleum Geologists.

Gosselet, J., 1885, Sur la structure geologique de l'Ardenne d'apies M. jan Lashnlx: Ann. Soc. Geol der Nord, v. 12, p. 195-202.

Harris, L. D., and R. D. Milici, 1977, Characteristics of thin-skinned style of deformation in the southern Appalachians, and potential hydrocarbon traps: U.S. Geol. Survey Prof. Paper 1018.

Hatcher, R. D., 1978, Tectonics of the western Piedmont and Blue Ridge, southern Appalachians; review and speculation: Am. Jour. Sci., v. 278, p. 276-404.

Hobbs, W. H., 1893, The geological structure of the Housatonic Valley lying east of Mount Washington: Jour. Geology, v. 1, p. 780-802.

——— 1894, Notes on the English equivalent of Schuppenstruktur: Jour. Geology, v. 2, p. 206.

Holmes, A., 1978, Principles of physical geology, 3d ed., revised by D. L. Holmes: New York, John Wiley and Sons.

Hossack, J. R., M. Garton, and R. P. Nickelsen, 1982, The geological section from the foreland up to the Jotun thrust sheet in the Valdres area, south Norway, *in* The Caledonide orogen—Scandinavia and related areas: Uppsala Caledonides Symp., 1982, New York, John Wiley and Sons.

Hsu, K. J., 1979, Thin skinned plate tectonics during Neo-Alpine orogenesis: Am. Jour. Sci., v. 279, p. 353-366.

King, P. B., and H. W. Ferguson, 1960, Geology of northeasternmost Tennessee: U.S. Geol. Survey Prof. Paper 311, 136 p.

McIntyre, D. B., 1954, The Moine thrust—its discovery, age and tectonic significance: Geol. Assoc. Canada Proc., v. 65, p. 203-223.

Milici, R. C., 1975, Structural patterns in the southern Appalachians; evidence for a gravity slide mechanism for Alleghanian deformation: Geol. Soc. America Bull., v. 86, p. 1316-1320.

Miller, R. A., W. D. Hardeman, and D. C. Fullerton, 1966, Geologic map of Tennessee: Tennessee Div. Geology, scale 1:250,000.

Milnes, A. G., 1974, Structure of the Pennine zone (central Alps); a new working hypothesis: Geol. Soc. America Bull., v. 85, p. 1727-1732.

——— and O. A. Pfiffner, 1980, Tectonic evolution of the central Alps in the cross section St. Gallen-Como: Eclogae Geol. Helvetiae, v. 73, p. 619-633.

Nevin, C. M., 1949, Principles of structural geology, 4th ed.: New York, John Wiley and Sons, 410 p.

Mudge, M. R., and R. L. Earhart, 1980, Lewis thrust fault and related structures in the Disturbed belt, northwestern Montana: U.S. Geol. Survey Prof. Paper 1174, 18 p.

Odom, A. L., and P. D. Fullagar, 1973, Geochronologic and tectonic relationships between the Inner Piedmont, Brevard zone and Blue Ridge belts, North Carolina: Am. Jour. Sci., v. 273-A, Cooper volume, p. 133-149.

Ollerenshaw, N. C., and F. R. Frey, 1975, Structural geology of the foothills between Savanna Creek and Panther River, S. W. Alberta, Canada: Canadian Soc. Petroleum Geologists Guidebook.

Oriel, S. S., 1950, Geology and mineral resources of the Hot Springs window: Madison County, North Carolina: North Carolina Div.

Mineral Resources Bull. 60.
Peach, B. N., et al, 1907, The geological structure of the north-west Highlands of Scotland: Mem. Geol. Survey Great Britain.
Perry, W. J., 1971, Structural development of the Nittany anticlinorium in Pendleton Co., West Virginia: PhD thesis, Yale Univ., 227 p.
——— 1978, Sequential deformation in the Central Appalachians: Am. Jour. Sci., v. 278, p. 518-542.
Rankin, D. W. , 1970, Stratigraphy and structure of Precambrian rocks in northwestern North Carolina, *in* G. W. Fisher et al, eds., Studies of Appalachian geology: central and southern: New York, Interscience, 460 p.
——— 1975, The continental margin of eastern North America in the southern Appalachians: the opening and closing of the Proto-Atlantic Ocean: Am. Jour. Sci., v. 275-A, p. 298-336.
——— G. H. Espanshade, and R. B. Newman, 1973, Geologic map of the west half of the Winston-Salem quadrangle, North Carolina, Virginia, and Tennessee: U. S. Geol. Survey Misc. Geol. Inv. Map I-709A, scale 1:250,000.
Rodgers, J. , 1953, Geologic map of east Tennessee with explanatory text: Tennessee Div. Geology Bull. 58, pt. 2, 168 p.
——— 1970, The tectonics of the Appalachians: New York, John Wiley and Sons, 271 p.
Roeder, D., O. C. Gilbert, and W. D. Witherspoon, 1978, Evolution and macroscopic structure of Valley and Ridge thrust belt, Tennessee and Virginia: Studies in Geology 2, Univ. Tennessee Dept. Geol. Sci., 25 p.
Roper, P. J., and D. E. Dunn, 1973, Superposed deformation and polymetamorphism, Brevard zone, South Carolina: Geol. Soc. America Bull., v. 84, p. 3373-3386.
Royse, F., M. A. Warner, and D. L. Reese, 1975, Thrust belt structural geometry and related stratigraphic problems, Wyoming, Idaho, northern Utah: Rocky Mtn. Assoc. Geologists Symp., p. 41-54.
Rutten, M. G., 1969, Geology of western Europe: Amsterdam, Elsevier Pub. Co., 520 p.
Sander, B., 1921, Zur Geologie der Zentralalpen: Geol. Staatsanst: Wien Jahrb. 6, v. 71, p. 171-224.
Shackleton, R. M., 1958, Downwards-facing structures of the Highland border: Geol. Soc. London Quart. Jour., v. 113, p. 361-392.
Spicher, A., 1972, Tektonische Karte der Schweiz: Schweizer. Geol. Kommission, scale 1:500,000.
Suess, E., 1904, The face of the earth, v. 1: Trans. from 1884 ed. by H. B. C. Sollas, Oxford, Oxford University Press, 604 p.
Termier, P., 1904, Les Nappes des Alpes Orientales et la synthese des Alpes: Soc. Geol. France Bull., v. 4, no. 3, p. 711-765.
Tollman, A., 1968, Die Grundbegriffe der deckentektonischen Nomenklatur: Geotektonische Forschungen, v. 29, p. 26-59.
Trumpy, R., 1960, Paleotectonic evolution of the central and western Alps: Geol. Soc. America Bull., v. 71, p. 843-908.
——— 1973, L'evolution de l'orogenese dans les Alpes Centrales; Interpretation des donnees stratigraphiques et tectoniques: Eclogae Geol. Helvetiae, v. 66, p. 1-10.
——— 1980, An outline of the geology of Switzerland, *in* Geology of Switzerland, a guidebook: 10th Internat. Geol. Cong., Basel, Wepf and Co.
Umbgrove, J. H. F., 1950, Symphony of the earth: The Hague, Martinus Nijhoff, 220 p.
Voll, G., 1976, Recrystallization of quartz, biotite, and feldspars from Erstfeld to the Leventina nappe, Swiss Alps, and its geological significance: Schweizer. Mineralogy u. Petrog. Mitt., v. 56, p. 641-647.
Uyeda, S., 1977, Some basic problems in the trench-arc-back arc system, *in* M. Talwani and W. C. Pitman, eds., Island arcs, deep sea trenches and back-arc basins: American Geophys. Union, Maurice Ewing Ser. 1, 470 p.
Willis, B., 1902, Stratigraphy and structure, Lewis and Livingston Ranges, Montana: Geol. Soc. America Bull., v. 13, p. 305-352.

Reprinted with permission from *Journal of Structural Geology*, v. 9, p. 207-219, W. R. Jamison, Geometric analysis of fold development in overthrust terranes, copyright 1987, Pergamon Press plc.

Geometric analysis of fold development in overthrust terranes

WILLIAM R. JAMISON

Amoco Production Company, P.O. Box 3385, Tulsa, OK 74102, U.S.A.

(*Received* 16 *July* 1985; *accepted in revised form* 30 *June* 1986)

Abstract—Fault-bend folding, fault-propagation folding, and detachment (or décollement) folding are three distinct scenarios for fold-thrust interaction in overthrust terranes. Simple kink-hinge models are used to determine the geometric associations implicit in each scenario. Bedding maintains constant thickness in the models except in the forelimb of the fold. The forelimb is allowed to thicken or thin without limit. The models address individual folds, and the calculated fold geometries are balanced structures.

Each mode of fold-thrust interaction has a distinct set of geometric relationships. Final fold geometry is adequate in itself to discern many fault-bend folds. This is not the case for fault-propagation and detachment folds. These two fold forms have very similar geometric relationships. Some knowledge of the nature of the underlying thrust or décollement zone is usually needed to distinguish between them. The geometry of a fold is altered, in a predictable fashion, by transport through an upper ramp hinge and by fault-parallel shearing of the structure. The shearing results in a tighter fold, whereas transport through the ramp hinge produces a broader fold.

The viability of the geometric analysis technique is demonstrated through its application to a pair of detachment folds from the Canadian Cordillera. The geometric analysis is also used to evaluate cross-sections through subsurface structures. In an example from the Turner Valley oil field, the analysis indicates how the interpretation should be altered so as to balance the cross-section. The analysis reveals hidden assumptions and specific inconsistencies in structural interpretations.

INTRODUCTION

ONE OF the basic functions of structural geology is the description and prediction of the geometry of naturally occurring structures. Prediction commonly consists of combining experience with some form of constraint, either conceptual, mechanical, or geometric. The constraint basically guides or justifies the predictor's intuition. This article examines a system of geometric constraints, based on the initial fold-thrust interaction, that may be used to examine individual folds in overthrust terranes.

The 'balancing' of cross-sections is certainly the most commonly utilized geometric constraint for overthrust structures. 'Balancing' is simply a test for conservation of volume. It is a check to assure that the total rock volume remains constant through the history of development of a geologic structure. Dahlstrom (1969) formally introduced the balancing procedure, and presented a set of tenets derived from his experience with structures in the southern Canadian Cordillera. The specific guidelines presented by Dahlstrom (1969) apply to regional cross-sections (e.g. Dixon 1982). They were not intended, and often are not appropriate, for the assessment of individual structures.

A construction procedure that is designed for the interpretation of individual structures as well as regional analyses is presented in Suppe (1983). Focusing on the fault-bend fold (Fig. 1a), or Rich-model mode of fold-thrust interaction (Rich 1934), Suppe derived a suite of curves relating the fold interlimb angle to upper and lower ramp angles. Similar relationships have also been derived for fault-propagation folds (Fig. 1b) (Suppe & Medwedeff 1984). While these analyses are not explicitly presented as balancing procedures, the basic tenet of balancing (conservation of volume) is the foundation behind the equations. As a consequence, a fold–thrust system that displays geometric relationships compatible with the curves and assumptions of Suppe (1983) or Suppe & Medwedeff (1984) is balanced.

In this article, a geometric analysis is presented that extends the approach of Suppe (1983) and Suppe & Medwedeff (1984). Certain of the constraints of their model are relaxed in order to permit variations in the amount or type of deformation through the fold. Detachment folds are added as a third type of thrust-associated

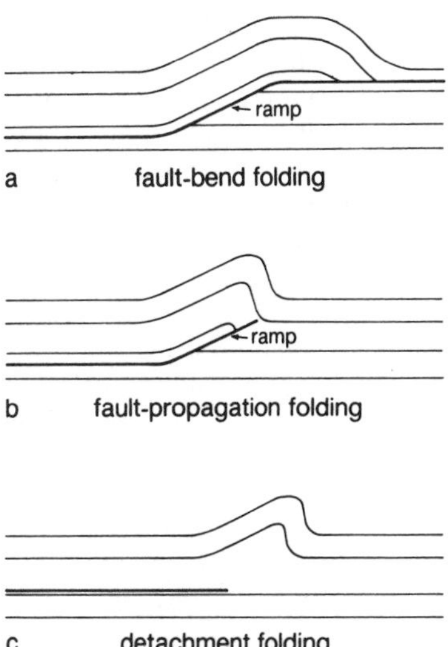

Fig. 1. Three types of fold-thrust interactions: (a) a fault-bend fold, (b) a fault-propagation fold, and (c) a detachment fold.

fold. Also, the effects of fold transport and shearing on fold geometry are considered. This analysis defines the range of geometries most probable for each mode of folding and demonstrates that the history of fold development and transport can, to a certain extent, be determined from final fold geometry. Unlike the procedures of Dahlstrom (1969), Suppe (1983) and Suppe & Medwedeff (1984), this analysis is not intended to be used for regional structural reconstructions (e.g. Suppe 1980). Rather, it is strictly a balancing guideline for individual folds in an overthrust terrane.

MODES OF FOLD-THRUST INTERACTION

Three fold forms are distinguished, viz. fault-bend, fault-propagation, and detachment (or décollement) folds (Fig. 1). For the geometric analysis, these three modes are defined simply in terms of geometry, with no kinematic or mechanical considerations. For the initial analysis, the structures are assumed to be in their early phases of development. A fault-bend fold (Fig. 1a) develops as the hangingwall of a thrust is transported through a ramp region on the thrust surface. This fold model was introduced by Rich (1934) to account for the geometry of the Pine Mountain overthrust in the southern Appalachians. The fold forms as a consequence of the movement of the hangingwall rocks through the ramp region. As a result, the forelimb of the fault-bend fold is always located on the foreland side of its associated ramp. While the fault-bend fold model has found widespread acceptance and application, in many cases it is simply, as Dahlstrom (1970, p. 361) noted, a 'convenient fiction'. Even though it is not universally appropriate, it is the easiest of the three fold-thrust forms to conceive, draw and model.

A fault-propagation fold (Fig. 1b) also has a direct association with the ramp region of the underlying thrust. Whereas the fault-bend fold develops subsequent to the ramp formation, the fault-propagation fold develops simultaneously with and immediately above the ramp. The displacement along the thrust diminishes progressively to zero along the ramp region beneath the fault-propagation fold. Stated otherwise, the fault-propagation fold develops at the termination of a thrust (Williams & Chapman 1983). The fold, in fact, is the geologic expression of the strain that implicitly must occur at a fault termination.

The detachment fold (Fig. 1c), like the fault-propagation fold, develops at the termination of a thrust. Unlike the fault-propagation fold, the detachment fold is not associated with a ramp in the underlying thrust. Rather, the detachment fold develops above a bedding detachment (or décollement), hence its name. Beneath the fold, the displacement along the detachment (bedding-parallel thrust) diminishes progressively to zero in a foreland direction.

A fold located at the tip of a blind thrust (Boyer & Elliott 1982) must be either a fault-propagation or detachment fold. The mechanical stratigraphy determines the probable fold-thrust relationship. A detachment fold might be expected where the thrust rides low in a ductile unit, that in turn underlies a more competent unit. In a layered sequence with more modest ductility contrasts, the fault-propagation fold may be more probable. (A hybrid is, of course, conceivable, maybe even common, but the geometry of that possibility is not considered here).

In both the fault-bend fold and the fault-propagation fold, the lower strata involved in the fold are truncated against the thrust. Consequently, both may be termed 'truncation anticlines'. The hangingwall truncation of the fault-bend fold lies foreland of the ramp, whereas in the fault-propagation fold it initially is on the ramp (Fig. 1). Thus the fault-bend and fault-propagation folds may be distinguished on the basis of the position, relative to the ramp, of the fold and the hangingwall truncation. However, as the fault in the fault-propagation fold grows and the fold is transported, both the hangingwall truncation and the fold are transported foreland of the associated ramp. As a result, the position of the fold and truncated beds no longer indicates the mode of fold genesis.

MODEL CONFIGURATIONS AND PARAMETERS

The analyses presented below are derived using structural models with simple kink-hinge geometries (e.g. Fig. 2b). The thickness of each stratigraphic unit remains constant throughout the structure, except in the forelimb of the fold. The forelimb can thicken or thin, but the thickness change must be constant throughout the forelimb. The kink-hinge configurations are used to simplify the geometric calculations discussed below, and are not intended to be strict representations of geological stuctures. Models with curvilinear fold hinges (e.g. Fig. 1), although aesthetically much more appealing than the angular forms, yield virtually the same angular relationships and are, geometrically, much more cumbersome.

The use of a variable forelimb thickness is a departure from most balancing procedures, wherein bedding maintains constant thickness throughout the entire structure (e.g. Dahlstrom 1969, Suppe 1983, Suppe & Medwedeff 1984). Detailed surface and subsurface studies of individual thrust-associated folds have shown, however, that the overall forelimb thickness can differ significantly from backlimb bedding thickness (e.g. Gallup 1951, Brown & Spang 1978, Williams & Chapman 1983). Small imbricate faults, minor folds or ductile flow of certain beds can produce this thickness contrast. Whereas the assumption of constant bedding thickness is probably adequate for regional sections, it can be inappropriate for the investigation of individual folds.

The solutions presented in the following section are derived for the dip cross-section of the structure. Plane strain is assumed, implying that there is no loss or gain of material into the plane of the cross-section. During fold development, geometric alterations occur only in the fold itself and in the hangingwall above the ramp. Out-

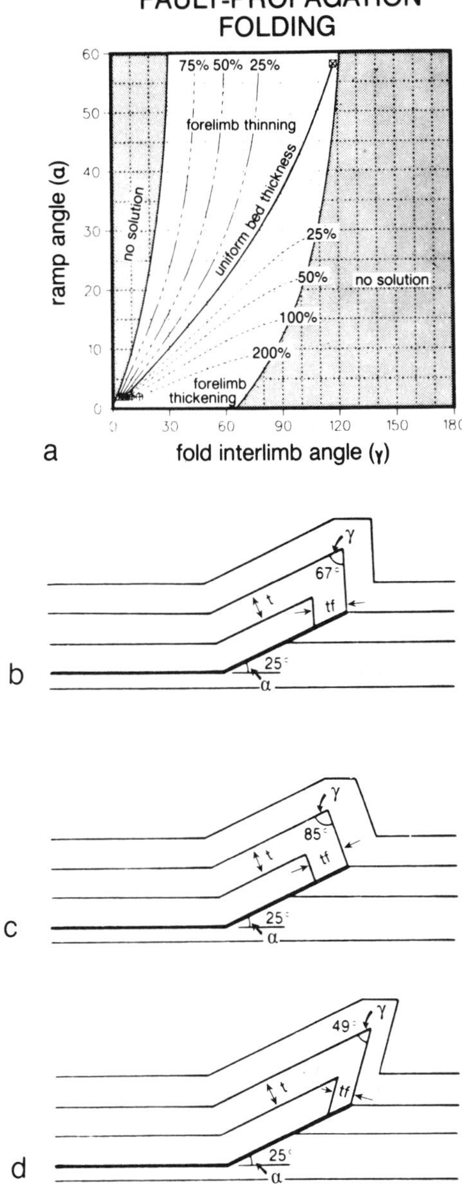

Fig. 2. Geometric analysis of fault-propagation folding. (a) Curves relating fold interlimb angle (γ) to ramp angle (α) for specified amounts of forelimb thickening and thinning. For any particular value of α, there exists a broad range of possible values for γ, depending on the amount of tectonic thickening or thinning of forelimb beds. For example, for $\alpha = 25°$, (b) a fold with uniform limb thickness has an interlimb angle of 67°, (c) a fold with 25% thickening of the forelimb has an interlimb angle of 85° and (d) a fold with 35% thinning of the forelimb has an interlimb angle of 49°.

side of this region, the hangingwall material is translated along the fault, but not altered in any other way. The footwall in all models is totally unaltered. The cross-sectional area of the material within the deformed region is equated with its undeformed area. This equality, combined with the plane strain assumption, ensures that rock volume is conserved during fold development and transport; i.e. the structures are balanced. The various geometric relationships presented in this analysis are simply the solutions to these balanced equations (see Appendix).

The interrelated parameters in the truncation anticlines are the ramp angle (α), the interlimb angle (γ), and the thickness change (tf/t) occurring in the forelimb (Figs. 2 & 3). In the detachment fold, backlimb dip (α_b) replaces ramp angle (Fig. 4). Additionally, the geometric analysis of the detachment fold requires specification of a fourth parameter, viz. the ratio of the fold amplitude (a) to the normal stratigraphic thickness (f) of the ductile unit infilling the core of the fold. As a result, the geometric relationships of the detachment fold are a direct function of the size (amplitude) of the structure. The angular relationships derived for the truncation anticlines have no dependence on fold amplitude.

Although the model configurations have been made simplistic to facilitate the calculations, the general forms are compatible with many folds observed in overthrust terranes. Faill (1969) noted that the dominant characteristics of folding in the Appalachians in Pennsylvania are planar limbs and angular hinges. Similar characteristics are evident in published examples of thrust-associated folds in Alberta (Brown & Spang 1978), British Columbia (Fitzgerald & Braun 1965), Tennessee (Serra 1977), England (Williams & Chapman 1983) and in the Jura of Switzerland (Laubscher 1977).

GEOMETRIC RELATIONSHIPS

Fault-propagation folding

Fold interlimb angle (γ) is a function of a ramp angle (α) and the amount of forelimb thickening or thinning (Fig. 2). The ramp angle for the fault-propagation fold is the angle that the ramp makes with the lower flat of the thrust. The curves relating α and γ (Fig. 2a) indicate that a fault-propagation fold with ramp angle of, say, 25° theoretically must have an interlimb angle of from 21° to 105°. If the beds maintain their original thickness throughout the fold (as in the analysis of Suppe & Medwedeff 1984), the interlimb angle is 67° (Fig. 2b). If the forelimb is thickened, the fold will have a larger interlimb angle. For the case of 25% thickening of the forelimb the interlimb angle is 85° (Fig. 2c). If the forelimb is attenuated, the fold will be tighter. A fold with 35% thinning of the forelimb beds should have an interlimb angle of 49° (Fig. 2d).

Fault-bend folding

Again, fold interlimb angle (γ) is a function of ramp angle (α) and the amount for forelimb thickening or thinning (Fig. 3). The ramp angle for the fault-bend fold is defined in terms of δ and β (Fig. 3b), the lower and upper ramp hinge angles, respectively, as

$$\alpha = \cot^{-1}\left(\cot\delta + 2\cdot\left(\tan\frac{\delta}{2} - \tan\frac{\beta}{2}\right)\right).$$

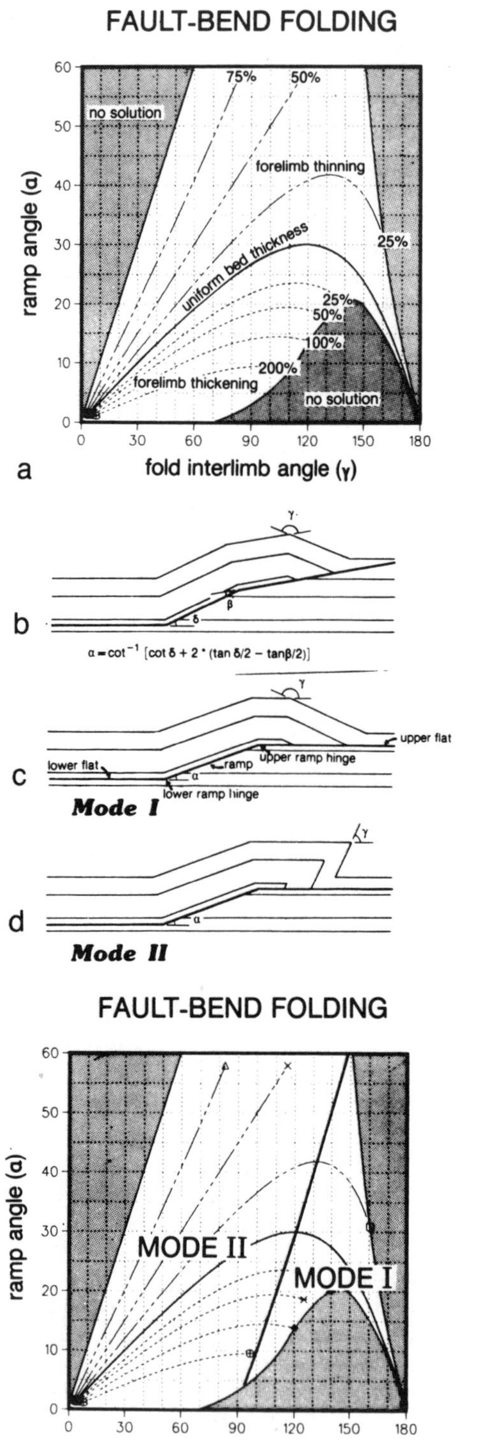

Fig. 3. Geometric analysis of fault-bend folding. (a) Curves relating fold interlimb angle (γ) to ramp angle (α) for specified amounts of forelimb thickening and thinning. Units maintain constant thickness except in the forelimb. (b) For the general case of unequal upper and lower ramp hinge angles (β and δ, respectively), an 'effective' ramp angle (α) is used to assess possible fold geometries. (c) Mode I and (d) Mode II fold geometries, following the distinction of Suppe (1983). The geometric analysis chart may be divided into domains (e) representing these two modes.

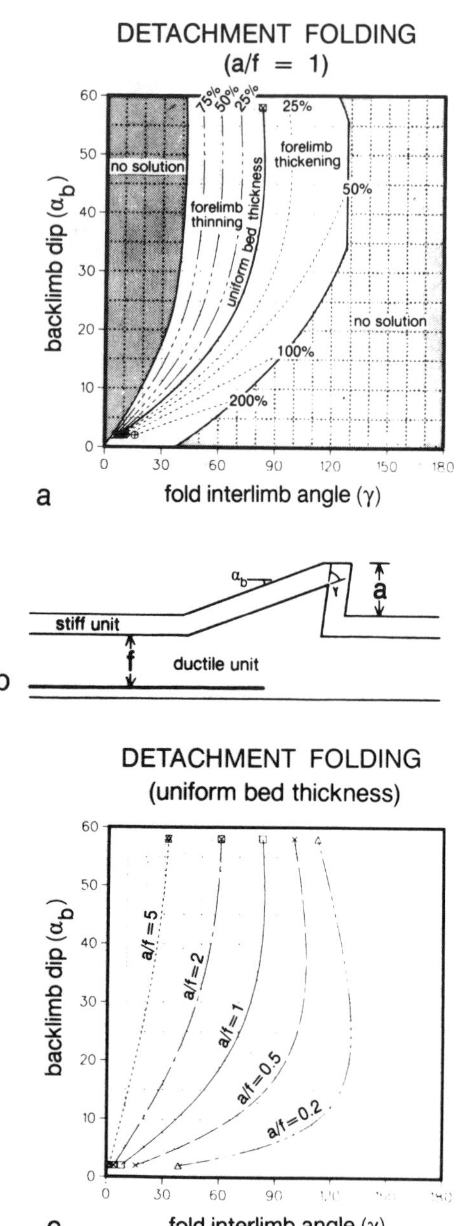

Fig. 4. Geometric analysis of detachment folding. (a) Curves relating fold interlimb angle (γ) to backlimb dip (α_b) for specified amounts of forelimb thickening and thinning, and $a/f = 1.0$. Bedding in the stiff layers maintains constant thickness except in the forelimb. (b) The detachment fold model is divided into 'stiff' and 'ductile' units. The fold form is expressed in the stiff units. The ductile units maintain constant thickness except in the region directly beneath the fold in the stiff units. (c) Curves relating γ to α_b for specified a/f values and the case of constant unit thickness in the stiff layers throughout the structure.

The most common values for both lower and upper ramp hinge angles observed in natural structures are 10–40° (Serra 1977, Boyer & Elliott 1982). Within this range $\alpha \approx \delta$. Commonly, both upper and lower 'flats' along the thrust are parallel to footwall bedding. In this case δ and β are equal, and $\alpha = \delta$.

The case of constant bed thickness through the fold was treated by Suppe (1983). He noted that there were two possible interlimb angles for any specified ramp angle (except $\alpha = 30°$, the maximum possible ramp angle in this case). To distinguish these two solutions, Suppe (1983) specified the larger interlimb angle solution to be a Mode I fold (Fig. 3c). The smaller interlimb angle fold is a Mode II fold (Fig. 3d). This distinction between Mode I and Mode II fault-bend folds can be extended through the range of solutions made in the current analysis (Fig. 3e).

Detachment folding

For the detachment anticline (Fig. 4), a suite of curves similar to those derived for the truncation anticlines can be generated for any specified *a/f* value (Fig. 4a). Note that interlimb angle (γ) is plotted against backlimb angle (α_b), where backlimb angle is the acute angle between projected backlimb strata and the décollement surface, which is assumed to be planar and parallel to footwall bedding.

Unlike truncation anticlines, the value of α (here, α_b) for a detachment anticline will generally change as the structure develops. Also, the fold amplitude increases as the structure grows and, thus, *a/f* increases (*f* does not change). The effect of *a/f* on interlimb angle is significant (Fig. 4c). In general, γ decreases as *a/f* increases; i.e. the fold gets tighter as amplitude increases.

For a given *a/f* value, the range of possible solutions (on the γ-α plots) for the detachment fold is more limited than for either of the truncation anticlines (compare Fig. 4a with Figs. 2a & 3a). However, because the swath of possible solutions shifts as *a/f* changes, there is actually a greater range of possible geometries for the detachment anticline than for either of the truncation anticline models.

SUBSEQUENT GEOMETRIC ALTERATIONS

Transport through an upper ramp hinge

A fault-propagation fold develops above the propagating ramp of its associated, underlying thrust. If the fault subsequently flattens as it continues to propagate in a foreland direction, an upper ramp hinge is formed (Fig. 5a). The geometry of the fold is altered if the fold is transported through this new ramp hinge. In all cases the fold becomes broader (Fig. 5b). Thickening or thinning of the forelimb is dictated by the original interlimb angle of the fold and the angle of this upper ramp hinge

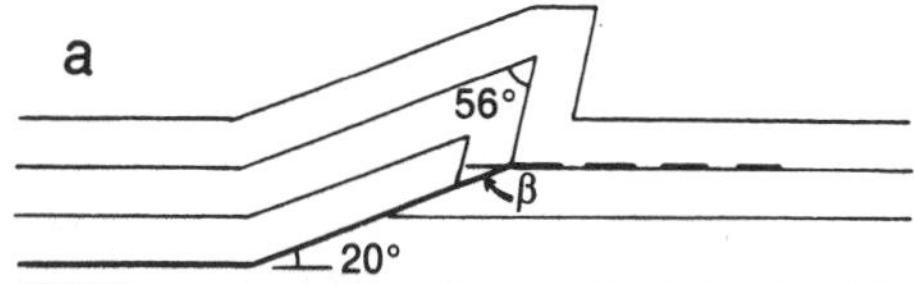

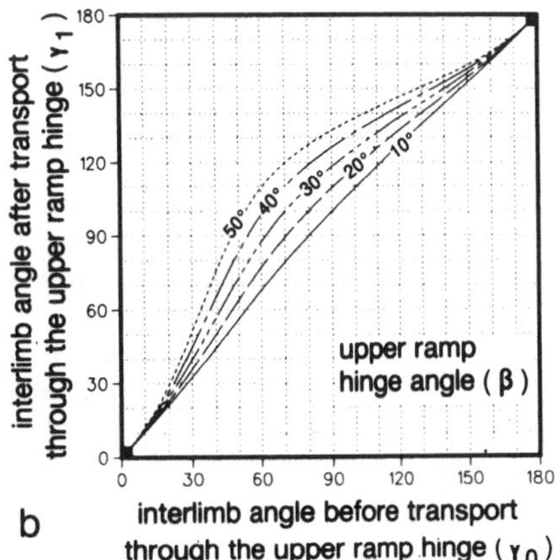

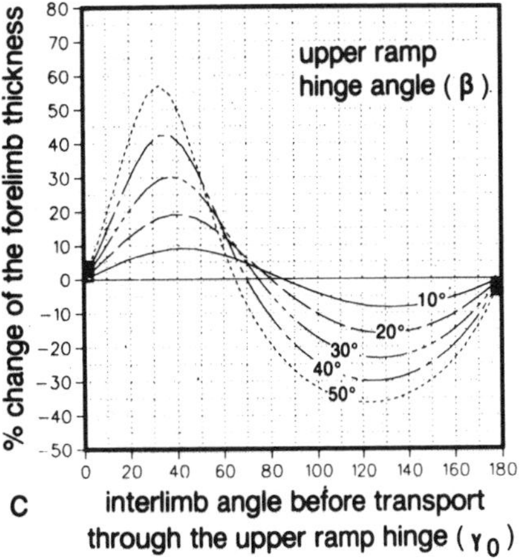

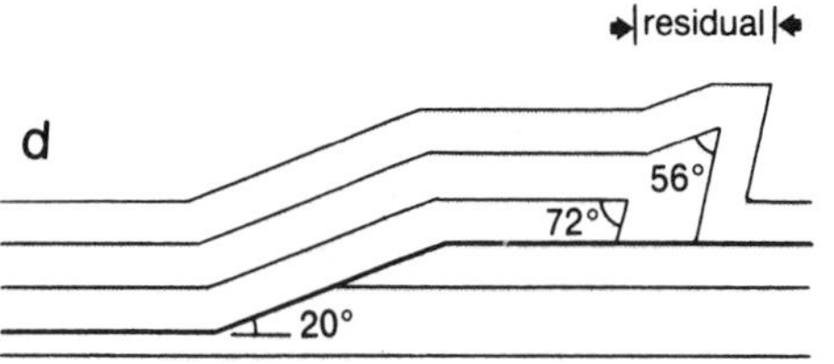

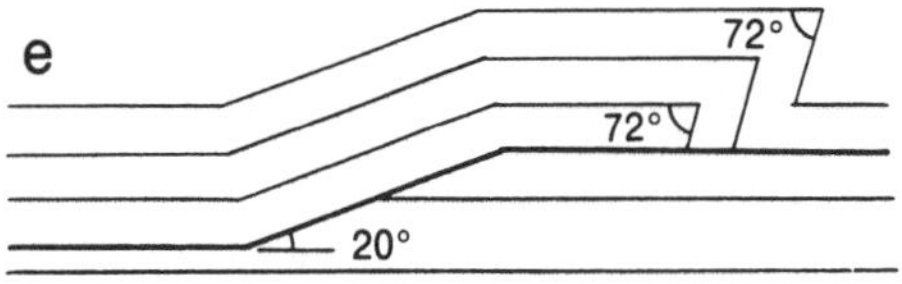

Fig. 5. (a) An upper ramp hinge develops beneath a fault-propagation fold if the underlying thrust flattens as it extends in a foreland direction. Geometric effects of transport through an upper ramp hinge upon (b) fold interlimb angle and (c) fold forelimb thickness. Transport through a lower ramp hinge would have the reverse effect. The interlimb angle of the original fold (a) is broadened by transport through the upper ramp hinge (d). The portion of the fold that does not move through the upper ramp hinge (the 'residual') maintains its initial geometry. The altered portion of the fault-propagation fold has the same geometry as a Mode II fault-bend fold with equivalent ramp angle (e).

(Fig. 5c). Only the portion of the fold that is transported through this hinge undergoes the geometric alteration. Any part of the fold that was already located foreland of the hinge is a 'residual' (Fig. 5d) that is transported without any associated geometric alteration. Roughly, the hangingwall beds that are truncated by the ramp undergo geometric alterations, whereas those beds not truncated form the residual. The residual will always have a smaller fold interlimb angle than that portion of tne fold that has been altered.

The portion of the fault-propagation fold that moves through the upper ramp hinge (Fig. 5d) is found to have a geometry that is identical to a Mode II fault-bend fold with the same ramp angle (Fig. 5e). The transport of the fold through the upper ramp hinge thus eliminates much of the geometric distinction between these two forms of truncation anticlines. For the general case, assuming that the fold is transported through an upper ramp hinge angle equal to the original ramp angle of the fold (i.e. its lower ramp angle), a suite of curves for transported fault-propagation folds may be generated (Fig. 6a). These new suites of curves coincide with the corresponding curves for the fault-bend fold model (Fig. 6b), but are effectively limited to the Mode II region of fault-bend folding.

A detachment fold may be converted into a truncation anticline if the detachment thrust, or an imbricate of this fault, cuts up-section through the forelimb of the fold. Only the portion of the fold that overlies this new ramp will be affected by subsequent transport of the fold through the upper ramp hinge. The geometric alterations can be calculated using the appropriate charts for fold transport (Figs. 5b & c). Because of the broad range of possible configurations for the original detachment fold, geometric similarities between a truncated and transported detachment fold and either a fault-propagation or fault-bend fold would be coincidental.

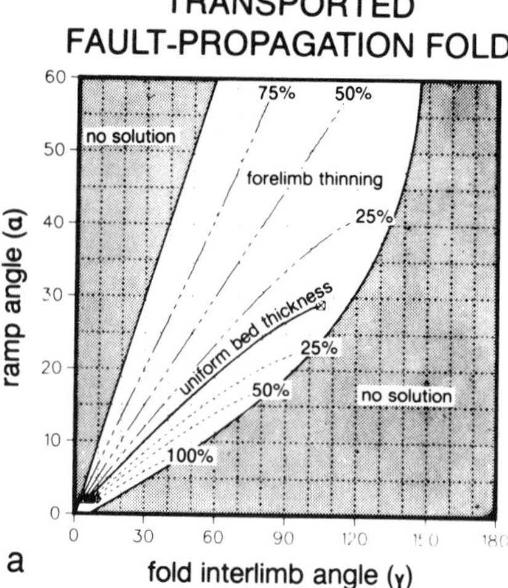

FAULT-BEND FOLDING

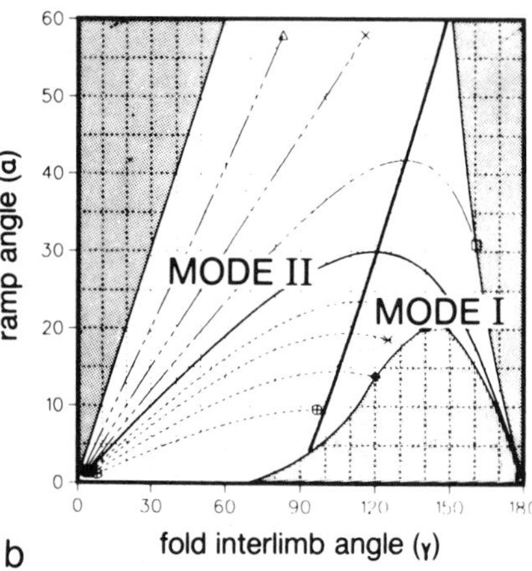

Fig. 6. Comparison of geometric relationships for (a) a fault-propagation fold transported through an upper ramp hinge of angle equal to the original ramp angle with those for (b) a fault-bend fold. The curves coincide, but the transported fault-propagation fold is essentially restricted to the Mode II domain of the fault-bend folding.

Fault-parallel shearing

In the truncation anticlines, the backlimb of the fold is parallel, or close to parallel, to the surface of the transporting thrust. In many thrust-associated folds there are suggestions that shear displacement has occurred along some of the backlimb bedding surfaces during fold development and/or transport. Such internal shearing will alter fold geometry. To assess these effects, movement on these surfaces is treated as fault-parallel shear occurring pervasively through the affected zone. Both the interlimb angle of the fold and the relative thickness of the forelimb are affected (Fig. 7).

Fault-parallel shear always reduces the fold interlimb angle (Fig. 7a). In general, this shearing will thin the forelimbs of fault-propagation folds, Mode II fault-bend folds, and most detachment folds, but will thicken the forelimbs of Mode I fault-bend folds. In nature, the existence of such bedding-plane slip is often difficult to demonstrate, and almost always impossible to measure. Thus, although the amount or even the existence of this fault-parallel shearing can seldom be directly assessed, it may be inferred indirectly via geometric analysis.

For a specified initial geometry and amount of shearing, the reduction in interlimb angle implies a change in forelimb thickness. However, the original charts for these fold models (Figs. 2–4) also associate a reduction in interlimb angle (maintaining a constant α) with a specific forelimb thickness change. Even so, the sheared fold has a geometry that is distinct from a fold that has become tighter without any associated fault-parallel shear. Specifically, for a given α and γ, the sheared fold has a thicker forelimb than an unsheared fold (compare Figs. 7d & f). This reflects the fact that the shearing process moves material into the hinge and forelimb regions of the fold.

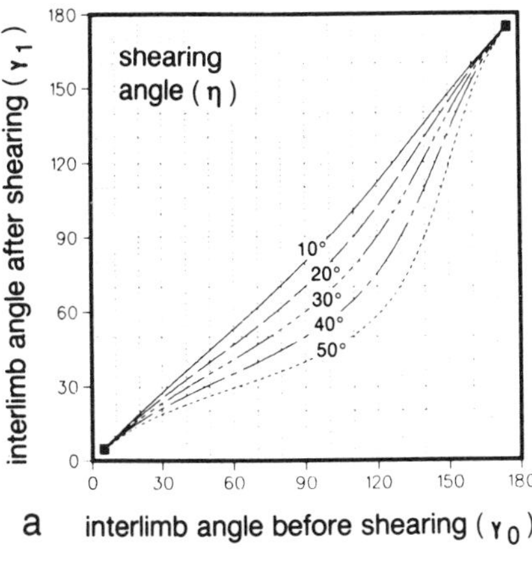

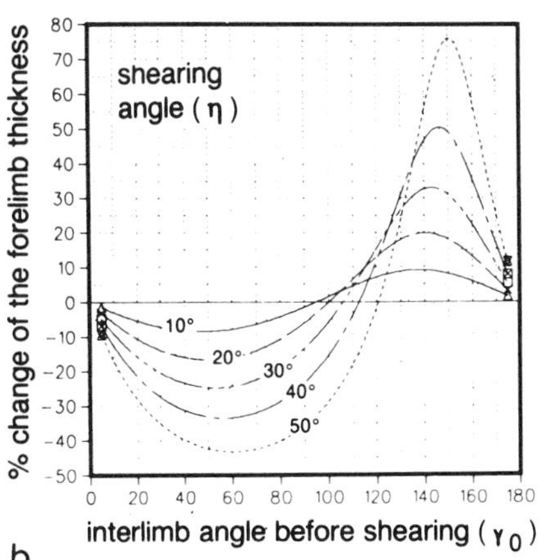

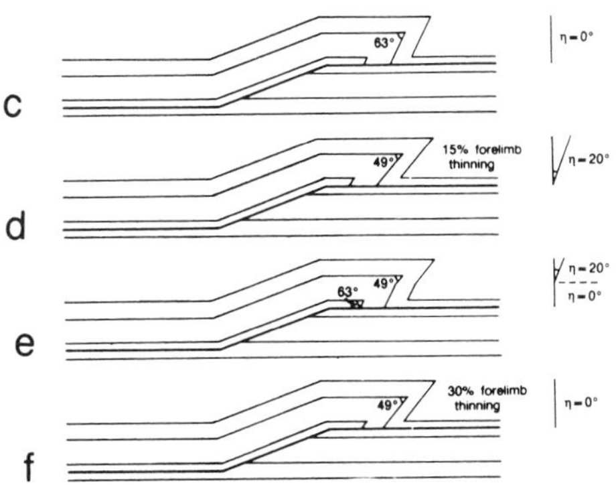

Fig. 7. Geometric effects of fault-parallel shearing on (a) fold interlimb angle, and (b) fold forelimb thickness. For the case of a Mode II fault-bend fold with $\alpha = 20°$ and uniform bed thickness, the original structure (c) may be sheared in its entirety (d) or only in part (e). An unsheared fold of similar interlimb angle (f) has a thinner forelimb than the corresponding sheared fold (d).

GEOLOGICAL INFERENCES

Although there is considerable overlap of the domains of possible solutions for the several modes of fold-thrust interaction (Figs. 2–4), some inferences regarding mode of development may be obtained directly from the final geometry of a structure. For example, most large folds with interlimb angles greater than 120° are Mode I fault-bend folds. Detachment folds that occur in this domain of large interlimb angles must have small *a*/*f* ratios (Fig. 4c), and fault-propagation folds are not geometrically viable. Subsequent tightening of a Mode I fault-bend fold, with or without fault-parallel shearing, will thicken the forelimb region. Folds with moderate interlimb angles ($80° \leqq \gamma \leqq 120°$) and substantial imbrication in the forelimb may be examples of modified Mode I fault-bend folds.

In the domain of folds with moderate to small interlimb angles, little can be inferred regarding mode of fold-thrust interaction simply from a knowledge of the interlimb angle itself. For example, the curves for a detachment fold with $a/f = 1.0$ (Fig. 4a) are very similar to those for a fault-propagation fold (Fig. 2a). Determination of interlimb angle, backlimb dip, and forelimb thickness change will not distinguish between these two forms of fault-tip folding. Only direct knowledge of the existence or nonexistence of a propagating ramp can resolve this case. On a positive note, many specific fold geometries do suggest a unique genesis. For example, a very tight, upright fold with little forelimb thinning is almost certainly a detachment fold with a high *a*/*f* ratio (Fig. 4c).

Additional inferences regarding fold genesis may be made using kinematic considerations. For the case of the fault-bend fold, the Mode I geometry is the form depicted by Rich (1934) and used in most other investigations, models, or analyses based upon the fault-bend fold model (e.g. Wiltschko 1979a,b, Morse 1977, Berger & Johnson 1980). The forelimb of the fold in Mode I is simply flexed downward as the hangingwall rock moves through the upper ramp hinge.

Conceptually, a Mode II fold must also initiate as a Mode I structure. The transition to the Mode II form occurs, in terms of the geometric charts, via a shift to the left along a constant ramp angle line (Fig. 3e). This shift requires the forelimb first to thicken and then to thin. This complex deformation path presumably would leave its mark in the form of substantial deformation of the forelimb beds. In fact, deformation might be so intense as to render bedding in the forelimb virtually unrecognizable.

If these speculations are valid, then it follows that a fold with a Mode II fault-bend fold geometry and only moderate forelimb deformation is more likely to be a transported fault-propagation fold (Fig. 6a) than a true fault-bend fold. Where specific structures are being assessed, features such as a slightly more appressed 'residual' in the non-truncated beds would add support to a fault-propagation interpretation.

APPLICATIONS

Geometric analyses of two sets of thrust-associated folds in the Canadian Cordillera serve to illustrate some of the uses and limitations of this technique. The first structure is a pair of folds, well exposed on a cliff face in northeastern British Columbia. The second is the Turner Valley Anticline, a subsurface structure whose configuration is defined by abundant well control.

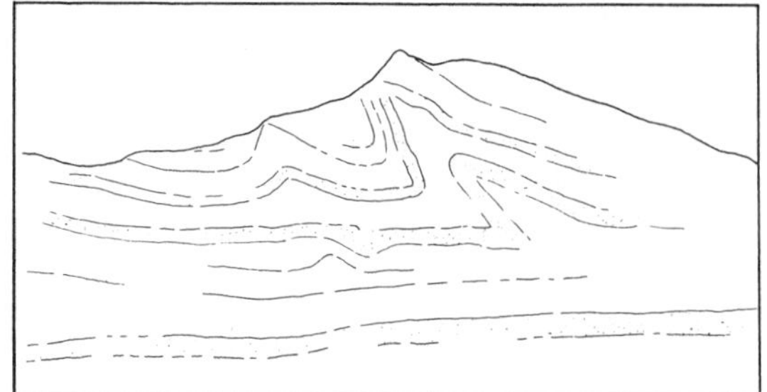

Fig. 8. Detachment anticlines in Besa River beds in northeastern British Columbia. After Fitzgerald & Braun (1965).

Outcrop example

These two folds (Fig. 8), located in the Main Ranges of the Canadian Cordillera in northeastern British Columbia, are developed in the Devonian–Mississippian Besa River Formation (Fitzgerald & Braun 1965). This formation is dominantly shale with some sandstone stringers. The folds occur above a bedding-parallel detachment surface. Direction of transport is from right to left in Fig. 8. There is no fault ramp associated with either fold; they are detachment folds in the context of this article.

The two folds appear to have developed above different detachment surfaces. The units used in a geometric evaluation must be located above zones of complex deformation that might occur in the core of the fold. In the case at hand, each fold is analyzed using the bedding units in the upper reaches of the outcrop. To assess the angular values needed in the geometric analysis, each limb of the fold, as well as the detachment surface, must be approximated by a single, straight line.

The right-hand fold (Fig. 8) is analyzed using horizon *m* (Fig. 9a). The fold interlimb angle (γ) at this horizon is 48°, the backlimb dip (α_b) measured relative to the detachment surface is 24° and the *a*/*f* value is 1.1. Note that *f* includes the entire interval between the detachment surface and horizon *m*. Consequently, the results of this particular analysis apply only to the units above horizon *m*. The geometric analysis indicates that a detachment fold with the measured angles and *a*/*f* value should exhibit a thinning in the forelimb units of about 50% (Fig. 9b). In the outcropping structure, the forelimb thickness (tf_r) is about 45% thinner than the presumed normal stratigraphic thickness (t_r).

In the left-hand fold set (Fig. 8), using horizon *k*, the fold interlimb angle (γ) is 82°, the backlimb dip (α_b) is 28°, and *a*/*f* is about 0.9 (Fig. 9c). Geometric analysis suggests a slight (5–10%) thickening should occur in the forelimb of the fold (Fig. 9d). In the outcropping structure, the upper units involved in this fold appear to have essentially uniform thickness across the fold.

The good correspondence between the geometry of naturally occurring folds and the model predictions supports the viability of the geometric analysis approach. The correlations should be good in this case, for the outcropping folds closely mimic the models used in the analysis (or vice versa). However, despite the simi-

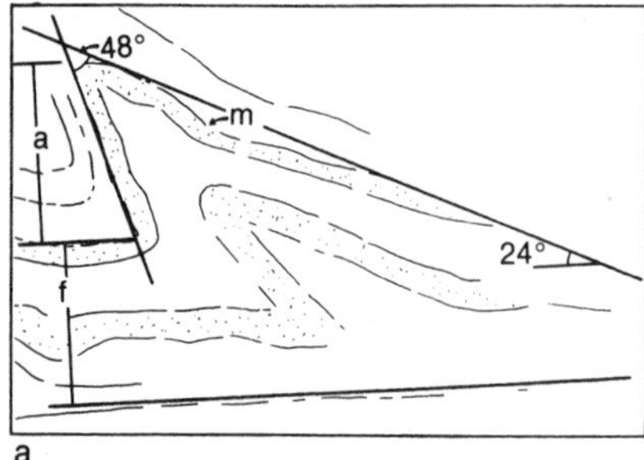

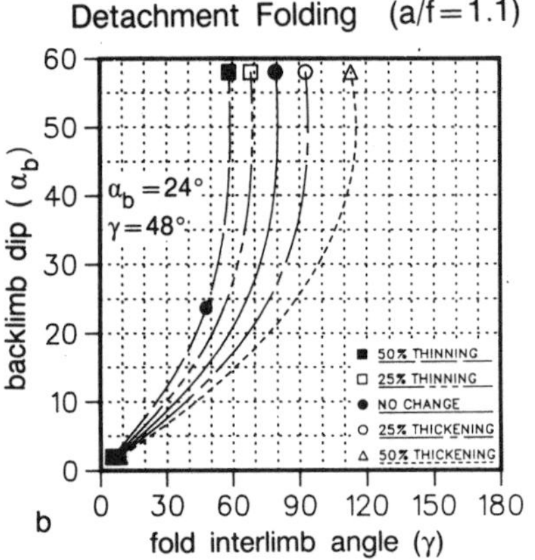

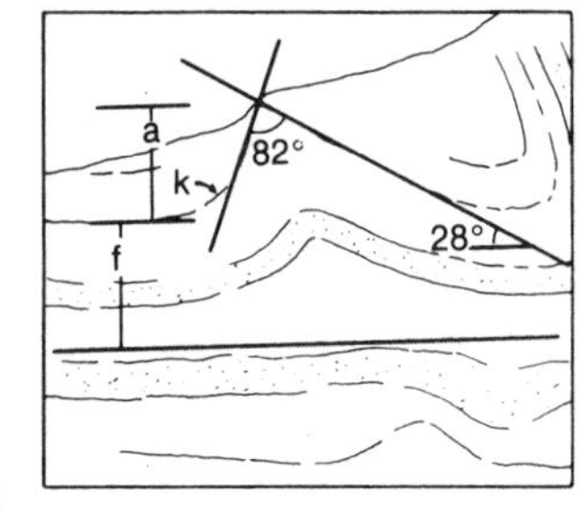

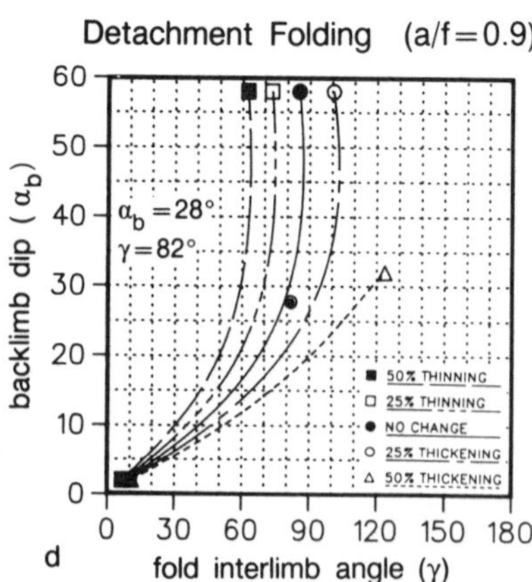

Fig. 9. Geometric analysis of Besa River folds. (a) The limbs of the right-hand fold of Fig. 8 are approximated by straight lines to determine the values needed for the geometric analysis. (b) Detachment fold charts for $a/f = 1.1$ suggest the outcropping fold should have forelimb thinning of 50% in the units immediately above horizon *m*. (c) Determination of the angular relationship in the left-hand fold of Fig. 8. (d) Geometric analysis indicates that this fold ($a/f = 0.9$) should have a slight thickening of forelimb beds at the level of horizon *k*.

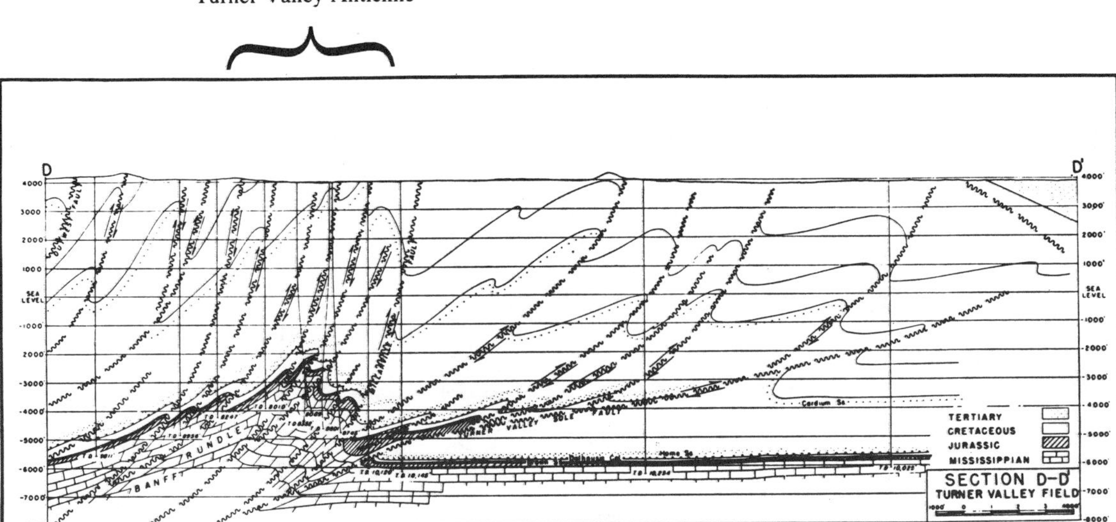

Fig. 10. Cross-sectional interpretation of the Turner Valley Anticline. Vertical to horizontal scales are 1:1. From Gallup (1951). (Reproduced with permission of American Association of Petroleum Geologists.)

larities, it should be noted that the model fold (Fig. 4b) becomes flat-topped upsection, and maintains a constant amplitude (a) throughout the folded interval. The outcropping folds do not become flat-topped, and they do not have identical amplitudes at different levels through the structure. Because of these dissimilarities, a single analysis is not appropriate for all units involved in the folding.

Subsurface example

The Turner Valley Anticline is located at the foreland limit of the Canadian Cordillera in southern Alberta (Fig. 10). Oil was discovered in the Paleozoic units of this structure in 1924, and subsequent drilling provided the data for construction of detailed cross-sections (Gallup 1951). Drilling of this structure predates the development of wire-line geophysical tools; therefore, the subsurface data consists of stratigraphic picks based on borehole cuttings.

The Turner Valley Anticline is a truncation anticline located above the ramp of the Turner Valley Sole Fault. The Mississippian units are truncated against the thrust ramp in both the hangingwall and footwall of the structure. Since both the fold and the truncated beds are located directly above the ramp of the transporting thrust fault, it can be concluded that this is a fault-propagation fold that has not been transported through the upper hinge of its associated ramp. Numerous imbricate faults are interpreted to occur throughout the Turner Valley structure (Fig. 10). One of the imbricates, the Millarville Fault, carries a separate Paleozoic fold along trend.

Geometric analysis of this structure requires more approximation than did the outcrop example simply because the structure, as interpreted, does not display smooth, planar limbs. In order to make the analysis, the forelimb, backlimb and fault ramp must be approximated by straight lines. Making these approximations (Fig. 11a), the fold interlimb angle at the top of the Mississippian carbonates is 92°, and at the top of the Cretaceous Cardium Sandstone it is 78°. Fault imbrications produce tectonic thickening in both limbs of the fold, but the thickening is greater in the forelimb. In the pre-Cardium Cretaceous units, the forelimb is about 10% thicker than the backlimb. In the Mississippian units, the relative forelimb thickening appears to be greater, although the lack of well control for basal Mississippian horizons prohibits a direct measurement of this value.

In the cross-section the ramp angle is 22°. For this ramp angle and the measured interlimb angles, the geometric analysis indicates that the Cretaceous beds should have been thickened 30% in the forelimb and the Mississippian units 45% (K_1 and M_1 in Fig. 11b). These are considerably larger amounts of forelimb thickening than depicted in the cross-section. In order to obtain folds with both the measured interlimb angles and the more modest forelimb thicknesses indicated in the cross-section, a ramp angle close to about 28° would be needed (K_2 and M_2 in Fig. 11b). As there is little control to constrain the depicted ramp angle, the section could just as easily be drawn with the larger (28°) ramp angle. Modifying the cross-section to be compatible with the geometric analysis will ensure that it is balanced.

CONCLUSIONS

Balancing techniques are the union of a belief in the conservation of matter and a suite of assumptions based upon geologic observations and simplifications. The

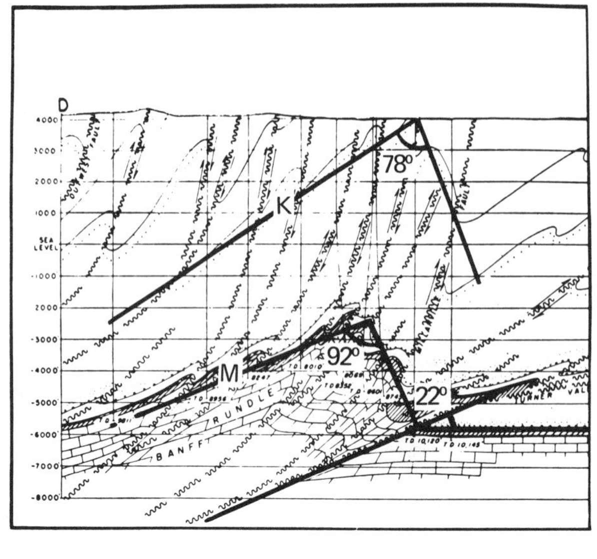

a

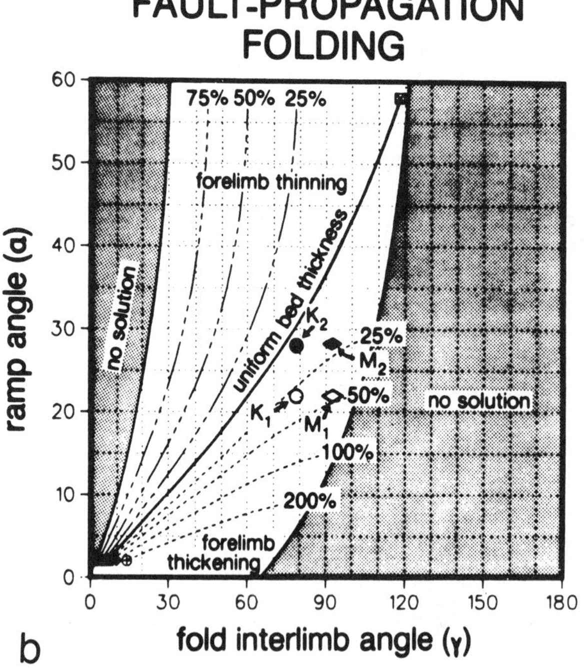

b

Fig. 11. Geometric analysis of the Turner Valley Anticline. (a) For the purposes of analysis, the fold forelimb and backlimb, the footwall bedding, and the fault are all approximated to straight lines. (b) The data from Mississippian (M_1 and M_2) and Cardium (K_1 and K_2) beds in the fold are plotted on fault-propagation fold charts relating γ and α to forelimb thickness. See text for discussion.

stated goal of the procedure dictates the chosen assumptions. The goal of this study is to relate fold geometry to the mode of fold-thrust interaction.

Fault-bend, fault-propagation and detachment folds each have a definable set of geometric associations. The most geometrically distinctive fold form is the Mode I fault-bend fold, which always has a relatively large interlimb angle. Also, tight, upright folds are more apt to be detachment folds than truncation anticlines. In general, however, there is considerable overlap and similarity of geometric solutions among the three modes of fold-thrust interaction. Consequently, determining the mode of fold development exclusively from final fold geometry is possible only for a limited suite of structures. However, the nature of the fold-thrust interaction can often be inferred by coupling the geometric analysis with kinematic considerations, details of the deformation, and familiarity with the local geology.

The geometry and geometric relationships of a fold are altered, in a predictable fashion, when it is transported foreland through a ramp hinge. For fault-propagation folds, the alteration is such that the geometry of the transported fold is identical to the geometry of a Mode II fault-bend fold with the same ramp angle. Fault-parallel shearing of the fold during or after fold development will tighten the fold and produce a forelimb that is unusually thick for the specified fold form. Excess material that is moved into the forelimb region by the shearing produces this effect.

Geometric analysis provides a means for balancing individual folds. Application of the analysis to existing cross-section interpretations can illuminate unstated assumptions and specific weaknesses of the construction. Structural interpretations inherently require an extrapolation from the reasonably well known into the unknown. The geometric analysis can provide guidelines for this extrapolation that have a geologic foundation.

Acknowledgements—My thanks to Amoco Production Company for permission to publish this paper, which is based on a 1982 Amoco Research Department report. I am especially grateful to Martha Withjack, Ron Nelson and Dave Wiltschko for their thoughtful criticism of the manuscript.

REFERENCES

Berger, P. & Johnson, A. M. 1980. First-order analysis of deformation of a thrust sheet moving over a ramp. *Tectonophysics* **70**, T9–T24.

Boyer, S. E. & Elliott, D. 1982. Thrust systems. *Bull. Am. Ass. Petrol. Geol.* **66**, 1196–1230.

Brown, S. P. & Spang, J. H. 1978. Geometry and mechanical relationship of folds to thrust fault propagation using a minor thrust in the Front Ranges of the Canadian Rocky Mountains. *Bull. Can. Soc. Petrol. Geol.* **26**, 551–571.

Dahlstrom, C. D. A. 1969. Balanced cross-sections. *Can. J. Earth Sci.* **6**, 743–757.

Dahlstrom, C. D. A. 1970. Structural geology in the eastern margin of the Canadian Rocky Mountains. *Bull. Can. Soc. Petrol. Geol.* **18**, 332–406.

Dixon, J. S. 1982. Regional structural synthesis, Wyoming salient of western overthrust belt. *Bull. Am. Ass. Petrol. Geol.* **66**, 1560–1580.

Faill, R. T. 1969. Kink band structures in the Valley and Ridge Province, central Pennsylvania. *Bull. Geol. Soc. Am.* **80**, 2539–2550.

Fitzgerald, E. L. & Braun, L. T. 1965. Disharmonic folds in Besa River Formation, northeastern British Columbia, Canada. *Bull. Am. Ass. Petrol. Geol.* **49**, 418–432.

Gallup, W. B. 1951. Geology of Turner Valley oil and gas field, Alberta, Canada. *Bull. Am. Ass. Petrol. Geol.* **35**, 797–821.

Laubscher, H. P. 1977. Fold development in the Jura. *Tectonophysics* **37**, 337–362.

Morse, J. D. 1977. Deformation in ramp regions of overthrust faults: experiments with small-scale rock models. *Wyoming Geol. Ass. Guidebook, 29th Ann. Field Conf.* 457–470.

Rich, J. L. 1934. Mechanics of low-angle overthrust faulting as illustrated by Cumberland thrust block, Virginia, Kentucky, Tennessee. *Bull. Am. Ass. Petrol. Geol.* **18**, 1584–1596.

Serra, S. 1977. Styles of deformation in the ramp regions of overthrust faults. *Wyoming Geol. Ass. Guidebook, 29th Ann. Field Conf.* 487–498.

Suppe, J. 1980. A retrodeformable cross-section of northern Taiwan. *Geol. Soc. China Proc.* **23**, 46–55.

Suppe, J. 1983. Geometry and kinematics of fault-bend folding. *Am. J. Sci.* **283**, 684–721.

Suppe, J. & Medwedeff, D. A. 1984. Fault-propagation folding. *Geol. Soc. Am. 1984 Ann. Mtg. Prog. With Abs.* **16**, 670.

Williams, G. & Chapman, T. 1983. Strain developed in the hangingwalls of thrusts due to their slip/propagation rate: a dislocation model. *J. Struct. Geol.* **5**, 563–571.

Wiltschko, D. V. 1979a. A mechanical model for thrust sheet deformation at a ramp. *J. geophys. Res.* **84**, 1091–1104.

Wiltschko, D. V. 1979b. Partitioning of energy in a thrust sheet and implications concerning driving forces. *J. geophys. Res.* **84**, 6050–6058.

APPENDIX

The basic premises behind the geometric analyses are discussed in the text. Each model begins by equating pre- and post-deformational area of the structure. The transformation of the initial equation into the relationships used to generate the curves shown in the text is an exercise in trigonometry. In this appendix, I show only the initial and final equations.

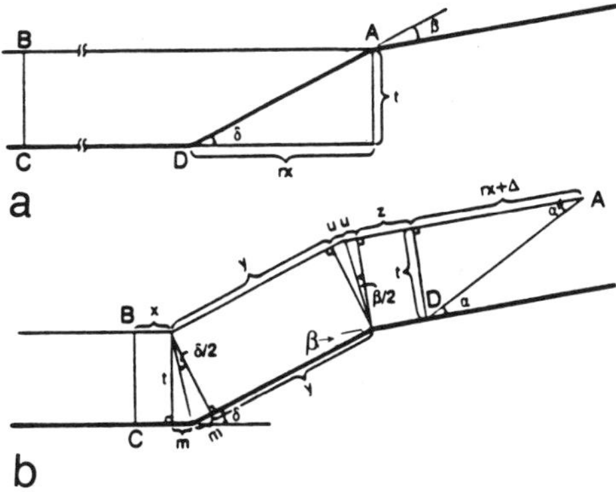

Fig. 12. Diagram for the evaluation of the effective ramp angle (α) of a fault-bend fold. (a) Fault with no offset of the hangingwall. (b) Fault after displacement of the hangingwall.

Fault-bend folding

A critical parameter in the fault-bend fold equations is the 'ramp angle' (α), which is a function of the upper and lower ramp hinge angles (β and δ, respectively). This relationship is determined by considering a transported, but unfolded, hangingwall (Fig. 12), where all units are assumed to have constant thickness. Constant cross-sectional area is maintained by keeping all bed lengths constant. From the pre-transport configuration:

$$\overline{AB} = \overline{CD} + rx. \quad (1)$$

In the post-transport configuration, this becomes:

$$x + y + 2u + z + rx + \Delta = x + 2m + y + z + rx, \quad (2)$$

or

$$\Delta = 2m - 2u, \quad (3)$$

which leads to:

$$\cot \alpha = \cot \delta + 2[\tan (\delta/2) - \tan (\beta/2)]. \quad (4)$$

The fault-bend fold is formed by allowing the leading-edge triangle of the hangingwall block to rest against the underlying thrust (Fig. 13). In so doing the forelimb of the fault-bend is formed. The bedding-normal thickness of this leading edge triangle (tf) is not required to equal the original bedding thickness (t). The pre- and post-folding geometries are equated as:

$$\text{area } \Delta ABD = \text{area } \Delta ABC + \text{area } \Delta ACE \quad (5)$$

which yields

$$\cot \alpha = \cot \gamma_1 + \sin \gamma_2/[\sin \gamma_1 \sin (\gamma_1 + \gamma_2)]. \quad (6)$$

Note that here, and in all ensuing models, the forelimb to backlimb thickness ratio may be expressed as:

$$tf/t = \sin \gamma_2/\sin \gamma_1. \quad (7)$$

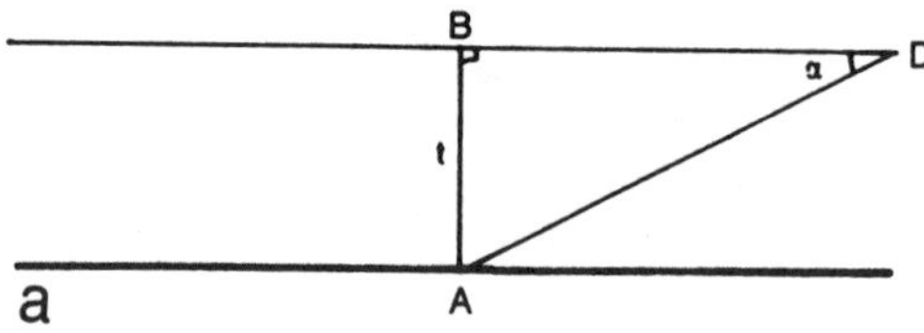

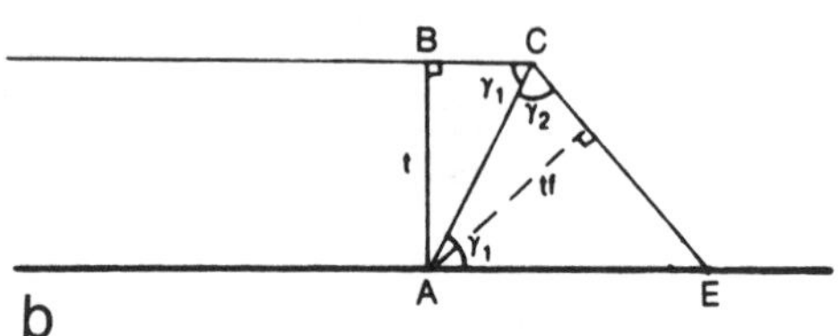

Fig. 13. Diagram for evaluation of fold interlimb angle of a fault-bend fold for a given α. (a) Prefolding and (b) post-folding geometries.

Fault-propagation folding

Separate geometries and equations are assigned to those units in contact with the fault versus those not in contact with the fault (Fig. 14). The ramp angle in this fold-thrust model is the lower ramp hinge angle. For those units in contact with the fault, the equated pre- and post-folding areas are expressed as:

$$\text{area } \Delta ABC = \text{area } \Delta ABF + \text{area } \Delta AFC, \tag{8}$$

which yields:

$$\cot \alpha = 2 \cot \gamma_1 - \tan \phi - \cot \gamma. \tag{9}$$

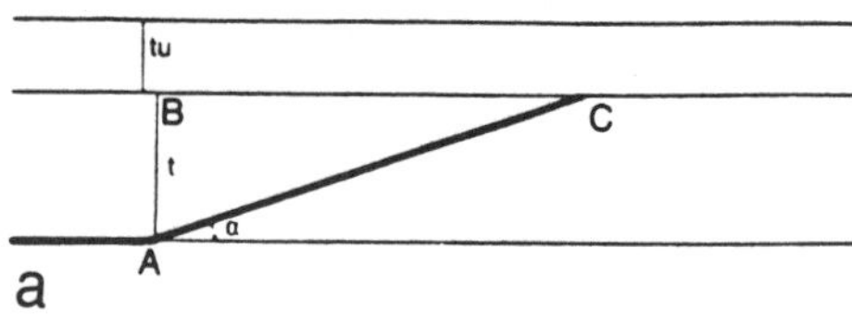

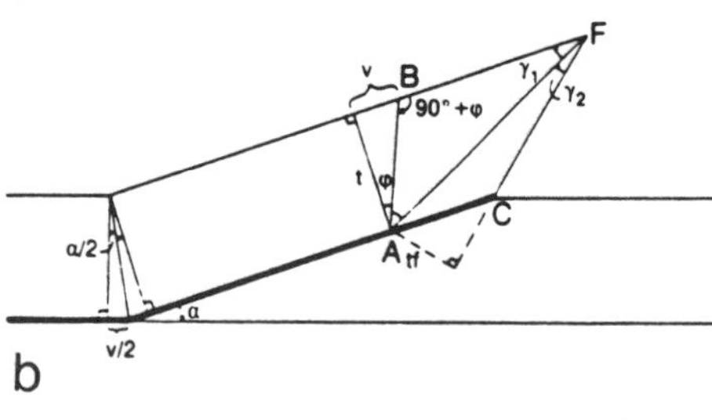

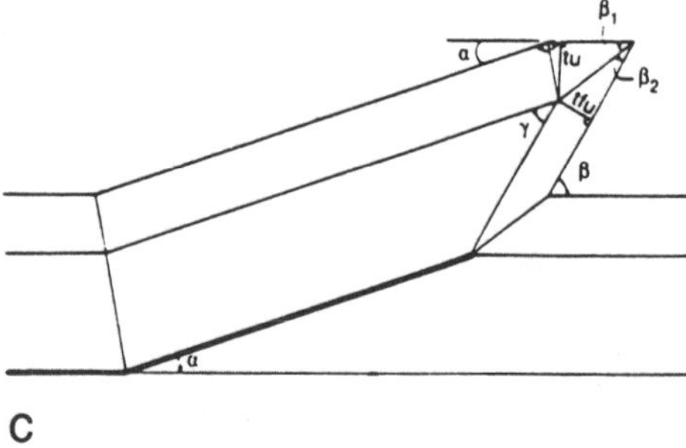

Fig. 14. Diagram for evaluation of fault-propagation fold. (a) Unfolded units. (b) Folded unit in direct contact with the propagating fault. (c) Folded unit not in contact with propagating fault.

Noting that:

$$\tan \phi = 2 \tan (\alpha/2), \tag{10}$$

the preceding equation may be written as:

$$\gamma = \gamma_1 + \gamma_2 + \tan^{-1} [1/(2 \cot \gamma_1 - (2 - \cos \alpha)/\sin \alpha)]. \tag{11}$$

For those units not in contact with the fault (Fig. 14c), the model is constructed such that:

$$\beta = \beta_1 + \beta_2 = \gamma + \alpha. \tag{12}$$

Using the assumption that:

$$tfu/tu = tf/t \tag{13}$$

leads to:

$$\beta_1 = \tan^{-1} [(2 \sin \alpha)/(2 \cos \alpha - 1)]. \tag{14}$$

The geometric relationships for the fault-propagation fold are defined by those units in contact with the fault.

Detachment fold

The chevron fold form used in this paper is a degenerative case of the more general box-fold geometry (Fig. 15). The basic geometric relationships come from analysis of the 'ductile' unit in the core of the fold. The pre-folding area (Fig. 15a) is:

$$A_1 = \text{area } ABCD = fL_o \tag{15}$$

where:

$$L_o = e_1(t/tf) + e_2 + e_3. \tag{16}$$

The post-folding area (Fig. 15b) is:

$$A_2 = \text{area } ABCD + \text{area } ADGH. \tag{17}$$

Equating these pre- and post-folding areas yields:

$$a/f = \frac{[t/(tf \cos \beta) + (1/\cos \delta) - \tan \beta - \tan \delta]}{[e_2/a + \frac{1}{2}(\tan \beta + \tan \delta)]}. \tag{18}$$

Allowing thickness changes to occur in the forelimb of the stiff unit, it is found that:

$$tf/t = \cos \beta \cot \xi_1 + \sin \beta. \tag{19}$$

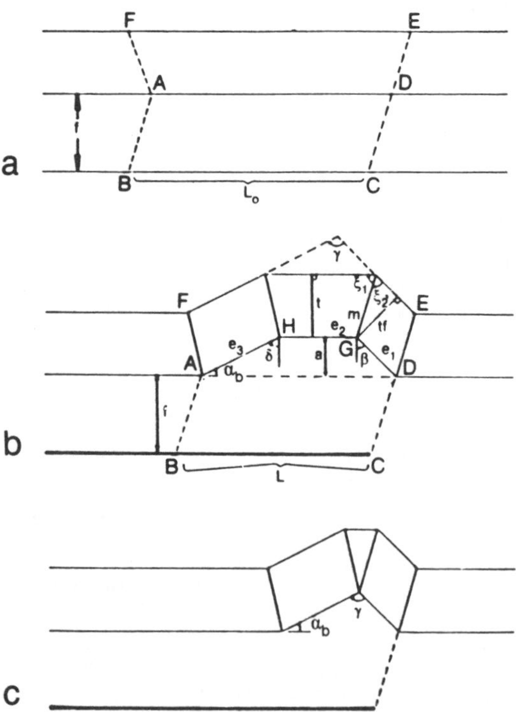

Fig. 15. Diagram for evaluation of detachment folds. (a) Units before folding. (b) General detachment fold (box fold) after folding. (c) Chevron fold form.

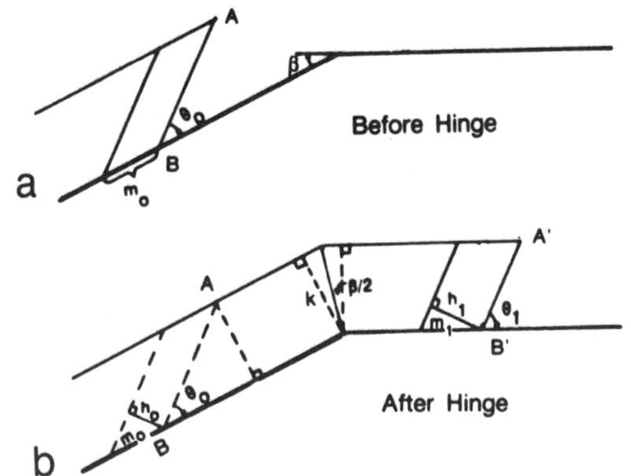

Fig. 16. Diagram for evaluation of the effect of upper ramp hinge on fold interlimb angle and forelimb thickness. (a) Pre-transport and (b) post-transport configuration.

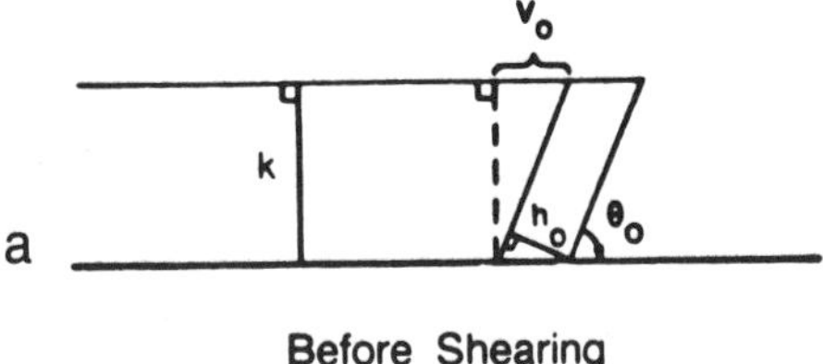

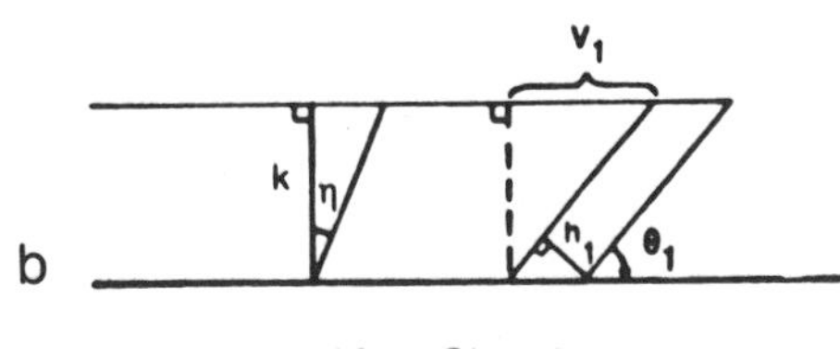

Fig. 17. Diagram for evaluation of the effect of fault-parallel shear on fold interlimb angle and forelimb thickness. (a) Pre-shearing and (b) post-shearing configuration.

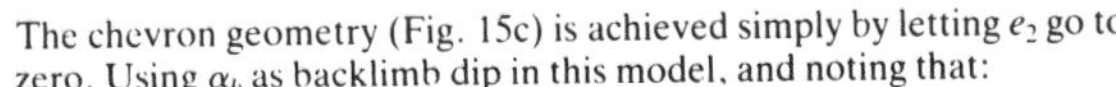

The chevron geometry (Fig. 15c) is achieved simply by letting e_2 go to zero. Using α_b as backlimb dip in this model, and noting that:

$$\alpha_b = 90° - \delta \quad \text{and} \quad \gamma = \beta + \delta \tag{20}$$

then:

$$a/f = 2(t/tf \sin \alpha_b/\sin \gamma + \cos \alpha_b + \sin \alpha_b \cot \gamma - 1). \tag{21}$$

Transport through an upper ramp hinge

For this calculation, it is assumed that there is no change in thickness of the structure, measured perpendicularly to the fault, during the transporting process (Fig. 16). Consequently, transport distances for the top and bottom surfaces of the structure are equal, i.e.:

$$\overline{AA'} = \overline{BB'}. \tag{22}$$

This observation allows pre- and post-transport angles to be related by:

$$\tan \theta_1 = \tan \theta_0/[1 - 2 \cdot \tan \theta_0 \cdot \tan (\beta/2)], \tag{23}$$

where the angles are measured relative to the fault surface. If θ_0 and θ_1 are bedding surface measurements, the post-transport thickness of this bed (h_1) relative to its initial thickness (h_o) is:

$$h_1/h_o = \sin \theta_1/\sin \theta_0. \tag{24}$$

Transport through a lower ramp hinge reverses the position of the variables.

Fault-parallel shearing

Uniform, fault-parallel shear through an angle η (Fig. 17) produces roughly the opposite effect as transport through an upper ramp hinge. Here:

$$\tan \theta_1 = \tan \theta_0/(1 + \tan \theta_0 \tan \eta) \tag{25}$$

and, as before:

$$h_1/h_o = \sin \theta_1/\sin \theta_0. \tag{26}$$

Reprinted by permission of the Canadian Society of Petroleum Geologists from *Bulletin of Canadian Petroleum Geology*, v. 20 (1972), p. 583-607.

BULLETIN OF CANADIAN PETROLEUM GEOLOGY
VOL. 20, NO. 3 (SEPTEMBER, 1972), P. 583-607

THOUGHTS ON OVERTHRUST FAULTING IN A LAYERED SEQUENCE

P. E. GRETENER[1]

ABSTRACT

This paper concentrates on some aspects of thrust faulting in a layered sequence. First, the fact that thrust faulting, as the name implies, constitutes duplication of the sequence is considered. The effects of loading due to an advancing thrust sheet are studied. In particular, the effect of the load on the fluid pressure in the underlying beds is reviewed. Second, the peculiar geometric attitude of thrust faults in layered sequence; i.e., the stepping, is discussed. Further, the apparent ease with which movement takes place along faults of this type is investigated, and previously advanced theories are supplemented. Such factors as the pore pressure, the existence of bedding planes (planes of weakness) and the unequal distribution of the stresses in a layered sequence under horizontal compression are all considered important when trying to understand the mechanics of overthrust faulting.

INTRODUCTION

Overthrusts have fascinated geologists for well over a hundred years. One of the classical locations in Canada is shown in Figure 1. The sheer wall of Mount Yamnuska on the Trans-Canada highway between Calgary and Banff consists of middle Cambrian limestone thrust over upper Cretaceous clastics along the McConnell fault. Thousands of travellers pass by this place every year, all of them admiring the scenic beauty, few of them realizing the profound geological significance of the picture. Older on younger — a defiance of the principle of superposition! A unique and unequivocal demonstration of the dynamics of the earth which only the insufficient yardstick of our lifespan makes seemingly 'dead' and static. This paper represents a tribute to this — one of our truly magnificent — geological location.

Figure 2 shows an overthrust in its most simple form. On the left of the figure the thrust is a sole fault stepping through the sedimentary sequence near the middle of the figure and becoming an actual overthrust (or toe) forward of the step. On the lower portion of Figure 2 are plotted the additional uneroded load (UL) due to the thrusting process as well as the stratigraphic displacement (SD) as observed along the thrust fault. A. Escher was the first to describe such structures from the Canton of Glarus in Switzerland as early as 1841 (Staub, 1954). The term *overthrust* is a straight translation from the German 'Ueberschiebung'. The term is widely used but recently some authors have objected to it because they feel it implies an absolute sense of motion or, in terms of Figure 2, that the left-hand side has moved over

---

[1]Department of Geology, The University of Calgary, Calgary 44, Alberta.

To my colleagues Drs. Simony and Spang my thanks for many stimulating discussions on the subject. Dr. D. W. Stearns read the manuscript and has contributed a number of valuable suggestions. This work was supported by a National Research Council of Canada Grant.

Fig. 1. Mount Yamnuska, Alberta. The cliff consists of middle Cambrian Elden lime-stone thrust over upper Cretaceous Belly River sandstones and shales forming the lower slopes.

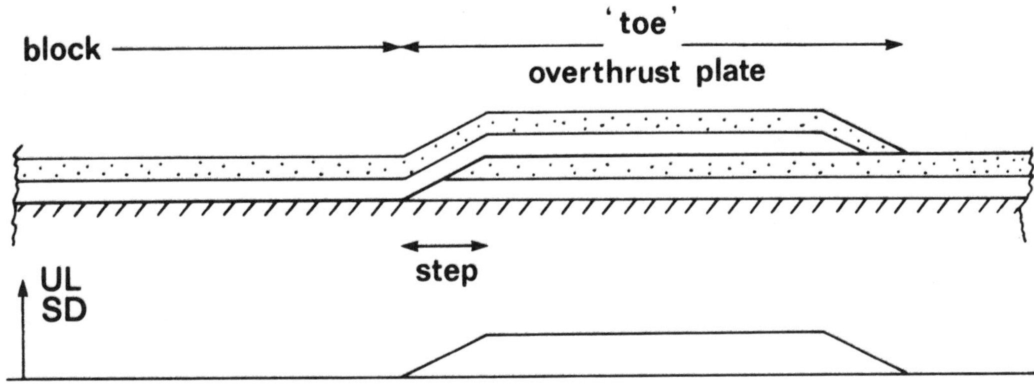

UL = uneroded load
SD = stratigraphic displacement

Fig. 2. Schematic diagram showing the major elements of an overthrust. The lower portion shows the additional uneroded load (UL) above, and the stratigraphic displacement (SD) along the thrust.

For convenient reference, certain notations have been adopted throughout the paper, as follows:

SS	step spacing (length of link)
SH	step height
TD	total displacement
SD	stratigraphic displacement
UL	uneroded load
p	pore pressure
S_z	total overburden pressure
σ_z	effective overburden pressure
σ_x	maximum effective horizontal stress
τ_0'	cohesion
τ_0	reduced cohesion on pre-existing fracture or bedding plane (may be zero)
n	coefficient of internal friction
μ	coefficient of sliding friction
δ	angle of fracture or bedding plane to axis of maximum compressive stress (σ_1)
λ	ratio of pore pressure to total overburden pressure (p/S_z)
E	Young's modulus
ν'	effective Poisson's ratio

1 bar = 1.02 kp/cm^2
1 bar = 14.5 psi
1 kp/cm^2 = 14.2 psi

the right-hand side, the latter remaining stationary. Of course, only a relative motion is indicated by Figure 2. To this author the question of overthrusting versus underthrusting is not of importance for the aspects discussed here, and when the term 'overthrust' is used, its connotation is simply a relative one.

Needless to say, Escher's interpretation was not readily accepted, (Heim, 1882). It was not until about the turn of the century that geologists were willing to acknowledge the existence of such fantastic structures (e.g., Staub, 1954, p. 143). The enigma of their formation remained: mechanically, these structures seemed to defy rational explanation. Only in 1959 did Hubbert and Rubey in two companion papers shed some light on the possible mechanics of these structures. Their ideas sparked much discussion, and it is the purpose of this paper to carry on the debate and point out some additional aspects which must be considered.

From Figure 2 it is apparent that the whole mass above and to the left of the fault plane has moved in regard to the sequence on the right. Only to the right of the area of the step does the 'toe' or overthrust plate exist. To the left of the step the fault plane is present, but it is no more than a bedding plane with no stratigraphic displacement. It is important to realize that only in the frontal part is the overthrust recognizable as such.

Figure 3 shows the area of the toe in more detail and clarifies the terminology used in this paper. Usually the thrust does not break to the surface in one giant step, but reaches it in a series of discrete minor steps. According to observations (Dahlstrom, 1970) the steps occur across the competent formations, whereas in the incompetent units the fault is frequently parallel or subparallel to bedding. In this paper that portion will be referred to as the link. This author accepts such behaviour of thrust faults as an observed fact (Rich, 1934; Douglas, 1950;

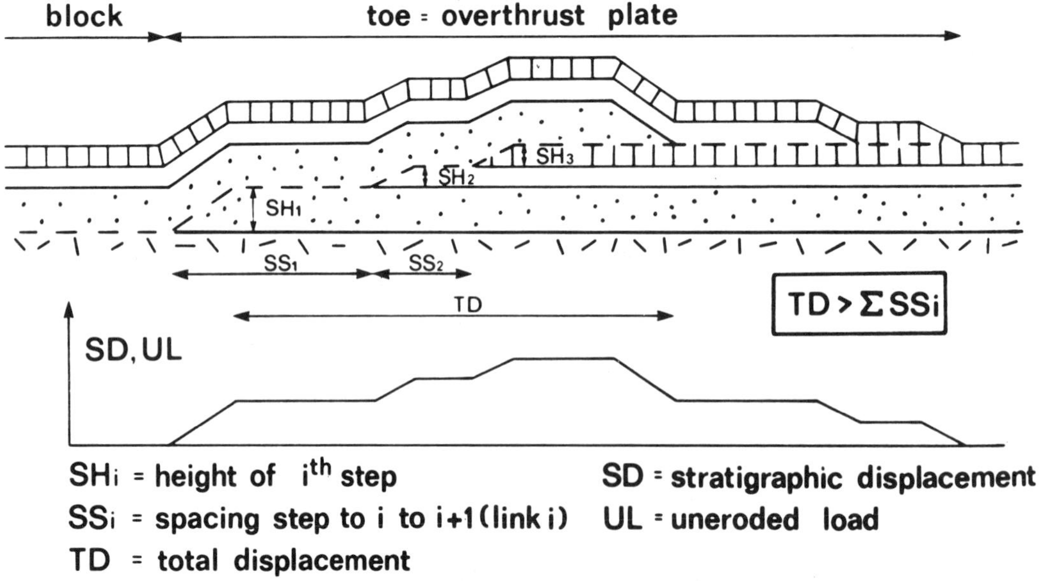

Fig. 3. Detail of the thrust plate for total displacement exceeding the sum of the links ($TD > \Sigma SS_i$).

Bally *et al.*, 1966; Dahlstrom, 1970). Thrusts become solefaults such as shown on the left in Figures 2 and 3, thus indicating that the shortening implied by thrust faulting has affected only an uppermost skin of the earth's crust in the area of observation. Thus, in one of the best explored thrust belts of the world, the southern Canadian Rockies, drilling and seismic work indicate that thrusting is confined to the sedimentary skin in the area east of the Rocky Mountain Trench and does not involve the Hudsonian basement (Bally *et al.*, 1966; Price and Mountjoy, 1969; Dahlstrom, 1970). The thrusts become observable forward from the deepest step. The thrust fault can be geometrically described by such parameters as the step height (SH), the step spacing or length of links (SS), the total displacement (TD) and the stratigraphic displacement (SD), all of which are explained in Figure 3. The uneroded load (UL) is directly proportional to the stratigraphic displacement. Uneroded load assumes that the thrust sheet remains essentially intact during the whole process of thrusting. It represents an extreme case and is not necessarily a reasonable assumption. The question of erosion during thrusting will be further discussed in the respective section.

From Figures 2 and 3 it is evident that the thrust motion has an appreciable vertical component only in the areas of the steps. On a link the motion is parallel or subparallel to bedding, so that deformation results only over the steps. This geometry was described by Rich as early as 1934 for the Appalachians. More recently it has become very important in the southern Canadian Rockies, where these structures are the loci of natural gas fields (Bally *et al.*, 1966).

The vertical extent of these structures is a function of the height of the respective buried step, whereas their lateral extent is given by the total displacement. In Figure 3 the total displacement is larger than the total length of the links (ΣSS_i). Under these conditions, maximum stratigraphic displacement is obtained and the structure formed is basically simple. In Figure 4 a case is shown where the total displacement is smaller than the sum of the links. In this case the result is a series of individual structures in the direction of transport. Areas of large stratigraphic displacement are separated by areas of small or zero displacement. Likewise, the uneroded load is concentrated immediately forward from the individual steps. Both stratigraphic displacement and uneroded load do not attain the maximum value corresponding to the total thickness involved in the thrusting, as in Figures 2 and 3.

It is important to note that Figures 2 through 4 all refer to the condition in an individual thrust sheet and not to that of the thrust belt as a whole. Individual sheets may advance on top of one another, leading to both larger stratigraphic displacements aand more severe conditions of loading.

Vital Statistics of Thrust Plates

It is not the purpose of this section to establish certain facts with a great degree of precision but rather to give orders of magnitude for the various terms explained in Figures 3 and 4. Dahlstrom (1970, p. 355) has most aptly coined the terms *fundamentals* and *incidentals*. This paper is concerned with some of the fundamentals of thrust faulting. It is thus permissible to look at the geological picture with a geophysicist's eye, filtering out the incidentals or the unimportant. (The

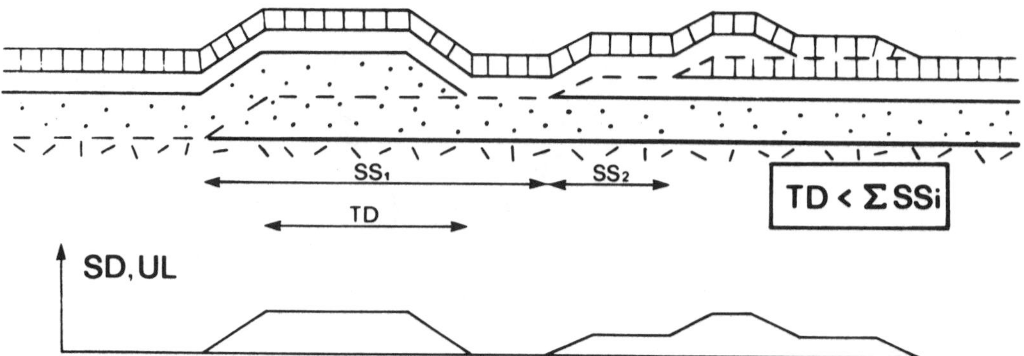

Fig. 4. Detail of the thrust plate for total displacement smaller than the sum of the links ($TD < \Sigma SS_i$).

latter term refers strictly to the objectives of this study. For those concerned with the detailed mapping or the economic exploration of such belts, the incidentals may well be all-important. For the present study they only obscure the fundamental picture or, in geophysical language they represent 'noise'.)

It may well be worth while to discuss briefly the definition of fundamentals and incidentals as used in this paper. To some extent it is of course a matter of scale, large features tending to be in the class of fundamentals whereas small ones belong to the incidentals. It may also be a matter of repetitiveness, fundamentals occurring more frequently than incidentals. More specifically, the fundamentals can be understood from a model but the incidentals cannot, being often the result of coincidence. Thus an original weakness in the rock may well locate a minor tear fault in a thrust sheet, or a facies change eliminating a thick competent unit will lead to a change in structural style.

There can be little question that in the Helvetide area of the Alps the incidentals play a much more prominent role than in the southern Canadian Rockies. The result is an apparent dissimilarity inducing some geologists to state: 'that there are no two mountain chains alike'. This statement is certainly unacceptable. Major structural provinces such as thrust belts must have certain basic features in common. It is suggested that the incidentals are largely responsible for masking those similarities.

In this paper an attempt is made to understand some fundamental features of thrust belts and, more precisely, individual thrust sheets. As a result, the details or incidentals shall not concern us or confuse the issue. It is for this reason that in the following subsection only very rough guides are established for the magnitude of certain features. Only the fact that they do exist, and to an appreciable degree, is of importance at this stage.

a) *Total Displacement*

This term is illustrated in Figure 3. According to the introduction to this section, no accurate values are sought but rather orders of magnitude. It is thus permissible to rely on rather simple techniques. Figure 5 shows the Lewis thrust in northwestern Montana just south of

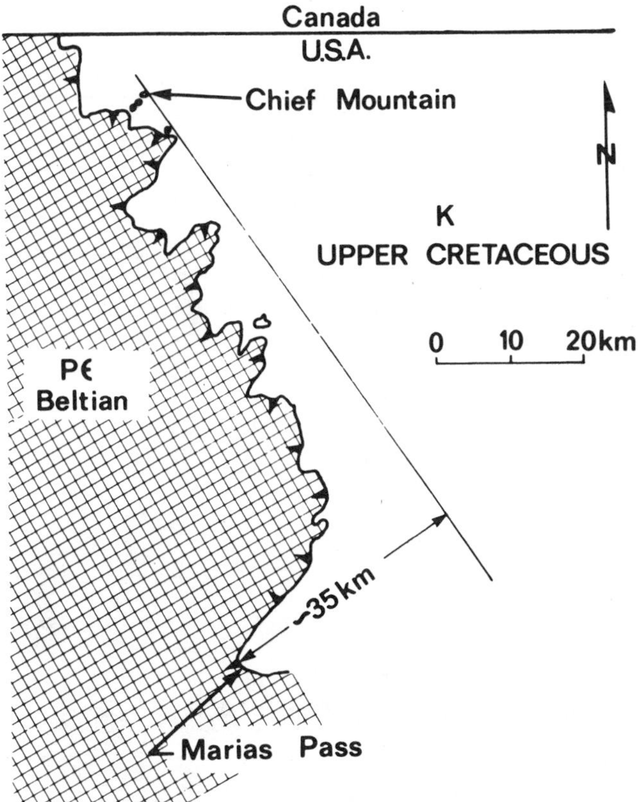

Fig. 5. The Lewis thrust plate in northwestern Montana. The minimum total displacement as read from the geological map.

the Canadian border, where pre-Cambrian Beltian sediments rest on Upper Cretaceous rocks. Near the international border Chief Mountain forms a spectacular Klippe, whereas at Marias Pass erosion has produced a deep re-entrant. Just a look at the map (Fig. 5) reveals a minimum total displacement of 35 km (22 mi). There is no interpretation involved in this estimate and it represents an absolute minimum.

Northwest of Calgary, Alberta, the McConnell thrust at the Panther River culmination has a visible displacement of 15 km (10 mi), where Cambrian rocks rest on Upper Cretaceous strata. The Glarus thrust in Switzerland shows Permian over Tertiary for a distance of 35 km (22 mi). Gansser's (1964) map of the Himalayas shows displacements of 50 km (31 mi) in several places. Maps of the Moine thrust show displacements of 10 km (6 mi).

All these displacements are simply taken from the geological maps in the manner shown for the Lewis thrust. They in no way reflect the actual total displacement and are undisputed minimum values. However, for the purpose of this study it is unimportant whether the true total displacement is one and a half, two or even three times the values stated above. It suffices to recognize that total displacements of major overthrust units in various parts of the world are measured in tens of kilometres.

b) *Stratigraphic Displacement*

At the Lewis thrust in Montana, pre-Cambrian Beltian rocks overlie Upper Cretaceous and the stratigraphic displacement is in the order of 6 km (4 mi). At Mount Yamnuska, west of Calgary (Fig. 1), the McConnell thrust brings middle Cambrian strata over Upper Cretaceous rocks with a stratigraphic displacement of more than 4 km (2.5 mi). Thickness of the Glarus thrust is estimated to have been 5 to 6 km (3 to 4 mi) at the time of emplacement (Trümpy, 1969).

Again it is not our intention to document in great detail the stratigraphic displacements on various thrusts. The above values indicate that on major thrusts the stratigraphic displacement is to be measured in kilometres.

c) *Observed Nature of Thrust Faults*

Not many detailed descriptions of thrust faults are available in the literature. However, modern mapping has led to many places where

Fig. 6. Detail of the McConnell thrust at Mount Yamnuska, west of Calgary, Alberta.

the thrusts themselves can be observed. Invariably the descriptions use such words as : 'knife sharp', 'drawn with a knife edge', etc. Figure 6 shows the McConnell thrust as exposed on Mount Yamnuska (detail, middle of Fig. 1). It is indeed possible to place one's hand on the surface separating the Middle Cambrian Eldon Formation from the Upper Cretaceous Belly River Formation. Descriptions in the literature (e.g., Gansser, 1964, photos 29 & 30, p. 132) and personal observations in both the Canadian Rockies and the Alps indicate that the picture shown may be termed 'typical' for thrust faults. The fact that this boundary is so sharp becomes more impressive when one considers that each point above the fault surface in Figure 6 has moved a minimum distance of 15 km (10 mi). It certainly points to an ease of movement and makes understandable the geologists' old postulate of a 'basal lubrication'. Until the classic companion papers of Hubbert and Rubey (1959) — that is, for over 100 years — this occurrence has remained a mystery. Today we begin to understand the mechanism of these faults and hopefully, the ideas presented further on in this paper wil contribute to that understanding.

While the sharp nature of the faults has been emphasized by many authors (Heim, 1919, p. 378, 387, 389; Holmes, 1964, p. 223, 1164; Shelton, 1966, p. 103, 323), one should not forget that the rocks above and below a thrust may show considerable deformation for distances of some hundreds of feet. Still the thrust plates as a whole have remained intact and are in no way comparable to the jumbled masses of landslides. Thus, the presumption underlying this paper — that we are dealing with virtually undistorted plates moving over the undisturbed sequence along sharp planes — is basically correct.

d) *Progession of Thrusting as Deduced from Field Evidence*

In the Canadian Rockies most authors agree that thrusting has progressed essentially from west to east; i.e., outward (Bally *et al.*, 1966; Dahlstrom, 1970). According to Rubey and Hubbert (1959, p. 190) and Armstrong and Oriel (1965), the same is true for the Idaho-Wyoming thrust belt. In the Helvetides of the Alps there is some evidence that in certain places thrusting has progressed outward, but the case cannot be as unequivocally stated as for the southern Canadian Rockies (Trumpy, *pers. comm.*)

As pointed out by Dahlstrom (1970) and Jones (1971), there are cases, particularly those of the so-called back-limb thrusts, that do not conform to this general rule. In our terminology they would fall into the class of the incidentals, their position and origin being determined by the preceding folding of a thrust plate due to an underlying step (Figs. 3, 4). These thrusts will only obscure the fundamental picture. As will be shown later, it seems mechanically very plausible that thrusting should progress basically outward. It is to be hoped that in future the field geologists will pay even more attention to this question.

e) *The Sedimentary Wedge*

The deformation in the outer thrust belts, under consideration in this paper, involves the sediments of a miogeocline. These sediments form a wedge and the case of the southern Canadian Rockies is particularly well documented. Even a casual field trip from Calgary to the Rocky Mountain Trench reveals the rapid thickening of the sedimentary section in this direction, as pointed out by Bally *et al.* (1966, Fig. 10).

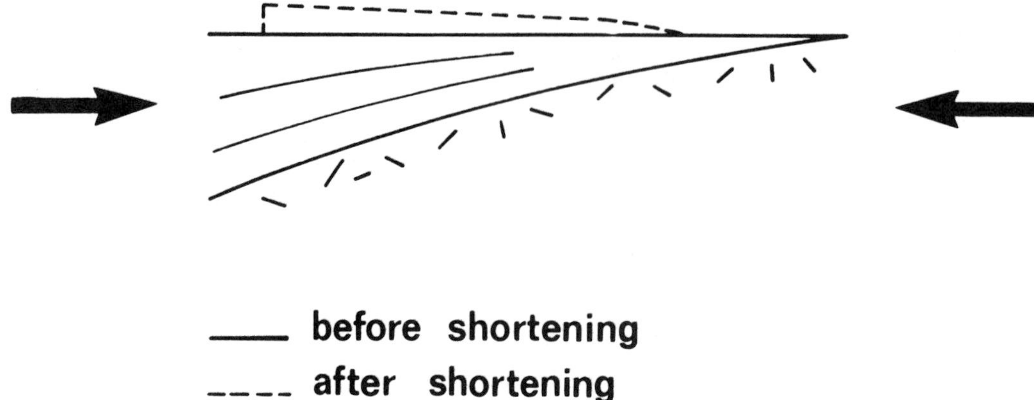

Fig. 7. The sedimentary wedge in compression.

The section deformed by thrusting is thus a wedge, a fact often not sufficiently appreciated. It necessarily means that more mass is shifted at the rear than is piled up at the front (Fig. 7). Furthermore, the bedding planes of such a sedimentary wedge are not completely parallel to the orientation of maximum horizontal compression during deformation.

Also, in the case of the southern Canadian Rockies, the tip of the wedge escapes deformation. In the Jura Mountains the very tip of such a wedge is strongly deformed (Laubscher, 1961), whereas the Swiss Plateau intervening between the Alps and Jura has experienced little deformation. A similar case seems to exist in the Mackenzie and Franklin mountains in the Northwest Territories of Canada. Clearly incidentals, such as the nature of the sediments involved, must play a decisive role. Thus Laubscher (1961) rightly attributes great significance to the presence of the Triassic evaporites in the subsurface of the Jura Mountains, the Swiss Plateau and the Alps proper. Incidentals may then determine to a large degree what happens to the tip of the wedge or, in other words, how far outward the deformation may reach. Whatever the case may be, certainly the fact that a wedge and not a block is deformed must be kept in sight.

The Question of the Rate of Advance

The formation of a total thrust belt is a process that must be counted in tens of millions of years. In the Austrian Alps, sizable thrust movements took place as early as pre-Coniacian and as late as post-Oligocene (Trümpy, *pers. comm.*) A total time span of at least 55 my is indicated. According to Dahlstrom (1970, p. 402) an even longer time span is indicated for the formation of the Canadian thrust belt. This is not, however, the question that is of concern in the present paper. The problem at hand is: how does an individual thrust sheet advance, what is the rate of movement? Is the process of thrusting a continuous one or does it proceed in spurts? Do individual sheets advance fast, followed by a pause until the next sheet becomes activated? It is these detailed questions of the kinematics in a thrust belt that are important in regard to the problems considered here. Not much has been written on these aspects of thrust faulting and it is certainly too early to present a final answer. The total time span covered by the formation of the thrust belts puts a certain upper time limit on the movements of single thrust sheets. Evidently these movements cannot be counted in anything higher than millions of years, and conceivably the sheets might have moved even faster.

Hsü (1969c) did study the Glarus overthrust in Switzerland in this respect. He estimates that it advanced at an average rate of 0.2 to 10 cm (0.8 to 3.9 in)/yr. For an advance of 40 km (24.9 mi), 0.4 to 20 my would be required. Thus an area 40 km wide and possibly 100 km (62 mi) long would be loaded with a plate about 3 km (1.9 mi) thick in the above time span. If we accept the minimum velocity we find that we load 3 km (1.9 mi) in 20 my, or an average of 0.15 mm/yr. For the maximum velocity we arrive at an average of 7.5 mm/yr. Certainly these values correspond to a fast rate of loading, compared with the values given for sedimentation rates by Rubey and Hubbert (1959, p. 171) and are in substantial agreement with their estimate for loading due to overthrusting in the Idaho-Wyoming thrust belt (Rubey and Hubbert, 1959, p. 192).

While it is not possible to give very precise answers to the question posed in this subsection, we may again be content; for our purposes it is sufficient to accept the advance of thrust sheets as a — geologically speaking — case of fast loading.

The Question of Contemporaneous Erosion

The thrust sheets as they present themselves today are strongly eroded remnants of initially much larger units. Since late uplift is a well-established fact in all major thrust belts (Trümpy, 1960; Bally *et al.*, 1966; Shaw, 1970), much of the erosion must postdate the thrusting and is therefore of no consequence with regard to the questions dealt with in this paper. Our problem requires the thrust sheets as they existed at the time of advance. It is almost certain that the uneroded load, as schematically given in Figures 2 to 4, does not provide the real situation. However, total erosion of the toe or thrust sheet, as assumed by Rubey and Hubbert (1959, p. 194, Fig. 9), is equally unrealistic in view of the total and stratigraphic displacements evident even today. The real question is thus to find the extent and thickness of the thrust sheets as they existed at the time of emplacement. An exact answer cannot be given at this time. Some evidence for more extensive loading than is evident today will appear in the following section. To establish the original extent of certain major thrust sheets is actually a challenging field problem that, hopefully, will be tackled in the near future.

Fortunately for this paper a precise answer is not required. It suffices to state that the evidence presently visible suggests that thrust sheets must have been extensive at the time of formation. For the Glarus thrust Trümpy (1969, p. 130) estimates the thickness of the sheet was 5 to 6 km (3.1 to 3.7 mi). Thus the shift of large loads on the earth's crust as a consequence of thrusting cannot be disputed — the only factor of real significance for our current considerations.

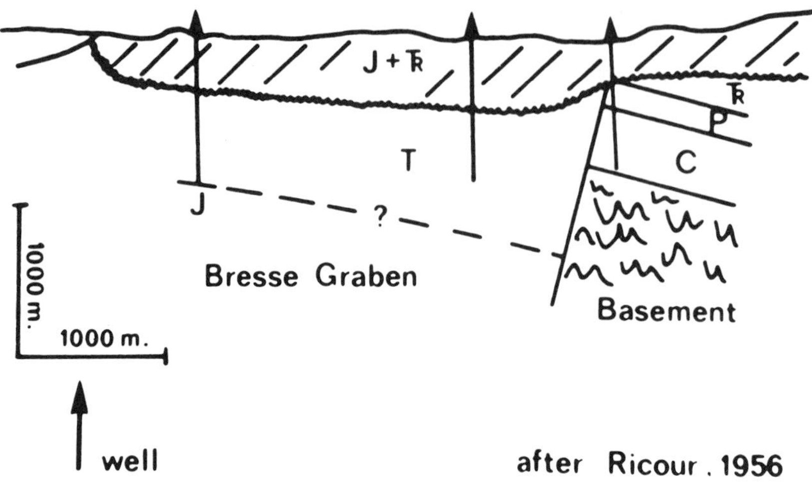

Fig. 8. Deformation of a thrust plane in the Jura Mountains. The downward deflection is interpreted as an effect of loading. The young Tertiary clastics of the Bresse graben were compacted under the load of the thrust sheet. (After Ricour, 1956).

The Effect of Loading

Loading due to thrusting must be accepted as a fact, even if it is not as severe as schematically shown in Figures 2 through 4. There are two examples that may be cited as additional evidence for loading due to thrust sheets.

The first concerns thrusting in the Jura Mountains. Recent work has shown that sizable thrusting is present. Figure 8 is after Ricour (1956) and shows one of the thrusts moving out over the Bresse Graben. According to Laubscher (1961), the graben structure was produced by pre-thrusting deformation of the basement. It is evident that the thrust plane drops about 200 m (650 ft) over the edge of the graben. This deformation of the fault plane is well documented by three wells. Laubscher (1961) attributes the deformation to the gouging action of the thrust. My preferred interpretation ascribes it to a loading effect. As the carbonates and evaporites of the thrust plate were loaded onto the Tertiary clastics of the Bresse Graben, the latter compacted under the additional load. The deformation of the thrust plane is thus a differential-compaction phenomenon and of post-thrusting age.

A second example relates to the McConnell thrust in the southern Canadian Rockies. A Cretaceous sandstone (ledge near Valley floor in Fig. 1) located under Mount Yamnuska capped by an erosional remnant of the McConnell plate was subjected to a crushing strength test. Six tests gave an average strength of 2450 kp/cm^2 with a coefficient of variation of 10 per cent. Cretaceous sandstones from more easterly positions never covered by thrust plates do not exhibit such high strength. While the idea requires further testing, the present results nonetheless suggest that Mount Yamnuska represents only a small remnant of a once much more extensive thrust sheet, and this tectonic loading may have produced the anomalously high strength in the underlying Cretaceous rocks.

It is impossible to escape the conclusion that thrust faulting leads to severe loading of the section in front of the thrust belt. It is therefore mandatory to look into the consequences such loading might produce. The isostatic behaviour of the earth's crust is well known and has been demonstrated in many parts of the world. Ice loading and unloading produces a response that is, furthermore, fast in geological terms. Thus, the Canadian Arctic has risen almost 300 m (1,000 ft) in some places since the unloading of the ice some 10,000 years ago. The crust also responds to small loads. The ponding of Lake Mead has led to a measurable depression of the crust (Longwell, 1954). The conclusion is therefore inescapable that loading due to thrust sheets must produce an isostatic depression. The effect is to create a foredeep that will act as a sediment trap, and the resulting clastic wedges will cause further passive subsidence. For the isostatic effect the question of contemporaneous erosion during thrusting is, therefore, of only secondary significance. Whether the load is in the form of thrust plates only or a combination of thrust plates and clastic wedges is immaterial. The foredeep is an essential yart of thrustbelts, an isostatic effect well explained by Price (1969) and Price and Mountjoy (1970, p. 16). The isostatic effects of thrusting may therefore be dismissed at this stage, and our attention turned to another consequence of the advancing thrust sheets more closely related to the actual progression of thrusting.

Discussion of the rate of advance has shown that thrusting represents a case of, geologically speaking, fast loading — a fact already briefly considered by Rubey and Hubbert (1959, p. 192). Under these circumstances we may expect high fluid pressures to develop under areas of maximum loading; i.e., in the front of major steps such as shown in Figure 9a. Under this condition the original fault may become inactive and a new thrust fault may develop in the area of high fluid pressure. As this new fault is activated, the old one is carried 'piggyback' and becomes folded in the process (Figure 9b). One postulates, therefore, thrusting to progress outward and deeper thrusts to be younger, for mechanical reasons. This general concept is well accepted in the Canadian Rockies (Bally *et al.*, 1966; Price, 1969; Dahlstrom, 1970). Exceptions or incidentals are the backlimb thrusts that may develop at a later stage on some thrust sheets. In the Helvetides of the Alps there is some evidence for the progression of thrusting in an outward direction, but the case cannot be as clearly stated as in the Canadian Rockies (Trümpy, pers. comm.). Theoretically one might, therefore, predict that thrusting is a somewhat selfperpetuating process (Gretener, 1969a). It is reassuring that the prediction that thrusting ought to progress in an outward direction is not at variance with present field observations.

On The Ease of Movement

The low basal friction that evidently must exist while the thrust sheets move into place has remained an enigma for many years. Only in 1959 did Hubbert and Rubey, in their companion papers discussing the role of the fluid pressure in this process, point the way to a possible solution. They effectively introduced the concept of effective stress into the geological literature. Possibly it is more accurate to say that they have put this concept in the limelight, since it had been already stated explicitly by Goguel (1962, p. 176, French edition 1952). The concept goes back to Terzaghi who developed it in 1923 (Skempton, 1960). In his classic paper (Terzaghi, 1950), he applied the idea to landslides, stating: 'that when the fluid pressure reaches a critical value the overburden may in fact be floated'. It is this principle that Hubbert and Rubey (1959) have applied to the mechanism of overthrusts.

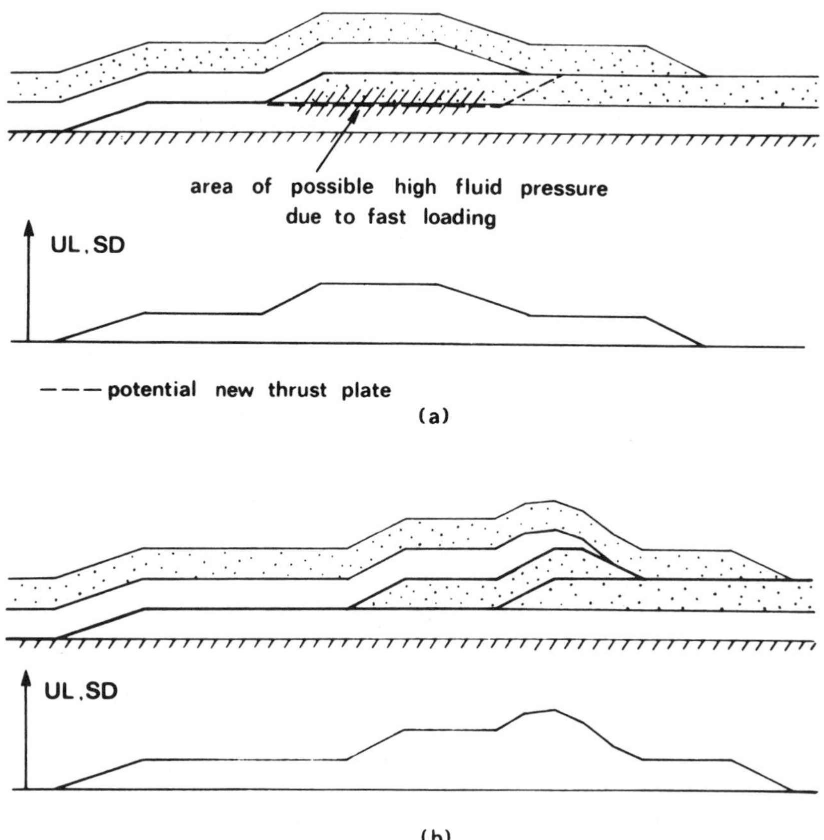

Fig. 9a. The effect of loading on the underlying sequence.

Fig. 9b. Activation of a lower thrust plane. The old thrust is carried piggyback and folded over the new step.

The concept may best be illustrated with the Mohr diagram. Figure 10a shows what happens in a porous medium under stress when the fluid pressure is raised by an amount Δp (Hubbert and Rubey, 1960). The Mohr circle describing the initial condition of stress is shifted to the left and approaches the failure envelope. As the pore pressure reaches a critical value, the Mohr circle will touch the failure envelope, and fracture of the material may be induced in this manner.

The significance of the Hubbert and Rubey papers (1959) is demonstrated by the intensive discussion that followed. Most recently Hsü

(1969a, b) criticized the authors' model of overthrust faulting. He claimed that the model works not so much because of the increased pore pressure — reduced normal stress across the failure plane — but rather since the authors chose to neglect the term cohesion (τ_0) in the Coulomb failure criterion (eq. 1) for the fracture plane:

$$\tau = \tau_0 + n\sigma \quad . \quad . \quad . \quad . \quad . \quad . \quad . \quad . \quad . \qquad (1)$$

where τ_0 is the cohesion; n the coefficient of internal friction; and σ the normal stress across the plane. Hubbert and Rubey (1959) did neglect this term, with the justification that the cohesion would be gradually broken; i.e., the fracture would process at a finite speed and the term should therefore not be retained.

It is well known that thrust faults follow bedding planes over long distances. It can also be said that bedding planes represent distinct discontinuities in the sedimentary sequence. The fact that during folding slippage is often concentrated on bedding planes points to the fact that at least some of them represent planes of weakness. It may thus be argued that the assumption of Hubbert and Rubey (1959) was justified on the grounds that many bedding planes are planes of low or zero cohesion. The discussion between Hsü (1969 a, b) and Hubbert and Rubey (1969), therefore, amounts to a dispute on the mechanical significance of the bedding plane.

The influence of pre-existing cracks on the strength of rocks has been investigated by a number of authors, notably Jaeger (1959), Jaeger and Rosengren (1969), and Handin (1969). The last-mentioned has not only added data of his own but reviewed the concept quite thoroughly. The idea has also been discussed in connection with thrust faulting by Price (1965). The effect of a pre-existing plane of weakness may be best understood when considering the Mohr diagram. Figure 10b shows such a diagram. For slippage to occur on a pre-existing fracture, the original Coloumb equation (1) must be replaced by:

$$\tau_0 = \tau_0' + \mu\sigma \quad . \quad . \quad . \quad . \quad . \quad . \quad . \quad . \qquad (2)$$

where τ_0' is the reduced (in the extreme zero) cohesion existing across the fracture; μ is the coefficient of sliding friction; and σ is the effective normal stress across the fracture plane. The upper failure line in Figure 10b represents the solid material (eq. 1) whereas the lower line applies to a plane of reduced cohesion (eq. 2). The two lines need not be parallel. The slope of the upper line is given by the coefficient of internal friction, while that of the lower line is the coefficient of sliding friction. It is quite evident from the inspection of Figure 10b that the effect of such a plane will be most pronounced for materials with high cohesion, and for the condition where the coefficient of sliding friction is smaller than the coefficient of internal friction.

In Figure 10b a Mohr circle is shown. The circle does not intersect the failure envelope for solid rock, and for an unbroken piece of rock this state of stress is stable. However, the circle does intersect the failure line for a plane of weakness in this rock. Thus, if the plane of weakness is so oriented that its angle (δ) with the σ_1-axis meets the condition:

$$\delta_{min} \leqslant \delta \leqslant \delta_{max} \quad . \quad . \quad . \quad . \quad . \quad . \quad . \qquad (3)$$

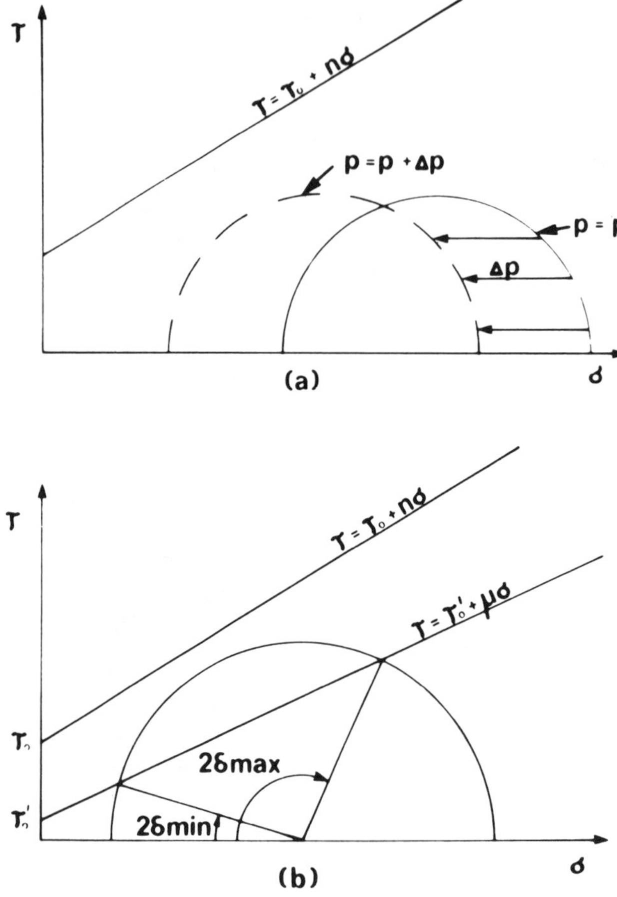

Fig. 10a. Shift of the Mohr circle due to an increase ($\triangle$p) in pore pressure.

Fig. 10b. The effect of a plane of weakness (reduced cohesion) as shown in the Mohr diagram.

sliding on this plane will be initiated and failure will occur (Handin, (1969). Many bedding planes qualify undoubtedly as planes of reduced or zero cohesion. It is most important to realize from Figure 10b that δ_{min} may be a very small angle; i.e. planes that are subparallel to the σ_1-axis may be activated.

Returning to the discussion between Hsü (1969a,b) and Hubbert and Rubey (1969), the following can be said: Taking the model as set out by Hubbert and Rubey (1959), we may plot the graph shown in Figure 11. The length of the block to be pushed without destruction is plotted versus the cohesion or the pore pressure - overburden ratio (λ) respectively. The thickness of the particular block is assumed to be 6 km (3.7 mi) — not an unreasonable value according to our previous considerations. The graph clearly shows that neither a reduction of the cohesion alone nor a simple increase of the pore pressure is sufficient to increase the permissible length of the block to any appreciable extent. Evidently an increase of pore pressure across a plane of low cohesion is necessary to make the model work. The latter may be called a plane of superweakness.

Figure 12a shows the concept of such a plane. The requirement is a zone of abnormally high fluid pressure containing at least one bedding plane with low cohesion (layer B in Figure 12a). This situation leads to a combination of the conditions shown in Figures 10a and 10b, given in Figure 12b. For zones A and C the Mohr circle is the one shown to the right. These zones are stable even if they do contain bedding planes. For zone C the Mohr circle is moved to the left, intersecting the failure line for the bedding plane. It can be seen that, provided the bedding plane forms an angle of 7° or more to the axis of maximum horizontal compression, failure will occur. Since the sedimentary sequence that is subject to deformation is wedge-shaped, such may well be the case. The angle δ—— in Figure 12b may be further lowered by increasing the fluid pressure to an even higher value.

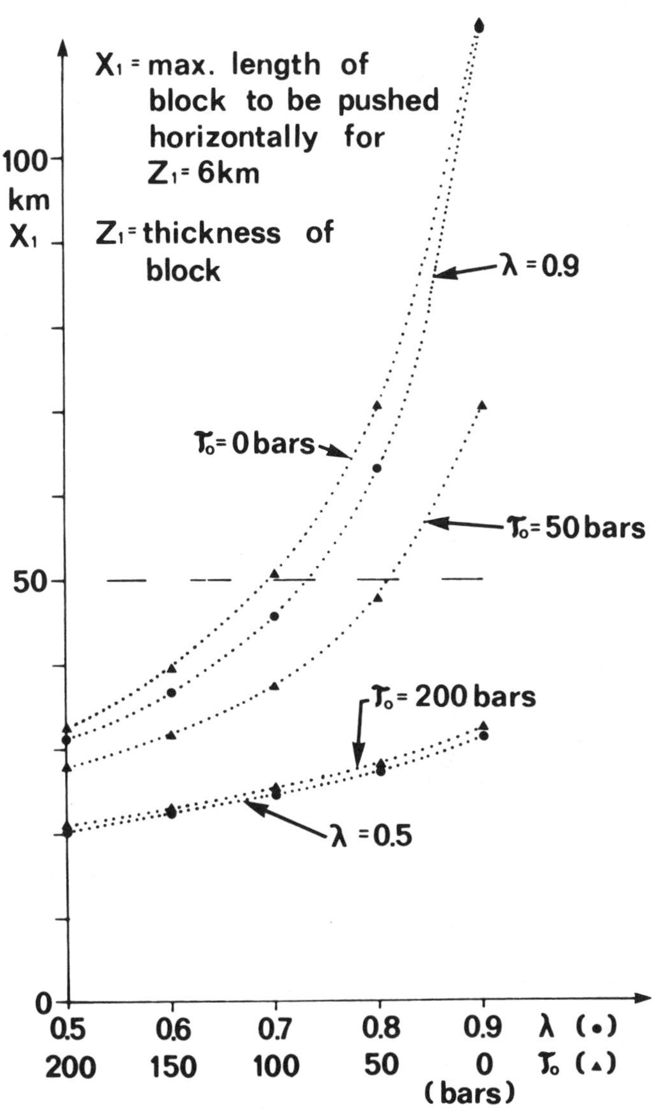

Fig. 11. The maximum length of a block 6 km (3.7 mi) thick that may be pushed as a function of the pore pressure and cohesion across the basal plane. (After Hsü, 1969a and Hubbert and Rubey, 1959).

The model of Figure 12 explains easy gliding along the links as proposed by Hubbert and Rubey (1959). It predicts the links to be concentrated in zones that are prone to high fluid pressures, such as shales, particularly when interbedded with evaporites. This is in accordance with published field observations (Douglas, 1950; Mostofi and Gansser, 1957; Dahlstrom, 1970). However, the plane of superweakness need not occur in this zone; it may well be located adjacent in the zone A or C of Figure 12a comprising carbonate or sandstones. In reality, the pore-fluid pressure in these rocks next to an overpressured shale zone will not be entirely normal but rather will drop off gradually. Thus we may well have the arbitrarily chosen conditions as shown in Figure 13. The pore pressure in the shale is 900 kp/cm², and in the dolomite adjacent to the shale 800 kp/cm². For the various mechanical parameters of importance we may assume the following numerical

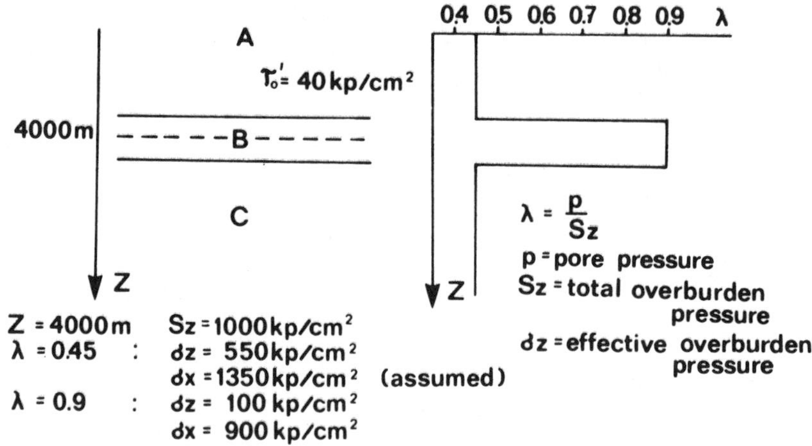

Fig. 12a. The concept of the plane of superweakness in cross-section.

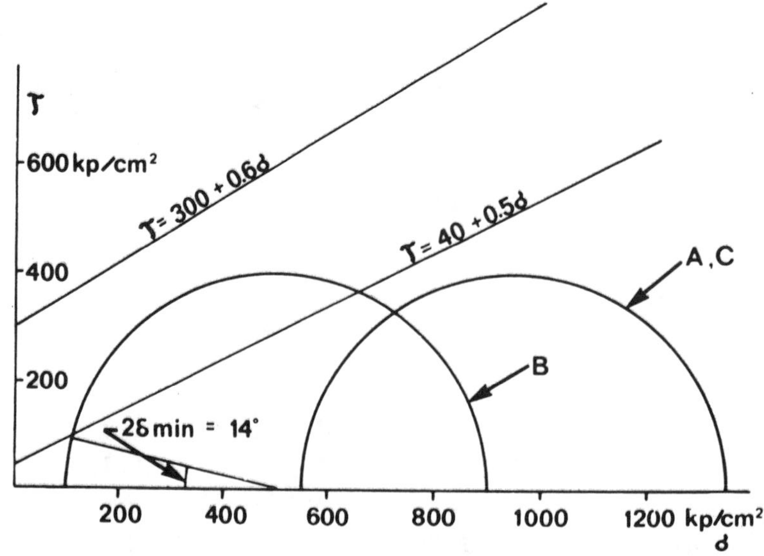

Fig. 12b. The concept of the plane of superweakness in the Mohr diagram.

values: On a bedding plane in the shale the cohesion (τ_0') is 80 kp/cm^2 whereas in the dolomite it is 40 kp/cm^2; the coefficient of gliding friction for the shale bedding plane is 0.6 and for the dolomite bedding plane 0.4. Under these conditions, horizontal compression may well induce gliding on a bedding plane in the dolomite next to the shale rather than in the shale itself. This explanation may well apply to the observation on the Heart Mountain thrust reported by Stearns *(pers. comm.)*.

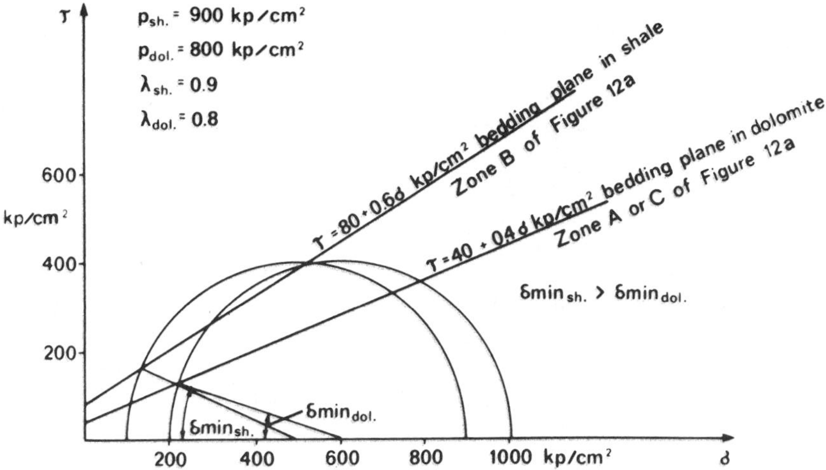

Fig. 13. Sliding on adjacent dolomite bedding plane in preference to a shale bedding plane.

In this case the detachment plane follows a bedding plane in the Bighorn dolomite about 8 to 10 ft above the Cambrian shales. In the terminology followed in this paper, this observation must be classified as an incidental due to the particular mechanical properties of the shale bedding planes versus the adjacent bedding planes in the dolomite. Also, the adjacent rock must be of low permeability in order to permit a gradual rather than abrupt drop of the fluid pressure.

So much for the links; what about the steps? Also, how can the high fluid pressures be maintained or re-created on the links, when the actual displacements along these faults must be measured in tens of kilometres?

The Pumping Effect

In order to attempt to understand thrust faulting in a layered sequence, there is yet another aspect that must be considered. The individual layers of such a sequence will undoubtedly differ markedly in their mechanical properties. Therefore, compression of such a layered medium must lead to a distribution of stress as schematically shown in Figure 14 (Gretener, 1969b). The competent layers such as massive carbonates, sandstones and, in particular, quartzites will act as beams and be highly stressed (Stearns, 1969). Until these layers are broken it is impossible to put any significant load on the intervening incompetent spacers (shale, evaporites). This automatically leads to the conclusion that the steps in the competent layers are the first to form. Only when a beam collapses (step formation) is the load suddenly transferred to the adjacent spacers and the link produced. This situation is illustrated in the Mohr diagram of Figure 15. Given are the failure envelopes for the solid rock as well as for the bedding planes. The large solid Mohr circle refers to the state of stress in the beam (e.g., dolomite) immediately before failure, whereas the small solid Mohr circle gives the state of stress in the adjacent incompetent bed (e.g., shale) at the same time. Note that the effective overburden stress (σ_z) is the same for both dolomite and shale. No allowance is made for an initially abnormal pore pressure in the shale. The effect of such an abnormal pressure would be to shift the Mohr circle for shale to the left and thereby further ease the process described below.

In Figure 15 the large solid Mohr circle (stress in dolomite) just touches the failure envelope for solid rock. A step forms, the stress in the beam is relieved (collapse of the large Mohr circle) and the load is rapidly transferred to the adjacent shale layer. Because of rapid loading, the additional load placed on the incompetent and often poorly permeable layer is carried by the fluid rather than the matrix. The rise in the fluid pressure results in a reduction of the confining (overburden) pressure. The small Mohr circle for the shale therefore grows in the manner shown in Figure 15. As it reaches a critical diameter, slippage on a bedding plane occurs and the link is formed. This is termed the pumping effect. It accounts for the initial formation and the peculiar geometry of the thrust faults.

The process may be better understood when considered in terms of a special triaxial test. A sample is in a pressure cell with confining pressure as well as axial load applied, and the pore pressure is moni-

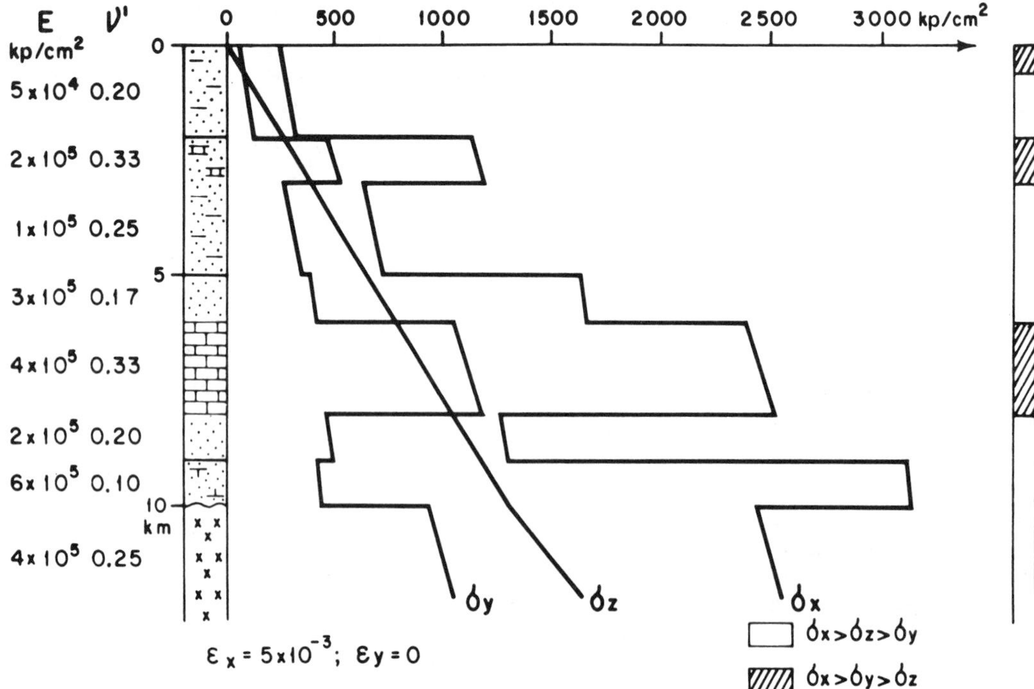

Fig. 14. Stresses in a layered sequence under horizontal compression (from Gretener, 1969b.)

tored. The axial load is slowly raised, with the result that both confining pressure and pore pressure remain constant while the effective axial load is increasing. At a given momen the rate of axial loading is drastically stepped up. Under these conditions, the fluid in the fully saturated sample is unable to move out of closing cracks and pore spaces. As a result, the fluid pressure will increase, with a corresponding drop in confining pressure, and the effective axial load will remain constant. This is the case for the shale as described above, the vertical stress component being the confining pressure and the horizontal compression being equivalent to the axial load.

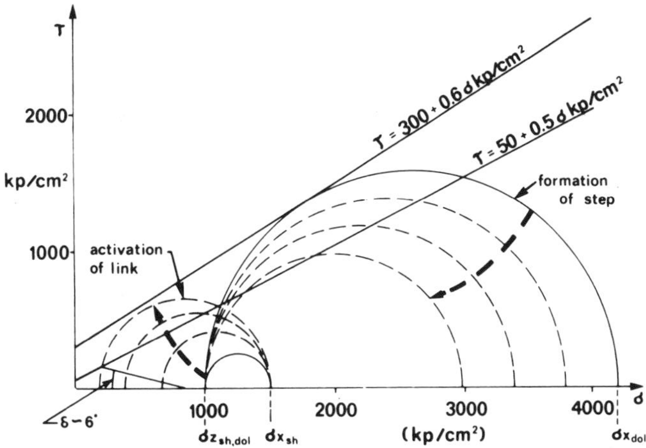

Fig. 15. The Mohr circle for a beam and the adjacent incompetent layer. Vertical stress is equal in both. Breaking the beam results in the collapse of the respective Mohr circle. The load is transferred to the spacer. Due to rapid loading only the pore pressure is increased. The result is to inflate the respective Mohr circle by decreasing the overburden pressure. A link (bedding plane) is activated as the circle reaches a critical diameter. As movement takes place, the latter Mohr circle collapses and motion on the fault is therefore limited.

This model describes what might happen during the initial formation of a thrust fault. However, we have seen that displacements along such faults are to be measured in tens of kilometres. It is not suggested that the total displacement take place at once for two reasons: First, as soon as movement occurs on the link, the respective Mohr circle must collapse and the fluid pressure decrease. Second, as is evident from Figures 3 and 4 the thrust sheet and, in particular, the beams must deform over every step. Therefore, displacement at any one time is limited, the link becoming 'stuck' because of a drop in fluid pressure, and the beam becoming locked over a step. Only segments of the fault are in motion at a given time, and the progress of the thrust sheet is much like the crawl of a caterpillar. This is the form of movement as envisaged by Barnes (1966), Hsü (1969c) and Price *(pers. comm.)*. Hsü has compared it to the stick-slip movement observed in experimental work (Brace and Byerlee, 1966). In the case of the thrust faults, sticking is related to the megaprotrusions on the fault plane — the steps — as well as to the drop in fluid pressure on the links as motion proceeds.

Thus, the process of rapid transfer of load from the beam to the incompetent layers may be repeated an unlimited number of times. High fluid pressures may be created on the links as the thrust plate moves along. A process that may further enhance the fluid pressure in the poorly permeable beds of the link is the pressurization due to shock waves. Every time the beam is unlocked, shock waves must be produced. Their effect would be to shift the Mohr circle for shale in Figure 15 even farther to the left, thereby facilitating movement on the links. The pumping effect is thus not limited to the initial condition, but repeats itself during the whole life span of the thrust fault. It represents an attempt to account for the evident ease of movement along such faults as suggested by the field observations. Clearly any acceptable mechanism must not only produce a favorable initial condition but must also account for maintenance of such a condition during the very sizable displacements.

It is interesting to note that, for this particular model, the jerky movement of the thrust plates is most important. The advance of a given point on a thrust sheet is shown in Figure 16. It is obvious that the average rate of advance is of no great importance except for the rate of loading of the underlying sequence. What is important is the form of the actual advance. Thus the old discussion of fast versus slow becomes irrelevant. The over-all movement is slow, but consists of a series of fast spurts. The average rate of progress tells only an incomplete story when the actual process consists of a series of fast, singular events (Gretener, 1967).

This explanation still leaves the positioning of the steps as an unresolved problem. Undoubtedly the step spacing will be related in some degree to the thickness of the beams. At least, in a statistical manner, one would expect the step spacing to increase with increasing beam thickness. Step spacing may also be related to gentle pre-thrust folding (Simony, *pers. comm.*), or the step formation may be entirely triggered by incidentals such as pre-existing fractures or lithological changes in the beam.

Conclusions

A number of aspects important to thrust faulting in a layered sequence have been considered. The role of the fluid pressure was introduced by Hubbert and Rubey in 1959 and the influence of the bedding plane was considered by Price in 1965. The combination of the two principles leads to the concept of the plane of superweakness: a plane with low or zero cohesion across it and situated in a layer of high fluid pressure, with the normal stress across the plane a very low value.

Field observations strongly suggest that any valid model of thrust faulting must account for a very low basal friction on the thrust plates. This suggestion appears particularly valid since Hubbert and Rubey (1959) have not considered the effect of the toe, and Raleigh and Griggs (1963) have done so only in a limited way (small toe). Extensive toes — or overthrust plates — are, however, an observed fact. Their existeice will tend to reduce the maximum lenth of plate that can be pushed without destruction, as published by Hubbert and Rubey (1959) and Raleigh and Griggs (1963). Or, as an alternative, the basal friction must be even lower than these authors assumed. Furthermore, this low

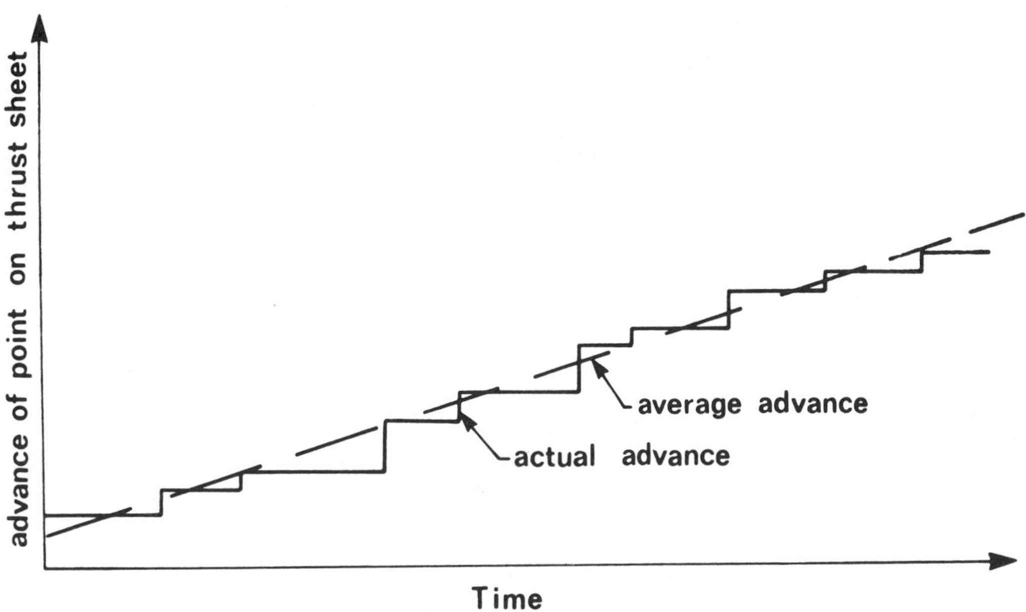

Fig. 16. Movement of a point on a thrust sheet. The actual movement is a series of jerky singular motions.

friction must be maintained in both time and space. Emplacement of a single major thrust sheet is a process that takes on the order of a million years and displaces rocks many tens of kilometres. Evidently, whatever the postulated causes of low basal friction, they must prevail during the whole time and over the total area involved.

In the present model it is suggested that the competent units called the beams carry a disproportionately high part of the horizontal load. When they break initially the load is suddenly transferred to the incompetent units, usually shales with low permeability. The effect of the sudden loading will be to raise the pore pressure in these units while reducing the effective overburden pressure, and the link is formed along a bedding plane of low or zero cohesion (Fig. 15). It is termed the pumping effect. Motion is limited because of the dissipation of the fluid pressure as movement takes place and also as the beams become locked in their movement over the steps. The process may thus be repeated many times. It is visualized that the plates move in the manner of caterpillars, with only a limited number of segments in motion at any time. Shock waves from the initial breaking of the beams or the breaking of the locks at the steps may well enhance the pore pressures in the links.

The shift of load that takes place during thrusting represents a case of fast loading, and the consequences of such action can be anticipated. High pore pressures will be induced in the impermeable layers of the overridden sequence. Thrusting is, therefore, predicted to progress outward, the lower and more forward thrusts being younger as shown in Figure 9b.

It is quite obvious that a full understanding of the mechanics of overthrust faulting is still a matter for the future. However, it is hoped that the foregoing discussion presents some ideas that may prove useful in the final analysis of these features.

REFERENCES

Armstrong, F. C. and Oriel, S. S., 1965, Tectonic development of Idaho-Wyoming thrust belt: Am. Assoc. Petroleum Geologists Bull., v. 49, no. 11, p. 1847-1866.

Bally, A. W., Gordy, P. L. and Stewart, G. A., 1966, Structure, seismic data, and orogenic evolution of southern Canadian Rocky Mountains: Bull. Can. Petroleum Geology, v. 14, no. 3, p. 337-381.

Barnes, W. C., 1966, Mechanics of overthrust faulting: Geol. Soc. America Ann. Mtg., San Francisco, [Abst.], p. 11-12.

Brace, W. F. and Byerlee, J. D., 1966, Stick-slip as a mechanism for earthquakes: Science, v. 153, no. 3739, p. 990-992.

Dahlstrom, C. D. A., 1970, Structural geology in the eastern margin of the Canadian Rocky Mountains: Bull. Can. Petroleum Geology, v. 18, no. 3, p. 332-406.

Douglas, R. J. W., 1950, Callum Creek, Langford Creek, and Gap Map-Areas, Alberta: Geol. Surv. Canada, Mem. 255, 124 p.

Gansser, A., 1964, Geology of the Himalayas: Regional Geol. Ser., L.U. de Sitter, *ed.*, New York J. Wiley and Sons Ltd., 289 p.

Goguel, J., 1962, Tectonics: San Francisco, W. H. Freeman & Co. [transl. from French ed., 1952, by H. E. Thalmann], 384 p.

Gretener, P. E., 1967, Significance of the rare event in geology: Am. Assoc. Petroleum Geologists Bull., v. 51, no. 11, p. 2197-2206.

——, 1969a, Fluid pressure in porous media — its importance in geology: a review: Bull. Can. Petroleum Geology, v. 17, no. 3, p. 255-295.

——, 1969b, On the mechanics of the intrusion of sills: Can. J. Earth Sci., v. 6, no. 6, p. 1415-1419.

Handin, J., 1969, On the Coulomb-Mohr failure criterion: J. Geophys. Res., v. 74, no. 22, p. 5343-5348.

Heim, A., 1882, Die Glarner Doppelfalte: Vierteljahrsch. Naturforschende Ges. Zurich, v. 27, no. 11.

——, 1919, Geologie der Schweiz: Cr. Herm. Tauchnitz, Leipzig, Band I & II, 1018 p.

Holmes, A., 1964, Principles of Physical Geology: London & Edinburgh, T. Nelson, 1288 p.

Hsü, K. J., 1969a, Role of cohesive strength in the mechanics of overthrust faulting and of landsliding: Geol. Soc. America Bull., v. 80, no. 6, p. 927-952.

——, 1969b, Role of cohesive strength in the mechanics of overthrust faulting and of landsliding: Reply: Geol. Soc. America Bull., v. 80, no. 6, p. 955-960.

——, 1969c, A preliminary analysis of the statics and kinetics of the Glarus overthrust: Eclogae Geol. Helv., v. 62, no. 1, p. 143-154.

Hubbert, M. K. and Rubey, W. W., 1959, Role of fluid pressure in mechanics of overthrust faulting: Geol. Soc. America Bull., v. 70, no. 2, p. 115-166.

——, 1960, Role of fluid pressure in mechanics of overthrust faulting: a reply: Geol. Soc. America Bull., v. 71, no. 5, p. 617-628.

——, 1969, Role of cohesive strength in the mechanics of overthrust faulting: Discussion: Geol. Soc. America Bull., v. 80, no. 6, p. 953-954.

Jaeger, J. C., 1959, The frictional properties of joints in rocks: Geofis, Pura e Appl., v. 43, p. 148-158.

—— and Rosengren, K. J., 1969, Friction and sliding of joints: Austr. I.M.&M. Proc., no. 229, p. 93-104.

Jones, P. B., 1971, Folded faults and sequence of thrusting in Alberta foothills: Am. Assoc. Petroleum Geologists Bull., v. 55, no. 2, p. 292-306.

Laubscher, H. P., 1961, Fernschubhypothese der Jurafaltung: Eclogae Geol. Helv., v. 54, no. 2, p. 221-282.

Longwell, C. R., 1954, Interpretation of levelling data, *in* Thomas, H. E., *ed.*, First Fourteen Years of Lake Mead: U.S. Geol. Surv., Circ. 346, p. 5.

Mostofi, B. and Gansser, A., 1957, The story behind the 5 Alborz: Oil Gas J., v. 55, no. 3, p. 78-84.

Price, R. A., 1965, Flathead Map-Area, British Columbia and Alberta: Geol. Surv. Canada, Mem. 336, 221 p.

——, 1969, The southern Canadian Rockies and the role of gravity in low-angle thrusting, foreland folding and the evolution of migrating foredeeps: Geol. Soc. America Ann. Mtg. [Abst.], pt. 7, p. 284-286.

Price, R. A. and Mountjoy, E. W., 1970, Geologic structure of the Canadian Rocky Mountains between Bow and Athabasca rivers, a progress report: Geol. Assoc. Canada, Spec. Pap. 6, p. 7-26.

Raleigh, C. B. and Griggs, D. T., 1963, Effect of the toe in the mechanics of overthrust faulting: Geol. Soc. America Bull., v. 74, no. 7, p. 819-830.

Rich, J. L., 1934, Mechanics of low-angle overthrust faulting as illustrated by Cumberland thrust block, Virgina, Kentucky, and Tennessee: Am. Assoc. Petroleum Geologists Bull., v. 18, no. 12, p. 1584-1596.

Ricour, J., 1956, Le Chevauchement de la Bordure Occidentale du Jura sur la Bresse dans la Région de Lons - Le Saunier: Bull. Ver. Schweizer Petrol. Geol. und Ing., v. 23, no. 64, p. 57-70.

Rubey, W. W. and Hubbert, M. K., 1959, Role of fluid pressure in mechanics of overthrust faulting: Geol. Soc. America Bull., v. 70, no. 2, p. 167-205.

Shaw, E. W., 1970, Vertical uplift in orogenic belts: Bull. Can. Petroeum Geology, v. 18, no. 3, p. 430-438.

Shelton, J. S., 1966, Geology illustrated: San Francisco & London, W. H. Freeman & Co., 434 p.

Skempton, A. W., 1960, Significance of Terzaghi's concept of effective stress. From theory to practice in soil mechanics: New York, J. Wiley & Sons, p. 42-53.

Staub, R., 1954, Der Bau der Glarneralpen und seine prinzipielle Bedeutung fuer die Alpengeologie: Glarus, Verlag Tschudi und Co., 187 p.

Stearns, D. W., 1969, Fracture as a mechanism of flow in naturally deformed layered rocks, *in* Baer, A. S. and Norris, D. K., *eds.*, Proceedings, Conference on Research in Tectonics: Geol. Surv. Canada, Paper 68-52, p. 79-90.

Terzaghi, K., 1950, Mechanism of landslides, *in* Application of Geology to Engineering Practice: Geol. Soc. America, Berkey Volume, p. 83-123.

Trümpy, R., 1960, Paleotectonic evolution of the central and western Alps: Geol. Soc. America Bull., v. 71, no. 6, p. 843-908.

——, 1969, Die helvetischen Decken der Ostschweiz. Versuch einer palinspastischen Korrelation und Ansätze zu einer kinetischen Analyse: Ecologae Geol. Helv., v. 62, no. 1, p. 105-142.

Reprinted by permission of the Australian Petroleum Exploration Association from *APEA Journal*, v. 26, pt. 1 (1986), p. 214-224.

A THIN SKINNED MODEL FOR THE PAPUAN THRUST BELT AND SOME IMPLICATIONS FOR HYDROCARBON EXPLORATION

D.M. Hobson
BP Australia Ltd, 1 Albert Road, Melbourne Vic. 3004

ABSTRACT

In the Papuan thrust belt the main risk in hydrocarbon exploration lies in identifying structural traps which are detached below the level of the primary reservoir. Because of the difficulty in obtaining usable seismic in the remote, inaccessible terrain, identification of prospects has been based on the interpretation of structural cross-sections drawn from geological maps. The structural models commonly used have been geometrically inadmissible, and as a result, some wells have penetrated thrusts at depth, before reaching the reservoir horizon.

More rigorously constrained regional cross sections through the thrust belt may be constructed using the principles of thin skinned tectonics. These may be used to identify structural provinces, characterised by different types of thrust pattern. In turn, the probability of reservoir involvement in a particular structure is directly related to its location in a particular province.

The cross-sections show that there has been as much as 100 km of shortening in the hinterland of the thrust belt. Restored versions of the cross sections may be used to reconstruct the shape of the basin and ideally to identify regions favourable for source and reservoir sediment distribution. In Papua New Guinea, palinspastic maps are currently feasible only for Miocene strata, but with the present level of exploration activity increased stratigraphic information should soon permit reconstructions for older formations.

INTRODUCTION

The predictive quality of geometrically consistent structural models for hydrocarbon exploration has only recently been fully appreciated. The primary application of the technique has been in the evaluation of thrust belts, in particular the Canadian Rocky Mountain foothills, to help assess whether a particular structure is detached above or below the stratigraphic level of a reservoir. In this paper, a series of restorable regional structural cross sections has been used to construct a model for the major thrust belt in Papua New Guinea, and an example of the use of structural modelling for ranking acreage and prospects is given.

In thrust belts, structural models are constructed on the basis of the principles of thin skinned tectonics, which have been described comprehensively by Boyer and Elliott (1982). The fundamental assumption made is that tectonic dip in folded sedimentary rocks is generated by the movement of strata over irregular fault surfaces, so that the attitude of folded beds may be used to infer the geometry of underlying faults.

An examination of published cross sections through the Papuan thrust belt shows that many do not balance and are geometrically impossible constructions. Even if such sections were corrected, they would imply that topographic elevation in the mountains is the result of predominantly vertical movements, and that the amount of horizontal shortening is at maximum a few kilometres. By contrast, in many foothills belts, shortening is of the order of tens or even hundreds of kilometres. Inductively, the Papuan foothills would be expected to be associated with comparable amounts of shortening. There are only a few published accounts of the Papuan thrust belt which suggest such a degree of horizontal shortening (e.g. Jenkins, 1974; St John, 1977).

In this paper a set of regional balanced cross sections through the Papuan thrust belt is presented. The sections show that the mountain belt has been built predominantly by horizontal shortening, involving displacements of the order of 100 km in the hinterland, which is comparable with that observed in other foothills regions. Although the quality of the data base is such that these sections cannot offer a unique solution for the regional structure, they are believed to be representative of the variations in tectonic style across the thrust belt.

Exploration for Hydrocarbons in Papua New Guinea

The initial impetus for exploration in Papua New Guinea was the discovery of numerous gas and oil seeps in the early part of the century (Rickwood, 1968). Many of the seeps are located near the front of the mountain belt that forms the New Guinea highlands.

The geological evaluation of these foothills has been slow and expensive, mainly due to the severe access difficulties in remote, deeply karstically eroded, jungle covered terrain. Even now, the available geological maps rely considerably on remote sensed data and must be regarded as provisional. There is also little regional seismic coverage to assist interpretation.

The first wells in the foothills were drilled near the front of the moutains where rig access by river was possible. Some of these wells encountered the Mesozoic reservoir target, whereas others such as Cecilia 1 and Mananda 1 penetrated thrusts and repeat Tertiary sequences, and were abandoned without reaching their target.

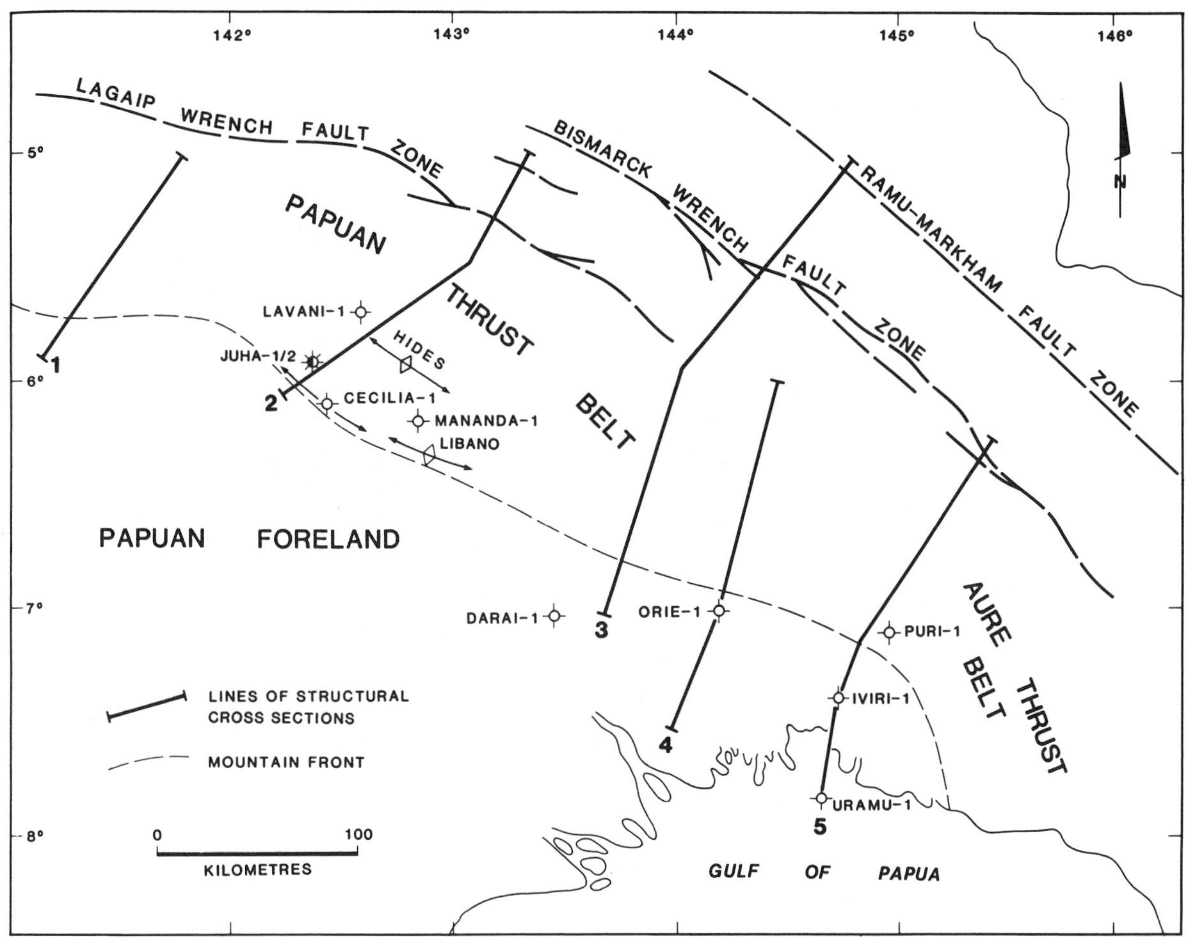

Figure 1 — Location map.

In recent years the advent of heliportable rigs has enabled wells to be drilled farther into the thrust belt. Some, such as Lavani 1, have been abandoned, due to structural complications. Others, such as the Juha wells and Mananda 3 have encountered hydrocarbons in the Mesozoic reservoir at depth. In planning future exploration programmes, reliable techniques for predicting valid structural traps are of obvious importance. It will be demonstrated below that cross section construction using the principles of thin skinned tectonics offers one rigorous means of assessing the risk on a particular trap.

THE PAPUAN THRUST BELT

The Papuan thrust belt trends in a southeasterly direction from the Irian Jayan border to the region of the Puri 1 well, where it swings north-south and is known as the Aure thrust belt (Fig. 1). The thrust belt formed in the Pliocene, in response to the oblique convergence between the Australian and Pacific plates. The obliquity of the collision has resulted in an important component of transcurrent faulting, which is largely confined to the northern part of the Papuan highlands, where the thrust belt is cut by the major, sinistral, Lagaip and Bismarck wrench fault zones.

Stratigraphy

The thrust belt is formed essentially of a Mesozoic clastic succession, which is overlain unconformably by a Late Tertiary sequence of limestones and clastics. A simplified stratigraphical succession is shown in Figure 2.

Mesozoic sediments were deposited on crystalline basement along the passive margin of the Australian continental plate. Because there has been little erosion of the foothills, there are few outcrops of the local Mesozoic succession, and the limited evidence for their facies distribution is based mainly on well data.

At the end of the Cretaceous, the foothills experienced a period of uplift and erosion, which was the local response to the opening of the Coral Sea ocean basin, to the southeast of the Gulf of Papua. On the regional scale, deposition did not recommence until the Miocene, when a thick limestone sequence was deposited across the Papua thrust belt. In the Aure thrust belt, and also near the eastern end of the Lagaip fault zone (Fig. 1), the

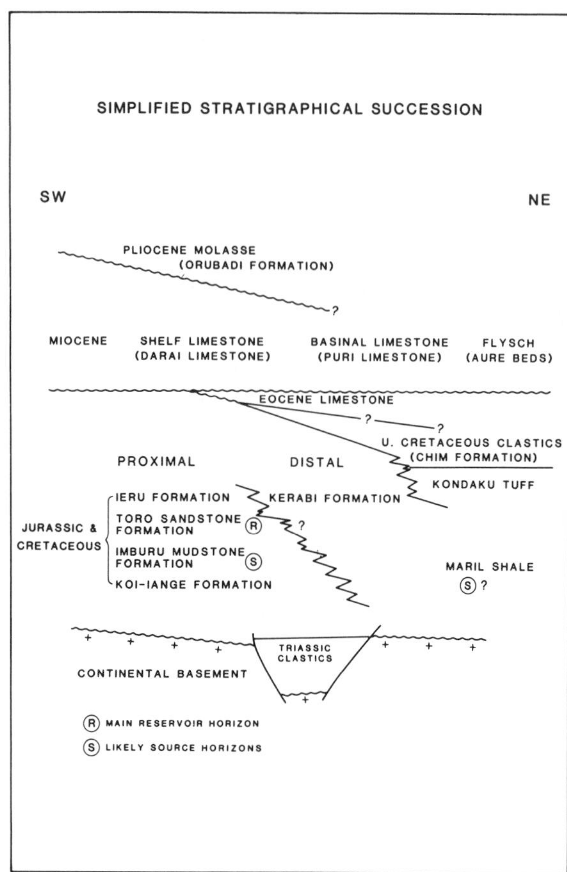

Figure 2 — Simplified stratigraphical succession.

Miocene consists of a flysch sequence, derived from a volcanic arc to the north.

During the Pliocene, when the arc collided with the Australian Plate, thrusting and uplift commenced and progressed from the northeast towards the southwest. Sediment eroded from the uplifted thrust sheets was redeposited as Pliocene molasse. At the front of the thrust belt, deformed Pleistocene-Recent volcaniclastic sediments are found, evidence for the youth of deformation there.

EVIDENCE FOR THIN SKINNED DEFORMATION IN THE PAPUAN THRUST BELT

In order to establish that a thrust belt is thin skinned, it is necessary to demonstrate that thrust fault surfaces possess irregular 'staircase' trajectories, with long 'flat' sections, parallel to bedding, separated by short 'ramps', which cut across stratal boundaries (Boyer and Elliott, 1982) (Fig. 3, a). It is also necessary to show that any folds observed could have formed by movement of hangingwall rocks over the irregular fault surface (Fig. 3, b & c).

In Papua New Guinea the evidence that the thrust belt is thin skinned comes from published geological maps, seismic sections, and also from well data:

(1) The traces of many thrusts have been mapped parallel to stratal boundaries in both the footwall and the hangingwall (Fig. 4). These observations have been traditionally interpreted that the outcropping fault is steep and reversed with limited throw (Fig. 4, a). However, the fact that faults such as the Soro thrust show the same mapped geometry over a distance of 80 km must imply that such limited throw is maintained over such distances, which is considered unlikely. The alternative interpretation of the map pattern is that the outcropping thrust surface is parallel to bedding in the footwall (Fig. 4, b). Changes in thrust displacement will result in variations in the width of the hangingwall ramp anticline, but the outcrop pattern across the thrust remains the same.

(2) Seismic sections, such as that across the Libano structure, show a prominent inclined reflector, interpreted as a thrust, parallel to bedding in the back limb of the hanging wall anticline (hangingwall flat); the thrust cuts across gently inclined reflectors in the footwall (footwall ramp).

(3) Some seismic sections in the thrust belt show deep reflectors which are gently inclined, parallel to the regional dip, and unaffected by faults or folds. These

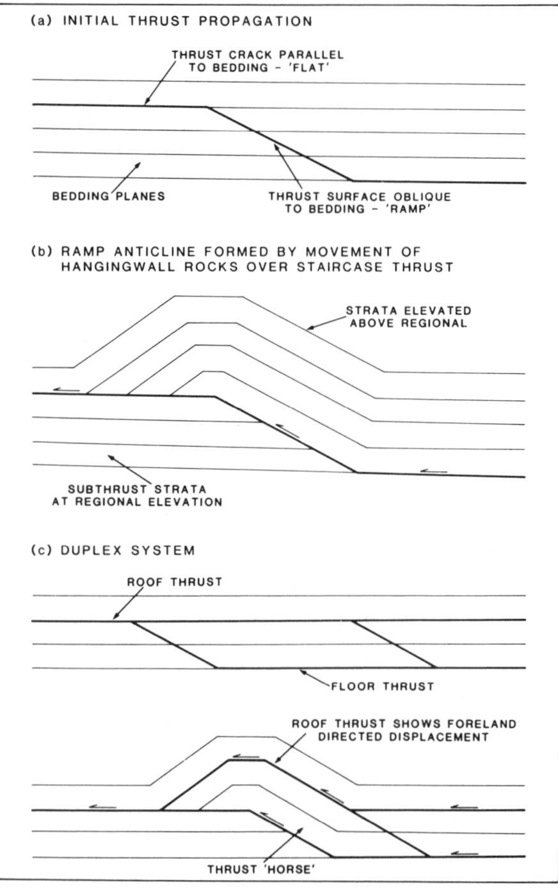

Figure 3 — Thrust surfaces and fold profiles in thin skinned belts.

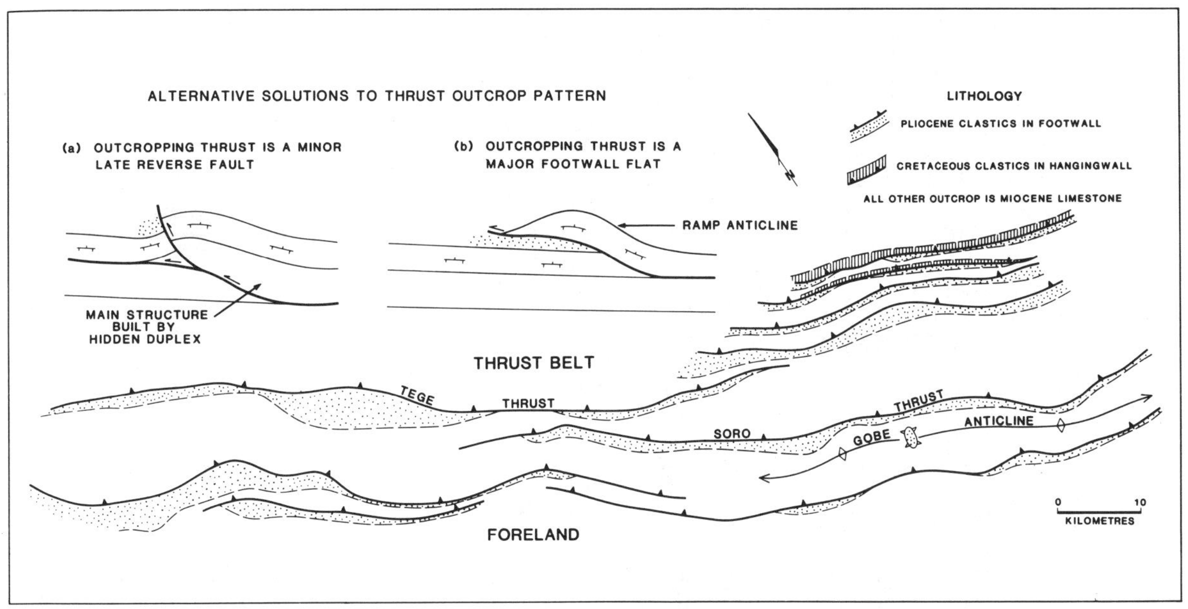

Figure 4 — Mapped thrust traces in the central section of the Papuan foothills.

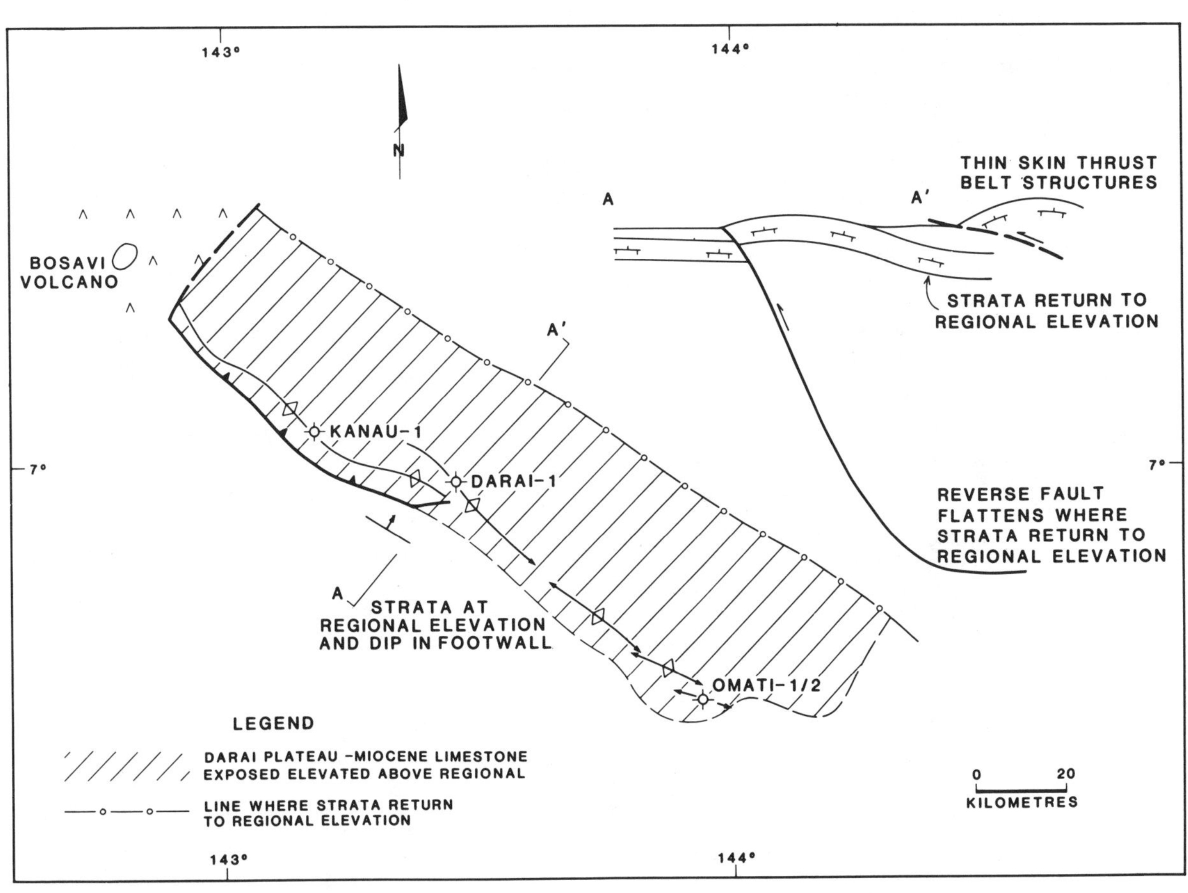

Figure 5 — The Darai Plateau inversion.

horizons are interpreted to lie beneath a decollement surface.

(4) Flat lying thrusts have been proven in the cores of the Gobe and Puri anticlines (by geological mapping and by well data respectively).

(5) The gravity data interpreted by St John (1977) suggest that some folds involve considerable overthrusting of the Miocene limestone sequence. The results of wells such as Mananda 1 and Cecilia 1 confirm that such repetition can occur.

(6) There is a strong positive correlation between topographic and structural relief in the Papuan thrust belt. Section construction shows that strata lie close to their regional elevation in the cores of synclines but are elevated above regional on anticlines. This observation is consistent with the principles of thin skinned tectonics.

It is important to note that there are also some structures in the mountain belt which have been formed by movement along steeply inclined faults which cut through basement. These structures are likely to be due to the inversion of faults which were initially active during Mesozoic sedimentation. The Darai Plateau, a large area of elevated limestone, is an example of one such feature (Fig. 5).

THE PAPUAN MOUNTAIN FRONT

Along the junction between the Papuan mountains and the foreland basin, there are changes in structural style which reflect different types of thrust tectonics within the foothills, in particular changes in thrust detachment levels. Vann *et al.* (1984) have reported comparable changes in the structure of mountain fronts in many parts of the world.

The conventional type of mountain front is an emergent thrust (Fig. 6, a), which occurs where the mountain and deformation fronts coincide. In Papua New Guinea this type of mountain front is only observed in the region of the Cecilia and Libano anticlines (see Fig. 8: Regional cross section 2), folds which are detached at shallow levels.

Elsewhere, the junction between the Papuan mountain belt and the foreland is rather more complex. Near the Darai Plateau the mountain front extends 50 km ahead of the thrust front, because parts of the foreland have been uplifted by movement along steeply dipping faults (Fig. 6, b). This movement is most probably an inversion of movement direction along faults which were originally extensional during sedimentation. It is likely that other inversion structures have been incorporated into the thrust belt as it has developed.

Near the Irian Jayan border and also close to the junction of the Papuan and Aure thrust belts, the mountain front is not defined by a fault, but by a foreland dipping panel of rocks, elevated above their regional position in a major monoclinal fold (Fig. 6, c). Vann *et al.* (1984) have observed mountain frontal monoclines in other foothills belts. They have noted that it is difficult to balance sections across such monoclines, because in the absence of outcropping thrusts it is difficult to see how

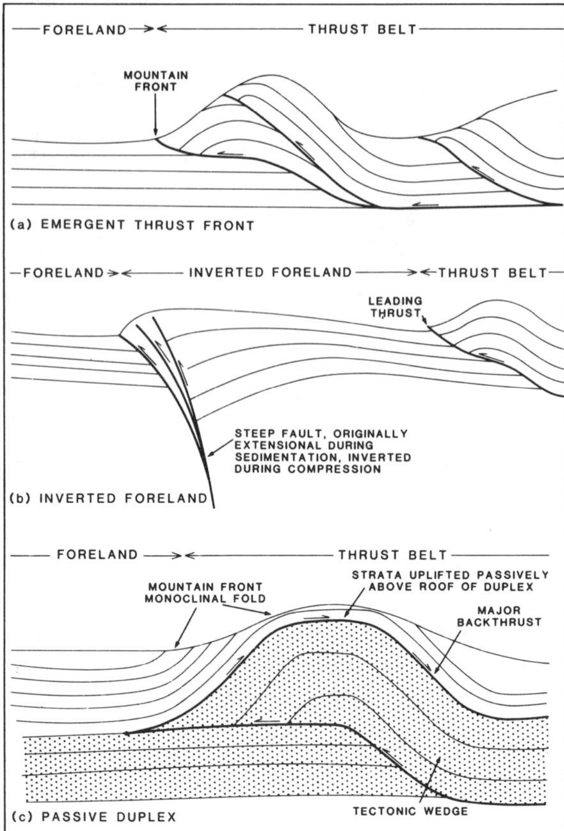

Figure 6 — Structural styles at mountain fronts.

the horizontal translations involved in building the mountain belt are transferred to the surface.

Vann *et al.* have proposed several solutions for this balancing problem. In Papua New Guinea the most likely explanation is that the frontal monocline is formed of a 'passive duplex' or 'intercutaneous wedge' (Charlesworth & Gagnon, 1985) which underthrusts the foreland. The upper detachment in such a system is a major backthrust (Fig. 6, c). Strata above the upper detachment have been uplifted and have folded to accomodate the shape of the underlying duplex, but they have not been translated towards the foreland.

THE REGIONAL STRUCTURE OF THE PAPUAN THRUST BELT

The foothills belt has evolved by forelandwards propagation of the thrust front. Consequently, the types of structure identified along the Papuan mountain front are likely to be typical of those within the thrust belt. These structural styles help constrain five regional restored cross sections which have been constructed between the foreland and the Lagaip-Bismarck wrench fault systems (Figs 1, 7-9). Because of the provisional nature of the geological maps on which they are based, the sections cannot provide unique solutions for the geological structure, but they do offer a geometrically admissible interpretation, and show the types of thrust related structures which are observed in other foothills-type belts around the world. These sections may be used to distinguish regions where structures are detached at shallow levels from those where deeper stratigraphic units are involved in thrusting. Thus, acreage where a Mesozoic reservoir is likely to be in structure may quickly be identified.

Cross Section 1

Regional cross section 1 cuts through the hinge of the Muller Anticline (Fig. 7), a major uplift with a core of granitic basement, exposed in the Strickland Gorge. The mountain front consists of a large monocline, and the Muller Anticline is interpreted as a major passive duplex, built of a horse of basement and Mesozoic cover.

The upper, backthrust detachment has been drawn at the base of a dominantly mudstone sequence of Cretaceous age (the Ieru Formation), where faults have locally been mapped around the Muller Anticline. As a geometrical consequence of the choice of this level for the top of the duplex, the lower detachment must lie within basement, about 14 km below sea level. The basement horse has been thrust over Jurassic rocks a distance of about 30 km southwest to build the Muller culmination. Distal facies Mesozoic sediments, which have been mapped behind the Muller Anticline, have also been displaced about 30 km southwest. The Late Cretaceous and Tertiary cover above the duplex has been uplifted and folded, but not appreciably thrust faulted.

The integrity of this interpretation depends on the identification of the correct stratigraphic position of the upper detachment in the passive duplex. In this instance the mapping is rather equivocal. There is an alternative interpretation, that the passive duplex involves only thrust slices of basement. In this case the displacement of the cover sequence towards the foreland would be negligible.

Just beyond the northeast end of line 1 the thrust belt is cut by the Lagaip wrench fault zone. The hinterland extension of the thrust belt has been translated by an unknown amount towards the west, into Irian Jaya.

Cross Section 2

Regional cross section 2 is drawn between the Cecilia frontal anticline and the Lagaip wrench fault zone (Fig. 8, upper). This interpretation is based on the assumption that the basement is inclined gently towards the hinterland, as is shown by Robertson Research Australia and Flower Doery Buchan (RRA/FDB) (1984, plate 7).

The mountain front is defined by the Cecilia ramp anticline, detached within Cretaceous clastics. This thrust surface forms the upper detachment of a major passive duplex of proximal Mesozoic sediments. Upper Cretaceous and Miocene sediments have been uplifted above the roof of the duplex. Northeast of the Sage thrust, distal Jurassic and possibly Triassic strata are inferred to form the core of the duplex.

The section has been restored as far back as the proximal/distal Jurassic boundary, which has been displaced 103 km towards the southwest. However, the edge of the shallow Miocene shelf (as located by reference to facies changes in the Miocene limestones) has been displaced only 15 km towards the southwest, as a consequence of its location above the roof of the passive duplex.

In this interpretation the passive Mesozoic duplex inferred along section 2 forms a structural transition between the deeper duplex involving basement, identified in section 1, and the shallow listric thrust belt of section 3 (see below).

St John (1977) has interpreted gravity data across the Lavani area differently from RRA/FDB (1984). He has identified a basement high in this area, essentially the eastern extension of the Muller Anticline. If this interpretation is correct, then the shortening of the Mesozoic and Tertiary sequence along section 2 would be minimal. However, there would be a dramatic increase in shortening towards the listric thrust belt in section 3 (see below) which would invoke a major and abrupt hangingwall lateral ramp. The diffuse nature of the lateral boundary separating the Lavani region from the listric thrust belt is inconsistent with the considerable change in shortening required by such an interpretation.

Cross Section 3

Regional cross section 3 (Fig. 8, lower) is drawn between the Darai Plateau and the Kubor Anticline, a large, basement-cored culmination. The Anticline is truncated by the Bismarck wrench fault zone. Beneath the thrust belt, basement is interpreted as inclined gently towards the hinterland as inferred by RRA/FDB (1984).

In this area the Darai Plateau foreland has been uplifted by movement, probably inversion, along steeply dipping faults, and the thrust front is located at the Gobe Anticline, behind the mountain front. Gobe is the frontal structure in a listric thrust belt detached close to the base of Cretaceous clastics.

Towards the hinterland, fewer thrusts have been mapped and there is a broad region where undulating folds have been identified at the surface. They are interpreted as accommodation structures, above a hidden Jurassic duplex, which is required to balance the shortening in the listric thrust belt.

In order to balance the Jurassic shortening, a major footwall flat, close to top basement level, is required beneath the Kubor Anticline. Findlay (1974) has identified a detachment close to top basement along the southern flank of the Kubor Anticline, which he believed to be a low-angle extensional fault. In this model, this fault is interpreted as the upper detachment of a basement duplex. As a geometrical consequence, the lower detachment must lie near the base of the crust, about 35 km below sea level.

There are facies and thickness changes in the Mesozoic succession across the Kubor Anticline (Bain *et al.*, 1975) which may be evidence for fault controlled sedimentation there. Thus it is possible that the Kubor Anticline is an inversion structure. Certainly, the lozenge-shape of the Kubor Anticline in plan could have been formed by inversion of an extensional fault, with an originally cuspate shape.

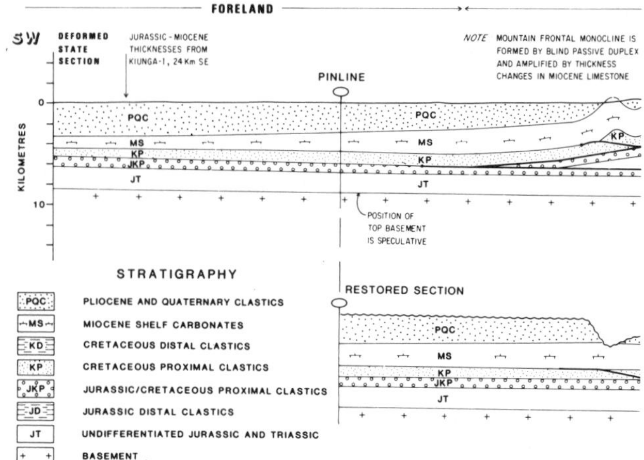

REGIONAL CROSS SECTION 1

SURFACE GEOLOGY FROM GEOLOGICAL SURVEY MAP SB 54-7 (BLUCHER RANGE) AND BP/APC INTERNAL REPORTS

Figure 7—Regional cross section 1. *To accompany:* Hobson, D.M., 1986 — A thin skinned model for the Papuan thrust belt and some implications for hydrocarbon exploration. *APEA Journal,* 26(1), 214-24.

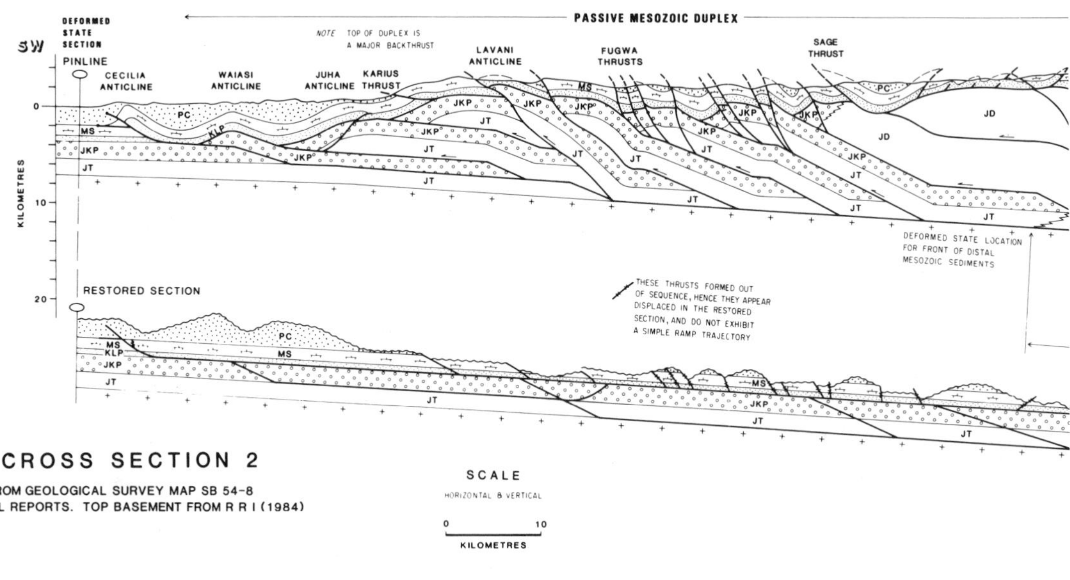

REGIONAL CROSS SECTION 2

SURFACE GEOLOGY FROM GEOLOGICAL SURVEY MAP SB 54-8 (WABAG) AND BP/APC INTERNAL REPORTS. TOP BASEMENT FROM R R I (1984)

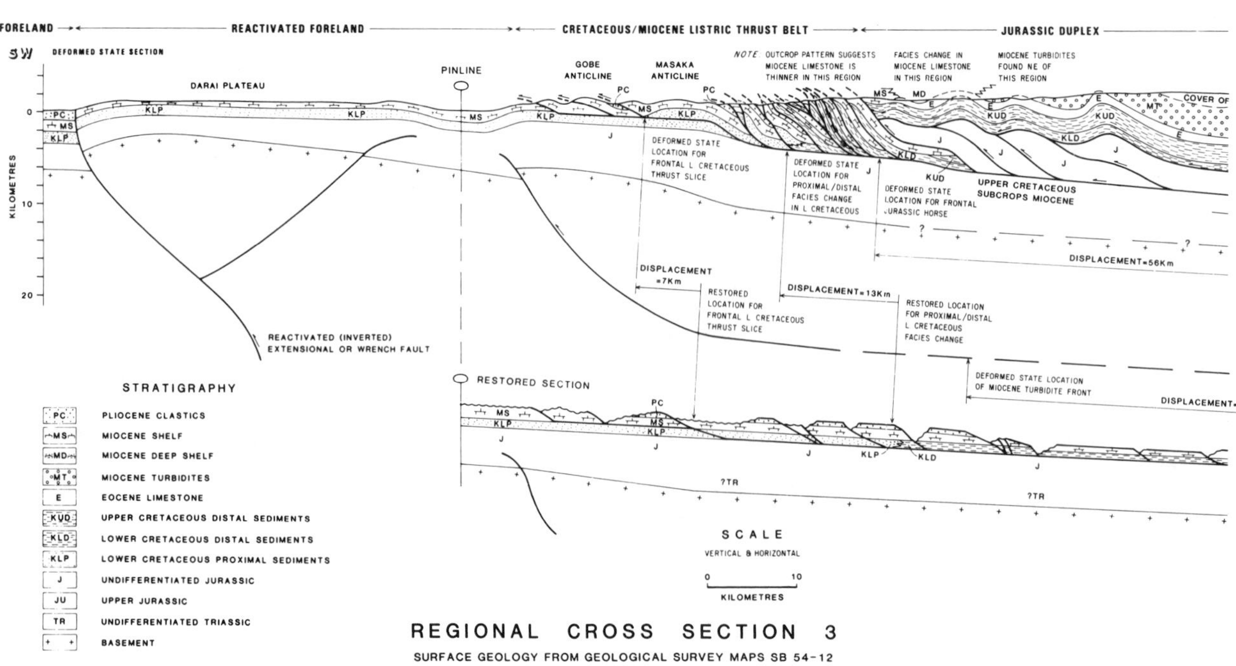

REGIONAL CROSS SECTION 3

SURFACE GEOLOGY FROM GEOLOGICAL SURVEY MAPS SB 54-12 (KUTUBU), SB 55-5 (RAMU) AND BP/APC INTERNAL REPORTS

Figure 8—Regional cross sections 2 and 3. *To accompany:* Hobson, D.M., 1986 — A thin skinned model for the Papuan thrust belt and some implications for hydrocarbon exploration. *APEA Journal,* 26(1), 214-24.

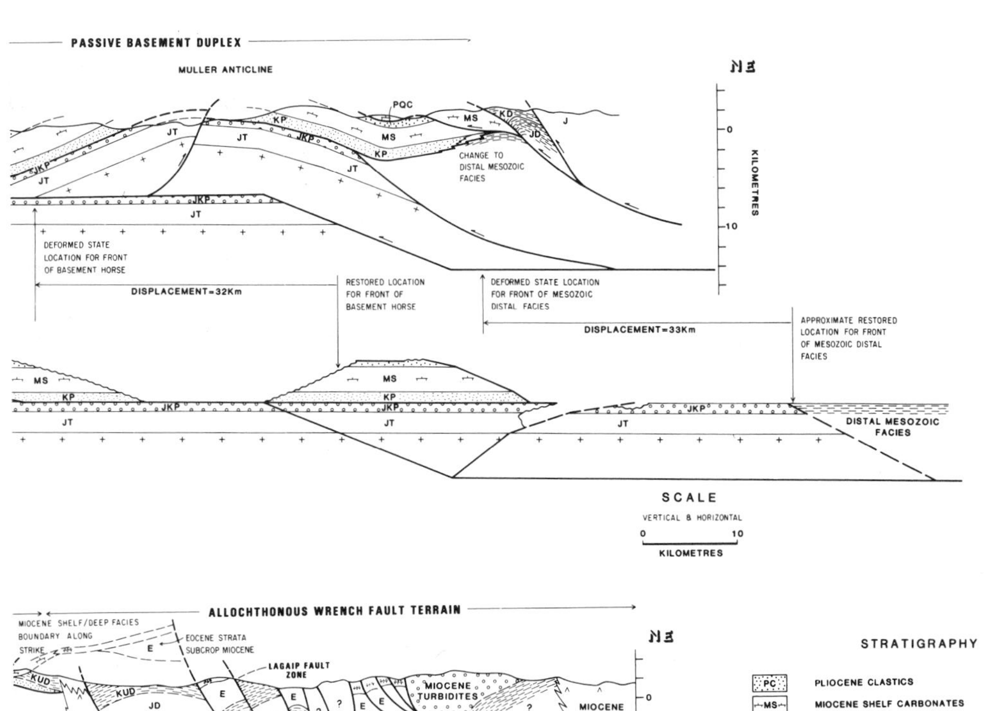
PASSIVE BASEMENT DUPLEX
MULLER ANTICLINE
NE
PQC
KP
MS
KD
JT
JKP
JD
J
CHANGE TO DISTAL MESOZOIC FACIES
KILOMETRES
0
10
DEFORMED STATE LOCATION FOR FRONT OF BASEMENT HORSE
DISPLACEMENT=32Km
RESTORED LOCATION FOR FRONT OF BASEMENT HORSE
DEFORMED STATE LOCATION FOR FRONT OF MESOZOIC DISTAL FACIES
DISPLACEMENT=33Km
APPROXIMATE RESTORED LOCATION FOR FRONT OF MESOZOIC DISTAL FACIES
DISTAL MESOZOIC FACIES
SCALE
VERTICAL & HORIZONTAL
0
10
KILOMETRES

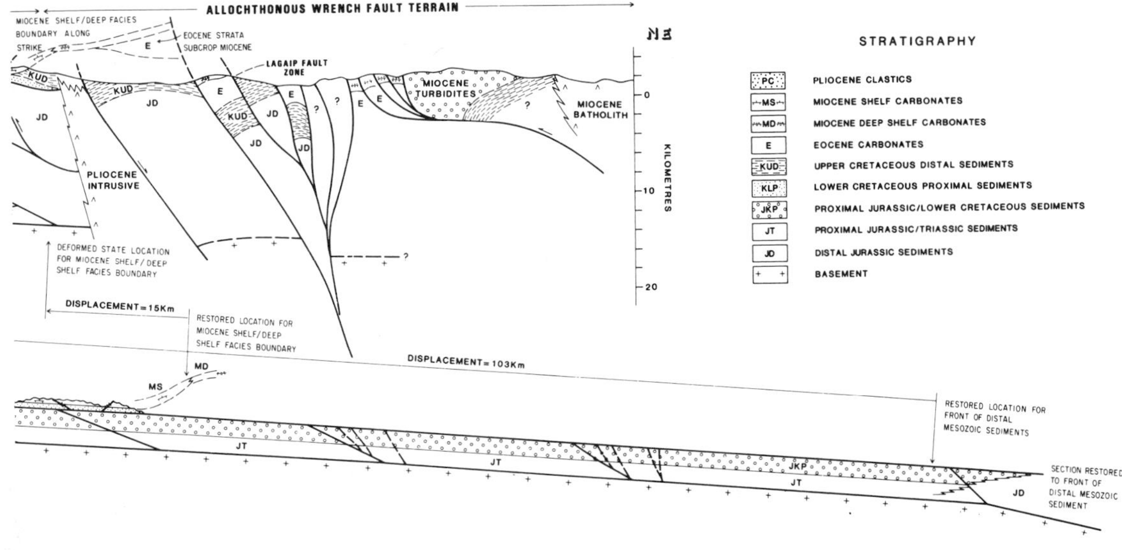
ALLOCHTHONOUS WRENCH FAULT TERRAIN
MIOCENE SHELF/DEEP FACIES BOUNDARY ALONG STRIKE
EOCENE STRATA SUBCROP MIOCENE
LAGAIP FAULT ZONE
NE
KUD
JD
E
MIOCENE TURBIDITES
MIOCENE BATHOLITH
PLIOCENE INTRUSIVE
KILOMETRES
0
10
20
DEFORMED STATE LOCATION FOR MIOCENE SHELF/DEEP SHELF FACIES BOUNDARY
DISPLACEMENT=15Km
RESTORED LOCATION FOR MIOCENE SHELF/DEEP SHELF FACIES BOUNDARY
DISPLACEMENT=103Km
MD
MS
JT
JKP
RESTORED LOCATION FOR FRONT OF DISTAL MESOZOIC SEDIMENTS
SECTION RESTORED TO FRONT OF DISTAL MESOZOIC SEDIMENT
STRATIGRAPHY
PC PLIOCENE CLASTICS
MS MIOCENE SHELF CARBONATES
MD MIOCENE DEEP SHELF CARBONATES
E EOCENE CARBONATES
KUD UPPER CRETACEOUS DISTAL SEDIMENTS
KLP LOWER CRETACEOUS PROXIMAL SEDIMENTS
JKP PROXIMAL JURASSIC/LOWER CRETACEOUS SEDIMENTS
JT PROXIMAL JURASSIC/TRIASSIC SEDIMENTS
JD DISTAL JURASSIC SEDIMENTS
BASEMENT

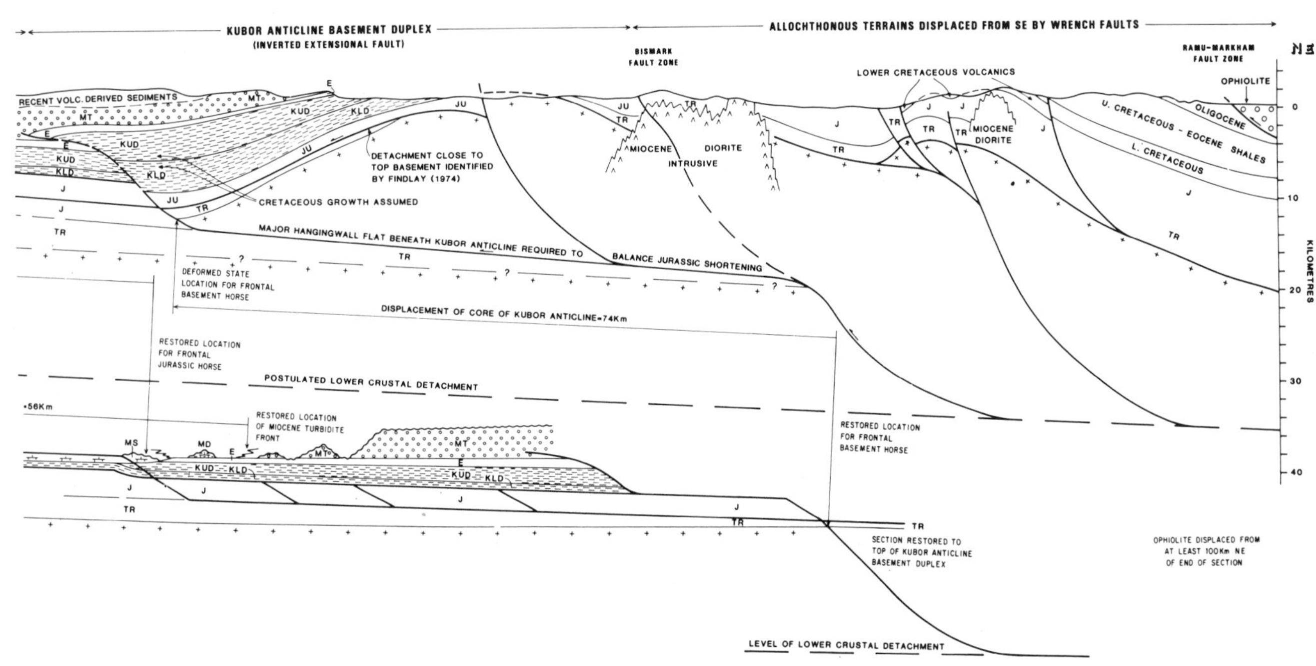
KUBOR ANTICLINE BASEMENT DUPLEX
(INVERTED EXTENSIONAL FAULT)
ALLOCHTHONOUS TERRAINS DISPLACED FROM SE BY WRENCH FAULTS
BISMARK FAULT ZONE
RAMU-MARKHAM FAULT ZONE
NE
LOWER CRETACEOUS VOLCANICS
OPHIOLITE
RECENT VOLC. DERIVED SEDIMENTS
MT
KUD
KLD
E
J
JU
TR
DETACHMENT CLOSE TO TOP BASEMENT IDENTIFIED BY FINDLAY (1974)
CRETACEOUS GROWTH ASSUMED
MIOCENE
DIORITE
INTRUSIVE
MIOCENE DIORITE
OLIGOCENE
U. CRETACEOUS - EOCENE SHALES
L. CRETACEOUS
MAJOR HANGINGWALL FLAT BENEATH KUBOR ANTICLINE REQUIRED TO BALANCE JURASSIC SHORTENING
DEFORMED STATE LOCATION FOR FRONTAL BASEMENT HORSE
DISPLACEMENT OF CORE OF KUBOR ANTICLINE=74Km
RESTORED LOCATION FOR FRONTAL JURASSIC HORSE
POSTULATED LOWER CRUSTAL DETACHMENT
KILOMETRES
0
10
20
30
40
56Km
RESTORED LOCATION OF MIOCENE TURBIDITE FRONT
MS
MD
RESTORED LOCATION FOR FRONTAL BASEMENT HORSE
SECTION RESTORED TO TOP OF KUBOR ANTICLINE BASEMENT DUPLEX
OPHIOLITE DISPLACED FROM AT LEAST 100Km NE OF END OF SECTION
LEVEL OF LOWER CRUSTAL DETACHMENT

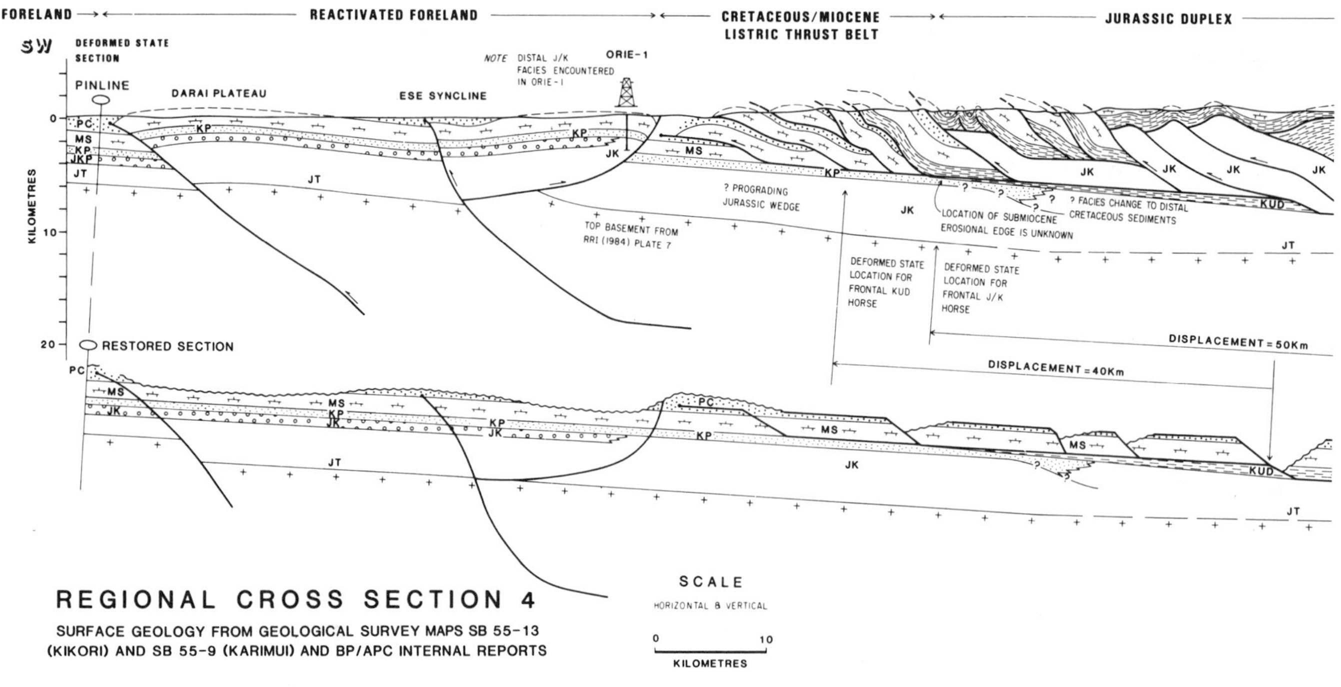

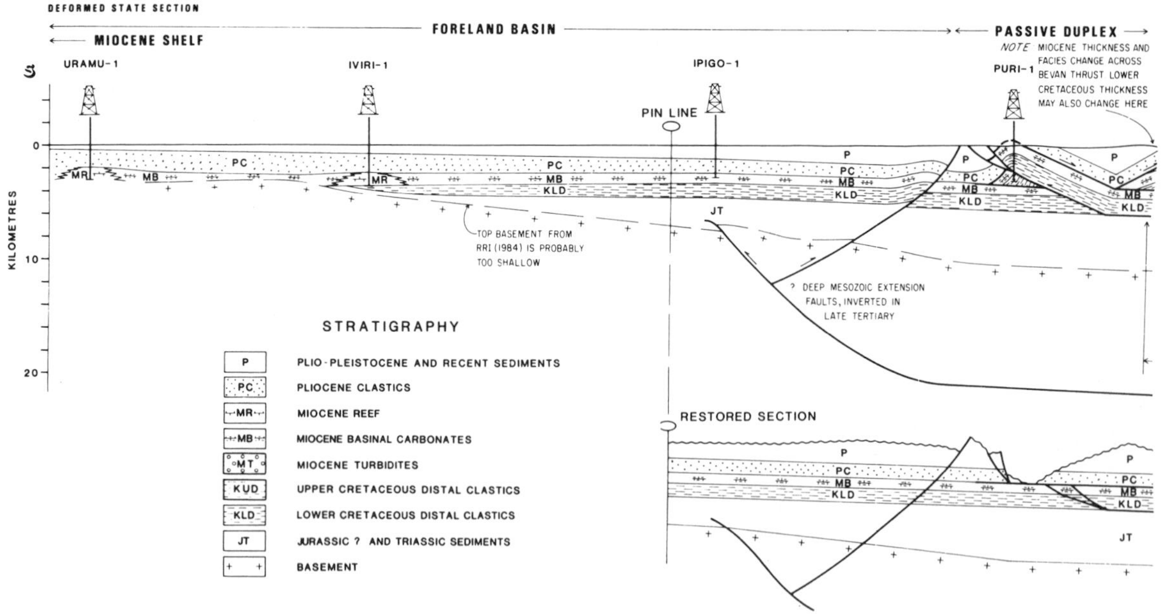

REGIONAL CROSS SECTION 5

SURFACE GEOLOGY FROM GEOLOGICAL SURVEY MAPS SB 55-13 (KIKORI) AND SB 55-9 (KARIMUI) AND BP/APC INTERNAL REPORTS

Figure 9—Regional cross sections 4 and 5. *To accompany:* Hobson, D.M., 1986 — A thin skinned model for the Papuan thrust belt and some implications for hydrocarbon exploration. *APEA Journal,* 26(1), 214-24.

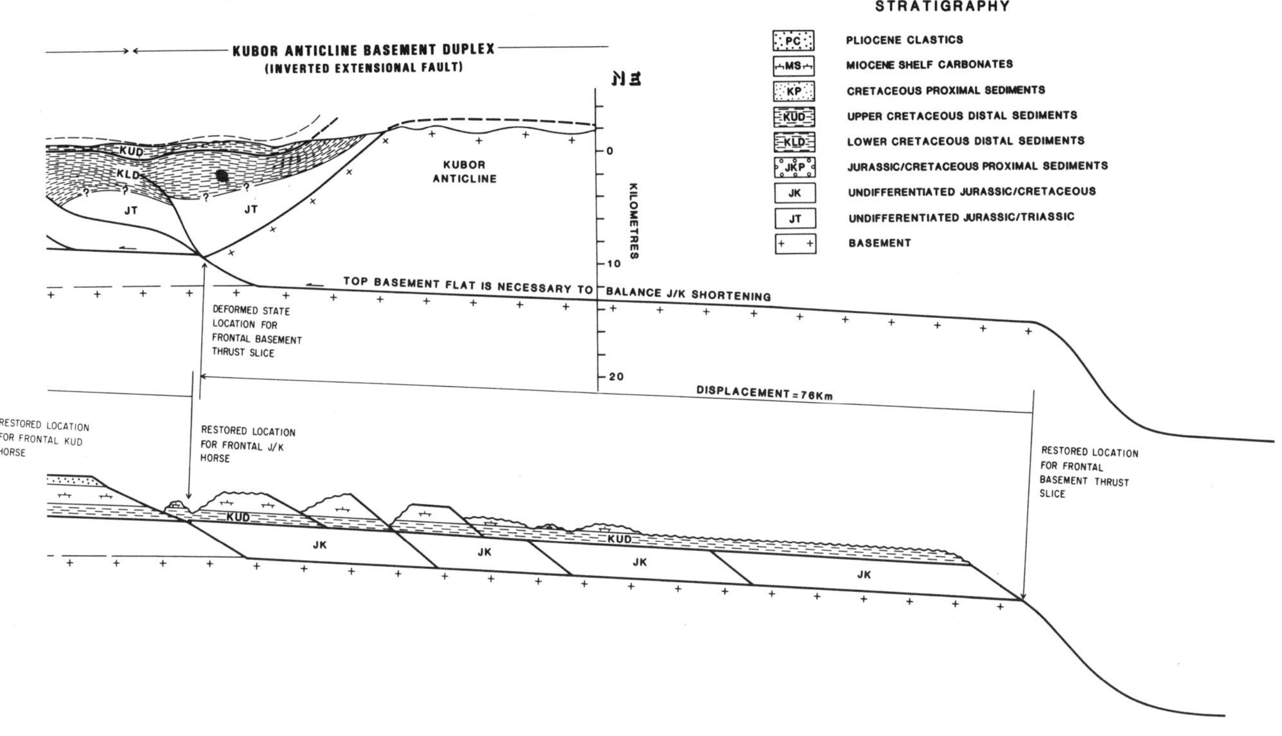
STRATIGRAPHY
PC PLIOCENE CLASTICS
MS MIOCENE SHELF CARBONATES
KP CRETACEOUS PROXIMAL SEDIMENTS
KUD UPPER CRETACEOUS DISTAL SEDIMENTS
KLD LOWER CRETACEOUS DISTAL SEDIMENTS
JKP JURASSIC/CRETACEOUS PROXIMAL SEDIMENTS
JK UNDIFFERENTIATED JURASSIC/CRETACEOUS
JT UNDIFFERENTIATED JURASSIC/TRIASSIC
BASEMENT
KUBOR ANTICLINE BASEMENT DUPLEX
(INVERTED EXTENSIONAL FAULT)
NE
KUBOR ANTICLINE
KILOMETRES
TOP BASEMENT FLAT IS NECESSARY TO BALANCE J/K SHORTENING
DEFORMED STATE LOCATION FOR FRONTAL BASEMENT THRUST SLICE
DISPLACEMENT=76Km
RESTORED LOCATION FOR FRONTAL KUD HORSE
RESTORED LOCATION FOR FRONTAL J/K HORSE
RESTORED LOCATION FOR FRONTAL BASEMENT THRUST SLICE

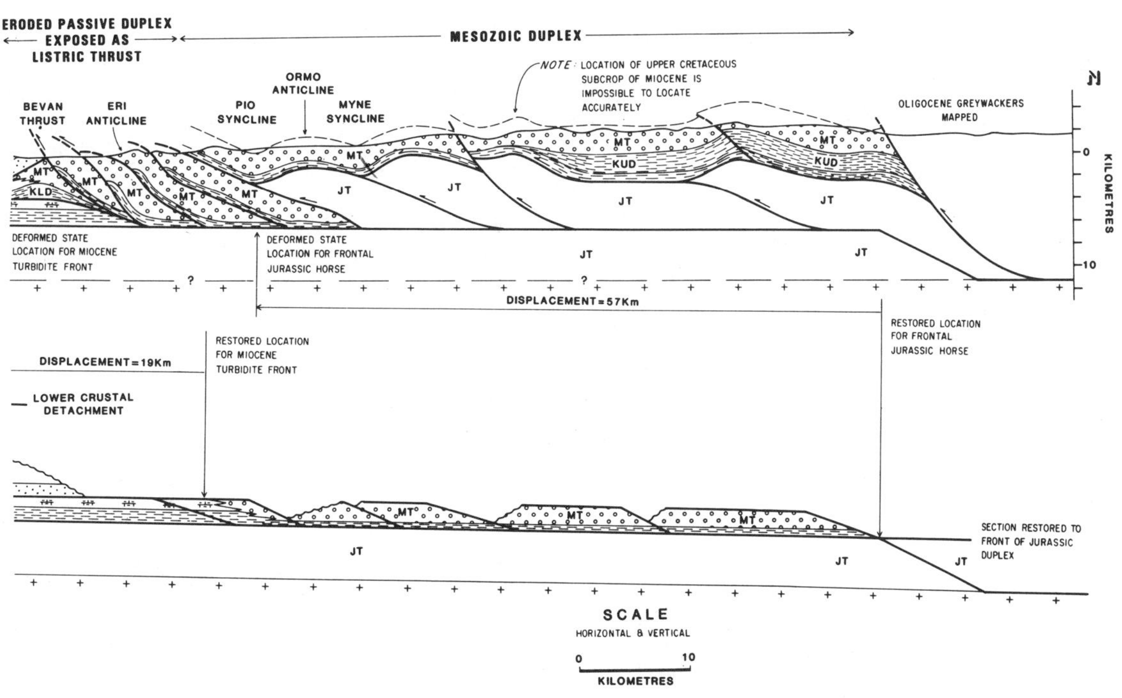
ERODED PASSIVE DUPLEX EXPOSED AS LISTRIC THRUST
MESOZOIC DUPLEX
NOTE: LOCATION OF UPPER CRETACEOUS SUBCROP OF MIOCENE IS IMPOSSIBLE TO LOCATE ACCURATELY
BEVAN THRUST
ERI ANTICLINE
PIO SYNCLINE
ORMO ANTICLINE
MYNE SYNCLINE
OLIGOCENE GREYWACKERS MAPPED
N
KILOMETRES
DEFORMED STATE LOCATION FOR MIOCENE TURBIDITE FRONT
DEFORMED STATE LOCATION FOR FRONTAL JURASSIC HORSE
DISPLACEMENT=57Km
RESTORED LOCATION FOR FRONTAL JURASSIC HORSE
DISPLACEMENT=19Km
RESTORED LOCATION FOR MIOCENE TURBIDITE FRONT
LOWER CRUSTAL DETACHMENT
SECTION RESTORED TO FRONT OF JURASSIC DUPLEX
SCALE
HORIZONTAL & VERTICAL
KILOMETRES

Section restoration shows that the frontal Jurassic thrust slice has been displaced 56 km towards the southwest, and that the basement core of the Kubor Anticline has been thrust some 74 km. Shallow and deeper shelfal Miocene carbonates, together with contemporareous turbidites, have been mapped along the line of section 3. The restored location of the shallow/deeper shelf facies lies 56 km northeast of its present position.

Section 3 has been constructed as far as ophiolitic ultramafic rocks in the Ramu-Markham fault zone. Because wrench fault displacements are likely to have added or subtracted strata from the plane of the section, it is not possible to reconstruct the section in this region rigorously. However, if it is assumed that there has been minimum area change due to wrench faulting, the ultramafic rocks must have been thrust from at least 100 km to the northeast.

Along the line of section 3, the Cretaceous and Miocene rocks have been actively shortened rather than simply uplifted as occurs to the west. There must be a hangingwall lateral ramp separating the two structural provinces. The trace of this ramp on the geological map is partially obscured by younger volcanic rocks from the Doma volcano, but it can be identified as a diffuse line which bounds the edge of the thrust belt (Fig. 10).

Along cross sections 1 and 2, the Miocene rocks are predominantly in a shallow shelfal facies (the Darai Limestone), to the south of the wrench fault zone. Along line 3, where shortening at Miocene levels is considerably greater, both shallow and deep shelf carbonate facies, and also turbiditic clastics have been mapped. This evidence suggests that the wrench fault zone has cut obliquely across the deformed Miocene shelf edge, and that the deep shelf carbonates and turbites have been displaced sinistrally towards Irian Jaya.

Cross Section 4

Regional cross section 4 (Fig. 9, upper) may be interpreted in a similar manner to section 3, in terms of a reactivated foreland, shallow listric thrust belt, Jurassic duplex, and basement duplex. The junction between reactivated foreland and thrust belt is more difficult to define in this area. Behind the Darai Plateau the Orie structure is thought to detach at or within basement, ahead of the leading thin skinned thrust slice. It is interpreted as an inversion structure.

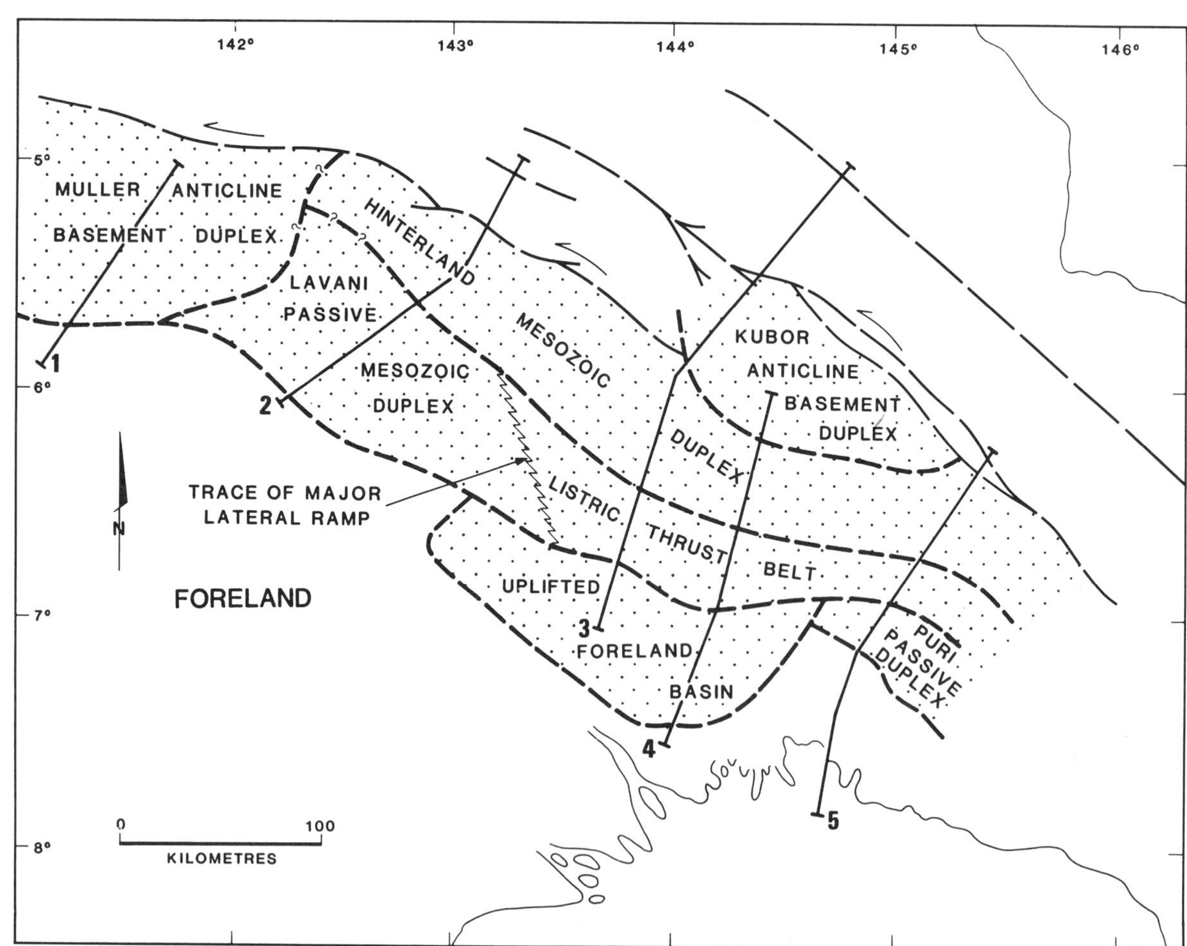

Figure 10 — Structural provinces.

Along section 4, Miocene rocks have only been mapped in shallow shelfal facies. The limestones are exposed to within 30 km of outcropping basement in the Kubor Anticline. Restoration of the section shows that the shallow shelf extended farther to the north west than along the line of section 3. It is likely that the Kubor Anticline marked the edge of the Miocene shelf, and that any deeper shelf carbonates and the turbidites facies were deposited farther to the northeast (now removed by wrench faulting). If the Kubor Anticline was a regional high throughout the Mesozoic, then the passive edge of the Australian Plate may have been irregular in shape throughout this time (see below).

Cross Section 5

Regional cross sections 5 (Fig. 9, lower) is drawn between the Gulf of Papua and the Bismarck fault zone. The top basement level has been based on RRA/FDB (1984, plate 7), but it may be drawn too shallowly, at least in the foreland region.

The Puri Anticline, which marks the front of the thrust belt, is a small passive duplex. To the east of section 5 this passive duplex becomes considerably larger, and is marked by a major mountain frontal monocline, visible on Landsat images. Behind Puri, Cretaceous and Miocene rocks are involved in other horses in the same duplex, although because the roof has been eroded, the structure is exposed as a listric thrust belt. North of the Pio Syncline, thrusts at surface are replaced by folds, interpreted as accommodation structures above a deeper, Mesozoic duplex. However, the geology of the region is poorly known and the maps here are imprecise.

In the foreland, the Miocene carbonate is mainly present in its basinal facies (the Puri Limestone). Across the Bevan Fault there is an important facies change to turbidites, which are predominantly found in the Aure thrust belt. The turbidite front has been thrust 19 km from the northeast.

Structural Provinces in the Papuan Thrust Belt

The regional cross sections through the thrust belt provide a basis for the identification of seven structural provinces (Fig. 10):

(i) the Muller Anticline passive basement duplex, which may be an inverted Mesozoic extensional fault;
(ii) the Lavani passive Mesozoic duplex;
(iii) the listric thrust belt, composed of Cretaceous-Miocene horses;
(iv) a hinterland Mesozoic duplex;
(v) the Kubor Anticline basement duplex, which may be an inverted Mesozoic extensional fault;
(vi) the uplifted part of the foreland basin;
(vii) the Puri passive duplex.

These provinces are now truncated by the Bismarck/Lagaip wrench fault zones.

There is a change in the amount of thrust shortening along the mountain belt. The Kubor Anticline basement duplex has been displaced about 75 km towards the foreland. However, the other basement high, the Muller Anticline basement duplex, has been displaced only 32 km southwest. Both duplexes are truncated by major wrench fault zones to the north.

The change in shortening along the thrust belt is probably due to location of the wrench fault zones. A major part of the thrust belt hinterland has been displaced along the Lagaip Fault, whereas more of the thrust belt has been preserved south of the Bismarck Fault. Alternatively, it is possible that cross section 1 (Fig. 7) is radically incorrect and that there is a major repetition of the whole sedimentary section beneath the Muller Anticline. It might be possible to identify such major thrust repetition by means of a regional gravity survey.

Because of the limited data base, it is not possible to arrive at a unique solution for the structure of the thrust belt. The sections in this paper have mainly been drawn assuming that the basement dips gently towards the hinterland at least in the frontal parts of the mountain belt. This is at variance with the interpretation of St John (1977), who has suggested that there are important basement highs, particularly in the Lavani area. If his interpretation is correct then there must be rapid changes in the amount of shortening in a lateral direction, which must be expressed on geological maps by means of hangingwall lateral ramps. The gradual change in outcrop style which is observed suggests a more gradational change in structural style from the Muller Anticline basement duplex to the listric thrust belt. The recognition of an intervening Lavani passive Mesozoic duplex provides a means of transition in structural styles without rapid lateral changes in shortening.

Inversion Structures

There is considerable evidence, especially in the Darai Plateau area, for inversion structures formed by uplift along earlier formed extensional faults. The Darai Plateau itself is a major elevated fault slice on the margin of the thrust belt (Fig. 5). The Iehi and Orie anticlines, both penetrated by wells, are examples of inversion structures now incorporated into the thrust belt. The Muller and Kubor anticlines are examples of major inversions, which have been thrust considerable distances towards the hinterland along major detachment surfaces. The considerable size and shape of these structures suggests that they were initiated as major cuspate extensional faults along the Mesozoic margin of the Australian Plate.

There are certainly other inversion structures in the thrust belt to be recognised. In this paper the Juha Anticline has been interpreted as a horse within a major passive duplex. It is also possible that Juha is an inversion strucuture detached deep in the basement. Unfortunately the existing seismic reflection data quality is insufficient to distinguish between these two alternative models.

PALINSPASTIC BASIN RECONSTRUCTION

The geometrical validity of the regional cross sections has been established by their restoration to a pre-deformation state. The restored parts of the sections show the

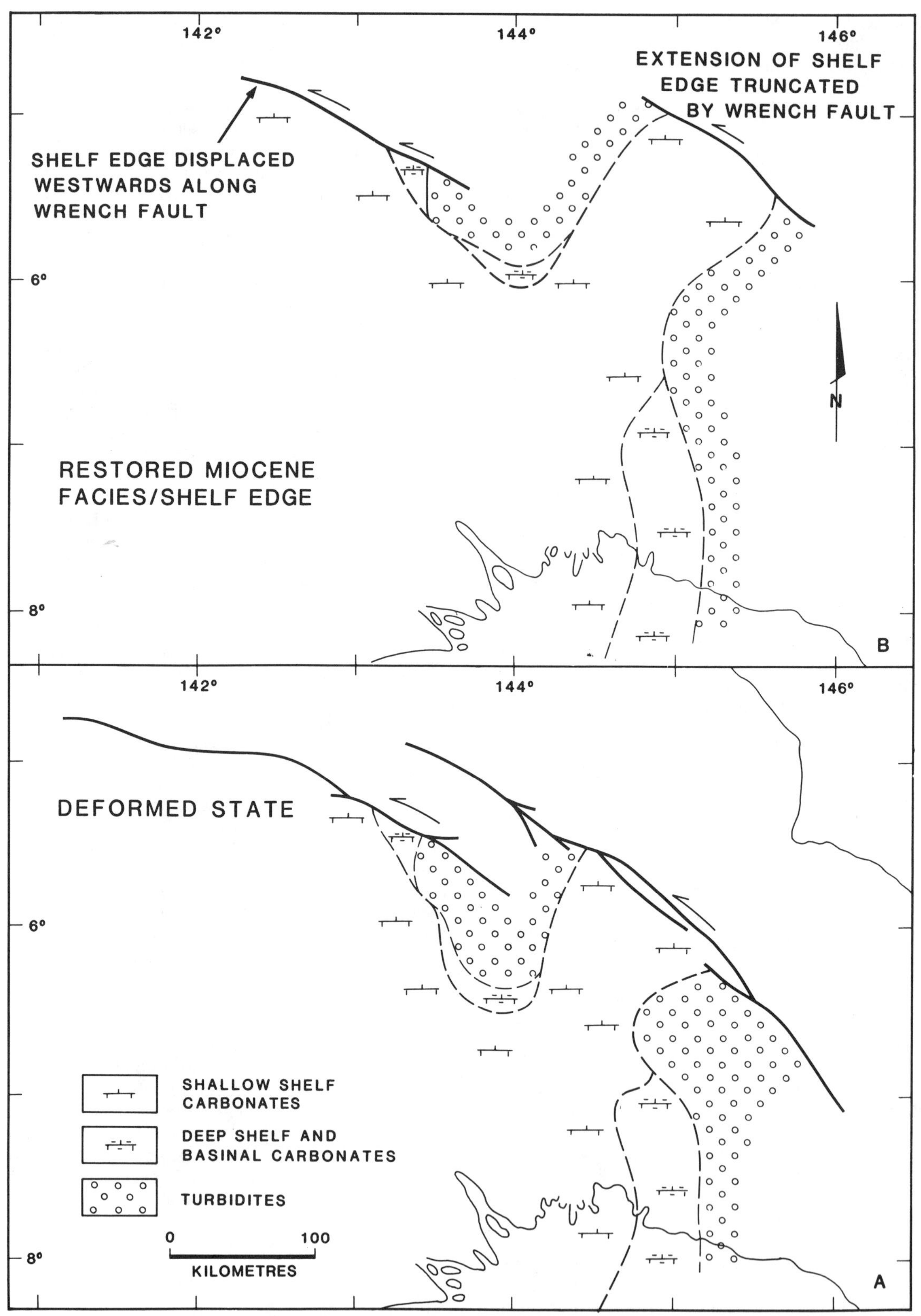

Figure 11 — Miocene facies palinspastic reconstruction.

original distribution of various facies and can be used to construct palaeogeographic maps which show the former shape of a basin. In turn, these maps may be used to predict the location of environments where favourable source and reservoir facies may be encountered.

Few of the published palaeogeographic reconstructions of the Papuan basin have been palinspastically corrected, probably because the structural model applied involves minimum shortening. The structural interpretation given in this paper involves considerable shortening, and thus the basin reconstructions would be expected to show significant differences to a deformed state version.

It is only practical to reconstruct the shape of the Papuan basin for the Miocene, where sufficient facies data are available. However, this reconstruction may serve as a basis for extrapolation of the geography of earlier periods.

It is possible to identify four facies types in the Miocene, from the available geological maps (Fig. 11, A):

(i) shallow shelf carbonates (the Darai Limestone),
(ii) a deeper shelf carbonate (the Nembi Limestone),
(iii) a basinal carbonate (the Puri Limestone),
(iv) a turbidite succession (the Aure Group).

The Miocene turbidites were derived from a volcanic arc, formed at the leading edge of the Pacific Plate. The deformed state facies relationships (Fig. 11, A) suggest that the turbidites were deposited on the deeper part of the carbonate shelf and basin. Locally they have accumulated up to the edge of the shallow part of the shelf. Two areas of turbidite deposition are currently preserved, on either side of the Kubor Anticline. It is likely that Miocene turbidites were also deposited to the north of the anticline, but this area has subsequently been displaced by the Bismarck wrench fault.

In the western part of Papua New Guinea, the shelf system has been truncated by the Lagaip wrench fault and the deeper water facies sediments are not now present.

The restored map for the Miocene shelf edge shows an irregular pattern, with a prominent northeasterly protruding ridge which corresponds to the Kubor anticlinal high (Fig. 11, B).

Because of the lack of facies data, it is not possible to restore the Australian shelf edge for pre-Miocene times. However, the observation that the Kubor Anticline was a high throughout the Mesozoic must imply that the north easterly trending ridge has been persistent over a long period of time. The Mesozoic shelf edge is likely to have been similarly irregular in plan, and shallow water sediments may have been deposited farther to the north than is conventionally believed.

IMPLICATIONS FOR HYDROCARBON EXPLORATION

In the Papuan thrust belt the main target has been identified as a Jurassic-Cretaceous clastic reservoir, the Toro Sandstone, involved in large anticlinal fold traps. The main exploration risk lies in identifying those structures which are detached above the reservoir level.

The deformed state sections may be used directly to identify regions where structures are likely to persist to Jurassic-Cretaceous levels. The integrity of structures in the Lavani passive Mesozoic duplex (Fig. 10) has been proven by wells at Juha and Mananda, where discoveries are being evaluated. However, structures in the adjacent listric thrust belt are likely to be detached at shallow levels and of little interest for this play. The precise location of the major lateral ramp separating the two provinces is of considerable economic importance, as it separates low and high risk acreage, from the point of view of trap development.

The deformed state sections may also be used to identify the regions where source rocks are likely to be mature. Fine grained Mesozoic clastics have been identified as the most likely source of the numerous seeps in the foothills, and RRA/FDB (1984) have argued that the distal Jurassic sediments (the Maril Shale) are an oil prone source. These sediments are certainly mature in the deeper parts of thrust slices, in particular within the hinterland Mesozoic duplex. Individual structures may be sourced from deeper levels in the same thrust slice.

CONCLUSIONS

Despite the provisional nature of geological mapping in Papua New Guinea, there is sufficient evidence to suggest that the structure should be interpreted in terms of thin skin tectonics. Five regional cross sections reveal the types of structure commonly encountered in thin skin belts: listric thrust slices, duplex and passive duplex systems, ramp anticlines, and inversion structures.

Conventional sections through the Papuan thrust belt imply a limited horizontal shortening, which is unusual in foothills regions. By comparison, in this paper the displacement of the Kubor Anticline is estimated at 75 km towards the foreland. If this shortening was accumulated during the Pliocene, a period of 3 Ma, this corresponds to an average displacement of 2.5 cm/y which is comparable with the westerly movement of the Pacific Plate at 1.5 cm/y.

The construction of regional cross sections such as these has important implications for hydrocarbon exploration. Despite 60 years of activity, exploration is still at an early stage in Papua New Guinea, and a major risk is perceived to be whether a surface structure is detached below reservoir. The deformed state sections may be used directly to identify high and low risk areas for any particular reservoir involvement, and to locate regions where a particular source rock is likely to be mature. In areas where there is a limited data base, such as around the Hides Anticline in PPL 27, a series of plausible models can be erected, and a suitable scheme to obtain the relevant data to distinguish them may be devised.

Ideally, the restored sections may be used to map the original location of facies belts and hence to identify favourable regions for source and reservoir sediment deposition. With the present state of knowledge it is not practical to rigorously restore the older parts of the succession in the Papuan thrust belt. However, there are now active exploration licences throughout the foothills, and the consequent expansion in stratigraphic knowledge should lead to the practicality of such reconstructions in the near future.

The use of rigorously restorable structural cross sections has considerable applications for regions of hydrocarbon exploration other than thrust belts. In both extensional basins offshore and also in inversion basins onshore, fault profiles may be precisely defined from a knowledge of the geometry of strata adjacent to the fault. Conversely, if the geometry of a fault is known, the attitude of strata may be computed for any given throw. Both approaches offer powerful techniques for testing the validity of interpreted seismic sections. A rapid appraisal of published cross sections through various onshore Australian basins shows that many are geometrically impossible constructions. Structural modelling techniques will allow geometrically admissible sections to be constructed, which can then be tested by specific geological and geophysical tools.

ACKNOWLEDGMENTS

The author would like to thank Dr P. J. Hill and Dr R. C. Williams for their editorial comments and the managements of the British Petroleum Company p.l.c. and the Australasian Petroleum Company Pty Ltd for permission to publish this paper.

REFERENCES

BAIN, J.H.C., MacKENZIE, D.E., & RYBURN, R.J., 1975 — Geology of the Kubor Anticline, central highlands of Papua New Giunea. *Bureau of Mineral Resources, Australia, Bulletin* 155.

BOYER, S.E., & ELLIOTT, D., 1982 — Thrust systems. *Bulletin of the American Association of Petroleum Geologists*, 66, 1196-1230.

CHARLESWORTH, H.A.K., & GAGNON, L.G., 1985 — Intercutaneous wedges, the triangle zone and structural thickening of the Mynheer coal seam at Coal Valley in the Rocky Mountain Foothills of Central Alberta. *Bulletin of Canadian Petroleum Geology*, 33, 22-30.

FINDLAY, A.L., 1974 — The structure of the Foothills south of the Kubor Range, Papua New Guinea. *APEA Journal*, 14, 14-20.

JENKINS, D.A.L., 1974 — Detachment tectonics in western Papua New Guinea. *Bulletin of the Geological Society of America*, 85, 533-48.

RICKWOOD, F.K., 1968 — The geology of Western Papua. *APEA Journal*, 8, 51-61.

ROBERTSON RESEARCH AUSTRALIA LTD & FLOWER DOERY BUCHAN LTD, 1984 — Petroleum potential of the Papuan basin, Papua New Guinea. (Consultants' report prepared for the Papua New Guinea Government.) Geological Survey of Papua New Guinea.

ST JOHN, V.P., 1977 — Detachment and basement tectonics in western Papua New Guinea and the shear-compressive development of the orogeny. *In: Carey Symposium — Orthodoxy and creativity at the frontiers of earth science*. University of Tasmania.

VANN, I.R., GRAHAM, R.H., & HAYWARD, A.B., 1984 — The structure of mountain fronts. (In press)

Reproduced by permission of the Geological Society from The nature and significance of large 'blind' thrusts within the northern Rocky Mountains of Canada by R. I. Thompson, in *Thrust and Nappe Tectonics*, Geological Society Special Publication 9, 1981, p. 449-462.

The nature and significance of large 'blind' thrusts within the northern Rocky Mountains of Canada

R. I. Thompson

SUMMARY: The northern Canadian Rocky Mountains comprise a rugged, structurally complex Foothills subprovince of large amplitude box and chevron folds, and a structurally diverse Rocky Mountain subprovince in which large mappable thrusts are rare. The boundary between them is, in some regions, defined by the unfaulted E-dipping limbs of an *en echelon* sequence of large mountain-front anticlines. The lack of thrusts, especially along the mountain front, contrasts with the well exposed linearly continuous thrusts of the Front Ranges structural subprovince within the southern Rocky Mountains, and leads to the impression that little lateral displacement has occurred.

Where deep cross-cutting valley erosion combines with increased fold plunge, it is apparent that the frontal anticlines are, in reality, large allochthonous sheets displaced many kilometres eastward relative to the craton on flat thrusts that separate Ordovician shales from underlying Devonian and Mississippian shales. The faults can be traced, in some places, eastward to the mountain front where they cut abruptly through the thick hanging wall successions of carbonate rocks; however, they cannot be mapped further eastward into surface exposures because they terminate within a décollement zone of Devonian and Mississippian shales, where the displacement on them is transformed into disharmonic folds and tectonic thickening of overlying units.

The Devonian and Mississippian shale succession is interpreted as a fundamental décollement zone of regional extent that separates a lower structural level of thrust-faulted carbonate rocks from an upper structural level characterized by folded late Palaeozoic and Mesozoic units. The shortening represented by Foothills folds is interpreted to equal the amount of shortening on 'blind' thrusts beneath the western margin of the Foothills structural subprovince.

A structural reinterpretation across the Muskwa Anticlinorium using the blind thrust interpretation demonstrates that the mountain-front Tuchodi Anticline may represent a large allochthonous thrust sheet folded over a large step in the blind thrust on which it was transported.

The northern Rocky Moutains, narrower and less foreshortened than the southern Rocky Mountains, are interpreted as a thin-skinned tectonic regime similar to but orogenically less mature than the southern Rocky Mountains.

The significant on-strike changes in structural style that affect the Canadian Rocky Mountain thrust and fold belt as it is traced northward beyond the well documented Waterton and Bow Valley regions (e.g. Bally *et al.* 1966; Price & Mountjoy 1970; Dahlstrom 1970) reflect a concomitant and gradual change in the stratigraphic character of the rocks that were deformed (Thompson 1979). The pattern of imbricate thrusts that typifies the structural fabric within the Foothills and Front Ranges in the S is less well developed northward where it is progressively replaced by a compensating pattern of folds and fold complexes in which thrusts are a subordinate tectonic element. Although thrusts and folds are never mutually exclusive, one may traverse some parts of the northern Rocky Mountains and not encounter a single thrust of consequence.

Without the well documented model of thin-skinned detachment tectonics so successfully applied to interpretation of the southern Rockies, and some appreciation for the gradual nature of stratigraphic and structural changes that occur northward along strike, one might be tempted to interpret the northern Rockies as a 'thick-skinned' terrain where vertical basement uplifts controlled the evolution of surface structures. The presence of broad mountain-front anticlines without visible thrusts along their eastern margins, a general lack of well defined imbricate thrust sheets involving large stratigraphic overlap, structural disharmony between major stratigraphic units, and an apparent minor amount of supracrustal shortening, constitute some of the reasons that influenced others (e.g. Taylor 1972; Taylor & Stott 1973) to speculate that the northern Rocky Mountains may differ in a fundamental way from its southern counterpart.

Much of the data and many of the ideas expressed here are taken from an earlier publication (Thompson 1979) that presented a thin-skinned detachment model of structural evolution for part of the northern Canadian Rocky Mountains. The theme of this paper remains unchanged, but its scope has been extended to include a structural reinterpretation across the Muskwa Anticlinorium (Figs 1 & 2), a large southward plunging anticlinorial closure cored by rocks of Middle Proterozoic age that is geologically similar to the Purcell Anticlinorium in the S (see Price, this volume).

FIG. 1. Location of study area in relation to the major tectonic subdivisions of the Canadian Cordillera.

Arguments will be presented that suggest the northern Rocky Mountains are tectonically less mature than the southern Rockies and as such represent a 'preview' of Canadian Rocky Mountain structural evolution prior to development of the spectacular thrust-faulted Front Ranges structural subprovince that now characterizes the eastern portion of the southern Rocky Mountains.

Geological framework

The northern Rocky Mountains consist of three major tectono-stratigraphic assemblages (Fig. 2). On the E, mechanically incompetent clastic rocks of Upper Palaeozoic and Mesozoic age dominate the complexly folded Foothills structural subprovince. Adjacent to the Foothills, a second assemblage of thrust faulted and folded Middle Proterozoic (Helikian) and Palaeozoic rocks is dominated by thick carbonate successions and forms the large southward plunging Muskwa Anticlinorium (Taylor & Stott 1973); southeastward, beyond the plunge-out if the anticlinorium, Palaeozoic carbonates form a narrow spine of topographically prominent peaks that defines the eastern edge of the Rocky Mountains structural subprovince. On the W is the third major assemblage comprising more penetratively deformed upper Proterozoic (Hadrynian) and Palaeozoic clastic facies (Taylor *et al.* 1979; Cecile & Norford 1979; Thompson 1976) in which continuity and scale of individual structures are less developed, and the structural styles more diverse.

It is usual to describe the Canadian Rockies in terms of structural subprovinces (North & Henderson 1954), each of which is characterized by distinctive topographic, stratigraphic and structural features separated by major thrusts or thrust systems. In the southern Rockies there are four: Foothills, Front Ranges, Main Ranges and Western Ranges, but as these subprovinces are traced northward to the latitude of the Peace River (56°N) the lateral limits and distinctive features of each become increasingly difficult to distinguish, and only two basic subdivisions remain: the Foothills structural subprovince on the E and the undivided Rocky Mountains structural subprovince on the W (Fig. 2); a small thrust may separate them or, alternatively, the Foothills subprovince may merge laterally with the Rocky Mountains subprovince across a topographically low, densely vegetated valley with no obvious evidence of a structural break between them (Fig. 2). This is in marked contrast with the southern Rockies where large thrusts such as the Lewis and McConnell, with displacements across them measured in tens of kilometres, place massive cliffs of Proterozoic or Palaeozoic carbonate rocks against topographically low Foothills terrain composed of Cretaceous shales.

Unlike the southern Foothills, the northern Foothills subprovince is a mountainous fold terrain comprising large-amplitude box and chevron folds that expose strata mainly of Mesozoic age (Fitzgerald 1968), although large folds of Mississippian through Permian strata occur along the western margin (Fig. 3). Some thrusts are present and are normally associated with tightly folded anticlinal complexes. Similarily, the undivided Rocky Mountains subprovince is in marked contrast with the Front and Main Ranges of the S. Rapid changes in structural style, both along and across the strike, disrupt the topographic and structural grain.

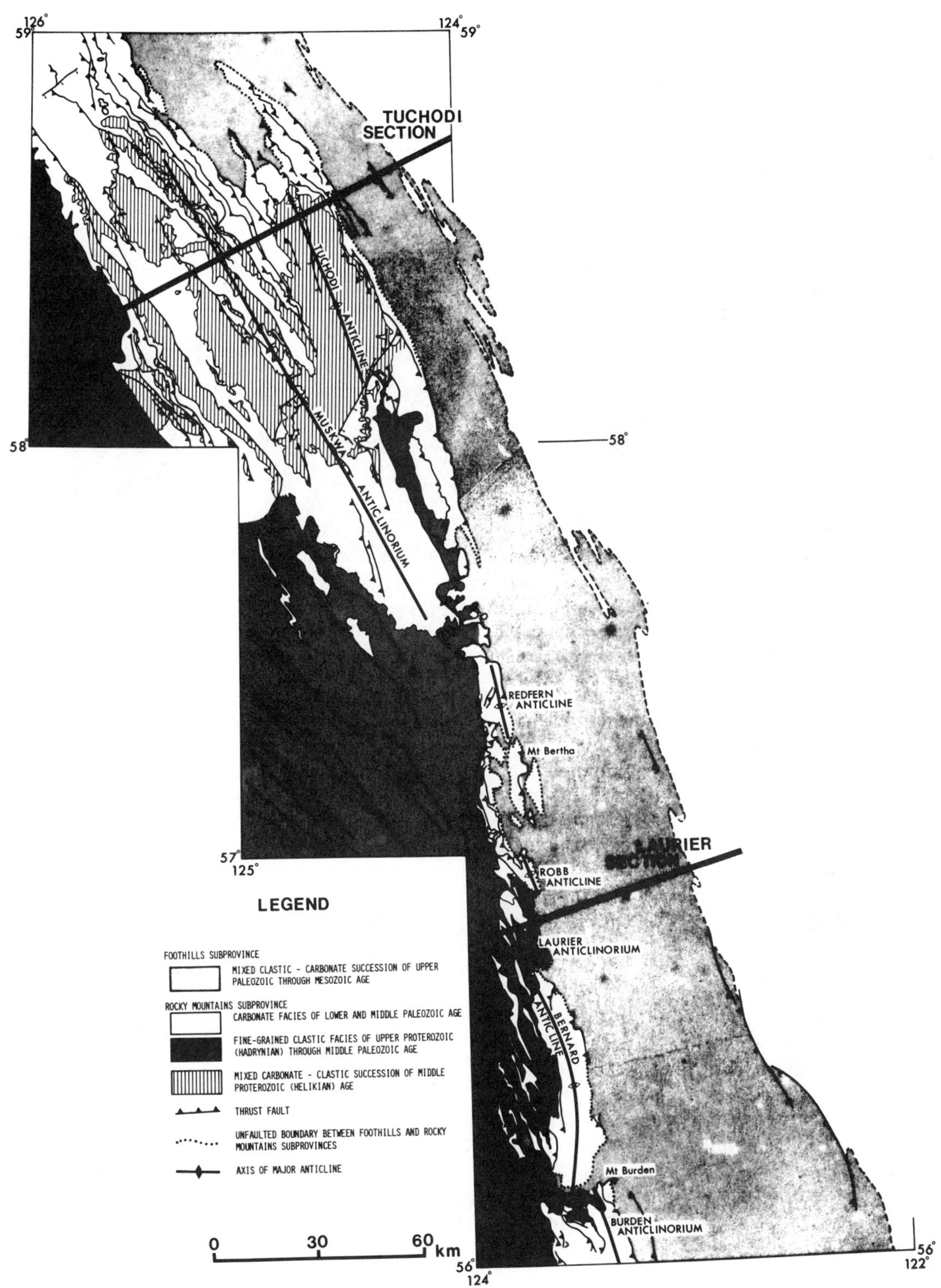

FIG. 2. Generalized tectonic-stratigraphic assemblage map showing the distribution of major thrust faults, the location of major mountain front anticlines, the Muskwa Anticlinorium, and the location of the Laurier and Tuchodi structure cross-sections.

FIG. 3. The eastern half of a large box fold of Mississippian carbonate rocks (as viewed from the NW) located along the western margin of the Foothills structural subprovince. White line highlights layering.

Two features are especially noticeable: (1) there are fewer thrusts, and those present lack the lateral continuity typical of major thrusts within the southern Rocky Mountains, and (2) the mountain front may be defined by the steep E-dipping limb of a large anticline that exposes Lower and Middle Palaeozoic carbonate rocks but with no obvious evidence of lateral transport of the anticline on a thrust (Figs 4 & 5).

Except for the folds within the Foothills, there is little evidence of significant lateral telescoping of the sedimentary prism—certainly nothing to suggest the 200 km or more of shortening calculated for the southern Rocky Mountains (Bally *et al.* 1966; Price & Mountjoy 1970). It will be argued that the lack of major thrusts is more apparent than real and that the large anticlinal structures within the northern Rocky Mountains have been displaced relatively eastward as much as 10 or more km on flat thrusts—termed blind thrusts here—that do not project into surface exposures.

The on-strike changes in structural style from S to N are closely linked to changes in the stratigraphic character of the miogeoclinal rock prism. Comparison of the cross sections in Fig. 11 illustrates the increased presence, in the N, of thick shale tongues, along with greater proximity of major carbonate to shale facies transitions to the eastern margin of the Rocky Mountain subprovince. Two of the incompetent units illustrated in Fig. 11 have played an important role in subsequent structural evolution: the unit of Cambro-Ordovician age (u€-lO) comprising thin bedded limestone, siltstone and shale is an important bedding glide zone into which thrusts flatten with depth; the Devonian–Mississippian shale succession ($D–M_{sh}$) is a stratigraphically higher décollement zone of regional extent into which thrusts may merge upward. A further complication is the oblique trend of major facies transitions to the structural strike, with the result that a thrust faulted carbonate succession may pass quickly along strike into a complex of stacked recumbent folds cut by subsidiary thrusts.

Blind thrusts

Blind thrusts played a critical role in the structural development of the northern Rocky Mountains, and are reponsible for some of the important differences in structural style between the northern and southern Rockies. Unlike thrusts which splay upward from a basal detachment zone to intersect the topographic surface, blind thrusts merge and flatten at some point in their upward trajectory into an 'upper' detachment zone(s). Consequently,

Fig. 4. View looking S along the eastern margin of the Bernard Anticline showing the unfaulted E-dipping fold limb which defines the eastern limit of the Rocky Mountain structural subprovince. Silurian and Devonian platform carbonate rocks (S–D) are overlain with apparent conformity by Devonian and Mississippian shales (M–D) of the Besa River Formation.

Fig. 5. View looking S along the eastern margin of the Laurier Anticlinorium showing the overturned limb of a near recumbent mountain front anticline consisting of Silurian (S) and Devonian (D) clastic facies rocks that are more distal clastic facies of the Silurian and Devonian platform carbonate rocks shown in Fig. 4. The white lines outline stratigraphic boundaries. The Laurier cross-section (Fig.9) passes through this structure.

they are rarely observed at surface and can influence the character of the surface geology to the point where one may traverse a low-angle thrust faulted terrain and not know it!

Geometric considerations require that the displacement on a blind thrust be transferred into the overlying (hanging-wall) beds through formation of a complex of disharmonic structures, otherwise balanced shortening from one stratigraphic level to another would not be possible (Dahlstrom 1969). This is illustrated in Fig. 6, a simplified diagrammatic representation of the blind thrust model.

As displacement on the blind thrust is initiated, a second detachment within the overriding sheet forms, and permits incompetent strata to behave independently and disharmonically with respect to the underlying rigid (hanging-wall) carbonates (Fig. 6a). If the amount of shortening caused by disharmonic folding keeps pace with the amount of displacement on the blind thrust, it will not extend laterally beyond the limits of the fold complex. If this balance is not maintained, then a splay from the blind thrust will develop, and break to surface as a kind of 'release valve' that accommodates the shortening imbalance. As displacement continues, the lateral limits of disharmonic folding are extended to keep pace with fault displacement (Fig. 6b).

By superimposing a hypothetical topographic surface onto Fig. 6c, it becomes apparent how misleading the surface geology may be. There is no structural break between the large surface anticline on the left and the fold complex on the right, and no way of determining that the large anticline is actually an allochthonous thrust sheet that is folded over a fault ramp or step at depth. Based on the surface evidence, one might argue in favour of vertical (basement) uplift as the agent responsible for the large anticline and that the fold complex adjacent to it is a consequence of gravitational sliding off the uplifted carbonate terrain.

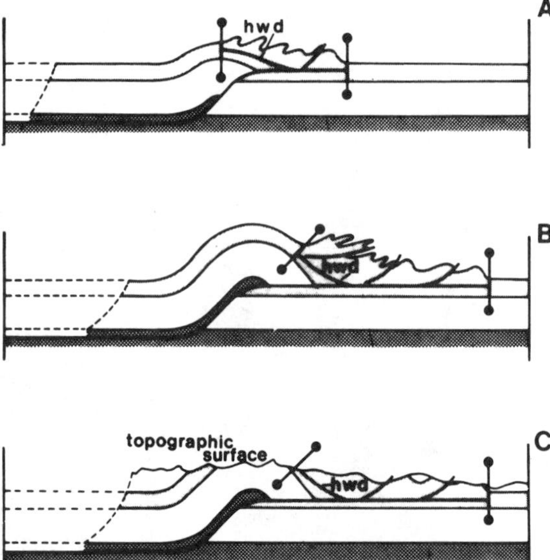

FIG. 6. A diagrammatic representation of the blind thrust model. The patterns define mechanically incompetent strata separated by a rigid carbonate unit with no pattern. (a) Illustrates the onset of displacement across the thrust accompanied by development of a hanging-wall detachment(s) (hwd) which allows the incompetent strata within the hanging-wall plate to deform disharmonically and absorb displacement on the underlying thrust. The thrust ceases to exist at the point where shortening due to folding in the hanging-wall equals displacement on the thrust—hence the pin on the right side of the section. In (b) continued thrust fault displacement increases the width of disharmonically deformed hanging-wall succession. (c) Illustrates the difficulty in deciphering the detached nature of the mountain front anticline using surface exposures. The major thrust remains 'blind', and much of the shortening within the disharmonically deformed incompetent unit may be difficult to assess unless good stratigraphic markers are present.

The field evidence in favour of the blind thrust model is present in those areas where the plunging portion of a mountain-front anticline is cut across-trend by a deep eroded valley in which lower structural levels are exposed. At Mount Burden, illustrated in Figs 7a and b, a flat thrust fault can be traced across the strike for 10 km. Over most of this length, the fault separates relatively flat lying Upper Cambrian–Lower Ordovician shales from underlying Devonian and Mississippian Besa River Shales; at its eastern end the complete Lower and Middle Palaeozoic carbonate succession in the hanging-wall dips eastward and is cut-off abruptly against the fault which then has Besa River shales in both the hanging-wall and footwall. Rather than continuing to cut up section, the fault is deflected into the Besa River Shale where it remains concealed from view along the strike of the mountain front. At the Mount Burden locality, some of these shales (and thin carbonates) are preserved within the frontal portion of the thrust plate, where the upper surface of the rigid (hanging-wall) carbonate succession is essentially undeformed and east-dipping whereas the overlying

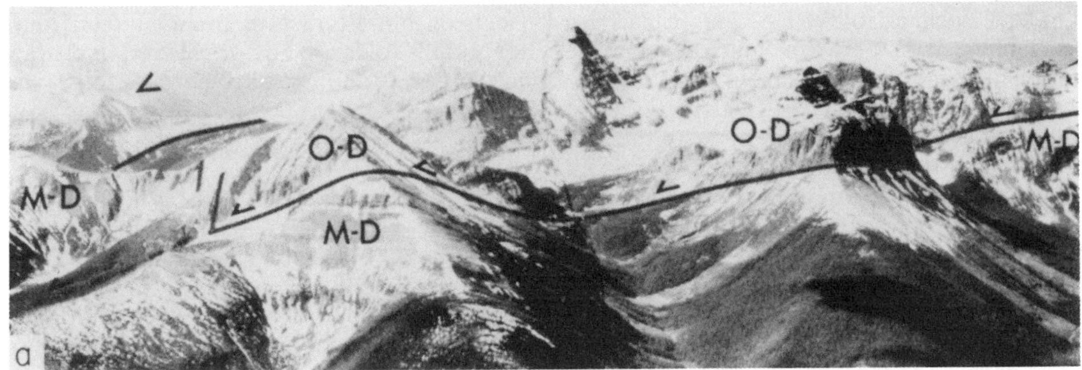

FIG. 7. (a) View toward the SW (oblique to the NW–SE structural grain) of the allochthonous Mt. Burden Anticlinorium. The thrust fault can be traced from W (right) to E (left) over 10 km but cannot be traced eastward beyond the hanging-wall cut-off of the Ordovician through Devonian carbonate succession (O—D) or along the trend of the mountain front. Below the fault trace are shales of Devonian and Mississippian age (M–D). Arrow head points to area shown in detail in Fig. 7b. The dark lines drawn above the fault trace approximate the stratigraphic boundary between Devonian platform carbonate rocks on the right and Devonian through Mississippian shales and argillaceous limestones on the left. (b) Detailed view toward the S, of structural disharmony within the toe region of the Mt. Burden thrust plate. The dark line adjacent to D represents the planar upper stratigraphic boundary of Middle Devonian dolostones; the black line within M–D outlines a succession of stacked recumbent folds within limestone and shale of the Devonian–Mississippian shale succession. This incompetent structural unit within the hanging-wall of the Mt. Burden thrust plate has deformed and shortened disharmonically with respect to the underlying more rigid dolostones (D).

shale succession forms a series of stacked recumbent isoclinal folds (Fig. 7b). Clearly, a detachment exists between the shales and the carbonates, and this detachment has served to allow displacement on the underlying blind thrust to be compensated, in part, by the fold complex above the hanging-wall detachment.

Further evidence for the existence of large blind thrusts is present along the northern limit of the Robb Anticline: the thrust over which it was displaced is clearly exposed along its northern limit as illustrated in Fig. 8a, yet the fault cannot be mapped southward along the strike of the Anticline, as illustrated in Fig. 8b.

Knowing the geometric characteristics of blind thrusts provides a more complete conceptual framework with which to interpret the surface and subsurface geology of the northern Rockies. Large surface anticlines of Lower and Middle Palaeozoic carbonate rocks such as the Burden and Redfern structures (Fig. 2) can be interpreted as allochthonous thrust sheets that have been displaced eastward several kilometres at least. The fold and fault complexes at higher structural and stratigraphic levels can be interpreted as large strain absorbers that compensate for the displacements on blind thrusts at depth; and the amount of bed-length shortening within the fold complex at surface should equal the amount of displacement across blind thrusts at depth—an important consideration when preparing balanced structural cross-sections using surface data (Dahlstrom 1969, 1970; Douglas 1950, 1958).

The generalizations outlined above can now be used to prepare structural interpretations of deeper features. To follow, are descriptions of two regional structure cross-sections (Fig. 9); one drawn across the Laurier Anticlinorium (termed the Laurier cross-section) where blind thrusts can be documented from surface geological relationships, and another drawn across the Tuchodi Anticline (termed the Tuchodi cross-section), a much larger structure that has the surface appearance of a broad upwarp. A third section drawn across the Bernard Anticline is described in Thompson (1979).

Two representative structure cross-sections

Laurier section

The Laurier section (Fig. 9) illustrates fold styles typical of the Foothills subprovince, the structural disharmony that exists across the Devonian and Mississippian Besa River Shale detachment zone, and the stacking of thrust sheets on blind thrusts at the mountain front.

Foothills folds are box or chevron in character, and conform in most respects to fold styles described by Faill (1969) from the Appalachian Plateau Province, and Laubscher (1977) from the Jura. This fold style is especially well developed in Triassic and younger strata but also typifies folds within the thick Mississippian carbonate succession (Fig. 3) exposed along the W margin of the Foothills subprovince. Some workers have interpreted such fold styles as representing the product of giant intersecting kink bands (e.g. Faill 1969). This approach is useful in a purely geometric sense to facilitate the balancing of bed length shortening within a single fold because no space problem is created by a decreasing radius of curvature, as occurs within concentric folds.

Deformation beneath the Foothills subprovince 'bottoms out' within the Besa River Shale (D–M_{sh}). The underlying Middle Devonian carbonate succession is represented as an essentially planar surface for two reasons: first, beds at the bottoms of large synclines are near the regional stratigraphic dip projected westward into the line of section from the Plains region on the E, and second, the Headstone Creek well, located on a large Foothills Anticline (Fig. 9), failed to intersect Middle Devonian strata despite having been terminated substantially below the point where Middle Devonian rocks should have been intersected had they been intimately involved in the folding. This assumption of an undeformed Middle Palaeozoic 'basement' beneath the Foothills should not be accepted too literally (Thompson 1979); it is probably cut by thrusts and is likely to be broadly folded, but the geometry and space constraints imposed by the overlying folded successions dictate that the scale and intensity of deformation is substantially less than that within the overlying rocks.

Middle Devonian and older rocks must become structurally involved in a major way beneath the W margin of the Foothills subprovince because synclines at surface are raised well above the regional stratigraphic gradient, with the result that a large 'hole' is generated between the 'undeformed' Middle Devonian carbonates projected westward and Mississippian carbonate rocks at surface. Additional evidence that a significant thrust repetition is present within this interval can be observed 30 km northward along strike at Mount Bertha (Fig. 2), where a slice of Middle Palaeozoic carbonate rocks has overridden Besa River

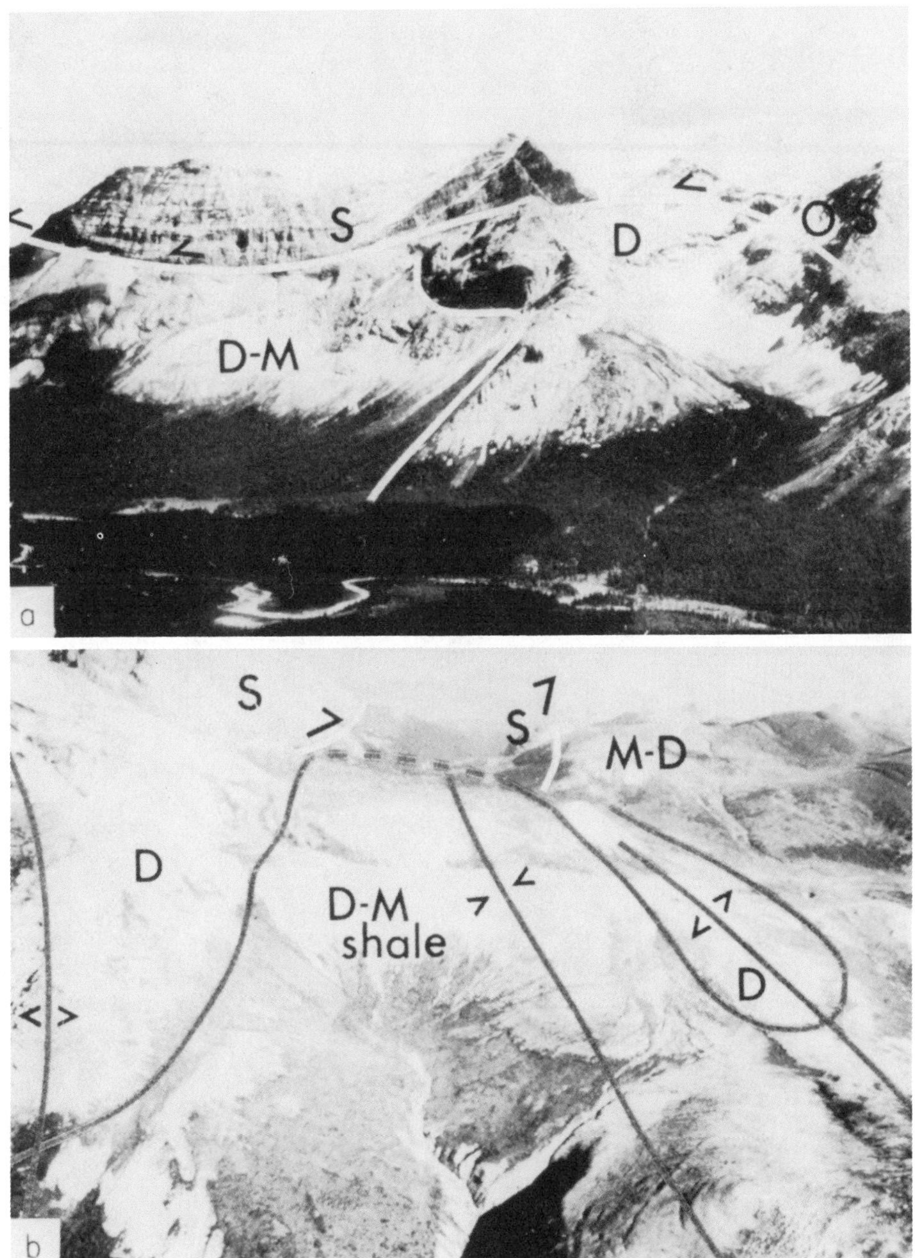

FIG. 8. (a) View toward the S of the up plunge limit of the Robb Anticline showing a thrust that places Ordovician and Silurian carbonate rocks (O, S) onto folded Middle Devonian dolostone (D) and Devonian–Mississippian shale (D–M); minimum lateral displacement is 5 km. (b) View from the S along the eastern flank of the Robb Anticline showing the abrupt termination of the thrust present in Fig. 8a 2 km S of that location. In the foreground are folds of Devonian dolostone (D) and Devonian–Mississippian shale (D–M) which appear unaffected by the major thrust immediately to the N; in the left foreground is the E dipping limb of the Robb Anticline which defines the mountain front. Without the more northerly up-plunge exposures, there would be little reason to suspect the allochthonous nature of the Robb Anticline.

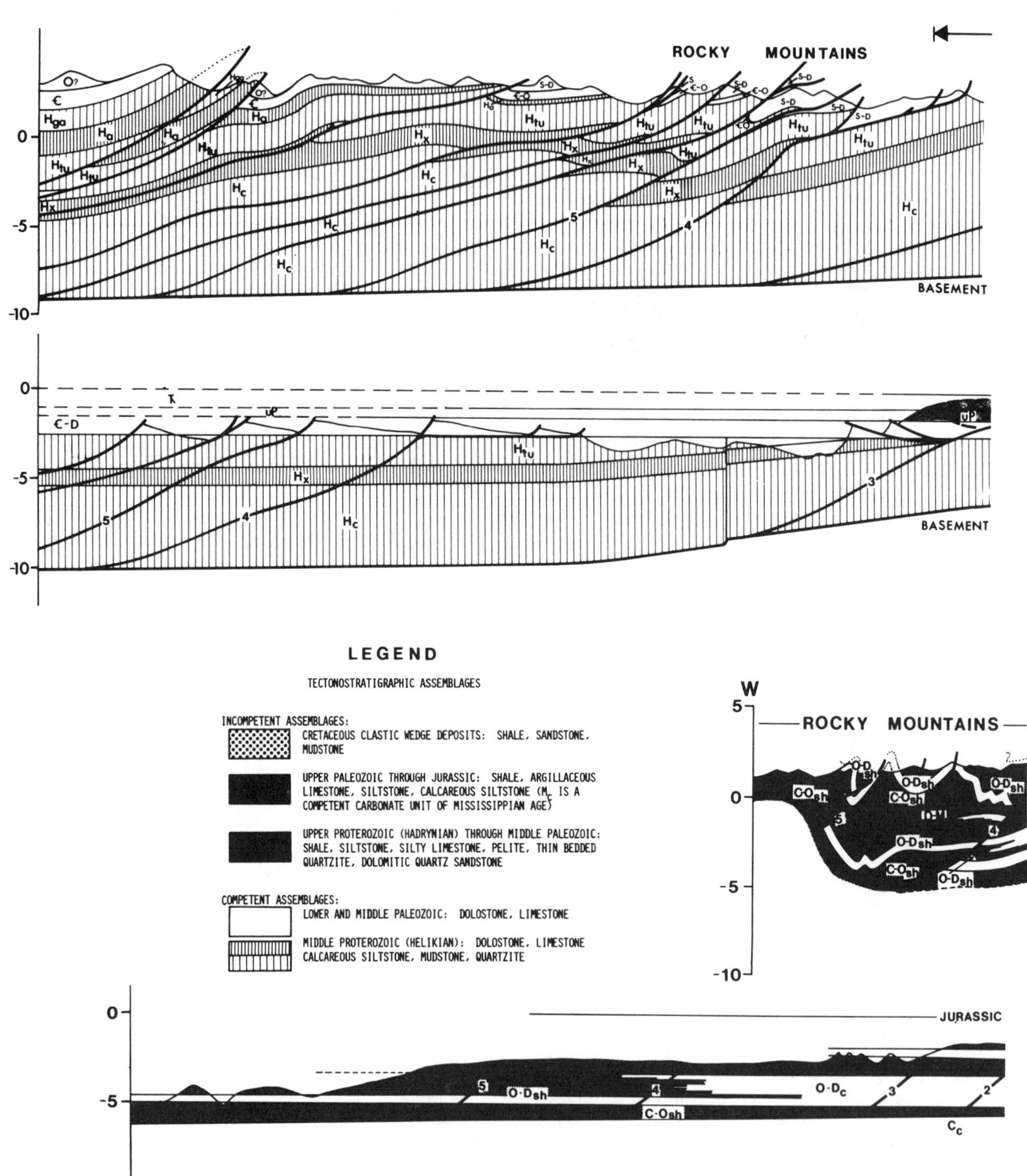

Fig. 9. The Tuchodi and Laurier structure cross-sections (see Fig. 2 for locations). Symbols on the Tuchodi section refer to the following stratigraphic units: H_c—Helikian Chischa Fm; H_x—Helikian Tetsa, George and Henry Creek Fms.; H_{tu}—Helikian Tuchodi Fm.; H_a—Helikian Aidia Fm.; H_{ga}—Helikian Gataga Fm.; C–O—Cambro-Ordovician Kechika Group; S–D—Silurian and Devonian Nonda, Muncho McConnell, Wokkpash, Stone and Dunedin Formations; uP–T—Upper Palaeozoic and Triassic Besa River, Kindle, Fantasque, Toad-Grayling, Liard and Luddington Formations; K–lower Cretaceous Fort St. John Group and Upper Cretaceous Dunvegan and Kotaneelee Formations (see Taylor 1973 for stratigraphic descriptions). Symbols on the Laurier

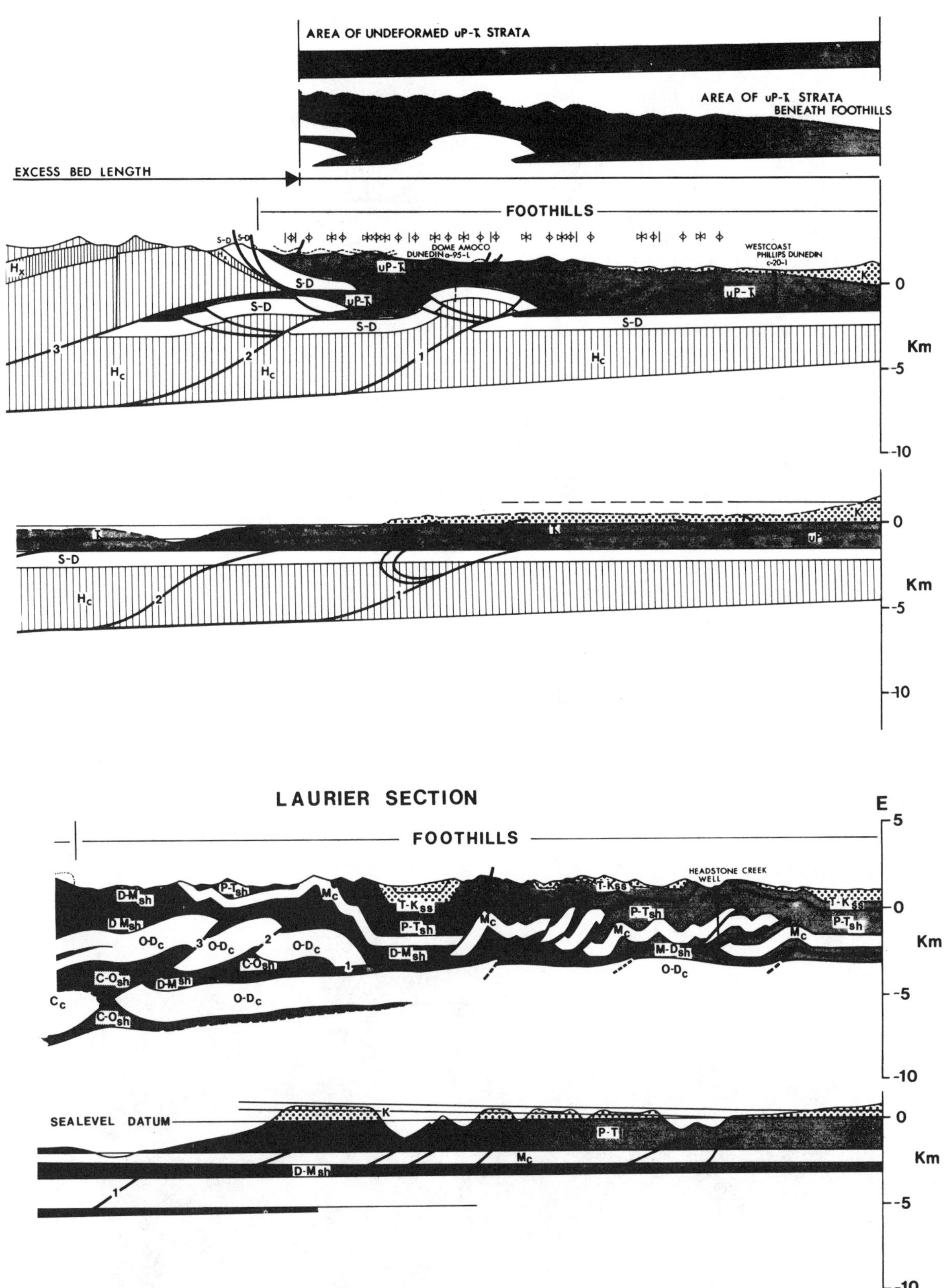

section refer to the following stratigraphic units: C_c—Cambrian carbonate unit; $C–O_{sh}$—Cambro-Ordovician Kechika Group; $O–D_{sh}$—Ordovician through Devonian unnamed clastic facies; $O–D_{sh}$—Ordovician through Devonian carbonate facies; $O–D_c$—Ordovician through Devonian carbonate facies; $D–M_{sh}$—Devonian and Mississippian Besa River Formation; M_c—Mississippian Prophet Formation; $P–T_{sh}$—Permian through Triassic Stoddart Group and Toad-Grayling Formation; $T–K_{ss}$—Triassic through lower Cretaceous Liard, Charlie Lake, Baldonnel, Pardonet Formations and Minnes, Bullhead and Fort St. John Groups (see Thompson 1979, 1976; Irish 1970 for stratigraphic descriptions).

Shales. The thrust cannot be traced laterally for any distance and it is interpreted here as a splay from a blind thrust.

The large volume of shale shown immediately in front of the Laurier Anticlinorium represents tectonically-thickened Besa River shale that has absorbed much of the displacement on the blind thrust over which the anticlinorium was displaced. An example of the style and disharmony of deformation within this interval of incompetent strata is illustrated in Fig. 10 (see also Fitzgerald 1965).

The Laurier Anticlinorium (Fig. 5) comprises a fold complex made up of fine-grained clastic facies that are equivalent in age to the Silurian and Devonian platform carbonate succession mapped to the N and S (Thompson 1976). The anticlinorium is thrust over the time-equivalent carbonate facies exposed within the southward plunging Robb Anticline necessitating that the major carbonate-to-shale facies transition of Silurian and Devonian age must occur within the thrust sheet that forms the Robb Anticline. It is speculated here that the position and geometry of the facies transition exerted an influence on the location of the footwall ramp over which the Laurier Anticlinorium was displaced and folded.

Structure W of the Laurier Anticlinorium consists of tightly folded, faulted and cleaved Upper Cambrian and Lower Ordovician shales and siltstone (€–O). This incompetent Cambro-Ordovician succession comprises an important detachment zone of regional extent. Like the Besa River Shale, it is a zone of structural disharmony between thrust faulted and folded Cambrian rocks below and thrust faulted middle Ordovician through Devonian carbonate rocks above.

Tuchodi section

The purpose of attempting a structural interpretation at the latitude of the Tuchodi section (Fig. 9) is to test the applicability of the blind thrust model as an alternative to the 'thick-skinned' interpretation of Taylor & Stott (1973). In their cross-sections, Taylor and Stott (op. cit.) show thrusts at surface which flatten down-dip into a regional detachment zone at the top of the Proterozoic sedimentary succession. Some of the fault displacement is then transferred onto steep W dipping (70–80°) contraction faults and onto curved convex-up contraction faults that presumably continue to steepen with depth to become subvertical basement controlled structures. The remainder of the displacement shown on faults above the regional detachment is unaccounted for within the underlying Proterozoic succession. The mountain front Tuchodi Anticline (Fig 2) is drawn as a simple upwarp, with no evidence of having undergone lateral W–E displacement, and folds within the adjacent

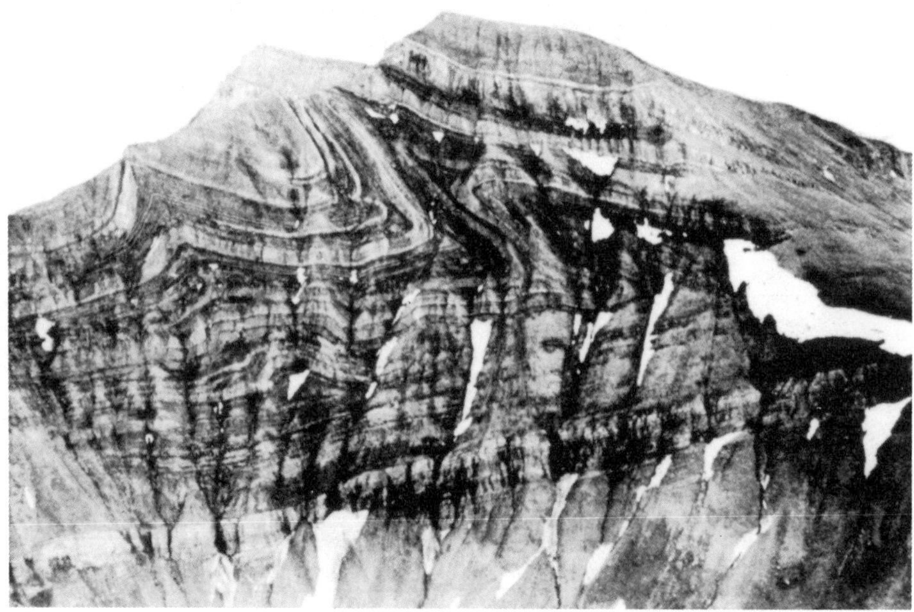

FIG 10. Disharmonic folds within argillaceous limestone and shales of Devonian and Mississippian age located adjacent (E) to the Mt. Burden Anticlinorium.

Foothills subprovince are portrayed as open concentric structures that affect the total Palaeozoic as well as the top of the Proterozoic stratigraphic assemblages. This geometric model is fundamentally different from the blind thrust model presented in the Laurier section (and in Thompson 1979) and I question that both 'thick- and thin-skinned' geometric solutions are compatible or even geologically plausible within the confined strike length distance between the Laurier and Tuchodi sections. Comparisons along strike reveal no changes in regional geological structural style that would indicate a fundamental change in the crustal processes that produced the surface structures.

Three important aspects of the Tuchodi section differ from the Laurier section. First, it was drawn with the built-in constraint of an uninvolved crystalline basement (Hudsonian) surface that dips gently westward; second, the dimensions of the Tuchodi Anticline are considerably larger than those of the Laurier Anticlinorium or Robb Anticline; and third, the mechanical character, age and thicknesses of stratigraphic assemblages that comprise the miogeoclinal prism are changed. The Upper Palaeozoic and Mesozic succession is thinner and less competent and lacks the thick Mississippian carbonate unit which separates the Besa River shale detachment zone below from Permian and Triassic shales above. The Lower and Middle Palaeozoic succession is much thinner and does not contain a thick Lower Ordovician shale detachment unit; and the Middle Proterozoic succession, which may or may not exist at the latitude of the Laurier section, is very thick and competent and has exerted an important influence on the structural style.

The depth to basement at the E limit of the Tuchodi section is estimated to be 5.4 km, and the westward dip of the basement surface is drawn to average 5°. These values are based on preliminary results from seismic reflection surveys in a region SE of the line of section (pers. comm. oil company geologists 1979). The depth and gradient estimates should be regarded as 'best guess' approximations that will undoubtedly be revised as the quality and quantity of seismic information improves.

The top of the Middle Palaeozoic carbonate succession (S–D) was projected into the line of cross-section from the undeformed Plains region to the E, and the regional slope maintained beneath the E portion of the Foothills subprovince to the position where thrust faulted Middle Devonian carbonate rocks are intersected above the regional dip by the Dome-Amoco Dunedin well. The thrust which cuts the Middle Devonian and older strata is interpreted as flattening into the detached and tectonically thickened succession of Upper Palaeozoic and Mesozoic shales.

The Tuchodi Anticline was produced by displacing and folding a thick coherent sheet of Proterozoic rocks over a large step on a major blind thrust. The remainder of the section was drawn by projecting surface faults downward into a major detachment at the top of the crystalline basement surface, making sure that bed length shortening was balanced from one stratigraphic level to another.

The essential surface geological constraints discussed by Taylor & Stott (1973) are satisfied by the cross-section. The Tuchodi Anticline appears as a broad surface fold; there is a detachment at the top of the Proterozoic sedimentary succession, and there is structural disharmony between the Rocky Mountains and Foothills structural subprovinces. In addition, the displacement shown at surface is accounted for at all stratigraphic levels down to basement, and there is no necessity to translate vertical uplift at depth into horizontal shortening at surface.

The Tuchodi section is geometrically possible only if the amount of displacement on the blind thrusts shown beneath the Foothills subprovince is compensated by an equal amount of bed length shortening within the disharmonically folded Upper Palaeozoic and Mesozoic rocks within the Foothills subprovince. A major disparity in values would indicate that one or more of the initial assumptions was incorrect. To check for balanced shortening, the cross-sectional area of deformed Upper Palaeozoic through Triassic (uP–T) strata was measured and compared with that of an undeformed strip of equal length (see upper right of Fig. 9). The difference in areas represents the additional amount of rock added to the cross-section as a result of folding and faulting during deformation. This excess area, if cast in terms of a normal stratigraphic thickness, has a length equal to the excess bed length due to shortening. In the Tuchodi section, it amounts to 21 km of Foothills shortening and agrees very well with the 19 km of displacement interpreted on the blind thrusts. It should be stressed that the computation was made after the section was completed and did not influence the initial assumptions used in any way.

Palinspastic restorations

The Laurier and Tuchodi cross-sections were partially restored to their original undeformed lengths (Fig. 9) by measuring along competent marker units, around folds and between faults. The thrust fault trajectories were plotted and keyed by number to faults in the structure sections to show the relative positions of ramps or steps, prior to eastward translation. The upward limit of preserved strata (i.e. the stretched-out surface topography) is also plotted to illustrate the relative effects of individual structures on the quantity of material removed by erosion.

Total shortening across Tuchodi section is 45 km: 20 km across the Foothills subprovince (including the kinematically-linked blind thrusts at depth), and 25 km on structures W of the Tuchodi anticline.

The Laurier section contains 28 km of shortening across the Foothills subprovince. Of this amount, 13 km occurs in Foothills folds W to the axis of the very large box syncline, and this has been balanced by placing an equal amount of displacement on blind thrust no. 1; the remaining 15 km is a minimum estimate of displacement on the blind thrusts at the mountain front.

Comparison of these net supracrustal shortening values, with values across similar segments of the southern Rockies, demonstrates a decrease in the amount of convergence between S and N. The Foothills subprovince together with the first Front Range Thrust (e.g. McConnell thrust) along the Bow Valley transect of the southern Rocky Mountains, has been shortened by 70 km. If this is extended westward to include the remainder of the Front Ranges, shortening exceeds 100 km (Bally *et al.* 1966; Price & Mountjoy 1970).

Despite the difference in amounts of net convergence, the basic structural style of the northern Rocky Mountains is consistent with that shown in cross-sections of the southern Rockies (e.g. Price, this volume; Price & Mountjoy, 1970; Bally *et al.* 1966; Keating 1966). The top of the crystalline basement surface can be interpreted as an undeformed passive planar element, at least to the western limits of the Laurier and Tuchodi sections. Thrust faults cut up stratigraphically from W to E and place older strata onto younger strata. Fault trajectories follow flat bedding glide zones except where they step or ramp upward from a lower to a higher glide zone, and fault displacements are on the scale of kilometres.

A comparison with the southern Rocky Mountains

The differences in structural style between the northern and southern Rocky Mountains reflect a change in the overall mechanical character of the layered rock sequence. In the N there is a greater proportion of incompetent, relative to competent, rock units that provided greater potential for the formation of décollement zones as well as structural disharmony between different stratigraphic levels. However, this explanation is incomplete because it fails to account for the reduction in supracrustal shortening from S to N, the narrower width of the northern Rockies, and the smaller volume of late Cretaceous and early Tertiary foredeep deposits adjacent to it. Changes northward in the lithostratigraphic character of the miogeocline should not have affected the width of the deformed belt or the amount of supracrustal shortening contained within it—only the surface or geometric expression of that shortening.

Comparison of the northern with the southern Rocky Mountains requires first that each belt be placed within a common frame of reference. In Fig. 11 (reproduced from Thompson 1979) the positions of individual structural subprovinces are plotted onto a restored cross-section of the miogeocline for both the northern and southern Rocky Mountains. Each cross-section is then compared graphically using the hinge lines as a common reference datum.

Three interdependent observations are apparent: (1) Deformation in the N did not progress as far craton-ward as in the S. A 20 km width (at least) of cratonic and clastic wedge deposits is deformed in the southern cross-section but remains undeformed in the northern cross-section: (2) the northern Rocky Mountains structural subprovince occupies a cross-section stratigraphic position within the miogeocline equivalent to the Main Ranges (and Western Ranges, not shown in Fig. 11) structural subprovince in the S; and (3) in the N there is no analogue for the closely spaced imbricated sheets of cratonic Palaeozoic carbonate rocks that characterize the southern Rockies Front Ranges structural subprovince because platform carbonate rocks occupying an equivalent cross-section position in the N remain essentially undeformed.

The E limit of deformation in each cross-section (Fig. 11) may be regarded as the preserved external limit of a strain front that passed progressively from W to E through the miogeocline. The narrower width and decreased amount of foreshortening in the N can be related directly to the observation that a narrower width of the sedimentary prism was deformed. If strain had persisted further eastward in the N, the platform carbonate succession that currently forms a basement to

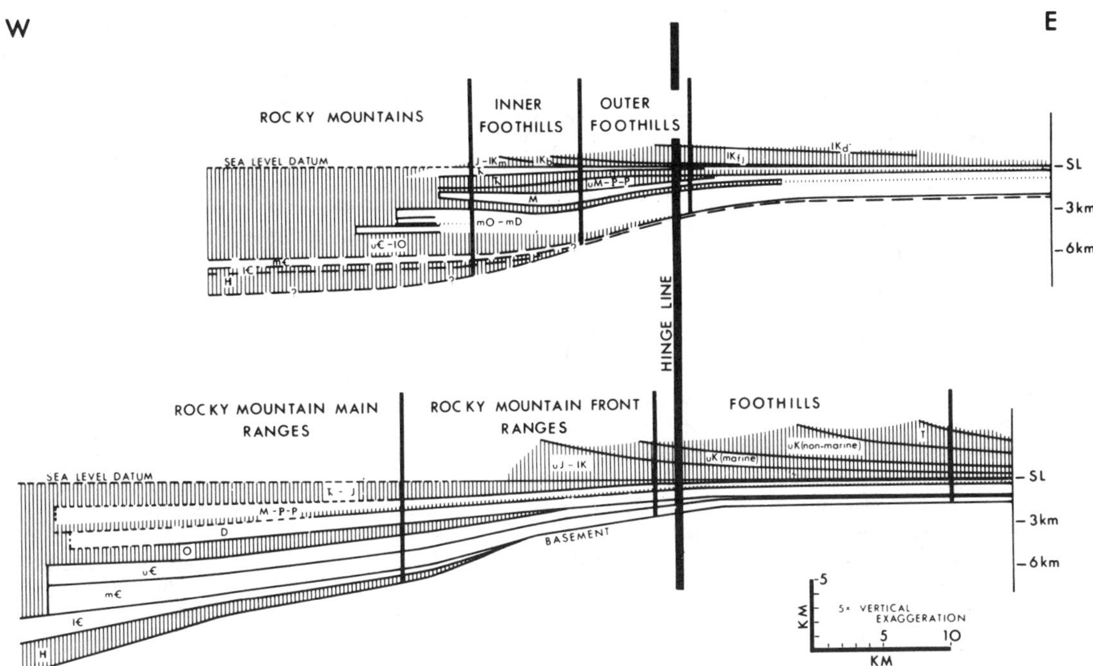

FIG. 11. A comparison of palinspastic reconstructions of stratigraphic sections across the miogeocline for the northern Rockies region near the Laurier section (above) and the southern Rockies (below; adapted from Price & Mountjoy 1970, Figs 2–3). The vertical line pattern is drawn across incompetent units, more rigid quartzites and carbonates are left open. The limits of structural subprovinces are indicated by solid vertical lines. The hinge line shown represents the approximate locus of accelerated thickening of the Lower and Middle Palaeozoic carbonate successions in each section. (Reproduced from Thompson 1979 with permission from the Canadian Journal of Earth Sciences.)

Foothills folds, would have become imbricated and uplifted as deeper décollement surfaces were extended eastward, resulting with a thrust-faulted structural subprovince similar in character to the Front Ranges of the southern Rocky Mountains. The thickness and lateral extent of the young portion of the adjacent clastic wedge deposits would also have been affected. Increased loading of the lithosphere by thrust sheets during late Cretaceous and early Tertiary time would have caused a deepening and enlargement of the adjacent foredeep basin (Price & Mountjoy 1970) and increased the potential for accumulation of a thicker more extensive clastic wedge sequence of that age. Instead, the northern Canadian Rocky Mountains represent a less complete stage in structural development of the foreland belt, a stage that preceeded development of a Front Ranges-type structural domain. This does not mean that deformation ended sooner. On the contrary, Lower Cretaceous foredeep deposits form a major component of the Foothills structural subprovince (Stott 1975) which was deformed in late Cretaceous (and early Tertiary?) time, synchronous with the major late Cretaceous–early Tertiary tectonic pulse to the S. However, the intensity of that pulse decreased northward.

It is not surprising then that a crude symmetry exists between the northern Rocky Mountains structural subprovince and the more internal zones of the southern Rockies. Each contains a large anticlinorium of Middle Proterozoic (Helikian) rocks—the Muskwa Anticlinorium in the N and the Purcell Anticlinorium in the S (see Price, this volume). Along their W margins is a thick, penetratively deformed and metamorphosed clastic sequence of Upper Proterozoic (Hadrynian) age, and along the E border of each is a cleaved succession of shales, siltstones and argillaceous limestones of Lower Palaeozoic age. The anticlinoria are cut by thrusts and high-angle faults that displace coherent rock slices, whereas the surrounding less competent assemblages are more intricately folded and faulted.

Conclusions

The northern Canadian Rocky Mountains are a thin-skinned detachment type structural domain that is fundamentally consistent with structural patterns documented for the southern part of the Rocky Mountain belt. Blind thrusts are an important structural element that disguise the presence of large allochthonous thrust sheets. The amount of supracrustal shortening is less within the northern Rockies and reflects the less intense nature of orogenic activity, especially during the late Cretaceous and early Tertiary.

ACKNOWLEDGMENTS This paper draws from important regional studies by Irish (1970), Taylor & Stott (1973), and Taylor (1979). My interpretation of their data does not necessarily reflect or support their views and sole responsibility for any errors of fact rest with me. I benefited from the comments of D. Cook and A. Okulitch who critically read the manuscript. This synthesis is a product of continuing regional geological studies of the northern Canadian Rocky Mountains by the Geological Survey of Canada. The prompt and efficient work of typist Claudia Thompson is appreciated.

References

BALLY, A. W., GORDY, P. L. & STEWART, G. A. 1966. Structure, seismic data, and orogenic evolution of southern Canadian Rocky Mountains. *Bull. Can. Pet. Geol.* **14,** 337–81.

CECILE, M. P. & NORFORD, B. S. 1979. Basin to platform transition, Lower Paleozoic strata of Ware and Trutch map areas, Northeastern British Columbia. *Pap. geol. Surv. Can.* **79–1A,** 219–26.

DAHLSTROM, C. D. A. 1969. Balanced cross sections. *Can. J. Earth Sci.* **6,** 743–57.

—— 1970. Structural geology in the eastern margin of the Canadian Rocky Mountains. *Bull. Can. Pet. Geol.* **18,** 332–402.

DOUGLAS, R. J. W. 1950. Callum Creek, Langford Creek, and Gap Map-areas, Alberta. *Mem. geol. Surv. Can.* **225,** 124.

—— 1958. Mount Head Map-area, Alberta. *Mem. geol. Surv. Can.* **291,** 241 p.

FAILL, R. T. 1969. Kink band structures in the Valley and Ridge Province, Central Pennsylvania. *Bull. geol. Soc. Am.* **80,** 2539–50.

FITZGERALD, E. L. 1968. Structure of British Columbia Foothills, Canada. *Bull. Am. Assoc. Petrol. Geol.* **52,** 641–64.

—— & BRAUN, L. T. 1965. Disharmonic folds in Besa River Formation, Northeastern British Columbia, Canada. *Bull. Alberta Soc. Petrol. Geol.* **49,** 418–32.

IRISH, E. W. J. 1970. Halfway River Map-area, British Columbia. *Pap. geol. Surv. Can.* **69,-11,** 154 p.

KEATING, L. F. 1966. Exploration in the Canadian Rockies and Foothills. *Can. J. Earth Sci.* **3,** 713–23.

LAUBSCHER, H. P. 1977. Fold development in the Jura. *Tectonophysics*, **37,** 337–62.

NORTH, F. K. & HENDERSON. G. G. L. 1954. Summary of the geology of the southern Rocky Mountains of Canada. *Alberta Soc. Petrol. Geol. Guidebook, 4th Annual Field Conference,* 15–81.

PRICE, R. A. & MOUNTJOY, E. W. 1970. Geologic structure of the Canadian Rocky Mountains between Bow and Athabasca Rivers, a progress report. *In:* WHEELER, J. O. (ed). *Structure of the Southern Canadian Cordillera.* Spec. pap. geol. Assoc. Can. **6,** 7–39.

STOTT, D. F. 1975. The Cretaceous System in-Noreastern British Columbia, *In*: CALDWELL W. G. E. (ed). *Cretaceous System in the Western Interior of North America.* Spec. pap. geol. Assoc. Can. **13,** 441–67.

—— & TAYLOR, G. C. 1972. Stratigraphy and structure, Rocky Mountains and Foothills of west-central Alberta and northeastern British Columbia. *24th Int. geol. Cong., Montreal, Guidebook Field Excursion,* **A10**.

TAYLOR, G. C. 1972. The influence of pre-Laramide tectonics on Rocky Mountain structures. *Geol. Assoc. Can. Cordilleran Section, Prog. & Abs.* 36–7.

—— 1979 Trutch (94G) and Ware East Half (94F, E 1/2) Map-areas, Northeastern British Columbia. *Geol. Surv. Can. Open File Report,* **606**.

——, CECILE, M. P., JEFFERSON, C. W. & NORFORD, B. S. 1979. Stratigraphy of Ware (East Half) Map-area, Northeastern British Columbia. *Pap. geol. Surv. Can.* **79–1A,** 227–31.

—— & STOTT, D. F. 1973. Tuchodi Lakes Map-area, British Columbia. *Mem. geol. Surv. Can.* **373,** 37 p.

THOMPSON, R. I. 1976. Some aspects of stratigraphy and structure in the Halfway River map-area (94B), British Columbia. *Pap. geol. Surv. Can.* **76–1A,** 471–77.

—— 1978. Geological maps and sections of Halfway River Map-area, British Columbia (94B). *Geol. Surv. Can., Open File Report,* **536**.

—— 1979. A structural interpretation across part of the northern Rocky Mountains, British Columbia, Canada.*Can. J. Earth Sci.* **16,** 1228–41.

R. I. THOMPSON, Geological Survey of Canada, 100 W Pender St., Vancouver, Canada

Reprinted by permission of the Geological Society of America from G. Eisenstadt and D. G. De Paor, *Geology*, v. 15 (1987), p. 630-633.

Alternative model of thrust-fault propagation

Gloria Eisenstadt, Declan G. De Paor
Department of Earth and Planetary Sciences, Johns Hopkins University, Baltimore, Maryland 21218

ABSTRACT

A widely accepted explanation for the geometry of thrust faults is that initial failures occur on deeply buried planes of weak rock and that thrust faults propagate toward the surface along a staircase trajectory. We propose an alternative model that applies Gretener's beam-failure mechanism to a multilayered sequence. Invoking compatibility conditions, which demand that a thrust propagate both upsection and downsection, we suggest that ramps form first, at shallow levels, and are subsequently connected by flat faults. This hypothesis also explains the formation of many minor structures associated with thrusts, such as backthrusts, wedge structures, pop-ups, and duplexes, and provides a unified conceptual framework in which to evaluate field observations.

INTRODUCTION

The mechanics of thrust faulting have intrigued geologists since the late 1800s. Initial investigative efforts by Cadell (1890) and Willis (1893) focused on duplicating the structures observed in the Scottish Highlands and the Appalachians by experimentally deforming layers of clay, plaster, and sand. Other researchers (e.g., Hayes, 1891) recognized that the formation of thrusts in sedimentary sequences was in some way associated with strata of contrasting competence. But until Rich's (1934) work on the Pine Mountain thrust, no coherent theory connected rheologic properties with thrust geometry. Rich identified the staircase trajectory of the Pine Mountain thrust, which previous workers had interpreted as a warped fault plane. He proposed that the thrust was formed by initial fracture on a weak shale bed. The overlying rock slid until frictional resistance became too great, then the thrust cut upsection through more competent strata until it reached another zone of easy gliding. In the current literature, the bedding-parallel areas of a thrust are called flats, and the parts that cut across bedding are called ramps.

Most major thrust systems display ramp-flat geometry (Douglas, 1950; Price, 1965; Dahlstrom, 1970; Boyer and Elliott, 1982), and there is general acceptance of Rich's explanation that thrusts begin as flats at depth and propagate toward the surface, cutting through more competent beds and again becoming parallel to bedding when they reach incompetent layers (Fig. 1). However, this model of ramp-flat formation is unsatisfactory for several reasons. It fails to explain the repetition of ramp-flat pairs; why and how does a fault cutting through a competent layer become bedding parallel again? Nor does it explain the occurrence of reverse faults in layered sequences of competent and incompetent rock that do not have a staircase trajectory. In addition, Rich's hypothesis predicts that thrust detachment horizons will form at the interface between the most competent and incompetent units. Instead, actual field observations present puzzling examples of faults that deviate from the preferred path. Miller (1973) illustrated this with major faults in Tennessee, Kentucky, and Virginia, explaining their anomalous behavior through a combination of increased fluid pressure and gravity sliding. Other examples include the Heart Mountain thrust of Wyoming (Pierce, 1957), where the major thrust horizon is between the Bighorn dolomite and the Grove Creek limestone (not in the shales above or below); two thrust systems in southern Nevada (Burchfiel et al., 1982), where the major decollement zones are within the dolostones rather than the underlying weaker rocks; and thrusts documented by Coleman and Lopez (1986) in the southern Appalachians, where the decollement horizon varies within the Knox Group dolomites. Finally, the current model does not accommodate actual field examples of thrusts that have flats that are subparallel to bedding rather than strictly parallel (Fox, 1959; Woodward, 1981).

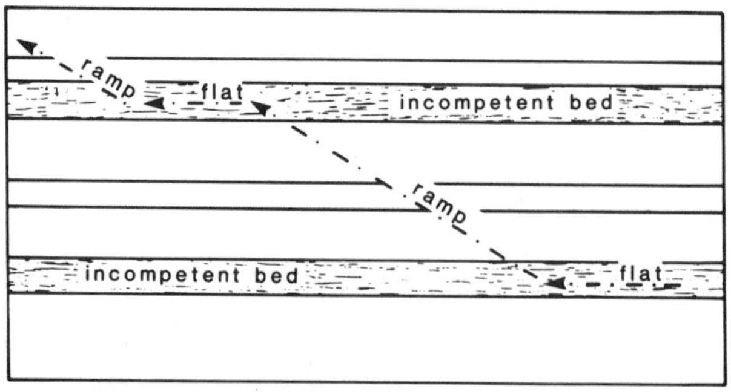

Figure 1. Rich's (1934) model of ramp-flat formation; thrusts begin as bedding-parallel faults (flats) at depth in incompetent layer and propagate toward surface, cutting through more competent layers as ramps. Dashed line and arrows indicate fault trajectory.

ALTERNATIVE MODEL

We suggest an alternative explanation for the formation of ramp-flat geometry. In our view, faults are most likely to begin in competent layers at shallow depths because lower layers cannot move until the brittle cover fractures. Gretener (1972) suggested that competent layers behave as beams and that movement occurs when the strongest beam fails, thereby transferring stress to weaker layers. Such failure is accompanied by increased pore pressure, and the thrust sheet moves along a flat surface until pore pressure drops. However, Gretener envisaged this process as being part of a model of stick-slip thrust movement and considered the final thrust geometry to be the result of ramp-flat pairs being formed as the thrust sheet propagated upsection. Recently, Bombolakis (1986) used a stick-slip model to predict conditions of frontal ramp formation during foreland thrust propagation. His analysis assumed that initial fault movement occurs along a basal fault, causing a buildup of strain in the potential

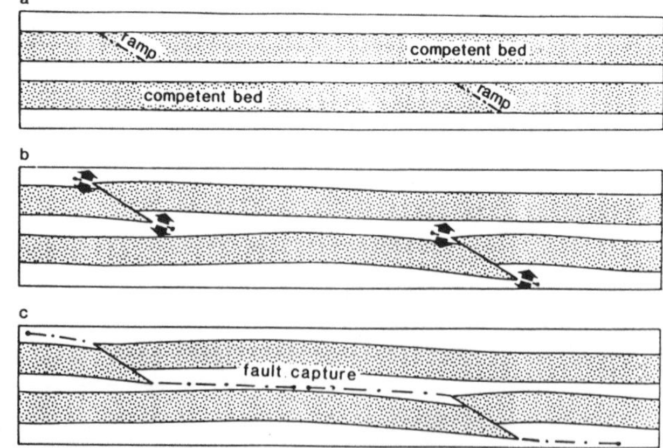

Figure 2. a: Ramp-flat formation by initial failure of competent units. b: Subsequent fault propagation by ripping of incompetent layers. Arrows indicate areas of tensional stress at fault tip. Flexure in this and following diagrams is exaggerated to enhance clarity. c: Fault capture as result of fault propagation in two directions. Flat fault is created between two ramps.

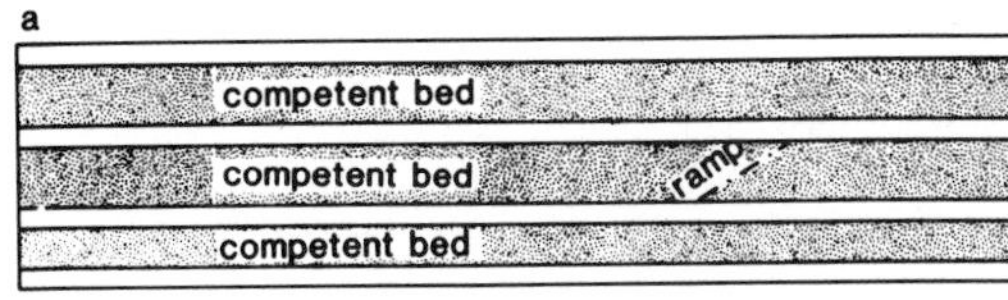

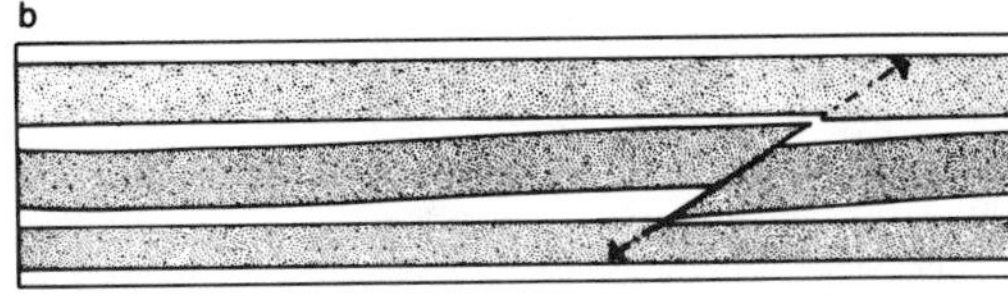

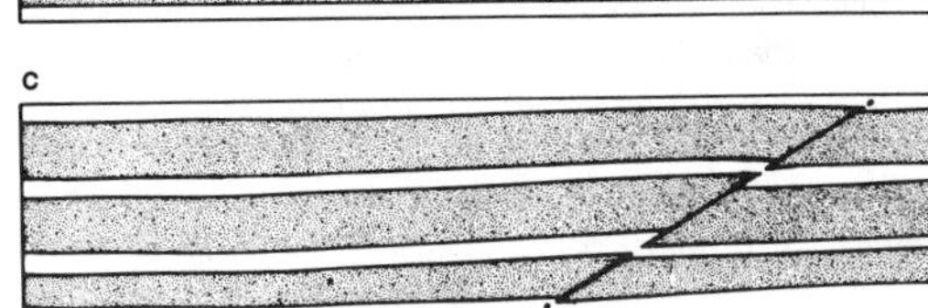

Figure 3. Fault propagation by chiseling. a: Initial ramp formation in competent layer. b: Impact on beds above and below causes new ramps to form. c: Thrust without ramp-flat geometry is created.

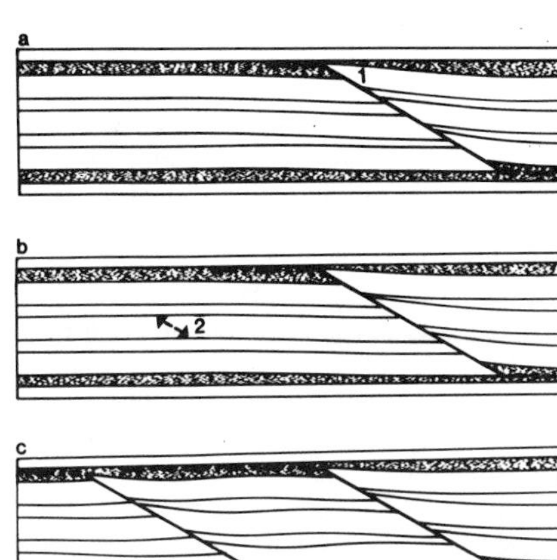

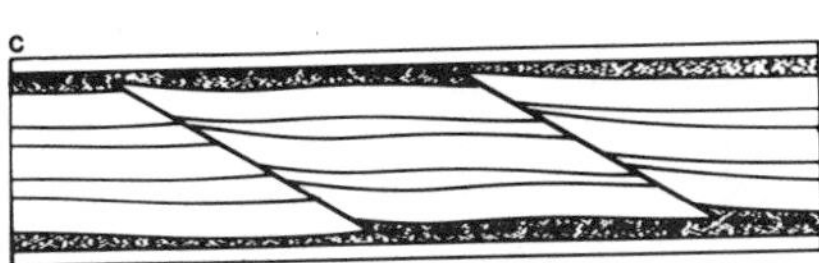

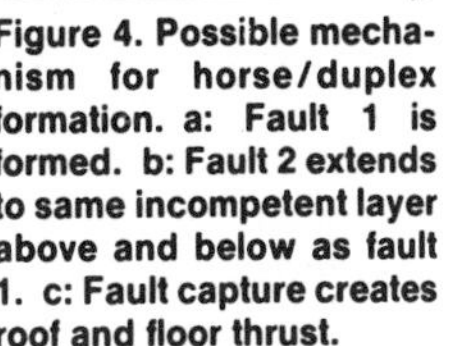

Figure 4. Possible mechanism for horse/duplex formation. a: Fault 1 is formed. b: Fault 2 extends to same incompetent layer above and below as fault 1. c: Fault capture creates roof and floor thrust.

ramp region and a subsequent fracturing of competent layers. In contrast, we propose that initial slip on ramps in the stronger layers creates tension between adjacent strata (Fig. 2); the flat fault propagates by ripping beds apart. If two flat faults propagating in opposite directions meet, they merge in a manner analogous to stream capture. The significant difference between this hypothesis and other models is the suggestion that ramps form first and that flats are the result of fault tip lines migrating both upsection and downsection from multiple source ramps. In other words, ramp-flat geometry is not the result of thrust movement but is already formed prior to significant thrust displacement. If thrusts did propagate from depth, their terminal accommodation structures (e.g., fault propagation folds) would soon grow very large. Our model requires only small structures at several localities along a thrust.

At high strain rate, with little cushioning between competent layers (Fig. 3), initial fracture and slip of the strongest layer may result in sudden impact on the beds above and below. The chisel-like impact may cause more ramps to form both upsection and downsection. A similar sequence could provide a mechanism for horse and duplex formation (Fig. 4). Initial fractures occur in an upper layer, and the fault propagates both upsection and downsection until it reaches an incompetent layer where it becomes flat. Later fractures occur at depth, and the migrating fault tips are captured by a decollement zone above and below, forming a roof and a floor thrust. According to our model, the primary difference between duplex and staircase geometry is whether later ramps form to the hinterland or to the foreland sides of earlier ramps. Staircase geometry only occurs when ramps lower in the stratigraphic sequence form toward the hinterland (Fig. 5). Conversely, duplex geometry may be the result of lower ramps forming in a foreland direction.

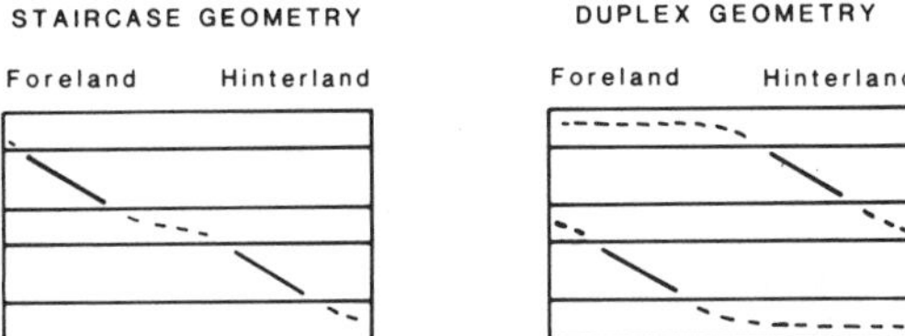

Figure 5. Staircase trajectory is result of ramp formation toward foreland in layers higher in stratigraphic sequence; duplex geometry may be result of converse sequence.

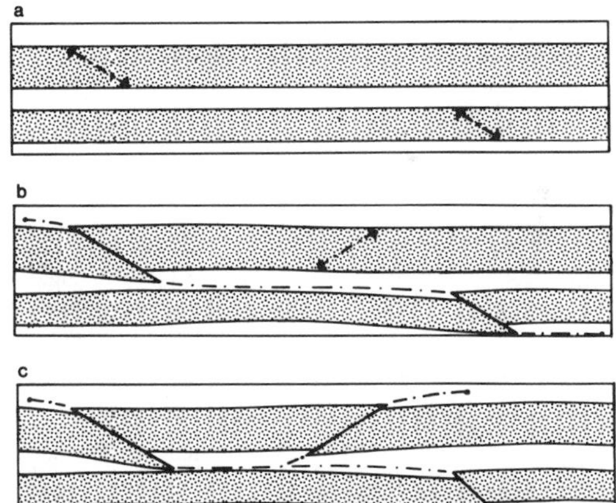

Figure 6. Backthrust (or pop-up structure) is due to formation of shear fractures in opposite directions. a: Initial ramp formation in two competent layers. b: Fault capture in incompetent layer and later fracturing within one of competent units in opposite direction. c: Propagating fault meets existing flat fault and creates backthrust.

PREVIOUS WORK

The fact that faults must propagate both upsection and downsection has been recognized by other researchers, though not in this context. In a little-known paper, Norris (1961) accepted the prevailing view that thrusts begin as glide surfaces in weak layers but recognized that there could be no significant displacement before failure occurred in the intervening strong layers. Newly formed ramps would link up existing glide surfaces up and down the stratigraphic section. We consider this to be the right approach, but we differ on the sequence of events. The propagation of thrust faults downsection was also suggested by Williams and Chapman (1983; Chapman and Williams, 1984), and by Wiltschko and Eastman (1983) who modeled stress concentrations resulting from preexisting basement warps and faults. However, failure was not achieved in the models, and hence the order of failure was not addressed.

Experimental studies to date, although answering many questions about deformation in ramp regions, have not addressed the question whether initial failure occurs first as a bedding-parallel fault in an incompetent layer or as a fracture of a competent layer. Morse (1978) used precut ramps in his experiments, whereas others have used unconsolidated sand and clay (Miyabe, 1934; Hubbert, 1951; Vincellette, 1964). Some experiments demonstrate the mechanical feasibility of our model. Chamberlain and Miller (1918) showed that in layered plaster of varying competency, the first fractures appear in the most competent layers and are followed by fracturing parallel to bedding. This result was later confirmed by Link (1928), who indicated that the initial break in his thrust experiments invariably occurred in the more brittle layers. Balicki and Spang (1975) deformed layers of clay of varying viscosities and found that ramps begin as closely spaced fractures that finally coalesce and cut the layer.

OTHER STRUCTURES

An attractive feature of our model is that it offers a unified explanation for thrust-related minor structures such as splays, wedge faults, pop-ups, duplexes, and backthrusts. These various structures can be created in thought experiments by invoking beam failure, ripping, chiseling, and fault capture in various sequences. Figure 6 shows the formation of a backthrust. Initially, two ramps form by beam failure in opposite directions. Resultant tip-line migration yields a forethrust and a backthrust. We would expect backthrusts to be a less frequent occurrence than forethrusts because most sedimentary sequences are wedge-shaped and thus are more easily telescoped in one direction. Pop-up structures may be the result of failure of a single layer on two such surfaces, as illustrated by the upper layer in Figure 6. Wedge faults (Cloos, 1961) can be explained by failure of a competent layer and subsequent movement on that ramp (Fig. 7).

On a larger scale, our model explains foreland and hinterland progression of thrust development. Because ductility increases with depth, initial failure occurs close to the surface on the most brittle competent layer. Stress is then transferred to less brittle layers at depth and to less lithified strata at shallower levels (Fig. 8), causing subsequent thrusts to reach the surface at some later time. Depending on the sequence of beam failure, the final configuration will represent either a foreland or a hinterland progression.

Finally, we present an example of how sequential ramp and flat faulting may be confused with a staircase trajectory. Figure 9 shows a set of ramps that formed in thick competent units, but initially the beds are still coherent across the fault. As the fault blocks rotate, the ramps rotate from a high to a low shear orientation. Eventually, as the hanging wall and footwall of the competent units no longer touch across the faults and shear stresses on bedding planes increase, a flat fault develops. This flat offsets the early formed ramps and creates a stratal repetition. Outside the ramp offset area there is no repetition of strata, and the flat fault may go unnoticed, leading to a staircase interpretation (other sequential geometries have been discussed by Bally et al., 1966; Fox, 1969; and Roeder et al., 1978). If such an interpretation is adopted, a cross section will not balance. Any attempt by the geologist to alter the section by adding missing area to the incompetent units may lead to grave errors in the final section.

DISCUSSION

The mechanisms proposed here offer a solution to the creation of multiple ramp-flat pairs through a stratigraphic sequence; a staircase fault trajectory is only an apparent, not a real, genetic unit. Instead, there may have been several source ramps whose connecting flats could have been formed in any number of different sequences. These mechanisms also offer an explanation for the variation of ramp-flat spacing in thrust sheets within a single thrust belt. A multiplicity of thrust-related structures can be explained by altering the sequence or direction of ramp formation. In addition, our hypothesis explains field examples of nonideal thrust faults. Where flats propagate by ripping of thinly bedded incompetent units, they are unlikely to be perfectly parallel to bedding. Instead, we predict that most flats form very low angle faults (less than 5°) over large distances. Where faults propagate by the chiseling of one competent unit into another, it is possible to form a thrust that does not ever become bedding parallel. A combination of multiple ramp formation and ripping may account for the occurrence of flats on bedding planes in competent units.

Recent work by Dunlap and Ellis (1986), using displacement-distance analysis on thrust faults, supports the model presented here. A detailed study of a pop-up structure shows that the fault had at least four nucleation sites along the two conjugate faults. Their evidence suggests that larger faults may be the result of multiple nucleation points and subsequent fault linkage. A test of the two hypotheses would be to perform similar analyses and an additional evaluation of the relative competency of the nucleation sites relative to the areas of fault linkage.

Another means of testing the two models of thrust-fault propagation

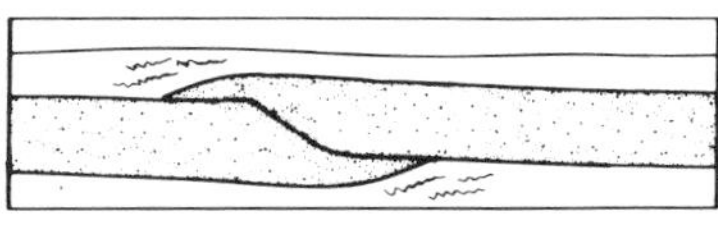

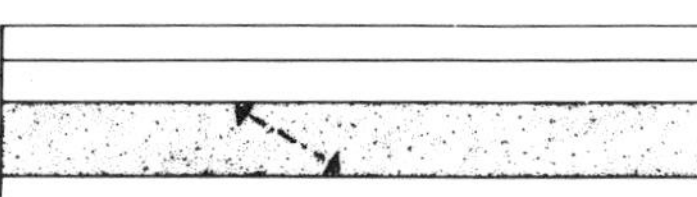

Figure 7. Diagram shows wedge fault formed by failure of competent layer. Photo shows example of wedge fault in interbedded limestones and shales, Kimmeridge Bay, Dorset, England (courtesy of Andrew Bell).

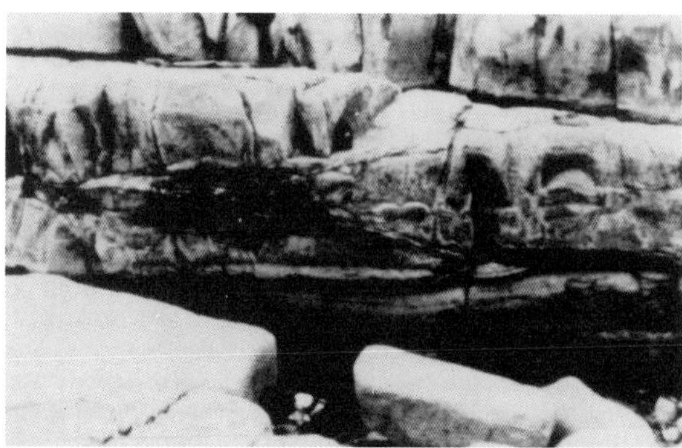

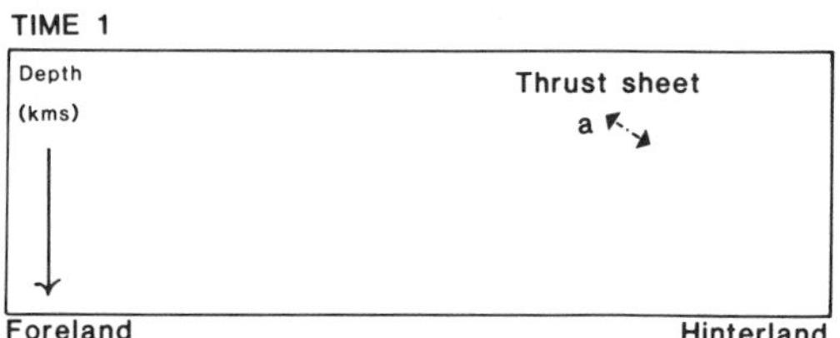

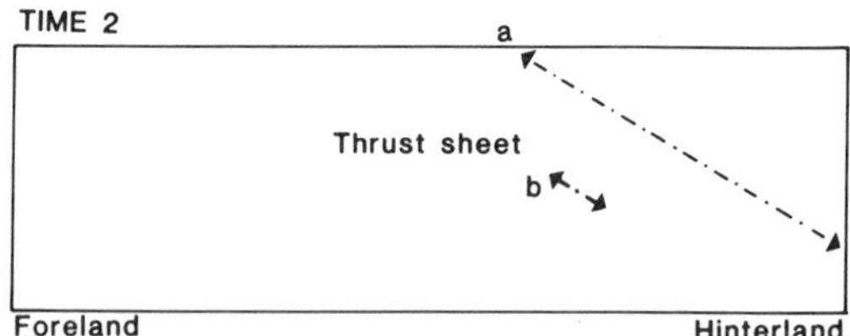

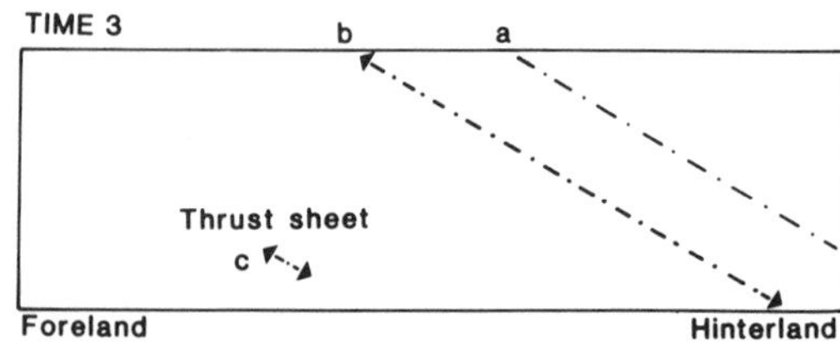

Figure 8. Example of foreland progression of thrusts. Thrust sheet a forms first because of initial ramp formation in brittle layer close to surface. Later fractures occur at depth, causing subsequent thrust sheets b and c to reach surface at later time.

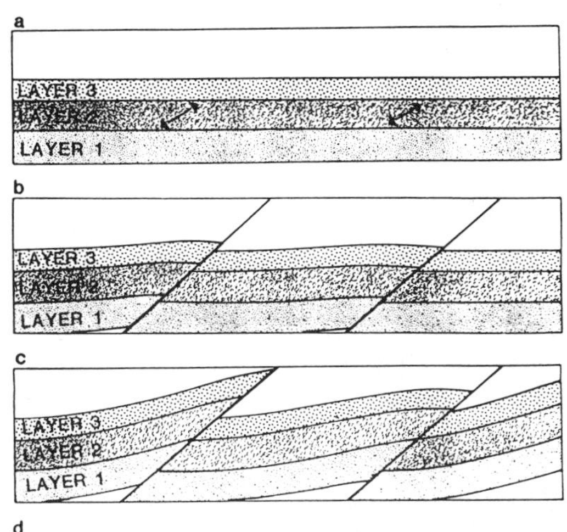

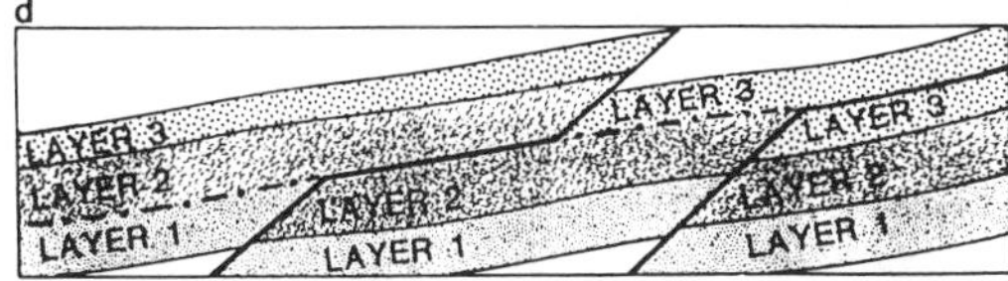

—·—·— unrecognized flat – no stratigraphic separation

Figure 9. Offset ramps may lead to false interpretation of ramp-flat geometry. a: Ramp formation in thick competent unit. b: Rotation of fault blocks causes decreasing shear stress on beds. c: Continued rotation eliminates overlap of competent layer on ramp fault, and flat fault develops. d: Apparent staircase geometry and unrecognized flat faults. Solid line indicates apparent ramp-flat geometry; dashed line indicates unrecognized flat.

is offered by Jamison's (1987) geometric analysis of thrust-associated folds. If thrust faults begin along basal decollement zones, then the final fold geometry should be that of a detachment fold which has been transported through at least one ramp region. On the other hand, if thrusts originate from multiple source ramps, then the final fold configuration along the length of the fault should be that of multiple fault-propagation folds that may also have passed through additional ramp regions. In Jamison's terms, one may be able to distinguish between a transported, altered detachment fold and a transported, altered fault-propagation fold.

Although we are not suggesting that the model presented here is the only mode of formation for thrusts that have ramp-flat geometry, we feel that casual acceptance of the current Rich-type mechanism has allowed geologists to comfortably simplyify some very real difficulties in our understanding of thrust initiation and propagation.

REFERENCES CITED

Balicki, M.A., and Spang, J.H., 1975, Multilayer clay models of thrust faults and associated structures: Geological Society of America Abstracts with Programs, v. 7, p. 586–587.

Bally, A.W., Gordy, P.L., and Stewart, G.A., 1966, Structure, seismic data and orogenic evolution of southern Canadian Rocky Mountains: Bulletin of Canadian Petroleum Geology, v. 14, p. 337–381.

Bombolakis, E.G., 1986, Thrust fault mechanics and origin of a frontal ramp: Journal of Structural Geology, v. 8, p. 281–290.

Boyer, S.E., and Elliott, D., 1982, Thrust systems: American Association of Petroleum Geologists Bulletin, v. 66, p. 1196–1230.

Burchfiel, B.C., Wernicke, B., Willemin, J.H., Axen, G.J., and Cameron, C.S., 1982, A new type of decollement thrusting: Nature, v. 300, p. 513–515.

Cadell, H.M., 1890, Experimental researches in mountain building: Royal Society of Edinburgh Transactions, v. 35, p. 337–357.

Chamberlain, R.T., and Miller, W.Z., 1918, Low-angle faulting: Journal of Geology, v. 26, p. 1–44.

Chapman, T.J., and Williams, G.D., 1984, Displacement-distance methods in the analysis of fold-thrust structures and linked-fault systems: Geological Society of London Journal, v. 141, p. 121–128.

Cloos, E., 1961, Bedding slips, wedges and folding in layered sequences: Société Géologique de Finlande, Extrait des Comptes Rendus, no. 33, p. 106–122.

Coleman, J.L., Jr., and Lopez, J.A., 1986, Dolomite decollements—Exception or rule? [abs.]: American Association of Petroleum Geologists Bulletin, v. 70, p. 576.

Dahlstrom, C.D.A., 1970, Structural geology in the eastern margin of the Canadian Rocky Mountains: Bulletin of Canadian Petroleum Geology, v. 18, p. 332–406.

Douglas, R.J.W., 1950, Callum Creek, Langford Creek, and Gap Map-Areas: Alberta: Geological Survey of Canada Memoir 255, 124 p.

Dunlap, W.J., and Ellis, M.A., 1986, Fault nucleation and propagation: Evidence from thrust displacement analyses: Geological Society of America Abstracts with Programs, v. 18, p. 590.

Fox, F.G., 1959, Structure and accumulation of hydrocarbon in Southern Foothills, Alberta, Canada: American Association of Petroleum Geologists Bulletin, v. 43, p. 992–1025.

——1969, Some principles governing interpretation of structure in the Rocky Mountain orogenic belt, *in* Kent, P.E., et al., eds., Time and place in orogeny: Geological Society of London Special Publication 3, p. 23–42.

Gretener, P.E., 1972, Thoughts on overthrust faulting in a layered sequence: Bulletin of Canadian Petroleum Geology, v. 20, p. 583–607.

Hayes, C.W., 1891, The overthrust faults of the southern Appalachians: Geological Society of America Bulletin, v. 2, p. 141–154.

Hubbert, M.K., 1951, Mechanical basis for certain familiar geologic structures: Geological Society of America Bulletin, v. 62, p. 355–372.

Jamison, W.R., 1987, Geometric analysis of fold development in overthrust terranes: Journal of Structural Geology, v. 9 (in press).

Link, T.A., 1928, Relationship between over- and under-thrusting as revealed by experiments: American Association of Petroleum Geologists Bulletin, v. 12, p. 825–854.

Miller, R.L., 1973, Where and why of Pine Mountain and other fault planes, Virginia, Kentucky, and Tennessee: American Journal of Science, v. 273-A, p. 353–371.

Morse, J.M., 1978, Deformation in ramp regions of thrust faults: Experiments with rock models [M.S. thesis]: College Station, Texas, Texas A&M University, 137 p.

Miyabe, N., 1934, Experimental investigations of the deformation of sand mass: Part IV: Tokyo University Earthquake Research Institute Bulletin, v. 12, p. 311–342.

Norris, D.K., 1961, An interstratal peel on Maverick Hill, Alberta: Journal of the Alberta Society of Petroleum Geologists, v. 9, p. 177–191.

Pierce, W.G., 1957, Heart Mountain and South Fork detachment thrusts of Wyoming: American Association of Petroleum Geologists Bulletin, v. 41, p. 591–626.

Price, R.A., 1965, Flathead map-area, British Columbia and Alberta: Geological Survey of Canada Memoir 336, 221 p.

Rich, J.L., 1934, Mechanics of low-angle overthrust faulting as illustrated by Cumberland thrust block, Virginia, Kentucky, and Tennessee: American Association of Petroleum Geologists Bulletin, v. 18, p. 1584–1587.

Roeder, D., Gilbert, E., and Witherspoon, W., 1978, Evolution and macroscopic structure of Valley and Ridge thrust belt, Tennessee and Virginia: University of Tennessee, Department of Geological Sciences, Studies in Geology 2, 25 p.

Vincellette, R.R., 1964, Structural geology of the Mt. Stirling quadrangle, Nevada, and related scale-model experiments [Ph.D. thesis]: Stanford, California, Stanford University, 141 p.

Williams, G., and Chapman, T., 1983, Strains developed in the hanging walls of thrusts due to their slip/propagation rate: A dislocation model: Journal of Structural Geology, v. 5, p. 563–571.

Willis, B., 1893, Mechanics of Appalachian structures: U.S. Geological Survey, 13th Annual Report, Part II, p. 222–223.

Wiltschko, D., and Eastman, D., 1983, Role of basement warps and faults in localizing thrust-fault ramps: Geological Society of America Memoir 158, p. 177–190.

Woodward, N., 1981, Structural geometry of the Snake River Range, Idaho and Wyoming [Ph.D. thesis]: Baltimore, Maryland, Johns Hopkins University, 261 p.

ACKNOWLEDGMENTS

Supported by Elf Aquitane (S.N.E.A.[P.]; Johns Hopkins University contract 6766), Sigma Xi, and the Geological Society of America. We thank Steve Boyer and Nick Woodward for their critical reviews of the manuscript.

Manuscript received September 29, 1986
Revised manuscript received March 23, 1987
Manuscript accepted April 8, 1987

The American Association of Petroleum Geologists Bulletin
V. 55, No. 2 (February 1971), P. 292-306, 10 Figs., 1 Table

Folded Faults and Sequence of Thrusting in Alberta Foothills[1]

P. B. JONES[2]
Calgary 1, Alberta

Abstract Thrust faults in the Alberta Foothills were folded by development of step-thrusts and other thrust-generated structures beneath them. Because step-thrusting is a pattern of faulting found in layered rocks of varying competence, folded faults may be expected in any thrust belt where the stratigraphic sequence is inhomogeneous and higher thrusts are emplaced before underlying ones. By invoking folded faults, several structures in the Alberta Foothills can be reinterpreted without the geologic and geometric inconsistencies inherent in existing interpretations.

Although a west-to-east sequence of thrust emplacement in the Alberta Foothills is implicit in the mechanics of folding of thrusts, some structural relations indicate an east-to-west sequence. The apparent contradiction may be resolved by the proposition that the time taken for a thrust to form, spreading both across and along strike, is greater than the time interval between initiation of successive thrusts.

Introduction

The Rocky Mountain orogeny (White, 1959) in the Alberta Foothills and eastern Rocky Mountains involved Precambrian to Paleocene sedimentary rocks (Table 1). Deformation consisted largely of overthrusting from the west along zones of *décollement* at various levels overlying the crystalline basement, which was not itself involved (Bally *et al.*, 1966; Keating, 1966). Faults other than thrusts are rare in the area under discussion. In this paper the term "fault" is used to denote thrust faults.

The presence of folded faults was demonstrated by Hake *et al.* (1942) in the central Foothills (Fig. 1). Folded faults also were described by Hage (1942) in the southern Foothills. Subsequent references to folded faults have been concerned mainly with those two areas. Scott (1951), MacKay (1943), and Erdman (1950) mapped other folded faults in the central Foothills: Douglas (1950; 1958a, b) described similar faults in the southern Foothills and Front Ranges of the Rocky Mountains. One purpose of this paper is to show that folded faults may be present anywhere in the Foothills and Rockies, and to present evidence for their presence in structures previously regarded as unaffected by folded faults.

Two explanations for the folding of faults have been given. Douglas (1950, Figs. 16, 17) and Bally *et al.* (1966, p. 37) regarded the folding as an integral part of the process of thrusting. Scott (1951, p. 2347), Dahlstrom *et al.* (1962, p. 394), and others attributed the folding to a discrete compressional phase which followed the period of thrusting in the disturbed belt.

In profile, thrust faults are curved and commonly are associated with tilted and folded strata, so that it is rarely possible to determine whether the curvature is the original form of the thrust plane or is the result of folding. It is therefore appropriate, before describing folded faults, to consider thrust faults that have not been folded.

Faults in Crimson Lake Area, Alberta

Killick (1954, p. 35) suggested that linear features in the vicinity of the sharp bend in the North Saskatchewan River (Fig. 1), east of the foothills belt, might be faults. Following photogeology by K. W. Roth, field checks by R. B. Sanders and the writer showed that two northwest-trending linears are southwest-dipping thrusts, the Sylvester Creek and Brewster Creek faults (Fig. 2). At the surface the Sylvester Creek fault dips 40° southwest. The footwall strata are horizontal. About 2,000 ft of beds above the fault also dip southwest, subparallel with the fault plane. In the subsurface, a seismic profile shows the fault plane flattening at depth, passing into a bedding-plane fault in the Alberta Group. The Brewster Creek fault appears to be similar. Wells drilled west of the fault trace penetrated thickened sections in the Alberta Group (Fig. 3). Farther west, wells drilled through both thrusts penetrated normal thicknesses in the Mesozoic sequence, with no

[1] Manuscript received, February 18, 1970; accepted, June 11, 1970. Published by permission of Amerada Hess Corporation.

[2] Amerada Hess Corporation.

The writer is indebted to R. L. Zell, F. G. Fox, P. Verrall, and E. L. Fitzgerald, who detected many errors and omissions in the manuscript and suggested improvements. Figures were drawn by the Amerada Hess drafting department. The typescript was prepared by Merlyn Trigg.

Table 1. Formations in Alberta Foothills
(N = Northern, C = Central, S = Southern)

System	*Formation*		*Lithology*	*Thickness (Ft)*
Tertiary	Paskapoo (N, C)	Porcupine Hills (S)	Nonmarine sandstone, shale, coal	2,000– 4,000
			Unconformity	
		Willow Creek (S)	Nonmarine sandstone, shale, mudstone	350– 2,700
Cretaceous, Upper	Brazeau (N, C)	Edmonton (S) Bearpaw (S) Belly River (S)	Mainly nonmarine sandstone, mudstone, shale, coal	2,000– 5,000
	Wapiabi		Marine shale and siltstone	1,100– 1,800
	Cardium		Marine sandstone, shale	30– 450
	Blackstone		Marine shale	400– 1,000
Cretaceous, Lower		Crowsnest (S)	Agglomerate, tuff	0– 1,800
	Mountain Park (N, C) Luscar (N, C)	Blairmore (S, C)	Nonmarine sandstone, shale	900– 6,500
	Cadomin (N, C)		Conglomerate	0– 200
			Unconformity	
Jurassic	Nikanassin (N, C)	Kootenay	Nonmarine sandstone, shale, coal	100– 4,000
	Fernie Group		Marine shale, limestone, sandstone	100– 1,300
			Unconformity	
Triassic	Spray River Group		Marine dolomite, siltstone	0–1,200
			Unconformity	
Permian	Ishbel Group		Marine sandstone, chert, sandy dolomite	0– 1,400
			Unconformity	
Pennsylvanian	Kananaskis		Marine silty dolomite	0– 300
Mississippian	Rundle Group		Marine limestone and dolomite	750– 3,000
	Banff		Marine argillaceous limestone	500– 1,100
	Exshaw		Marine black shale	0– 40
			Unconformity	
Devonian	Palliser		Marine limestone, dolomite	650– 1,200
	Alexo		Silty limestone and dolomite	20– 600
	Fairholme		Marine limestone, shale, dolomite, local basal clastic unit	950– 1,600
			Unconformity	
Cambrian			Marine limestone and dolomite, sandstone and shale	700– 2,500
			Unconformity	
Precambrian	Purcell Group (S) Mountains only		Dolomite, argillite, limestone	7,000–15,000

indication of the existence of faults in the section. The Stolberg structure, the most easterly Foothills structure in which Paleozoic strata are involved in thrusting, is present 20 mi southwest of the outcrop of the Sylvester Creek fault. Two fault slices of Mississippian carbonates overlie the autochthon. Figure 3 shows the correlation of these two faults with the Sylvester Creek and Brewster Creek faults, via bedding-plane faults, as suggested by Thomas (1958, p. 127).

The Sylvester Creek and Brewster Creek faults are not folded. The footwall strata are horizontal and the tilting of the hanging wall strata is due to fault movement only. According to Fox (1959, p. 1024), the arcuate profile of many faults in the disturbed belt is the result of folding or rotation of thrusts having a uniform low dip. If this were true of the Sylvester Creek fault, footwall and hanging wall strata would be affected equally. They are not, hence the arcuate profile must be related to the stress pattern causing the thrusting (Hafner, 1951, p. 385), modified by the vertical anisotropy of the sedimentary sequence.

The Sylvester Creek and Brewster Creek

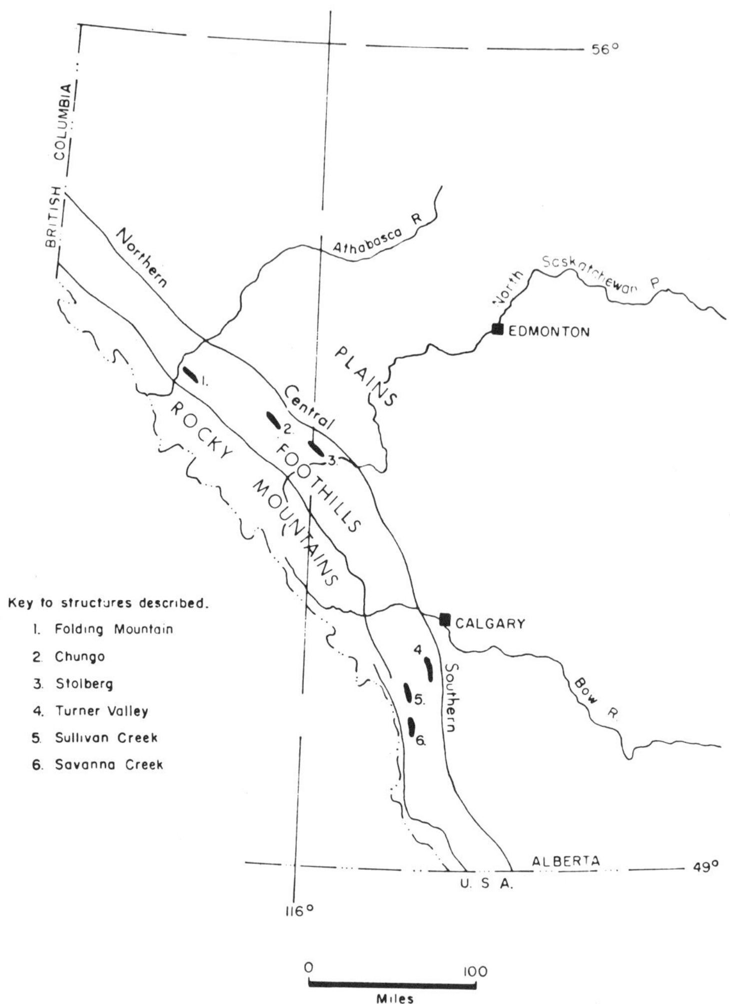

Fig. 1.—Location map, Alberta Foothills.

faults transect the Paleozoic in the Stolberg structure at low angles, parallel the bedding of the Alberta Group for 16 mi, and curve upward to the surface cutting the footwall strata at up to 40°. They are step faults, and their profiles in cross section (Fig. 3) are probably typical of a thrust fault before folding.

Fault Structures in Central Foothills of Alberta

Stolberg Structure (Fig. 3)

Three wells have been drilled on the Stolberg structure, and extensive seismic work has been conducted by several oil companies. The warping of the Ancona thrust over the subsurface structure is shown both in plan (Fig. 2) and cross section (Fig. 3). Although the Ancona thrust and adjacent strata are folded, the underlying faulted slices of Mississippian rocks are barely folded. The Ancona thrust appears to have been folded by the emplacement of the Sylvester and Brewster Creek faults beneath it, as well as by the imbrications in the intervening Upper Cretaceous strata. The lower of the two faults that displace the Mississippian is not folded, presumably because no structure was developed beneath it to cause folding. If folding were due to a discrete compressional phase following thrusting, the entire sequence would have been folded. The slight folding of the overthrust slices of Mississippian rocks can be attributed entirely to thrust movement. Apparently, the original profile of the Ancona thrust was similar to the Sylvester Creek fault, curving upward to the surface in the same way. The present anticlinal and synclinal profile is believed to have resulted from upwarping of the near-horizontal sector of the thrust plane by emplacement of the thrusts beneath it.

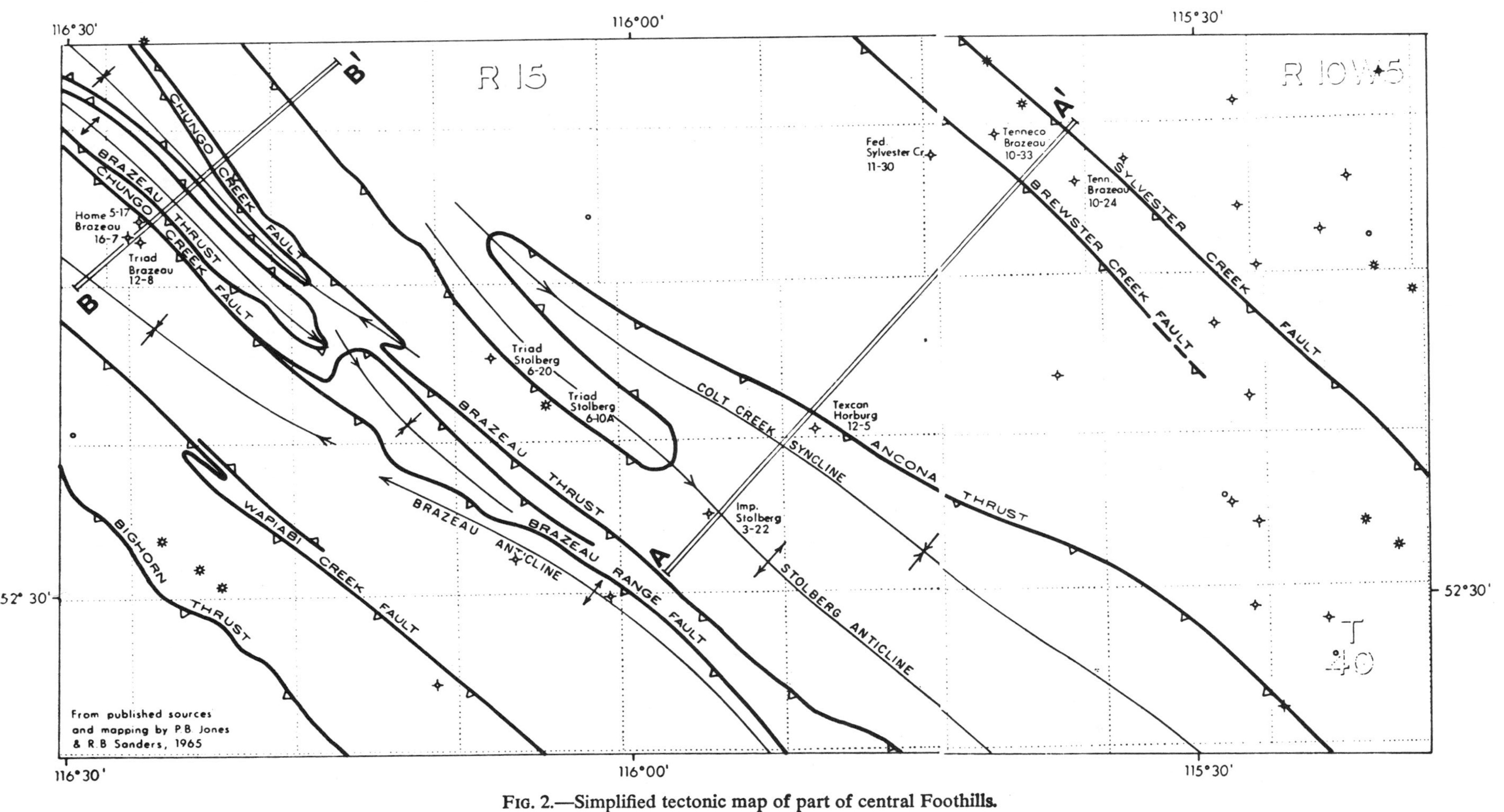

FIG. 2.—Simplified tectonic map of part of central Foothills.

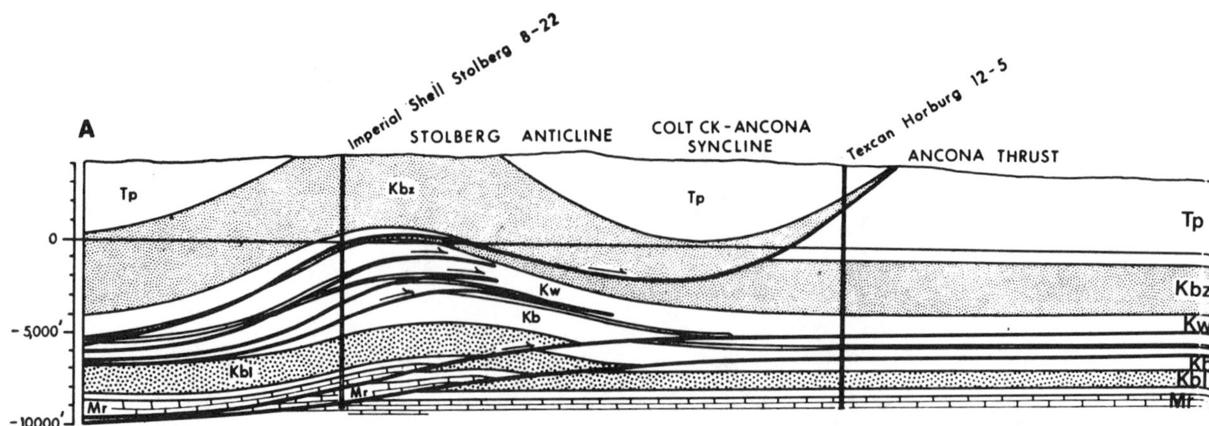

FIG. 3.—Structural cross-section A-A′, Stolberg structure to plains.

Chungo Structure (Fig. 4)

The Chungo structure is similar to the Stolberg structure, although more complex. The Brazeau and adjacent faults are strongly folded over a subsurface structure which involves faulting only. As at Stolberg, folding of the overlying thrusts is believed to be due to the development of thrust structures in the underlying Paleozoic and Mesozoic strata. Scott (1951, p. 1347) concluded that the order of tectonic events in the area is thrusting followed by folding on a regional scale. However, this mechanism fails to explain the absence of folding at depth, even after allowance has been made for disharmonic folding.

Correlation of the sole faults of the Chungo structure with the outcropping Ancona and adjacent fault (Fig. 4) is less certain than the correlation of the sole faults of the Stolberg structure with the Sylvester Creek and Brewster Creek faults. It is the simplest correlation, however, and avoids the practice commonly used in the construction of structural cross sections of projecting sole faults into the nearest incompetent part of the overlying Mesozoic section, where they are conveniently lost. Sole faults involving a net slip of several miles must extend for long distances across strike unless the slip is taken up by extensive folding or imbrication in the hanging wall.

The Chungo and Stolberg structures are adjacent and closely related, though their culminations are not directly across strike from each other (Fig. 2). Their formation can be explained by a west-to-east sequence of development of step thrusts, each thrust being folded by development of the underlying thrust toward the east (Fig. 5). The lowest and most easterly thrust, the Sylvester Creek fault, is not folded because there was no subsequent step fault developed beneath it to cause folding.

Thrust Faults as Guides in Subsurface Exploration in Foothills

The lack of direct correspondence between surface and subsurface structure limits the use of surface geology for locating subsurface structural traps for petroleum. The Stolberg and Chungo structures are typical examples of this situation. The following suggestions may be useful guides in subsurface exploration in the Alberta Foothills.

1. Faults are folded by development of structures beneath them. Anticlinally folded faults are indicators of underlying anticlines and/or thrust structures.
2. Study of a sole fault at the surface may provide information on the subsurface structure on the west where the fault intersects the Paleozoic. For example, there appears to be a rough correspondence between the south-plunging end of the Stolberg structure and the southern limits of the Sylvester Creek and Brewster Creek faults. The northern limit of the Stolberg structure may be determined by finding the northern limits of the same faults. The strike extent of the Chungo structure probably corresponds to the extent along strike of the Ancona thrust. The culminations of subsurface structures should lie approximately downdip from the points of maximum displacement of their sole faults at the surface.

Not all sole faults of subsurface structures involving overthrust Paleozoic have been recognized at the surface. This may be due to (1) Quaternary and possibly Tertiary cover of fault traces; (2) lack of well control, or misin-

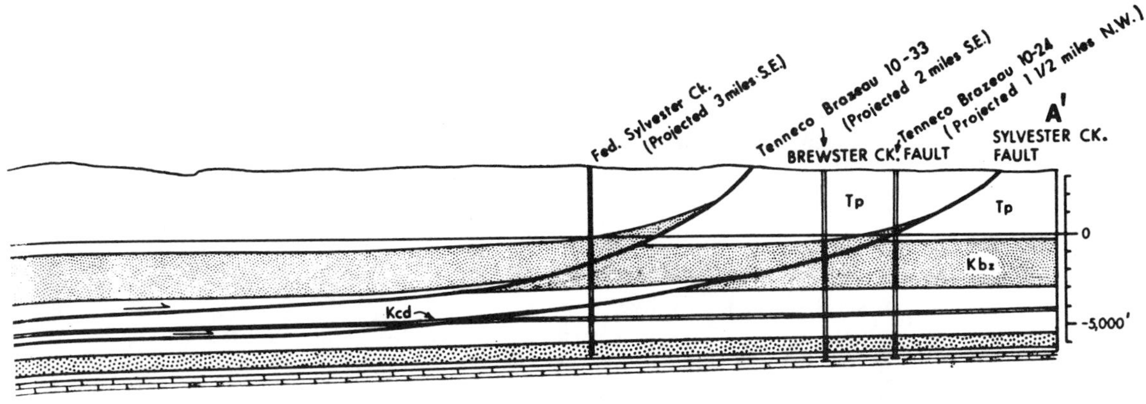

0 1 2
Miles

Tp Paskapoo
Kbz Brazeau
Kw Wapiabi
Kcd Cardium
Kb Blackstone
Kbl Mountain Park } Blairmore
Luscar
Jfn Fernie-Nikanassin
Mr Rundle

terpretation of well data; (3) passage of sole faults into bedding plane faults which are unrecognizable, and result in the faults reaching the surface many miles east of the structures they have generated; and (4) bifurcation of major thrusts into numerous smaller ones whose aggregate slip is equal to the net slip of the major thrusts.

Origin of Folded Faults

Hypotheses of formation of folded faults fall into two groups. First, it is supposed that the development of the Foothills belt took place by a period of thrusting, followed by a period of folding during which the thrusts were folded. Scott (1951, p. 2347) concluded that the sequence of tectonic events in the Chungo area involved a period of thrusting followed by an episode of compressional folding on a regional scale. Similarly, Dahlstrom *et al.* (1962, p. 394) attributed the synclinal form of the Lewis thrust in southern Alberta and British Columbia to a second cycle of deformation. In both cases, however, the structures underlying the folded faults are simple fault structures with little or no folding. If folding of the Lewis plate had taken place after thrusting, the footwall section would also have been involved. Bally *et al.* (1966, cross section E-E′) show the Lewis thrust folded over the Sage Creek structure in the west and the Waterton structure in the east. Both these structures are the result of faulting, and the syncline in the Lewis thrust plate is not present in the underlying strata.

Douglas (1950, Figs. 16, 17) showed that if thrusting took place along a stepped fault plane and if the compressive stress was not relieved completely by the thrusting, folding of the fault plane would take place, accentuating any existing irregularities. Hume (1957, p. 407) stated:

> It is inevitable, therefore, as shown by Douglas, that at the point where the dip of the fault changes upward from a steeper to a gentler slope, an anticline will develop and produce not only a fold in the fault itself but in the beds above and below it.

Figure 6 represents a process of folding of thrusts described by Verrall (1968, p. 114). It shows that the folding of fault planes involves neither the separate episode of folding proposed by Scott nor accentuation of the stepped profile in the manner shown by Douglas. The only essential mechanism is the emplacement of a step-thrust beneath the fault concerned. Figure 6 shows that (1) movement of the fault along a stepped thrust plane may not affect the underlying strata but inevitably causes folding of the overlying sequence, including any thrusts —thus, fault A, which is not a step-thrust, is folded by movement of fault B, a step-thrust, beneath it; (2) emplacement of successive step-thrusts B and C, if superimposed on one another, accentuates the folding higher in the sec-

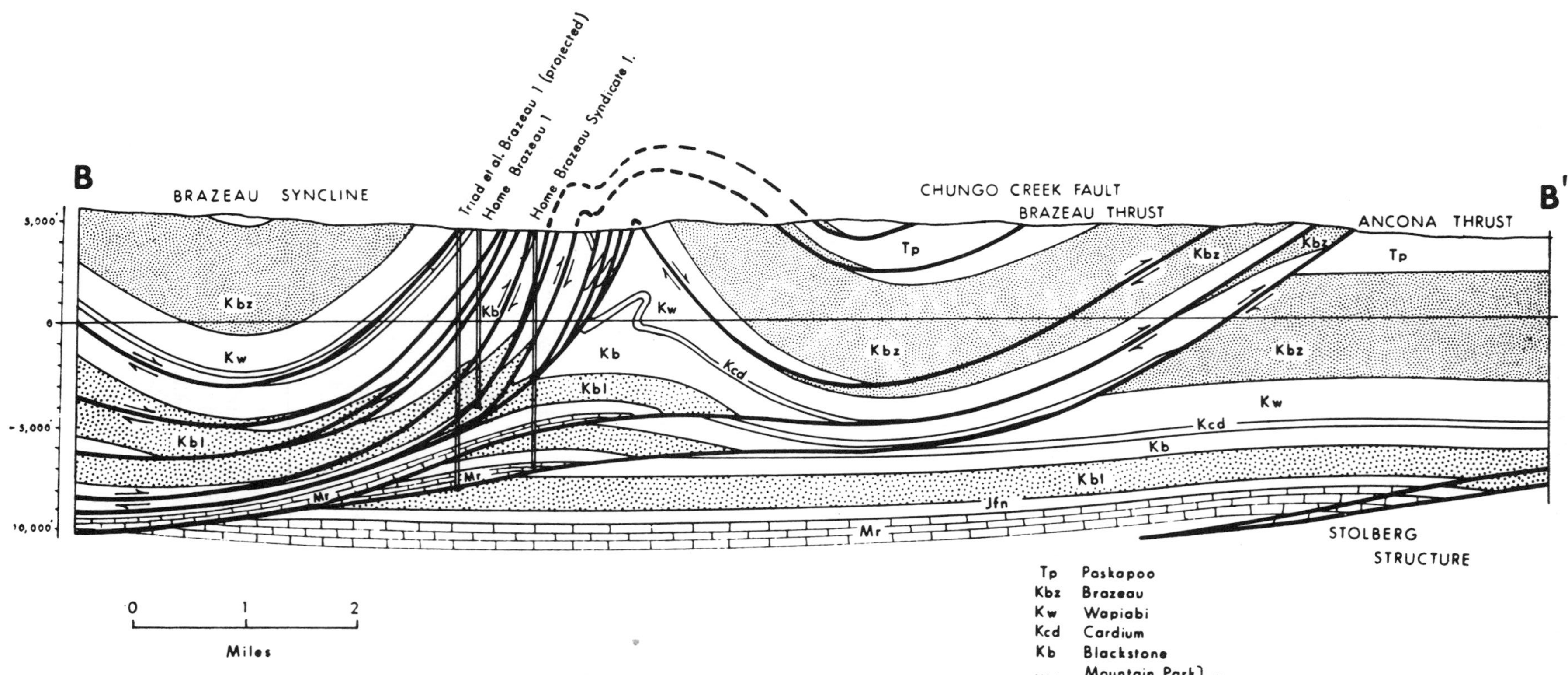

FIG. 4.—Structural cross-section B-B′, Chungo structure. Modified after Douglas (1958b).

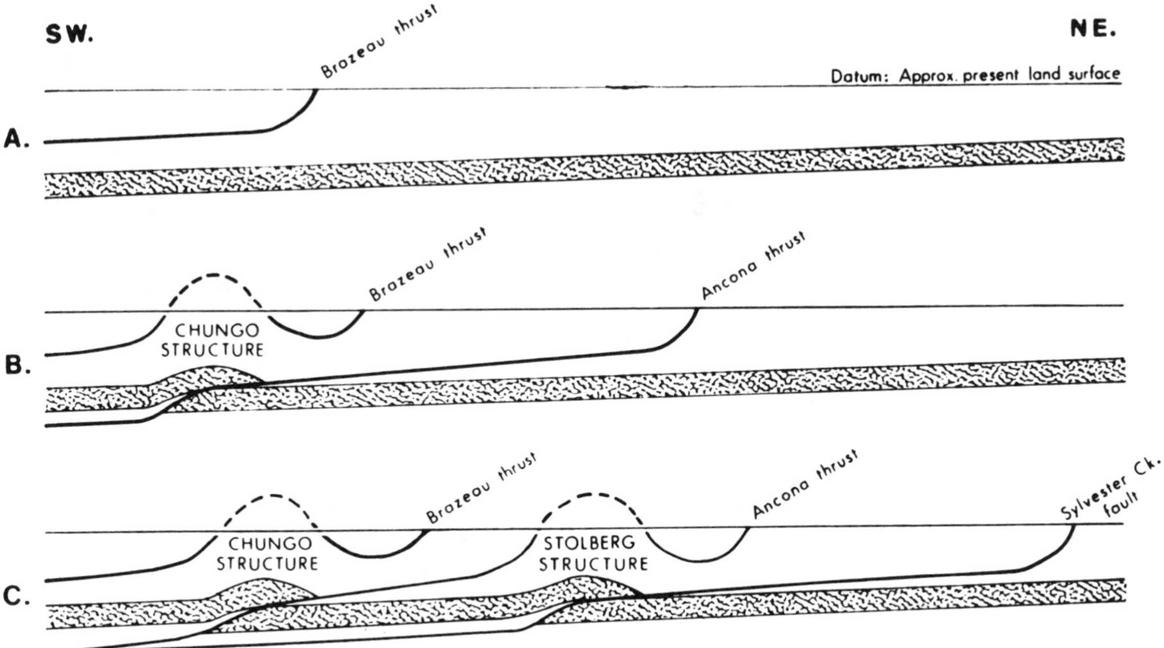

FIG. 5.—Development of Stolberg and Chungo structures, diagrammatic, not to scale. Brewster Creek and other intervening faults have been omitted for clarity. **A.** Emplacement of Brazeau thrust. **B.** Emplacement of Ancona thrust, folding of Brazeau thrust, and formation of Chungo structure. **C.** Emplacement of Sylvester Creek fault, folding of Ancona thrust, and formation of Stolberg structure

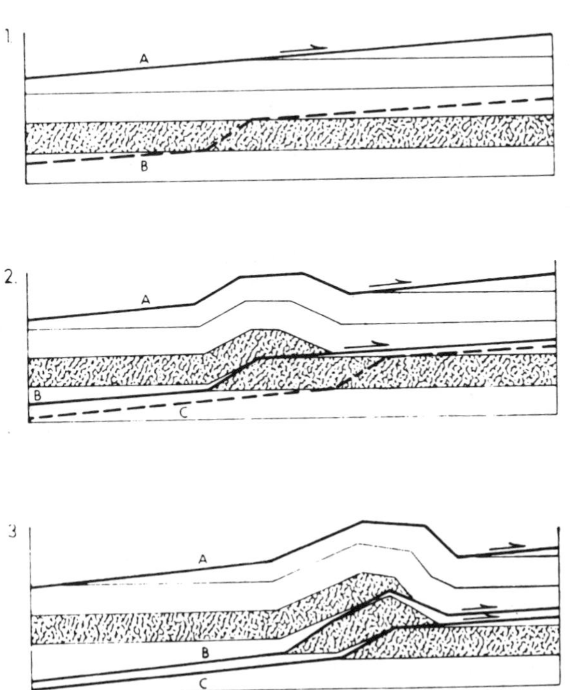

FIG. 6.—Diagram showing development of folded faults. **1.** Planar fault A, incipient step-fault B. **2.** Movement on step-fault B folds overlying sequence, including fault A. Incipient step-fault C. **3.** Movement on C folds fault B and increases folding of fault A.

tion; and (3) thrusts are concentrated in certain parts of the section, in which they appear to be of little significance, because they are bedding-plane or near bedding-plane thrusts. Only two conditions are required for this process to take place. At least one fault, preferably a lower one, need be a step-thrust, and the sequence of emplacement of thrusts is from higher to lower, *i.e.,* from west to east in the Foothills belt of Alberta.

The formation of structures such as the Savanna Creek structure in the southern Alberta Foothills can be explained simply by this mechanism. Fox (1959, p. 1013) pointed out that there is no evidence that the folded faults penetrated by wells are step faults. However, the autochthon was not penetrated and there are several major faults above it that were not penetrated by drilling, any of which may be a step fault. Similarly, the folding of the Dyson Mountain thrust in the Sullivan Creek structure (Fig. 7) can be explained in terms of a stack of stepped thrust plates.

Step faults are common in the Alberta Foothills. Geologic maps of the area and well records show that faults are much more common in some formations than others (Fox, 1959, p. 994). If faults cut all formations at the same

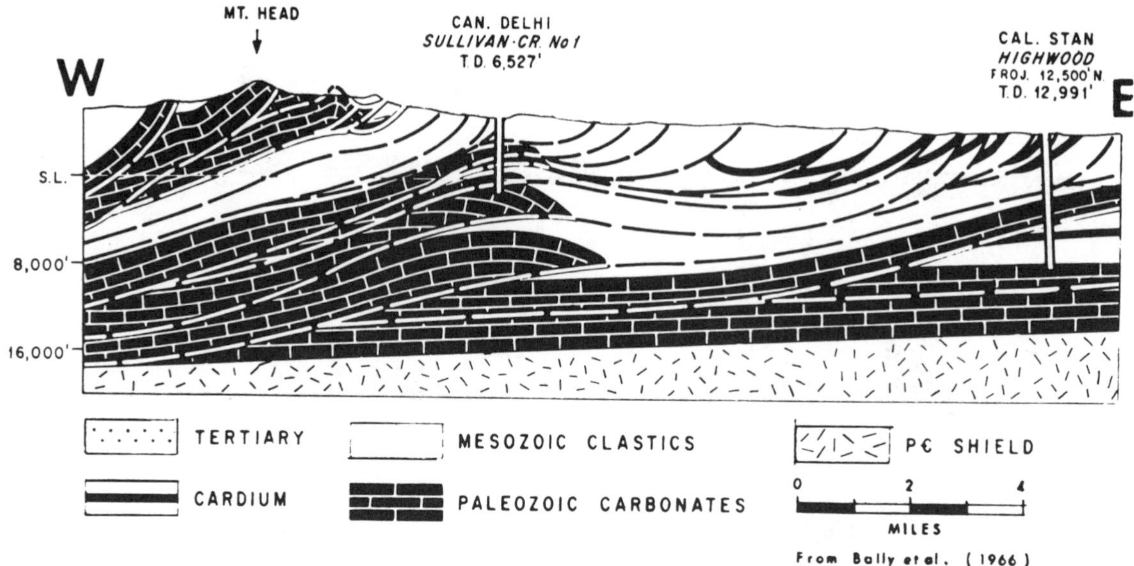

Fig. 7.—West-east cross section through Sullivan Creek structure. Compare with Figure 6, stage 3.

angle, their stratigraphic distribution would be random. Their concentration in less competent formations requires that they cut those formations at much lower angles than the others—*i.e.*, they are step faults. The step faulting is due to variations in competence through the stratigraphic column. Thus, folded faults are to be expected in the development of any thrust belt where the stratigraphic column is inhomogeneous and the order of thrusting is from higher to lower.

The mechanism described explains the formation of folded fault structures, such as Savanna Creek and Sullivan Creek, but it may not account completely for the more extreme folding of higher thrusts in the Stolberg and Chungo structures. In these structures there is considerable thickening of the incompetent shales of the Alberta Group. Over double the normal thickness of Blackstone shale was penetrated in Triad Stolberg 6–20. Fox (1959, p. 996) suggested that the thickening is due to flowage of shale or to repetition of the Blackstone by faulting. The shale-sandstone-shale sequence of the Alberta Group is an ideal sequence for the generation of step-thrusts. Given sufficient faults, even the extreme folding of the Brazeau thrust over the Chungo structure (Fig. 4) could have been produced by the process shown in Figure 6.

Reinterpretation of Foothills Structures

The formation and occurrence of folded faults may be more common in the Alberta Foothills and Rocky Mountains than are indicated by existing geologic maps and reports. Much of the early mapping of the Foothills took place before folded thrusts were recognized, or when they were regarded as exotic phenomena. There are many structures that should be reexamined. Two examples follow.

Folding Mountain Structure (Fig. 8)

The Folding Mountain structure is close to the western edge of the Foothills belt (Fig. 1). It is an overthrust anticline with a core of Paleozoic carbonate rocks, a moderately dipping west limb, and a vertical east limb. The underlying sole thrust was penetrated by the Jasper Syndicate No. 1 well, which passed from Devonian shale into sandstone and shale of the Lower Cretaceous Blairmore Group. Stratigraphic displacement (minimum net slip) is about 10,000 ft. Northeast of the well, the Folding Mountain thrust, which cuts the east flank of the structure, has been correlated with the sole thrust penetrated by the well (Webb, 1955; Mountjoy, 1959). However, net slip of the Folding Mountain thrust at outcrop near the well is not over 2,400 ft (Mountjoy, 1959, cross-sections A-B, C-D). Mapping by the writer indicates a slip of not more than 400 ft just updip from the well. If this is the sole thrust, there is a tremendous amount of "stretching" of the Blackstone beneath the thrust. The earlier cross section also implies that there is a complete reversal of dip direction only a few feet beneath the surface, for

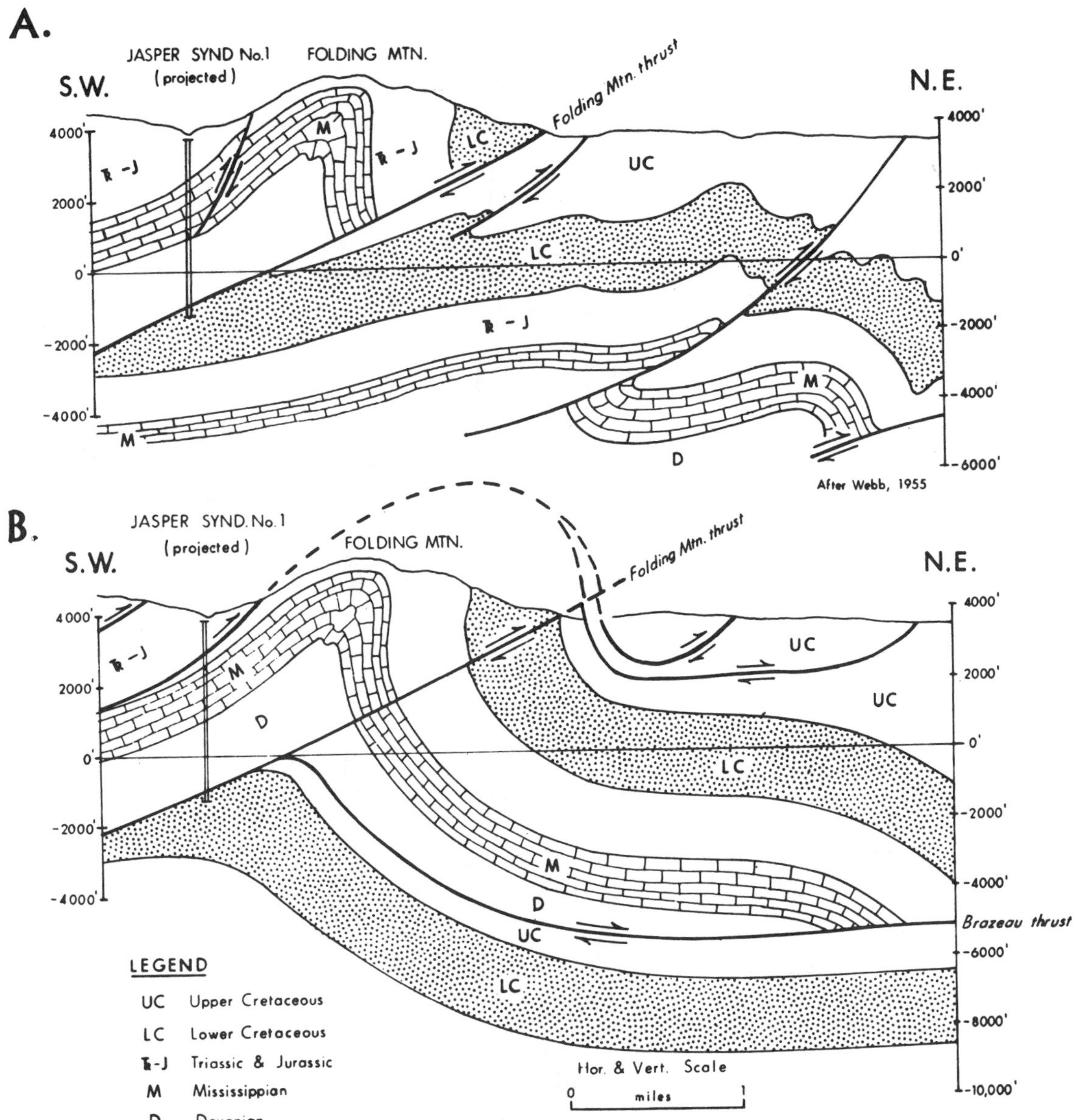

FIG. 8.—Cross sections through Folding Mountain structure. **A.** Interpretation without folded faults. **B.** Interpretation involving folded faults.

which there is no evidence. A simpler alternative explanation is offered.

In the Livingstone and Brazeau ranges of the Foothills, Douglas (1950, 1956) showed that faults cutting east limbs of similar anticlinal structures are minor imbrications of major sole faults which themselves crop out several miles farther east. This interpretation can be applied to the Folding Mountain thrust. This thrust, which cuts the east limb of the fold, is believed to be an imbrication of a folded major sole fault that was formed at or near the end of the folding. The sole fault itself is folded parallel with the overlying strata beneath the Brazeau syncline, as shown in Figure 8, emerging on the east flank of the syncline as the Brazeau thrust (Mountjoy, 1959).

Turner Valley Structure (Fig. 9)

Turner Valley is the oldest oil and gas field in Alberta. The overthrust anticline was drilled in 1913, and gas was found in the Lower Cretaceous Blairmore Group. Later drilling proved the presence of oil beneath the gas cap and of

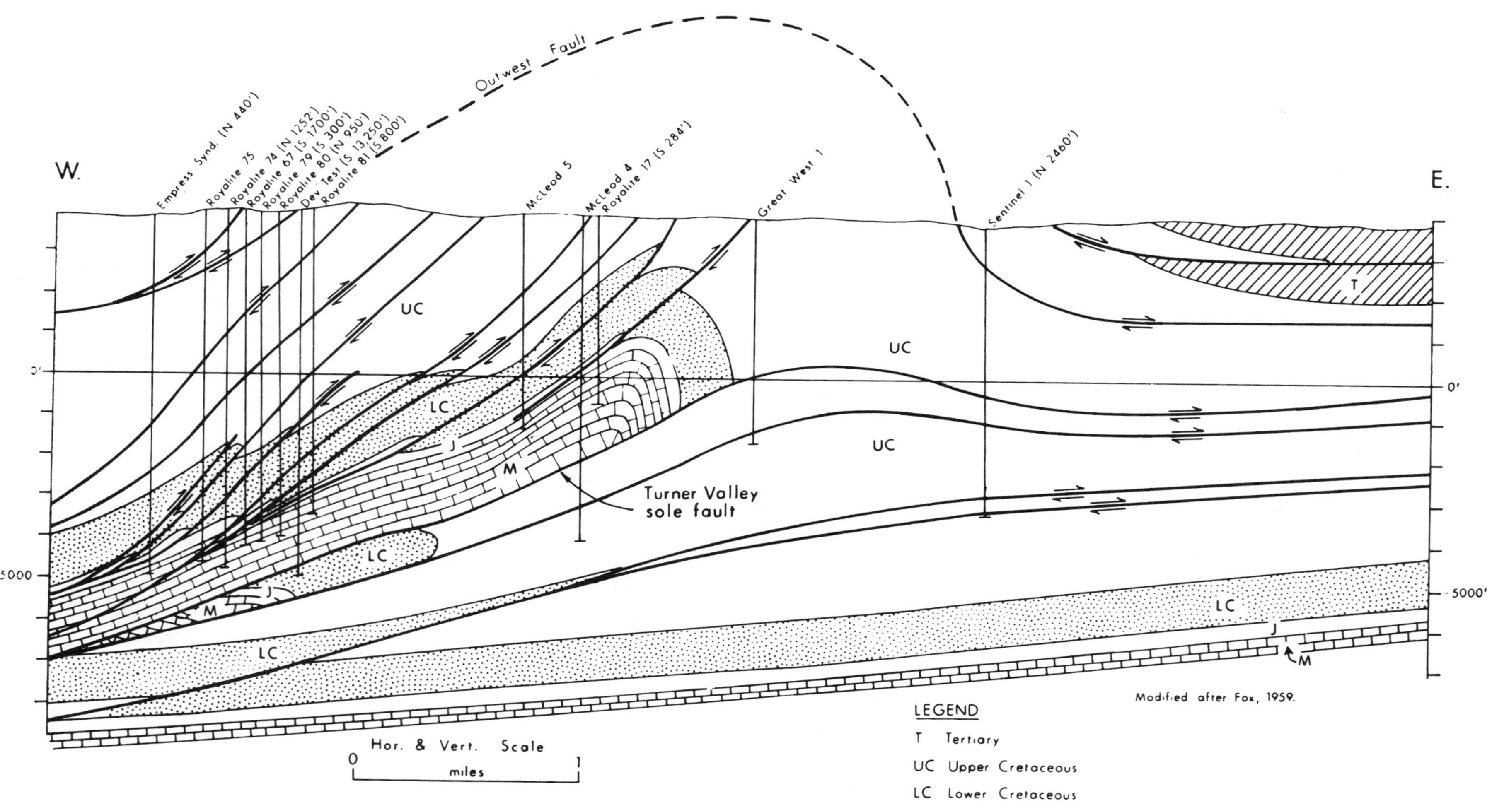

FIG. 9.—West-east cross section through Turner Valley structure, interpretation involving folded faults.

another reservoir in Mississippian carbonates, underlain by a major thrust fault. Almost 400 wells have been drilled on the structure; at least 11 penetrated the sole fault.

Geologic maps and cross sections of the structure by Hume (1931), Gallup (1951, Plate I), and Fox (1959, Fig. 4; 1969, Plate 5b) illustrate two problems that are inherent in current structural interpretations. The first is that, although the Turner Valley sole fault, which was penetrated by wells in the central part of the field, has a minimum net slip of 2 mi, the fault outcropping in the east flank of the structure, mapped as the Turner Valley sole fault (Gallup, 1951, Plate I), has little or no net slip. Hume (1939, p. 9) suggested that the sole fault bifurcates upward and is represented at the surface by numerous small faults, each having small net slip, whose aggregate slip is equal to the slip of the sole fault at depth. Fox (1969, Plate 5b) showed the sole fault dying out before it reaches the surface. The second problem is that the width of outcrop of the east-dipping Cretaceous Edmonton and Belly River Formations exposed on the east flank of the Turner Valley anticline is as much as four times greater than the width calculated from their true formation thicknesses and observed dips. Hume (1931, p. 9) suggested that this was the result of repetition by west-dipping faults. However, although thrusting causes repetition of beds, the outcrop width is increased only if the thrusts dip in the same direction as the involved beds. If they dip in the opposite direction, the result is a reduction of the outcrop width. Conventional interpretation of the Turner Valley structure fails to explain why there is no surface fault having a displacement comparable to that of the sole fault, and why the strata on the steep east flank are thickened instead of thinned.

Figure 9 shows an alternative interpretation, involving folded faults. Strata in the east flank are shown to be thickened by east-dipping thrusts, folded roughly parallel with the underlying anticline in the Paleozoic carbonates. It is probable that the numerous smaller thrusts in the west flank also extend in folded form into the east flank. The Turner Valley sole fault does not steepen and reach the surface in the east flank of the structure, but extends eastward as a low-angle fault. This interpretation explains both the excessive outcrop width and the absence of a major west-dipping thrust in the east flank. Fox (1959, p. 998) described faults in the Mesozoic section several miles southeast of Turner Valley, underlying horizontal Tertiary strata. The Turner Valley sole fault may be one of these.

The Outwest fault has a slip of 6 mi (Bally *et al.*, 1966, cross-section C-C′). Its folding over the Turner Valley structure is suggested by Gallup's (1951, Plate I) geologic map, which shows the Outwest and adjacent faults in the west flank of the Turner Valley anticline converging on a similar fault in the east flank, as they are traced toward the south-plunging anticlinal nose. It is assumed that the Outwest fault maintains its character as a near-bedding-plane thrust and follows the strata across the fold axis into the east flank. The absence of faults south of Turner Valley that are big enough to be correlated with the Outwest fault suggests that it does not continue southward.

The Folding Mountain and Turner Valley structures are not unique. Similar structures are present in other parts of the Foothills and Rocky Mountains. The examples given show how structural interpretations involving folded thrusts can reduce geologic and geometric inconsistencies. Foothills structures should be examined carefully for evidence of folded thrusts. Structural cross sections of some geologic maps of the Foothills attempt to reconcile the surface geology with well data by showing abrupt reversals of observed dip direction of strata within a few feet of the surface. Many of these dubious constructions can be eliminated by a structural interpretation involving folded faults. Such interpretations may allow reevaluation of the petroleum prospects of some areas and indicate the presence of underlying structures that are suitable for petroleum accumulation.

Sequence of Thrusting in Canadian Rockies and Foothills

Several authors have discussed the sequence of thrusting in the disturbed belt, both on a local and on a regional scale. The temporal relation between folding and thrusting—*i.e.*, whether folding preceded, followed, or was contemporaneous with thrusting—is also involved. Douglas (1950, Fig. 26) showed how back-limb thrusts form in sequence from east to west within a major overthrust sheet in the southern Alberta Foothills. Hume (1957, p. 411) follows Douglas in stating that back-limb thrusts are younger than their associated sole faults and formed from east to west, whereas the sole faults themselves formed in the reverse sequence, from west to east.

It is difficult to apply Douglas' argument because of the problem of distinguishing back-limb thrusts from major sole faults. Most major thrusts pass along strike into minor back-limb thrusts within the next major thrust plate below and on the east. An example is the McConnell thrust, underlying the eastern Rockies through much of southwest Alberta. It passes southward into a back-limb thrust within the Livingstone major overthrust sheet of the Foothills. Similarly, a small back-limb thrust within the McConnell overthrust sheet passes southward into the Lewis thrust, which is a major thrust in southern Alberta. Thus, strict application of Douglas' (1950) principles of back-limb thrusting leads to the conclusion that the Lewis thrust is later than the Livingstone thrust, the next major thrust to the east, and that the entire disturbed belt was formed in an east to west sequence.

Many geologic maps of the Alberta Foothills show major thrusts intersecting thrusts east of them. This relation led Choquette (1959) and Hunt (1956) to conclude that the order of thrusting was from east to west and that some Foothills structures existed long before they were overridden by thrusts developed west of them. However, many structures have steep to vertical flanks, and it is difficult to visualize an overthrust mass moving over such relief.

Working in the disturbed belt of Montana, Deiss (1943, p. 256) concluded that, because the Lewis thrust truncates the westernmost "high-angle" thrust of the Sawtooth Range to the east, it is the younger structure. Hurley (1959, Figs. 6, 7) also stated that the Lewis thrust is younger than the structures which it overrode. Both these authors favor an east-to-west sequence of thrusting. In the same area, Childers (1963, p. 160–162) proposed a west-to-east sequence of major thrusting, combined with a variable sequence of thrusting within each major thrust plate. He also related the folding of the Brazeau and associated thrusts in Alberta to development of thrust structures beneath them.

Fitzgerald (1962, Fig. 5) explained the warping of the McConnell thrust in the Ghost River area of Alberta by subsequent emplacement of the Panther River thrust beneath and to the east of it. The Panther River thrust itself is not shown as a folded thrust.

Bally *et al.* (1966, Figs. 6, 13) discussed the overriding and cutting of thrusts by those developed west of them. In an attempt to reconcile these relations with a west-to-east sequence of thrusting, they suggested that the relation is surficial only, and is reversed at depth. Early formed thrusts were bedding-plane slides from which imbrications cut upward to the surface. Subsequent thrusts formed below and on the east, cutting steeply upward through both the bedding-plane slides and their imbrications, and reaching the surface west of their predecessors. This explanation is unacceptable to the writer because it implies a different manner of emplacement for the earlier and later thrusts, for which there is no evidence. Moreover there is no example in the area of a low-angle thrust being displaced by a high-angle one having the same dip direction.

Hypothesis of Thrust Development

There is evidence for both east-to-west and west-to-east sequences of thrust development. A hypothesis concerning the sequence must be based on the premise that items of evidence presented are complementary and not mutually exclusive.

The structural relations between thrusts in the Alberta Foothills and Rockies are clearly shown on maps and cross sections. It is from these relations that the conflicting conclusions regarding thrust sequences have been drawn. It should be realized, however, that field relations show only the sequence of *termination* of events. They do not show the sequence in which those events started. The longer the time interval involved in the deformation of the Foothills and Rockies, the greater is the probability of lack of correspondence between sequence of initiation of thrust emplacement and sequence of termination of thrust movement. Furthermore, study of maps and cross sections may lead to a subjective impression that the relations show a sequence of discrete events. There is no reason why this should be so.

The mechanism of thrusting envisaged by the writer involves three assumptions, the first of which was stated by Douglas (1958b, p. 130). They are:

1. A thrust develops as an initial break at a given point and spreads along strike in both direction.
2. The amount of slip of a thrust is more or less proportional to the duration of its movement. A thrust that has a net slip of 5 mi moved over a much longer period than one whose net slip is 1 mi. Within one thrust, the sector where the net slip is greatest is the sector that moved longest. Although it is unlikely that thrusts moved at the same speed, the range of velocities is probably small.
3. The time interval between emplacement of successive thrusts is considerably smaller than the time required for full development of a single thrust.

If these assumptions are true, particularly

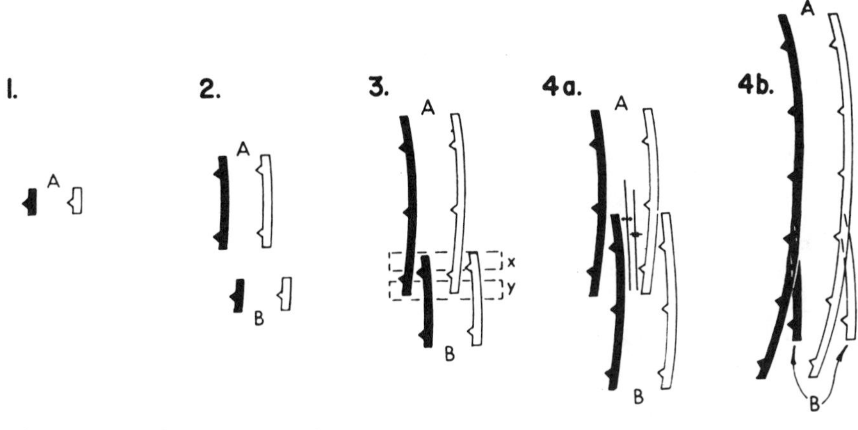

FIG. 10.—Diagram of show time and space relations between successive thrusts (plan view). **1.** Initial break of thrust A. **2.** Movement of A, initial break of thrust B. **3.** Movement of A and B. At X, thrust A is older than B. At Y, thrust A is younger than B. **4a.** Movement of thrust A ceases. Continued movement of B causes folding of thrust A. **4b.** Movement of B ceases. Continued movement of A causes it to override B.

the third, then in many places the sequence of initiation of emplacement is of no significance, and the final relations are largely independent of it. A major thrust such as the Lewis thrust, moving for a long period, would cut and override a minor one east of it, moving for a short period, regardless of which was initiated first (Childers, 1963, p. 160). The cutting of eastern by western thrusts would depend on the difference in duration of their movement or magnitude, not the sequence in which they were initiated.

Figure 10 shows how structural relations indicating an east-to-west sequence of thrusting may have been caused by a west-to-east sequence of thrust initiation. The folding of thrusts cannot be explained by an east-to-west sequence of thrust initiation. It appears, therefore, that at least all the larger thrusts were initiated in a west-to-east sequence, although locally the sequence may have been reversed.

During the Tertiary deformational history of the disturbed belt, major thrusts such as the McConnell and Lewis thrusts were active for much longer periods than lesser faults. The sequence of thrusting was predominantly from west-to-east, but because major thrusts moved for a longer period, they overrode smaller ones on the east that started to move later and stopped earlier.

Conclusions

1. Folded faults are believed to have been formed by the development of new thrusts and associated structures beneath them. In the Alberta Foothills and Rockies this means that faults were folded by subsequent emplacement of thrusts cropping out farther east.

2. Step-thrusting results in the folding of both the overlying strata and any pre-existing faults present in the hanging wall sequence.

3. Folded faults may be expected to occur in any thrust belt if the stratigraphic column is inhomogeneous and the sequence of thrusting is from higher to lower.

4. Folded faults overlie deeper structures that may contain hydrocarbons. Sole faults, where exposed, may reveal significant information on the buried structures which they generated several miles down-dip.

5. Thrusting in the Alberta Foothills and Rocky Mountains was initiated mainly in a west-to-east sequence. In many cases, however, order of termination of thrust movement was from east to west. This apparent paradox is explicable if it is accepted that (a) thrusts develop from an initial break and spread along strike; (b) amount of slip is dependent on duration of fault movement; and (c) the time interval between initiation of successive thrusts is considerably less than the time involved in the emplacement of a single thrust.

References Cited

Bally, A. W., P. L. Gordy, and G. A. Stewart, 1966, Structure, seismic data, and orogenic evolution of southern Canadian Rocky Mountains: Bull. Canadian Petroleum Geology, v. 14, p. 337–381.

Childers, M. O., 1963, Structure and stratigraphy of the southwest Marias Pass area, Flathead County, Montana: Geol. Soc. America Bull., v. 74, p. 141–164.

Choquette, A. L., 1959, Theoretical approach to foothills and mountain deformation of western Canada: Alberta Soc. Petroleum Geologists Jour., v. 7, no. 10, p. 234–237.

Dahlstrom, C. D. A., R. E. Daniel, and G. G. L. Henderson, 1962, The Lewis thrust at Fording Mountain, British Columbia: Alberta Soc. Petroleum Geologists Jour., v. 10, no. 7, p. 373–395.

Deiss, C. F., 1943, Stratigraphy and structure of southwest Saypo quadrangle, Montana: Geol. Soc. America Bull., v. 54, p. 205–262.

Douglas, R. J. W., 1950, Callum Creek, Langford Creek and Gap map-areas, Alberta: Geol. Survey Canada Mem. 255, 124 p.

——— 1956, George Creek, Alberta (geologic map with marginal notes): Canada Geol. Survey Paper 55–39, scale 1 in. = 1 mi.

——— 1956, Nordegg, Alberta: Canada Geol. Survey Paper 55–34, 31 p.

——— 1958a, Chungo Creek map-area: Canada Geol. Survey Paper 58–3, 45 p.

——— 1958b, Mount Head map-area, Alberta: Canada Geol. Survey Mem. 291, 241 p.

Erdman, O. A., 1950, Alexo and Saunders map-areas, Alberta: Canada Geol. Survey Mem. 254, 100 p.

Fitzgerald, E. L., 1962, Structure of the McConnell thrust sheet in the Ghost River area, Alberta: Alberta Soc. Petroleum Geologists Jour., v. 10, no. 10, p. 553–574.

Fox, F. G., 1959, Structures and accumulation of hydrocarbons in southern Alberta Foothills, Alberta, Canada: Am. Assoc. Petroleum Geologists Bull., v. 43, no. 5, p. 992–1025.

——— 1969, Some principles governing interpretation of structures in the Rocky Mountain orogenic belt: London Geol. Soc. Spec. Pub. 3, p. 23–41.

Gallup, W. B., 1951, Geology of Turner Valley oil and gas field, Alberta, Canada: Am. Assoc. Petroleum Geologists Bull., v. 35, no. 4, p. 797–821.

Hafner, W., 1951, Stress distributions and faulting: Geol. Soc. America Bull., v. 62, p. 373–398.

Hage, C. O., 1942, Folded thrust faults in Alberta Foothills west of Turner Valley: Royal Soc. Canada Trans., 3rd ser., sec. 4, v. 36, p. 67–78.

Hake, B. F., R. Willis, and C. C. Addison, 1942, Folded thrust faults in the Foothills of Alberta: Geol. Soc. America Bull., v. 53, p. 291–334.

Hume, G. S., 1931, Turner Valley, Alberta: Canada Geol. Survey Map 257A, scale, 1 in. = 1 mi.

——— 1939, Preliminary report on the stratigraphy and structure of Turner Valley, Alberta: Canada Geol. Survey Paper 39–4, 19 p.

——— 1957, Fault structure in the Foothills and eastern Rocky Mountains of southern Alberta: Geol. Soc. America Bull., v. 68, p. 395–412.

Hunt, C. W., 1956, Panther dome: a minor orogen of the Canadian cordillera: Alberta Soc. Petroleum Geologists 6th Ann. Field Conf. Guidebook, p. 44–55.

Hurley, G. W., 1959, Overthrust faulting and Paleozoic gas prospects in Montana's disturbed belt: Billings Geol. Soc. 10th Ann. Field Conf. Guidebook, p. 98–108.

Keating, L. F., 1966, Exploration in the Canadian Rockies and Foothills: Canadian Jour. Earth Sci., v. 3, p. 713–723.

Killick, G. H., 1954, Surface faults in the Alberta syncline: Western Miner (Feb.), p. 35–37.

MacKay, B. R., 1943 (Preliminary Map) Foothills belt of central Alberta: Canada Geol. Survey Paper 43–3.

Mountjoy, E. W., 1959, Miette, Alberta: Canada Geol. Survey Map 40–1959, scale, 1 in. = 1 mi.

North, F. K., and G. G. L. Henderson, 1954, Summary of the geology of the southern Rocky Mountains of Canada: Alberta Soc. Petroleum Geologists 4th Ann. Field Conf. Guidebook, p. 15–81.

Scott, J. C., 1951, Folded faults in Rocky Foothills of Alberta, Canada: Am. Assoc. Petroleum Geologists Bull., v. 35, no. 11, p. 2316–2347.

Thomas, A. N., 1958, Note on a geological cross-section through the Nordegg area: Alberta Soc. Petroleum Geologists 8th Ann. Field Conf. Guidebook, Aug., p. 121–127.

Triad Oil Co. Ltd., 1958, Geological map of Nordegg area, *in* Alberta Soc. Petroleum Geologists 8th Ann. Field Conf. Guidebook, Aug., scale 1 in. = 4 mi., 203 p.

Verrall, P., 1968, Observations on geological structure between the Bow and North Saskatchewan rivers: Alberta Soc. Petroleum Geologists 16th Ann. Field Conf. Guidebook, p. 106–118.

Webb, J. B., 1955, Cross-section of Foothills belt east of Jasper Park: Alberta Soc. Petroleum Geologists 5th Ann. Field Conf. Guidebook, p. 130a.

White, W. H., 1959, Cordilleran tectonics in British Columbia: Am. Assoc. Petroleum Geologists Bull., v. 43, no. 1, p. 60–100.

Reprinted by permission of the Canadian Society of Petroleum Geologists from *Bulletin of Canadian Petroleum Geology*, v. 17, no. 3 (1969), p. 326-346.

BULLETIN OF CANADIAN PETROLEUM GEOLOGY
VOL. 17, NO. 3 (SEPTEMBER, 1969), P. 326-346

THE UPPER DETACHMENT IN CONCENTRIC FOLDING[1]

C. D. A. DAHLSTROM[2]
Calgary, Alberta

ABSTRACT

During concentric folding the deformed rocks are divided into two or more discrete structural units. In the Fernie Basin area the internal structure and both boundaries of one concentrically folded panel can be studied. This structural-lithic unit is separated from the overlying rocks by a decollement and from the underlying rocks by a major thrust which probably was a decollement during folding. Recognizing these relationships requires the application of the principle of down-plunge projection.

INTRODUCTION

The purpose of this paper is to show from geometry and to illustrate by example that both the upper and lower boundaries of a concentrically folded sequence of rocks are detachments. The necessity for a lower detachment is generally appreciated, but the upper parts of the structures are ordinarily eroded so that little thought is given to whether an upper detachment existed or not. Illustrations were selected from the Fernie Basin area in southeastern British Columbia (Fig. 1) because here it is possible to recognize the upper detachment and to see the major thrust plane which was the lower detachment. Deriving these conclusions from the available Fernie Basin data requires an understanding and application of several geometric concepts which many geologists have not had occasion to use in their normal work. Consequently, this paper begins with a brief description of these concepts.

THE FOOTHILLS FAMILY

The Cordillera that forms the backbone of the North American continent is flanked on its eastern side by a marginal belt of thrusts and folds which adjoin relatively undeformed rocks lying in normal stratigraphic succession on the Canadian Shield basement. The segment of this marginal belt that contains the Fernie Besin extends unbroken from the Liard River in British Columbia to the Idaho Batholith in Montana, a distance of 1000 mi. Much of the basic understanding of this segment derives from government and industrial geological work undertaken in connection with the search for oil and gas in the "Foothills" and "Front Ranges" of Alberta. Physiographically these are two separate provinces, but structurally they

---

[1]Manuscript received June 5, 1969.

[2]Chevron Standard, Calgary, Alberta. The writer is indebted to Chevron Standard for permission to publish this paper and for the opportunity to participate in their foothills mapping program. Chevron's basic mapping of the Fernie Basin was done during the summers of 1956 and 1957 by parties under the direction of G. G. L. Henderson. All the oblique photographs were taken by K. T. Hyde, aerial photographer of Calgary.

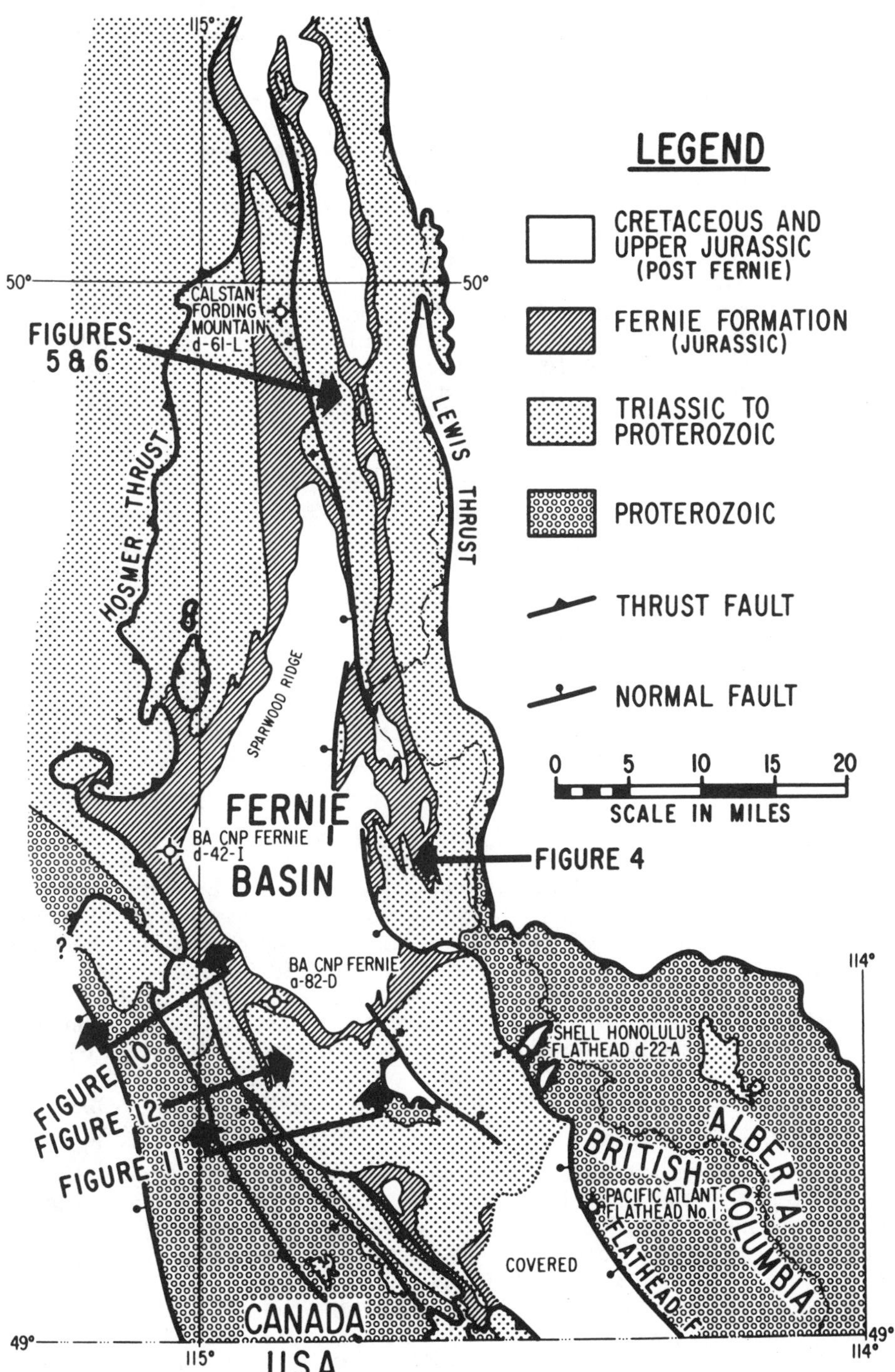

Fig. 1. Regional geology of the Fernie Basin area.

are a single entity characterized by a specific assemblage of structures, the "foothills family."

The word "family" is intended to emphasize the concept that a particular structural province can contain only a *specific and limited* suite of structural types. This idea is not new, since it is fundamental to Bucher's (1933) concept of "natural laws" and inherent in De Sitter's (1956) treatment of "comparative structural geology." Nevertheless, reiteration is justified because the idea of structural families is so useful in structural interpretation when data is incomplete and there are several conceivable solutions. Here the family concept is helpful because it permits one to eliminate those solution that require structures not appropriate to that particular structural province. It also permits effective use of well-exposed foreign samples as models for interpretation by analogy, because structural provinces have the same family of structures regardless of the geographic location. The foothills family, for example, is characteristic of the margin of an orogen whether it be the Canadian Rockies, the Appalachians or the Andes. The "foothills family" of structures consists of:

1. concentric folds,
2. detachments (decollements),
3. low-angle thrust faults (often folded),
4. tear faults,
5. late normal faults.

This paper is primarily concerned with the detachment horizons, but their existence is contingent on concentric deformation.

Concentric Folding

Concentric folds are those wherein the thickness of individual units, measured perpendicular to the bedding planes, is constant in all parts of the fold. Acording to Goguel's "law of conservation of volume," beds have essentially the same volume before and after deformation, except for the rather negligible amount of compaction that occurs during deformation. If total volume and bed thickness are both constant during concentric folding, it necessarily follows that the surface area of individual bedding planes is also constant. In transverse cross-sections this means that the bedding planes have the same length after folding as they had when deposited.

Figure 2 shows a simple concentric fold. In the central beds (2 to 5) of the folded panel there is an increase in the circumference of the synclines downwards which is exactly compensated by a decrease in the circumference of the anticlines. Bed lengths in units 2 to 5 are constant between the anticlinal and synclinal axial planes. This neat balance between synclines and anticlines fails beyond the centres of curvature, where geometric problems of excess bed length and volume develop (unit 6) that are resolved by complex crenulations or faulting or both (unit 1). Figure 2 depicts the geometric problems in the synclines and the alternative geometric solutions in the anticlines, but this is purely for explanatory purposes. The model is, of course, bilaterally symmetrical about the

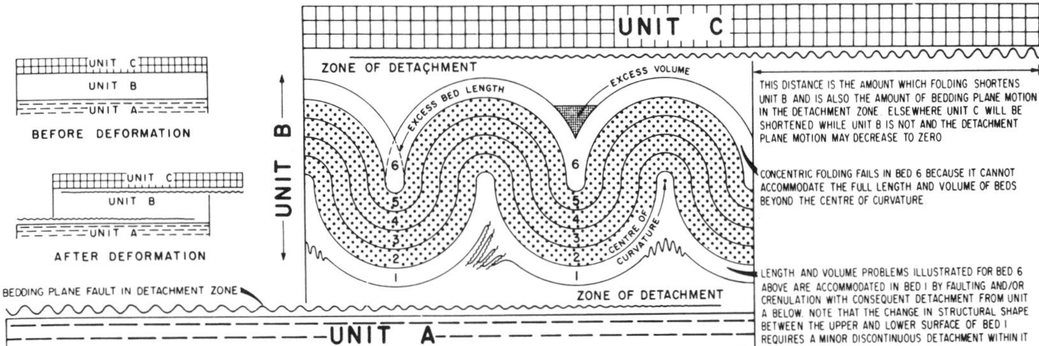

Fig. 2. The ideal concentric fold (from Dahlstrom 1960).

horizontal plane, so that in nature both the synclinal troughs and the anticlinal crests must contain contorted or faulted beds. This structural thickening (units 1 and 6 of Fig. 2) abruptly diminishes fold amplitude, thereby restricting the vertical range of concentric folding and limiting the thickness of sedimentary rock that can fold together as a simple unit. Therefore, a concentrically folded "panel" or structural-lithic unit (Currie *et al,* 1962) must be separated from the rocks above and below it by a detachment horizon. The lower detachment always exists, but sometimes the upper detachment does not and never did exist because the folded panel extended all the way to surface during deformation, so that the upper "detachment" was the interface between rock and air.

Concentric folding is characterized by substantial changes of structural shape with depth, which means that "structural style" is a function of the erosion level. Figure 3 shows four distinctly different tectonic styles derived from one concentrically folded panel. Recognizing these tectonic styles as different aspects of the same basic model is helpful because each is a statement of depth to detachment (or height to an eroded detachment). This may be quite obvious when fold plunges are steep, but it is not at all obvious when fold plunges are subhorizontal because the same tectonic style then exists over very large areas. The Fernie Basin area is a particularly convenient example because several tectonic styles are exposed in a relatively small area, making their interrelation more apparent.

Figure 4 is from a study by R. A. Price (1964) of the geometric characteristics of concentric folding. Several of his carefully mapped field examples are in the Fernie Basin area and show that concentric folding is characteristic of the district.

Plunge Projection

Figures 5 and 6 illustrate the geology of the Wisukitsak Range in the northern part of the Fernie Basin. The features of particular interest are several relatively small concentric folds in the Permo-Penn Rocky Mountain, the Mississippian Rundle and the Triassic Spray River formations, which do not seem to extend through the overlying Jurassic Fernie Formation and into the Jurassic Kootenay Formation. The folds are well exposed

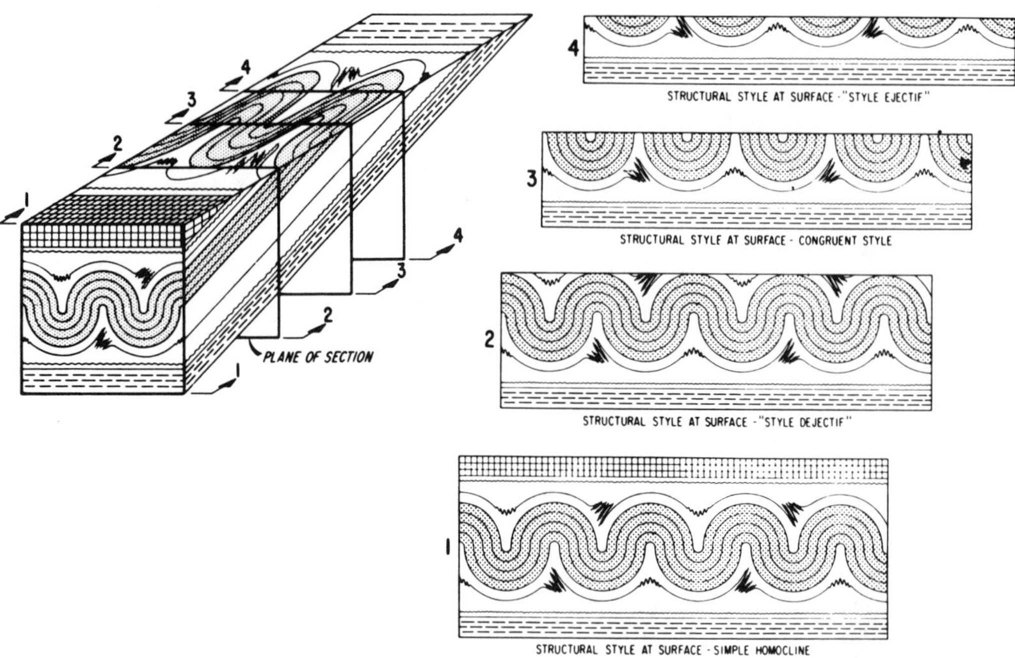

Fig. 3. Relationship between structural style and depth of erosion in a concentrically folded terrane.

in the cliff face and their crests can be traced as they plunge down the back slope of the Wisukitsak Range toward the Jurassic exposures. From the photographs in Figure 6 it appears that there is a structural discordance between the folded Triassic and the overlying Jurassic, of the kind shown in the composite cross-section of Figure 5. However, photographs are two-dimensional representations that juxtapose objects in the foreground with those in the background. Structurally, this is equivalent to taking the fold shapes in the foreground and projecting them, unchanged, several miles to a distant cross-sectional plane in the background (Fig. 5). Alternatively one might postulate that the folds die out along plunge, and that they are not reflected in the Jurassic units because they do not extend that far horizontally. Either of these alternatives could be true, depending on whether the folds are cylindrical or conical.

Cylindrical folds maintain a constant shape, so that successive cross-sections show the same fold form. In contrast, conical folds do not maintain their shape, so that successive cross-sections show a progressive change in fold form. Imagine a line moving parallel to itself in such a way that both its ends are describing the same curved form. The surface formed by such a moving line is a cylindrical surface. If the line moves so that only one end describes a curved form, while the other pivots about a fixed vertex, then the surface formed would be conical. If a cardboard mailing tube is cut in half longitudinally, the result is a cylindrical fold with a semicircular cross-section. If one end of the tube is pinched, then the fold becomes conical because one end has a semicircular cross-section while

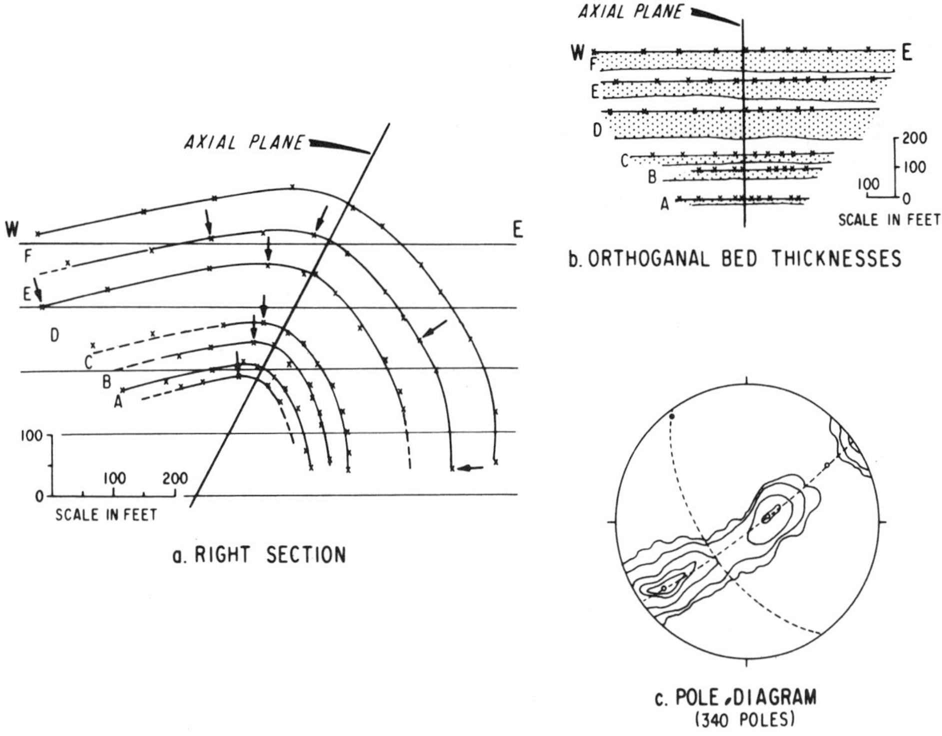

Fig. 4. A concentric fold from the Fernie Basin (from Price, 1964, reproduced with permission).

the pinched end has a parabolic cross-section (see Wilson, 1967, and the papers cited in his bibliography).

There are two basic ways of determining whether a fold is cylindrical or conical:

1. By direct observation. If a series of natural cross-sections is available, one can determine whether the shape is essentially constant (cylindrical fold) or whether it changes (conical fold). This is the simplest method when exposure permits and it is the one used in this paper.

2. By statistical analysis of bedding-plane attitudes. When poles to bedding attitudes are plotted on a stereonet they form girdle patterns, which lie along great circles if the fold is cylindrical or along small circles or other curved shapes if the fold is conical (Dahlstrom 1954). For example, in Figure 4a the contoured poles to bedding define a great circle and therefore the fold is cylindrical. Figures 4b and 4c also establish that the fold is concentric, so it is, then, a cylindrical, concentric fold, which is the normal kind in the "foothills family."

Whether a fold is conical or cylindrical is important because it determines how map data should be projected to the plane of section.

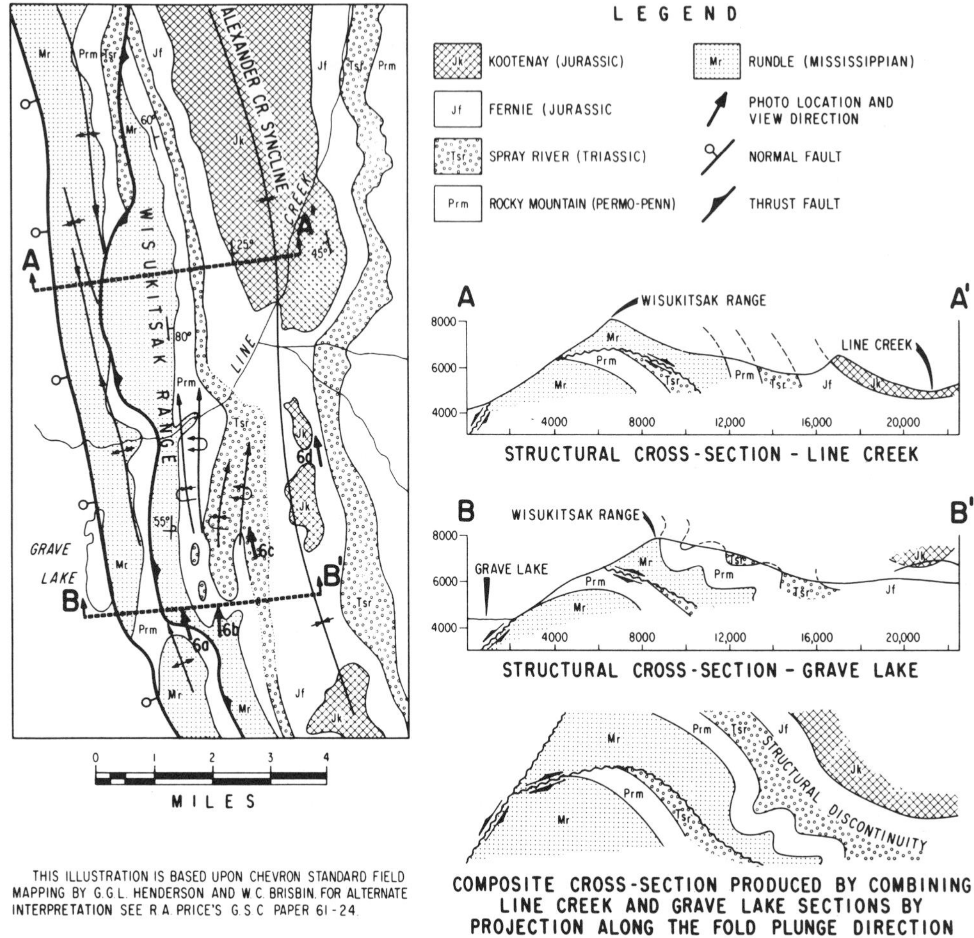

Fig. 5. Geology of the Wisukitsak Range, in the northern part of the Fernie Basin.

Imagine an actual cylinder like a simple piece of pipe oriented in space with a distinct plunge. If that pipe is cut by a horizontal plane, the resulting "map view" shows an elliptical form. If it is cut by a vertical plane, the resulting "cross section" is another ellipse. Both these ellipses are sections of the same piece of pipe, so that given one of the ellipses (e.g., the map shape) and the plunge of the pipe, it is a relatively simple problem in descriptive geometry to construct the other (i.e., the cross-sectional shape) by projecting data parallel to the pipe axis (Fig. 7a). Similarly structural data pertaining to a cylindrical fold can be transferred from a map to its appropriate location on a cross-section by projecting it parallel to the fold axis (Stockwell 1950).

Fig. 6. Oblique aerial views of the Wisukitsak Range. Looking north, 6a is on the west and 6d on the east.

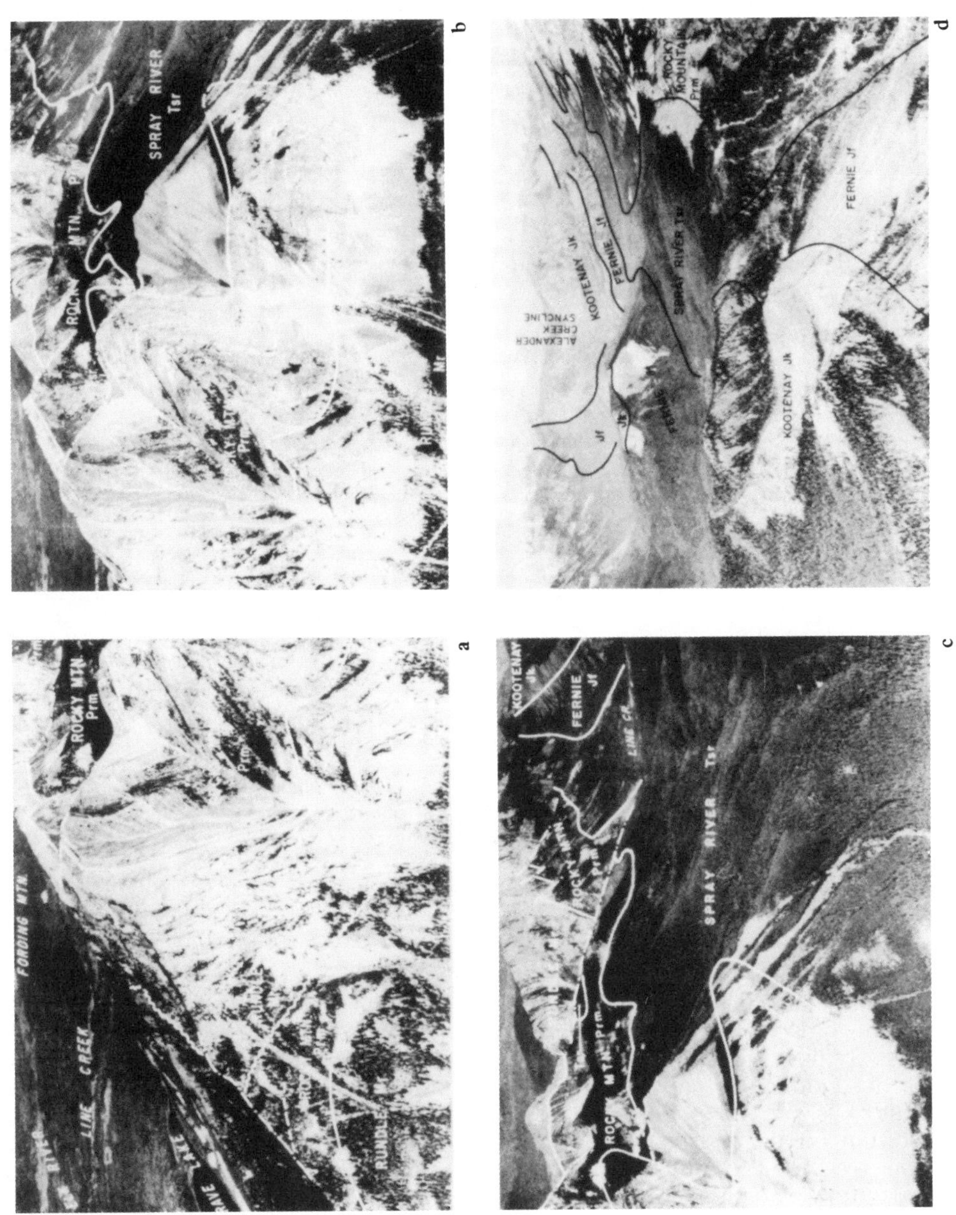
a
FORDING MTN.
LINE CREEK
ROCKY MTN.
Prm
RUNDLE
b
ROCKY MTN.
SPRAY RIVER
Tsr
c
KOOTENAY
Jk
FERNIE
Jf
LINE CR.
SPRAY RIVER Tsr
ROCKY MTN. Prm
d
ALEXANDER
CREEK
SYNCLINE
KOOTENAY Jk
FERNIE Jf
SPRAY RIVER Tsr
ROCKY
MOUNTAIN
Prm
KOOTENAY Jk
FERNIE Jf

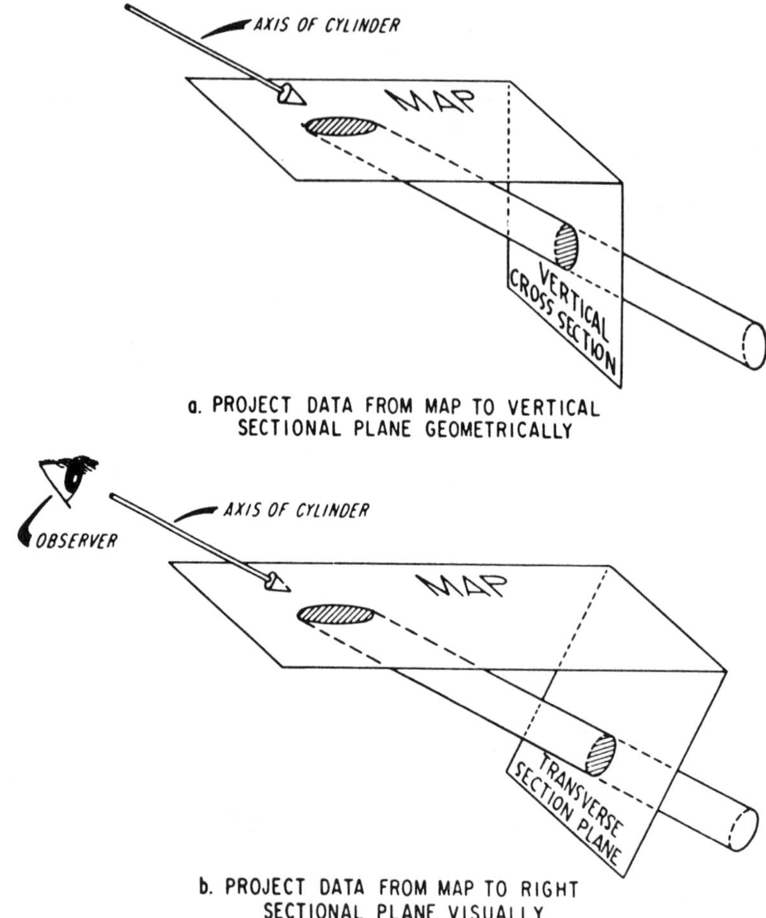

Fig. 7. Down-plunge projection and viewing.

If the pipe is cut by a plane perpendicular to the pipe axis, the "cross section" shows a circular form which is a true representation of the shape of the pipe. It is the same "end-on" view that one can get by looking directly down the pipe axis (Fig. 7b). This idea of end-on viewing can be used to optically convert geological maps into cross-sections. The procedure is simply to hold the map in a position such that the observer is looking obliquely into it down the plunge direction of the fold. Closing one eye helps, by eliminating the sense of depth perception. The result is an optical foreshortening of the map to produce a cross-section (Mackin 1950). The reader is invited to apply this technique to the geological map in Figure 9, particularly if he is the kind of fellow who peeks at the latter pages of "whodunnits."

Similar but more complicated analytical and projection techniques are available to handle conical folds. These will not be discussed because most concentric folds (including those in the Fernie Basin) are cylindrical over

TABLE OF FORMATIONS

MODIFIED AFTER DOUGLAS (1952) CRABB (1957) AND PRICE (1962)

AGE	FORMATION	PRINCIPAL LITHOLOGY	THICKNESS
TERTIARY	KISHENEHN	CONGLOMERATES, MUDSTONES & MARLS	0 - 6600?
	UNCONFORMITY		
UPPER CRETACEOUS	BELLY RIVER	NON-MARINE SANDSTONE & SHALE	3000 - 4200
	WAPIABI	MARINE GREY SHALE	1000 - 1600
	CARDIUM	MARINE SANDSTONE	30 - 300
	BLACKSTONE	MARINE GREY SHALE & SILTSTONE	450 - 1000
UPPER CRETACEOUS ?	CROWSNEST	VOLCANIC AGGLOMERATE & TUFF	0 - 1000
LOWER CRETACEOUS	BLAIRMORE	NON-MARINE SANDSTONE, SHALE & CONGLOMERATE	1000 - 2200
	DISCONFORMITY		
LOWER CRETACEOUS -JURASSIC	KOOTENAY	NON-MARINE SANDSTONE, SHALE & COAL	600 - 2400
JURASSIC	FERNIE	MARINE BLACK SHALE	480 - 1500
	DISCONFORMITY		
TRIASSIC	SPRAY RIVER	MARINE LAMINATED SILTSTONE	0 - 1800
PERMO-PENN	ROCKY MOUNTAIN	ORTHOQUARTZITE & ARENACEOUS DOLOMITE	100 - 1000
MISSISSIPPIAN	RUNDLE GROUP		2000
	ETHERINGTON	SILTY DOLOMITE	
	MOUNT HEAD	THIN BEDDED LIMESTONE	
	LIVINGSTONE	MASSIVE CRINOIDAL LIMESTONE	
	BANFF	DARK, ARGILLACEOUS CHERTY LIMESTONE	800
MISSISSIPPIAN ?	EXSHAW	BLACK, FISSILE SHALE	15 - 40
	DISCONFORMITY		
DEVONIAN	PALLISER	CLIFF-FORMING MOTTLED LIMESTONE	900 - 1000
	FAIRHOLME	DARK GREY LIMESTONE, REEFOID DOLOMITE	1600
	UNCONFORMITY		
CAMBRIAN	DOLOMITE UNIT	LIGHT GREY DOLOMITE	280 - 350
	SHALE UNIT	GREEN SHALE & LIMESTONE	215
	QUARTZITE UNIT	LIGHT YELLOWISH GREY QUARTZITE	140
	UNCONFORMITY		
PRECAMBRIAN	KINTLA	RED & GREEN ARGILLITE & QUARTZITE	600 - 1600
	SHEPPARD	DOLOMITE, QUARTZITE, SILTSTONE	150
	PURCELL LAVA	AMYGDALOIDAL ANDESITE	320
	SIYEH	GREY DOLOMITE	1150
	GRINNELL	RED ARGILLITE	350
	APPEKUNNY	GREEN ARGILLITE	1700
	ALTYN	GREY ARGILLACEOUS LIMESTONE	1500
	WATERTON	BANDED LIMESTONE & DOLOMITE	300 +

Fig. 8. Table of formations, Fernie Basin area.

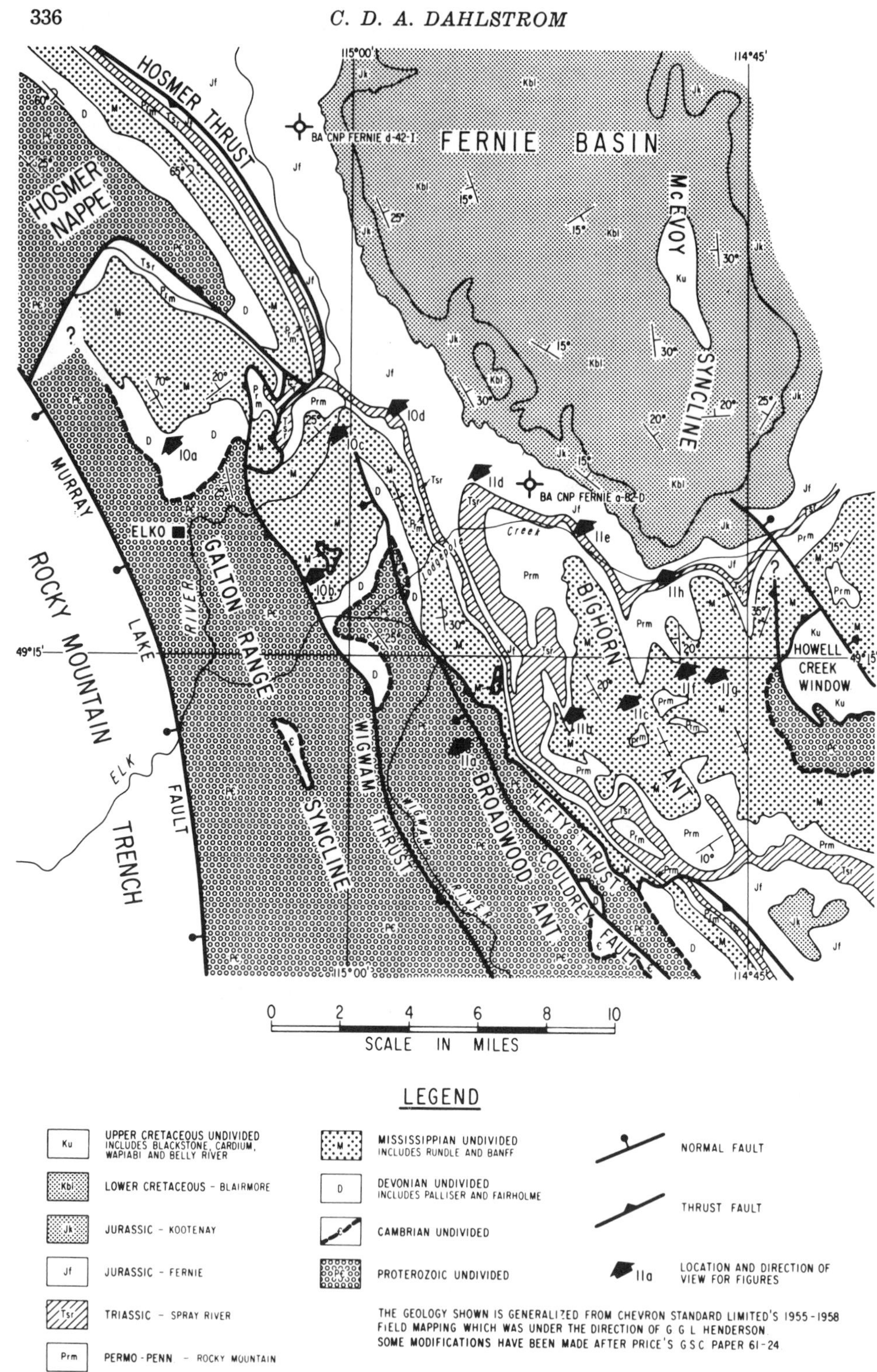

Fig. 9. Geology of the south end of the Fernie Basin area.

most of their length. This does not mean that folds continue forever. Obviously they have to end somewhere, and one of the ways in which they can end is to become conical. Even then the aberrant conical section is a relatively small proportion of the fold's total length.

However, if the fold does become conical then obviously the simple methods of cylindrical projection no longer apply. This raises the question of just how far it is legitimate to project a cylindrical structure beyond the point where one was last able to prove it cylindrical. Clearly one's judgment must, in some way, be related to scale, so that small folds are projected short distances whereas large folds are projected longer distances. Aside from this rather platitudinous generality, the best guide is the observed structural pattern of the area itself. If the fold trends are long and straight, then so can the projections be, but if the fold trends are complex *en echelon* features, then projections must necessarily be short.

Geology of the Fernie Basin

The term "basin" is a misnomer. Although the Fernie area is structurally a broad open syncline at Mesozoic level, it is not a "depositional basin" at any level. The term derives from the extensive coal deposits in the Jurassic Kootenay Formation and the common usage of "basin" as a term to describe almost any significant coal-mining area.

The sediments in the Fernie Basin area belong to the shelf and miogeosynclinal sequences. The Precambrian rocks are unaltered carbonates and clastics, the Palaeozoics are carbonates and the Mesozoics are clastics. As is usual in shelf and miogeosynclinal sequences, there are several disconformities within the succession which represent substantial time gaps without appreciable angularity. Figure 8 briefly describes the stratigraphic column.

The photographs in Figure 10 in conjunction with the maps in Figures 1 and 9 illustrate the structure of the western edge of the Fernie Basin. (In viewing these and subsequent photographs the reader should bear in mind that the terrain is very rugged. Although the photographs look along strike, they were taken at a constant elevation and therefore do not look down plunge. Because they are rarely the kind of end-on view shown in Figure 7b, they do contain some optical illusions). The western edge of the Fernie Basin is composed of three structural elements:

1. The uppermost element is the great recumbent nappe overlying the Hosmer thrust. The overturned limb and "brow" of this nappe retains its shape over 15 mi of gentle north plunge (Henderson and Dahlstrom 1959).
2. The intermediate element is a panel of rock bounded by the Hosmer and Wigwam thrusts. It contains the Galton Range syncline, which retains its basic form over 25 mi of subhorizontal plunge from the United States border to the Elk River, where it begins to plunge abruptly to the north, bringing the anticline in the Palaeozoics on the hanging wall of the Wigwam fault into view. The anticline-syncline pair retain their form over 7 mi of substantial north plunge until they disappear beneath the Hosmer Nappe.

3. The lower element is the Broadwood anticline, which retains its fundamental form over 25 mi of subhorizontal plunge from the U.S. border to the Wigwam River. At that point it begins to plunge abruptly northward. Except for complications to be discussed, the Broadwood anticline at Palaeozoic level retains its form over 5 mi of pronounced north plunge. The changes from subhorizontal to pronounced north plunge in the Galton Range syncline and the Broadwood anticline are an expression of the North Kootenay Pass monocline (Price 1965).

Because Figure 10c is practically a plunge view down the axis of the Broadwood anticline it is, except for perspective distortion, a cross-section which displays the relationship between these three structural elements quite clearly. The folds within these elements retain their basic form over long distances and are essentially cylindrical folds. The object in showing these structures is to establish the existence of cylindrical folds and thereby to justify the projection of data along plunge to produce geometric, optical or mental cross-sections like Figure 10c.

Figures 11a and b are additional views of the Broadwood anticline which, taken in conjunction with Figure 10c, provide graphic evidence that structures can and do change laterally. At the north-plunging end (Figure 10c) the Mississippian beds wrap unbroken around the nose of the Broadwood anticline and into the overturned eastern limb (Figures 10c and 10d). Farther south (Figures 9 and 11a), that same overturned limb is cut by the Hefty fault, which thrusts Precambrian rocks from the anticlinal core eastward over the overturned Mississippian forelimb. Because Mississippian rocks are involved in both instances this is a true lateral change in structure from north to south, caused by an increase in the amount of horizontal displacement on the Hefty thrust. Lateral changes like this preclude simple plunge projection, because the anticlinal axis above the thrust and the syncline axis below have different plunge directions. Fortunately in such instances, attempts at simple projection usually produce contradictions which show that something is wrong. When it is recognized that there are actually two plunges involved, it is not particularly difficult to project structural data from the anticlinal segment above the fault along the anticlinal plunge, and the data from the syncline along the synclinal plunge. However, this example does sound a warning that wholesale projection of large areas along a single plunge direction may be misleading, if even structures as closely related as an anticline-syncline pair can have divergent plunges.

The Couldrey fault is a late normal fault which displaces the anticlinal part of the Broadwood fold. It is a small feature by comparison with the other structures, and it simply offsets them because it appears after the folds and thrusts were fully developed.

Figures 11a, b, c, d, f, and g provide a photographic traverse from the forelimb of the Broadwood anticline (which is visible in the background of Fig. 11a) eastward to the major culmination of the Howell Creek window. In this distance there are three anticlinal axes plunging north towards the Fernie Basin. Figure 11c is a view looking parallel to the vertical limb of

---

Fig. 10. Oblique aerial views of the structures in the Paleozoic and Proterozoic rocks on the western margin of the Fernie Basin. Looking north, 10a is on the west and 10d is on the east.

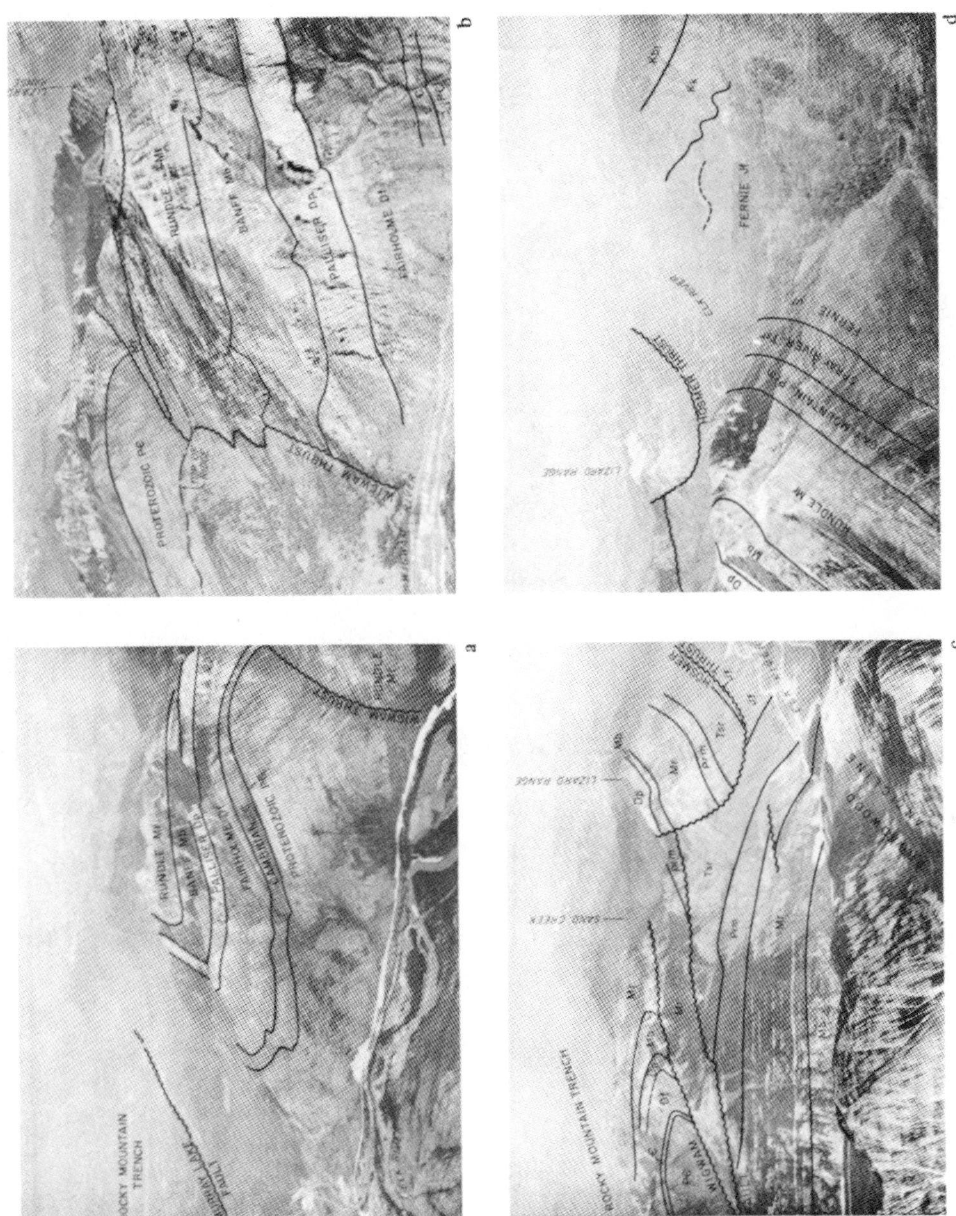
a
ROCKY MOUNTAIN TRENCH
RUNDLE Mr
BANFF Mb
PALLISER Dp
FAIRHOLME Df
CAMBRIAN Є
PROTEROZOIC PЄ
WIGWAM THRUST
b
PROTEROZOIC PЄ
WIGWAM THRUST
RUNDLE
BANFF Mb
PALLISER Dp
FAIRHOLME Df
c
ROCKY MOUNTAIN TRENCH
SAND CREEK
LIZARD RANGE
HOSMER THRUST
d
LIZARD RANGE
HOSMER THRUST
FERNIE Jf
RUNDLE M
ROCKY MOUNTAIN Prm
SPRAY RIVER Trs

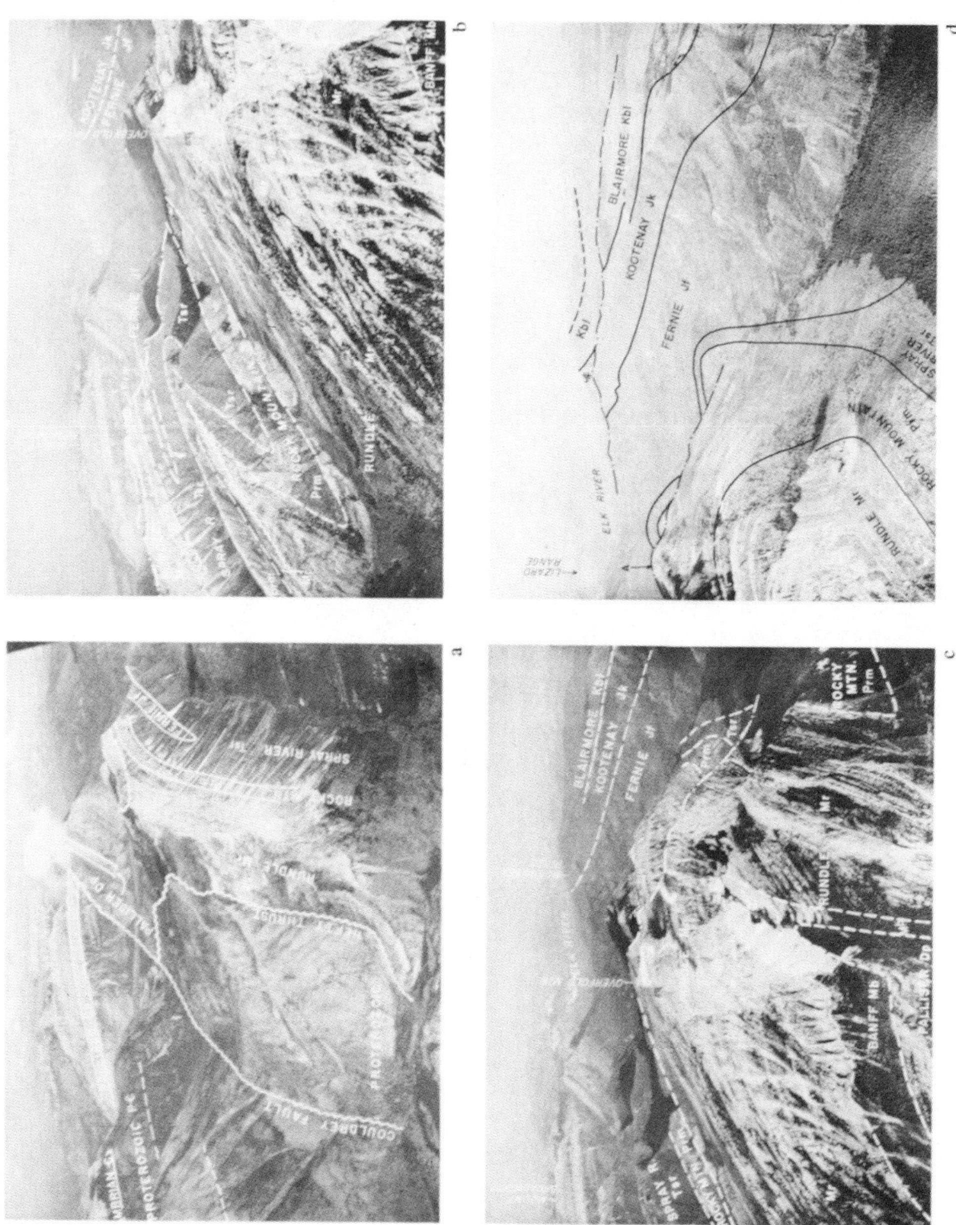
a
b
c
d
BLAIRMORE Kbl
KOOTENAY Jk
FERNIE Jf
ELK RIVER
LIZARD RANGE
ROCKY MOUNTAIN Prm
RUNDLE Mr
COULDREY FAULT
PROTEROZOIC
SPRAY RIVER Trs
BANFF Mb

the Bighorn anticline, the central and largest of these three. According to one of the basic rules of map interpretation, the vertical dips on a fold strike in the direction of plunge (Wegmann 1929). Therefore, this view is in the direction of the plunge, but it is not quite a down-plunge view because the point of observation is about a thousand feet too low. This produces an optical illusion wherein the Permo-Penn and Triassic units disappear on the plunge end of the anticline, although they can be seen to wrap around the nose of the anticline in Figure 11d. Despite the cross-sectional imperfections of this photograph, it does demonstrate that the tight asymmetric Bighorn anticline is plunging directly toward the gently folded Jurassic of the Fernie Basin, and that there is no reflection of the existence of the Bighorn anticline in the attitude of the Jurassic beds (Figs. 11c, d and e). In Figure 11f the syncline east of the Bighorn anticline can be seen to plunge beneath the Fernie Basin without affecting the simple structure of the Jurassic. The very tight asymmetric anticline in Figures 11g and h also plunges northward beneath the Fernie Basin. This is perhaps the most striking discordance of all, because the anticlinal axis in the Palaeozoics plunges directly toward the principal synclinal axis in the Fernie Basin (Figs. 9 and 11h).

From the photographs and from down-plunge inspection of the geological map in Figure 9 it is evident that the tightly folded structures in the Palaeozoic are not reflected in the open folds of the Fernie Basin. Since the folded structures plunge directly toward the Basin, it is necessary either to introduce a structural discontinuity between the Mesozoic and the Palaeozoics or to claim that the folds in the Paleozoic are conical and do not reach the Fernie Basin. As shown on the geological map of Figure 9, the Bighorn anticline is cylindrical because it maintains its basic shape over an observed length of 12 miles. One would expect it to continue the two more miles that will bring it under the Jurassic of the Fernie Basin. That it should continue is also supported by the structural continuity which was earlier demonstrated for the structures on the western margin of the basin. Therefore, it is concluded that these structures do extend northward in the subsurface, and that there is a structural discordance between the Palaeozoics and the Mesozoics at the south end of the Fernie Basin.

Discordant Palaeozoic-Mesozoic relationships exist in several parts of the Fernie Basin (e.g., Figs. 5 and 6). That this phenomenon is regionally recurrent was recognized by R. A. Price (1961, pages 52 and 53), who stated:

"The geometry of the margin of the Fernie Basin, as defined by the base of the Kootenay Formation, appears to be structurally discordant relative to some of the folds developed within the Paleozoic rocks around the periphery of the basin. Folds in Paleozoic rocks south of Lodgepole Creek and west of Sparwood Ridge are characterized by relatively high structural relief, yet where they pass under the margin of the basin they have little or no expression in the Kootenay-Fernie contact. The relief of the upper surface of the Paleozoic succession is largely obliterated at higher levels by deformation within the relatively incompetent rocks of the Fernie Group. Folding of the Kootenay and Blairmore strata in the basin appears to have proceeded, in part, independently of that in the

---

Fig. 11. Oblique aerial views of the structures in the Paleozoic and Mesozoic rocks at the south end of the Fernie Basin. Looking North, 11a is on the west and 11h on the east.

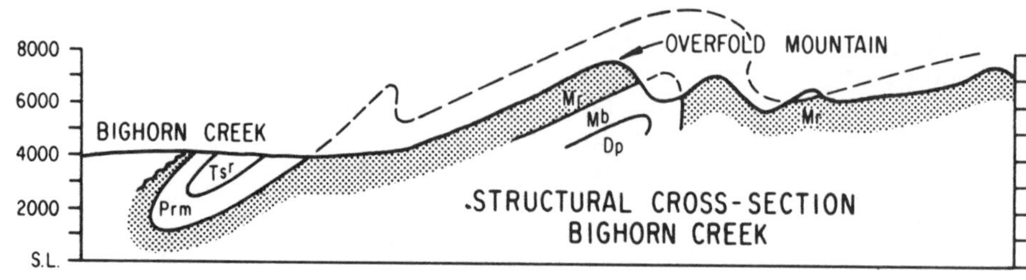

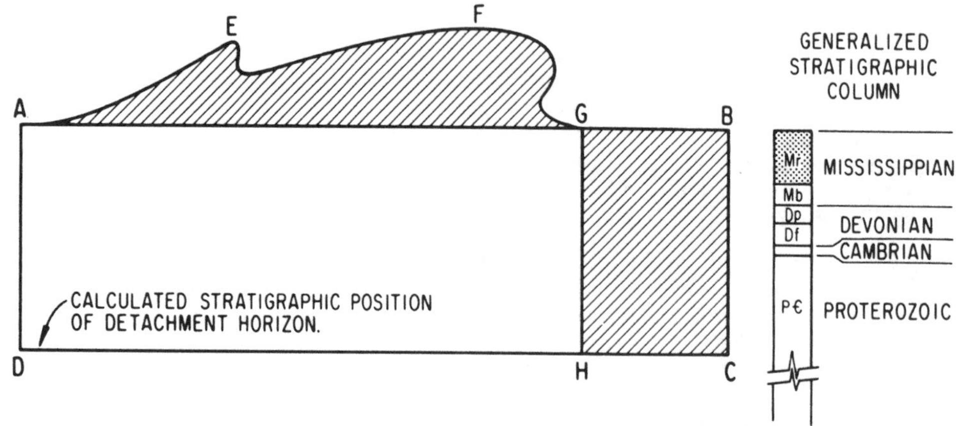

Fig. 12. Depth to detachment calculation for the Bighorn Anticline. Deformation has transformed the area ABCD to the area AEFGHD. Area AEFG can be measured directly on the section and it is equal to the area GBCH. GB is the amount of shortening which can be determined by subtracting the distance AG from the distance AEFG. From the area GBCH and the distance GB, the depth to the detachment, BC, can be calculated. For further discussion see Goguel (1962).

Paleozoic strata beneath the basin. Furthermore, the geometry of the axial surface of the McEvoy syncline, and an apparent offset between it and the folds in the northern part of MacDonald Range, provide some measure of the apparent relative eastward translation of the Kootenay and Blairmore strata of the Fernie Basin over the underlying Paleozoic structures along the incompetent Fernie interval."

Unconformity or Detachment

A discordance between folded beds below and flat-lying beds above can be explained by either a structural or a stratigraphic discontinuity (i.e., detachment or unconformity). To discriminate with certainty between them it is necessary to see whether there has been erosional truncation at the surface of discontinuity (Craddock, 1960). If the discontinuity is an unconformity, there will be units preserved in the synclinal sections which have been eroded from the anticlines. If the discontinuity is a detachment, the stratigraphic succession will be the same on both syncline and anticline but the synclinal axis will be filled with a thick "wad" of folded and faulted beds (Fig. 2).

In the Fernie Basin area the discontinuity is primarily within the Jurassic Fernie Formation. Because these shales do not provide good outcrops, simple direct evidence of truncation or lack thereof is not available. However, several lines of indirect evidence indicate that the surface is a detachment.

1. The B.A. C.N.P. Fernie d-42-I well spudded within the Jurassic Fernie Formation and drilled 9500 ft without leaving the Fernie, although the true thickness of this shale unit should be less than 1500 ft. This abnormal thickness was due to a structurally thickened "wad" of Fernie in the syncline east of the down-plunge end of the Broadwood anticline (Fig. 9).

2. The Fernie Formation is a black marine shale with very little sand. This kind of lithology hardly suggests a major angular unconformity. Regionally the Fernie Formation is sandy in the east and shaly in the west (Frebold, 1957), quite the inverse of what one would expect if there were an angular unconformity within the western Fernie shale. One might also entertain the possibility that the folds in the Palaeozoics were formed during the deposition of the Fernie without ever rising above sea level, so that there was thick deposition (over 10,000 ft) in the synclinal troughs and thin deposition on the anticlinal crests (1000 to 1500 ft). Such drastic changes in thickness should be reflected by extra members and general difficulties in correlation within the Fernie Formation. These, however, do not seem to exist as the stratigraphic sequence is apparently rather straightforward.

3. In the Fernie Basin there generally is a structural discontinuity within the Jurassic shales, but there are some folds which do affect both Mesozoic and Palaeozoic beds. To explain these folds, two generations of structure (pre- and post-Fernie shale) are needed if the discontinuity is an unconformity, whereas only one generation is needed if it is a detachment.

4. The appearance of a pronounced angular unconformity would not be consistent with the pattern of historical geology within the Fernie "Basin." In an active orogenic belt there are usually three parts of the depositional basin. On the cratonic or foreland side the sequence is incomplete, being interrupted by epeirogenic uplifts which produce numerous disconformities. In the center, deposition is more or less complete. On the active or hinterland side of the basin, orogeny is taking place and there are numerous major angular unconformities. In the regional sense, the Fernie "Basin" area is on the cratonic side, at the pont where some of the cratonic unconformities are fading out into the continuous deposition of the central basin. The clastic lithology of the Mesozoics shows that the active side of the Western Cordilleran orogen did move eastward, but it was not until early Tertiary time that the active side of the basin had migrated far enough east for one to expect major angular unconformities in the Fernie area.

From these indirect lines of evidence it is concluded that the discontinuity within the Fernie shale is a detachment which permitted the pre-Fernie rocks to deform independently of the post-Fernie rocks.

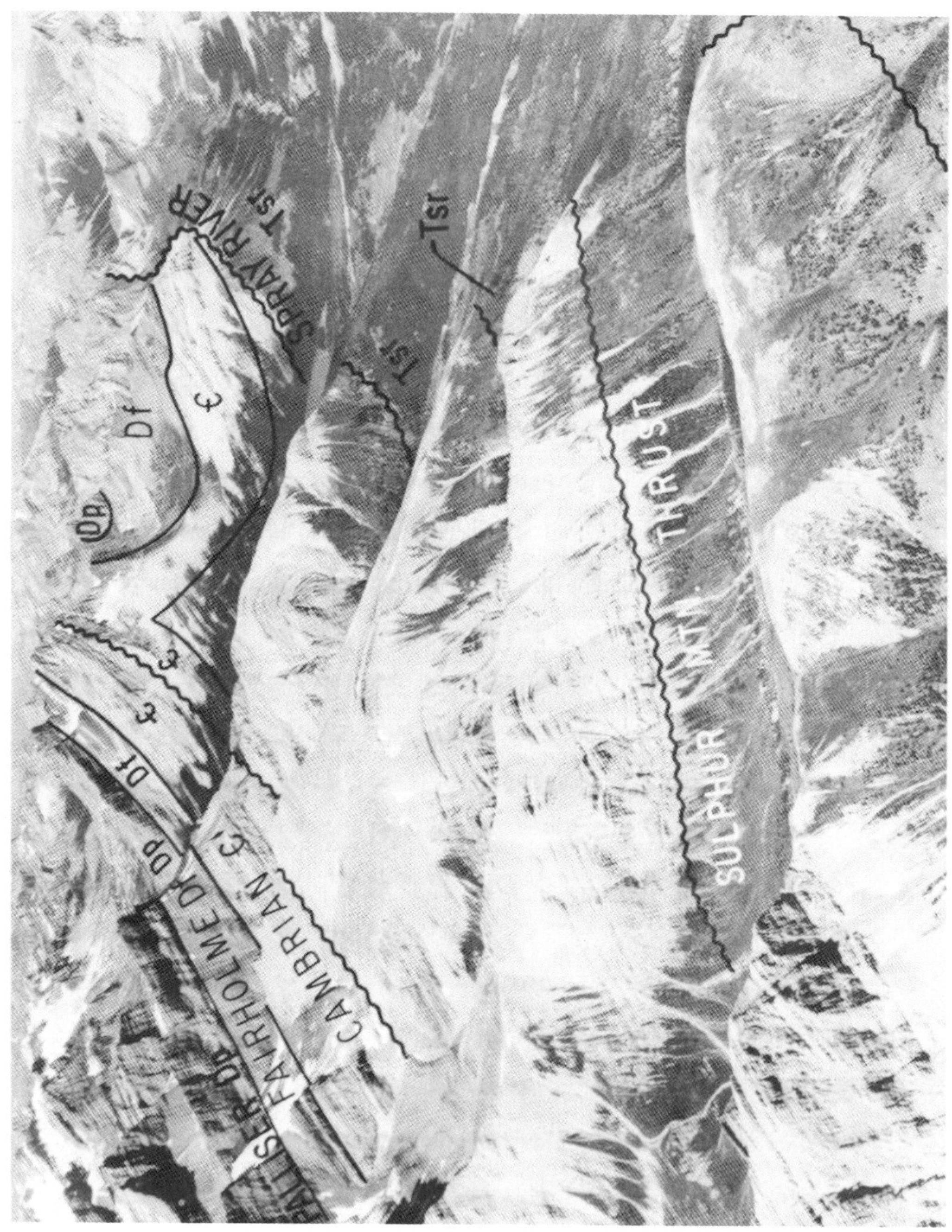
PALLISER
Dp
FAIRHOLME
Df
CAMBRIAN
SPRAY RIVER
Tsr
SULPHUR MTN. THRUST

The Lower Detachment

The existence of a lower detachment in the Fernie Basin area cannot be demonstrated with photographs as can the upper. However, its existence can be inferred with confidence from rather straightforward geometric reasoning.

The Bighorn anticline is a concentric fold (Figs. 11c and d) which necessarily implies the existence of a lower detachment. The stratigraphic position of this detachment can be determined roughly by a simple calculation (Fig. 12) which shows that it is in the Proterozoic sediments. The Howell Creek window exposes a major fault which thrusts Proterozoic rocks upon Upper Cretaceous and dips westward beneath the Bighorn anticline (Fig. 1). Such major thrusts can and do function as detachment horizons, as illustrated in Figure 13. Therefore one would conclude that the major thrust in the Howell Creek window probably was the lower detachment during the folding of the Bighorn anticline and related structures. Saying that this fault plane was both a detachment and a thrust does not imply that folding and thrusting were synchronous: the geometric relationships could have been developed by folding before, during or after the thrusting.

Conclusions

The Fernie Basin area provides a good illustration of a concentrically deformed panel of rock which is bounded by detachments both above and below. The upper detachment is a decollement and the lower a major thrust fault. In concentrically folded terrane it is quite possible to have simple structures at surface which conceal more complicated structures at depth, or to have complicated structures at surface which conceal simple structure at depth. An adequate understanding of such vertical changes in structural form can be useful in mineral exploration.

References

Bucher, W. H., 1933, The deformation of the earth's crust: Princeton, N.J., Princeton Univ. Press, 518 p.

Crabb, J. J., 1957, A summary of the geology of the Crowsnest Coal Fields and adjacent areas: in Alberta Soc. Petroleum Geologists, Guidebook, 7th Ann. Field Conf., Sept., p. 77-85.

Craddock, C., 1969, The origin of the Lincoln fold system, southeastern New Mexico: Intern. Geol. Cong. 21st, Copenhagen part XVIII, p. 34-44.

Currie, J. B., Patnode, H. W., and Trump, R. P., 1962, Development of folds in sedimentary strata: Geol. Soc. America Bull., v. 73, p. 655-674.

Dahlstrom, C. D. A., 1954, Statistical analysis of cylindrical folds: Canadian Institute Mining and Metallurgy Trans., Vol. LVIII, p. 140-145.

——, 1960, Concentric folding, *in* Edmonton Geol. Soc. 2nd Ann. Field Trip Guide Book, p. 82-84.

De Sitter, L. V., 1956, Structural geology: New York, McGraw-Hill Book Co. Inc., 552 p.

Fig. 13. The Sulphur Mountain thrust near Miette, Alberta. An anticline-syncline pair can be seen at Devonian level in the background. In the foreground the contorted Cambrian rocks in the anticlinal core show that the thrust was a decollement. The rocks below the thrust are Triassic in age.

Douglas, R. J. W., 1952, Waterton, Alberta: Geol. Surv. Canada, Paper 52-10.

Frebold, Hans, 1957, The Jurassic Fernie group in the Canadian Rocky Mountains and foothills: Geol. Surv. Canada, Mem. 287.

Goguel, J., 1962, Tectonics, translation of 1952 edition: San Francisco, Freeman and Company, 384 p.

Henderson, G. G. L., and Dahlstrom, C. D. A., 1959, First-order nappe in Canadian Rockies: Am. Assoc. Petroleum Geologists Bull., v. 43, no. 3, p. 641-653.

Mackin, J. H., 1950, The down structure method of viewing geologic maps, Jour. Geology, v. 58, p. 55-72.

Price, R. A., 1961, Fernie map-area, east half, Alberta and British Columbia: Geol. Surv. Canada, Paper 61-24.

——, 1962, Geologic structure of the central part of the Rocky Mountains in the vicinity of Crowsnest Pass: Alberta Soc. Petroleum Geologists Jour., v. 10, no. 7, p. 341-351.

——, 1964, Flexural-slip folds in the Rocky Mountains, southern Alberta and British Columbia: Geol. Surv. Canada, Reprint 78, 16 p.

——, 1965, Flathead map area, British Columbia and Alberta: Geol. Surv. Canada, Mem. 336, 221 p.

Stockwell, C. H., 1950, The use of plunge in the construction of cross sections of folds: Proc. Geol. Assoc. Canada, v. 3, p. 97-121.

Wegmann, C. E., 1929, Beispiele tektonischer analysen des grundgebirges in Finnland: Bull. Commn. geol. Finl. no. 87, p. 98-127.

Wilson, G., 1967, The geometry of cylindrical and conical folds: Proc. Geol. Assoc. London, v. 78, p. 179-210.

Reprinted with permission from *Journal of Structural Geology*, v. 8, p. 229-237, C. J. Banks and J. Warburton, 'Passive-roof' duplex geometry in the frontal structures of the Kirthar and Sulaiman mountain belts, Pakistan, copyright 1986, Pergamon Press plc.

Journal of Structural Geology, Vol. 8, Nos. 3/4, pp. 229 to 237, 1986
Printed in Great Britain

0191-8141/86 $03.00 + 0.00

'Passive-roof' duplex geometry in the frontal structures of the Kirthar and Sulaiman mountain belts, Pakistan

C. J. Banks and J. Warburton

B.P. Petroleum Development of Spain, S.A.

(*Received* 1 *July* 1984; *accepted in revised form* 30 *August* 1985)

Abstract—Exploration for hydrocarbons over the past few years has greatly improved our understanding of the geometry of frontal mountain belt structures. In this study we introduce and discuss the concept of the 'Passive-roof duplex', using as the main example the Kirthar and Sulaiman Ranges in the Baluchistan Province of Pakistan. Structures similar to those described here have been recognized previously in other mountain belts, and they appear to exist as a common feature in many more frontal regions of mountain belts. Our example of a Passive-roof duplex which we describe from Pakistan is compared briefly with similar structures reported by others.

The *Passive-roof duplex* is here defined as a duplex whose roof thrust has backthrust sense (*Passive-roof thrust*) and whose *roof sequence* (those rocks lying above the roof thrust) remains relatively 'stationary' during foreland directed piggy-back style propagation of horses within the duplex.

GEOLOGICAL SETTING OF THE KIRTHAR AND SULAIMAN THRUST BELTS

FIGURE 1 shows the geographical and tectonic setting of the Kirthar and Sulaiman thrust belts. The thrust belts of western and north-western Pakistan were initiated during the collision of the Indian and Eurasian plates throughout the Tertiary. The details of plate configuration and motion are omitted from this study and the reader is referred to several publications (Molnar & Tapponier 1975, Bordet 1978, Acharyya 1979, Bingham & Klootwijk 1980). A major contribution to understanding of the geology of Pakistan was provided by a substantial report and maps published by the Hunting Survey Corporation in 1960. A collection of twenty four papers compiled by the Geological Survey of Pakistan (1979) also provides a broad overview of the geodynamics of Pakistan.

The Kirthar thrust belt and the structures within it strike approximately N–S and are bounded to the west by a zone of steep to vertical left lateral transcurrent faults, named the Chaman Fault Zone (Lawrence & Yeats 1979, Lawrence *et al.* 1981). The relationship between the westerly dipping thrust surfaces of the Kirthar thrust belt and the steep Chaman faults is poorly understood.

The Sulaiman Range forms a continuation of the Kirthar Range around the tight arc of the Sibi Trough. Like the Kirthar thrust belt, the Sulaiman Range is formed from imbricate slices which here developed during southerly propagating piggy-back thrusting. Figure 2 shows a geological map, and Fig. 3 a simplified stratigraphic column for the Kirthar and Sulaiman Ranges. Rocks exposed at the surface, which are involved in thrusting, range in age from Carboniferous to Recent, although older rocks may be involved at depth. The most likely major detachment horizons are indicated in Fig. 3.

The Sibi Trough (Figs. 1 and 2) is a molasse basin which developed as a foredeep ahead of the developing Kirthar and Sulaiman thrust complexes in Miocene to Recent time, with the accumulation of at least 7000 m of sediment. In the region between Quetta and Sibi the molasse basin is tightly constricted and structures are extremely complex. It is in this area that the molasse sediments are affected by structures which propagated eastward and southward and which, in Fig. 2, display obvious interference patterns. The majority of coarse clastic sediment supplied to the Sibi Trough was locally derived from the uplifting mountain belts, and has been continuously reworked during development of the thrust belts. A similar relationship between tectonics and sedimentation has been detailed for the Siwalik molasse basin further north along the southern margin of the Himalaya and Hindu Kush (Burbank & Reynolds 1984).

Duplex geometry

Figures 4 and 5 show two vertical cross-sections, one WNW–ESE across the Kirthar thrust belt in the vicinity of the Bolan Pass, and one N–S across the Sulaiman Range. The lines of section are shown in Figs. 1 and 2. The structure to the east of the Bolan Pass and external part of the Sulaiman Range is supported in part by fair to good quality seismic reflection data, but the overall interpretation is based mostly on field and airphoto/Landsat image mapping. In the Sulaiman Ranges (Fig. 5) the floor thrust is suspected to be within a sequence of evaporites of Eo-Cambrian age. The roof thrust is within the Lower Eocene Ghazij Formation shales in most of the range but appears to be in Lower Cretaceous Goru Formation shales in the northernmost horses of the duplex. All of these units are relatively incompetent and the roof and floor thrusts are separated by a thick sequence (up to 8 km) composed dominantly of Jurassic, and in the south also, Palaeocene limestones.

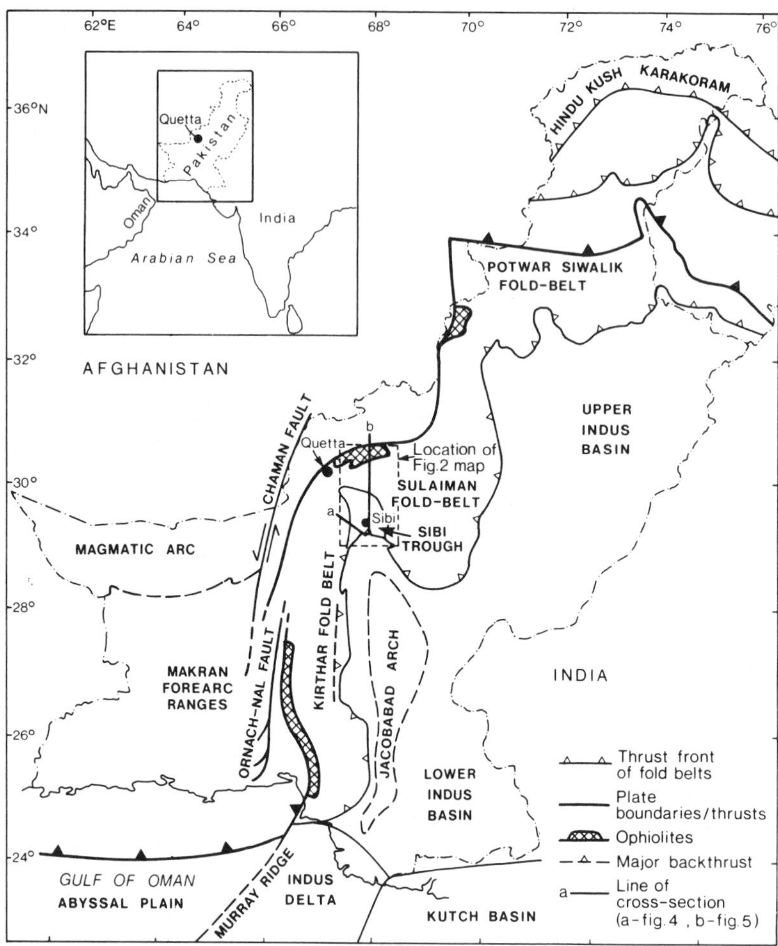

Fig. 1. Geographic and tectonic setting of the Kirthar and Sulaiman Ranges, with the locations of two cross-section lines for Figs. 4 and 5.

In the Bolan Pass area of the Kirthar Range (Figs. 2 and 4) there is a distinct difference in the style of folding above and below the Ghazij Formation. Below the Ghazij Formation, folds involving the thick limestone sequence are generally open, whilst above, they are commonly close to tight and occasionally isoclinal and strongly asymmetric. This demonstrates some detachment within the Ghazij Formation. By analogy with the Sulaiman Range it is likely that the regional sole thrust in the Kirthar Range is within Eo-Cambrian evaporite or equivalent section. However, as yet there are no data to confirm or refute this assumption.

The most important features of the geological structure in the frontal Kirthar and Sulaiman Ranges are as follows:

(1) There is a broad zone of steep foreland dip (and occasional overturning) which bounds and involves molasse sediments of the Sibi Trough at the mountain front (Figs. 2, 4 and 5).

(2) Jurassic (Chiltan Formation) limestones are elevated almost 9 km above the regional level on the internal side of the steep zone, yet no major displacement thrusts outcrop (Figs. 4 and 5).

(3) Only minor displacement thrust structures occur within the molasse sediments of the Sibi Trough on the external side of the steep zone. These intra-molasse thrusts, reflected by the Dezgat and Bannh anticlines at the surface (Figs. 2 and 4), are interpreted as out-of-syncline thrusts developed during formation of the major steep zone.

The seismic data are of limited value in the interpretation of the frontal mountain structures because lines do not extend across the steep zone. The main conclusions to be drawn, are that the pre-molasse stratigraphic section is essentially undeformed and that the Kirthar (limestone) Formation can be extrapolated from outcrop in the steep zone (unfaulted), directly into a flat-lying stratigraphic section at a regional level below the molasse.

We interpret the steep zone as a major frontal culmination wall (see Butler 1982) in the roof sequence of a duplex developed in the thick limestone units beneath a roof thrust in the Ghazij (shale) Formation. Since the outcropping Jurassic limestones are elevated about 9 km above their regional level within the duplex and undeformed stratigraphic sections occur immediately on the

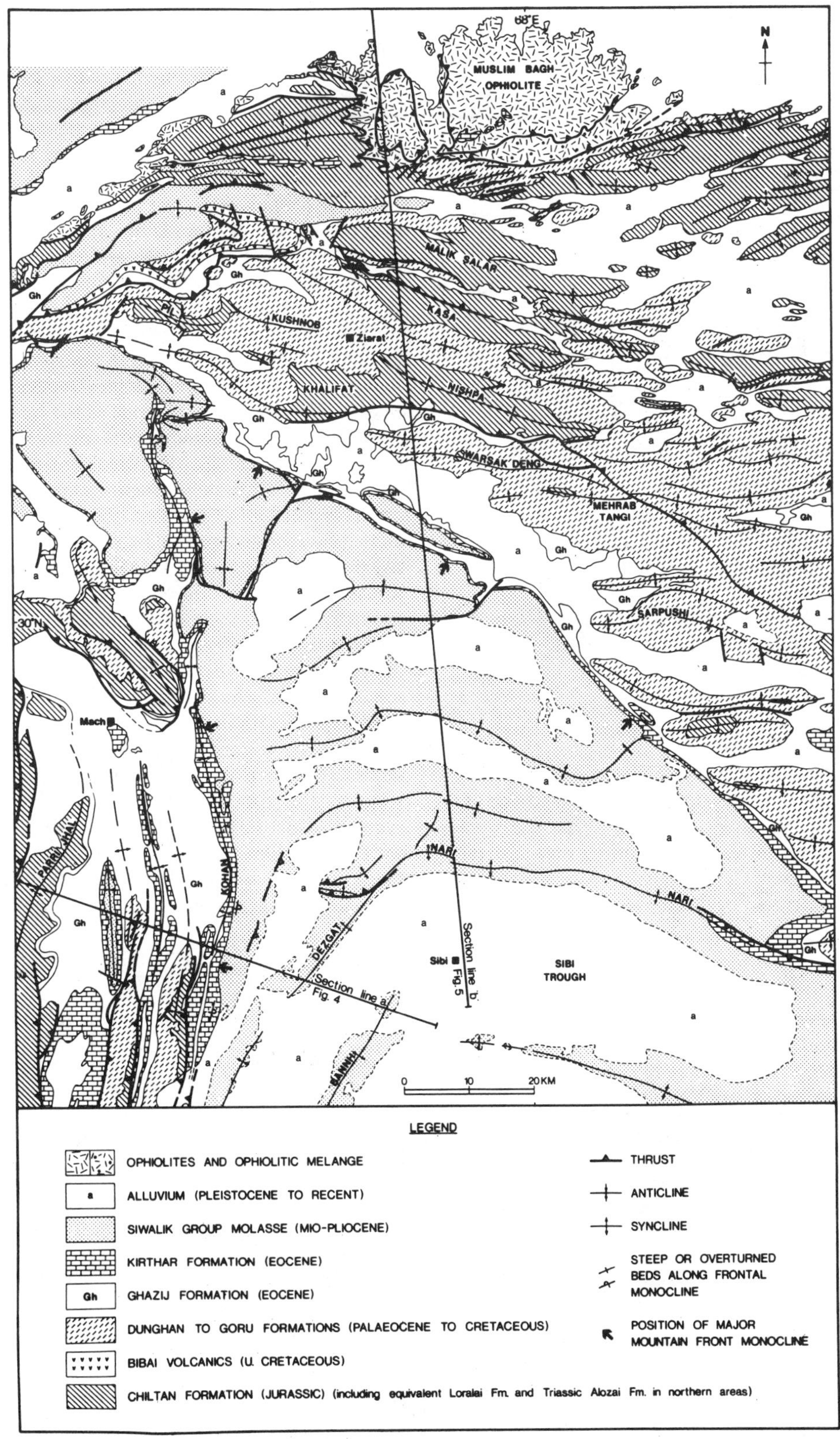

Fig. 2. Geological map of part of the Kirthar and Sulaiman thrust belts and the north-western part of the Sibi Trough. The location is shown as an inset on Fig. 1. The names on the map refer to the major surface anticlines, some of which appear on the cross-section in Fig. 5.

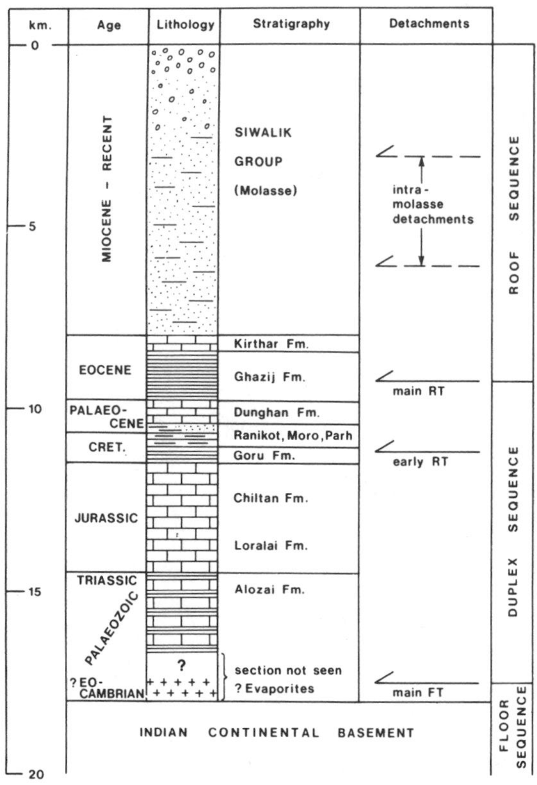

Fig. 3. Simplified stratigraphic column for the Kirthar and Sulaiman Ranges indicating the most likely detachment horizons.

foreland side of the steep culmination wall, then a blind thrust must exist immediately on the foreland side of the most external hangingwall cutoff within the duplex. Similar mechanisms of blind thrusting on the external side of the Northern Canadian Rocky Mountain foothills have been described by Thompson (1981). This means that shortening within the duplex could not have been accompanied by foreland-directed transport of the roof sequence. Instead, the roof sequence must have a back-thrust sense of displacement relative to the foreland propagating imbrication of the limestones in the duplex (Fig. 6). This type of structure we refer to as a *passive-roof thrust*, which is equivalent to the "upper detachment" described by Jones (1982) in the Alberta syncline.

The Geological Survey of Pakistan have recognized a long backthrust structure in the vicinity of the mountain front along part of the length of the Kirthar thrust belt (Fig. 1) (Kazmi & Rana 1982). Our own studies in the Bolan Pass area and Sulaiman Range suggest that a passive backthrust exists along most of the length of the mountain front, forming the roof thrust of a *passive-roof duplex*. In a conventional duplex model the roof sequence would be displaced toward the foreland and we would expect to see some major structure caused by the accommodation of that displacement above the roof thrust. Instead, the only structures are relatively minor displacement out-of-syncline thrusts in the molasse above undeformed regional stratigraphy. The present day frontal culmination wall in the Kirthar and Sulaiman thrust belts projects up into the air and has been eroded to supply molasse sediment to the Sibi Trough.

It is likely that this thrust process also occurred at intervals during the development and propagation of the mountain front. As a result of such erosion, the preserved bed length in the roof sequence is considerably less than the restored bed length of rocks within the duplex. In the Bolan Pass area and in the Sulaiman Ranges, the roof sequence is preserved mainly in synclines which lie between the ramp anticlines of horses within the duplex (Figs. 2 and 4). One or both limbs of these synclines are usually steeply dipping (Fig. 4) and these steep zones may represent the successive positions

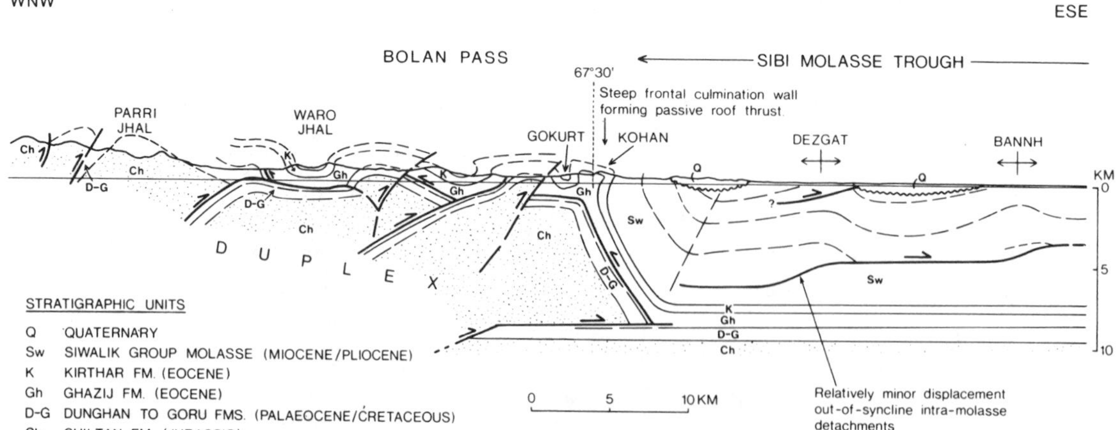

Fig. 4. WNW–ESE geological cross-section across the Kirthar thrust belt in the vicinity of the Bolan Pass. The line of cross-section is labelled "a" on Figs. 1 and 2. The cross-section is constructed from field data along with air photo and Landsat image interpretation. Seismic data over the Sibi Trough does not extend westwards as far as the steep zone, but demonstrates flat lying undeformed stratigraphy below the molasse, with relatively minor displacement on out-of-syncline intra-molasse detachments. Imbrication and shortening within the duplex causes about 9 km of uplift of the Jurassic Chiltan Formation above its regional level. Since no equivalent amount of shortening has been transferred into the foreland (undeformed sub-molasse stratigraphy and small intra molasse displacements only) then a blind thrust must be present at depth within the Ghazij décollement and a passive roof thrust must pass over the Jurassic Limestone imbricates.

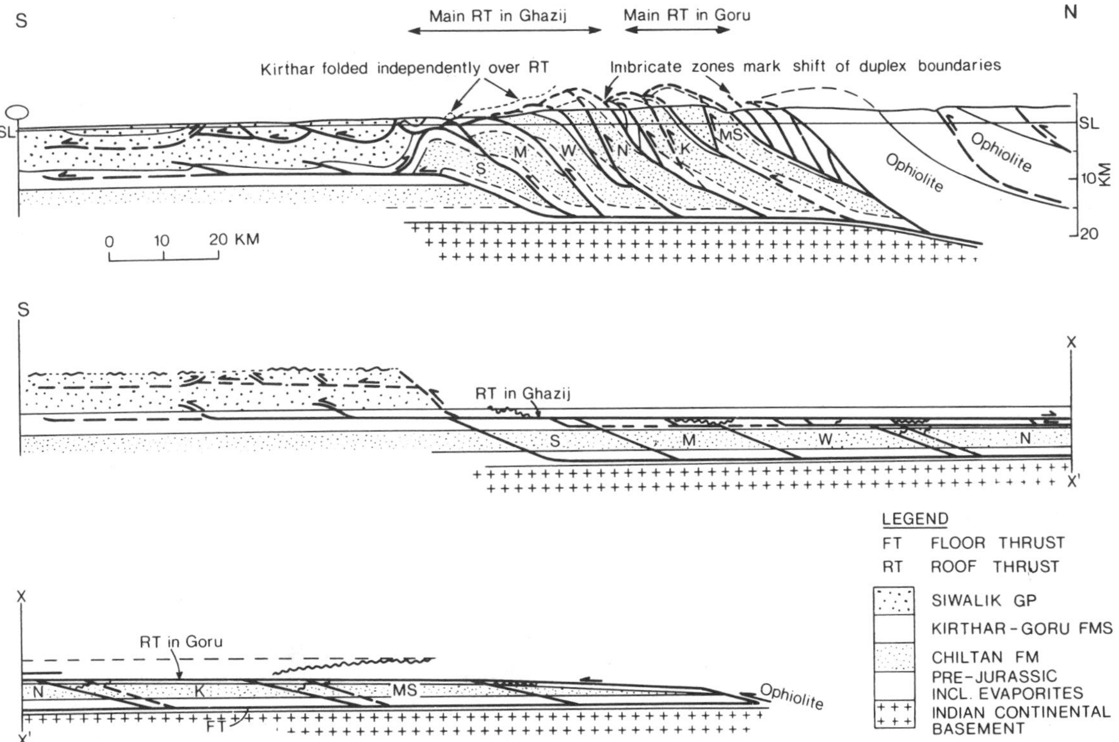

Fig. 5. Actual and restored N–S geological cross-sections across the Sulaiman Range. The line of cross-section is labelled "b" on Figs. 1 and 2. The same comments apply as those outlined in Fig. 4 caption. Letters identifying the major duplex horses are from individual mountains (shown on Fig. 2) formed from the ramp anticlines. From south to north these are: S, Sarpushi; M, Mehrab Tangi; W, Warsak Deng; N, Nishpa; K, Kasa; MS, Malik Salar.

during thrust propagation at which the passive roof sequence became emergent.

In the Sulaiman Range cross-section (Fig. 5), conventional duplexes on a relatively minor scale occur where the levels of the roof or floor thrust change. The transition from Lower Cretaceous (Goru) to Eocene (Ghazij) roof levels is marked by a minor Goru to Ghazij duplex which is overridden by the Jurassic slice to the north, i.e. it is a conventional duplex.

DISCUSSION: POSSIBLE STRUCTURES IN THE ROOF SEQUENCES OF PASSIVE-ROOF DUPLEXES

During development of a duplex, excess bed length rapidly accumulates in the roof sequence relative to the highly shortened duplex. This excess bed length may be eroded close to the mountain front, as in the Sulaiman Range, but if the roof sequence is not eroded, it is likely to become shortened by deformation occurring within it, not necessarily related to the structure of the underlying duplex. In the Bolan Pass the excess bed length in the roof sequence is taken up by hinterland-facing isoclinal folds. Another mechanism, not identified in this area in the field, but possibly present, is the development of passive backthrusts in the roof sequence. For a duplex forming purely by the passive-roof mode we may expect an overstep sequence (Butler 1982) of foreland-dipping passive backthrusts to develop in the roof sequence (Fig. 7). Such passive backthrusts have been recognized in the Alberta foothills of the Rocky Mountains (see Fig. 8b).

If passive roof duplexes are mainly restricted to moun-

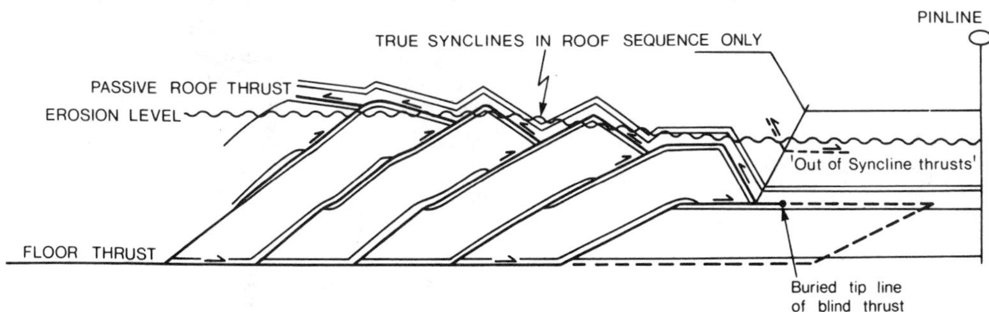

Fig. 6. Passive-roof duplex model: the roof sequence has backthrust sense relative to the forelandward propagated duplex.

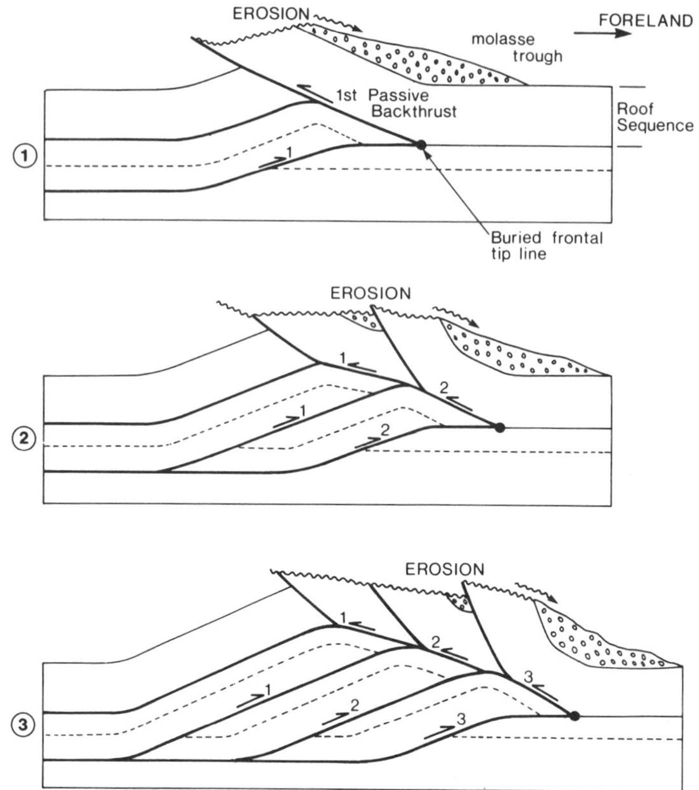

Fig. 7. Development of an overstep sequence of foreland dipping passive backthrusts in the duplex roof. 1–3 represent successive stages in the development of the duplex. Thrusts are numbered in order of development.

tain fronts it is not surprising that the geometry of their roof sequences is poorly documented. The structures there are becoming uplifted and emergent and a duplex roof is unlikely to be preserved for long. The key feature identifying the structure is the monoclinal forelandward dip at the mountain front with little major thrusting at outcrop.

Similar structures in other mountain belts

The concept of a backthrust overlying a frontal mountain duplex is not new. The term "triangle zone" was introduced by Gordy *et al.* (1977) to describe the region of opposed dip thrusts which extends along a great length of the Alberta foothills of the Rocky Mountains. Figure 8a is of a deformed and partially restored section across the Alberta syncline (after Price 1981), which demonstrates the passive-roof duplex geometry. The main features are the frontal mountain monocline (the Alberta syncline), the passive-roof thrust (the Waldron fault) and the foreland-thrusted passive-roof duplex which completes the "triangle zone".

Figure 8b is a modification of an original cross-section by Ziegler (1969) produced by Jones (1982) for the Athabasca Valley foothills. An important feature of that cross-section is the existence of foreland-dipping passive backthrusts within the roof sequence; a set of structures which we suggest in Fig. 7 but for which we have seen little evidence in the field in Pakistan.

As explained earlier, in the case of the Kirthar and Sulaiman examples, seismic data are of limited value in demonstrating the overall structure of the passive-roof duplex. Some impressive seismic data however, are presented by Jones (1982) for the Alberta syncline structure and demonstrate the buried frontal tip and passive-roof thrust ("upper detachment" of Jones) above the duplex. Further examples of passive-roof duplexes are discussed by Jones (1982).

A passive-roof duplex exists in the western frontal ranges of the Taiwan thrust belt, as illustrated by Suppe & Namson (1979), Suppe (1980a,b, 1981), Davis *et al.* (1983) and Suppe (1983). Suppe (1980a) demonstrated that no deformation extended west of the Taiwan mountain front and that the upper detachment of the duplex must be a passive backthrust. Microseismic data from the Taiwan thrust belt (Wu *et al.* 1979, Suppe 1981) has shown that the upper part of the stratigraphic section rides passively over a presently imbricating duplex wedge (Fig. 9). Numerous microearthquakes have occurred within the duplex, whilst the rocks of the roof sequence, and those beneath the sole thrust remain relatively aseismic.

Continuity of a passive-roof sequence

A problem of passive-roof duplex development is the

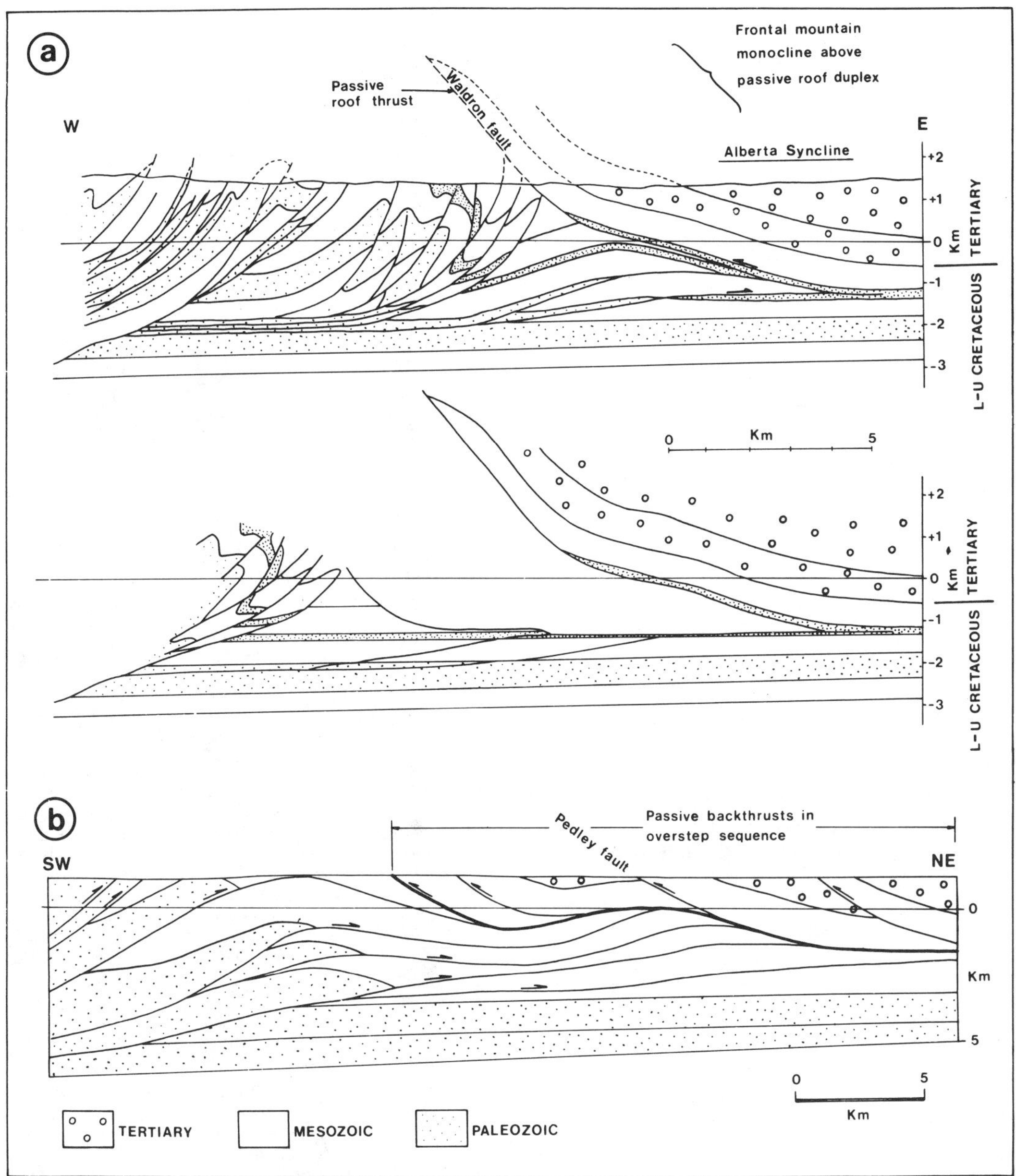

Fig. 8. Cross-sections across the eastern Rocky Mountain Foothills. (a) Deformed and partially restored E–W geological cross-sections across the Alberta Syncline at latitude 49°45′N (modified after Price 1981). (b) NE–SW cross-section across the Athabasca Valley Foothills showing backthrust imbricate stack in duplex roof sequence (modified by Jones 1982 after Ziegler 1969).

distance over which the upper backthrust (or roof sequence) can extend. In examples where the roof sequence is preserved, for example in the Athabasca Valley, Alberta (Fig. 8 cross-section *b*), it is dissected by several foreland-dipping backthrusts, each with a present length of up to 7.5 km (e.g. the Pedley Fault). It is conceivable that only this length of passive backthrust acted at any one time during duplex development, as indicated by Fig. 7. In contrast, in the Taiwan example, the length of continuous passive-roof sequence is considerably greater, up to 14 km (Fig. 9). Restoration of the western limb of the Alberta Syncline by Price (1981, see Fig. 8) indicates a single passive-roof backthrust extending some 13 km over hinterland-dipping horses. Jones (1982) predicts that 50 km of passive-roof sequence may have extended across the Alberta foothills. With such a limited number of well-documented examples it is difficult to estimate the dimensions of a passive-roof sequence in sections parallel to the dominant thrust movement direction. However, our feeling (based mainly on the analysis of the Pakistan structures and previously cited examples) is that the passive-roof mode of duplex deformation is most common in the vicinity of the mountain front.

Rather than inferring a single continuous passive-roof

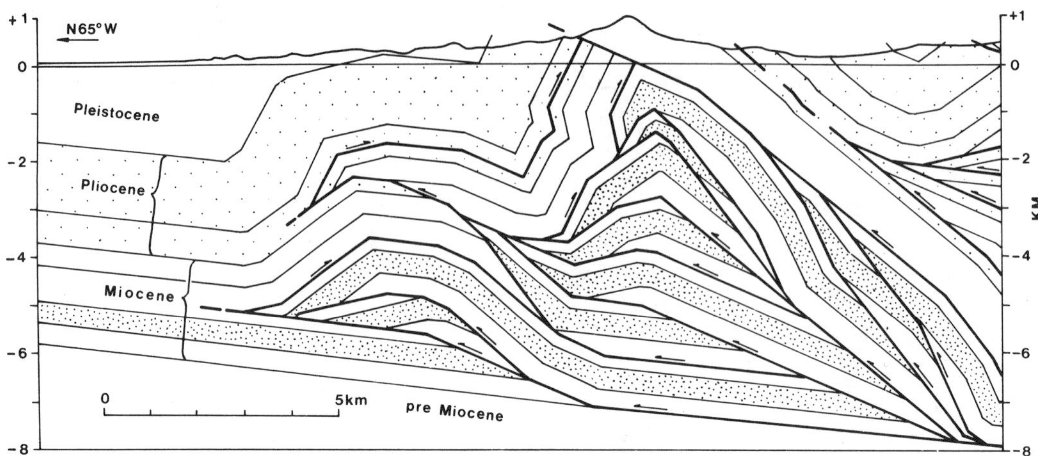

Fig. 9. Cross-section across part of the southern Taiwan thrust belt showing passive-roof duplex at the mountain front (modified after Suppe 1980a).

sequence which extends back over a large number of duplex horses, it is likely that the roof sequence becomes imbricated (as in Fig. 7) such that 'older' (more internal) segments of the passive-roof sequence become 'inactive' earlier, and are transported with earlier formed duplex horses toward the foreland. However, examples described by others, in particular the Brooks Range of Alaska (I. R. Vann pers. comm. 1983) suggest that continuous passive-roof sequences may extend several hundreds of kilometres across regional strike.

The common occurrence of backthrust structures hindward of mountain fronts and within the internal zones of mountain chains may reflect the further importance of passive backthrusting across entire orogenic belts.

CONCLUSIONS

(1) The passive-roof duplex is a duplex whose roof sequence has backthrust sense and remains relatively stationary during piggy-back style thrust propagation within the underlying duplex.

(2) Passive-roof duplexes can be inferred as a mechanism of orogenic contraction where the mountain front is marked by a forelandward-dipping monocline rather than a thrust, and only blind thrusts exist on the foreland side of the last mountain belt imbricate slice.

(3) To conserve bed length equality in and above a passive-roof duplex, there may be erosion of emergent imbricate slices of roof sequence. The passive-roof mode may therefore be of major importance near surface in the vicinity of the mountain front. However some examples appear to have a passive-roof sequence which extends for large distances towards the hinterland. The common existence of backthrusts in the internal zones of mountain belts possibly indicates a more general importance of the passive-roof mode of deformation.

(4) The emergence of passive-roof sequences in foredeep molasse basins (e.g. the Sibi Trough in Pakistan) may rework molasse sediments several times and produce multiple unconformities. This feature is not however restricted to emergent backthrusts, but occurs widely in areas where frontal imbricate fans emerge at surface.

(5) Overstep backthrust roof sequences with foreland dip may be a characteristic feature of passive-roof duplexes.

Acknowledgements—The authors would like to thank the management of BP Petroleum Development Ltd. London for permission to publish the work, which results from an exploration programme in Pakistan carried out by BP in conjunction with the Oil and Gas Development Corporation. We also thank S. G. Abbas of the Geological Survey of Pakistan for his participation in the project. David Elliott worked with us in Karachi for a week in May 1982 and we are indebted to him for imparting some of his knowledge and enthusiasm. Samir Hanna, Chris Morley and Ian Vann are thanked for discussions on frontal mountain structures, and the comments of Mike Coward and two referees are gratefully acknowledged.

REFERENCES

Acharyya, S. K. 1979. India and Southeast Asia in Gondwanaland Fit. *Tectonophysics* **56**, 261–276.

Bingham, D. K. & Klootwijk, C. T. 1980. Palaeomagnetic constraints on Greater India's underthrusting of the Tibetan Plateau. *Nature, Lond.* **284**, 336–338.

Bordet, P. 1978. The western border of the Indian plate: implications for Himalayan geology. *Tectonophysics* **51**, T71–T76.

Burbank, D. W. & Reynolds, R. G. H. 1984. Sequential Late Cenozoic structural disruption of the northern Himalayan foredeep. *Nature, Lond.* **311** (13), 114–118.

Butler, R. W. H. 1982. The terminology of structures in thrust belts. *J. Struct. Geol.* **4**, 239–245.

Davis, D., Suppe, J. & Dahlen, F. A. 1983. Mechanics of fold-and-thrust belts and accretionary wedges. *J. geophys. Res.* **88**, 1153–1172.

Farah, A. & De Jong, K. A. (Editors) 1979. *Geodynamics of Pakistan*. Geological Survey of Pakistan, Quetta, Pakistan.

Gordy, P. L., Frey, F. R. & Norris, D. K. 1977. Geological guide for the C.S.P.G. and 1977 Waterton–Glacier Park Field conference: Canadian Society of Petroleum Geologists, Calgary.

Hunting Survey Corp. Ltd. 1960. Reconnaissance geology of part of West Pakistan, a Colombo Plan cooperative project, Toronto. (A report published for the Government of Pakistan by the Government of Canada.)

Jones, P. B. 1982. Oil and gas beneath east-dipping underthrust faults in the Alberta Foothills. In: *Geological Studies of the Cordilleran*

Thrust Belt (edited by Powers, R. B.). Rocky Mountain Association of Petroleum Geologists **1**, 61–74.

Kazmi, A. H. & Rana, R. A. 1982. Tectonic Map of Pakistan. Government of Pakistan Geological Survey of Pakistan Special Publication.

Lawrence, R. D. & Yeats, R. S. 1979. Geological reconnaissance of the Chaman Fault in Pakistan (edited by Farah, A. & De Jong, K. A.). Geological Survey of Pakistan, Quetta, 351–357.

Lawrence, R. D., Khan, S. H., De Jong, K. A., Farah, A. & Yeats, R. S. 1981. Thrust and strike-slip fault interaction along the Chaman transform zone, Pakistan. In: *Thrust and Nappe Tectonics* (edited by McClay, K. R. & Price, N. J.). *Spec. Publs geol. Soc. Lond.* **9**, 363–370.

Molnar, P. & Tapponnier, P. 1975. Cenozoic tectonics of Asia: effects of a continental collision. *Science, Wash.* **189**, 419–426.

Price, R. A. 1981. The Cordilleran foreland thrust and fold belt in the southern Canadian Rocky Mountains. In: *Thrust and Nappe Tectonics* (edited by McClay, K. R. & Price, N. J.). *Spec. Publs geol. Soc. Lond.* **9**, 427–448.

Suppe, J. & Namson, J. 1979. Fault-bend origin of frontal folds of the Western Taiwan fold-and-thrust belt. *Petrol. Geol. Taiwan* **16**, 1–18.

Suppe, J. 1980a. Imbricated structure of Western Foothills Belt, South-Central Taiwan. *Petrol. Geol. Taiwan* **17**, 1–16.

Suppe, J. 1980b. A retrodeformable cross section of northern Taiwan. *Proc. geol. Soc. China* **23**, 46–55.

Suppe, J. 1981. Mechanics of mountain building and metamorphism in Taiwan. *Mem. geol. Soc. China* **4**, 67–89.

Suppe, J. 1983. Geometry and kinematics of fault-bend folding. *Am. J. Sci.* **283**, 684–721.

Thompson, R. I. 1981. The nature and significance of large "blind" thrusts within the northern Rocky Mountains of Canada. In: *Thrust and Nappe Tectonics* (edited by McClay, K. R. & Price, N. J.). *Spec. Publs geol. Soc. Lond.* **9**, 449–462.

Wu, F. T., Yeh, Y. H. & Tsai, Y. B. 1979. Seismicity in the Tsengwen reservoirs area, Taiwan. *Bull. seism. Soc. Am.* **69**, 1783–1796.

Ziegler, P. A. 1969. The development of sedimentary basins in Western and Arctic Canada. Alberta Society of Petroleum Geologists.

Reprinted by permission of the Geological Society of America from G. Mulugeta and H. Koyi, *Geology*, v. 15 (1987), p. 1052-1056.

Three-dimensional geometry and kinematics of experimental piggyback thrusting

Genene Mulugeta, Hemin Koyi
Hans Ramberg Tectonic Laboratory, Institute of Geology, University of Uppsala
Box 555, S-751 22 Uppsala, Sweden

ABSTRACT

The three-dimensional geometry and kinematics of piggyback stacks of imbricate thrust sheets are illustrated and discussed using a single model shortened in a squeeze box. Strike-parallel geometric elements simulated include lateral ramps, eyed sheath folds, splays, and thrust/thrust interference. Fine details of these structures were exposed by eroding a shortened wedge of sand using a newly developed vacuum-eroding technique. A kinematic analysis of the model shows a stepwise increase in imbricate thrust spacing and/or a decrease in rate of nucleation of imbricate thrusts in the direction of thrust transport. Despite the steady forward advance of a rear wall, the piggyback wedge accreted episodically, recording different strain domains in longitudinal cross sections. Strain partitioning in single layers by bed-length balancing showed an increase in layer shortening with volume loss and a corresponding decrease in imbricate thrusting and ramp folding with depth.

INTRODUCTION

Imbricate thrust systems comprise a series of thrust sheets that overlap one another like roof tiles, all dipping in the same general direction. Suess (1904) initially recognized this special type of thrust system as schuppen structure. Geological and geophysical mapping and drilling in recent years have revealed fine details of the three-dimensional geometry of thrust systems (Bally et al., 1966; Dahlstrom, 1969; Boyer and Elliot, 1982; Coward, 1983). Boyer and Elliot (1982) explained how individual thrust faults spread into thrust systems, either as imbricate families or duplexes. Moreover, they emphasized the importance of establishing space-time interconnectivity between thrusts for a correct understanding of their three-dimensional geometry.

The piggyback mode of thrust propagation is well established in continental collision zones (Boyer and Elliot, 1982). However, the interplay of gravity (Elliot, 1976; Ramberg, 1977) and a push from the rear (Chapple, 1978; Davis et al., 1983) in generating such structures is not well understood. In this paper, we examine the three-dimensional geometry and kinematics of imbricate thrust sheets in models pushed from the rear at a steady rate above a rigid decollement.

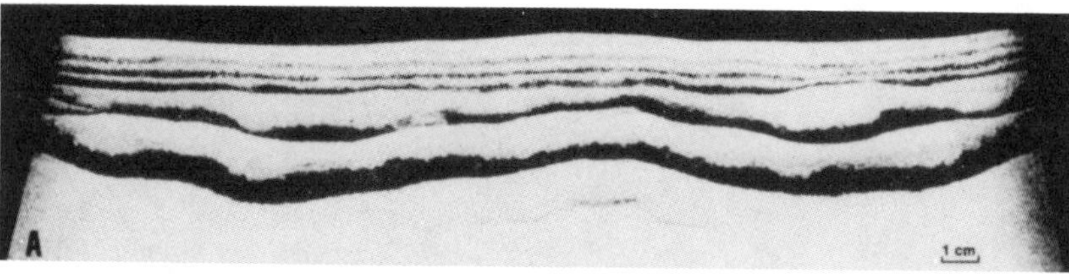

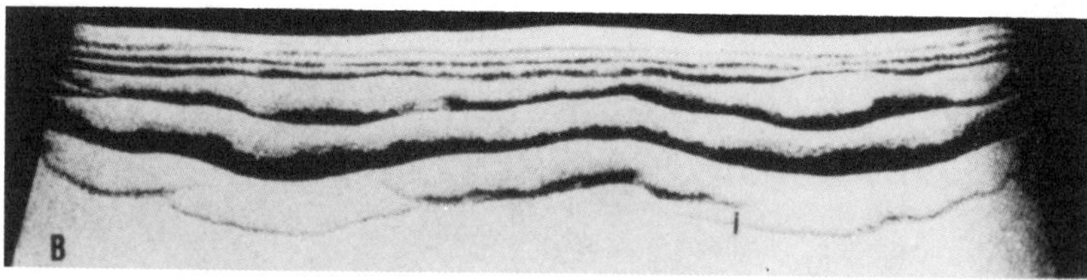

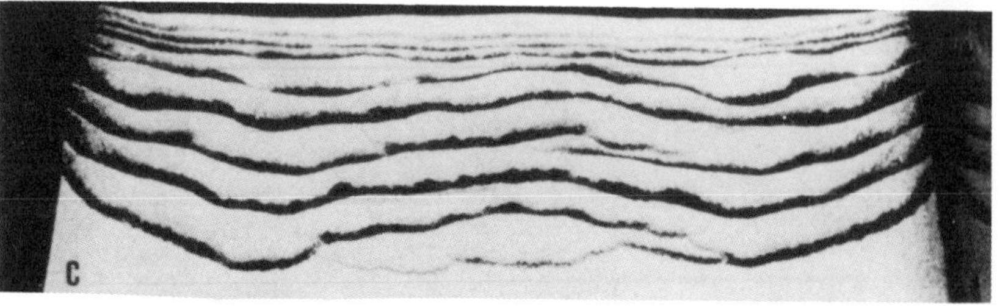

Figure 1. Front perspective view of thickening and imbricating wedge at percentage bulk shortenings of 15% (A), 25% (B), and 40% (C).

EXPERIMENTAL PROCEDURE

A single model, which mimicked all essential geometric features of piggyback thrusting observed in thrust terranes, is discussed in detail here (Figs. 1 and 2). Finely laminated sand layers were constructed in a squeeze box with overall dimensions of 20 × 20 × 10 cm. Individual layers 0.1–0.2 mm thick were prepared by pouring sand of different colors in alternate layers. Only one material was used: quartz sand having grain diameters in the range 0.08–0.18 mm with some 5% mica flakes, 10% subangular feldspar grains, and <1% subangular tourmaline, hematite, magnetite, and rutile. Coulomb rheology of the sand material used is $\tau_f \sim 10$ (Pa) + 0.4 σ. The shear stress necessary to cause sliding of the bed of sand particles along a plexiglas detachment is $\tau_s \sim 0.3\ \sigma$.

The use of sand in analogue modeling of thrust mechanics is not new. Davis et al. (1983), for example, used such a Coulomb material in order to constrain the bulk geometry of thrust wedges at the scale of foreland fold-thrust belts and accretionary wedges. However, the purpose here is to examine three-dimensional geometry and kinematics of piggyback accretion in models using such materials, not in analogue models of a specific thrust terrane. With this in mind, the field of stability of piggyback thrusting was investigated in several tests by varying coefficients of basal friction prior to the test described here. Scaling using a model ratio of cohesive strength meant that 1 cm in the model represents ~0.8 km in nature.

After 40% bulk shortening (Fig. 2), sections both along strike and parallel to transport direction were eroded by means of a low-power vac-

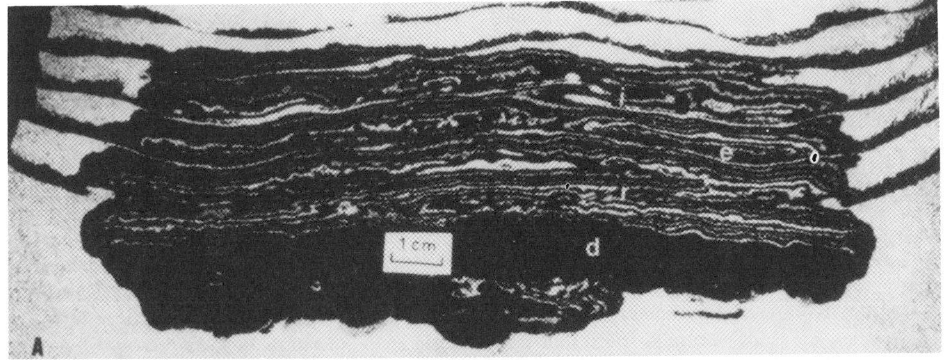

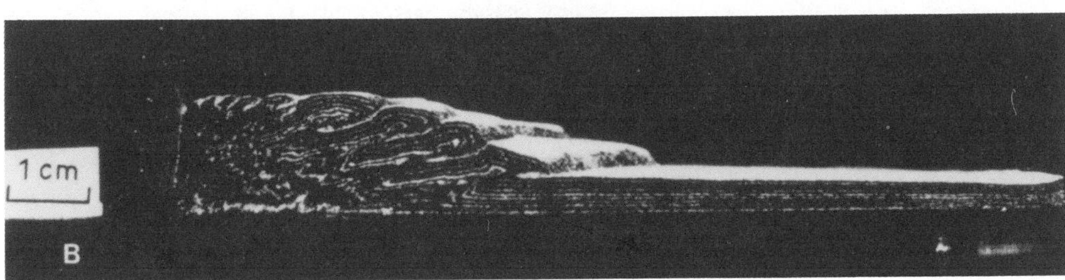

Figure 2. A: General perspective of model shown in Figure 1 at 40% bulk shortening; vacuumed excavations partially expose internal three-dimensional geometries through nappe pile; r = lateral ramp, i = interference zone, d = decollement, e = eyed sheath fold. B: Cross section of same model exposing piggyback stack of imbricate sheets.

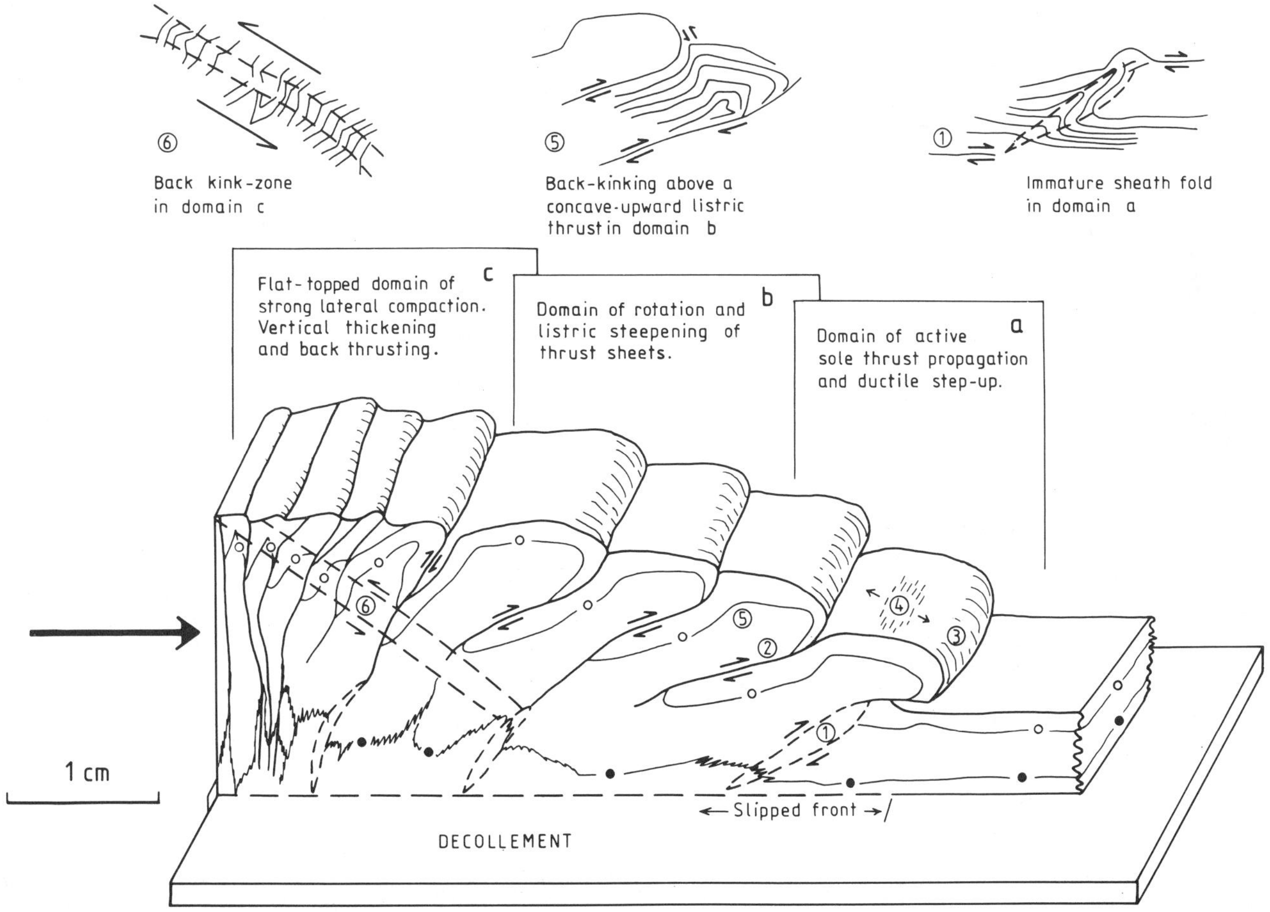

Back kink-zone in domain c

Back-kinking above a concave-upward listric thrust in domain b

Immature sheath fold in domain a

Flat-topped domain of strong lateral compaction. Vertical thickening and back thrusting. c

Domain of rotation and listric steepening of thrust sheets. b

Domain of active sole thrust propagation and ductile step-up. a

1 cm

← Slipped front →

DECOLLEMENT

Figure 3. Tracing of thrust wedge in Figure 2B, showing different deformation regimes (a–c), and three close-ups of characteristic internal geometries; 0 = single layers at different depths showing different modes of compression, 1 = ductile thrust in evolving ramp sheath fold, 2 = listric upward imbricate thrust, 3 = slump surface, 4 = small-scale extension faults, 5 = back-kink folds, and 6 = back thrusting.

uum cleaner with a 0.5-cm-diameter nozzle. The new technique of stepwise erosion exposed the three-dimensional geometry of a piggyback stack of imbricate sheets (Fig. 2).

MECHANICS OF PIGGYBACK ACCRETION

Figure 1 illustrates, in perspective top view, the initiation and propagation of thrust sheets to generate a three-dimensional piggyback network. The imbricate sheets, which developed prior to 10% shortening, were remarkably straight (Fig. 1A). Later imbricate thrusts anastomosed and propagated sideward from one or more nucleation sites and interfingered with one another to create a mechanically interlocked system of thrust faults (e.g., i in Fig. 1B) which also showed a repetition of stratigraphy in the zone of overlap (i in Fig. 2A). The anastomosing geometry of thrust faults in map view is well documented (e.g., Elliot, 1976).

In longitudinal cross section, the model shows a piggyback stack of imbricate sheets (Fig. 2B). The mechanics of piggyback imbrication involved episodes of thrust propagation and slip along the basal detachment. These were punctuated by step-up of a ductile ramp from the frontal tip of a sole thrust. Serial repetition of the cycle outlined above generated a piggyback stack of imbricate sheets (Fig. 2B).

DEFORMATION DOMAINS

The piggyback wedge recorded different deformation domains in longitudinal cross section during sequential accretion (Fig. 3). For example, the deformation front (domain a in Fig. 3) delineates a zone of alternating active sole-thrust propagation and step-up of a ramp from the decollement. Each ramp stepped up by ductile thrusting (shear-zone localization), narrowed down to a discrete thrust in upper layers, and subsequently propagated downward, cutting forelimbs of highly noncylindrical folds.

Domain b in Figure 3 exhibits a zone of active rotation and steepening of the imbricate thrusts. In this zone, thrusts in upper layers changed into concave-upward listric geometry during episodes of forward migration of the zone of active shortening and during episodes of climbing of older imbricate sheets on structurally lower and younger ones. In deeper layers the initially horizontal beds rotated and steepened into parallelism, the thrusting taking place at inflection points (Fig. 2B). Small-scale extension features developed on top of ramp anticlines during climb and collapse of each of the imbricate sheets (see 4 in Fig. 3). Moreover, the successive imbricate sheets, which developed subsequent to ~15%–20% bulk shortening, generated internal back folds, which developed in response to differential loading by the older and structurally higher thrust sheets during forward translation. These are likely to develop to back thrusts in strongly compacted sediments, which would separate triangular zones across serial imbricate sheets (see 5 in Fig. 3).

The area under the flat-topped domain c (Fig. 3) is strongly compacted: in this area the stack of imbricate sheets, which developed prior to ~10% shortening, steepened to a high angle (~80°) and locked, thus inhibiting differential forward transport along each imbricate thrust. This favored strong lateral compaction and vertical thickening, and formation of back thrusting (6 in Fig. 3). The piggyback wedge strained internally by different mechanisms and recorded a rapid change in deformation style during its episodic forward advance. These stages include sole-thrust propagation with consequent nonuniform thickening above the slipped zone of decollement. This was followed by step-up involving ductile rather than discrete thrusting at the frontal tip of a sole thrust during a pause in its forward advance (domain a, Fig. 3). Progressive forward migration of the zone of active shortening, coupled with movement on structurally lower and younger thrust sheets, led to rotation and steepening of older thrusts (domain b, Fig. 3), which eventually locked and ceased working (domain c, Fig. 3).

STRAIN PARTITIONING

Strain partitioning in single layers, by bed-length balancing at 40% bulk shortening (Fig. 4), showed that the experiment has undergone ductile strain with volume loss during sequential accretion. Bulk shortening in single layers was partitioned into three components; ramp folding, layer shortening with volume loss, and imbrication. However, such a procedure neither portrayed the true reversal of the kinematic sequence nor retained true angular relations between thrusts and layering. Nevertheless, it gave

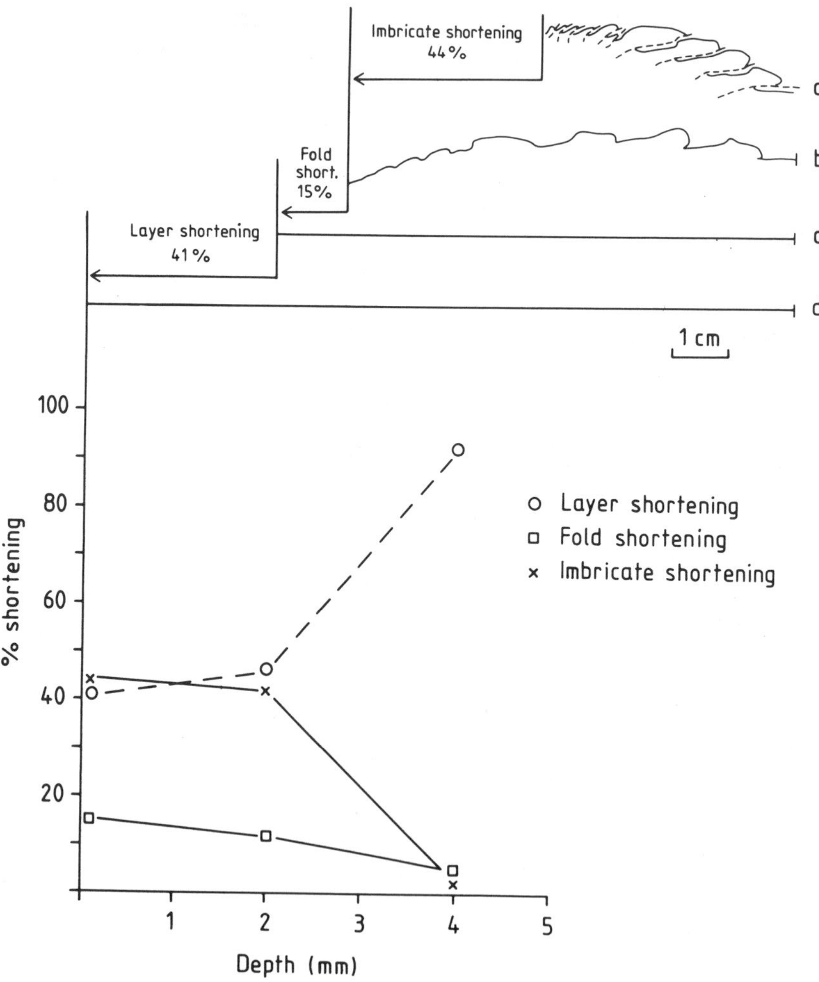

Figure 4. Successive restoration of top layer of Figure 2B at 40% bulk shortening; a = percentage shortening by imbrication (44%); i.e., after removal of imbricate shortening; b = percentage shortening by ramp folding (15%) after unfolding of layer; c = percentage layer shortening and volume loss (41%), estimated from difference in length between true original length and apparent original length after restoration.

an idea of the depth dependency of brittle vs. ductile strain involved in the shortening. This may in turn provide a qualitative insight into deformation partitioning in natural thrust sheets, which are usually inaccessible for direct observation. Bed-length balancing, using some selected layers at different depths, shows an increase in layer shortening with volume loss and a decrease in fold and imbricate shortening with increase in confining pressure, for the same total bulk shortening (Fig. 4).

INTERNAL THREE-DIMENSIONAL GEOMETRIES

Sheath Folds

The ductile shear-zone regimes associated with step-up of decollement to the surface define the limbs of immature sheathlike folds. These folds nucleated at the lateral termination of zones of nonuniform thickening above slipped areas of decollement. Initially nucleated with variable kink-band geometry, they became increasingly asymmetric and noncylindrical with time and developed strongly curved fold hinges (Fig. 5). Such folds present elliptical eye shapes in sections normal to the transport direction (Fig. 5).

During progressive shortening, the ductile shear zones forming the limbs to sheath folds narrowed first to zones of intense shear before localization of a sharp thrust. The thrust surfaces propagated both downward and laterally and increased in length by cutting through forelimbs of these folds (A–C, Fig. 5). Where shear gradients are very high, as along the base of imbricate sheets, the layers became intensely elongated and thinned (e.g., Fig. 2, A and B). Quinquis et al. (1978) have discussed models of development of sheathlike folds in a progressive simple-shear-deformation regime. Experimental simulation of such structures in viscous fluids was carried out by Cobbold and Quinquis (1980). Lacassin and Mattauer (1985) have emphasized the importance of deformation in crustal ductile shear-zone regimes in generating sheath folds in, for example, the internal Alps. However, these works did not consider the mutual relations between sheath folds and the continually growing and adjusting thrust surfaces into which they mature.

Lateral Ramps

Some of the structurally lower and younger imbricate thrusts did not propagate continuously along strike, but rather stepped upward and joined structurally higher and older thrust sheets as lateral ramps (r in Fig. 2A). Such structures are well documented in natural thrust sheets (Boyer and Elliot, 1982; Butler, 1982). The experimental lateral ramps developed as a result of spontaneous nucleation and propagation; they were not preexisting static features. If the lateral three-dimensional propagation space of a newly formed thrust is inhibited by a structurally higher and older one—for example, due to differential movement along strike—the new underlying thrust ramps laterally upward to join the higher and older thrusts. Such lateral ramps provide an efficient means of connecting different generations of thrust systems to shorten a sequence (r in Fig. 2A).

KINEMATICS OF IMBRICATE SPACING

The kinematics of imbricate spacing during piggyback accretion have been studied in detail (Fig. 6, A–C). The serial imbricates exhibited increased spacing with progressive shortening (Fig. 6A). As imbricate spacing increased toward the transport direction, the rate of nucleation of imbricates decreased. For example, the initial four imbricates, which were narrow, nucleated at a rate of about 0.17 imbricates/s (Fig. 6A). The nucleation rate decreased to 0.03 imbricates/s after a bulk shortening of ~10% (Fig. 6A). This has to be true if the wedge grows during accretion. Spacing will remain constant if the toe keeps a constant distance from the back.

This may be understood from the critical wedge taper equation of Davis et al. (1983). A wedge that loses volume by erosion will not slide and broaden at a constant taper along its base; rather, it will shorten and thicken into a wedge until the critical taper is attained. Such a reduction in wedge volume during accretion lowers the serial spacing of imbricates. This was confirmed in another erosion/accretion model by removing the wedge after every successive imbrication (Fig. 6B). In contrast to the total accretion model, the nucleation rate of imbricates in the erosion/accretion model remained nearly constant for the same convergence rate (0.05 cm/s). Moreover, sequential removal of the wedge increased the rate of nucleation of imbricates by a factor of ~32 times as compared to the total accretion model (cf. Fig. 6A and 6B). This implies that erosion during accretion lowers serial spacing of imbricates in the transport direction. This may, in turn, suggest that the qualitative finite-element elastic analysis of Mandle and Shippam (1981), which suggests that ramp spacing decreases in the transport direction, may not be generally applicable.

However, the above discussion concerns the role of wedge volume in the kinematics of imbricate spacing in a homogeneous Coulomb material with initial constant layer thickness, when detached and totally accreted upon a smooth rigid decollement. In natural settings, a systematic variation in ramp spacing—increasing or decreasing toward the toe—may be perturbed by variable friction along the sole and/or by control of preexisting basement features. Other likely factors that may control imbricate spacing are facies changes in the sedimentary layering and variations in layer thickness above a decollement. Therefore, further analysis of this important problem must be carried out before all the factors that control ramp spacing can be adequately understood. For example, Bombalakis (1986) cited arguments for both an increasing and a decreasing sequence of ramp spacing in the Utah-Idaho-Wyoming thrust belt. (See also Dixon, 1982; Wiltschko and Dorr, 1983; and Bombalakis, 1986, for additional discussions related to the ramp spacing problem.)

SIDEWARD PROPAGATION RATES OF IMBRICATES

The sideward propagation rates of imbricate thrusts from their initial nucleation site was 20 to 100 times faster than the push rate (Fig. 6C). Another interesting feature of the kinematics of piggyback imbrication pushed at a constant rate

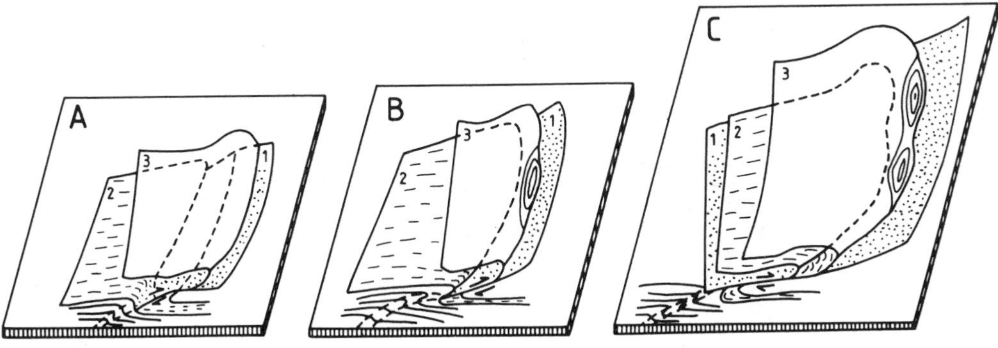

Figure 5. Schematic illustration of development of sheathlike folds (2) in relation to propagating thrust surface (1, stippled). A, B, and C indicate different shape changes of folds and thrust surface with progressive shortening. B shows sheathlike fold exposing elliptical eye; C shows eyes-within-eye fold, which develops due to differential movement within sheath folds.

is that forward propagation and step-up of successive imbricates was about 15 to 30 times slower than the rate at which they propagated to the side. This might suggest that in a prototype system emplaced under similar boundary conditions and having similar ratios of push vs. gravity force, the along-strike propagation rate of imbricates should be much faster than the rate of forward propagation and step-up between successive imbricates. This may account for thrust sheets being longer than their width.

CONCLUSIONS

1. Vacuum erosion unroofed three-dimensional thrust-sheet geometries involving sheath folds, eye folds, and lateral ramps very similar to those observed in natural fold/thrust belts.

2. The classical ramp/flat geometry shown by many workers is a transient phenomenon in the tests. Each ramp begins as a ductile shear zone (= sheath fold), tears to a discrete thrust, but then rotates and steepens with progressive accretion.

3. Internal geometry and mode of compression in the model thrust sheets were related to serial and episodic forward migration of a piggyback wedge during a steady forward advance of a rear wall.

4. Stepwise increase in imbricate spacing with progressive shortening was a direct geometric consequence of a progressively thickening and accreting wedge.

5. Sideward propagation rates of imbricates decreased as serial imbricates nucleated with increasing distance from the rear wall.

6. The strain partitioning of single layers showed that discrete thrusting gives way to more penetrative strain and compaction with depth.

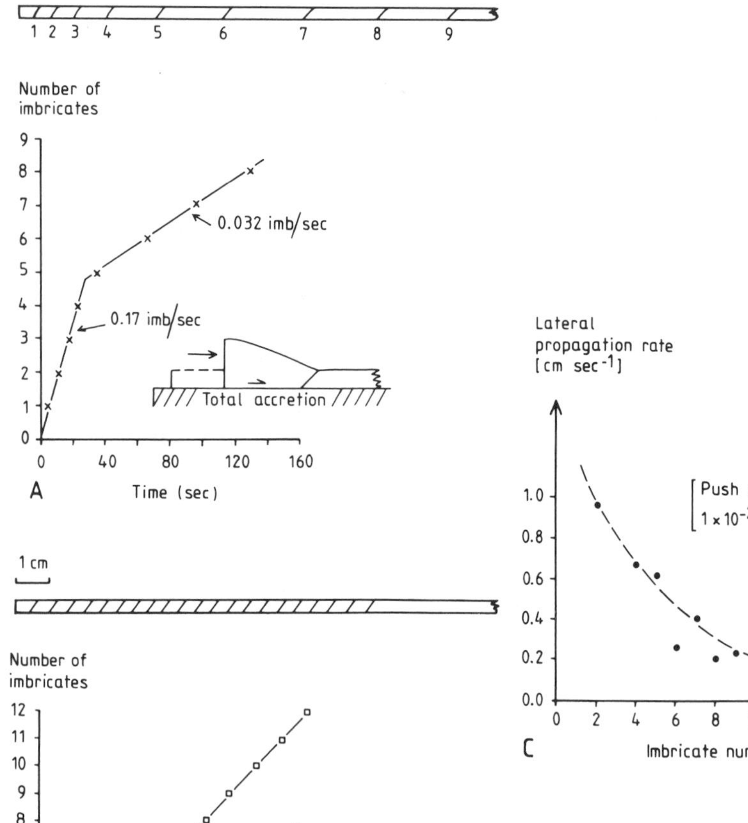

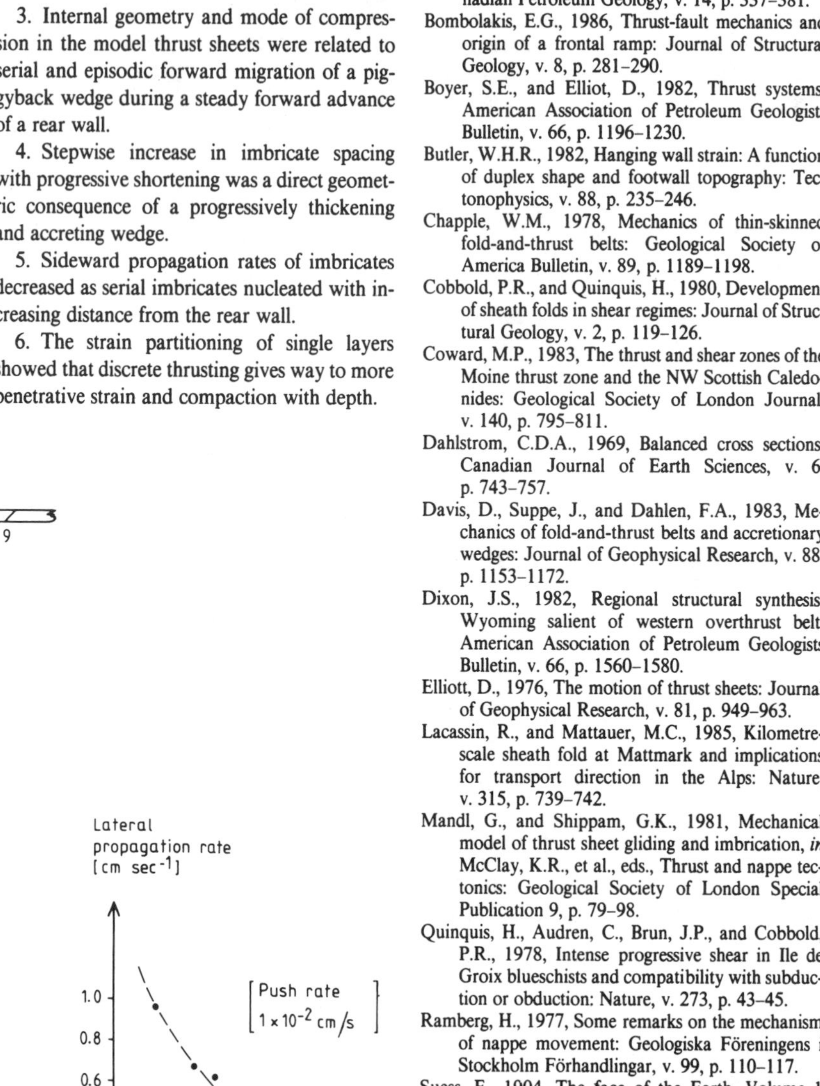

Figure 6. Kinematics of imbricate spacing. A: "Total accretion wedge" with stepwise increase in spacing. B: "Erosion-accretion" wedge with nearly constant spacing. C: Decrease in sideward propagation rate with serial imbrication.

REFERENCES CITED

Bally, A.W., Gordy, P.L., and Steward, G.A., 1966, Structure, seismic data and orogenic evolution of the southern Canadian Rockies: Bulletin of Canadian Petroleum Geology, v. 14, p. 337–381.

Bombolakis, E.G., 1986, Thrust-fault mechanics and origin of a frontal ramp: Journal of Structural Geology, v. 8, p. 281–290.

Boyer, S.E., and Elliot, D., 1982, Thrust systems: American Association of Petroleum Geologists Bulletin, v. 66, p. 1196–1230.

Butler, W.H.R., 1982, Hanging wall strain: A function of duplex shape and footwall topography: Tectonophysics, v. 88, p. 235–246.

Chapple, W.M., 1978, Mechanics of thin-skinned fold-and-thrust belts: Geological Society of America Bulletin, v. 89, p. 1189–1198.

Cobbold, P.R., and Quinquis, H., 1980, Development of sheath folds in shear regimes: Journal of Structural Geology, v. 2, p. 119–126.

Coward, M.P., 1983, The thrust and shear zones of the Moine thrust zone and the NW Scottish Caledonides: Geological Society of London Journal, v. 140, p. 795–811.

Dahlstrom, C.D.A., 1969, Balanced cross sections: Canadian Journal of Earth Sciences, v. 6, p. 743–757.

Davis, D., Suppe, J., and Dahlen, F.A., 1983, Mechanics of fold-and-thrust belts and accretionary wedges: Journal of Geophysical Research, v. 88, p. 1153–1172.

Dixon, J.S., 1982, Regional structural synthesis, Wyoming salient of western overthrust belt: American Association of Petroleum Geologists Bulletin, v. 66, p. 1560–1580.

Elliott, D., 1976, The motion of thrust sheets: Journal of Geophysical Research, v. 81, p. 949–963.

Lacassin, R., and Mattauer, M.C., 1985, Kilometrescale sheath fold at Mattmark and implications for transport direction in the Alps: Nature, v. 315, p. 739–742.

Mandl, G., and Shippam, G.K., 1981, Mechanical model of thrust sheet gliding and imbrication, *in* McClay, K.R., et al., eds., Thrust and nappe tectonics: Geological Society of London Special Publication 9, p. 79–98.

Quinquis, H., Audren, C., Brun, J.P., and Cobbold, P.R., 1978, Intense progressive shear in Ile de Groix blueschists and compatibility with subduction or obduction: Nature, v. 273, p. 43–45.

Ramberg, H., 1977, Some remarks on the mechanism of nappe movement: Geologiska Föreningens i Stockholm Förhandlingar, v. 99, p. 110–117.

Suess, E., 1904, The face of the Earth, Volume 1 (Translation from 1884 [Sollas, H.B.C., ed.]): Oxford, England, Oxford University Press, 604 p.

Wiltschko, D.V., and Dorr, J.A., Jr., 1983, Timing of deformation in overthrust belt and foreland of Idaho, Wyoming and Utah: American Association of Petroleum Geologists Bulletin, v. 67, p. 1304–1322.

ACKNOWLEDGMENTS

Funding was provided by the Swedish Natural Science Research Council. C. J. Talbot criticized various versions of the manuscript. A. A. AlDahan determined the mineral contents of the sand material.

Manuscript received April 9, 1987
Revised manuscript received July 27, 1987
Manuscript accepted August 12, 1987

Printed in U.S.A.

Reprinted by permission of the Royal Society from *Philosophical Transactions of the Royal Society*, v. A283 (1976), p. 289-312.

Phil. Trans. R. Soc. Lond. A. **283**, 289–312 (1976) [289]
Printed in Great Britain

The energy balance and deformation mechanisms of thrust sheets

By D. Elliott
Department Earth and Planetary Sciences, The Johns Hopkins University, Baltimore, Maryland 21218, *U.S.A.*

[Plates 12 and 13]

The total energy involved in emplacing a thrust sheet is expended in initiation and growth of the thrust surface, slip along this surface, and deformation within the main mass of the sheet. This total energy can be determined from potential energy considerations knowing the initial and final geometry from balanced cross sections after defining the thrust's thermodynamic system boundaries. Emplacement of the McConnell thrust in the Canadian Rockies involved *ca.* 10^{19} J of gravitational work, an order of magnitude greater than any possible work by longitudinal compressive surface forces. A new theory for the initiation and growth of thrusts as ductile fractures is based on a demonstration that thrust displacement is linearly related to thrust map length and that fold complexes at the ends of thrusts are constant in size for a given metamorphic grade. Much of the total work is dissipated within the body of the sheet. Field observations show which mechanisms of dissipation are most important at various positions within the thrust sheet, and it is found that only the top 5 km of the McConnell was dominated by frictional sliding. A novel type of sliding along discrete surfaces is pressure solution slip, in which obstacles are by-passed by diffusive mass transfer. Fibres and pressure solution grooves are diagnostic features of this sliding law, in which slip velocity is linearly related to shear stress. Pressure solution slip is widespread at depths greater than about 5 km, but at this depth penetrative whole rock deformation by pressure solution becomes dominant – marked by cleavage and stretching directions – and accounts for much of the finite strain within the thrust sheet. The McConnell thrust has an outer layer which deformed by frictional sliding and this overlies a massive linearly viscous core responsible for much of the energy dissipation and gross mechanical behaviour.

Symbols

a, b thickness and length of ductile zone at end of thrust
A amplitude of sinusoidal surface
C, C', K, K' dimensionless constants
d diameter of perfectly plastic zone at tip of ductile crack and also used for mean diameter of grains
D diffusion coefficient
g gravitational acceleration
h, $\bar{h}$ thickness of thrust sheet and mean thickness averaged over dip length p
H, H_j, H_0, H', ΔH centre of gravity in general, of volume element, of original state, of final state, and change in centre of gravity of thrust sheet
k Boltzmann constant
l, p strike length measured on map and dip length measured on cross section of a thrust
L arc length of thrust dislocation within a particular stratigraphic interval
M metamorphic quantity
n_j component of outwards surface normal

r grain axial ratio
S surface area
t time
T absolute temperature
u, u_j, $\bar{u}$, u^* displacement, displacement component, mean displacement, transitional displacement
V, V_j volume and volume element
V_c, V_p volume of ductile bead and volume swept out by bead during growth of thrust
W deformation work in general, when primed represents work calculated on a cross section of unit thickness
W_g, W_s, W_i work of gravitational and surface forces on the thrust system, and deformation work within interior of sheet
W_c, W_p, W_b work to create, propagate and slide along on thrust fault surface V, V_j
x_i rectangular coordinates
α dip of topographic surface of thrust sheet
γ general measure of finite shear strain intensity, a bar indicates average value, a dot indicates shear strain rate
$\bar{\gamma}_c$, γ_f average shear strain and shear strain at fracture within volume of ductile rock V_c
γ_t shear strain in wall rock adjacent to thrust
δ thickness of grain boundary or solution film
η coefficient of shear viscosity
λ ratio of pore pressure to normal stress on the surface
μ coefficient of sliding friction
ρ density
σ_{ij} elements of stress tensor
τ general shear stress intensity, a bar indicates average value
τ_b, $\bar{\tau}_i$ basal shear stress and mean shear stress within interior
$\bar{\tau}_s$ average shear stress on boundary due to surface forces
τ_y yield or flow stress for deforming rock
τ_n, σ_n shear and normal stress on surface whose normal is n
τ', τ^0 constants with dimensions of shear stress
ϕ slope of Mohr–Coloumb yield envelope on Mohr Diagram
Ω atomic volume

Introduction

Three principal aims of this paper are to calculate the mechanical energy consumed in the emplacement of a thrust sheet, to see where in the thrust sheet this energy is dissipated, and to find out how, or by what process, the energy dissipation occurs. In addition I wish to present some ideas on how the thrust fracture surface arises.

As in so many problems in structural geology the key lies in proper deciphering of the geometry and sequences of structural events within a thrust complex. Many of these 'rules' have been developed in the Canadian Rockies in a literature scattered through several journals and guidebooks, in part summarized by Dahlstrom (1970).

In addition to the powerful methods for determining the structural geometry and sequence of events, some progress into understanding the emplacement of thrusts can be made purely from considerations of static equilibrium. For example, an entire thrust sheet is so long in cross section (*ca.* 100 km) and contains so large a volume of rock that, compared to the effect of gravitational forces, any conceivable compressive surface force exerted from the hinterland is of negligible importance in effecting forward motion. Consequently, it is possible to write a simple equation for the basal shear stress operating along a thrust fault (Elliott 1976*a*). But only limited progress can be made from statics alone and further progress is hindered by our lack of knowledge of the constitutive equations appropriate for deforming rocks.

It is frequently suggested that the detailed stress distribution and motion along the thrust may be controlled by a dislocation type mechanism. The importance of such dislocation mechanisms along thrust fault surfaces was first suggested in a general way by Oldham (1921) and Hubbert & Rubey (1959). Barnes (1966) and Price (1973) have emphasized the importance of this point of view. Creeping-type slip for various kinds of faults has been modelled in a number of ways with groups of smeared-out dislocations (e.g. Nason & Weertman 1973).

But all these dislocation models require an explicit statement of the appropriate constitutive equations for the sliding surfaces, the tectonic breccia, or the wall rock, as the case may be. There seems to be little advantage in using bold assumptions about ancient sliding surfaces until the constitutive equations can be brought at least into sharper focus, and it is pointed out in this paper that progress is possible in this direction purely from field observations.

Some of these difficulties may be overcome by looking at the energy balance of a thrust sheet and seeing how it is partitioned among the various processes. If surface forces are insignificant compared to gravitational forces, then the work expended in emplacing a thrust sheet must be provided by a decrease in the potential energy of the thrust mass. By comparison of the initial and final potential energy of the thrust we may measure the amount of energy expended in mechanical deformation. Such an approach was pioneered by Goguel in 1948 and applied by him to a number of thrust complexes in the French Alps, but does not seem to have been used since then. Unfortunately, few techniques for working out the structural geometry were available at the time and several complications in Alpine geology made determination of the initial and final states of these thrust sheets rather imprecise.

The McConnell thrust system

The first objective is to define the thermodynamic system as it existed between the original state just before motion along the particular thrust until the final state following a thrust's activity. This thrust system is specified by the four boundaries listed below (figure 1).

1. The top boundary lies along the surface where all elements of the stress tensor σ_{ij} are zero.

2. Thrust sheets lose mass by erosion, but it has been shown that during motion of a thrust most of this erosion occurs from the vicinity of the toe (Price & Mountjoy 1970).

Thrust sheets also gain mass by successive imbrication into the footwall rocks below and in front of the current outcrop of the thrust. In order to avoid problems with this large mass transfer occurring in the vicinity of the toe, we will define the front boundary as the material surface in the original state which will eventually become the topographic front surface in the final state. This type of boundary is quite general, yet has the advantage of defining a closed system – or one of constant mass.

3. The lower boundary lies within the footwall below the thrust surface.

Below this boundary are undeformed rocks structurally attached to the foreland, and above is a thin rind of footwall rocks deformed by passage of the thrust sheet. Along this lower boundary, all strain and displacement components are zero ($u_j = 0$).

4. In the original state the trailing edge boundary is a vertical material line chosen to enclose a volume of rock of interest. Since the thrust system is assumed to be undergoing plane incompressible flow, a vertical line in the final state chosen to enclose the same volume as the initial state will give the average position of the trailing edge boundary but not its curved shape.

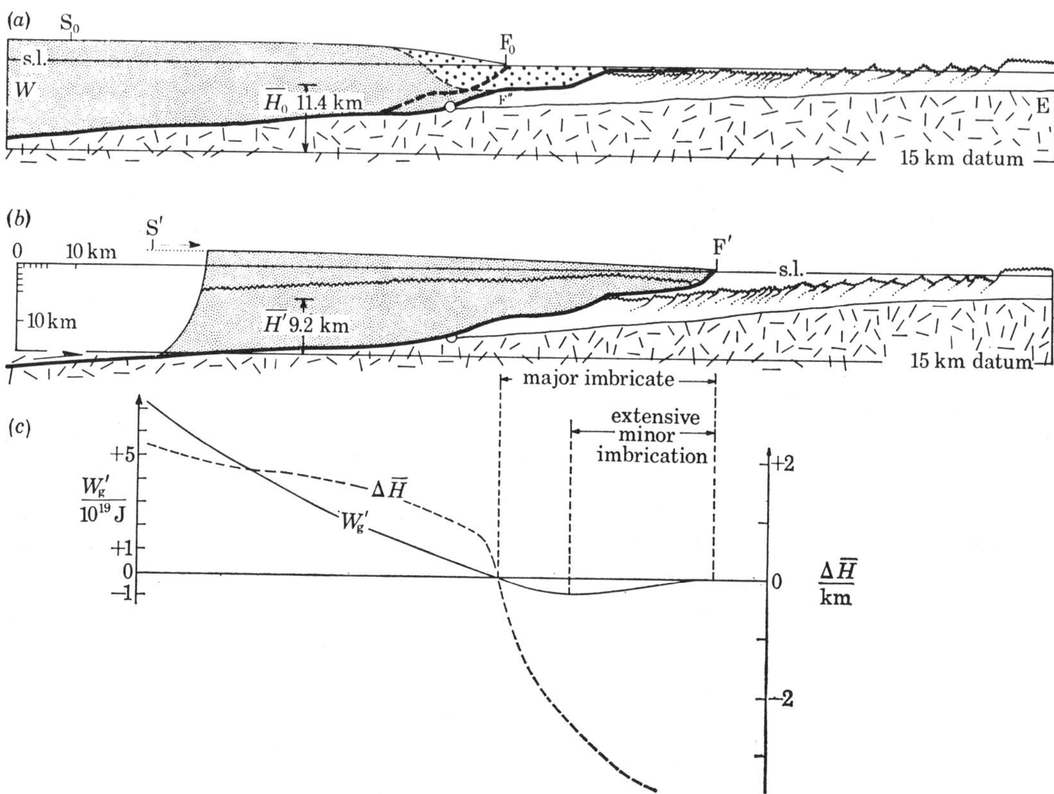

FIGURE 1. (*a*) Bow Valley cross section from the planes into the Western Main Ranges in the Early Maestrichian 'original state' before motion along the McConnell thrust. F_0 and S_0 are the mountain front and 3500 m topographic summit. F″ is the original position of the hanging wall beds which will be at the mountain front F′ in the final state. Height H_0 of the centre of gravity is 11.4 km above the –15 km horizontal datum. The present erosion surface is shown as a serrated line. Location of section shown on figure 4.

(*b*) The final state in the latest Maestrichian after motion on the McConnell. F′ and S′ are the mountain front and the 2500 m topographic summit. The centre of gravity H' is 9.2 km above the horizontal datum.

(*c*) Graph of W'_g and ΔH with horizontal position in the final state. The major imbricate slice in figure 1*b* coincides almost exactly with $W'_g \simeq 0$. The front toe has its centre of gravity raised. The toe portion is dominated by extensive minor imbrication, but this is not shown.

However, a substantial component of simple shear parallel to bedding is a direct geometric consequence of a thrust surface cutting up section in the direction of motion (figure 2). This effect imparts a strong curvature to the trailing-edge boundary and will be discussed later in the paper.

With our thrust system defined, we can now look at the forces which produce changes in shape and position of this system. The external forces are of two possible sorts.

1. Gravitational forces acting on the mass of the thrust system. Positive gravitational work W'_g is accomplished on the system whenever the centre of gravity is decreased from the initial to the final state.

2. Surface forces may accomplish work W'_s on the boundaries of the thrust system.

The total work by all the external forces is the sum of two terms,

$$W' = W'_g + W'_s \tag{1}$$

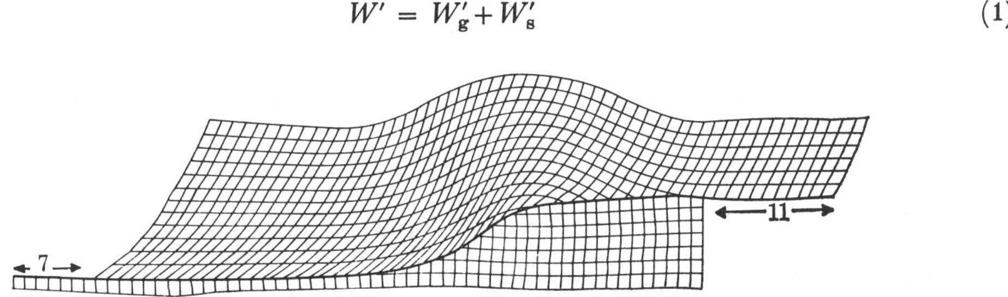

FIGURE 2. Originally rectangular grid stamped onto a stack of paper and cut by a thrust fault. Trailing edge moves forwards 7 units along thrust and necessarily assumes a concave shape. Leading edge along thrust moves forward 11 units. Note the widespread distribution of simple shear strain as a result of movement over this footwall step. (Drawn from photograph.)

Both of these external forces act very slowly, therefore inertial terms are negligible and mechanical equilibrium is attained between the system and the surroundings. Consequently, the total work by all the external forces is exactly balanced by the work of internal forces which resist this motion and deformation of the thrust sheet. This internal work is used up in shearing motions completely and irreversibly and is dissipated as heat, i.e. shear-strain heating.

We will now calculate the gravitational work W'_g for the McConnell thrust in the Canadian Rockies, then we will consider the effects of work by surface forces.

The section westwards from the Bow River passes through the central part of the McConnell thrust (figure 4). Along this line, original and final state cross sections were constructed (figure 1) based upon the restored section by Bally *et al.* (1966, plate 12), in turn supported by published seismic sections.

The thrusts making up the Foothills, Front and Main ranges were developed during deposition over a 50 Ma time span from Campanian to Oligocene of the synorogenic Belly River-Paskapoo assemblage, the second major molasse sequence deposited in the Canadian Rockies (Eisbacher *et al.* 1974). Foothill thrusts cut the Paskapoo Formation but are overlain by the Porcupine Hills (Carrigy 1971); this dates the Foothill thrusting between 65–60 Ma ago. The Foothill thrusts deform and therefore post-date the McConnell and must have ceased activity about 65 Ma ago. The McConnell thrust at Black Rock has a good portion of the lower Brazeau Formation (Lower Maestrichian) as footwall and the thrusting here must be after about 72 Ma ago. Consequently the toe of the McConnell thrust was active over at most an 8 Ma time span. Its displacement is about 40 km, giving a minimum mean velocity of 5 mm/a or 1.6×10^{-10} m/s.

The effects of Foothill thrusting may be removed by constructing a restored section (Bally *et al.* 1966, plate 12) to establish the undeformed trajectory of the McConnell thrust. The eroded portion was reconstructed by allowing the toe of the McConnell to climb in the lower Brazeau Formation with the characteristic stairstep fashion shown by nearby Foothill thrusts.

As a thrust sheet advances so much mass is transferred laterally along the crust that a large isostatic depression spreads far out in front of the thrust outcrop; such isostatic depressions are a dominant feature in controlling molasse basins (Bally *et al.* 1966; Price & Mountjoy 1970; Price 1974). The depoaxis of successive molasse foredeeps migrate cratonwards, driven by the isostatic response to a moving thrust mass.

Sealevel at a given time can be estimated from the shallow water, alternatively fresh and marine, formations. Restored sections for various times can then be drawn using the different estimated sealevels as horizontal datum. This has been done for the Early Maestrichian (*ca.* 74 Ma), before motion on the McConnell (figure 1 *a*), and for the latest Maestrichian (*ca.* 62 Ma), not long after motion ceased (figure 1 *b*). The sealevel lines must be projected for considerable distances without any preserved evidence – a procedure which is probably the chief source of error. Other sources of error are numbered below. The current erosion level is close to the Early Maestrichian time line so that this section is much more soundly based than the latest Maestrichian one. The isostatic depression is observed to increase rapidly westwards; but this cannot be maintained forever and the angle of dip of the décollement to the west is taken to gradually approach its current dip.

The thrust induced isostatic depression becomes strikingly clear on these restored sections (figure 1). By creating a depression ahead, thrusts make it much easier to move themselves forwards; indeed it is doubtful if the McConnell thrust could have moved at all without this isostatic adjustment. Since the horizontal datum can be fixed for both original and final states, this isostatic response is implicitly accounted for in the gravitational work calculations.

In the original state, the mountain front was assumed to start at the thrust which outcrops immediately west of the McConnell. The mountains were assumed to have a relief of 3500 m, no higher than the relative relief today between the Andes and its eastern craton. The surface slope was drawn at an average of 2.5°, steeper at first and then tapering off. In the final state, the mountain starts at the surface outcrop of the McConnell and rises to 2500 m with an average surface slope of 1.5°. Such reconstructed surfaces are clearly a source of error (2), although judging by contemporary active mountain belts there is not too much room for a significantly different surface topography.

The geometry of the thrust sheet is subject to increasing error (3) west of the Continental Divide, where listric normal faults become important and probably post-date development of the McConnell. For this reason, the system is arbitrarily terminated about 6 km east of the Trench. The concave shape of the trailing edge boundary in the final state is an estimate also subject to error (4). The trailing edge boundary to the original state is drawn so as to enclose a volume equal to that of the final state of 1.1×10^3 km^3. We assume that both original and final state cross sections are 1 km thick perpendicular to the plane of section.

Prior to being displaced 40 km along the McConnell thrust, point F′ in the final state was once at position F″ in the original state. The position in the original state of the material points in the deformed final state which make up the rest of the topographic surfaces S′ F′ may be estimated and the uneroded mass below and to the west of S_0 F″ is the original position of that volume of rock which became the McConnell sheet. This front boundary again is subject to error (5).

The thin layer of the footwall which is part of the thrust system has a very small volume in comparison to the rest of the system, nor does this volume undergo much change of its centre of gravity. Its contribution to the gravitational work may be safely neglected.

The gravitational work of the rest of the thrust system may now be calculated by the classical methods. Draw a horizontal datum line 15 km below the reconstructed sealevel on each section (figure 1). The cross sections can now be subdivided into convenient 10 km long elements and the volume V_j, centre of gravity H_j of each element measured.

Then starting from the toe and moving towards the hinterland the cumulative volume V,

$$V = \sum_0^j V_j$$

has a centre of gravity H,

$$H = \sum_0^j V_j H_j \Big/ V. \tag{2}$$

The difference in the centre of gravity between the original H_0 and final H' states is for a particular volume of rock

$$\Delta H = H_0 - H'.$$

This ΔH is plotted with respect to position in the final state (figure 1*c*).

The work accomplished by gravity W'_g for an equivalent volume V in the original and final states is

$$W'_g = \rho g V \Delta H. \tag{3}$$

This gravitational work is also plotted with respect to position in the final state (figure 1*c*). Note that from equation (3) gravitational work per unit volume (W'_g/V) varies directly with change in centre of gravity, so that ΔH is a measure of energy density.

How accurate are these results, particularly in view of the five numbered geological sources of error indicated in the text above? The largest error undoubtedly arises during restoration to sealevel. Although exact limits to this and the other errors cannot be given, trials with various 'worse possible' cross sections did not change the order of magnitude of the gravitational work. Note also that for a western boundary anywhere within the last 60 km of the thrust sheet, the gravitational work is in the order of 10^{19} J. It appears as if errors might not give rise to wildly fluctuating results, and the value of 10^{19} J would be attained even after moderate alterations to the cross sections.

Errors of measurement from the cross sections are comparatively small, and it is in this sense that the gravitational work is reported to two significant figures.

The work of surface forces can only be applied at the boundaries of the thrust system. At such a boundary

$$W_s = \int_0^S u_j \sigma_{ij} n_j \, dS \tag{4}$$

where the displacement component is u_i, stress component σ_{ij}, the outward normal n_j, and the surface area of the boundary is S.

Now consider the 4 bounding surfaces of the thrust system. The top surface is stress free $\sigma_{ij} = 0$, the lower boundary is deformation free $u_i = 0$, and the front boundary is close to the surface so that the stress σ_{ij} and displacement u_i are small. The only part of the boundary of the thrust system on which surface forces may accomplish work is the trailing edge boundary. In other words this is the 'principal' boundary condition because the state of the surface forces on this boundary specifies the prime agency generating the motion of the thrust (see, for example, Batchelor 1965, p. 217). We represent these surface forces by a stress whose magnitude is zero at the surface and might reach a maximum value limited by the shear strength of the rock of 2.5×10^7 Pa at the thrust fault. This stress is largely a deviatoric shear stress. The maximum long term average regional shear stress $\bar{\tau}$ for an entire sheet is in the order of 10^7 Pa (Elliott 1976*a*). Equation (4), when applied to the trailing edge boundary, may be expressed as

$$W'_s = u\bar{\tau}h\, dx_3. \tag{5}$$

Here the area of the trailing edge boundary is the height h times the arbitrary thickness dx_3 normal to the plane of section and u is displacement of the top with respect to the bottom of the trailing edge boundary.

Recall that the total work expended is the sum of gravitational work and the work accomplished by the surface forces. This work W' must be positive:

$$W' = W'_g + W'_s. \tag{1}$$

There are three special cases of this exact equation, and we will now see that each of these cases applies to natural subdivisions of the thrust system found on the cross sections and graphs (figure 1).

(1) One of the main theories for thrusting is that it occurs as a result of *longitudinal compressive surface forces*. One form of this theory is expressed as

$$W' = W'_s - W'_g > 0. \tag{6}$$

In this case the centre of gravity is raised so gravitational work is negative. Much of the work exerted by surface forces is consumed in overcoming this gravitational resistance. This special case clearly applies to the front 150 km^3 toe region of the McConnell thrust, where the centre of gravity is raised as a result of deformation, and the gravitational work is negative with a value of -7.5×10^{18} J (figure 1*c*). The magnitude of the gravitational energy density reaches higher values here than anywhere else in the thrust system.

The toe of the McConnell thrust almost certainly resembled the swarm of minor imbricates marking this stratigraphic level in the foothills. This extensive minor thrusting proves that the thrust sheet is losing coherency or unity, suggesting that the stress $\bar{\tau}_s$ is greater than 10^7 Pa in this region; in fact if $\bar{\tau}_s \approx 2.5 \times 10^7$ Pa then the work of surface forces acting on this front toe region would be 8×10^{18} J, which is enough to overcome gravitational resistance.

(2) Another form of the *longitudinal compressive surface force theory* is

$$\left.\begin{aligned} W' &\approx W'_s \\ W'_g/W'_s &\ll 1. \end{aligned}\right. \tag{7}$$

For the 230 km^3 front portion of the thrust sheet the gravitational work W'_g is approximately zero. The decrease in local centres of gravity $\bar{H}_j$ of volume elements V_j (see equation (2)) at the back of this region are enough to raise the centre of gravity of the toe, and no net gravitational work is produced or consumed.

The major imbricate of 230 km^3 is about 11 km thick at its trailing edge. The maximum possible work by the surface forces here, from equation (4), is about 5.6×10^{18} J.

(3) The final special case of equation (1) is the *gravity spreading theory* which may be stated as 'the driving force for thrust emplacement is entirely a consequence of the sheet's weight'. For this case

$$\left.\begin{aligned} W' &\approx W'_g \\ W'_s/W'_g &\ll 1. \end{aligned}\right\} \tag{8}$$

The main body of the sheet, 1.1×10^4 km^3 is about 15 km thick at its trailing edge, so that the maximum possible work of the surface forces on this system is 6×10^{18} J.

From the volume of about 300 km^3 upwards, gravitational work is expended by the motion of the sheet and is of the order of 10^{19} J for the entire western 60 km of this sheet. For the sheet as a whole (1.1×10^3 km^3) the centre of gravity fell through 2.2 km and $+7.3 \times 10^{19}$ J of gravitational work was expended, an order of magnitude greater than any possible work of surface forces. In summary, this special case of gravity spreading will apply for either the whole of the McConnell thrust system or the frontal portion, so long as it is greater than about 300 km^3. Small portions near the front involve substantial longitudinal compressive surface forces. What seems to be happening is that the surface forces acting on the toe region and on the major imbricate are themselves provided by the gravitational forces acting on the main mass. The prime agency is gravity, but on a smaller scale the longitudinal compressive surface force theory is also clearly applicable. It all depends on the volume of material considered within the thrust system.

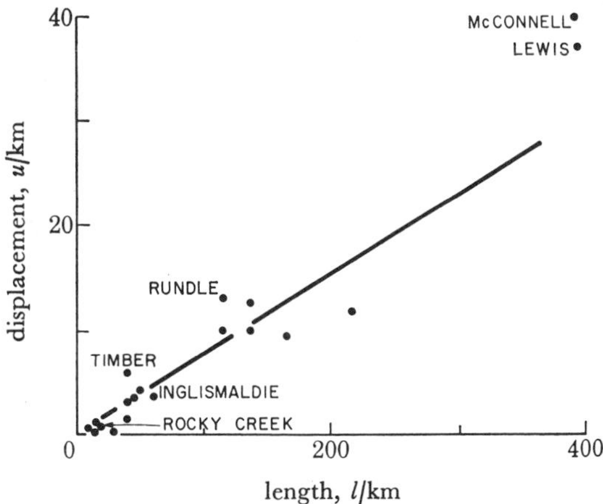

FIGURE 3. Maximum displacement u in central part of a thrust versus map length l for 20 thrusts from the foothills and front ranges. A linear relation is suggested. The named thrusts are referred to in the text and subsequent figures. Further data, statistical analysis, and a detailed bibliography are presented elsewhere (Elliott 1976*b*).

THE INITIATION AND GROWTH OF A THRUST SURFACE

I have plotted in figure 3 the maximum stratigraphic separation observed along thrusts in cross section versus their map length for 20 thrusts in the foothills and front ranges of the Canadian Rockies. The main criteria in choice of a particular thrust to plot were (*a*) the existence of fairly recent cross sections and maps and (*b*) thrusts without complex splays. Faults whose map pattern showed extensive branches proved awkward to handle and they seemed to plot much more irregularly. Nevertheless, I believe the particular thrusts chosen are reasonably repesentative. There are several implications from this graph.

(1) It has been suggested for some time that 'long thrusts are strong thrusts'. We see here that in fact not only do the longest thrusts have the largest displacements, but there is a roughly linear relation between strike length l on the map and displacement u,

$$u \simeq Kl \tag{9}$$

where $K \simeq 7\%$.

(2) Presumably thrusts in all different stages of growth are present on this graph. As Douglas (1958, p. 130) pointed out, a thrust starts as an initial break at a given point and

spreads out sideways in both directions. Consequently this relation (9) must apply to the development in time of a given thrust.

$$du/dt = K\,dl/dt.$$

(3) The initial break is roughly in the centre of the map length and the maximum displacement must be at this point. This is clearly demonstrated by the McConnell thrust (figure 4) where the displacement is 0 at its northern termination, reaches a maximum, then loses displacement towards its southern termination.

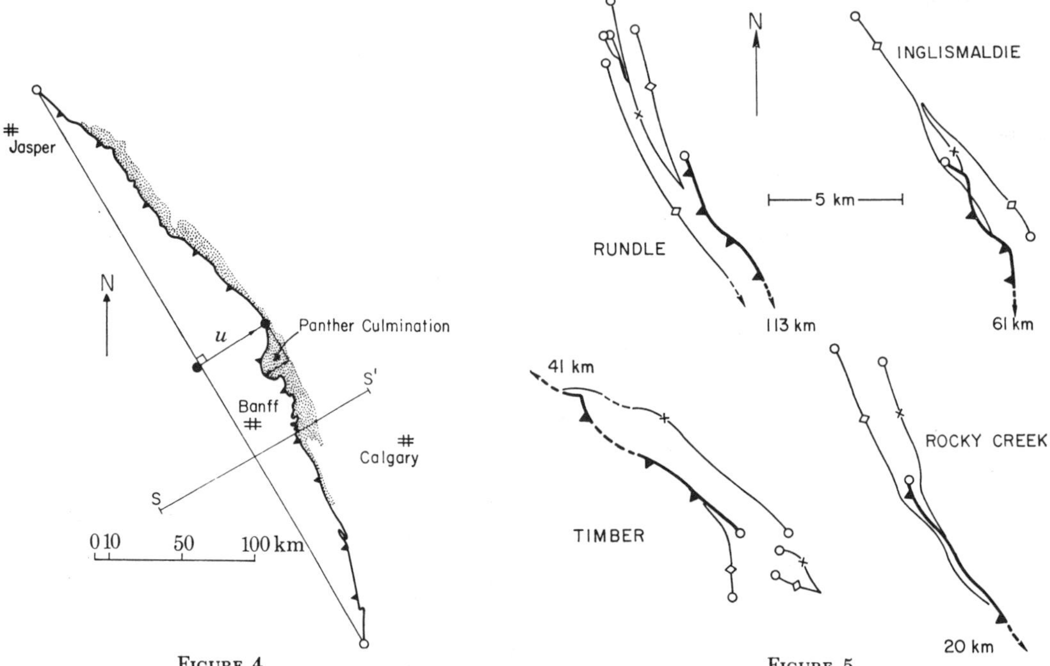

FIGURE 4 FIGURE 5

FIGURE 4. Map of the McConnell thrust, which has a strike length of 410 km. The re-entrant at the Panther culmination gives an absolute minimum displacement of 20 km. The balanced cross section of figure 1 is through S–S′ and along this line displacement $\simeq$ 40 km. The chord joining the 2 ends of the thrust has a normal bisectrix, and along this bisectrix, by the 'bow-and-arrow' rule, lies the maximum displacement $u \simeq$ 45 km of the thrust. In the central portion of the McConnell thrust the hanging-wall is Middle Cambrian shelf carbonate facies and the footwall is Cretaceous molasse (stippled). Towards the ends one goes up section in the hanging-wall and down section in the footwall, and the terminations of the thrust are in Devonian–Permian shelf carbonates. Fold complexes at the ends of the thrust are not shown.

FIGURE 5. Maps of fold crest and trough surface traces near the ends of 4 thrusts in the Foothills and Front Ranges. All the fold complexes die out about 8 km (= a) from the end of the thrusts. The ends of the thrusts are about 0.5 km (= $\frac{1}{2}b$) from adjacent crest and trough surface traces. Note that a and b are about the same size regardless of size of thrust.

(4) The direction of this maximum displacement vector u on a map is the normal bisector of the straight line joining the two ends of the particular thrust. I call this the 'bow and arrow' rule.

(5) Stratigraphic evidence (such as given earlier for the McConnell) indicates that thrusts move between 10^{-10} and 10^{-9} m/s, from equation (9) the rate of lateral propagation of the thrust $dl/dt \approx 10^{-11}$ to 10^{-12} m/s. In other words, *thrust fracture surfaces propagate very slowly*.

This process of slow, lateral growth of the thrust fault can be clarified by studying the way thrusts die out, summarized in the following points.

(6) It has been known at least since Heim (1921) that thrusts die out along strike into folds.

(7) These folds are strongly non-cylindrical with both converging and diverging hinges. Fold pairs are linked in the sense that as antiform and synform converge the folds become tighter and tighter towards a singularity. Thrusts usually pass through such singularities (Ollerenshaw 1968).

(8) These fold complexes have characteristic dimensions, regardless of the length of the thrust. For example figure 5 shows fold complexes dying out about 5–10 km from the end of the thrusts (dimension b) and a typical distance from the fold hinges to the end of the thrust is about 0.5 km (dimension $\frac{1}{2}a$).

(9) Where thrusts die out in the overturned short limbs of asymmetric folds ('forelimb thrusts') finite simple shear strains are fairly high, $\gamma_f \approx 1$.

When thrusts die out in the right-way-up long limbs ('backlimb thrusts') the finite simple shear near the fracture is considerably less, $\gamma_f \simeq 0.1$. This strain γ_f is estimated assuming that the asymmetric folds are of flexural-slip type, and regardless of the length of the thrust and its maximum displacement the shear fracture strain γ_f appears to be roughly constant.

(10) The magnitudes of a, b and γ_f cited above are for the very low grade rocks of the Foothills and Front Ranges. As the metamorphic grade increases, the values of a, b and γ_f probably change, all increasing as the ductility of the rocks increases.

From these last 6 observations (6 to 11), we may conclude that a non-cylindrical fold complex travels just ahead of a sideways propagating thrust. The rocks are folded, the folds grow and tighten, and then the thrust fracture extends laterally into this strained mass. The folds indicate that ductile deformation of substantial magnitude affects the rock before the crack slowly propagates into it. Presumably, as the strain progresses, *damage* accumulates. Preliminary field work suggests that in some instances this damage is caused by minor tension fractures which become steadily larger, more numerous, and record higher strains towards the thrust. Eventually these flaws reach a stage where they suddenly run together forming the large scale thrust fracture. The propagation of thrust faults appears to be a case of *ductile fracturing*, and the most widely used ductile fracture criteria is that of a limiting finite strain above which a fracture appears as the coalescence of flaws (*see* MacClintock & Argon 1966, for a good review).

Odé (1960, p. 300) defined purely ductile faults to be those for which there is no stored elastic strain energy; an assumption precisely opposite to what one would make for a Griffith-type purely brittle fault. Odé suggested thrusts as examples of purely ductile faults; his plastic analysis is rather different from the approach presented below.

A key relation is that the distortional work W due to a particular rock deformation mechanism affecting a volume V of rock can be expressed in terms of the octahedral shear stress τ and strain γ as

$$W = \frac{3}{2}\int_0^V \int_0^\gamma \tau \,\mathrm{d}\gamma \,\mathrm{d}V. \tag{10}$$

This equation expresses the fact that the work is dependent upon the path of finite strain γ.

Throughout the rest of this paper we shall assume simple shear paths for ease of discussion, but this in no way restricts the physical reasoning. Further, deformation paths are a measurable quantity in naturally deformed rocks. For example, the fold complex at the thrust tip may be observed to be of flexural slip type so that the deformation path is one of simple shear along bedding. Alternatively, in penetratively deformed rocks the deformation path can be measured from fibres (Elliott 1972; Durney & Ramsay 1974).

The local shear stress must have been higher in the ductilely deformed fold complex than its regional value in the much less deformed rocks outside the immediate vicinity of the thrust termination. This local stress would be reached by the stress concentrating abilities of the fracture. If we assume that a characteristic flow stress τ_y had to be achieved to produce the ductile flow within these fold complexes, some interesting conclusions can be drawn. Note that use of such a flow shear stress may be justified in three different ways:

(1) In an exact sense, as a shear stress averaged over both time and through the volume of ductilely deformed material,

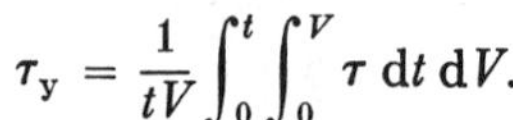

$$\tau_y = \frac{1}{tV}\int_0^t \int_0^V \tau \, dt \, dV.$$

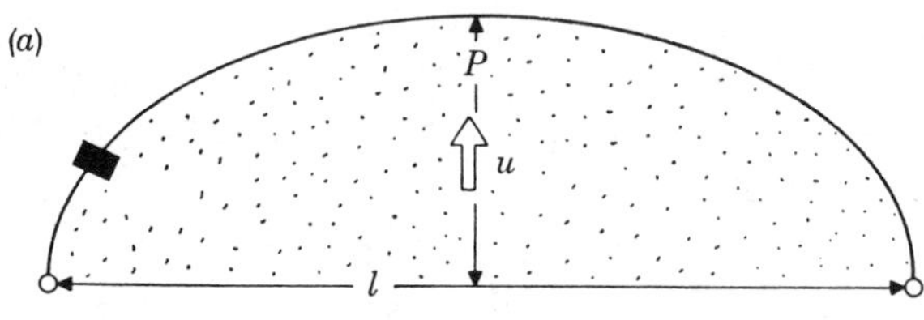

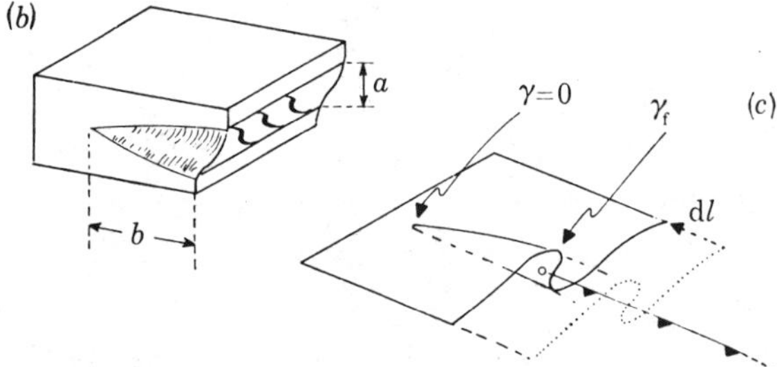

FIGURE 6. (*a*) Plan view of thrust. The thrust surface is stippled. Displacement reaches a maximum u in the central region. The perimeter of the thrust surface is the thrust dislocation, and outside of this perimeter the formation is unfaulted. The thrust dislocation is assumed to be a half ellipse enclosing an area $(\frac{1}{4}\pi)\,pl$ where p and l are the dip and strike lengths of the thrust surface within the formation. (*a*) and (*b*) examine various detailed aspects of the black rectangle.

(*b*) Schematic view of portion of formation just beyond thrust. Ductile deformation zone dies out in distance b and is of thickness a. Strain damage suggested by sigmoidal tension cracks, shown black within deformation zone. This type of ductile bead lies all along thrust dislocation.

(*c*) Folds in formation at end of thrust shown by thin marker layer. Thrust has lengthened by moving distance dl into previously deformed rock. Finite strain at fracture $\gamma_f \approx 0.6$ here where the interlimb angle $\approx 60°$. Fold dies out at point of zero, strain $(\gamma = 0)$ over length a.

(2) In an exact sense again, as the shear yield stress in a perfectly plastic rock.

(3) In an approximate sense, as the yield stress of the perfectly plastic approximation to a power law material, $\dot{\gamma} \propto \tau^n$. This approximation is good whenever $n > 3$.

For a simple shear deformation path, with τ_y and γ measured in the simple shear direction,

$$W = \tfrac{1}{2}\tau_y \gamma V. \tag{11}$$

The tip which surrounds an active thrust separates faulted from unfaulted rock and is a type of fracture dislocation (figure 6). This fracture or 'thrust dislocation' has edge character in the central region and screw character at both sides. The central tip of the thrust dislocation eventually climbs out of the formation and may reach the ground surface, splitting the thrust dislocation into two parts each of essentially screw character. The screw type component affects a particular formation as if it were being torn like a sheet of paper.

We shall now see that the region of deformed and folded rocks at the ends of thrusts provides insight into how a thrust within a particular formation is created, propagated and slowly grows.

(*a*) *Creating the thrust fracture*

We have seen field evidence which indicates a lip or 'bead' of ductile rock all along the thrust dislocation. This bead has thickness a, breadth b, and cross sectional area $\frac{1}{4}\pi ab$ (figure 6). With arc length L of thrust dislocation within the formation, then the volume V_c of the ductile bead is

$$V_c = \tfrac{1}{4}\pi abL.$$

The work W_c necessary to produce a shear strain $\bar{\gamma}_c$ averaged over this ductile volume V_c, assuming a simple shear deformation path

$$W_c = \tfrac{1}{4}\pi\tau_y\bar{\gamma}_c abL.$$

The work W_c' to create the part of the ductile bead within the cross section slice which is dx_3 thick is

$$W_c' = \tfrac{1}{4}\pi\tau_y\bar{\gamma}ab\,dx_3.$$

For the entire McConnell thrust system, $\tau_y \simeq 2\times10^7$ Pa, $\bar{\gamma} \simeq 0.25$, $a \simeq 1$ km, $b \simeq 8$ km (from figure 5), $dx_3 = 1$ km, so $W_c' \simeq 1.6\times10^{16}$ J.

(*b*) *Propagating the thrust fracture*

Over its active life the dislocation bounding the edge of the thrust surface sweeps out an area of $\frac{1}{4}\pi\, pl$ (figure 6). But in order to move this thrust dislocation forwards or sideways, the ductile bead of thickness a must move ahead of the propagating fracture and sweeps out a volume V_p

$$V_p = \tfrac{1}{4}\pi apl.$$

This entire volume of rock reached the finite strain necessary to induce ductile fracture γ_f; assuming a simple shear deformation path the work W_p necessary to push this ductile zone ahead is

$$W_p = \tfrac{1}{8}\pi\tau_y\gamma_f apl. \tag{13}$$

For a cross-section slice of thickness dx_3 the work W_p' is

$$W_p' = \tfrac{1}{2}\tau_y\gamma_f ap\,dx_3. \tag{14}$$

For the McConnell thrust system, if the finite strain before fracture $\gamma_f \simeq 0.5$ (figure 5), $p \simeq 90$ km, and for a slice of 1 km thick, $W_p' \simeq 8.8\times10^{17}$ J.

Increase in ductility of the rock would be reflected in higher values of a, b, $\bar{\gamma}$, γ_f, V_c and V_p, and in a lower value of the yield stress τ_y. This increase in ductility affects the work both to create and to propagate a thrust.

We saw that the sides of a thrust dislocation are essentially of screw character, and in fact resemble the 'tearing' or 'mode III' type problem in fracture mechanics. Fortunately this is one of the few problems in ductile fracturing for which analytical solutions exist. This problem is briefly discussed, for example, in MacClintock & Argon (1966, p. 409) where assuming a perfectly plastic ductile bead which just touches the crack tip and is of circular diameter d, they derive that $d \propto \tau_y^{-2}$ and $\gamma_f \propto \tau_y^{-1}$. Thus a and $b \propto \tau_y^{-2}$ and $W_c \propto \tau_y^{-4}$, $W_p \propto \tau_y^{-2}$, and a small increase in ductility causes a large increase in the work necessary to create and propagate thrusts. In zones of high grade metamorphism one should expect to see less thrusting. Field work seems to bear this out; in fact, structural mapping in higher grade rocks often reveals that any thrusting which occurred did so at an early stage under very low grade metamorphic conditions.

(c) *Sliding on the thrust surface*

The work W_b involved in sliding along the base of area $\frac{1}{4}\pi lp$ is

$$\tfrac{1}{4}\pi\tau_b pl\bar{u}.$$

Note that we have defined this basal sliding work in such a way that it occurs only along the thrust, a two dimensional surface of zero thickness.

Evaluation of the mean displacement $\bar{u}$ is important. Consider first the average along the dip length p.

Study of figure 2 reveals that the trailing edge has less thrust displacement than the leading edge. The central portion of the model thrust cuts steeply up section in the hanging wall and in this region simple shear from within the body of the sheet is transferred to displacement along the base.

We seem to have a method whereby simple shear within the body of the sheet can be converted into displacements along the base, so that the thrust surface gathers slip as it cuts up section. A large proportion of the observed 40 km displacement observed at the leading edge of the McConnell could die out down dip.

Another feature which we must take into account is that the displacement along the leading edge of a thrust reaches a maximum value u in the central part and dies out at each end over a strike length l (figure 4). Assuming that the variation in displacement with strike length traces a semiellipse, the value of the displacement averaged over the strike length is $\frac{1}{4}\pi u$.

Averaging over the dip and strike lengths,

$$\bar{u} = C\tfrac{1}{4}\pi u.$$

Here $C < 1$, and accounts for the thrust gaining displacement as it cuts up section. Therefore

$$W_b = C(\tfrac{1}{4}\pi)^2\,\tau_b plu. \tag{15}$$

On a cross section the thrust surface has an area $p\,\mathrm{d}x_3$ and the work of basal sliding W_b' is

$$W_b' = C\tau_b up\,\mathrm{d}x_3. \tag{16}$$

It was demonstrated in a previous article that the shear stress acting on the base of a thrust sheet τ_b is (Elliott 1976*a*).

$$\tau_b \simeq \rho gh\alpha \tag{17}$$

where α is the surface slope. This basal shear stress τ_b represents an average value over a distance along the base several times the local thickness h of the sheet. A typical value for the basal shear stress along the sole thrust in the southern and central Canadian Rockies was estimated at about 5×10^6 Pa, probably higher at the beginning of thrusting and dropping towards the end.

For the McConnell thrust (figure 1) C is unknown; a value of 0.5 might be reasonable and in this case $W_b' \approx 9\times10^{18}$ J.

(d) Growth of a thrust

When a thrust is first starting to grow within a formation the relative importance of the work terms W_b, W_p, W_c may be quite different from the later, large displacement stage.

The conditions under which $W_b = W_p$ is found by equating equations (15) and (13). The transitional displacement u^* is

$$u^* = 2/\pi C(\tau_y/\tau_b)\,\gamma_f a. \tag{18}$$

By equating equations (11) to (13), and using equation (9), the conditions under which $W_c = W_p$ is

$$u^* = K(\bar{\gamma}_c/\gamma_f)\,(L/p)\,b. \tag{19}$$

Substituting in appropriate values for the Foothills and Front ranges we find that the transitional displacements u^* are about 1–2 km. For small thrusts with displacements below these values the work required to create the fracture surface is a dominant term, $W_c > W_p > W_b$. For displacements above these transitional displacements, $W_b > W_c > W_p$, and in the previous section we found that for the entire McConnell thrust system with $u \approx 40$ km that $W'_b > W'_p > W'_c$.

(e) Deformation within the main body of the sheet

A simple way of visualizing the resistance set up within the mass of the thrust is to imagine a situation where the base of the sheet was perfectly lubricated so that there could be no basal terms W_c, W_p, W_b. Nevertheless, everytime the thrust went over a substantial bump or obstacle in the base, the entire sheet would have to be deformed into conformity with the bump so that no voids would open up. At low metamorphic grades such deformation would be essentially flexural slip folding (parallel to kink folds) with sliding on a large number of discrete surfaces set up throughout this portion of the thrust sheet.

Another and possibly greater source of internal deformation is the simple shear within the thrust sheet which must be transferred to the base as the thrust cuts up section (figure 2).

There are, then, two types of deformation within the sheet which must proceed in order for basal sliding to occur without continuity being destroyed or holes opening up. During basal sliding internal deformation is therefore a necessary process, or in other words, basal sliding and internal deformation are *dependent processes* and the overall rate determining step is the slowest of the two. It seems that the internal slip which must be activated over the huge area of internal surfaces such as bedding within a thrust sheet makes this internal deformation the rate limiting step.

On balancing the external work provided to the internal work expended within the system, assuming gravity spreading (equation 8),

$$W' \simeq W'_g = W'_i + W'_b + W'_c + W'_p.$$

For the McConnell thrust system the last two terms W'_c, W'_p are two or three orders of magnitude less than the second term W'_b, and by subtraction $W'_i \simeq 6.4 \times 10^{19}$ J. But is it generally true that $W'_i > W'_b$? The evidence is at least suggestive, but information on the strain distribution within thrust sheets is essential. Much of the resistance to forwards motion of the sheet may lie here.

Now we wish to determine by what physical processes this deformation occurs. It is useful to classify the mechanisms of deformation operating within thrust systems depending upon whether slip on discrete surfaces or continuous deformation of some finite volume is involved. In the next sections we will look at the mechanical processes involved in these kind of deformation, using the 'flow chart' (figure 3) as a guide.

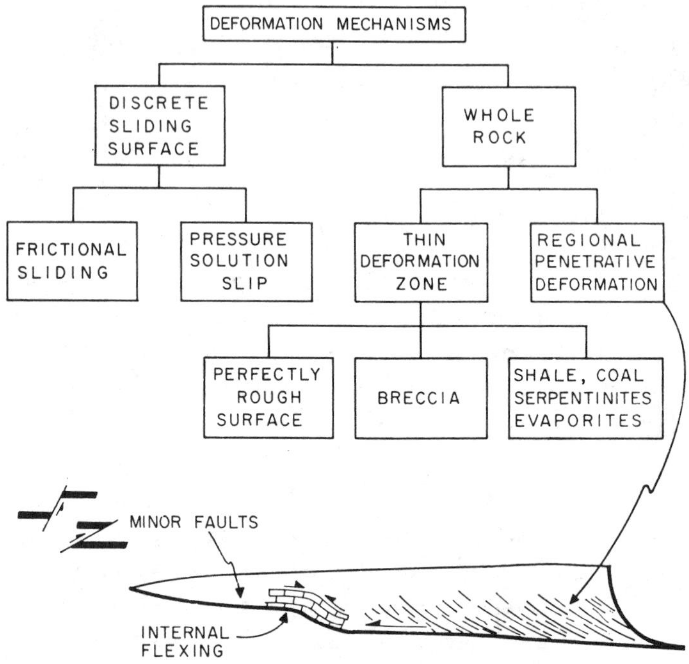

FIGURE 7. One manner of classifying the various kinds of deformation mechanisms occurring in a thrust system. Discrete sliding surfaces include bedding planes and minor faults. Slip is activated as the thrust climbs section.

DEFORMATION BY SLIP ON DISCRETE SURFACES

These discrete sliding surfaces appear to the eye to have no thickness, and the rock in between two such sliding surfaces undergoes little deformation.

Essentially undeformed fossils and oolites occur within a few metres of major thrusts such as the Pulaski and Saltsville in the central Appalachians. These observations were made within individual bedding units, and simply proves that internal deformation within the thrust sheet must have occurred by slip along discrete surfaces, and deformation of the rock between such sliding surfaces is below the limit of detection. There are two different sliding mechanisms which are distinguishable in the field.

(a) *Stress criterion for frictional sliding*

Many workers assume outright that the slip associated with thrusting obeys the linear law of sliding friction between solids (e.g. Hubbert & Rubey 1959; Carlisle 1965; Hsu 1969). This type of sliding law states that slip cannot occur unless the shearing stress τ_n along the sliding surface reaches a certain value determined by the coefficient of sliding friction μ, the normal stress on the sliding surface σ_n, and the ratio λ of pore pressure to normal stress on the surface:

$$\tau_n = \tau' + \mu(1-\lambda)\,\sigma_n. \tag{20}$$

Here τ' is a constant shear stress below which sliding cannot occur even with $\sigma_n = 0$ or $\lambda = 1$.

By considering the details of the physical processes which operate along the sliding surface it is possible to derive this sliding law from first principles. For example, asperities project from one surface towards the other (figure 8*a*). These asperities slide along, indenting and penetrating, and eventually break off producing wear particles. Assuming that the asperities break off in a brittle-elastic fashion when their finite shear fracture strength is reached, Byerlee (1967) was able to derive the linear frictional sliding law (equation (20)).

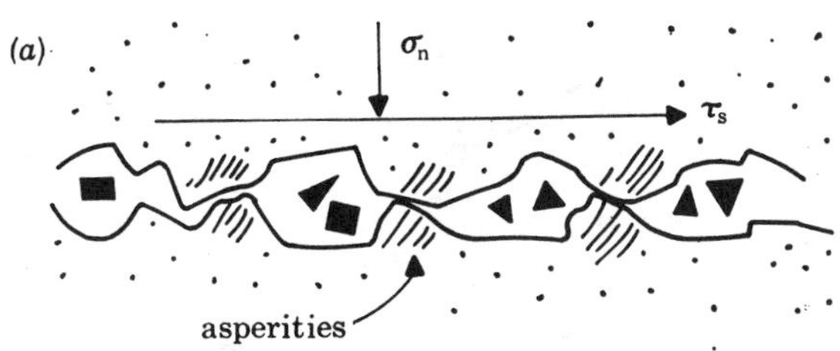

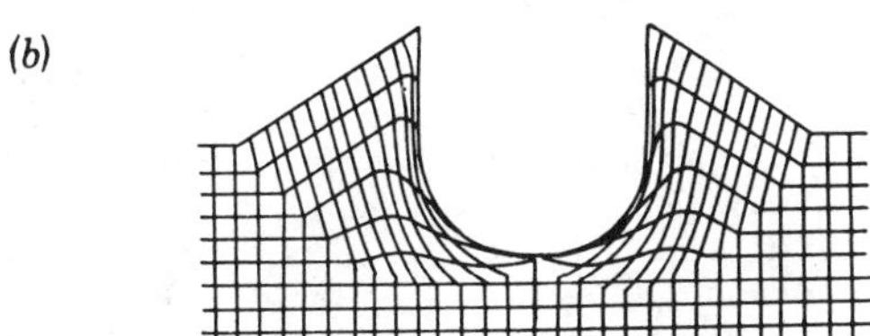

FIGURE 8. (*a*) Longitudinal section through a sliding surface undergoing shear and normal stresses τ_s, σ_n. Behaviour of the asperities is the key to deriving the linear sliding criteria by either brittle-elastic fracturing or plastic yielding. Black wear particles (tectonic breccia) are accumulating.

(*b*) Cross section through such a sliding surface showing the plastic deformed zone on both sides of a track left by a rigid asperity. Compare with figure 13*c*.

This linear sliding law can also be derived if the microscopic processes involve perfectly plastic yielding rather than elastic-brittle fracturing. In this case the asperities plough through the surface as a cutting or scratching process (figure 8*b*), forming junctions or welds which are then sheared off by plastic yielding (Bowden & Tabor 1964).

These physical mechanisms produce diagnostic minor features on the sliding surfaces. For example, figures 9, 10 and 11, demonstrate a sliding surface which has an observed magnitude and sense of slip. This surface shows striations, raised steps, chevron marks, flutes, and crescentic gouge marks (Wegmann & Schaer 1957; Dzulynski & Kotlarczyk 1965; Tjia 1968). These minor structures may be used as a guide to the processes going on, and whenever a particular sliding surface is covered by this type of feature the linear frictional sliding criteria was in force. These sliding surface markings are virtually identical to some of those formed beneath glaciers; they must be produced in the same way and I see no reason for a distinct terminology.

(*b*) *Pressure solution slip*

Many of the sliding surfaces in the central Appalachians are covered by 'fibrous mineral growth' (Cloos 1971, p. 52), often arranged into imbricated 'fibrous shingles' or 'accretion steps' (figure 13). These fibre coated sliding surfaces have also been recognized in the Western Alps (Durney & Ramsay 1973), and they are widespread in the Canadian Rockies (figures 11,

12, 14, 15). Striations and grooves are also found, produced by pressure solution around resistant and sliding tools (figure 13*c*). All scales of styolites abound in the wall rock around the slip surfaces.

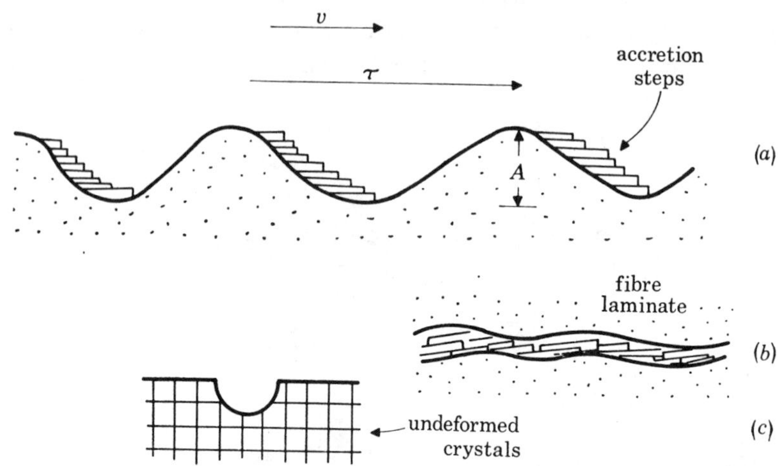

FIGURE 13. (*a*) Longitudinal section through a sinusoidal surface with amplitude A undergoing pressure solution slip with velocity v under a shear stress τ_n. Accretion steps produced on lee side of bumps. If fibres terminate with euhedral crystal faces and into vuggy fillings, this is evidence of a fluid-filled void and these are occasionally found on the lee side of accretion steps.

(*b*) Fibres between two sliding surfaces build up into a 'fibre laminate', in which fibres form thin layers, often with different orientation, grain size, and composition in each layer.

(*c*) Cross section through a groove left by a tool which has presolved its way across surface. Crystals adjacent to groove are essentially undeformed (compare with figure 8*b*).

Such sliding surfaces, covered with fibres, grooves, and accretion steps, are promptly pronounced to be 'slickensided' by most geologists who see them. Unfortunately the term slickenside often has genetic connotations of abrasive scratching produced by sliding friction (equation (20)). I will use slickenside in a purely descriptive and non-genetic fashion in this article.

It has been shown elsewhere that pressure solution and fibrous growth are diagnostic of diffusion creep, for which an explicit flow law may be written (Elliott 1973). We shall now show that a diffusion controlled *sliding law* may be inferred from a slip surface covered with fibre growth and pressure solution features.

An ideal surface exactly planar down to an atomic scale could slide without the necessity of transport of material along the surface. There is, of course, no such thing as a perfectly smooth surface and, with protrusions, sliding is possibly only if material is transported from one part of the surface to another. One way of moving such material about is by diffusion from the stressed surface, into the diffusion 'pipe', and precipitation on a less stressed part of the surface. Define a 'metamorphic quantity' M,

$$M = \Omega\delta D/kT. \tag{21}$$

The diffusion coefficient D at absolute temperature T is that of the rate limiting ionic species with atomic volume Ω travelling in a grain boundary or solution film of thickness δ (see Elliott (1973) for further discussion of these quantities and Fisher & Elliott (1974, Fig. 2), for possible values of D). This metamorphic quantity M has the dimensions of volume flow rate of material per unit shear stress. At a given shear stress a large flow of material (a high value of M) would allow a high rate of sliding.

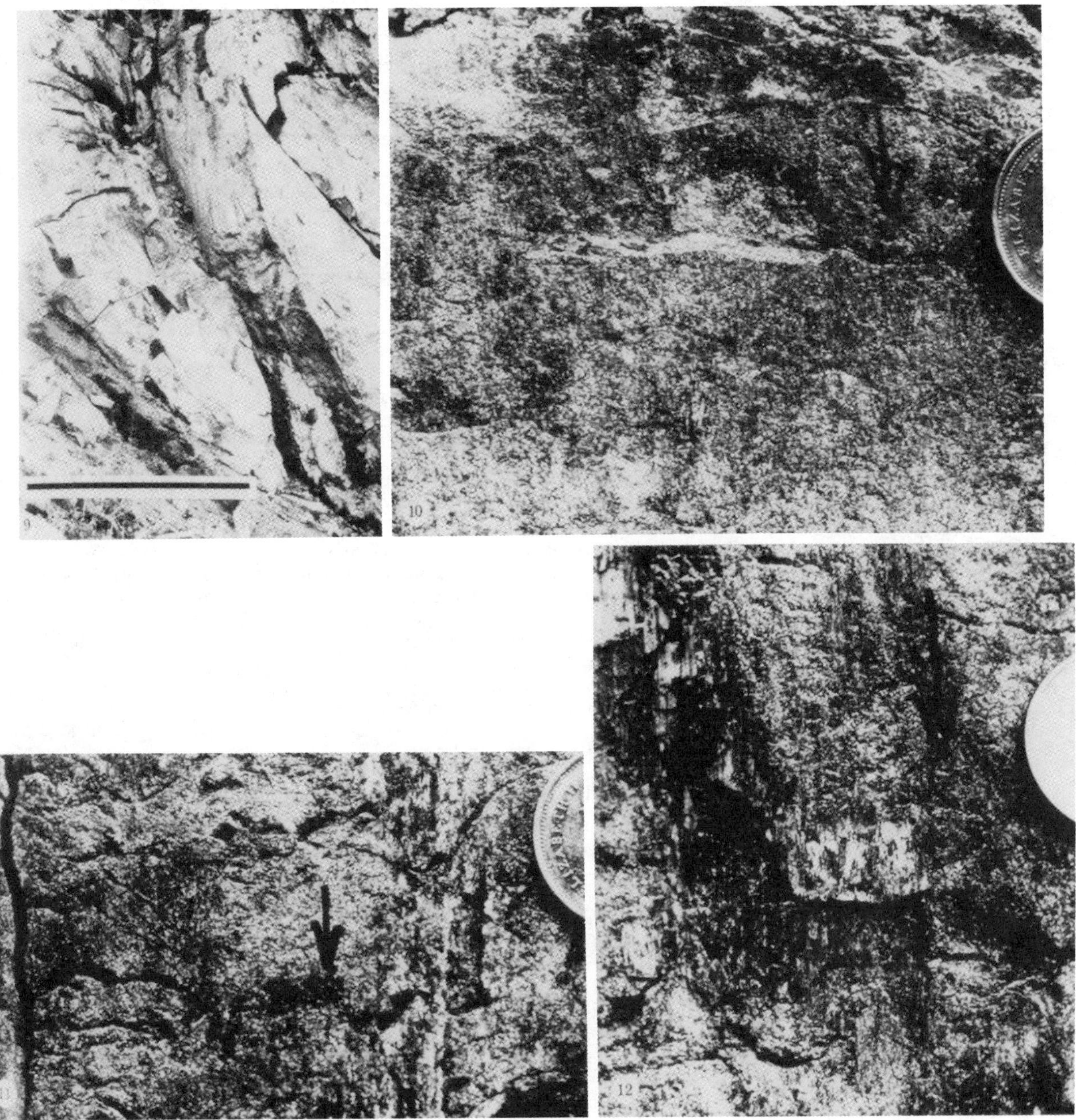

FIGURE 9. Minor contraction fault. On the exposed right face are a number of sliding features which indicate the physics of the sliding process. In photos (figures 10, 11 and 12) of these sliding features arrows indicate downward slip of the left side over this exposed right face. Scale is 1 m, note that displacement on the fault is a few metres. Upper Cretaceous Brazeau formation along Sheep Creek in the Panther Culmination (figure 4). Thickness of overburden $\approx$ 5 km.

FIGURE 10. Stepped surface with steps facing opposite sense of motion of left side of fault. Frictional sliding; a tool might have burst, leaving a step (Lindström 1974), or it may be a Hertzian crack.

FIGURE 11. Tectonic chevron marks, horns point in sense of motion of left side of fault. Caused by frictional sliding of indentor which is not rolling, Hertzian cone cracks form behind this tool (MacClintock 1953): note fibre between coin and arrow.

FIGURE 12. Fibrous quartz in accretion steps facing in the sense of motion of left side of fault. Evidence of pressure solution slip such as this covers far less surface area at this outcrop than minor features produced by frictional sliding.

(*Facing p.* 306)

12, 14, 15). Striations and grooves are also found, produced by pressure solution around resistant and sliding tools (figure 13*c*). All scales of styolites abound in the wall rock around the slip surfaces.

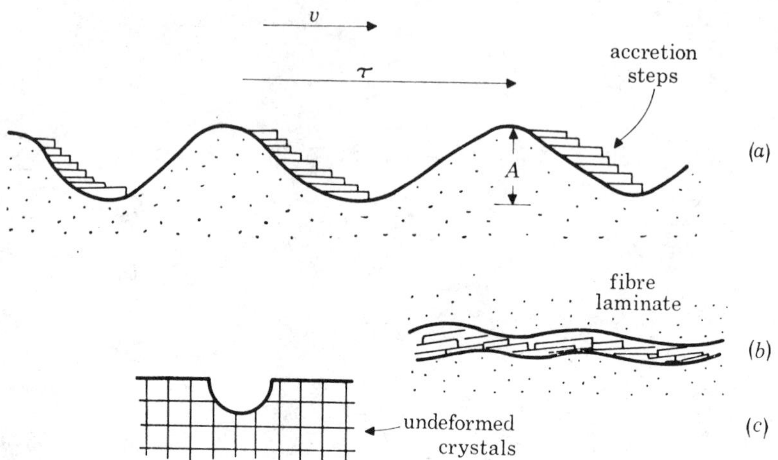

FIGURE 13. (*a*) Longitudinal section through a sinusoidal surface with amplitude A undergoing pressure solution slip with velocity v under a shear stress τ_n. Accretion steps produced on lee side of bumps. If fibres terminate with euhedral crystal faces and into vuggy fillings, this is evidence of a fluid-filled void and these are occasionally found on the lee side of accretion steps.

(*b*) Fibres between two sliding surfaces build up into a 'fibre laminate', in which fibres form thin layers, often with different orientation, grain size, and composition in each layer.

(*c*) Cross section through a groove left by a tool which has presolved its way across surface. Crystals adjacent to groove are essentially undeformed (compare with figure 8*b*).

Such sliding surfaces, covered with fibres, grooves, and accretion steps, are promptly pronounced to be 'slickensided' by most geologists who see them. Unfortunately the term slickenside often has genetic connotations of abrasive scratching produced by sliding friction (equation (20)). I will use slickenside in a purely descriptive and non-genetic fashion in this article.

It has been shown elsewhere that pressure solution and fibrous growth are diagnostic of diffusion creep, for which an explicit flow law may be written (Elliott 1973). We shall now show that a diffusion controlled *sliding law* may be inferred from a slip surface covered with fibre growth and pressure solution features.

An ideal surface exactly planar down to an atomic scale could slide without the necessity of transport of material along the surface. There is, of course, no such thing as a perfectly smooth surface and, with protrusions, sliding is possibly only if material is transported from one part of the surface to another. One way of moving such material about is by diffusion from the stressed surface, into the diffusion 'pipe', and precipitation on a less stressed part of the surface. Define a 'metamorphic quantity' M,

$$M = \Omega\delta D/kT. \tag{21}$$

The diffusion coefficient D at absolute temperature T is that of the rate limiting ionic species with atomic volume Ω travelling in a grain boundary or solution film of thickness δ (see Elliott (1973) for further discussion of these quantities and Fisher & Elliott (1974, Fig. 2), for possible values of D). This metamorphic quantity M has the dimensions of volume flow rate of material per unit shear stress. At a given shear stress a large flow of material (a high value of M) would allow a high rate of sliding.

FIGURE 14. Fibrous accretion steps on fault surface in Cambrian carbonates of Mid-Chancellor formation 0.8 km west of Field on Trans-Canada Highway. Rocks are also intensely cleaved by pressure solution processes. Overburden $\simeq$ 8 km.

FIGURE 15. Thin light layers are made up of vertically oriented fibrous growth between beds which are drifting apart. Intense pressure solution cuts beds and also is affecting previously deposited fibres, recycling them. Penetrative pressure solution is the dominant deformation process at this outcrop, which is near that of figure 14.

It is possible that much of the diffusion along sliding surfaces occurs in a discrete, hydrous film rather than by grain boundary diffusion. Evidence of open spaces and a hydrous film are fibre terminations with euhedral crystal faces and open-space drusy fillings behind bumps and accretion steps.

Approximate a sliding surface by a sinusoidal one with amplitude A (figure 13*a*). It can be shown that if a shear stress τ is applied parallel to the surface an exact solution for the velocity (after Raj & Ashby 1971),

$$\nu = (8M/\pi A^2)\tau. \tag{22}$$

It is interesting that the wavelength does not appear in this equation. A sinusoidal surface is a convenient approximation for any clearly periodic bumps and differs by not more than a factor of two from more exact solutions. Diffusional redistribution quickly removes local irregularities and reveals the basic periodic pattern.

(*c*) *Relative importance of the discrete sliding mechanisms*

We have two processes of slip on discrete surfaces: frictional sliding (equation (20)) and pressure solution slip (equation (22)). Each of these processes does not need to wait for the other to happen; they can occur either simultaneously or in any time order. In other words frictional sliding and pressure solution slip are *independent* processes. If a sliding surface is entirely covered by fibrous growth and pressure solution features the sliding mechanism is essentially pressure solution slip, and in a similar way we can identify frictional sliding. Most sliding surfaces I have observed are clearly in one category or the other, but the mixed case could be handled by assuming that the ratio of frictional to pressure solution slip is equal to the ratio of surface areas covered by pressure solution versus frictional sliding features. For example, at the Sheep Creek locality (figures 9, 10, 11 and 12) frictional sliding features cover a far higher percentage (*ca.* 80) of the observed surface area than do pressure solution features. The work W by either frictional or pressure solution sliding of m such sliding surfaces of a particular type in a volume of rock is

$$W = \sum_{0}^{m} u_m \tau_m S_m \tag{23}$$

where u_m, τ_m is the displacement and shear stress on the mth surface whose surface area S_m is covered by minor structures of a particular type.

Whole rock deformation

Rather than being restricted to a discrete surface, slip may be distributed over an intensely deformed layer of finite thickness bounded by zones of much less deformed material. These zones of whole rock deformation range in thickness up to thick zones of penetrative whole rock deformation with strain gradients extending over several kilometres.

(*a*) *Thin deformation zones*

These are low grade versions of the ductile shear zones described by Ramsay & Graham (1971) from high grade rocks. There are three different ways for these thin (< 100 m) deformation zones to originate.

(i) *Perfectly rough surfaces*

It is possible that the asperities interlock to such an extent that the surfaces can be described as 'perfectly rough' and the wall rock on both sides deforms as a mass. This could arise along sliding surfaces if the pore pressure became sufficiently low. Clearly the concept of sliding friction is no longer useful when we reach 'full' or 'sticking' friction, as it is known in plasticity. Pressure solution slip could also get locked up in a similar fashion. These cases are recognized. In the field they appear as discrete surfaces with a definite but small amount of slip compared to the displacement accounted for the ductile shear zones with continuous strain gradients in the adjacent wall rock; and they are occasionally identified as 'feather' or 'pinate' joints or gash veins.

(ii) *Tectonic breccia*

After a certain amount of frictional sliding along discrete surfaces, wear particles accumulate into a continuous layer of breccia separating the two moving surfaces. Within this tectonic breccia the strain is accomplished by ductile deformation in the particles and sliding of the particles over each other accompanied by progressive reduction in size. The deformation may be distributed throughout this deformation zone and the finite strain recorded by such breccia could be enormous.

Such granular aggregates have been extensively studied in soil mechanics. In bulk, the breccia may be described as an ideally plastic solid with a Mohr–Coulomb yield surface,

$$\tau_y = \tau^0 + \tan\phi(1-\lambda)\,\sigma_n. \tag{24}$$

Here σ_n is the normal stress on the boundaries of the deformation zone and the slope ϕ of the yield envelope is determined from a Mohr diagram. τ^0 is a constant.

This breccia is frequently observed. Even major thrust surfaces may have a layer of tectonic breccia only a few centimetres thick, although there are a few reports of breccia layers along thrusts which are as thick as 20 m (Stanley & Morse 1974). But tectonic breccia does not appear to cover the majority of sliding surfaces within the body of the thrust sheet.

(iii) *Shale, coal, evaporites, serpentinite*

These easily deformed rocks cover a substantial percentage of the surface area of many thrust faults. Such rocks could even be produced by the thrust motion itself, by breaking down the tectonic breccia to clay size particles as suggested by Wilson (1970) and Engelder (1974). Serpentinite and gypsum could have a high pore pressure effect (λ near 1) as a result of dehydration reactions (Raleigh & Patterson 1965; Heard & Rubey 1966). Again, an ideally plastic solid with Mohr–Coulomb yield surface (equation (14)) is favoured.

It is interesting that two of the equations discussed above are *formally* identical but have quite different *physical* interpretations; this has produced some confused discussion. The first equation (20) is a stress condition for frictional sliding of two surfaces in direct contact. The second equation (24) is a plastic yield condition for a finite volume of material, which could be (1) the deforming wall rock, (2) tectonic breccia, or (3) shale, coal, evaporites or serpentinite.

Special and rather unique rock types such as shale, coal, gypsum, and serpentinite cannot provide a general solution, for although all of these rocks can be found along various thrust surfaces, much of the energy involved in the emplacement of a thrust sheet may be dissipated within the main mass itself rather than by basal slip. Recall also that thrusts become more difficult to create and propagate (W_e and W_p) in ductile rocks. Such rocks tend to deform and flow rather than fracture.

(b) Regional penetrative deformation

Penetratively deformed rocks are immediately recognized by throughgoing metamorphic fabrics such as cleavage and stretching direction. Finite strains are large, and strain gradients occur over distances of the order of kilometres. We must use the constitutive equations appropriate for a three dimensional volume. Flow laws relating stress to strain rate depend exponentially upon temperature and on an inverse power of the grain size.

In low grade rocks pressure solution (or grain boundary diffusion) creep is often the dominant process,

$$\dot{\gamma} = (1/\eta)\tau.$$

Grain boundary diffusion creep is a Newtonian flow with viscosity η,

$$\eta = d^3 r^{\frac{3}{2}}/K'\pi M.$$

In this equation d is grain diameter, r is grain axial ratio, and K' is a constant which may have a value between 21 and 8γ (Elliott, 1973).

Many of the formations in the Western Main Ranges show a penetrative cleavage (Cook 1975). This cleavage is of pressure solution type – indicating Newtonian flow (figure 15).

The rate with which deformation work is expended per unit volume in a simple shear path with steady flow is

$$\dot{\omega} = \tfrac{1}{2}\tau\dot{\gamma} = \left(\frac{\eta}{2}\right)\dot{\gamma}^2,$$

therefore

$$W = \left(\frac{\eta}{2t^2}\right)\gamma^2.$$

We see that deformation work depends upon a *power* of the finite shear strain.

At higher temperatures and coarse grain sizes dislocation creep is important, with the shear strain rate $\dot{\gamma}$ dependent upon the shear stress τ to some power n,

$$\dot{\gamma} \propto \tau^n.$$

The work W due to each of the whole rock deformation mechanisms, such as pressure solution and dislocation creep, must be separately evaluated.

Which were the main energy dissipating processes within the mass of the thrust sheet? Information from several sources can be patched together onto a cross section of the McConnell thrust (figure 16).

(1) In the Central Appalachians penetrative cleavage and finite strain – almost entirely a result of massive pressure solution – is observed to become a dominant process when the metamorphic temperatures reach ≥ 250 °C (Elliott 1973, and in preparation). With a thermal gradient between 25 °C km^{-1} and 50 °C km^{-1} this temperature would be reached at depths between 9.2 and 4.6 km. The 250° isotherm is sketched onto the cross section (figure 16).

(2) The overburden during deformation of the outcrops illustrated in this paper can be estimated (figure 16). Processes observed in these outcrops are related to adjacent thrusts, but they can be placed in the structural positions they might occupy if they had deformed within the moving McConnell thrust mass.

We find that as the depth of overburden increases (and metamorphic temperature) there is a transition from frictional sliding to pressure solution slip, but that shortly after the zone of extensive pressure solution slip is reached the rock becomes penetratively cleaved (figure 15). About half the volume of the McConnell thrust system is stamped with this penetrative pressure solution cleavage, and much of the finite strain within the thrust system occurred here.

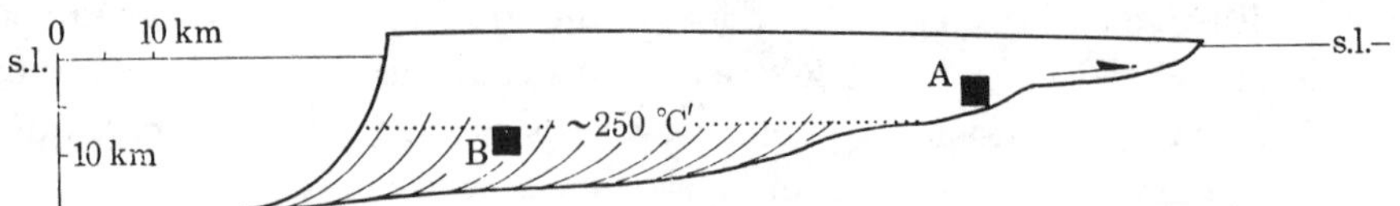

Figure 16. Outcrops associated with different thrusts may be interpreted as if they were at corresponding geological positions within the same thrust sheet. In this case A represents the outcrop of figures 9, 10, 11 and 12 and B the outcrops of figures 14 and 15. The position of the 250 °C isotherm is sketched in.

Conclusions

Knowing the initial and final geometry of the thrust fault system it is possible to calculate the gravitational work involved in thrusting. For the entire McConnell thrust sheet, gravity spreading is the appropriate theory because the gravitational work of 7.3×10^{19} J is much greater than the maximum possible work by any surface forces. But if one considers smaller portions of the McConnell, longitudinal compressive surface forces are important. For example if one considers only the major imbricate slice negligible gravitational work is accomplished; on a still smaller scale the front toe portion has its centre of gravity raised and gravity is a resisting rather than a driving force.

A new theory for the initiation and growth of thrust fault surfaces is outlined. It is found that the maximum displacement along thrusts is linearly related to their map length. Fold complexes at the lateral ends of thrusts appear roughly constant in size for a given metamorphic grade. These observations suggest that thrust fault surfaces are formed as ductile fractures, and the theory predicts increasing difficulty in forming thrusts as the ductility rises, such as would be produced by increase in metamorphic grade. Whenever thrust displacement is greater than a few kilometres, then the work involved in basal sliding is larger than that required to propagate the thrust.

Finite simple shear within the body of the sheet is necessarily transferred into displacement along the thrust fault surface wherever the thrust cuts up section in the hanging wall. Bending the thrust over obstacles in the footwall sets up simple shear and flexural slip folding within the body of the sheet. A substantial amount of work is expended in this internal deformation, and even for major thrusts like the McConnell this work of internal deformation may exceed that of basal sliding.

This picture of the growth of thrust faults can be checked by field work, and it is also possible to establish the physical mechanisms of rock deformation by field observations. A novel type of sliding along discrete surfaces is pressure solution slip, in which obstacles are bypassed by diffusive mass transfer. Fibres and pressure solution grooves are diagnostic features of this sliding law, in which the velocity of sliding is linearly related to the applied shear stress. It appears that roughly a 5 km thick surface layer of the McConnell thrust system was dominated by linear frictional sliding. Pressure solution slip only becomes significant deeper than about 5 km, but at this depth penetrative whole rock deformation – marked by cleavage and stretching directions –

is becoming dominant with regional strain gradients of the order of kilometres. This penetrative whole rock deformation occupies roughly half the volume of the McConnell thrust system, and accounts for most of the finite strain and deformation work. Pressure solution is the most important kind of whole rock deformation, suggesting that the McConnell thrust has a frictional sliding surface layer overlying a massive linearly viscous core.

The National Science Foundation supported the work reported in both this and a companion paper (Elliott 1976*a*) on thrust faults. My colleagues and students at Johns Hopkins criticized this paper and suggested several improvements. The Geological Survey of Canada provided both an office in Calgary and incisive discussion.

References (Elliott)

Bally, A. W., Gordy, P. L. & Stewart, G. A. 1966 Structure, seismic data, and orogenic evolution of southern Canadian Rocky Mountains. *Bull. Can. Petrol. Geol.* **14**, 337–381.

Barnes, W. C. 1966 Mechanics of overthrust faulting. *Abstract Geol. Soc. Am. Mtg.*

Batchelor, G. K. 1965 *An introduction to fluid dynamics.* Cambridge University Press.

Bowden, F. P. & Tabor, D. 1964 *The friction and lubrication of solids* Vol. 2. Oxford University Press.

Byerlee, J. D. 1967 Theory of friction based on brittle fracture. *J. Appl. Phys.* **38**, 2928–2934.

Carlisle, D. 1965 Sliding friction and overthrust faulting. *J. Geol.* **73**, 271–291.

Carrigy, M. A. 1971 Lithostratigraphy of the uppermost Cretaceous (Lance) and Palaeocene strata of the Alberta plains. *Res. Council Alberta, Bull.* **27**.

Cloos, E. 1971 *Microtectonics along the western edge of the Blue Ridge, Maryland and Virginia.* Baltimore: The Johns Hopkins Press.

Cook, D. G. 1975 Structural style influenced by lithofacies, Rocky Mountain Main Ranges, Alberta–British Columbia. *Geol. Survey Can. Bull.* **233**, 73 p.

Dahlstrom, C. D. A. 1970 Structural geology in the eastern margin of the Canadian Rocky Mountains. *Bull. Can. Petrol. Geol.* **18**, 332–406.

Douglas, R. J. W. 1958 Mount head map-area, Alberta. *Geol. Surv. Can. Mem.* **291**, 241 pp.

Durney, D. W. & Ramsay, J. G. 1973 Incremented strains measured by syntectonic crystal growths. In *Gravity and tectonics* (ed. A. D. Kees & R. Scholten). New York: John Wiley and Sons.

Dzulynski, S. & Kotlarczyk, J. 1965 Tectoglyphs on slickensided surfaces. *Bull. Acad. Polonaise Sci. Géolog.* **13**, 149–154.

Eisbacher, G. H., Carrigy, M. A. & Campbell, R. B. 1974 Paleodrainage pattern and Late Orogoenic basins of the Canadian Cordillera. *Tectonics and sedimentation* (ed. W. R. Dickinson). Soc. Econ. Paleo. and Min., Special Pub. 22, 143–166.

Elliott, D. 1972 Deformation paths in structural geology. *Geol. Soc. Am. Bull.* **83**, 2621–2635.

Elliott, D. 1973 Diffusion flow laws in metamorphic rocks. *Geol. Soc. Am. Bull.* **84**, 2645–2664.

Elliott, D. 1976*a* The motion of thrust sheets. *J. geophys. Res., Red.* **81**, 949–963.

Elliott, D. 1976*b* Thrust length, displacement, and breeding behaviour. In preparation.

Engelder, J. T. 1974 Cataclasis and the generation of fault gouge. *Bull. Geol. Soc. Am.* **85**, 1515–1522.

Fisher, G. W. & Elliott, D. 1974 Criteria for quasi-steady diffusion and local equilibrium in metamorphism. In *Geochemical transport and kinetics* (ed. A. W. Hofmann, B. J. Giletti, H. S. Yoder & R. A. Yund). Washington: Carnegie Inst.

Goguel, J. 1948 (2nd edn) Introduction a l'étude méchanique des déformations de l'écorce terrestre. *Mém. Carte Geol. France.* 530 pp.

Heard, H. C. & Rubey, W. R. 1966 Tectonic implications of gypsum dehydration. *Bull. Geol. Soc. Am.* **77**, 741–760.

Heim, A. 1921 Geologie der Schweiz. II. Die Schweizer Alpen. Tauchnitz, Leipzig.

Hsü, K. J. 1969 Role of cohesive strength in the mechanics of overthrust faulting and land sliding. *Geol. Soc. Am. Bull.* **80**, 927–952.

Hubbert, M. K. & Rubey, W. W. 1959 Role of fluid pressure in mechanics of overthrust faulting. *Bull. Geol. Soc. Am.* **70**, 115–166.

Lindström, M. 1974 Steps facing against the slip direction, a model. *Geol. Mag.* **111**, 71–74.

MacClintock, P. 1953 Crescentic crack, crescentic gouge, friction crack, and glacier movement. *J. Geol.* **51**, 186.

McClintock, F. A. & Argon, A. S. 1966 *Mechanical behavior of materials*. Reading, Massachussetts: Addison-Wesley.

Nason, R. & Weertman, J. 1973 A dislocation theory analysis of fault creep events. *J. geophys Res., Red* **78**, 7745–7751.

Odé, H. 1960 Faulting as a velocity discontinuity in plastic deformation. In Griggs, D. & Handin, J. (editors), *Geol. Soc. Am. Mem.* **79**, 293–321.

Oldham, R. C. 1921 Know your faults. *Q. J. Geol. Soc. Lond.* **78**, 78–92.

Ollerenshaw, N. C. 1968 Preliminary account of the geology of limestone mountain map-area, southern foothills, Alberta. *Geol. Surv. Can. Pap.* 68–24, 37 pp.

Price, R. A. 1973 The mechanical paradox of large overthrusts. *Abstract Geol. Soc. Am. Ann. Mtg.*

Price, R. A. 1974 Large scale gravitational flow of supracrustal rocks southern Canadian Rockies. In *Gravity and tectonics* (ed. A. J. Kees & R. Scholten). New York: John Wiley and Sons.

Price, R. A. & Mountjoy, E. W. 1970 Geologic structure of the Canadian Rocky Mountains between Bow and Athabasca rivers – a progress report. *Geol. Ass. Can., Spec. Pap.* **6**, 7–25.

Raj, R. & Ashby, M. F. 1971 On grain boundary sliding and diffusional creep. *Metallurg. Trans.* **2**, 1113–1127.

Raleigh, C. B. & Patterson, M. S. 1965 Experimental deformation of serpentinite and its tectonic implications. *J. Geophys. Res. Red* **70**, 3965–3985.

Ramsay, J. G. & Graham, R. H. 1970 Strain variation in shear belts. *Can. J. Earth Sci.* **7**, 786–813.

Stanley, R. S. & Morse, J. D. 1974 Fault zone characteristics of two well exposed overthrusts: the Muddy Mountain Thrust, Nevada, and the Champlain Thrust at Burlington, Vermont. Abstracts NE Sect. *Geol. Soc. Am. Ann. Mtg.*

Tjia, H. D. 1968 Fault-plane markings. 23*rd Int. Geol. Congress* **13**, 279–284.

Wegmann, E. & Schaer, J. P. 1957 Lunules tectoniques et traces de mouvements dans les plis du Jura. *Eclogae Geologicae Helvetiae*, **50**, 492–496.

Wilson, R. C. 1970 Mechanical properties of the shear zone of the Lewis overthrust, Glacier National Park, Montana. Ph.D. Thesis, Texas A and M, 89 p.

Salt Tectonics

BULLETIN OF THE AMERICAN ASSOCIATION OF PETROLEUM GEOLOGISTS
VOL. 44, NO. 9 (SEPTEMBER, 1960), PP. 1519-1540, 23 FIGS.

MECHANISM OF SALT MIGRATION IN NORTHERN GERMANY[1]

F. TRUSHEIM[2]
Hannover, Germany

ABSTRACT

The object of this paper is to describe and explain the formation of salt stock structures in Northern Germany and to contribute in this way to a better understanding of similar phenomena in other parts of the world.

The majority of the structures in the North German basin can be directly or indirectly attributed to "halokinesis." This term, proposed by the writer (1957), designates the formation of salt structures, and their structural and stratigraphic implications, which are essentially the result of the autonomous movements of salt under the influence of gravity. Phase-bound tectonic forces play only a minor part.

The Permian salt structures of Northern Germany are classified into salt pillows, salt stocks, salt walls, and extrusions along fissures. They are accompanied by primary, secondary, and third-order peripheral sinks. The halokinetic movements have taken place in an essentially continuous and autonomous fashion from the Triassic to the present day. Reckoned throughout long periods of geological time, the absolute rate of flow of the salt averages 0.3 mm. per year.

INTRODUCTION

Since 1945, the German petroleum industry has greatly increased its efforts in exploration. Thousands of deep wells[3] and a very extensive network of reflection seismic measurements have yielded much information, which has led to new views being formed and old ones revised. This applies especially to the salt stock region of Northern Germany, the classic salt stock area, which has been the birthplace of some fundamental theories of salt geology.

For the last 50 years, Northern Germany has been considered the type example of geodynamic salt structures, *i.e.*, structures essentially produced by tangential compressive pressure, to be precise, by "Saxonian tectonics" as defined by H. Stille. The salt was considered to have played only a passive part in the periodic folding.

The observations of recent years, and in particular the decisive impetus which geophysics has given to interpretation of the deeper levels, have, however, shown that in Germany, too, the majority of the salt structures belong to the *geostatic* type, being essentially gravity phenomena. This theory, which was first put forward in 1910–1912 by Lachmann and Arrhenius, and elaborated in the United States by Barton (1933) and Nettleton (1934 and 1936) in particular, now is rehabilitated as far as Germany is concerned. The writer (1957) proposed the term "halokinesis"[4] as a collective term for all processes connected causally with the autonomous, isostatic movement of salt. Halo-kinetic structures are contrasted with halo-tectonic structures, which originated predominantly as a result of compressive tectonic forces. Every conceivable transition between the two types is to be found in the world.

Throughout the greater part of Northern Germany geological conditions are such that all zones, down to the underlying pre-Permian foundation, lie within the recording range of modern reflection seismic methods, and can be penetrated by deep wells.[5] It is therefore possible to record the history of the salt structures from their origin throughout all stages of development, and thus to obtain models, which, in principle, might also be valid for other regions of the earth, in which a cover of overlying strata "floats" on thicker salt layers, but where the salt layers lie at inaccessible depths.

[1] Manuscript received, January 23, 1960.

[2] Exploration manager of Gewerkschaft Brigitta.

The writer is indebted to many of his colleagues in Gewerkschaft Brigitta for helpful discussion and inspiring criticism, Dieter Sannemann in particular. He designed most of the illustrations, interpreted most of the seismic profiles, and gave considerable assistance in other respects in the preparation of this report.

Some of the enclosed profiles have been made available for publication by the following oil companies: Gewerkschaft Brigitta, Gewerkschaft Elwerath, Itag, Mobil Oil AG, Wintershall AG, to which the writer expresses his gratitude.

Gewerkschaft Brigitta have kindly given permission for publication of the paper.

[3] So far more than 13,500 wells have been drilled for petroleum and natural gas in Western Germany, of which more than 3,300 are in production.

[4] From the Greek words hals = salt, kinein = to move.

[5] The deepest well in Germany to date is DEA's Haya Z, which is 4,779 m. deep.

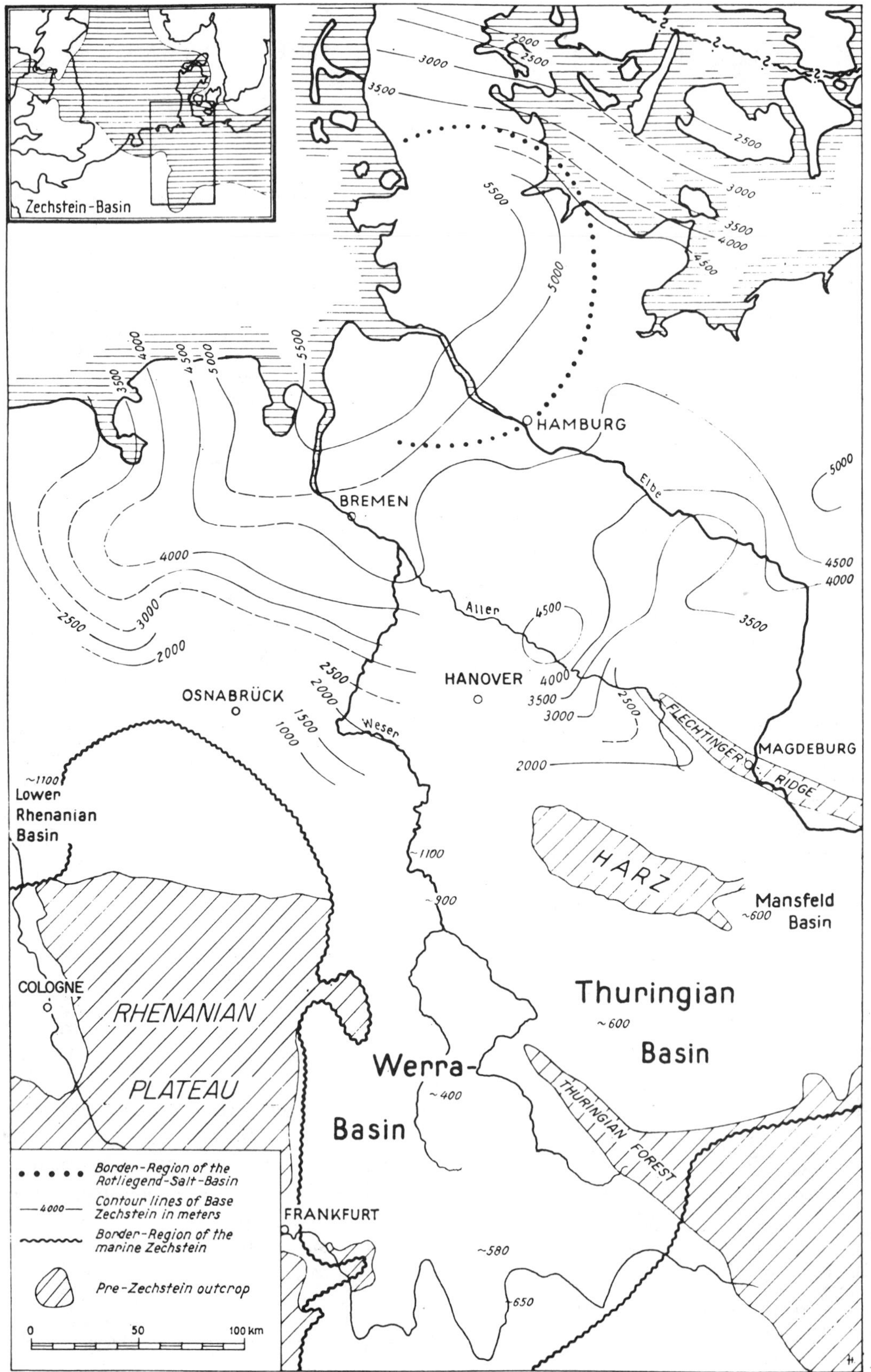

Fig. 1.—Northwest German Zechstein basin and its border region. Contours of the Zechstein base in meters. (Border region of Rotliegendes salt after Schott, 1942.) Designed by D. Sannemann.

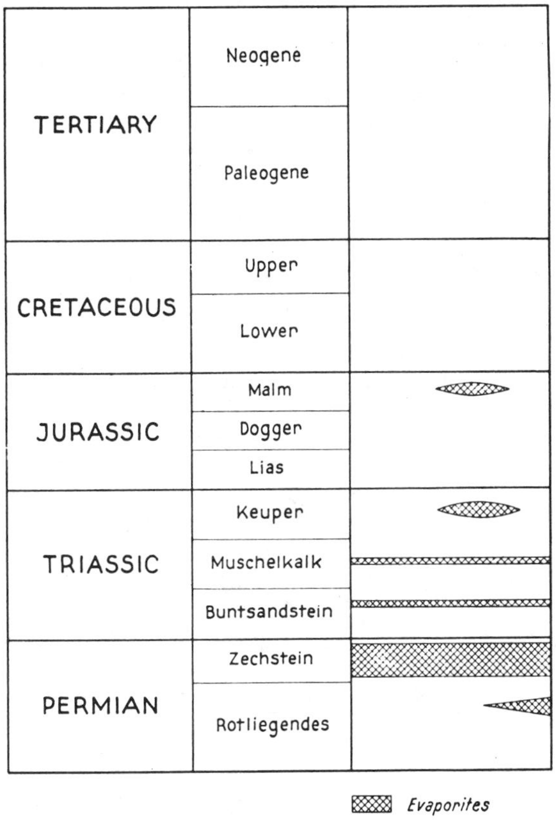

TERTIARY	Neogene	
	Paleogene	
CRETACEOUS	Upper	
	Lower	
JURASSIC	Malm	
	Dogger	
	Lias	
TRIASSIC	Keuper	
	Muschelkalk	
	Buntsandstein	
PERMIAN	Zechstein	
	Rotliegendes	

Evaporites

FIG. 2.—Stratigraphic column of North Germany and distribution of evaporites.

STRATIGRAPHY AND PALEOGEOGRAPHY

The Northwest German basin (Fig. 1) already existed during Paleozoic. For our purposes, only its history since the Permian is of interest (Fig. 2). The "mother salt" formation of most of the German salt stocks is Permian in age. The salt of the Rotliegendes (=Lower and Middle Permian) was deposited only in the deepest part of the basin (Fig. 1). The salt layers of the Zechstein (=Upper Permian), precipitated in several cycles, were deposited throughout the greater part of the basin, including its bays at the edges. From the edges toward the central part of the basin, the primary thickness of the salt increases to more than 1,000 m. Anhydritic, pelitic, and carbonate deposits, as well as potash beds, are intercalated in the salt layers. These intercalations are generally thickest at the edges of the basin.

The cover of the salt formation is formed by Triassic beds consisting of thick clastic and carbonate sediments averaging about 1,500 m. in total thickness. These are overlain by Jurassic, Cretaceous, and Tertiary. The glacial Pleistocene masks all older structures. The subsidence from the beginning of the Mesozoic amounted to about 3,000–4,000 m. in the central part of the North German basin, and was epeirogenic and not geosynclinal in character.

REGIONAL ARRANGEMENT OF SALT STRUCTURES (Fig. 3)

So far, more than 200 salt stocks have been found in Northwestern Germany. Some of them are elongate or vermiform, meandering structures. These *salt walls* average 4–5 km. in breadth and may reach a length of more than 120 km. On an average, they are 8–10 km. apart. They have risen from the deepest part of the basin (approximately the area between the Danish frontier and Bremen), that is, from the SSW.-NNE. striking depression which contains also the Permian Rotliegendes salt (Fig. 1). Here, the base of the Permian salt might lie deeper than 5,000 m.

The region of salt walls is adjoined by that of the *salt stocks*. They can be circular or oval in outline, as well as elliptical or elongate. The round salt stocks are 2–8 km. in diameter in their upper part. The salt stocks are in many places linked together like strings of pearls, in parallel-striking straight or winding lines. In this region, the pre-saline base is estimated as lying at 3,000–5,000 m.

The region of salt stocks is surrounded by a girdle of *salt pillows*. Evidently the primary thickness of the salt and the depth of subsidence were not sufficient to bring about diapirs in these regions. Accordingly, these "undernourished" salt structures remained at depth, arrested in an embryonic stage.

Genuine salt pillows and salt stocks are missing in the exterior parts of the basin.

The salt structures of Northern Germany therefore seem to be largely dependent, in both shape and size, on the primary thickness of the mother salt formation, and on the weight of the overlying strata (Fig. 4).

DEVELOPMENT OF SALT STRUCTURES

PRE-SALINE RELIEF

In large areas of Northwestern Germany, seismic surveys reveal one or more distinct reflectors, lying at depths ranging from 3,000 to more than 5,000 m., and also continuing below many salt structures. According to drilling results, these reflectors are situated at the base of

Fig. 3.—Salt structures in Northwest Germany. Numbers refer to structures mentioned in text: 1. Nusse, 2. Barrien-Thedinghausen, 3. Cismar, 4. Nienhagen-Hänigsen, 5. Eilte, 6. Lichtenhorst, 7. Niendorf II, 8. Gorleben, 9. Jaderberg, 10. Schwedeneck, 11. Düste-Twistringen, 12. Töps-Asendorf-Bahlburg, 13. Reitbrook, 14. Mölme. Designed by D. Sannemann.

the Permian salt layer, and thus correspond approximately with the top of the pre-saline basement. In general, the relief of this basement, upon which the marine Permian transgressed, is very subdued and flat (Fig. 1).

This mature configuration of the pre-saline basement is in contrast to the often strongly disturbed post-saline zones, whose structural features (anticlines, synclines, transgressions, downthrown faults, overthrusts, etc.) are largely independent of the underlying basement.

It is a remarkable fact that the structures become constantly simpler with depth, with decreasing number and throw of the disturbances.

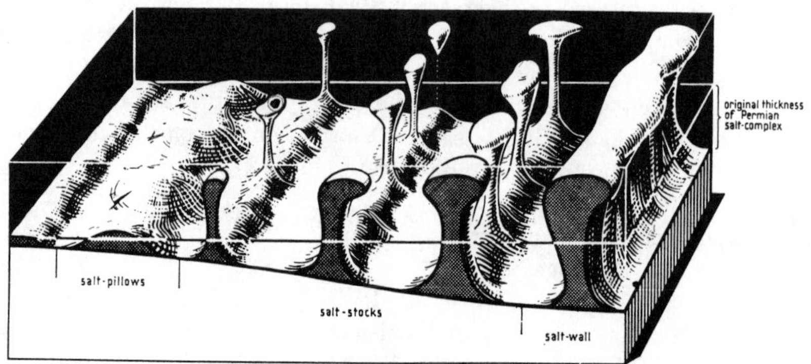

Fig. 4.—Diagram of different types of salt structures in relation to original thickness of Permian salt complex of Northwest Germany.

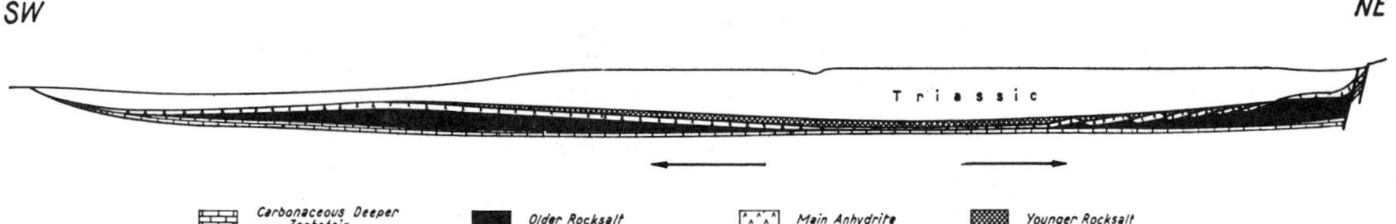

FIG. 5.—Cross section of Mansfeld basin (after Fulda). Arrows indicate direction of salt migration.

It is impossible to doubt that this pronounced disharmony between the pre-saline and the post-saline levels is due almost exclusively to the mobility of the Permian salt.

CAUSE AND COMMENCEMENT OF SALT MOVEMENTS

Salt can begin to flow only when it has been buried under sufficient load to cross the boundary between the elastic and the plastic condition. The lateral fluid-like plastic flow is facilitated if the basement dips at an angle of more than 1°, and if the salt layer possesses a certain thickness and is not too strongly mixed by competent intercalations. Experiences with the German Zechstein basins showed that an overburden of about 1,000 m. (sandstones, limestones, pelites of the Triassic and Jurassic) and a thickness of at least 300 m. of salt were necessary to initiate the process of flowing. If those conditions have become effective, the overlying strata "floated" on the layer of highly viscous Zechstein salt.

Because the critical boundary between the elastic and the plastic condition is lower in potash salt than in rock salt, the *potash beds* intercalated in the salt must start flowing even before the rock salt. Experience in the German potash mines has shown that potash salts advance much more rapidly into the mine galleries than the rock salt. The Werra Basin (Fig. 1) is a good example of potash deposits being intensively folded in an earlier stage of halokinesis; the rock salt in the hanging and footwalls still lies almost horizontal. Accordingly, the potash beds form, so to speak, the advance guard in halokinesis, and often yield exellent illustrations of salt migration processes.

The first cause initiating salt movement is beyond the range of observation. The various theoretical possibilities are not discussed here. The initial impulse to movement can be sought in inhomogeneities, either in the basement under the salt, or in the salt itself, or even in the roof of the salt layer. Of course, stresses already present may have been converted into the initial movement by a tectonic event or by an earthquake. According to mechanical considerations put forward by Dürschner (1957), the presence of a sufficiently deep sedimentation trough, in the shape of a shallow saucer, was sufficient to start the salt moving.

The requisite instabilities may have been so small that it may never be possible to define them with certainty. But it seems that with the first movement performed by the salt, a chain reaction was set in motion, which led to all stages of development of salt structures observed in nature.

The origin of the regional pattern of salt structures, with the virtual equi-distance of homologous salt structures, is an unsolved problem. There are two main opinions on this point.

On the one hand, the view is held that the salt pattern can be interpreted as result of a *large-scale rhythmical phenomenon* of salt migration, as the experiments of Parker and McDowell (1955) indicate. According to this view, the salt migration would have begun in the deepest part of the basin, in waves travelling centrifugally like ripples from a stone thrown into water. It is interesting to note in this connection that structural salt trends in great parts of Northern Germany run approximately parallel with the contours on the Zechstein base.

Another case of regional salt migration is indicated in those basins from whose central area all the salt has travelled out into the border regions. Accordingly, in the central parts of such basins the covering layers lie almost directly on top of the basement. The salt has accumulated in the border areas, in structural trends parallel with the rims of the basins. An example of these in Germany is the Mansfeld basin (Fig. 5), in Spain the Ebro basin (Lotze, 1957, p. 344) and in Rumania the Transylvanian basin.

This regional behavior of salt migration in

Fig. 6.—Diagram of salt-dome family (after D. Sannemann, 1960).

different basins can perhaps be explained by the differences in configuration of the basin floor. In subsiding basins with flat bottoms salt will reach the plastic limit and be activated more or less simultaneously over large areas, whereas in saucer-shape basins salt will begin to move in the more central parts. In addition, the thickness of the salt, the size of the basin, the rate of subsidence, etc., all play a part.[6]

Another view proceeds from the hypothesis that the salt pattern reflects a network of *faults in the pre-saline basement.* The existence of such disturbances has actually been proved, or at least shown to be probable, by evidence from boreholes drilled near the edge of the basin. In addition, the salt lines along the Aller River and between the Aller and Elbe rivers, linking together various elongate salt domes and running transversely to the old relief, could be connected with disturbances in the basement, following the northwest-striking tectonic lineaments which predominate throughout Central Europe.[7]

These few proved or probable fracture lines might represent the skeleton from which the salt migration began. The "mother" structures arising here could then have initiated "daughter," "grandchild," etc., generations parallel with and on both sides of the mother structures (Fig. 6).

It is to be hoped that the problem of the origin of regional salt structures in Northern Germany will soon approach solution. A prerequisite to the eventual solution is an exact dating of the origin and development of each structure. This will reveal whether the origin of the salt movements was monophyletic (from *one* ancestor structure only), or polyphyletic (with more than one structural ancestor).

[6] Underground migrations of salt masses are directly noticed. According to Lotze (1957) and Borchert (1959), in certain German potash mines the entire salt body is moving toward the exploited terrane, so that some adits have been pushed out of their original position.

[7] It is still a moot point whether the long salt walls in the northern part of the basin are connected, at least in part, with disturbance systems in the basement.

PILLOW STAGE AND PRIMARY PERIPHERAL SINKS

It is postulated that the development of a salt structure normally began with the *salt pillow* stage. The salt pillow consists of a salt accumulation (Figs. 7–9), which is plani-convex, at first hourglass-shape, later dome-shape, and usually almost symmetrical. The migration of salt into these pillows took place predominantly from the sides inward. This mass displacement of the salt caused the cover at the periphery of the salt pillow to subside but this subsidence was compensated at the surface by levelling sedimentation. This is the origin of the "*primary peripheral sinks*," which reflect the salt migrations at depth and are indicated as local maxima of sedimentation with decrease in thickness toward the salt pillow.

The primary peripheral sinks are thus typical attributes of salt pillow formation. Vertically they show a progressive shift toward the salt pillow throughout long periods, corresponding with the continuous upward doming of the pillow. This shift is an important criterion for judging the salt migration phenomena.

In each stage of salt migration, the excess rock volume of the primary peripheral sinks corresponds with the volume of the migrated salt. The peripheral sinks can therefore be used as indicators of the salt migration processes.

It also appears, when the supply of mother salt was interrupted, that the potential energy stored in the salt pillow was not sufficient to enable it to break through the covering beds. Such an undernourished salt pillow was then "frozen" for a variable period of time (Figs. 8–9).

DIAPIR STAGE AND SECONDARY PERIPHERAL SINKS

Under a continuous supply of mother salt the swelling of the salt pillow increased. Its flanks became constantly steeper. The extended sedimentary roof over the swelling salt pillow was gradually forced to break along shearing cracks. The salt began to break through. The *diapir stage* commenced, characterized as the stage of pre-

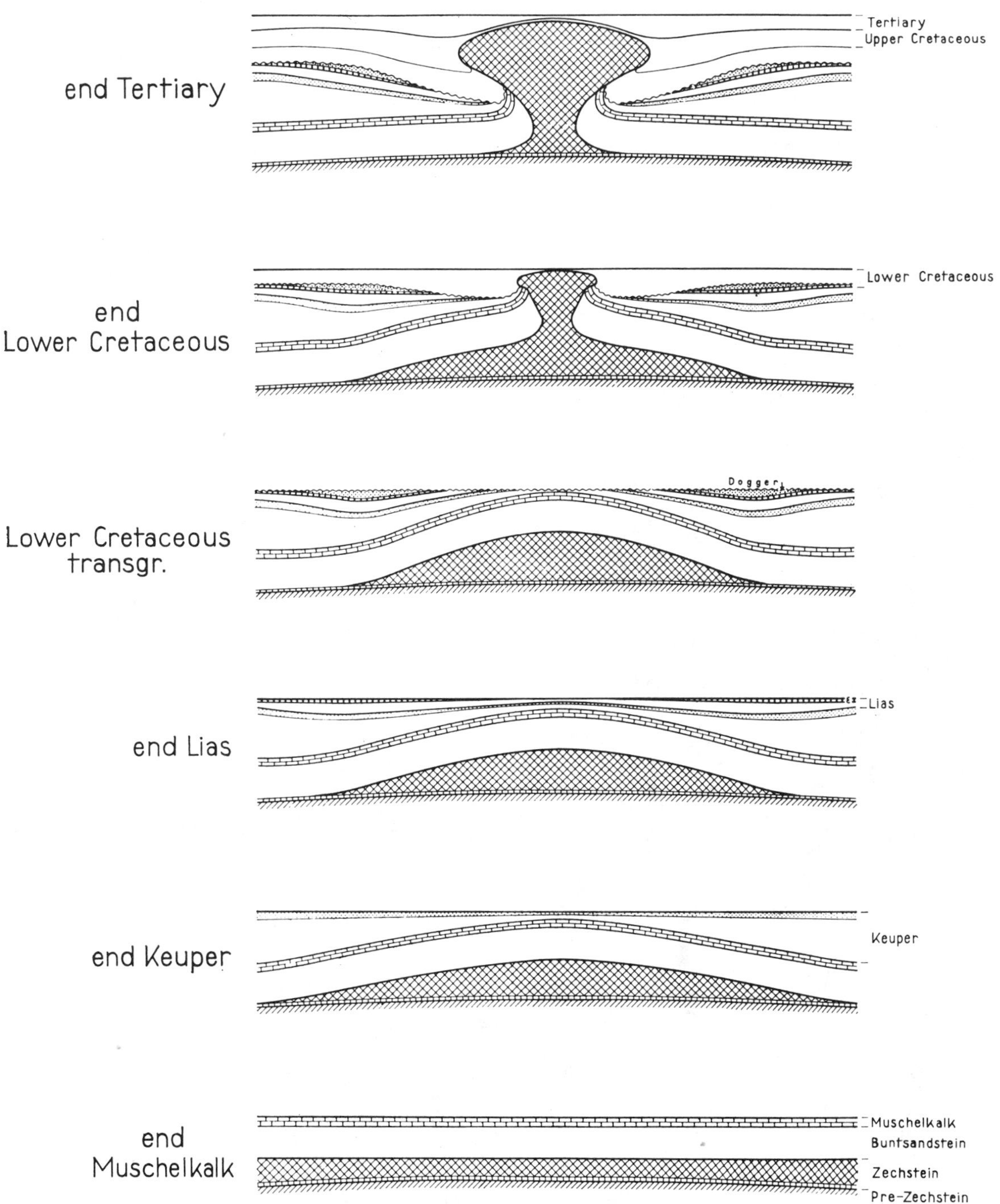

FIG. 7.—Diagrammatic development of Zechstein salt stock.

dominantly vertical salt migration (Figs. 7, 10).

Since salt was able to move upward against the confining force of overburden, the head of the diapir could also break through strata whose specific gravity was less than that of the salt. The diapir could rise for a period until the gravitative equilibrium of the rock system surrounding the salt was reached. If the cover of lighter sediments was very thick, and the salt replenishment was limited, the diapir could remain stationary at a certain level for any length of time. On the other hand, in the presence of a

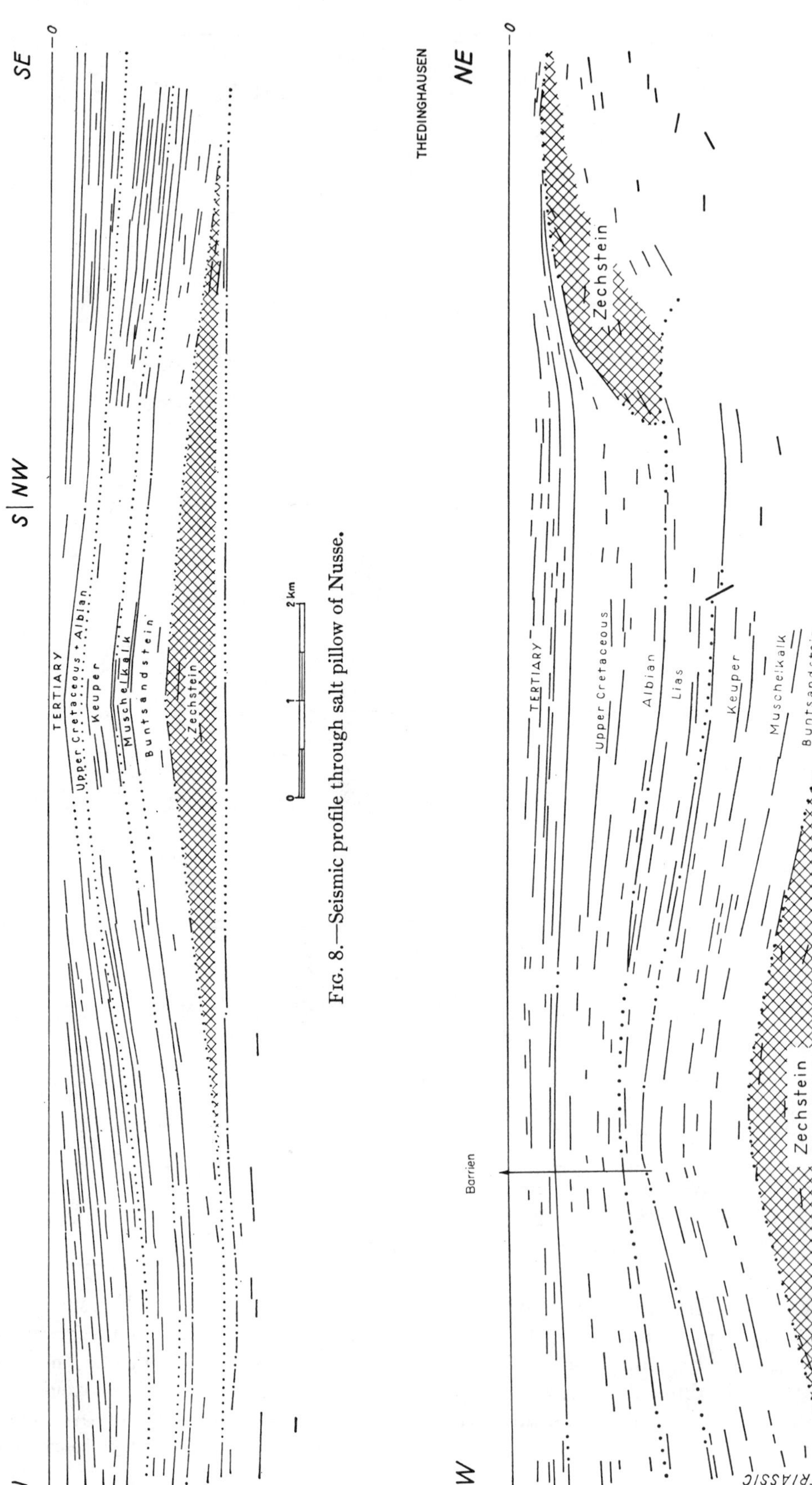

FIG. 8.—Seismic profile through salt pillow of Nusse.

FIG. 9.—Seismic profile through salt pillow of Barrien and salt stock of Thedinghausen.

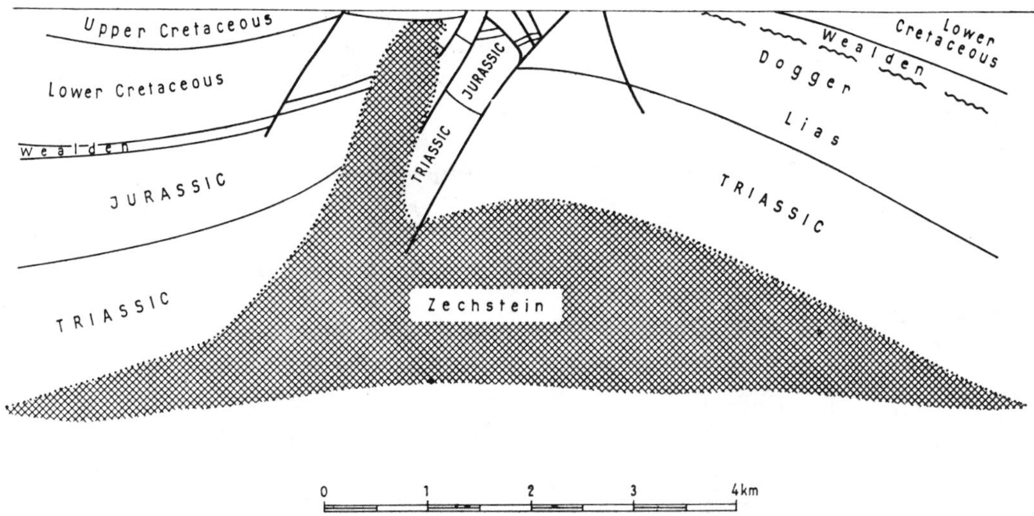

FIG. 10.—Cross section of salt dome of Mölme. Example of early stage in diapir (after O. Heermann, 1948).

thin cover of lighter material the salt could reach the surface and flow out.

Such *extrusions* are known in both fossil and recent form. In Northern Germany these outflows seem to have taken place predominantly under water. Presumably, the greater part of the flap or mushroom-shape overhangs (Fig. 11) must be regarded as the remains of "salt glaciers," which extruded below a veneer of unconsolidated bottom sediments (Gripp, 1958, p. 255), rather than as protrusions of the diapir stage.

In connection with these outflows breccias have also been observed. These coarse masses of detritus, with blocks more than half a meter in diameter, partly lie on top of the pierced salt and have partly been intercalated as rock flows in the flank sediments. It seems that the debris of the broken roof is entrained by the salt paste in process of extrusion and remains as "*saline moraines*," after the salt has been leached out. These extrusive breccias represent a special type of solution breccias.

The presence of extrusions of Permian salt during subsequent geologic periods may also be demonstrated indirectly by microfossils. In both Triassic and Upper Jurassic salt layers, Permian spores have been detected side by side with those of Triassic and Upper Jurassic age, respectively. Such evaporite deposits, which may become several hundreds of meters thick and are characterized by unconventional stratigraphic sequences, may have been derived from extruded Permian salt, which was dissolved, and subsequently redeposited together with the Permian spores.

The size, shape, and development of the diapirs are largely dependent on the thickness of the mother salt layer in relation to the composition and thickness of the overburden (Fig. 4). However, the old idea that the North German salt stocks were cylindrical in form, with predominantly vertical walls, has been largely refuted by seismic surveying and drilling results.

Many salt structures in Northern Germany are *asymmetric* especially in the deeper levels (Figs. 11–15). The salt has been forced obliquely upward, through a half-open "trapdoor" so that in cross section the salt body has assumed the appearance of a duck's head. The salt paste, during the upward thrust, probably intruded along the inclined shearing planes, and consequently initiated the asymmetric development of the diapir (Fig. 14). In this way one of the German salt plugs even becomes detached from its source of supply, and rises into higher levels as a rootless, drop-shape salt body (Fig. 15). In general, it may be stated that even neighboring salt stocks can be of very different construction. No two salt stocks are alike.

All these regional and local salt migrations necessarily result in strong modification of the original saline columns. Extremely chaotic conditions are known, especially in the diapirs. As a result of the flowing salt, complicated changes of the original sequence may even arise below extrusion zones, especially where rigid, competent members are intercalated in the salt layer (Fig. 16). The more mobile the rock salt, the longer its path of migration; and the narrower

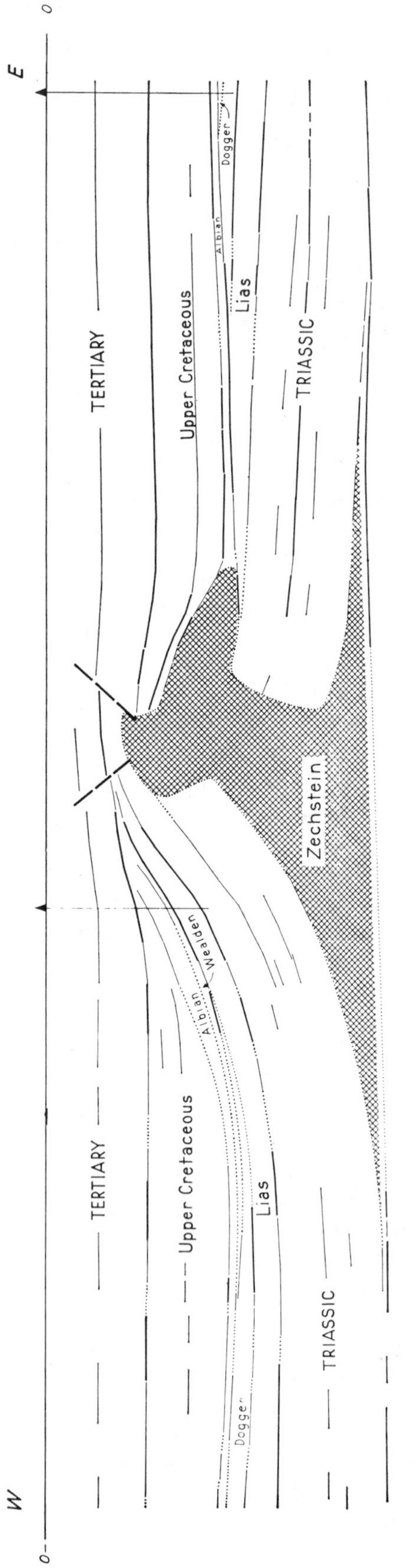

FIG. 11.—Section through salt stock of Niendorf II, showing asymmetric salt extrusion, after interpretation of seismic profile by D. Sannemann.

the passages, the more the primary sequence must have been fluidly deformed by being squeezed out, heaped up, injected, and folded. This also explains the difficulties of correlating well logs in the saline Permian of Northern Germany.

The extrusion of a diapir from a salt pillow resulted in subsidence above the shrinking pillow and compensating sedimentation on the surface. This was the cause of the *secondary peripheral sinks* (=rim synclines). Their most important characteristic is an increase in thickness toward the salt stock, or toward the zone of extrusion (Figs. 7, 12, 13, 14, 17) in contrast to the increase in thickness away from the salt pillow in the primary peripheral sinks.

The change from pillow to diapir stage is also reflected in the *facies* of the infilling sediments, namely, as a transition from the prevailing shallow-water uplift-facies of the pillow stage to a deeper-water basin-facies in the rim syncline of the diapir stage. The "break," which, in extreme cases, has the character of an emersion or transgression plane, is commonly a good seismic reflector.

The more seismic surveying and drilling in Northwestern Germany progress, the more intricate the *paleogeographic* picture of the post-Permian formation becomes. The old idea of great uniform sedimentation basins, surrounded by landmasses supplying detritus, is being abandoned in favor of a region with strongly varied relief. The paleogeographic picture is determined by local basins, isolated troughs, subaqueous bars and islands, in horizontal and vertical alternation. The facies vary correspondingly. Besides the well known regional sources of sediments, local influxes from neighboring regions took place to an ever-increasing extent, and these are especially important in oil geology. Most of the sedimentary iron-ore deposits in Northern Germany are the result of such local detrital influxes.

Due to the great difference in the character of the salt structures, the *size and shape* of the peripheral sinks in Northern Germany also differ widely. On the sides of the elongate salt ridges, the peripheral sinks have the shape of troughs, more than 100 km. long.

It has not yet been possible to ascertain the exact *geometrical relations* between the diameter of a salt stock and the diameter of the correlative peripheral sinks, especially since the volume of the primary peripheral sinks and the amount of

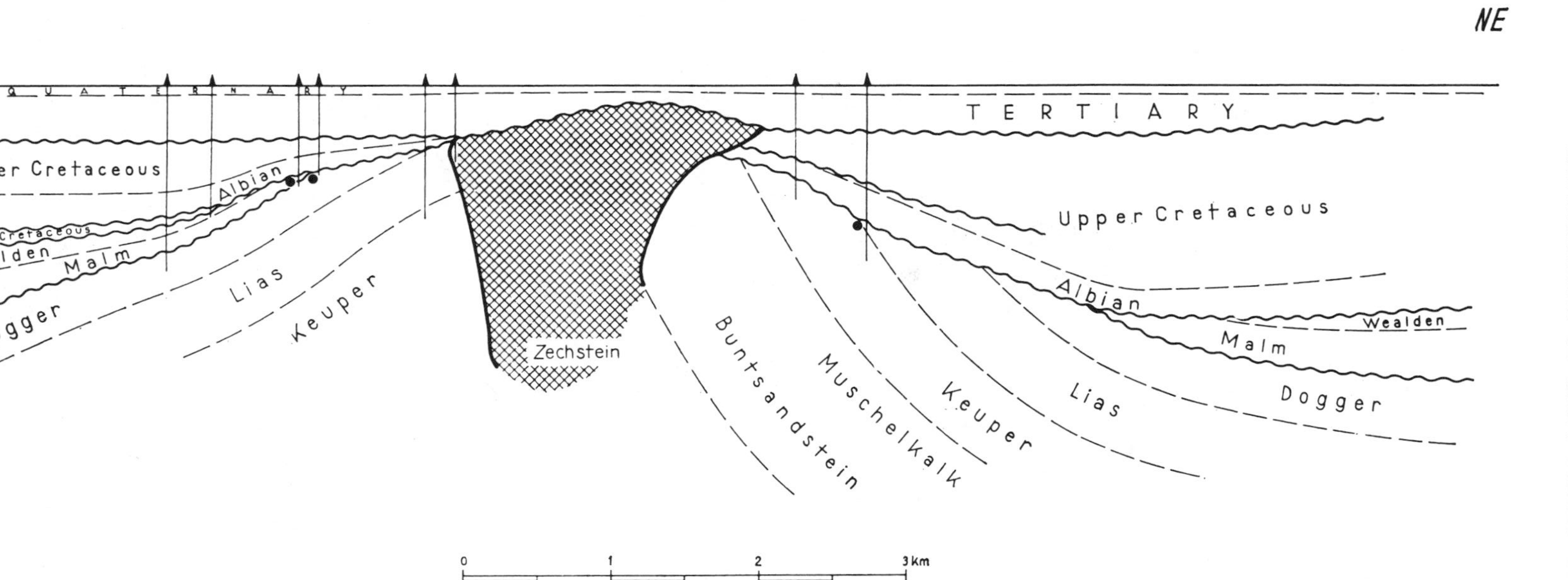

FIG. 12.—Cross section of asymmetric salt stock of Lichtenhorst. Moderately dipping southwest flank, steeply inclined pre-Albian series on northeast flank, strong secondary peripheral sink in Upper Cretaceous in northeast. Interpretation by E. Plein.

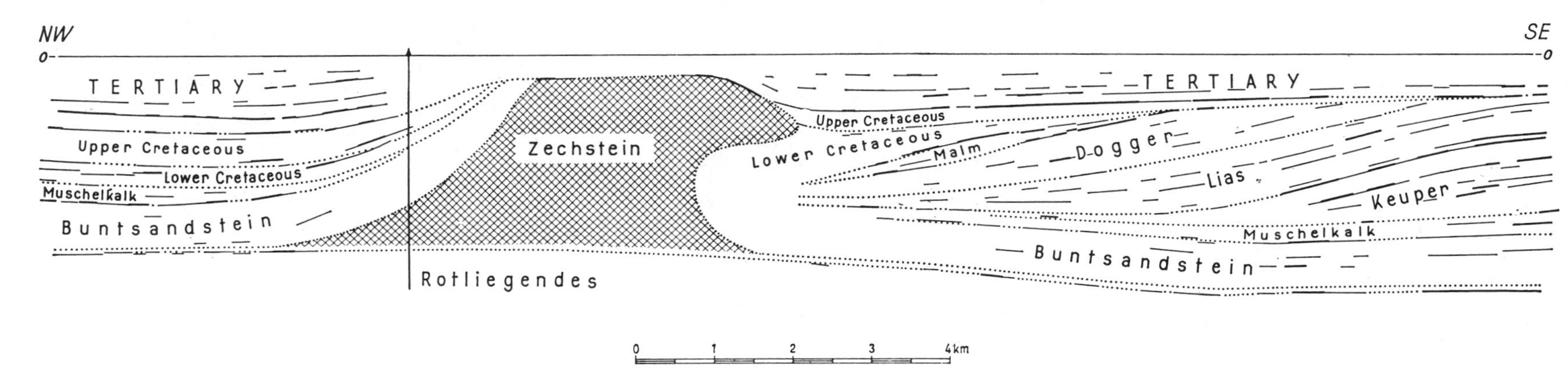

FIG. 13.—Seismic profile through asymmetric salt stock of Gorleben. Primary peripheral sink: Keuper to Malm. Secondary peripheral sink formed during Lower and Upper Cretaceous (at southeastern flank only). Interpretation by D. Sannemann

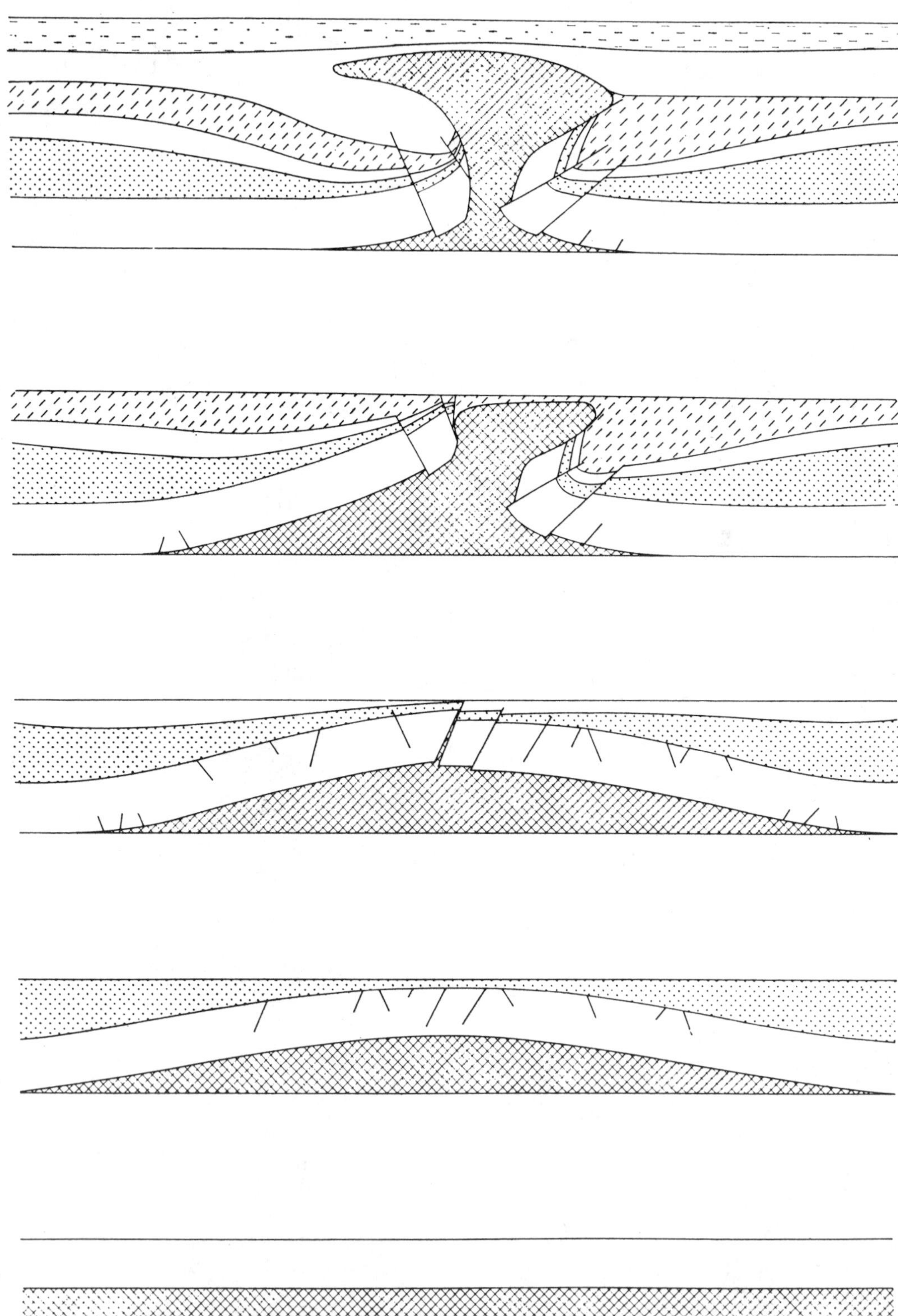

FIG. 14.—Diagrammatic development of asymmetric Zechstein salt stock.

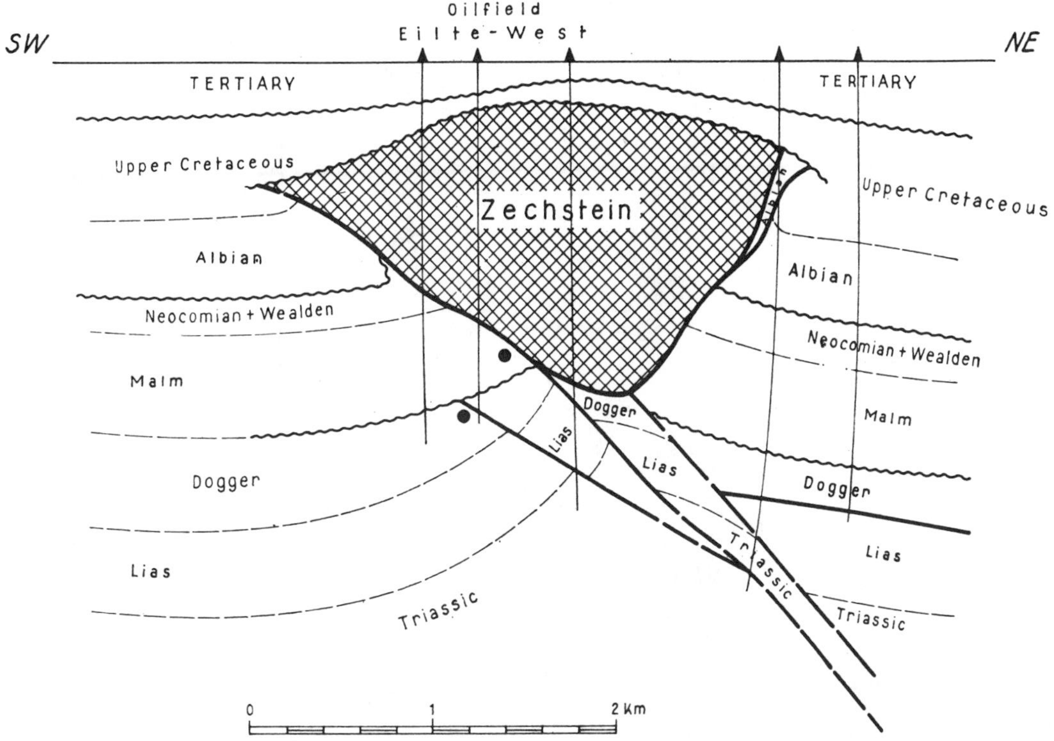

FIG. 15.—Cross section of drop-like salt stock of Eilte (with oil field Eilte-West below salt body). Ascent of salt might be associated with fault zone below salt body. Interpretation by E. Plein.

the regional epeirogenic subsidence have to be taken into account, and the quantity of salt that has escaped or been dissolved can not be exactly calculated.

The positive and negative changes in level caused by moving salt can be so drastic that the picture of epeirogenic development may be modified and obscured to the point of unrecognizability. In the Mesozoic of Northern Germany the *increase in thickness* between the sediments on the ridges and those in the correlative basins is in some places of the order of 1:10 and more, commonly over rather short distances.[8]

In such areas, it is difficult to ascertain the *average thickness* of the individual formations. If the maximum values were added together, the post-Permian sediments would have a thickness of more than 13 km. On the other hand, the average total thickness of the post-Permian sediments in northwestern Germany is only about 4 km. The explanation of this discrepancy is that thickness maxima are shifting and are compensated by the lesser thickness in older or younger formations.

If the movement of the salt is considered as a motive force in the genesis of the structures, a better interpretation of many *structural peculiarities* of Northern Germany can be given. Almost all the larger structures are limited to the post-saline strata. In general, the intensity of folds and faults decreases with depth. As a rule, they do not continue into the pre-saline basement. In contrast to tectonic seesaw movements, the halokinetic structural development is almost invariably irreversible, since the salt must usually adhere to the direction of flow that has been once imposed on it. The *transformation of structural relief* is a typical consequence of the salt migration. Old swells are buried under younger troughs, and, conversely, older trough-fills may be converted into younger anticlines. These phenomena occurred because the salt that had accumulated in the core of the existing swells broke

[8] Up till now, free gliding of unconsolidated sediments from the ridges toward the neighboring basins has not been observed in Northern Germany. Because of the very slow rate of ascent of the salt bodies, and the synchronous sedimentation that completely compensated the rise, the angle of inclination necessary to enable such free gliding to take place probably was not achieved.

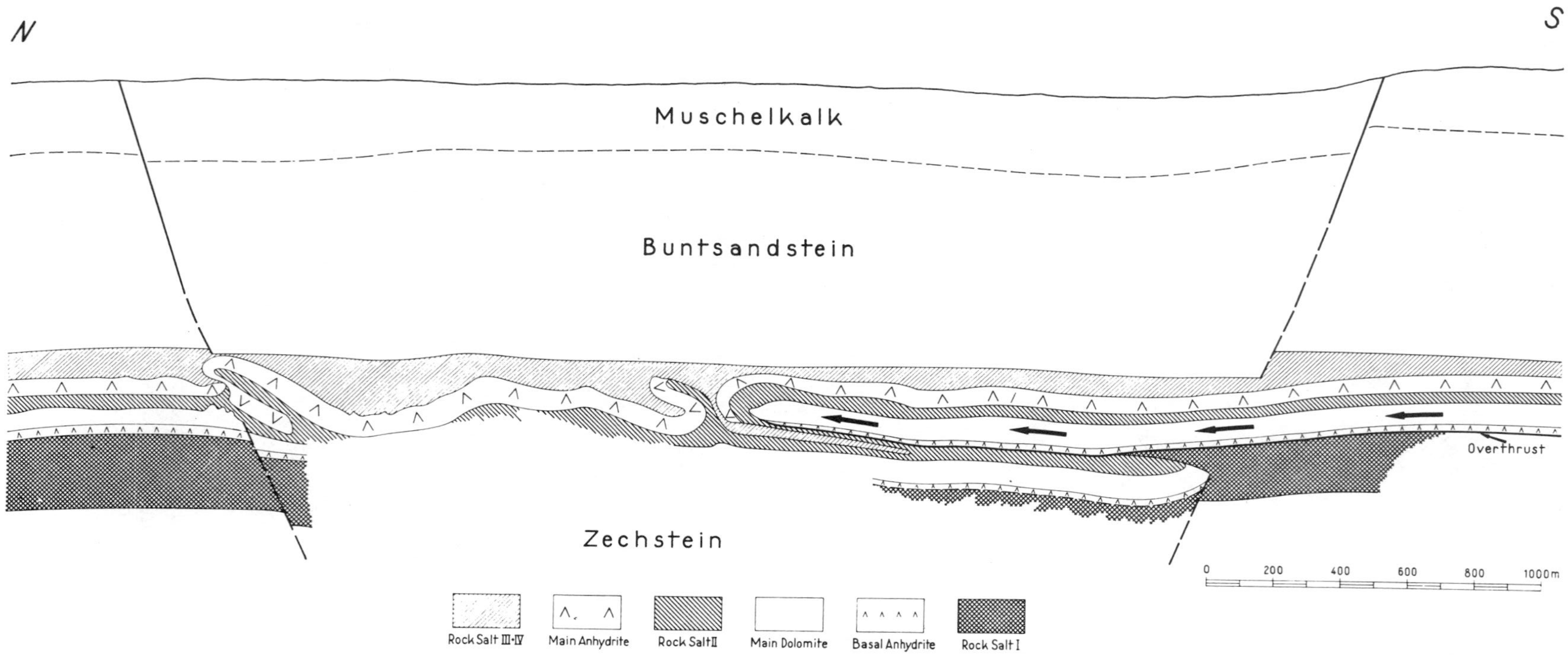

Fig. 16.—Cross section of Schlotheimer graben in Thüringian basin, according to observations in mines and in wells (after Knieg, 1958). Displacements in Zechstein layer, competent main dolomite pushed 1.4 km. into graben (arrows), in form of nearly horizontal plate.

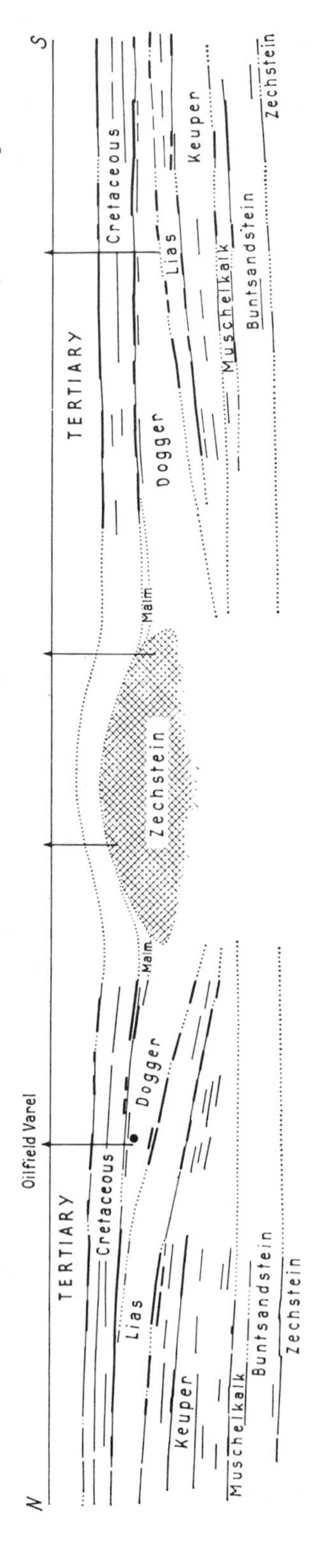

Fig. 17.—Seismic profile through salt dome of Jaderberg. Primary peripheral sink: Keuper and Lias, secondary: Dogger to Malm (after Mobil Oil AG and Gewerkschaft Brigitta).

through upward, and as a result, the roof of the swell collapsed. The depressions arising over the existing swells were continuously filled with basin sediments, whose volume more or less equals that of the salt which has moved away. Simultaneously, the trough-fill already in existence was transformed into a pseudo-anticline, because the flanks of the existing trough flattened out more and more. Accordingly, such *turtle-shape structures* owe their vaulting not to a later, "updoming" of the trough-fill, but exclusively to the collapse of the swell regions. It is therefore typical of these pseudo-anticlines that they become steadily more gentle with depth (Figs. 18, 19).

The greater part of the *downthrown faults* observed are also connected with the salt movement processes in the regions in which halokinesis was active. They can not be attributed to orogenic forces which might have caused or influenced the rise of the salt. The fault systems on salt structures are caused by salt accumulation. They are related only to the former or present roof of the structures, and consequently do not continue into the pre-saline basement. In the case of collapsed salt structures, they are overlain by younger trough sediments (Fig. 18).

The *arrangement* of these faults was predetermined by the shape of the salt structures (long, oval, round). Their number and throw decrease downward. Their *age* corresponds with the age of the salt movement.

Where the roof of the salt accumulation collapsed, the *extension* phase (during the accumulation) was followed by a *compression* phase (during the collapse), a sequence which has been observed in many structures in Northern Germany. This transition is in many places connected with a reversal of relative movement tendencies. Hence, at the same fault, downthrow movements were active at one time, and upthrow movements at another. This "schizoid" behavior on the part of numerous disturbances is indicative of their halokinetic origin.

POST-DIAPIRIC STAGE

After conclusion of the diapir stage proper, the majority of the North German salt stocks were covered by Upper Cretaceous and (or) Tertiary sediments. As a result of this fresh load the salt rose again, though usually on a smaller scale (Figs. 12, 13, 15, 17). From the seismic reflections of the younger strata which were uplifted, it can be seen, however, that the process persisted over long periods of time. It is known to be continuing to a certain extent today.

In this later upward movement of the salt plugs, the compensating movements of the adjoining formations must have brought about (*third-order*) *peripheral sinks*, which were only very small and shallow (Figs. 7, 17).

EXTRUSIONS ALONG FISSURES

Salt stocks are absent from the shallower areas of the German Zechstein basin. But here, too, the moving Zechstein salt has influenced the structural development of the overlying strata. Upward salt movements were induced by generally regional fault zones, through which the salt was extruded. Accordingly, the structural picture of these regions is dominated by long ridges, more or less collapsed now, and accompanied by long, gully-like troughs. Their axes are commonly sigmoidal or are arranged *en échelon*.

The salt migration processes in many of these shallower regions can be clearly reconstructed with the help of reflection seismic surveys (Fig. 18). This is the *locus classicus* of moving trough axes, in which the maxima of sedimentation migrate regularly and continuously toward the ridges.

Here too, as a rule, a salt accumulation corresponding with the pillow stage associated with primary peripheral sinks was formed first. Then fissure extrusions associated with secondary peripheral sinks followed. In many places the salt could only partly escape *via* the fracture zones, and a larger part was left behind at depth (Fig. 18).

END OF SALT MIGRATION

Seen as a whole, halokinetic development reached a certain maturity in Northern Germany at the end of Lower Cretaceous. Throughout most of the region, the Permian mother salt originally present in layers had migrated into the salt structures. The only parts left were those which were not capable of movement, because of intercalation of thick series of anhydrite and carbonate. Seismic surveying has shown that the Triassic in many places directly overlies the pre-saline in the areas between the salt stocks (Figs. 11, 13, 19). The mobile salt, together with entrained intercalations of thinner zones of anhydrite and carbonate,

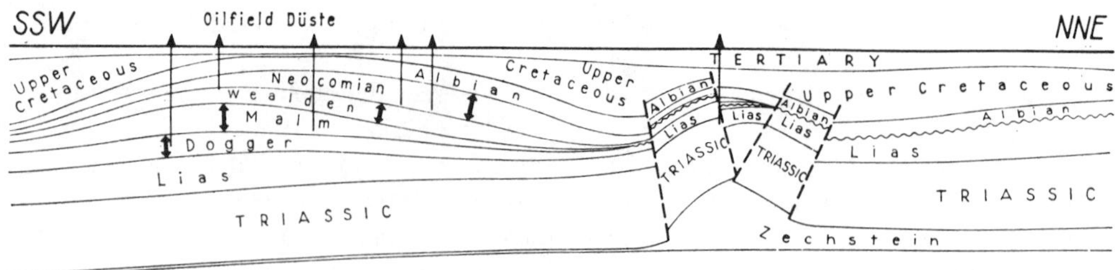

Recent

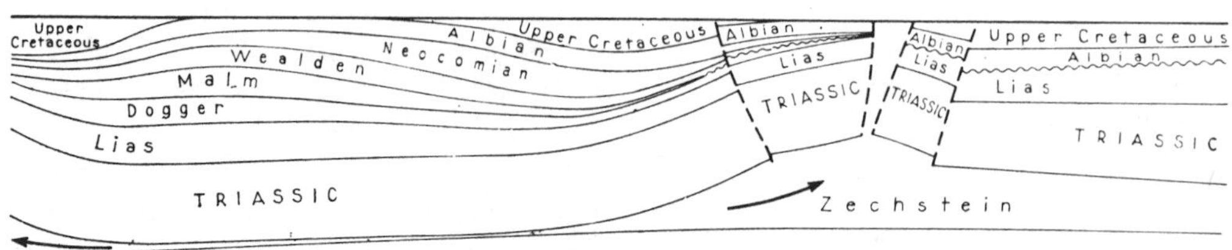

Pre-Santonian

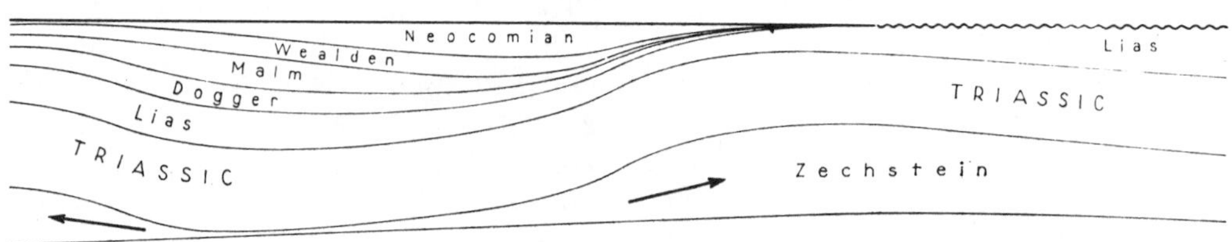

Pre-Albian

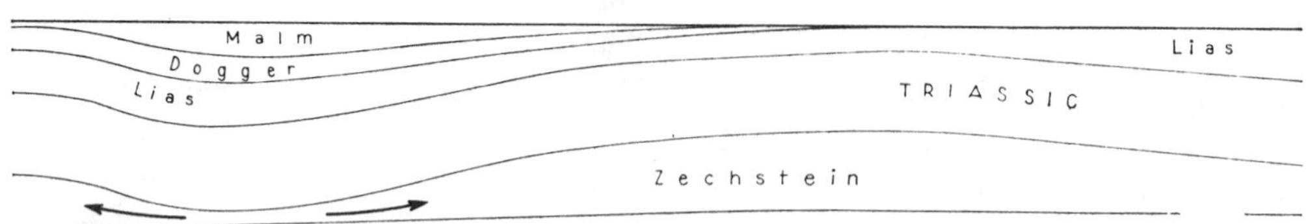

Pre-Wealden

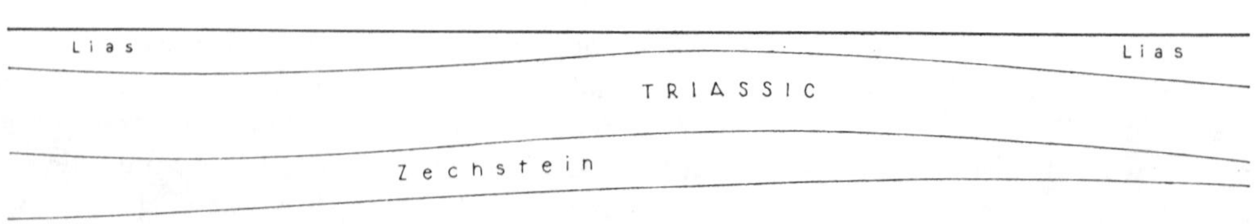

Pre-Dogger

FIG. 18.—Development of pseudo-anticline and extrusion zone in Lower Saxonian basin (Düste-Twistringen). Arrows: direction of salt migration. Double arrows: maximum thicknesses in rim syncline.

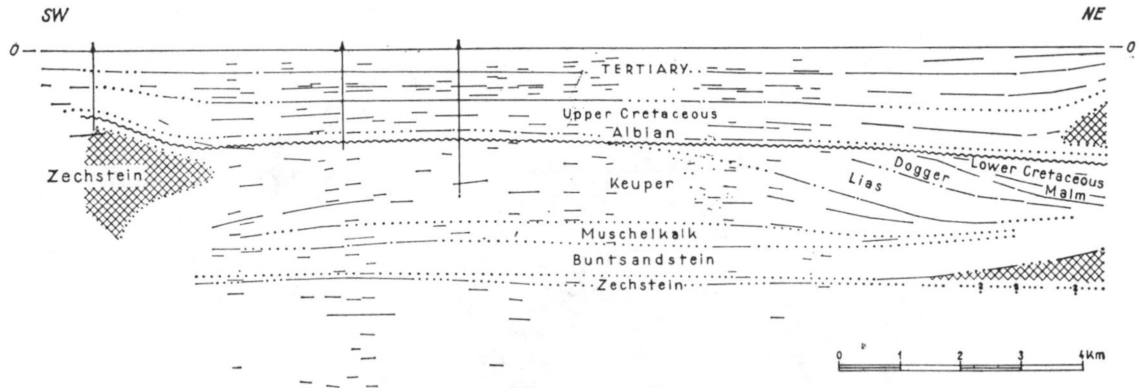

FIG. 19.—Seismic profile through turtle-shape structure at Asendorf, between salt domes of Töps and Bahlburg.

was, for the most part, concentrated in the salt structures, or had extruded in the course of geological development, or was dissolved by sea water or ground water. Accordingly, it was hardly possible for any more large-scale regional salt movements to take place. Since the movements occurred over a flat basement, a sedimentation level was produced, in general parallel with the old basement. The halokinetically deformed zone lies between these two parallel planes (e.g., Figs. 13, 17, 19). In general, only the salt concentrated in the salt stocks was able to rise again in a post-diapiric phase.

In the salt pillows, on the other hand, this state of maturity was not achieved. Salt pillows are known to exist (e.g., Fig. 9) which have remained virtually quiescent since the Upper Cretaceous or the Tertiary. The forces of buoyancy in the accumulated salt mass were not sufficient to cause a break-through. In other places, the growth of the salt pillow continued into the Tertiary or even beyond (e.g., Fig. 8).

Mobile salt is a sediment restricted to the upper layers of the earth's crust. In those regions of the world which are not, or not strongly, subject to orogenic stresses, thicker salt masses rise directly (as salt stocks) or indirectly (as newly sedimented derivatives), ever higher in accordance with progressive covering and the gravitative potentials resulting therefrom. The salt is increasingly expelled from the deeper levels, especially when basins subside so deep that they approach the front of regional metamorphism (and migmatization), by which salt is completely destroyed and consumed.

More or less complete expulsion of salt takes place in the orogens. This is in accordance with the fact that the earth's great salt deposits tend to have arisen immediately after the major orogenies (Lotze, 1957, p. 189). The writer believes that salt, the oldest sediment of our globe, is involved in an eternal cycle, as a result of its special physical properties, without its becoming consumed anywhere to any considerable degree, and without its ever coming finally to rest (permanence of salt), since the light salt can never achieve geostatic equilibrium amidst heavier sediments. By the standards of geological time, halokinesis is only an episodically recurring stage in the history of salt.

AGE OF SALT MOVEMENTS

In Northern Germany the first movements of the Permian salt began before the end of the Lower Triassic. From the Upper Triassic onward, throughout the Mesozoic period, great regional salt movements took place, which partly continued in the Cenozoic. These movements began in the individual structures at very different times, and terminated at different times. Each salt stock has its own history, and closely adjoining structures may have developed quite differently. The section from a subcrop geological map of Northwestern Germany (Fig. 20) shows the different ages of neighboring salt stocks, deduced from the age of their peripheral sinks. An analysis of the structural development (D. Sannemann, 1960) shows how the individual stages of development gradually merged into each other (Fig. 21). No dependence on orogenic phases can be discerned as regards either the beginning or the end of the movement.

The concept, contradictory to this, of a tectonically phase-bound development dates from

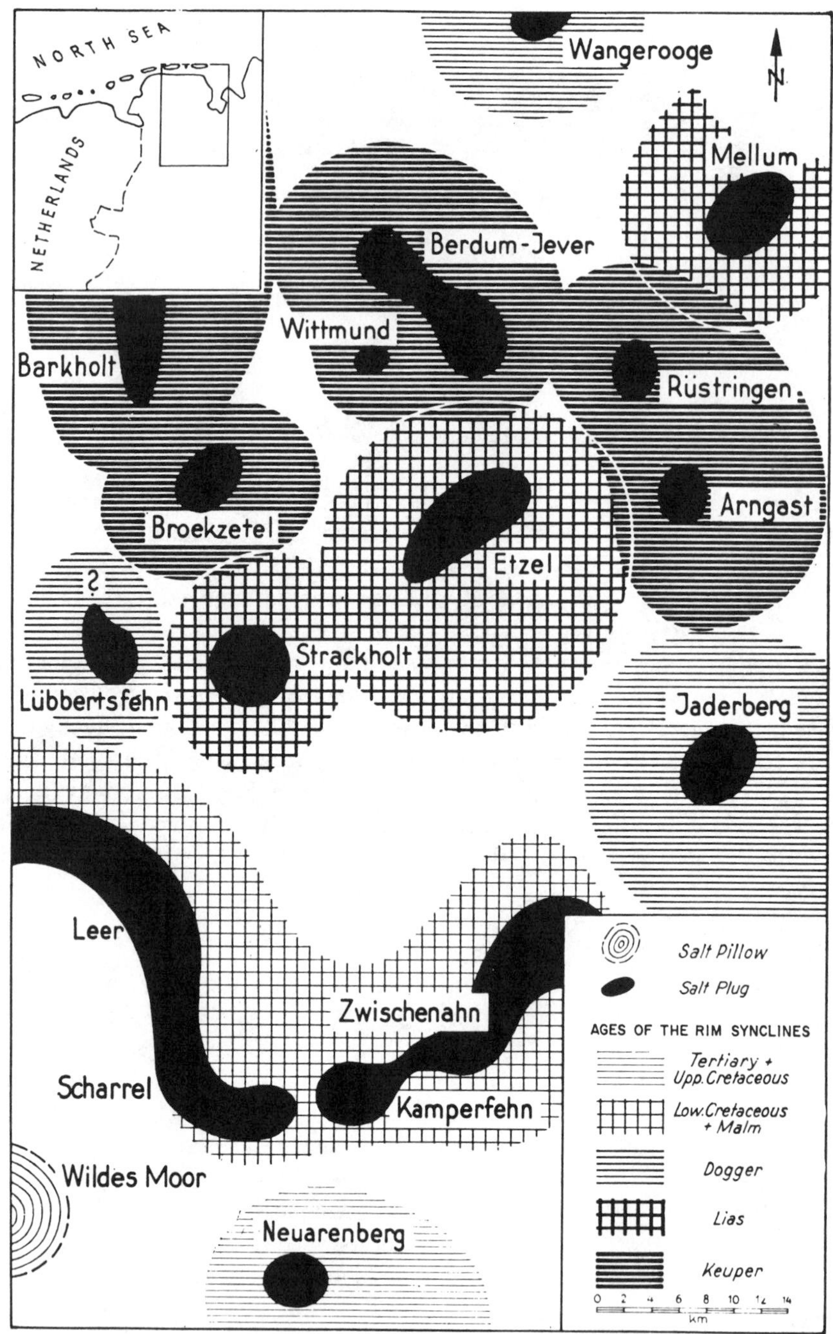

Fig. 20.—Part of salt dome family in Northwest Germany, showing different ages of secondary peripheral sinks. Oldest salt stocks (secondary peripheral sink in Keuper) surrounded by younger diapirs (after D. Sannemann, 1960).

the time in which reflection seismic surveying had not yet been applied, and only relatively few wells had been drilled in and around the salt stocks. Meanwhile, many regions have been explored by closely spaced wells. These have proved in many places that unconformities occur only in the upper parts of salt structures, and are absent in the accompanying peripheral sinks. These unconformities were not caused by one brief orogenic event but are local ingressions continually progressing throughout long periods of time, in the course of which any stratigraphic unit of the overlying series may overlap any of the older beds.

However, in contrast to these discordances re-

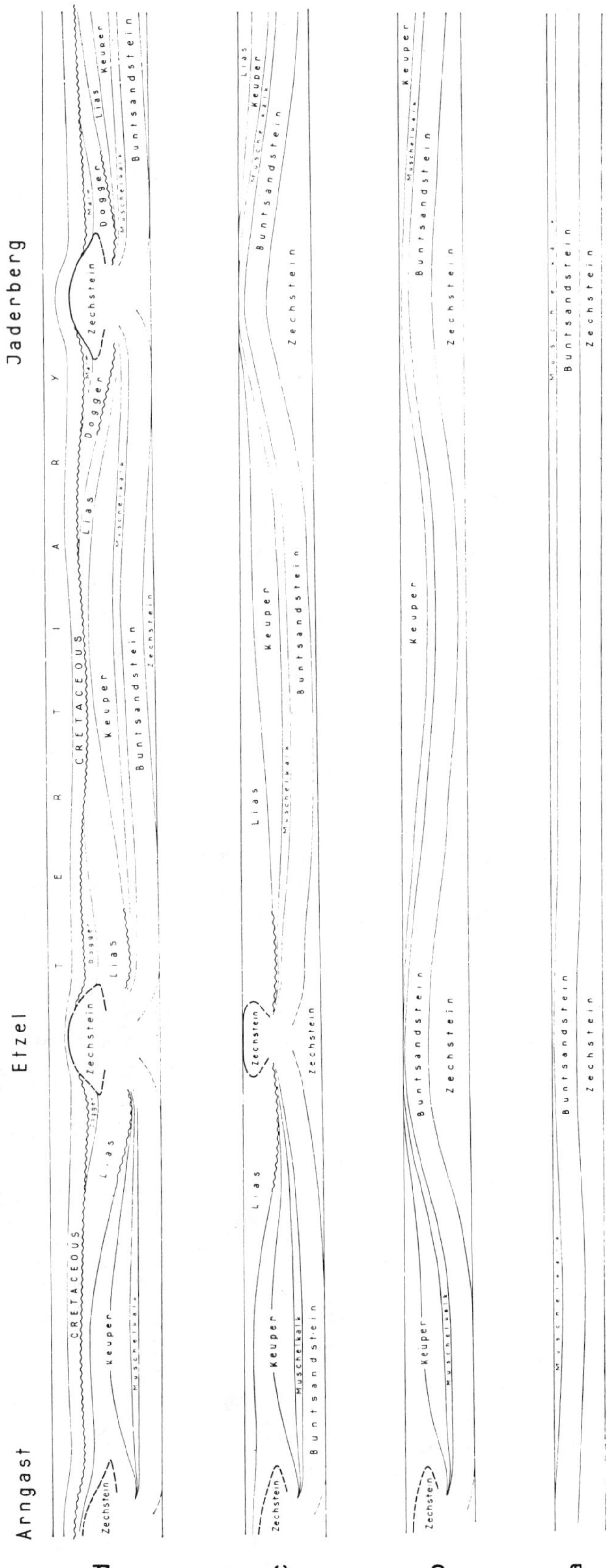

FIG. 21.—Development of three adjacent salt domes of different age (see Fig. 18) (after D. Sannemann, 1960).

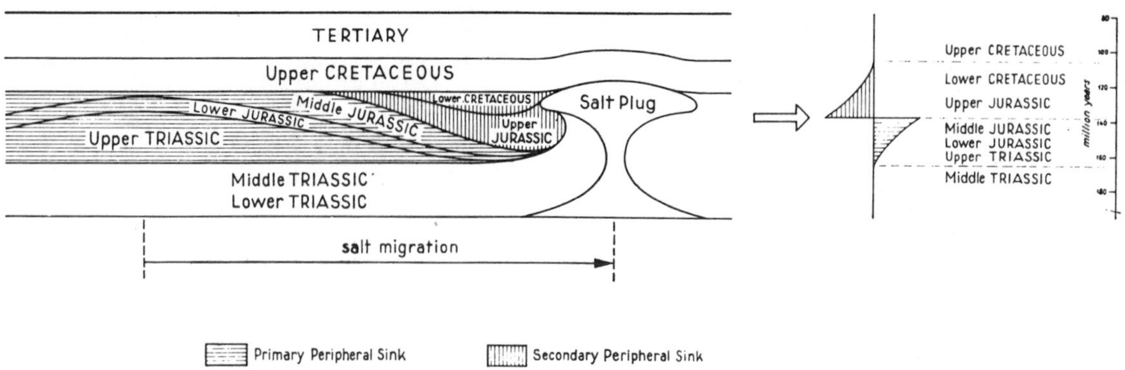

FIG. 22.—Diagram illustrating method, by which speed of salt migration can be ascertained. Designed by D. Sannemann.

stricted to the salt structures, the Albian transgression is obviously of continental dimensions. It is probable that this world-wide transgression was caused by a eustatic rise of the sea.

The broad features of the salt movements can best be recognized with the help of systematic analysis and comparison of the salt structures over large regions, on the basis of reflection seismic surveys and well records. The completeness of sedimentation in the peripheral sinks, the relation between peripheral sinks and the apparently uniform shift of their axes through time and space, clearly demonstrate that the salt movements proceeded in an *essentially continuous, autonomous* mode throughout the Mesozoic. Theoretically, it is possible that stresses present in times of regional tectonic unrest might have accelerated the flowing movement of the salt. However, we believe that the daily disturbance of the earth's solid crust by lunar forces should have a certain influence on the initiation and maintenance of salt flow. This source of energy, disregarded till now, consists in a constant jarring, working like a vibrating engine, and reducing friction. Small effects of tidal forces have been known for a long time (Ellenberger, 1954).[9]

SPEED OF SALT MOVEMENTS

The lateral migration of the salt, in a direction toward the salt structures, is reflected by the progressive shifting with time of the primary peripheral sinks, or, as indicated stratigraphically, by the shifting of the corresponding thickness maxima. In many places this migration took place continually throughout several epochs. Since the absolute duration of these epochs is known to a certain extent, the speed of the salt movement must be approximated (Fig. 22). The peripheral sinks can therefore be used to measure the order of magnitude of the salt movement.

The writer's associate, D. Sannemann, has analyzed the development in time of numerous peripheral sinks on the basis of this principle, and has reached the interesting conclusion (Sannemann, 1960) that the rate of flow of the salt on the geological time scale *averages 0.3 mm. per year.* This value, which is more or less of the order of magnitude of epeirogenic movements, is certainly only a rough approximation. Very likely the rate of flow fluctuated both in the unit of time observed and also within the salt body. Also the petrographic nature of the salt may modify the rate. However, surprisingly enough, it was found that this mean rate of flow of the salt of about 0.3 mm. per year was constant, both in the more lateral flow during the salt pillow stage and in the more vertical flow during the diapir stage; in other words, a specific salt pillow took about as long to form as the diapir developing from it (Fig. 23). In addition, the time analysis of the salt stocks so far investigated permits the conclusion that, in general, the geologically older salt structures required a shorter period of development than the geologically younger structures. The number of salt stocks analyzed is, however, not yet sufficient to permit a final verdict.

These statements contradict the view prevailing up till now, i.e., that, because of relief of pressure, the diapir stage was consummated more quickly than the pillow stage.[10] It is suggested

[9] Personal communication by H. Dürschner.

[10] On the other hand, calculations of Dobrin (1941, quoted from Scheidegger, 1958, p. 260) have already

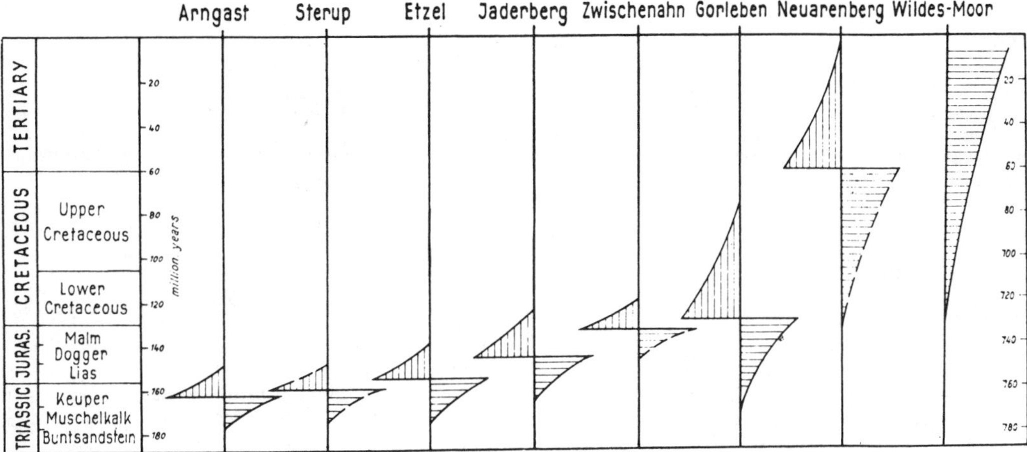

FIG. 23.—Time chart of development of some German salt domes by D. Sannemann. Vertical-line pattern: diapiric stage. Horizontal-line pattern: pillow stage.

that the speed of the salt in both cases was the maximum speed, which is mainly dependent on the elastic constants of the salt (interior friction, viscosity).

The average value of 0.3 mm. per year, was found for movements which lasted over long periods of geological time. Moreover, results of isolated measurements, carried out by precision levelling on recently rising diapirs are available. For instance, a relative rise of 1–2 mm. per year was measured on salt stocks of the Caspian depression (Lotze, 1957, p. 355). To what extent shallow heaving processes in the zone of the ground water may also play a part is not known. According to geological studies, the post-diapiric rise of the salt stock of Segeberg, in Holstein (Northern Germany) has amounted to about 2 mm. per year in the last 20,000 years (Teichmüller, 1958). Recent movements in Iraq (which, in the opinion of the writer, were halokinetic) have reached a rate of 2.4 mm. per year (calculated from observations by Lees and Falcon, 1952). These higher values may be maximum values, in contrast to the aforementioned average values over long time intervals.

It is known that salt can penetrate much more quickly into artificially created cavities (boreholes, mine galleries). There are examples from German potash mining (Spackeler, 1957), according to which salt has been driven into open mine galleries at a rate of 72, 88, even 1,250 mm. a year.

shown that, outside its initial stage, the rate of rise of a salt body is constant. Air bubbles arising in water at a constant speed constitute a comparable example.

Moreover, the rate of flow might probably be greater in outflowing salt glaciers than in the formation of salt pillows and diapirs.

REFERENCES

ARRHENIUS, S., AND LACHMANN, R., 1912, "Die physikalisch-chemischen Bedingungen der Bildung der Salzlagerstätten," *Geol. Rundschau*, Vol. 3.

BARTON, D. C., 1933, "Mechanics of Formation of Salt Domes, with Special Reference to Gulf Coast Salt Domes of Texas and Louisiana," *Bull. Amer. Assoc. Petrol. Geol.*, Vol. 17, pp. 1025–83.

BEHRMANN, R. B., 1949, "Geologie und Lagerstätte des Ölfeldes Reitbrook bei Hamburg," *Erdöl und Tektonik von NW.-Deutschland*. Hanover.

BENTZ, A., 1949, "Ergebnisse der erdölgeologischen Erforschung NW.-Deutschlands 1932–1947," *ibid.*

BORCHERT, H., 1959, *Ozeane Salzlagerstätten*. 184 pp. Gebrüder Borntraeger, Berlin.

BRAND, E., 1956, "Ergebnisse neuerer Aufschlusstätigkeit im Raum Rehden-Aldorf," *Erdöl und Kohle*, Vol. 9, pp. 2–9.

CAREY, S. W., 1954, "The Rheid Concept in Geotectonics," *Jour. Geol. Soc. Australia*, Vol. 1, pp. 67–117.

DE GOLYER, E., 1934, "Origin of North American Salt Domes," *Problems of Petroleum Geology*, Amer. Assoc. Petrol. Geol., pp. 629–78.

DOBRIN, M. B., 1941, "Some Quantitative Experiments on a Fluid Salt-Dome Model, and Their Geological Implications," *Trans. Amer. Geophys. Union*, Vol. 22, Pt. 2, pp. 528–42.

DÜRSCHNER, H., 1957, "Einige physikalische Überlegungen zum Problem der Halokinese," *Zeits. Deutsch. Geol. Ges.*, Vol. 109, pp. 152–58.

ELLENBERGER, H., 1954, "Die Erdgezeitenforschung unter besonderer Berücksichtigung des deutschen Beitrages," *Sonderhefte der Zeits. für Vermessungswesen*, Vol. 2. Stuttgart. (With references.)

FULDA, E., 1938, *Steinsalz und Kalisalze*. Ferdinand Enke, Stuttgart.

GRIPP, K., 1958, "Salzspiegel und Salzhut auf dem Lande und unter dem Meere," *Abh. Naturw. Ver. Bremen*, Vol. 35, pp. 249–58.

HANNA, M. A., 1934, "Geology of the Gulf Coast Salt Domes," *Problems of Petroleum Geology*, Amer. Assoc. Petrol. Geol., pp. 629–78.

Heermann, O., 1948, "Application of Seismic Reflection Methods to German Salt Dome Structures," *Reprint, Intern. Geol. Congr. 18th. Sess.*, Pt. V. London.

Heim, A., 1958, "Beobachtungen über Diapirismus," *Eclog. Geol. Helvetia*, Vol. 51, pp. 1–32. Basel.

Kent, P. E., 1958, "Recent Studies of South Persian Salt Plugs," *Bull. Amer. Assoc. Petrol. Geol.*, Vol. 42, pp. 2951–72.

Knieg, N., 1958, "Zur Tektonik des westlichen Schlotheimer Grabens (Thüringer Becken)," *Abh. Deutsch. Akad. Wiss.*, Berlin, Jahrg. 1956, No. 6.

Lachmann, R., 1911–1913, "Der Salzauftrieb," *Kali.* Halle (Saale).

Lees, G. M., and Falcon, N. L., 1952, "The Geographical History of the Mesopotamian Plains," *Geogr. Jour.*, Vol. 118, Pt. 1, pp. 24–39.

Lotze, F., 1957, *Steinsalz und Kalisalze*, Pt. I. 466 pp. Gebrüder Borntraeger, Berlin.

Nettleton, L. L. 1934–1936, "Fluid Mechanics of Salt Domes," *Bull. Amer. Assoc. Petrol. Geol.*, Vol. 18, No. 9 (1934), pp. 1175–1204; *Gulf Coast Oil Fields*, Amer. Assoc. Petrol. Geol. (1936), pp. 79–108.

Parker, T. J., and McDowell, A. N., 1955, "Model Studies of Salt-Dome Tectonics," *Bull. Amer. Assoc. Petrol. Geol.*, Vol. 39, pp. 2384–2470.

Roll, A., 1949, "Die stratigraphische Entwicklung und die Geschichte der Salzstockbildung im Hannoverschen Becken," *Erdöl und Tektonik von NW.-Deutschland.* Hanover.

Sannemann, D., 1960, "Zur Entwicklung von Salzstock-Familien" (in preparation).

Scheidegger, A. E., 1958, *Principles of Geodynamics.* Berlin-Göttingen-Heidelberg.

Schott, W., 1942, "Das Perm und seine Erdölführung," *Jahrb. Reichsamt Bodenforschung*, Vol. 63, pp. 567–89.

Schreiber, A., 1956, "Zum tektonischen Werdegang des Erdölfeldes Georgsdorf (Emsland)," *Geol. Jahrbuch*, Vol. 71, pp. 675–700.

Spackeler, G., 1957, *Lehrbuch des Kali- und Steinsalzbergbaus.* Halle.

Stille, H., 1924, *Grundfragen der vergleichenden Tektonik.* Berlin.

———, 1925, "The Upthrust of the Salt Masses of Germany," *Bull. Amer. Assoc. Petrol. Geol.*, Vol. 9, pp. 417–41.

———, 1932, "Asymmetric Folds with Reference to German Salt Bodies, *ibid.*, Vol. 16, pp. 169–77.

Teichmüller, R., 1948, "Das Oberflächenbild des Salzdomes von Segeberg in Holstein," *Zeits. Deutsch. Geol. Ges.*, Vol. 98, pp. 7–29.

Trusheim, F., 1957, "Über Halokinese und ihre Bedeutung für die strukturelle Entwicklung Norddeutschlands," *ibid.*, Vol. 109, pp. 111–51 (with references).

Reprinted by permission of the Geological Society of America from M. P. A. Jackson and C. J. Talbot, *Geological Society of America Bulletin*, v. 97 (1986), p. 305-323.

External shapes, strain rates, and dynamics of salt structures

M.P.A. JACKSON *Bureau of Economic Geology, The University of Texas at Austin, Austin, Texas 78713*
C. J. TALBOT *Institute of Geology, University of Uppsala, S-751 22 Uppsala, Sweden*

ABSTRACT

Salt structures continue to attract attention as petroleum traps and as storage vessels for wastes or fuels. Drawing on field studies, experiment, and theory, we examine the megascopic structure and large-scale dynamics of salt structures.

Salt flowage can transform a tabular salt body into nondiapiric rollers, anticlines, and pillows; into diapiric walls, stocks, massifs, and nappes; and into extrusive domes and salt glaciers. These structures distort at widely variable strain rates of 10^{-8} s^{-1} to 10^{16} s^{-1}, with a comparatively restricted range of dominant wavelengths of 7-26 km.

Buoyancy is an ineffective diapiric mechanism unless the salt structure has pre-existing relief of at least 150 m beneath a denser overburden of terrigenous clastics. Differential loading is a far more effective mechanism in the early stages of diapirism and commonly results in asymmetric salt structures. Gravity spreading modifies the shape of salt structures with negative buoyancy. With a heat-induced density inversion, thermal convection may lead to internal circulation and stirring of a still-tabular salt body. At least four mechanisms may form broad bulbs on mature salt stocks. Consideration of the effective viscosity contrast between salt and its cover suggests that mature stocks may have stems much more narrow than is commonly envisaged.

Salt tectonics is classified here on the basis of change of gravity potential energy that promotes or retards salt flow. Halokinetic movements can be initiated, succeeded, retarded, or accelerated by regional tangential forces that stretch, wrench, or compress sedimentary basins.

INTRODUCTION

A polycrystalline aggregate of halite has unusual physical properties. Rock salt is remarkably soluble, weak, and easily mobilized by solid-state flow. Its density is anomalously low. Most other sediments originate with densities less than that of salt, but they compact, dehydrate, or cement to greater densities as they are buried. Salt, however, develops a polycrystalline microstructure of low permeability during accumulation and remains almost incompressible throughout subsequent burial and metamorphism.

Most rocks require the temperatures and pressures of orogeny near plate margins to regionally metamorphose and penetratively deform. Salt can flow at such low temperatures and differential stresses that it is usually metamorphosed and expelled from sedimentary sequences in compactional and thermobaric regimes, while other sediments are merely diagenetically altered. By the time sediments are involved in the "nonmetamorphosed" fold and thrust belts that fringe or precede orogens, salt has commonly been lost from the sequence by subsurface dissolution or by squeezing toward the surface as gneissose diapiric intrusions, or even as mylonitic extrusions, followed by solution in the meteoric regime.

Diapiric stocks of salt and their domed envelopes of country rocks have been intensively studied because they trap hydrocarbons. Comprehensive reviews of salt deformation on all scales nevertheless are rare. Gera (1972) reviewed selected aspects of salt tectonics. Halbouty (1979) reviewed the morphology of many Gulf Coast domes. Internal structures of salt bodies were described by Kupfer (1968) and Richter-Bernburg (1980).

Petroleum exploration around salt structures and, to a leser extent, mining for rock salt, potash, and sulfur within them have provided the main stimulus for salt dome studies. New potential uses for salt domes have created a surge of interest in their geology at all scales. Salt stocks provide relatively pure rock in which it is comparatively inexpensive to mine or dissolve underground caverns for long-term storage of valuable or dangerous materials (Cohen, 1977). Such materials include high- and low-level radioactive wastes and other poisons as well as petroleum crude and its refined products for economic and strategic purposes. The pressure difference between these deep caverns and the surface also has potential for the temporary storage of energy in the future as compressed air or electrolytically derived fuel.

Programs to study the possibility of storing radioactive wastes in salt stocks in the United States and northwestern Europe have yielded considerable data on the rheology and rock mechanics of salt, summarized, for example, by Baar (1977), Gevantman (1981), Carter and Hansen (1983), and Hardy and Langer (1984). Such studies deform salt relatively rapidly in samples usually of centimetre scale. Extrapolation of results to larger scale requires the absence of mesoscopic heterogeneity and discontinuities in the rock mass (Handin and Logan, 1981). Extrapolation to slower strain rates by five to eight orders of magnitude relies on demonstrating that the deformation mechanisms operating during the experiment are identical to those operating during natural strain (Carter, 1976).

New information is also available on salt structures at the megascopic scale (>1 km) as geophysical techniques such as seismic reflection and refraction and gravity surveys continue to revolutionize tectonic studies (for example, Martin, 1980; Bally, 1983). Borehole and seismic data from explorationally mature basins also provide improved estimates of strain rates of rising salt structures (Seni and Jackson, 1984; Labao and Pilger, 1985). Sophisticated mathematical and material modeling in the laboratory has simulated large-scale salt diapirism.

Dixon and Summers (1985) showed that very thin (13 per mm) active or passive marker layers in scaled centrifuge models can explore the little-studied mesoscopic scale (1–1,000 m) of natural deformation. No such studies of diapirs have yet been published, however, and we depend on old field work in domal salt mines for data. Recent field studies of salt extrusions in Iran have also shed light on flow within diapirs.

This paper describes and attempts to explain the macroscopic evolving shape of the contact between salt and its overburden in the form of rock, surficial deposits, water, or air, as seen from the viewpoint of structural geologists. After outlining the various gross shapes developed in

Geological Society of America Bulletin, v. 97, p. 305–323, 11 figs., 1 table, March 1986.

natural salt, we consider the rates at which such structures form and then discuss the dynamics of their growth and relation to tectonics. In a forthcoming companion paper, we examine the mesoscopic and microscopic internal fabric of salt structures.

A progress report should distill what is known but should also acknowledge what is unknown. The failure of many underground spaces in salt by unanticipated, rapid creep closure, seismic rock bursts, and sudden flooding (for example, Baar, 1977) suggests that we still have much to learn about rock salt. Our approach has been to exploit the three tools of theory, experiment, and observation of nature. In so doing, we try to reconcile two different viewpoints: first, that of the practical petroleum geologist or geophysicist with everyday experience of salt domes, the full shape and growth history of which are often imperfectly known; second, that of the theoretician or experimentalist working with simplified models, the geometry, kinematics, and dynamics of which are far better known than those of real salt domes. As we show, some differences in our concepts of natural and model domes remain.

EXTERNAL SHAPES OF SALT STRUCTURES

A salt body is approximately tabular after its deposition by evaporation. Facies changes cause the salt body to feather out laterally and vertically into mixed zones of evaporites and clastics. Salt thickness varies if the original basin floor was irregular or if subsidence of the floor was not uniform. Most of these geometric and compositional variations can focus later salt movement. Despite these primary variations within and around it, the salt sequence is basically stratiform and subhorizontal before salt flow.

Salt is the type example of a rock flowing in the solid state by gravity alone, because it has such low density and negligible yield strength. A salt layer of adequate thickness can flow plastically under minute stress differences due to the body force of gravity, to imposed tectonic forces, or to a combination of these two. The effectiveness of differential stress in driving deformation depends on such factors as temperature, confining pressure, and the presence of impurities such as water, which increases diffusive flow, or nonevaporite minerals, which reduce plasticity.

Salt flowage can transform a tabular body into a wide variety of salt structures. Subsurface bodies are grouped into diapiric and nondiapiric structures. In describing natural structures, we use the term "diapir" strictly to refer to a body that has pierced, or appears to have pierced, shallower overburden (O'Brien, 1968). Natural diapirs are separated from their cover by a strain discontinuity; nondiapiric structures are in conformable contact with their overburden. In referring to mathematical and material models, we use diapir in a looser sense, applying it to all structures that resemble natural salt diapirs in their external form, regardless of whether or not the model "salt" contact is a discontinuity. This is the case because in many models the overburden is so ductile that even mature diapirs deform it as a thin stretched skin rather than actually piercing broken overburden. Such extreme thinning in relatively incompetent model overburdens is of uncertain relevance to nature. We know virtually nothing about the strains and rheology of the overburden envelope around natural salt stocks. In both natural and simulated structures, the salt or its equivalent is referred to as the source layer. "Salt dome" is a general term applied to a salt pillow or stock and its overlying and surrounding aureole of arched strata.

Salt structures generally evolve from concordant, low-amplitude structures to discordant, high-amplitude intrusions and thence to extrusions, but they can stop growing at any stage (Fig. 1). Gentle, immature salt structures concordant with their cover include salt anticlines, salt rollers, and salt pillows. Salt anticlines (salt ridges, salt waves, rides salifères) have approximately symmetrical cross sections with a planar base and arched roof. Salt rollers (Bally, 1981) are also ridgelike but are asymmetrical in cross section; a long, gentle dip slope is in stratigraphic contact with the cover, whereas a short, steep scarp slope is in normal fault contact with the cover. Salt pillows (Salzkissen, intumescences salifères) are periclinal subsurface domes; plan views are circular to moderately elliptical, and their bases are generally subplanar.

Diapiric intrusion results in the formation of salt walls, salt stocks, salt nappes, and possibly detached teardrop-shaped blobs of salt. Salt stocks (salt plugs, Salzstöcke, noyau de sel, núcleo de sal) vary in shape from squat to columnar and can be conical or barrel-shaped. The top of a diapir may swell sideways to form what we refer to as a "bulb" (cap, hat, head) on a stem (Fig. 1). The part of the bulb that is wider than the stem is the overhang (Salzüberhang, sel surplombent). Overhangs also result from tilting of cylindrical stocks. Regardless of diapir shape, the uppermost surface of salt is referred to as the crest. Some salt diapirs shaped like inverted teardrops may have pinched off from their source layers (for example, Eilte Dome, northern Germany; Trusheim, 1960). Without exceptionally closely spaced drilling over the entire stock, however, such diapirs would be difficult to distinguish from diapirs with broad bulbs and narrow stems.

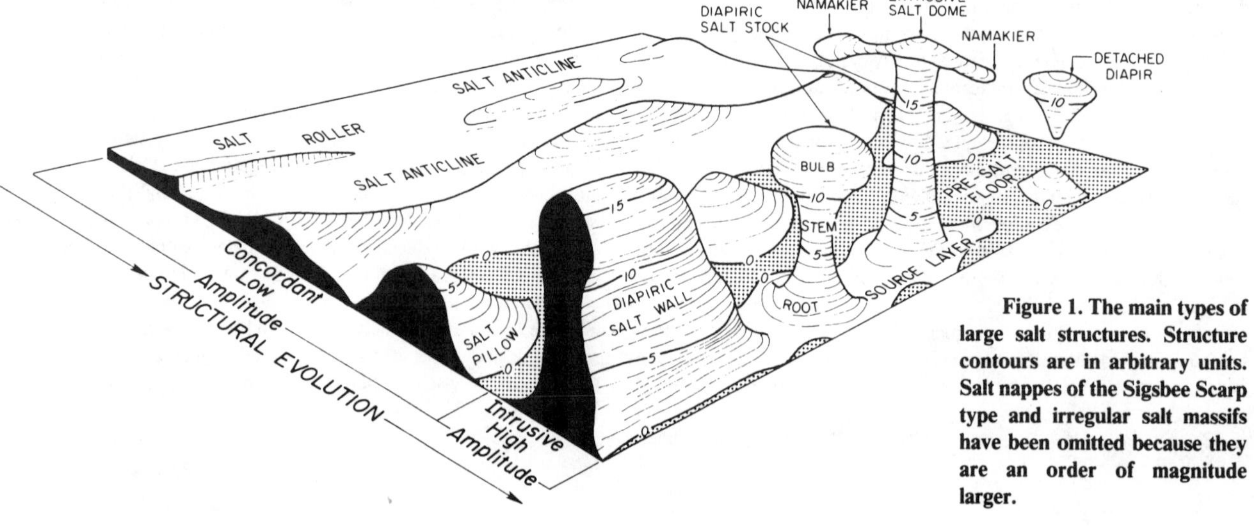

Figure 1. The main types of large salt structures. Structure contours are in arbitrary units. Salt nappes of the Sigsbee Scarp type and irregular salt massifs have been omitted because they are an order of magnitude larger.

Salt walls (Salzmauern) are elongated like salt anticlines but are intrusive and of much greater amplitude (Fig. 1). Salt ridges can evolve directly into intrusive walls, or their growth can be focused at particular sites that evolve into periclinal pillows and thence into columnar stocks. In the Zechstein basin of northwestern Europe, both walls and stocks grew in the deeper parts of the basin. Traced upward and horizontally, parallel-sided walls can develop pinch-and-swell plan forms and separate into elliptical to circular stocks. The direction of elongation in plan is parallel to base-of-salt contours, salt isopachs, and reactivated basement faults. Salt nappe complexes, which are too large to show in Figure 1, move laterally tens of kilometres by climbing up the stratigraphic section on thrust ramps. Salt diapirs can bud off the leading edge and roof of the spreading wedges of salt, as appears to have occurred at the Sigsbee Scarp in the northern Gulf of Mexico (Buffler and others, 1978; Humphris, 1979). The salt massif is another large complex: an uplift of salt with outward-dipping flanks and irregular plan shape, with a crest consisting of smaller pillows or stocks. Massifs are generally deep seated and resemble the early growth stage of experimental diapir systems.

Evolving from the concordant and intrusive types of salt structure, there is a third type, which is extrusive. Bun-shaped extrusive domes of salt represent the exposed crests of diapiric stocks. Thin, broad flanges of salt flow from some of these over the land surface or sea floor, in the rarest and most bizarre form of salt tectonics. These extrusive sheets are known as "salt glaciers" (Salzgletscher, glaciers de sel, glaciars de sal) or, as recently proposed by Talbot and Jarvis (1984), "namakier," from *namak* (Farsi for salt) and glac*ier*. Salt extrusions can be confined to valleys or spread over alluvial plains as piedmont namakiers.

The characteristics and dimensions of salt anticlines, salt pillows, salt walls and stocks, and salt extrusions are described below, using specific examples that emphasize recent findings.

Salt Anticlines

The Mississippi delta and its fan on the continental slope represent the current depocenter in the northern Gulf of Mexico. Anticlines and walls of Jurassic Louann Salt are rising through the distal apron of the fan in response to differential loading (Watkins and others, 1978). The plan shape and wavelength of the walls are assumed to reflect those of the precursor salt anticlines. The anticlines and walls are straight or sinuous, are as much as 260 km long, are parallel to regional strike, and have mean wavelengths of ~20 km. Anticlines away from the fan (and that of the Rio Grande) are less regular in shape and trend. Immature structures in the lower Sigsbee slope have a characteristic wavelength of 10 km, whereas more mature structures in the upper slope have a wavelength of 27 km (Lindsay, 1977), possibly due to the landward increase in cover and source-layer thickness. Paleoanticlines originally formed in the Late Jurassic were palinspastically reconstructed from the present distribution of diapirs in the East Texas Basin; these had a mean wavelength of 18 km and a maximum length of 140 km (Jackson and Galloway, 1984).

Salt Pillows

What appear to be pillows are growing beneath the Great Kavir salt desert of central Iran, a triangular area between the Elburz Mountains in the north and the Zagros Mountains in the southwest. Their folded cover is exposed by an almost horizontal erosional section through periclinal domes and basins in the Miocene-Pliocene Upper Red Formation consisting of red beds, salt, gypsum, and salt-bearing clays and marls that have flowed since the early Tertiary (Stöcklin, 1968). The broad shapes, closely packed arrangement, and complete lack of visible discordance around these structures suggest that they are evaporite pillows with a wavelength of ~10 km. Immediately to the west, salt diapirs described below are exposed.

Subsurface reconstruction of Jurassic-Cretaceous salt tectonics in the East Texas Basin indicates that pillows grew either along salt anticlines or at their junctions (Jackson and Galloway, 1984). Such anticlines coincide with areas of salt that presumably were too thin to form diapirs. The constructive interference of two merging anticlines formed a local culmination that focused pillow growth there, however. The mean wavelength of 10 pillows in the western side of the basin is 7 km. The mean wavelength of 7 pillows in the North Louisiana basin is 15 km (Labao and Pilger, 1985).

Salt Diapirs

The width of salt stocks varies greatly. In a field of immature salt structures, there is a complete gradation between individual, mature, jet-like stocks and irregular, immature massifs having mean diameters as much as 40 km (Martin, 1980). Mature stocks are those the flanks of which are vertical or inward dipping. Structure-contour maps of >120 salt domes in South Louisiana have been published (New Orleans Geological Society, 1960, 1962, 1983). Most of these shapes are conical, with sides sloping outward to depths of 3,000–4,000 m, the limits of well data (Fig. 2A). A minority are tilted cylinders (Fig. 2B). The mean wavelength of United States Gulf Coast diapirs is 26 km (Bishop, 1978).

Mature salt stocks having the largest indisputable dimensions are probably those of southern and central Iran. Those in southern Iran are of Precambrian Hormuz Salt and are commonly obscured by flanges of extrusive salt, but the diameters of the stocks themselves have been estimated to range to at least 10 km (Kent, 1979). Those in central Iran consist of much younger lower-middle Tertiary salt and are as much as 11 km in diameter (aerial photographic interpretation by Jackson and Talbot).

The symmetry of salt diapirs is highly variable, even within a single basin. They can be axially symmetrical about a straight, vertical axis; orthorhombic (Figs. 2C, 2F); monoclinic (Fig. 2B); or nonsymmetrical about curved axes (Fig. 2E). Many diapiric stems are tilted beneath asymmetric bulbs with subhorizontal crests. Ground-water solution apparently has truncated the crest of many stocks, creating a dissolution table (salt mirror, Salzspiegel).

Salt Extrusions

Although currently confined largely to Iran, salt extrusions, which were first described by de Böckh and others (1929), are important to understanding salt dynamics, including diapirism. They are tangible evidence of geologically rapid flow of salt at surface temperatures and provide large field demonstrations of the internal structures generated.

In southern Iran, Precambrian Hormuz Salt has risen diapirically from depths of 5 to 10 km through an almost continuous upper Paleozoic and Mesozoic carbonate overburden, breaching the surface in >200 salt stocks (Ala, 1974; Kent, 1979). The Hormuz Salt provides a décollement zone for the Zagros folding and thrusting, which began only 15 Ma, much later than the Mesozoic diapirism. Anticlines along the northeast coast of the Persian Gulf are still rising at rates between 1.4 and 1.9 mm · a^{-1} (per year) and shortening at rates estimated between 14 and 18 mm · a^{-1} (Vita-Finzi, 1979). Along the coast, the bellows action of Zagros folding reactivates pre-existing salt diapirs and pumps salt up to and over the surface. Farther inland, the declining ratio of extrusion to erosion results initially in decreasing volumes of surviving extrusions and eventually to

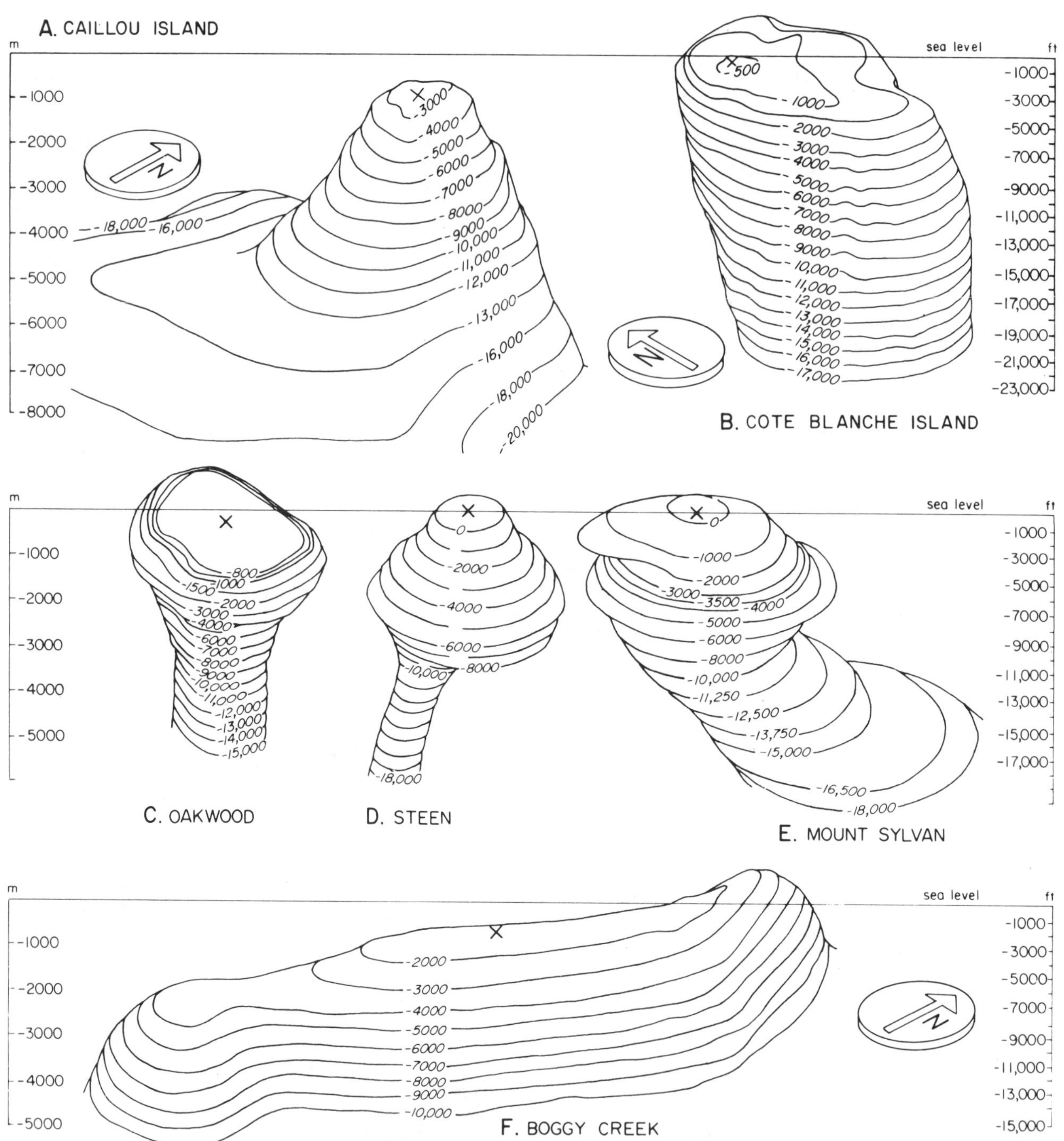

Figure 2. Isometric block diagrams of six salt diapirs in south Louisiana (A, B prepared by W. F. Mullican from data in New Orleans Geological Society, 1983) and East Texas (C-F adapted from Jackson and Seni, 1984). Contours show depths (ft) below mean sea level; unlabeled contours are interpolated. A. Caillou Island Dome, a largely immature, elliptical, conical stock with small size (13 km^2); asymmetric, farside overhang; tilted axis; no symmetry. The stock rests on the junction of two deep salt walls. Data from ~1,500 wells. B. Cote Blanche Island Dome, a mature, cylindrical, elliptical stock with small size (15 km^2), asymmetric overhang, tilted axis, and monoclinic symmetry. Data from 44 wells. C. Oakwood Dome, a mature, subcircular stock with small size (10 km^2), symmetrical overhang, vertical axis, and orthorhombic symmetry. Data from 82 wells, a seismic profile, and gravity modeling. D. Steen Dome, a mature, subcircular stock with small size (10 km^2) and an inclined, slender stalk. Data from 5 wells and gravity modeling. E. Mount Sylvan Dome, an irregular, slightly elliptical, strongly tilted mature stock with small size (6 km^2), multiple overhangs, and no symmetry. Data from 4 wells, a seismic profile, and gravity modeling. F. Boggy Creek Dome, a structurally immature, highly elliptical wall with large size (>50 km^2) and no overhang. Data from 5 wells.

salt-free cirques or craters in which the former orifices are choked by insoluble debris. Present-day salt extrusion is indicated by several lines of evidence: the necessary maintenance of the extrusive dome against continual erosion and solution, sounds of movement (Kent, 1966), the burial of its own insoluble moraine by the namakier (Harrison, 1930), and measured flow rates (Talbot and Rogers, 1980).

One of the most active salt intrusions—Kuh-e-Namak (Dashti province)—in southern Iran feeds two extrusive sheets. The larger, northern one is 50–100 m thick and 2,000 m wide. It flows almost 3,000 m, and its average upper-surface slope is 8° to the plain. There it spreads over alluvium and its own insoluble moraine.

The thickness and density of the overburden suggest that the uneroded and uncollapsed height of the extruded salt plug would be 2,800 m through buoyancy alone (Talbot and Jarvis, 1984); the tectonic effects of the Zagros shortening would increase this height by an unknown amount. On account of gravity spreading, solution, and erosion of the extruding salt, however, the summit is only ~1,600 m above sea level, 1,400 m above the surrounding plains, and 1,000 m above the country-rock orifice.

Kuh-e-Namak and its two namakiers owe their existence mainly to the high rate of salt extrusion caused by Zagros fold pressurization. The dry climate helps, but an estimated 670 mm · a^{-1} rainfall would be required to dissolve salt as fast as it rises (Talbot and Jarvis, 1984). The climate of 51% of the contiguous United States would be dry enough for at least part of this salt glacier to be preserved (Geraghty and others, 1973). A third factor aiding preservation is a patchy veneer of Quaternary marine sediments on distal and static parts of the northern namakier. These sediments provide a minimum estimated age of 30,000 to 300,000 yr for the start of current extrusion.

RATES OF SALT FLOW

Strain rate is the proportional change in length per second of a deforming body. As a geologic datum, the mean strain rates for flow of the asthenosphere (Carter, 1976) and for crustal orogeny (Pfiffner and Ramsay, 1982) have both been estimated at ~10^{-14} s^{-1} (per second).

By comparison, strain rates for *in situ* deformation of rock salt vary enormously over 8 orders of magnitude from 10^{-8} s^{-1} to 10^{-16} s^{-1} (Table 1). This wide range reflects the diverse conditions of flow. The most rapid peak rates are those of borehole closure during accelerating creep (10^{-8} s^{-1}), whereas those of mine closures and steady-state borehole closures (10^{-9} to 10^{-11} s^{-1}) and namakiers (10^{-8} to 10^{-11} s^{-1}) are not far behind. Estimates of 10^{-13} s^{-1} for the rate of diapiric extrusion assisted by folding are slower, and those for the most active phase of growth of diapirs driven by gravity alone are still slower at 10^{-14} to 10^{-15} s^{-1}.

The timing of the evolutionary history of subsurface salt structures in the Gulf of Mexico is elucidated by two approaches.

The first approach averages plan shapes of hundreds of salt structures in relation to the age of diapirism. They can be grouped into four main zones, regardless of whether they are anticlines, walls, pillows, or stocks: (1) continental slope, (2) continental shelf, (3) coastal plain, and (4) interior basin. Both the area and the degree of ellipticity in plan decrease with age (Fig. 3). The domes become smaller and more circular in plan as they mature. Statistically, these salt domes reach structural maturity after 40 Ma.

A second approach is to examine the growth history of individual structures (Seni and Jackson, 1984; Labao and Pilger, 1985). Here, the difference between gross and net rates of growth is of fundamental importance. Gross rates are a function of the volume of salt evacuated from the withdrawal basin and risen in the diapir. Net rates are a function of this process and all other processes that affect diapir height and vertical growth rate, such as salt dissolution, extrusion, and lateral intrusion. Gross rates of growth thus approximate the true rate of salt flow of the diapir as a whole, regardless of the independent motion of the diapir crest. Net rates of growth approximate the actual movement of the diapir crest relative to surrounding strata.

Studies of this type in the East Texas Basin indicate that the durations of pillow and diapir stages of growth of the same structure are similar: 10 to 60 Ma, with a mean of 20 Ma for pillows and 25 Ma for diapirs during periods of significant growth. After diapiric breakthrough, these diapirs typically show a surge of maximum vertical growth lasting 10 to 15 Ma. Strain rates have been calculated on the basis of changes in the relief of salt diapirs over different stratigraphic time intervals in East Texas by Seni and Jackson (1984). The diapir rates are summarized in Table 1; the pillow rates were derived in similar fashion from the vertical growth rates published in the same study.

The combined evidence from these two approaches suggests that although salt flow started in different places at different times, the average

TABLE 1. NATURAL STRAIN RATES IN THE DEFORMATION OF ROCK SALT.

Environment		Strain rate* (per second)
Glacial salt		
Direct measurement of flow†		1.1×10^{-11} to 1.9×10^{-9}
Comparison of theoretical profile with actual profile§		6.7×10^{-13} to 9.0×10^{-13}
Estimated from namakier morphology**		2×10^{-13} to 2×10^{-8}
Diapiric salt		
Direct measurement of closure of mined cavity††		1×10^{-9} to 9×10^{-12}
Direct measurement of peak borehole closure§§		3×10^{-8}
Direct measurement of long-term borehole closure***		7.4×10^{-11} to 3.5×10^{-9}
Measurement of topographic mound†††		2×10^{-14}
Comparison of theoretical profile with actual profile§		8.4×10^{-13}
Estimates from stratigraphic thickness changes around diapirs		
Over-all rates:	range§§§	1.1×10^{-15} to 1.1×10^{-16}
	mean§§§	6.7×10^{-16}
Fastest rates:	range§§§	3.7×10^{-15} to 6.2×10^{-16}
	mean§§§	2.3×10^{-15}
Average growth for Zechstein domes****		2×10^{-15}

*Conventional strain rate $\dot{e} = e/t$, where elongation e = change in length/original length and t = duration of strain in seconds.

†Talbot and Rogers (1980). Survey of markers distorted by glacial flow of Hormuz Salt, Ku!.-e-Namak namakier, Iran. Duration of strains = 292 days (2.5×10^7 s). Maximum flow after 5-mm rainfall. Calculated shear stress = $\tau < 0.25$ MPa.

§Talbot and Jarvis (1984), observed longitudinal profile of Kuh-e-Namak exposed stock and namakier compared with profile of numerical model of viscous fluid extruding from a narrow orifice. Stock height = 6.4 km; namakier length = 6 to 8 km.

**Wenkert (1979). Average shear strain rates for 5 Iranian namakiers of Hormuz Salt, assuming steady-state equilibrium between extrusion and wasting and based on erosion rates of 0.08–0.25 cm·a^{-1}. Calculated shear stress = $\tau \sim 0.03$ MPa.

††Serata and Gloyna (1959), Reynolds and Gloyna (1960), and Bradshaw and McClain (1971). Peak rates in Grand Saline Dome (East Texas Basin) and bedded rock salt (Hutchinson and Carey Mines, Kansas). Upper limit corresponds to wall temperatures of ~100 °C. Estimated stress difference $\Delta\sigma \sim 10$ MPa. Duration of strains = 10 to 30 a (3.2×10^8 s to 9.5×10^8 s).

§§Martinez and others (1978), Vacherie Dome (Louisiana Interior Basin). Duration of strain = 90 days (7.8×10^6 s).

***Thoms and others (1982), Vacherie Dome (Louisiana Interior Basin). Slowest at 100 °C 351 m deep under $\Delta\sigma$ = 4.2 MPa; fastest at 160 °C 1,509 m deep under $\Delta\sigma$ = 18.1 MPa. Duration of strain = 890 days (7.7×10^7 s).

†††Ewing and Ewing (1962), Sigsbee Knolls (Gulf of Mexico abyssal plain). Calculation based on salt-stock height of 1,300 m (seismic section *in* Martin, 1980). Duration of strain = 11,000 a (3.5×10^{11} s).

§§§Seni and Jackson (1984). Based on salt diapirs in the East Texas Basin. Over-all rates are those during entire known history, duration = 30 to 56 Ma (9.5×10^{14} s to 1.8×10^{15} s) of 16 diapirs; net rates calculated by maximum rate of deposition in salt-withdrawal basins. Fastest rates are those during stratigraphic periods characterized by the most rapid diapirism. Duration = 1 to 13 Ma (3.2×10^{13} s to 4.1×10^{14} s); gross rates calculated by dividing the volume of sediments in a salt-withdrawal basin by the product of the maximum cross-sectional area of the diapir and the duration of the stratigraphic interval the sediments of which constitute the salt-withdrawal basin. See Seni and Jackson (1984) for further details.

****Sannemann (*in* Trusheim, 1960), based on stratigraphic-thickness data and salt-stock height of 4 km. Duration of strain = 35 to 130 Ma (1.1×10^{15} s to 4.1×10^{15} s).

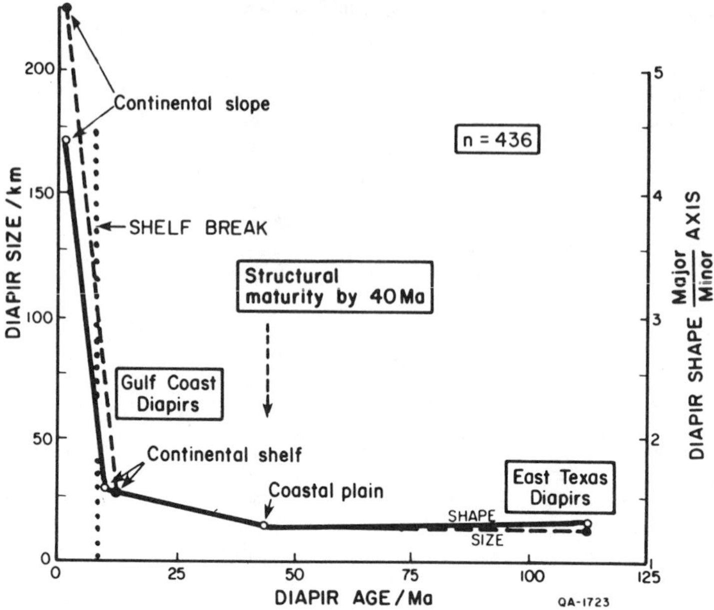

Figure 3. Structural evolution of United States Gulf Coast and East Texas Basin salt structures, which become smaller (dashed curve) and more circular (solid curve) in plan view with increasing age and distance landward, reaching structural maturity some 40 Ma after salt flow began. (Data for 436 salt structures collected by S. J. Seni from the map of Martin, 1980.)

United States Gulf Coast salt structure developed to maturity at rates shown in Figure 4. Curves are derived by synthesizing the data on average growth history outlined in the three preceding paragraphs. The growth-rate curve for East Texas domes forms the basis. The time scale is based on a 40-Ma growth to maturity. Strain rates are those calculated for East Texas domes from their growth rates. Although vertical growth rates for Gulf Coast domes may be generally higher than those of East Texas, the greater heights of the former lower their conventional strain rates. For example, a salt dome on the Texas-Louisiana continental shelf studied by Wilson (1985) using a dense grid of 592 km of seismic-reflection lines yielded a rapid Pleistocene growth rate of 760 m Ma^{-1}. Owing to its estimated height of 12 km, however, the strain rate of this dome is only a

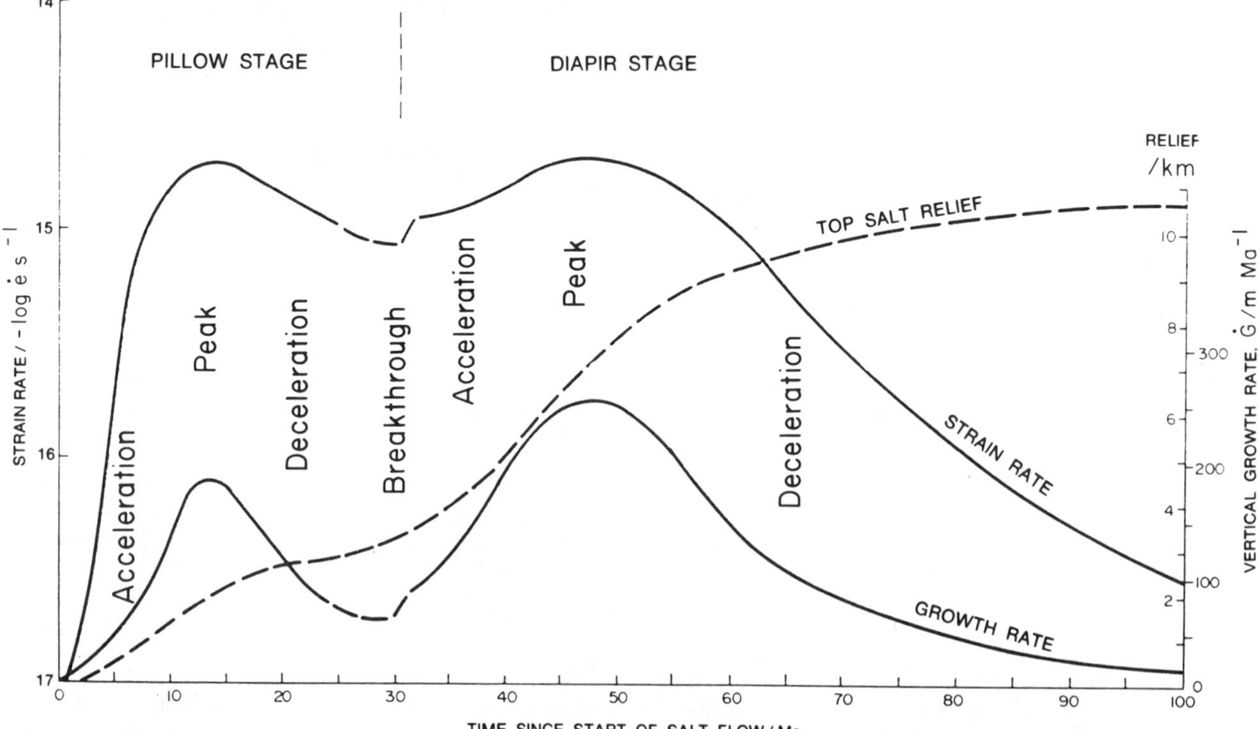

Figure 4. The life of a typical United States Gulf Coast salt structure in the pillow (including anticline) stage and diapir (including wall) growth stages. See text for derivation.

moderate 2×10^{-15} s^{-1}. The top-salt-relief curve in Figure 4 is calculated from the growth-rate curve based on a final overburden thickness of 10.6 km, typical of the continental shelf here (Buffler and others, in press).

We cannot overstress the fact that the graph represents the average behavior of a large number of salt structures and thus disguises large individual variations, such as multiple surges of growth in the same growth stage. Both pillow and diapir stages have their own times of accelerating, peak, and decelerating growth. Decelerating pillow growth may be related to increased resistance of the cover by diagenesis before salt breakthrough or to the decline in gravity potential as the pillow penetrates low-density near-surface strata. The much longer decelerating stage of diapiric growth is presumably caused by the effects of low-density, partly consolidated sediments near surface when the diapir grows by downbuilding.

How representative these rates are globally remains to be seen, but similar durations for each stage characterize the diapirism of the Zechstein Salt in northwest Germany (pillow stage, 16 Ma; diapir stage, 12 Ma; Jaritz, 1973; Hunsche, 1978) and Denmark (pillow stage, 20 Ma; diapir stage, 20 Ma; Madirazza, 1975).

Once a diapir extrudes into air or water, it can deform much faster. This is true partly because of the softening effects of water (the Joffee effect) and partly because frictional drag of country rocks is restricted to the base of spreading tongues of extrusive salt. By assuming steady-state conditions (zero net growth), gross rates of flow for the Kuh-e-Namak (Dashti province) extrusive dome and namakier have been calculated (Talbot and Jarvis, 1984). The reported rainfall of 280 mm · a^{-1} can potentially dissolve as much as 470 mm of vertical thickness. To balance this solution requires vertical salt extrusion of ~110 mm · a^{-1} over the entire 9.4-km^{-2} area of the elliptical orifice. The top free surface of the bun-shaped extrusive salt dome is a parabolic fountain. To maintain a steady net height, salt in this fountain must rise at a rate of 170 mm · a^{-1}, assuming an effective viscosity of 10^{16} Pa · s, or at 17 mm · a^{-1}, assuming a less likely viscosity of 10^{17} Pa · s (Talbot and Jarvis, 1984). We have seen how 110 mm of the 170 mm · a^{-1} rise of salt is potentially removed in solution, leaving 60 mm · a^{-1} vertical thickness of the broad extrusive dome to spread sideways and feed 2 dip-slope namakiers. Salt enters the head of the more active northern namakier at an estimated average rate of ~2,000 mm · a^{-1}. Most of this flow rate, which is much greater than the diapiric rate, is dissipated by episodes of rapid extrusive flow during the short winter rainy season, when brief flows of 500 mm per day have been measured along the namakier's flanks (Talbot and Rogers, 1980).

SALT DYNAMICS

Buoyancy Halokinesis

First proposed by Arrhenius (1912) and further developed by many subsequent workers, the simplest natural case of salt dynamics involves the buoyant rise of salt into a laterally uniform overburden (Fig. 5A). Compositional density inversions in horizontal layers of viscous fluids are inherently unstable, because the low-density buoyant source layer is overlain by a denser viscous overburden. The dominant theoretical view of halokinesis is that it involves gravity-driven motion stabilizing the system.

The style of deformation of a particular interface depends on the ratios of the densities, thicknesses, and viscosities of the layers involved, their boundary conditions, and the time following beginning of stabilizing motions (Ramberg, 1981). Comparatively simple linear mathematics using Rayleigh-Taylor theory (Chandrasekhar, 1982) can describe the very early stages of motion in the interface, when the shape of the upwellings is still identical to the shape of the downwellings in cross section. At these minute strains, the interface deforms exceedingly slowly, with a wide range of potential wavelengths. Biot and Odé (1965) demonstrated that in a simple two-layer system of viscous source and cover, as the viscosity of the cover increases, the dominant wavelength increases, and the rate of amplification (vertical growth) decreases. A relatively thin cover results in slow, weakly selective amplification of small wavelengths; a thick cover encourages the reverse. Their simulation of time-dependent thickness and density increases in the cover due to continued deposition and compaction showed that the wavelength selectively amplified also increases. Salt structures at this stage would be conformable with their overburden. Multiple-wavelength structures in layers of mixed properties were calculated in general terms for a hypothetical salt-bearing sequence (Ramberg, 1968, 1981), but there is ample scope for more specific study along these lines.

Linear mathematical descriptions become inadequate as the gentle early structures amplify beyond amplitude/wavelength ratios of 0.015–0.09, depending on the viscosity contrast (Woidt, 1978). Numerical or material models (for example, Ramberg, 1966, 1981; Fletcher, 1972; Whitehead and Luther, 1975; Heye, 1978, 1979; Hunsche, 1978; Woidt, 1978, 1980) become necessary to follow the accelerating flow and the narrowing of the range of amplifying wavelengths as the rising source anticline and sinking cover syncline begin to differ in profile. Narrowing of the dominant wavelength has also been reported for some models (Woidt, 1980). The steady rise of a diapir being fed by a thin source layer can be approximated by Stokes' law (Dobrin, 1941; Berner and others, 1972). According to Dobrin, the rise rate is controlled by the viscosity of the thick cover, rather than by that of the source layer. Statistical analysis of his Table 1, however, indicates that it is the viscosity of the stiffer layer—regardless of whether this is thin source or thick cover—that is rate controlling.

Although it has been known for decades (for example, Grout, 1945; Goguel, 1948) and has been incorporated in numerical models, it is still worth emphasizing that diapirism must involve lateral as well as vertical motion of the source and cover. Indeed, if, as is commonly the case, the wavelength of diapirism is more than twice the thickness of the sequence involved, the horizontal component of the motion must exceed the vertical.

The end result of overturn in viscous fluids is inversion. The light layer overlies the heavy layer, separated by a plane, horizontal interface. Such complete inversion, however, currently appears to be rare in rocks, because they are not Newtonian fluids and generally have yield strengths. After partial overturn, buoyancy forces are eventually too weak to overcome the yield strength of rocks, and diapiric overturn ceases.

Geologically realistic model domes, in which overturns of the source layer and its overburden are only partial, develop where model materials—particularly the overburden—are non-Newtonian and have yield points, as do compacted sediments (Parker and McDowell, 1955; Talbot, 1974, 1977; Dixon, 1975; Schwerdtner and others, 1978; Schwerdtner and Tröeng, 1978; and many examples in Ramberg, 1981). Motion ends when the layers reach a geometry that is gravitationally stable. This can occur before the source layer is exhausted, so that any further compaction of the overburden (or the application of lateral forces) can renew the movement. Diapirs can rise until they reach their level of neutral buoyancy, generally above the level where the cover-rock density is equal to that of the salt. In sedimentary basins, this level is usually within 1 km of the surface (see below) but is above the topographic surface in cover rocks of density greater than that of salt ($\rho > 2{,}200$ kg · m^{-3}).

Gravity-driven motion does not necessarily occur at a constant rate throughout a region. The instability around a mother diapir can propagate laterally outward to form successive generations of daughter and granddaughter diapirs. At least three generations of such salt-dome families are well documented in seismic profiles of the Zechstein evaporites in northern

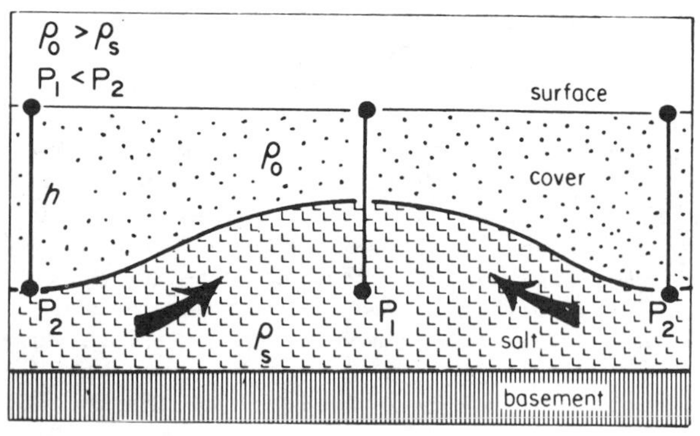

A. BUOYANCY HALOKINESIS

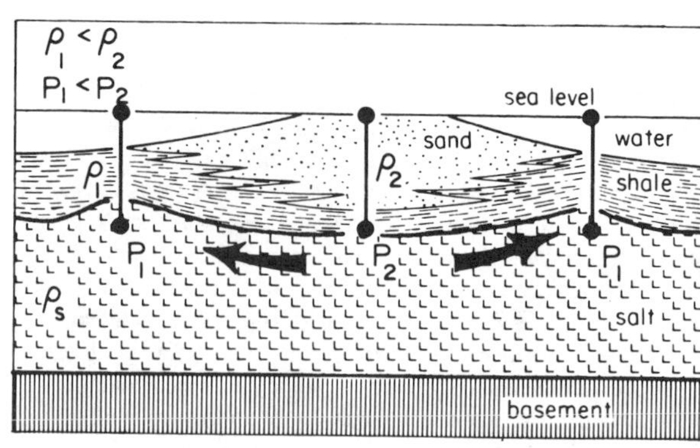

B. DIFFERENTIAL LOADING HALOKINESIS

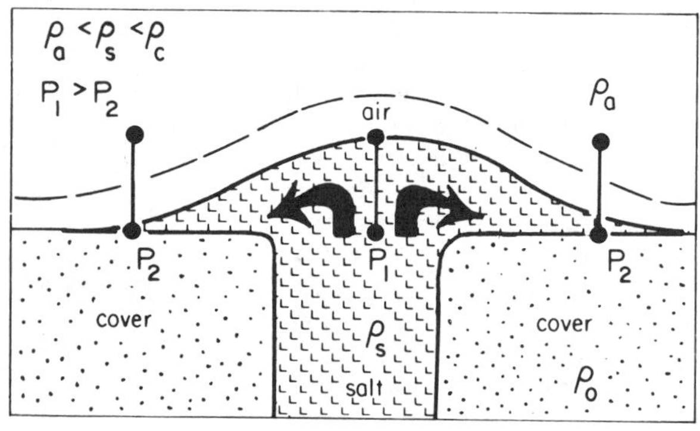

C. GRAVITY SPREADING HALOKINESIS

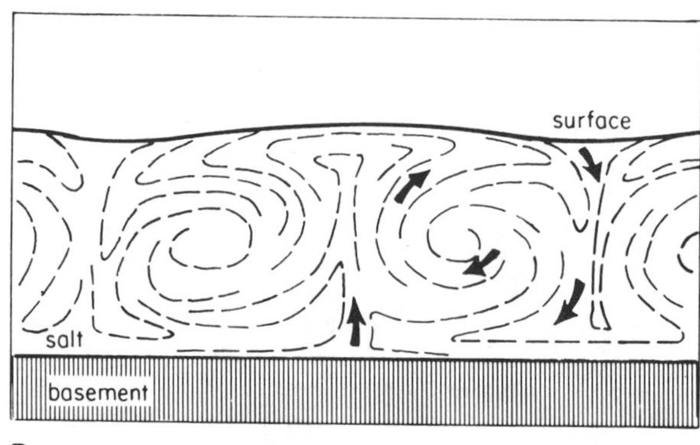

D. THERMAL CONVECTIVE HALOKINESIS

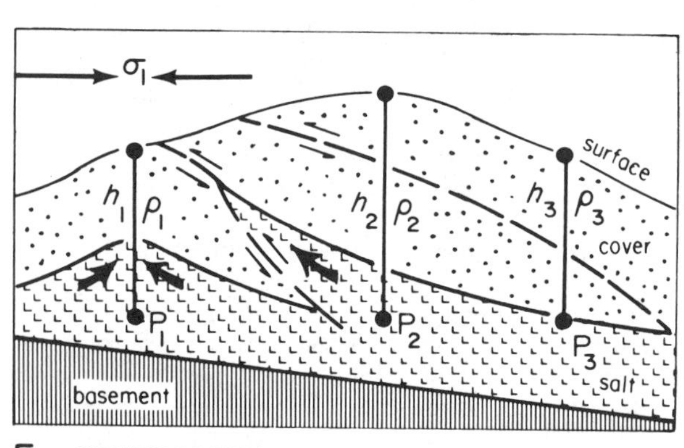

E. CONTRACTION $h_2 > h_1 = h_3$

Stable: $\rho_s > \rho_o \therefore \rho_1 > \rho_2 > \rho_3$ and $P_3 < P_1 < P_2$

Unstable: $\rho_s < \rho_o \therefore \rho_1 < \rho_2 < \rho_3$ and $P_1 < P_3 < P_2$

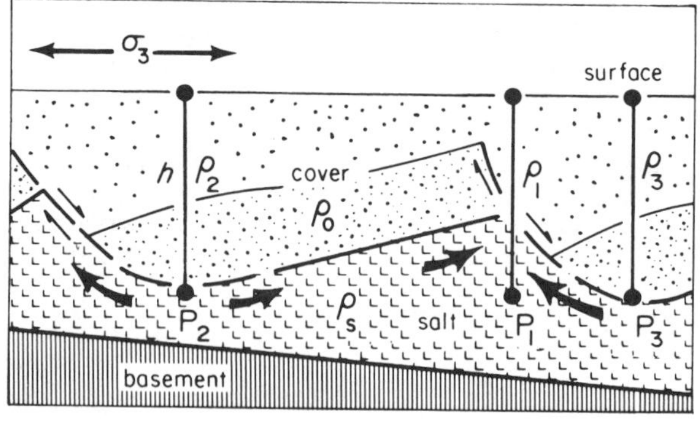

F. EXTENSION

Stable: $\rho_s > \rho_o \therefore \rho_1 > \rho_2 > \rho_3$ and $P_1 > P_2 > P_3$

Unstable: $\rho_s < \rho_o \therefore \rho_1 < \rho_2 < \rho_3$ and $P_1 < P_2 < P_3$

Figure 5. Six principal mechanisms of salt tectonics. All types can combine. P refers to a point or to the lithostatic pressure at that point, based on thickness and density of the overburden; ρ refers to the mean bulk density of a unit where the symbol is isolated or to the mean bulk density of a complete crustal section where the symbol is adjacent to a vertical line defining the section. A. Buoyancy halokinesis. B. Differential loading halokinesis. The salt does not need to be overlain by denser cover, as in case A. C. Gravity spreading halokinesis. An extrusive dome spreads sideways, completing the cycle of overturn. Dashed line shows hypothetical profile of extrusive salt without erosional attrition. D. Thermal convective halokinesis. The convection cell has overturned several times; dashed lines represent partly homogenized layering. E. Contraction halotectonics. Salt acts as a detachment zone for an overlying fold-and-thrust belt; lithostatic pressures can retard the shortening (stable contraction) with normal density stratification or augment it (unstable contraction) if densities are inverted. F. Extension halotectonics. Salt rollers form a decoupling zone below listric normal faults in extending cover; lithostatic pressures can retard or augment this flow, depending on whether a density inversion is absent (stable extension) or present (unstable extension), respectively.

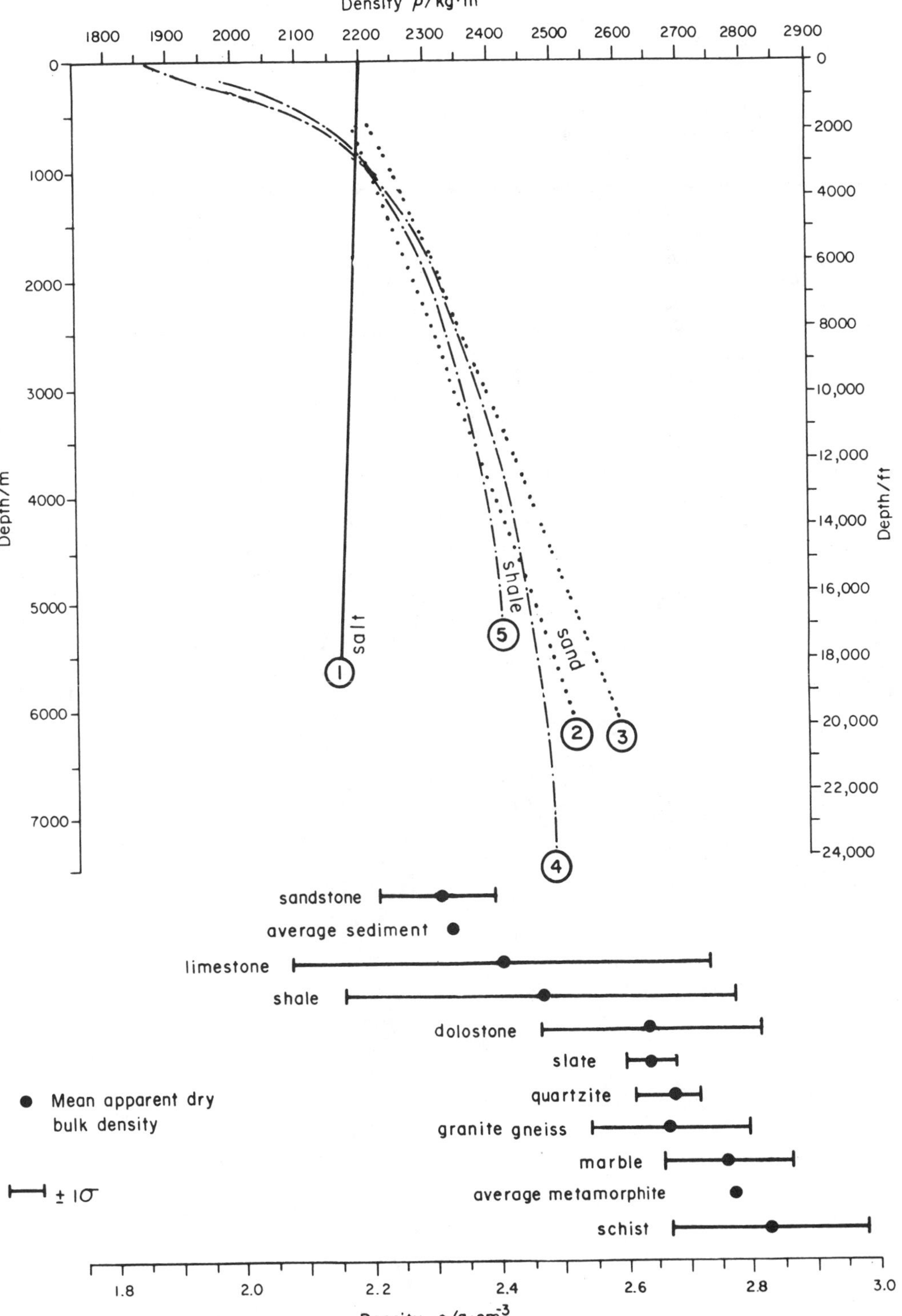

Figure 6. Relation between bulk density and depth in salt and associated terrigenous clastics in the United States Gulf Coast. 1. Rock salt with a few percent anhydrite has a density close to 2,200 kg · m^{-3}, but on burial with a geothermal gradient of 30 °C · km^{-1}, it expands by heat more than it contracts by confining pressure, thus lowering the density slightly (Clark, 1966; Gussow, 1968). 2, 3. 4,393 and 3,171 specimens of brine-filled nonmatrix and matrix lower Tertiary sandstone, calculated from porosity data of Loucks and others (1979) by equation 1 of Chapman (1974), assuming grain density of 2,650 kg · m^{-3} and brine density of 1,070 kg · m^{-3}. 4. Brine-filled Tertiary shale (Dickinson, 1953). 5. Brine-filled shale (Gardner and others, 1974). Mean apparent dry bulk densities of common sedimentary and metamorphic rocks (impermeable mass/bulk volume) ± 1 standard deviation. Sample size of individual rock types averages 184 (Touloukian and others, 1981).

Germany (Sannemann, 1968, 1983). Each generation started at different times—210 Ma, 180 Ma, and 140 Ma—and ended at different times, growth having partly overlapped to produce imbricated withdrawal basins. Other natural examples are scarce, but diapir families are a feature of many experiments (Parker and McDowell, 1955; Heye, 1978, 1979; Hunsche, 1978; Woidt, 1978, 1980). Diapir families are initiated wherever perturbations are strong and localized. A local disturbance triggers successive generations migrating outward in concentric rings. Alternatively, an abrupt lateral boundary can trigger successively weaker generations inward to the basin center across a source layer of constant thickness (Hunsche, 1978).

Density inversions are common but not ubiquitous wherever salt (ρ = 2,200 kg · m^{-3} with anhydrite impurities) is buried. All terrigenous clastics have to be buried before they compact, dehydrate, or cement to densities in excess of salt. The critical depth at which a density inversion between the salt and its cover becomes established varies from basin to basin. Average United States Gulf Coast Tertiary shales and sands saturated with brine must be buried deeper than 800–900 m and 450–650 m, respectively, before their densities exceed that of salt (Fig. 6). At such depths, overpressuring (λ >0.46, where λ is the ratio of pore pressure to lithostatic pressure) is absent or slight. In the thick Cenozoic shale sequences of South Texas, overpressure begins only at depths of 1,430 m (Berg and Habeck, 1982). Here, shale overpressured to λ = 0.9 requires burial to 2,100–3,000 m to compact more densely than salt. Gypsum (ρ = 2,300 kg · m^{-3}) generally dehydrates to anhydrite (ρ = 2,960 kg · m^{-3}) below depths of ~800 m, assuming an open system with respect to fluids (Braitsch, 1971). Well-cemented carbonates (mean ρ = 2,600 to 2,800 kg · m^{-3} (Clark, 1966; Touloukian and others, 1981) are denser than salt, even at the surface.

A density inversion provides the instability necessary for salt flow but does not guarantee that a salt pillow will form. For now, we shall ignore mechanical properties of the cover that may prevent the rise of salt beneath it and into it. Inherent in the rheology of rock salt itself, there are two factors that may retard or prevent growth of a salt structure. First, a yield point in salt, if present, would require a threshold of differential stress to be reached before salt could flow plastically. Second, the power-law nature of salt flow means that at very low stresses and strain rates, the effective viscosity of salt is extremely high, perhaps too high to allow the growth of a salt diapir in a geologically reasonable time.

Opinions are divided as to whether salt has a true yield point. Confusion derives partly from the phenomenon of work hardening and partly from the difficulty of interpreting stress-strain curves. Baar (1977) quoted a wide range of yield points measured in experimentally deformed salt ranging from 4 to 23 MPa (1 MPa = 10 bars). He pointed out that most of these values are not true yield points but have been artificially elevated by repeated loading, which results in work hardening. Borchert and Muir (1964) and Baar (1977) believed that the true yield point of rock salt is ~ 1 MPa or less. Others stated in general terms that a yield point exists (Fischer, 1984; Hardy and others, 1984).

To identify a yield point, one must subjectively interpret a stress-strain curve where it is generally very steep and asymptotic to the stress axis. In some cases, all that can be deduced is a technical yield point, arbitrarily defined as the stress at which permanent extension is 2% (Langer, 1984). Even minute creep strains below the technical yield point become important, however, if cumulative over geologic time, so that the concept of a technical yield point is spurious in tectonics. Enough stress-strain curves (for example, Gera, 1972; Heard, 1972; Farmer and Gilbert, 1984; Hansen and Carter, 1984; Hansen and others, 1984; Horseman and

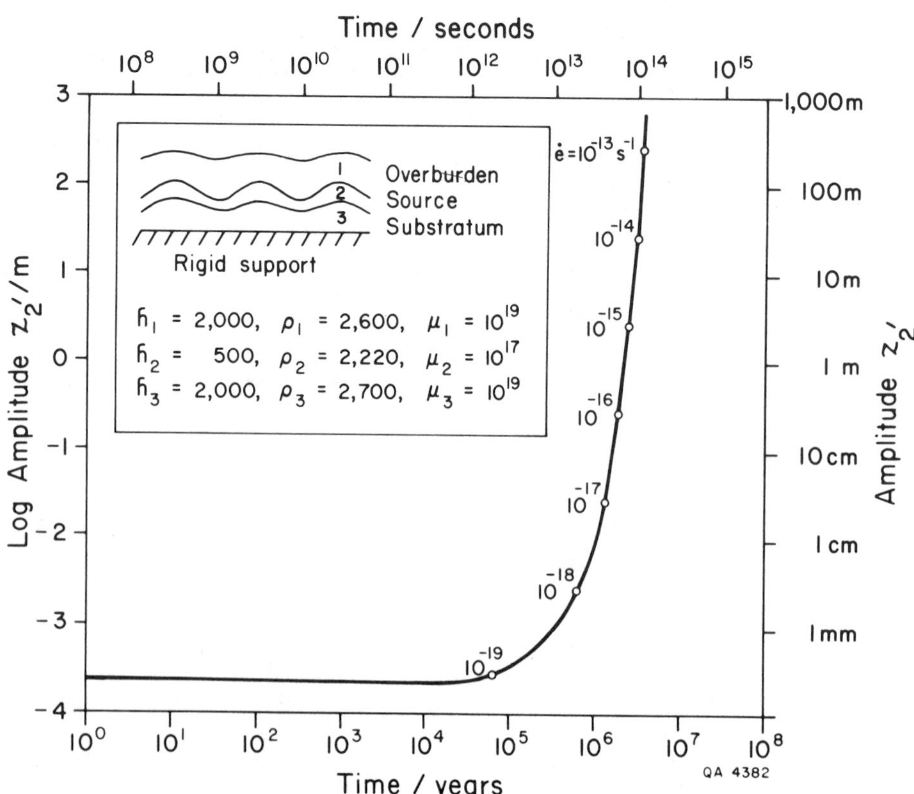

Figure 7. Accelerating strain rates (ė) during growth of a viscous pillow numerically modeled by Ramberg (1981, his Fig. 7.20). Inset shows model geometry, layer thickness (h in metres) density (ρ in kg · m^{-3}), and viscosity (μ in Pa · s).

Passaris, 1984) show no evidence of a yield point for us to believe that no minimum stress is required for salt flow, probably on a human time scale and certainly on a geologic time scale.

After disposing of the problem of yield point, we turn to the flow behavior of salt at very low stresses and strain rates. The following equations are relevant to the early growth stages of a buoyant mass like a salt pillow:

$$\Delta\sigma = (z_2 \cdot \rho_1 \cdot g \cdot 10^{-6}) - (z_2 \cdot \rho_2 \cdot g \cdot 10^{-6}) \quad (1)$$

$$z_2 = \Delta\sigma/(\Delta\rho \cdot g \cdot 10^{-6}) \quad (2)$$

$$\dot{e} = (\Delta\sigma \cdot 10^6)/\mu \quad (3)$$

where $\Delta\sigma$ = differential stress (MPa); z'_2 = relief of buoyant source layer after growth of pillow (m); z_2 = original relief of buoyant layer before growth (m); h_1 = original thickness of cover (m); h_2 = original thickness of buoyant layer (m); h_3 = original thickness of substratum below buoyant layer (m); ρ_1 = density of cover ($kg \cdot m^{-3}$); ρ_2 = density of buoyant layer ($kg \cdot m^{-3}$); ρ_3 = density of substratum ($kg \cdot m^{-3}$); $\Delta\rho = \rho_1 - \rho_2$; $\dot{e}$ = steady-state conventional strain rate (s^{-1}); μ = dynamic viscosity or effective dynamic viscosity ($Pa \cdot s$); g = free-fall acceleration due to gravity = $9.81\ m \cdot s^{-2}$.

What is a geologically realistic strain rate for a growing pillow? There is no fixed strain rate for a dome. The data for salt domes in Figure 4 show that active salt structures have vertical strain rates between 10^{-14} and $10^{-16}\ s^{-1}$. The threshold of movement was arbitrarily taken at $\sim 10^{-17}\ s^{-1}$. Is it possible to determine the lowest strain rate required to produce a salt structure in, say, 20 Ma, the active growth span of most United States Gulf Coast pillows?

Let us examine the growth history of a pillow in the numerical model of Ramberg (1981, his Fig. 7.20). The buoyant layer is a Newtonian fluid with constant viscosity of $10^{17}\ Pa \cdot s$ and originally 500 m thick. Figure 7 shows the growth history of the rising pillow, based on Ramberg's data and equations. Over 4 Ma the structure grows from an initial relief of 0.22 mm to a final relief of 600 m (beyond which the analysis becomes unrealistic). Growth is explosive after 0.1 Ma, and the structure reaches geologic dimensions (amplitude ~30 m) after 3 Ma. This growth is fast because of the high density contrast ($380\ kg \cdot m^{-3}$) chosen for the model.

For selected round-number strain rates, the corresponding differential stress can be calculated by equation 3, knowing the viscosity. These stresses can be converted to amplitudes using equation 2. This enables the round-number strain rates to be plotted on the growth curve shown in Figure 7. The pillow reaches a strain rate of $10^{-15}\ s^{-1}$ after 2.3 Ma or 1.7 Ma before it reaches 600 m high. Clearly, a pillow a few metres high must grow with strain rates of at least $10^{-15}\ s^{-1}$ to attain an amplitude of 600 m within 1.7 Ma. To grow this large in 17 Ma (a typical Gulf Coast pillow requires ~ 20 Ma to reach the diapir stage of growth), a strain rate of at least $10^{-16}\ s^{-1}$ is necessary. This strain rate is minimal, because this model assumes a flexible base to the source layer, which aids amplification.

We can estimate the actual behavior of salt—a non-Newtonian fluid—in a growing pillow on the basis of a current flow law for dry polycrystalline rock salt. Equations 19 and 9 of Carter and Hansen (1983) and Hansen and Carter (1984), respectively, reduce to the following for a temperature of 100 °C:

$$\dot{e} = (3.8 \times 10^{-13})\ \Delta\sigma^{4.5}. \quad (4)$$

The differential stresses, strain rates, and equivalent viscosities for salt pillows of various amplitudes were calculated using equations 2, 4, and 3, respectively. The results shown in Figure 8 are based on a density contrast of $100\ kg \cdot m^{-3}$, corresponding to a Gulf Coast depth of 1,640 to 2,250 m (Fig. 6). Seismic-reflection evidence indicates that many Gulf Coast salt pillows were initiated early in their burial history beneath overburdens considerably less than 1,000 m, perhaps less than 600 m (for example, Rosenkrans and Marr, 1967; Hughes, 1968; Jackson, 1982). Some of this cover consisted of Upper Jurassic carbonates, which are expected to be denser than the terrigenous clastics shown in Figure 6, and so a density contrast of $100\ kg \cdot m^{-3}$ seems reasonable.

Figure 8 shows that with low relief of the salt pillow, strain rates are extraordinarly low and equivalent viscosities are very high when compared with those of an equivalent Newtonian fluid. Figure 7 showed that a minimum strain rate of $10^{-16}\ s^{-1}$ is necessary to form a salt pillow 600 m high within 17 Ma from an initial relief of a few metres. Figure 8 shows that this minimum strain rate is attainable when relief is at least 163 m. Any smaller pillow will continue accelerating but will rise too slowly to form a pillow of typical dimensions in the normal time frame for the Gulf Coast. On the basis of this flow law and the density contrast quoted, buoyancy is thus ineffective in pillow growth unless the pillow is at least 150 m high (or 80 m with twice the density contrast).

This minimum amplitude could be reduced if the salt were dampened by formation waters. At strain rates less than $\sim 10^{-8}\ s^{-1}$, wet salt appears to deform by diffusion processes, rather than by crystal plasticity, which lower the stress exponent from 4.5 to as low as 1.5 (Spiers and others, 1984). Accordingly for a given stress and pillow amplitude, the viscosity would be lower, and the strain rate would be higher than shown in Figure 8. Only when the appropriate deformation mechanism for salt at geologic strain rates has been agreed upon can we determine the true minimum relief for buoyancy to be effective in forming a salt dome. As long, however, as the power-law creep described in equation 4 is accepted as realistic for natural salt at geologic strain rates, we must also accept that buoyancy

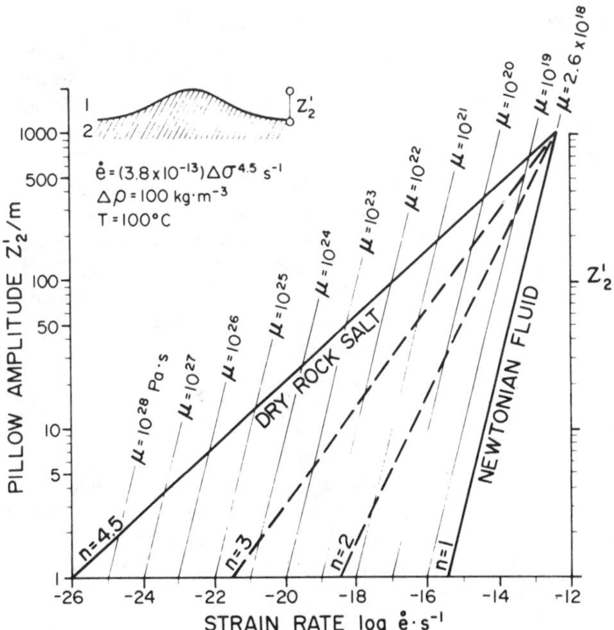

Figure 8. Relation between pillow amplitude (or relief on top of salt), strain rate, and effective viscosity (μ) for dry rock salt, based on the flow law (Carter and Hansen, 1983), density contrast, and temperature shown. Relations for equivalent model fluids with other stress exponents (n) are given for comparison.

can become effective only after 150 m of relief in the salt upper surface has already been formed by some other mechanism.

We have so far entirely ignored the mechanical constraints imposed by an overburden of real (non-Newtonian) sediments. We have insufficient data to evaluate whether a partly consolidated overburden has a yield point sufficiently high to prevent further growth of a low-amplitude salt pillow. If, furthermore, the cover is of higher effective viscosity than is the salt source layer, it, not the source layer, will control the rate of pillow growth. Salt strain rates will then be lower than those in Figure 8, and even greater relief on the top of salt will be necessary before buoyancy is effective.

Buoyancy becomes increasingly powerful as the pillows grow in amplitude and as a thickening cover gradually compacts to greater densities. The logic presented above, however, suggests that buoyancy alone is unlikely to initiate the growth of salt pillows. To initiate a pillow, we must either call upon special original relief of salt bodies or, as seems more reasonable, examine the role of other mechanisms of salt flow. This leads to the concept of differential loading, which is a far more efficient machine for near-surface, gravity-driven salt flow than is the simple density inversion of horizontal layers with laterally uniform properties.

Differential Loading Halokinesis

Material (Talbot, 1977) and numerical (for example, Woidt, 1980) models show that as long as a dome is not affected by boundary conditions, salt domes rising from a horizontal source layer through a thicker overburden that is laterally uniform will have high symmetry about a vertical axis or plane. In practice, however, the more that is known about individual salt structures, the less often they are found to be either symmetrical or vertical. Such structures are probably the result of differential loading. Centrifuge experiments have demonstrated that any asymmetry about the vertical in the starting configuration of a Rayleigh-Taylor instability (that is, differential loading built in from the start of the experiment) is exaggerated during overturn, so that diapirs move up and over the least stable sector (Talbot, 1977). On the basis of centrifuge experiments, asymmetry in natural salt structures therefore suggests that differential loading was probably involved in their growth.

Lateral variations in thicknesses (including surface slopes), densities, strengths, or strain rates in a layer of salt or its cover can all result in differential loading of the salt. These factors are commonly related because of sedimentary facies changes, and so they often work in concert. Examples are prograding deltas—as originally proposed by Rettger (1935) and shown in Figure 5B—alluvial or turbiditic fans, coral reefs, oolitic or clastic shoals, volcanic accumulations, and thrust and ice sheets.

Surface relief is particularly efficient in initiating flow in a soft substrate, because the density contrast between unconsolidated sediments and either air or water is greater than between most rock types (R. O. Kehle, 1983, personal commun.). At densities of 2,000 $kg \cdot m^{-3}$ for unconsolidated deltaic sediments and 1,000 $kg \cdot m^{-3}$ for sea water, equation 2 and Figure 8 indicate that a bathymetric relief of only 16 m in the delta provides enough differential stress to drive flow in a tabular salt layer below. This relief is only one-tenth of that required by buoyancy (Fig. 8). A salt pillow having an amplitude of only 16 m, growing by buoyancy alone, would grow at strain rates of $\sim 10^{-21}$ s^{-1}, or 10^5 times as slowly as by differential loading (Fig. 8).

The effects of lateral variations in loading diminish with depth if the salt is overlain by weak, plastic sediments like overpressured shale, the lateral flow of which to positions of lower gravity potential, can reduce the pressure gradients set up by differential loading. If, however, the overlying weak layer is compartmentalized by growth faults or rising salt structures, lateral flow is restricted. Differential loading is accentuated with time as the compartment continues to deepen with respect to its surroundings. Structural configurations and shale-sand rheologies are highly variable, and loading, growth, density, and viscosity are time dependent. We thus cannot quantify the effectiveness of differential loading in causing salt flow beneath overburdens of varying thickness. Differential loading can initiate halokinesis even where the overburden is less dense than salt. This is not to claim that mature salt domes develop solely because of differential loading by less dense cover—the process provides the initial salt relief required for buoyancy to subsequently amplify.

As they rise toward the surface, salt structures can influence the depositional facies of their own overburden. Depocenters thus can become established over their own areas of salt withdrawal, reinforcing the supply of salt to the relatively rising structure. Differential loading can be perpetuated wherever sedimentary facies stack vertically, such as in deltas on rapidly subsiding basin margins like the Gulf Coast; this type of stacking is further accentuated by the uplap associated with growth faulting (Brown and Fisher, 1982, Figs. I-70 and II-19).

Gravity Spreading Halokinesis

Halokinesis is traditionally envisaged as salt moving up and through denser lithologies, rather than sinking below less dense rocks, sediments, water, or air. Gravity, however, can also dissipate relief in the top of any salt body that is above its level of neutral buoyancy—whether this is above or below the surface. The primary relief in the salt illustrated in Figure 5C thus will tend to dissipate if the overburden was less dense than salt and if any yield strengths are exceeded by lateral differential stress set up by the surface slope of the salt body.

Like any other soft material, salt on the surface can become unstable and flow by gravity spreading down slopes as low as 3° (Wenkert, 1979). In Iran, salt is driven above its level of neutral buoyancy, forming extrusive domes that spread under their own weight. Even the carbonate-rich surface rocks there are denser than salt, and so the resulting salt fountains are subaerial.

Thermal Convective Halokinesis

Clean, dry rock salt has unusual thermal properties, for it is relatively translucent to energy radiating in the visible-to-radar electromagnetic spectrum and also has a high thermal conductivity compared with most other crystalline rocks. In a thick, exposed, tabular body of salt, two thermal regimes are likely. A shallow zone heated by the sun is subject to diurnal ($\leqslant 1$ m deep), seasonal (< 10 m deep), and secular temperature fluctuations. A deeper zone is continually warmed by the Earth's heat flow. Rock salt also has a high thermal expansivity (and is relatively incompressible under confining pressure), and so both sources of heat induce thermal gradients that result in density gradients. The short-term density gradients induced by the Sun's heat in the shallow levels are mechanically stable. In contrast, the long-term density gradients due to the much lower heat flow from the Earth are mechanically unstable, because light, hot salt is overlain by dense, cold salt. This could result in thermal convection, which increases the rate of heat flow upward through the salt. The thermal gradient, not the absolute temperature, is the destabilizing agent combined with gravity.

Crude calculations (along the lines of Talbot, 1978) taking monomineralic salt to be a simple Newtonian fluid with a kinematic viscosity of 10^{13} $m^2 \cdot s^{-1}$ (dynamic viscosity = $2.2 \cdot 10^{16}$ $Pa \cdot s$, after Biot and Odé, 1965) suggest that salt layers thicker than 4,500 m could be undergoing cyclic convection in regions with a thermal gradient as low as 30 °C km^{-1}. Carter and Hansen's (1983) equation predicts that such a low viscosity requires strain rates on the order of 10^{-11} s^{-1} even at several hundred degrees

Celsius in dry rock salt. Such high rates have been recorded only in damp salt (Table 1), and simple thermal convection seems possible in nature only in water-softened salt.

The geometry of convection due to secondary, thermally induced (Rayleigh-Bénard) density inversions appears to be broadly similar to that due to primary compositional (Rayleigh-Taylor) instabilities. Major differences are that thermally induced motions are normally considered to be within a layer of uniform composition and that, rather than being single overturns, they are cyclic as long as the conditions are suitable (Fig. 5D). Unlike halokinesis due to compositional density inversions, that arising from thermally induced density inversions could theoretically operate without any overburden to the salt. Wholesale cyclic convection may be possible in a tabular salt body beneath either a rigid roof or a free surface of the sea or land.

Most rising salt structures must act as thermal plumes to some extent. The tendency for salt to cool at shallow levels and to sink may contribute to the circulation within bulbs that we discuss later, but in general this circulation is unnecessary for efficient heat transfer. Salt has two to three times the thermal conductivity of its typical cover, so that when salt structures penetrate a thermally insulating cover, they can leak the subsalt heat flow by conduction without further mass movement. The regional heat flow reaching the salt from below thus tends to be preferentially channeled up through the sedimentary pile via the salt diapirs. This is why the thermal gradients measured in salt structures are usually much lower than in the surrounding country rocks, even though their absolute temperatures are commonly higher (Selig and Wallick, 1966; Law Engineering Testing Company, 1983). O'Brien and Lerche (1984) suggested that such thermal conduction diminishes the general hydrocarbon maturity index near the roots of salt diapirs and could increase it above them; however, the extra heat flow conducted to diapir overburdens can be either dissipated or amplified by hot (~140 °C) fluid flow convecting up through the permeable cover in the fractured aureole around the intrusion (Keen, 1983).

Richter-Bernburg (1980) listed six tabular salt sequences in Germany, North America, and Brazil in which isoclinal recumbent folds are clearly visible in mines or drill cores (see his Figs. 4 through 7). Such structures commonly involve potash minerals and have usually been interpreted as soft-sediment slumping. Another tabular body with intense internal structures can be added to the list, however: that at Boulby potash mine, the deepest mine in England. Here, various lines of evidence can be used to argue that a multiwavelength cellular movement pattern within part of the Zechstein sequence was due not to slumping but to thermal convection during burial to 2 to 3 km (Talbot and others, 1982). These complex structures record various stages of thermal convection of a sylvinite-shale multilayer up to the stage where several cycles of overturn dispersed the shales as microscopic clasts in a uniform gneissose sylvinite. A quantified thermal convective model was used to predict an unexpectedly low viscosity for damp sylvinite (10^7–10^8 Pa · s in the temperature range 170–300 °C at confining pressures of 50–72.5 MPa), but, as far as we know, these figures have not yet been tested in the laboratory.

Thermal convection in salt is still only a hypothesis, not yet convincingly demonstrated in nature. Nevertheless, because heat can soften rocks and thermal gradients can accentuate primary density inversions, the regional heat flow is, at the very least, an extra, mechanically destabilizing factor that is largely overlooked in basins containing salt.

Summary of Halokinesis Mechanisms

Buoyancy is ineffective in raising salt below terrigenous clastic cover less than ~700 m thick, or at whatever depth sediments in a particular basin compact to a density greater than that of the salt sequence. Buoyancy can be effective below thinner cover if the cover contains appreciable thicknesses of well-cemented carbonates, certain evaporites and other chemical sediments like iron formation, or volcanics if the salt can deform such cover. Current flow laws for steady-state creep of salt suggest that buoyancy can amplify only pre-existing relief; it is effective only when a relief of tens to hundreds of metres is present in the upper surface of the salt unit, depending on the cover density. This relief may conceivably be primary but is more likely to be induced by other mechanisms of halokinesis or by tectonic faulting or folding.

Differential loading is probably the main mechanism when salt is buried less than a few hundred metres. Its effectiveness is enhanced and can be perpetuated and propagated to several kilometres if sedimentary facies are being vertically stacked and structurally compartmentalized, as in deltas on rapidly subsiding basin margins like the Gulf Coast. The effectiveness of differential loading is reduced if weak strata such as overpressured shale overlie the salt, for their lateral flow can reduce the lateral pressure gradients set up by differential loading. However, diapirs of pressurized shale and salt flowing as a composite source layer are probably possible.

Gravity spreading becomes the most important mechanism of salt flow once a salt dome reaches a level of neutral buoyancy or becomes extrusive. Gravity spreading also distorts salt structures formed by gravitationally stable tectonics once regional stresses decay. Gravity spreading may also drive salt masses basinward beneath a prograding continental margin.

Thermal convection is at present merely a possibility based on order-of-magnitude calculations and is likely only in thick salt bodies with steep thermal gradients if the effective viscosities are orders of magnitude lower than existing data on dry salt predict.

Bulbs and Effective Viscosity Contrast

Salt structures can potentially form swollen bulbs on their growing stems by four different mechanisms. Of these, three are associated with upper boundaries to the diapir, and the fourth might develop before the relatively rising structure reaches such a boundary (Fig. 9). Space is available for lateral spreading of the diapir bulb by concomitant sideways and downward creep of the cover into space vacated by the depleted source layer. Local topographic uplift over the diapir can also create space.

First, salt structures could spread beneath a resistant barrier (Fig. 9B). The resulting bulbs have symmetries dependent on the orientation of the barrier in relation to the direction of approach by the rising structures; other factors being equal, bulbs spread fastest parallel to the azimuth of the approach direction in a plane parallel to the barrier (Talbot, 1977). Any pendant peripheral lobes (Fig. 9C) tend to keep rising until they catch up with the rest of the retarded bulb; the lobes are squeezed flat by the pressure gradient and can rotate to recumbent attitude against the base of the bulb (P. Heeroma, unpub. data on centrifuge experiments at the University of Uppsala. The bulbs may eventually penetrate the roof as their plan areas increase (Bishop, 1978) or the overburden thickness increases, compacting the overburden and increasing the buoyancy pressure.

Second, the rising salt structures can spread sideways at a much more subtle barrier, that of neutral buoyancy. If they mature to the stage where they can lift unconsolidated, less dense cover (like bulbous dikes feeding sills), salt intrusions will spread sideways (Fig. 9D). At the neutral level, the density of the cover (ρ_o) equals that of the salt (ρ_s). Negative buoyancy above the neutral level causes spreading. Where the cover remains denser than salt all the way to the surface, salt can emerge, possibly by erosion of the cover above the rising salt, and spread as an extreme, extrusive bulb with or without namakiers.

A third mechanism for the formation of salt overhangs is the subsequent burial of salt extrusions (Fig. 9E). A probable example of this type

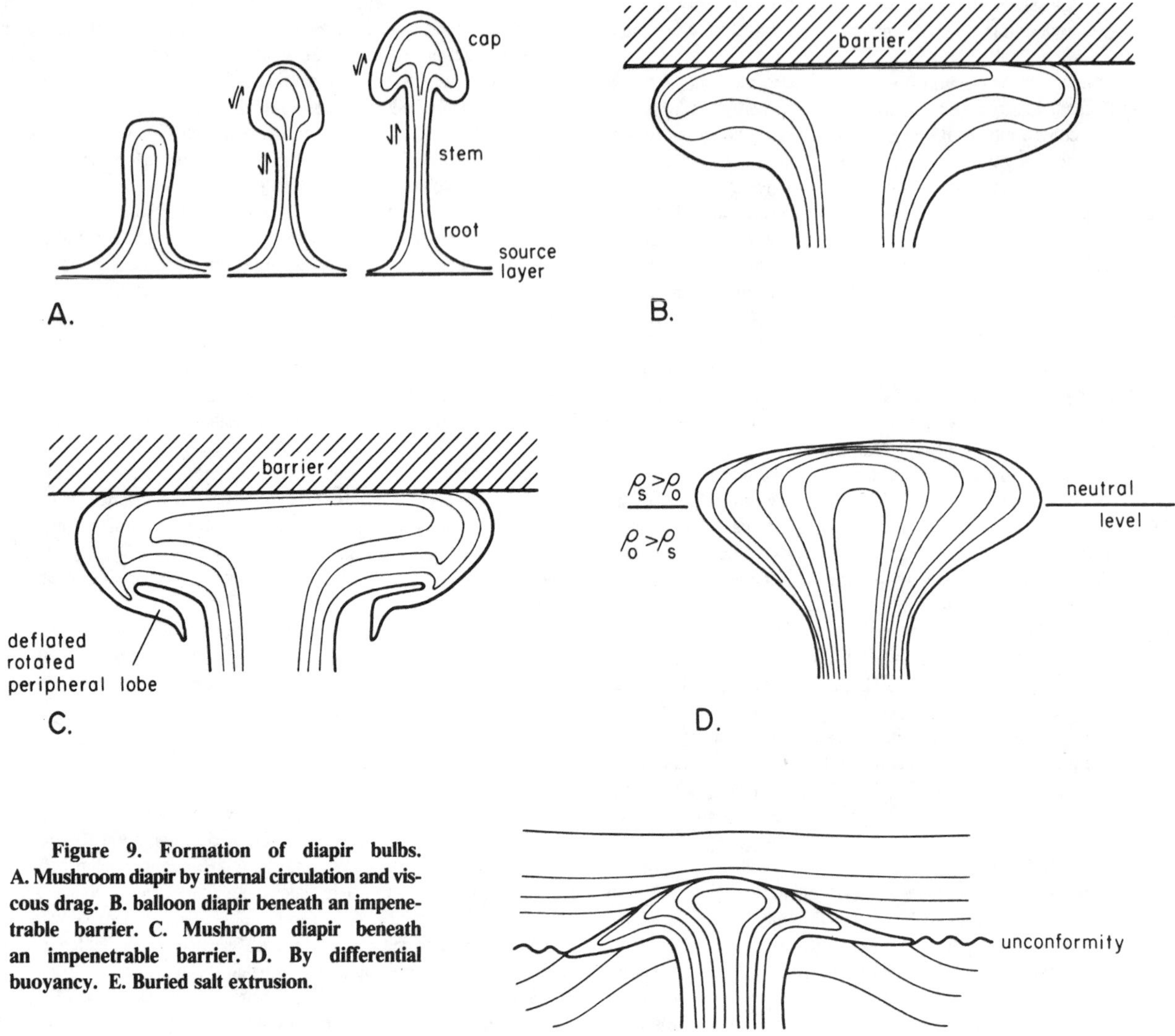

Figure 9. Formation of diapir bulbs. A. Mushroom diapir by internal circulation and viscous drag. B. balloon diapir beneath an impenetrable barrier. C. Mushroom diapir beneath an impenetrable barrier. D. By differential buoyancy. E. Buried salt extrusion.

of extreme bulb is a Triassic salt stock in the offshore Aquitaine basin, France. Its 12-km-wide crest dwarfs its 1-to-2-km-wide stem (Curnelle and Marco, 1983). Other examples proposed, such as the Wienhausen-Eicklingen Dome, West Germany (Bentz, 1949), are less convincing as buried extrusions.

A fourth possible mechanism in which the bulb forms by internal circulation and viscous drag between the sinking cover and rising salt is suggested by theoretical and model studies (Fig. 9A). The spacing and general shape of gravity structures maturing well below an upper boundary appear to depend largely on the parameter m, the ratio of the viscosity of the overburden divided by the viscosity of the source (Berner and others, 1972; Whitehead and Luther, 1975; Woidt, 1978; Heye, 1978, 1979). We recognize three morphological types, differentiated qualitatively as "types A, B, and C." Where $m << 1$, relatively widely spaced type A diapirs swell only slightly to thumblike shapes (Fig. 10A); where $m \approx 1$, a pronounced type B mushroom-shaped bulb trailing a downward-facing peripheral lobe develops because of friction between the sinking cover and the rising source (Fig. 10B). This is a mushroom diapir (pilzförmiger Diapir, diapir pédoncule). Where $m >> 1$, large type C blobs trailing only tenuous stems (like balloons on strings) rise from pronounced withdrawal basins in the deep source layer (Fig. 10C). Experimental evidence suggests that a detached blob rising through a non-Newtonian overburden with a shear-thinning viscosity will form a pointed tail on its lower side (Van Dyke, 1982, his Figs. 182 and 183).

Salt is one of the most intensively studied materials in rock mechanics. Estimates of the effective viscosity of dry rock salt nonetheless currently range from 10^{14} to 10^{20} Pa · s (LeCompte, 1965; Odé, 1968; Heard, 1972; Carter and Hansen, 1983). The effects of water on salt rheology have not yet been adequately explored by experiment, but because water lowers the viscosity of halite, this factor would increase the probability of type C diapirs. Unfortunately, we are even more ignorant of the effective viscosities of most other rocks. For this reason, the possibility of the effective viscosities of some other rocks being less than that of salt cannot be ruled out. Many unconsolidated and overpressured sediments certainly have effective viscosities less than that of crystalline rock salt; however, beds of nearly all other lithologies (other than those rich in

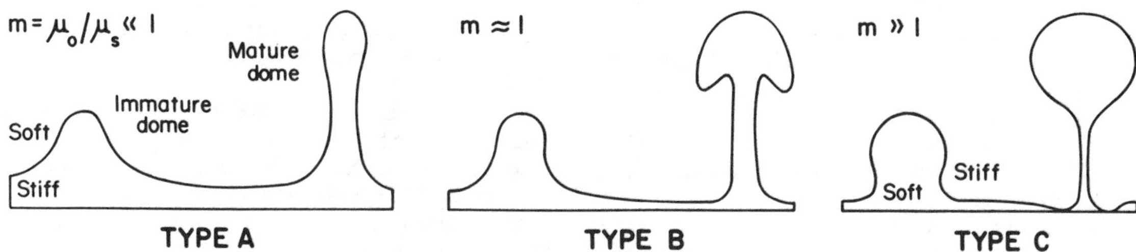

Figure 10. Schematic effects of viscosity contrast on shapes of immature and mature domes based on material and numerical modeling. Overburden viscosity is μ_0; source-layer viscosity is μ_s. Immature domes show subtle but recognizable differences that magnify with maturity.

potash minerals and bischofite) buckle or display boudinage when incorporated in deforming salt, indicating that, by the time they are deformed together, rock salt has an effective viscosity lower than that of its country rocks. Scaling calculations by Dobrin (1941) and Hunsche (1978), based on experimental diapirs in viscous fluids, furthermore indicate that United States Gulf Coast and north German country rocks have viscosities on the order of 10^{20} Pa · s and 10^{19} Pa · s, respectively. Our own unpublished scaling calculations based on centrifuge experiments suggest that the sediments around salt domes in the East Texas Basin and offshore United States Gulf Coast have viscosities of 3×10^{20} and 6×10^{19} Pa · s, respectively.

As a working hypothesis, then, we assume that the effective viscosity of nonevaporites is similar to or higher than that of salt. Accordingly on theoretical grounds, we should expect diapir bulbs like those of type B or C. A common criticism of models that involve viscous behavior of the cover is that fracturing of natural cover may affect the shape of the ductile intrusion. This is true in the immediate vicinity of the fault (for salt dikes are exposed in the Danakil depression, Ethiopia; Holwerda and Hutchinson, 1968). The concept of brittle versus ductile behavior, however, is scale dependent (ductile creep can be resolved into discontinuities on the intragranular scale). Even in experimental models incorporating brittle cover (for example, Parker and McDowell, 1955), the over-all shape of the diapir conforms to one of the three types we have designated A, B, and C. However important faults are in initiating pillows or diapirs, it appears that the over-all behavior of salt masses and their cover is approximately viscous on a tectonic time scale (Carey, 1954). This is evidenced by the rounded forms of salt stocks, suggesting that their form is controlled primarily by flow, not fracturing of the cover (Nettleton, 1934). The relation between dome shape and viscosity, moreover, does not appear to be affected by whether the modeled dome growth is syndepositional (the process of downbuilding proposed by Barton, 1933) or, as more commonly modeled, postdepositional (Jackson and Talbot, 1985).

For these reasons, we believe that the experimental and theoretical relations between dome shape and viscosity also apply to natural salt diapirs.

Our perception of the shape of salt structures is based largely on seismic-reflection profiles, gravity profiles, and drilling data. Apart from the shallow crest, the contact of a mature salt structure is not clearly defined by routine seismic-reflection profiling, for it is masked by seismic noise and curved diffractions from its steep sides (Tucker and Yorston, 1973, example 13). A drill passing through the overhang of a bulb is rarely sufficiently deflected inward to reach any stem beneath the overhang; thus the broad overhangs drilled through, for example, in Bethel Dome (Halbouty, 1979) and San Felipe Dome (Halbouty and Hardin, 1954), may be much broader than generally supposed because this drilling limits only the maximum width of the stem. Most of the well-explored salt domes in South Louisiana are not mature enough to have formed bulbs (New Orleans Geological Society, 1960, 1962, 1983). Even some of these apparently simple conical forms, however, disguise overhangs that require determined drilling through 1–2 km of salt (in the case of Calcasieu Lake Dome) to prove their existence.

The gross shape of mature salt structures below any bulbs therefore will remain largely speculative until the results of recent improvements in horizontal seismic-reflection profiling and mapping are published and methods such as vertical seismic-reflection profiling, downhole seismic refraction, radar probing (Holser and others, 1972), and magnetotellurics (Losecke, 1972) are applied in earnest. Slender stems have already been predicted by gravity modeling (Fig. 2D). German geologists, who have had access to data accrued during a century of intensive mining in deformed evaporites, have long tended to depict salt structures with broad bulbs and narrow stems (Trusheim, 1960; Sannemann, 1968).

If salt diapirs actually do have larger overhangs and more slender stems than is generally envisaged, substantial hydrocarbon reserves may still await discovery in overhang traps.

Salt Flow within the Tectonic Framework

The role of lateral tectonic forces in the flow of salt has been debated as long as salt domes have been studied. In the early part of this century, lateral squeezing of the crust was held to be the engine for all salt intrusion, on the basis of diapirs originally described in the Romanian Carpathians (Mrazec, 1907). Subsequent gravity measurements over many United States Gulf Coast salt domes provided the evidence for Nettleton's (1934, 1943) treatise, which firmly established the role of buoyancy in driving diapiric growth. In a third stage of conceptual evolution, we are beginning to realize the importance of crustal extension in initiating gravity-driven rise of salt structures in petroleum basins such as the North Sea (for example, Jenyon, 1985).

Recognizing that gravity alone could initiate and amplify salt structures, Trusheim (1957, 1960) defined this process of gravity-driven "autonomous" salt flow as "halokinesis." He defined salt movement driven mainly by lateral compression as "halotectonics." As defined, these terms are unsatisfactory on several counts. First, the Greek roots of halokinesis (*hals* and *kinein*) merely signify salt movement without reference to the driving force. On etymological grounds, halokinesis could be applied to all types of salt flow. Second, halotectonics was defined with reference only to lateral contraction of the crust, thus omitting the probably more common type of salt basin characterized by lateral extension. Third, the two terms halokinesis and halotectonics are too vaguely defined to determine at what stage one process grades into the other.

Despite these limitations, the term "halokinesis" is now entrenched in the literature. In attempting to define the role of salt movement in tectonics, we have thus taken the conservative approach of expanding and more precisely defining Trusheim's concepts and terminology rather than introducing completely new terms.

The classification scheme in Figure 11 can be applied to all forms of salt tectonics. The x axis is kinematic, and the y axis is dynamic. Gravity-enhanced and gravity-retarded deformation (in the -y and y halves of Fig. 11) involve loss and gain, respectively, of gravity-potential energy. A diapiric system loses gravity potential by lowering the center of gravity of a denser unit and raising the center of gravity of a less dense unit (Ramberg, 1981, p. 9–27). Regional horizontal extension characterizes the x halves, and regional horizontal contraction characterizes the -x halves. Along the y axis, deformation is either nonexistent in the vertical plane or is of the strike-slip type, between the fields of transpression (horizontal convergence combined with wrench-type shear) and transtension (horizontal divergence combined with wrench-type shear) (Harland, 1971; Sanderson and Marchini, 1984). True halokinesis is restricted to the -y axis, where regional strain is absent.

Conventional lateral extension and contraction of sequences without density inversions or other gravitational instabilities characterize the (x,y) and (-x,y) quadrants. Once regional lateral stresses decline, such gravity-retarded structures remain unchanged or decay by isostasy. In terms of fluid dynamics, these structures are stable after their generating lateral forces have declined. Extension and contraction by lateral forces encouraged by gravity characterize the (x,-y) and (-x,-y) quadrants, respectively. After decline of the lateral forces, structures in these quadrants could continue to grow by the ever-present acceleration of gravity if further gravity potential can be lost. In fluid dynamics, these active, gravity-aided structures are known as "instabilities" and are therefore termed "unstable" in Figure 11.

Halotectonics as originally defined by Trusheim (1957) is restricted to quadrant (-x,y), which is here termed "stable contraction." "Salt tectonics," synonymous with halotectonics, has grown to embrace all types of salt movement, a usage we follow here.

Salt is likely to have been already expelled from metamorphic zones where crystalline basement shortens by penetrative ductile strains. The contraction of salt-bearing sedimentary sequences is therefore likely to be thin-skinned and concentrated above a décollement in the salt. The long-wavelength folds and imbricated thrusts in the unmetamorphosed sediments marginal to many orogens thus are thought to die out beneath the salt, as in the French Jura, the Spanish Pyrenees, and the Iranian Zagros Mountains (Fig. 5E). Thrust flats and sole thrusts commonly follow salt layers over huge areas, as in the Zagros, where only a few of the many earthquakes originate beneath the salt layers or have structural expression as surface faults (Berberian, 1981). There, the still-active Alpine folding and thrusting that started 15 Ma ago as Arabia collided with central Iran (Stöcklin, 1968; Berberian, 1981) were superposed on an older pattern of gravity-driven salt structures (Kent, 1979). The original salt diapirs of Mrazec (1907) are composed of Miocene salt deformed in the Alpine fold-and-thrust zones of the Carpathian foredeep in Romania (Burchfiel and Royden, 1982).

Salt is also an important tectonic component in regions of deep-seated crustal extension. An example is the North Sea, beneath which a

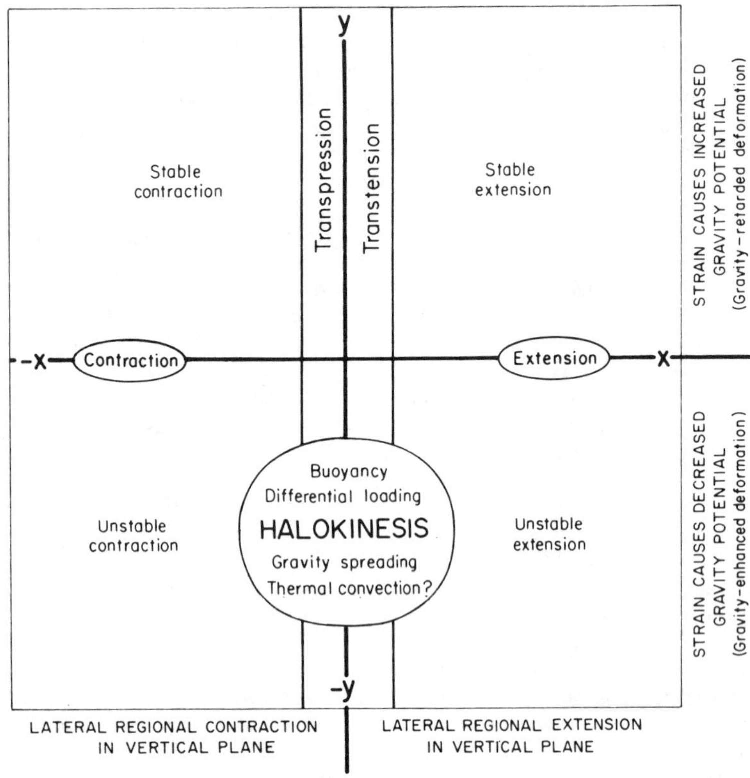

Figure 11. A classification of salt tectonics (halotectonics) based on the type of regional lateral strain and on increase or decrease of gravity potential during deformation. The -y axis is the domain of halokinesis. Other axes and fields involve regional extension or contraction.

triple junction rifted a Permian sag in Pangea and nearly developed into a spreading ocean basin in the Triassic and Jurassic (Ziegler, 1982). The Triassic extension also reactivated old faults in the pre-Carboniferous Caledonian basement. These faults influenced the trend of salt structures already growing in the Permian Zechstein and almost certainly initiated more salt structures. Later, near the Cretaceous-Tertiary time boundary, major transcurrent faults were reactivated in the basement and propagated upward through the cover to be modified and partially absorbed by pronounced flowage within the salt (Talbot and others, 1982).

Some salt flow may be initiated by lateral forces. For example, in the margins of United States Gulf Coast basins, brittle extension of the postsalt Jurassic carbonate cover created asymmetric salt rollers (Bally, 1981) that may have seeded larger salt structures where the salt was thick enough (Fig. 5F). The buoyant rise of the Hormuz Salt in the Persian Gulf region was delayed until its largely carbonate cover was ~5 km thick ~180 Ma ago, some 450 Ma after deposition of the salt (Kent, 1979). Extrusion of these salt stocks began only ~120 Ma ago (Kent, 1979). A strong density contrast was clearly available, and so this delay must have been caused by other factors. One of these could be strength. Carbonates are relatively strong at laboratory strain rates to depths of at least 10 km if water-saturated (Handin and others, 1963). This strength could have resisted the buoyancy stresses set up by low-amplitude salt mounds until the system was destabilized by extensional halotectonics (Kent, 1979). Between 200 and 140 Ma ago, the Zagros sedimentary pile was a passive margin subsiding and extending as shelf carbonates accumulated (National Iranian Oil Company, 1977; Berberian and King, 1981). Normal faulting and perhaps gravity-glide folding may have accentuated the relief in the upper surface of the Hormuz Salt, increasing buoyancy stresses enough to pierce the carbonate overburden fractured in the same tectonic episode.

Whether the northeast grain of the salt structures in and beneath the Great Kavir in central Iran is due to lateral forces between the flanking Elburz and Zagros Mountains is unclear. This basin may have originated as a back-arc basin floored by stretched continental crust (Bally and Snelson, 1980; Berberian and King, 1981). Trapped in the megasuture between the colliding Arabian, Indian, and Eurasian plates, central Iran is being laterally compressed by means of folding and mountain-bordering reverse faults (Berberian, 1981); domes growing in strata under lateral compression are expected to be elongated normal to the axis of maximum compressive stress. On the other hand, linear grains can occur even in simple halokinesis because of slopes, thickness variations, or basement-fault reactivation.

Sedimentary sequences already overturning because of gravity are certainly vulnerable to modification by later lateral forces. The effects of gravity-driven deformations, however, are so dominant on salt that gravity is always a significant factor at map scale (>1 km), regardless of the tectonic setting. Complications such as salt acting as a decoupling zone for both cover and substrate during combinations of compressive, tensile, and transcurrent shear stresses can be expected before, during, or after any gravity-driven salt deformation.

SYNTHESIS

The different aspects of salt tectonics that we have discussed are all part of a geochemical cycle in which salt crystallizes in beds, which may then mature as diapirs that ultimately return to solution. A tabular salt body in an evaporite basin is sufficiently mobile to flow. This flow is powered by gravitational potential energy released by the change of internal geometry as a rock assemblage is rearranged into a more stable configuration. Tabular salt bodies transform into the shapes described here as anticlines, pillows, rollers, walls, stocks, nappes, massifs, or namakiers.

A water-softened, thick salt layer exposed to a moderately high geothermal gradient typical of the early stages of basin formation by crustal stretching can theoretically become unstable without any overburden. A heat-induced density inversion can potentially lead to internal circulation and stirring of a still-tabular salt body. Complex internal recumbent folds in tabular salt bodies may be due to soft-sediment slumping or metamorphic volume changes, but they may also record convective stirring.

Salt must be buried beneath other rocks or sediments to form diapirs. Even with a primary density inversion, a tabular, horizontal salt body overlain by a uniform, dense overburden is stable if the relief on the interface between salt and cover is insufficient to induce flow at geologic strain rates. Natural salt is rarely overlain by such uniform overburden, however, for lateral variations in thickness, density, yield strength, and effective viscosity abound.

Salt structures are commonly asymmetric and do not necessarily rise vertically. They are driven by buoyancy (which requires a density inversion in the rock sequence) or differential loading (which does not) at speeds about one-tenth those of orogeny. The structures cease growing (1) as the diapir penetrates a layer of lower density, or (2) by encountering resistant overburden, or (3) by detachment of the diapir stem from its root, or (4) after effective exhaustion of the source layer. At least four mechanisms may form broad bulbs in mature diapirs, which may have stems much narrower than is commonly envisaged. Diapiric growth can be reactivated by further burial and compaction of the cover.

Halokinetic movements can be initiated, succeeded, retarded, or accelerated by the tangential forces that stretch, wrench, or compress sedimentary basins. The type of halokinesis that occurs depends on the relative timing of the imposed forces and whether or not density inversions are present.

If the overburden is denser than salt all the way to the surface, salt intrusions can extrude over the land surface or sea floor as slow fountains of salt with parabolic upper profiles. These extrusive domes collapse under their own weight and can spread in air or water as namakiers at some of the fastest known natural strain rates for crystalline rocks. As few overburdens are denser than salt right to the surface, most shallow diapirs lift their cover like ductile laccoliths and rarely extrude. Salt starts dissolving when it penetrates unsaturated aquifers, although a largely protective armor of cap rock can form. The survival of extrusive domes and namakiers requires the rate of supply of salt from depth to exceed the rate of attrition from above and can be aided by mantles of insoluble sediment.

The solution of salt at or below the surface and its transport and precipitation in either the same basin or the world ocean complete the geochemical cycle, leaving behind clogged craters, breccia chimneys, or collapsed folds in a partly evacuated and subsided sedimentary basin that long before hosted desiccating seas or saline lakes.

ACKNOWLEDGMENTS

Research was funded by the U.S. Department of Energy Office of Nuclear Waste Isolation under contract no. DE-AC97-83WM46615 (MPAJ) and by the Swedish Natural Science Research Council under contract nos. G-GU-1652-105 & 6 (CJT). We thank Steven Seni for supplying unpublished data; Reinold Cornelius, William Mullican, and Patricia Bobeck for assistance; and the following for critical review: Ruud Weijermars, Ray Fletcher, Fred Donath, Matt Werner, Peter

Murphy, Don Kupfer, William Muehlberger, Gerald O'Brien, Kirk Hanson, and Thomas Fails.

REFERENCES CITED

Ala, M. H., 1974, Salt diapirism in southern Iran: American Association of Petroleum Geologists Bulletin, v. 58, p. 1758–1770.

Arrhenius, Svante, 1912, Zur Physik der Salzlagerstätten: Meddelanden Vetenskapsakademiens Nobelinstitut, v. 2, no. 20, p. 1–25.

Baar, C. A., 1977, Applied salt-rock mechanics 1: The in-situ behavior of salt rocks: Amsterdam, Elsevier, 294 p.

Bally, A. W., 1981, Thoughts on the tectonics of folded belts, *in* McClay, K. R., and Price, N. J., eds., Thrust and nappe tectonics: Oxford, England, Geological Society of London, p. 13–32.

——— 1983, Seismic expression of structural styles—A picture and work atlas, Volume 2: American Association of Petroleum Geologists Studies in Geology Series 15, 357 p.

Bally, A. W., and Snelson, S., 1980, Realms of subsidence, *in* Miall, A. D., ed., Facts and principles of world petroleum occurrence: Canadian Society of Petroleum Geologists Memoir 6, p. 9–94.

Barton, D. C., 1933, Mechanics of formation of salt domes with special reference to Gulf Coast salt domes of Texas and Louisiana: American Association of Petroleum Geologists Bulletin, v. 17, p. 1025–1083.

Bentz, A., 1949, Ergebnisse der erdölgeologischen Erforschung Nordwestdeutschlands 1932–1947, ein Überlick, *in* Erdöl und Tektonik in Nordwestdeutschland: Amtliche Bodenforschung Hanover-Celle, p. 7–18.

Berberian, Manuel, 1981, Active faulting and tectonics of Iran, *in* Gupta, H. K., and Delany, F. M., eds., Zagros–Hindu Kush–Himalaya geodynamic evolution: American Geophysical Union Geodynamics Series, v. 3, p. 33–69.

Berberian, Manuel, and King, G.C.P., 1981, Towards a paleogeography and tectonic evolution of Iran: Canadian Journal of Earth Sciences, v. 18, no. 2, p. 210–265.

Berg, R. R., and Habeck, M. F., 1982, Abnormal pressures in the lower Vicksburg, McAllen Ranch field, South Texas: Gulf Coast Association of Geological Societies Transactions, v. 32, p. 247–253.

Berner, H., Ramberg, Hans, and Stephansson, Ove, 1972, Diapirism in theory and experiment: Tectonophysics, v. 15, p. 197–218.

Biot, M. A., and Odé, Helmer, 1965, Theory of gravity instability with variable overburden and compaction: Geophysics, v. 30, p. 213–227.

Bishop, R. S., 1978, Mechanism for emplacement of piercement diapirs: American Association of Petroleum Geologists Bulletin, v. 62, p. 1561–1583.

Borchert, Hermann, and Muir, R. O., 1964, Salt deposits: The origin, metamorphism and deformation of evaporites: London, Van Nostrand, 338 p.

Bradshaw, R. L., and McClain, W. C., 1971, Project Salt Vault: A demonstration of disposal of high-activity solidified wastes in underground salt mines: Oak Ridge, Tennessee, Oak Ridge National Laboratory, ORNL-4555.

Braitsch, O., 1971, Salt deposits: Their origin and composition: Berlin, Springer-Verlag, 297 p.

Brown, L. F., Jr., and Fisher, W. L., 1982, Seismic stratigraphic interpretation and petroleum exploration: American Association of Petroleum Geologists Continuing Education Course Note Series 16, 183 p.

Buffler, R. T., Worzel, J. L., and Watkins, J. S., 1978, Deformation and origin of the Sigsbee Scarp—Lower continental slope, northern Gulf of Mexico: Annual Offshore Technology Conference, 10th, OTC 3217, p. 1425–1438.

Buffler, R. T., Winker, C. D., Rosenthal, D. B., Viele, G. W., Suter, Sherman, Lillie, R. J., Miles, A. E., Pilger, R. H., Jr., Nicholas, R. L., Watkins, J. S., Martin, R. G., and Sawyer, D. S., in press, Continent-ocean transect F1: Ouachitas to Yucatan: Geological Society of America Decade of North American Geology Program.

Burchfiel, B. C., and Royden, L., 1982, Carpathian foreland fold and thrust belt and its relation to Pannonian and other basins: American Association of Petroleum Geologists Bulletin, v. 66, p. 1179–1195.

Carey, S. W., 1954, The rheid concept in geotectonics: Geological Society of Australia Journal, v. 1, p. 67–117.

Carter, N. L., 1976, Steady state flow of rocks: Reviews of Geophysics and Space Physics, v. 14, p. 301–360.

Carter, N. L., and Hansen, F. D., 1983, Creep of rocksalt: Tectonophysics, v. 92, p. 275–333.

Chandrasekhar, S., 1982, Hydrodynamic and hydromagnetic stability: New York, Dover, 704 p.

Chapman, R. E., 1974, Clay diapirism and overthrust faulting: Geological Society of America Bulletin, v. 85, p. 1597–1602.

Clark, S. P., Jr., ed., 1966, Handbook of physical constants (revised edition): Geological Society of America Memoir 97, 587 p.

Cohen, B. L., 1977, The disposal of radioactive wastes from fission reactors: Scientific American, v. 263, p. 21.

Curnelle, R., and Marco, R., 1983, Reflection profiles across the Aquitaine basin (salt tectonics), *in* Bally, A. W., Seismic expression of structural styles—A picture and work atlas, Volume 2: American Association of Petroleum Geologists Studies in Geology Series 15, p. 2.3.2-11–2.3.2-17.

de Böckh, H., Lees, G. M., and Richardson, F.D.S., 1929, Contribution to the stratigraphy and tectonics of the Iranian Ranges, *in* Gregory, J. W., ed., The structure of Asia: London, Methuen, p. 58–176.

Dickinson, George, 1953, Geological aspects of abnormal reservoir pressures in Gulf Coast Louisiana: American Association of Petroleum Geologists Bulletin, v. 37, p. 410–432.

Dixon, J. M., 1975, Finite strain and progressive deformation in models of diapiric structures: Tectonophysics, v. 28, p. 89–124.

Dixon, J. M., and Summers, J. M., 1985, Recent developments in centrifuge modelling of tectonic processes: Equipment, model construction techniques and rheology of model materials: Journal of Structural Geology, v. 7, p. 83–102.

Dobrin, M. B., 1941, Some quantitative experiments on a fluid salt-dome model and their geological implications: American Geophysical Union Transactions, v. 22, pt. 2, p. 528–542.

Ewing, Maurice, and Ewing, J. I., 1962, Rate of salt-dome growth: American Association of Petroleum Geologists Bulletin, v. 46, p. 708–709.

Farmer, I. W., and Gilbert, M. J., 1984, Time dependent strength reduction of rock salt, *in* Hardy, H. R., Jr., and Langer, Michael, eds., The mechanical behavior of salt: Proceedings of the First Conference, Pennsylvania State University, November 1981: Houston, Texas, Gulf Publishing Co., p. 3–18.

Fischer, F. J., 1984, An axisymmetric method for analyzing cavity arrays, *in* Hardy, H. R., Jr., and Langer, Michael, eds., The mechanical behavior of salt: Proceedings of the First Conference, Pennsylvania State University, November 1981: Houston, Texas, Gulf Publishing Co., p. 661–680.

Fletcher, R. C., 1972, Application of a mathematical model to the emplacement of mantled gneiss domes: American Journal of Science, v. 272, p. 197–216.

Gardner, G.H.F., Gardner, L. W., and Gregory, A. R., 1974, Formation velocity and density—The diagnostic basics for stratigraphic traps: Geophysics, v. 39, p. 770–780.

Gera, Ferruccio, 1972, Review of salt tectonics in relation to disposal of radioactive wastes in salt formations: Geological Society of America Bulletin, v. 83, p. 3551–3574.

Geraghty, J. J., Miller, D. W., Van der Leeden, Frits, and Troise, F. L., 1973, Water atlas of the United States: Port Washington, New York, Water Information Center, 120 pls.

Gevantman, L. H., ed., 1981, Physical properties data for rock salt: U.S. National Bureau of Standards Monograph 167, 282 p.

Goguel, Jean, 1948, Introduction à l'étude méchanique des déformations de l'écorce terrestre, (2nd edition): Mémoires de la carte géologique détaillée de la France: Paris, Ministère de l'Industrie et du Commerce, 530 p.

Grout, F. F., 1945, Scale models of structures related to batholiths: American Journal of Science, v. 243A, p. 260–284.

Gussow, W. C., 1968, Salt diapirism: Importance of temperature, and energy source of emplacement, *in* Braunstein, Jules, and O'Brien, G. D., eds., Diapirism and diapirs: American Association of Petroleum Geologists Memoir 8, p. 16–52.

Halbouty, M. T., 1979, Salt domes, Gulf region, United States and Mexico: Houston, Texas, Gulf Publishing Co., 561 p.

Halbouty, M. T., and Hardin, G. C., 1954, New exploration possibilities established by thrust fault at Boling Dome, Wharton County, Texas: American Association of Petroleum Geologists Bulletin, v. 38, no. 8, p. 1725–1740.

Handin, John, and Logan, J. M., 1981, Experimental tectonophysics: Geophysical Research Letters, v. 8, no. 7, p. 647–650

Handin, John, and others, 1963, Experimental deformation of sedimentary rocks under confining pressure: Pore pressure tests: American Association of Petroleum Geologists Bulletin, v. 47, p. 717–755.

Hansen, F. D., and Carter, N. L., 1984, Creep of Avery Island rocksalt, *in* Hardy, H. R., Jr., and Langer, Michael, eds., The mechanical behavior of salt: Proceedings of the First Conference, Pennsylvania State University, November 1981: Houston, Texas, Gulf Publishing Co., p. 53–69.

Hansen, F. D., Mellegard, K. D., and Senseny, P. E., 1984, Elasticity and strength of ten natural rock salts, *in* Hardy, H. R., Jr., and Langer, Michael, eds., The mechanical behavior of salt: Proceedings of the First Conference, Pennsylvania State University, November 1981: Houston, Texas, Gulf Publishing Co., p. 71–83.

Hardy, H. R., Jr., and Langer, Michael, 1984, The mechanical behavior of salt: Proceedings of the First Conference, Pennsylvania State University, November 1981: Houston, Texas, Gulf Publishing Co., 901 p.

Hardy, H. R., Jr., Bakhtar, K., Mrugala, M., and Kimble, E. J., Jr., 1984, Development of laboratory facilities for evaluating the creep behavior of salt, *in* Hardy, H. R., Jr., and Langer, Michael, eds., The mechanical behavior of salt: Proceedings of the First Conference, Pennsylvania State University, November 1981: Houston, Texas, Gulf Publishing Co., p. 85–117.

Harland, W. B., 1971, Tectonic transpression in Caledonian Spitsbergen: Geological Magazine, v. 108, p. 27–42.

Harrison, J. V., 1930, The geology of some salt plugs in Laristan (southern Persia): Geological Society of London Quarterly Journal, v. 86, p. 463–522.

Heard, H. C., 1972, Steady-state flow in polycrystalline halite at pressure of 2 kilobars, *in* Heard, H. C., and others, eds., Flow and fracture of rocks: American Geophysical Union Geophysical Monograph Series, v. 16, p. 191–210.

Heye, Dietrich, 1978, Experimente mit viskosen Flüssigkeiten zur Nachahmung von Salzstrukturen: Geologisches Jahrbuch, v. E12, p. 31–51.

——— 1979, Modellversuche zum Salzdiapirismus mit viskosen Flüssigkeiten: Geologisches Jahrbuch, v. E16, p. 39–51.

Holser, W. T., and others, 1972, Radar logging of a salt dome: Geophysics, v. 37, no. 5, p. 889–906.

Holwerda, J. G., and Hutchinson, R. W., 1968, Potash-bearing evaporites in the Danakil area, Ethiopia: Economic Geology, v. 63, p. 124–150.

Horseman, Stephen, and Passaris, Evan, 1984, Creep tests for storage cavity closure prediction, *in* Hardy, H. R., Jr., and Langer, Michael, eds., The mechanical behavior of salt: Proceedings of the First Conference, Pennsylvania State University, November 1981: Houston, Texas, Gulf Publishing Co., p. 119–157.

Hughes, D. J., 1968, Salt tectonics as related to several Smackover fields along the northeast rim of the Gulf of Mexico basin: Gulf Coast Association of Geological Societies Transactions, v. 18, p. 320–330.

Humphris, C. C., Jr., 1979, Salt movement on continental slope, northern Gulf of Mexico: American Association of Petroleum Geologists Bulletin, v. 63, no. 5, p. 782–798.

Hunsche, Udo, 1978, Modellrechnungen zur Entstehung von Salzstockfamilien: Geologisches Jahrbuch, v. E12, p. 53–107.

Jackson, M.P.A., 1982, Fault tectonics of the East Texas Basin: The University of Texas at Austin Bureau of Economic Geology Geological Circular 82-4, 31 p.

Jackson, M.P.A., and Galloway, W. E., 1984, Structural and depositional styles of Gulf Coast Tertiary continental margins: Application to hydrocarbon exploration: American Association of Petroleum Geologists Continuing Education Course Note Series 25, 226 p.

Jackson, M.P.A., and Seni, S. J., 1984, The domes of East Texas, *in* Presley, M. W., ed., The Jurassic of East Texas: Tyler, Texas, East Texas Geological Society, p. 163–239.

Jackson, M.P.A., and Talbot, C. J., 1985, The internal structure of model and natural salt domes (Final report prepared for U.S. Department of Energy under contract no. DE-AC97-83WM-46651): Austin, Texas, The University of Texas at Austin Bureau of Economic Geology, 57 p.

Jaritz, Werner, 1973, Zur Enstehung der Salzstrukturen Nordwestdeutschlands: Geologisches Jahrbuch, v. A10, p. 3–77.

Jenyon, M. K., 1985, Fault-associated salt flow and mass movement: Geological Society of London Journal, v. 142, p. 547–553.

Keen, C. E., 1983, Salt diapirs and thermal maturity: Scotian Basin: Canadian Petroleum Geology Bulletin, v. 31, no. 2, p. 101–108.

Kent, P. E., 1966, Temperature conditions of salt dome intrusions: Nature, v. 211, p. 1387.

——— 1979, The emergent Hormuz salt plugs of southern Iran: Journal of Petroleum Geology, v. 2, p. 117–144.

Kupfer, D. H., 1968, Relationship of internal to external structure of salt domes: American Association of Petroleum Geologists Memoir 8, p. 79–89.

Labao, J. J., and Pilger, R. H., Jr., 1985, Early evolution of salt structures in the North Louisiana salt basin: Gulf Coast Association of Geological Societies Transactions, v. 35, p. 189–198.

Langer, Michael, 1984, The rheological behaviour of rock salt, *in* Hardy, H. R., Jr., and Langer, Michael, eds., The mechanical behavior of salt: Proceedings of the First Conference, Pennsylvania State University, November 1981: Houston, Texas, Gulf Publishing Co., p. 201–240.

Law Engineering Testing Company, 1983, Geothermal studies of seven interior salt domes: Columbus, Ohio, Battelle Memorial Institute, Office of Nuclear Waste Isolation, ONWI-289, 41 p.

LeCompte, P., 1965, Creep in rock salt: Journal of Geology, v. 73, p. 469–484.

Lindsay, J. F., 1977, Salt tectonism and the evolution of the Sigsbee Scarp, Gulf of Mexico: Tectonophysics, v. 39, p. 607–619.

Losecke, W., 1972, Über die Bestimmung von Salzstockgrenzflächen mit Hilfe der Magnetotellurik: Zeitschrift für Geophysik, v. 38, p. 959–962.

Loucks, R. G., Dodge, M. M., and Galloway, W. E., 1979, Sandstone consolidation analysis to delineate areas of high-quality reservoirs suitable for production of geopressured geothermal energy along the Texas Gulf Coast: U.S. Department of Energy, Division of Geothermal Energy Contract No. EG-77-5-05-5554, 97 p.

Madirazza, Ivan, 1975, The geology of the Vejrum salt structure, Denmark: Geological Society of Denmark Bulletin, v. 24, p. 161–171.

Martin, R. G., 1980, Distribution of salt structures in the Gulf of Mexico: Map and descriptive text: U.S. Geological Survey Miscellaneous Field Studies Map MF-1213, 8 p., scale 1:2,500,000.

Martinez, J. D., and others, 1978, An investigation of the utility of Gulf Coast salt domes for the storage or disposal of radioactive wastes, Volume 1: Louisiana State University Institute for Environmental Studies Contract Report EW-78-C-05-5941/53, 390 p.

Mrazec, Ludovic, 1907, Despre cute cu simbure de strapungere [On folds with piercing cores]: Soc. Stiite Bulletin Romania, v. 16, p. 6–8.

National Iranian Oil Company, 1977, Explanatory notes to geological map, cross sections and tectonic map of south-central Iran, sheet 5: Tehran, National Iranian Oil Company Exploration and Production, 1 p., scale 1:1,000,000.

Nettleton, L. L., 1934, Fluid mechanics of salt domes: American Association of Petroleum Geologists Bulletin, v. 18, p. 1175–1204.

——— 1943, Recent experimental and geophysical evidence of mechanics of salt-dome formation: American Association of Petroleum Geologists Bulletin, v. 27, p. 51–63.

New Orleans Geological Society, 1960, Salt domes of South Louisiana: Volume I, 145 p.

——— 1962, Salt domes of South Louisiana: Volume II, 107 p.

——— 1983, Salt domes of South Louisiana: Volume III, 142 p.

O'Brien, G. D., 1968, Survey of diapirs and diapirism, *in* Braunstein, Jules, and O'Brien, G. D., eds., Diapirism and diapirs: American Association of Petroleum Geologists Memoir 8, p. 1–9.

O'Brien, J. J., and Lerche, Ian, 1984, The influence of salt domes on paleotemperature distributions: Geophysics, v. 49,

no. 11, p. 2032–2043.
Odé, Helmer, 1968, Review of mechanical properties of salt relating to salt dome genesis: Geological Society of America Special Paper 88, p. 543–595.
Parker, T. J., and McDowell, A. N., 1955, Model studies of salt-dome tectonics: American Association of Petroleum Geologists Bulletin, v. 39, p. 2384–2470.
Pfiffner, O. A., and Ramsay, J. G., 1982, Constraints on geological strain rates: Arguments from finite strain states of naturally deformed rocks: Journal of Geophysical Research, v. 87, no. B1, p. 311–321.
Ramberg, Hans, 1966, Experimental and theoretical study of salt-dome evolution, *in* Rau, J. L., and Dellwig, L. F., eds., Symposium on salt, 3rd: Cleveland, Ohio, Northern Ohio Geological Society, p. 261–270.
——— 1968, Instability of layered systems in the field of gravity, II: Physics of the Earth and Planetary Interiors, v. 1, p. 448–474.
——— 1981, Gravity, deformation and the Earth's crust in theory, experiments and geological application (2nd edition): London, Academic Press, 452 p.
Rettger, R. E., 1935, Experiments on soft-rock deformation: American Association of Petroleum Geologists Bulletin, v. 19, p. 271–292.
Reynolds, E., and Gloyna, E. F., 1960, Reactor fuel waste disposal project: Permeability of rock salt and creep of underground cavities, final report: Austin, Texas, University of Texas, 40 p.
Richter-Bernburg, Gerhard, 1980, Salt tectonics, interior structures of salt bodies: Centres de Recherches Exploration-Production Elf-Aquitaine Bulletin, v. 4, p. 373–393.
Rosenkrans, R. R., and Marr, J. D., 1967, Modern seismic exploration of the Gulf Coast Smackover trend: Geophysics, v. 32, p. 184–206.
Sanderson, D. J., and Marchini, W.R.D., 1984, Transpression: Journal of Structural Geology, v. 6, no. 5, p. 449–458.
Sannemann, D., 1968, Salt-stock families in northwestern Germany: American Association of Petroleum Geologists Memoir 8, p. 261–270.
——— 1983, Migration of salt-induced structures, *in* Bally, A. W., ed., Seismic expression of structural styles, Volume 2—Tectonics of extensional provinces: American Association of Petroleum Geologists Studies in Geology Series 15, p. 2.3.2-1–2.3.2-2.
Schwerdtner, W. M., and Tröeng, Björn, 1978, Strain distribution within arcuate diapiric ridges of silicone putty: Tectonophysics, v. 50, p. 13–28.
Schwerdtner, W. M., Sutcliffe, R. H., and Tröeng, Björn, 1978, Patterns of total strain within the crestal region of immature diapirs: Canadian Journal of Earth Sciences, v. 15, p. 1437–1447.
Selig, F., and Wallick, G. C., 1966, Temperature distribution in salt domes and surrounding sediments: Geophysics, v. 31, p. 346–361.
Seni, S. J., and Jackson, M.P.A., 1984, Sedimentary record of Cretaceous and Tertiary salt movement, East Texas Basin: Times, rates, and volumes of salt flow and their implications to nuclear waste isolation and petroleum exploration: The University of Texas at Austin Bureau of Economic Geology Report of Investigations 139, 89 p.
Serata, S., and Gloyna, E. F., 1959, Development of design principle for disposal of reactor fuel waste into underground salt cavities, Reactor Fuel Waste Disposal Project: Austin, Texas, University of Texas, 173 p.
Spiers, C. J., Urai, J. L., Lister, G. S., and Zwart, H. J., 1984, Water weakening and dynamic recrystallization in salt: Geological Society of America Abstracts with Programs, v. 16, no. 6, p. 665.
Stöcklin, Jovan, 1968, Salt deposits of the Middle East, *in* Mattox, R. B., ed., Saline deposits: Geological Society of America Special Paper 88, p. 157–181.
Talbot, C. J., 1974, Fold nappes as asymmetric mantled gneiss domes and ensialic orogeny: Tectonophysics, v. 24, p. 259–276.
——— 1977, Inclined and asymmetric upward-moving gravity structures: Tectonophysics, v. 42, p. 159–181.
——— 1978, Halokinesis and thermal convection: Nature, v. 273, p. 739–741.
Talbot, C. J., and Jarvis, R. J., 1984, Age, budget and dynamics of an active salt extrusion in Iran: Journal of Structural Geology, v. 6, p. 521–533.
Talbot, C. J., and Rogers, E. A., 1980, Seasonal movements in a salt glacier in Iran: Science, v. 208, p. 395–397.
Talbot, C. J., Tully, C. P., and Woods, P.J.E., 1982, The structural geology of Boulby (Potash) Mine, Cleveland, United Kingdom: Tectonophysics, v. 85, p. 167–204.
Thoms, R. L., Mogharrebi, M., and Gehle, R. M., 1982, Geomechanics of borehole closure in salt domes, *in* Annual Convention, 61st, Proceedings: Dallas, Texas, Gas Processors Association, p. 228–230.
Touloukian, Y. S., Judd, W. R., and Roy, R. F., eds., 1981, Physical properties of rocks and minerals, McGraw-Hill/CINDAS Data Series on Material Properties, Volume II-2: New York, McGraw-Hill, 502 p.
Trusheim, Ferdinand, 1957, Über Halokinese und ihre Bedeutung für die strukturelle Entwicklung Norddeutschlands: Zeitschrift Deutsche Geologische Gesellschaft, v. 109, p. 111–151.
——— 1960, Mechanism of salt migration in northern Germany: American Association of Petroleum Geologists Bulletin, v. 44, p. 1519–1540.
Tucker, P. M., and Yorston, H. J., 1973, Pitfalls in seismic interpretation: Society of Exploration Geophysicists Monograph Series 2, 50 p.
Van Dyke, Milton, 1982, An album of fluid motion: Stanford, California, Parabolic Press, 176 p.
Vita-Finzi, C., 1979, Rates of Holocene folding in the coastal Zagros near Bandar Abbas, Iran: Nature, v. 278, p. 632–633.
Watkins, J. S., and others, 1978, Occurrence and evolution of salt in deep Gulf of Mexico: American Association of Petroleum Geologists Studies in Geology 7, p. 43–65.
Wenkert, D. D., 1979, The flow of salt glaciers: Geophysical Research Letters, v. 6, p. 523–526.
Whitehead, J. A., Jr., and Luther, D. S., 1975, Dynamics of laboratory diapir and plume models: Journal of Geophysical Research, v. 80, p. 705–717.
Wilson, C. H., 1985, Depositional, structural, and thermal evolution of a Pleistocene oil-productive area: Texas-Louisiana continental shelf [M.A. thesis]: Austin, Texas, The University of Texas at Austin, 93 p.
Woidt, W.-D., 1978, Finite element calculations applied to salt-dome analysis: Tectonophysics, v. 50, p. 369–386.
——— 1980, Analytische und numerische Modellexperimente zur Physik der Salzstockbildung: Institut für Geophysik und Meteorologie der Technischen Universität Braunschweig GAMMA 38, 151 p.
Ziegler, P. A., 1982, Geological atlas of Western and Central Europe: The Hague, Shell Internationale Petroleum Maatschappij B.V., 130 p.

Manuscript Received by the Society February 8, 1985
Revised Manuscript Received September 19, 1985
Manuscript Accepted September 19, 1985
Publication Authorized by the Director, Bureau of Economic Geology, The University of Texas

The American Association of Petroleum Geologists Bulletin
V. 67, No. 8 (August 1983), P. 1219-1244, 22 Figs.

Evolution of Salt Structures, East Texas Diapir Province, Part 1: Sedimentary Record of Halokinesis[1]

S. J. SENI and M. P. A. JACKSON[2]

ABSTRACT

Post-Aptian (post-112 Ma) strata in the East Texas basin were strongly influenced by halokinesis and therefore record the evolution of associated salt structures. Dome-induced changes in patterns of sandstone distribution, depositional facies, and reef growth indicate that thickness variations in strata surrounding domes were caused by syndepositional processes rather than by tectonic distortion.

Salt domes in the East Texas basin exhibit three stages of growth: pillow, diapir, and post-diapir, each of which affected surrounding strata differently. Pillow growth caused broad uplift of strata over the crest of the pillows; the resulting topographic swell influenced depositional trends and was susceptible to erosion. Fluvial channel systems bypassed pillow crests and stacked vertically in primary peripheral sinks on the updip flanks of the pillows. Diapir growth was characterized by expanded sections of shelf and deltaic strata in secondary peripheral sinks around the diapirs. Lower Cretaceous reefs on topographic saddles between secondary peripheral sinks now host major oil production at Fairway field. Post-diapir crestal uplifts and peripheral subsidence affected smaller areas than did equivalent processes during pillow or diapir stages.

Documented facies variations over and around domes at different stages of growth enable prediction of subtle facies-controlled hydrocarbon traps. Facies-controlled traps are likely to be the only undiscovered ones remaining in mature petroliferous basins such as the East Texas basin.

INTRODUCTION

This paper, Part 1 of a larger report, describes the stages of diapir growth in the East Texas basin (Fig. 1) and documents the influence of salt movement on adjacent and overlying strata. Part 2 of this report summarizes patterns of dome growth through time and space and presents quantitative data on the rates and volumes of salt flow.

Dome growth creates a wide range of subtle traps for migrating petroleum, such as stratigraphic, unconformity, and paleogeomorphic types (Halbouty, 1980). The early formation of subtle traps enables oil to be trapped at the onset of migration. These subtle traps are especially significant for future exploration in highly mature areas such as the Gulf Interior and Gulf Coast basins. Using approximately 2,000 wells (Fig. 2) in the East Texas basin, we document specific stages of salt-dome growth, each characterized by different combinations of subtle traps, as well as more obvious structural ones. Understanding this evolution and its lithologic and structural effects allows prediction of subtle traps in both mature basins and in other, less explored salt basins.

The East Texas basin is one of several inland Mesozoic salt basins in Texas, Louisiana, and Mississippi that flank the northern Gulf of Mexico (Fig. 1). The general stratigraphy (Fig. 3) and structure of the East Texas basin have been summarized in many articles (e.g., Eaton, 1956; Granata, 1963; Bushaw, 1968; Nichols et al, 1968; Kreitler et al, 1980, 1981; Wood and Guevara, 1981). The evolution of this basin in relation to opening of the Gulf of Mexico is summarized by Jackson and Seni (1983).

The Jurassic Louann Salt was deposited on a planar angular unconformity across Triassic rift fill and Paleozoic basement (Fig. 3). The early post-Louann history of the basin was dominated by slow progradation of platform carbonates and minor evaporites during Smackover to Gilmer deposition (Fig. 3). After this phase of carbonate-evaporite deposition, massive progradation of Schuler-Hosston siliciclastics took place in the Late Jurassic–Early Cretaceous. Subsequent sedimentation comprised alternating periods of marine carbonate and siliciclastic accumulation. By the Oligocene, subsidence in the East Texas basin had ceased and major depocenters shifted to the Gulf of Mexico. Paleocene and Eocene strata crop out in most of the basin, indicating that net erosion characterized the last 40 Ma (millions of years ago).

Salt in the East Texas basin first moved during the early period of basin formation, defined as Jurassic to Early Cretaceous, prior to 112 Ma (Hughes, 1968; Jackson et al, 1982). We have limited this report to diapirism in the middle and late periods of basin evolution (112 to 48 Ma) because insufficient subsurface information on the early period prevents rigorous analysis of salt movement at that time. Consequently, this report does not cover the initial stage of movement of most East Texas diapirs. However, it includes the full growth history of the younger diapirs, so that all growth stages are represented. All 16 shallow-

[1]Manuscript received, November 12, 1982; accepted, February 28, 1983.

Published with permission of the Director, Bureau of Economic Geology, The University of Texas at Austin, Austin, Texas 78712.

[2]Bureau of Economic Geology, The University of Texas at Austin, Austin, Texas 78712.

This research was supported by the U.S. Department of Energy, Contract Number DE-AC97-80ET46617.

L. F. Brown, Jr., W. E. Galloway, N. Tyler, T. E. Ewing, and R. T. Budnik critically reviewed the manuscript; their comments are gratefully acknowledged. We thank E. Bramson, R. Conti, S. Ghazi, S. Lovell, B. Richter, J. Smith, and D. Wood for their help in data collection and processing. J. Ames, M. Bentley, M. Evans, R. Flores, J. Horowitz, and J. McClelland drafted figures under the supervision of D. Scranton. G. Zeikus and K. Bonnecarrere typed original manuscripts. Twyla J. Coker word processed the manuscript under the direction of L. Harrell, and S. Doenges supervised manuscript editing.

and intermediate-depth (<2,000 m; <6,500 ft) diapirs were studied.

EVOLUTIONARY STAGES OF SALT MOVEMENT

The evolution of salt from planar beds to nearly vertical subcylindrical stocks involved pillow, diapir, and post-diapir stages in the North German salt basin (Trusheim, 1960). Data presented here indicate that this three-stage model of dome growth is also appropriate for the East Texas basin. Each stage had distinctive effects on depositional facies, lithostratigraphy, and thickness of surrounding sediments (Fig. 4).

The evolution of natural salt structures has received much attention in the literature (e.g., Bornhauser, 1958; Atwater and Forman, 1959; Trusheim, 1960; Bishop, 1978; Halbouty, 1979) because of their status as obvious structural traps for petroleum. Controversy surrounds the emplacement history of diapirs, and hinges on whether the dominant processes were intrusion (favored by DeGolyer, 1925; Barton, 1933; Nettleton, 1934; Trusheim, 1960; Sannemann, 1968; Kupfer, 1970, 1976; Smith and Reeve, 1970; O'Neill, 1973; Stude, 1978; Kent, 1979; Woodbury et al, 1980) or extrusion (favored by Loocke, 1978; Turk, Kehle, and Associates, 1978; Jaritz, 1980; R. O. Kehle, personal commun., 1982). Bishop (1978) theorized that diapirism typically occurs by extrusion or alternates between intrusion and extrusion. Barton (1933), Bornhauser (1958, 1969), and Johnson and Bredeson (1971) emphasized the role of sediment "downbuilding" around salt structures whose crests remain more or less stationary and relatively close to the depositional surface. Bishop (1978) emphasized the importance of understanding the depositional history of surrounding sediments in order to interpret dome-growth history, an approach followed here.

Regardless of the mechanism responsible for salt movement and diapirism, flow of salt into a growing structure

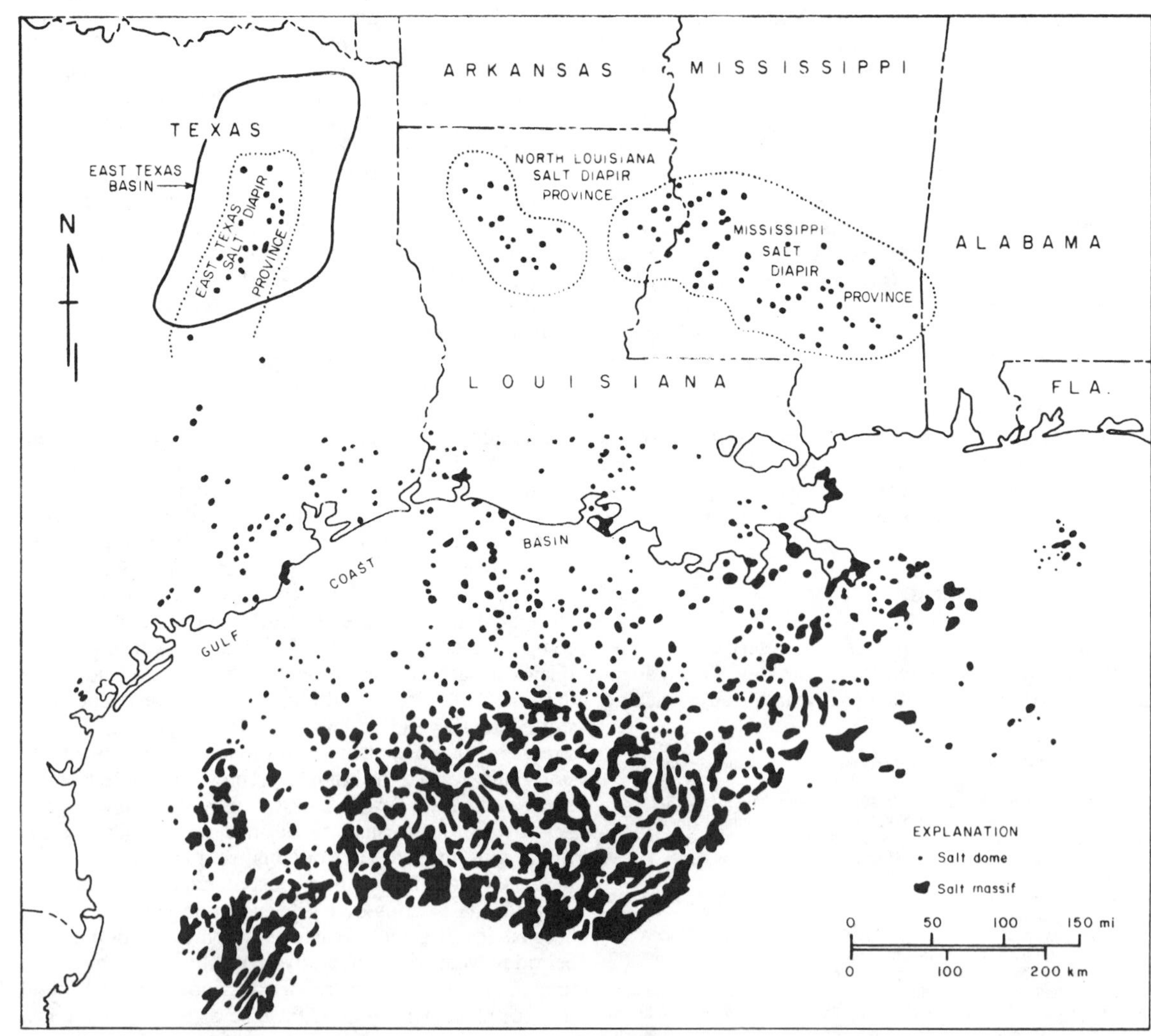

FIG. 1—Map showing East Texas basin, location of inland salt-diapir provinces, and salt domes. After Martin (1978).

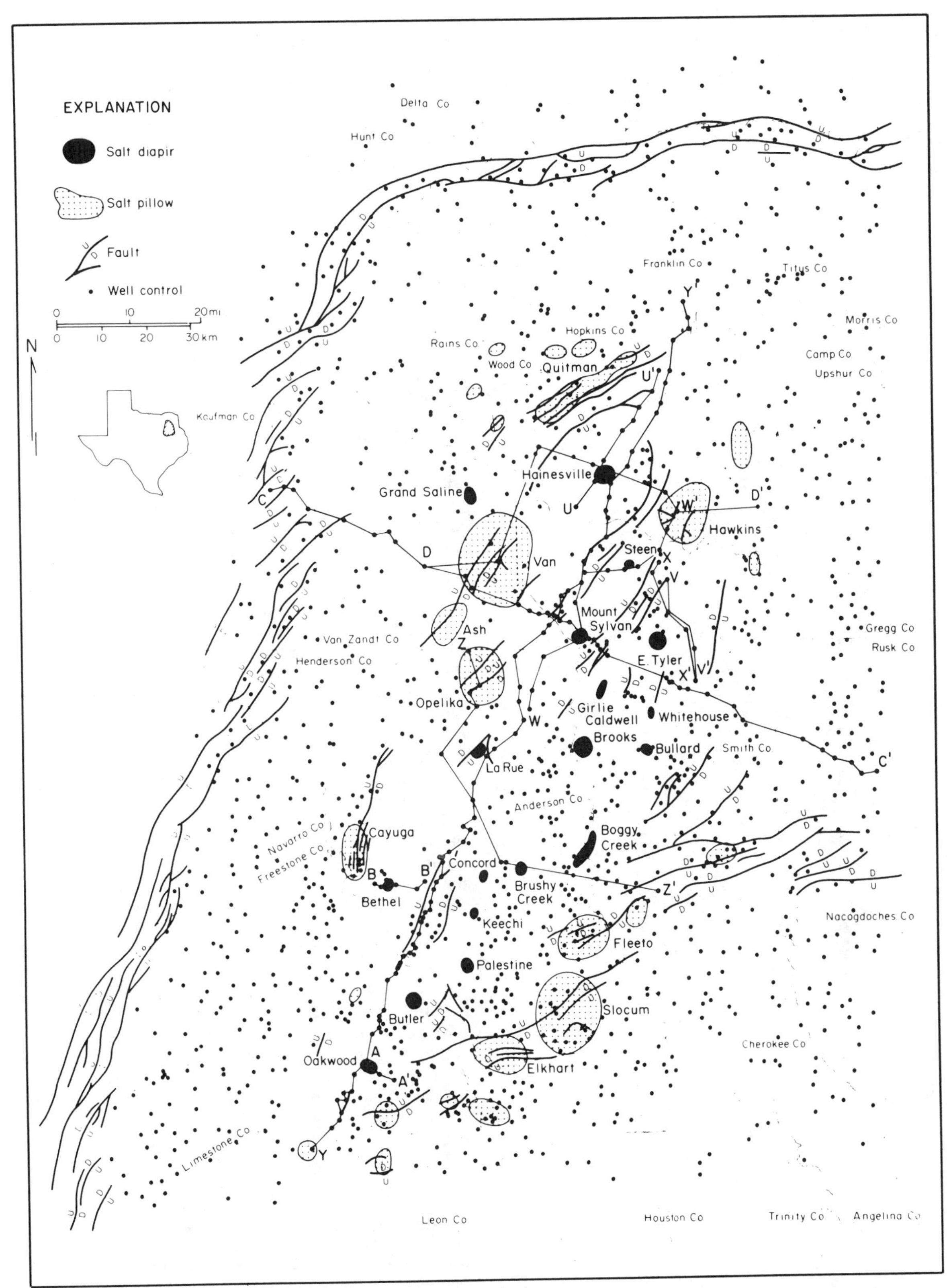

FIG. 2—Index map of East Texas basin showing locations of well logs used in interval isopach mapping and cross-section lines.

creates a withdrawal basin that is a structural low and an isopach thick. "Withdrawal basin" is a general term that includes rim syncline (a geometric term) and primary, secondary, and tertiary peripheral sink (genetic terms) (Fig. 4). Trusheim (1960) defined primary peripheral sinks as forming during pillow growth, secondary peripheral sinks or rim synclines during diapir growth, and tertiary peripheral sinks during post-diapir growth. We define secondary peripheral sinks as containing units at least 50% thicker than adjacent units unaffected by salt withdrawal; tertiary peripheral sinks have thickening less than 50% because of much slower rates of salt movement at this later stage. The term "sink" is used in a structural sense. Ramberg (1981, p. 286) pointed out that in terms of fluid dynamics, the rim syncline is actually the source of the flow, whereas the dome is the true sink.

The following sections present lithostratigraphic effects of the three stages of dome growth observed in the East Texas basin (Fig. 4).

ERA	SYSTEM	SERIES	STAGE	GROUP		FORMATION or MEMBER	HORIZON	AGE (Ma)	DURATION
CENOZOIC	TERTIARY	EOCENE		CLAIBORNE		COOK MOUNTAIN			NOT STUDIED
						SPARTA			
						WECHES			
						QUEEN CITY			
						REKLAW			
						CARRIZO			
		PALEOCENE		WILCOX		UNDIFFERENTIATED	TOP WILCOX	48	8 Ma
				MIDWAY		UNDIFFERENTIATED	TOP MIDWAY	56	10 Ma
MESOZOIC	CRETACEOUS	UPPER CRETACEOUS	SENONIAN	NAVARRO		UPPER NAVARRO CLAY			
						UPPER NAVARRO MARL	TOP NAVARRO	66	7 Ma
						NACATOCH SAND			
						LOWER NAVARRO			
				TAYLOR	UPPER				
						PECAN GAP CHALK	TOP L TAYLOR	73	13 Ma
					LOWER	WOLF CITY SAND			
				AUSTIN		GOBER CHALK			
						AUSTIN CHALK			
						ECTOR CHALK			
			TURONIAN	EAGLE FORD		EAGLE FORD	BASE AUSTIN	86	NOT STUDIED
			CENOM-ANIAN	WOODBINE		LEWISVILLE	TOP WOODBINE	92	6 Ma
						DEXTER			
		LOWER CRETACEOUS	ALBIAN	WASHITA		MANESS SHALE			
						BUDA LIMESTONE	TOP WASHITA	98	6 Ma
						GRAYSON SHALE			
					GEORGETOWN	MAIN STREET LS			
						WENO-PAWPAW LS			
						DENTON SHALE			
						FORT WORTH LS			
						DUCK CREEK SHALE			
						DUCK CREEK LS			
				FREDERICKSBURG		KIAMICHI SHALE		104	NOT STUDIED
						GOODLAND LIMESTONE			
				TRINITY		PALUXY/WALNUT	TOP PALUXY	104	1 Ma
					UPPER GLEN ROSE	UPPER GLEN ROSE	TOP GLEN ROSE	105	7 Ma
			APTIAN		LOWER GLEN ROSE	MASSIVE ANHYDRITE			
						RODESSA LS			
						JAMES LS	TOP JAMES	112	
						PINE IS. SH			
						PETTET LS			
						HOSSTON [TRAVIS PEAK]			
	JURASSIC	UPPER JURASSIC		COTTON VALLEY		SCHULER		135	
						BOSSIER			
						GILMER		138	
						BUCKNER			
						SMACKOVER		140	
						NORPHLET			
		MIDDLE JURASSIC				LOUANN SALT		143	
						WERNER		156	
	TRIASSIC?					EAGLE MILLS			
PALEOZOIC						OUACHITA			

FIG. 3—Stratigraphic column of East Texas basin (after Wood and Guevara, 1981). Right column shows duration of isopach intervals used in this report. Mapped horizons were selected on basis of ease of regional subsurface correlation rather than exact equivalence to group boundaries. Geochronology based on van Hinte (1976a) and van Eysinga (1975).

Pillow Stage

Salt pillows are defined here as concordant anticlinal or laccolithic salt structures, with any amplitude/wavelength ratio. Their growth is initiated and maintained by factors such as depositional rate and erosional rate of post-salt deposits on the pillow crest, uneven sediment loading, salt buoyancy, downdip creep of salt, and subsalt discontinuities. Although the relative importance of these various factors is poorly understood, evidence of early (pre-Gilmer) salt movement under thin sedimentary cover of less than 600 m (2,000 ft) (Hughes, 1968; Jackson et al, 1982) suggests that uneven sediment loading and rate of deposition are the principal controlling factors early in the history of salt movement (Bishop, 1978).

Sediments deposited during pillow growth are characterized by: (1) thinning toward the axis of salt uplift; (2) only minor thickening into relatively distant primary peripheral sinks; and (3) lithostratigraphic variations over the crest of the pillow and in primary peripheral sinks.

Geometry of overlying strata.—Syndepositional thinning of sediments over the crest and flanks of growing salt pillows is the most diagnostic feature of salt movement during this stage. Quitman, Van, and Hawkins salt pillows are at similar elevations, about −3,650 m (−12,000 ft), but show differing patterns in sediment thinning over the pillow crests (see Figs. 7, 8, 9 in Seni and Jackson, 1983, pt. 2). Accordingly, drape and differential compaction of sediments over the salt structures had less effect on thinning than did rate of salt movement.

A 100 to 400 km^2 (40 to 155 mi^2) area over each of four Paluxy salt pillows—Van, Hawkins, Hainesville, and Bethel domes—contains stratigraphic intervals thinned from 10 to 100%; thinning is typically about 25%. The areas of strata thinned by salt uplift are stacked vertically over the crest of each pillow (Fig. 5).

Hainesville dome provides the best example of the geometry of strata around a growing pillow (Fig. 6). Lower Cretaceous strata onlap and pinch out toward the dome, indicating syndepositional sedimentation and erosion around a growing swell during pillow stage growth (Loocke, 1978, p. 40-46).

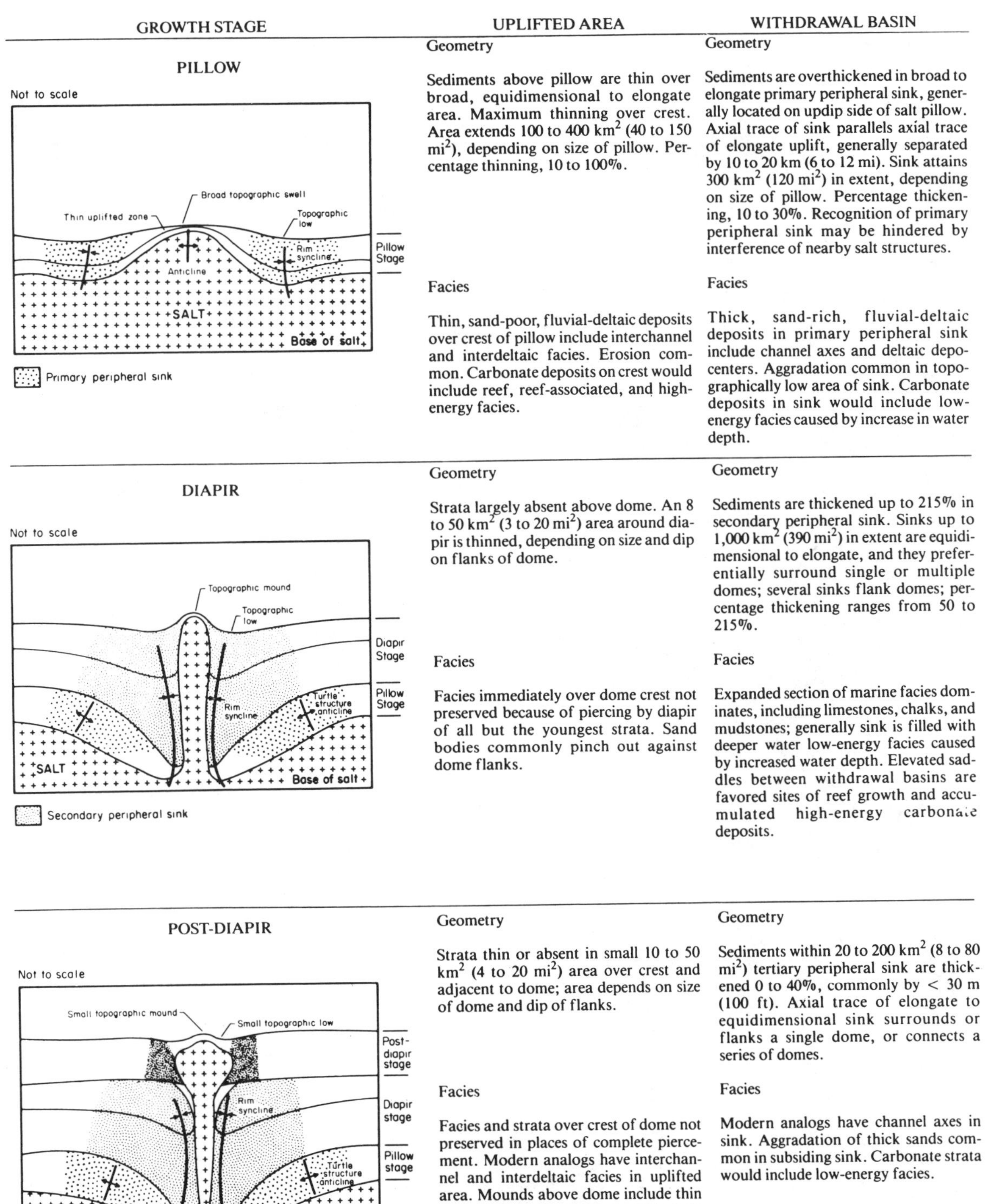

GROWTH STAGE	UPLIFTED AREA	WITHDRAWAL BASIN
PILLOW	**Geometry** Sediments above pillow are thin over broad, equidimensional to elongate area. Maximum thinning over crest. Area extends 100 to 400 km^2 (40 to 150 mi^2), depending on size of pillow. Percentage thinning, 10 to 100%.	**Geometry** Sediments are overthickened in broad to elongate primary peripheral sink, generally located on updip side of salt pillow. Axial trace of sink parallels axial trace of elongate uplift, generally separated by 10 to 20 km (6 to 12 mi). Sink attains 300 km^2 (120 mi^2) in extent, depending on size of pillow. Percentage thickening, 10 to 30%. Recognition of primary peripheral sink may be hindered by interference of nearby salt structures.
	Facies Thin, sand-poor, fluvial-deltaic deposits over crest of pillow include interchannel and interdeltaic facies. Erosion common. Carbonate deposits on crest would include reef, reef-associated, and high-energy facies.	**Facies** Thick, sand-rich, fluvial-deltaic deposits in primary peripheral sink include channel axes and deltaic depocenters. Aggradation common in topographically low area of sink. Carbonate deposits in sink would include low-energy facies caused by increase in water depth.
DIAPIR	**Geometry** Strata largely absent above dome. An 8 to 50 km^2 (3 to 20 mi^2) area around diapir is thinned, depending on size and dip on flanks of dome.	**Geometry** Sediments are thickened up to 215% in secondary peripheral sink. Sinks up to 1,000 km^2 (390 mi^2) in extent are equidimensional to elongate, and they preferentially surround single or multiple domes; several sinks flank domes; percentage thickening ranges from 50 to 215%.
	Facies Facies immediately over dome crest not preserved because of piercing by diapir of all but the youngest strata. Sand bodies commonly pinch out against dome flanks.	**Facies** Expanded section of marine facies dominates, including limestones, chalks, and mudstones; generally sink is filled with deeper water low-energy facies caused by increased water depth. Elevated saddles between withdrawal basins are favored sites of reef growth and accumulated high-energy carbonate deposits.
POST-DIAPIR	**Geometry** Strata thin or absent in small 10 to 50 km^2 (4 to 20 mi^2) area over crest and adjacent to dome; area depends on size of dome and dip of flanks.	**Geometry** Sediments within 20 to 200 km^2 (8 to 80 mi^2) tertiary peripheral sink are thickened 0 to 40%, commonly by $<$ 30 m (100 ft). Axial trace of elongate to equidimensional sink surrounds or flanks a single dome, or connects a series of domes.
	Facies Facies and strata over crest of dome not preserved in places of complete piercement. Modern analogs have interchannel and interdeltaic facies in uplifted area. Mounds above dome include thin sands. Carbonate strata would include reef or high-energy deposits; erosion common.	**Facies** Modern analogs have channel axes in sink. Aggradation of thick sands common in subsiding sink. Carbonate strata would include low-energy facies.

FIG. 4—Schematic stages of dome growth and variations in associated strata above and around salt structures.

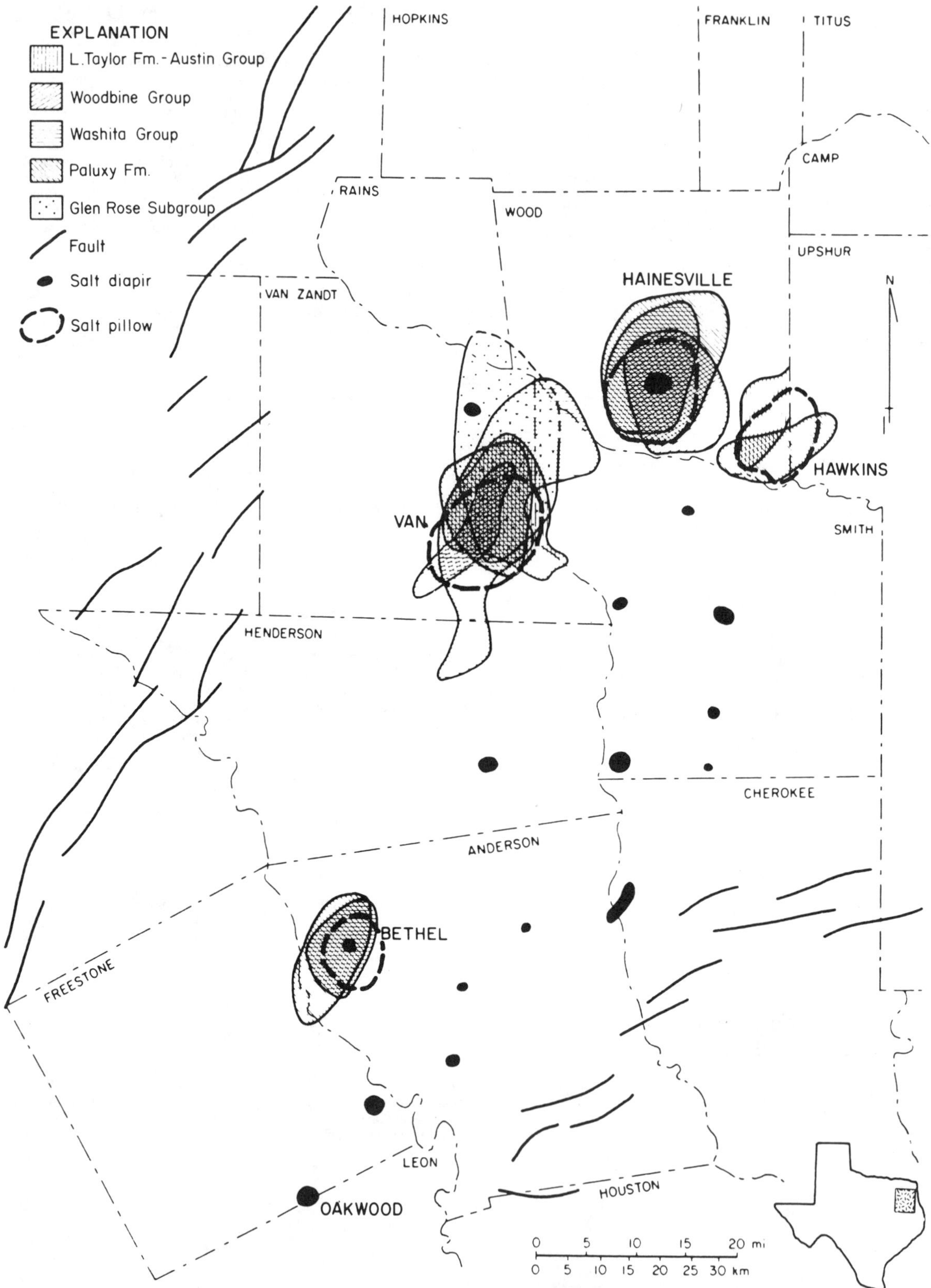

FIG. 5—Mapped areas of stratal thinning in five isopach intervals over crests of salt pillows in East Texas basin.

Geometry of surrounding strata.—A second but less diagnostic characteristic of pillow growth is primary peripheral sinks. Primary peripheral sinks are typically broad, shallow basins that are 10 to 30% overthickened with respect to adjacent strata unaffected by salt flow. The axial traces of these basins are located 10 to 20 km (6 to 12 mi) from the crest of the pillows in the Van, Hawkins, Hainesville, and Bethel domes (Fig. 7). The axial traces are either subparallel to crest lines of pillows or partially concentric to them, as in a rim syncline. Sinks are equidimensional or elongate in plan and are concentrated on the updip side of the salt structures, as exemplified by Bethel, Van, and Hainesville domes (Fig. 7). In the Zechstein basin of northern Germany, the primary peripheral sinks migrated toward the growing salt pillows with time as the flanks of the salt pillows continually steepened (Trusheim, 1960). This migration of primary peripheral sinks was not observed near East Texas pillows, but secondary and tertiary sinks are nearer the domes.

Depositional facies and lithostratigraphy.—Depositional facies and sandstone distribution patterns in the Paluxy Formation (Lower Cretaceous) illustrate the influence of syndepositional salt movement on surrounding strata. The Paluxy is typical of relatively thin (generally less than 150 m, 500 ft) Cretaceous siliciclastics around the margin of the basin that interfinger with carbonates of the basin center (Caughey, 1977; Seni, 1981).

A net-sandstone map of the Paluxy Formation (Fig. 8) documents sandstone distribution in fluvial and deltaic deposits around three salt pillows—Van, Hawkins, and Hainesville. Pillow growth is shown by decreased net and percentage sandstone in strata deposited over these structures. Dip-oriented trends of net sandstone outline fluvial axes that bypassed the pillows. Sediments in the primary peripheral sinks are significantly richer in sand than are deposits over the pillows (Fig. 9); F- and t-statistical tests indicate that, based on the boreholes shown in Figure 9, the crestal areas contain between 5 and 20% less sand at the 95% confidence level. The response of facies trends in other environments is summarized in Figure 4. A natural example is provided by Cretaceous rudist reefs that grew on swells over salt pillows in northern Mexico (Elliot, 1979).

Diapir Stage

A primary peripheral sink is synclinal during the pillow stage (Fig. 10B, C). During the subsequent diapir stage, the flanks of the pillow deflate because of salt withdrawal into the central growing diapir. Pillow deflation results in a secondary peripheral sink in which the originally uplifted and thinned strata collapse (Fig. 10D). The thickened primary peripheral sink remains unaffected by collapse, thereby forming an anticlinal structure cored by unde-

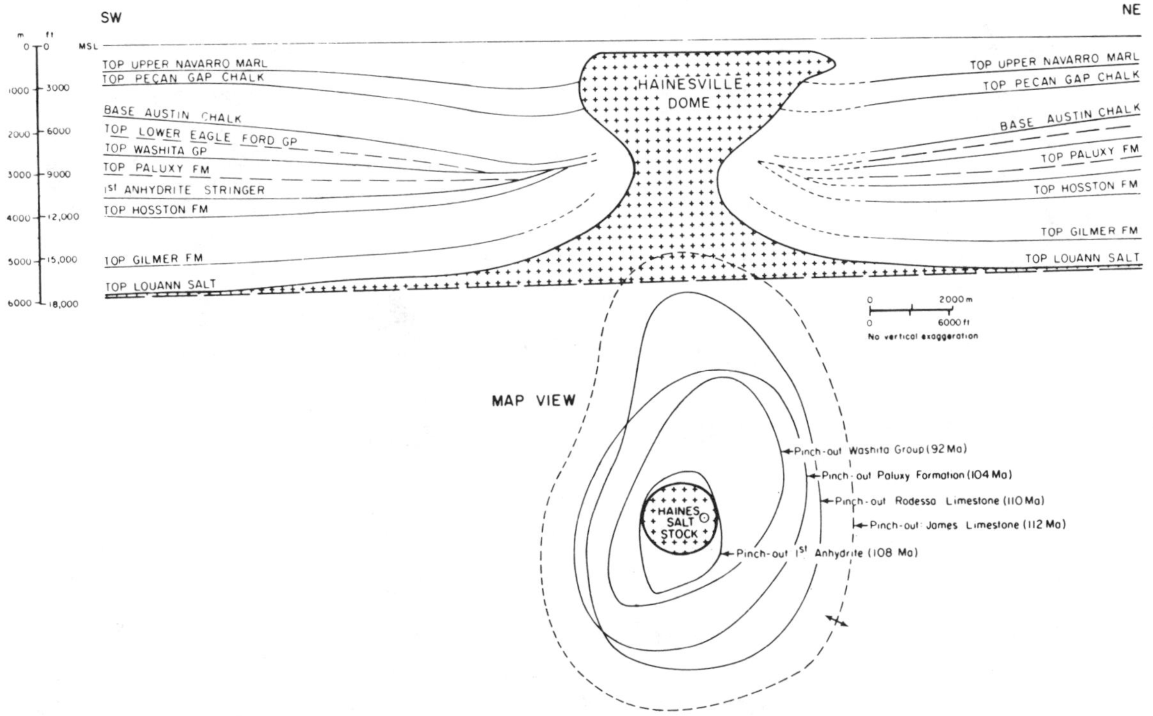

FIG. 6—Southwest-northeast cross section and map, Hainesville dome, East Texas basin, showing structure of surrounding strata. Lower Cretaceous strata onlap, offlap, and pinch out around dome. Both syndepositional and postdepositional erosion was active. Dome growth evolved from pillow stage to diapir stage between end of Washita time and end of Eagle Ford deposition. Cross section and map of interpretation of seismic data by Loocke (1978).

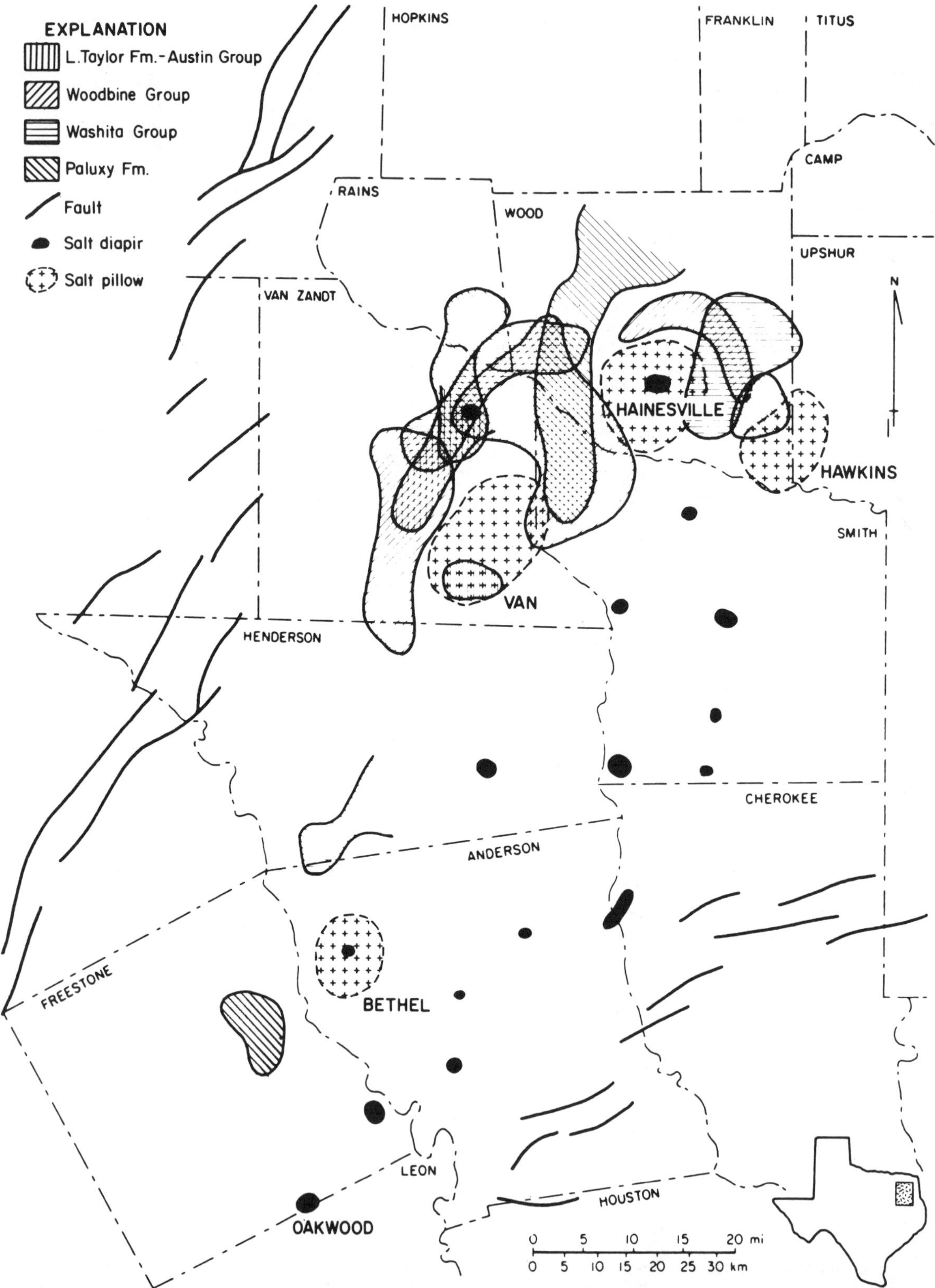

FIG. 7—Primary peripheral sinks in East Texas basin based on isopach maps of four stratigraphic units. Strata in primary peripheral sinks thickened mainly by salt flow from areas updip of salt pillow, with subordinate lateral flow into growing pillow.

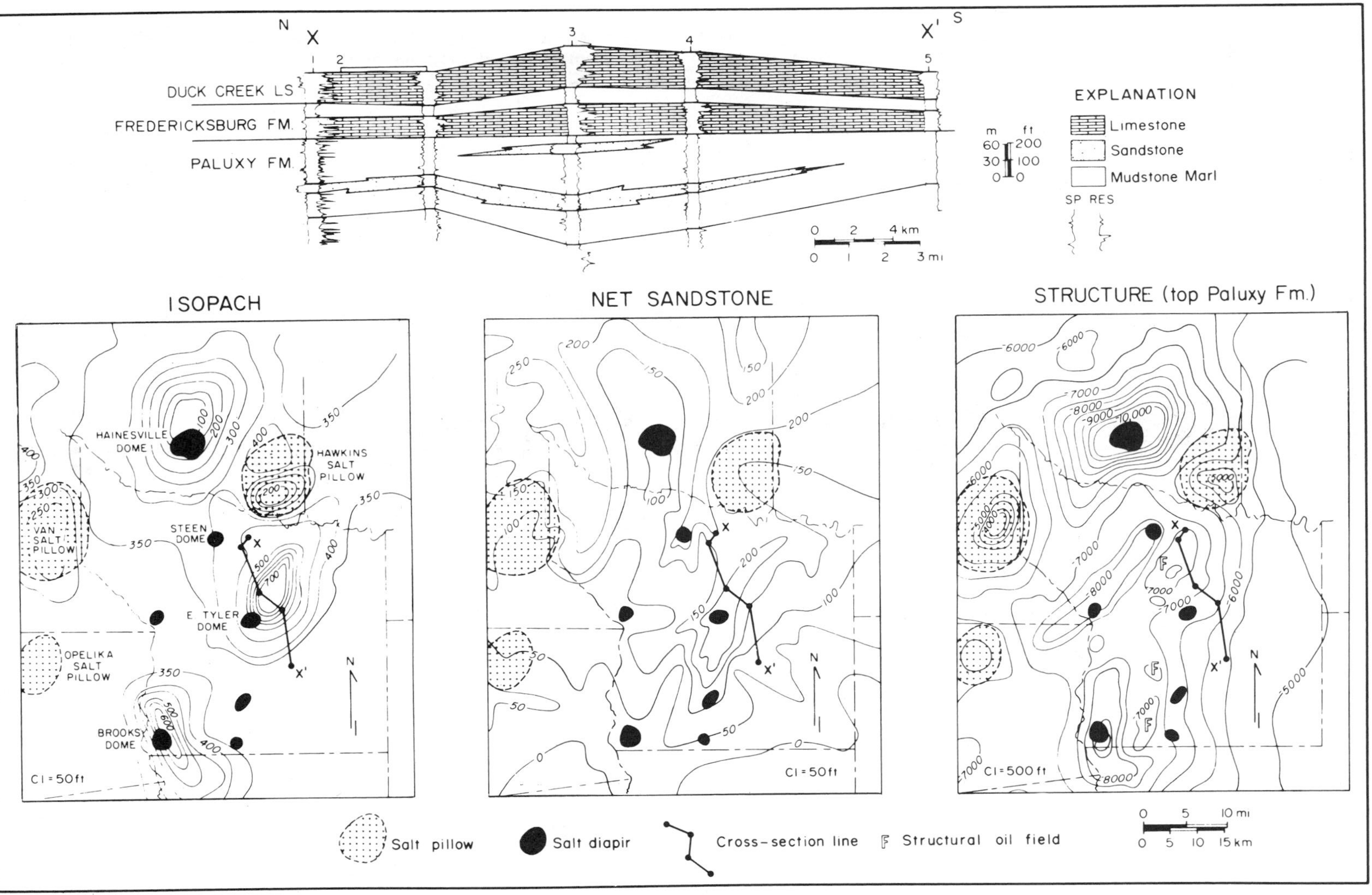

FIG. 8—Cross section XX′ and maps of isopach, net sandstone, and structure (top Paluxy) of Paluxy Formation, northern part of East Texas basin. Effects of both pillow (Van, Hawkins, and Hainesville) and diapiric (Brooks and East Tyler) salt movement are shown. Sand body in withdrawal basin around East Tyler dome (cross section and net sandstone map) was isolated from sandstone feeder system between Hainesville and Hawkins pillows by subtle structural saddle east of Steen dome (structure map). This saddle was also a topographic high during Paluxy deposition (isopach map).

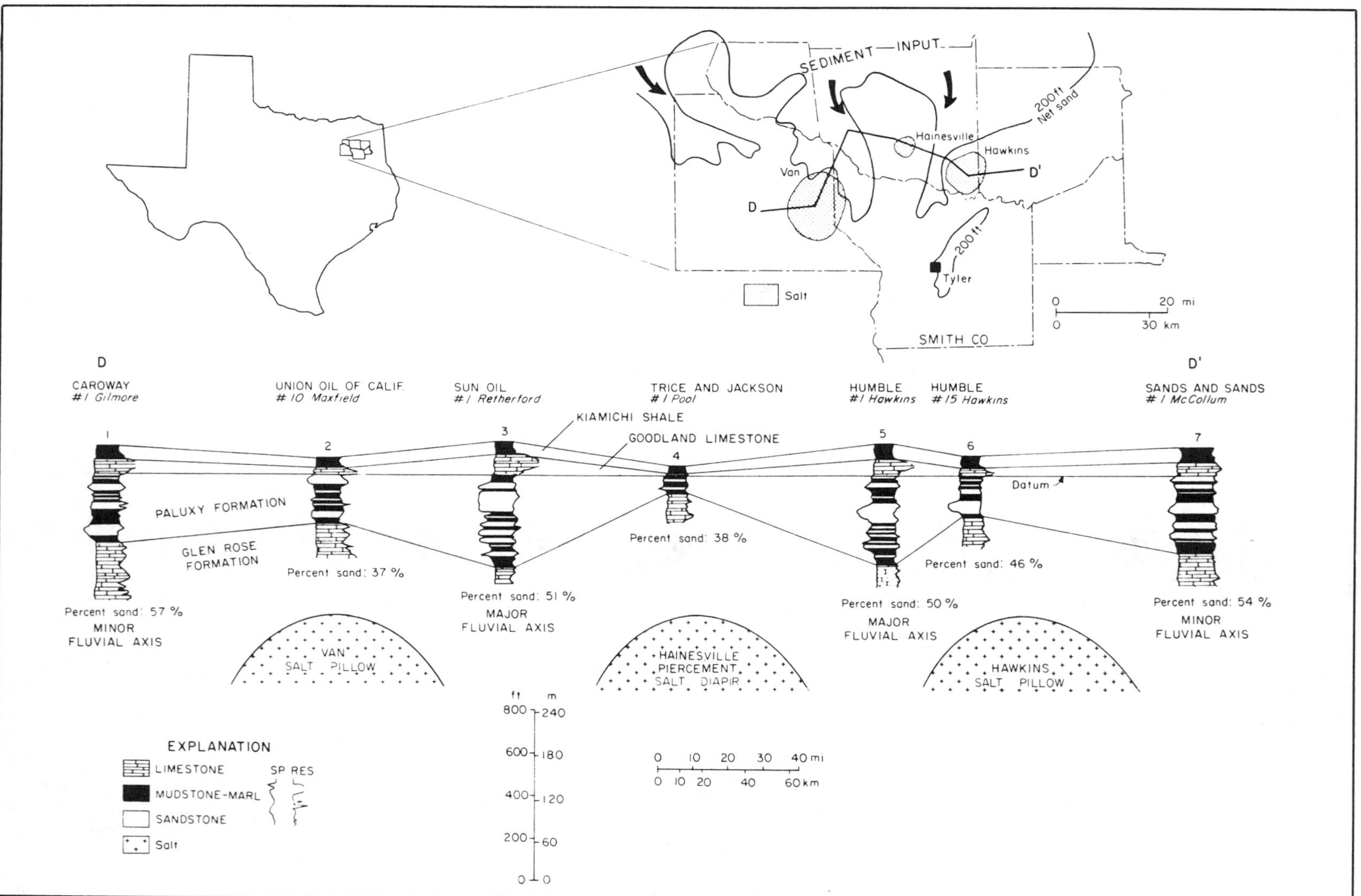

FIG. 9—Cross section DD′ across Van, Hainesville, and Hawkins salt structures, northern part of East Texas basin. Decreased thickness and sand percentage over each structure indicate that fluvial systems bypassed topographic swells over salt structures during Paluxy deposition.

formed, overthickened sediments and flanked by collapsed, thinned sediments (Fig. 10E). Interdomal strata thereby undergo structural reversal from synclines to anticlines creating a turtle-structure anticline (Trusheim, 1960), whereas the reverse takes place for strata immediately adjacent to diapirs. Turtle-structure anticlines are economically important because they have yielded 363 million bbl of oil, or 22% of the cumulative oil production, from the central part of the East Texas basin (Wood and Giles, 1982).

The diapir stage of salt movement is therefore characterized by deep sediment-filled sinks that surround or flank the salt dome as rim synclines (Fig. 4). Secondary peripheral sinks contain thicker sediment accumulations and are of greater area than the primary or tertiary peripheral sinks. Diapiric uplift exposes overlying strata to erosion, thereby destroying the sedimentary record over the diapir. We can only speculate on the nature of these sedimentary environments (Fig. 4). Units thin near the dome crests, commonly abruptly; this thinning may be either syndepositional or postdepositional.

Geometry of surrounding strata.—Seven secondary peripheral sinks are recognized around Bethel, Brooks, Boggy Creek, East Tyler, Hainesville, La Rue, and Steen domes (Fig. 11). These basins vary from equidimensional to elongate in plan. The axial traces of most of the secondary peripheral sinks intercept the associated domes. The remaining two traces are within 6 km (4 mi) of the associated domes.

Axial traces of these secondary peripheral sinks are aligned in two dominant directions, northwest and northeast. Possible controls of this alignment are orientation of early salt anticlines (northeast) and their crestal depressions (northwest), interference folding of salt, or regional faulting (Jackson, 1982). The orientation of salt-withdrawal basins may in turn partly control similar orientations of surface lineaments in the East Texas basin (Dix and Jackson, 1981).

Secondary peripheral sinks are up to 215% thicker than the strata unaffected by salt movement. The maximum observed thickness increase is 1,347 m (4,420 ft) in the fine-grained terrigenous clastics and carbonates of the Austin through Midway Groups around Hainesville dome. In Figures 12 and 13 the effects of this thickening are shown for the lower Taylor and Austin Groups. The time of maximum withdrawal-basin subsidence was different around different domes, even adjacent domes (Figs. 5, 11, 14; also see Seni and Jackson, 1983, pt. 2).

Depositional facies and lithostratigraphy.—Marine and deltaic strata (mostly limestone and fine-grained terrige-

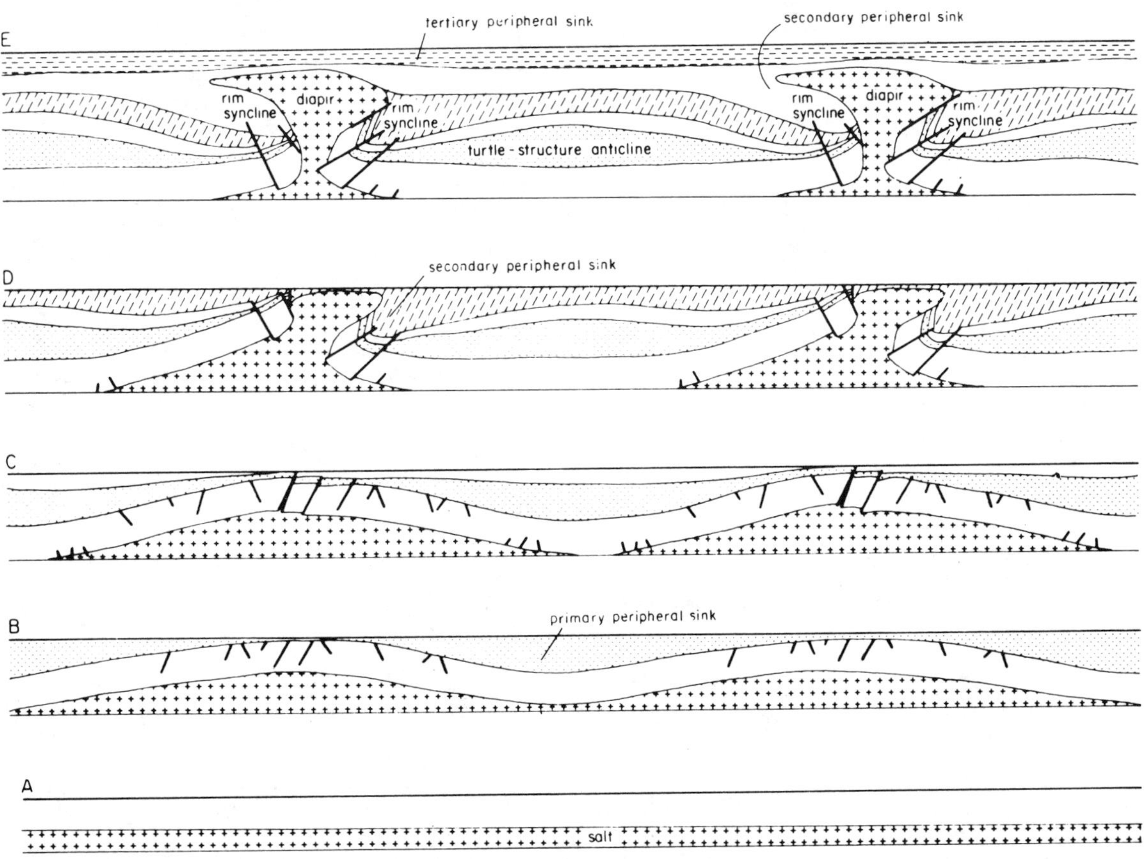

FIG. 10—Schematic cross sections showing inferred evolution of salt structures from pillow stage (B and C), through diapir stage (D), to post-diapir stage (E). Modified from Trusheim (1960).

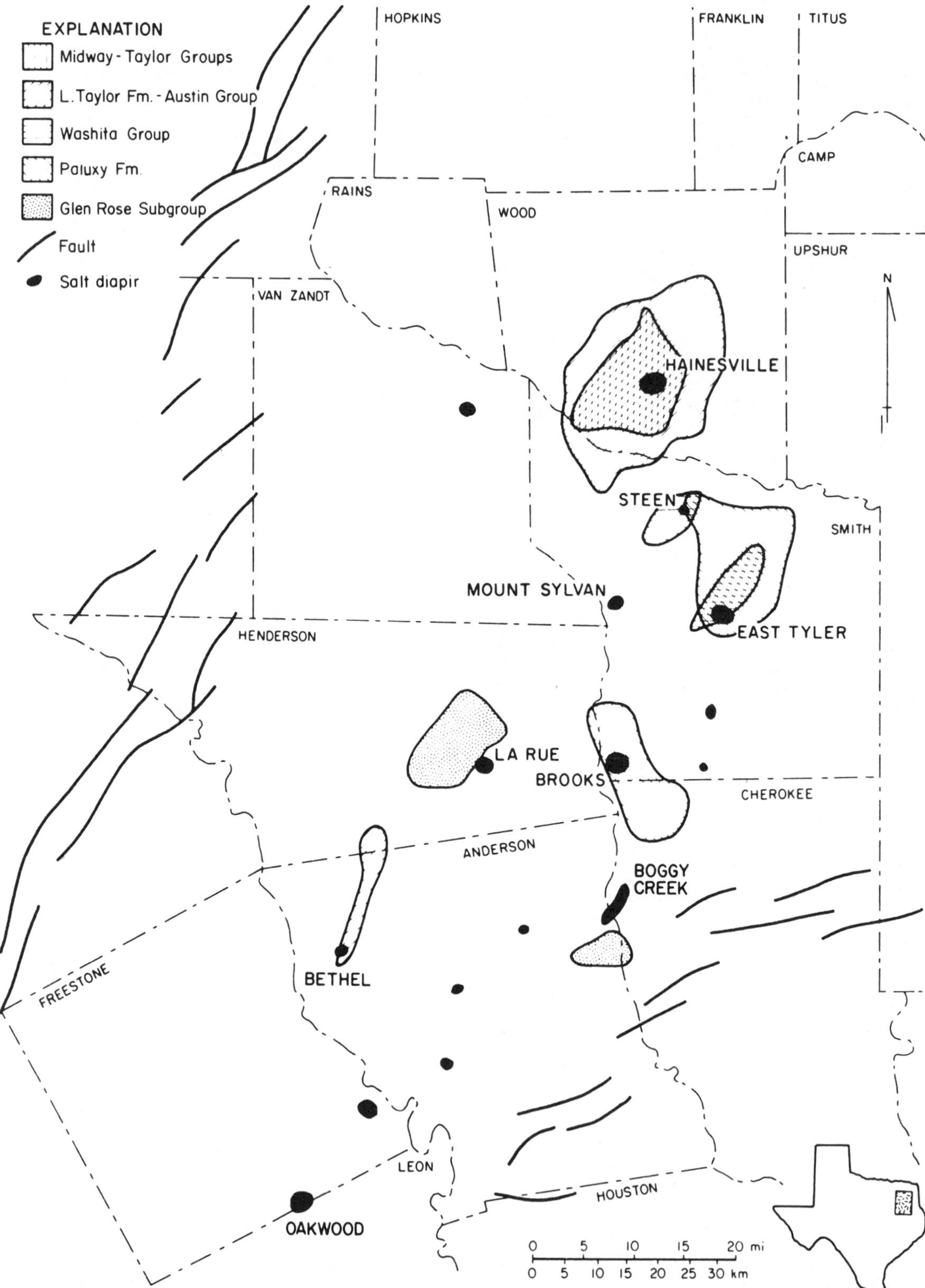

FIG. 11—Secondary peripheral sinks in East Texas basin from isopach maps of five stratigraphic units. Only sinks thickened greater than 50% with respect to regional thickness are shown. Actual area affected by salt withdrawal is much greater than secondary peripheral sinks shown here (compare Figs. 13, 14, 15).

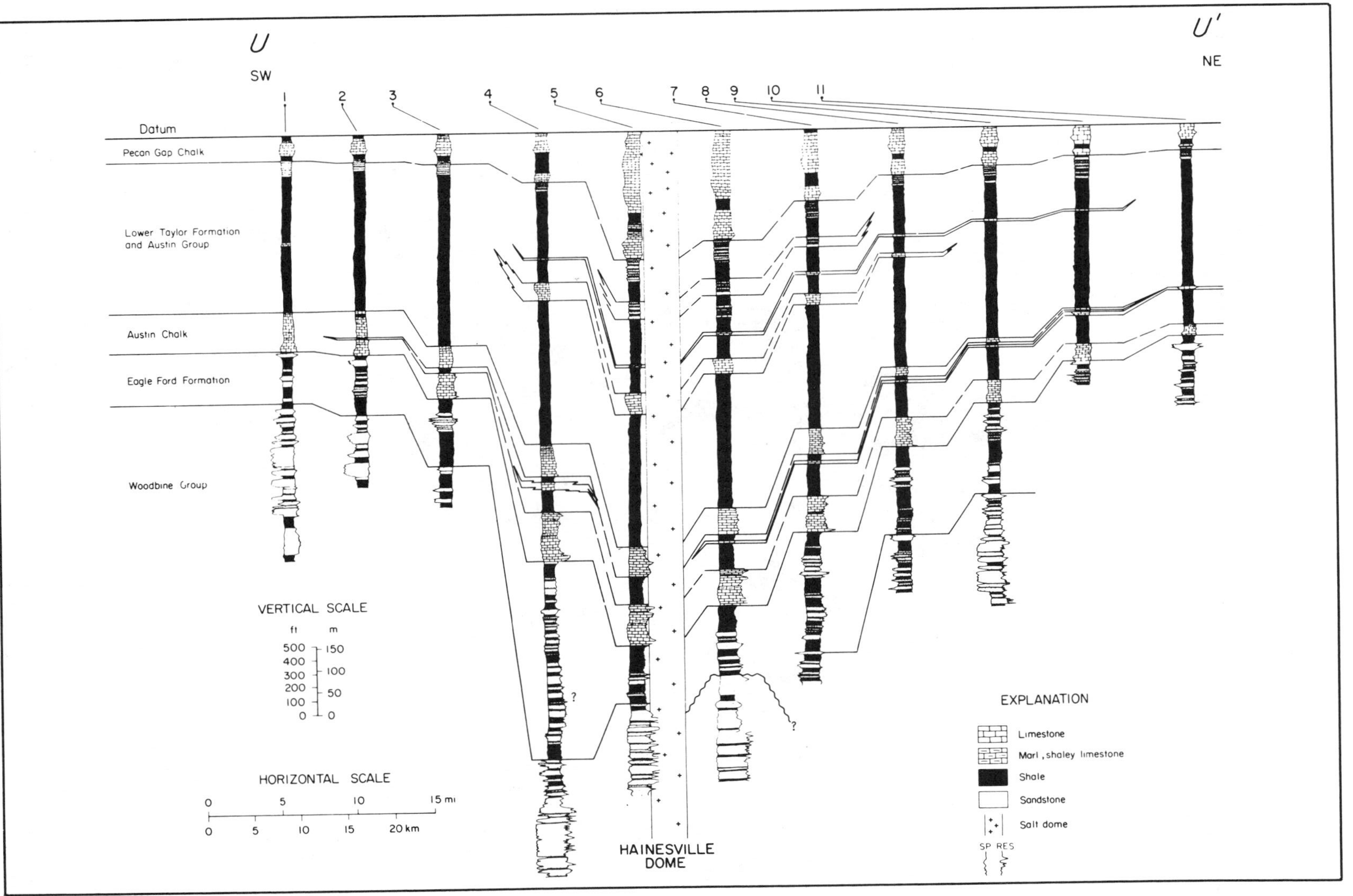

FIG. 12—Lithostratigraphic cross section UU′, pre-Pecan Gap Chalk (Upper Cretaceous) strata, in secondary peripheral sink around Hainesville dome. Stratigraphic section expands up to 215%, but lithic variations are minor. Cross section located on Figures 2 and 13.

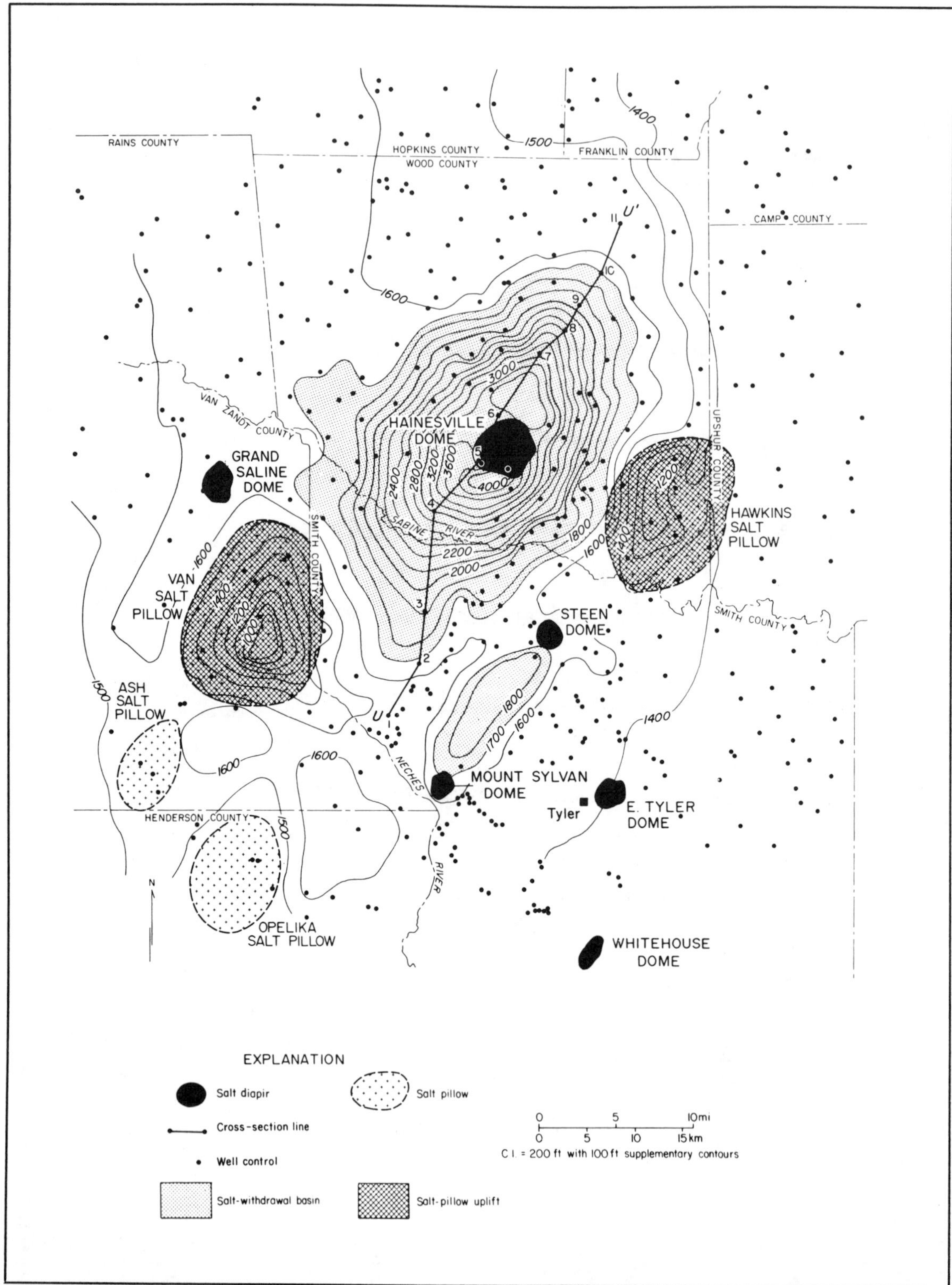

FIG. 13—Isopach map, lower Taylor-Austin Group, around Hainesville dome, northern part of East Texas basin. Axial trace (approximately cross-section line UU′, Fig. 12) of secondary peripheral sink intercepts Hainesville dome.

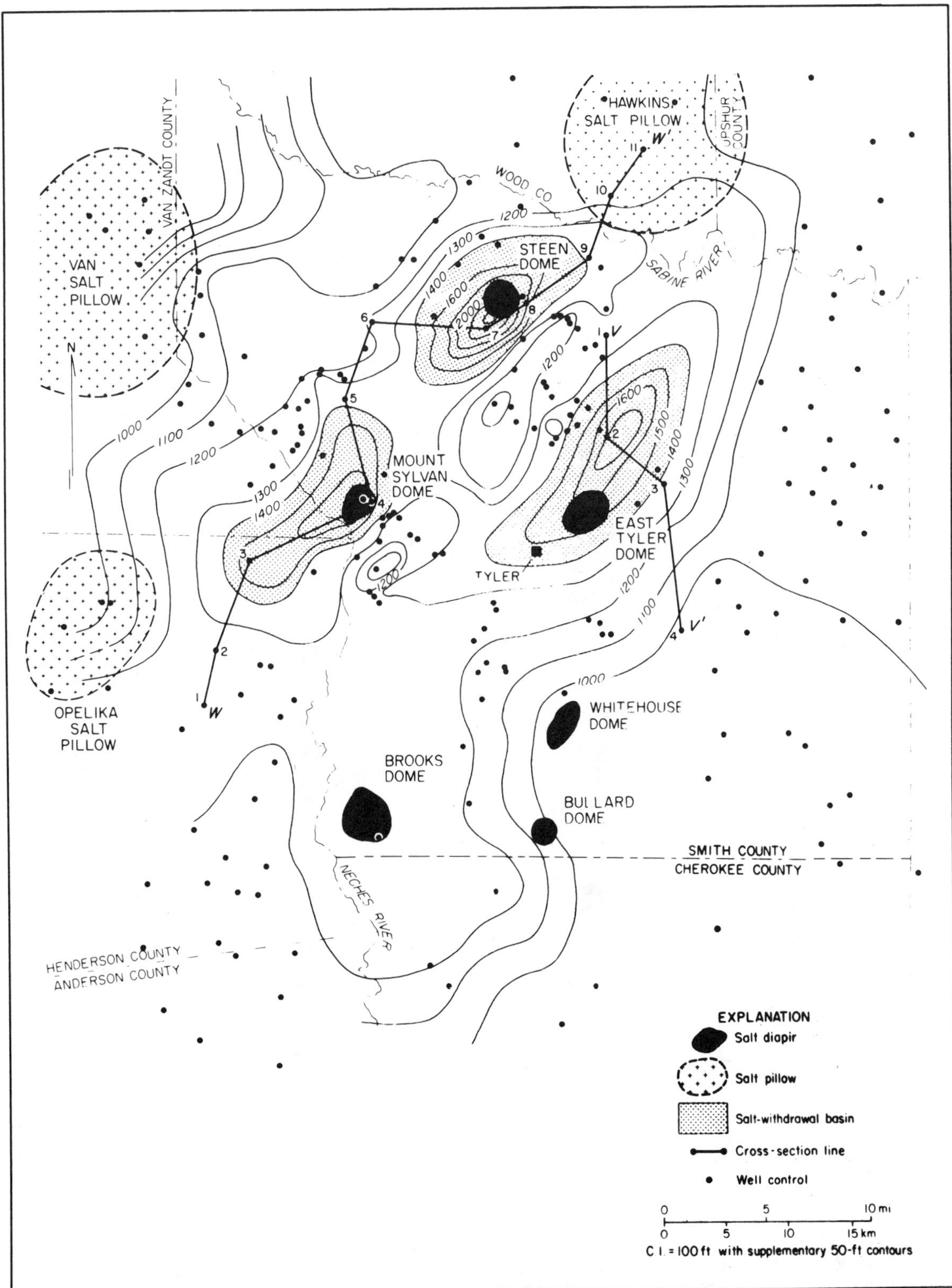

FIG. 14—Isopach map, Washita Group, central East Texas basin, showing overthickened strata in salt-withdrawal basins around Mount Sylvan, Steen, and East Tyler domes.

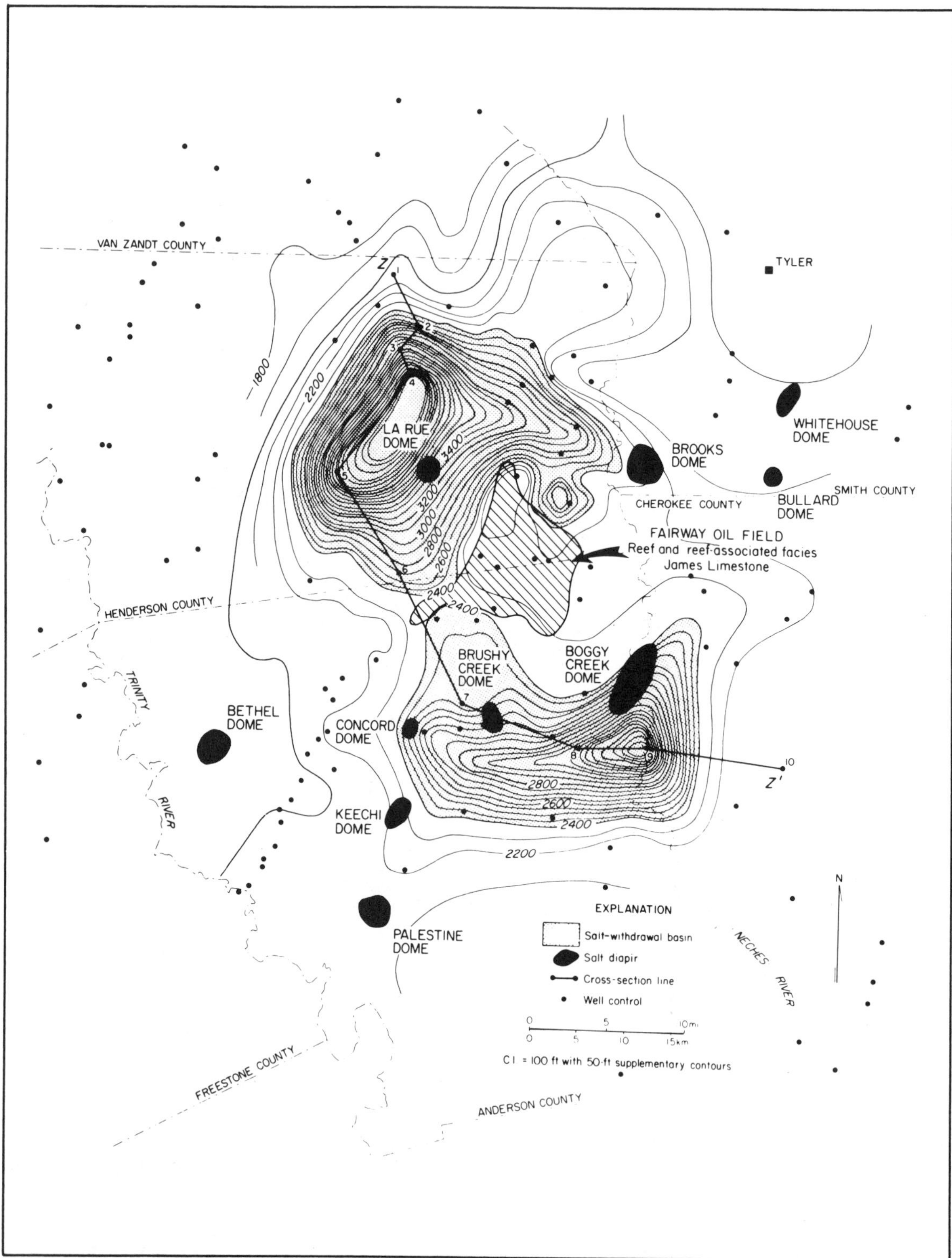

FIG. 15—Isopach map, Glen Rose Subgroup, central East Texas basin near La Rue, Brushy Creek, and Boggy Creek domes. Domes are flanked by large secondary peripheral sinks indicating rapid dome growth during Glen Rose deposition. Fairway field (lined pattern) is located in reef and reef-associated facies on elevated saddle between withdrawal basins. Cross section ZZ′ is shown in Figure 16.

nous clastics) dominate the expanded stratigraphic section within secondary peripheral sinks in the East Texas basin. Uplift and erosion occurred over the diapir during subsidence and deposition of deeper water facies in the adjacent peripheral sinks.

La Rue and Brushy Creek domes are surrounded by prominent salt-withdrawal basins, containing 152 km^3 (37 mi^3) of overthickened strata (Fig. 15). The region between these large basins was an elevated saddle, which favored growth of reefs during deposition of the Lower Cretaceous James Limestone (Fig. 16). Today these reefs and reef-associated facies host oil production of the Fairway field (Terriere, 1976) (Fig. 15).

Diapiric growth of Steen and East Tyler domes affected distribution of sand and mud in nearshore deposits of the Paluxy Formation (Fig. 8). A strike-oriented trend of aggregate sandstone thickness is isolated in the mudstone fill of a withdrawal basin around East Tyler dome (Fig. 8). Greater subsidence in this basin preserved what is interpreted to be a barrier bar or shelf-sand body produced by delta destruction.

Post-Diapir Stage

Post-diapir growth can be viewed as the waning phase of salt movement after rapid growth during the diapir stage. The post-diapir stage is generally the longest stage of salt

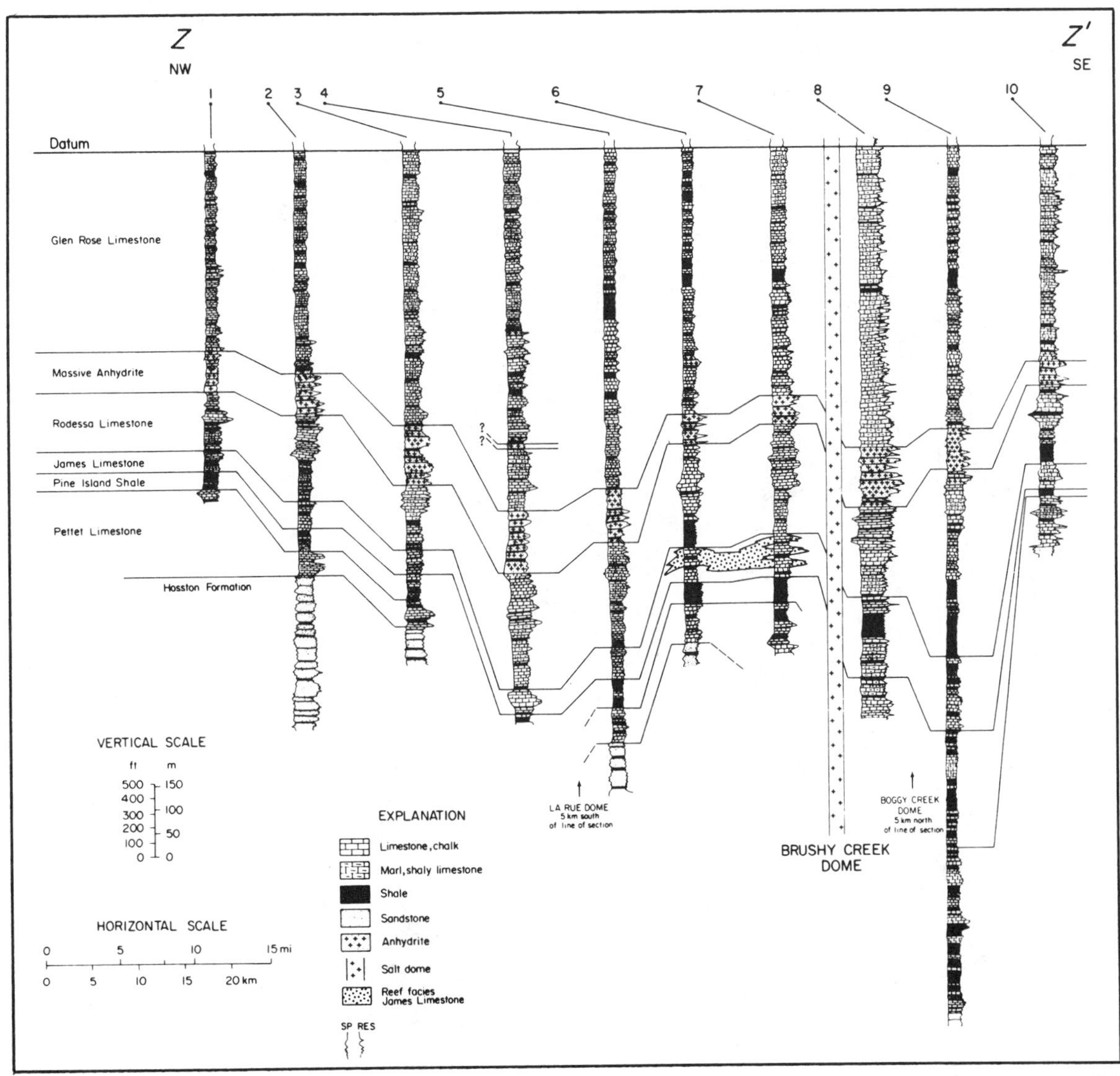

FIG. 16—Lithostratigraphic cross section ZZ′ near La Rue and Brushy Creek domes, East Texas basin, showing overthickening of Glen Rose strata in secondary peripheral sinks, and existence of reef facies in James Limestone in elevated saddle between sinks. Cross section located on Figures 2 and 15.

flow. Over geologic time this movement is steady-state compared with the relatively brief surge of diapirism. During the post-diapir stage, domes stay at or near the sediment surface despite continued regional subsidence and deposition.

Post-diapir salt movement is characterized by tertiary peripheral sinks (Trusheim, 1960). These sinks surround or flank domes, and some are characterized by lithologic variations in fluvial deposits that encase the diapirs. Given the contour interval used in this study (30 m, 100 ft), changes in thickness may be too subtle to reveal some tertiary peripheral sinks.

All diapirs examined here show evidence of post-diapir growth. All but five of these domes are within 600 m (2,000 ft) of the surface (the exceptions being Boggy Creek, Brushy Creek, Concord, Girlie Caldwell, and La Rue domes). The post-diapir rise of these five deep domes failed to keep pace with sedimentation and subsidence of the salt source layer in the center of the East Texas basin.

The influence of post-diapir salt flow on thickness and geometry of surrounding strata, on depositional systems, and on lithostratigraphy of the Eocene Wilcox Group was studied for three reasons: (1) post-diapir flow is minor, with little influence on surrounding strata, thus its effects are best revealed in the youngest units, which have been less complicated by differential subsidence and compaction; (2) domes have not completely "pierced" the Wilcox Group in East Texas so that supra-dome strata can also be investigated; and (3) sand-body geometry and depositional systems of the Wilcox in Texas are well known

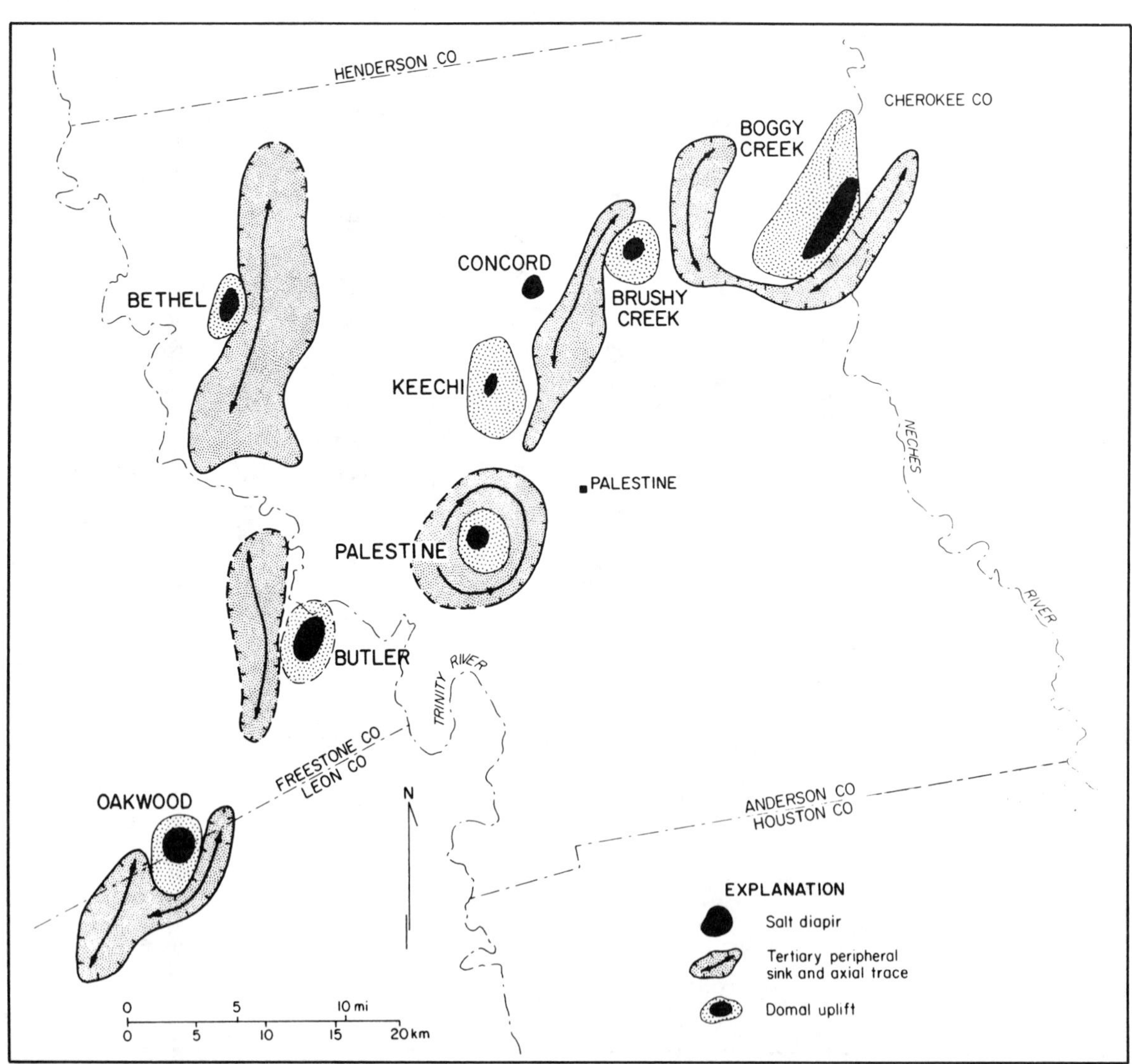

FIG. 17—Map of tertiary peripheral sinks and uplifted areas in Wilcox Group over domes in southern part of East Texas basin. Subsidence in sinks affects greater areas than does uplift over domes. Strata are well preserved in withdrawal areas and are poorly preserved in uplifted areas.

(Fisher and McGowen, 1967; Kaiser, 1974; Kaiser et al, 1978, 1980).

Geometry of surrounding strata.—In the southern part of the East Texas basin, eight diapirs were active in the early Tertiary and are flanked by tertiary peripheral sinks 8 to 40% thicker than areas unaffected by salt flow (Fig. 17). The sink areas range from 20 to 100 km^2 (8 to 39 mi^2). The tertiary peripheral sink with the largest volume is on the eastern flank of Bethel dome. The uplifted and thinned areas over the dome crests cover 10 to 50 km^2 (4 to 19 mi^2) (Fig. 17), but rarely extend more than 3 km (2 mi) beyond the salt stock.

Depositional systems and lithostratigraphy.—Post-diapir growth produced mounds over the domes that locally influenced distribution of sand and mud in Wilcox fluvial deposits. Aggrading fluvial channels were preferentially localized by greater subsidence in tertiary peripheral sinks. Deflection of fluvial channels away from the domal mounds allowed deposition of fine-grained, floodplain sediments over the dome. These relationships are well illustrated in the southern part of the East Texas basin by eight domes in the interaxial areas between major sand belts of the Wilcox Group (Fig. 18).

Sand-body distribution in the Wilcox around Bethel (Fig. 19) and Oakwood domes (Fig. 20) illustrates the effect of dome growth on coeval sedimentation. The tertiary peripheral sink east of Bethel dome (Fig. 19) includes four stacked channel-fill sands, each more than 15 m (50 ft) thick. In contrast, uplifted strata over Bethel dome are thinned and include only one sand body greater than 15 m

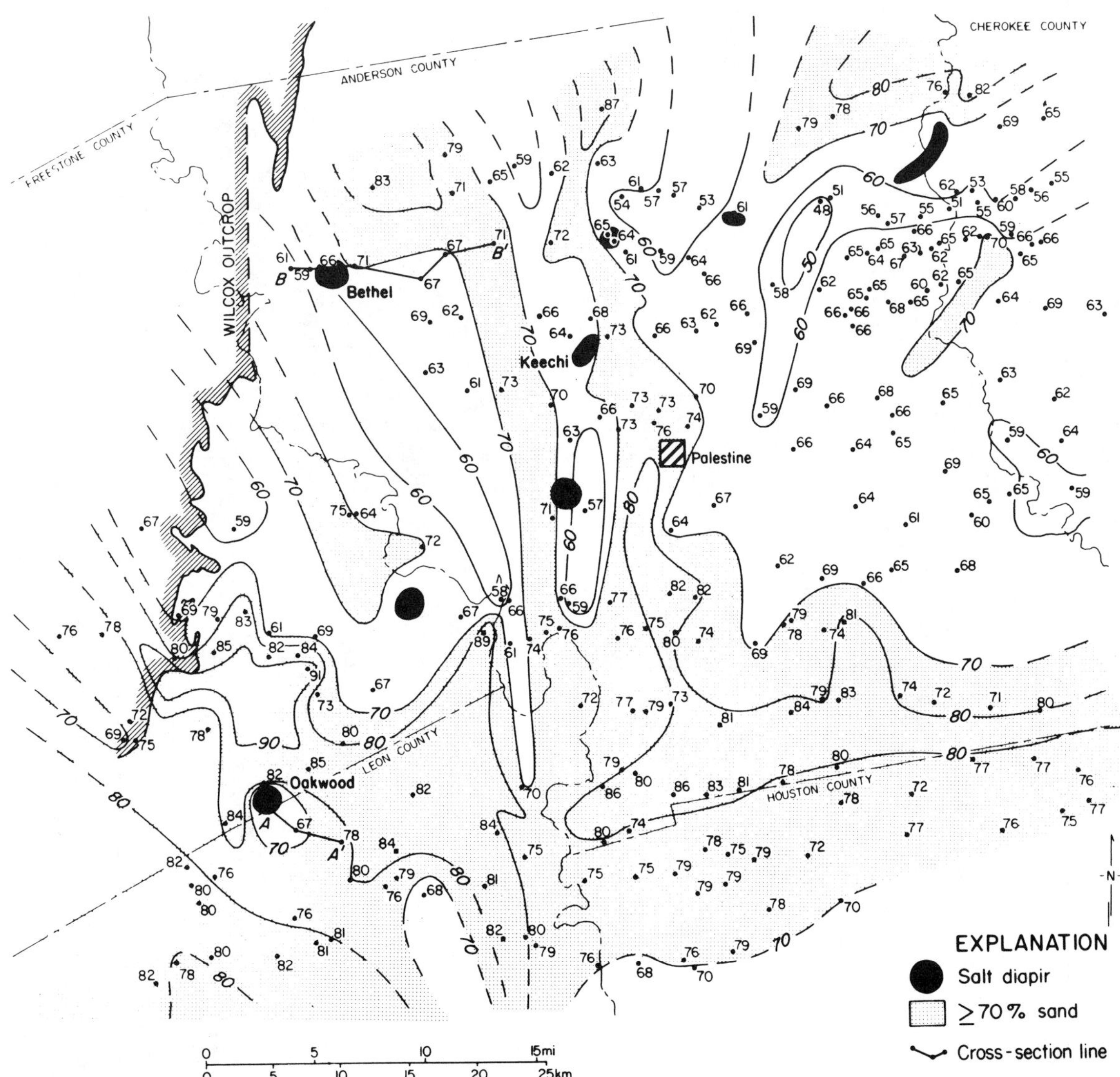

FIG. 18—Percentage sandstone map, Wilcox Group, southern part of East Texas basin. Eight diapirs in this area lie in interaxial areas containing lower percentage of sand.

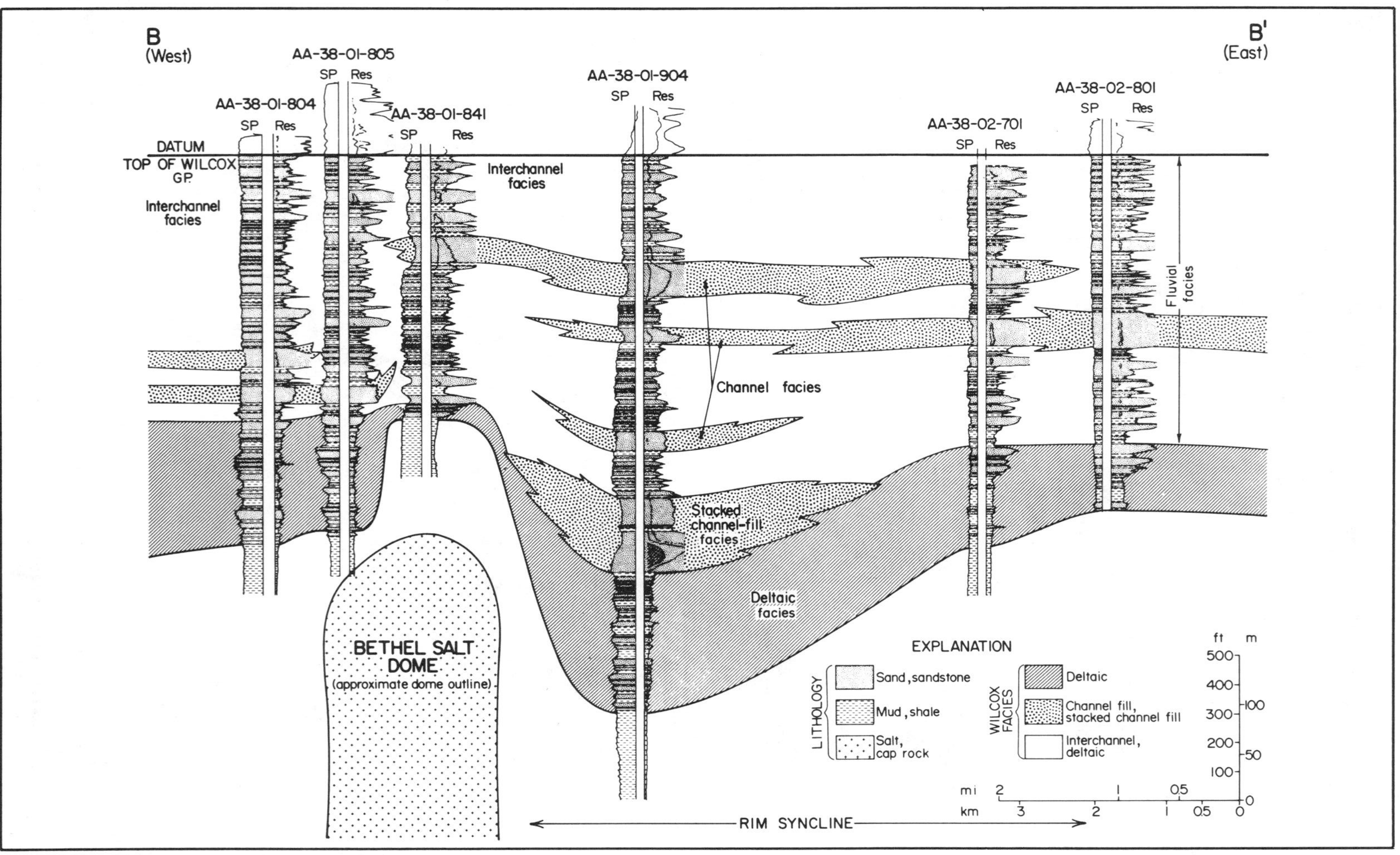

FIG. 19—Cross section BB′, Wilcox Group around Bethel Dome (cross section located on Figs. 2 and 18). Four channel-fill sandstones 15 m (50 ft) thick occupy the tertiary peripheral sink east of Bethel Dome. Five of the six sandstones pinch out over the diapir.

(50 ft) thick, although the percentage of sand is only slightly lower there than in the peripheral sink. Vertically stacked, channel-fill sands also dominate the tertiary peripheral sink 3 to 10 km (2 to 6 mi) southeast of Oakwood dome (Fig. 20). Muddy sediments dominate the flood plain over the dome and are interbedded with crevasse-splay sands 0.3 to 4.0 m (1 to 13 ft) thick. F- and t-statistical tests indicate that, at the 95% confidence level, strata over the diapirs contain 7 to 18% less sand than do strata in nearby channel axes.

Holocene analogs.—Surface mapping of the Texas coastal zone (Fisher et al, 1972, 1973; McGowen et al, 1976; McGowen and Morton, 1979) provides valuable information on Holocene topography and surficial-

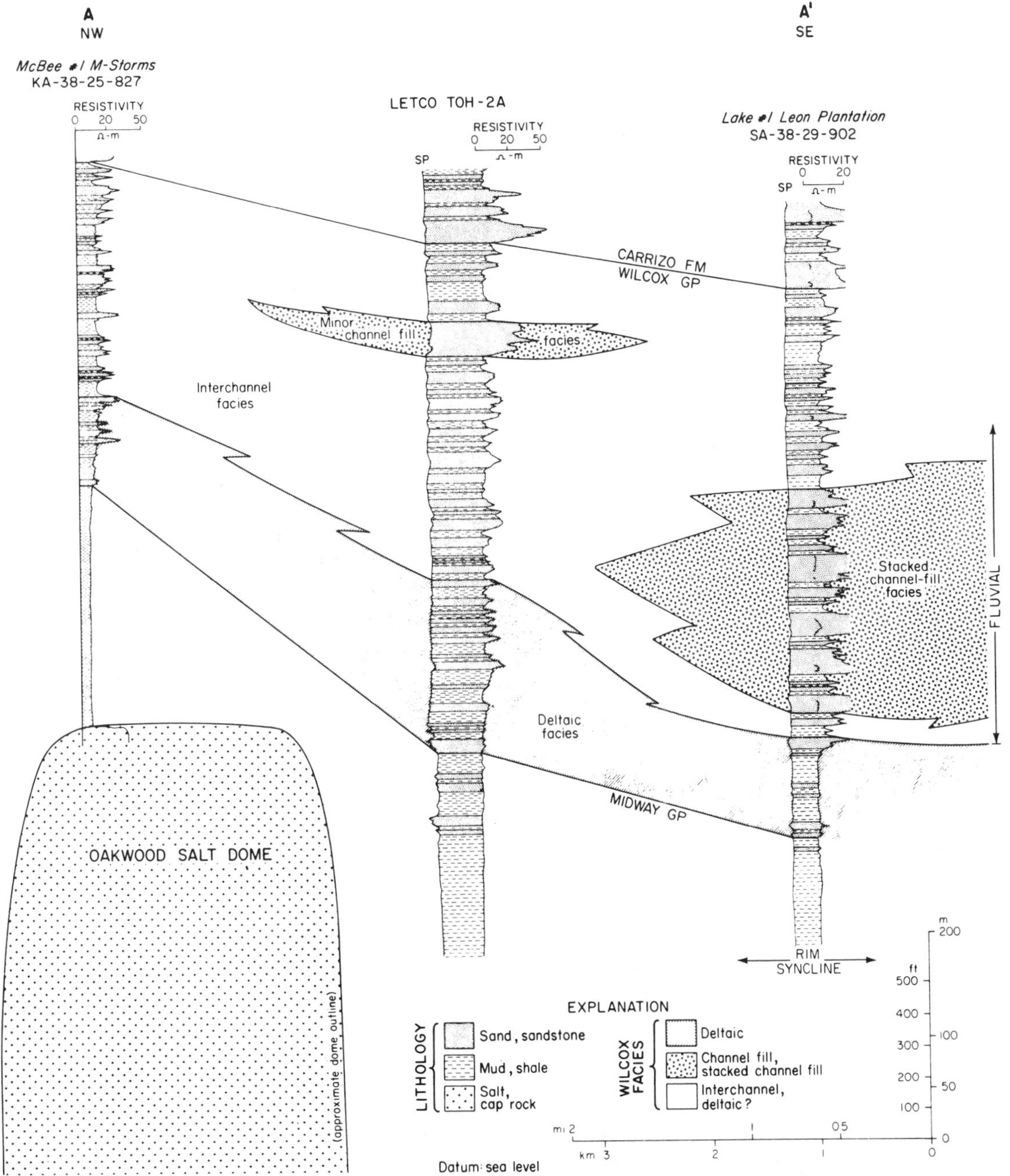

FIG. 20—Cross section AA′, Wilcox Group around Oakwood Dome (cross section located on Figs. 2 and 18). The tertiary peripheral sink contains sand-rich facies. A paleotopographic mound over the dome deflected Wilcox fluvial systems, so that thinned strata over the crest of the dome comprise mud-rich, floodplain facies.

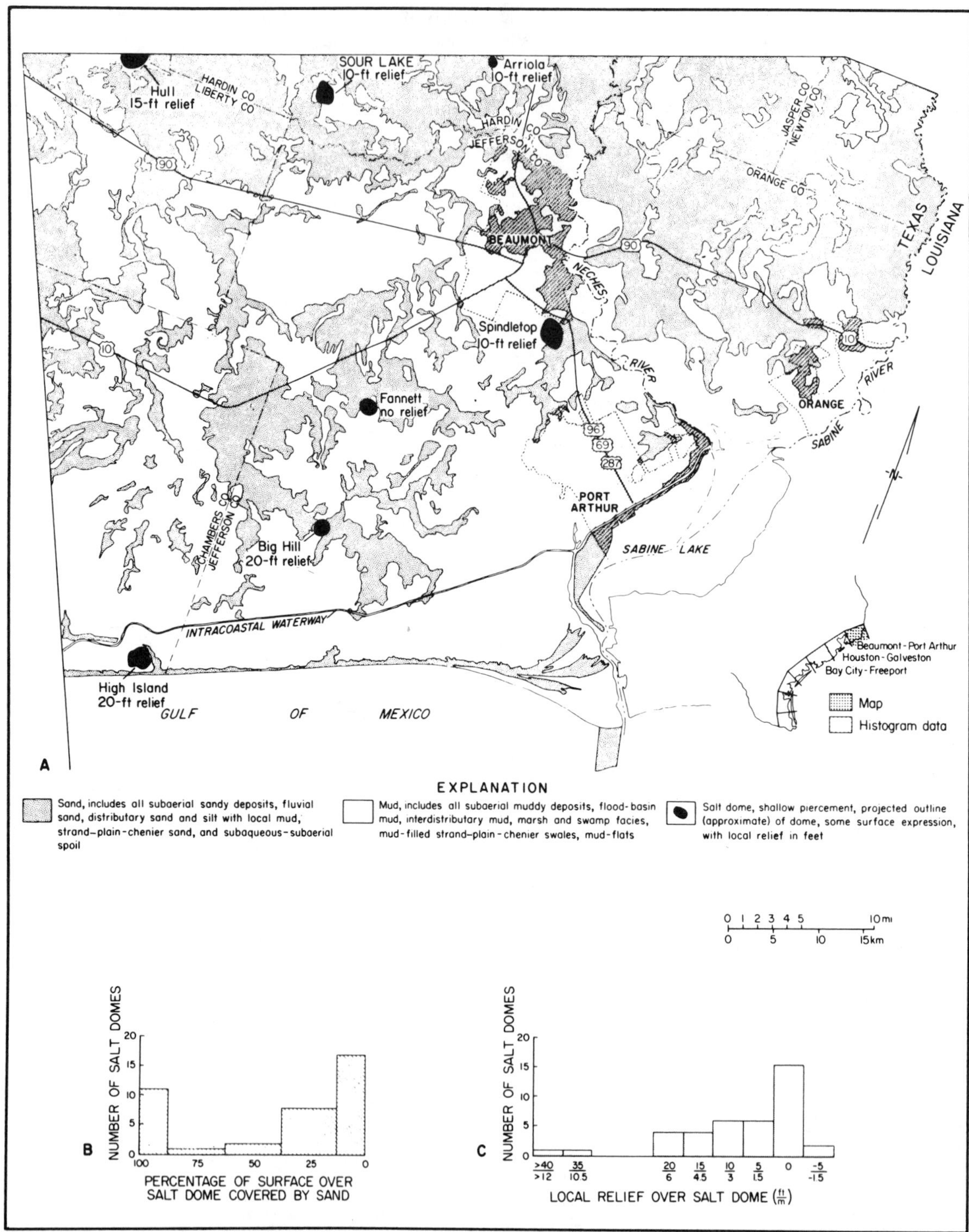

FIG. 21—Map of shallow salt domes and surficial sand and clay, uppermost Texas coast (Beaumont-Port Arthur area). Coastal diapirs are located preferentially along margins of dip-oriented sand belts or in muddy, interaxial areas. Histograms show relief over domes and percentage of surface over domes covered by sand on upper Texas coast south to the Bay City–Freeport area. Abundant surficial sand (including eolian) blankets most of upper Texas coast in the Bay City–Freeport area. Domes in this anomalously sandy area constitute about half the domes in the 75-100% sand class. (Histograms after Fisher et al, 1972, 1973; McGowen et al, 1976; map after Fisher et al, 1973.)

sediment distribution over shallow domes in a post-diapir stage for analogy with the East Texas domes during early Tertiary deposition (Fig. 21).

Fifty-six percent of the diapirs on the upper Texas coast have greater than 1.5 m (5 ft) positive relief over the domes (Fig. 21C). This relief has apparently influenced the distribution of Holocene surficial sediments. Texas coastal diapirs generally occur in sand-poor areas or along sand-belt margins. Since the Tertiary, fluvial facies and environments of the Texas coastal zone have tended to stack vertically owing to rapid subsidence. For example, sandstones are vertically stacked in the upper Pliocene and Pleistocene fluvial-deltaic sequences in the Houston-Galveston area (Kreitler et al, 1977). Thus, the present association of coastal diapirs in mud-rich surficial deposits indicates a high probability that older deposits encasing the diapirs are also mud rich. The lack of relief over some salt structures is related to greater depth of burial, to cessation of upward growth, or to dissolution.

The modern Persian Gulf is a shallow epicontinental sea with many similarities to the East Texas basin during the Mesozoic. Holocene sediments in the Persian Gulf are primarily carbonates similar to the Washita Group and Glen Rose Subgroup. Shallow salt domes form mounds on the sea floor, and particularly active diapirs form islands with salt exposed at the surface (Purser, 1973; Kent, 1979). Some of these salt-dome islands (Fig. 22) are flanked by arcuate depressions inferred to be the surface expression of rim synclines (Purser, 1973). A zone of coral-algal reefs fringes many salt-cored islands and sea floor mounds,

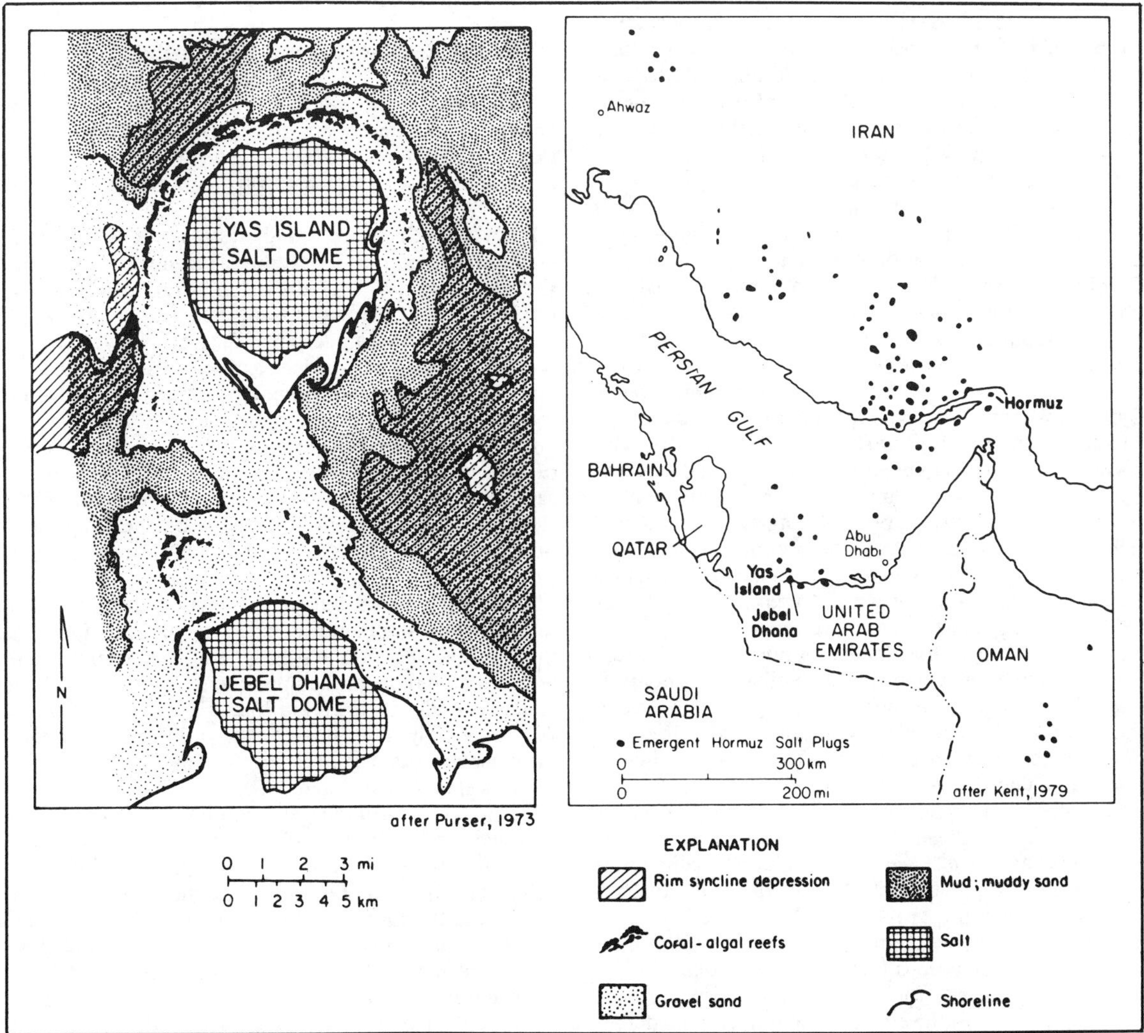

FIG. 22—Bathymetry and the distribution of modern carbonate sediments in the Persian Gulf are strongly controlled by salt flow. Yas salt dome forms an island north of Trucial Coast, United Arab Emirates (righthand map), flanked by coral-algal reefs and carbonate sand and gravel. Area between Yas and Jebel Dhana salt dome on Trucial Coast mainland also contains coarse carbonate clastics and patch reefs. Farther offshore from Yas, rim synclines are expressed as topographic depressions on sea floor in which carbonate mud and muddy sand are accumulating. Modified from Purser (1973).

whereas mud and muddy carbonate sand accumulate in topographic depressions of rim synclines located 1 to 5 km (0.6 to 3 mi) offshore (Fig. 22). The sea floor around Hormuz Island, the most spectacular salt-dome island, is littered with exotic blocks of the Hormuz Formation that have been rafted up by the salt, indicating that, in the past, salt was extruded on the surface and sea bottom (Kent, 1979).

DISCUSSION

Syndepositional lithostratigraphic variations in response to salt flow highlight the interdependence among sediment accumulation, dome evolution, and potential for petroleum accumulation. These lithostratigraphic variations are primarily a function of paleotopography. Salt uplift formed swells and mounds over salt pillows and diapirs, respectively. Concurrently, topographic and structural basins formed over zones of salt withdrawal, a process that formed saddles with residual elevation between the basins. This salt-related topography influenced sedimentation patterns, which, in turn, enhanced continued salt flow by increased sedimentary loading in the basin.

In the East Texas basin, growth of salt pillows was responsible for uplift and thinning in areas of 100 to 400 km^2 (40 to 150 mi^2), whereas diapir growth caused uplift and thinning in areas of only 10 to 50 km^2 (4 to 20 mi^2). Continued domal "piercement" commonly destroyed the uplifted strata by erosion or by shoving the uplifted units aside in trapdoor manner. In contrast, much of the very broad, thinned zone over pillow crests was preserved after pillow collapse when diapirism buried the thinned region deep below secondary and tertiary peripheral sinks.

Dome and pillow uplifts influenced net-sandstone trends because fluvial systems bypassed mounds. Uplifted areas, therefore, tend to be thin and sand poor. Subsidence of the peripheral sinks, in turn, promoted aggradation of sand-rich fluvial-channel facies. These variations are commonly illustrated in nonmarine facies deposited both in pillow stage (Paluxy Formation) and post-diapir stage (Wilcox Group) sinks, but are rare in marine facies deposited in diapir-stage sinks. Under marine conditions, sand can accumulate by winnowing on bathymetric shoals, so that salt domes with sufficient surface expression, such as those in the modern Persian Gulf, are overlain by sand-rich sediments, in direct contrast to diapirs in fluvially dominated depositional environments. Small reefs might also be expected on topographic highs over dome crests, but these have not been found in East Texas. Such dome-crest reefs have been recognized in Oligocene sediments of the Texas Gulf Coast (Cantrell et al, 1959), in Holocene strata in the northwestern Gulf of Mexico (Bright, 1977; Rezak, 1977), and in the modern Persian Gulf (Purser, 1973). Lower Cretaceous reefs have been found in East Texas on saddles between salt-withdrawal basins.

During diapirism, the effect of the topographic depression in the peripheral sink far overshadows the effect of uplift over the dome. Diapir growth is characterized by enormous secondary peripheral sinks. In East Texas the largest secondary peripheral sink encloses 1,000 km^2 (390 mi^2), around Hainesville dome. Low-energy marine facies characteristically dominate the fill of secondary peripheral sinks. In contrast to secondary sinks, primary and tertiary peripheral sinks are usually difficult to map because they are only slightly thicker than surrounding strata, and because of interference from other active salt structures nearby.

Locations of sinks are related to evolutionary stage and regional dip. Axial traces of primary peripheral sinks are 10 to 20 km (6 to 12 mi) from the crest of the associated pillows, and tend to be located updip of the structure. In contrast, secondary and tertiary peripheral sinks commonly encircle the diapir. This shifting of the peripheral sinks through time reflects the changing character of salt migration through the various stages of dome growth. The primary change was from predominantly downdip lateral flow in the pillow stage to a combination of centripetal and vertical flow in the diapir and later stages.

SIGNIFICANCE TO SUBTLE PETROLEUM TRAPS

Variations in thickness and syndepositional facies characterize near-dome strata during salt flow. These variations enable inference of dome-growth stages and provide a framework with which to predict subtle hydrocarbon traps.

Pillow growth caused broad crestal uplift so that syndepositionally and postdepositionally thinned strata overlie the pillow crest. Fluvial and deltaic strata deposited over the crests of salt pillows are sand-poor but are likely to be flanked by stratigraphic pinch-outs of sandy reservoirs. Sand-rich fluvial channel systems bypassed pillow crests and occupied adjacent primary peripheral sinks. Under marine conditions, paleotopographic swells over pillows are potential reservoirs because they were preferred sites of reef growth, high-energy grainstone deposition, and sand concentration by winnowing. Primary peripheral sinks formed preferentially updip of the salt pillows because of greater salt flow into the pillow from the updip side.

Structural reversal during diapirism transforms a primary peripheral sink into a turtle-structure anticline. Thus the location of a primary peripheral sink establishes the position of the core of the subsequent turtle-structure anticline, generally 10 to 20 km (6 to 12 mi) updip from the dome crest. This is a valuable exploration guide for one of the most important salt-related structural traps, especially at the deeper, less explored horizons.

During diapirism, large secondary peripheral sinks enclosed or flanked the diapir. In East Texas, marine strata dominate the fill of secondary peripheral sinks and represent expanded but otherwise normal, low-energy sequences. Because secondary peripheral sinks were local sites of greater subsidence and, hence, topographic lows, they were more likely to preserve marine sand bodies formed during transgressive reworking. These pinch-outs of marine sand bodies may subsequently act as subtle hydrocarbon traps.

Sea-floor mounds over diapirs may become petroleum reservoirs because they were sites for reef growth, grainstone deposition, and sand concentration by winnowing, as in the case of pillows. In contrast, however, these supradomal mounds were much smaller than analogous suprapillow swells and were almost invariably destroyed by further uplift, erosion, and salt emplacement.

Another important stratigraphic trap may be formed during diapirism. Raised saddles between secondary peripheral sinks allowed reef growth in James Limestone (Lower Cretaceous Glen Rose Subgroup); both the structure and lithology of these saddles favored petroleum accumulation, such as the giant Fairway field in Henderson and Anderson Counties.

Post-diapir growth had only a minor effect on surrounding strata. Mounds over domes undergoing post-diapir growth deflected Wilcox fluvial channel systems around supradome areas, so that mud-rich interaxial sediments were deposited over the diapir. Differential subsidence caused Wilcox fluvial channel sandstones to stack vertically in tertiary peripheral sinks. Subtle petroleum traps formed during this stage are expected to be much smaller than those formed during earlier stages of diapirism.

REFERENCES CITED

Atwater, G. I., and M. J. Forman, 1959, Nature of growth of southern Louisiana salt domes and its effect on petroleum accumulation: AAPG Bulletin, v. 43, p. 2592-2622.

Barton, D. C., 1933, Mechanics of formation of salt domes with special reference to Gulf Coast salt domes of Texas and Louisiana: AAPG Bulletin, v. 17, p. 1025-1083.

Bishop, R. S., 1978, Mechanism for emplacement of piercement diapirs: AAPG Bulletin, v. 62, p. 1561-1583.

Bornhauser, M., 1958, Gulf Coast tectonics: AAPG Bulletin, v. 42, p. 339-370.

——— 1969, Geology of Day dome (Madison County, Texas)—a study of salt emplacement: AAPG Bulletin, v. 53, p. 1411-1420.

Bright, T. J., 1977, Coral reefs, nepheloid layers, gas seeps, and brine flows on hard banks in the northwestern Gulf of Mexico: Third International Coral Reef Symposium Proceedings, no. 3, v. 1, p. 39-46.

Bushaw, D. J., 1968, Environmental synthesis of the East Texas Lower Cretaceous: Gulf Coast Association of Geological Societies Transactions, v. 18, p. 416-438.

Cantrell, R. B., J. C. Montgomery, and A. E. Woodard, 1959, Heterostegina reef on Nash and other piercement salt domes in northwestern Brazoria County, Texas: Gulf Coast Association of Geological Societies Transactions, v. 9, p. 59-62.

Caughey, C. A., 1977, Depositional systems in the Paluxy Formation (Lower Cretaceous) northeast Texas—oil, gas and ground water resources: University of Texas Bureau of Economic Geology Geological Circular 77-8, 59 p.

DeGolyer, E., 1925, Origin of North American salt domes: AAPG Bulletin, v. 9, p. 831-874.

Dix, O. R., and M. P. A. Jackson, 1981, Statistical analysis of lineaments and their relation to fracturing, faulting, and halokinesis in the East Texas basin: University of Texas Bureau of Economic Geology Report of Investigations, no. 110, 30 p.

Eaton, R. W., 1956, Resume on subsurface geology of northeast Texas with emphasis on salt structures: Gulf Coast Association of Geological Societies Transactions, v. 5, p. 78-84.

Elliot, T. L., 1979, Deposition and diagenesis of carbonate slope deposits, Lower Cretaceous, northeastern Mexico: PhD dissertation, University of Texas at Austin, 330 p.

Fisher, W. L., and J. H. McGowen, 1967, Depositional systems in the Wilcox Group of Texas and their relationship to occurrence of oil and gas: University of Texas Bureau of Economic Geology Geological Circular 67-4, p. 105-125.

——— L. F. Brown, J. H. McGowen, and C. G. Groat, 1973, Environmental geologic atlas of the Texas coastal zone—Beaumont-Port Arthur area: University of Texas Bureau of Economic Geology, 93 p.

——— J. H. McGowen, L. F. Brown, and C. G. Groat, 1972, Environmental geologic atlas of the Texas coastal zone—Galveston-Houston area: University of Texas Bureau of Economic Geology, 91 p.

Giles, A., and D. H. Wood, 1981, Petroleum accumulation patterns in the East Texas salt dome area, *in* Geology and geohydrology of the East Texas basin, a report on the progress of nuclear waste isolation feasibility studies (1980): University of Texas Bureau of Economic Geology Geological Circular 81-7, p. 67-77.

Granata, W. H., 1963, Cretaceous stratigraphy and structural development of the Sabine uplift area, Texas and Louisiana, *in* Report on selected north Louisiana and south Arkansas oil and gas fields and regional geology: Shreveport Geological Society, Reference Volume V, p. 50-95.

Halbouty, M. T., 1979, Salt domes Gulf region, United States and Mexico, second edition: Houston, Gulf Publishing Company, 561 p.

——— 1980, Methods used, and experience gained, in exploration for new oil and gas fields in highly explored (mature) areas: AAPG Bulletin, v. 64, p. 1210-1222.

Hughes, D. J., 1968, Salt tectonics as related to several Smackover fields along the northeast rim of the Gulf of Mexico basin: Gulf Coast Association of Geological Societies Transactions, v. 18, p. 320-339.

Jackson, M. P. A., 1982, Fault tectonics of the East Texas basin: University of Texas Bureau of Economic Geology Geological Circular 82-4, 31 p.

——— and S. J. Seni, 1983, Geometry and evolution of salt structures in a marginal rift basin of the Gulf of Mexico, east Texas: Geology, v. 11, p. 131-135.

——— ——— and M. K. McGowen, 1982, Initiation of salt flow, east Texas basin (abs.): AAPG Bulletin, v. 66, p. 584-585.

Jaritz, W., 1980, Einige Aspekte der Entwicklungsgeschichte der nordwestdeutschen Salzstocke: Zeitschrift der Deutschen Geologischen Gesellschaft, v. 131, p. 387-408.

Johnson, H. A., and D. H. Bredeson, 1971, Structural development of some shallow salt domes in Louisiana Miocene productive belt: AAPG Bulletin, v. 55, p. 204-226.

Kaiser, W. R., 1974, Texas lignite near surface and deep basin resources: University of Texas Bureau of Economic Geology Report of Investigations no. 79, 70 p.

——— W. B. Ayers, Jr., and L. W. La Brie, 1980, Lignite resources of Texas: University of Texas Bureau of Economic Geology Report of Investigations no. 104, 52 p.

——— J. E. Johnston, and W. N. Bach, 1978, Sand-body geometry and the occurrence of lignite in the Eocene of Texas: University of Texas Bureau of Economic Geology Geological Circular 78-4, 19 p.

Kent, P. E., 1979, The emergent Hormuz salt plugs of southern Iran: Journal of Petroleum Geology, v. 2, p. 117-144.

Kreitler, C. W., E. Guevara, G. Granata, and D. McKalips, 1977, Hydrogeology of Gulf Coast aquifers, Houston-Galveston area, Texas: University of Texas Bureau of Economic Geology Geological Circular 77-4, 18 p.

——— O. K. Agagu, J. M. Basciano, E. W. Collins, O. R. Dix, S. P. Dutton, G. E. Fogg, A. B. Giles, E. H. Guevara, D. W. Harris, D. K. Hobday, M. K. McGowen, D. Pass, and D. H. Wood, 1980, Geology and geohydrology of the East Texas basin, a report on the progress of nuclear waste isolation feasibility studies (1979): University of Texas Bureau of Economic Geology Geological Circular 80-12, 112 p.

——— E. W. Collins, E. D. Davidson, Jr., O. R. Dix, G. W. Donaldson, S. P. Dutton, G. E. Fogg, A. B. Giles, D. W. Harris, M. P. A. Jackson, C. M. Lopez, M. K. McGowen, W. R. Muehlberger, W. D. Pennington, S. J. Seni, D. H. Wood, and H. V. Wuerch, 1981, Geology and geohydrology of the East Texas basin, a report on the progress of nuclear waste isolation feasibility studies (1980): University of Texas Bureau of Economic Geology Geological Circular 81-7, 207 p.

Kupfer, D. H., 1970, Mechanism of intrusion of Gulf Coat salt, *in* D. H. Kupfer, ed., The geology and technology of Gulf Coast salt, a symposium: School of Geoscience Louisiana State University, Baton Rouge, Louisiana, p. 25-66.

——— 1976, Times and rates of salt movement in north Louisiana, *in* J. D. Martinez and R. L. Thoms, eds., Salt dome utilization and environmental considerations: Symposium Proceedings, Louisiana State University, Baton Rouge, Louisiana, p. 145-170.

Loocke, J. E., 1978, Growth history of the Hainesville salt dome, Wood County, Texas: Master's thesis, University of Texas at Austin, 95 p.

Martin, R. G., 1978, Northern and eastern Gulf of Mexico continental margin: stratigraphic and structural framework *in* Framework, facies, and oil-trapping characteristics of the upper continental margin: AAPG Studies in Geology 7, p. 21-42.

McGowen, J. H., and R. A. Morton, 1979, Sediment distribution, bathymetry, faults, and salt diapirs, submerged lands of Texas: University of Texas Bureau of Economic Geology, 31 p.

——— L. F. Brown, Jr., T. J. Evans, W. L. Fisher, and C. G. Groat, 1976, Environmental geologic atlas of the Texas coastal zone—Bay City-Freeport area: University of Texas Bureau of Economics Geology, 98 p.

Nettleton, L. L. 1934, Fluid mechanics of salt domes: AAPG Bulletin, v. 18, p. 1175-1204.

Nichols, P. H., G. E. Peterson, C. E. Wuestner, 1968, Summary of subsurface geology of northeast Texas *in* Natural gases of North America AAPG Memoir 9, p. 982-1004.

O'Neill, C. A., III, 1973, Evolution of Belle Island salt dome, Louisiana: Gulf Coast Association of Geological Societies Transactions, v. 23, p. 115-135.

Purser, B. H., 1973, Sedimentation around bathymetric highs in the southern Persian Gulf, *in* B. H. Purser, ed., The Persian Gulf: Holocene carbonate sedimentation and diagenesis in a shallow epicontinental sea: New York, Springer-Verlag, p. 157-177.

Ramberg, H., 1981, Gravity, deformation and the earth's crust in theory, experiments and geological application, second edition: London, Academic Press, 452 p.

Rezak, R., 1977, West Flower Garden Bank, Gulf of Mexico: AAPG Studies in Geology 4, p. 27-35.

Sannemann, D., 1968, Salt-stock families in northwestern Germany, *in* Diapirism and diapirs: AAPG Memoir 8, p. 261-270.

Seni, S. J., 1981, Depositional systems of the Lower Cretaceous Paluxy Formation, East Texas basin *in* Geology and geohydrology of the East Texas basin, a report on the progress of nuclear waste isolation feasibility studies (1980): University of Texas Bureau of Economic Geology Geological Circular 81-7, p. 53-59.

——— and M. P. A. Jackson, 1983, Evolution of salt structures, east Texas diapir province, part 2: patterns and rates of halokinesis, AAPG Bulletin, v. 67 (following this paper).

Smith, D. A., and F. A. E. Reeve, 1970, Salt piercement in shallow Gulf Coast salt structures: AAPG Bulletin, v. 54, p. 1271-1289.

Stude, G. R., 1978, Depositional environments of the Gulf of Mexico South Timbalier Block 54: salt dome and salt dome growth models: Gulf Coast Association of Geological Societies Transactions, v. 28, p. 627-646.

Terriere, R. T., 1976, Geology of Fairway Field, East Texas, *in* North American oil and gas fields: AAPG Memoir 24, p. 157-176.

Trusheim, F., 1960, Mechanism of salt migration in northern Germany: AAPG Bulletin, v. 44, p. 1519-1540.

Turk, Kehle, and Associates, 1978, Tectonic framework and history, Gulf of Mexico region: Report prepared for Law Engineering Testing Company, Marietta, Georgia, 28 p.

van Eysinga, F. W. B., 1975, Geological time table: Amsterdam, Elsevier. One sheet.

van Hinte, J. E., 1976a, A Jurassic time scale: AAPG Bulletin, v. 60, p. 489-497.

——— 1976b, A Cretaceous time scale: AAPG Bulletin, v. 60, p. 498-516.

Wood, D. H., and A. B. Giles, 1982, Hydrocarbon accumulation patterns in the East Texas salt dome province: University of Texas at Austin Bureau of Economic Geology Geological Circular 82-6, 36 p.

——— and E. H. Guevara, 1981, Regional structural cross sections and general stratigraphy, East Texas basin: University of Texas Bureau of Economic Geology, 21 p.

Woodbury, H. O., I. B. Murray, Jr., and R. E. Osborne, 1980, Diapirs and their relation to hydrocarbon accumulation, *in* A. D. Miall, ed., Facts and principles of world petroleum occurrence: Calgary, Canadian Society of Petroleum Geologists, p. 119-142.

The American Association of Petroleum Geologists Bulletin
V. 67, No. 8 (August 1983), P. 1245-1274, 22 Figs., 3 Tables

Evolution of Salt Structures, East Texas Diapir Province, Part 2: Patterns and Rates of Halokinesis[1]

S. J. SENI and M. P. A. JACKSON[2]

ABSTRACT

The effects of salt mobilization on Aptian and younger (post-112 Ma) strata in the East Texas basin can be used to illustrate patterns of dome growth through time, and to estimate rates and amounts of salt movement.

Pre-Aptian domes grew in three areas around the margin of the diapir province, apparently in pre-Aptian depocenters. Maximum dome growth along the basin axis coincided with maximum regional sedimentation there during the middle Cretaceous (Aptian, Albian, and Cenomanian). In the Late Cretaceous the sites of maximum diapirism migrated to the periphery of the diapir province. Diapirism began after pillows were erosionally breached, leading to salt extrusion and formation of peripheral sinks.

The duration of pillow and diapir stages of growth was subequal, ranging from 10 to 30 Ma. Post-diapiric stage of growth continued for more than 112 Ma in some cases. Diapirs grew fastest in the Early Cretaceous, when peak growth rates ranged from 150 to 530 m/Ma (490 to 1,740 ft/Ma), declining in the Early Tertiary to 10 to 60 m/Ma (30 to 200 ft/Ma). Assuming steady-state conditions over periods of 1 to 17 Ma, strain rates for the rise of the East Texas diapirs averaged 6.7×10^{-16}/sec; peak gross rate of growth averaged 2.3×10^{-15}/sec, similar to slow orogenic rates. The evolution of East Texas salt domes essentially ended in the early Tertiary with uplift rates less than 30 m/Ma (100 ft/Ma).

INTRODUCTION

In part 1 (Seni and Jackson, 1983a), we described how each growth stage since 112 Ma (million years ago) of salt structures in the East Texas basin left a characteristic imprint on surrounding strata in terms of syndepositional facies and thickness changes; we also recognized specific types of subtle hydrocarbon traps at each growth stage. In part 2, we show how this sedimentary record of halokinesis can be used as a powerful tool for quantifying patterns, timing, volumes, and rates of salt flow. Subsurface data down to the depth of the salt source layer are not necessary for reconstruction of the Cretaceous to Tertiary history of salt movement.

On the basis of these patterns it is possible to anticipate the occurrence of stratigraphic traps in different areas and at different stratigraphic levels. Knowing the form of individual salt structures at different evolutionary stages is also vital for reconstruction of the history of petroleum migration and pooling in structural traps.

Quantitative studies of this type are also increasingly necessary for evaluating the geologic stability of salt domes for storage of nuclear waste and hydrocarbons. This study uses simple concepts and techniques such as determination of the volumes of salt and sediments by planimetry, gross versus net rates of dome growth, strain rates based on regional subsurface data, cumulative probability analysis and standard deviation maps of sediment accumulation rates.

ASSUMPTIONS

Syndepositional thickness variations in surrounding strata not only allow recognition of the timing and patterns of halokinesis, but also the measurement of volumes and rates of salt flow. All previous estimates of the growth rate of pillows and diapirs (e.g., Trusheim, 1960; Ewing and Ewing, 1962; Sanneman, 1968; Kupfer, 1976; Netherland, Sewell and Associates, 1976; Kumar, 1977; Jaritz, 1980) rely on certain basic assumptions, and the present study is no exception.

In assessing the times, rates, and volumes of salt movement, certain propositions can be made. They can be divided into three proven propositions, three unproven propositions (assumptions), and two incorrect propositions.

Proven Propositions

1. *Upper surfaces of mapped units were originally horizontal and planar.*—Absence of deep-water (more than 200 to 300 m, 650 to 980 ft) post-Aptian deposits indicates that on a regional scale the depositional surface was virtually horizontal.

2. *Contour intervals smaller than 100 ft (30 m) provide little increase in accuracy of computed volumes.*—Closed isopach contours around salt structures delineate the area influenced by local salt flow in different stratigraphic

---

[1]Manuscript received, November 12, 1982; accepted, February 28, 1983. Published with permission of the Director, Bureau of Economic Geology, The University of Texas at Austin, Austin, Texas 78712.

[2]Bureau of Economic Geology, The University of Texas at Austin, Austin, Texas 78712.

This research was supported by the U.S. Department of Energy, Contract Number DE-AC97-80ET46617.

L. F. Brown, Jr., W. E. Galloway, T. E. Ewing, and R. T. Budnik critically reviewed the manuscript; their comments are gratefully acknowledged. We thank E. Bramson, R. Conti, S. Ghazi, S. Lovell, B. Richter, J. Smith, and D. Wood for their help in data collection and processing. J. Ames, M. Bentley, M. Evans, R. Flores, J. Horowitz, and J. McClelland drafted figures under the supervision of D. Scranton. G. Zeikus and K. Bonnecarrere typed original manuscripts. Twyla J. Coker word processed the manuscript under the supervision of Lucille Harrell, and S. Doenges supervised manuscript editing.

intervals. The total volume of a salt-withdrawal basin is based on planimetry of hand-drawn isopach maps and calculated by the technique shown in Figure 1. Reducing the contour interval below 100 ft (30 m) has little effect on the calculated volume. In this type of integration, errors due to approximation cancel out except at the boundary of the basin (perimeter of area 1 in Figure 1). Here the maximum vertical error is ± 100 ft. Because the actual basin edge is likely to lie on either side of the outermost contour, errors will also tend to nullify each other.

Errors may be classed as relative (compared with the volume of an individual salt-withdrawal basin) or absolute (compared with the total volume of all salt-withdrawal basins). Maximum relative error for the volume of an individual salt-withdrawal basin is greatest in the smallest basins. However, because such basins constitute a small proportion of the total volume of all salt-withdrawal basins, the absolute error is small.

3. *Deformation around and above salt structures resulted directly from gravity-induced salt tectonics (halokinesis).*—The tectonic setting of the evolving East Texas basin is that of a subsiding passive margin or aulacogenic reentrant. Regional crustal shortening in such a setting is improbable, and there is no evidence for its occurrence. Nearly 500 km (300 mi) of seismic-line data indicate that the Louann Salt is a décollement zone separating deformed overburden from the virtually planar upper surface of the pre-Louann basement (Jackson and Harris, 1981; Jackson and Seni, 1983; McGowen and Harris, 1983).

Unproven Propositions

4. *Geochronology is reasonably accurate.*—Acknowledging that no single time scale is universally accepted (Baldwin et al, 1974), we use the Jurassic-Cretaceous time scale of van Hinte (1976a, b) which includes radiometric, paleomagnetic, and faunal data. A single age from van Eysinga (1975) was used for the top of the Eocene Wilcox Group.

5. *Upper surfaces of mapped units are isochronous.*—Variations in rates of progradation and transgression result in small age variations along the upper surface of a stratigraphic unit because of dipping bedding surfaces. Considering the relatively small study area and the long depositional history of mapped units, these age variations are negligible.

6. *Stratigraphic record is sufficiently complete to recognize long-term trends.*—We recognize that the geologic record includes periods of nondeposition, erosion, and removal of stratigraphic section. Local angular unconformities as a result of dome uplift are present around Hainesville and Butler domes. Only one regional unconformity (base of Austin Chalk) truncates a significant thickness of stratigraphic section. Halbouty and Halbouty (1982) see evidence for an additional regional unconformity (base of Woodbine Group) in the eastern part of the East Texas basin. Even in a highly explored and relatively small area like the East Texas basin it is doubtful that all unconformities, especially disconformities, have been recognized. We have diminished the potential problem associ-

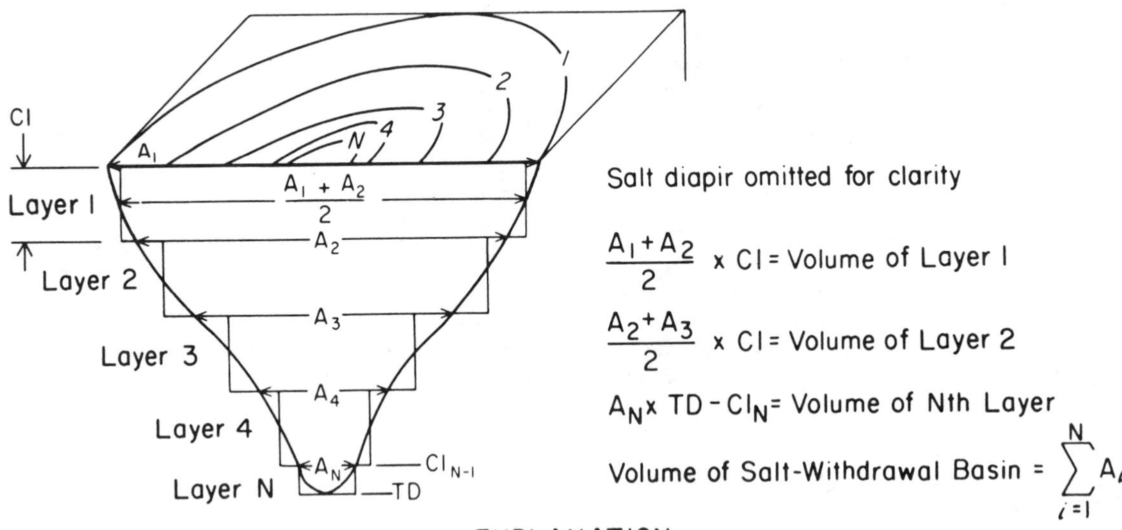

FIG. 1—Technique for calculating volume of an individual salt-withdrawal basin.

ated with intermittent dome growth by averaging the time intervals. The Cretaceous-Tertiary growth history was divided into seven consecutive time intervals, each with a duration of 1 to 17 Ma. We recognize that the time intervals for which the mean dome growth rates were calculated may include periods of growth and nongrowth. In situations of intermittent sedimentation and dome growth, therefore, these mean rates are less than actual rates over shorter durations. Nevertheless, long-term growth trends are clearly recognizable when graphically compared with the mean rates of the seven consecutive intervals.

Incorrect Propositions

7. *Effects of compaction were negligible.*—During burial, compaction progressively reduces volume and porosity of sediments by expulsion of pore fluids. Our calculations of volumes and rates of salt flow are based on the present (compacted) volumes of sediments, rather than on their original (uncompacted) volumes. Compactional effects on shale can be assessed for various burial depths. Depths of 152 m (500 ft), 1,524 m (5,000 ft), and 3,048 m (10,000 ft) represent the range of burial for strata studied in this paper. Using averaged porosity-depth data from Magara (1980), volume loss for shale at depths of 152 m (500 ft), 1,524 m (5,000 ft), and 3,048 m (10,000 ft) is 5%, 35%, and 39%, respectively. Fine-grained clastics are the most common sediment in the East Texas basin, and probably compacted the most. The calculated volume losses therefore represent the probable maxima for the entire stratigraphic section studied. Based on these maximum volume losses, true rates of dome growth calculated on original, "decompacted" sediment thicknesses (1/100-volume loss), are estimated to exceed "compacted" rates by, at most, 1.05, 1.54, and 1.64 for depths of 152 m (500 ft), 1,524 m (5,000 ft), and 3,048 m (10,000 ft).

8. *The volume of sediments in a salt-withdrawal basin is equivalent to the volume of salt moved into the diapir during filling of that basin.*—This proposition is based on propositions 5, 6, and 7, together with the principle of conservation of volume. This proposition forms a cornerstone of our treatment of dome growth. Under this proposition the evacuation of salt from a particular zone allows the accumulation of an equivalent volume of extra-thick sediments (salt-withdrawal basin) above the zone. Other studies have equated the volume of sediments in a salt-withdrawal basin to the volume of salt that migrated into the diapir (Trusheim, 1960; Crowe, 1975; Kupfer, 1976; Reese, 1977). The concept is also used for various purposes in the oil industry. To the authors' knowledge, the present study is the first to use volumes measured by planimetry to calculate volumes and rates of dome growth.

Despite acceptance of this proposition in the literature, three factors in addition to compaction reduce its reliability: (1) structural thickening by folding, (2) dissolution below the withdrawal basin, and (3) centrifugal salt flow (flow away from diapir).

The influence of structural thickening by folding in volume calculations is thoroughly treated in the following section. This treatment indicates that *theoretical* effects of fold thickening are much less than the *observed* thickening in withdrawal basins. Dissolution and collapse of the salt source layer below a withdrawal basin, could potentially yield erroneously high rates of diapir growth if, instead of migrating to feed the diapir, the salt dissolved and entered the basinal ground-water regime. Average salinity values of ground water increase with depth in the East Texas basin; for example, the average salinity of the Hosston group is 200,000 mg/L (sea water is 35 to 40 mg/L). There is no evidence for differential salt dissolution in withdrawal basins. Thus, we cannot estimate the degree to which sedimentary volume calculations might be affected by dissolution of salt before it migrated into the diapir. In contrast, the dissolution of salt at the diapir crest and flank is registered by volumetric changes in the withdrawal basin if further diapir growth replaced the lost salt.

Centrifugal flow of salt from below a withdrawal basin and away from a diapir could cause erroneously high estimates of diapir-growth rates. Centrifugal salt flow has been produced experimentally in diapir models by Dixon (1975, Figure 21). Reflection seismic data indicate no rings of salt structures around salt-withdrawal basins; salt is largely absent between diapirs. Such peripheral salt rings would be expected if centrifugal flow had occurred. It cannot be argued that the salt rings have subsequently dissolved; their removal would be recorded by anomalously thickened sediments above them, and there is no such evidence.

Distinguishing Between Syndepositional and Postdepositional Thickness Variations

The problem.—This report deals extensively with thickness variations in strata around and above growing salt structures. The origin of these variations is debatable (Bishop, 1978); they may arise either syndepositionally or postdepositionally. Recognition of a specific origin is critical because only syndepositional thickness variations can document the flow of salt during deposition of a particular stratigraphic unit, thereby providing a time frame for the history of diapirism.

Thickness variations of strata over and around salt structures can be ascribed to four processes: (1) local rise or fall of the sedimentation surface during deposition (i.e., syndepositional thickness variations); (2) postdepositional erosion; (3) structural distortion of strata due to stresses imposed by growth of nearby salt structures; or (4) increased compaction of sediments which drape an incompressible salt body (i.e., postdepositional thickness variations). The effects of these processes are broadly similar. For example, thinning of strata over a salt pillow can be caused by syndepositional rise of the pillow, erosion, or increased drape compaction over the rigid, relatively incompressible salt body, analogous to draping in varvite laminae over dropstones. Thickening of strata in a rim syncline around a diapir can be caused by syndepositional deepening of the syncline or by postdepositional ductile folding of poorly consolidated sediments around the salt stock.

Postdepositional strain of originally planar layers overlying a rising buoyant mass has been documented by scaled centrifuge modeling (Dixon, 1975; Ramberg, 1981). In the case of a low-viscosity, low-density body such as rock salt, the overburden above the crest of a circular rising dome distorts by oblate flattening and by horizontal extension. This distortion is identical to nontectonic compactional strain and cannot be differentiated from compaction above the crest. Furthermore, if traced laterally, tectonic flattening diminishes because the rise of underlying salt in the flanks is less than in the dome axis. Compactional strain also diminishes laterally because the sediments are not directly overlying the incompressible salt. Differentiating between tectonic and compactional thinning over a salt dome is therefore difficult. There are practical problems also in detecting lateral changes in strain where strain gradients are minute. Figure 2 shows the proportional thickness changes, Δh, in sedimentary units near East Texas salt domes. Maximum thinning averages only about −0.75%/100 m (−0.75%/330 ft) of lateral distance. It would be difficult for strain analysis to identify these small changes in strain and to filter out background variations.

Centrifuge modeling also shows that overburden is tectonically thickened in rim synclines around both low-viscosity (salt) and high-viscosity (e.g., gneiss) diapirs. The combination of vertical shortening (induced by compaction) and oblique shortening (induced by folding) of planar layers of constant lateral thickness, would be extremely difficult to differentiate from an analogous combination of compaction and folding of syndepositionally thickened layers of variable lateral thickness. Apart from this inherent problem of similarity, the practical problem of low strain gradients exists; maximum thickening averages only +1.5 to 1.0%/100 m (1.5 to 1.0%/330 ft) in rim synclines (Fig. 2).

If syndepositional thickness changes cannot be differen-

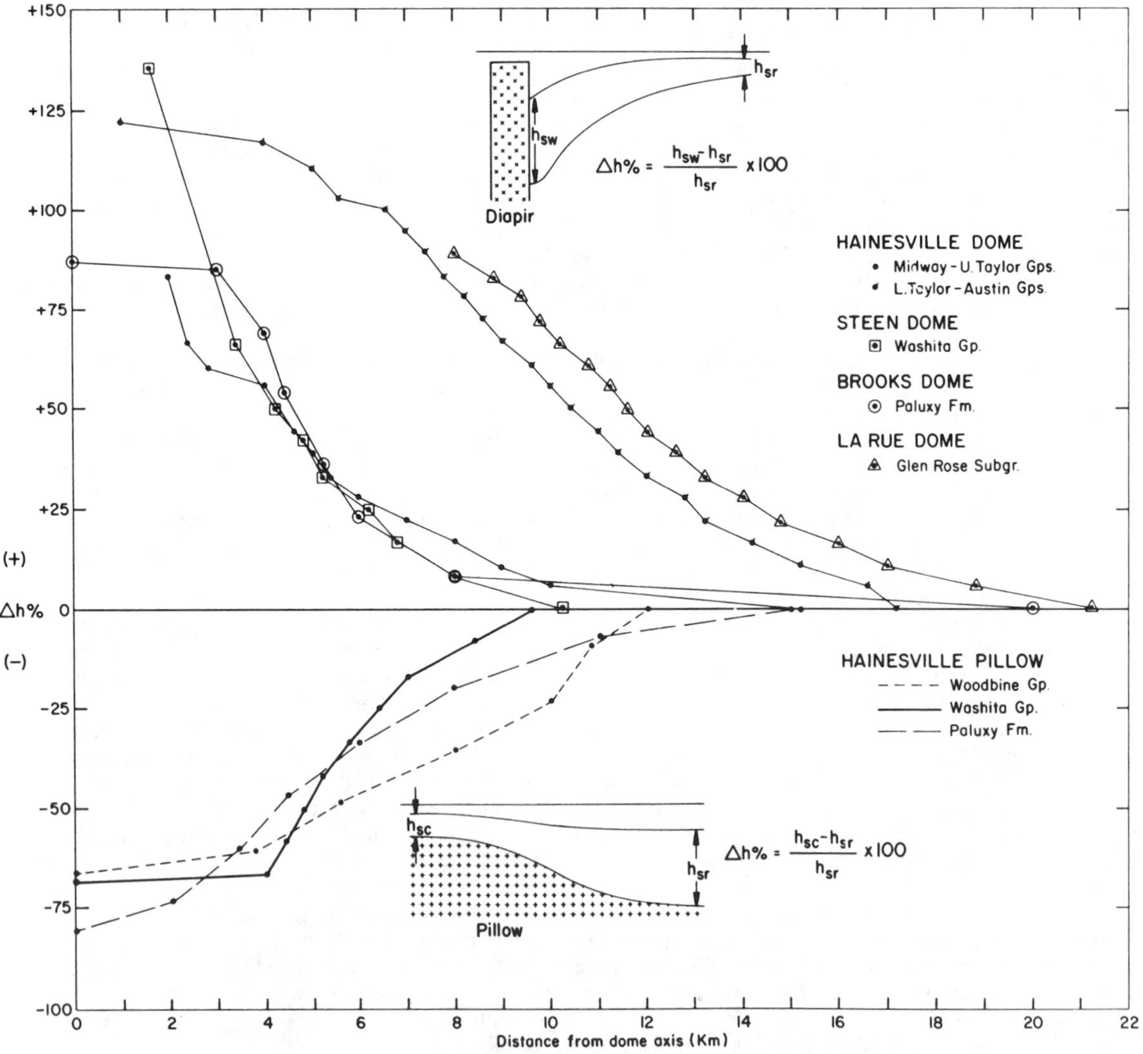

FIG. 2—Percentage thickness changes, Δh, in stratigraphic units versus lateral distance from axes of salt pillows and diapirs.

tiated from postdepositional thickness changes by strain analysis, are other means available? We propose the changes can be differentiated by structural and sedimentologic criteria.

Structural evidence.—The structural criteria are based on comparison of the geometry of folds around natural salt domes with that around experimental models of domes with comparable limb dips. The maximum large-scale dip of post- 112 Ma strata is 10°; at deeper levels the maximum dip of the Louann Salt (apart from salt stocks) is 25°.

Thickness changes induced by folding were measured from 14 models (toward rim synclines of diapirs) and 11 models (toward crests of pillows); the results are shown in Figure 3. Both the mean and median values of thickness change in rim synclines are less than 2% thickening, with a maximum of 7% thickening and a minimum of −12% thinning. The median value is identical to the maximum

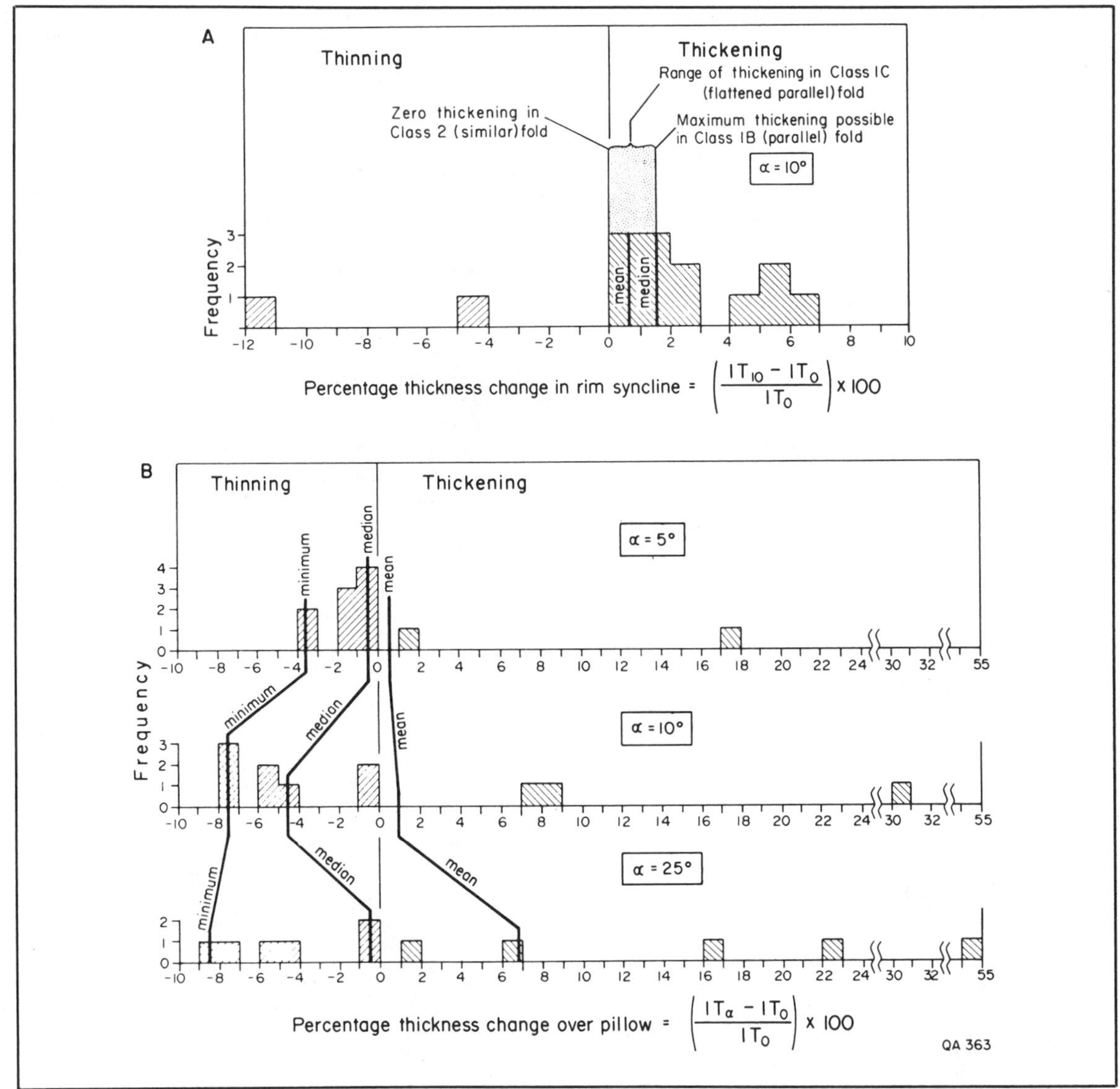

FIG. 3—Histograms showing frequency distribution of thickness changes measured in model diapiric rim synclines (14 layers) and over model pillow crests (11 layers). Models are illustrated in Dixon (1975, his Figures 6A, 21A) and Ramberg (1981, his Figures 11.13, 11.17, 11.38, 11.46, 11.58, 11.80, 11.81, 11.83, 11.93, 12.1). (A) The median percentage change of 1.5% thickening is identical to maximum thickening theoretically possible at this limb dip in a fold formed by buckling, as calculated by the equation $T'_\alpha = \sec \alpha$ (from Ramsay, 1967, equation 7-6). Maximum thickening is 7%. (B) Increased thinning with increased steepness of pillow flanks is reflected by decreasing minimum curve; maximum thinning (minimum thickness) at 10° dip is −8%. The mean and median curves are skewed rightward with increasing dip because increasing dip causes increasing difference between vertical thickness and orthogonal (stratigraphic) thickness.

thickening theoretically possible in a limb dipping 10° in a fold formed by buckling ($T'_{\alpha} = \sec \alpha$, Ramsay, 1967, p. 367). Higher measured values suggest either errors in model construction or measurement, or the action of a different folding mechanism. In thickness changes over a pillow (Fig. 3), the maximum thinning increases from −3.5% change at 5° flank dip to −8.5% change at 25° flank dip; the maximum thinning at 10° dip is −7.5%.

In summary, the maximum thickness changes induced by growth of experimentally modeled domes at comparable maximum dips of 10° are (1) 7% thickening in rim synclines of diapirs, and (2) −8% thinning above pillows. These thickness changes are much less than those actually measured around the salt dome of the East Texas basin (cf. Figs. 2, 3). Equivalent thickness changes in Figure 2 are (1) 135% thickening, and (2) −80% thinning. These findings suggest that because of the comparatively low dips within the basin, folding cannot account for more than a small fraction of the observed thickness changes. Rise and fall of the depositional surface induced by flow of underlying salt, remains the only reasonable explanation for huge thickness changes induced over wide areas of gentle dip.

Additional structural evidence that thickness changes around the East Texas salt domes are largely syndepositional, rather than postdepositional, results from comparison of the positions of axial traces with those of experimental models. The axial traces of experimental dome models are shown in Figure 4. Their shapes vary widely, but they share one important characteristic—all of the axial traces, except one, curve away from the diapir as they ascend through higher layers. This trend of inward curvature in model domes is different from the pattern of the axial traces of the East Texas salt domes (Fig. 4). The axial traces of the primary, secondary, and tertiary peripheral sinks migrate progressively toward the diapirs, so that axial traces in younger units are closer to their diapirs. This migration is caused by shrinkage of a broad salt pillow as salt was evacuated from it up a central diapir; it is an inherent characteristic of the model shown in part 1, Figure 4 (Seni and Jackson, 1983a).

Sedimentological evidence.—Syndepositional thickness changes are favored also by sedimentologic evidence. Changes in sandstone distribution patterns, depositional facies, and localized reef growth near domes in East Texas are documented in part 1 (Seni and Jackson, 1983a). These changes reflect the influence of syndepositional topographic variations that were salt controlled. In other basins, ancient examples are provided by facies variations associated with salt structures, including sandbody pinchouts, changes in sandbody geometry, and preferential growth of reefs over domes and pillows (Halbouty and Hardin, 1951; Cantrell et al, 1959; O'Neill, 1973; Stude, 1978; Elliot, 1979; Trippet, 1981). Holocene examples of analogous topographic, lithologic, and faunal variations associated with salt structures have also been recorded (Ewing and Ewing, 1962; Fisher et al, 1972, 1973; Purser, 1973; McGowen et al, 1976; Bright, 1977; Rezak, 1977; Kent, 1979).

On the basis of these structural and sedimentological observations, we have concluded that *most* thickness variations are syndepositional in origin. For the purposes of estimating dome-growth rates, *all* thickness variations are assumed to be syndepositional in origin.

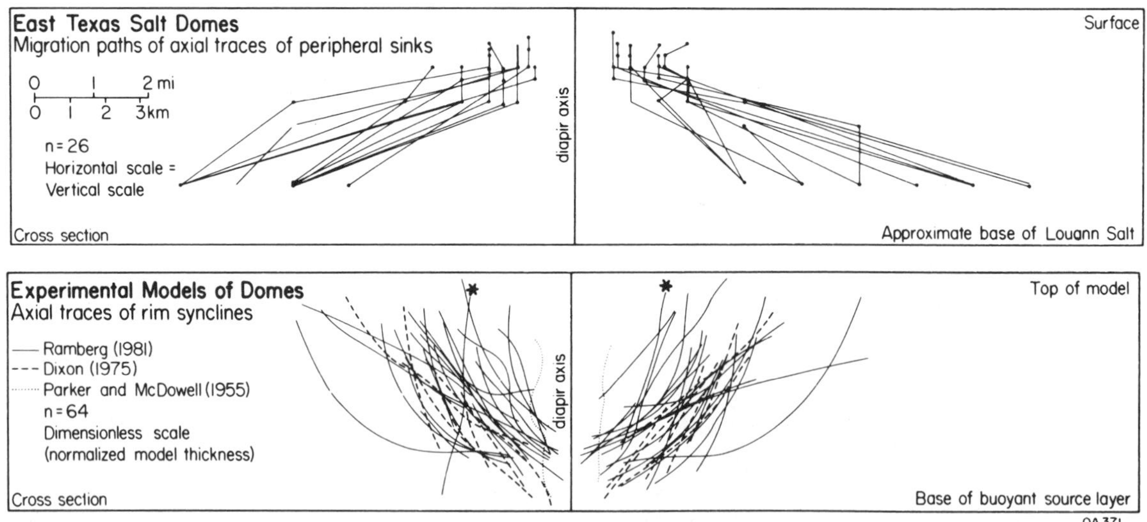

FIG. 4—Comparison of axial-trace position in vertical cross sections through 14 of 16 salt domes (Bullard and Whitehouse Domes have a measurable sink at only one level) of the East Texas basin and 23 experimental domes (illustrated in Parker and McDowell, 1955, their Figure 21; Dixon, 1975, his Figures 2B, 3B, 4B, 5B, 21A, 21B, 21C; Ramberg, 1981, his Figures 11.2, 11.13F, 11.16, 11.17, 11.19, 11.25, 11.32, 11.38, 11.39, 11.45, 11.50, 11.58B, 11.93, 11.95, 12.1B). The axial trace (measured from maps) migrates progressively closer to diapirs through evolutionary stages of primary, secondary, and tertiary peripheral sinks. In contrast, axial traces in model-dome cross sections curve away from the dome as they ascend, with the one exception of the unrepresentative model marked by asterisks, which is cored by dense "basement."

PATTERNS OF SALT MOVEMENT IN TIME AND SPACE

Sixteen salt diapirs in the East Texas basin constitute three groups defined by timing (Fig. 5) and location (Fig. 6) of diapirism: group 1—pre-Glen Rose Subgroup (pre-112 Ma), periphery of diapir province; group 2—Glen Rose Subgroup to Washita Group (112 to 98 Ma), basin axis; and group 3—post-Austin Group (86 to 53 Ma), periphery of diapir province. The following sections present lithostratigraphic evidence for this grouping. Two deep diapirs—Concord and Girlie Caldwell domes—were not considered in this report, because they do not intrude the Glen Rose Subgroup, the deepest unit for which adequate subsurface data exist. The stratigraphic framework and regional setting of the study area were described in part 1 (Seni and Jackson, 1983a).

Group 1: Pre-Glen Rose Subgroup (pre-112 Ma)—Periphery of Diapir Province

The seven diapirs in group 1 (Figs. 5, 6) can be divided spatially into three subgroups: Grand Saline dome, in the northwest of the province; Whitehouse and Bullard domes, on the eastern margin; and Oakwood, Butler, Palestine, and Keechi domes, in the southwest (Fig. 6).

Group 1 diapirs are the smallest (mean volume = 21 km^3, 5.0 mi^3) of the dome groups in East Texas basin. Diapir volumes were calculated by means of gravity-derived diapir models and a structure contour map of the top of the James Limestone in the Glen Rose subgroup. Thus, all quoted volumes refer to parts of the diapirs above the top of the James Limestone (see part 1, Seni and Jackson, 1983a, Fig. 3). The crests of all these diapirs are less than 300 m (1,000 ft) deep (maximum depth, 268 m, 880 ft, for Oakwood dome; minimum depth, 37 m, 121 ft, for Palestine dome; mean depth, 144 m, 472 ft). No primary or secondary peripheral sinks surround these domes in strata younger than 112 Ma, because these diapirs had attained postdiapir stage by Glen Rose deposition. Effects on sedimentation and thickness range from effects so small that they cannot be detected with a contour interval of 100 ft (30 m), to minor tertiary peripheral sinks in post-112 Ma strata. For example, Bullard and Whitehouse domes have no discernible tertiary peripheral sinks in sediments that accumulated since 112 Ma. In contrast, sediments around Grand Saline dome have small tertiary peripheral sinks in Paluxy and Walnut Formations and lower Taylor Formation and Austin Group strata (Figs. 7, 8).

The three subgroups of group 1 diapirs each consist of a cluster of coeval structures (Fig. 6). In at least one of the

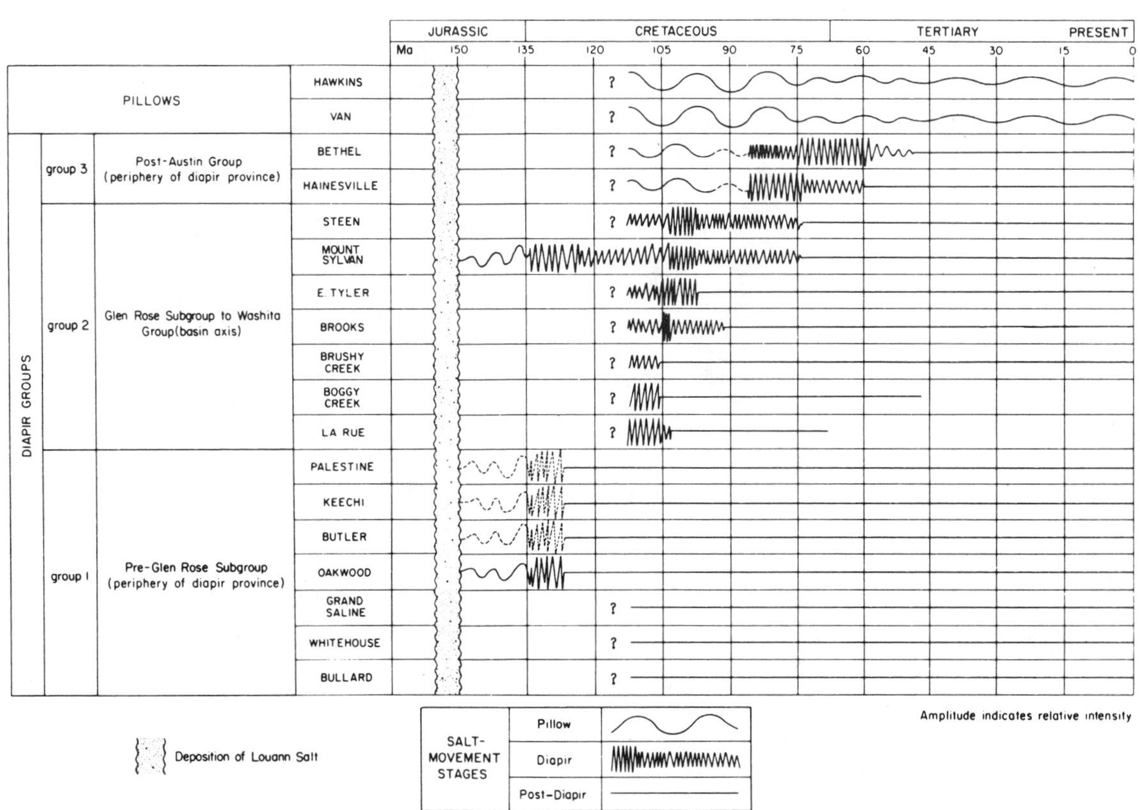

FIG. 5—Three groups of domes based on timing of salt movement stages. Only the final evolutionary stage of the oldest (group 1) diapirs is preserved in Glen Rose and stratigraphic units younger than 112 Ma. Where available, pre-Glen Rose history is based on interpretation of seismic sections.

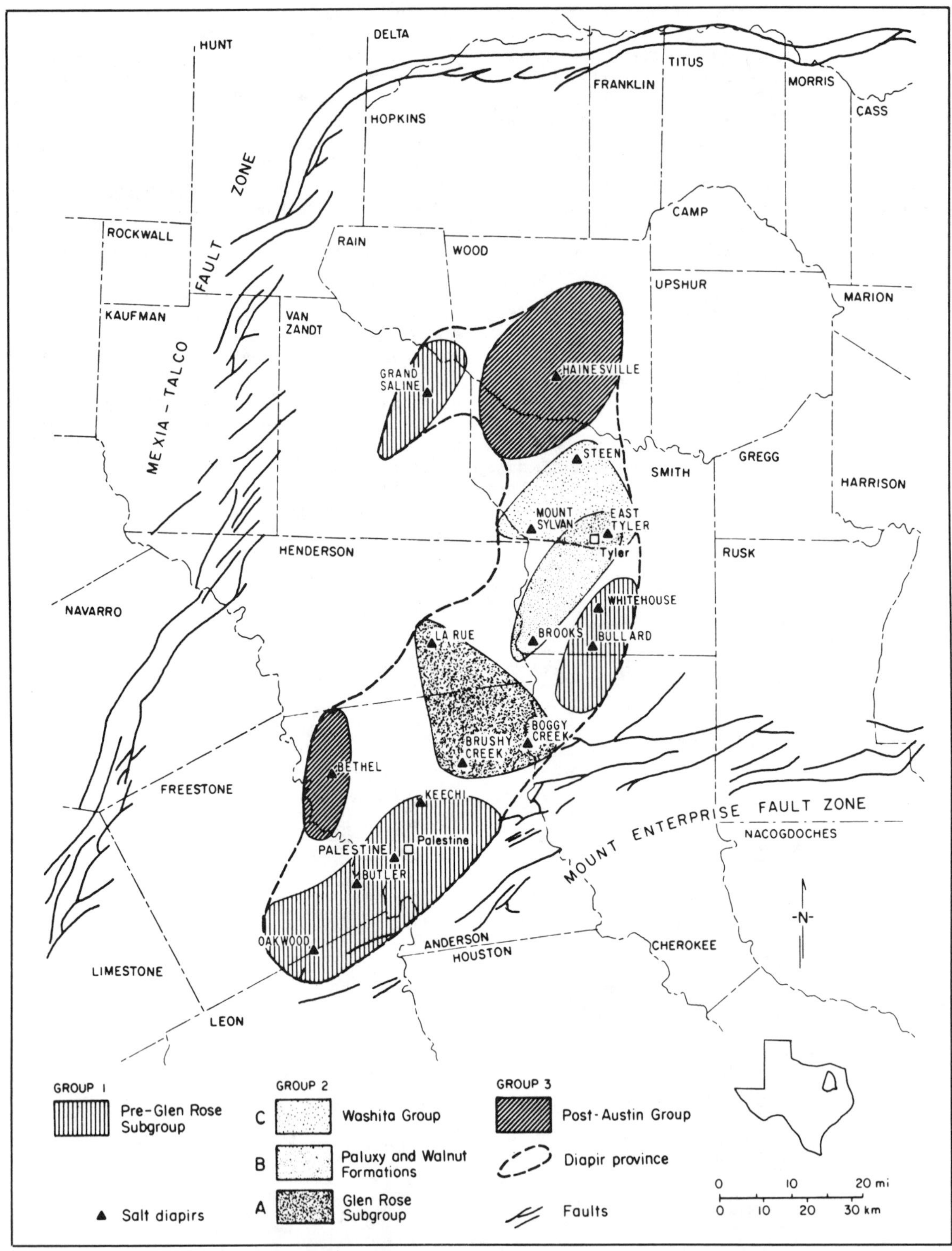

FIG. 6—Map showing spatial distribution of three age groups of salt diapirs in East Texas basin. Ornamented areas around diapirs correspond to outer limits of secondary peripheral sinks where known. Note gradual migration of group 2 subgroups toward group 3 area. The Mexia-Talco fault system defines northern and western margin of basin; dashed line defines diapir province.

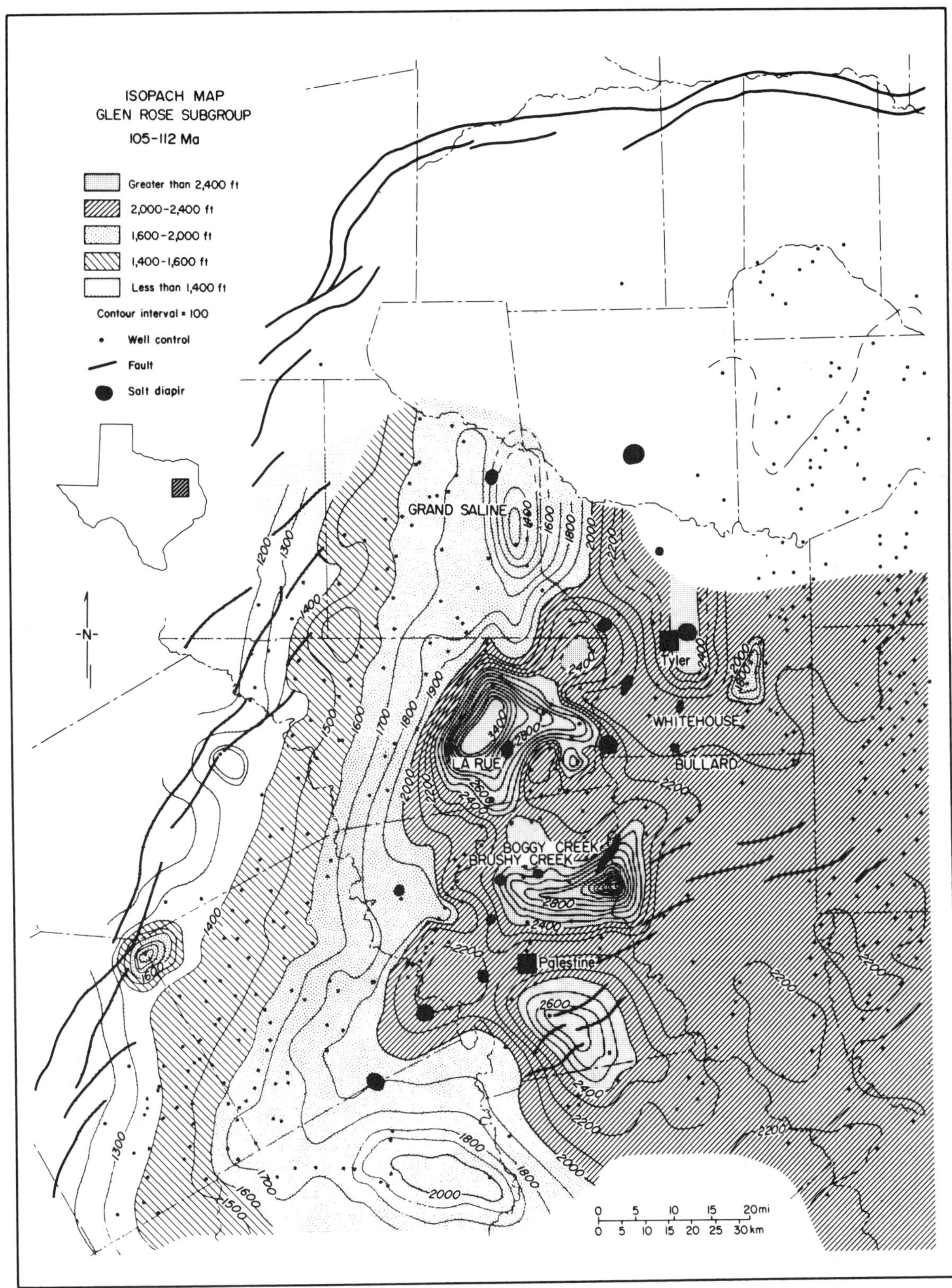

FIG. 7—Regional isopach map, Glen Rose Subgroup, East Texas basin. Large secondary peripheral sinks in central part of basin are associated with diapiric growth of La Rue, Brushy Creek, and Boggy Creek domes.

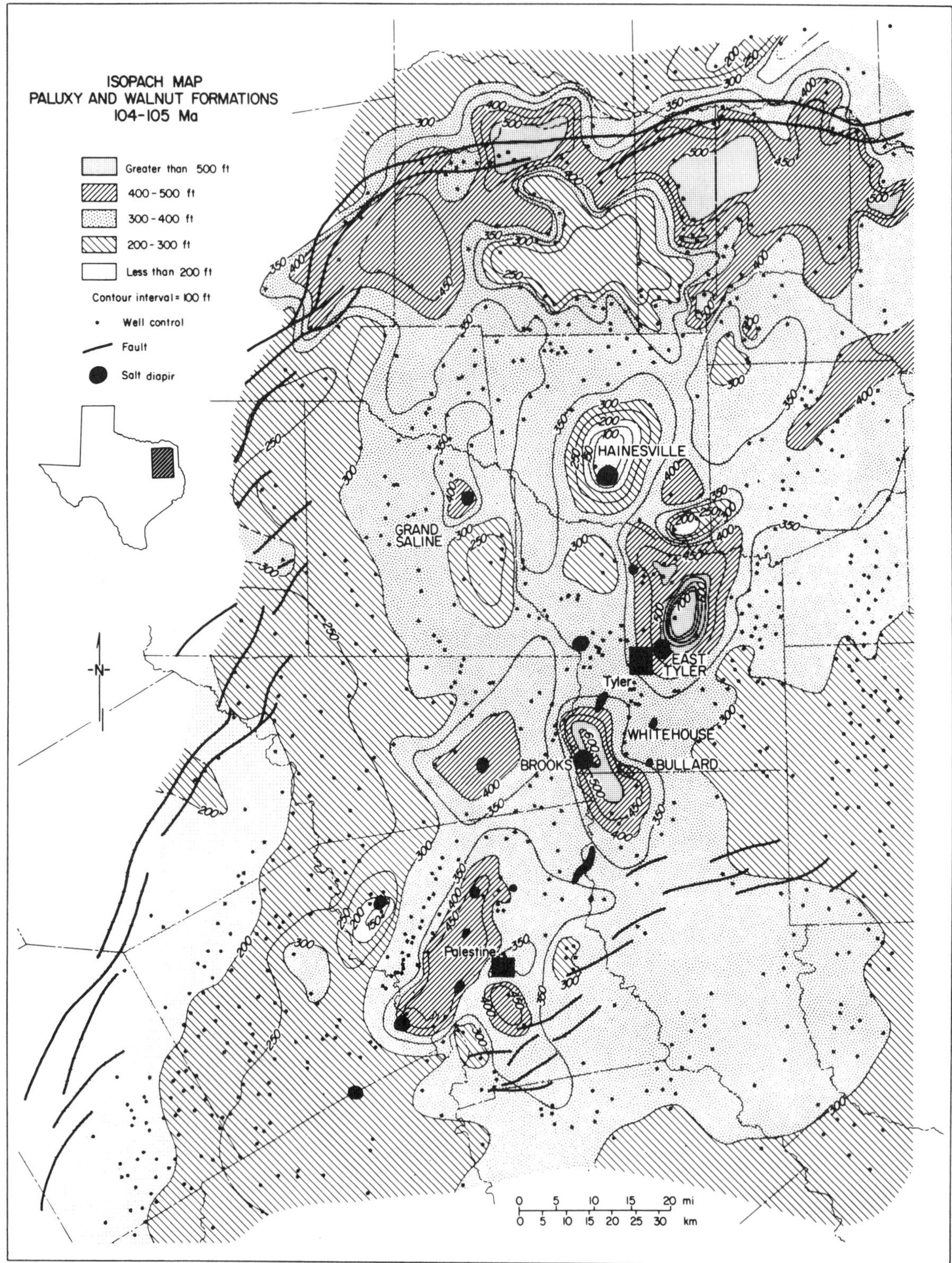

FIG. 8—Regional isopach map, Paluxy and Walnut Formations, East Texas basin. The Walnut Formation is an offshore, time-equivalent facies of Paluxy Formation. Small secondary peripheral sinks around Brooks and East Tyler domes are in central part of basin; large thin areas in northern part of basin formed over the crest of growing salt pillows such as those around Hainesville dome.

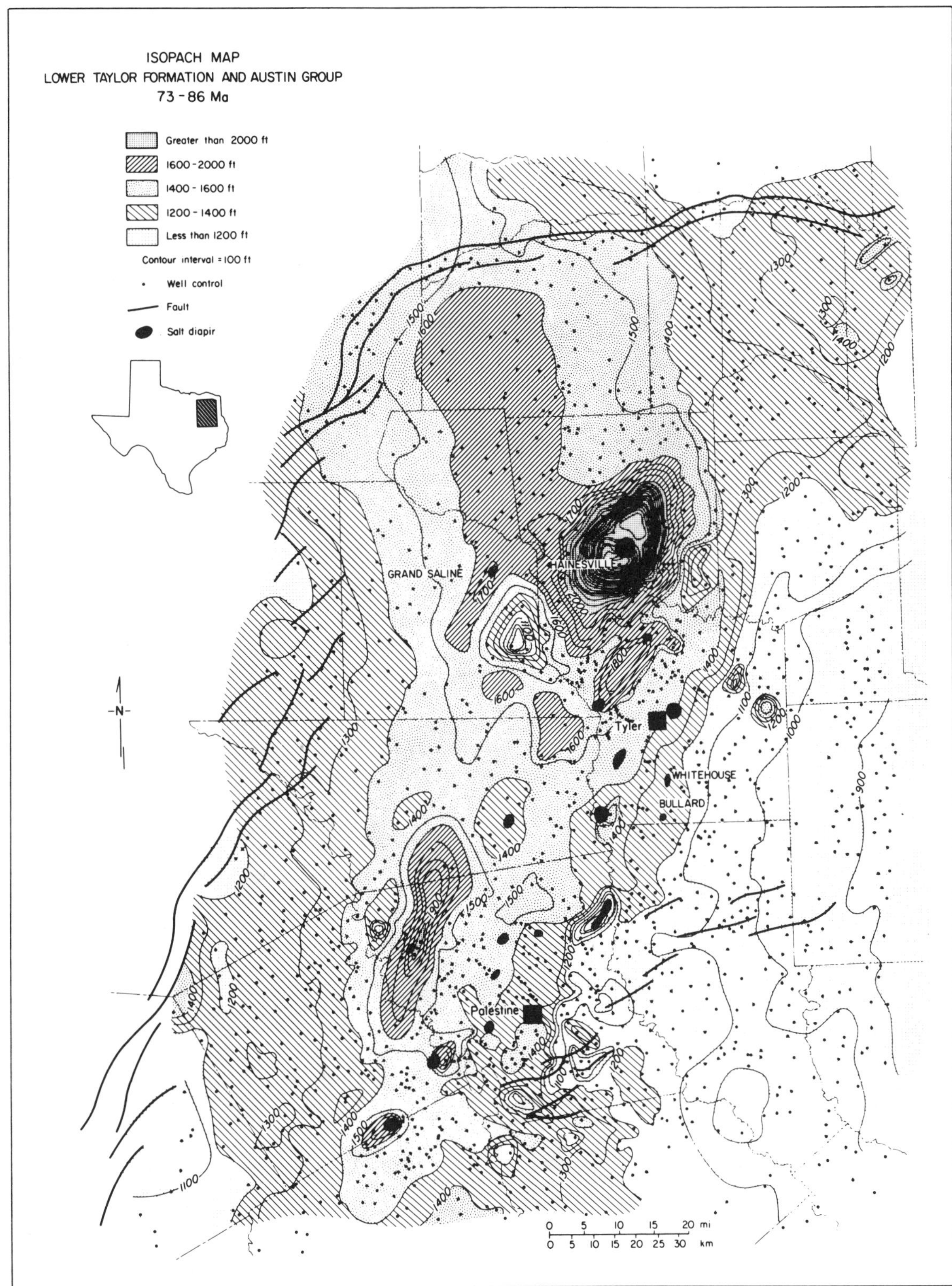

FIG. 9—Regional isopach map, lower Taylor Formation and Austin Group, East Texas basin. A large secondary peripheral sink around diapiric Hainesville dome dominates northern part of basin.

Method

A. Net Growth by Sediment Thinning

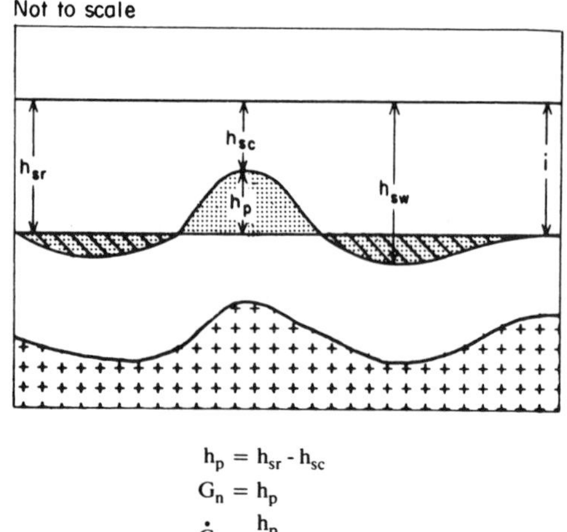

$$h_p = h_{sr} - h_{sc}$$

$$G_n = h_p$$

$$\dot{G}_n = \frac{h_p}{t_i}$$

h_p = Height of pillow

h_{sr} = Regional mean sediment thickness

h_{sc} = Minimum sediment thickness over crest of structure

h_{sw} = Maximum sediment thickness in salt-withdrawal basin

G_n = Net growth of pillow

$\dot{G}_n$ = Net rate of growth of pillow

t_i = Duration of stratigraphic interval (i)

Net growth calculated by sediment thinning will equal net growth calculated by sediment thickening only when

$$h_{sc} = 0$$

$$\text{and } h_{sw} = h_{sr}$$

$$\therefore G_n = h_{sr} - h_{sc} = h_{sw}$$

Application, Assumptions, Restriction, Advantages

Application: Pillow stage, postdiapir stage (only for non-pierced strata)

Assumptions:

(1) Sediment thinning is syndepositional

(2) Sediment thinning is due to uplift of crest of structure

Restriction: Only records extension greater than shortening caused by extrusion or dissolution

Advantages:

(1) Simple quantitative methodology

(2) Can be measured from single cross section

(3) Applicable to youngest strata not pierced by diapir, thus provides rates of most recent growth

FIG. 10—Methods for calculating net rate pillow growth (Fig. 10A) and net (Fig. 10B) and gross (Fig. 10C) rates of diapir growth, and applications, assumptions, restrictions, and advantages of each method.

clusters, the diapirs appear to have evolved from a single parental structure, thus forming a "family" of related diapirs. The linear alignment and similar post-112 Ma growth histories of Oakwood, Butler, Palestine, and Keechi domes suggest evolution from a single, parental, salt-cored anticline trending northeast. Seismic control around Oakwood dome discloses growth history prior to 112 Ma for this dome, which may also be applicable to the three similar domes in this southern subgroup. Giles (1981) recognized domeward thinning of Smackover and Gilmer (Cotton Valley lime) carbonates and therefore inferred pillow growth during Late Jurassic. Domeward thickening of post-Gilmer to pre-Pettet terrigenous clastics indicates that Oakwood dome grew diapirically during the Late Jurassic to Early Cretaceous, from 143 to 112 Ma.

The post-Early Cretaceous histories of Butler, Keechi, and Palestine domes are broadly similar to that of Oakwood dome. Butler, Keechi, and Palestine domes had slightly higher growth rates during deposition of the Lower Cretaceous Paluxy and Walnut Formations, and slightly slower rates during subsequent deposition of the Washita Group.

Group 2: Glen Rose Subgroup to Washita Group (112 to 98 Ma)—Basin Axis

The seven diapirs of group 2 (Figs. 5, 6) constitute three subgroups, all of which straddle the basin axis: La Rue, Brushy Creek, and Boggy Creek domes in the basin center; Brooks and East Tyler domes north of basin center; and Mount Sylvan, Steen, and East Tyler domes farther north (Fig. 6). These subgroups are characterized by progressively younger diapirism. The site of maximum diapirism shifted sequentially northward along the basin axis toward the margin of the diapir province from 112 to 98 Ma. During this period the rate of sediment accumulation was fastest along the basin axis, in contrast to the earlier stage when sediments accumulated fastest around the periphery of the diapir province during group 1 diapirism. Group 2 diapirs (mean volume = 31 km^3, 7.4 mi^3) are larger than group 1 diapirs.

In the center of the basin, La Rue, Brushy Creek, and Boggy Creek domes of subgroup A (see Figure 6) are associated with two very large secondary peripheral sinks (Fig. 7; part 1, Seni and Jackson, 1983a, Fig. 15) that indicate

Method	Application, Assumptions, Restriction, Advantages

B. Net Growth by Sediment Thickening

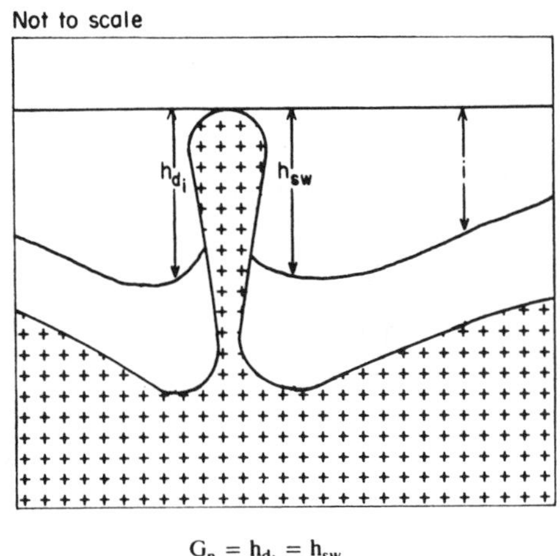

$$G_n = h_{d_i} = h_{sw}$$

$$\dot{G}_n = \frac{h_{sw}}{t_i}$$

h_{d_i} = Height of diapir

h_{sw} = Maximum thickness of stratigraphic interval measured in salt-withdrawal basin

G_n = Net growth of diapir

$\dot{G}_n$ = Net rate of growth of diapir

t_i = Duration of stratigraphic interval (i)

Application: Pillow stage, diapir stage, postdiapir stage

Assumptions:

(1) Diapir remains near sediment surface during deposition

(2) Rate of deposition controls or is controlled by diapir growth

Restriction: Only records net extension greater than shortening caused by diapir extrusion or dissolution

Advantages:

(1) Simple quantitative methodology

(2) Can be measured from single cross section

Method	Application, Assumptions, Restrictions, Advantage

C. Gross Diapir Growth

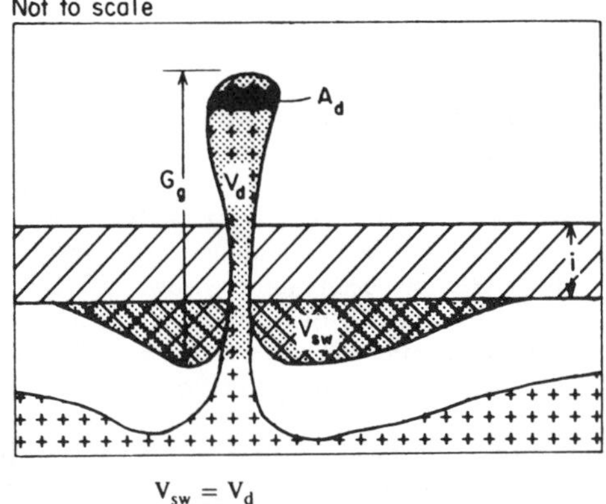

$$V_{sw} = V_d$$

$$G_g = \frac{V_{sw}}{A_d}$$

$$\dot{G}_g = \frac{V_{sw}/A_c}{t_i}$$

V_{sw} = Volume of sediments in salt-withdrawal basin

V_d = Volume of diapir

A_d = Maximum cross-sectional area of diapir

G_g = Gross growth of diapir

$\dot{G}_g$ = Gross rate of growth of diapir

t_i = Duration of stratigraphic interval (i)

Application: Diapir stage only

Assumptions:

(1) Present cross-sectional area of diapir equals cross-sectional area of diapir during filling of withdrawal basin

(2) Volume of withdrawal basin equals volume of salt mobilized during filling of withdrawal basin

(3) All salt mobilized during filling of withdrawal basin migrated into diapir

Restrictions:

(1) Requires measuring volume of withdrawal basin and area of diapir, which requires close well spacing for map construction

(2) Growth by tear-drop detachment of diapir base is not measurable

Advantage: Records total extension independent of possible dissolution or extrusion

FIG. 10—Continued

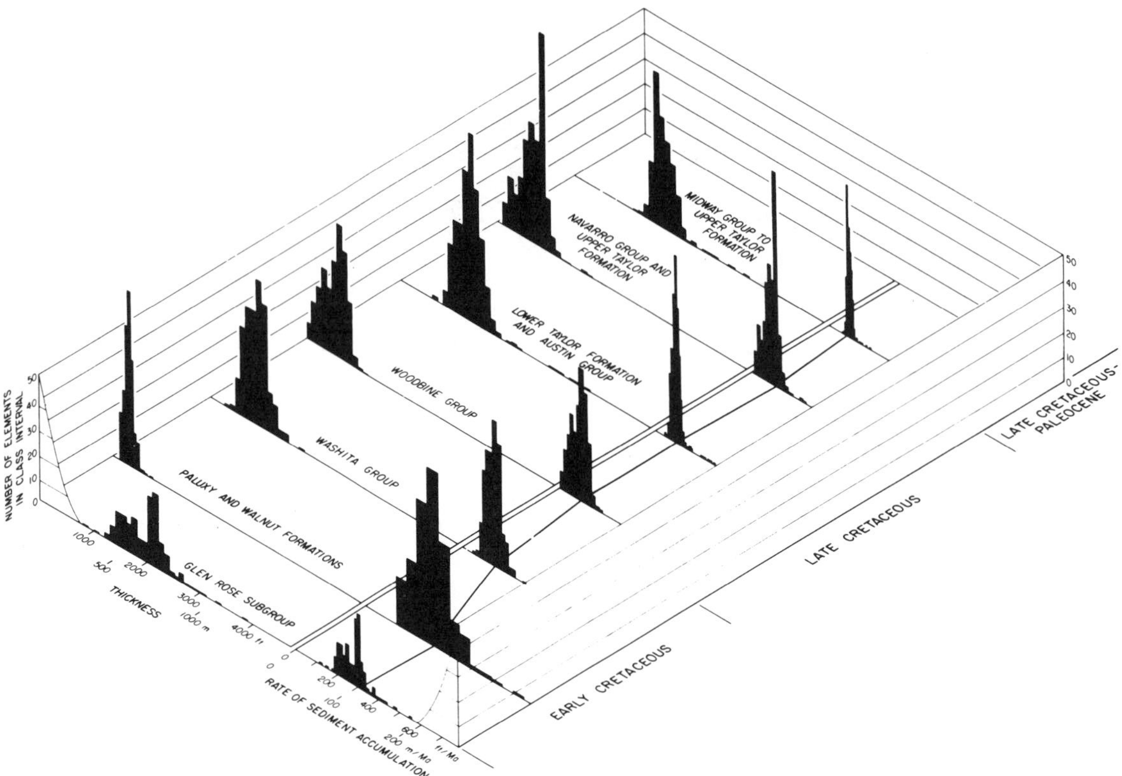

FIG. 11—Histograms of thickness and rate of regional sediment accumulation of principal stratigraphic units in the East Texas basin from 112 to 56 Ma. Sediment-accumulation rate equals vertical thickness divided by duration of each unit. Solid horizontal line connects modes of sediment-accumulation rates and reveals a systematic decline in accumulation rates with time. Salt-influenced values of increased thickness and rate, form tails on the right side of each histogram.

diapir growth during deposition (112 to 105 Ma) of the Glen Rose package (mean regional thickness = 571 m, 1,873 ft) of shelf carbonates and thin evaporites. The Glen Rose Subgroup exceeds regional thickness by as much as 610 m (2,000 ft) in the secondary peripheral sinks. The area affected by salt withdrawal is 1,900 km^2 (734 mi^2). La Rue dome is surrounded by the largest salt withdrawal basin (112 km^3, 27 mi^3) around group 2 diapirs.

The crests of La Rue, Boggy Creek, and Brushy Creek domes are presently 1,356 m (4,450 ft), 557 m (1,830 ft), and 1,088 m (3,570 ft) deep, respectively. The net rate of postdiapiric growth for these domes in the basin center was less than regional depositional rates, thus resulting in burial. This lag in upward dome growth may be related either to the deposition of massive, resistant carbonate strata around and above the diapirs, to rapid deposition in the center of the basin, to depletion of the salt source layer, or to some combination of these factors.

The diapirs of subgroup B (see Figure 6), Brooks and East Tyler domes, were active during deposition of the Paluxy and Walnut formations from 105 to 104 Ma. These secondary peripheral sinks (486 km^2, 188 mi^2; and 727 km^2, 281 mi^2 in extent, respectively) are filled with Paluxy to Walnut strata (Fig. 8). The Paluxy-Walnut sequence is relatively thin, being only 88 m (289 ft) in mean thickness and 219 m (720 ft) in maximum thickness. This thinness and the lack of evidence of pillow-phase thinning during the preceding Glen Rose deposition, indicate that Brooks and East Tyler domes probably did not evolve from a pillow stage during deposition of Paluxy-Walnut strata and are thus inferred to have been diapirs prior to this time. A renewed surge of diapiric growth, initiated by unknown causes during Paluxy and Walnut deposition, is indicated by localized, massive overthickening of the secondary peripheral sinks.

Mount Sylvan, East Tyler, and Steen domes constitute subgroup C and are surrounded by secondary peripheral sinks filled with Washita Group carbonates deposited from 104 to 98 Ma (see part 1, Seni and Jackson, 1983a, Fig. 14). Salt-withdrawal basins in Washita strata are similar in size and style to basins formed during Paluxy-Walnut deposition, suggesting that Washita dome growth was also caused by rejuvenation of preexisting diapirs. East Tyler dome is included in subgroups B and C because it showed rapid rates of growth during Paluxy deposition (subgroup B) and Washita deposition (subgroup C). The similarity between subgroups B and C also includes the shallow depth of their diapir crests. The average depth of the crests of subgroup B and C diapirs are 168 m (550 ft) and 183 m (600 ft), respectively.

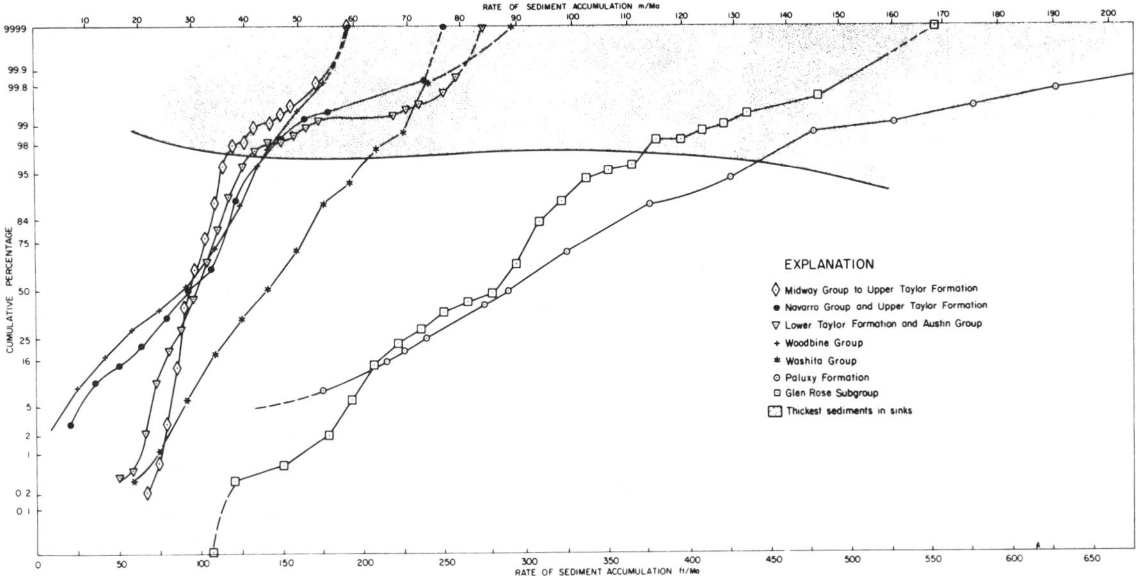

FIG. 12—Cumulative-probability curves of sediment-accumulation rate versus cumulative percentage. Declining regional accumulation rates are evidenced by displacement of curves of younger units to the left. The increase in slope of curves of younger strata compared with older strata reflects less variable sediment accumulation rates. Salt-induced skewness is clearly evident by decreasing slopes above the 95th or 98th percentile (stippled). These thickest parts of each unit accumulated in peripheral sinks. Salt-induced thinning over pillows is not apparent at lower parts of curves in this diagram because of small number of post-Glen Rose pillows compared with post-Glen Rose diapirs. Wilcox data are omitted because the Wilcox is restricted to southernmost part of basin.

Group 3: Post-Austin Group (86 to 56 Ma)—Periphery of Diapir Province

Hainesville and Bethel domes are group 3 diapirs (Figs. 5, 6); their crests are 366 m (1,200 ft) and 488 m (1,600 ft) deep, respectively. Because of their young age, these are the only domes in the East Texas basin with a complete history of salt movement preserved in strata younger than 112 Ma (Fig. 9). Group 3 diapirs (mean volume = 47 km^3, 11 mi^3) are by far the largest diapirs in the East Texas basin. According to Loocke (1978), approximately 78 km^3 (19 mi^3) of salt constitute the Hainesville salt stock.

Seismic control (Loocke, 1978) and well data indicate pillow growth of Hainesville dome from 112 to 92 Ma during Glen Rose to Woodbine deposition. Diapir growth occurred from 86 (possibly 92) to 56 Ma during post-Woodbine to Midway deposition.

The history of the late growth and location on the periphery of the diapir province (Fig. 6) suggests that group 3 diapirs have a growth history significantly different from that of groups 1 and 2. Group 1 and 2 diapirs grew during rapid regional sediment accumulation (to be discussed), but group 3 diapirs grew fastest when regional rates of sedimentation had declined significantly. The significance of local angular unconformities over Hainesville dome during pillow growth is also discussed later in this article.

Overview of Dome History

In the East Texas basin the location of diapirism varied through time. Viewed against a background of basin infilling, the shifting areas of dome movement followed an orderly progression. Diapirism was concentrated first along the edge of the diapir province (group 1) and then in basin center and northward along the basin axis (group 2) in response to the shifting of depocenters from the basin margin to the center. Basin-edge tilting and erosion over pillow crests localized the final episode of diapiric activity (group 3) on the updip margin of the diapir province.

The pre-Glen Rose history of diapirism is described elsewhere (Jackson and Seni, 1983; McGowen and Harris, 1983), but it is summarized here to provide an overview of salt movement in the East Texas basin. Rapid peripheral filling of the previously starved basin during Hosston–Glen Rose deposition may account for the location of the three subgroups of group 1 diapirs around the margin of the diapir province. The distribution and timing of group 1 diapirs suggest that uneven loading by thick terrigenous clastics of the Schuler and Hosston Formations, prograding toward the basin center, triggered diapirism in the Jurassic and Early Cretaceous. Group 1 diapirs grew in sites of maximum regional sedimentation. Group 2 diapirs underwent diapirism along the basin axis as sedimentation and subsidence rates peaked in Early Cretaceous (112 to 98 Ma). We are uncertain whether diapirs of groups 1 and 2 "pierced" their overburden by subsurface intrusion or by erosional breaching. But the growth of group 2 diapirs during rapid regional sedimentation and subsidence suggests that erosion was not the prime cause. Rather, uneven (in terms of both thickness and lithofacies) sediment loading was probably responsible for initiating diapirism.

Group 3 diapirs grew on the northern and western margin of the diapir province when regional sedimentation rates were significantly reduced from former levels. Loading of the lithosphere in the center of basins commonly causes basin-edge tilting, making locally elevated areas like pillow crests prone to erosion between episodes of sedimentation (Boillet, 1981). Erosion could therefore have exposed and breached salt pillows on the updip margin of the East Texas diapir province. This breaching initiated diapirism in at least one dome. Structural evidence shows that Hainesville dome reached the surface by erosional exposure of the salt pillow (Loocke, 1978; part 1, Seni and Jackson, 1983a, Fig. 6). Salt extrusion, probably forming salt-cored islands, allowed massive diapirism because of the lack of vertical constraint to the rise of salt (Bishop, 1978). The minor amount of cap rock supports the conclusion that salt was removed by extrusion and erosion, rather than by ground-water dissolution.

RATES OF SALT MOVEMENT AND DOME GROWTH

Syndepositional thickness variations in surrounding strata not only allow recognition of timing and patterns of halokinesis, but also measurement of volumes and rates of salt flow. All previous estimates of the growth rate of pillows and diapirs (e.g., Trusheim, 1960; Ewing and Ewing, 1962; Sanneman, 1968; Kupfer, 1976; Netherland, Sewell and Associates, 1976; Kumar, 1977; Jaritz, 1980) rely on certain basic assumptions, and the present study is no exception; these assumptions are discussed at the beginning of this paper. Methods vary widely and include some based entirely on structural uplift through time; such approaches have complicating factors such as the possibility of collapse of marker beds due to salt dissolution.

The concept of gross rate of growth versus net rate of growth is fundamentally important but almost always ignored. Gross rates are a function of the volume of salt evacuated from the withdrawal basin and mobilized up the diapir. Net rates are a function not only of this process, but also of all other processes that affect diapir height and growth rate, such as salt dissolution, extrusion, and lateral intrusion. Thus, gross rates of growth approximate the true rate of salt flow regardless of the independent motion of the diapir crest. However, net rates of growth approximate the actual movement of the diapir crest.

Stratigraphic data were employed in four types of calculations; the first is self explanatory and the other three are reviewed in Figure 10.

1. The volume of sedimentary fill in a salt-withdrawal basin equals the volume of salt that migrated during filling of that basin (Fig. 1). This volume, divided by the duration of withdrawal-basin activity, quantifies the rate of salt movement in terms of volume through time.

2. Net rates of salt-pillow uplift are calculated by the rate of stratigraphic thinning through time over the crest of salt pillows (Fig. 10A). The upward growth of nonpiercement salt pillows resulted in thinning of strata over the pillow. Rate is obtained by dividing growth by duration.

3. Net rate of dome-crest uplift (or net rate of dome growth) is equated with the maximum rate of sediment accumulation in the associated withdrawal basin (Fig. 10B). This technique assumes that salt remained at or near the depositional surface through most of its growth history, representing a balance between basin subsidence, ground water dissolution, and extrusion on one side, and upward movement of the diapir crest on the other. Again, rate is obtained by dividing growth by duration.

4. Gross rate of diapir elongation is calculated by dividing the volume of salt moved (estimated by method 1) by the maximum cross-sectional area of the diapir (Fig. 10C). This technique assumes that all salt migrated from below the withdrawal basin into the diapir and rose through a constriction defined by the cross-sectional area of the stock, thereby lengthening the stock. Rate is obtained by dividing growth by duration.

Distinguishing Between Regional and Salt-Related Thickness Variations

Regional isopach maps (Figs. 7, 8, 9) and statistical analysis of thickness data were used to differentiate between changes in interval thickness caused by basin-wide subsidence as opposed to salt-related local subsidence. In the East Texas basin, salt-related thickness changes are evident on regional isopach maps as isolated perturbations of regional thickness trends. This contrasts sharply with isopach trends in the Tertiary section of the upper Texas Gulf Coast, where thickness trends are strongly influenced by salt migration or growth faulting (Bebout et al, 1978; Galloway et al, 1982). Isopach maps of strata in the East Texas basin clearly show that local thickness anomalies around salt domes shifted their positions with time (Figs. 7, 8, 9). For example, pre-Austin Group strata are thin (Figs. 7, 8) around Bethel and Hainesville domes, whereas younger strata are massively thickened (Fig. 9) around these domes. Whitehouse and Bullard domes have little effect on thickness of post-112 Ma strata.

Statistical thickness data from wells in each isopach interval are documented in Seni and Jackson (1983b). Each isopach interval was treated as a separate population and analyzed by the procedures described in the following paragraph.

The effect of salt tectonics on thickness and rate of sediment accumulation is illustrated by histograms, cumulative-probability curves, and contour maps showing sediment-accumulation rate and standard deviation of accumulation rate. Figure 11 is a histogram of measured thicknesses and calculated rates of sediment accumulation for various isopach intervals. Increased thickness and rates of sediment accumulation on the right side of the histograms are especially evident for the Glen Rose Subgroup and Paluxy Formation. These skewed values are known to be salt-induced, for they are yielded by areas adjacent to salt diapirs. Cumulative-probability curves (Fig. 12) were constructed using the procedures of Folk (1980) for grain-size analysis. These curves approach a straight-line (normal or Gaussian) distribution over the central 16th to 84th percentiles. The tails of the distribution above the 95th to 98th percentiles (stippled) are associated with much greater rates of accumulation in withdrawal basins. The thickness and rates of sediment accumulation in diapir-stage withdrawal basins are typically three standard deviations (3σ) more variable than regional values (Fig. 13).

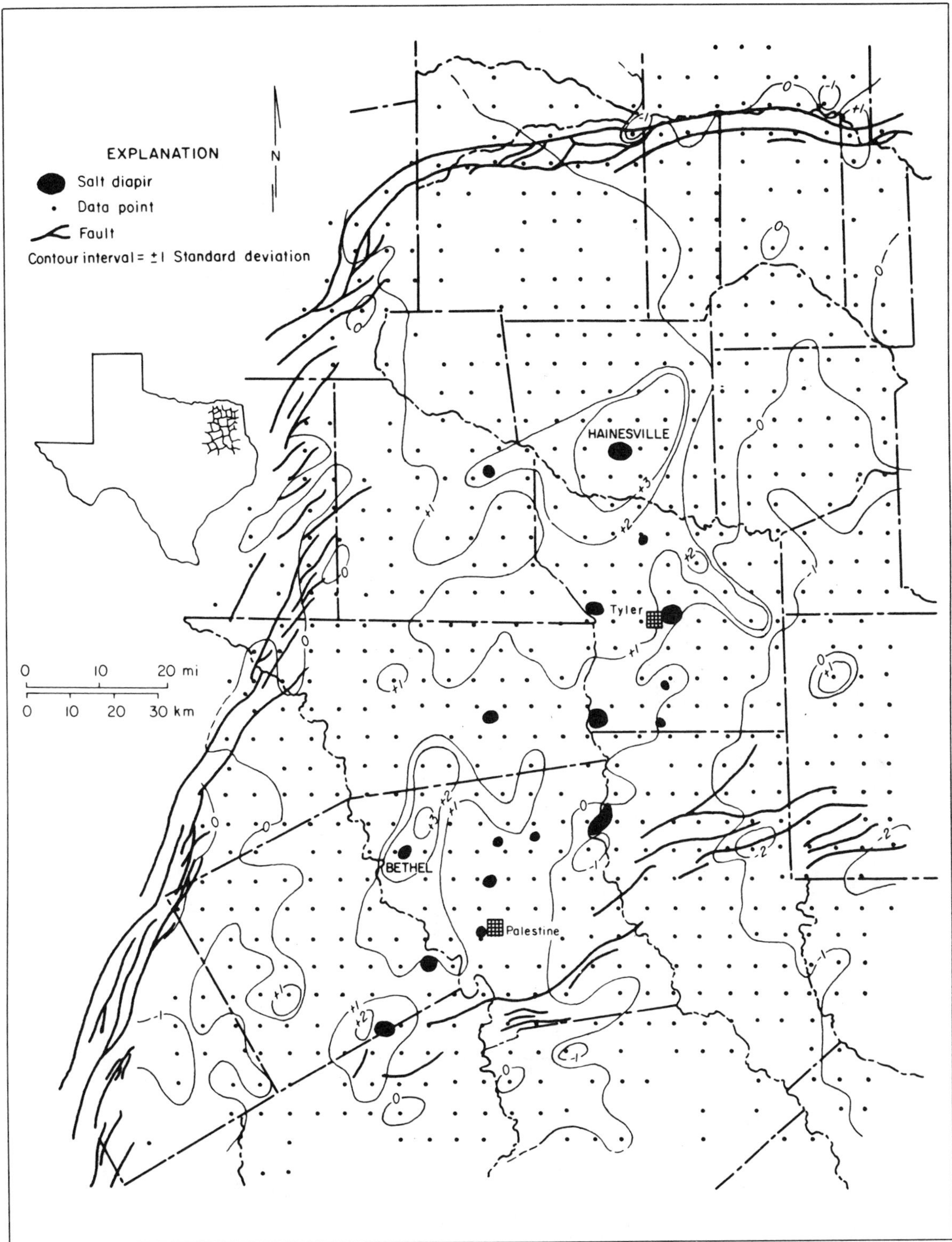

FIG. 13—Contour map showing sample grid spacing and standard deviation (σ) of sediment accumulation rate for the lower Taylor Formation–Austin Group. Secondary peripheral sinks around Hainesville (north) and Bethel domes (west) exhibit rates of sediment accumulation 3 σ more variable than regional values indicating high degree of local variability induced by salt flow. Compare with isopach map, Figure 9.

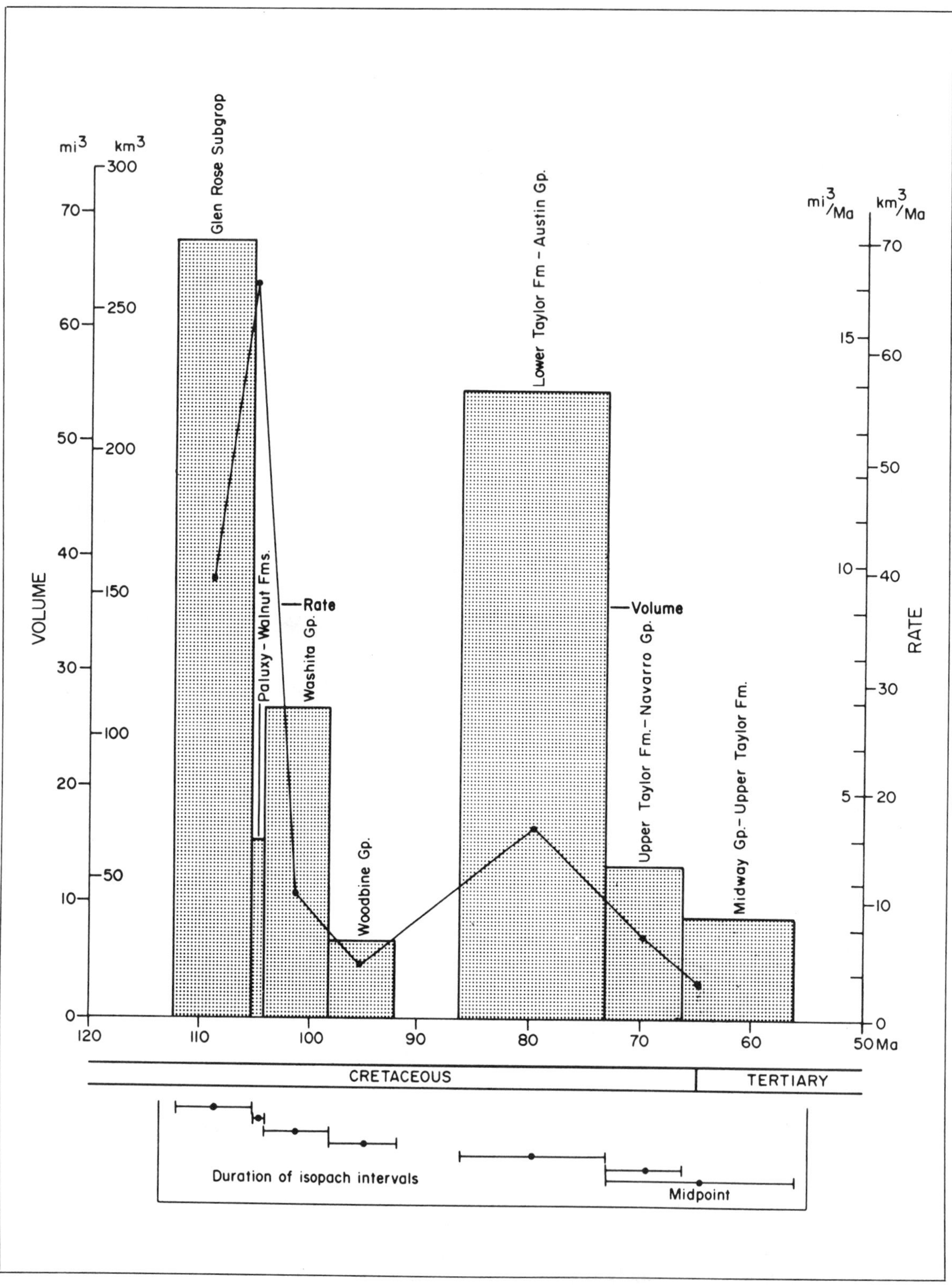

FIG. 14—Histogram of the volume of salt-withdrawal basins compared with the volumetric rate of regional basin infilling. The rate and volume of sedimentary fill in salt-withdrawal basins are equal to rate and volume of salt flow. The volume is a function of both duration and rate.

Volume of Salt Mobilized and Estimates of Salt Loss

Basin-wide summation of the volumes of salt-withdrawal basins (Table 1) shows a general decline in the rates of salt movement from Early Cretaceous to the Eocene (Fig. 14). The volume of mobilized salt peaked during deposition of the Lower Cretaceous Glen Rose Subgroup, when 276 km^3 (67 mi^3) of sediment accumulated in withdrawal basins. The maximum rate of salt movement occurred during deposition of the Paluxy and Walnut Formations at about 105 Ma (Fig. 14). This estimate of about 65 km^3/Ma (15 mi^3/Ma) appears high, probably because the duration of Paluxy deposition is so short that small errors are magnified. Actual rates of salt movement were probably similar to rates during deposition of the Glen Rose Subgroup, about 40 km^3/Ma (10 mi^3/Ma).

A subsidiary peak of salt moment coincided with filling of the large salt-withdrawal basin (Mineola basin) around Hainesville dome in the Late Cretaceous. The proportion of the volume of salt mobilized to the total volume of sediments accumulated during the Eocene is 2.0×10^{-3}, one order of magnitude lower than the equivalent proportion in the Early Cretaceous (2.0×10^{-2}).

The volumes of withdrawal basins for respective diapirs in the East Texas basin are shown in Figure 15. Hainesville and Bethel domes have a complete growth history from pillow to postdiapir stages preserved in post-Glen Rose strata; thus, large primary peripheral sinks can be mapped. Only the latest stage of the growth history of older domes is preserved in post-Glen Rose strata. Because this part of their history is characterized by slow growth, their salt-withdrawal basins are small (less than 15 km^3, 4 mi^3).

The volume of all salt-withdrawal basins is much greater than the total volume of salt in stocks projecting above the James Limestone in the Glen Rose Subgroup (Table 1). The volume of known and probable withdrawal basins indicates that approximately 800 km^3 (190 mi^3) of salt migrated during deposition of Lower Cretaceous to Eocene strata. Subtracting from this the volume of salt in stocks above the James Limestone indicates approximately 380 km^3 (90 mi^3) of salt have been lost by groundwater dissolution, extrusion, or erosion. This loss of salt and the abundance of anhydrite and calcite cap rock over most east Texas salt domes, support a residual origin for cap-rock formation by salt dissolution. Hainesville dome is surrounded by the largest withdrawal basin (243 km^3, 58 mi^3) in the East Texas basin. Subtracting the volume of the Hainesville salt stock (40 to 78 km^3; 10 to 19 mi^3) from the volume of the withdrawal basin indicates that 165 to 203 km^3 (40 to 49 mi^3) of salt are missing from Hainesville

Table 1. Volume of Withdrawal Basins (km^3)

Domes	Wilcox Group	Midway and Navarro Groups and Upper Taylor Formation	Navarro Group and Upper Taylor Formation	Lower Taylor Formation and Austin Group	Woodbine Group	Washita Group	Paluxy and Walnut Formations	Glen Rose Subgroup	All Intervals	Volume of Salt in Diapirs Above Glen Rose Subgroup (km^3)
Oakwood	-	-	-	3.4	-	9.3	-	-	12.7	18.2
Butler	-	-	-	1.2	-	-	3.9	-	5.1	17.8
Palestine	-	-	-	-	3.9	7.8	3.9	-	15.6	22.0
Keechi	-	-	-	-	-	7.8	3.9	4.6	16.2	55.4
Bethel	4.5	15.6	12.9	19.6	-	-	-	-	52.6	e16.4
Boggy Creek	-	-	-	3.2	-	-	-	73.5	76.7	71.3
Brushy Creek	-	-	-	3.2	-	-	4.6	9.2	17.0	e13.1
Whitehouse	-	-	-	-	-	-	-	-	-	10.2
Bullard	-	-	-	-	-	-	-	-	-	8.4
La Rue	-	-	-	12.1	-	22.0	6.3	111.8	152.2	e12.3
Brooks	-	-	-	-	5.1	4.6	12.0	12.4	34.1	e41.0
Mt. Sylvan	-	-	-	2.1	1.1	8.0	-	6.4	17.6	28.0
East Tyler	-	-	-	-	-	10.6	20.1	-	30.7	e28.4
Steen	-	-	-	4.1	4.5	15.6	3.6	-	27.8	e20.5
Grand Saline	-	-	-	3.5	-	-	-	-	3.5	e16.4
Hainesville	-	73.3	40.8	169.6	12.8	11.4	5.9	-	272.9	e41.0
Other basins	7.0	-	-	-	-	7.8	2.3	4.6	14.8	-
Basins of unknown affinity	-	-	-	-	-	-	3.5	53.2	-	-
Unknown basins and domes	11.5	88.9	53.7	222.0	27.4	104.9	66.5	222.5	749.4	420.4
Total	11.5	88.9	53.7	222.0	27.4	104.9	70.0	275.7	806.1	420.4

e = Estimated by extrapolation of residual gravity data.
[= Domes that share a single withdrawal basin.

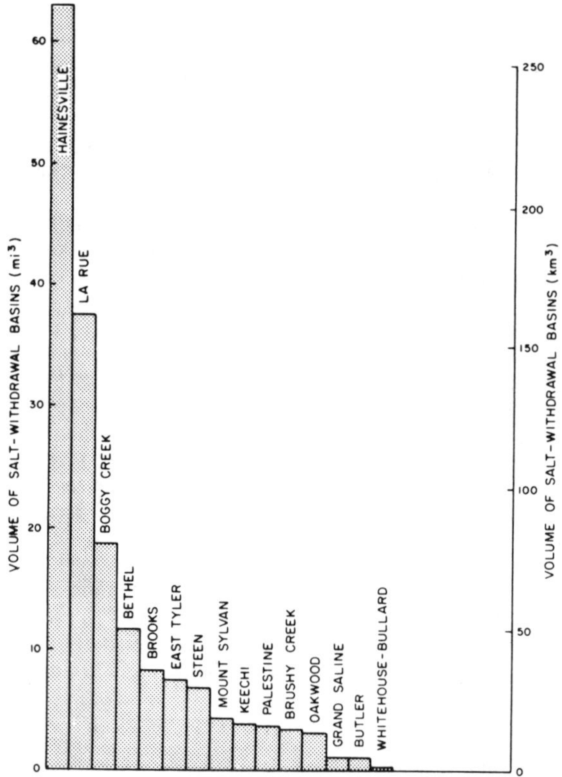

FIG. 15—Histogram of volume of salt-withdrawal basins around individual diapirs for basins formed since 112 Ma (early Glen Rose deposition).

dome. Loocke (1978) estimated salt loss by a seismic interpretation of original and residual salt volumes and concluded that 92 to 133 km^3 (22 to 32 mi^3) of salt were missing. Extrusion and erosion of salt are the most probable explanations for the large salt loss around Hainesville dome because the cap rock is too thin to have formed from ground-water dissolution of large volumes of salt.

Net and Gross Rates of Dome Growth

Methods of calculating rates of dome growth are based on particular inferred mechanisms of dome growth. Thus, different methods yield different estimates of growth rate for the same dome in a particular time interval. The reliability of these methods depends on the validity of the inferred mechanism of dome growth and on the accuracy of the estimates of volume, area, thickness, and duration.

Gross rates of dome growth are based on a minimum of mechanistic assumptions, but they require reliable estimates of volumes and areas (Fig. 10C). Net rates of dome growth assume the diapir crest remains at or near the sediment surface.

Net rates of pillow growth.—Net rates of uplift were calculated both by the maximum rate of deposition in withdrawal basins and by the rate of sediment thinning over pillow crests (Figs. 10A, B, 16). Maximum net rates of 40 to 100 m/Ma (130 to 330 ft/Ma) for pillow uplift are calculated by the rate of sediment thinning. In comparison, net rates of uplift range from 100 to 130 m/Ma (330 to 427 ft/Ma) if calculated by rate of sedimentation in primary peripheral sinks. The two methods of calculation yield the same growth rate only when sediments over the pillow crest are thinned to zero thickness and where the thickness of the primary peripheral sink is approximately equal to the regional thickness.

Generally, the growth rate of salt pillows that later evolved into diapirs exceeded the growth rate of pillows that are still in a pillow stage. Perhaps a certain threshold value of geologic momentum (a function of growth rate and size) must be exceeded for a pillow to evolve into a diapir).

Net rates of diapir growth.—Net rates of diapir growth, as calculated from the maximum rate of sediment accumulation in withdrawal basins (Figs. 10B, 17A, 17B), show trends similar to those of salt movement and regional sediment accumulation rates (Fig. 18). Maximum net rates of dome growth ranged from 150 to 230 m/Ma (490 to 755 ft/Ma) during deposition of the Lower Cretaceous Glen Rose Subgroup and Paluxy and Walnut Formations. Growth rates then generally declined into the Eocene. Mean net rates of growth were 30 m/Ma (100 ft/Ma) from 73 to 56 Ma. Except for Hainesville and Bethel domes, maximum rates of net dome growth were associated with high rates of regional sediment accumulation (Figs. 10, 17A).

Subtracting regional rates of sediment accumulation from maximum rates of sediment accumulation in withdrawal basins provides a method of examining net rates of dome growth independently from background effects of regional sedimentation. Figure 19 shows that most domes have an initially rapid growth rate in the Early Cretaceous even with removal of the effects of rapid sediment accumulation during that time.

Gross rates of diapir growth.—Dome growth rates calculated with this method (Figs. 10C, 20) are generally the highest of the three methods used. Most domes show a rapid decline in rates of growth with time. Brooks, Steen, and Hainesville domes thus show peak rates of 530 m/Ma (1,740 ft/Ma), 420 m/Ma (1,380 ft/Ma), and 460 m/Ma (1,510 ft/Ma), respectively. Minimum growth rates range from 10 to 20 m/Ma (33 to 66 ft/Ma); these estimates are probably too low because some salt-withdrawal basins escaped detection, thereby effecting low estimates of salt volume mobilized. The growth rates given here are based on the maximum cross-sectional area of the stock, a procedure that yields conservative growth rates.

The hypothetical gross heights of six east Texas diapirs are shown on Figure 21. Gross heights of Brooks, Mount Sylvan, and Oakwood domes are less than the height of the enclosing strata above the James Limestone datum surface because some withdrawal basins were thin and remain unrecognized. The shortness of the gross length of these domes also indicates little salt loss by dissolution or extrusion since Glen Rose deposition. In contrast, the gross heights of Hainesville, Bethel, and Steen domes are 48 to 163% greater than the thickness of enclosing strata (equivalent to net column length), indicating abundant loss of salt during deposition of post-Glen Rose strata.

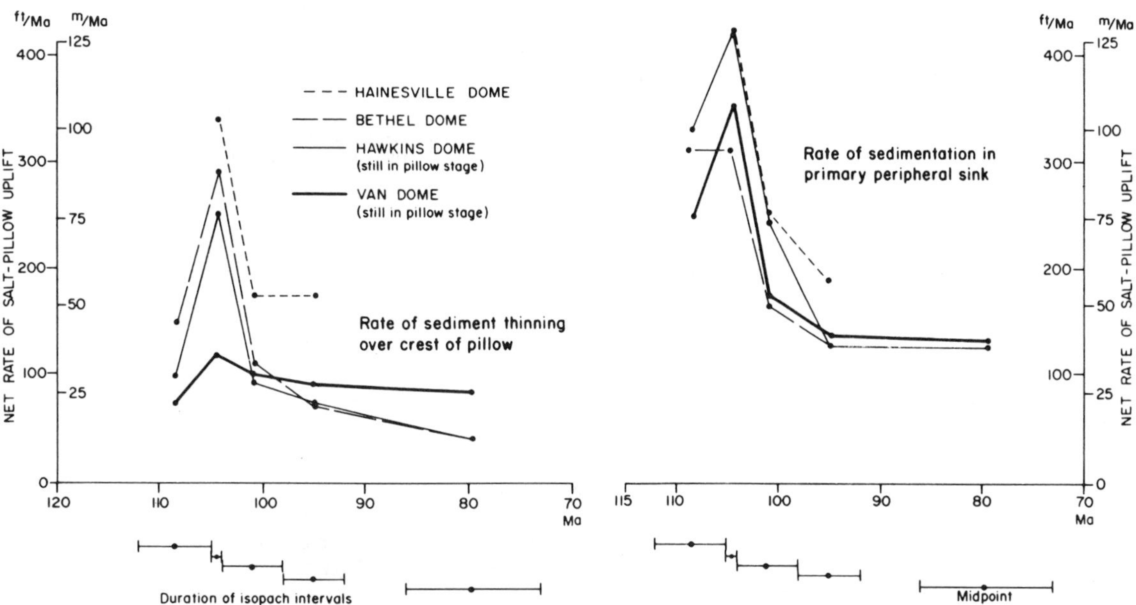

FIG. 16—Net rate of salt-pillow uplift calculated by rate of sediment thinning over crest of pillow (cf. Fig. 17). Pillows grew fastest during Early Cretaceous.

Significance of Growth Rates and Strain Rates

With the current state of knowledge, rates of dome growth are at best, semiquantitative. None of these methods can determine absolute movement of dome crests—that is, the movement of the crest relative to the topographic surface and the geoid. For example, the dome crest can remain stationary relative to the geoid (no absolute movement), while undergoing relative upward movement with respect to strata subsiding around the dome. Geologic examples of absolute rise of salt structures are known; for example, uplift of abyssal plain sediments by dome growth in the Gulf of Mexico was described by Ewing and Ewing (1962).

Methods of estimating rates of diapir growth are of two types: those estimating net rates, and those estimating gross rates. The advantages of each method depend on the intended purpose for these estimates. Gross rates of growth are especially appropriate for feasibility studies of nuclear waste repositories in salt domes because they provide estimates of rates of salt flow within a diapir. Diapiric salt around a repository could conceivably carry a repository upward (gross growth), while the diapir crest remains stationary (zero net growth) owing to salt dissolution.

Published values for rates of dome growth (Fig. 22, Table 2) vary widely. A comparison of growth rates for salt domes in Gulf interior basins shows that maximum growth rates ranged from 12 to 540 m/Ma (40 to 1,770 ft/Ma). Despite this 45-fold spread and despite the variety of techniques used to measure growth rates, the trends are qualitatively similar. Dome-growth rates were initially high in the Early Cretaceous or Jurassic (150 to 100 Ma). All growth rates show a general decline into the Tertiary. The spread in growth rates for the most recent growth episodes is very low (less than two-fold) because of an exponential decline (growth rate, $\dot{G} = a^{-t}$, where t = duration and a is a variable). Growth rates in the Tertiary are the lowest and range from < 10 to 20 m/Ma (33 to 66 ft/Ma). The similarity of growth histories for domes in the east Texas and north Louisiana diapir provinces suggests that dome growth depended primarily on regional factors, including basin evolution, rather than on local factors.

Growth rates calculated in this study provide an excellent means of estimating the strain rate of rock salt undergoing natural, nonorogenic, gravity-driven deformation. The relation between growth rate and strain rate is simple: growth rate represents the absolute lengthening per unit time (m/Ma), whereas strain rate ($\dot{e}$) represents the proportional change in length per unit time of 1 sec (one second). Methods of converting diapiric growth rate to strain rate are outlined in Seni and Jackson (1983b).

Converting growth rates to strain rates takes into account the differences in height of the diapirs; obviously a 1-m (3-ft) rise of a 10-m (30-m) high diapir is a far larger strain (elongation, e = 10%) than a 1-m (3-ft) rise of a 1,000-m (3,300-ft) high diapir (e = 0.1%), and this is reflected in a higher strain rate in the shorter diapir. Another useful attribute of the strain rate is that it enables comparison with all other types of deformation, ranging from meteoritic impacts ($\dot{e} = 10^{+3}$/sec) to compaction or isostatic rebound ($\dot{e} \simeq 10^{-16}$/sec) (Price, 1975). Estimation of the strain rates of natural deformation is still in its infancy and continues to be vigorously pursued. The problem hinges on how to obtain a reasonable approximation of the time during which a measured strain occurred. This problem is lessened in dealing with very slow strain rates because the need for accurate dating is less. Accordingly, strain rates for salt diapirs can be obtained to the nearest order of magnitude with a high degree of confidence.

Parameters such as isostatic rebound, displacement rates on the San Andreas fault, sea-floor spreading, and inferred flow rate of the asthenosphere show remarkable agreement that a "representative" geologic strain rate is 10^{-14}/sec (Heard, 1963, 1976; Carter, 1976). By comparing known finite strains, based on strain analysis in orogenic zones, with estimated durations of young orogenies (≤ 5 Ma), Pfiffner and Ramsay (1982) bracketed conventional strain rates for orogeny between the limits of 10^{-13}/sec and 10^{-15}/sec. Diapiric strain rates based on the growth rates in Figures 16, 19, and 20 are shown in Table 3. Overall rates refer to the rate during the entire known history of diapirism; the mean overall rate over about 53 Ma is 6.7×10^{-16}/sec. The fastest rates refer to the rates during a particular stratigraphic period characterized by the most rapid diapirism; the mean strain rates over approximately 5 Ma, based on gross growth and net growth, are 2.3×10^{-15}/sec and 9.8×10^{-16}/sec, respectively. These values agree closely with the lower limit for orogeny of

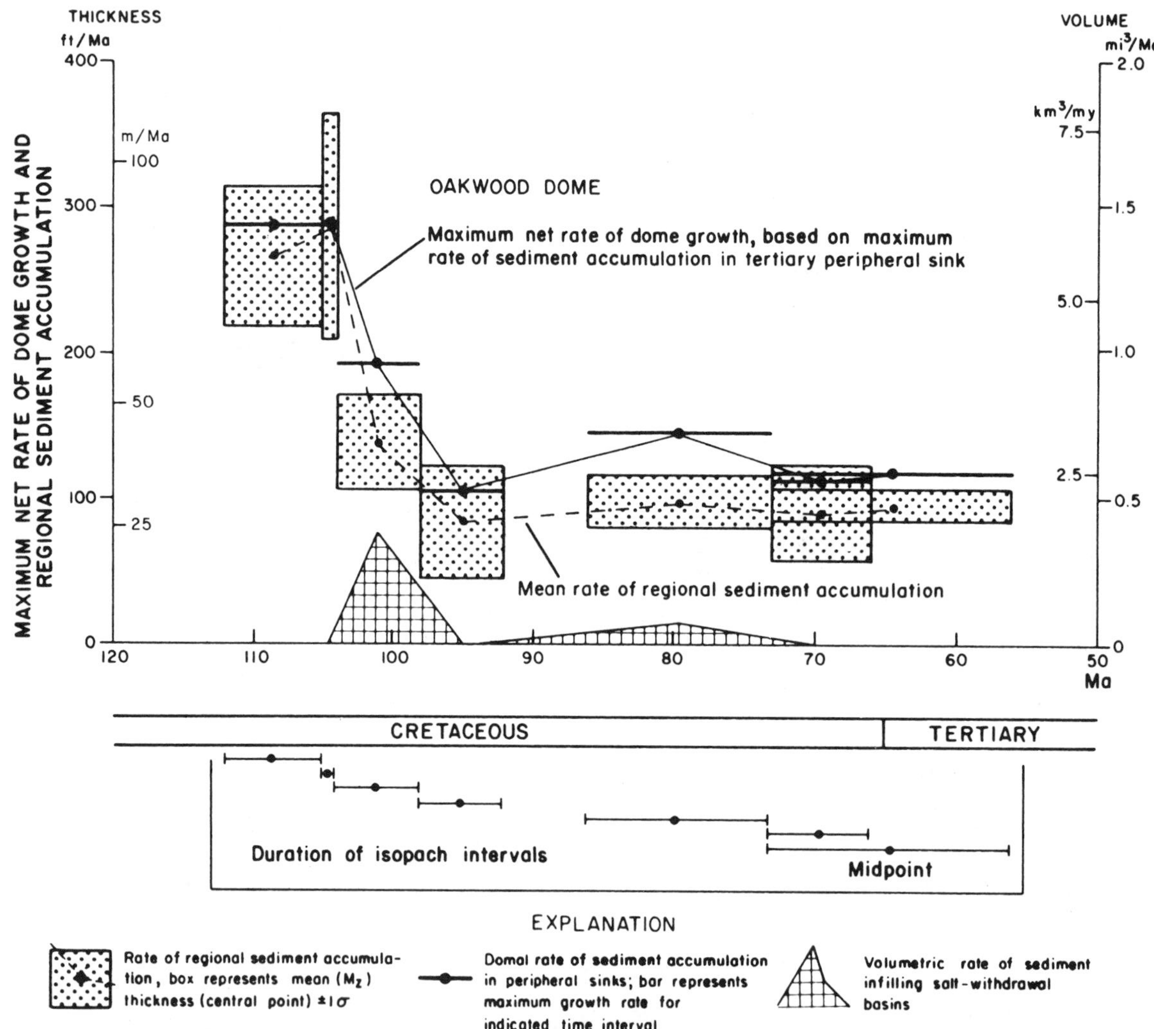

FIG. 17—Net rate of dome growth from 112 to 56 Ma for two diapirs calculated by equating maximum rate of sediment accumulation in peripheral sinks to net rate of dome growth in East Texas basin. Mean regional rate of sediment accumulation (dashed line) is shown. Volumetric rate of sedimentary infilling of salt withdrawal basins (hatched area) is inferred to equal volumetric rate of salt flow into salt structure. Volumetric rate calculated by dividing planimetered volume of salt withdrawal basin (Fig. 1) by duration of isopach interval. A. Oakwood dome is a group 1 diapir showing little difference between regional rates of sediment accumulation and net rates of dome growth. This reflects absence of thick secondary peripheral sinks in stratigraphic units younger than 112 Ma that would indicate large volumes of salt flow and rapid rates of dome growth. Therefore, the period of greatest diapiric growth for Oakwood dome predated 112 Ma (Figs. 5, 6). B. In contrast, the younger Hainesville dome is a group 3 diapir showing a large difference between regional rates of sediment accumulation and net rates of dome growth. Hainesville dome is surrounded by the largest secondary peripheral sink in East Texas basin, filled from 86 to 56 Ma (Figs. 9, 15).

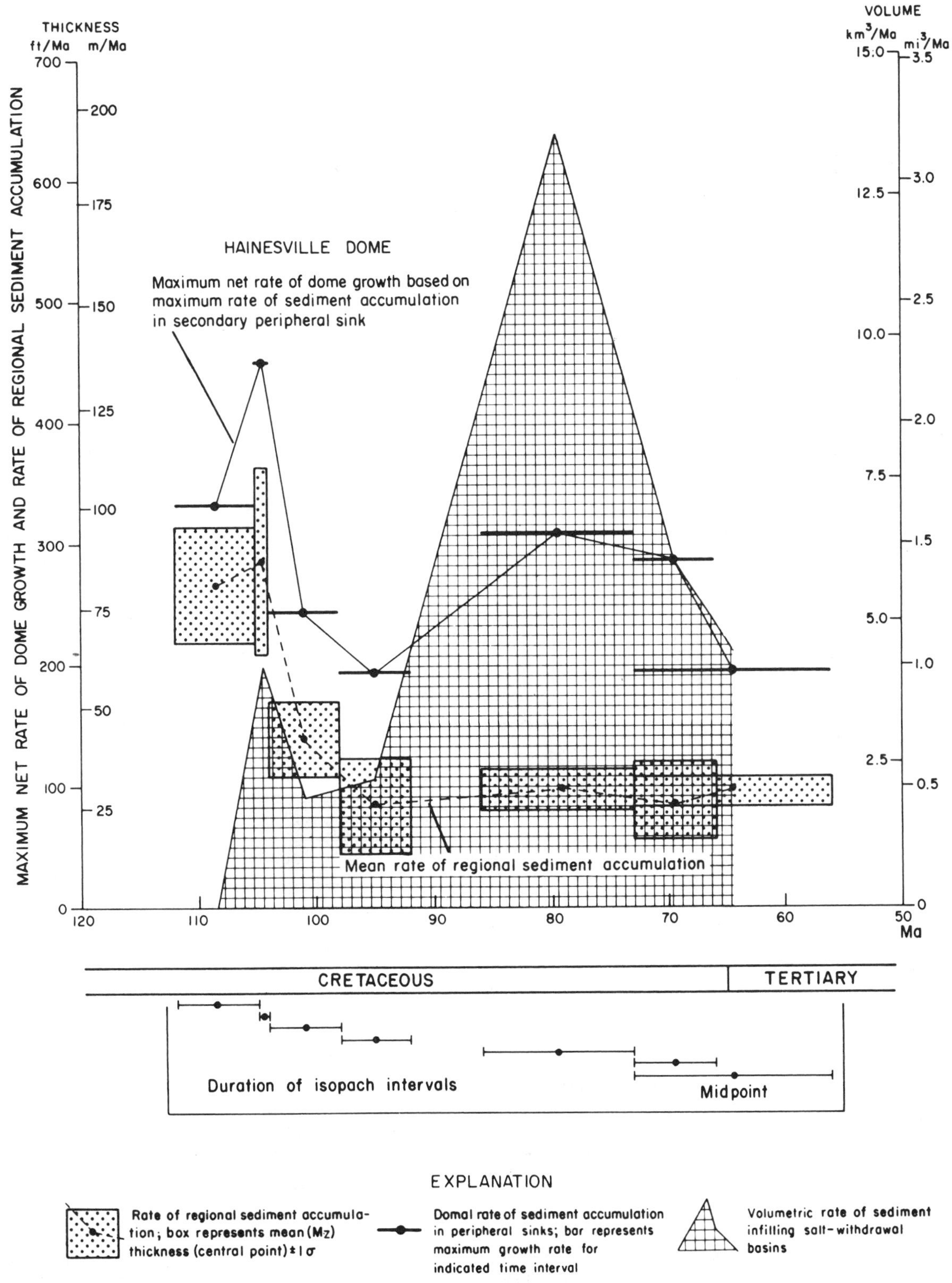

FIG. 17—Continued

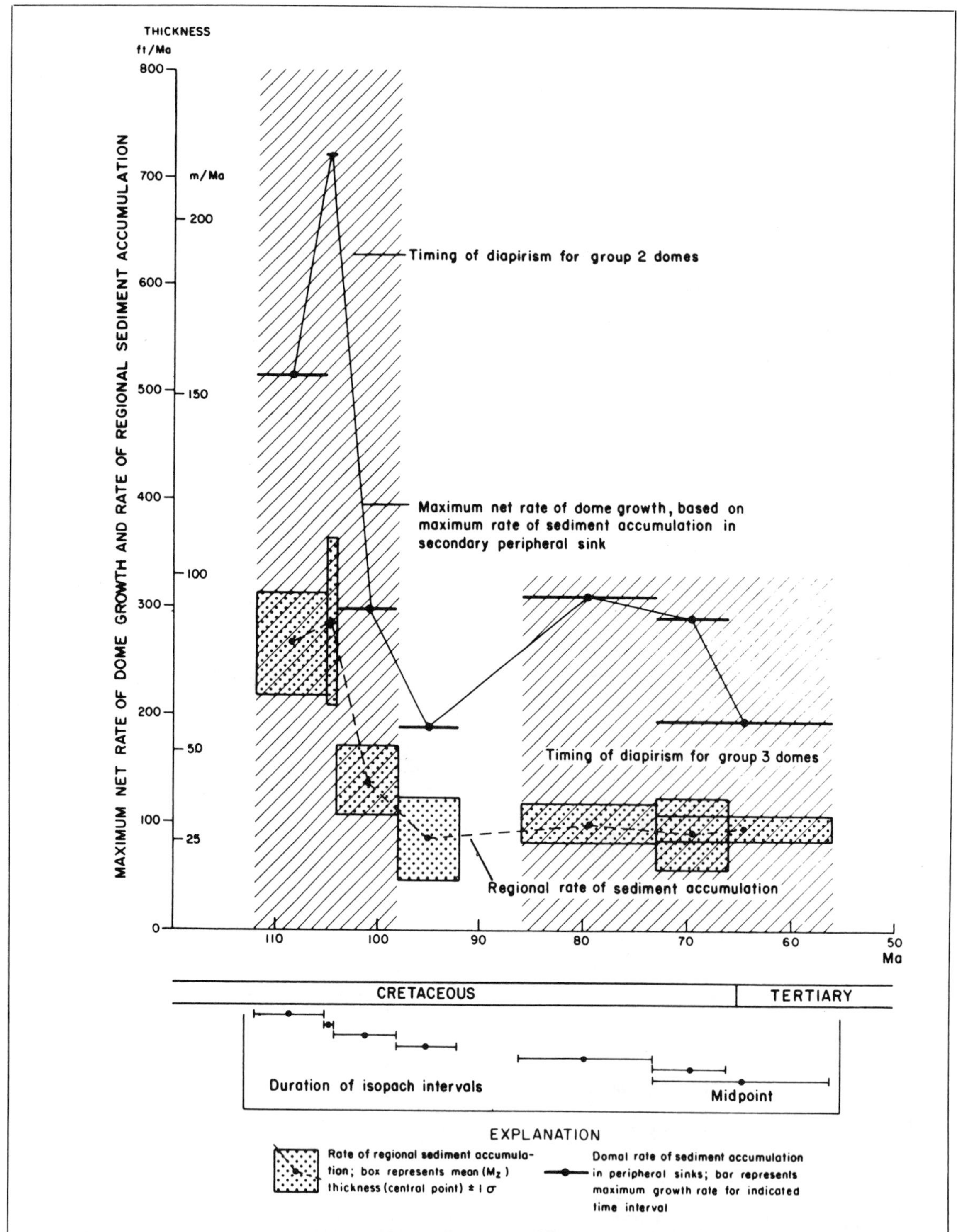

FIG. 18—Maximum net rate of dome growth (solid line represents growth rate of most rapidly growing diapir during deposition of any given stratigraphic unit) and mean regional sediment accumulation rate in East Texas basin from 112 to 56 Ma. Dome-growth rates peaked in Early Cretaceous during a time of high regional sediment accumulation. In contrast, a subsequent diapiric episode associated with growth of Hainesville and Bethel domes in Late Cretaceous was accompanied by low rates of regional sediment accumulation.

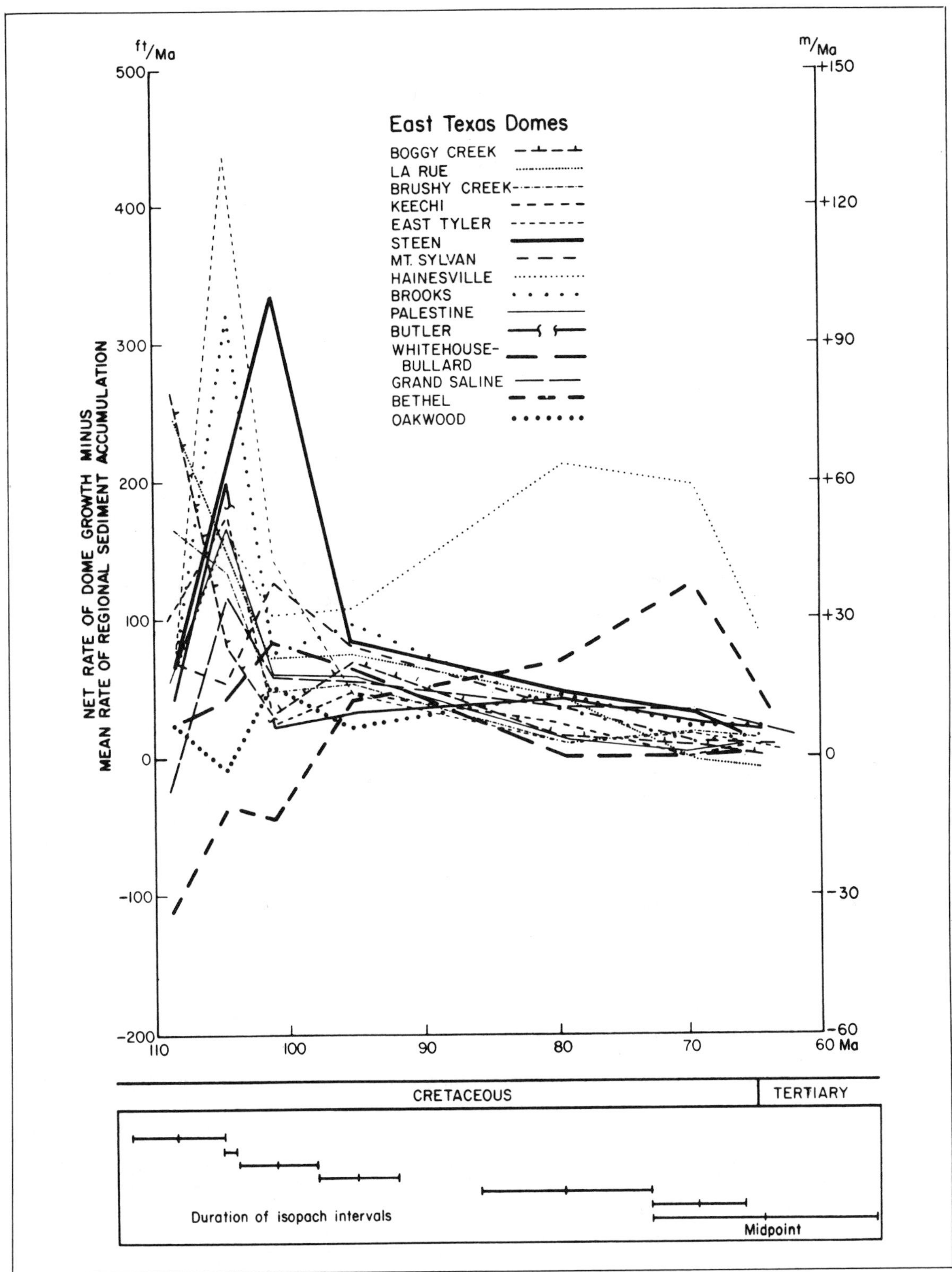

FIG. 19—Net rate of dome growth (calculated by rate of sediment accumulation in peripheral sinks) minus mean regional rate of sediment accumulation for 16 domes from 112 to 56 Ma. Even with removal of high rates of regional sedimentation (cf. Fig. 18), most domes grew fastest during Early Cretaceous. Negative values result from sediment thinning over pillow crests, as in Bethel dome.

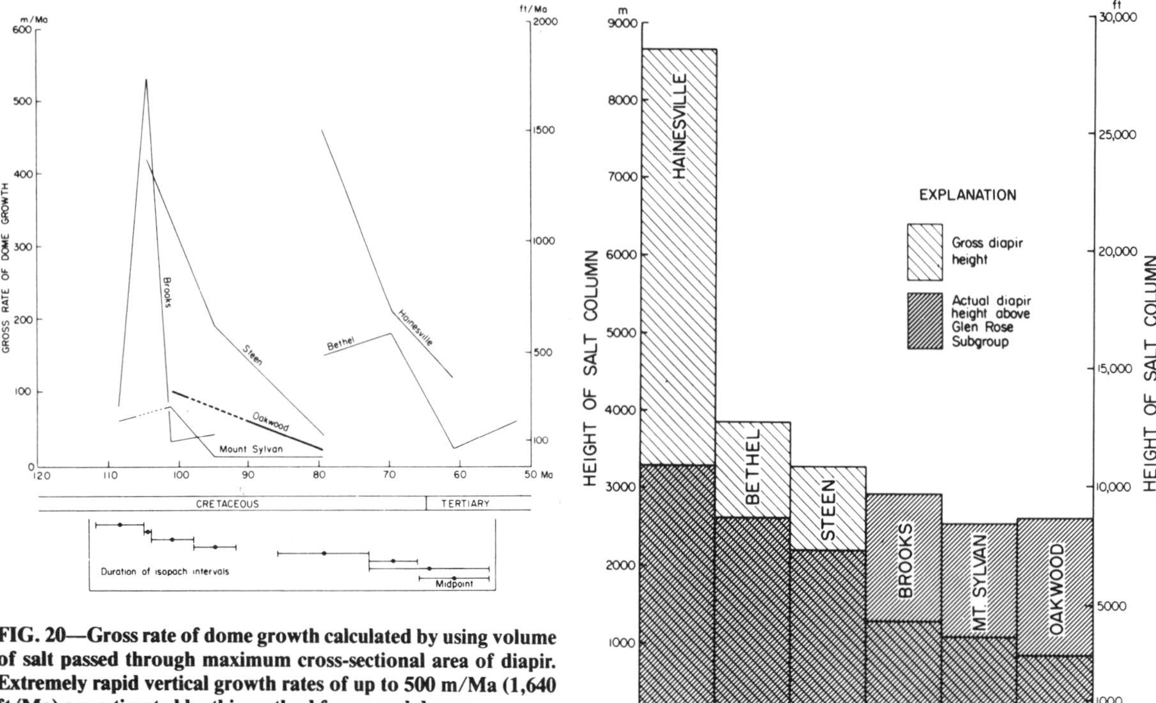

FIG. 20—Gross rate of dome growth calculated by using volume of salt passed through maximum cross-sectional area of diapir. Extremely rapid vertical growth rates of up to 500 m/Ma (1,640 ft/Ma) are estimated by this method for several domes.

FIG. 21—Gross height of post-James Limestone (lower part of Glen Rose Subgroup) salt columns for various domes, calculated by dividing volume of salt-withdrawal basin by maximum cross-sectional area of diapir. Present net height of post-James Limestone salt columns is shown for comparison. Hypothetical gross height exceeds actual net height of Hainesville, Bethel, and Steen domes because their crests were continually removed by dissolution or extrusion. For Oakwood, Mount Sylvan, and Brooks, the calculated gross height is less than actual height of salt column above the top of the Glen Rose Subgroup. This is because growth rates during postdiapir stage were too low to be fully accounted for by contour interval used in this method.

10^{-15}/sec given by Pfiffner and Ramsay (1982). A passive, gravity-driven process such as halokinesis would be expected to develop at a slower pace than orogeny. Of course, our estimates refer to a cumulative elongation over millions of years. The actual strain rate of the deforming rock salt is likely to be much higher, probably in the range of $10^{-12 \text{ to } -14}$/sec, for two reasons. First, deformation is likely to be spasmodic rather than the steady state used in our calculations. Second, deformation is likely to be concentrated in specific parts of the salt stock at any one time, with the largest strains and strain rates being in ductile shear zones between more massive tongues of rising rock salt.

CONCLUSIONS

As in the Zechstein salt basin (Trusheim, 1960), the durations of the pillow and diapir stages of salt dome growth in the East Texas basin are subequal and range from 10 to 30 Ma. The duration of postdiapir growth is commonly the longest, exceeding 112 Ma in some places. Approximately 800 km^3 (190 mi^3) of salt migrated during the Early Cretaceous to Eocene in the East Texas basin. Approximately 380 km^3 (90 mi^3) of salt that migrated during this time were lost subsequently by erosion and dissolution.

Group 1 diapirs, represented by Grand Saline, Butler, Oakwood, Palestine, Keechi, Bullard, and Whitehouse domes, are the oldest. These domes ceased diapirism before deposition of the Lower Cretaceous Glen Rose subgroup. They were initiated around the eastern, northwestern, and western periphery of the diapir province because these were the clastic depocenters for the Lower Cretaceous-Jurassic Hosston Formation and Cotton Valley Group.

Group 2 diapirs are La Rue, Boggy Creek, Brushy Creek, Brooks, East Tyler, Steen, and Mount Sylvan domes. These domes underwent diapirism from 112 to 98 Ma, with the center of diapiric activity migrating northward along the basin axis. They grew in the area of maximum sediment accumulation during a period of rapid regional subsidence.

Group 3 diapirs are Hainesville and Bethel domes. These domes underwent diapirism from 86 to 56 Ma, a period of low regional sediment accumulation. Local unconformities over Hainesville dome and the absence of 165 to 203 km^3 (40 to 49 mi^3) of salt from the Hainesville area indicate that erosion of strata above a salt pillow probably triggered diapirism by exposing the pillow, thereby inducing rapid sediment accumulation in an enormous secondary peripheral sink.

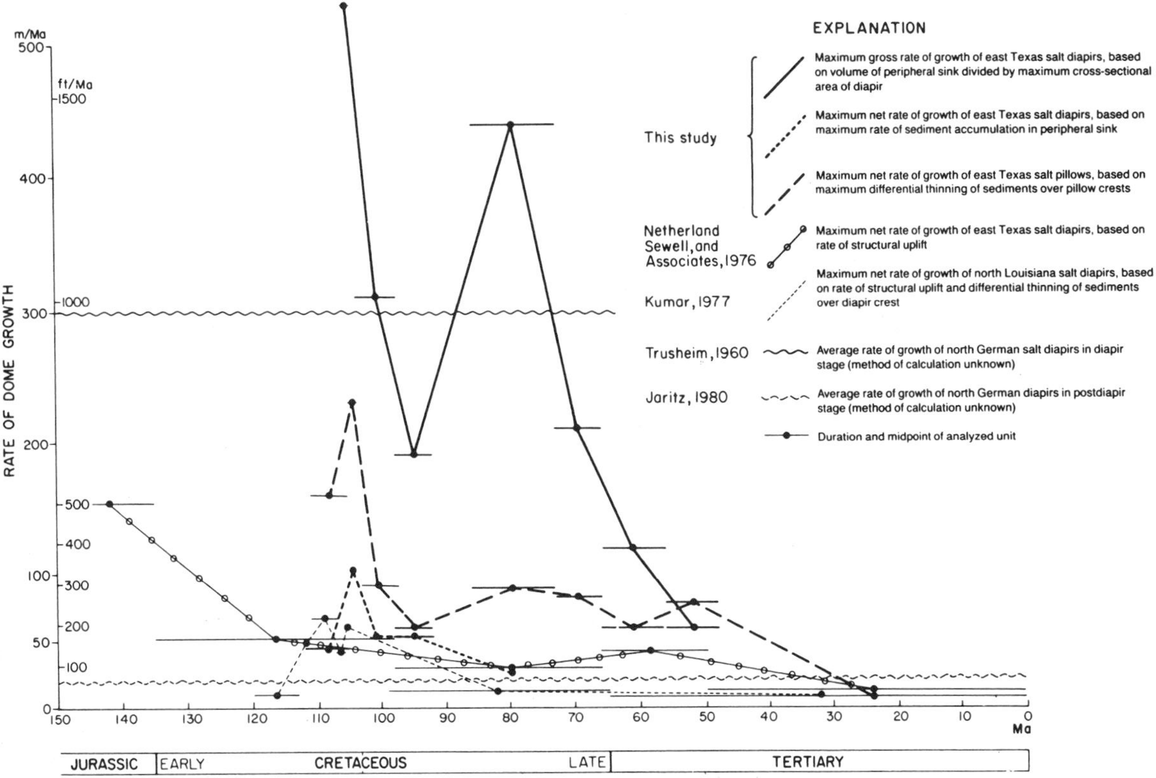

FIG. 22—Comparison of published dome growth rates with those of present study. Shapes of curves of growth rate for north Louisiana and east Texas diapirs are similar. Estimated dome-growth rates based on youngest strata decline exponentially and converge to a low value of approximately 20 m/Ma (66 ft/Ma). A similar terminal growth rate for north German Zechstein salt domes was estimated by Jaritz (1980).

In terms of the basin as a whole, peak rates of salt movement were 39 to 65 km^3/Ma (9 to 16 mi^3/Ma) during deposition of Lower Cretaceous Glen Rose Subgroup and Paluxy and Walnut Formations. A subsidiary peak of salt movement was 17 km^3/Ma (4.0 mi^3/Ma) during deposition of the Upper Cretaceous lower Taylor and Austin Groups. Rates of salt movement and dome growth declined exponentially into the Tertiary.

Calculated rates of dome growth vary according to techniques used, but long-term trends are similar. Dome-growth rates are highest when calculated by dividing the volume of salt moved by the cross-sectional area of the diapir because these are gross rates. Net rates of dome growth were based on maximum rates of sediment accumulation in peripheral sinks, and by the rates of sediment thinning over the crest of salt pillows. Net growth rates (which have been affected by dissolution or erosion) of pillows are lower when calculated by the rate of thinning. Maximum gross rates of dome growth (400 to 530 m/Ma; 1,310 to 1,740 ft/Ma) coincided with maximum regional rates of deposition in the Early Cretaceous from 112 to 104 Ma. Rapid gross rates of dome growth (180 to 460 m/Ma gross; 590 to 1,510 ft/Ma) recurred along the northern and western margins of the east Texas diapir province in the Late Cretaceous, from 86 to 56 Ma, with growth of Hainesville and Bethel domes.

Strain rates for growth of the East Texas salt diapirs, treated as steadily rising homogeneous bodies, averaged 6.7×10^{-16}/sec throughout the recorded diapiric history; strain rates peaked at a mean values of 2.3×10^{-15}/sec. These values are equivalent to the slower orogenic rates estimated in the literature. However, strain rates within the diapirs are likely to have been much higher in rock salt undergoing spasmodic and inhomogeneous strain.

REFERENCES

Baldwin, B., P. J. Coney, W. R. Dickinson, 1974, Dilemma of a Cretaceous time scale and rates of sea floor spreading: Geology, v. 2, p. 267-270.

Bebout, D. G., R. G. Loucks, and A. R. Gregory, 1978, Frio sandstone reservoirs in the deep subsurface along the Texas Gulf Coast: University of Texas Bureau of Economic Geology Report of Investigations 91, 92 p.

Bishop, R. S., 1978, Mechanism for emplacement of piercements diapirs: AAPG Bulletin, v. 62, p. 1561-1583.

Boillet, G., 1981, Geology of the continental margins: New York, Longman Inc., 115 p., translated by A. Scarth.

Bright, T. J., 1977, Coral reefs, nepheloid layers, gas seeps, and brine flows on hard banks in the northwestern Gulf of Mexico: Proceed-

Table 2. Comparison of Published Methods and Rates of Salt-Dome Growth with Those of Present Study

Author	Growth Rate $\dot{G}$(m/Ma)	Location	Method	Comments
Ewing and Ewing, 1962	1,000 (net)	Sigsbee Knolls, Gulf of Mexico	Uplift of dated marker bed	Abyssal-plain seamounts project above flat floor. Pleistocene sediments have been uplifted 150 to 300 m.
Trusheim, 1960	300 (average)	Zechstein salt, Germany	Unknown	Average.
Sanneman, 1968				Consistent long-term rate, same for pillows and domes.
Jaritz, 1980	100 to 500–diapir stage ~ 20–later stages			
Netherland, Sewell, and Associates, 1976	6 (low average) 44 (high average)	Various domes, East Texas basin	Uplift/time	General decrease in growth rates for individual domes through time. Maximum calculated rate of growth is 54 m/Ma for Mt. Sylvan dome in Early Cretaceous. Minimum calculated rate of growth is 3 m/Ma for Mt. Sylvan and Bullard domes in post-Eocene. Maximum estimated growth rate is 153 m/Ma in Late Jurassic.
Kupfer, 1976	5 to 10	Vacherie dome, N. Louisiana salt basin	Volume of salt/time	General decrease in dome growth rate through time. Maximum growth in Early Cretaceous (0.4 km^3/Ma), sharp decline in Late Cretaceous to 0.04 km^3/Ma.
Kumar, 1977	1 to 42 (net)	Various domes, N. Louisiana salt basin	Stratigraphic thinning and uplift/time	General decrease in dome growth through time. Maximum growth in Early Cretaceous is 42 m/Ma, maximum growth in Late Cretaceous is 19 m/Ma and minimum growth rate in Cenozoic is 1 m/Ma.
This study	13 to 104 (net)	Various domes, East Texas basin	Net rate of pillow growth—rate of sediment thinning	
	26 to 222 (net)	Various domes East Texas basin	Net rate of dome growth—maximum rate of sedimentation in withdrawal basin	
	10 to 530 (gross)	Various domes, East Texas basin	Gross rate of diapir growth—volume of salt divided by cross-sectional area of salt stock	

ings, Third International Coral Reef Symposium, no. 3, v. 1, p. 39-46.

Cantrell, R. B., J. C. Montgomery, and A. E. Woodward, 1959, Heterostegina reef on Nash and other piercement salt domes in northwestern Brazoria County, Texas: Gulf Coast Association of Geological Societies Transactions, v. 9, p. 59-62.

Carter, N. L., 1976, Steady state flow of rocks: Reviews of Geophysics and Space Physics, v. 14, no. 3, p. 301-360.

Crowe, C. T., 1975, The Vacherie salt dome, Louisiana and its development from a fossil salt high: Master's thesis, Louisiana State University, 172 p.

Dixon, J. M., 1975, Finite strain and progressive deformation in models of diapiric structures: Tectonophysics, v. 28, p. 89-124.

Elliot, T. L., 1979, Deposition and diagenesis of carbonate slope deposits, Lower Cretaceous, northeastern Mexico: PhD dissertation, University of Texas at Austin, 330 p.

Ewing, M., and J. Ewing, 1962, Rate of salt-dome growth: AAPG Bulletin, v. 46, p. 708-709.

Fisher, W. L., L. F. Brown, Jr., J. H. McGowen, and C. G. Groat, 1973, Environmental geologic atlas of the Texas coastal zone—Beaumont-Port Arthur area: University of Texas Bureau of Economic Geology, 93 p.

——— J. H. McGowen, L. J. Brown, Jr., and C. G. Groat, 1972, Environmental geologic atlas of the Texas coastal zone-Galveston-Houston area: University of Texas Bureau of Economic Geology, 91 p.

Folk, R. L., 1980, Petrology of sedimentary rocks: Austin, Texas, Hemphill Publishing, 182 p.

Galloway, W. E., D. K. Hobday, and K. Magara, 1982, Frio Formation of Texas Gulf coastal plain: depositional systems, structural framework, and hydrocarbon distribution: AAPG Bulletin, v. 66, p. 649-688.

Giles, A. B., 1981, Growth history of Oakwood salt dome, East Texas, *in* C. W. Kreitler et al, Geology and geohydrology of the East Texas basin, a report on the progress of nuclear waste isolation feasibility studies (1980): University of Texas Bureau of Economic Geology Geological Circular 81-7, p. 39-42.

Table 3. Strain Rates in East Texas Salt Diapirs*

	t_i (Ma)	Overall Mean Net $\dot{e}_i$ (on $\dot{G}_n$)	t_i (Ma)	Fastest Net $\dot{e}_i$ (on $\dot{G}_n$)	t_i (Ma)	Fastest Gross e_i (on G_n)
La Rue	56	1.1×10^{-16}	7	1.5×10^{-15}		
East Tyler	56	1.1×10^{-15}	1	1.5×10^{-15}		
Boggy Creek	56	6.8×10^{-16}	7	1.4×10^{-15}		
Brooks	56	7.4×10^{-16}	1	1.3×10^{-15}	1	3.7×10^{-15}
Steen	56	8.3×10^{-16}	1	1.1×10^{-15}	1	3.2×10^{-15}
Brushy Creek	56	5.9×10^{-16}	7	1.1×10^{-15}		
Butler	56	5.9×10^{-16}	1	1.0×10^{-15}		
Keechi	56	6.0×10^{-16}	1	9.6×10^{-16}		
Palestine	56	5.7×10^{-16}	1	9.4×10^{-16}		
Grand Saline	56	6.5×10^{-16}	1	9.0×10^{-16}		
Oakwood	56	5.2×10^{-16}	7	8.2×10^{-16}		
Mt. Sylvan	56	9.7×10^{-16}	1	7.9×10^{-16}	6	6.2×10^{-16}
Whitehouse	56	6.7×10^{-16}	1	6.5×10^{-16}		
Bullard	56	6.0×10^{-16}	1	6.5×10^{-16}		
Hainesville	30	6.5×10^{-16}	13	6.2×10^{-16}	13	3.0×10^{-15}
Bethel	30	7.9×10^{-16}	7	4.0×10^{-16}	7	1.1×10^{-15}
Mean	53	6.7×10^{-16}	4	9.8×10^{-16}	6	2.3×10^{-15}
Standard deviation		2.1×10^{-16}		3.3×10^{-16}		1.4×10^{-15}

*Strain rates ($\dot{e}_i$ = elongation per second) for given durations (t_i) based on gross growth rates ($\dot{G}_g$) and net growth rates ($\dot{G}_n$). Domes are listed in order of fastest net growth rate.

Halbouty, M. T., and J. J. Halbouty, 1982, Relationships between East Texas field region and Sabine uplift in Texas: AAPG Bulletin, v. 66, p. 1042-1054.

——— and G. C. Hardin, Jr., 1951, Types of hydrocarbon accumulation and geology of South Liberty salt dome, Liberty County, Texas: AAPG Bulletin, v. 35, p. 1939-1977.

Heard, H. C., 1963, Effects of large changes in strain rates in the experimental deformation of Yule marble: Journal of Geology, v. 71, p. 162-195.

——— 1976, Comparison of the flow properties of rocks at crustal conditions: Royal Society of London Philosophical Transactions, Series A, v. 283, p. 173-186.

Jackson, M. P. A., and D. W. Harris, 1981, Seismic stratigraphy and salt mobilization along northwestern margin of the East Texas basin, *in* C. W. Kreitler et al, Geology and geohydrology of the East Texas basin, a report on the progress of nuclear waste isolation feasibility studies (1980): University of Texas Bureau of Economic Geology Geological Circular 81-7, p. 28-32.

——— and S. J. Seni, 1983, Geometry and evolution of salt structures in a marginal rift basin of the Gulf of Mexico, East Texas: Geology, v. 11, no. 3, p. 131-135.

Jaritz, W., 1980, Einige Aspekte der Entwicklungsgeschichte der nordwestdeutschen Salzstocke: Zeitschrift der Deutschen Geologischen Gesellschaft, v. 131, p. 387-408.

Kent, P. E., 1979, The emergent Hormuz salt plugs of southern Iran: Journal of Petroleum Geology, v. 2, p. 117-144.

Kumar, M. B., 1977, Growth rates of salt domes of the North Louisiana salt basin, *in* J. D. Martinez, R. L. Thoms, C. G. Smith, Jr., C. R. Kolb, E. J. Newchurch, and R. E. Wilcox, An investigation of the utility of Gulf Coast salt domes for the storage or disposal of radioactive wastes: Louisiana State University Institute of Environmental Studies, p. 225-229.

Kupfer, D. H., 1976, Times and rates of salt movement in north Louisiana, *in* J. D. Martinez and R. L. Thoms, eds., Salt dome utilization and environmental considerations: Louisiana State University Proceedings of a Symposium, p. 145-170.

Loocke, J. E., 1978, Growth history of the Hainesville salt dome, Wood County, Texas: Master's thesis, University of Texas at Austin, 95 p.

Magara, K., 1980, Comparison of porosity-depth relationship of shale and sandstone: Journal of Petroleum Geology, v. 3, no. 2, p. 175-185.

Martin, R. G., 1978, Northern and eastern Gulf of Mexico continental margin: Stratigraphic and structural framework, *in* Framework, facies, and oil-trapping characteristics of the upper continental margin: AAPG Studies in Geology 7, p. 21-42.

McGowen, J. H., L. F. Brown, Jr., T. J. Evans, W. L. Fisher, and C. G. Groat, 1976, Environmental geologic atlas of the Texas coastal zone—Bay City-Freeport area: University of Texas Bureau of Economic Geology, 98 p.

McGowen, M. K., and D. W. Harris, 1983, Cotton Valley (Upper Jurassic)-Hosston (Lower Cretaceous) depositional systems and their influence on salt tectonics in the East Texas basin: University of Texas Bureau of Economic Geological Circular (in press).

Netherland, Sewell, and Associates, 1976, Geologic study of the interior salt domes of northeast Texas salt dome basin to investigate their suitability for possible storage of radioactive waste material as of May, 1976: Report prepared for the Office of Waste Isolation, Energy Research and Development Administration, Union Carbide Corp., Nuclear Division, 57 p.

O'Neill, C. A., III, 1973, Evolution of Belle Island salt dome, Louisiana: Gulf Coast Association of Geological Societies Transactions, v. 23, p. 115-135.

Parker, T. J., and A. N. McDowell, 1955, Model studies of salt-dome tectonics: AAPG Bulletin, v. 39, p. 2384-2470.

Pfiffner, O. A., and J. G. Ramsay, 1982, Constraints on geological strain rates: Arguments from finite strain states of naturally deformed rocks: Journal of Geophysical Research, v. 87, no. B1, p. 311-321.

Price, N. J., 1975, Rates of deformation: Geological Society of London Journal, v. 131, p. 553-575.

Purser, B. H., 1973, Sedimentation around bathymetric highs in the southern Persian Gulf, *in* B. H. Purser, ed., The Persian Gulf Holocene carbonate sedimentation and diagenesis in a shallow epicontinental sea: New York, Springer-Verlag, p. 157-177.

Ramberg, H., 1981, Gravity, deformation and the earth's crust in theory, experiments and geological application, 2d ed.: London, Academic Press, 452 p.

Ramsay, J. G., 1967, Folding and fracturing of rocks: New York, McGraw-Hill, 568 p.

Reese, R. J., 1977, Salt kinematics of southern Bienville Parish, Louisiana: Master's thesis, Louisiana State University, 136 p.

Rezak, R., 1977, West Flower Garden bank, Gulf of Mexico, *in* Reefs and related carbonates—ecology and sedimentology: AAPG Studies in Geology 4, p. 27-35.

Sannemann, D., 1968, Salt-stock families in northwestern Germany, *in* Diapirism and diapirs: AAPG Memoir 8, p. 261-270.

Seni, S. J., and M. P. A. Jackson, 1983a, Evolution of salt structures, east Texas diapir province, part 1: sedimentary record of halokinesis: AAPG Bulletin, v. 67 (precedes this article).

——— ——— 1983b, Sedimentary record of Cretaceous to Tertiary salt movement, East Texas basin: times, rates, and volumes of salt flow, implications for nuclear-waste isolation and petroleum exploration: University of Texas Bureau of Economic Geology Report of Investigations (in press).

Stude, G. R., 1978, Depositional environments of the Gulf of Mexico South Timbalier Block 54: salt dome and salt dome growth models: Gulf Coast Association of Geological Societies Transactions, v. 28, p. 627-646.

Trippet, A. R., 1981, Characteristics of diapirs on the outer continental shelf-upper continental slope boundary, northwest Gulf of Mexico: Gulf Coast Association of Geological Societies Transactions, v. 31, p. 391-397.

Trusheim, F., 1960, Mechanism of salt migration in northern Germany: AAPG Bulletin, v. 44, p. 1519-1540.

van Eysinga, F. W. B., 1975, Geologic time table: Amsterdam, Elsevier, 1 sheet.

van Hinte, J. E., 1976a, A Jurassic time scale: AAPG Bulletin, v. 60, p. 489-497.

——— 1976b, A Cretaceous time scale: AAPG Bulletin, v. 60, p. 498-516.

American Association of Petroleum Geologists Studies in Geology 7. *Framework, Facies, and Oil-Trapping Characteristics of the Upper Continental Margin*, edited by A. H. Bouma, G. T. Moore, and J. M. Coleman, copyright 1978, p. 69-85.

Salt Movement on Continental Slope, Northern Gulf of Mexico[1]

C. C. HUMPHRIS, JR.[2]

Abstract Regional sparker lines across the continental slope of the northern Gulf of Mexico demonstrate the close relation between salt movement and sediment deposition.

Salt features on the outer slope are not as well developed as those near the shelf because sedimentation has been much less on the slope. Salt-generated structures in the eastern part of the gulf are more mature than those in the western gulf because of higher rates of sedimentation. The youngest salt features on the outer slope are much larger than domes on the shelf.

Seismic data from the outer slope suggest that salt dome growth in this area was initiated by southward salt flowage caused by sediment loading updip. The Sigsbee Escarpment appears to be a salt scarp (formed by this gulfward salt flowage) that has extruded over younger sediments for a considerable distance.

Areas of salt diapirs in the Gulf of Mexico, with the exception of diapirs on much of the lower continental slope, are considered to be areas of original thick salt deposition. It is suggested that these areas of thick salt were deposited in one central rift in Jurassic time, and have moved to their present position by seafloor spreading. The present Red Sea is a model for the Gulf of Mexico at the time of Mesozoic breakup.

Introduction

The origin of the salt in and around the Gulf of Mexico, and the transformation of this originally thick-bedded salt into swells and diapirs, long has been of great interest to geologists. The age of the salt was discussed by a number of writers, including Hazzard et al (1947), Jux (1961), Halbouty (1967), Viniegra (1971), and Kirkland and Gerhard (1971). Barton (1933), Nettleton (1934), and Parker and McDowell (1955) contributed to the various concepts concerning the initiation and growth of salt domes in the gulf.

Extent of the various salt deposits in and around the gulf was described by Andrews (1960), Antoine and Bryant (1969), and Garrison and Martin (1973). This paper illustrates the close relation between sediment deposition and salt movement on the continental slope of the northern Gulf of Mexico. Concepts involving the initiation of salt dome growth on the slope, the origin of the Sigsbee Escarpment, and a suggested model for the formation of the Gulf of Mexico are presented.

Much of the data presented are derived from sparker surveys and core holes on the continental slope completed in 1966 by Chevron, Exxon, Gulf, and Mobil. This program consisted of approximately 19,300 km of sparker profiles along with gravity and magnetic coverage, and 42 core holes at 36 locations. These core holes penetrated up to 305 m of sediment in water depths up to 1,525 m.

[1] Manuscript received, January 31, 1977; accepted, June 23, 1977.

[2] Chevron, U.S.A., Inc., New Orleans, Louisiana 70112.

Modified from a paper read before the Association at New Orleans, May 24, 1976 which received the AAPG Matson Award for the best technical paper presented at the 1976 meeting. Published with permission of Chevron, U.S.A., Inc.

Article Identification Number: 0149-1377/78/SG07-0004/$03.00/0.

Relation of Sediment Deposition to Salt Movement on Continental Slope

The continental slope of the northern Gulf of Mexico (also known as the Texas-Louisiana Slope) extends from the shelf edge (133 m water depth) gulfward across the Sigsbee Escarpment. It extends from the Mexico-Texas boundary on the southwest, to the Mississippi Fan on the northeast. This area essentially is the same that Lehner (1969) covered in his study of salt diapirs and Pleistocene stratigraphy of the continental slope.

Figure 1 is an index map of the Gulf of Mexico and surrounding areas. Locations of the Sigsbee Escarpment, the Sigsbee Knolls, and the Florida and Campeche Escarpments are shown.

Figure 2 is a regional north-south sparker profile extending from the Mississippi Delta southward across the Sigsbee Escarpment for a distance of 210 km (Line 90, Fig. 1). The vertical exaggeration is a factor of six on this and all subsequent sparker profiles. Seven core holes drilled on (interpreted) diapiric salt features found caprock material, salt crystals in the sediment, or high salinity in interstitial waters, indicating that these tests bottomed near salt. Lehner (1969), reported that salt was recovered from 10 core holes drilled on similar diapiric features.

This sparker profile shows a series of individual salt features, separated by areas of thick sediment. These structures, at this stage in their growth history, appear to grow mainly by

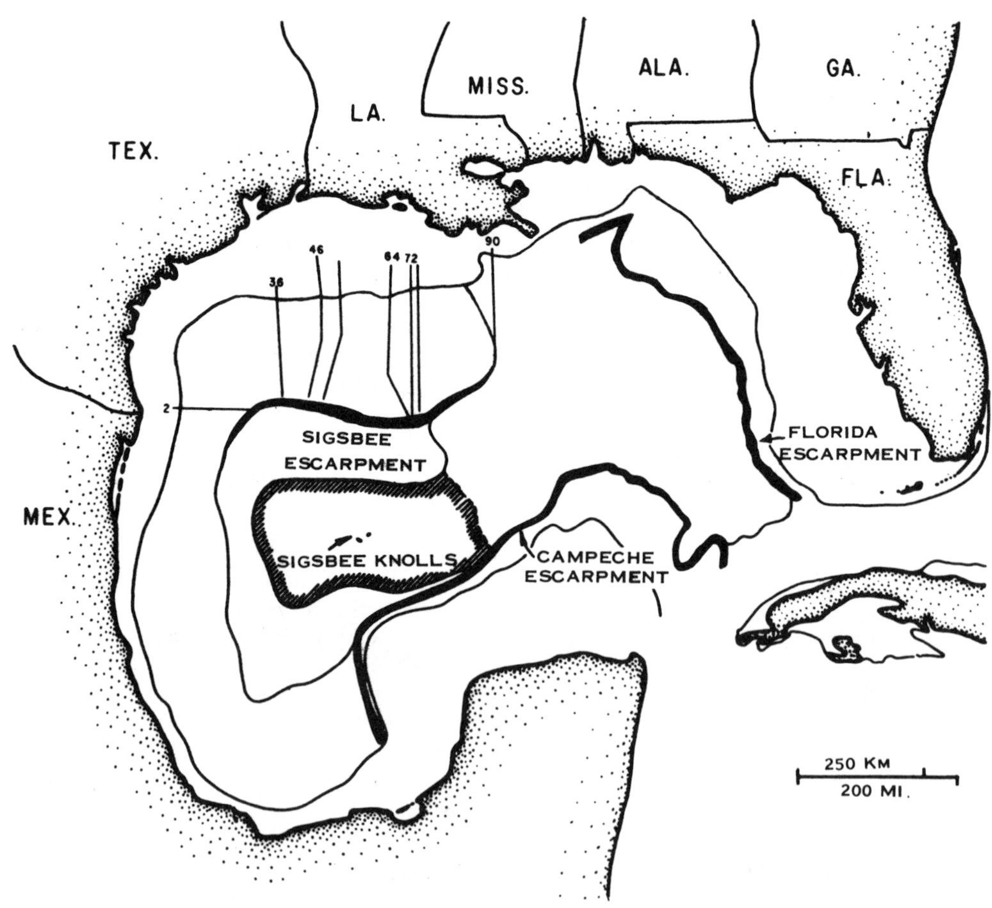

FIG. 1—Regional map of Gulf of Mexico and vicinity. Locations of sparker and seismic profiles discussed in text are indicated.

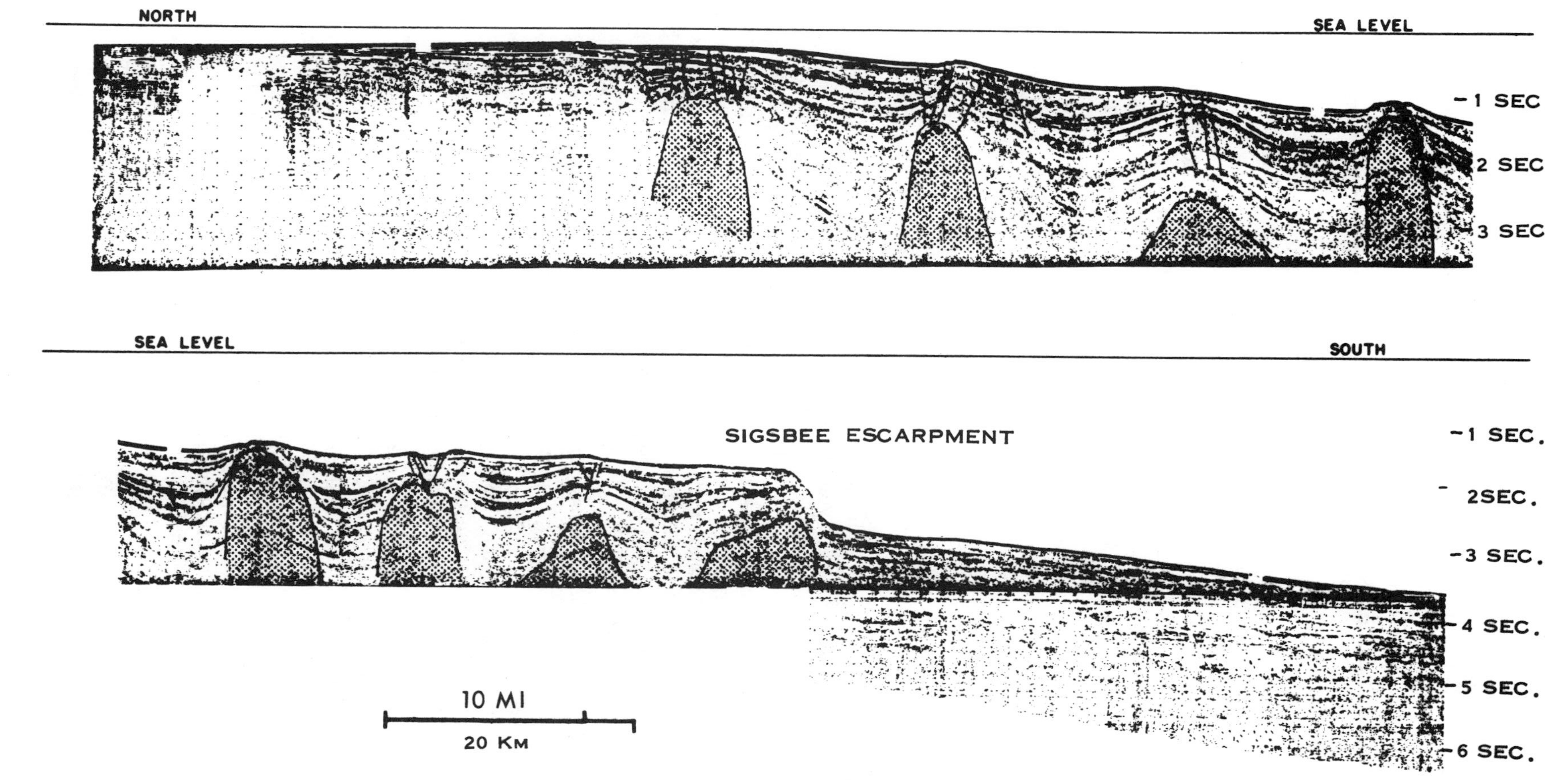

FIG. 2—North-south sparker profile, line 90, Figure 1. Profile in two parts, left end of lower panel is continuation of right end of upper panel. Interpreted salt features shown by pattern. Prominent scarp in bottom panel is Sigsbee Escarpment. Vertical exaggeration is approximately 6 on this and other sparker profiles.

a downbuilding of the sediments around the diapir. This and other regional sparker profiles indicates that domes are close to the water bottom throughout their growth, and maintain position relative to the bottom as sediments build downward around them.

A north-south sparker profile, about 145 km west of the previous profile, (Fig. 3; Line 72, Fig. 1) also extends from the shelf edge, across the slope and Sigsbee Escarpment, and onto the continental rise. Diapirs similar to the discrete features observed on the profile shown in Figure 2 developed on the shelf and upper slope where the greatest amount of deposition occurred. Features on the lower slope are much larger than those in Figure 2 because they have not received sufficient sediment to develop discrete diapirs. These larger features on the lower slope are interpreted as being in an early stage of diapiric development compared to the diapirs in Figure 2.

Figure 4 is the westernmost regional sparker profile, commencing south of the High Island area and continuing south across the Sigsbee Escarpment (Line 36, Fig. 1). Salt features on the upper slope have become individual diapirs. On the lower slope, no individual salt diapirs are present for a distance of 65 km north of the escarpment. Apparently, sediment loading in this area was not adequate to develop individual diapirs. Salt domes on this lower slope are in an early or youthful stage of development, similar to that of salt features observed on the lower slope in Figure 3.

A panel of structural cross sections developed from regional sparker profiles plus two common-depth point (CDP) seismic profiles (CDP seismic profiles penetrate much deeper than sparker profiles) is shown in Figure 5 (see Fig. 1 for locations). The point of reference for all lines is the Sigsbee Escarpment, on the extreme right-hand side of each section. The top section, which runs west to east, is located on the Texas-Mexico border, and the remaining sections are arranged in clockwise sequence around the northern Gulf of Mexico.

Stages in the evolution of domes from an initial salt mass can be observed on these sections. The first or youthful stage (to right of wavy line), typified by the lower slope east of Texas, is an area tens of kilometers in width where salt is overlain by a thin veneer of sediments. A second and later stage, left of the wavy line, developed after additional sedimentation. Individual diapiric features are recognizable at this stage; this later stage extends much farther seaward in the eastern gulf than in the western part because of the greater Pleistocene sedimentation in the east. A later, more mature stage brought about the formation of salt spines occurring along the present shelf and onshore area where the thickest amount of Gulf Coast sediments are present. The loci of these three stages correspond in a general way to the structural provinces described by Woodbury et al (1973).

Suggested Mechanism for Initiation of Salt Dome Growth on Continental Slope

Various mechanisms have been suggested for the initiation of salt domes, including a pre-salt irregular surface, variation of thickness and density of the overburden, faulting of the bedded salt, and externally applied compressive stresses (Parker and McDowell, 1955).

Regional mapping of salt features (from sparker profiles on the slope) indicates no linearity which would reflect basement involvement. The northern Gulf of Mexico generally is considered to have been in an extensional regime from the Mesozoic to the present with no compressional activity.

A substantial thickness variation in the overlying Tertiary and younger sediments would be required to initiate salt dome growth. Density logs from offshore wells indicate that a thickness difference of 1,220 to 1,525 m of Gulf Coast Tertiary or younger sediments is needed before the sediment density surpasses that of salt, and growth could be initiated. Where a salt feature begins to grow, the density contrast between the salt and the heavier sediments is sufficient to maintain growth.

Several minor salt swells are on a part of a sparker profile from the lower slope (Fig. 6). A maximum sediment thickness of 900 m is interpreted to overlie the salt in the lows, with approximately 350 m of sediment occurring over the structural highs. Contemporaneous growth is indicated by the positive bathymetric relief and the structural attitude of the reflectors. Because the observed 550-m differential in sediment thickness is not enough to initiate salt movement, it is suggested that this structural growth must be caused by lateral

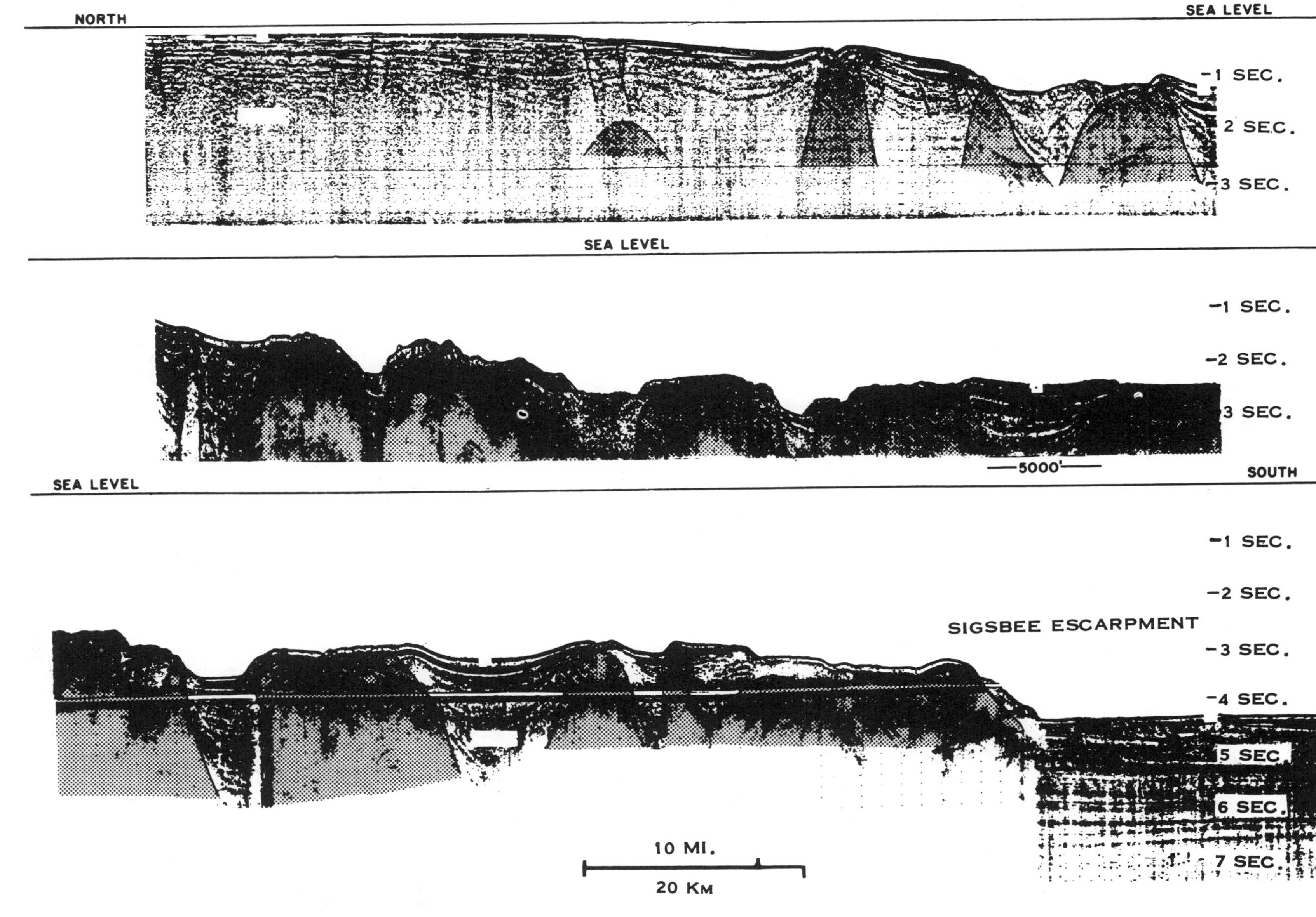

FIG. 3—North-south sparker profile, line 72, Figure 1. Profile in 3 parts. Note larger size of salt features on lower slope, as compared to salt features in Figure 2. Sigsbee Escarpment shown near right end of lower panel. Sparker penetration insufficient to observe top of salt between diapirs.

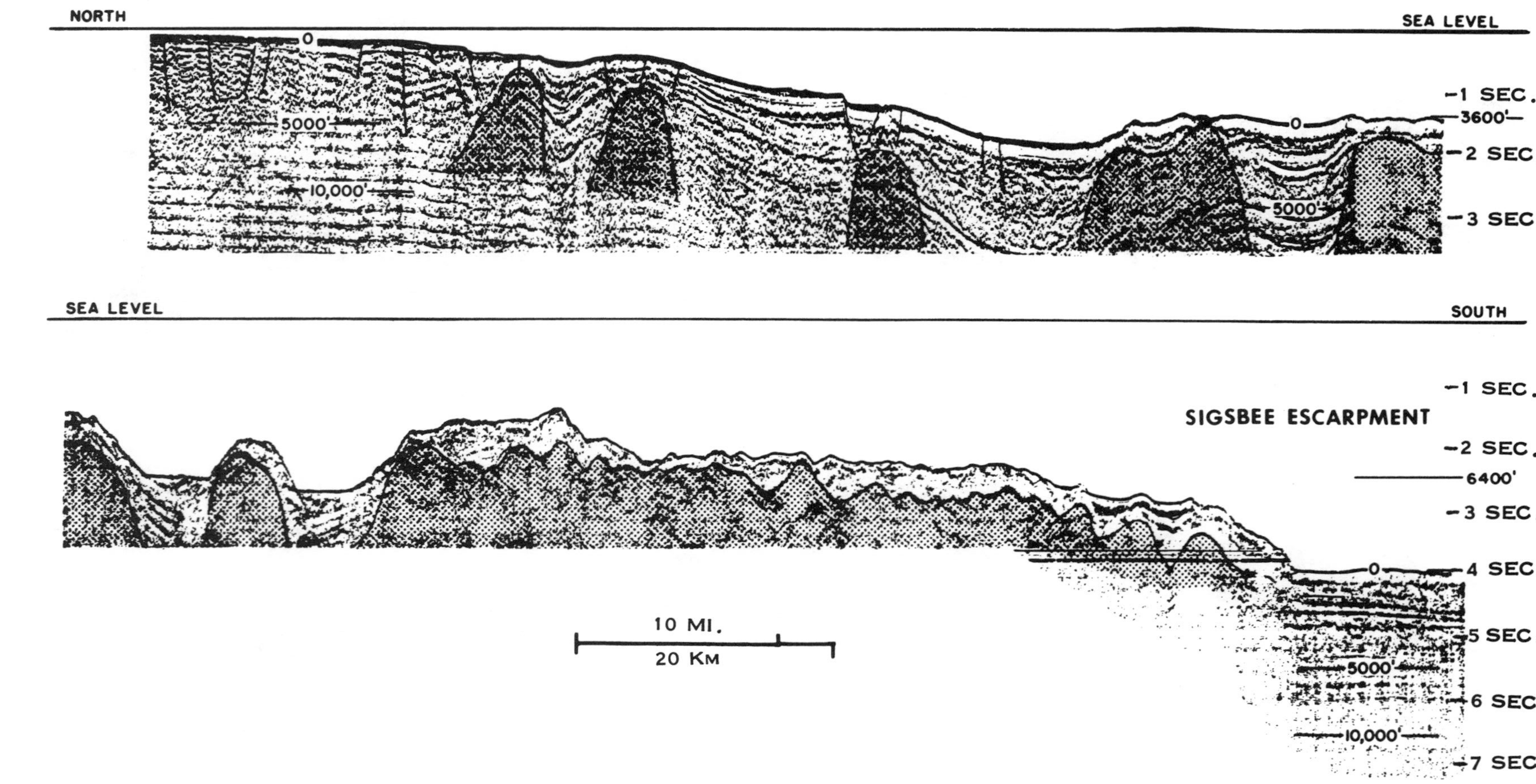

FIG. 4—North-south sparker profile, line 36, Figure 1. Profile in 2 parts. At least 3,000 m of flat lying reflectors south of Sigsbee Escarpment. No individual salt diapirs have developed for 65 km north of scarp because of insufficient sedimentation.

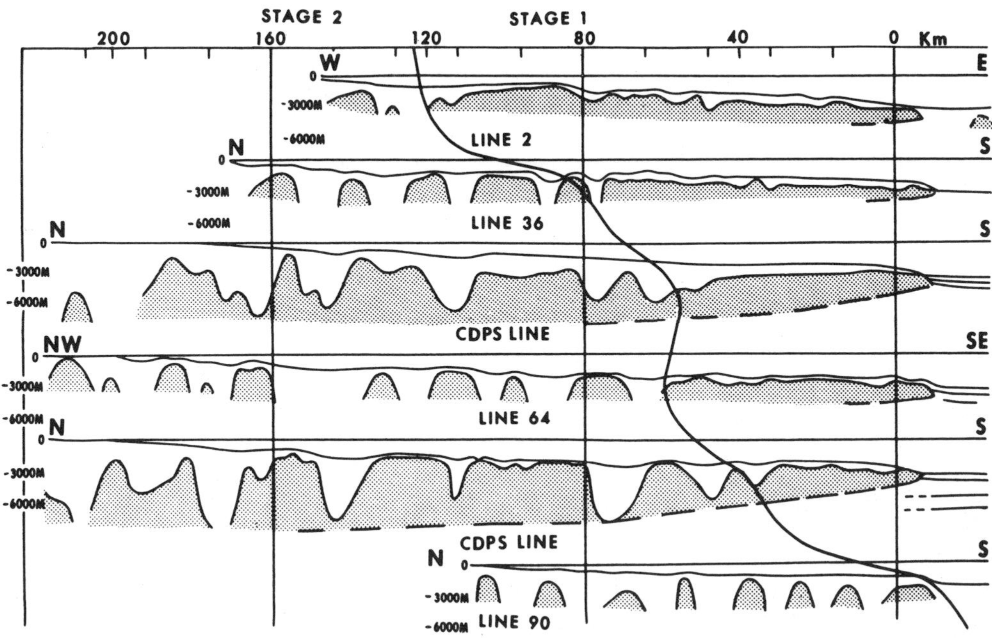

FIG. 5—Panel of structural cross sections across continental slope showing interpreted salt features, developed from sparker and seismic profiles. Location of profiles indicated on Figure 1. Sections demonstrate stages in development of salt diapirs on slope. Stage 1 is youngest, diapirs have not formed because of insufficient sedimentation. Base of salt has been interpreted at southern end of two seismic profiles, which have much deeper penetration than sparker profiles. Sigsbee Escarpment shown near right edge. Vertical exaggeration 2.6.

salt flowage resulting from stresses exerted by sediment loading updip. A diagrammatic representation of this mechanism is illustrated in Figure 7. Salt flowage as depicted here might be analogous to the extrusion flow of ice in a glacier (Demorest, 1943). Continued growth of these features would later produce bathymetric lows, in which pods of sediment (of sufficient thickness to produce a density contrast and continued domal development) could accumulate. This formation of salt features due to lateral salt flowage, resulting from the sediment loading on the shelf, is probably a major mechanism for the initiation of salt domes on the continental slope.

Origin of Sigsbee Escarpment

The Sigsbee Escarpment is the most prominent topographic feature on the lower slope of the northern Gulf of Mexico (Fig. 1). De Jong (1968) and Amery (1969) were among the first to demonstrate that the Sigsbee Escarpment was a wall of salt that is extruding southward over younger sediments. The scarp appears to be the leading edge of a salt mass of probable Jurassic age that extruded seaward over Quaternary and Tertiary sediments for a considerable distance, possibly tens of kilometers.

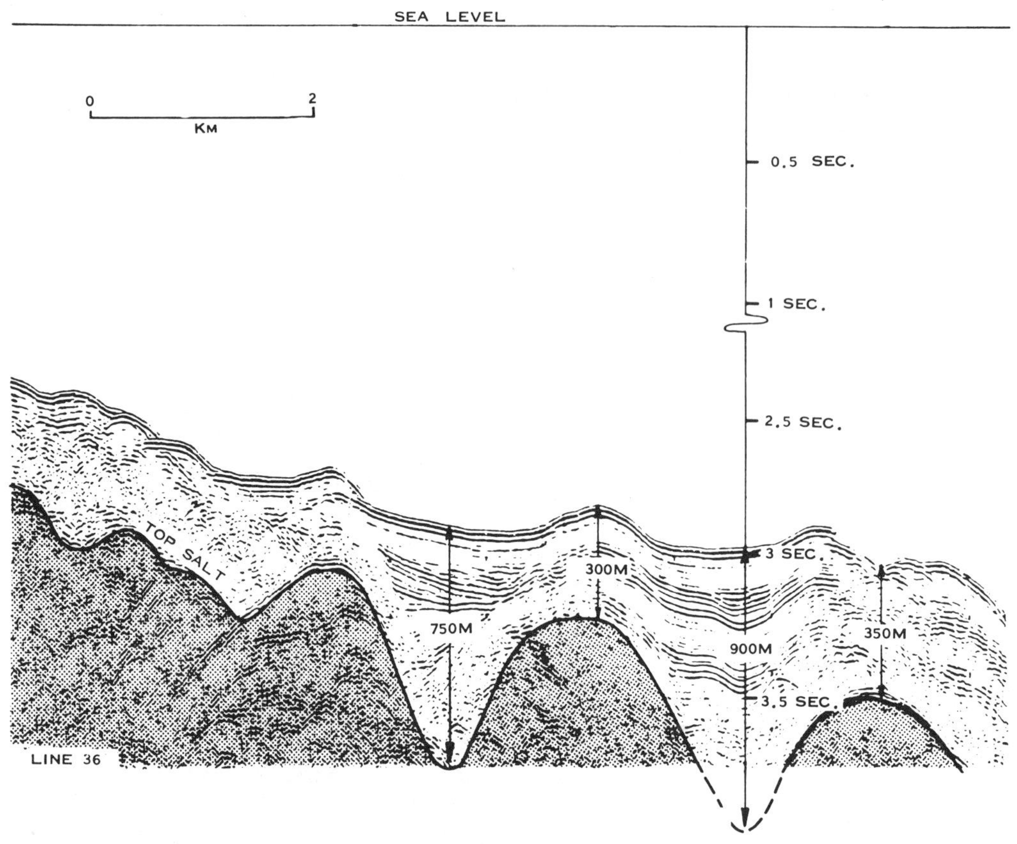

FIG. 6—Part of sparker profile 36, near Sigsbee Escarpment (Fig. 4). Interpreted salt indicated by pattern. Bathymetric highs over salt features demonstrate there has been salt movement with sediment thickness differential of 550 m. Vertical exaggeration 6.

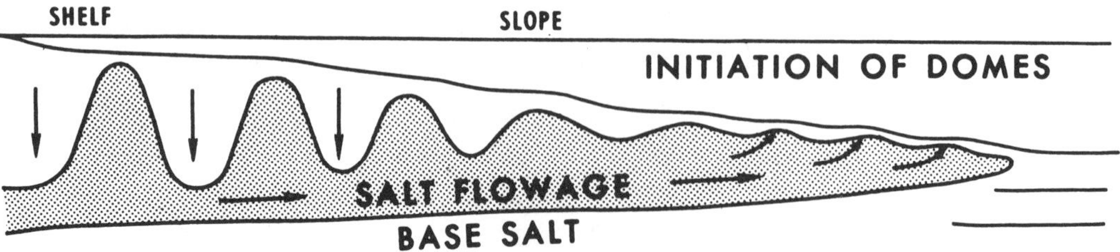

FIG. 7—Diagrammatic representation of initiation of salt dome growth on continental slope as result of sediment loading on shelf and upper slope.

Part of a true scale north-south CDP seismic depth section across the leading edge of the scarp is shown in Figure 8. This line was recorded by the Marine Science Institute of the University of Texas (Watkins et al, 1976). A top salt and a base salt reflector are interpreted in the section north of the scarp; near-horizontal reflectors can be traced to the scarp edge where they turn up abruptly; and noncontinuous reflectors can be observed beneath the salt on this section for a distance of 14.5 km north of the escarpment. Several continuous reflectors appear to be truncated against the base salt reflector.

Processing of this depth interpretation used a sediment velocity for the entire section. It is suggested that if a salt velocity (instead of the slower sediment velocity) had been applied to the salt mass, reflectors beneath the salt would appear as a continuation of the near-horizontal reflectors from the continental rise. Amery (1969) demonstrated this on a sparker profile, and showed that reflectors beneath the salt could be traced at least 9.6 km north of the scarp. De Jong (1968) analyzed velocities from CDP seismic lines some distance north of the Sigsbee Escarpment. He demonstrated that beneath the high velocity layer, which he considered to be salt, a sequence of lower velocity strata existed which could be correlated with the abyssal section of similar low velocity in front of the scarp.

Another suggestion that salt flowed laterally for a considerable distance can be observed on the panel of sparker and seismic profiles (Fig. 5). The two seismic lines (which have the deeper penetration) have been interpreted to show north dip on the base of salt near the scarp. This north dip has been interpreted from the delineation of the base of the salt at the scarp (Fig. 8), and from the observation that sediment accumulations separating the salt features demonstrate a progressive thickening shoreward. The base of salt is interpreted here to be a north-dipping surface that truncates the underlying abyssal sediments.

The Pleistocene thickness map (Fig. 9) can be interpreted to indicate a definite relation between the area of thickest Pleistocene deposition and the southward bulge of the Sigsbee Escarpment. It is suggested that the salt that forms this bulge has been displaced gulfward by lateral flowage (Wilhelm and Ewing, 1972), resulting from the Pleistocene loading on the shelf. The original area of salt deposition in the northern Gulf of Mexico must have been much smaller than the area underlain by salt features today.

Regional Cross Section

A generalized north-south structural cross section depicts the continental slope in relation to the entire Gulf of Mexico (Fig. 10). Gulfward from the coast the salt features are progressively larger and younger due to less post-salt sedimentation. The Sigsbee Escarpment marks the southern limit of the extrusive salt that flowed southward over continental rise or abyssal sediments during sediment loading on the shelf.

Published seismic lines indicate the presence of at least 6,000 m of horizontal sediments underlying the abyssal plain south of the scarp, thought to be mainly Tertiary in age and younger, with minor thicknesses of Cretaceous and Jurassic sediments.

Sediments underlying the abyssal plain appear to have a different source from those of the slope. They were transported across the shelf and deposited as turbidites in the abyssal area at the same time that deposition was occurring on the shelf and upper slope. This deposition on the shelf and upper slope produced gulfward salt flow, which, having broken through a thin layer of Mesozoic sediments, has continued to flow out over Tertiary and younger abyssal sediments.

Areas of Diapiric Salt

A generalized map showing areas of diapiric salt in the Gulf of Mexico and vicinity is shown in Figure 11. These areas include the various interior basins, the Gulf basin (including diapirs of East Mexican Shelf and Slope, and Texas-Louisiana Shelf and Slope) and the Sigsbee Knolls–Campeche Knolls–Isthmian basin trend. Salt in all of these basins is thought to be of approximately the same age, Middle to Late Jurassic (Kirkland and Gerhard, 1971).

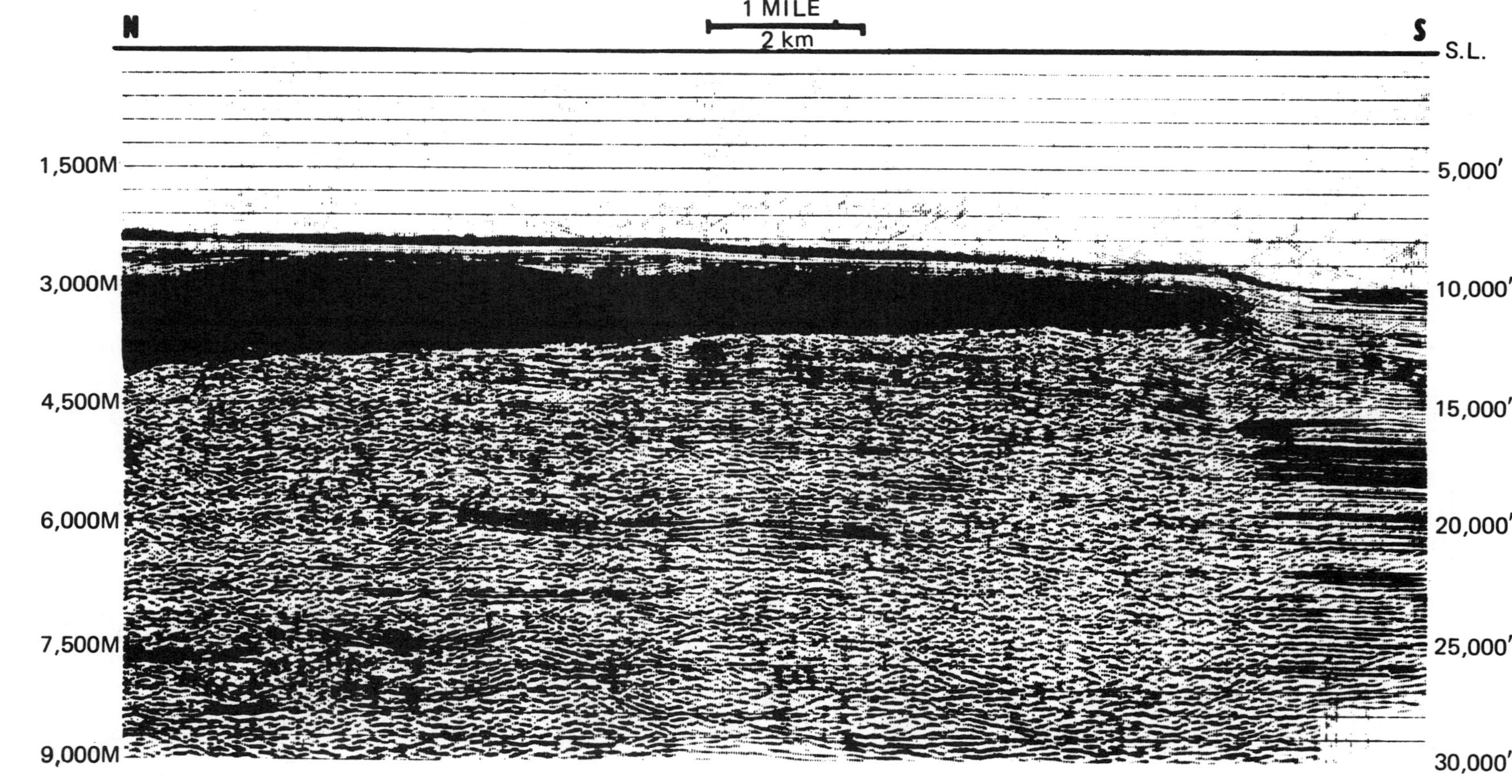

FIG. 8—Seismic depth profile across Sigsbee Escarpment (near right edge) shot by UTMSI (Watkins et al, 1976). Salt layer indicated by pattern. North dipping reflector that truncates flatter reflectors interpreted to be base of salt. Salt velocity was not applied when constructing this depth section. If faster salt velocity were applied to interpreted salt layer, salt would be shown to be thicker. Pull up of reflectors beneath leading edge of salt observed here would flatten to appear as northward continuation of flat-lying abyssal reflectors toward south.

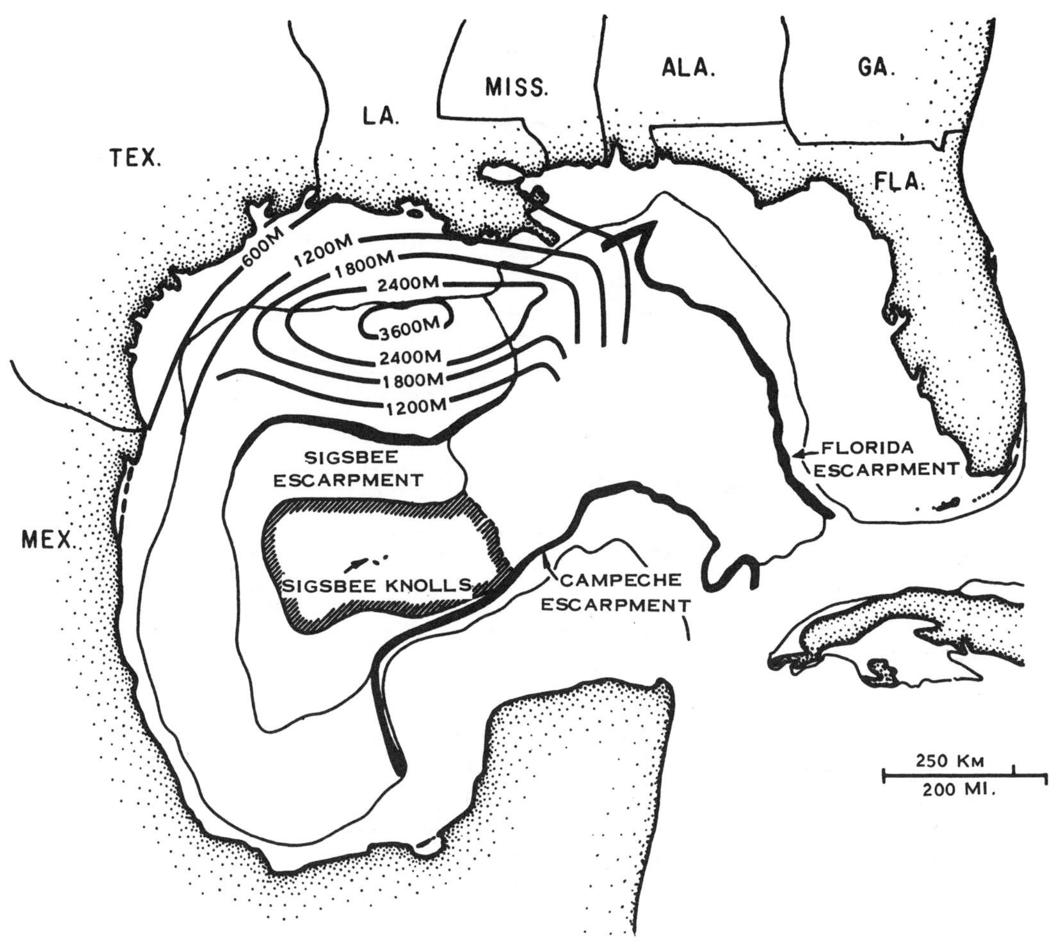

FIG. 9—Generalized Pleistocene thickness map (Woodbury et al, 1973), showing relation of area of thick sediment to bulge of Sigsbee Escarpment.

Evidence of diapiric salt features appears (on published sparker profiles) along the western margin of the Gulf as far south as 22° N lat. (Garrison and Martin, 1973). This diapiric trend is interpreted to extend farther south, based on the evidence of isolated diapiric features on published profiles. The Mexican Ridges, previously thought to be salt-generated, now are considered to be probably nonsalt cored, based on recent seismic data (Watkins et al, 1976). These ridges are just east of the diapiric salt along the western margin of the gulf.

Salt diapirs that have been mapped in the Sigsbee and Campeche Knolls areas extend onshore into the Isthmian salt basin of Mexico (Garrison and Martin, 1973). No diapiric features have been reported between the west-northwest boundary of the Sigsbee Knolls–Campeche Knolls trend, and salt diapirs along the western margin of the Gulf or between the Sigsbee Knolls and the Sigsbee Escarpment. Published sparker profiles indicate that this nondiapiric area probably has received more sedimentation than the area around the Sigsbee Knolls diapirs. It appears that if salt had been deposited over the entire Gulf in a thickness equal to that of the Sigsbee Knoll diapiric trend, salt diapirs would have developed in a more regional fashion. The absence of diapiric features between the Sigsbee Knolls diapiric trend and the Sigsbee Escarpment suggests that any salt present is too thin to produce diapirs.

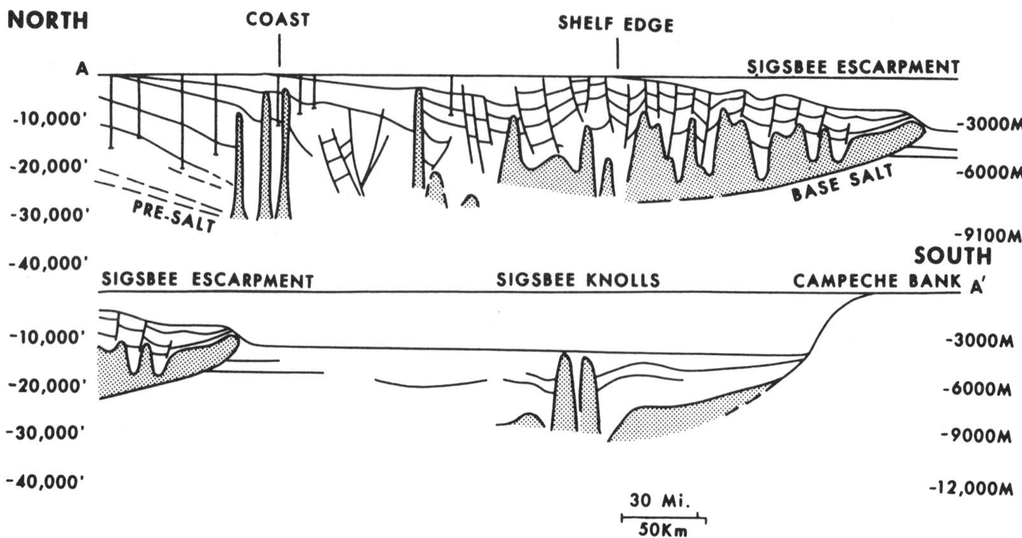

FIG. 10—Generalized north-south cross section of Gulf of Mexico in two parts showing relation of continental slope and related salt features to Gulf of Mexico basin. (Line of section A-A′, Fig. 11). Based on well control, seismic, and published sparker profiles. Demonstrates concept of gulfward movement of salt caused by sediment loading on shelf and upper slope. Salt extrudes over younger sediments that have been deposited in abyssal area contemporaneous with shelf and upper slope sedimentation. Vertical exaggeration: 10.

An area of diapiric features just west of the Florida Escarpment is indicated from sparker profiles shot by the U.S.N.S. *Elisha Kane* in 1969. Well-developed diapirs are only detectable (on these profiles) in a narrow trend close to the escarpment. Absence of such diapirs between the Sigsbee Knolls and this narrow diapiric trend near the Florida Escarpment can be interpreted to indicate an original salt thickness too thin to have allowed them to form. Additional seismic coverage will be necessary to map the extent of these diapiric features near the Florida Escarpment.

Areas of salt diapirs in the Gulf of Mexico region may represent areas of original thick salt deposition, with the exception of the laterally displaced salt of much of the continental slope. Two general trends of original thick salt are suggested, one that parallels the present coast from Louisiana to Mexico (the Gulf basin), and one that trends from the Sigsbee Knolls to the Isthmian basin of Mexico. It also is suggested that the area between the Sigsbee Scarp and the Sigsbee Knolls, which is probably underlain by oceanic crust, has little (if any) underlying salt.

These two salt diapir trends, the Gulf basin (excluding the area of interpreted lateral salt movement) and the Sigsbee Knolls–Campeche Knolls–Isthmian basin, suggest that the present Gulf of Mexico was formed by rifting, probably in Jurassic time. Opening of the Gulf by some form of rifting has been proposed by many, including Dietz and Holden (1970), Kupfer (1974), and Burgess (1976). These areas of suggested thick salt deposition are considered to have been originally deposited in a central rift basin, then moved to their present position by seafloor spreading. This interpretation would be analogous to the present salt basins of approximately the same age off the east coast of Canada and west coast of Portugal and North Africa. These modern salt basins, on both sides of the Atlantic, can be interpreted to have been deposited in a central rift basin at the time of the Mesozoic breakup of continents now bordering the Atlantic (Evans, 1974; Kinsman, 1974).

Mesozoic Reconstruction of Gulf of Mexico

Figure 12 is a map showing part of the present North Atlantic and Gulf of Mexico. The location of the Mid-Atlantic Ridge and a series of magnetic anomalies also is shown. Areas of diapiric salt are indicated off the coast of Canada and the Gulf of Mexico, with the exception of the area of interpreted displaced salt which underlies the outer continental slope of the northern Gulf. These areas of diapirs are suggested to be sites of original thick salt deposits.

Figure 13 is a suggested pre-drift reconstruction of the North Atlantic region, with the continental fit based largely on the relation of magnetic anomalies and JOIDES core holes. Areas of diapiric salt off the coast of Canada are shown, as well as salt diapiric areas of approximately the same age off Portugal and off the northwestern coast of Africa. This diapiric salt on either side of the North Atlantic is believed to be of Late Triassic–Early Jurassic age, slightly older than the Louann Salt of the Gulf of Mexico. Salt diapirs of probable Early Cretaceous age are present off the western coast of Africa and the eastern coast of South America. Salt from which these diapirs have developed along both the eastern and western margins of the North and South Atlantic is interpreted to have been originally deposited in a system of rift basins, then moved to their present position on opposite sides of the Atlantic by seafloor spreading.

Thick salt deposits in the Gulf of Mexico may have been deposited in the same manner as those surrounding the margins of the Atlantic (Pautot et al, 1970). The suggested reconstruction of the Gulf (Fig. 13) rotates Yucatan counterclockwise into the Gulf, to bring the Sigsbee Knolls–Campeche Knolls–Isthmian basin salt diapir trend in juxtaposition with salt diapirs of the Gulf basin. No indication of magnetic anomalies (anticipated from seafloor spreading) has been reported in the Gulf. Absence of such anomalies might be explained by the deep burial of the source of these anomalies, which can attenuate the anomalies to be unrecognizable (Kupfer, 1974). Another possible explanation for this lack of anomalies can be that spreading occurred during a magnetic quiet period in which no anomalies were produced.

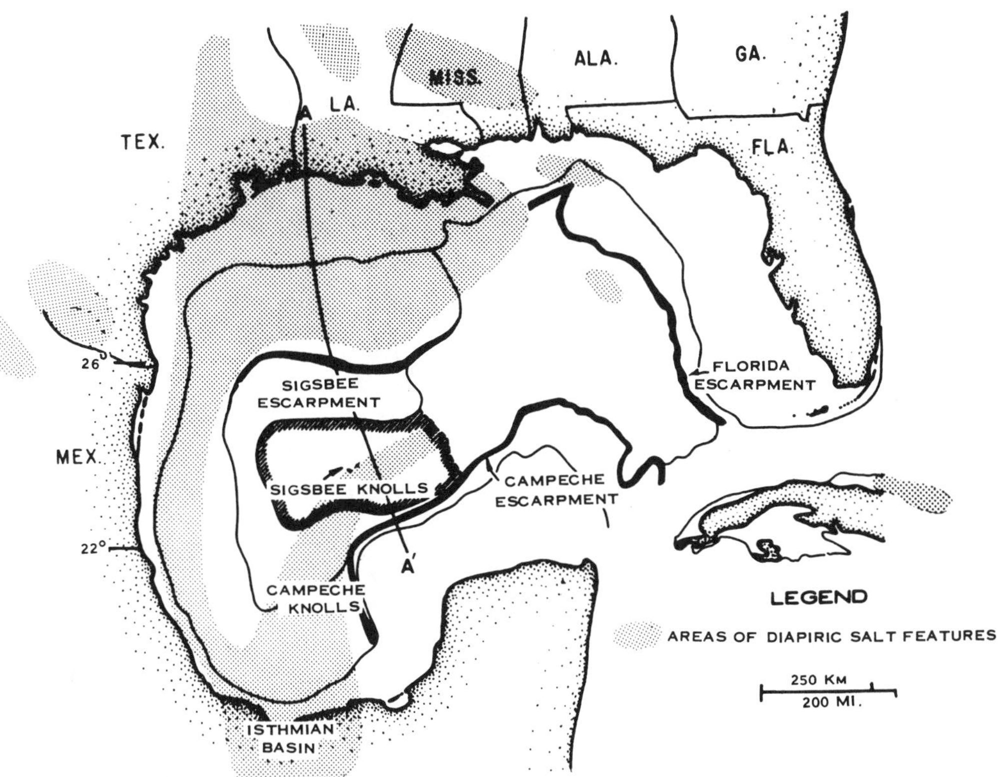

FIG. 11—Areas of diapiric salt features in Gulf of Mexico (modified from various sources, including Antoine and Bryant, 1974). Diapiric salt extended south of 22° along Mexican coast, based on interpretation of published sparker profiles. Area of diapirs indicated near Florida Escarpment, based on USNS *Elisha Kane* sparker profiles. No diapirs have been reported in area between Sigsbee Knolls and Sigsbee Escarpment. With exception of salt forming Sigsbee Escarpment and salt features of lower slope, these areas of diapiric salt are considered to be areas of thick salt deposition.

Red Sea Model for Gulf of Mexico

A model for salt deposition in the Gulf of Mexico and later opening of the gulf is the modern Red Sea. A structural cross section across the Red Sea, on the same scale as the Gulf of Mexico section (Fig. 10), is shown in Figure 14. The presence of more than 5,000 m of Miocene salt is reported based on seismic work (Lowell and Genik, 1972). Magnetic lineations in the central Red Sea indicate the presence of oceanic crust. In this model, sediment loading near the coast caused diapirs to begin development. If a major source of sediments (such as the Mississippi River) was depositing on the salt, and spreading continued with turbidite deposition in the deep water where oceanic crust was forming, it is believed that lateral salt flow would develop on a major scale similar to the one suggested for the gulf.

Possible Sequential Evolution of the Gulf

A possible sequential development of the gulf is shown in Figure 15 based on the regional cross section (Fig. 10). The amount of salt is held constant for each stage.

End of Louann Deposition (Middle-Late Jurassic)

The Gulf is considered to have originated by rifting in Jurassic time. The Louann Salt, probably 3,050 m or more thick, accumulated rapidly in a central rift similar to the modern Red Sea. Shallow-water conditions prevailed at the end of Louann deposition.

End of Cretaceous Time

The original salt basin was spread apart by the formation of new oceanic crust. Shallow-water sediments (deposited on the shelves as the gulf region subsided) thin gulfward by starvation. Some salt movement may have occurred due to the loading on the shelf, but probably not enough to generate diapiric features.

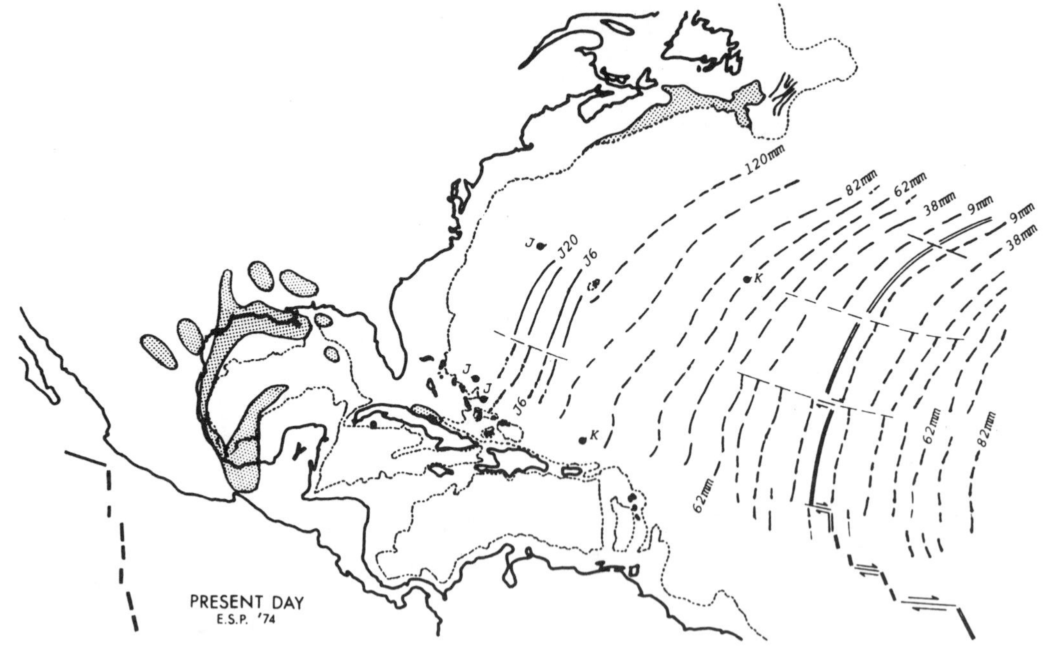

LEGEND
AREAS OF DIAPIRIC SALT (EXCEPTION OF SIGSBEE SCARP AND VICINITY)

FIG. 12—Map of present Gulf of Mexico and part of North Atlantic. Areas of salt diapirs off Canada and in Gulf of Mexico indicated, with exception of laterally displaced salt in vicinity of Sigsbee Escarpment. Mid-Atlantic Ridge (double lines) and magnetic lineations in North Atlantic shown.

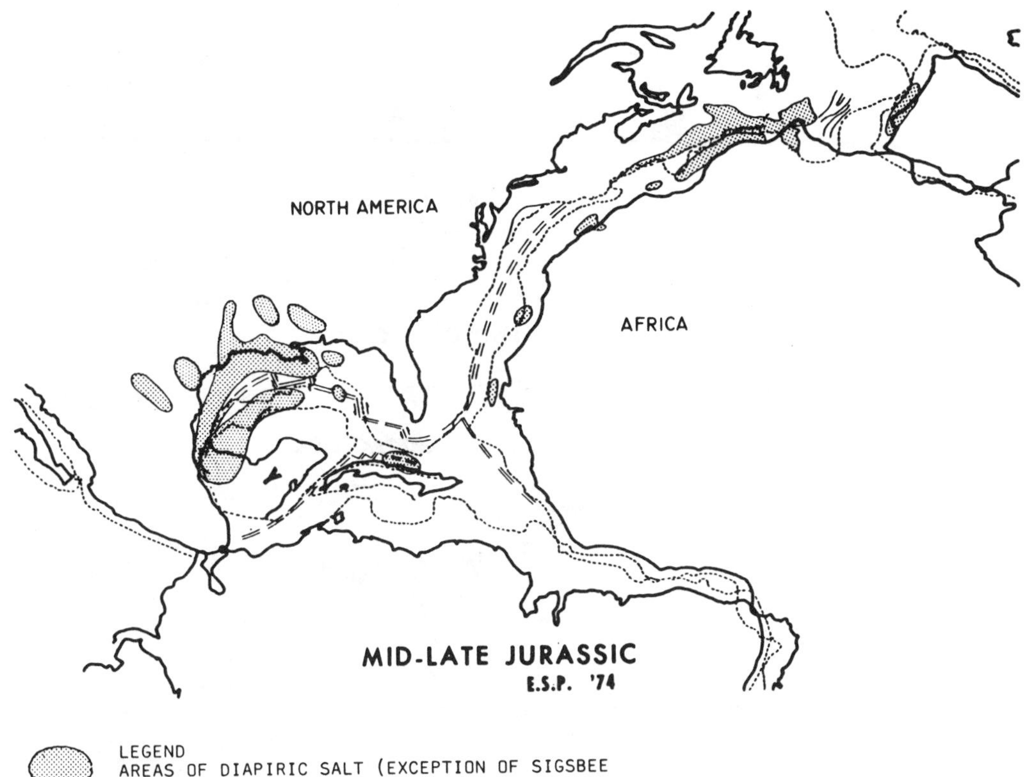

FIG. 13—Jurassic reconstruction of North Atlantic area based on magnetic lineations and JOIDES core holes. Shows relation of salt diapiric areas off Canada to those off Portugal and northwest Africa, and Gulf Coast basin to Sigsbee Knoll–Isthmian basin. Yucatan (Y) has been moved counterclockwise into Gulf of Mexico to suggest two salt trends were originally deposited in one central rift, similar to the salt basins around the North Atlantic.

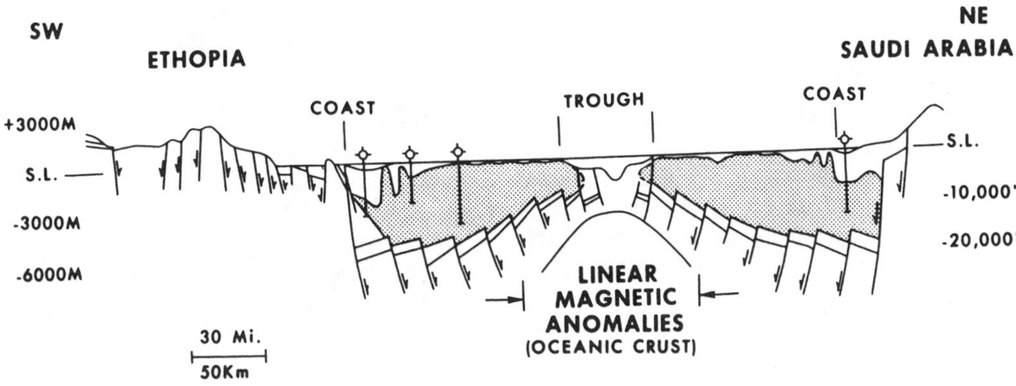

FIG. 14—Generalized cross sections across southern part of Red Sea, same scale as Gulf of Mexico section (Fig. 10). Modified from various sources, including Lowell and Genik (1972). Seismic information indicates more than 5 km of salt. Vertical exaggeration 10.

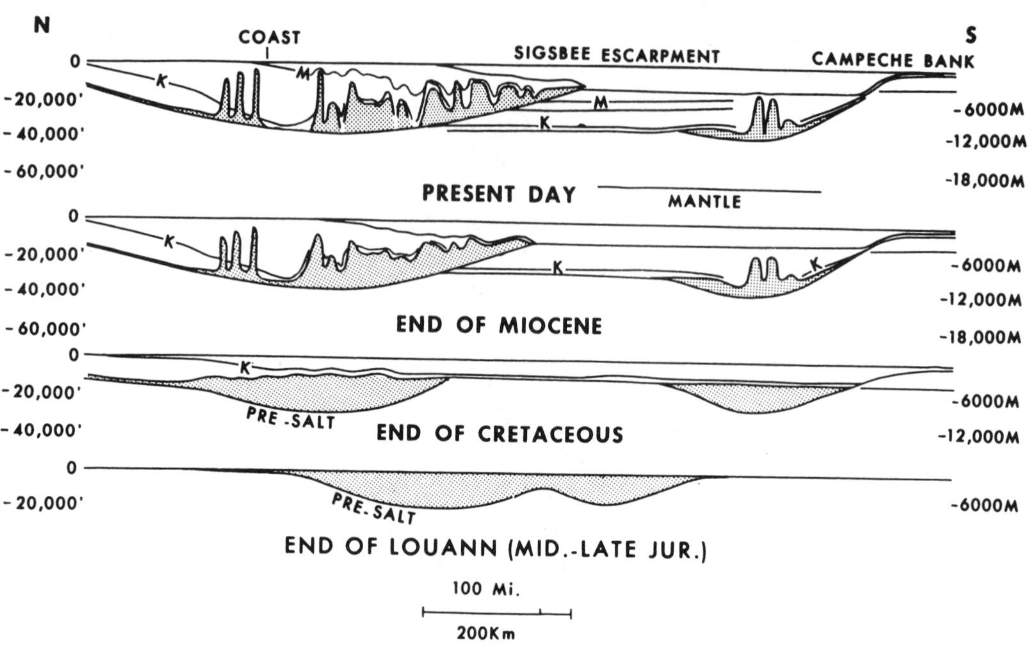

FIG. 15—Suggested sequential development of Gulf of Mexico. Developed from section shown in Figure 10. Area of salt similar for all stages.

End of Miocene Time

Thick Tertiary deposits moved progressively seaward. Salt diapirs formed, and loading by deposition on the shelf and upper slope generated gulfward salt flow. The forward edge of the salt is visualized to have broken through the thin overlying Mesozoic beds and to have extruded over younger abyssal sediments. These abyssal sediments are considered to have been deposited as turbidites, contemporaneous with sedimentation on the shelf and upper slope.

Present

Additional diapiric features have formed seaward of the Pleistocene depocenter. Salt flow continues to the south, extruding over the younger sediments.

Summary

Regional sparker profiles across the continental slope of the northern Gulf of Mexico demonstrate that salt movement has occurred in response to sediment loading. Salt diapirs on the slope have been initiated by lateral salt flow as a result of sediment loading on the shelf. The Sigsbee Escarpment is interpreted to have been formed by lateral salt flow, with salt extruding for a considerable distance over flat-lying continental rise or abyssal sediments. It is suggested that the present Gulf was formed by rifting in Jurassic time. Evidence for this are the two separate salt diapir trends, interpreted to have formed in one central rift but since spread apart.

References Cited

Amery, G. B., 1969, Structure of Sigsbee Scarp, Gulf of Mexico: AAPG Bull., v. 58, no. 12, p. 2480-2482.

Andrews, D. I., 1960, The Louann Salt and its relationship to Gulf Coast salt domes: Gulf Coast Assoc. Geol. Socs. Trans., v. 10, p. 215-240.

Antoine, J. W., and W. R. Bryant, 1969, Distribution of salt and salt structures in Gulf of Mexico: AAPG Bull., v. 53, no. 12, p. 2543-2550.

——— et al, 1974, Continental margins of the Gulf of Mexico, *in* C. A. Burk and C. L. Drake, eds., The geology of continental margins: New York, Springer-Verlag, p. 683-693.

Barton, D. C., 1933, Mechanics of formation of salt domes with special reference to Gulf Coast domes of Texas and Louisiana: AAPG Bull., v. 17, p. 1025-1083.

Burgess, W. J., 1976, Geologic evolution of the Mid-Continent and Gulf Coast areas—a plate tectonics view: Gulf Coast Assoc. Geol. Socs. Trans., v. 26, p. 132-143.

de Jong, A., 1968, Stratigraphy of the Sigsbee Scarp from a reflection survey (abs.): Soc. Exploration Geophysicists Prog., Fort Worth Mtg., p. 51.

Demorest, M., 1943, Ice sheets: Geol. Soc. America Bull., v. 54, p. 363-400.

Dietz, R. S., and J. C. Holden, 1970, Reconstruction of Pangaea—breakup and dispersion of continents, Permian to present: Jour. Geophys. Research, v. 75, no. 26, p. 4939-4956.

Evans, R., 1974, The significance of the evaporites in the basins around the Atlantic margin (abs.), *in* A. H. Coogan, ed., Fourth symposium on salt, v. 1: Northern Ohio Geol. Soc., p. 271.

Garrison, L. E., and R. G. Martin, Jr., 1973, Geologic structures in the Gulf of Mexico basin: U.S. Geol. Survey Prof. Paper 773, 85 p.

Halbouty, M. T., 1967, Salt domes—Gulf region, United States and Mexico: Houston, Gulf Publishing Co., 425 p.

Hazzard, R. T., W. C. Spooner, and B. W. Blanpied, 1947, Notes on the stratigraphy of the formations which underlie the Smackover Limestone in South Arkansas, Northeast Texas, and North Louisiana: Shreveport Geol. Soc. (1945) Reference Report, v. 2, p. 483-503.

Jux, U., 1961, The palynologic age of diapiric and bedded salt in the Gulf Coastal province: Louisiana Geol. Survey Geol. Bull. 38, 46 p.

Kinsman, D. J. J., 1974, Evaporite deposits of continental margins, *in* A. H. Coogan, ed., Fourth symposium on salt, v. 1: Northern Ohio Geol. Soc., p. 255-259.

Kirkland, D. W., and J. E. Gerhard, 1971, Jurassic salt, central Gulf of Mexico, and its temporal relation to circum-gulf evaporites: AAPG Bull., v. 55, p. 680-686.

Kupfer, D. H., 1974, Environment and intrusion of Gulf Coast salt and its probable relationship to plate tectonics, *in* A. H. Coogan, ed., Fourth symposium on salt, v. 1: Northern Ohio Geol. Soc., p. 197-213.

Lehner, Peter, 1969, Salt tectonics and Pleistocene stratigraphy on continental slope of northern Gulf of Mexico: AAPG Bull., v. 53, no. 12, p. 2431-2479.

Lowell, J. D., and G. J. Genik, 1972, Sea-floor spreading and structural evolution of southern Red Sea: AAPG Bull., v. 56, p. 247-259.

Nettleton, L. L., 1934, Fluid mechanics of salt domes: AAPG Bull., v. 18, p. 1175-1204.

Parker, T. J., and A. N. McDowell, 1955, Model studies of salt-dome tectonics: AAPG Bull., v. 39, no. 12, p. 2384-2470.

Pautot, G., J.-M. Auzende, and X. LePichon, 1970, Continuous deep-sea salt layer along North Atlantic margins related to early phase of rifting: Nature, v. 227, July 25, p. 351-354.

Viniegra, O. F., 1971, Age and evolution of salt basins of southeastern Mexico: AAPG Bull., v. 55, p. 478-494.

Watkins, J. S., et al, 1975, Multichannel seismic reflection investigation of the western Gulf of Mexico: Offshore Technology Conference, p. 797-804.

——— J. L. Worzel, and J. Ladd, 1976, Deep seismic reflection investigation of occurrence of salt in Gulf of Mexico, *in* Beyond the shelf break: AAPG Marine Geology Comm. Short Course, v. 2, p. G-1 to G-34.

Wilhelm, O., and M. Ewing, 1972, Geology and history of the Gulf of Mexico: Geol. Soc. America Bull., v. 83, p. 575-600.

Woodbury, H. O., et al, 1973, Pliocene and Pleistocene depocenters, outer continental shelf, Louisiana and Texas: AAPG Bull., v. 57, p. 2428-2439.

Reprinted by permission of the Gulf Coast Association of Geological Societies from *Transactions of the Gulf Coast Association of Geological Societies*, v. 18, 1968, p. 320-330.

SALT TECTONICS AS RELATED TO SEVERAL SMACKOVER FIELDS ALONG THE NORTHEAST RIM OF THE GULF OF MEXICO BASIN

DUDLEY J. HUGHES
Hughes & Hughes
Jackson, Mississippi

ABSTRACT

Along the north rim of the Mississippi Salt Basin, the Louann salt was deposited as a basinward thickening wedge. Shallow piercement domes are found only in the deeper portion of the basin where the salt has attained a considerable thickness. Around the rim of the basin, between the shallow piercement dome area and the updip limits of the salt, more obscure structures are formed which take the shape of salt ridges, pillows and anticlines. Salt structures having similar characteristics are developed in areas of similar salt thickness. In the area studied along the northeast rim of the Mississippi Salt Basin, these salt structures can be classified as four common types.

The farthest updip structures are along the wedgeout of the salt. These are formed by flowage of the salt toward the wedgeout, which developed a salt ridge parallel to the wedgeout and immediately basinward from it. The suggested name for this type structure is "peripheral salt ridge." Reverse dip is exhibited from the crest of this ridge to the wedgeout. Associated with this type structure is a complex graben system in overlying sediments. This graben system is caused by stresses created by salt movement adjacent to a stable floor updip from the salt wedgeout. Flowage of salt in the peripheral salt ridge has been slow and continuous since deposition of first overburden.

Immediately downdip from these peripheral structures, the mother salt is still relatively thin and forms "low relief salt pillows." These exhibit low relief anticlinal closures aligned in ridge-like patterns. Salt movement in the pillows was rapid during early deposition of overburden but showed little structural movement after deposition of the Haynesville.

Basinward from the "low relief salt pillows," the mother salt attains greater thickness and structures developed have the appearance of "intermediate salt anticlines." These attain a fair size and show major growth during deposition of Jurassic rocks but little movement during Cretaceous and younger deposition. The differential salt uplift on these structures is usually sufficient to form a primary graben over its crest. This graben is sometimes spread apart at Smackover and Norphlet horizons to form a "drawbridge" pattern. On structures exhibiting the "drawbridge" pattern younger sediments are deposited directly on salt within the graben area.

Basinward from the intermediate salt anticlines, the salt attains greater thickness and results in formation of "high relief salt anticlines." These structures exhibit major differential growth from deposition of earliest overburden through Upper Cretaceous time. Commonly a primary graben was developed during the Jurassic which terminates upward in Lower Cretaceous rocks. Continued movement of the salt has caused development of a secondary graben within the primary graben which is most prominent in Upper and Lower Cretaceous sediments. A pronounced "drawbridge" effect at Smackover and Norphlet horizons is common on the high relief salt anticlines.

A study of the thin salt area around the north rim of the Mississippi Salt Basin indicates that salt flowage and structural deformation has taken place even in areas of very thin mother salt. As the salt thickens from its wedgeout basinward, the structures formed progressively farther downdip have more salt available and thereby develop a progressively thicker salt core. Salt flowage began very early after deposition of a small amount of overburden. Most salt structures exhibited a fast rate of growth in the early stages which continued as long as abundant salt was available from the mother salt bed. Faulting is confined to strata above the salt.

INTRODUCTION

There is a notable lack of information published about the areas of thin Louann salt around the rims of the salt basins of the Gulf Coast. This paper will describe structures which are formed from thin mother salt in a portion of the Mississippi Salt Basin, between the updip Louann salt wedgeout and the downdip shallow piercement dome areas found in the interior of the basin. The area studied is underlain by salt too thin to form shallow piercement salt domes, but does form salt ridges, pillows and anticlines. These low relief salt structures are currently the object of an intensive exploration effort in a search for Jurassic fields.

Since approximately 1960 much new information has become available through deep drilling and improved seismic methods utilizing multiple coverage (Rosenkrans and Marr, 1967). These structures can be interpreted with reasonable accuracy even to the base and configuration of the salt itself. Many test wells have been drilled into the Louann salt.

To illustrate the characteristics of the different types of structures in this area of thin salt, four Smackover fields are used as examples.

REGIONAL SETTING

The study area is in the northeast Mississippi Salt Basin in the northeastern Gulf of Mexico Basin (See Figure 1). The structures studied are considered characteristic of the types of salt structures which produce oil and gas from Jurassic rocks. Structures studied are: Quitman Field and Nancy Field, Clarke County, Mississippi; Cypress Creek Field, Wayne County, Mississippi; and Pool Creek Field, Jones County, Mississippi. These particular structures were chosen because they illustrate a progression from structures

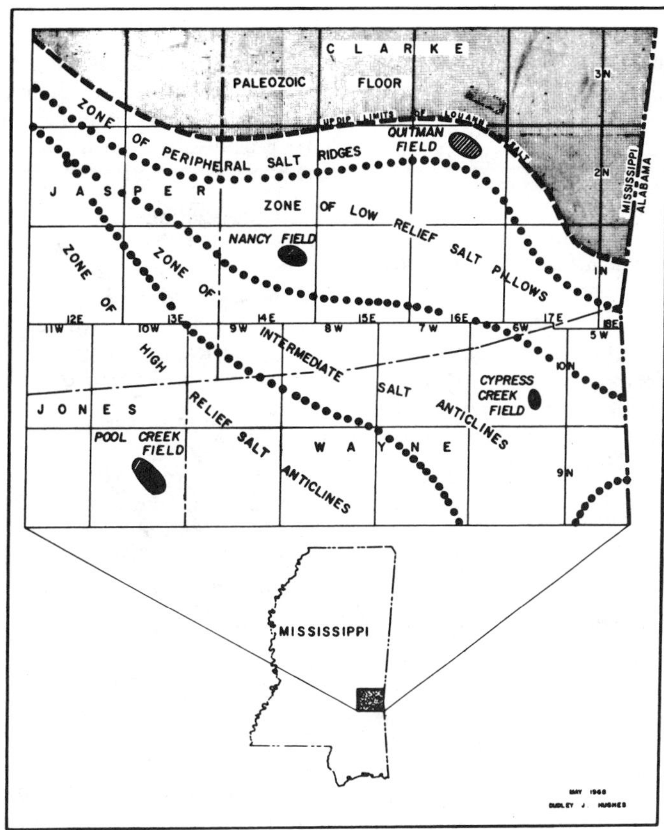

Figure 1. Area of study along the north rim of the Mississippi Salt Basin. The Louann salt wedges out against eroded Paleozoic floor to the north, and thickens progressively southward. Salt structures having similar characteristics are developed in areas of similar salt thickness. A zone of "peripheral salt ridges" parallels the wedgeout; adjacent to these is a zone of "low relief salt pillows"; basinward from these, is a zone of "intermediate salt anticlines"; and in thicker salt basinward is a zone of "high relief salt anticlines." The mother salt thickens more rapidly basinward on western side of area studied than on eastern side.

in areas of thin salt near the updip limit of the salt (Quitman and Nancy) to structures in areas of thicker salt farther basinward (Cypress Creek and Pool Creek).

The basin floor (Paleozoic) on which the Louann salt was deposited shows little indication of structural movement since Louann time except southward tilting as deposition of Mesozoic and Cenozoic sediments occurred. Seismic control indicates the base of the salt is presently a fairly flat basinward dipping surface. The original thickness of the Louann salt varied from a feather edge along its updip limit in Clarke and Jasper Counties to several thousand feet basinward in Wayne and Jones Counties. It is probable that nowhere in the area studied does the salt retain its original thickness for a sustained distance. The salt is thicker in the structures and thinner in the withdrawal areas adjacent to the structures.

Deposition of overlying younger Mesozoic sediments was more widespread, completely overlapping the salt. Immediately overlying the Louann salt is the Norphlet Formation of Upper Jurassic age. The Norphlet in the study area is porous quartz sandstone and red shale of variable thickness. Overlying the Norphlet is the Smackover Formation which is primarily limestone and dolomite. A porous zone at the top of the Smackover produces oil and gas in the fields discussed. Unconformably overlying the Smackover is the Haynesville Formation which consists of anhydrite, dolomite and limestone interbedded with sandstone and shale. A major unconformity at the top of the Haynesville is overlain by the Cotton Valley sandstones of Upper Jurassic age. A sequence of Lower Cretaceous sandstones and shales overlies the Cotton Valley. The Upper Cretaceous section is predominantly sandstone and shale in the lower portion and chalk in the upper portion. Above the Upper Cretaceous are undifferentiated Cenozoic beds.

EXPLORATION HISTORY

Prior to the "Jurassic play," two shallower trends were actively explored in the area studied. During the 1940's and early 1950's, hundreds of wells were drilled for production from Eutaw sandstones (Upper Cretaceous) at depths of 3000 to 5000 feet. From this first regional subsurface control, it was determined that a complex graben system averaging four miles in width (at the Eutaw horizon) is present around the northern periphery of the basin. This system, which trends from northwest to southeast through central Jasper and Clarke Counties, is known as the Pickens-Gilbertown-Pollard fault zone and is a portion of the peripheral graben belt found along the updip margins of the salt basins of the Gulf Coast.

Basinward from the Pickens-Gilbertown fault zone, isolated grabens were found which are associated only with individual structures and which exhibit different characteristics than in this peripheral graben belt (Hughes, 1960).

In the mid 1950's and early 1960's many wells were drilled to depths of 9000 to 12,000 feet in search of Lower Cretaceous oil fields. These wells provided deeper control which showed more structural relief and isolated some structures not apparent on Eutaw beds. On some of the more pronounced uplifts in the area, domal salt was found to have penetrated Lower Cretaceous beds at the apex of the structure. Prior to 1963, few wells had been drilled to the Jurassic.

Since the discovery of Jurassic oil in Bienville Forest, Smith County, Mississippi, in 1963, the primary objective of exploration in the area has been for production from Jurassic rocks. Most early drilling for Jurassic objectives was in the deep portion of the Mississippi Salt Basin on the more obvious and pronounced salt uplifts. Drilling on these structures had a discouraging effect on Jurassic exploration for a short time because many of the earlier tests encountered domal salt in the Lower Cretaceous or Cotton Valley and the primary objectives were missed. One of these larger downdip features which did not produce from the Smackover was Pool Creek Field, found in 1965 (Minihan and Oxley, 1966).

Emphasis on the areas of thin salt was accelerated in 1966 by the discovery of Quitman Field in Clarke County, Mississippi. This field is near the updip limit of the salt in the Pickens-Gilbertown-Pollard fault zone where Jurassic beds are found at relatively shallow depths (11,500'). The Nancy Field was discovered downdip from Quitman in 1967, in an area showing only regional south dip on shallower Upper and Lower Cretaceous beds. The Cypress Creek Field was discovered early in 1968 basinward from the Nancy Field. This structure also had little structural expression on the shallower Upper and Lower Cretaceous beds.

The rapid unfolding of geological information which has taken place in the past few months prompted the writer to arrive at the following interpretations.

STRUCTURE TYPES

Salt structures in the area studied generally exhibit a flat base (tilted basinward) and an anticlinal upper surface. Four structure types are recognized with the principal distinguishing characteristics being thickness of salt within the structure, its influence on overlying sediments, and its position in the basin. These structure types are classified as "peripheral salt ridges," "low relief salt pillows," "intermediate salt anticlines," and "high relief salt anticlines" (modified after Halbouty, 1967, p. 69). Figure 1 is a map of the area studied showing the portion of the area in which each type structure develops. The following discussion of each type structure is accompanied with a cross section and structure map contoured on top of the Louann salt of the field representing that type structure.

Peripheral Salt Ridge (Quitman Field)

A salt ridge is developed near the updip limit of the Louann salt which parallels the wedgeout of the salt. This ridge has reverse dip (in this area—north dip) from its crest line to the salt wedgeout. Low relief anticlinal closures are developed along this ridge. Such a closure resulted in the Quitman Field (Figures 2 and 6). These closures occur most commonly where the crest line of the ridge runs parallel to regional strike.

Associated and parallel to the peripheral salt ridge is the Pickens-Gilbertown-Pollard fault zone, a complex graben system developed in beds above the salt. The principal faults comprising this graben system converge downward and terminate on the north flank of the peripheral salt ridge near the salt wedgeout.

In the Quitman Field, both seismic and subsurface control indicate that the faulting is confined to beds above the salt. Subsurface control shows the throw of the faults to increase with depth to mid-Lower Cretaceous (Paluxy) and then to rapidly decrease with depth. The faults die out in Jurassic beds above the Louann salt. This same pattern of fault behavior can be seen on other peripheral salt ridge structures.

The origin of the peripheral salt ridge and its associated peripheral graben belt is explained by simple salt movement contemporaneous with deposition. The salt was deposited as a wedge with gradual thickening to the south from its updip limit. As the basin floor tilted southward, the wedgeout edge of the salt became the highest portion of the salt mass. Overburden was deposited contemporaneous with tilting to create differential forces which caused the salt to flow toward the wedgeout. This produced the salt ridge immediately south of the wedgeout.

The peripheral graben system was caused by stresses created in overlying beds by the salt movement. The complexity and throw of the faulting is much greater than would be expected by the low magnitude of salt movement alone. This is explained, however, by the contrast in underlying rock. The younger sediments which overlap the wedgeout of the salt rest on a stable Paleozoic floor north of the wedgeout and rest on an unstable salt mass south of the wedgeout. These beds are subject to stresses by plastic flowage of the salt south of the wedgeout while anchored to the unyielding floor to the north. The stresses thus created in overlying beds have strong shear and tension components so that a complex graben system is developed even with minor salt movement.

It is probable that most of the faulting in the Pickens-Gilbertown-Pollard fault zone is caused only by salt movement adjacent to a stationary floor. There is less evidence to attribute this faulting to "regional basement faults" (Oxley, Minihan, and Ridgeway, p. 27).

The source of the salt which forms the peripheral salt ridge was from the basin to the south and was never depleted. For this reason the peripheral salt ridges appear to have undergone a slow, steady growth even to present. Also, as the peripheral salt ridge builds southward from the wedgeout zone, it is likely that when the salt mass became higher south of the wedgeout new differential forces are created to cause a secondary flowage of the salt from the wedgeout toward the crest line of the ridge. This would increase the tensional stresses causing faulting in the shallow beds. If this is the case, at Quitman the maximum "backflow" would probably have begun during Lower Cretaceous times. Cloos (1968, p. 428-437) produced grabens in clay models which are strikingly similar to those found in the peripheral graben belt. He concluded, however, that the graben formed was a result of "creeping sediments" which are "torn loose from the stationary area," rather than salt movement.

Characteristics of the peripheral salt ridge are: (a) north

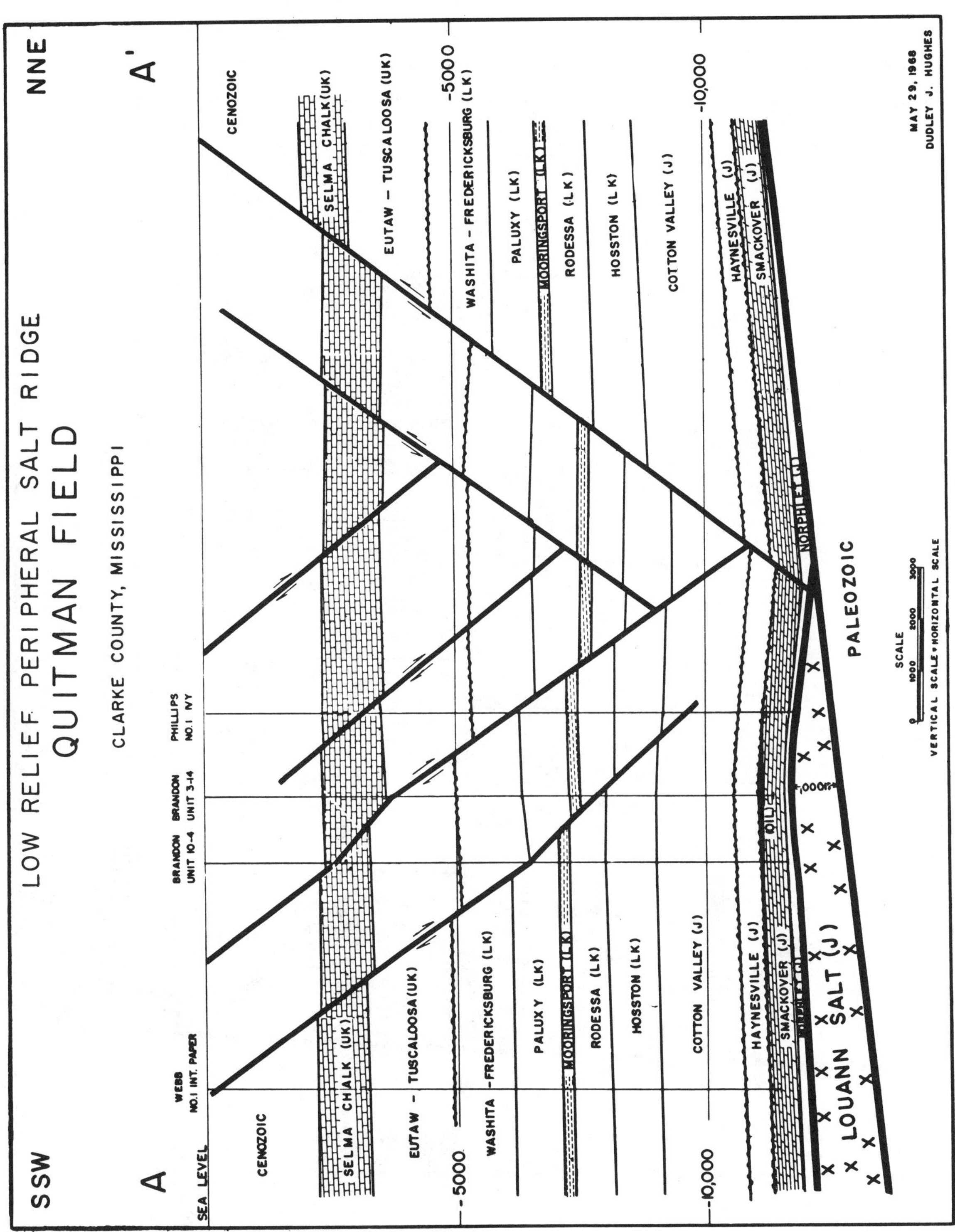

Figure 2. Cross section of Quitman Field, Clarke County, Mississippi, illustrating a "peripheral salt ridge" type structure. Flowage of the salt toward wedgeout produces a salt ridge immediately dowdip from wedgeout. Stresses induced in overlying beds by salt movement adjacent to stationary floor produce a complex graben system. This may explain origin of the Pickens-Gilbertown-Pollard fault zone.

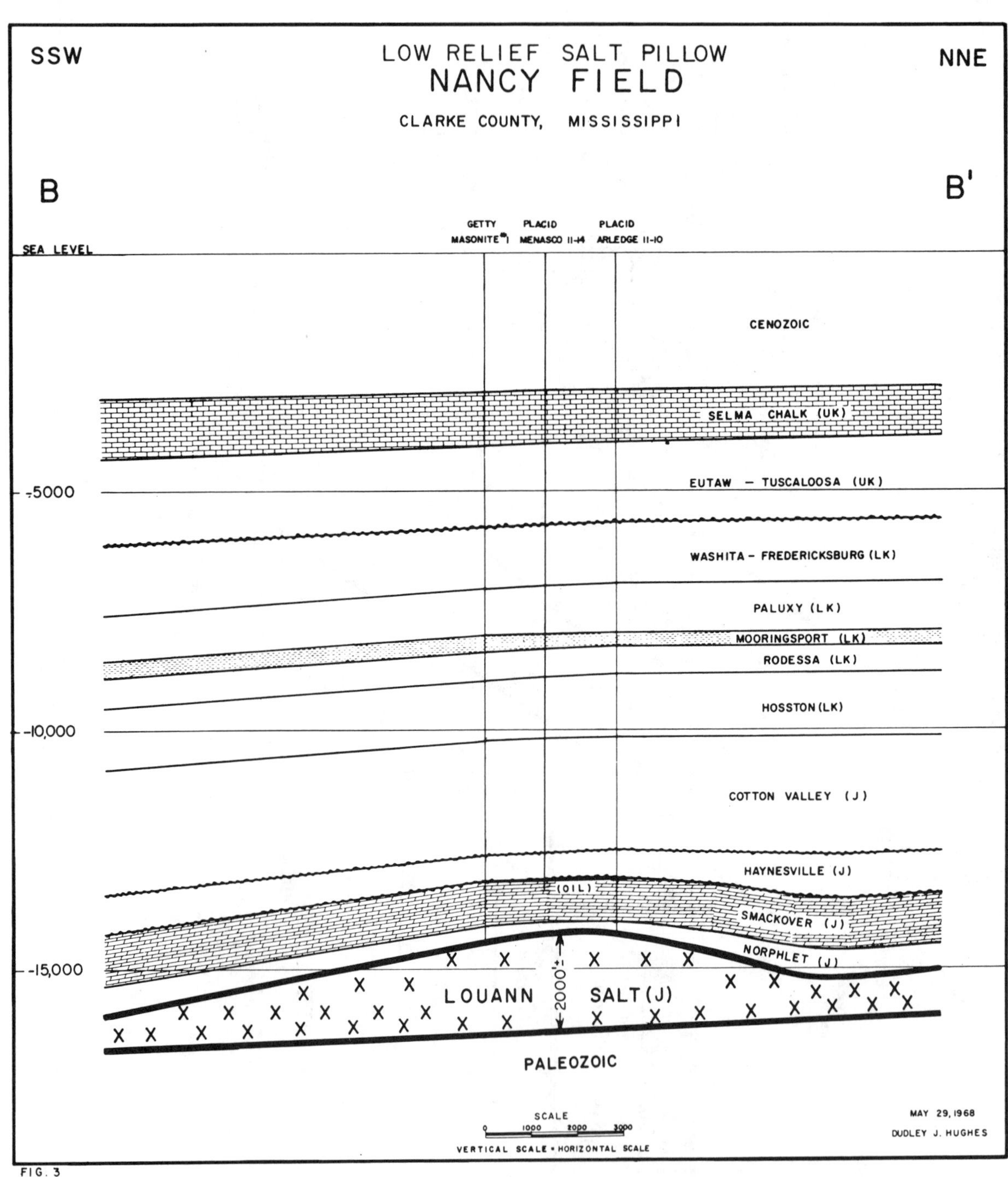

Figure 3. Cross section of Nancy Field, Clarke County, Mississippi, illustrating "low relief salt pillow." Most salt movement took place during Norphlet, Smackover, and Haynesville times. There is little structure above Haynesville.

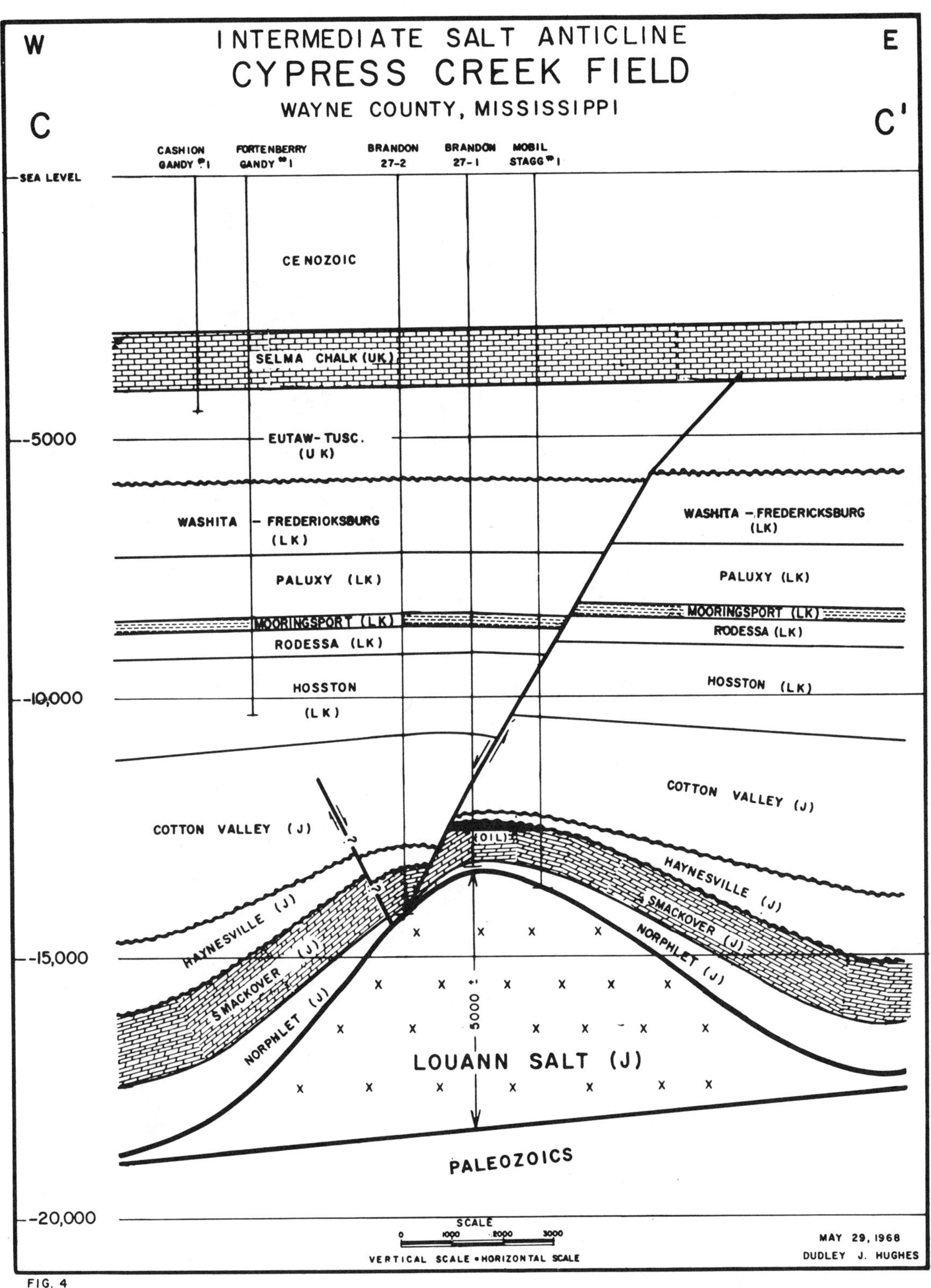

Figure 4. Cross section of Cypress Creek Field, Wayne County, Mississippi, illustrating an "intermediate salt anticline." Salt movement was predominant during Norphlet, Smackover, Haynesville, and Lower Cotton Valley deposition. A "primary" graben is developed near crest of structure.

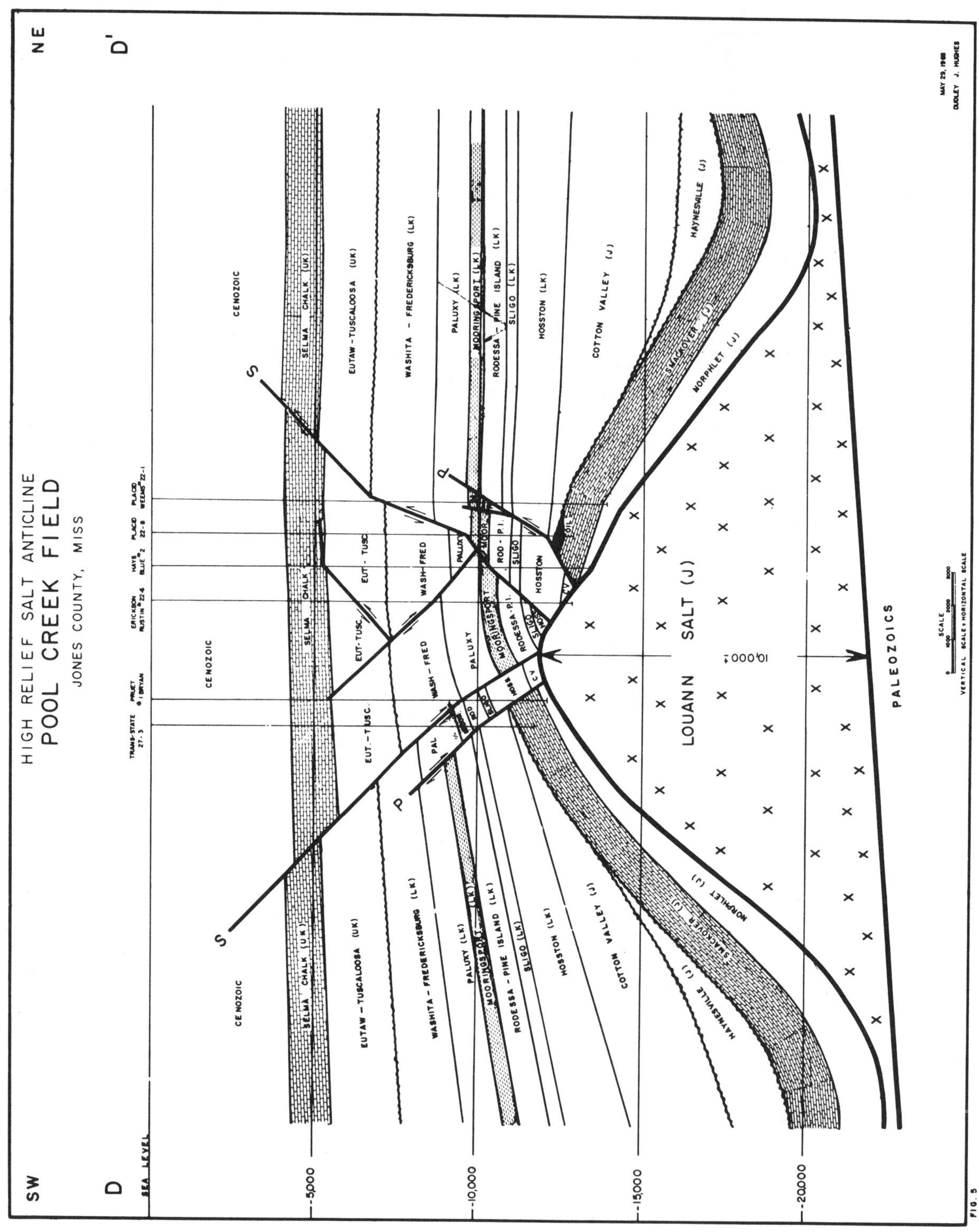

Figure 5. Cross section of Pool Creek Field, Jones County, Mississippi, illustrating a "high relief salt anticline." Faults of the "primary" graben are labeled "P." Faults of the "secondary" graben are labeled "S." Limbs of Smackover structure exhibit a pronounced "drawbridge" pattern. Salt movement continued a fast rate through Upper Cretaceous time.

dip from the top of the ridge to the wedgeout of the salt, (b) location of this ridge along the south side of the regional peripheral graben belt, (c) the north flank of the ridge exhibits steeper dip than the south flank, (d) the overlying sediments are not pierced by the salt, (e) salt movement began early but there appears to be nc periods of rapid growth of the salt feature, (f) the salt withdrawal area south of the ridge is not pronounced, (g) structure on the Smackover Formation generally conforms with the salt, (h) the faults in the beds above the salt dip approximately 50 to 60 degrees and die out prior to cutting the salt, (i) salt thickness at the apex of the peripheral salt ridge usually ranges from 500 to 2000 feet.

Low Relief Salt Pillow (Nancy Field)

Downdip and farther basinward from the peripheral salt ridge the mother salt layer is relatively thin for a number of miles though gradually thickening southward. A different type salt structure is developed in these areas. These are the "low relief salt pillows" (Figure 1). The pillows are commonly arranged in ridge fashion approximately parallel to the peripheral salt ridges. Nancy Field is an example of this type salt structure (Figures 3 and 6). The pillows generally have little structural expression above the Haynesville, and usually have not caused faulting in the overlying sediments. In contrast to the peripheral salt ridge, the salt forming the low relief salt pillows exerts less stress on the overlying beds because the beds are underlain by a common supporting rock (salt). The presence of a common supporting rock explains the lack of faulting or a lesser amount of faulting associated with these low relief salt features, even where the overall salt movement was greater than in the peripheral salt ridge.

The growth of these structures appears to have begun during Norphlet time; and relatively rapid growth continued through Smackover-Haynesville times. There was little structural growth after Haynesville time. Closure over these structures forms ideal traps for Smackover oil accumulations.

The characteristics of the low relief salt pillow are: (a) lack of structural expression above the Haynesville, (b) little or no faulting associated with the feature, (c) no piercement of the overlying Mesozoic beds, (d) ridge-like alignment of salt pillows, (e) salt thickness at the apex of the pillows generally ranges from 1000 to 3500 feet.

Intermediate Salt Anticline (Cypress Creek Field)

Farther downdip from the salt ridges and pillows, the mother salt layer is sufficiently thick to form "intermediate salt anticlines" (Figure 1). Salt available to these structures is sufficient to develop a salt core from approximately 3500 feet to 7000 feet in thickness. Major salt growth took place on these structures from Norphlet through mid-Cotton Valley times, with only minor growth during late Cotton Valley and Cretaceous deposition. Usually a primary graben (Hughes, 1960) is developed in sediments over the crest of the intermediate salt anticlines. The faults exhibit a maximum throw near the top of the Smackover, but from this maximum the throw decreases upward and downward. The faults terminate upward in the Cretaceous and downward in the Norphlet. A nose, or faulted nose, is the only indication of structure on the Cretaceous beds. The Cypress Creek field is an example of an intermediate salt anticline (Figures 4 and 7).

On intermediate salt anticlines (though not on the Cypress Creek Structure) the large uplift of the salt often causes spreading apart of the graben at Smackover and Norphlet horizons. This creates a zone barren of Smackover and Norphlet between the faults in which younger beds (Haynesville or Cotton Valley) are deposited over salt. The gap between limbs of the Smackover is usually no more than three-quarters of a mile in width on an intermediate salt anticline. This pattern is called the "drawbridge" effect because the two limbs of the Smackover structure on each side of the graben have been raised and in cross section appear as an open drawbridge. This effect is common only on the larger salt features.

The key characteristics of the intermediate salt anticlines are (a) major growth during Jurassic time, but only minor growth during later deposition, (b) the common presence of a primary graben over the crest of the salt, which sometimes exhibits a drawbridge pattern on the Smackover and Norphlet beds, (c) Smackover oil accumulation occurs on the upthrown side of the faults, (d) the salt anticline thickness will range from 3500 to 7000 feet at its apex.

High Relief Salt Anticline (Pool Creek Field)

Basinward from the intermediate salt anticlines, the salt becomes thicker and much larger "high relief salt anticlines" are formed (Figure 1). These structures had sufficient salt available to maintain continuous fast growth from Norphlet into Upper Cretaceous times. A primary graben normally developed above the salt during the Jurassic deposition and secondary grabens may have developed within the primary graben during Lower Cretaceous deposition. The faults of the primary graben usually terminate upward in Cotton Valley or Lower Cretaceous rocks. The faults of the secondary grabens carry upward into the Upper Cretaceous or younger beds. The drawbridge pattern may be very pronounced at Smackover level with the Smackover terminated at the upthrown sides of the primary graben. Smackover production is found in these fault closures on the upthrown sides of the primary graben. The gap present between the two limbs of a Smackover structure with a drawbridge pattern may range from three-quarters to two miles in width on high relief salt anticlines. On large uplifts Lower Cretaceous or Cotton Valley beds rest on the salt within the grabens.

The Pool Creek field is an example of a high relief salt anticline with a drawbridge pattern on Norphlet and Smackover (Figures 5 and 7).

Characteristics of the high relief salt anticline are: (a) major salt growth from Jurassic into Upper Cretaceous; (b) usual presence of a primary graben which does not carry into Upper Cretaceous, and presence of a secondary graben (or grabens) within the primary graben, which is prominent in Lower Cretaceous and Upper Cretaceous beds; (c) pronounced "drawbridge" pattern on Smackover and Nor-

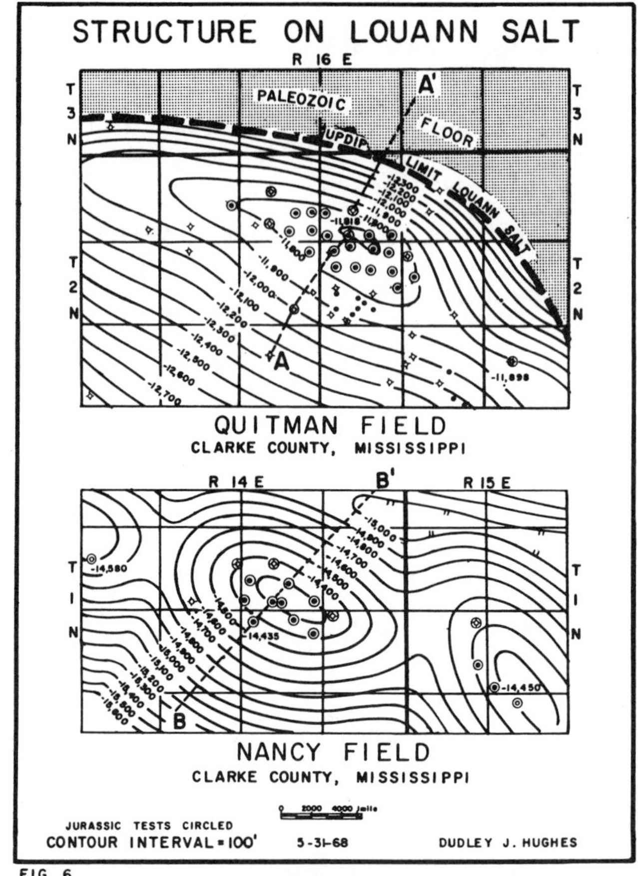

Figure 6. Structure contours on Louann salt of Quitman Field, Clarke County, Mississippi, (peripheral salt ridge), and Nancy Field, Clarke County, Mississippi (low relief salt pillow). Figure 2 is cross section A-A' of the Quitman structure; and Figure 3 is cross section B-B' of the Nancy structure.

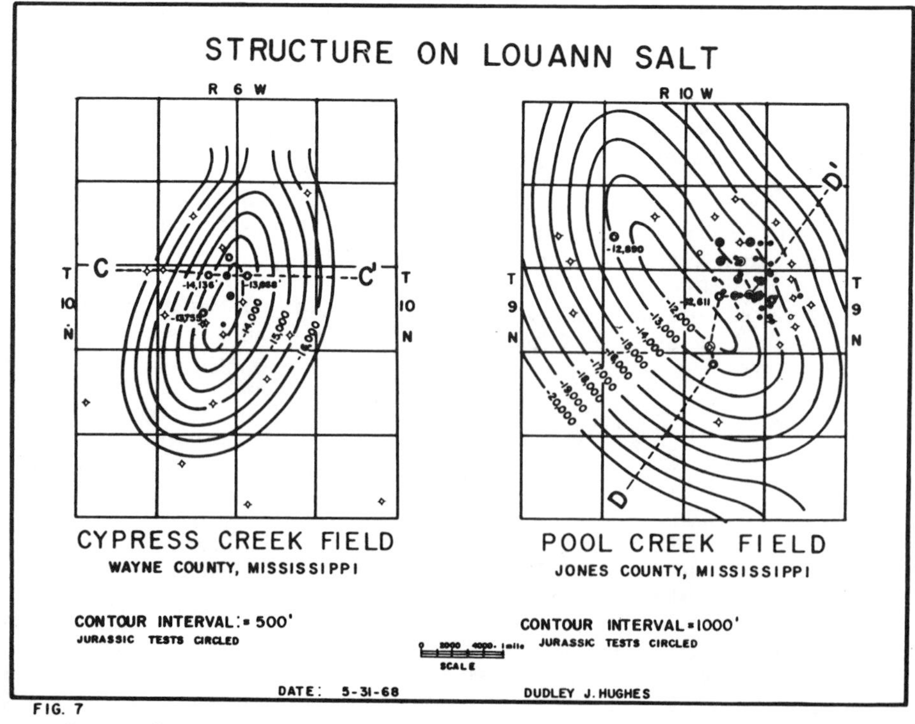

Figure 7. Structure contours on Louann salt are shown for the Cypress Creek Field, Wayne County, Mississippi, (intermediate salt anticline), and the Pool Creek Field, Jones County, Mississippi (high relief salt anticline). Note change in contour intervals. Figure 4 is cross section C-C' of the Cypress Creek structure, and Figure 5 is cross section D-D' of the Pool Creek structure.

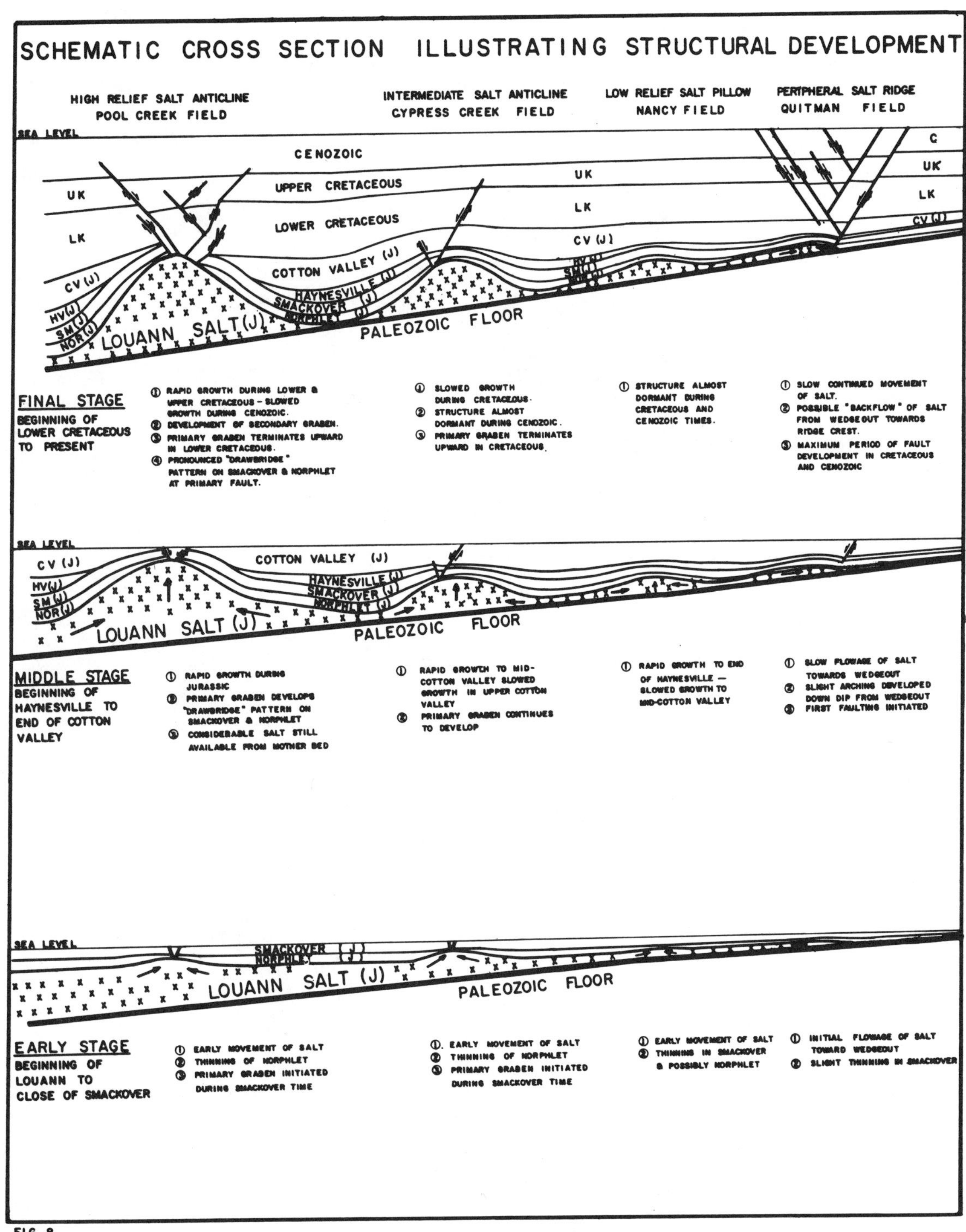

Figure 8. Schematic cross section illustrating evolution of the salt structures discussed in this paper. Louann salt was initially deposited as a basinward thickening wedge. Movement of salt began with deposition of Norphlet. Type structure developed was governed principally by thickness of mother salt bed.

phlet Formations, (d) salt thickness at apex of anticline 7000 to 15,000 feet.

Figure 8 illustrates the evolution of the four structure types discussed.

CONCLUSION

The following generalizations appear to be valid in the area studied: (1) Salt does not require any particular thickness before flowage and structural deformation will take place. Flowage has taken place and structures are formed in very thin salt near the salt wedgeout. (2) The size of the salt core in salt anticlines is proportional to the amount of salt available from the mother bed (this relationship may not be true of residual salt highs). (3) Salt flowage began soon after deposition of a small amount of overburden. In the area studied salt flowage began after only a few hundred feet of Norphlet was deposited on top of the Louann salt. (4) The fast rate of growth of most salt structures in their early stages continued until most of the salt available from the mother salt layer had flowed into the structure. The growth rate of the mature salt structure was very slow and probably was the result of reshaping of the salt already isolated in the uplift.

REFERENCES CITED

Cloos, E., 1968, Experimental analysis of Gulf Coast fracture patterns; Am. Assoc. Petroleum Geologists, v. 52/3, pp 420-444.

Halbouty, M. T., 1967, Salt domes, gulf region, United States and Mexico: Gulf Publishing Co., pp 1-86.

Hughes, D. J., 1960, Faulting associated with deep-seated salt domes in the northeast portion of the Mississippi salt basin: Trans G.C.A.G.S., v. 10, pp 154-173:

Minihan, E. D. and Oxley, M. L., 1966, Pre-Cretaceous geology of Pool Creek Field, Jones County, Mississippi: Trans. G.C.A.G.S., v: 16, pp 35-43:

Oxley, M. L., Minihan, E. and Ridgeway, J. M., 1967, a study of the Jurassic sediments in portions of Mississippi and Alabama: Trans. G.C.A.G.S., v: 17, pp 24-48:

Rosenkrans, R. R. and Marr, J. D., 1967, Modern seismic exploration of the Smackover trend: Journal of Geophysics, v. 32, n. 2, pp 184-206.

Reprinted with permission from *Journal of Structural Geology*, v. 6, p. 521-533, C. J. Talbot and R. J. Jarvis, Age, budget and dynamics of an active salt extrusion in Iran, copyright 1984, Pergamon Press plc.

Journal of Structural Geology, Vol. 6, No. 5, pp. 521 to 533, 1984
Printed in Great Britain

0191–8141/84 $03.00 + 0.00

Age, budget and dynamics of an active salt extrusion in Iran

C. J. Talbot

Institute of Geology, Department of Mineralogy and Petrology, University of Uppsala, Box 555, S-751 22 Uppsala, Sweden

and

R. J. Jarvis

Department of Mathematics, The University, Dundee DD1 4HN, Scotland, U.K.

(*Received* 21 *March* 1983; *accepted in revised form* 3 *October* 1983)

Abstract—The Hormuz salt of Kuh-e-Namak, Iran began rising through its Phanerozoic cover in Jurassic times and had surfaced by Cretaceous times. In Miocene times, the still-active Zagros folds began to develop and the salt is still extruding to feed a massive topographic dome and two surface flows of salt which have previously been called salt glaciers but are here called *namakiers*.

Two crude but independent estimates for the rate of salt extrusion and loss are shown to balance the salt budget if the current salt dynamics are assumed to be in steady state. First, to replace the extrusive salt likely to be lost in solution in the annual rainfall, the salt must rise at an average velocity of about 11 cm a^{-1}. Second, the foliation pattern shows that the extruding (and partially dissolved) salt column spreads under its own weight. The maximum height of the salt dome is consistent with a viscous fluid with a viscosity of 2.6×10^{17} poises extruding from its orifice at a rate of almost 17 cm a^{-1}. Both estimates are consistent in indicating that salt can extrude onto the surface 42–85 times faster than the average long term rate at which salt diapirs rise to the surface.

The structure, fabrics, textures and deformation mechanisms of the impure halite all change along the path of the extrusive salt from the dome down the length of both namakiers. Such changes tend to occur when the flowing salt encounters changes in its boundary conditions, and the recognition of buried namakiers is discussed in the light of such observations. Episodes of salt flow at a rate of 0.5 m per day have been measured along the margin of the N namakier after significant rain showers. Such brief episodes of rapid flow alternate with long periods when the namakier is dry and stationary. The shape of the colour bands cropping out on the N namakier indicate that the flow over the surface of impure salt with a mylonitic texture obeys a power law with $n \sim 3$. Although the reported annual rainfall has the potential of dissolving both namakiers in about 2000 years, a superimposed thin marine cover may protect static parts of them for as long as 30,000 to 300,000 years.

INTRODUCTION

This report describes one of the many bodies of Hormuz (Eocambrian–Cambrian) salt extruding through a thick sequence of Phanerozoic sedimentary rocks as they shorten along a NNE–SSW axis in the Zagros fold belt of Iran.

Most previous studies of bodies of deformed salt have been at structural levels at which the salt is confined to an intrusive pipe. Even if such diapirs reach the surface they are either inactive or the salt in them is dissolved as fast as it rises so that the salt ends upwards in a solution surface beneath a cap rock of relatively insoluble residues.

The majority of the 150 or so emergent diapirs of Hormuz salt in Iran are marked by topographic domes of extrusive salt indicating that their rise has been faster than their solution near or at the surface (Kent 1979). Sheets of salt have flowed from many of these domes down dip slopes of their country rocks to spread as piedmont lobes over the surrounding plains of recent sands and gravels.

The diapirs of Hormuz salt and their associated extrusives have an extensive literature and the following list is not exhaustive: de Böckh *et al.* (1929), Lees (1927, 1931), Harrison (1930, 1956), O'Brien (1957), Kent (1958, 1966, 1970, 1979), Gansser (1960), Gussow (1966, 1968), Gera (1972), Ala (1974), Fürst (1976), Barr (1977), Wenkert (1979), Talbot (1979, 1981), Talbot & Rogers (1980). The extrusive salt sheets have so far been known as salt glaciers but this is an inappropriate and unwieldy term and it is proposed here that a sheet or lobe which is predominantly of halite and which has flowed over the surface is called a *namakier*. This term is derived from *namak* which is Farsi for salt.

Descriptions of the salt structures and fabrics of such salt extrusives is long overdue in view of the questions as to whether the overhangs common to many buried salt structures have formed as intrusives or extrusives. Even quite thin sub-horizontal sheets of salt can be almost opaque to routine seismic investigations (but see Amery 1969) and might overlie and obscure far larger volumes of unsuspected hydrocarbon reservoirs than the sub-surface overhangs developed in association with asymmetric or non-vertical diapirs (Talbot 1977).

Deformation maps (Verall *et al.* 1975, Arieli *et al.* 1980) which summarize several decades of theoretical and laboratory based salt mechanics do not yet allow for the possibility of namakiers flowing over the surface. It has therefore been suggested that the namakiers of Iran could only have extruded hot like lavas (Gussow 1966—but see Kent 1966 and Gera 1972) or that they only

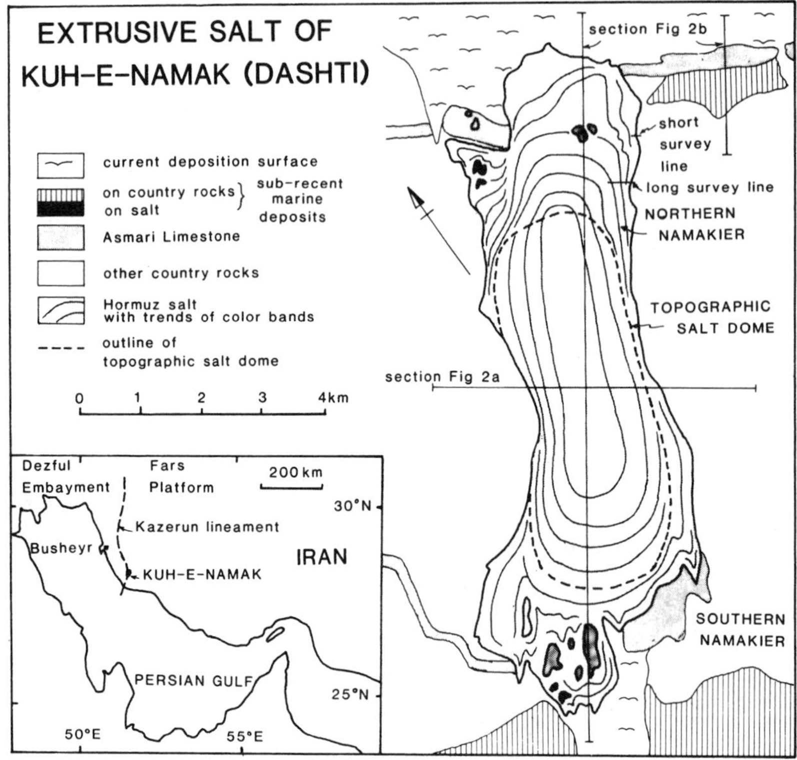

Fig. 1. Geological sketch map of the salt extrusives at Kuh-e-Namak (Dashti) showing locations of profiles in Fig. 2.

appear to move by the recementation of fallen blocks (Gussow 1968, Gera 1972).

The research programme which led to this report was cut short by the political situation in Iran and cannot be definitive; it therefore emphasizes problems by inference rather than solves them by measurements. The problems considered are the age of the salt extrusives, the budget of supply and loss of salt and the rate at which salt extrudes at the surface.

GEOLOGICAL SETTING OF KUH-E-NAMAK (DASHTI)

The Hormuz salt of Kuh-e-Namak (Dashti) has risen from depth where the south end of the Kazerun lineament crosses an anticline of typical NW–SE Zagros trend 120 km southeast of Busheyr about 40 km east of the coastline (Fig. 1). Kuh-e-Namak means 'mountain of salt' in Farsi and there are so many of these in the Zagros that the name of the province is usually added to identify a particular example; however, this convention will not be continued in this work which deals with the same Kuh-e-Namak throughout.

The Kazerun lineament is a 250 km N–S line of en-échelon wrench faults and local thrusts which are discontinuous on the surface but probably continuous along a major fault in the Precambrian basement (Kent 1979). This deep fault has been reactivated during several phases of movement since the Permian and appears to have been the site of several depocentres in Fars (Miocene–Pliocene) times (Kent 1979).

The Kazerun lineament separates the Fars platform from the Dezful embayment in the simply folded zone of the Zagros mountains. The Permian to Asmari (Eocene–early Miocene) cover in the Dezful embayment is thicker than it is on the Fars platform and, presumably as a consequence, the Zagros folds differ in style on either side of the Kazerun lineament. The anticline through which Kuh-e-Namak emerges crosses the Fars platform and is not alone in terminating close to the lineament.

The normal displacement across the Kazerun lineament reaches 5000 m further north (Kent 1979), but at the level of the Hormuz salt source layer beneath Kuh-e-Namak is only about 1000–1500 m down to the NW (Templeton pers. comm. 1977), and at the level of the Jurassic is only about 100 m (Figs. 2 and 3).

The surface expression of salt diapirs along the northern portion of the Kazerun lineament tends to be craters clogged with insoluble Hormuz debris. This suggests that they may be comparatively inactive, although the rainfall is higher there (Kent 1979); alternatively they may have been decapitated from their source by one or more thrusts spreading southwards as the basement beneath the simply folded Zagros moved northwards. Further south the salt extrusives become increasingly obvious and reach a climax at Kuh-e-Namak, which is probably one of the most active salt structures in Iran. Kuh-e-Darang, 20 km south-southwest and on the same

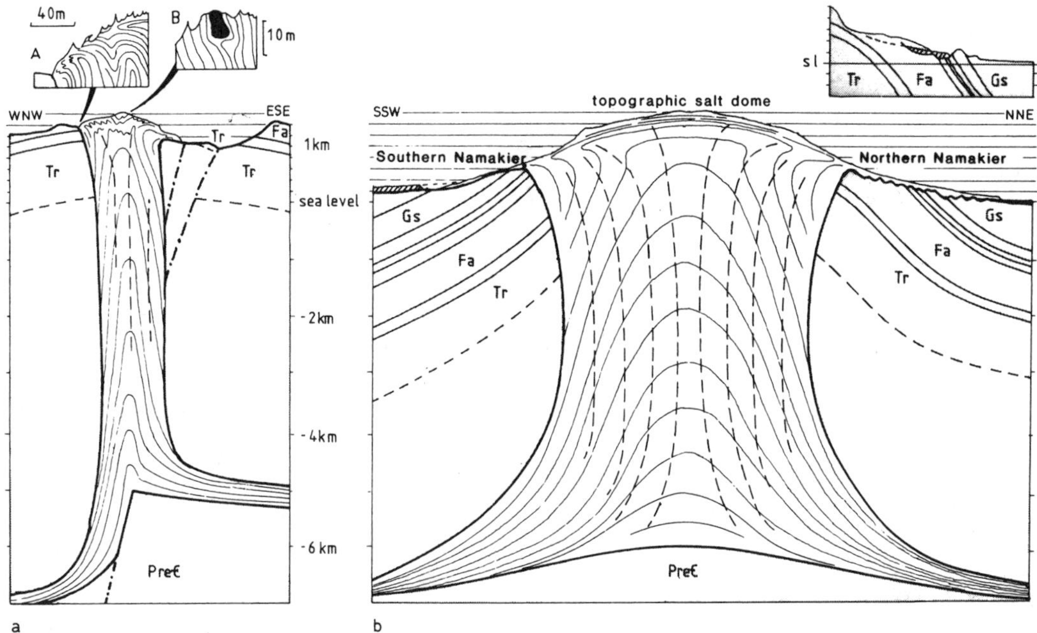

Fig. 2. Vertical cross (a) and longitudinal (b) sections of the Hormuz (= Eocambrian) salt mass showing the geometry of the flow profiles inferred from surface exposure. Locations of profiles are shown on Fig. 1. The colour banding (= bedding?) is shown in solid lines while the trace of the foliation (= stream lines) is indicated by dashed lines. Country rocks are: Pre€, Precambrian; Tr, Trias; Fa, Fahliyan-Surmeh Formation (Cretaceous); Grs, Gachsaran Formation (Miocene).

Inserts on Fig. 2(a): (A) The subvertical axial surface of mature folds inherited from depth curl to parallel the top free surface of the salt dome on its flank. These are joined or refolded by immature folds with axial planes which also curl to parallel the free surface but are too small-scale to illustrate (see Talbot 1979, fig. 8). (B) Occasional large blocks of insoluble Hormuz material (e.g. black limestone) are incorporated in the salt emerging in the dome. These are milled and dispersed in the halite if they are carried towards the distal portions of the namakiers in (rather than on) the salt.

Insert on Fig. 2(b): Cross section 1.5 km east of N namakier (see Fig. 1 for location) to show recent marine deposits (oblique ruling) and marine bevel (dashed) which define a recent surface (33,000–283,000 BP?) already showing significant folding. The same deposits and bevel can be seen to define the south limb of the recent increment of the Kuh-e-Namak anticline on the main longitudinal profile (Fig. 2b).

fault, has a salt core which is probably active but has not only not yet emerged, but may not even have yet reached the piercement stage (Kent 1979). Both Kuh-e-Namak and Kuh-e-Darang have developed from the same broad N–S trending salt pillow associated with attenuation of the Jurassic rocks exposed nearby (Kent 1979). The deep salt pillow at the south end of the Kazerun lineament is the oldest known indication of movement of the Hormuz salt in Iran (Kent 1979). Hormuz debris in Cretaceous sediments near Kuh-e-Namak indicate that, like others elsewhere in the Zagros, this particular diapir had already surfaced by Cretaceous times (Player pers. comm. 1976) long before the Zagros folding began. There can be little doubt that the rise of Hormuz salt in this structure was reactivated when the Zagros folds started to form in or around middle Miocene times (Kent 1979) and that both the folding and the salt extrusion are still active.

MORPHOLOGY OF KUH-E-NAMAK (DASHTI)

For descriptive purposes the salt body at Kuh-e-Namak may be considered to consist of the following components: two portions of a source layer at different structural levels, a sub-vertical pipe of elliptical cross-section, an extrusive topographic dome with a short extrusive flange along its southeast margin and two namakiers extending down to the plains just above sea level (Fig. 3). No piedmont lobes spread from either of these namakiers now, although moraines suggest that this may have happened in the past and such lobes are still common in other examples further east (Harrrison 1930).

If the orifice of the pipe is correctly inferred to underlie the steepest flanks of the extrusive salt dome (Fig. 1) then the orifice is a 6000 × 1000 m ellipse elongate NNE–SSW, along the fault bounding the southeast flanks of the salt bodies beneath both Kuh-e-Namak and Kuh-e-Darang.

Where the salt emerges in the extrusive dome it is coarsely crystalline (~4 cm in grain diameter), pale in colour and incorporates discrete blocks of insoluble Hormuz material up to several metres across (see insert B on Fig. 2a). Slight variations in the grain size and muted colours of the halite (pink, honey, pale green, grey and white) combine with the nature of entrained insoluble Hormuz debris (black limestone and sulphur, specular haematite, anhydrite, etc.) to define broad colour bands metres to tens of metres wide. These bands almost certainly represent the original bedding after it has flowed tens of kilometres sideways along the source layer and then several kilometres more up the pipe. These colour bands are generally steep and strike along the length of the extrusive salt dome and are also folded about doubly plunging fold axes with the same NNE–SSW orientation (see Fig. 2). A slight grain shape fabric in the halite, and the orientation of both the halite grains and any entrained spicules of insoluble materials define a foliation which is axial planar to folds exposed in the dome and both namakiers (Talbot 1979). Continual regeneration of this foliation ensures that it has only a short memory of its strain history (Talbot 1979). Increasing in intensity down the lengths of the namakiers, this flow foliation is interpreted as recording the stream lines of the salt mass (Talbot 1981).

The surface relief of the extrusive salt dome at Kuh-e-Namak is minor (~40 m) compared to its overall dimensions (6000 × 2000 × 1000 m). The salt emerging in the upper parts of the salt dome has incorporated within it mature (high amplitude/wavelength ratio) major folds which presumably generated at depth (Fig. 2a). The axial planes of these folds, and the foliation parallel to them, tend to be subvertical with a SSW–NNE strike along the SSW–NNE midline of the dome where the axes are doubly plunging and parallel the top free surface of the salt (and the shallow colour bands indicated on Fig. 2b). The axial surfaces splay outwards in the SSW–NNE trending slopes of the dome, curl through the horizontal just short of where any salt flows over the country rock, and thereafter dip gently downslope (Fig. 2a).

The colour bands inferred beneath the surface on Fig. 2 are probably grossly oversimplified in terms of the complications in known diapirs (Richter-Bernberg 1980) and ignore the possibility of thermal convection within the diapir (Talbot 1980).

In a vertical cross-section along the length of the topographic salt dome (Fig. 2b) the flow foliation, the colour bands and the fold axes broadly parallel the top free surface of the dome and the deeper stream lines can be inferred to spread beneath the summit. The emergent salt has been less confined in this section than in Fig. 2(a) and feeds two dip-slope namakiers.

Although the folds inherited from depth must extend into the upper reaches of both namakiers, none were recognized as such. Either any such folds were so isoclinal to be inconspicuous or the transition occurs beneath the surface (Fig. 2b). The colour bands and the foliation are close to being parallel throughout most of the exposed salt mass. Only where the porphyritic salt encounters abrupt changes in boundary conditions do the colour bands act as passive markers and fold around the stream lines indicated by the foliation (Talbot 1979).

The boundaries of the salt mass fall into two categories: those limited by rigid country rocks and those bounded by free air. The stream lines (flow foliations) of most previously described salt bodies are truncated upwards by an erosion surface. This is not the case at Kuh-e-Namak where Fig. 2 illustrates that the stream lines tend to parallel not only the rigid country rock boundaries but also the free boundaries of most of all the extrusive salt.

ACTIVE FOLDING OF THE COUNTRY ROCKS AND THE AGE OF THE NAMAKIERS

Remnants of an unsorted clastic deposit without any noticeable Hormuz components lie on the south namakier up to a height of about 300 m above sea level. A similar deposit lies up to about 365 m above sea level on the north namakier where it is dammed by a ridge of Asmari limestone. These deposits, only a few metres thick, are unbedded and consist of unsorted angular to sub-rounded clasts, up to a few centimetres in diameter and of at least three different types of fossiliferous limestone, supported by a fine grained calcareous matrix. These sediments contrast with the remainder of the insoluble material on the salt which are recognizably Hormuz in derivation. The exotic sediments are interpreted as the remnants of a formerly continuous cover of marine sediments which were deposited on the lower portions of both namakiers when they were below sea level and then jumbled as they settled because of the loss by solution of some of the underlying halite. It is noticeable that the remnants only survive on what are interpreted as inactive parts of the extrusive Hormuz salt.

Kent (1979) refers to a Bakhtyari conglomerate free of Hormuz debris on top of part of the Kamarj North salt plug 160 km north of Kuh-e-Namak (Dashti). He considered (1979, p. 138) that such conglomerates were deposited on a modern erosion plane cut across a dead intrusion emplaced before the Bakhtyari, probably in Miocene–Pliocene times. This may be so for the Kamarj North plug, but the present writers consider the exotic deposits on both namakiers at Kuh-e-Namak to be younger; indeed, they will be interpreted as contemporaneous with a sub-recent depositional surface recognizable over large parts of the area surrounding Kuh-e-Namak.

Only a tiny strip of this surface survives (up to a height of about 400 m above sea level) behind a protecting ridge of Asmari limestone east of the snout of the north namakier (see insert on Fig. 2b). What is assumed to be the same surface on the south side of Kuh-e-Namak towers above the distal feather edge of the south namakier and is bypassed by rivers currently draining off the salt. This surface extrapolates up to a height of close to 400 m above sea level on both namakiers where it is represented by the exotic deposits but is interrupted by the extrusive salt dome. However, the same surface can be recognized with few disturbances southwards from Kuh-e-Namak across a gentle syncline in the topography

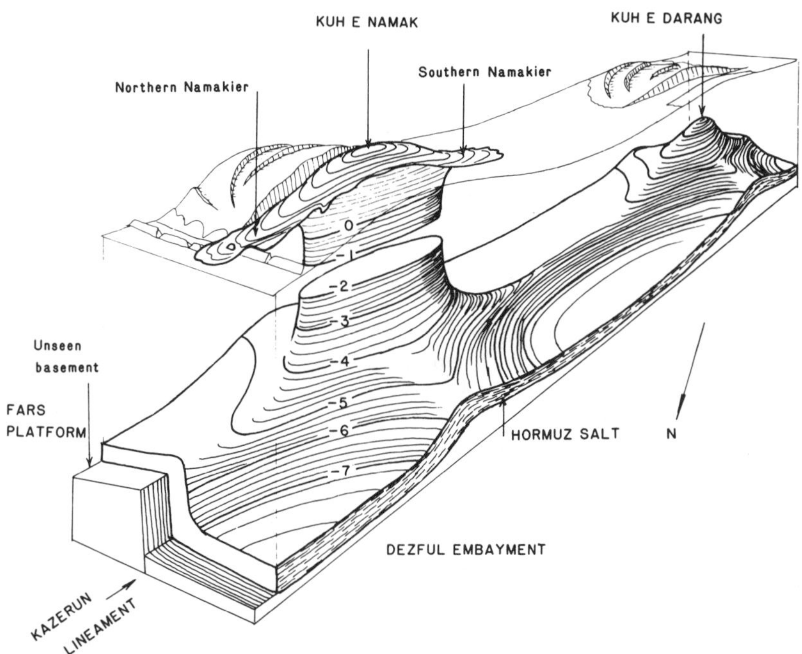

Fig. 3. Three-dimensional sketch of the salt bodies at Kuh-e-Namak and Kuh-e-Darang rising from the same salt pillow which developed at the southern end of the Kazerun lineament in Jurassic times. The recent surface mentioned in the caption to Fig. 2(b) and text can be seen to define an anticline through the rift across Kuh-e-Darang and a syncline between Kuh- e-Namak and Kuh-e-Darang.

to a smooth anticline which can be traced along a central rift through Kuh-e-Darang where it rises to 395 m above sea level (Fig. 3).

Vita-Finzi (1979) has dated the rise of similar recent surfaces and demonstrated that the next anticline inland of that through which Kuh-e-Namak rises is shortening between 14 and 18 mm a^{-1} where it meets the Gulf coast west of Banda Abbas. The rate at which this anticline has risen over the last 7000 years is spasmodic but averages between 1.4 and 1.9 mm a^{-1} using the eustatic factor of Flint (Vita-Finzi 1979). This is not as fast as the rate of rise of 7.4 mm a^{-1} calculated for folds east of Banda Abbas (Vita-Finzi 1979) or 12 mm a^{-1} for the Shaur anticline near the head of the Gulf (Lees & Falcon 1952). If the exotic deposits on the salt and what is interpreted as the contemporaneous depositional surface round about are taken as having formed near sea level, and if their maximum heights are attributed to uplift along Zagros anticlines at rates of between 1.4 and 12 mm a^{-1} then they are likely to be between 33,000 and 283,000 years old.

SALT BUDGET

Unlike a glacier, which is replenished with snow from above and loses ice by melting and ablation mainly in its lower reaches, a salt dome and any namakiers flowing from it are fed through a diapiric pipe from below and can lose salt by solution in rain falling on and draining off all exposed surfaces. On Kuh-e-Namak the topographic surface of the salt truncates former flow planes to an increasing degree from the dome down to the distal ends of both namakiers. However, minor refolding of the colour bands about axial surfaces parallel to the current free surface indicates that the direction of salt flowage can adapt to changes in its free boundaries which can only have been comparatively recent (see insert A on Fig. 2a).

We will approach the problem of the salt budget by attempting to calculate the volume of salt required to rise up the pipe to replenish the potential loss of salt by solution in the run-off of the rainfall. Both namakiers have been thicker and more extensive in the past than now but this analysis will assume that the current budget of supply and loss is approximately in balance.

All the extrusive salt is proud of its immediate surroundings so that all rain falling on it will be assumed to drain off it. The surface area of the exposed salt is huge because of the small-scale relief of the solution surface (pinnacles, fluting, etc.). However, because rainfall is measured in vertical cm we are interested in the area of salt in plan. The area of the salt dome (minus the flange along its ESE lip) is ~8.4 km^2. The area of the extrusive salt is about 23 km^2; that of the salt which is judged to be currently active about 20 km^2 (see later). The rainfall is confined to December and January and in 1945 was recorded as having a mean of 28 cm a^{-1} (Anon 1945).

At 25°C, 100 cc of pure water can dissolve 36 g of NaCl of density 2161 kg m^3. One cm of rain can therefore potentially dissolve 0.1667 cm of salt and the annual 28 cm rainfall on Kuh-e-Namak can dissolve a vertical thickness of 4.7 cm. The 100 m thick namakier could therefore potentially dissolve in 2150 years and the 1000 m high salt dome could disappear in 21,000 years if they were not replenished by more salt rising from below.

It will be assumed that the evaporation is low during the short rainy season and that all the falling rain runs off fully saturated with NaCl. The rain falling on the dome alone could then dissolve 39.2×10^{10} cm^3 of salt per year. To replace such a volume of salt requires a rise of salt of 4.6 cm a^{-1} averaged over the complete area of the orifice. However, such a calculation ignores the finding that a large volume of salt leaves the dome by flowing into the namakiers. It is therefore the area of all the active salt extrusives which is relevant.

The rain falling on all the exposed salt judged to be still flowing is $20 \times 10^{10} \times 28 = 560 \times 10^{10}$ cm^3. At 25°C this could dissolve approximately

$$\frac{560 \times 10^{10} \times 36}{2.161 \times 100} = 93.3 \times 10^{10} \text{ cm}^3 \text{ of NaCl.}$$

To replace this volume, the salt needs to rise out of the 8.4 km^2 (8.4×10^{10} cm^2) orifice at a rate of just over 11 cm a^{-1}.

This figure is an average for the complete area of the orifice and can be expected to be higher than average in the centre of the orifice and rather less at the margins. This figure is also likely to be a maximum for, although some rain undoubtedly evaporates while still on the salt, and, like the insoluble halite debris, some of the halite may be carried off as a solid load, it is unlikely that all of the actual run-off carries its full potential load of NaCl.

Even a rate of rise of 11 cm a^{-1} would mean that only 0.02% of the area of the extrusive salt dome in Fig. 2(b) is actually passing through the dome each year. Another way in which such a rate of rise can be put in perspective is to consider that a rainfall of 67 cm a^{-1} could dissolve the salt as fast as it rises (by this calculation). This implies that an emergent salt body as active as Kuh-e-Namak would be extrusive in most areas of the world.

THE DYNAMICS OF THE EXTRUSIVE SALT DOME

Another, independent, approach to the problem of how fast the salt extrudes from the orifice in the bedrock beneath Kuh-e-Namak is to consider the topography in terms of the fluid dynamics of the salt as a viscous fluid. If the extrusive column of salt is driven only by the weight of the more dense cover rocks resting on the source layer (e.g. Lees 1931, see also Gera 1972), then the height of the salt column (from source to summit) multiplied by its density (2200 kg m^{-3}) should equal the average density of the country rocks (2500–2600 kg m^{-3}) multiplied by their thickness.

Mapping, gravity, seismic and drilling surveys over the last few decades suggest that a stratigraphic thickness of at least 9 km of Phanerozoic sediments overlies the Hormuz salt in the Zagros fold belt as a whole and that the present cover to the salt round about Kuh-e-Namak is about 5000 m on the Fars platform and about 7000 m in the Dezful embayment (Kashfi written pers. comm. 1981). A salt column supported by a cover 5500 m thick could theoretically rise 2200 m above the surrounding plains while a cover 7000 m thick could extrude a salt column 2800 m high. Such calculations of the potential height of the summit are almost certainly underestimates because the still active folding of the country rocks is also likely to be pressurizing the salt and, in effect, driving it upwards.

In fact the summit of Kuh-e-Namak is just over 1600 m above sea level and about 1400 m above the surrounding plains. However, the splaying of the stream lines visible in the top free surface (which can be inferred from Fig. 2) demonstrates that the salt column at Kuh-e-Namak cannot support (and is spreading under) its own weight. The height of the dome is therefore not a simple measure of the forces expelling the salt from its vent. Nevertheless, we will use the height of the dome to approximate the rate of extrusion of the salt beneath it.

As Figs. 2 and 3 demonstrate that the vast majority of the salt leaves the dome in the NNW–SSE section (Fig. 2b), the extrusion dynamics will be treated in two dimensions. Ignoring, at this stage, the loss of salt at the surface by solution, the top free surface of the salt dome will be considered to be in a steady state and to owe its shape to the gravitational spread over a horizontal rigid surface of a viscous fluid rising out of a line source (aligned perpendicular to Fig. 2(b), see Fig. 4). We will neglect the inertial forces because the viscosity of salt is so high. The stream function ψ of a viscous fluid extruding vertically out of a narrow orifice in xy space has the form

$$\psi = \left(\frac{m}{\pi}\right)\left(\tan^{-1}\frac{y}{x} + \frac{xy}{(x^2+y^2)} - \frac{16}{3}\frac{x^2y}{a^3} + \frac{10}{3}\frac{x^3y}{a^4}\right) \quad (1)$$

where m is the mass/unit time/unit distance and a is the maximum height the fluid reaches above its orifice (i.e. x at $y = 0$).

At the top free surface $\psi = 0$, in which case

$$\tan^{-1} y/x + xy(x^2+y^2) = \frac{16}{3}\frac{x^2y}{a^3} - \frac{10}{3}\frac{x^3y}{a^4} \quad (2)$$

and substituting

$$x = \Gamma\cos\theta, \qquad y = \Gamma\sin\theta \text{ in (2)}$$

we find

$$\theta + \sin\theta\cos\theta = \frac{16}{3}\frac{x^3}{a^3}\cdot\tan\theta - \frac{10}{3}\frac{x^4}{a^4}\tan\theta$$

and therefore

$$\frac{x^3}{a^3} - \frac{5}{8}\frac{x^4}{a^4} = \frac{3}{16}\cos\theta\left(\frac{\theta}{\sin\theta} + \cos\theta\right). \quad (3)$$

The left hand side of expression (3) $= f(\theta)$, is independent of the viscosity, and has been solved by iteration to yield Fig. 4(a). This is the non-dimensional shape of the top free surface of a viscous fluid extruding onto a

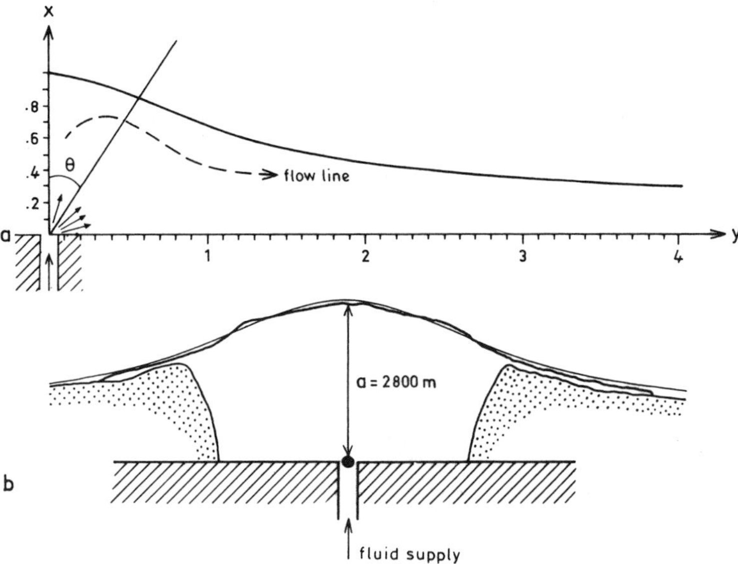

Fig. 4. (a) Shape of the top free surface (where the stream function, $\psi = 0$) of a viscous fluid extending out of a narrow linear orifice (an xz crack perpendicular to page) onto a plane yz (horizontal) surface. (b) The top free surface of the salt in longitudinal profile (Fig. 2b) scaled to fit the curve of (a) reflected about $y = 0$. Notice that 'a', the maximum height the free surface reaches above its line source (normal to the page and indicated by black dot) is 2800 m.

horizontal plane rigid surface through a narrow orifice.

Figure 4(b) illustrates a dimensionalized version of this curve superimposed on the longitudinal profile of Kuh-e-Namak and shows that the fit is good if the line source is considered to be 2800 m below the summit. This is the height 'a' expected if the salt column extruded because of the weight of the overburden acting on the source layer at depth and if it did not spread under its own weight. However, this fit is coincidental, particularly over the lengths of the namakiers, because the approach we have used is only designed to fit the top surface of the fluid extrusion closely near $y = 0$. The orifice at Kuh-e-Namak has finite width and the rigid boundaries over which the salt extrudes are neither plane nor horizontal. More important, Fig. 4(a) takes no account of the very substantial loss of the extruded salt in solution. Thus Fig. 4(a) shows significant thicknesses of viscous fluid large distances from the orifice along y whereas neither of the namakiers extend beyond 6 km from $y = 0$. However, the correspondence between the theoretical curve and the topography of Kuh-e-Namak emphasizes what the geology already suggests that, despite the considerable loss of salt by solution, the flow lines adapt to changes in the boundary conditions on a short time scale.

We know too little about the rate of salt flow in the dome to attempt to adapt the shape of Fig. 4(a) to take account the loss of 4.7 cm thickness of salt per year. Instead we notice by inspection of Fig. 2(b) that the geology suggests that 'a' is in fact about 1000 m.

In the appendix it is argued that, where $y = 0$, the mass flow/unit time/unit distance of a viscous fluid extruding from a narrow orifice, has the relationship

$$m = 3\pi\rho g a^3/112\mu$$

where ρ is the density, g the acceleration due to gravity and μ the viscosity. Substituting in (4) the following realistic values $\rho = 2200$ kg m^{-3}, $g = 9.81$ m s^{-2}, $a = 1000$ m and $\mu = 2.6 \times 10^{13}$ m^2 s^{-1} ($=2.6 \times 10^{17}$ poise for dry rock salt at NTP from Griggs 1929) we find

$$m \sim 0.07 \text{ kg s}^{-1}\text{ m}^{-1}.$$

We approximate the rate of extrusion $\dot{E}$ through an orifice 6000 m wide by $\dot{E} = m \times$ seconds in a year (3.16×10^7)/width of orifice $\times$ density $\simeq 17$ cm a^{-1}.

As in the earlier calculation for the rate of salt extrusion required to replace the potential loss of NaCl in solution, so the rate of extrusion calculated as necessary to account for the maximum height the salt reaches above its rigid orifice is averaged over the complete elliptical area of the orifice. Notice that our use of a narrow orifice is likely to be an underestimate for what is in fact extrusion through an orifice of significant width.

We have already estimated that, of the annual average of 17 cm of salt extruded through the orifice, 11 cm can be dissolved from its surface by the annual rainfall. If we take the value of $a = 2800$ m suggested by the best fit of the topography of the salt dome with our model of an extruding viscous fluid and remove from it 11/17 of its height,

$$2800 - \left(\frac{2800 \times 11}{17}\right) = 2800 - 1800 = 1000 \text{ m}$$

we find the actual value of 'a' suggested by the geology.

By taking note of the effect the solution of salt has on the maximum height that the salt reaches above its orifice in this simple manner we find that our budget of salt supply and loss is balanced.

THE SOUTHERN NAMAKIER AT KUH-E-NAMAK

About 2000 m long and 3500 m wide, the south namakier may exceed a thickness of 50 m in places. This sheet consists of almost pure white and pink halite and its former extent is marked by soil-free country rocks rather than the lateral and terminal moraines of the north namakier. Kent (1979) considers that this namakier may still be active and he may well be correct for salt periodically avalanches from a cliff backed by successive curling bergschrunds in the SE corner of the topographic dome. (This was the only area of routine avalanches encountered; elsewhere the salt movement was predominantly by dynamically recrystallizing flow.) However, the eroded relief of the south namakier is of the same order as its thickness, its distal margin has a feathered edge and five or six significant windows of country rock are visible in it (see Fig. 1); it is therefore considered relatively inactive compared to the north namakier.

THE NORTHERN NAMAKIER AT KUH-E-NAMAK

Down the length of both namakiers the fabric and texture of the salt undergo episodic but cumulative changes reflecting changes in the mechanisms of deformation at successive changes in the rigid boundary conditions (Talbot 1981). The sequence of changes is now abbreviated in the south namakier but is well represented down the length of the more active north namakier. This namakier is now 3000 m long, 2000–2500 m wide and different parts of the same ridge of Asmari limestone form a window in the lower part of the main salt stream and a dam to a triangular area of almost static salt on the western margin (Fig. 5 & Talbot 1979). This ridge is one of perhaps 15 buried scarp and dip slopes encountered by the salt as it flows down to the plains only 30–40 m above sea level.

Rather than eroding its bedrock channel the north namakier tends to partially fill in and smooth out irregularities in its bottom and side boundaries and shear over or past such static infills by sliding on the face-parallel plane of halite grains (Talbot 1981). Wherever the salt mass decelerates and thickens to surmount bedrock obstructions, the stream lines marked by the flow foliation diverge and become axial planar to trains of asymmetric folds which progressively tighten to flow up and over the obstruction (Talbot 1979). Each train of folds dramatically thins and repeats several times the colour bands inherited from upstream. Such folds are only obvious where they are generating for, in between obstructions, they are isoclinal and become very inconspicuous. Each bedrock obstruction is of course stationary and the moving salt picks up and then carries with it internal slides and folds generated at the obstruction.

All the halite of the dome is coarsely crystalline and clear and transparent. Diffusion processes, which probably characterize the salt flow at depth, appear to persist into the dome. However, dynamic recrystallization of tiny sub-grains around the grain boundaries within each zone of folding (or refolding) increases the proportion of fine grained translucent salt at the expense of porphyroclasts of transparent halite surviving from the dome. More and more of the salt assumes a mylonitic texture down the lengths of both namakiers until perhaps only 0.1% of the salt in the 30 m high snout of the north namakier, and perhaps 10% of the distal feather edge of the south namakier consist of clear porphyroclasts greater than 1 cm across.

As well as the internal banding being thinned and repeated, the colours of the namakier also intensify at every obstruction. This is partly due to the conversion of transparent coarse-grained halite to translucent fine-grained halite and partly due to the comminution and dispersion of insoluble entrained minerals. Progressive solution of halite down the length of the namakiers occurs but is considered to be a largely superficial phenomenon (resulting in surface moraine) rather than an internal process.

RECOGNITION OF BURIED NAMAKIERS

Buried namakiers are more likely to be associated with salt diapirs which have surfaced in arid climates (or saline seas) rather than those in temperate or tropical climates or less saline bodies of water. Where a namakier surmounts an obstacle in its path the colour bands and any grain shape or orientation foliation in it may be steep; however, along most of its length such features are likely to dip at low angles away from a steep feeder with a steep grain shape fabric.

The colour bands and any foliation may also be sub-horizontal in a deeper source layer but a formerly extrusive namakier may be distinguishable from its source layer by the following differences:

(1) stratigraphic and structural levels,

(2) the colours are likely to be more intense in thinner bands,

(3) a higher proportion of insoluble components which are finer grained and more dispersed and

(4) the smaller grain size and porphyroclastic texture can be expected to survive annealing until quite deep burial.

Fossil namakiers may not spread from the top of a salt plug as the salt in the pipe (and even parts of the namakier) may rise further during subsequent dome-building. Spreads of insoluble Hormuz material deep in the stratigraphic succession close to salt diapirs in Iran could indicate the moraines of former (but dissolved) namakiers if they are not too reworked by superimposed fluvial or marine processes.

THE DYNAMICS OF A NAMAKIER

Measurements were made of movements along two lines of markers on the east flank of the north namakier

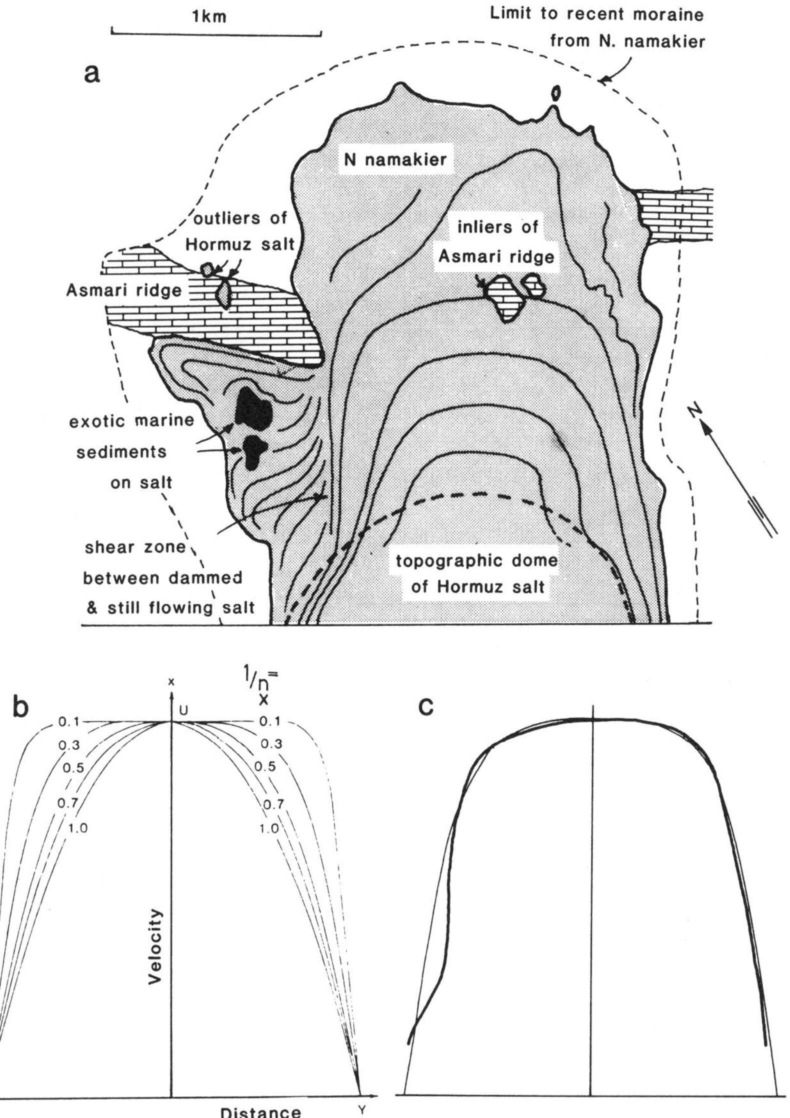

Fig. 5. (a) The outcrop of the colour bands on the top free surface of the N namakier are interpreted as recording the flow profile of a fluid flowing between two rigid walls upstream of the ridge of Asmari limestone. (b) Velocity profiles of Newtonian ($n = 1$) and strain-thickening fluids in a channel of width Y, U is the maximum velocity in the x direction. Labels on curves are $1/n$. (c) Superimposition of the colour band crossing the inliers of the Asmari ridge from (a) with the strain-thickening fluid profile with $n = 3$ from (b) suggesting that mylonitized rock salt flows over the surface with $n \sim 3$.

(see Fig. 1). When the extrusive salt was dry on the surface (as it is most of the year) the markers moved both up and downstream a few cms in a daily cycle with small movements occurring in immediate response to small temperature changes hundreds of metres away (Talbot & Rogers 1980). During most of the year the dry salt in the lower portions of the north namakier deforms as an elastic solid. However, within a day of a significant rain shower the salt softened and no longer transmitted thermal forces elastically. Instead some of the markers moved downstream at a rate of 0.5 m day^{-1}. The strain along both lines of markers was differential and persisted longer where a stream flowed over the surface. As the surface dried, the namakier shrank for four days and recovered parts of its advance. There can be little doubt that the wet salt flowed with an effective viscosity of about 10^{12} poises ($\sim 10^8$ m^2 s^{-1}) with a basal shear stress close to 2.5 kg cm^{-2} for about 5 days of the monitoring session (Talbot & Rogers 1980). If a large portion of the north namakier only flows when it is sufficiently damp then such flow is episodic and only likely for a few weeks each year.

A minimum estimate of the long term rate of flow of the north namakier is provided by the absence from the main stream of what is supposed to have been a continuous cover of marine sediments deposited an estimated 283,000–33,000 years ago. The highest remnants of this former cover survive 1730 m behind the present snout (Figs. 1 and 5a). The main stream needs only to have advanced at an average rate of between 6 mm a^{-1} and

5 cm a^{-1} to shed entirely any former superimposed marine cover.

Figures estimated earlier suggest a more likely, but very approximate, value for the average annual rate of flow of at least the upper reaches of the north namakier. Of the ~17 cm a^{-1} estimated to extrude out of the country rock orifice, ~5 cm a^{-1} are dissolved from the surface of the dome itself. This leaves ~12 cm a^{-1} to flow into the heads of two namakiers. As the flow of the north namakier is judged to be considerably more vigorous than that to the south, we will rather arbitrarily assume that ~10 cm a^{-1} of the extruded salt enters the upper reaches of the north namakier. As these have a cross-sectional area of ~2.2×10^5 m^2, the average rate of flow of the salt entering the top of the north namakier is something like 2 m a^{-1}.

The scenario of a slow (continuous?) input of salt into the head of the north namakier with brief episodes of much faster flow punctuating long stationary periods further down its length can be reconciled by various lines of evidence.

Firstly, the bulk material properties of the salt mass can be inferred to change continually from the deep source, up the pipe, through the dome and in each of several fold generation zones down the length of the namakier to its snout. Any potential slowing of flow by the dispersion of entrained impurities (see e.g. Verrall *et al.* 1975, p. 11) must be outweighed by the decrease in the average grain size of the halite. This is illustrated by the change of deformation mechanisms and structures down the length of the namakier (Talbot 1981). The internal slides and then a crenulation fail to develop upstream of obstructions of much the same size. Eventually all internal folding ceases to develop and the (superplastic?) salt mass flows over the last few bedrock ridges without internal folding—a situation attributed to a lack of a significant proportion of halite porphyroclasts in the salt near the snout (Talbot 1979). These halite porphyroclasts demonstrate that the grain boundary diffusion in the dome is largely replaced in the namakier by dynamic recrystallization together with tensional separation and sliding along the face parallel planes of the halite grains.

The softening of halite by water has been known since 1908 as the Joffée effect (Odé 1968). The theory assumes that halite is inherently ductile but, by absorbing gases (e.g. ozone) from the atmosphere, becomes 'case hardened' so that intercrystalline flaws driven by strain cannot reach the poisoned surface unless they are wet. The Joffée effect was originally described in single crystals but also appears to hold for the north namakier at Kuh-e-Namak. Wenkert (1979), in the theoretical approach which so far comes closest to accounting for the behaviour of a namakier, attributes the flow of wet polycrystalline halite to the intercrystalline diffusion of ions down chemical gradients through an interstitial brine. However, a major difficulty remaining is to visualize how an elastic salt sheet 50–100 m thick can start flowing within a day (if not a few hours) of a few mm of rain falling on it (Talbot & Rogers 1980). Brittle fractures exist in the salt but are tightly closed during the day. Even though they open when the temperature drops, we have to imagine a few mm of rain softening a relatively large volume of salt surprisingly rapidly.

The outcrop of the colour and dirt bands on the top surface of the south namakier (Fig. 1) reflects its current relief rather than a flow pattern. The top free surface has been dissolved progressively deeper into the stream lines down the length of the north namakier but the arcuate outcrop of the bands can be treated in a manner analogous to studies of glaciers (Nye 1965, Paterson 1969).

The thickness of the north namakier is variable and unknown in detail but is probably between 50 and 100 m over most of its area. The half-width of the main salt stream is 1000 m. The ratio W (half-width divided by the thickness) is about 10–20, far higher than W for most valley glaciers which ranges between 1 and 4. Compared to most glaciers therefore the north namakier is almost infinitely wide although the current active salt sheet flow is confined laterally by parallel rigid walls (Fig. 1). Unlike valley glaciers in which the surface flow patterns are complicated by variations in thickness and the influence of the bottom boundary across their widths, the influence of the bottom boundary of the north namakier is expected to be more or less uniform across its width. The surface flow pattern is therefore only likely to be controlled significantly and consistently by the rigid sides.

In effect this means that the flow profile on the top surface of the namakier can be treated as though the salt were flowing between two walls. Figure 5(b) illustrates the flow profiles of fluids which obey different power laws flowing in such a channel (Coward 1980). For Newtonian fluids $n = 1$, $n > 1$ for dilatant or strain-thickening fluids like ice (or salt) and $n < 1$ for pseudo-plastic fluids. A comparison between these curves and the shape of the colour bands cropping out on the surface of the north namakier (Fig. 5c) suggests that n for impure salt flowing over the surface is close to 3. This is at the lower end of the range determined experimentally for dry rock salt where $3 < n < 12$ (Tsianbaos 1978).

SUMMARY

The Hormuz salt at Kuh-e-Namak is known to have started rising in the Jurassic period and to have reached the surface by Cretaceous times, long before the current Zagros folding began. Whether salt extrusion was continuous from Cretaceous times to the present is unknown. However, the survival on both namakiers of what are interpreted as parts of a marine cover inferred to be between 33,000 and 283,000 years old implies that namakiers can survive beneath a thin but relatively impermeable cover for such intervals. The distortion of a depositional clastic surface interpreted to be of the same age close to Kuh-e-Namak up to a height of 365 m above sea level over the crest of the rising salt body beneath Kuh-e-Darang indicates that the local Zagros folding is still vigorous.

A commonly accepted figure for the rate of rise of the top surface of salt diapirs averaged over geological periods is about 2–4 mm a^{-1} although local spines have been known to rise 8–15 mm a^{-1} for brief intervals (Gera 1972, Kupfer 1976). Gera (1972) recognized that salt is likely to rise fastest when it has just disrupted the surface for then the pressure differential on the source layer due to the weight of the overburden is at its greatest value. At Kuh-e-Namak we have the additional factor that the salt beneath the orifice is pressurized to an unknown degree by active folding of the Phanerozoic cover. If it were rigid the extrusive salt column could theoretically rise to 2800 m above the surrounding plains even if the only driving force were the weight of the cover rocks. However, anticlines near Kuh-e-Namak are rising at rates of between 1.4 and 1.2 mm a^{-1} and such tectonic forces are likely to extrude a salt column of even greater height. Nevertheless, despite such potential, the summit of Kuh-e-Namak is only about 1600 m above sea level, 1400 m above the surrounding plains and 1000 m above its rigid rock orifice. This is because the extruding salt is dissolved from its surface and because the salt column spreads under its own weight as it rises. The height and shape of the longitudinal profile of the extrusive salt dome has been shown to be close to that expected if the extrusive salt deforms with a Newtonian viscosity of 2.6×10^{17} poise and rises up the elliptical pipe with an average velocity of almost 17 cm a^{-1}. About 11 cm of this extruded salt could be dissolved each year.

These rates of extrusion and solution are independent order-of-magnitude approximations based on simple models applied assuming that the extrusive salt dynamics are now at a steady state. Nonetheless they combine to suggest that, at 17 cm a^{-1}, the salt beneath the summit of Kuh-e-Namak is currently extruding 42–85 times faster than the generally accepted figure for the rate of rise of salt diapirs averaged over geological periods.

The mobility of the Hormuz salt sequence may decrease as it rises from depth because of a fall in temperature; although this effect may be outweighed by strain softening and (or) shear heating in such rapidly moving salt (Gera 1972). The supply of salt to the base of the dome is likely to be continuous and the salt volume is likely to take a long time to adjust to any changes in the mass balance of the dome; the flow of the north namakier is likely to have a much shorter response time and at least part of it is known to be episodic on a time scale of days and weeks. Furthermore, the seasonal nature of the mobility of the namakier can be inferred to intensify down its length. Thus the strain softening visible as a decrease in grain size down its length appears to overcome any poisoning effect of the entrained insoluble minerals. Such strain softening accounts for the change in deformation structures, mechanisms of flow, and measured or inferred flow velocities from the salt dome down the north namakier. The Hormuz salt is inferred to rise continuously up the pipe as a Newtonian fluid at an average rate of about 17 cm a^{-1} and to flow into the head of the north namakier at about 2 m a^{-1}, and yet temporary flows at 0.5 m day^{-1} have been measured along parts of its margins. Such a velocity, which probably implies flows at over a metre per day along the centre line of the namakier, raises the problem of how a substantial thickness of dirty mylonitized halite ($n \sim 3$) can be softened by small volumes of water in a matter of days if not hours.

Extrusive salt can be softened by rain or sea water and buried salt can be softened by groundwater or by water released by dehydration reactions (e.g. gypsum → anhydrite). The height of the summit of Kuh-e-Namak could be constant despite the throughput of salt beneath; just as the movements of a table tennis ball supported on a water fountain are no measure of the velocity of the water jet beneath. The buried top surface of pipe-like or tabular bodies of salt may not move much whereas the underlying salt could be convecting vigorously (Talbot 1980, Talbot *et al.* 1982).

A general understanding of salt deformation becomes increasingly significant as salt rock engineering grows more sophisticated. Not only are mined and solution cavities in salt becoming larger and deeper but their uses are multiplying. Engineering on, in or near salt bodies which are either as active as Kuh-e-Namak, or could become as active in the life of the engineered facility, could have difficulties if the potential dynamics of the salt mass are not fully appreciated.

Acknowledgements—Considerable thanks are due to Ernie Rutter, Stan Murrel, Rick Sibson, Dick Templeton, Bob Stonely, Eric Rogers, R. Player and Sir Peter Kent for useful comments, corrections and discussions at various stages in this work.

Invaluable logistic support in the field from the Iranian Geological Survey and the Oil Service Consortium of Iran is gratefully acknowledged together with financial support from the Royal Society and the Carnegie Trust.

REFERENCES

Ala, M. A. 1974. Salt diapirism in southern Iran. *Bull. Am. Ass. Petrol. Geol.* **58**, 1758–1770.

Amery, G. B. 1969, Structure of Sigsbee Scarp, Gulf of Mexico. *Bull. Am. Ass. Petrol. Geol.* **53**, 2480–2482.

Anon, 1945. *Geographical Handbook.* Persia BR525, London.

Arieli, A., Heard, H. C. & Mukherjee, A. K. 1980. Deformation modelling in NaCl at intermediate and elevated temperatures. UCRL offprint 85066, Lawrence Livermore Laboratory.

Barr, C. A. 1977. *Applied Salt-Rock mechanisms—I. The in-situ Behaviour of Salt Rocks.* Elsevier, Amsterdam, 294.

Coward, M. P. 1980. The analysis of flow profiles in a basaltic dyke using strained vesicles. *J. geol. Soc. Lond.* **137**, 605–615.

de Böckh, H., Lees, G. M. & Richardson, F. D. S. 1929. Contribution to the stratigraphy and tectonics of the Iranian ranges. In: *The Structure of Asia* (edited by Gregory, J. W.). Methuen, London.

Fürst, M. 1976. Tektonic und Diapirismus der Östlichen Zagrosketten. *Z. dt. geol. Ges.* **127**, 183–225.

Gansser, A. 1960. Über Schlammvulkane und salzdome. *Vjschr. Naturf. Ges. Zürich* **105**, 1–46.

Gera, F. 1972. Review of salt tectonics in relation to disposal of radioactive wastes in salt formations. *Bull. geol. Soc. Am.* **83**, 3551–3574.

Griggs, D. 1929. Creep of rocks. *J. Geol.* **47**, 225–251.

Gussow, W. C. 1966. Salt temperature: a fundamental factor in salt dome intrusion. *Nature, Lond.* **210**, 518–519.

Gussow, W. C. 1968. Salt diapirism: importance of temperature and energy source of emplacement. In: *Diapirism & Diapirs* (edited by Braunstein, J. & O'Brien, A.) *Mem. Am. Ass. Petrol. Geol.* **8**, 16–52.

Harrison, J. V. 1930. The geology of some salt-plugs in Laristan (southern Persia). *Q. Jl. geol. Soc. Lond.* **86**, 463–522.

Harrison, J. V. 1956. Comments on salt domes. *Verh. K. Ned. geol-mijnb. Genoot.* **14**, 1–8.

Kent, P. E. 1958. Recent studies of south Persian salt plugs. *Bull. Am. Ass. Petrol. Geol.* **42**, 2951–2973.

Kent, P. E. 1966. Temperature conditions of salt dome intrusions. *Nature, Lond.* **211**, 1387.

Kent, P. E. 1970. The salt plugs of the Persian gulf region. *Trans. Leicester lit. phil. Soc.* **441**, 56–88.

Kent, P. E. 1979. The emergent Hormuz salt plugs of southern Iran. *J. Petrol. Geol.* **2**, 117–144.

Kupfer, D. H. 1976. Shear zones inside gulf coast salt stocks help to delineate spines of movement. *Bull. Am. Ass. Petrol. Geol.* **60**, 1434–1447.

Lees, G. M. 1927. Salzgletscher in Persien. *Mitt. geol. Ges. Wien* **22**, 29–34.

Lees, G. M. 1931. Salt-dome depositional and deformation problems. *J. Instn Petrol. Technol.* **17**, 259–280.

Lees, G. M. & Falcon, N. L. 1952. The geographic history of the Mesopotamian plains. *Geogrl Jl* **118**, 24–39.

Nye, J. F. 1965. The flow of a glacier in a channel of rectangular, elliptical or parabolic cross-section. *J. Glaciol.* **5**, 661–690.

O'Brien, C. A. E. 1957. Salt diapirism in south Persia. *Geologie Mijnb.* **19**, 357–376.

Odé, H. 1968. Review of mechanical properties of salt relating to salt dome genesis in saline deposits. *Spec. Pap. geol. Soc. Am.* **88**, 543–596.

Paterson, W. S. B. 1969. *The Physics of Glaciers.* Pergamon Press, Oxford.

Richter-Bernberg, G. 1980. Salt tectonics—Interior structures of salt bodies. *Bull. Cent. Rech. Exploration–Production: Elf Aquitaine* **4**, 373–393.

Talbot, C. J. 1977. Inclined and upward moving gravity structures. *Tectonophysics* **42**, 159–181.

Talbot, C. J. 1979. Fold trains in a glacier of salt in southern Iran. *J. Struct. Geol.* **1**, 5–18.

Talbot, C. J. 1980. Halokinesis and thermal convection. *Nature, Lond.* **273**, 739–741.

Talbot, C. J. 1981. Sliding and other deformation mechanisms in a glacier of salt, S. Iran. In: *Thrust and Nappe Tectonics* (edited by McClay, K. R. & Price, N. J.). *Spec. Publs geol. Soc. Lond.* **9**, 173–183.

Talbot, C. J. & Rogers, E. A. 1980. Seasonal movements in a salt glacier in Iran. *Science, Wash.* **208**, 395–397.

Talbot, C. J., Tully, C. P. & Woods, P. J. E. 1982. The structural geology of Boulby (Potash) Mine, Cleveland, U.K. *Tectonophysics* **85**, 167–204.

Tsianbaos, G. 1978. Mechanical behaviour of rock salt under uniaxial compression. Unpublished M.Sc. thesis, Imperial College, University of London.

Verall, R. A., Fields, R. J. & Ashby, M. F. 1975. Deformation-mechanism maps for LiF and NaCl. Tech. Report 1. Div. Engineering Applied phys., Harvard University, U.S.A.

Vita-Finzi, C. 1979. Rates of Holocene folding in the coastal Zagros near Bandar Abbas, Iran. *Nature, Lond.* **278**, 632–633.

Wenkert, D. D. 1979. The flow of salt glaciers. *Geophys. Res. Lett.* **6**, 523–526.

APPENDIX

The flow of a viscous, incompressible fluid issuing from a long straight crack in a horizontal plane boundary

We choose axes $0xyz$ with $0x$ vertically upwards and such that the plane boundary is given by $x = 0$ and the crack by $x = y = 0$. Figures 4(a) & (b) in the main text illustrate the orientation of these axes in relation to the longitudinal profile of the salt mass.

We assume that the flow is that due to a line source of strength m per unit length along the crack, that it is two-dimensional, and that the effects of inertia are negligible. The vertical and horizontal components of velocity may then be found from $u = \partial\psi/\partial y$, $v = -\partial\psi/\partial x$, where the stream-function ψ satisfies Stokes's equations

$$\mu \frac{\partial}{\partial y}(\nabla^2\psi) = \frac{\partial p}{\partial x} + \rho g,$$

$$\mu \frac{\partial}{\partial x}(\nabla^2\psi) = -\frac{\partial p}{\partial y}.$$

Here $\nabla^2\psi = \partial^2\psi/\partial x^2 + \partial^2\psi/\partial y^2$, μ is the viscosity and ρ the density of the fluid, p is the excess of the pressure in the fluid above atmospheric pressure and g is the gravitational acceleration.

The boundary conditions to be satisfied are that $u = v = 0$ on $x = 0$ and that the upper surface of the fluid, which may be taken to be the streamline $\psi = 0$, should be stress free. This zero-stress condition leads to the equations

$$p \cos 2\theta = 2\mu \frac{\partial^2\psi}{\partial x\,\partial y},$$

$$p \sin 2\theta = \mu\left(\frac{\partial^2\psi}{\partial y^2} - \frac{\partial^2\psi}{\partial x^2}\right),$$

where θ is the downward inclination of the surface to the horizontal, to be satisfied on $\psi = 0$. In addition the symmetry of the flow requires that $\psi(x, -y) = -\psi(x, y)$.

It is convenient to rewrite the equations in non-dimensional form by setting $x = ax'$, $y = ay'$, $\psi = (m/\pi)\psi'$, $p = (\mu m/\pi a^2)p'$, where a is the maximum height of the free surface, which is attained on $y = 0$. Then on dropping the primes we obtain

$$\frac{\partial}{\partial y}(\nabla^2\psi) = \frac{\partial p}{\partial x} + \lambda,$$

$$\frac{\partial}{\partial x}(\nabla^2\psi) = -\frac{\partial p}{\partial y}, \tag{A1}$$

where $\lambda = \pi\rho g a^3/\mu m$, with the zero-stress conditions

$$p \cos 2\theta = 2\frac{\partial^2\psi}{\partial x\,\partial y}, \tag{A2}$$

$$p \sin 2\theta = \frac{\partial^2\psi}{\partial y^2} - \frac{\partial^2\psi}{\partial x^2}, \tag{A3}$$

to be satisfied on the free surface.

The general solution of equations (A1) which also satisfies the boundary conditions on $x = 0$ may be written

$$\psi = \tan^{-1}\frac{y}{x} + \frac{xy}{x^2 + y^2} + \sum_{n=0}^{\infty}\frac{(-1)^n(n+1)x^{2n+2}}{(2n+2)!}f^{(2n)}(y) + \sum_{n=0}^{\infty}\frac{(-1)^n(n+1)x^{2n+3}}{(2n+3)!}g^{(2n)}(y), \tag{A5}$$

$$p = k - \lambda x + \frac{2(x^2 - y^2)}{(x^2 + y^2)^2} + \sum_{n=0}^{\infty}\frac{(-1)^n x^{2n+1}}{(2n+1)!}f^{(2n+1)}(y) - G(y) + \sum_{n=0}^{\infty}\frac{(-1)^n x^{2n+2}}{(2n+2)!}g^{(2n+1)}(y), \tag{A6}$$

where f and g are arbitrary functions, $G(y) = \int_0^y g(z)\,dz$, $f^{(r)}$ denotes the rth derivative of f, and k is a constant.

The exact determination of the functions f and g is not possible due to the non-linearity of (2) and (3), but assuming the curvature of the free surface is small we may obtain an approximate solution as follows.

When the curvature of the free surface is small f and g may be approximated by $f(y) = Ay$, $g(y) = By$, the terms omitted being of order y^3, and the free surface may be approximated by $x = h(y) = 1 - \alpha y^2$. On expanding the condition $\psi(h(y), y) = 0$ in powers of y and equating to zero the coefficients of y and y^3 we obtain

$$\frac{1}{2}A + \frac{1}{6}B + 2 = 0,$$

$$\alpha A + \frac{1}{2}\alpha B - 2\alpha + \frac{4}{3} = 0,$$

whence

$$A = -16 + \frac{8}{3\alpha}, \quad B = 36 - \frac{8}{\alpha}.$$

Similarly expanding p, $\cos 2\theta$, and $\partial^2\psi/\partial x \partial y$ on $x = h(y)$ in powers of y and equating constants and coefficients of y^2 in (A2) we obtain

$$k - \lambda + 4 - \frac{4}{3\alpha} = -\frac{8}{3\alpha},$$

$$\alpha\lambda - 16\alpha - \frac{56}{3} + \frac{4}{\alpha} = \frac{104}{3}(1 - 2\alpha),$$

whence

$$\lambda = \frac{160}{3}\left(\frac{1}{\alpha} - 1\right) - \frac{4}{\alpha^2},$$

$$k = -\frac{172}{3} + \frac{156}{3\alpha} - \frac{4}{\alpha^2}.$$

Likewise equating coefficients of y in (3) we find $\alpha = \frac{1}{2}$.

Thus $A = -32/3$, $B = 20$, $\lambda = 112/3$ and $k = 92/3$.

It follows that the maximum height a and the mass flux m are related approximately by

$$\frac{\pi\rho g a^3}{\mu m} = \frac{112}{3}$$

which gives

$$m = \frac{3\pi\rho g a^3}{112\mu} = 0.084\rho g a^3/\mu.$$

BULLETIN OF THE AMERICAN ASSOCIATION OF PETROLEUM GEOLOGISTS
VOL. 50, NO. 1 (JANUARY, 1966), PP. 108-158, 36 FIGS.

OIL AND GEOLOGY IN CUANZA BASIN OF ANGOLA[1]

GEORGES P. BROGNON[2] AND GEORGES R. VERRIER[2]
Brussels, Belgium

ABSTRACT

The Cuanza basin is in northwestern Angola on the Atlantic Coast of West Africa. This basin is about 300 km. long north-south and 170 km. wide east-west, and contains an Early Cretaceous carbonate-evaporite sequence and a Late Cretaceous and Tertiary argillaceous-arenaceous sequence. The Precambrian crystalline basement is partly covered by extrusive rocks and granite-wash type sediments. Surface and subsurface sediments of the basin consist of Lower and Upper Cretaceous, Paleocene, Eocene, and Miocene strata.

Occurrences of oil and gas have been reported in almost all of the stratigraphic units in the Cuanza basin, and there is major production from the Cretaceous rocks. Study of these hydrocarbon occurrences and of the geological history of the basin shows that close relationships exist between sources, migration, and entrapment of oil, and environment of deposition controlled by the basement and salt tectonics.

During Early Cretaceous time, subsidence of the central part of a restricted basin determined the regional cyclical deposition of a carbonate-evaporite sequence providing a favorable situation for genesis and entrapment of oil. Thus, the deposition during Aptian time of a very fine crystalline limestone, interbedded with argillaceous limestone and overlain by an oölitic sandy calcarenite, itself underlying evaporites, had an important influence on the subsequent extent of oil accumulations in the Binga Formation. During Aptian-Albian time, differential subsidence on the western margin of the basin caused lateral interfingering of back-reef calcarenite, argillaceous carbonate, and evaporite. This interfingering is believed to be related closely to oil accumulations in this area. Very important vertical development of reef deposits in the Longa area is related to lateral migration of the underlying Massive Salt, which flowed with the help of the excess of weight introduced by the growing reef. On the eastern margin, upper Albian reef buildups capped by marine shale also provided a favorable situation for generation and accumulation of oil.

During Late Cretaceous and Tertiary time, a major basement flexure or fault zone appears to have been associated genetically downdip with deposits that accumulated with greater thickness than elsewhere. This flexure and the loci of maximum deposition moved eastward during Late Cretaceous and Paleocene, then westward during Eocene and Miocene. These thick formations, which are mainly argillaceous-arenaceous and which were deposited partly in deltaic and lagoonal environments, grade westward into thinner marine deposits and eastward into thinner continental deposits. During each particular epoch corresponding with a stabilization of this moving flexure, favorable conditions for genesis of hydrocarbons seem to be related to these transitional environments.

Oil production is located above the Massive Salt at the crest of salt anticlines, and one small oil field has been discovered below the Massive Salt along a ridge of the Basement Complex in a pinch-out of sandstones between Precambrian mica-schist below, and salt above.

INTRODUCTION

The Cuanza basin is a Cretaceous and Tertiary sedimentary basin which extends about 315 km. along the northwestern coast of Angola, on both sides of the mouth of the Cuanza River, between 8° and 10° S. Lat. The surface of this basin—22,000 sq. km. in area—is covered in its central part by Pleistocene sands, whereas, along the flanks, formations of Aptian to Miocene age are exposed (Fig. 1). This study is based on geophysical data (chiefly reflection-seismic) and subsurface information from wells (Fig. 2).

The sedimentary formations of the basin may have a total thickness greater than 4,000 m. (3,525 m. of section was drilled in the deepest well). The stratal section is divisible into three units of unequal economic importance, readily differentiated into mappable groups both on the surface and in the subsurface (Fig. 3).

1. The basal unit, a terrigenous detrital sequence of pre-Aptian to possibly early Aptian age, was deposited in continental-deltaic-lagoonal and marine littoral to sub-littoral environments.

2. The middle unit, a carbonate marine-lagoonal sequence with evaporites of Aptian-Albian age, was deposited during a marine transgression. At the base is a thick member of salt called the "Massive Salt." This grades westward into marine carbonate facies and eastward into continental arenaceous facies.

[1] Manuscript received, May 17, 1965. Publication authorized by Petrofina, Brussels.

[2] Petrofina, S. A. The writers are indebted to Sherman A. Wengerd, Department of Geology, University of New Mexico, for his critical reading of the manuscript.

3. The upper unit, consisting of marine formations ranging in age from Late Cretaceous to Miocene with a major gap during Oligocene time, grades eastward into transitional deltaic and lagoonal deposits.

In general, units 1 and 2 are separated by a widespread unconformity indicating emergent to subaerial conditions, but no such regional discontinuity exists between units 2 and 3.

Occurrences of oil and gas in all stratigraphic units in the Cuanza basin have been reported. The reservoirs range in age from pre-Aptian to Miocene (Fig. 4). Oil pools have been discovered above the Massive Salt at the crests of salt anticlines in the Binga and Cabo Ledo Formations, and below the Massive Salt in a pinchout of the Cuvo Formation. Gas and condensate have been found in volcanic ash beds lying between igneous rocks and the Massive Salt. Outcrops containing asphaltic material in the Rio Lifune area are part of the Binga Formation, and outcrops of the Cuvo Formation contain bedded "asphaltic coal."

This study outlines the relations between the occurrence of oil and the sedimentary environment, both in great part controlled by the architecture and tectonics of the Cuanza basin; therefore, the writers first discuss the tectonic framework which influenced the sedimentation.

Tectonic Control of Sedimentation

Architecture of floor of basin.—Initial field geological studies of the basin indicated the presence of anticlines and faults related to movements and deformations of the Massive Salt. This salt layer, ranging up to 600 m. in original thickness, is situated near the base of the stratigraphic sequence, beneath 1,500–3,000 m. of younger sediments. Geophysical results and drilling data permit, to some extent, reconstruction of the actual architecture of the infra-saliferous basin. This reconstruction is depicted by the structural map, contoured with the top of pre-Aptian strata as datum (Fig. 5). This map shows the boundaries of the following structural units.

1. A homocline along the eastern limit of the basin, with an average dip of 5°-6°, reaches to a depth of 3,000 m. or more below sea-level in the central part of the basin.

2. The floor of the basin itself reaches a depth of 3,600 m. (subsea) in the center near Galinda but might be as deep as 4,000 m. (subsea) or more in the graben of Quenguela.

3. The Cacuaco and Morro Liso ridges, respectively, on the northern and southern ends of the basin, interrupt the eastern homocline at a depth of about 2,500 m. (subsea).

4. The Cabo Ledo uplift, undoubtedly the most ancient feature of the pre-salt architecture, extends along the present coast for more than 100 km. above the subsea depth of 3,000 m. Its culmination at Tobias is at 2,083 m. (subsea); consequently, it is more than 1,200 m. above the basin floor which extends toward the east. The bottom of the basin is limited westward by this Cabo Ledo uplift and farther south by the Longa ridge, which also extends along the present coast.

The pre-Aptian paleogeological map (Fig. 6) shows the pre-Aptian age of these tectonic features, which doubtless were contemporaneous with the beginning of the subsidence which initiated the Cuanza basin.

These major structural features, which existed before Aptian time, continually influenced sedimentation during the successive basinal subsidences and continued more or less during the remainder of geologic time until the Oligocene.

Evolution of subsidence.—The zones of maximum subsidence and the hinge flexures which separate them from the less negative areas were at different positions, separated by large distances, between Aptian and Miocene time.

The evaporitic Albian-Aptian basin lies in a relatively western position, with its eastern limiting flexure being the Cacuaco and Morro Liso ridges (Fig. 7). Concomitantly there existed also a western Albian-Aptian flexure, which is called the Cabo Ledo uplift. This uplift is partly responsible for the isolation of the basin and for the development of evaporitic facies within the interior lagoonal basin. These evaporitic facies grade eastward into terrigenous detrital facies on the eastern shelf and westward into bioclastic carbonates on the Cabo Ledo uplift. This facies gradation is accompanied by a marked thinning of both the terrigenous clastic and bioclastic carbonate equivalents of the central evaporite facies.

During Cenomanian time, the eastern flexure was considerably farther east, encompassing the Cacuaco and Morro Liso ridges. The axis of maximum Cenomanian subsidence also was shifted

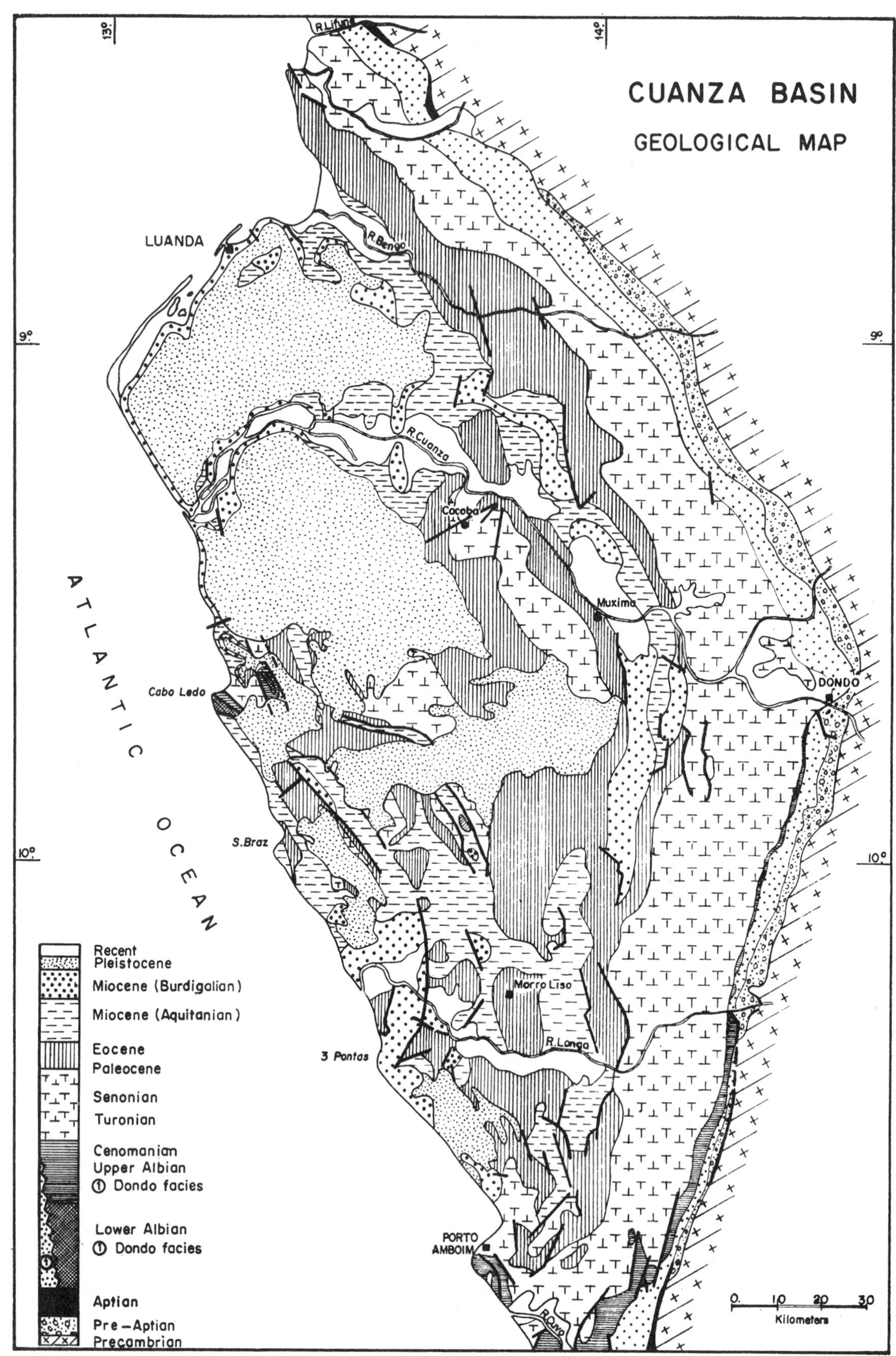

Fig. 1 – Geological map of the Cuanza basin.

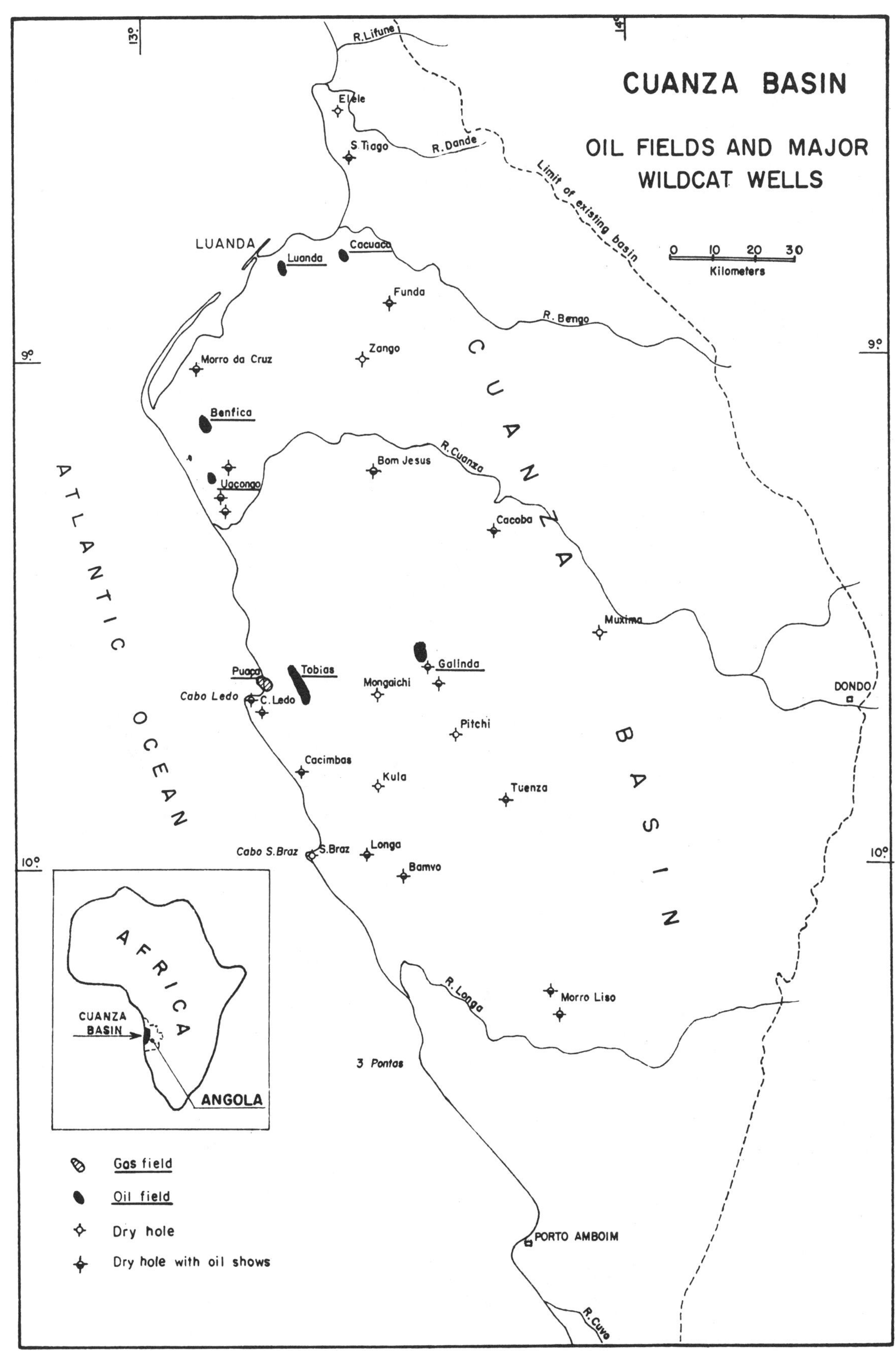

Fig. 2—Oil fields and wells of the Cuanza Basin.

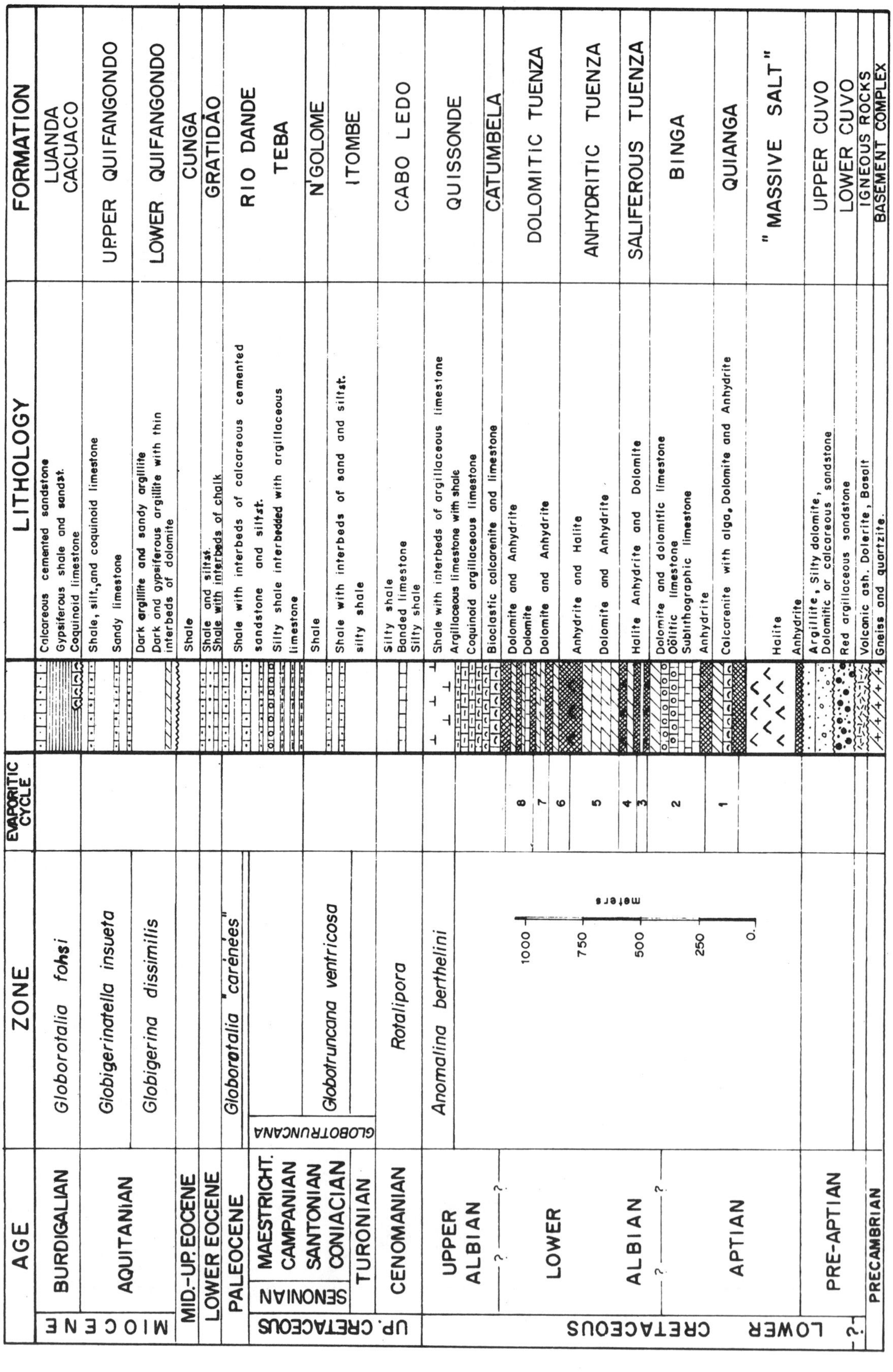

	AGE	ZONE	EVAPORITIC CYCLE	LITHOLOGY	FORMATION
MIOCENE	BURDIGALIAN	*Globorotalia fohsi*		Calcareous cemented sandstone Gypsiferous shale and sandst. Coquinoid limestone	LUANDA CACUACO
MIOCENE	AQUITANIAN	*Globigerinatella insueta*		Shale, silt., and coquinoid limestone Sandy limestone	UPPER QUIFANGONDO
MIOCENE	AQUITANIAN	*Globigerina dissimilis*		Dark argillite and sandy argillite Dark and gypsiferous argillite with thin interbeds of dolomite	LOWER QUIFANGONDO
	MID.-UP. EOCENE			Shale	CUNGA
	LOWER EOCENE			Shale and siltst. Shale with interbeds of chalk	GRATIDÃO
	PALEOCENE	*Globorotalia "carénées"*		Shale with interbeds of calcareous cemented sandstone and siltst.	RIO DANDE
UP. CRETACEOUS	SENONIAN: MAESTRICHT.	GLOBOTRUNCANA		Silty shale interbedded with argillaceous limestone	TEBA
UP. CRETACEOUS	SENONIAN: CAMPANIAN, SANTONIAN	GLOBOTRUNCANA: *Globotruncana ventricosa*		Shale	N'GOLOME
UP. CRETACEOUS	SENONIAN: CONIACIAN; TURONIAN	GLOBOTRUNCANA: *Globotruncana ventricosa*		Shale with interbeds of sand and siltst. silty shale	ITOMBE
UP. CRETACEOUS	CENOMANIAN	*Rotalipora*		Silty shale Banded limestone Silty shale	CABO LEDO
LOWER CRETACEOUS	UPPER ALBIAN	*Anomalina berthelini*		Shale with interbeds of argillaceous limestone Argillaceous limestone with shale Coquinoid argillaceous limestone	QUISSONDE
LOWER CRETACEOUS	UPPER ALBIAN			Bioclastic calcarenite and limestone	CATUMBELA
LOWER CRETACEOUS	— ? — ? — LOWER ALBIAN		8, 7	Dolomite and Anhydrite Dolomite Dolomite and Anhydrite	DOLOMITIC TUENZA
LOWER CRETACEOUS	LOWER ALBIAN		6, 5	Anhydrite and Halite Dolomite and Anhydrite	ANHYDRITIC TUENZA
LOWER CRETACEOUS	LOWER ALBIAN		4, 3	Halite Anhydrite and Dolomite	SALIFEROUS TUENZA
LOWER CRETACEOUS	— ? — ? — APTIAN		2	Dolomite and dolomitic limestone Oölitic limestone Sublithographic limestone Anhydrite	BINGA
LOWER CRETACEOUS	APTIAN		1	Calcarenite with alga, Dolomite and Anhydrite	QUIANGA
LOWER CRETACEOUS	APTIAN			Halite Anhydrite	"MASSIVE SALT"
LOWER CRETACEOUS	PRE-APTIAN			Argillite, Silty dolomite, Dolomitic or calcareous sandstone	UPPER CUVO
LOWER CRETACEOUS	PRE-APTIAN			Red argillaceous sandstone	LOWER CUVO
?				Volcanic ash. Dolerite, Basalt	IGNEOUS ROCKS
	PRECAMBRIAN			Gneiss and quartzite.	BASEMENT COMPLEX

Scale: 0. – 250 – 500 – 750 – 1000 meters

Fig. 3 — Stratigraphic and lithologic column for the Cuanza basin

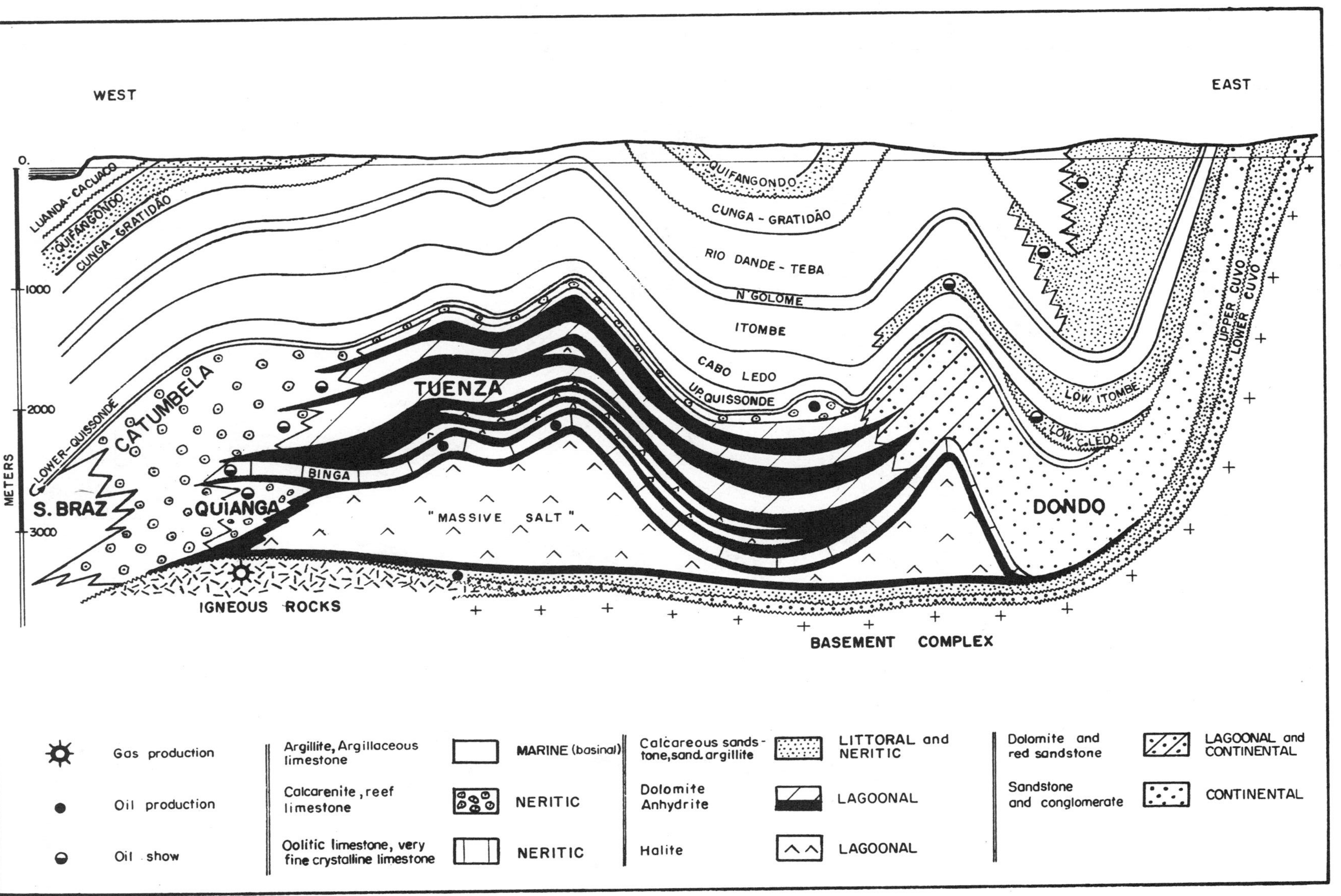

Fig.-4. Diagrammatic cross-section in Cuanza Basin. The West-East cross-section shows the variations of thickness and variations of facies of the different formations, with oil occurrences.

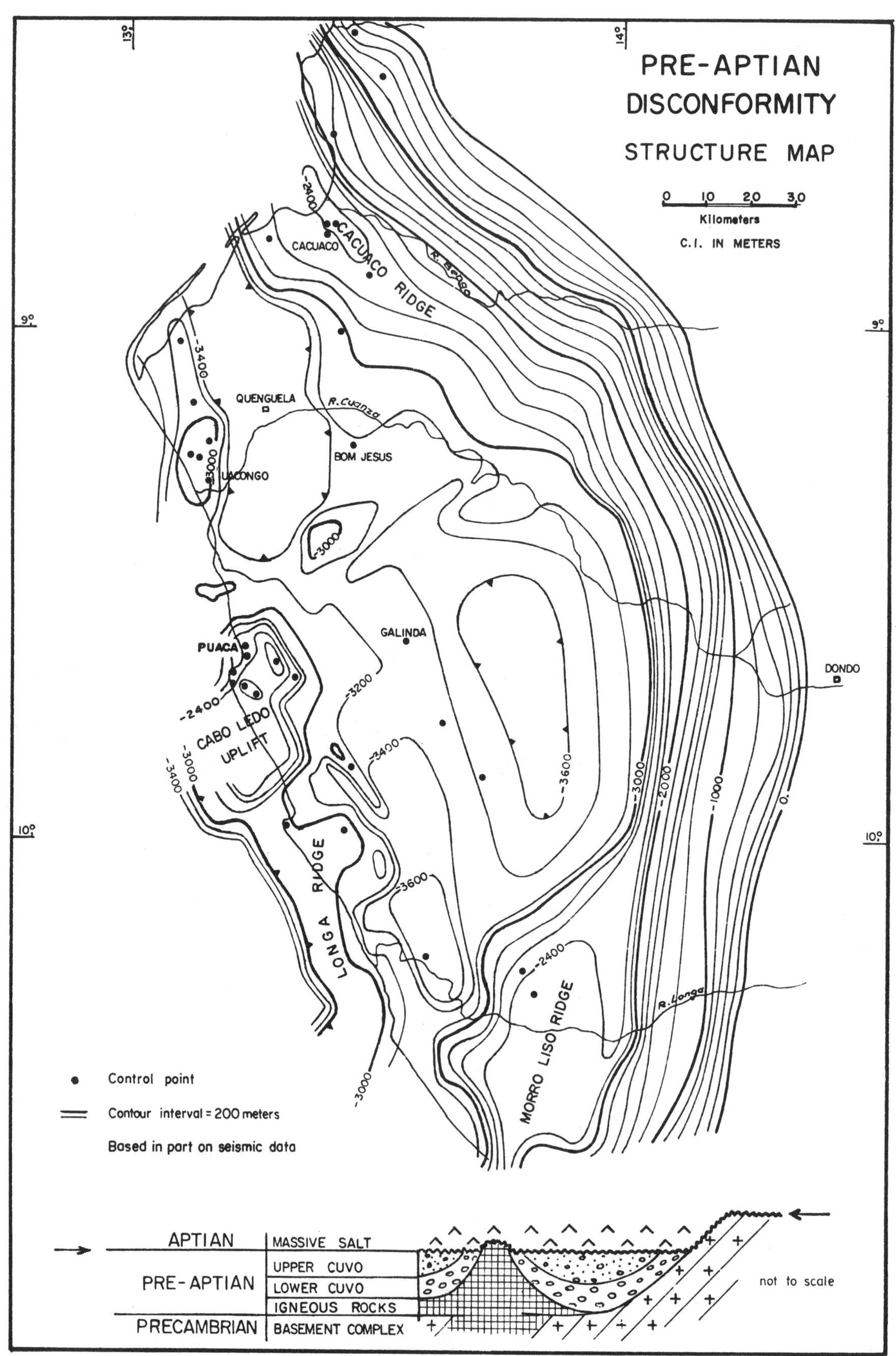

Fig. 5 – Present structure, top of pre-Aptian disconformity.

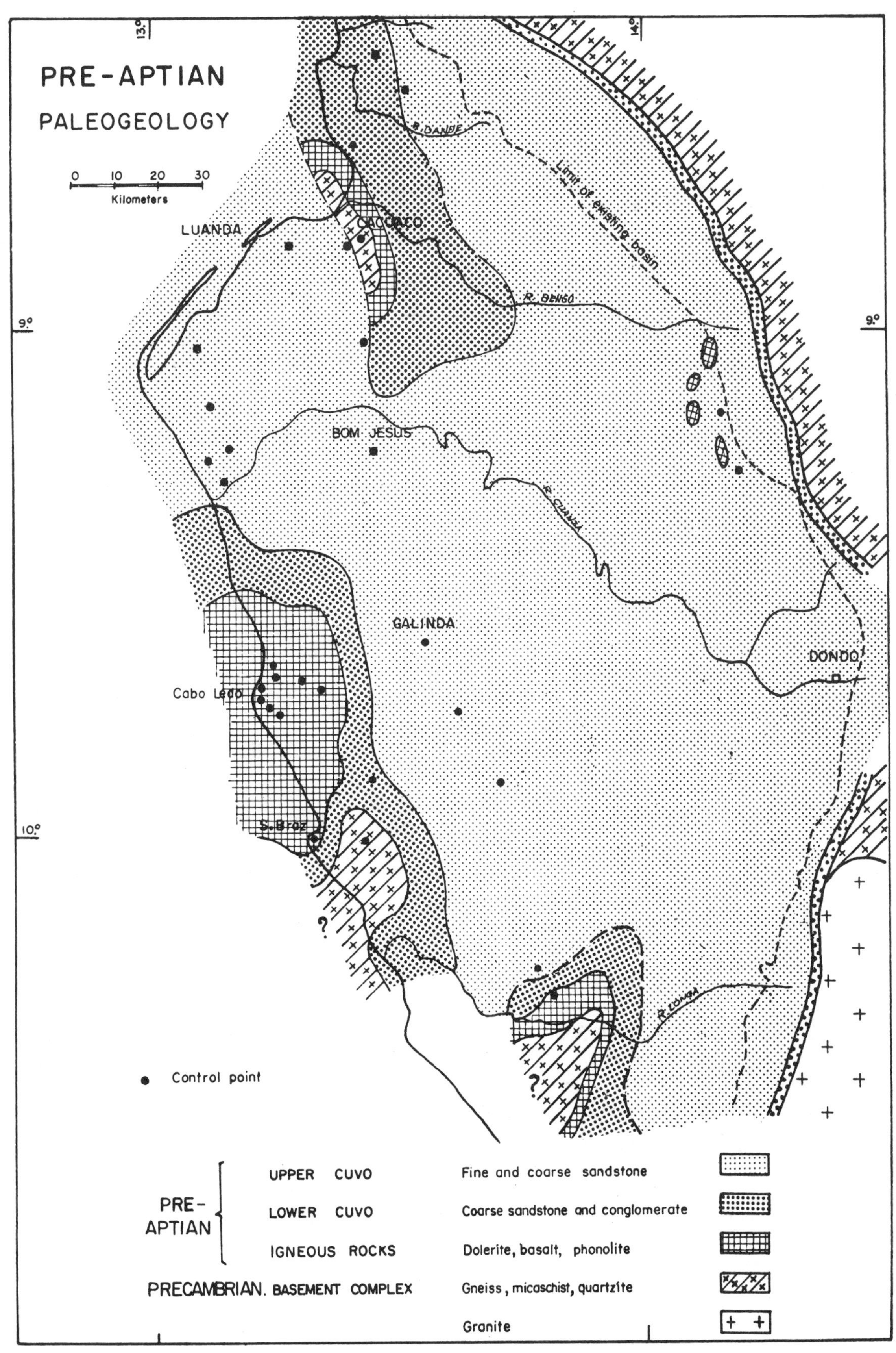

Fig. 6– Pre-Aptian paleogeology. The map shows age and distribution of strata immediately underlying Aptian or Albian formations.

eastward, and along this depositional axis were deposited chiefly transitional strata of the sublittoral, deltaic, and lagoonal facies. This transitional system of sediments grades westward into true marine facies and eastward into continental facies.

These structural units continued to be shifted farther east during late Senonian time (Fig. 7), but finally reoccupied more westerly positions during Eocene time and principally during Miocene time (Fig. 8).

Saliferous tectonics.—Salt tectonics, through extensive continued activity from Albian to Miocene time, also was an important controlling factor on the concomitant sedimentation. The isopachous map (Fig. 9) of the Massive Salt near the base of the sedimentary sequence gives some idea of the magnitude of these tectonic effects of the salt. Tectonic flow-accumulations of non-diapiric salt are 1,400 m. thick in some of the drilled structures. This map illustrates the relatively constant northwest-southeast orientation of the salt domes caused by the influence of faulting in the underlying basement (Figs. 5, 6). Thus, basement faults along the Cabo Ledo uplift impart an elongate shape to salt structures (*e.g.,* Tobias), in contrast to the massive bulbous central structures, such as Galinda, where the basement is known to have been relatively free of tectonic movements (Figs. 10, 11).

The first salt migrations may have begun near the end of Albian time, in an area encompassing the present coast from the mouth of Rio Longa to Uacongo (Fig. 9). These first salt migrations probably were related to basement tectonics and later accelerated by differential loading of superjacent sediments. From this initial area, it appears that the salt deformations spread from west to east, reaching a line extending from Galinda to Luanda by Cenomanian-Turonian time. The saliferous tectonic movements reached the Cacoba area by Senonian time, and the Muxima area by Eocene time, or possibly as late as Miocene time (Fig. 9).

The salt migrations between Albian and Miocene time have greatly affected the thickness of the overlying formations. They have instigated the development of some Lower Cretaceous reef trends (Fig. 12) and, perhaps, are related to the extension toward the west of the Upper Cretaceous and lower Tertiary clastic transitional facies (Fig. 32). During Miocene time, salt migration seems to have been primarily responsible for the thick accumulation of sandstone, siltstone, and shale in the Quenguela and Tres Pontas troughs. Miocene deposits also exist in grabens developed along anticlinal axes related to salt domes.

Basal Arenaceous Formations

At the base of the stratigraphic sequence of the Cuanza basin, and overlying the gneisses and other rocks of the Basement Complex or the igneous rocks which, in places, cover the basement, there are detrital terrigenous strata called the Lower and Upper Cuvo Formations. The age generally attributed to these formations is pre-Aptian Neocomian, although the uppermost Cuvo might be of early Aptian age.

The Lower Cuvo is composed of coarse, conglomeratic, red sandstone, with elements coming from the underlying gneisses or igneous rocks, cemented by red silty claystone, rich in kaolin, derived from pre-Cuvo weathering of the basement rocks. The sandstone and conglomerate, mixed in places with volcanic ash beds, grade upward into the Upper Cuvo sandstone; the same gradation probably takes place laterally from east to west.

The Upper Cuvo consists of fine to coarse quartz sandstone, with some interbeds of coquinoidal, more or less dolomitic, limestone, in places containing abundant ostracods, with a calcareo-dolomitic and kaolinitic cement. Carbonaceous fragments are present, and, in some places, thin coal layers. In places such as Calucala, the sandstone is bituminous.

These data lead the writers to believe that the Cuvo downwarp of the continent, which began the subsidence history of the Cuanza basin, was accompanied by volcanism, and that the early Cuvo phase of fluvio-continental filling took place at the same time. As the Early Cretaceous sea came into the basin for the first time, terrigenous sedimentation was dominant, but the marine Upper Cuvo almost reached the limits of the entire basin. The relatively small thickness of 200–300 m. of Upper Cuvo suggests that the initial subsidence was relatively slight over this wide area.

The isopachous map of the Upper Cuvo (Fig. 13) shows the areas of non-deposition and erosion

(Cacuaco, Morro Liso, and Cabo Ledo) which were responsible for basinal isolation. The zone of interfingering of deltaic, lagoonal, and continental facies known at Calucala and, more generally, in the eastern part of the basin, corresponds with the isopachous maximum.

These transitional facies, in which were probably located the source environments of the oil in the Cuvo, grade westward into white and gray quartz sandstone. At Luanda and Cacuaco, this sandstone is well-sorted and relatively clean, probably reworked and washed out by current and wave action along the shores of Cacuaco, Morro Liso, and Cabo Ledo islands.

Carbonate Formations and Evaporite Equivalents of Early Cretaceous Age

The Aptian-Albian (Lower Cretaceous) strata of the Cuanza basin are a succession of evaporite cycles, beginning with the initial Massive Salt covering about four-fifths of the actual sedimentary basin. The *average* original sedimentary thickness of this Massive Salt is about 350 m., with projected and reconstructed original thicknesses of about 600 m. along the axis of maximum subsidence (Fig. 14). This unit has been called the "Massive Salt" in order to distinguish it from other, thinner, salt layers present in the Tuenza section higher in the sequence (Figs. 2, 3).

This succession of Aptian evaporitic cycles is transgressive, inasmuch as this first saline body is overlain by other evaporitic strata developed as thinner, less saline, and less extensive lentils than the Massive Salt. This evolution of the basin during Aptian time progressed first toward a saline-penesaline stage during which halite and anhydrite were deposited, and subsequently to a stage of penesaline anhydrite deposition, with each cycle separated from the next one by a phase of carbonate-dolomite deposition.

Above, and partly laterally equivalent to the Massive Salt, are two well-developed cycles corresponding, from top to bottom, to the Quianga and Binga Formations (carbonate units). The others are grouped into the Tuenza Formation, which is divided into three members: (1) the Saliferous Tuenza (moderate saline-penesaline conditions); (2) the Anhydritic Tuenza (saline-penesaline conditions present only locally); and (3) the Dolomitic Tuenza (penesaline to almost normal marine conditions) (Figs. 2, 3).

Finally, at the top of this sequence also are the Catumbela and Quissonde Formations. The Catumbela consists of oölitic and bioclastic calcarenite, containing local carbonate buildups. In some places, the Catumbela succeeds, but in other places is equivalent to, the evaporitic cycles which pre-date the Quissonde Formation. The Catumbela is an argillaceous limestone of late Albian age. The age of the overlying Quissonde is easily determined by ammonites and Foraminifera, and these normal marine deposits indicate that evaporitic conditions were terminated by late Albian marine transgression into the Cuanza basin.

Although the Quissonde and Catumbela Formations themselves include no evaporites (by mappable definition), these formations are included in the Aptian-Albian evaporitic sequence because: (1) they overlie it conformably in the interior basin without discernible submarine or subaerial disconformity; and (2) on the western uplift or platform of Cacimbas-Cabo Ledo, the lower parts of these mappable facies are lateral equivalents of evaporite lentils in the central basin.

On the eastern shelf, the entire Aptian-Albian sequence grades into continental facies of the Dondo clastic facies (Fig. 15).

Evaporite members.—The study of variations in thickness and lithology of the different saline and penesaline members is of double interest: (1) it permits separation of the 2,000 m. of Aptian-Albian sediments into isochronous slices corresponding with cycles of deposition; and (2) this division provides knowledge of the structural evolution of the basin and its subsidence history.

In the central part of the evaporitic basin, sections in drilled tests indicate the following cyclical succession.

1. Dolomite 2. Anhydrite and dolomite 3. Anhydrite	penesaline member
4. Anhydrite and halite	saline-penesaline member
5. Anhydrite 6. Anhydrite and dolomite 7. Dolomite	penesaline member

Several wells show that the same succession which is found vertically in units 1–2–3–4 also can be found laterally. (Figs. 18, 22, 24). This suggests that the sulfate and chloride precipitation, which at one place in the basin created vertical variations in concentrations controlling the de-

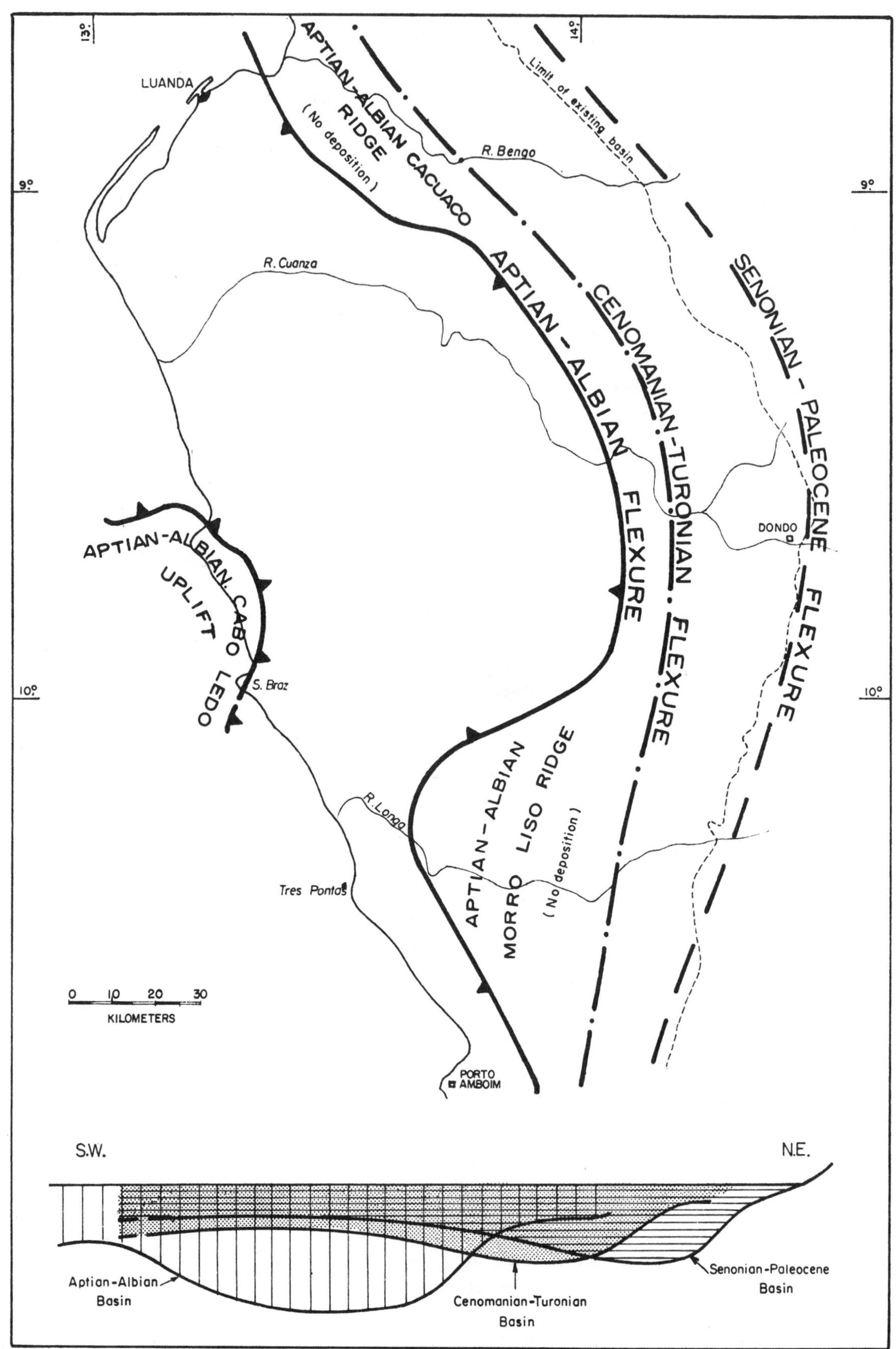

Fig. 7—Tectonic features controlling sedimentation during Cretaceous time.

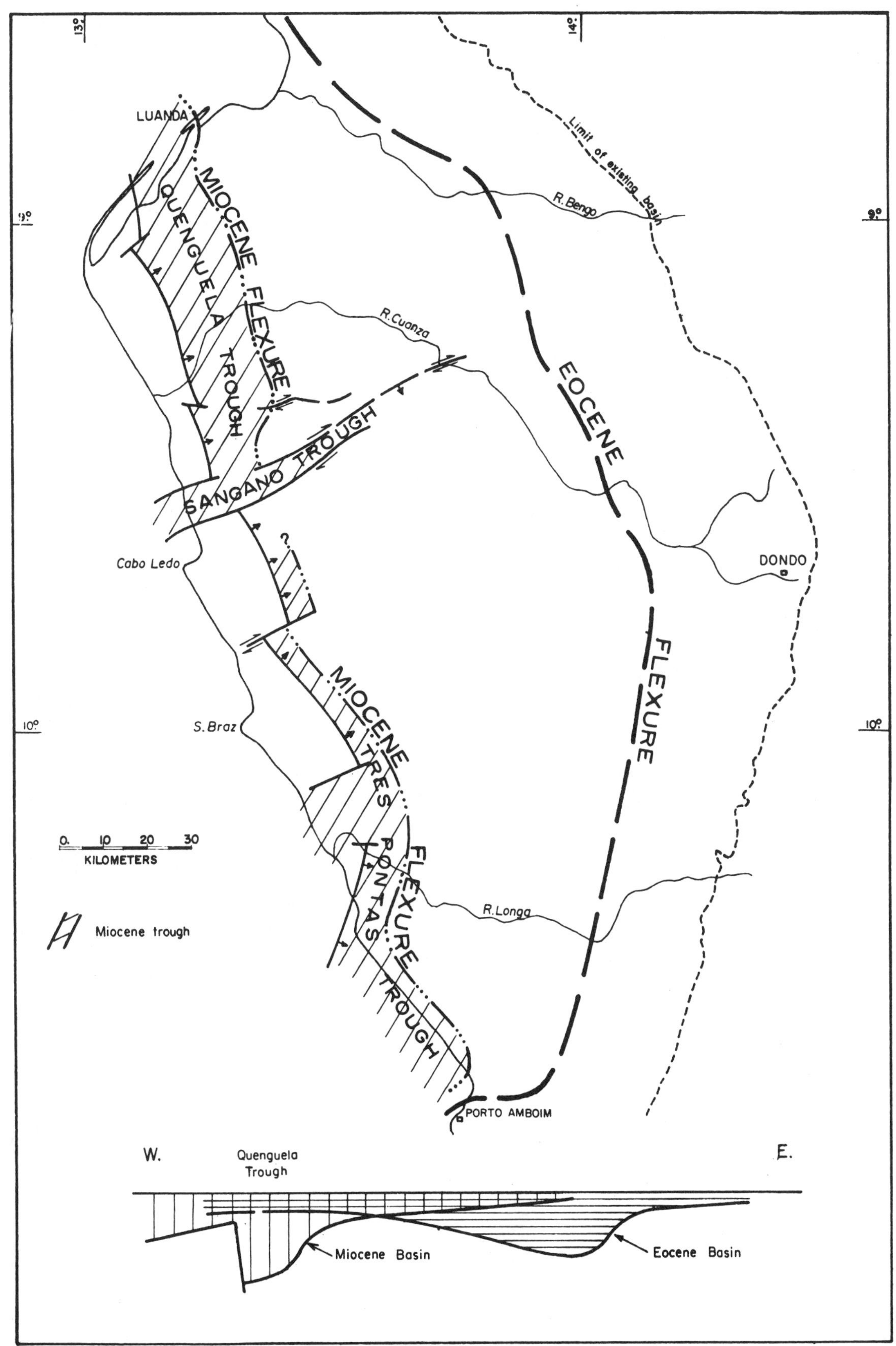

Fig. 8–Tectonic features controlling sedimentation during Tertiary time.

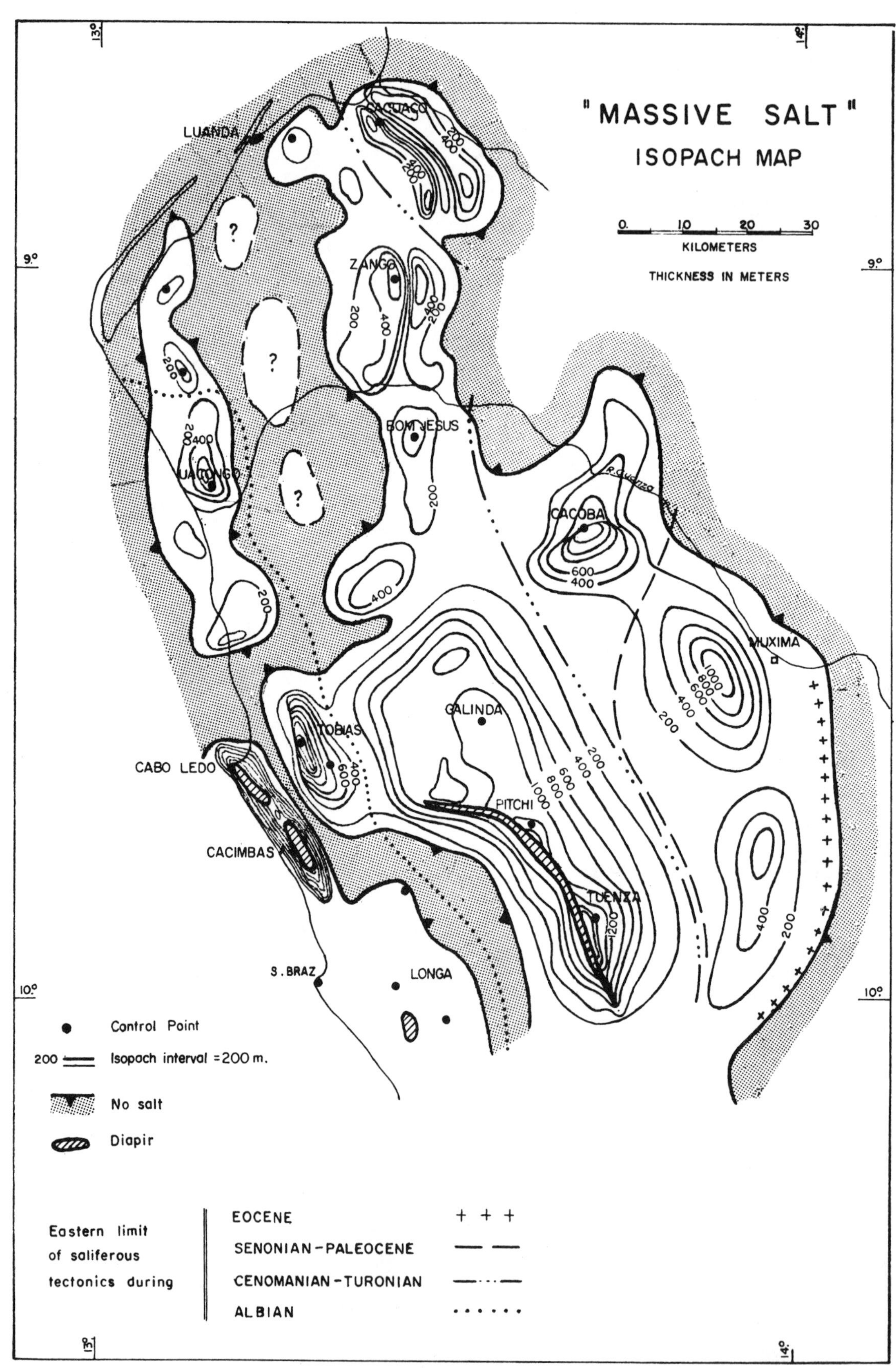

Fig. 9— Present thickness and age of deformation of the "Massive Salt"

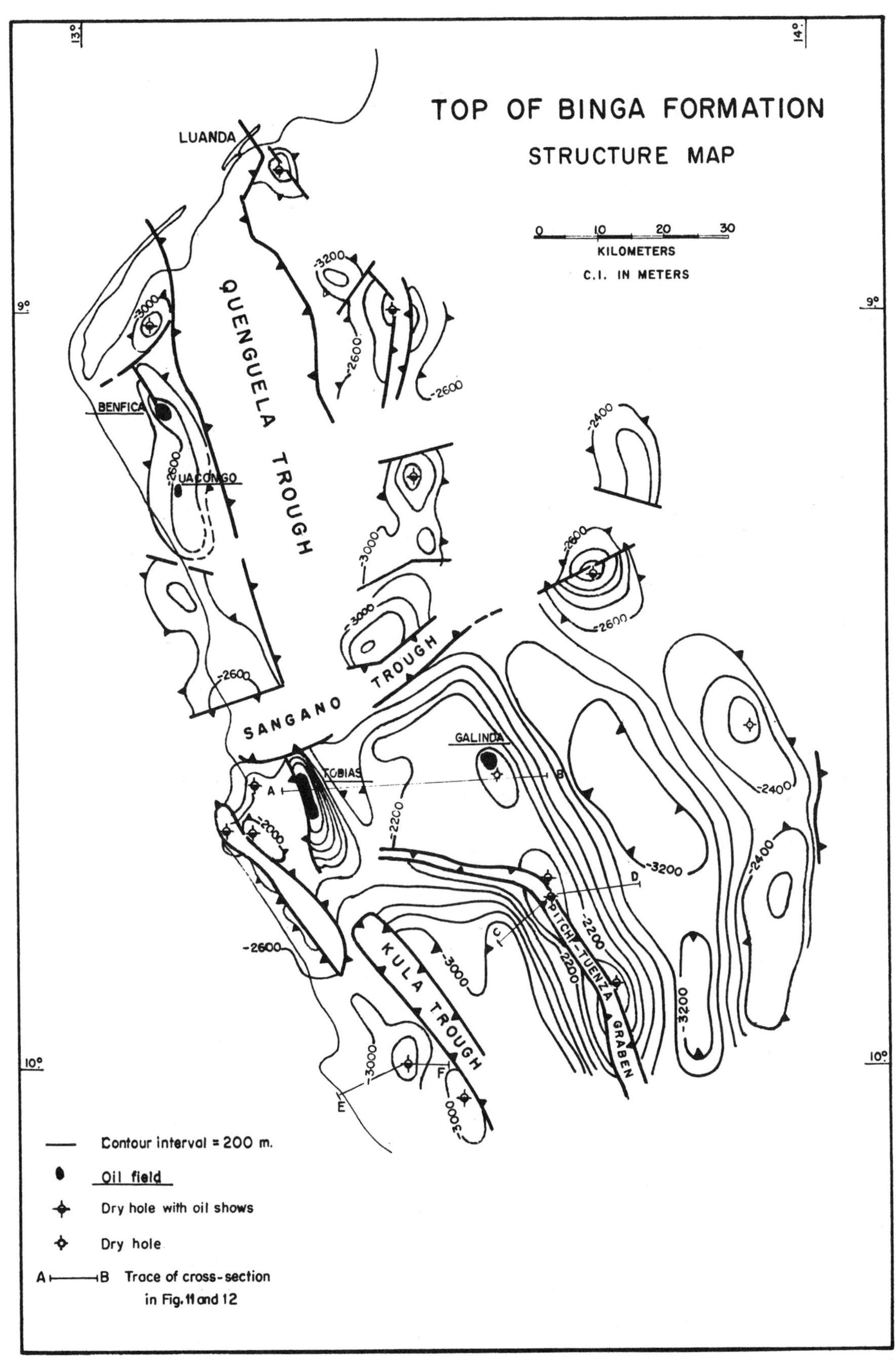

Fig. 10 – Present structure, top of Binga formation.

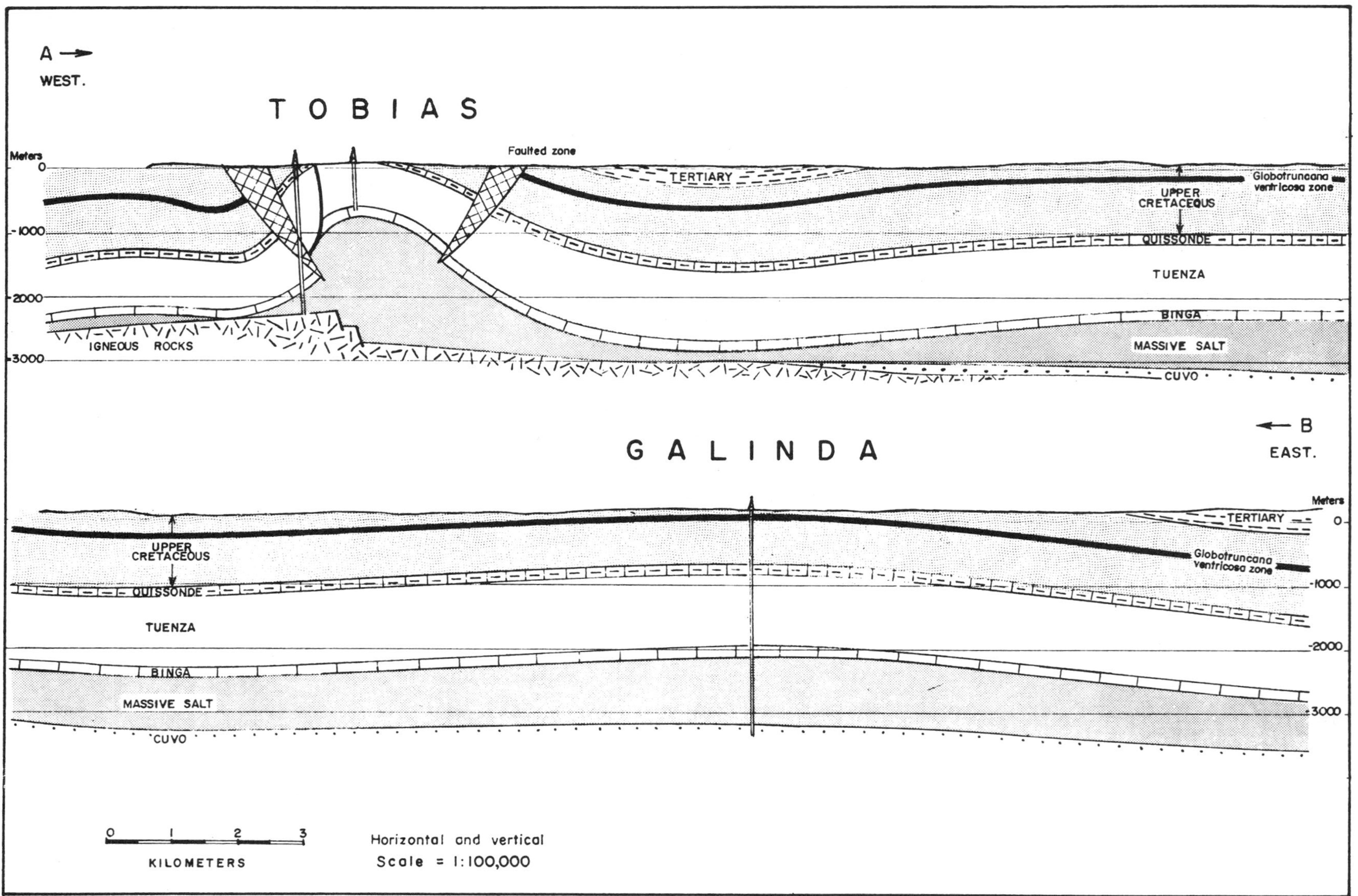

Fig.11 — Cross-section A–B through the Tobias and Galinda anticlines.

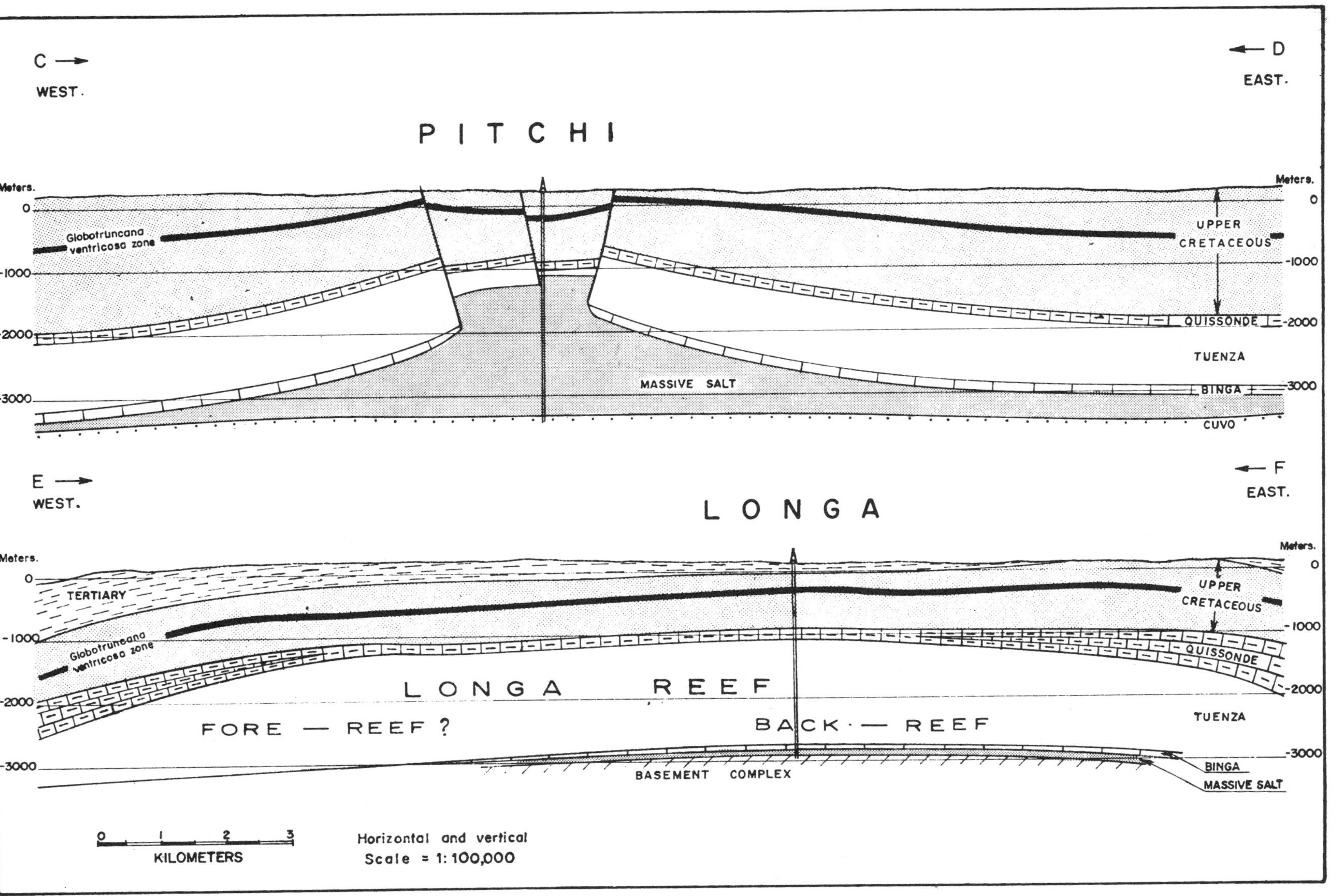

Fig.12 — Cross-sections C—D and E-F through the Pitchi and Longa anticlines. The thin "Massive Salt" section penetrated in the Longa well section suggests that a migration of the salt towards the Pitchi, Galinda or Tobías zone is responsible, in part for the build-up of the Longa Reef.

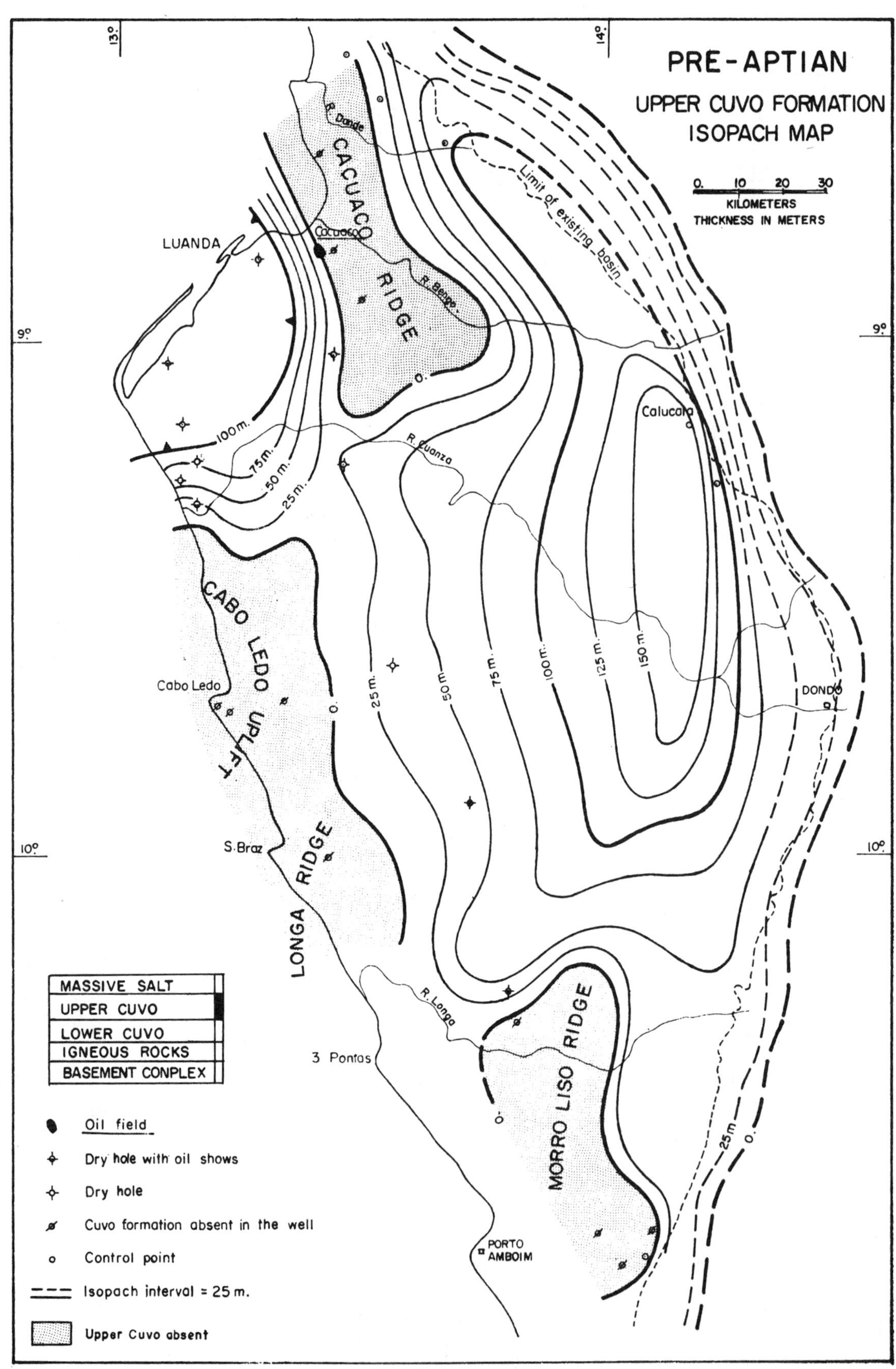

Fig. 13—Thickness of Upper Cuvo formation. The Cacuaco oil field which produces from this formation is indicated.

posits of anhydrite or halite, also created lateral variations during Aptian-Albian evaporite deposition. Thus, evaporitic bodies generally are included in a dolomitic mass and represent a basinal milieu; in turn, the halite core is enveloped by anhydrite which is overlain and underlain by dolomite lentils as parts of the same cycle (Fig. 16).

The concept of an evaporitic cycle then becomes a valid correlation tool. Indeed, the appearance and disappearance of one of the saline units in a vertical section corresponds with an event which is an acceleration, then a slowing down, of the subsidence, yielding a recognizable maximum of subsidence in the tectonic evolution of the basin.

During study of the wells, a marker was selected on the gamma ray logs. This marker was picked in the center of the saline unit, if present, or, if not present, in the central penesaline lentil. The marker was chosen to correspond with the maximum of subsidence for each cycle (Figs. 18, 22, 24). The writers were thus able to divide the 2,000 m. of evaporitic sediments of the Aptian-Albian basin into mappable isochronous slices, each corresponding with a complete cycle in which the oldest cycles proceed from saline to saline, and the youngest ones from penesaline to penesaline. This division into slices presents another advantage in that it makes possible a study, cycle by cycle, of the facies variations for the intra-member dolomitic carbonates of the central basin, as well as for the biocalcarenites of the equivalent platform facies along the western side of the Cuanza basin.

From a structural point of view, the outlines of the feather edges of the saline bodies show the unstable central part of the basin downwarped by a concentric subsidence (Fig. 15). They also show the more local zones of instability such as the Bamvo and São Braz troughs, which separated the more stable zones such as the Kula and Longa axes, whereas the limits of the penesaline bodies delineate the extension of the Cabo Ledo uplift and the extension of certain reef trends.

Carbonate members.—The lithologic succession intercalated between two saline bodies, the Massive Salt below and the Tuenza evaporites above, serves as an example of the economically important carbonate sections in the Cuanza basin. Cycle No. 2, or Binga cycle, includes, from top to bottom, the following lithologic associations, with the corresponding depositional environments.

1. Halite and anhydrite—lagoonal saline-penesaline.
2. Massive anhydrite overlain by interbedded anhydrite, dolomite, microcrystalline limestone (more or less dolomitized), or by the calcareous mud type of limestone—lagoonal penesaline to marine (central bank).
3. Oölitic and bioclastic calcarenite, more or less sandy—marine (bank margin).
4. Sublithographic limestone, with *Calpionella,* in places black and argillaceous, showing an euxinic tendency—marine (central bank).
5. Microcrystalline dolomite and anhydrite, overlain by marine anhydrite—lagoonal penesaline.
6. Halite—lagoonal saline.

Data from wells have proved that the same succession of units 1–5 that exists vertically also exists laterally. However, it is necessary to define briefly the enumerated environments which come from American and Canadian publications, based in great part on studies of actual sedimentation in the Bahama Islands.

According to Illing (1954), the Bahama Banks constitute a large submarine "plateau" of about 100,000 sq. km., where the depth of the water is less than 6 fm. The banks have numerous islands on them, and the sides of the banks plunge steeply toward oceanic depths. These numerous islands, in places covered by dunes, occupy one-tenth of the total surface of the Bahama Banks. The submarine bottoms are covered by cryptocrystalline aragonite grains which have been and are being precipitated from the sea water. On the margins of the banks, in the areas of maximum turbulence, coral reefs are associated with bioclastic and oölitic sands. Calcareous sands transported by tidal currents are deposited on the bank surface forming local shoals, but the amount of coarser detrital material decreases toward the interior of the central bank where the calcareous mud facies is predominant.

The writers believe that the deposition of lithologic units 3 and 4 of Cycle No. 2 (Binga) occurred under conditions similar to those observed now on the Bahama Banks. Those conditions were introduced by an important slowing-down of the subsidence, which may have permitted a long period of reworking of the unit 3-type

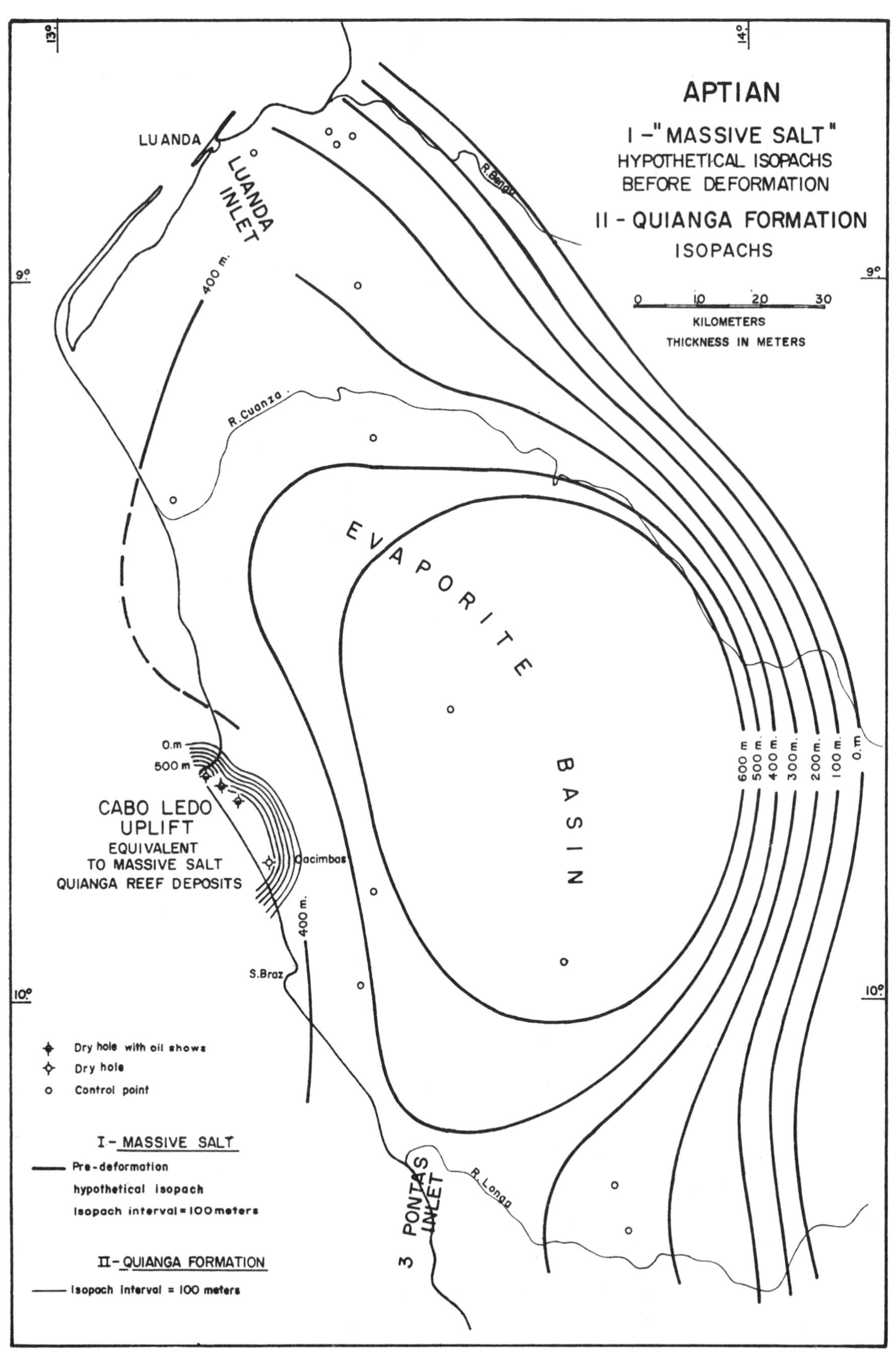

Fig.14 - Hypothetical thickness of "Massive Salt" before deformation, and thickness of equivalent massive salt reef deposits of the Quianga formation.

sediments. Thus, it is logical to assume that all the Aptian-Albian evaporitic cycles of the Cuanza basin were controlled chiefly by periods of maximum subsidence during evaporite deposition separated by periods of stability, allowing the formation of more normal marine carbonates.

The western area of more permanent stability was less affected by this cycle subsidence; this allowed the maintenance of carbonate depositional conditions of the bank-margin type during most of Aptian-Albian time in the Cabo Ledo area. Toward the landmass east of the eastern Aptian-Albian flexure, a lagoonal-continental environment was maintained to form the eastern detrital shelf (Fig. 16).

Within this general framework of the Aptian-Albian sequence, it is possible to examine in greater detail the successive units (Figs. 2, 3).

Massive Salt and Quianga Cycle (Cycle No. 1).—Overlying the terrigenous deposits of the Upper Cuvo, this unit corresponds with a complete change of sedimentation conditions caused by an abrupt increase in subsidence and also by a change to a hot, arid climate.

This subsidence was not accompanied by an increased supply of terrigenous detrital sediments, a fact which suggests that no notable rejuvenation of relief occurred in the landmass on the east. This allowed the transgressing sea to invade and almost completely cover the alluvial Cuvo plain. During the beginning of the transgression, the highlands, which dominated this Cuvo plain on the north (Cacuaco) and south (Morro Liso), as well as on the west (Cabo Ledo platform), were the sites of volcanic activity that created a puzzling volcanic ash sequence whose exact relation to the Cuvo detritals is not yet fully understood.

The westernmost uplifts, such as Cabo Ledo, probably caused the initial isolation of the Cuanza basin, as it was transformed into a large evaporite lagoon. The water lost by evaporation probably was replaced by normal marine waters entering through the Luanda and Tres Pontas sags or access-ways. With the accentuation of subsidence, the evaporitic basin onlapped the limits of the ancient Cuvo alluvial plain step by step, slowly erasing the existing relief which included those highlands that may have been present at the end of Cuvo time. As the Massive Salt was deposited, the remaining relief features were buried by the salt. Thus this salt is evidence of marine transgression, covering the lower and upper Cuvo deposits, the Basement Complex in Cacuaco, and the eruptive rocks in the Morro Liso and Cabo Ledo areas (Fig. 6).

The presence of the Cabo Ledo uplift on the ocean side at the west, during the basinal subsidence, favored the development of the reef build-ups and reef-barrier trends discovered in the wells drilled at Cabo Ledo and Cacimbas. In these locales, below the sediments of the Binga cycle, were found several hundred meters of bioclastic and oölitic calcarenites associated with algal limestones and corals surrounded by anhydrite. These deposits correspond with a pre-Binga cycle, the Quianga cycle, which in part is a lateral equivalent of the Massive Salt (Figs. 14, 18). This lateral gradation from halite to calcarenitic limestone occurs in a distance of less than 10 km. from east to west.

Some oil shows have been found in the Cabo Ledo (CL-2b) and Petroleo (PL-1) wells in the porous calcarenite member of the Quianga cycle, but numerous gas and oil shows with asphalt were observed in the anhydritic member below the carbonate sequences (Fig. 18). The anhydritic member is a lateral equivalent of the carbonate sequence and corresponds with a transitional environment between reef and saline. The anhydrite is in fact interbedded with numerous beds of thin, dark, very finely crystalline, argillaceous dolomite and dark shale. Immediately after cores were obtained, numerous bubbles of gas and oil were visible along bedding and slumping planes, asphalt being in the fractures and vugs in the anhydrite. There is no doubt that these argillaceous dolomite and dark shale lentils are the sources for this oil, which originated *in situ,* as well as for the oil observed in the adjacent dolomitic members.

The gas and condensate of the Puaca field (Fig. 2), discovered in volcanic ash lentils lying between the argillaceous anhydrite member and igneous rocks at the top of the Cabo Ledo uplift, also might have originated in the same source environments (Fig. 4).

Binga Cycle (Cycle No. 2).—In all of the Aptian-Albian evaporite basin, except the Cabo Ledo area, the Massive Salt was overlain directly by the Binga carbonate succession, previously described as an example of a typical carbonate cycle.

The vertical change from halite to anhydrite to

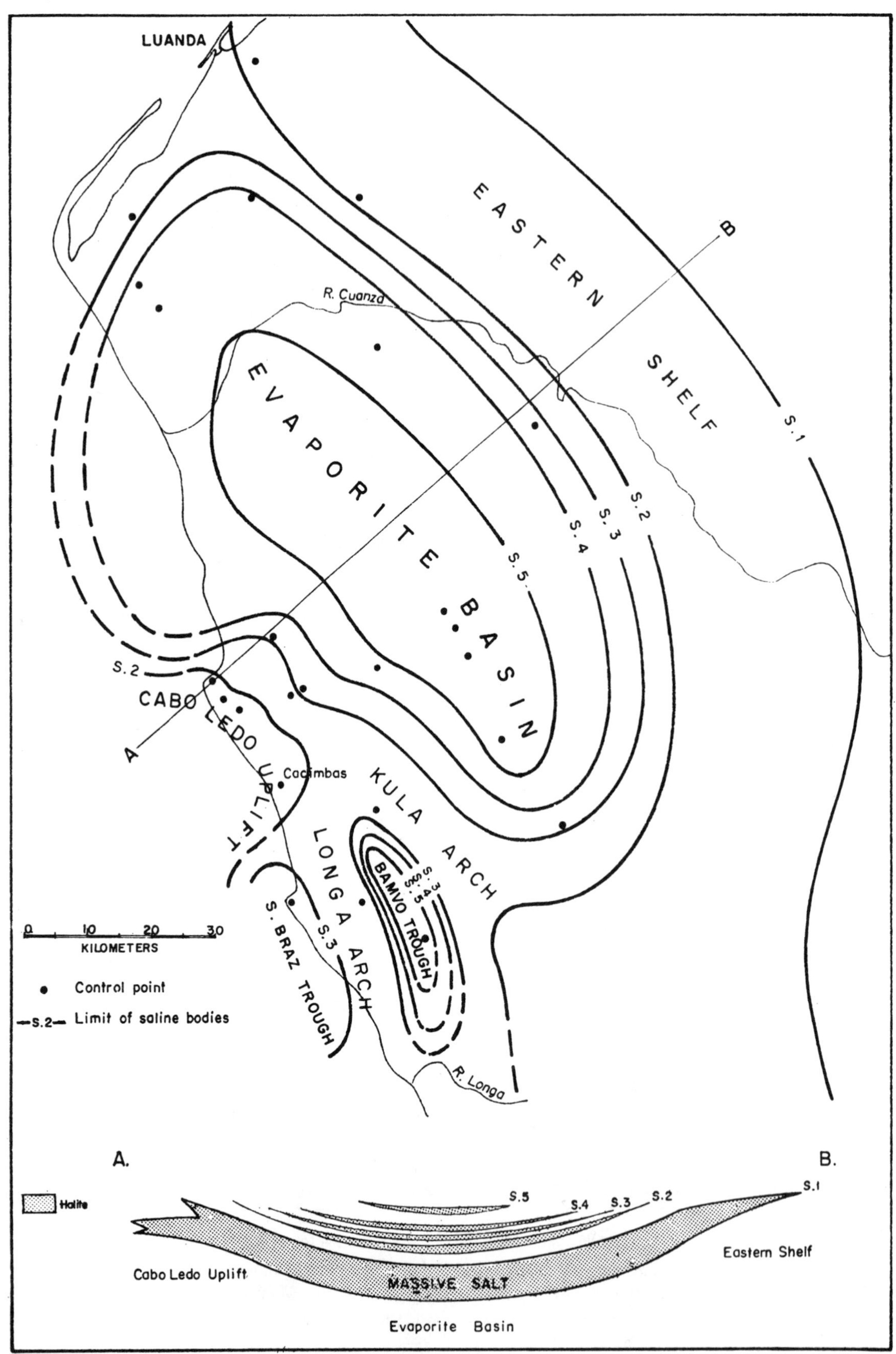

Fig. 15 – Limits of Aptian – Albian saline bodies

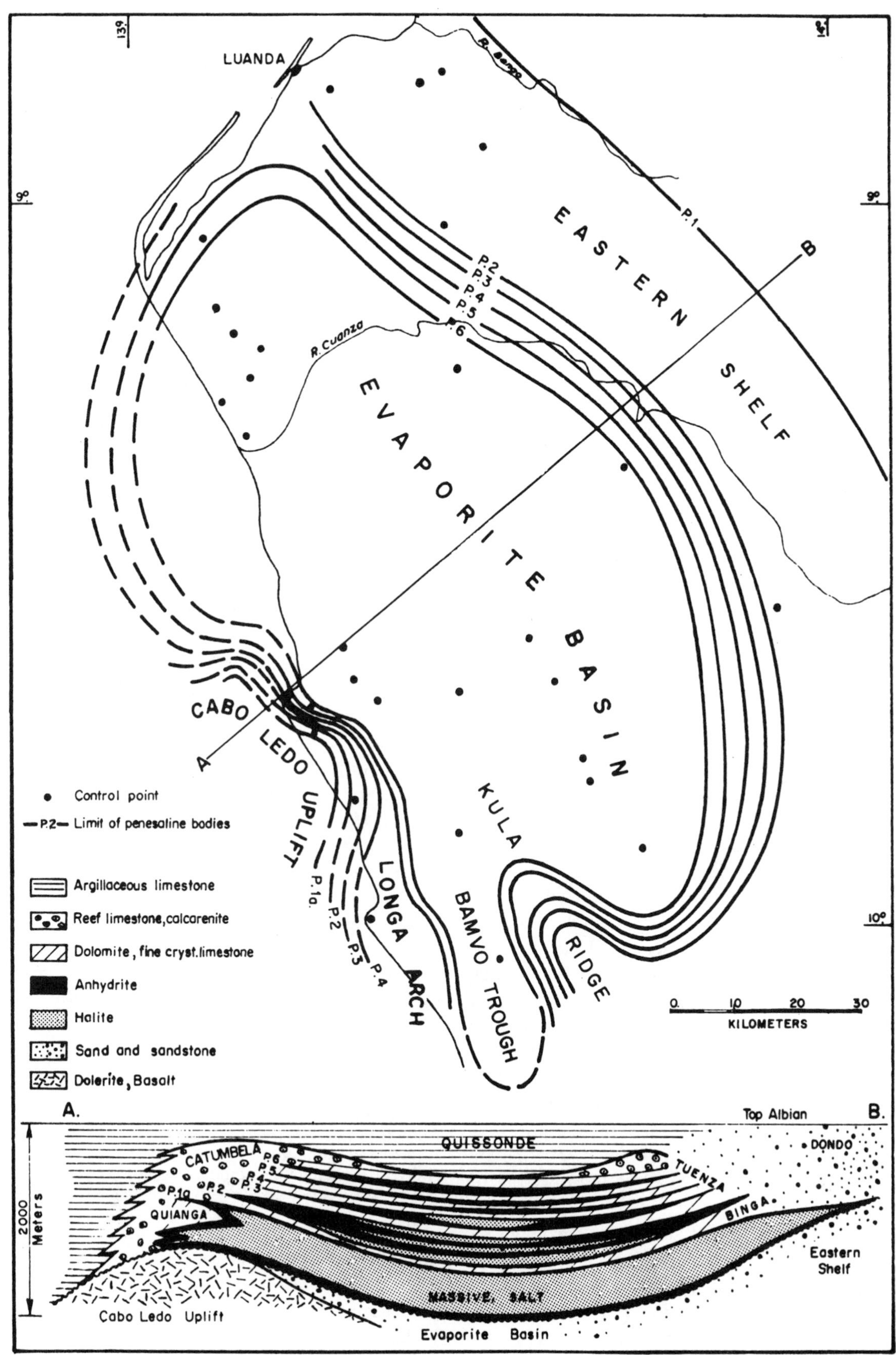

Fig. 16 – Limits of Aptian-Albian penesaline bodies

sublithographic limestone and calcarenite corresponds, after the deposition of the Massive Salt, with a slowing down of subsidence. This stability transformed the saline basin into a large bank or shoal area, covered by about 2 fm. of water in which were deposited sediments of calcareous mud type, grading westward into oölitic and bioclastic calcarenite on the bank-margin of the Cabo Ledo uplift. This calcareous mud provided an early seal which protected the Massive Salt from widespread solution by the transgressing normal marine waters.

The isopachous map of the Binga cycle (Fig. 17) shows the extension of the central part of the basin, where the calcareous mud sediments were deposited; conditions were saline-penesaline both at the beginning and at the end of the cycle. Zones of more local instability, such as the Bamvo and São Braz troughs, are surrounded by two deep structural axes flanking the Longa and Kula arches. The Cacuaco and Morro Liso ridges remained emergent, separating, respectively, the Aptian Cuanza basin from the Lifune basin on the north and Quilonga basin on the south, the latter being situated outside of the area of Figure 15. These two small subsidiary basins correspond with the areas where the Binga is now well exposed on the surface (Fig. 1).

The depth of the water then decreased as the slowing of the subsidence was accentuated. Under the action of coastal currents, the calcarenite was reworked, transported, and mixed with some terrigenous elements coming from the continent on the east. These sediments finally were deposited by currents into localized banks and shoals, perhaps in places which emerged episodically; in general, these sediments have a widespread distribution on the bank surface. This reworking led finally to the deposition of sandy calcarenite lentils (Figs. 18, 19), in many places dolomitized, particularly on the bank margins and near the Luanda inlet threshold. This "Sandy Calcarenite" comprises oölitic and bioclastic fragments, mixed with quartz and feldspar grains, cemented by large crystals of clear, sparry calcite.

After this period of stability and maximum reworking, the subsidence began again, but in a distinct way, in the southern part of the basin, where halite lies directly above the Sandy Calcarenite. In the northern part of the basin, however, between halite and calcarenite, some transitional beds appear, comprising interbedded thin layers of anhydrite, sublithographic limestone with ostracods, microcrystalline dolomite, and more or less dolomitized bioclastic calcarenite. The first layer of halite is present only 30–40 m. higher in the sequence (Fig. 18).

The isopachous map of these "Transitional Beds" (Fig. 20) shows clearly the influence of the Luanda inlet-threshold, prolonging the marine influence in the northern part of the basin, while the Tres Pontas inlet access-way was closed to the entry of normal marine waters.

The Binga Formation is the main petroliferous formation of the Cuanza basin. The oil is entrapped either in the Transitional Beds where they exist in the northern part of the basin, or in the Sandy Calcarenite where it is directly covered by the saline or penesaline unit of the regressive part of the Binga cycle.

Lithologic aspects and geochemical analyses suggest that the source of this oil is the massive, very finely crystalline limestone interbedded with dark shale and argillaceous limestone underlying the Sandy Calcarenite lentils. The broad extent of this source-reservoir association explains why the Binga is petroliferous wherever it is found in wells and outcrops. Good primary porosities and permeabilities, improved by secondary dolomitization, are classically encountered along the seaward margin of the shoals or near the Luanda inlet, but commercial oil accumulations have been localized chiefly by salt folding and fracturing of the formation.

Saliferous Tuenza Cycles (Cycles Nos. 3 and 4).—After the Binga cycle, which ended with the deposition of a saline-penesaline unit, two other cycles were deposited which pass directly into the saline-penesaline stage of evaporite development (Figs. 21, 22). From top to bottom, the following lithologic associations are observed.

Halite and anhydrite	Cycle No. 4
Anhydrite and microcrystalline dolomite	
Halite and anhydrite	
Anhydrite and microcrystalline dolomite	Cycle No. 3
Halite and anhydrite	

The writers believe that the Saliferous Tuenza member corresponds in time with the evaporitic basin at the end of the Binga cycle, but minor oscillations permit separation of these two cycles within the evaporitic mass.

The isopachous map of the Saliferous Tuenza

cycle illustrates much the same tectonic framework which controlled the Binga cycle sedimentation; that is, the bottom of the basin lies athwart the broad low valley of the present-day Rio Cunaza, whose location corresponds with the area of maximum deposition of the underlying saline-penesaline bodies. On the Cabo Ledo uplift, shoal- and intershoal-type carbonate sedimentation continued as in pre-Tuenza time.

In the south, the Longa arch was still in existence as indicated by the absence of saline-penesaline bodies S3 and S4. The Longa arch was flanked by two local unstable zones of subsidence called the Bamvo and São Braz troughs in which saline-penesaline bodies S3 and S4 are well developed (Fig. 22).

A few barrels of oil, probably generated *in situ,* were produced in the Bamvo well on the Longa arch from dark gray, fractured dolomite with black bituminous shale interbedded in massive anhydrite.

Anhydritic Tuenza Cycle (Cycle No. 5).— This cycle represents, in the interior basin, a slowing of subsidence, a condition that again favored the appearance of a large broad bank above the large evaporitic lagoon of the Saliferous Tuenza (Fig. 23). However, the carbonate phase of this cycle is composed essentially of microcrystalline dolomite, derived from calcareous mud sediments, interbedded with anhydrite associated with more or less argillaceous dolomite. These lithologic types imply that water salinity remained relatively high to deposit anhydrite, but that the bank was deep enough that current action could not spread over its surface the oölitic or bioclastic calcarenites which were forming on the western margin of the basin.

This carbonate phase is surrounded by the penesaline bodies P2 (below) and P3 (above), the latter representing a return to conditions of sufficient subsidence where evaporitic deposition could resume. The maximum subsidence and associated thicker evaporites occur in the Galinda-Pitchi area, where the saline-penesaline body S5 is well developed (Fig. 24).

The isopachous map of the Anhydritic Tuenza cycle (Fig. 23) shows the continued permanence of the interior basin, flanked by the stable shelf on the east and by the Cabo Ledo uplift on the west. In addition, there developed in the southern part of the interior basin, inside the unstable zone, a reef buildup along the deep Longa arch. This arch already was in existence as a tectonic spur during the deposition of the earlier Cycles 2, 3, and 4. The Longa arch, during the time of marine sedimentation, was an almost emergent high, permitting currents to accumulate oölitic and bioclastic calcarenites associated with quartz and feldspar sandstone. In addition, true colonies of bivalves and echinoderms built a reef with sufficient vertical development to resist wave action along the flanks of the Longa arch.

In the Bamvo well located on the Longa arch, very good oil shows, consisting of a few barrels of recovered oil, have been observed in this calcarenite just below the anhydrite of the penesaline body P3. It is the only place in the Cuanza basin where oil has been reported in the Anhydritic Tuenza. No possible oil-source facies is known in anhydritic Tuenza beds, but the impregnated calcarenites might grade eastward into anhydrite interbedded with dark shale and argillaceous dolomite similar to those in the Saliferous Tuenza below.

Dolomitic Tuneza Cycles (Cycles Nos. 6, 7, and 8).—The same general remarks apply to these three cycles as to the previous Tuenza cycles, especially regarding the subsidence accelerations in the interior basin, bringing back periodically the evaporitic phases. However, Cycles 6, 7, and 8, from oldest to youngest, are progressively less developed both in thickness and areal extent, indicating that the normal transgressive character of the sequence became increasingly accentuated despite the decreasing areal extent (Figs. 25, 26).

The carbonate phases consist of microcrystalline dolomite, deposited in a central-bank environment, with extremely porous saccharoidal dolomite lentils which originally might have been shoal-type calcarenites episodically developed on the bank surface.

Cycle No. 7, however, similar to the earlier Cycle No. 5, can be distinguished from the others by the presence of a reef buildup developed on the Longa arch, which was still in existence at the end of Cycle No. 6. The saline-penesaline bodies existing in Bamvo and São Braz are absent in Longa. This critical development, more than 300 m. thick, of oölitic and bioclastic calcarenite (pelecypods and echinoderms) contains quartz and feldspar sand grains cemented by coarse-grained,

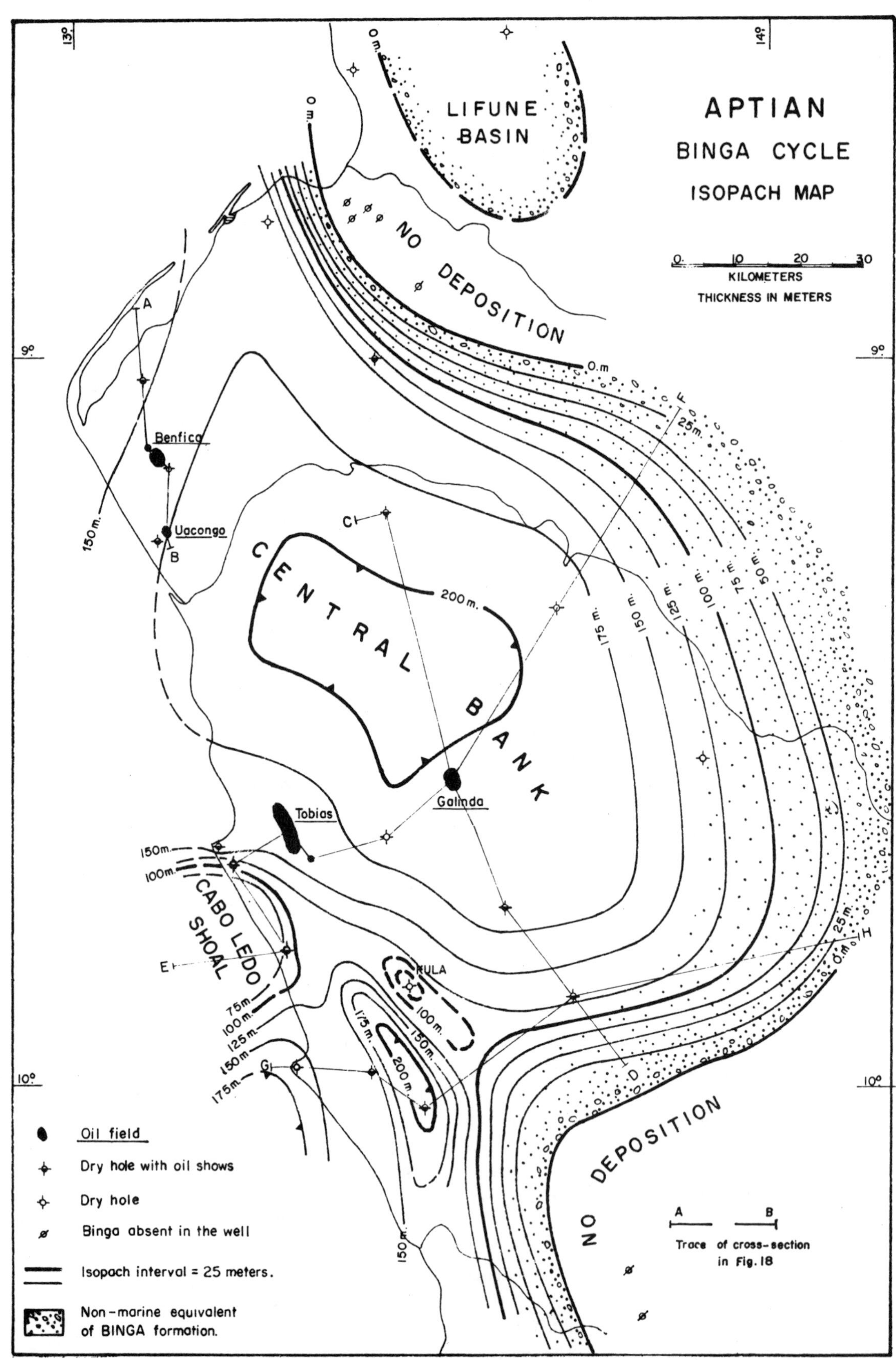

Fig. 17 — Thickness of Aptian Binga Cycle deposits. Oil fields producing from this interval are indicated.

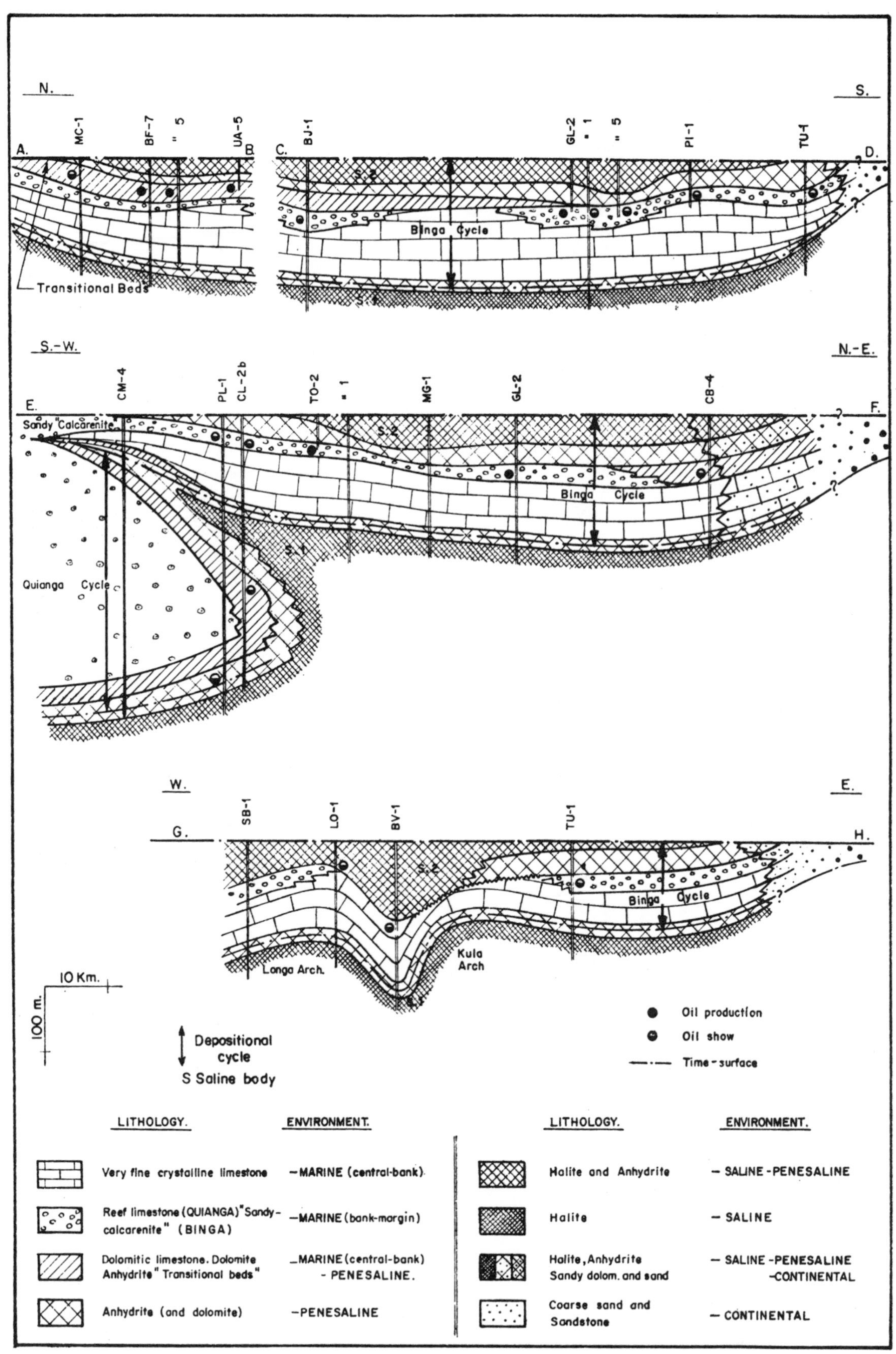

Fig. 18 — Cross sections of Aptian Binga and Quianga Cycle deposits.

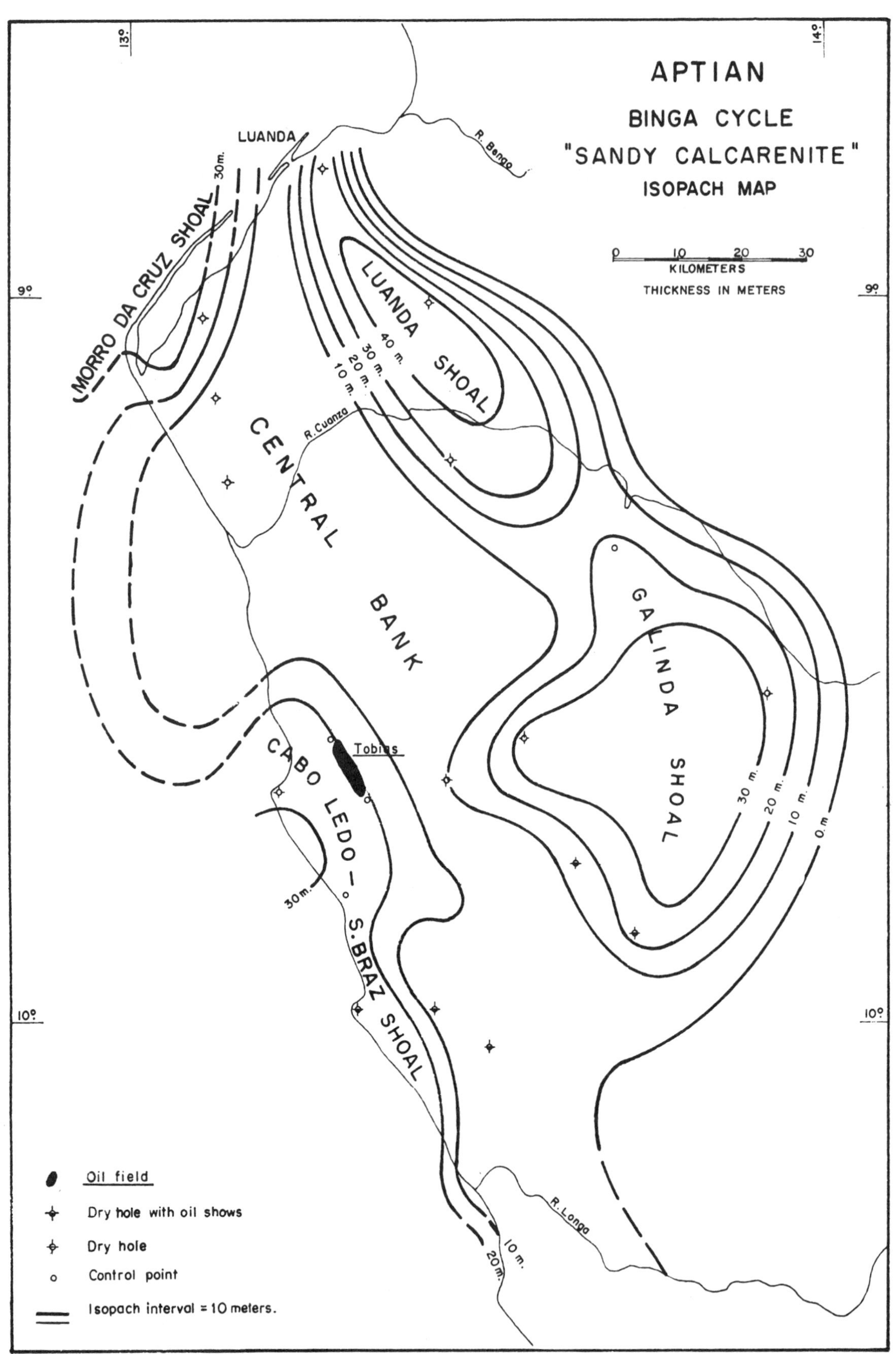

Fig. 19 – Thickness of the "Sandy calcarenite" of Aptian Binga Cycle.
The Tobias oil field which produces from this formation is indicated.

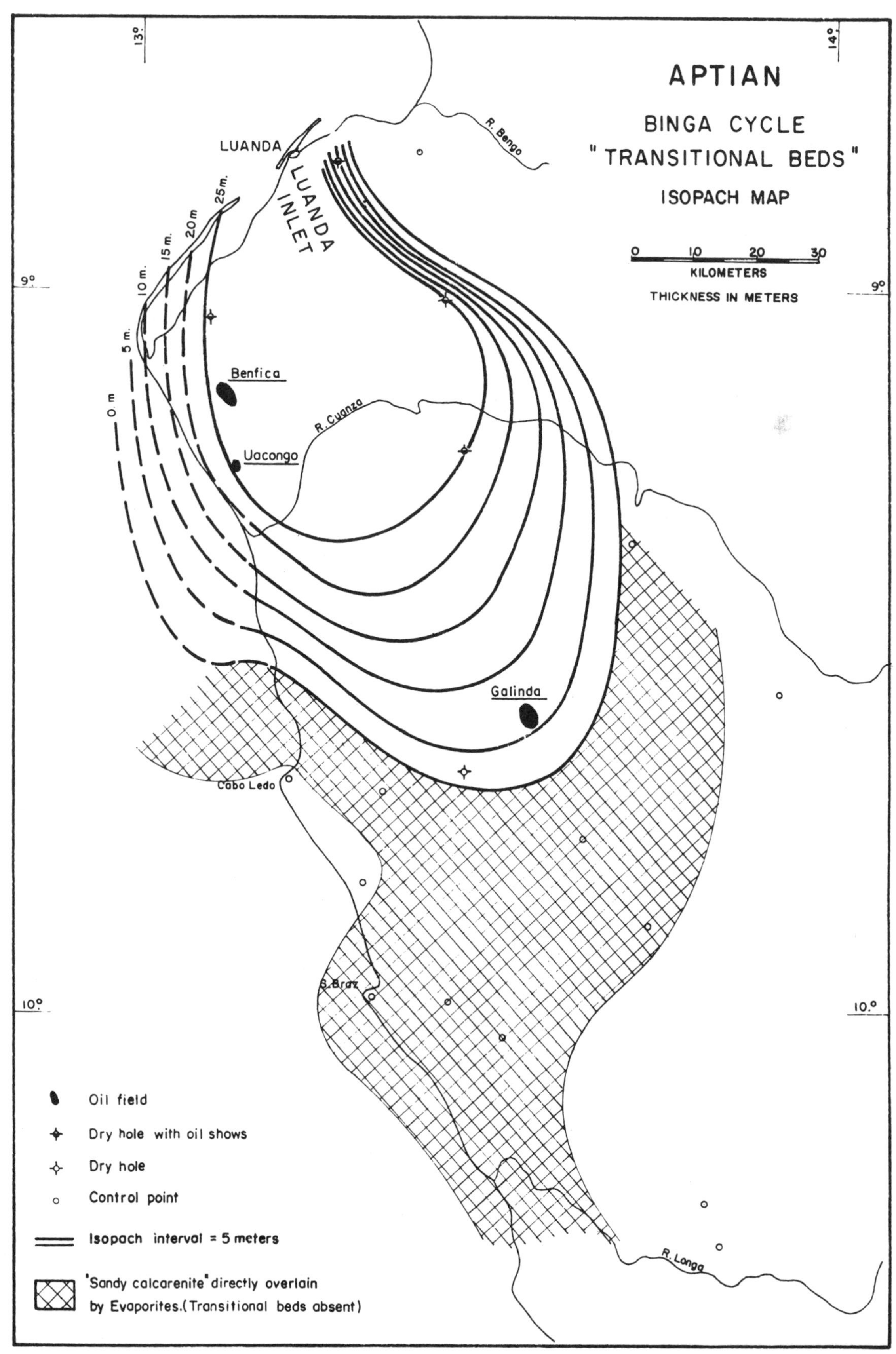

Fig.20—Thickness of the "Transitional beds" of Aptian Binga Cycle. Oil fields which produce from this formation are indicated.

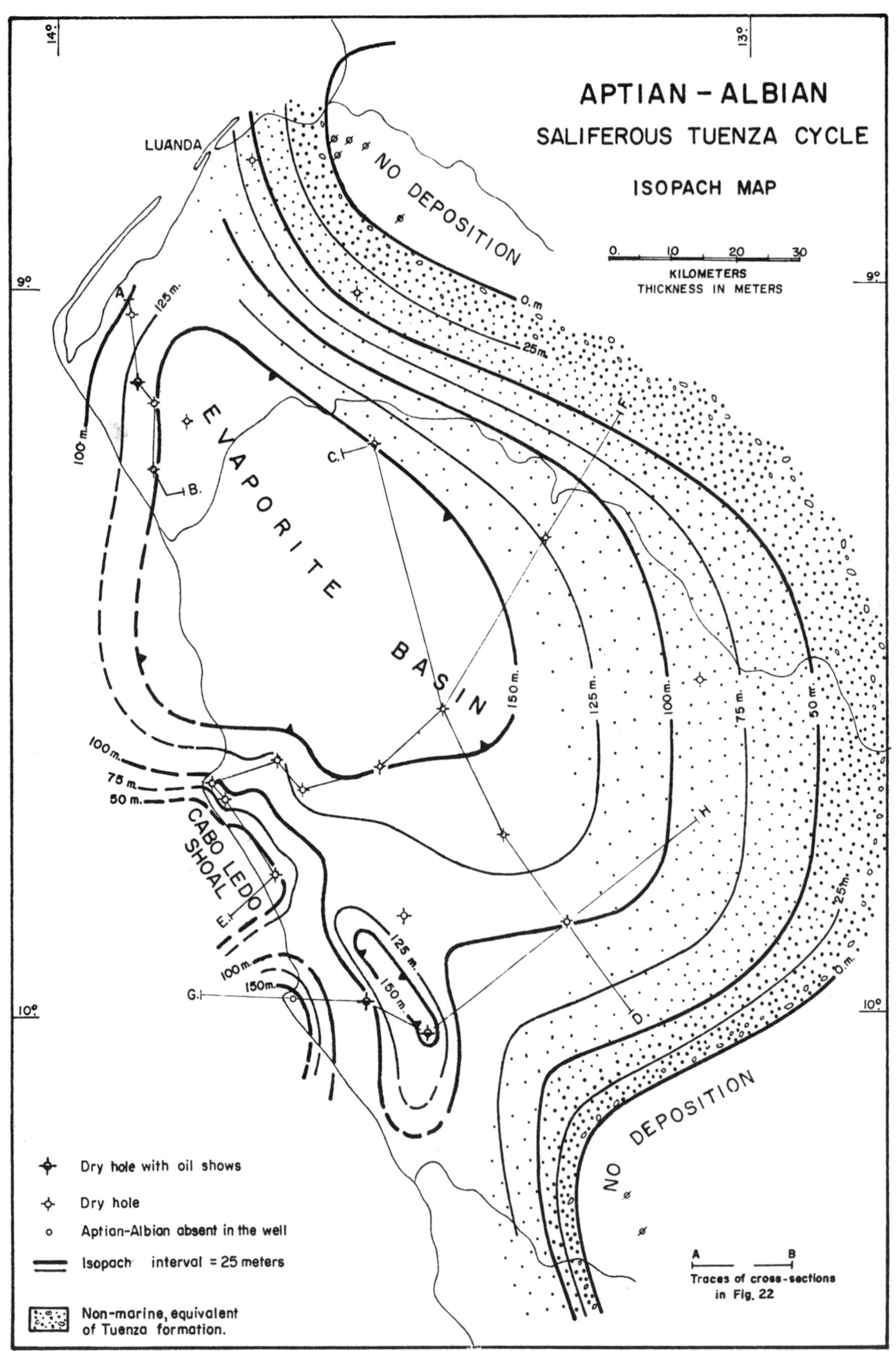

Fig.21 — Thickness of Aptian-Albian Saliferous Tuenza Cycle deposits. Oil shows from this interval are indicated.

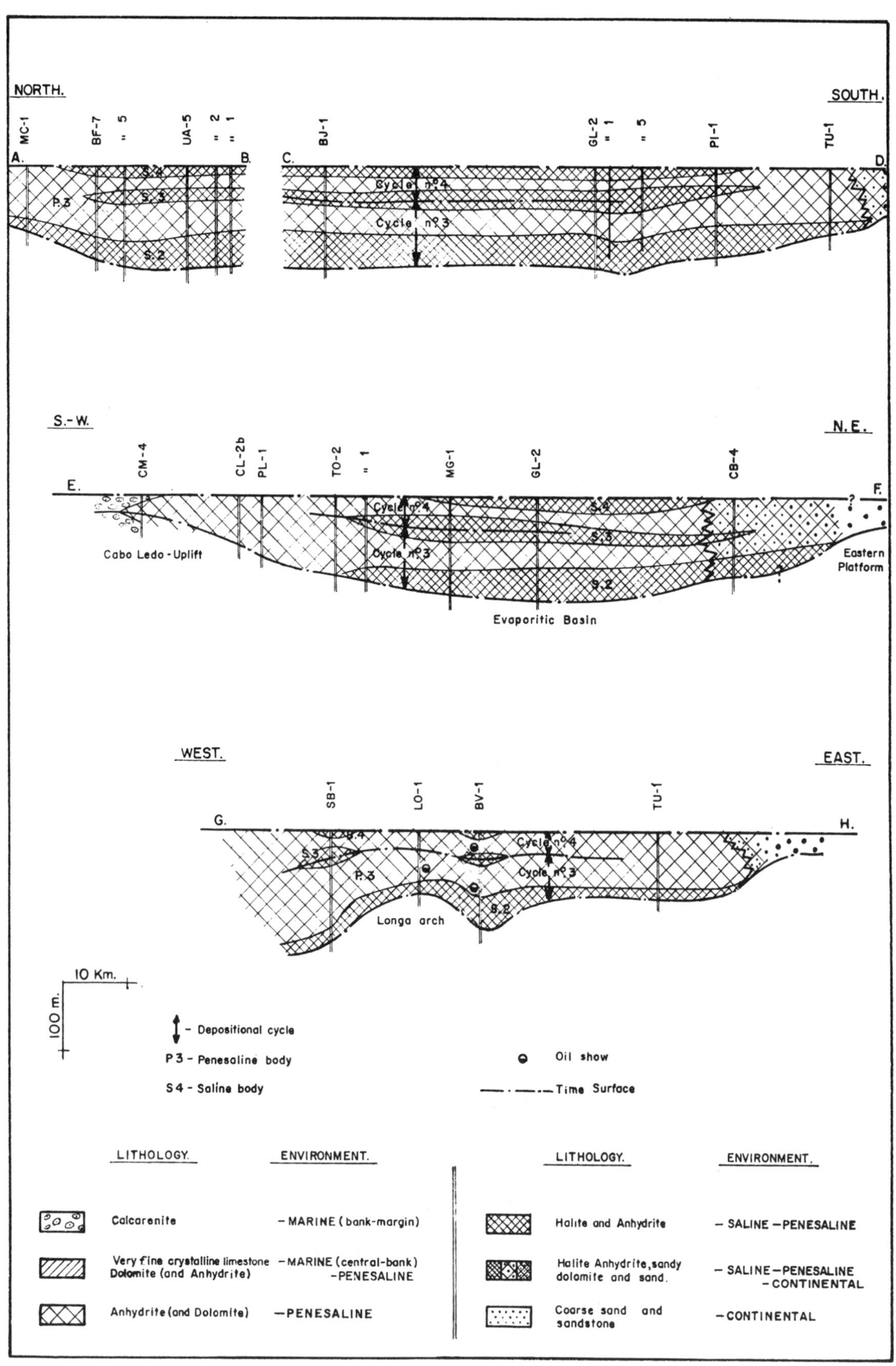

Fig.22 — Cross-sections of Aptian-Albian Saliferous Tuenza Cycles.

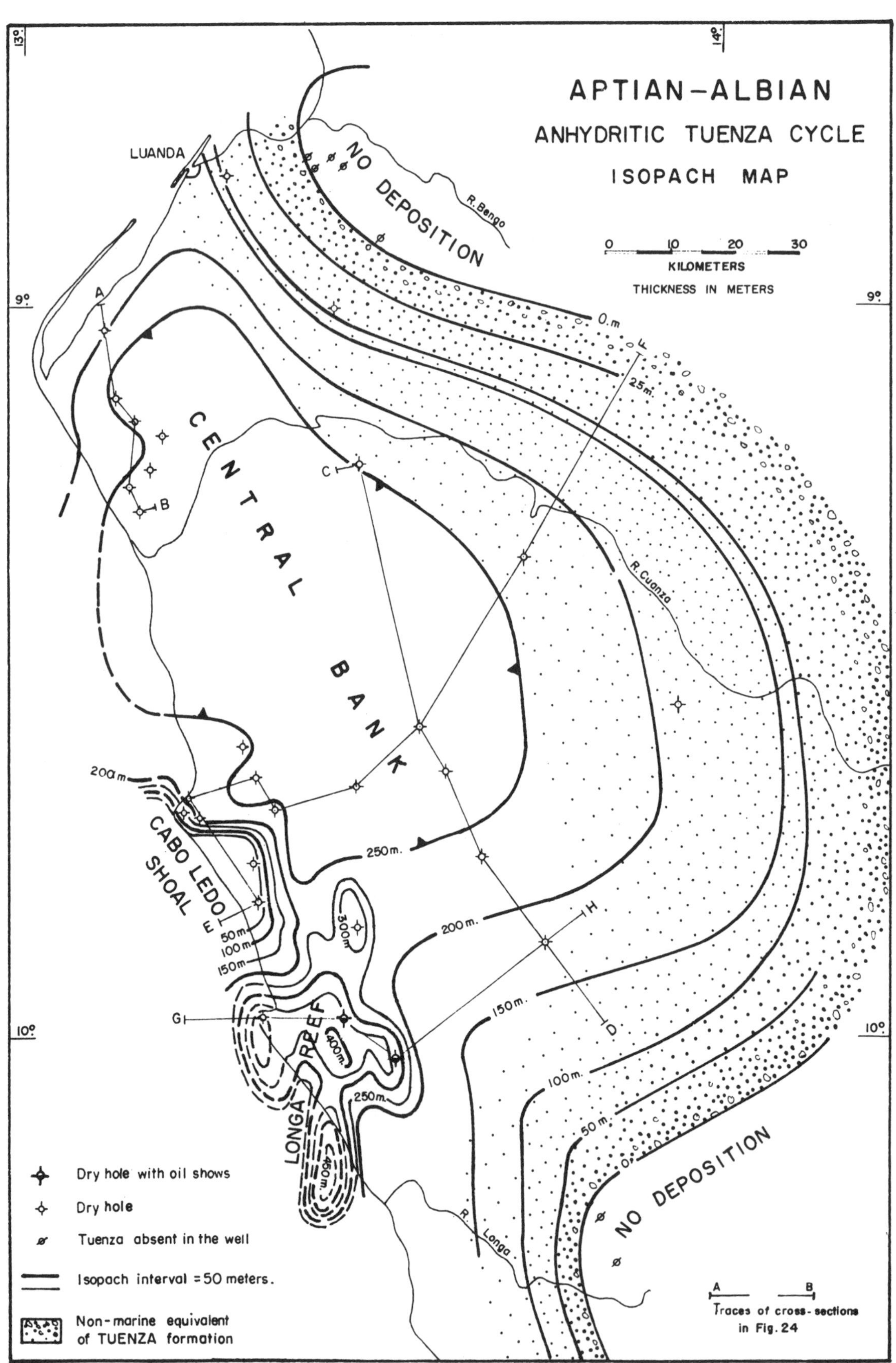

Fig.23—Thickness of Aptian-Albian Anhydritic Tuenza Cycle. Oil shows from this interval are indicated.

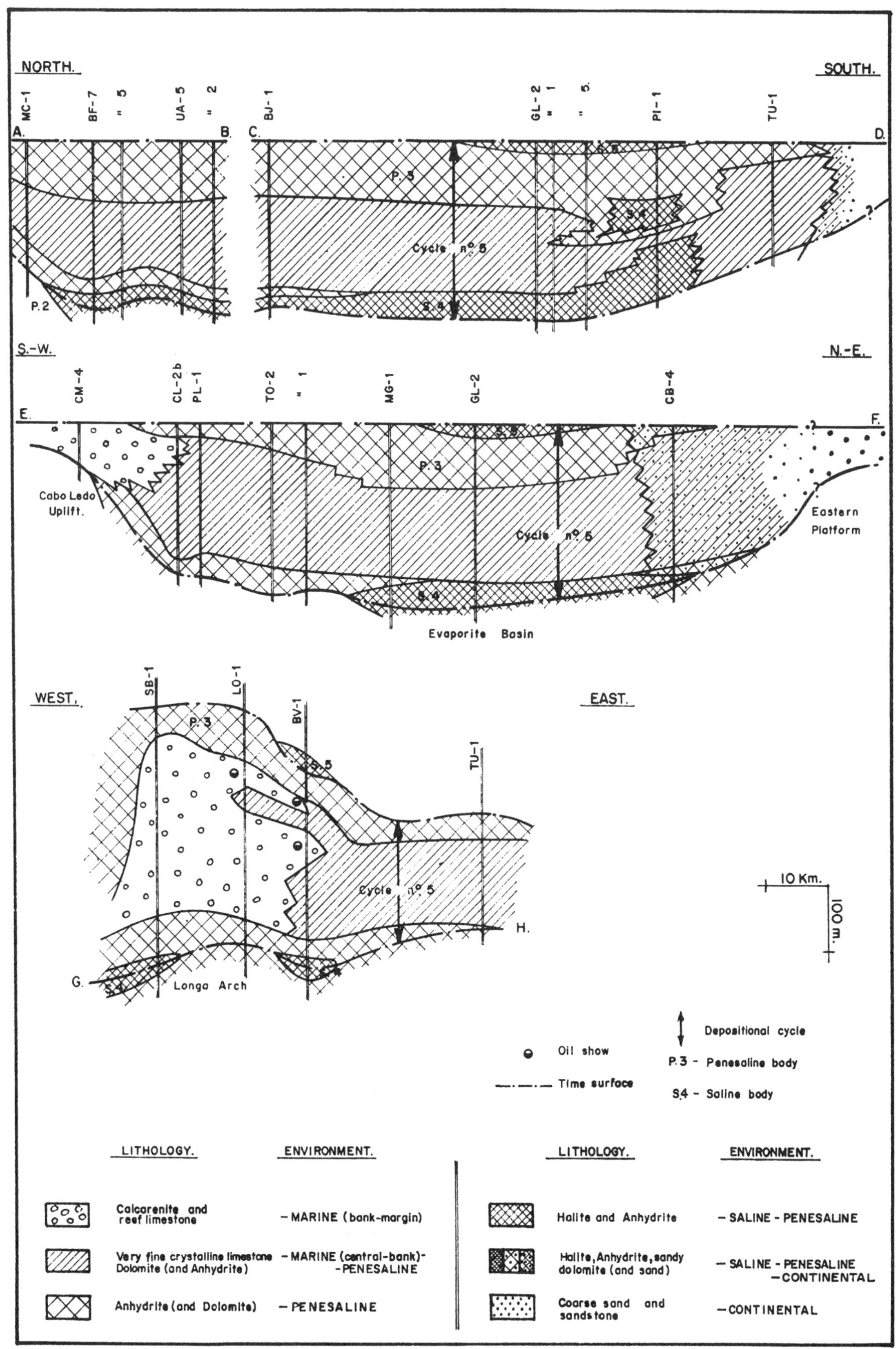

Fig. 24 – Cross-sections of Aptian-Albian Anhydritic Tuenza Cycle.

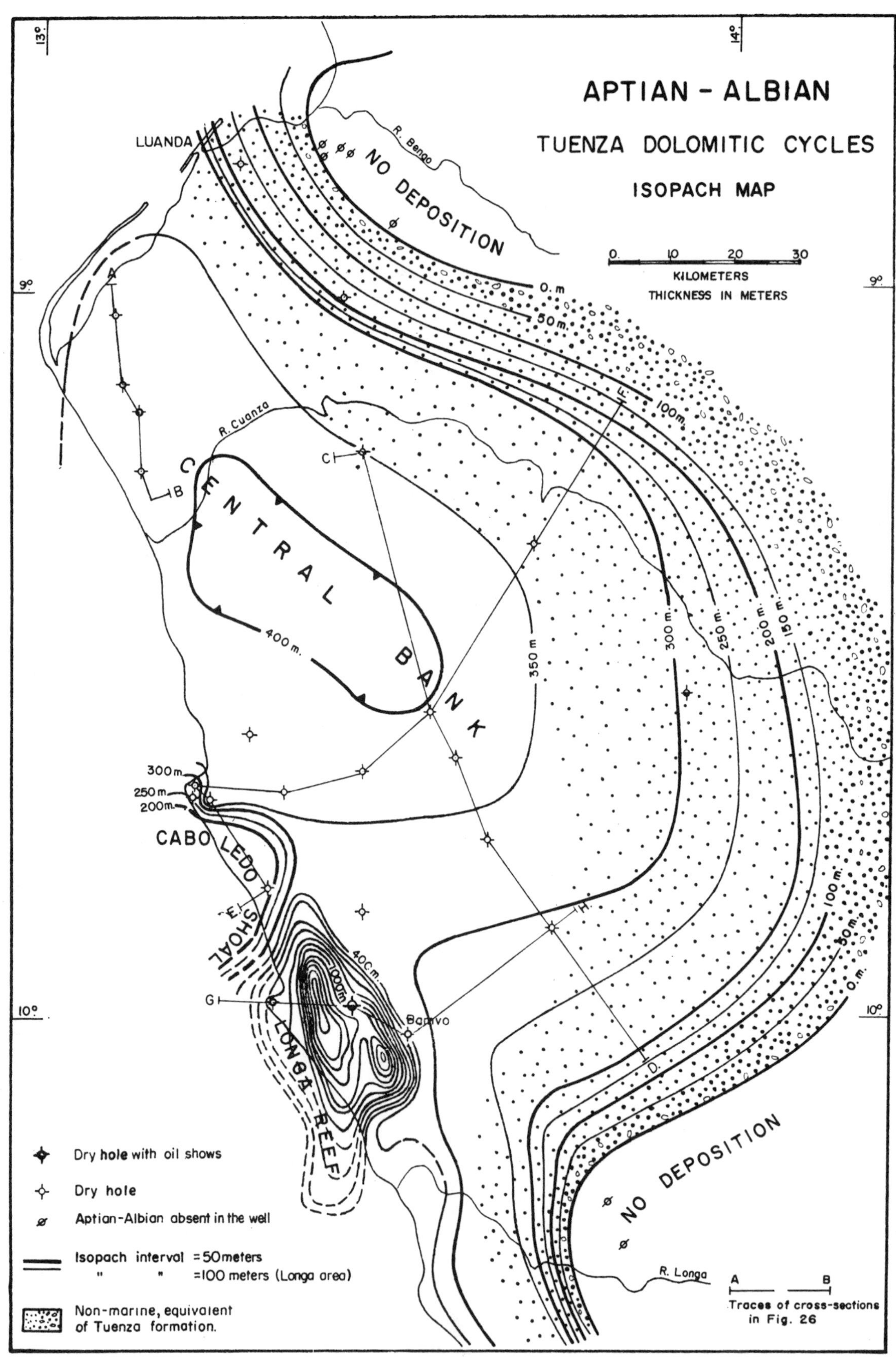

Fig. 25—Thickness of Aptian-Albian lower and middle Dolomitic Tuenza Cycles
Oil shows from these formations are indicated.

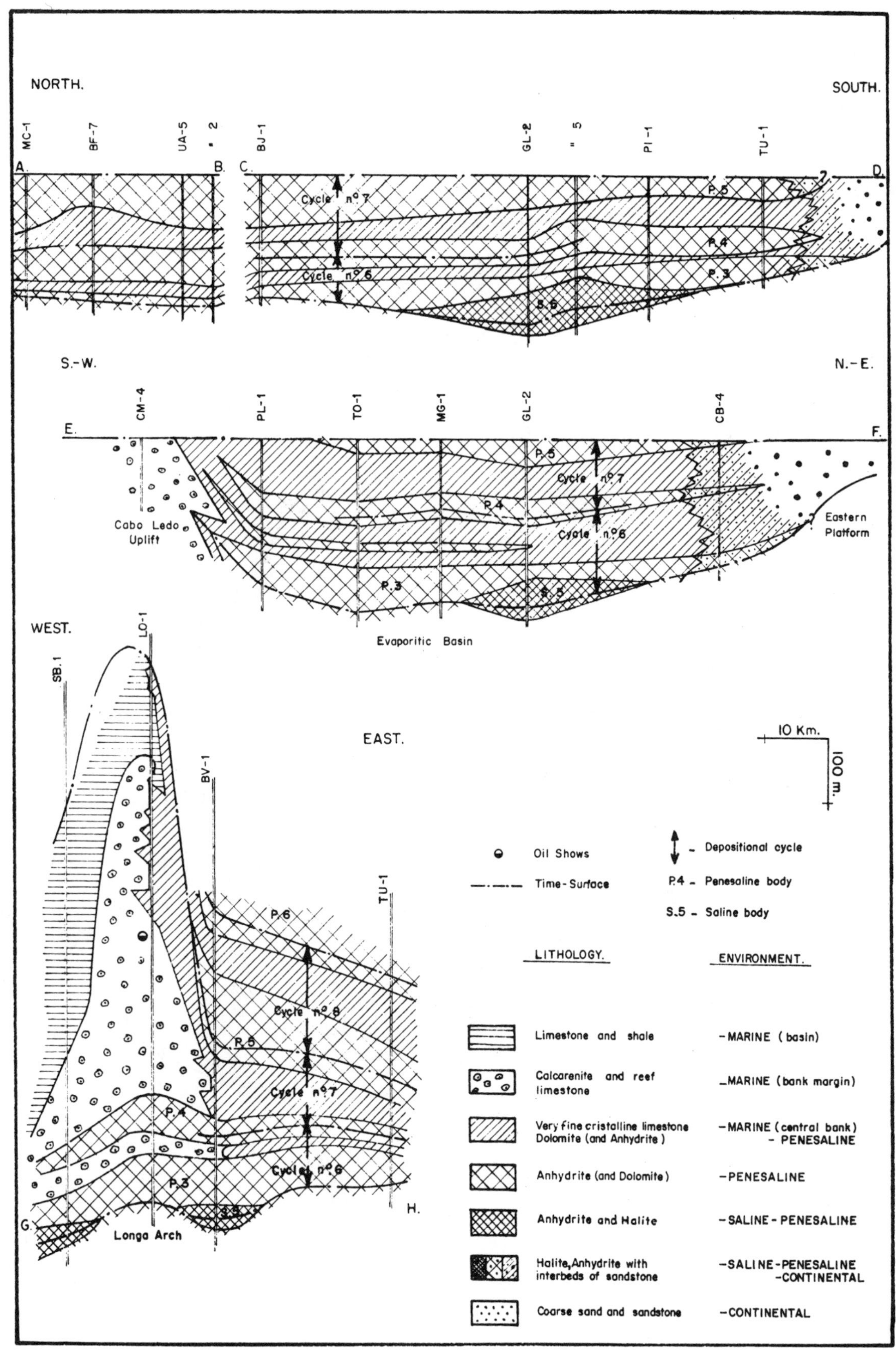

Fig. 26 — Cross-sections of Aptian-Albian lower and middle Dolomitic Tuenza

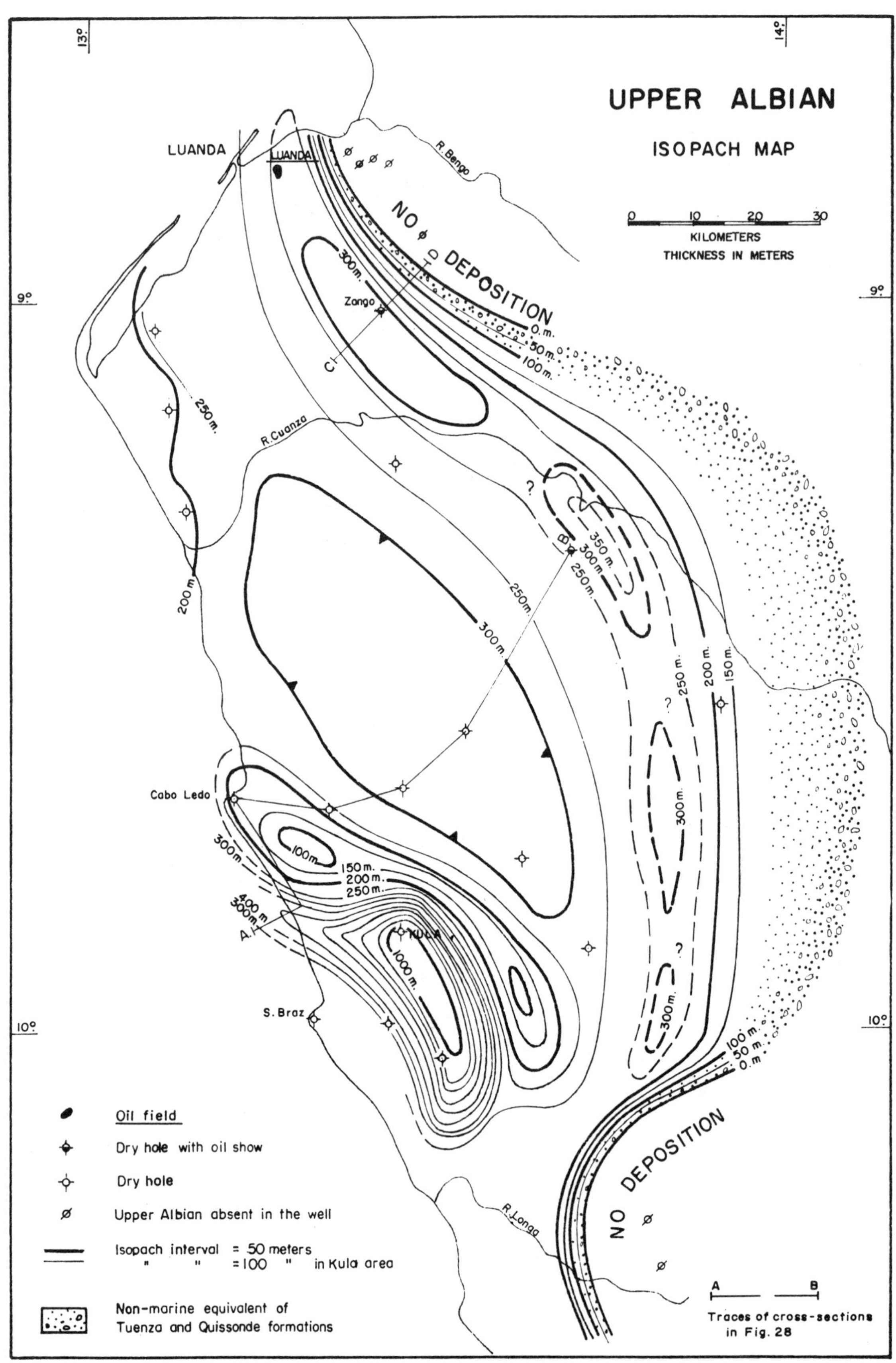

Fig. 27–Thickness of upper Albian strata. The Luanda field producing from rocks of this age is shown.

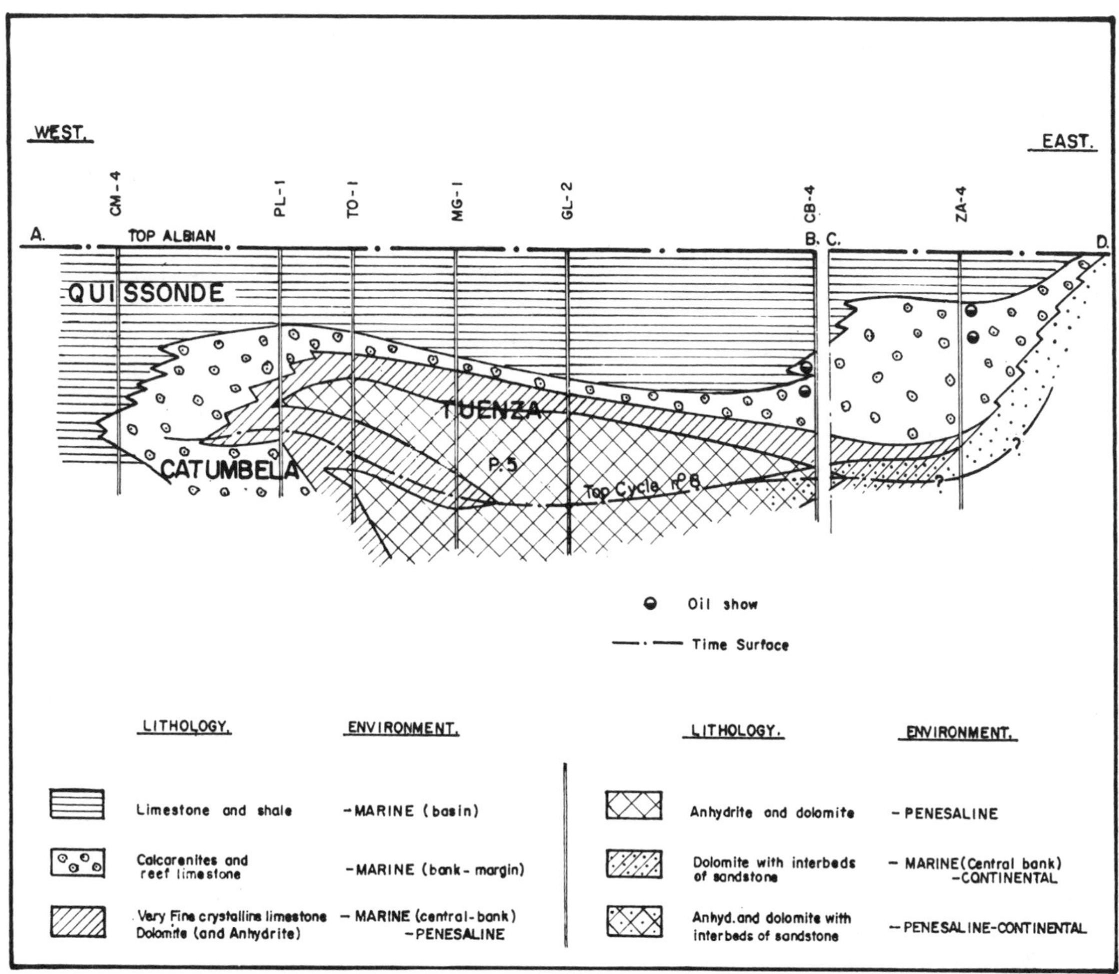

Fig. 28 – Cross-section of upper Albian strata.

clear, crystalline sparry calcite. Thus, Cycle No. 7 has a total thickness which seems exaggerated in relation to the thickness of the equivalent non-reef marine sediments of the interior basin. It is also notable that the larger reef buildup zone corresponds exactly with the zone where the thickness of the Massive Salt (averaging 100 m.) is considerably less than normal (Fig. 9). This suggests a thinning of the Massive Salt by loading of about 400 m. in relation to the original depositional thickness.

The fact was mentioned before that, after Albian time, important salt movements took place from west to east, from the Longa area toward the Galinda area, continuing during Late Cretaceous and even into Tertiary time, finally resulting in the enormous saliferous core of the Galinda anticline (Fig. 9). It may be assumed that lateral salt migration may have begun near the end of Aptian time with the help of excess weight introduced above the Longa arch by the reef buildup. Salt migration, accentuated by loads on the west, probably caused further development of the Longa reef buildup. This buildup attains a vertical relief difficult to explain solely by differential subsidence between the Longa arch and the surrounding areas (Fig. 12).

As in the Anhydritic Tuenza, oil shows consist of several barrels of oil produced in a well on the Longa arch. Oil also impregnates the coarse calcarenite lentils just below a thick anhydrite bed.

Upper Dolomitic Tuenza, Catumbela, and Quissonde.—This synchronous interval overlies the youngest Tuenza evaporitic Cycle No. 8, and corresponds with the complete invasion of the Aptian-Albian evaporitic basin by normal marine waters of an epicontinental sea (Fig. 27). The upper limit of this interval is defined by the appearance in the wells of *Anomalina berthelini.* This interval, in an average central part of the basin, from top to bottom, consists of the following lithologic succession:

1. Upper Quissonde: argillaceous limestone with *Lagena* sp. and shell fragments.
2. Lower Quissonde: argillaceous limestone with shell fragments.
3. Catumbela: bioclastic calcarenite with pisolites of large rounded fragments; caprinoidal limestone with algae and corals.
4. Upper Dolomitic Tuenza: dolomite and dolomitic limestone; anhydrite and dolomite.

In all of the area corresponding with the ancient Aptian-Albian evaporitic basin, the Catumbela is relatively thin (10–20 m.). It thickens, however, near the western margin of the basin, where it grades laterally westward into the Quissonde and eastward into continental clastic sediments.

This rather abrupt transition from Tuenza evaporites to almost normal-salinity sediments is good evidence for the invasion of the interior basin by oceanic water from the west. Indeed, at the end of Cycle No. 8, a remnant evaporitic penesaline lagoon was still in existence for a short time within the smaller interior basin, but almost surrounded by an arcuate shelf and a western arc of low semi-barrier uplifts. Because of a slowing of the subsidence, a broad bank or welt was formed as a final stage. On the surface of the welt, the latest Catumbela bioclastic shoals were formed and were exposed to a long period of reworking by the advancing Quissonde sea.

When subsidence was again initiated, the climatic conditions were both cooler and more humid. Thus, eventually the paleogeographic conditions changed and bank sedimentation changed gradually upward to normal epicontinental marine deposits. With this accentuation of subsidence, the littoral or sub-littoral limestones at the base of the Quissonde Formation progressively grade upward into fine marine facies of the upper Quissonde. However, a fringing reef developed during this time along the eastern flexure which limited the late Albian epicontinental sea. Thus the Quissonde Formation grades into two Catumbela reef trends which are not synchronous; the older one developed along the western platform of the interior basin just before the widespread invasion by the sea, and the second one, younger, along the eastern flexure (Fig. 28).

One well produces oil from this interval. In addition, several good oil shows have been found, all of them associated with the Catumbela facies of the eastern trend. The origin of this oil is not clear, but possible oil-source environments might have existed, represented by transitional strata intercalated between the Catumbela facies and the last lagoonal formations of the Dolomitic Tuenza. The oil also might have come from overlying uppermost Albian strata, where dark silty shale with some Foraminifera and vegetal debris is known to be present in argillaceous limestone. The trap for the oil of the Luanda field generally is related to a faulted zone in this formation.

Argillaceous-Arenaceous Formations of Late Cretaceous and Tertiary Age

Upper Cretaceous and Tertiary strata of the Cuanza basin are essentially a marine sequence, deposited in a bathyal to neritic environment, which overlapped eastward during Late Cretaceous time to the limits of the Albian basin and occupied nearly all the earlier basinal area.

The westward extent of Upper Cretaceous clastic terrigenous deposits was determined by the subsidence of the continental shelf and by the rejuvenation of the emerged relief, which controls the amount, type, and distribution of detrital material transported to and within the marine domain. However, the terrigenous material never extended greatly westward except during the late Miocene, and was deposited generally just west of the eastern flexure zone (Figs. 7, 8).

Thus, true infraneritic to bathyal marine facies were deposited during Late Cretaceous to Eocene time through most of the actual basin, suggesting that practically continuous sedimentation took place from latest Albian to the end of Eocene time. However, some local emergence occurred as a result of localized intra-basin saliferous tectonics. The transitional facies, littoral or more generally deltaic or lagoonal, were confined until Eocene time to the eastern part of the basin, where numerous stratigraphic lacunae appear, corresponding with brief emergences that affected only the eastern high shelf of the basin. During Miocene time, in contrast, the westward shift of the flexure (Fig. 8) which already existed during Eocene time allowed the transitional facies, mainly deltaic, to be deposited in the western part of the basin.

The detailed Upper Cretaceous and Tertiary stratigraphy is based essentially on micropaleontologic studies which made possible the establishment of several mappable subdivisions that can be grouped as follows: (1) Cenomanian-Turonian-lower Senonian; (2) upper Senonian-Paleocene; (3) Eocene; and (4) Miocene (Fig. 3).

Cenomanian-Turonian-lower Senonian.—This interval, extending from the top of the *Anomalina berthelini* zone upward to the top of the *Globotruncana ventricosa* zone (top of the Santonian), is represented in the western part of the basin by a monotonous series of brown, foraminiferal, more or less silty shale, overlying the upper Albian calcareous shales.

The Cenomanian Cabo Ledo Formation is distinguished from the surrounding formations because it corresponds almost exactly with the *Rotalipora* sp. zone (Fig. 3). The Cenomanian sequence represents a general transgression over the entire basin, and, in the east, it successively oversteps and truncates continental Albian, the Massive Salt, igneous rocks, and the Basement Complex. Only in the São Tiago area did transgression begin with deposition of a basal detrital deposit.

At Cacoba, the Itombe Formation is fine- to coarse-grained sandstone and little-consolidated, asphaltic sandstone interbedded with red claystone. This deltaic sequence is overlain by a coquinite of *Plicatula* sp. (caprinoidal limestone) of late Turonian age, indicating a new transgression. This section is succeeded by silty shale of Coniacian age, and covered in turn by the N'Golome shale of Santonian age.

The interval corresponding to these three formations, Cabo Ledo, Itombe, and N'Golome, extending from Cenomanian through Santonian time, begins and ends with two easily recognized transgressive phases separated by a Turonian regressive phase that corresponds with the maximum extension of the Cacoba delta. The eastern flexure, which limited the interior Albian basin on the east, was shifted far toward the east of the Cacuaco and Morro Liso ridges; the area of maximum subsidence also was shifted in the same direction, becoming situated west of the new position of the flexure.

The isopachous map of this interval (Fig. 29) shows some local thickness variations in the western half of the basin which were caused by salt tectonics; in the eastern half thickness variation results from gradations of marine into transitional facies.

The thickness maxima of this section, found in the wells Morro Liso No. 1 (1,486 m.) and Cacuaco No. 6 (1,082 m.), comprise black and gray, thin-bedded, carbonaceous and bituminous, silty shale, with Foraminifera, interbedded with black, silty, and bituminous claystone containing vegetal debris (Fig. 32C). Effective permeability exists in fractures which extend in all directions, mostly vertically, and several hundred barrels of high-gravity oil have been produced from the Morro Liso well. This formation is underlain by Massive Salt and covered by gray marine shale. It is reasonable to believe that the Cabo Ledo and Itombe Formations in this area are petroleum source beds.

Upper Senonian-Paleocene.—Sedimentation continued during late Senonian-Paleocene time, conditions being much the same as during Cenomanian-early Senonian time. The central part of the basin received a monotonous sequence of more or less silty shales with Foraminifera, the highest occurrence of *Globorotalia* sp. enabling identification of the Paleocene-Eocene contact.

Toward the east several lithologic criteria make it possible to distinguish formations above the Santonian N'Golome Formation. These include the lower and middle Teba Formations (Campanian) and the upper Teba Formation (lower Maestrichtian) which is characterized by the presence of *Inoceramus* fragments, suggesting the return to more neritic conditions. The Rio Dande Formation (upper Maestrichtian-Paleocene), composed of marly siltstone and limestone, indicated a return to deeper-water sedimentation.

Like the Cenomanian-lower Senonian, the upper Senonian and Paleocene strata indicate a new accentuation of the eastward-shifting flexure and the area of maximum subsidence which remained close to and just west of the flexure (Figs. 30, 32B). The terrigenous transitional facies are now incomplete in outcrop, because of continual uplift and destruction by erosion along this eastward-shifting flexure. Near Dondo, for example, on both sides of the Cuanza River, the upper Senonian and the Paleocene are thick sequences of unfossiliferous yellow to tan sandstone and red to tan argillite which may represent a fluvio-deltaic environment. Toward the north,

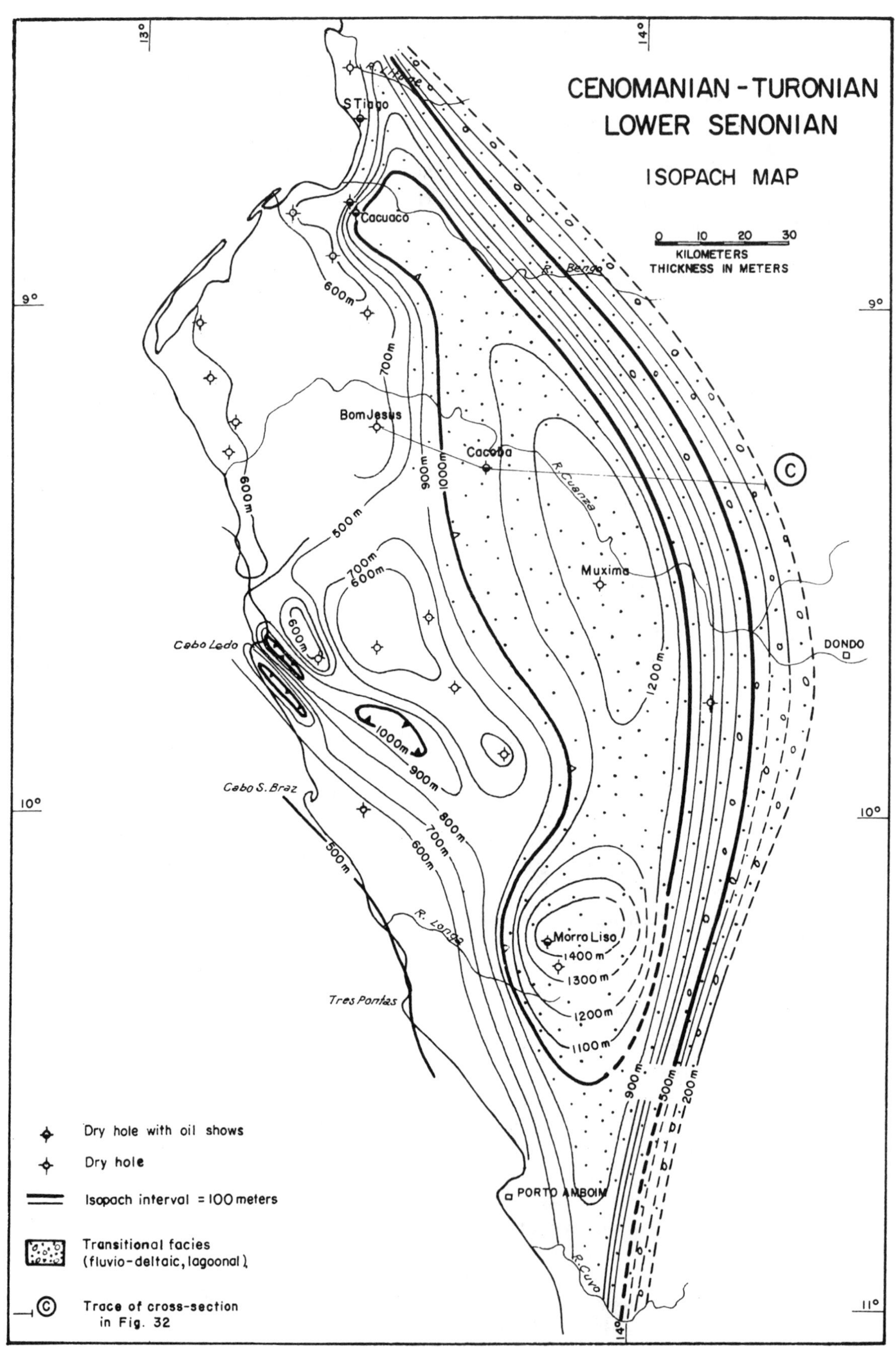

Fig.29 — Thickness of Cenomanian, Turonian and lower Senonian Strata.
Oil shows from rocks of this age are indicated.

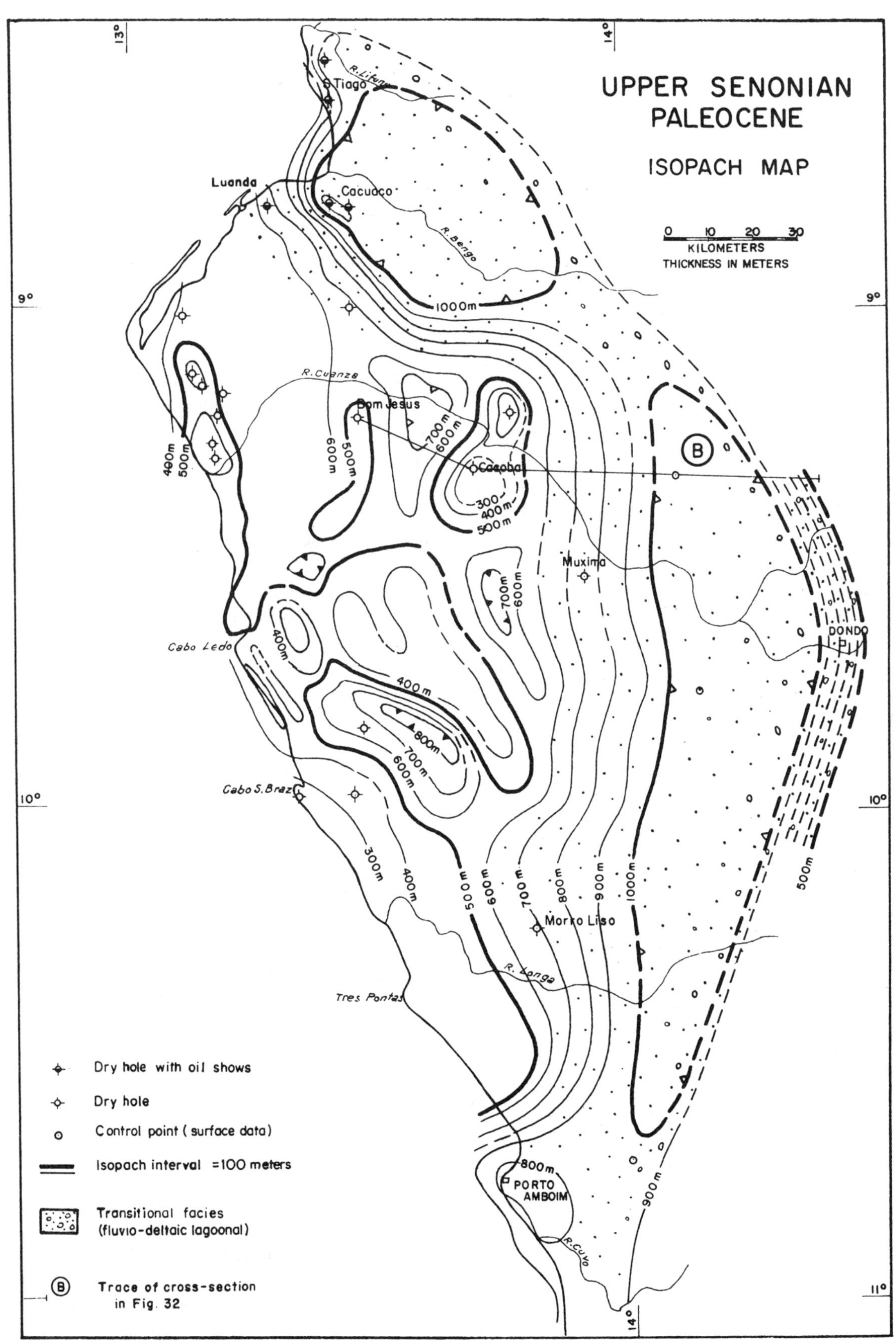

Fig. 30— Thickness of upper Senonian and Paleocene strata.
Oil shows from rocks of this age are indicated.

at Cacuaco and also in Luanda, the wells encounter black unfossiliferous shale, argillite, and finely laminated interbeds of dark marl, containing numerous Foraminifera filled with free oil. It is believed that this oil originated *in situ* in these upper Senonian-Paleocene strata.

Eocene.—Attributed to the Eocene are the Gratidão (lower Eocene) and Cunga (middle and upper Eocene) Formations, but the base of the Gratidão may in places be of late Paleocene age (Fig. 3). The Cunga Formation in other places may be only of early and middle Eocene age, late Eocene sediments having been eroded and being present in very few places.

Inside of the basin, the Eocene, whose average thickness does not exceed 200 m., is composed of shale rich in Foraminifera, radiolarians, and fragments of very small unidentified organisms. The Eocene can be distinguished from the surrounding formations, especially from the Paleocene, only on the basis of its microfossil content. This is true because in the greater part of the basin sedimentation was continuous from Paleocne through Eocene time, except on the top of the growing salt structures where Eocene strata are preserved in very few localities (Fig. 31).

Near the basin margin, in outcrop, the Eocene can be recognized fairly easily by its whitish chalk aspect, by the rich chert occurrences in some places, and by the presence of silicified limestone beds in the Gratidão. Its thickness increases considerably (up to 1,000 m.). This thickening is coincident with the appearance within the marl of extremely fine detrital material and also of numerous organic and coaly fragments, as at the Morro Liso outcrops. Interbeds of black bituminous marl and argillite containing fine laminae of black microcrystalline dolomite are found at São Tiago cliff.

The axis of maximum subsidence of this euxinic-type sequence was shifted westward from Paleocene through Eocene time (Figs. 7, 8), reaching approximately the same geographic location as that of Cenomanian-Turonian time. During the Oligocene emergence, this stratal section was removed from many locales by subaerial erosion, adding to the difficulty of mapping.

In the Luanda and Cacuaco areas, oil impregnates silty shales about 100 m. thick, apparently deposited in an euxinic environment. Bubbles of oil and gas were observed along bedding and slumping planes in Cacuaco wells. It is difficult to visualize this as other than *in situ* oil.

Oligocene.—Despite the fact that no Oligocene sediments are present, the Oligocene is a most important sedimentational epoch during which active salt tectonics took place as the entire area of the Cuanza basin stood above sea-level.

Miocene.—The Miocene of the Cuanza basin is represented only by sediments of the Aquitanian and Burdigalian stages. It corresponds with a new, and the last, sedimentary cycle after the Oligocene emergence.

At the base of the Miocene section is the lower Quifangondo Formation, which, in most places, is black or variegated, highly gypsiferous, unfossiliferous shale, resting unconformably on the upper, middle, or lower Eocene or, in some places, Campanian strata. Above this is black to brown shale with Foraminifera of the *Globigerina dissimilis* zone. Following this the upper Quifangondo was deposited, consisting of silty shale, very rich in Foraminifera, which grades upward into coquinoidal limestone with sandy limestone interbeds. This member corresponds with the *Globigerinatella insueta* zone (Fig. 3).

The next younger Luanda Formation has brown shale at its base, containing Foraminifera of the *Globorotalia fohsi* zone, abruptly transitional upward into littoral and deltaic sands with sandstone interbeds that close the cycle.

The Miocene of the Cuanza basin seems thus to correspond with one complete sedimentary cycle, beginning with a mild marine transgression during early Aquitanian time, covering the older lowlands of the Oligocene relief. The transgression intensified and reached its maximum during late Aquitanian and early Burdigalian time, when the Miocene seas occupied the greater part of the sedimentary basin (Figs. 33, 34).

Adjacent to the Miocene flexure, great thicknesses of sediments were deposited which comprise marine shale and dark azoic argillite with lensing interbeds of sandstone; however, on the platform east of the flexure, sedimentation took place in much shallower water where limestone with algae, pelecypods, and echinoderms of the Cacuaco Formation was deposited (Fig. 35). This development came to a maximum in early Burdigalian time, followed by a period of fairly active tectonism, which was succeeded by an intense period of erosion that allowed the ancient Cuanza River to bring into the Quenguela trough more than 1,500 m. of poorly consolidated sand and clay in a fluvio-deltaic and littoral environment. The Miocene sedimentary cycle came to an abrupt close, ending Tertiary sedimentation in the Cuanza basin. As the subsidence slowed in relation to the amount of available sediments, the sea was expelled from the area of the presently known Cuanza basin.

Drilling in the Quenguela and Funda troughs indicates the existence of a very thick Aquitanian section in which the base of the unit shows the greatest thickening (the *Globigerina dissimilis* zone). The thickening is accomplished by the appearance of intercalations of dark, unfossiliferous, micaceous, silty argillite beds containing fine interbeds of microcrystalline dolomite and relatively thick white sandstone. The lithologic types indicate the presence of a transitional facies related to the flood plain and delta of the Miocene Rio Cuanza.

Some oil and gas occurrences are known in these Aquitanian strata, which are petroleum source beds as well as reservoirs.

Oil Fields and Areas of Potential Productive Hydrocarbons

Petroleum occurrences are known in almost all of the formations from Lower Cretaceous to Miocene, but each formation is productive, or potentially productive, in different localized areas of the basin. The geographical extent of these areas is related to regional tectonic features, which control favorable conditions for reservoir and source-bed developments. The distribution of these productive zones and descriptions of their respective oil occurrences are set forth in the following paragraphs.

Oil in basal igneous and arenaceous pre-Aptian formations.—In the stratigraphic sequence, these formations are located below the Massive Salt. Therefore, they are not affected by salt tectonics, and oil and gas in stratigraphic traps are related to old positive structural features.

Gas and condensate have been discovered at 2,129 m. (subsea) at Puaca (Fig. 2) on the Cabo Ledo uplift (Fig. 5) in lentils of volcanic ash between basalt below, and the Quianga-Binga anhydrite above. The offshore extension of this gas field is not known.

The oil of the Cacuaco field is trapped at 2,425 m. (subsea) along a ridge of the Basement Complex (Fig. 5) in clean sandstone pinching out between mica-schist below and the basal anhydrite of the Massive Salt above. The sandstones are cemented by kaolinite and dolomite, and a downdip wedge-out of permeability is related to a downdip increase of the kaolinite cement. The clean sandstone zone of the updip permeability wedge-edge may correspond with an ancient shoreline of the Cuvo sea. Oil may have originated in transitional facies deposited downdip, similar to the facies presently known to crop out at Calucala (Fig. 13).

Good oil shows have been found in the Upper Cuvo Formation on the flanks of such old structural highs as in the Morro Liso well No. 1, where a pinch-out of the upper Cuvo exists. Therefore, the flanks of these positive zones, which emerged before the deposition of the salt (Cabo Ledo uplift, the Longa, Cacuaco, and Morro Liso ridges; Figs. 5, 6), are potential productive zones where oil might be trapped in pinch-outs of the Upper Cuvo standstones.

Oil in Aptian Binga Formation.—Hydrocarbon occurrences in the Binga are known everywhere in the Aptian Cuanza basin (Fig. 17) either in the subsurface or at the surface where the asphaltic Binga limestone crops out in the Rio Lifune Valley (Fig. 1).

This regional distribution of oil occurrence in Aptian Binga deposits might be explained in two ways. (1) Over the whole basin, carbonate muds were deposited in a restricted environment. These were in turn overlain by porous sandy calcarenite or dolomite shoal deposits. A regional and progressive decrease in the rate of subsidence after the deposition of the Massive Salt appears to have caused this widespread close association between source and reservoir rocks. (2) The anhydrite cap overlying the calcarenite and dolomite corresponded with a new and sudden beginning of subsidence of the filled basin.

The Binga Formation is potentially productive only within the areal limits of this anhydrite (penesaline body P2) and where it overlies the reservoir beds (Fig. 16). The best primary porosity and permeability, improved by secondary dolomitization and related to a bank-margin environment, are found at the periphery of this potential productive zone (Fig. 19). Inside of this peripheral area, however, the Binga oil fields (Tobias, Galinda, Benfica, and Uacongo) appear to occur on the top of salt anticlines, where fracturing generally compensates for a possible lack of primary permeability.

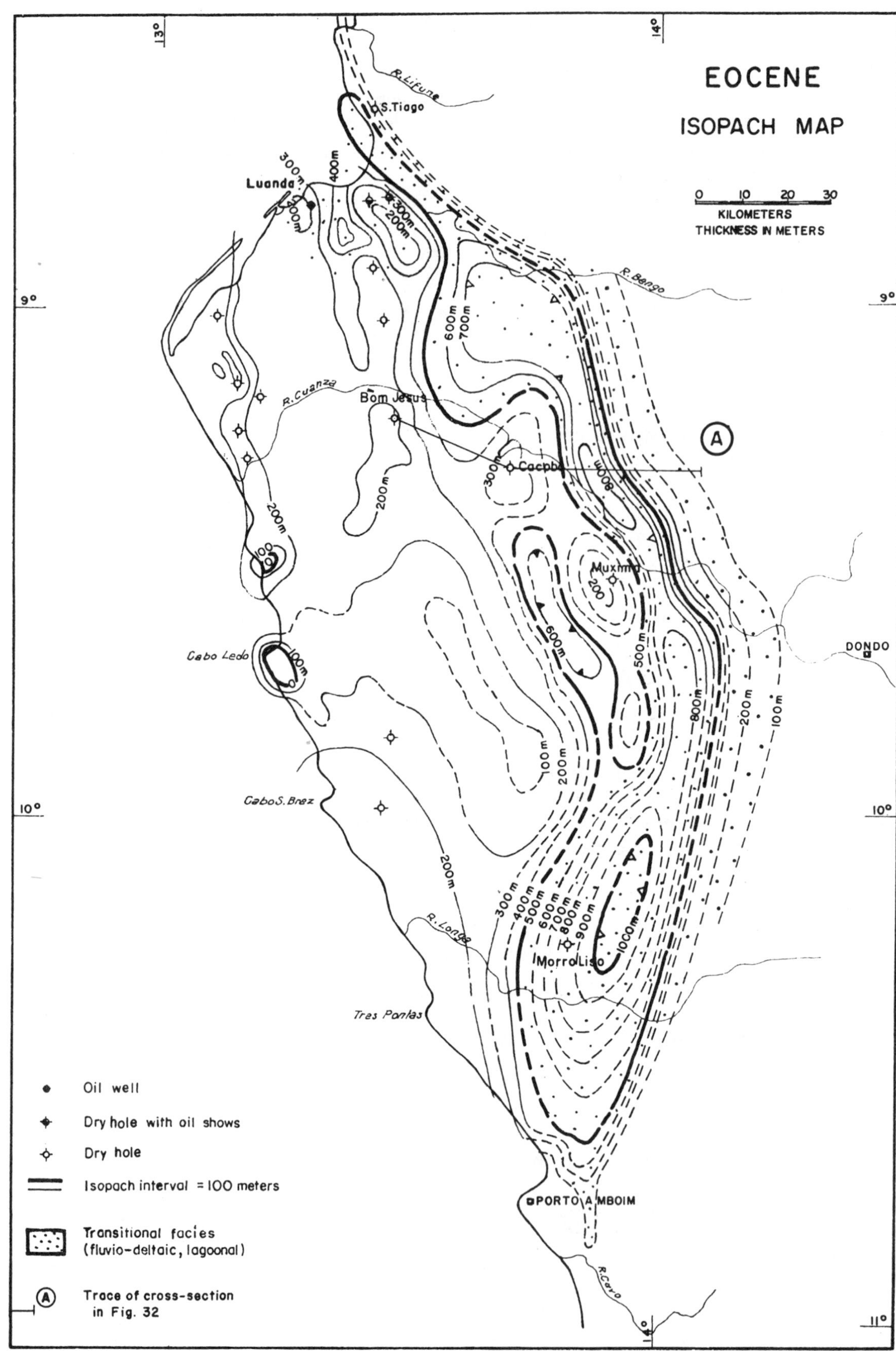

Fig. 31—Thickness of Eocene strata. Oil shows from rocks of this age are indicated.

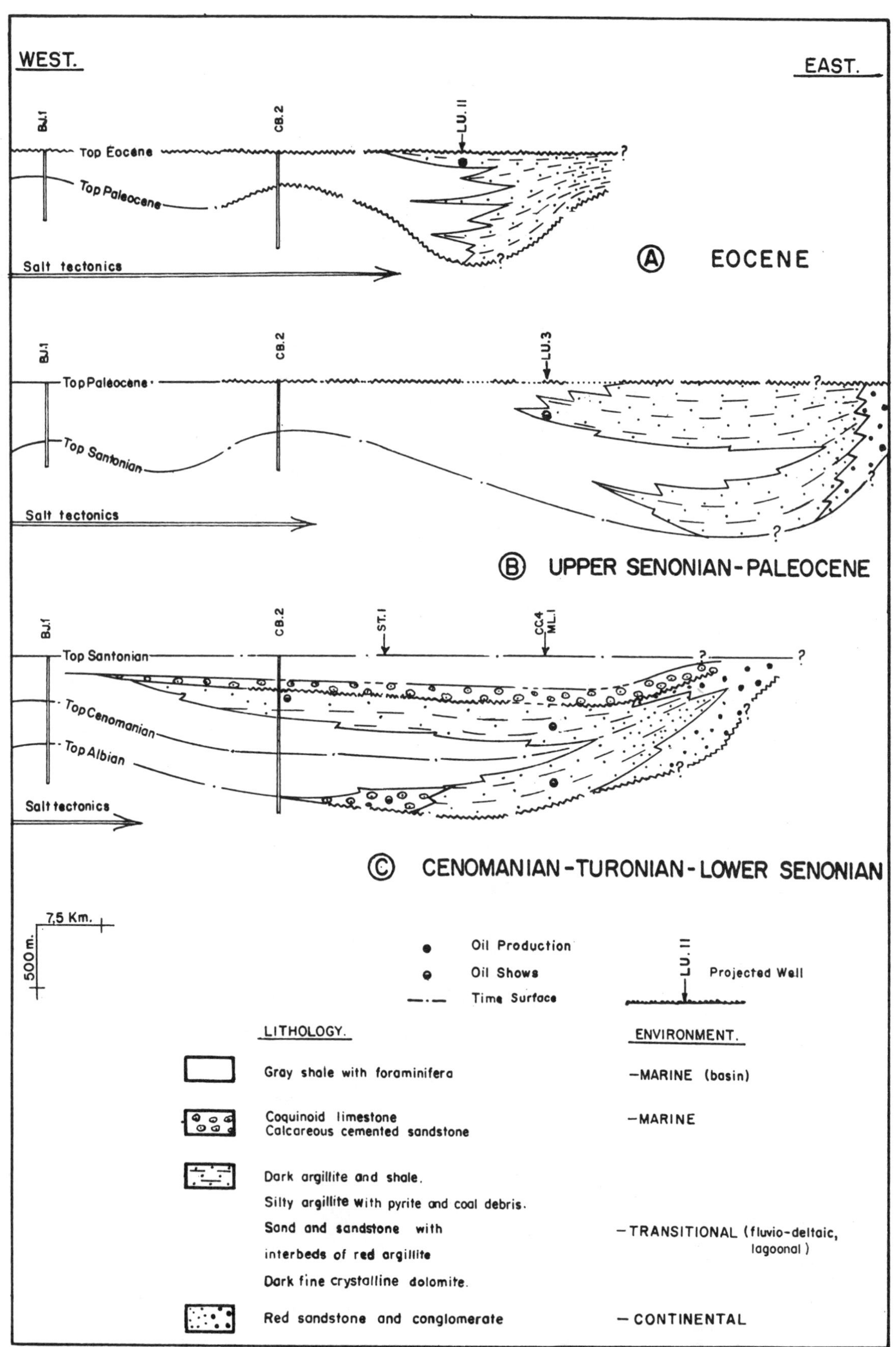

Fig.32 – Cross-sections of Upper Cretaceous, Paleocene and Eocene beds. The eastern limit of salt tectonics coincides with the western limit of the transitional facies.

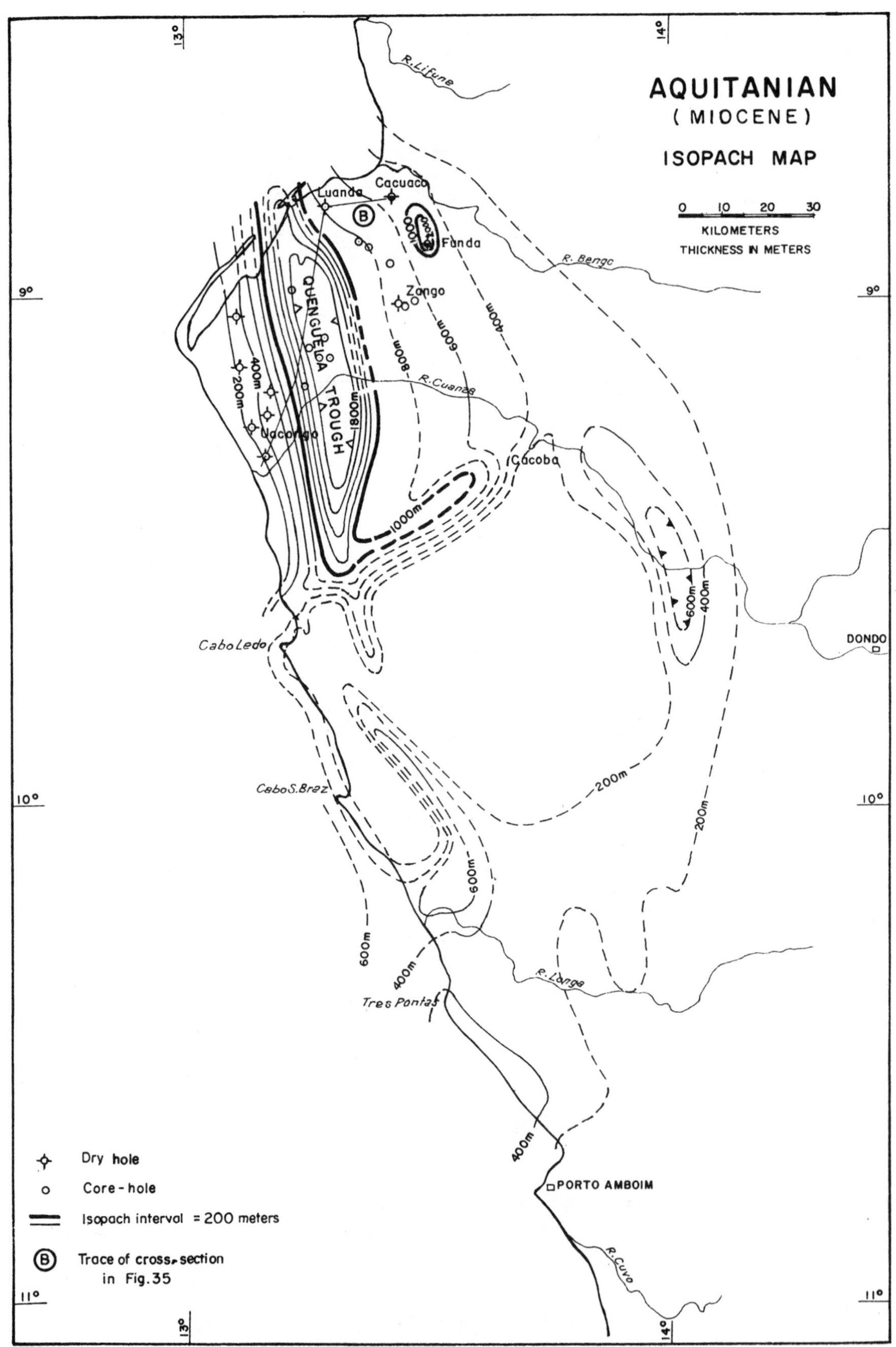

Fig.33—Thickness of Miocene Aquitanian beds.

In the Cuanza basin, the various types of salt structures (expanded or residual) represent all the stages of salt deformation and evolution (Figs. 9–12). The stages include: (1) non-piercement salt expanded into a flat anticlinal fold (*e.g.*, Galinda); (2) non-piercement salt expanded into a high, ridge-like anticlinal fold with adjacent strata strongly faulted on the flanks (*e. g.*, Tobias); (3) piercement salt dome (*e. g.*, Cacimbas); (4) piercement salt domes with an axial graben affecting the overlying layers (*e. g.*, Pitchi); (5) "relic" piercement salt structure where salt was largely extruded and destroyed (*e. g.*, Cabo Ledo); and (6) residual interdomal salt features (*e.g.*, Luanda and Benfica anticlines).

Oil fields in the Binga Formation are related solely to expanded non-piercement or interdomal residual salt structures, and are never related to piercement salt, which seems to have destroyed pre-existing oil accumulations. Thus, the few barrels of oil produced with salt water from the Binga reservoir on the flank of the Pitchi salt diapir, and the strong oil shows observed in the salt itself, suggest that intrusion of the salt destroyed a pre-existing oil pool trapped before the end of Late Cretaceous time.

The Binga reservoirs related to expanded salt domes are the most fractured and the most prolific. The largest oil field of the Cuanza basin, the Tobias field (Fig. 36), corresponds with a maximum expansion of the Massive Salt just before piercement.

Characteristics of the Tobias field are the following.

Producing zone: oölitic sandy limestone and very fine crystalline limestone (both highly fractured) of the Binga Formation
Thickness of productive zone: 100 m.
Matrix porosity: 4-12 per cent (oölitic limestone)
Oil column: 350 m.
B. O. P. D.: 18,000
Cumulative oil production through December 31, 1964: 15,746,430 barrels (88.7 per cent of Angola production)

The original fluid pressure in the Binga Formation at Tobias is abnormally low (948 psi, at 2,343 ft.) in contrast to the excess reservoir pressure of other Binga oil pools. This pressure is also *less* than that in the shallower formations of the Tobias anticline.

Oil in Lower Cretaceous formations (except Binga Formation).—Oil shows have been found in the Aptian-Albian carbonate formations penetrated by wells drilled along the western margin (Quianga and Tuenza) of the basin, where carbonate platform sediments persisted on a western stable area. Along the Cabo Ledo uplift and the Longa ridge, the interfingering of source beds with reef or reef-like calcarenite and anhydrite provided the source and reservoir environments for the oil and gas occurrences in Aptian and Albian reservoirs (Figs. 24, 26).

Oil occurrences in upper Albian deposits are known along the eastern flexure of the Albian basin in the Catumbela reef facies, capped by upper Albian marl. The source of oil in this formation could be the upper Albian shale or the transitional lagoonal deposits which underlie the Catumbela facies (Fig. 28).

In the Cacoba well No. 1 and the Zango well No. 1, the Catumbela Formation was found to be impregnated with heavy oil, asphalt, and salt water. The Zango No. 1 is on the flank of a large saliferous anticline with an axial graben, the structural picture resembling very much the Pitchi anticline (Fig. 12). This similarity between the two structures suggests that a piercement salt dome might have formed in the Zango graben and destroyed an oil pool trapped before Late Cretaceous time. One well of the Luanda field, related to a residual salt anticline, is producing from the Catumbela Formation; the other wells produce from a fault zone in the Quissonde Formation, overlying the Catumbela Formation.

In the Albian formations, potentially productive zones can be related to the Cabo Ledo uplift and the Longa arch in the west, and to the Albian flexure (Fig. 7) in the east.

Oil in Upper Cretaceous and lower Tertiary Formations.—In Upper Cretaceous and Tertiary formations, oil is found in deposits of transitional, lagoonal-deltaic environments. The depositional profile of these sediments, which resembles the classic delta, is demonstrated in electric logs. Depocenters (thickness maxima) are everywhere located west of a major flexure which was shifted eastward during Cretaceous and westward during Tertiary time. Visual observation and chemical analyses suggest that the black, very fine, silty, argillaceous, calcareous deposits are representative of oil-generating environments. This may be the only way to explain oil impregnations observed in sections 100 m. or more in thickness along bedding and slumping planes of impermeable rocks found in the Cenomanian, Senonian, Paleocene, and Eocene strata of the Cacuaco and

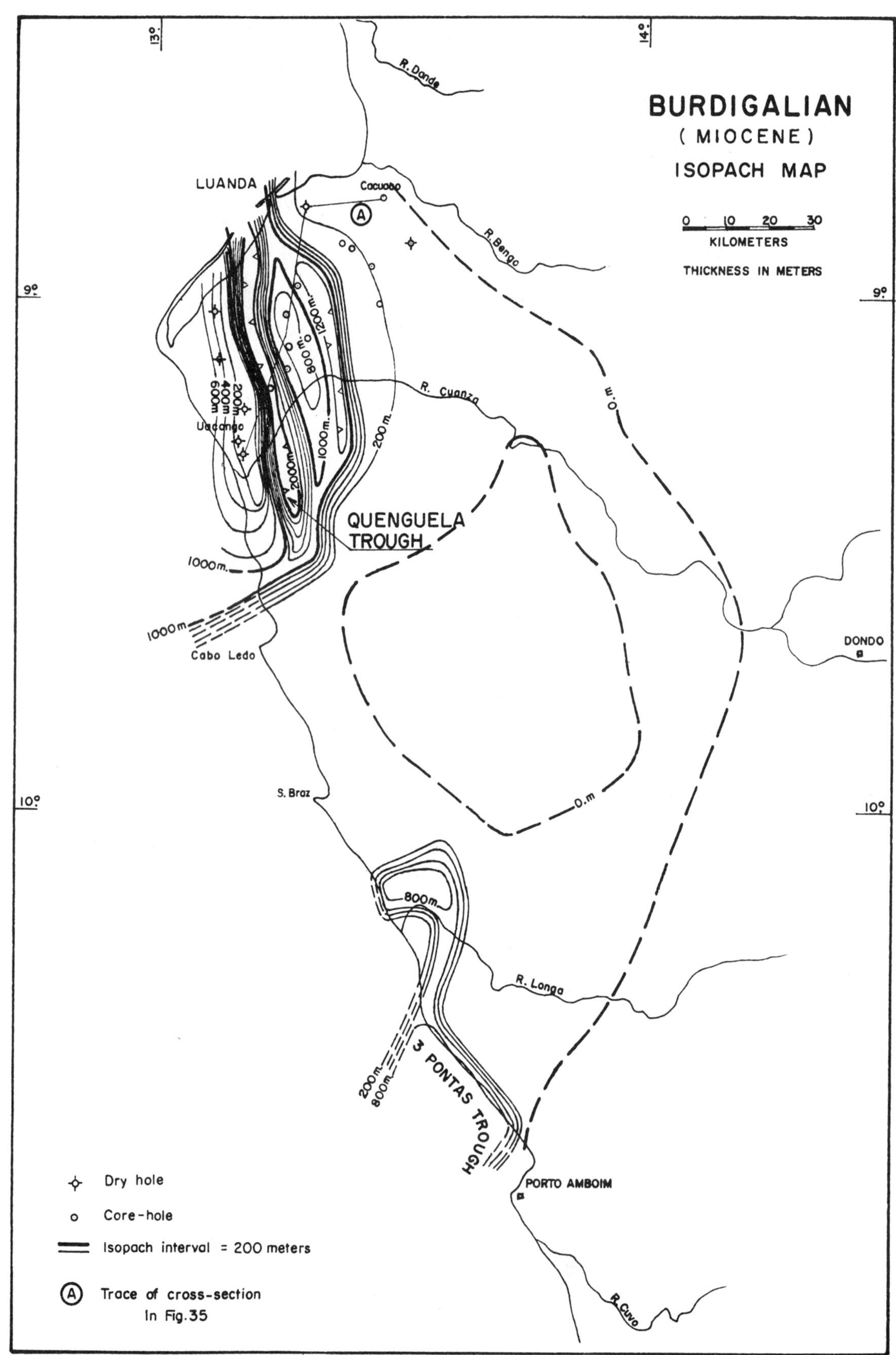

Fig. 34—Thickness of Burdigalian beds.

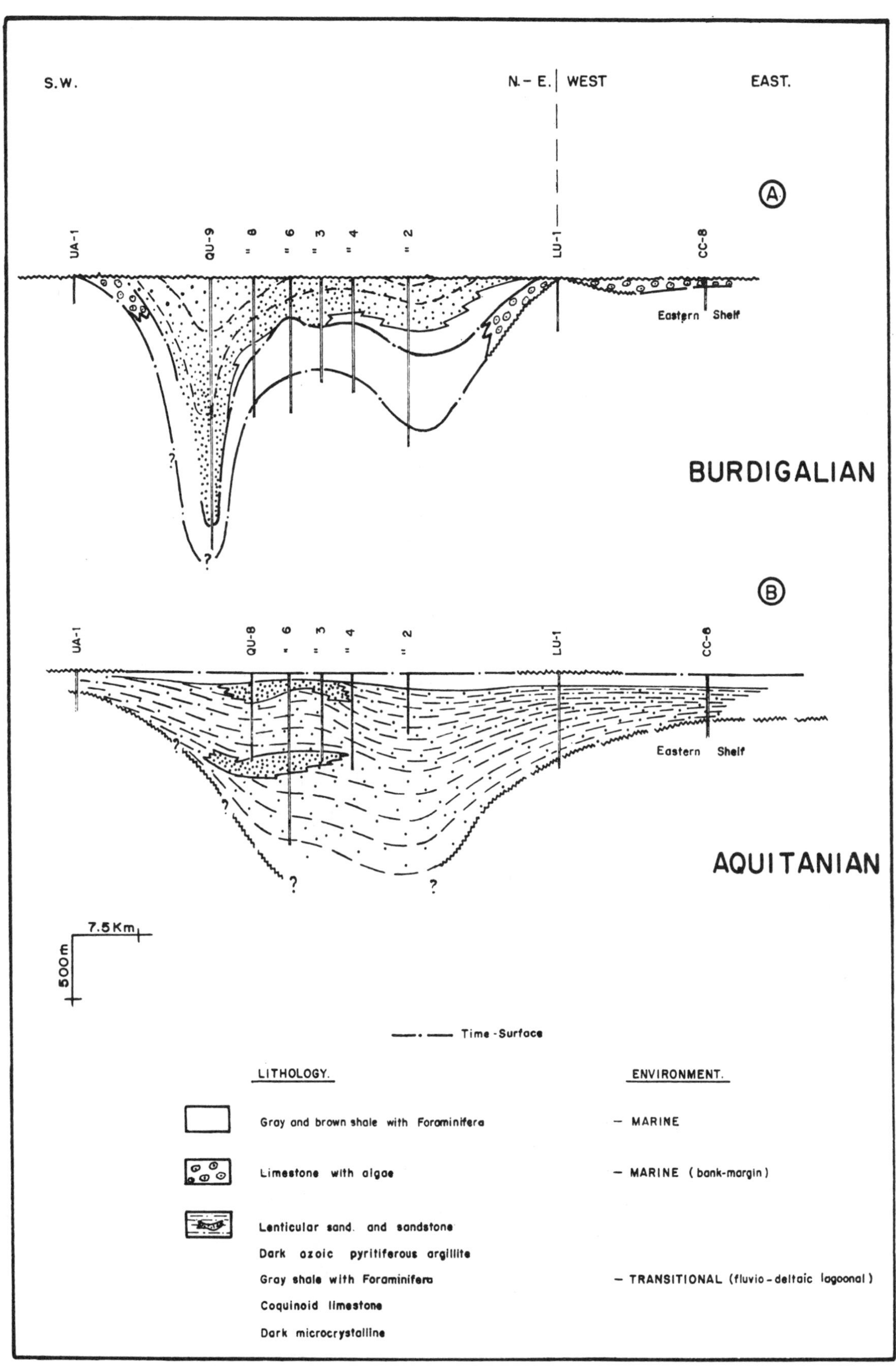

Fig. 35 – Cross-sections of Miocene Quenguela trough.

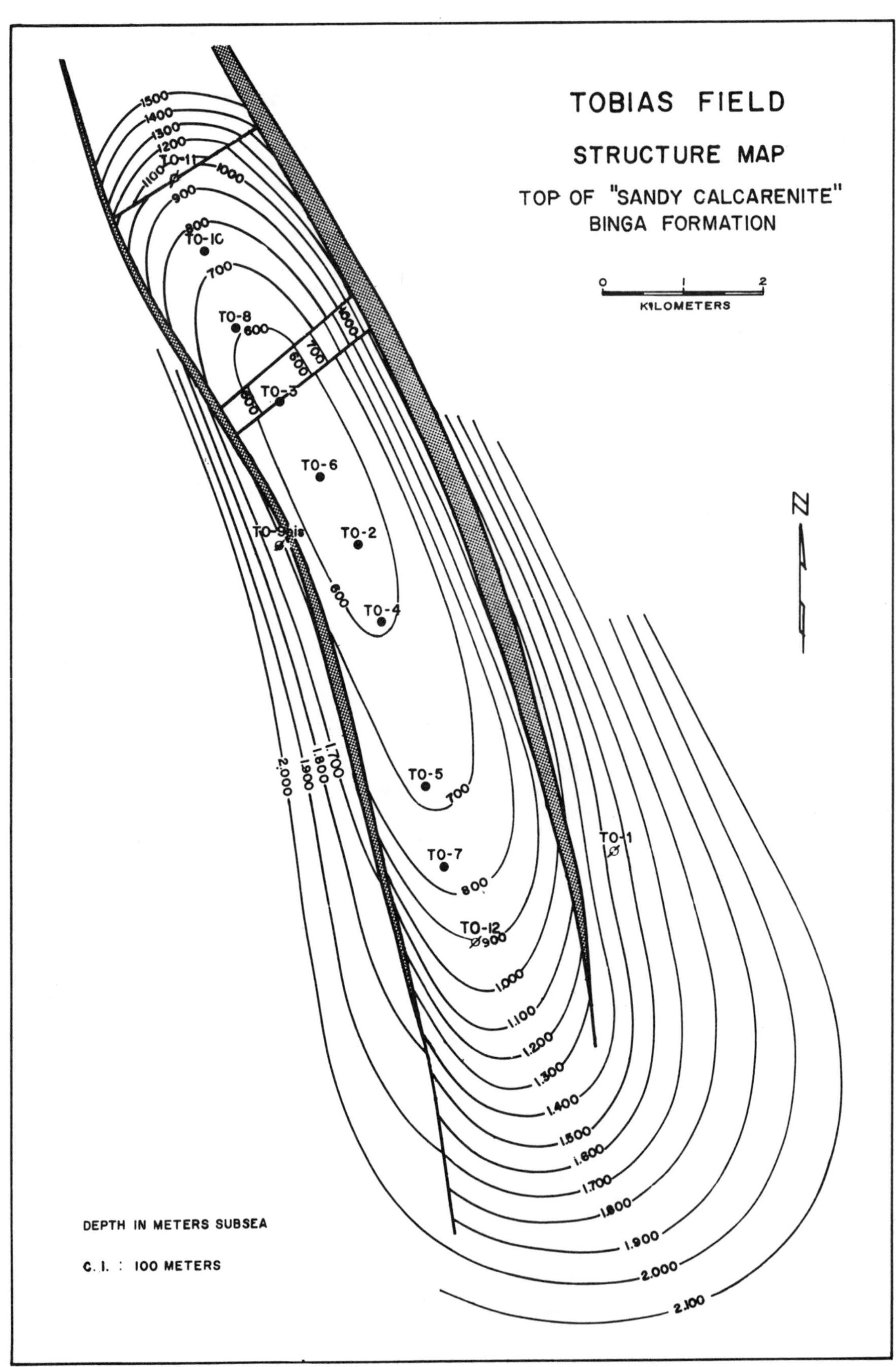

Fig. 36—Structure map of Tobias Field

Morro Liso areas, where rapid deposition and burial of a large volume of sediments with organic matter created optimum conditions for the generation of hydrocarbons.

Several barrels of oil have been produced from fractured, silty, indurated shale of Cenomanian-Turonian age (Morro Liso well) and of Senonian-Paleocene age (Cacuaco well). On the Luanda anticline, one well has been completed as an oil producer from the sandy Eocene facies. In the Cacoba No. 1, heavy oil shows and asphalt have been found in the regressive wedge of sediments which correspond with the maximum westward extent of the Turonian Cuanza River delta (Fig. 32).

It has been pointed out earlier that, during Late Cretaceous and early Tertiary time, the eastern limit of the area affected by salt tectonics coincided approximately with the front of the thick transitional beds (Fig. 9). This suggests that in front of deltas or delta-associated depositional units, clastic-sediment loading might have formed salt structures which may have made possible early entrapment of the oil generated in Upper Cretaceous and lower Tertiary formations.

The geographical extent of the potential productive area of the Upper Cretaceous and lower Tertiary is relatively small because: (1) a high rate of subsidence downdip from the flexure prevented large westward extension of the transitional deposits, and relatively thin, marine, shaly formations were deposited in the largest (western) part of the basin; and (2) the flexure of Late Cretaceous-early Tertiary time (Figs. 7, 8), and the transitional beds which were deposited close behind it, now crop out along the uplifted basin homocline.

Restricted potential productive zones exist in the northern Luanda-Cacuaco area and in the southern Morro Liso area where Upper Cretaceous and lower Tertiary formations were not wholly removed by erosion and are still well protected (Figs. 29–31).

Oil in Miocene formations.—During Miocene time, and after a fairly active period of tectonism, the continent on the east, adjacent to the Cuanza basin, became a major source of sediments. The ancient Cuanza River brought into the western part of the basin a great supply of terrigenous material and deposited it in transitional deltaic and marine neritic environments.

The subsiding part of the basin, west of the Miocene flexure (Fig. 8), already had been affected since the end of Early Cretaceous time by well-developed salt tectonic features that consisted of non-piercement and piercement domes. Miocene clastic loading in pre-existing synclines, rims, or grabens started a strong acceleration of salt tectonism. The concomitant outflow of salt accompanying the filling of these rims and grabens caused numerous salt structures of the northern part of the basin to reach a final stage of evolution approaching complete destruction. The major zones of faulting, which limit the Quenguela trough on the eastern and western sides (Fig. 10), might correspond with such "relic" structures.

Some oil and gas shows have been found in Miocene beds of the Funda graben and of the Quenguela trough which, until now, have been investigated by only a few shallow wells. The Quenguela and Tres Pontas troughs with their offshore extensions may be considered good prospective oil areas for the following reasons. (1) A thick section of deltaic and neritic sediments related to basinal subsidence (and outflow of salt) was deposited. (2) This thick section of shale, clay, and silt, in many places rich in organic material and interbedded with clean sand, was deposited and buried rapidly. (3) Oil could occur in regressive wedges of deltaic sediments which extend westward between beds of marine shale. (4) Oil could have been trapped in large anticlines related to interdomal residual salt structures.

Conclusions

The Cuanza basin is a Cretaceous and Tertiary structural depression affected by the eastward displacement of a boundary flexure during the Cretaceous and by westward displacement during the Tertiary. Other major features are the volcanic Cabo Ledo uplift and the Longa, Cacuaco, and Morro Liso ridges, which were active contemporaneously with the first subsidence that created the basin, and which were still active at the end of Albian time. Salt tectonics, through extensive continued activity from Albian through Miocene time, also were important in controlling contemporaneous sedimentation.

Deposition of a pre-Aptian terrigenous sequence was initiated by a transgressive sea entering the basin during and after strong volcanic activity. During Early Cretaceous time and after general emergence, the sea transgressed again into a restricted basin. The sea in part submerged the existing uplifts and ridges. Cyclical subsidence in the central part of the basin caused cyclic carbonate-evaporite deposition behind carbonate buildups along the Cabo Ledo uplift and the Longa ridge. Upper Cretaceous and Tertiary sediments are represented principally by a marine sequence, deposited in a bathyal to neritic environment. However, thick transitional deltaic and littoral facies, accumulated as lenticular sedimentary masses, grade, respectively, from continental deposits on the east to thinner marine deposits on the west. The loci of maximum deposition and the associated flexure moved eastward during Cretaceous time and westward during Tertiary.

The reconstruction of the geologic history of the Cuanza basin is only a first approach in the understanding of origin, migration, and accumulation of oil, on the one hand, and environment of deposition related to tectonic history, on the other. However, the following preliminary conclusions may be listed.

1. The numerous oil occurrences found in the vertical stratigraphic sequence from Lower Cretaceous through Miocene appear to be related to abundant source rocks existing in each of these units. Lithologic aspects and (or) geochemical analyses suggest that source rocks were deposited during almost all periods from Early Cretaceous to Miocene, and at every time, the lateral distribution of the possible source rocks corresponds generally with a specific, tectonically controlled, depositional environment.

2. Old positive areas, probably contemporaneous with the beginning of the subsidence which initiated the Cuanza basin, provided a favorable situation for entrapment of oil below the Massive Salt in pinch-outs of the Upper Cuvo littoral sandstones.

3. Production of oil from the Binga Formation (Lower Cretaceous), the most prolific oil-producing formation of the basin, appears to be related primarily to a broad occurrence throughout the basin of a close association of source and reservoir related to a progressively decreasing rate of subsidence after the deposition of the Massive Salt. Production from this formation also is controlled by the broad extent of the anhydrite overlying the Binga calcarenite which corresponds with a renewed and sudden beginning of subsidence. Commercial accumulations in the Binga have been localized secondarily by salt folding and by fracturing of the Binga reservoir.

4. Development of Albian reef trends along the western and eastern stable marginal areas of the Albian evaporitic basin provided favorable situations for the generation and accumulation of oil.

5. From the geological history of the Cuanza basin, it is concluded that oil accumulations in the Lower Cretaceous producing formations possibly were completed before the end of Cretaceous time.

6. Commercial accumulation of oil in Upper Cretaceous and Tertiary strata is not yet known. However, the thick argillaceous-arenaceous sequences associated genetically with major flexures or fault zones and deposited in transitional environments could be productive of oil. They contain both source beds and reservoir rocks.

References

Brognon, G., and Verrier, G., 1955, Contribution à la géologie du bassin du Cuanza en Angola: 4th World Petroleum Cong., Rome, Sec. 1, June, 1955, p. 251–265.

——— Verrier, G., and Masson, P., 1959, La tectonique salifère du bassin du Cuanza en Angola: 5th World Petroleum Cong., New York Sec 1, paper 6, p. 109–122.

——— and Verrier, G., 1964, Tectonique et sedimentation dans le bassin du Cuanza (Angola): XXII Internatl. Geol. Cong., New Delhi, 1964.

——— *et al.*, 1965, Salt tectonics of the Cuanza basin, Angola (abs.): Am. Assoc. Petroleum Geologists Bull., v. 49, no. 3, p. 336.

Dartevelle, Ed., 1945, Résumé de quelques observations sur les collections de Service Géologique et des Mines de l'Angola et sur la géologie du littoral de l'Angola: Unpub. rept., Maio, Luanda (Angola).

——— 1953, Echinides fossiles du Congo et de l'Angola: Ann. Mus. Roy. Congo Belge, ser. in-8°, sc. geol., v. 13, p. 71.

Haas, O., 1942, Ammonites from Angola: Am. Mus. Nat. History Bull., v. 81, no. 1, p. 1–224, pl. I–XLVII.

Hedberg, H. D., 1964, Geologic aspects of origin of petroleum: Am. Assoc. Petroleum Geologists Bull., v. 48, no. 11, p. 1755–1803.

Illing, L. V., 1954, Bahaman calcareous sands: Am. Assoc. Petroleum Geologists Bull., v. 38, no. 1, p. 1–95.

Mouta, F., 1951, Carte géologique de l'Angola, Notice explicative: XVIII Internatl. Geol. Cong., pt. XIV, London, 1948, p. 118–130.

Rennie, J. V. L., 1945, Lamelibranquios e gastropodos do Cretacio superior de Angola: Junta das Miss. Geogr. Inv. Colon., Memorias, Ser. Geologica, 1, 67 p., Lisbon (original in English).

Scruton, P. C., 1953, Deposition of evaporites: Am. Assoc. Petroleum Geologists Bull., v. 37, no. 11, p. 2498–2512.

Spath, L. F., 1951, Preliminary notice on some Upper Cretaceous ammonite faunas from Angola: Com. Serv. Geol. Portugal, t. 32, p. 123–130, Lisbon.

Reprinted by permission of the Saskatchewan Geological Society from C. G. Carlson and J. E. Christopher, eds., *Fifth International Williston Basin Symposium*: Saskatchewan Geological Society Special Publication 9, 1987, p. 169-177.

STRUCTURAL HISTORY OF POPLAR DOME AND THE DISSOLUTION OF CHARLES FORMATION SALT, ROOSEVELT COUNTY, MONTANA

DAVID M. ORCHARD
BHP Petroleum (Americas) Inc.
5613 DTC Parkway, Suite 600
Englewood, CO 80111

ABSTRACT

Poplar Dome, in western Roosevelt County, Montana, is a major structural feature of the western Williston Basin. Approximately 150 feet of structural closure exists at the Mississippian Ratcliffe horizon. Minor structural movement occurred in the area of Poplar Dome during Mississippian through Cretaceous time. Much of the present structural relief is the result of Laramide uplift.

Beds of salt in the Mississippian Charles Formation (Poplar interval) abruptly pinch out at Poplar Dome. Over 300 feet of salt from regionally extensive beds is missing on and near the structure. Nonsalt beds, however, extend across the dome with little thickness change and few facies changes. The stratigraphic evidence along with the presence of breccia in cores indicate that the salt beds were originally present on Poplar Dome but were subsequently dissolved. The dissolution was probably a Tertiary event caused by the introduction of fresh water along fractures induced by uplift. The removal of the salt caused a collapse over the dome which affected the structural configuration of beds younger than the Charles Formation.

In addition to the dissolution of salt beds, pore-filling salt has been dissolved from limestone of the Charles "A" and "B" units. The dissolution of this salt from Charles "A" porosity is related to the removal of bedded salt, while the flushing of salt from the Charles "B" porosity is related to a regional dissolution front encroaching from the southwest.

Poplar Dome is the site of a major oil field which has yielded more than 45 million barrels of oil, primarily from the Charles "B" porosity. The removal of pore-filling salt from the Charles "B" was a critical factor in creating this oil accumulation. The removal of salt beds also allowed the upward migration of oil into reservoirs of the Charles "A" and the Pennsylvanian Tyler Formation.

INTRODUCTION

The Poplar Dome is located in western Roosevelt County, Montana, at the eastern end of a structural trend that extends eastward along the Wolf Creek nose from Bowdoin Dome (Fig. 1). The northeast-southwest trending Brockton-Froid fault system lies to the east of Poplar Dome. Poplar Dome is a major structural feature in the Montana portion of the Williston Basin and is host to a large oil field. More than 45 million barrels of oil have been produced at Poplar Dome, primarily from reservoirs in the Mississippian Charles Formation. Unit outlines, field names, well control, and a cross-section index are shown on Figure 2. Stratigraphic terminology as used in this paper is shown on a type log in Figure 3.

Poplar Dome is shown as a closure on structure maps drawn on the base of the Upper Cretaceous Colorado Shale (Dobbin and Erdman, 1955) and the Cretaceous Greenhorn Limestone (Colton and Bateman, 1956). The discovery and development of the oil reserves has lead to structural and stratigraphic discussions by Cox (1953), Beekly (1956), Moore (1958), Brunson (1985), and Diehl (1985).

An anomalous feature of Poplar Dome is the thinning or absence of beds of halite in the Charles Formation on and near the structure. This anomaly was noted by Beekly (1956) and was briefly discussed by Cook (1976). This paper describes the manner in which the Charles salt beds pinch out at Poplar Dome. Evidence from stratigraphic relationships and core observation suggest that the absence of salt is caused by post-depositional dissolution. Related aspects of salt-plugged and salt-free porosity are of particular economic importance in the presence of reservoir-quality carbonate rock in the Charles Formation.

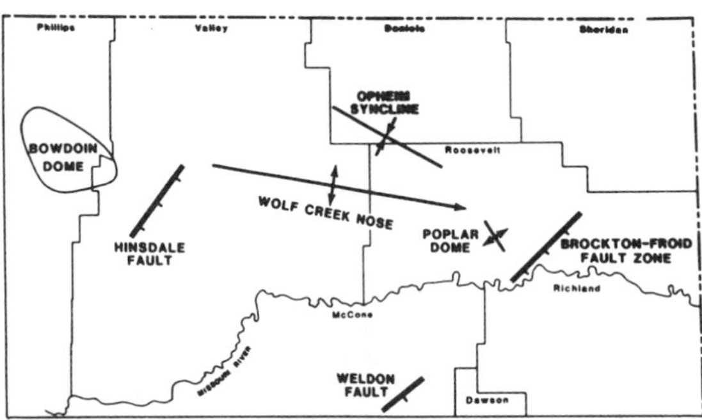

Fig. 1: Location of Poplar Dome in relation to other tectonic features of northeastern Montana.

STRUCTURE

Minor structural movements occurred along the area of the Wolf Creek Nose and Poplar Dome during the Mississippian. Northeasterly-trending anticlines were formed as shown on the Devonian Nisku to Mississippian Ratcliffe isopachous map (Fig. 4). The dome remained gently positive through the Cretaceous as shown on the Cretaceous Greenhorn to Mississippian Kibbey Limestone isopachous map (Fig. 5).

Structure on the top of the Mississippian Ratcliffe interval (base of the Green Point anhydrite) is illustrated on Figure 6. Northeasterly trending anticlines are present along the Wolf Creek Nose west of Poplar Dome, and an anticline with this trend is present at Northwest Poplar Field. The Ratcliffe structure at West Poplar Field and the East Poplar Unit is not uniformly domal but rather consists of three broad closures separated by narrow synclines. The maximum closure on the overall structure at the Ratcliffe horizon is 150 feet. The steepest dip occurs along a flexure on the northeastern and eastern sides of the dome.

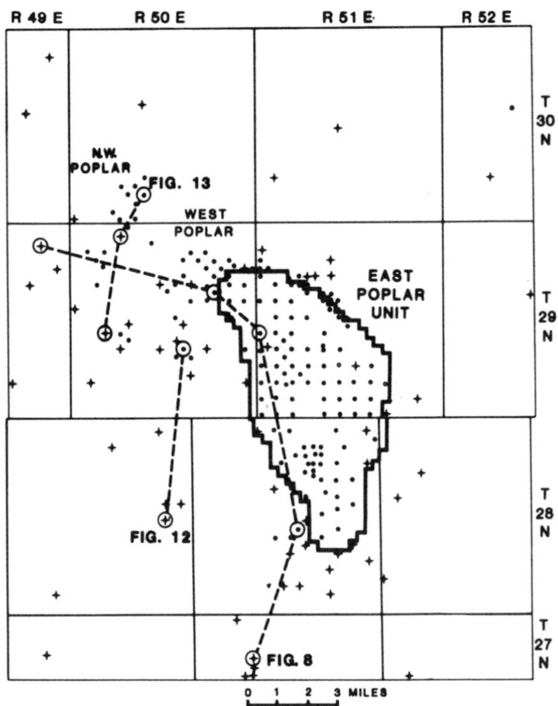

Fig. 2: Well control, unit and field locations, and cross-section index for study area.

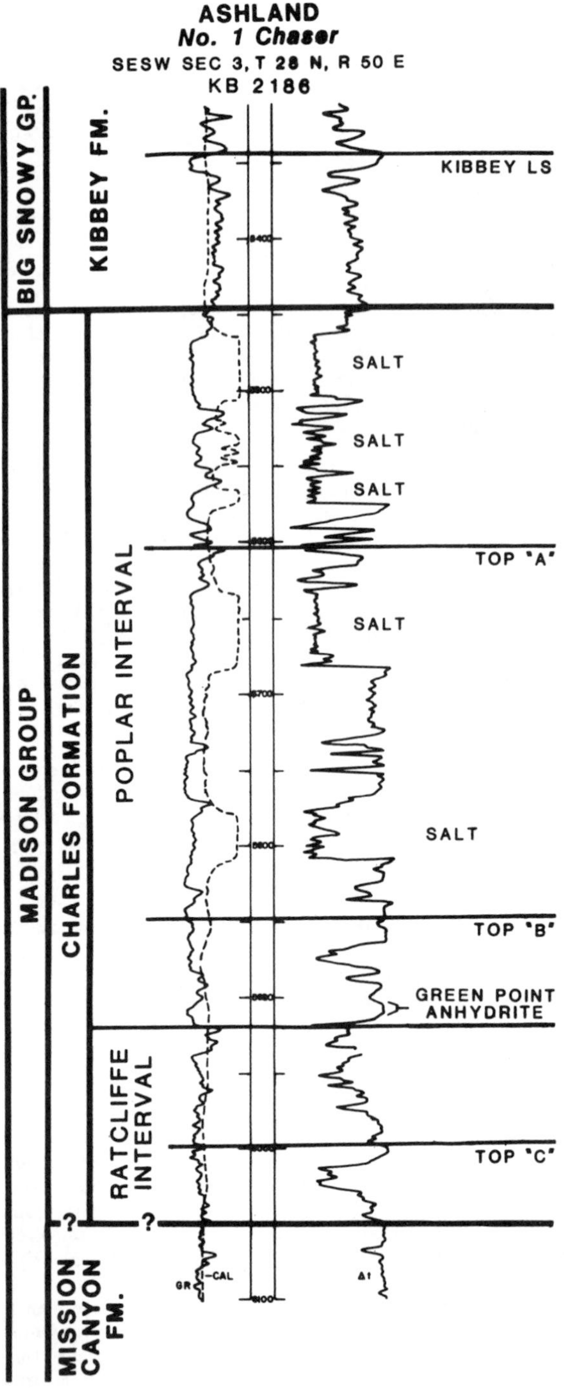

Fig. 3: Type log for Charles Formation with the stratigraphic terminology as used in this paper.

The structure on the Mississippian Kibbey Limestone (Fig. 7) and younger horizons has a different configuration than that on the Ratcliffe. The highest parts of the structure still occur at West Poplar Field and East Poplar Unit. However,a prominent ridge is present trending northwest-southeast to the southwest of the producing areas. This ridge also appears on the Greenhorn structure map of Colton and Bateman (1956). The relationship of this ridge to the presence of salt beds in the Charles Formation is discussed later.

The major uplift leading to the present-day structural configuration occurred in post-Paleocene time. The Paleocene Fort Union Formation is eroded off Poplar Dome and the Wolf Creek nose, and the Cretaceous Fox Hills and Bearpaw Formations are exposed (Colton and Bateman, 1956). Thomas (1974) discussed the involvement of Poplar Dome and the Wolf Creek Nose in Laramide tectonics. Clement (1986) stated that the major uplift on the Cedar Creek Anticline also occurred in post-Paleocene time, with regional mid-Tertiary epeirogenic uplift.

Fracturing of Charles carbonates is described from cores by Beekly (1956) and was also observed by the present author. The configuration of the Ratcliffe structure map (Fig. 6) suggests that the broad closures may be separated by narrow grabens with faulted boundaries trending north-northeast. The overall structure trend of Poplar Dome is northwest to southeast. The influence of both northwest-southeast and northeast-southwest trends can be discerned on the isopachous map of the net Charles salt (Fig. 8).

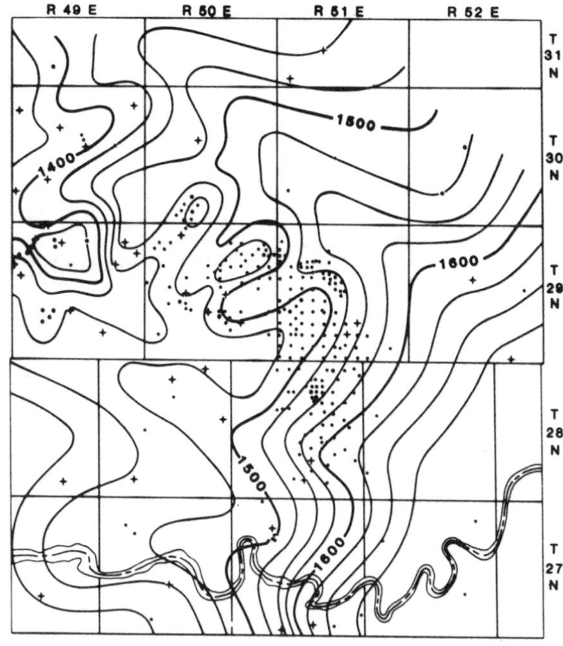

Fig. 4: Devonian Nisku to Mississippian Ratcliffe isopachous map of Poplar Dome area. Contour interval is 25 feet. Devonian penetrations shown by well symbols; shallower penetrations shown by open circles.

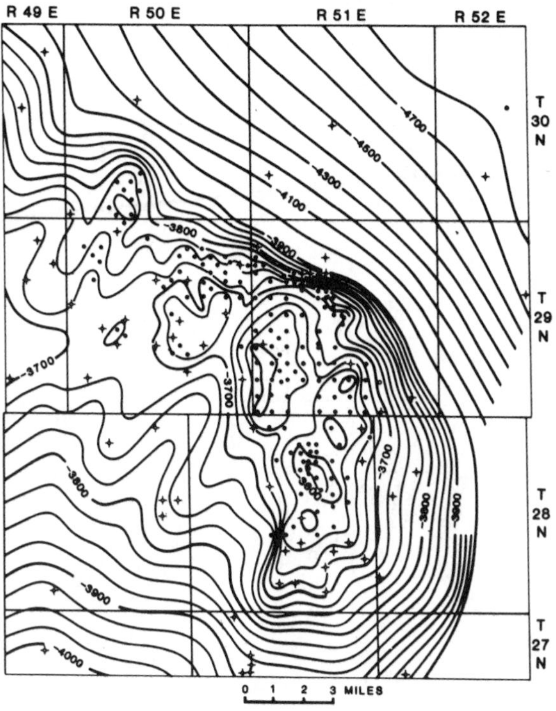

Fig. 6: Structural map of top of Ratcliffe interval, Poplar Dome area. Contour interval is 25 feet.

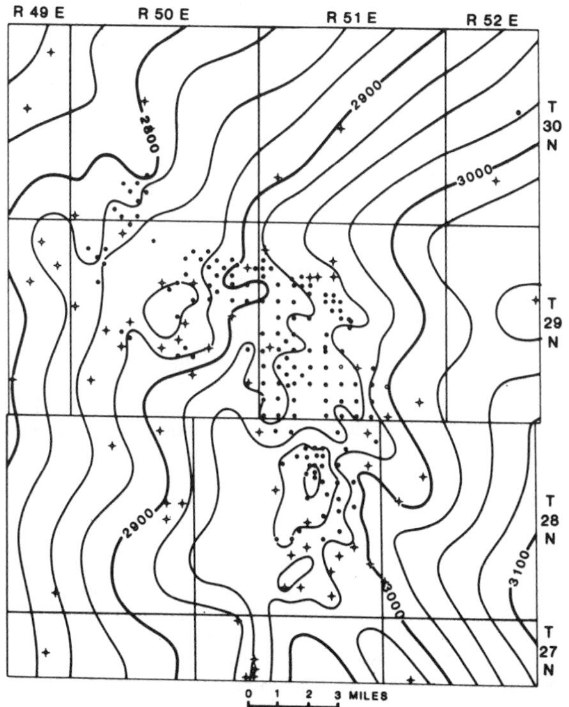

Fig. 5: Mississippian Kibbey Limestone to Cretaceous Greenhorn isopachous map of Poplar Dome area. Contour interval is 25 feet.

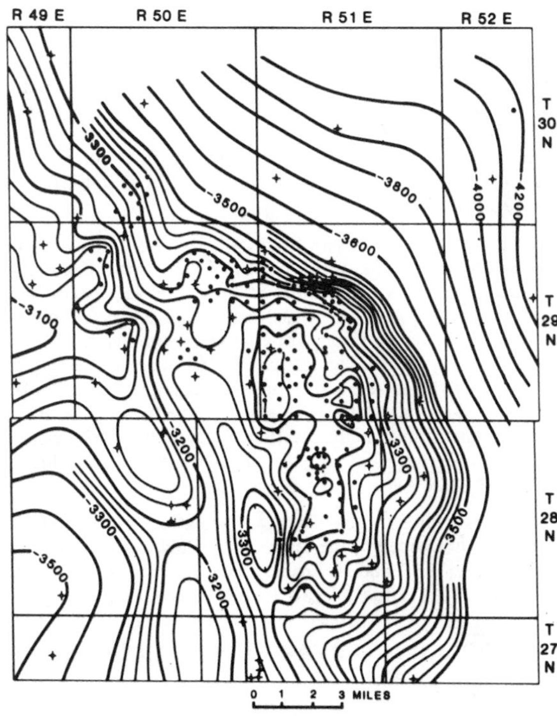

Fig.7: Structure map on top of Mississippian Kibbey Limestone, Poplar Dome area. Contour interval is 25 feet.

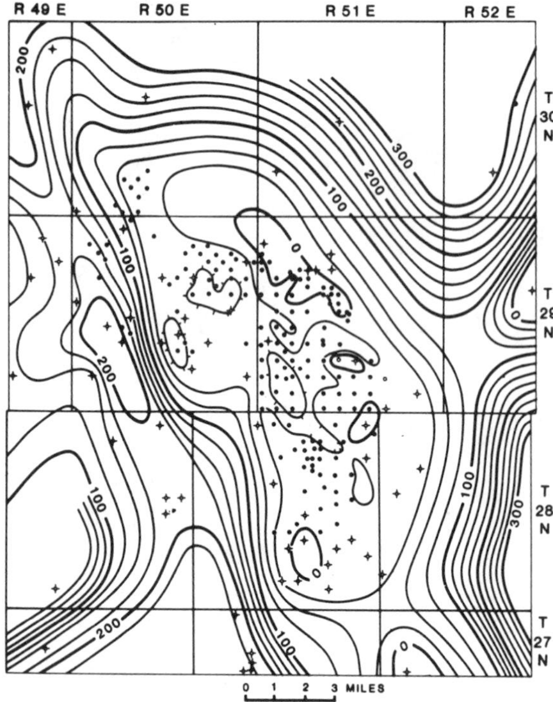

Fig.8: Isopachous map of net salt in Charles Formation, Poplar Dome area. Contour interval is 25 feet.

POPLAR INTERVAL

The Poplar interval (Carlson and Anderson, 1965; Cook, 1976) is characterized by the presence of several beds of halite separated by carbonate and anhydrite (Fig. 3). The Poplar interval overlies the Ratcliffe interval, and the contact is considered by the author to be the base of the Green Point anhydrite. This corresponds to the top of the Charles B-4 of Brunson (1985).

Individual salt beds can be traced over wide areas of the Williston Basin. The western limit of salt in the Charles is shown in Figure 9. The extent of individual salt, carbonate, and anhydrite beds in western North Dakota is presented by Cook (1976). In the vicinity of Poplar Dome, five major salt beds were deposited. The salt at 5775 feet on the type log of Figure 3 is the lowest salt deposited throughout the Poplar Dome area. A thin salt was locally deposited at the top of the Green Point anhydrite. In addition, a regionally developed salt below the Green Point anhydrite appears immediately to the east of the study area.

The reservoirs in the Charles Formation have been informally subdivided by the operators at Poplar Dome into the Charles "A," "B," and "C" zones (Beekly, 1956; Moore, 1958; Brunson, 1985). These interals are identified on Figure 3.

POPLAR INTERVAL AT POPLAR DOME

On and near Poplar Dome, many of the salt beds are missing. A map of the net Charles salt thickness (Fig.8) illustrates that the salt ranges from zero to 218 net feet. Figure 10 is a cross-section which illustrates the manner of disappearance of the salt beds onto Poplar Dome. As the dome is approached along the line of cross-section, the upper salt beds disappear first. Wells with thin salt (less than 30 feet) invariably have salt present only in the lowest salt beds. Significantly, the carbonate and anhydrite beds between the salts do not change facies or appreciably thin as the salt beds pinch out. Individual non-salt beds can be traced by log character across the structure. With neutron-density crossplots it can be demonstrated that very few lithologic changes (i.e. limestone to dolomite or carbonate to anhydrite) occur. No occurrences of non-salt to salt transition are present. Some relict insoluble sediment is present, however, at the stratigraphic position of missing salts, particularly those salts that are stratigraphically highest in the section. Units above and below a salt converge across a salt edge with no related thickness changes.

At the base of the illustration on Figure 10, values are given for net salt and net non-salt thicknesses within the Poplar interval. While the salt thickness ranges from nearly zero to over 200 net feet, the thickness of on-salt rocks varies only from 297 to 314 net feet.

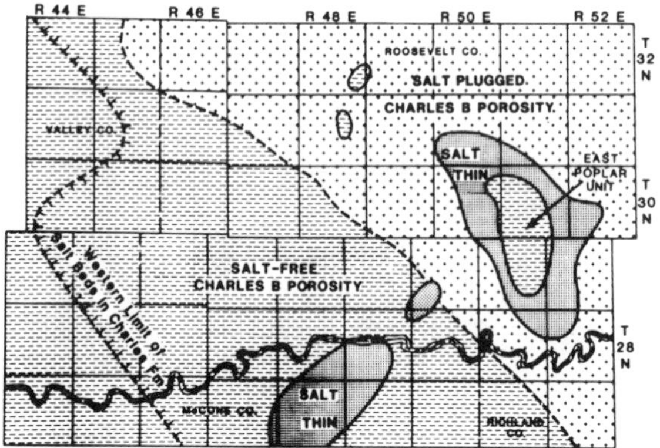

Fig.9: Regional map of salt in the Charles Formation showing western limit of bedded salt, areas of thin or absent bedded salt, and areas of salt plugging in Charles "B" porosity.

SALT DISSOLUTION

The missing salt on Poplar Dome has been previously noted in the literature. Beekly (1956) stated that the question of whether these salts were not deposited or were deposited and subsequently removed had not been resolved. Cook (1976) noted Charles salt thins over the Nesson Anticline, "Divide High," and Little Knife Anticline, and attributed these thins to syndepositional structural growth. For Poplar Dome, however, he suggested that the absence of salt beds and the lack of salt plugging in certain reservoir intervals might be the result of groundwater solution.

The question of non-deposition versus dissolution can be approached by looking at pre-Poplar interval isopachs, the character of non-salt beds across the dome, core evidence, and other instances of salt absence in the area. Non-deposition of salt on Poplar Dome would imply the presence of an underlying or syndepositional structure during Charles deposition onto which the salt beds could pinch out. Dissolution of the salt would require entry of fresh or brackfish water into the formation at some time subsequent to the deposition of the salt.

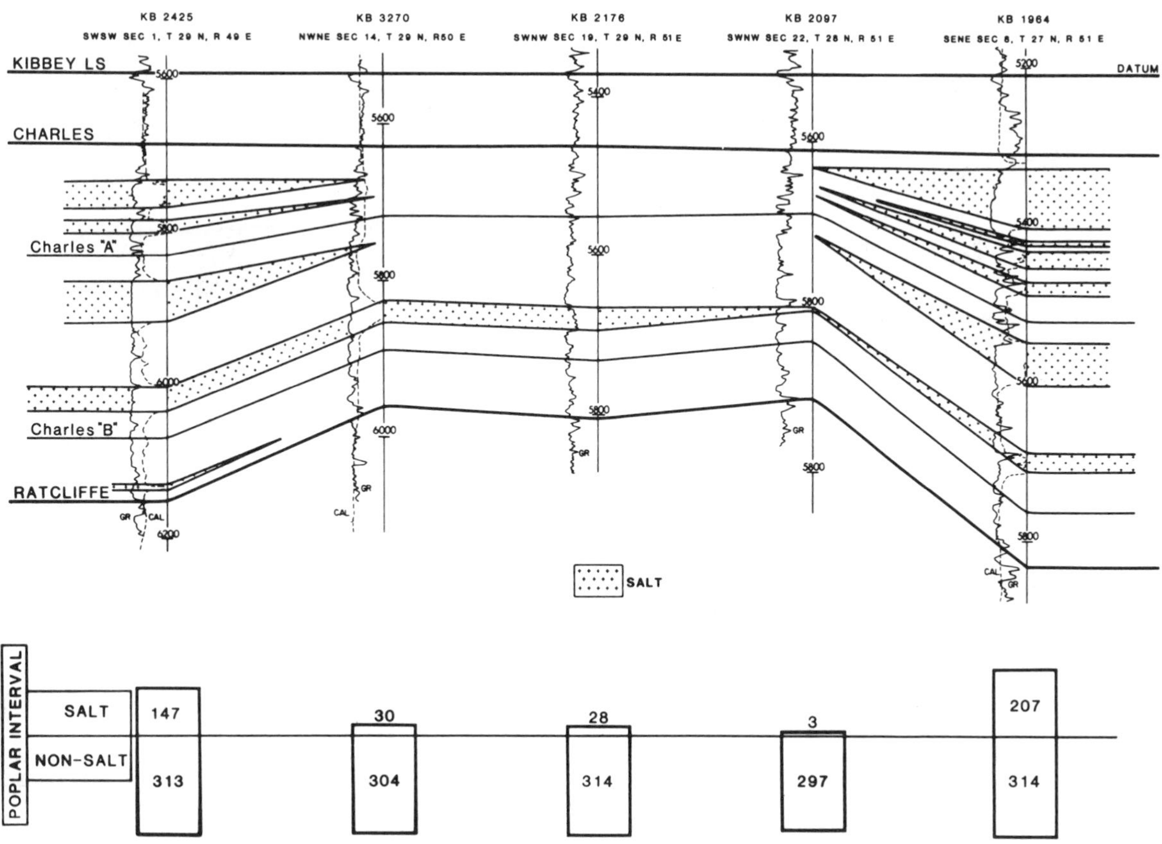

Fig. 10: Cross-section showing manner of salt pinch out onto Poplar Dome. Note the relative persistance of the lowest salt (immediately above Charles "B") and the persistence of non-salt beds. Datum is Kibbey Limestone. Values for net salt and net non-salt thicknesses in Poplar interval given at base of illustration. Locations on figure 2.

An argument against an underlying pre-existing structure can be made from the Nisku-Ratcliffe isopach (Fig. 4). On this map, the Poplar Dome is not present as a closure. Rather, northeast-southwest trending anticlines are present similar to those in the area of the Wolf Creek Nose. The magnitude and configuration of the pre-Poplar structure would not preclude deposition of salt across the dome.

Syndepositional structural growth is argued against by the character of the non-salt beds across the dome. As discussed above, these carbonate and anhydrite beds do not thin appreciably and exhibit few facies change as the salt beds pinch out. Also, the contact between the Charles and overlying Kibbey Formations is conformable, and the Kibbey Limestone to Charles interval is uniformly thick across the structure (Fig.10). To explain these features, a model of non-deposition would involve episodic uplift coincident only with periods of salt deposition and ceasing at the onset of Kibbey deposition.

Anomalously thin salt in the Charles Formation is not limited to Poplar Dome. Salt is also completely missing along the steeply dipping northeastern and eastern flanks of the structure, with the thin extending into T28-29N, R52E. Other salt thins which together form a northeast-southwest trend are present to the southwest of Poplar Dome (Fig.9).

Salt solution implies the collapse of the strata above and between the salt beds. This could lead to breccias which would be observable in cores. Most of the cores taken during the development of East Poplar Unit were through Charles "B" intervals below the lowest extensive salt bed. During a recent search, no cores could be found that sampled the exact interval from which a salt bed was missing.

A core through Charles "A" limestone at 5636-5643 feet in the Murphy #44 East Poplar Unit (SW SW Sec, 24, T28N-R51E) was examined. This section consists of dark gray lime mudstone with fractures and brecciation. Two views (horizontal and vertical faces) of a sample at 5636 feet (Fig. 11) show closely set vertical fractures and horizontal fractures and stylolites. Some of the fractures are partly or entirely healed by anhydrite. A large vug in the center of the sample is formed by the angular fit of a number of limestone fragments. This sample comes from an interval approximately 75 feet above the horizon from which 30 feet of the salt are missing. In combination with tectonic fracturing, the deformation observed in the core may be the result of the collapse of the formation when the underlying salt was dissolved.

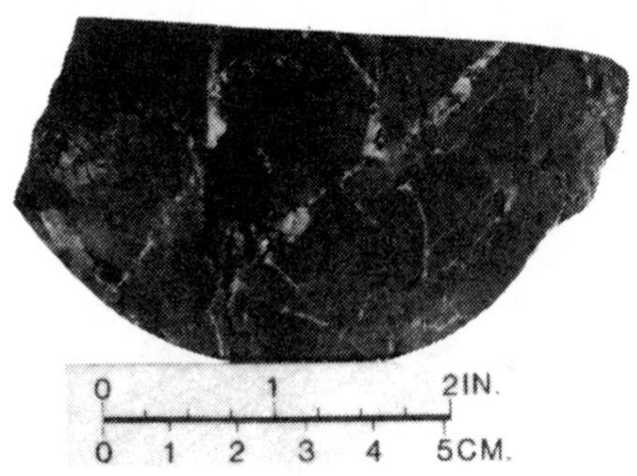

Fig. 11: Brecciated and fractured Charles "A" limestone from depth of 5636 feet in the Murphy #44 East Poplar Unit. Views of horizontal (top) and vertical (bottom) cuts through same sample.

AGE OF SALT DISSOLUTION

The age of a salt dissolution event can be recognized by identifying depositional thickening in some stratigraphic interval above the salt equal to the thickness of the missing salt. Up to 218 feet of salt is missing from the Charles Formation on Poplar Dome. No depositional compensations for this salt can be seen on the Kibbey Limestone to Greenhorn isopach (Fig. 5). From Figure 10 it can be seen that no compensation occurred in the interval between the Kibbey Limestone and the Charles Formation.

The Cretaceous Bearpaw and Fox Hills Formations are exposed at the surface at Poplar Dome, and the Paleocene Fort Union Formation has been erosionally removed (Colton and Bateman, 1956). Tertiary sediments are not present that could demonstrate depositional thickening to determine the exact time of salt removal, and that occurrence can only be stated as post-Fox Hills and probably post-Fort Union. The salt dissolution therefore appears to be a Tertiary event and may be related to fractures induced by Laramide uplift.

STRUCTURAL EFFECTS OF SALT DISSOLUTION

It was previously noted that the structural configuration on the Kibbey Limestone and other horizons above the missing salt differs from that on the Ratcliffe (below the salt). The large ridge that appears to the southwest of the dome on the horizons above the Charles is due to the preservation of salt under the ridge with removal of salt on the dome.

A comparison of two wells, Banner #1-24 Tribal (NW NW Sec. 24, T28N-R50E) and the Century #22-15 Robbins (SW SE Sec.22, T29N-R50E), illustrates the structural effect of the dissolution of the salt (Fig. 12). The salt is 163 feet thicker in the Banner well than in the Century well. At the Ratcliffe horizon, the Banner well is 103 feet low, but it is 44 feet high at the Kibbey Limestone. A 147-foot structural change has therefore occurred coincident with the removal of 163 feet of salt. The remaining 16 feet of structural difference is attributed to minor thickness variation in the non-salt beds.

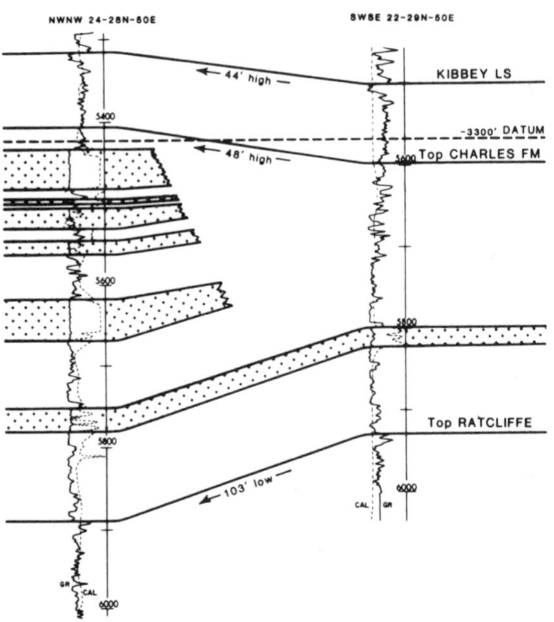

Fig. 12: Structural cross-section illustrating reversal of dip on tops of Charles Formation and Kibbey Limestone versus top of Ratcliffe interval caused by the removal of salt beds. Locations on Figure 2.

The Poplar Dome is expressed on the surface as an extension of the Wolf Creek Nose, but neither topography nor the pattern of outcropping formations clearly express domal closure (Colton and Bateman, 1956). The model of salt dissolution implies that the dome would have approximately 200 feet greater relief had the salt remained, and the surface expression of the surface would have been more pronounced.

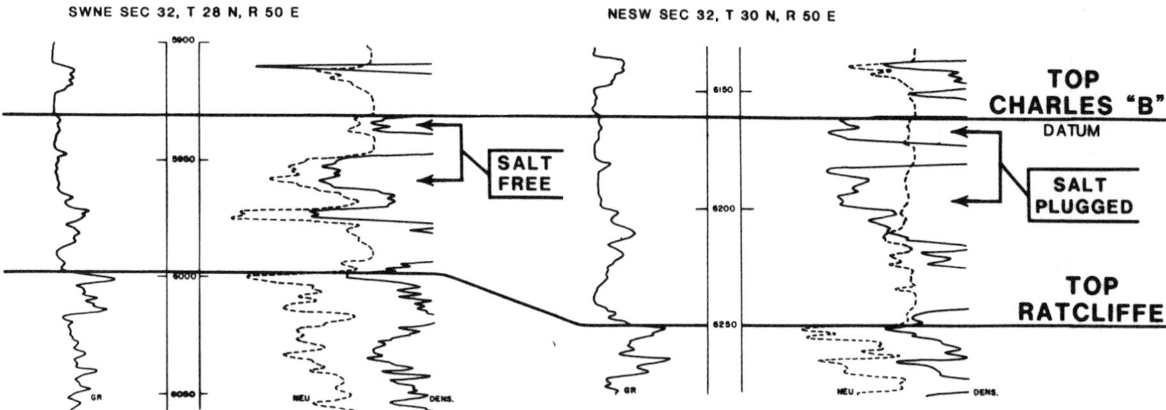

Fig.13: Comparison of neutron-density log response for salt-free and salt-plugged porosity in Charles "B" limestone.

SALT PLUGGING IN POPLAR INTERVAL RESERVOIRS

Porous zones of the Charles "A" and Charles "B" units are regionally salt-plugged but locally salt-free on Poplar Dome (Fig.9). The widespread presence of pore-plugging salt in these units in the central basin suggests that salt-plugging is the normal condition for these rocks over a wide area. The absence of salt-plugging towards the margins of that area as in the present study may be attributed to flushing.

Figure 13 compares the neutron-density log response of salt-free porous Charles "B" limestone to the same section with salt plugging. The neutron-density cross-over indicative of salt-plugging has been illustrated for other Williston Basin reservoirs by Dean (1982) and Weinzapfel and Neese (1986). Figure 14 is a cross-section showing the pattern of dissolution of the pore-plugging salt for these zones. This cross-section also reinforces the observations that the upper salt beds disappear first, the non-salt beds show continuity across the structure, and the variation is much greater for net salt than for net non-salt thicknesses within the Poplar interval (Fig. 8).

Porous zones in the Charles "A" become salt free in relation to the dissolution of stratigraphically higher salt beds. As salt beds progressively deeper in the section were removed, pore-filling salt was also flushed from porous carbonate. In Figure 14, two porous salt-free beds in the Charles "A" have lateral boundaries with salt-plugged porosity. Each of these units becomes salt-free when no bedded salt remains higher in the section.

In contrast to the Charles "A," porous Charles "B" units are variously salt-plugged or salt-free in relation to a regional boundary rather than to the dissolution of overlying salt beds. This regional boundary (Fig. 9) trends northwest to southeast, with salt-free rocks to the southwest and salt-plugged rocks to the northeast. This boundary is abrupt, with salt in equivalent beds disappearing within a lateral distance of less than one mile. Outliers of salt-free porosity to the northeast of the regional boundary are present, including a large area under the East Poplar Unit. Diehl (1985) mapped the limits of salt-plugging in Charles "B" reservoirs along the edge of the outlier through West Poplar Field. The regional salt-plugging boundary has a trend approximately parallel to the western limit of salt beds in the Charles (Fig. 9). At Poplar Dome, the area of salt-free Charles "B" porosity is coincident with, but smaller than, the area affected by dissolution of bedded salt.

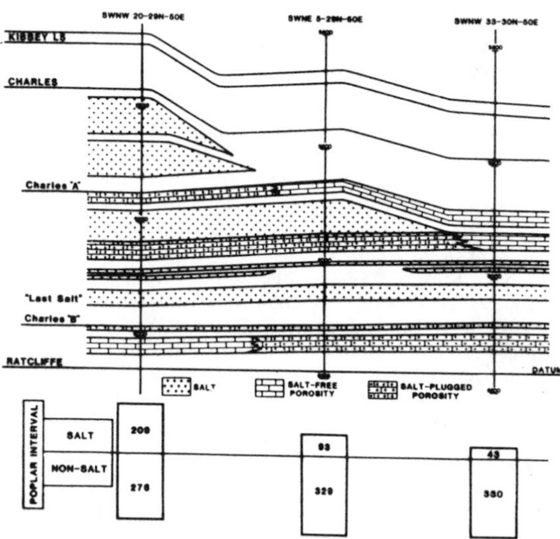

Fig. 14: Cross-section showing salt pinch outs and salt-plugged to salt-free transitions in Charles "A" and Charles "B" reservoirs. Datum is top Ratcliffe. Value for net salt and net non-salt thicknesses in Poplar interval given at base of illustration. Locations on Figure 2.

MECHANISMS OF SALT DISSOLUTION

The Madison Group is a major aquifer with hydrodynamic flow to the northeast into the Williston Basin (Feltis, 1980; Downey, 1984; Maccary, 1984). In addition, Ordovician, Pennsylvanian/Permian and Jurassic/Cretaceous aquifers flow in the same direction (Downey, 1984).

Beekly (1956) reported a chloride concentration of 182,004 ppm with sodium of 117,620 ppm for produced water at Poplar Dome. The interval from which this water was produced was not stated, but it was most probably Charles "B." Diehl (1985) provides values for total dissolved solids, at Northwest Poplar Field of 250,000 ppm for the Charles "B" and 40,000 ppm for the Charles "C." This fresher water for the Charles "C" zone, the lowest producing zone in the Charles Formation, may be related to an influx of water in the underlying Mission Canyon Formation. Maccary (1984) mapped apparent water resistivities for the Mission Canyon and identified fresher water (higher Rwa) approaching Poplar Dome from the west.

The regional salt-plugging boundary in the Charles "B" is oriented approximately along strike with the hydrodynamic gradient for the Madison aquifer. It is probable that the salt-free porosity to the southwest of that boundary is an expression of the advancing fresh water.

Dissolution of bedded salt and salt in Charles "A" porosity appears related to water from some overlying aquifer. This is indicated by the top-down pattern of salt removal, and the relatively extensive occurrence of the lowest salt bed (5775 feet, Fig. 3) across Poplar Dome. On the net salt isopachous map (Fig. 8), northwest-southeast and northeast-southwest structural elements are expressed, suggesting that water may have been introduced into the Charles Formation along fractures which those trends.

SIGNIFICANCE OF SALT SOLUTION TO OIL PRODUCTION

The bulk of the oil produced at Poplar Dome comes from Charles "B" reservoirs. Some production has been attained from Charles "A" zones above the stratigraphic position of the lowest salt beds. Also, approximately 300,000 BO have been produced from Pennsylvanian Tyler Formation sandstones above the Charles.

Dow (1974) proposed that the Charles salt beds provide a regional seal to vertical migration of oil generated in the Bakken Formation. This model was reviewed and supported by Meissner (1978) and Webster (1984). Dow (1974) recognized that oil produced in the Tyler sandstones above Poplar Dome was Bakken oil that had been able to migrate vertically along fractures above the Charles due to the absence of salt. Other instances of upward migration cited are oil in Spearfish reservoirs in the northeastern Williston Basin and oil in Kibbey reservoirs along the Weldon Fault. The current work would add Charles "A" reservoirs to the Tyler as being charged by vertical migration through the areas where salt is totally absent.

From the regional distribution of salt-plugged porosity in the Charles "B" reservoirs (Fig. 9), it appears that the major reservoirs at Poplar Dome were originally plugged. The removal of this salt is a major factor in the creation of the large oil field at Poplar Dome. In addition, removal of pore-filling salt from Charles "A" reservoirs has allowed a small amount of production from those zones.

ACKNOWLEDGEMENTS

This paper is presented with the permission of BHP Petroleum (Americas) Inc., and the author is grateful for that permission and support. The manuscript was critically reviewed by Nels E. Voldseth, Jeffrey M. Dunleavy, Robert G. Lindsay, and Suzanne E. Reeves. The illustrations were prepared by Barry Perow, and the manuscript was typed by Kathryn Portus.

REFERENCES CITED

Beekly, E.K., 1956, Poplar-type field, Montana: North Dakota Geol. Soc. and Saskatchewan Geol. Soc.: First International Williston Basin Symposium, pp. 61-65.

Brunson, T., 1985, Poplar east field In: Montana Oil and Gas Fields Symposium, Montana Geol. Soc., Billings.

Carlson, C.G. and Anderson, S.B., 1965, Sedimentary and tectonic history of North Dakota part of Williston Basin: AAPG Bull., Vol. 49, No. 11, pp. 1833-1846.

Clement, J.H., 1986, Cedar Creek: a significant paleotectonic feature of the Williston Basin; In: Peterson, J.A., ed., Paleotectonics and Sedimentation in the Rocky Mountain Region, United States: AAPG Mem. 41, pp. 213-240.

Colton, R.B., and Bateman, A.F., Jr., 1956, Geologic and structure contour map of the Fort Peck Indian Reservation and vicinity, Montana, USGS Misc. Geol. Invest. Map I-225.

Cook, C.W., 1976, A mechanical well log study of the Poplar interval of the Mississippian Madison Formation in North Dakota; North Dakota Geol. Surv. Rept. of Invest. #52, 20 pp.

Cox, H.M., 1953, Williston Basin - Mississippian reservoir characteristics and proved reserves: AAPG Bull., Vol. 37, No. 10, pp. 2294-2302.

Dean, K., 1982, Devonian Dawson Bay Formation in northwestern North Dakota: Fourth International Williston Basin Symposium, Saskatchewan Geol. Soc. and North Dakota Geol. Soc., pp. 89-92.

Diehl, L.A., 1985, Poplar Northwest Field; In: Montana Oil and Gas Fields Symposium, Montana Geol. Soc., Billings.

Dobbin, C.E., and Erdman, C.E., 1955, Structure contour map of the Montana plains: USGS Oil and Gas Inv. Map #OM-178A.

Dow, W.G., 1974, Application of oil-correlation and source-rock data to exploration in the Williston Basin: AAPG Bull., Vol. 58, No. 7, pp. 1253-1262.

Downey, J.S., 1984, Geohydrology of the Madison and associated aquifers in parts of Montana, North Dakota, South Dakota, and Wyoming: USGS Prof. Paper #1273-G.

Feltis, R.D., 1980, Potentiometric surface map of water in the Madison Group, Montana: Montana Bur. Mines and Geol., Hydrogeologic Map #2.

Jensen, F.S., and Varnes, H.D., 1964, Geology of the Fort Peck area, Garfield, McCone,and Valley Counties, Montana: USGS Prof. Paper #414-F, 49 pp.

Maccary, L.M., 1984, Apparent water resistivity, porosity, and water temperature of the Madison Limestone and underlying rocks in parts of Montana, Nebraska, North Dakota, South Dakota, and Wyoming: USGS Prof. Paper #1273-D.

Meissner, F.F., 1978, Petroleum geology of the Bakken Formation, Williston Basin, North Dakota and Montana, In: Montana Geol. Soc., Williston Basin Symposium: Billing, pp. 207-230.

Moore, W.J., 1958, Poplar field east, In: Nordquist, J.W. and Johnson, M.C., eds., Montana Oil and Gas Fields Symposium (revised): Billings Geol. Soc., pp. 196-197.

Thomas, G.E., 1974, Lineament-block tectonics, Williston-Blood Creek Basin: AAPG Bull., Vol. 58, No. 7, pp. 1305-1322.

Webster, R.L., 1984, Petroleum source rocks and stratigraphy of the Bakken Formation; In: Woodward, J., Meissner, F.F., and Clayton, J.L., eds., Hydrocarbon source rocks of the greater Rocky Mountain region: Rocky Mtn. Assoc. of Geol., pp. 57-81.

Weinzapfel, A.C., and Neese, D.G., 1986, Gooseneck Field, northern Williston Basin; In: Rocky Mountain Oil and Gas Fields, Wyoming Geol. Soc., pp. 61-82.